Process Analysis by Statistical Methods

Another Wiley book by the author

Process Analysis and Simulation: Deterministic Systems
David M. Himmelblau and Kenneth B. Bischoff
1968

Process Analysis by Statistical Methods

David M. Himmelblau
Professor of Chemical Engineering
University of Texas

John Wiley & Sons, Inc.
New York · London · Sydney · Toronto

Library of Congress Catalog Card Number: 73-89684
SBN 471 39985 X
Printed in the United States of America.

10 9 8 7 6 5 4

To Betty

Preface

Successful process analysis involves both deterministic and statistical techniques. The estimation of coefficients in process models, the development of empirical models, and the design of efficient experiments should be tools as familiar to the scientist and engineer as are the techniques of solving equations and using computers. However, many students and practicing scientists and engineers, even while employing quite sophisticated mathematical techniques, treat their process calculations as if the processes were deterministic. Such an approach can be quite misleading and, in the design of equipment, results in the use of large safety or "fudge" factors to accommodate the reality of uncertainty. While the introduction of statistical techniques into process calculations may not always reduce the uncertainty, it can lead to more precise statements about the uncertainty and hence to better decision making.

In their discovery and application of the laws of nature, engineers and scientists are concerned with activities such as experimentation, operation of processes, design of equipment, trouble shooting, control, economic evaluation, and decision making. The concepts and statistical techniques incorporated in this book have been selected from the viewpoint of their pertinence to these activities. This text differs from others that describe the applications of statistics primarily in the direction of its emphasis rather than the specific statistical techniques discussed. The emphasis here is on process model building and evaluation rather than on statistical theory or the applications of theory to pseudorealistic experiments. The term process analysis as used in this text does not refer to the theory of random walks. Brownian motion, Markov processes, queuing theory, or similar random phenomena. Instead it refers to the analysis by statistical techniques of continuous industrial processes typified by the chemical, petroleum, and food industries, or of continuous natural processes such as river flows and biological growth and decay.

This book is divided into three major parts. Because the book is designed for the reader who has had minimal initial contact with the theory or application of statistics, Part I reviews the necessary background material underlying the other two parts. Part I is not a broad exposition of statistics but simply a description of the terminology and tools of analysis. Part II treats the topics of how to build and evaluate empirical models and how to design experiments effectively. Part III is concerned with the estimation of model parameters and the identification of process models that are based on transport phenomena principles. Certain of the later chapters adopt the view that a digital, or perhaps hybrid, computer is available to relieve the analyst of much of the tedious detailed calculations, permitting him to concentrate on the more productive role of evaluation and interpretation.

Since the level of the text is aimed at the college junior or senior engineer, it is assumed that the reader has a firm grasp of calculus and differential equations, as well as an elementary acquaintance with simple matrix algebra and operational transforms. Nevertheless, the appendices summarize the essential aspects of these latter two topics inasmuch as many college students seem not to have encountered them. Several topics are first introduced without matrix notation and then repeated in matrix notation because the redundancy has proved pedagogically more effective than either approach alone.

The objective of the book is to enable the engineer or scientist to orient his views in the context of what he knows about deterministic design and analysis so as to accommodate the concept of randomness in process variables. Consequently, many topics encountered in other texts, topics with considerable intrinsic merit, have been omitted. The choice of topics has been governed primarily by one question: Is the information or technique of any practical use in process analysis? Special attention has been paid, insofar as possible, to discussing what happens if the assumptions made about the process model are not fulfilled in practice and to illustrating some nonideal experimental data encountered in practice.

It is my hope that this book will prove of value to undergraduates who have the opportunity to take just one course in statistics as well as to research workers and engineers who would like to apply statistical techniques in their work but are unable to labor through numerous books and technical articles in order to do so.

David M. Himmelblau
Austin, Texas
1969

Contents

PART I STATISTICAL BACKGROUND FOR PROCESS ANALYSIS

PART II DEVELOPMENT AND ANALYSIS OF EMPIRICAL MODELS

APPENDIX

PART I

Statistical Background for Process Analysis

The purpose of Part I is to describe certain statistical techniques that are useful in process analysis. These initial chapters are by no means a comprehensive introduction to statistical analysis. Instead, they are intended to present primarily those facets of analyzing experimental data that are needed to understand the subsequent material on the design of experiments and empirical modelling.

CHAPTER 1

Introduction

Techniques of process analysis which take into account the existence of error in the process variables and coefficients can be implemented separately or in conjunction with techniques which ignore the error. To make correct decisions in the face of uncertainty, the analyst must be able to choose rationally from among alternatives. Hence, when process error exists, scientific decision making requires additional skills on the part of the analyst. The objective of the analysis may be to test a hypothesis, to develop a suitable relationship among variables, or perhaps to arbitrate a disputed decision. But no matter what the objective of experimentation and subsequent analysis, the tools of analysis to a large extent make use of the discipline of statistics.

There is no doubt that modern developments in digital computers have made the analysis of data considerably less tedious and enhanced the ability of the analyst to treat complex problems. Recent developments in data processing, display techniques, and pattern recognition suggest even more revolutionary things to come. If the analyst is to take advantage of these events, he must acquire a dual capability. First, and most obvious, he must command a sound and versatile background in engineering and mathematics. Second, he must be perceptive enough to find where the techniques described in this book can be effectively employed. The latter is by no means an unimportant attribute.

In this introductory chapter we shall briefly define some of the terminology to be used, examine how stochastic processes differ from deterministic ones, and classify the mathematical models used to represent real processes. This chapter is designed to demonstrate how real processes intermesh with their more formal representations, and to indicate under what circumstances statistical techniques can be introduced.

1.1 TERMINOLOGY AND CLASSIFICATION OF MODELS

By *process analysis* we refer to the application of scientific methods to the recognition and definition of problems and to the development of procedures for their solution. In more detail, this means: (1) mathematical specification of the problem for the given physical situation, (2) detailed analysis to obtain mathematical models, and (3) synthesis and presentation of results to ensure full comprehension. The *process* denotes an actual series of operations or treatments of materials as contrasted with the *model*, which is a mathematical description of the real process.

Models are used in all fields—biology, physiology, engineering, chemistry, biochemistry, physics, and economics. It is probably impossible to include under one definition all the varied connotations of the word model, but here we are concerned with mathematical descriptions of processes that aid in analysis and prediction.

Deterministic models or elements of models are those in which each variable and parameter can be assigned a definite fixed number, or a series of fixed numbers, for any given set of conditions. In contrast, in *stochastic* or *random* models, uncertainty is introduced. The variables or parameters used to describe the input–output relationships of the process and the structure of the elements (and the constraints) are not precisely known. Stochastic variables and models will be examined in more detail in Section 1.2.

Three very general types of models (and their combinations) can be written for a process

1. *Transport phenomena models*—use of physicochemical principles.
2. *Population balance models*—use of population balances.
3. *Empirical models*—use of empirical data fitting.

Examples of transport phenomena models are the phenomenological equations of change, that is the continuum equations describing the conservation of mass, momentum, and energy. Residence time distributions and other age distributions are examples of population balance models. Finally, examples of typical empirical models are polynomials used to fit empirical data.

Table 1.1-1 classifies transport phenomena models from the viewpoint of the complexity of the physical

TABLE 1.1-1 CLASSIFICATION OF MODELS BASED ON TRANSPORT PHENOMENA ACCORDING TO THE DEGREE OF PHYSICAL DETAIL

Stratum of Physico-chemical Description	Extent of Use by Engineers	Topical Designations	Typical Parameters for Analysis
Molecular and atomic	Fundamental background	Treats discrete entities; quantum mechanics, statistical mechanics, kinetic theory	Distribution functions; collision integrals
Microscopic	Applicable only to special cases	Laminar transport phenomena; statistical theories of turbulence	Phenomenological coefficients; coefficients of viscosity, diffusion, thermal conduction; Soret coefficient
Multiple gradient	Applicable only to special cases	Laminar and turbulent transport phenomena; transport in porous media	"Effective" transport coefficients
Maximum gradient	Used for continuous flow systems; "plug flow"	Laminar and turbulent transport phenomena, reactor design	Interphase transport coefficients; kinetic constants
Macroscopic	Very widely used	Process engineering; unit operations; classical kinetics and thermodynamics	Interphase transport coefficients; macroscopic kinetic constants; friction factors

detail drawn into the model; the degree of detail about a process decreases as we proceed down the table. Examples of specific models can be found in the tables and examples in Part III.

Table 1.1-2 is an alternate classification of transport phenomena models made from the viewpoint of the nature of the equations appearing in the model; hence, it is oriented toward the *solution* of models. As a rough guide, the complexity of solving the mathematical model roughly increases as we go down Table 1.1-2. *Steady state* means that the accumulation term (time derivative) in the model is zero. A *lumped parameter* representation means that spatial variations are ignored; the various properties and the state (dependent variables) of the system can be considered homogeneous throughout the entire system. A *distributed parameter* representation, on

TABLE 1.1-2 CLASSIFICATION OF DETERMINISTIC TRANSPORT PHENOMENA MODELS BASED ON MATHEMATICAL STRUCTURE

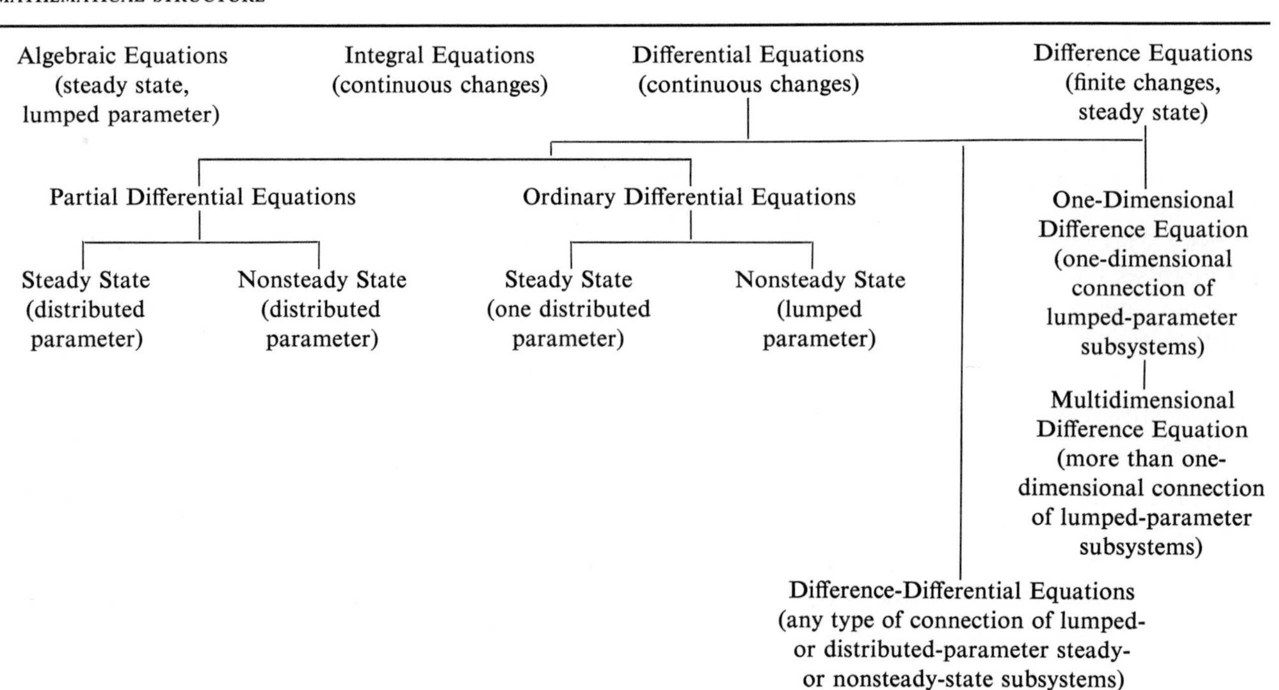

the other hand, takes into account detailed variations in behavior from point to point throughout the system. All real systems are, of course, distributed in that there are some variations throughout them. Often, however, the variations are relatively small, so they may be ignored and the system may then be "lumped."

As used in this text, a *system* is a process or part of a process selected by the engineer for analysis; subsystems (or elements) are further subdivisions of the system. The concept of a system does not necessarily depend upon the apparatus in which a process takes place nor upon the nature of the process itself. Instead, the concept is an arbitrary one used by the engineer to isolate a process, or a part of a process, for detailed consideration. For example, a packed distillation column is usually treated as a system, whereas a plate distillation column is treated as a system composed of subsystems of individual stages. There is nothing inviolable about this treatment, because a packed column can be considered as a staged process if desired, and a plate column can be considered as a continuous entity.

If the output y of a subsystem is completely determined by the input x, the parameters of the subsystem, and the initial and boundary conditions, in a general sense we can represent the subsystem symbolically by

$$y = \mathscr{H}x \tag{1.1-1}$$

The operator $\mathscr{H}$ represents any form of conversion of x into y. Suppose now two separate inputs are applied simultaneously to the subsystem so that

$$y = \mathscr{H}(x_1 + x_2) = \mathscr{H}(x_1) + \mathscr{H}(x_2) = y_1 + y_2 \tag{1.1-2}$$

Operator $\mathscr{H}$ is then, by definition, a linear operator, the properties of which are described in more detail in Appendix B. A system is termed linear if its operator $\mathscr{H}$ is linear, and the model of a linear system, which is represented by linear equations and boundary conditions, is called a linear model. Otherwise, the model is nonlinear.

Further details on the classification and application of process models can be found in the text by Himmelblau and Bischoff.†

1.2 STOCHASTIC VARIABLES AND MODELS

In real life most measurements or experiments result in values of the measured variables which vary from one repetition of the experiment to another. These outcomes are termed *random*, *stochastic*, chance, statistical, or probabilistic, depending upon the author and his particular emphasis; the associated variables are termed *random* or *stochastic variables*.

† D. M. Himmelblau and K. B. Bischoff, *Process Analysis and Simulation*, John Wiley, New York, 1967.

Many reasons exist why observations or measurements obtained by experiment are random rather than deterministic. In some cases the randomness rests on physical phenomena, such as the decay of a radioactive species or the emission of electrons from a thermionic cathode, processes which take place on a molecular or atomic scale but which are measured with macroscopic devices. In other cases there is insufficient information about the variable or a lack of techniques to gather the required information, so only certain manifestations are observed. Often the observer is just negligent or careless. Under actual plant conditions, process noise, cycling, signal noise, and other phenomena interfere with all measurements. Figure 1.2-1 illustrates the record of a feedwater flow transmitter as the time travel of the pen is speeded up. The top figure is a typical industrial recorder; in the middle figure the signal becomes clearer; in the bottom figure the 60-cycle noise inherent in the apparatus becomes evident but intrinsic variability still remains. Finally, uncertainty exists because the process models do not adequately represent the physical process. In general, basic indeterminacy in measurements is a phenomenon the analyst faces in all of his work.

The "true" value of a variable is that value which would be obtained on measurement if there were no stochastic feature associated with the measuring. Hence the true value of a process variable is in one sense a hypothetical value which is postulated as existing. Tied in with the concept of a true value is the concept of error, because an "error" represents the difference between the actual measurement and the true value. Therefore, a *random error* is an error which represents the difference between a random variable and its true value.

Random outcomes obtained by experiment thus incorporate error or uncertainty. This type of error must be distinguished from: (1) a large, one of a kind, isolated error which might be called a "blunder," and (2) an error introduced continuously and due, say, to faulty calibration of an instrument or to a preconceived idea of the expected data. This latter type of error causes *bias* or lack of accuracy and is termed a *systematic error*. *Accuracy* refers to how close the average value of the experimental data is to the "true" value; *precision* refers to how widely the individual data points are scattered about their average value. Systematic errors cannot be treated by the methods presented in this book.‡

Thus, experiments can be viewed as having different outcomes, ζ_i; each experiment can be assigned a function

‡ For treatment of such errors refer to: W. J. Youden, *Technometrics* **4**, 111, 1962; and W. J. Youden, *Physics Today* **14**, 32, Sept. 1961. An extended discussion of accuracy and precision can be found in an article by C. Eisenhart, *J. Res. Nat. Bur. Standards* **67C**, 161, 1963.

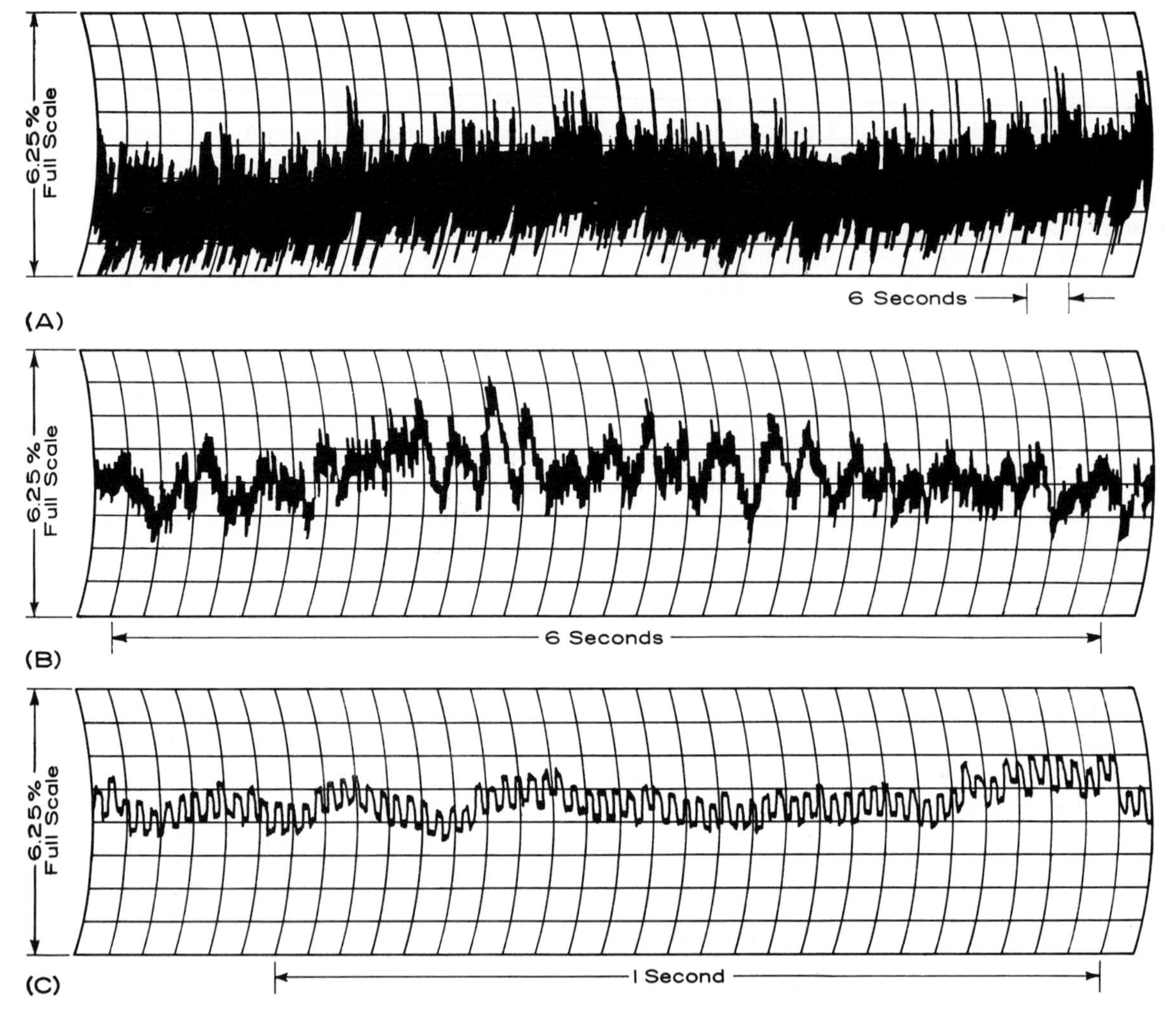

FIGURE 1.2-1 Field data taken from a differential pressure transmitter on three different time scales. (From B. D. Stanton, *ISA J.*, p. 77, Nov. 1964.)

of time, $X(t, \zeta)$, which may be real or complex. The family (collection) of all possible functions $X(t, \zeta)$ is commonly termed a stochastic or random process. In this text, however, a *stochastic process* will refer to a physical operating process which demonstrates stochastic characteristics because it includes a random input, output, coefficient, initial or boundary conditions, or any combination thereof. The term *ensemble* will be given to the family of functions $X(t, \zeta)$ which are the collection of all possible time records of experiments. Figure 1.2-2 portrays three *sample functions* (sample records) from the ensemble for the same variable observed over a finite time interval. The graphs may represent repeated runs on the same apparatus or simultaneous runs on identical apparatus.

The ensemble itself is a random variable as is a single time record and as is a group of experiments at one time. Some stochastic variables can be expressed as explicit functions whereas others can be defined only by graphical or tabular data. In what follows we shall suppress the notation of ζ in the argument of X and simply use $X(t)$ to denote both:

1. The ensemble (the collection of time functions).
2. A single function for one experiment in time, in general.

A subscript number will be used to distinguish one variable (either deterministic or stochastic) from another, and occasionally to distinguish one time record from another such as $X_2(t)$ in Figure 1.2-2. The particular meaning will be clear from the text. The random variable at a given time will be denoted by a subscript on t, such as $X(t_1)$, or by the absence of (t) as the argument of X if the variable is independent of t. In many instances it will be necessary to distinguish between the random variable itself, X, and the value of the variable by using lower case letters for the value. Random variables for the most part will be designated by capital letters taken from the latter half of the alphabet. However, some well-accepted symbols for the random variables, such as for

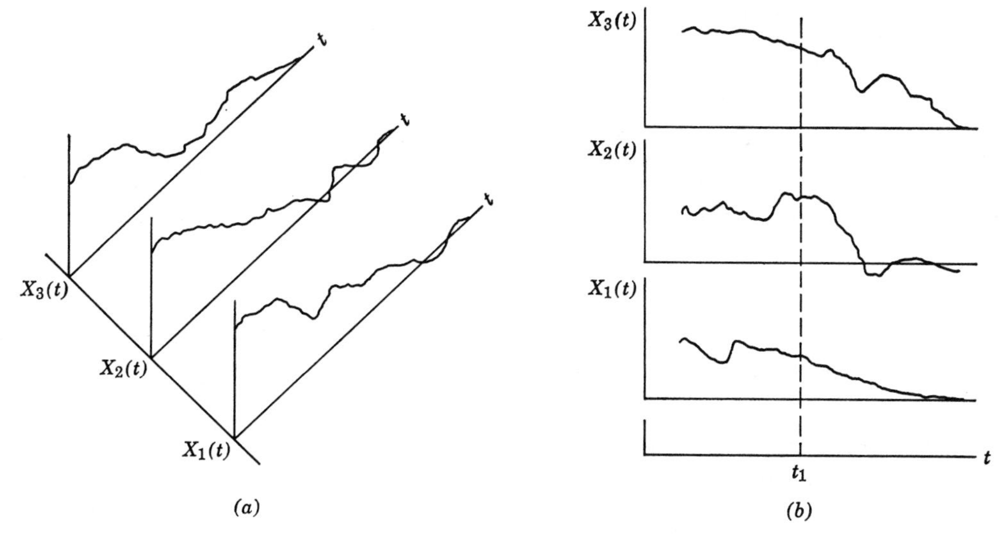

FIGURE 1.2-2 Sample random functions from an ensemble showing part of the ensemble $X(t)$: (a) three-dimensional representation and (b) two-dimensional representation.

the estimated variance, will be lower case. Deterministic variables will in general be lower case except for special engineering symbols such as the absolute temperature. The above description, like all other descriptions of a random variable, gives very little insight as to its nature or to the kinds of calculations that can be carried out on stochastic processes. Such insight and analytic knowledge can only come with further experience.

A *stochastic model* is nothing more than a mathematical representation of a stochastic process. Figure 1.2-3 illustrates the information flow for two simple stochastic models. In Figure 1.2-3b, a random error is added to the output of the deterministic model of Figure 1.2-3a to give a random output. In Figure 1.2-3c, a random input is introduced into the model to yield a random output. It would be quite possible for the differential equation(s) in the model to be stochastic because of a random coefficient. The process dependent and independent variables may be either *continuous* or *discrete*. Most, but not all, of the variables associated with continuous processes are continuous variables such as temperature, pressure, and composition—variables that can assume any values within an interval. A discrete variable can take on only distinct values in an interval.

Stochastic models can be classified in an arrangement similar to that shown in Table 1.1-2 or Figure 1.2-4. The terms in Figure 1.2-4 are discussed in Chapters 2 and 12. While the stochastic model may only be an abstraction of the real process, it presumably represents the process with reasonable faithfulness for the variable(s) of interest. As long as the model represents the real situation sufficiently well so that the conclusions deduced from mathematical analysis of the model have the desired precision, the model is adequate. The advantages of working with the model rather than with the experimental results directly are:

1. Relationships in the model can be precisely stated and manipulated mathematically; in the actual world the relationships among the process variables hold only approximately.

2. The model concentrates attention on relevant features of the process while removing from consideration many perplexing and unimportant features not subject to rigorous analysis.

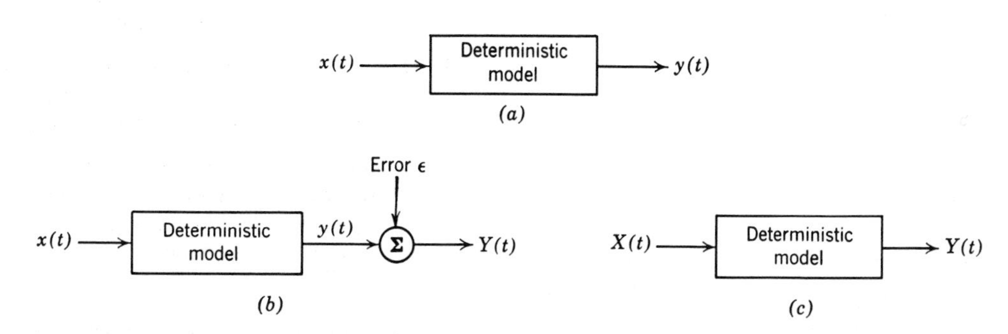

FIGURE 1.2-3 Block diagram representation of stochastic models.

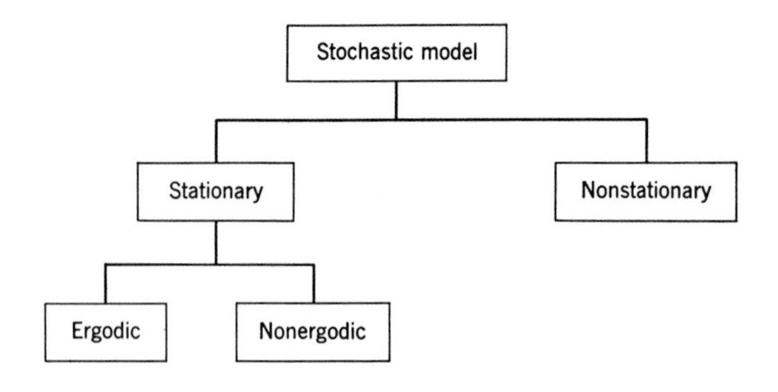

FIGURE 1.2-4 Alternate classification of stochastic models.

3. The model can be used to infer what will happen in the domain in which experimental observations are lacking.

Assuming that only random errors and not systematic errors are present in measurements of a process variable, the analyst is most interested in determining on the basis of a finite number of measurements: (1) the central tendency of the observations of a given variable, (2) the dispersion of the observations about a central value, and (3) the uncertainty in these estimates. The central value is usually characterized by the ensemble mean and estimated by the sample mean or by a time average. The dispersion is characterized by the ensemble variance, which can be estimated from a sample variance or a suitable time average. In the next chapter we shall consider these descriptive statistics that enable the experimenter to reduce a mass of information into a compact form.

Supplementary References

Formby, John, *An Introduction to the Mathematical Formulation of Self-organizing Systems*, D. Van Nostrand, Princeton, N.J., 1965.

Himmelblau, D. M. and Bischoff, K. B., *Process Analysis and Simulation*, John Wiley, New York, 1968.

Papoulis, A., *Probability, Random Variables, and Stochastic Processes*, McGraw-Hill, New York, 1965.

Petersen, E. L., *Statistical Analysis and Optimization of Systems*, John Wiley, New York, 1961, Chapters 1–3.

Problems

1.1 Indicate the appropriate classification of models in terms of Table 1.1-1 for each of the following cases.

(a) Laminar flow through a circular tube

$$\frac{1}{r}\frac{d}{dr}r\frac{dv_z}{dr} = -\left(\frac{\Delta p}{L}\right)$$

where:

r = radial direction

v_z = velocity in axial direction

$\frac{\Delta p}{L}$ = pressure drop

(b) Heat conduction in an infinite cylinder

$$\frac{\partial T}{\partial t} = \frac{\alpha}{r}\frac{\partial}{\partial r}r\frac{\partial T}{\partial r}$$

where:

T = temperature

r = radial direction

α = constant

(c) Heat transfer in a jacketed kettle

$$q = UA\,\Delta T$$

where:

q = heat transfer

U = constant

A = area for heat transfer

ΔT = temperature difference

1.2 What kind of model (lumped or distributed parameter) is represented by the following cases.

(a) Heat transfer with flow

$$\frac{\partial T}{\partial t} + u\frac{\partial T}{\partial z} = a\frac{\partial^2 T}{\partial z^2} + Q$$

(b) Mass transfer in tank

$$\frac{dc}{dt} + ac = w(t)$$

(c) Dispersion in packed tube

$$\frac{d^2 z}{dx^2} + a\frac{dz}{dx} + bz = w(x)$$

1.3 Classify each equation in Problem 1.1 (or 1.2) in one of the categories listed in Table 1.1-2.

1.4 In carrying out measurements on a process variable, how is it possible to ascertain whether or not the variable is stochastic or deterministic?

1.5 Helgeson and Sage† reported the data in Table P1.5 for the heat of vaporization of propane. Does the average deviation have a bearing on the accuracy or the precision of the reported data?

† N. L. Helgeson and B. H. Sage, *J. Chem. Eng. Data* **12**, 47, 1967.

TABLE P1.5

Author	Number of Data Points Used	Temperature Range Min	Temperature Range Max	Average Deviation of All Points for Heat of Vaporization
A	14	100	135	1.12
B	16	103	167	1.43
C	4	100	190	0.98

1.6 Explain two ways in which the deterministic process input $x(t) = a \cos \omega t$ could be made into a stochastic input.

1.7 What is one additional way in which error can be introduced into a process model besides the methods illustrated in Figure 1.2-3?

1.8 Is the error introduced by a numerical integration scheme for the solution of a model represented by a differential equation a stochastic (random) error? Is the truncation error introduced by the partial difference approximation to the differential equation in the process model a stochastic error?

1.9 A thermocouple is placed in a tank of water and the leads attached to a potentiometer. List some of the random errors that will appear in the observed voltage.

1.10 The following figure represents the relative frequency distribution of measurements of a presumed random variable. Can you tell from the graph whether or not the measurements are biased? Explain.

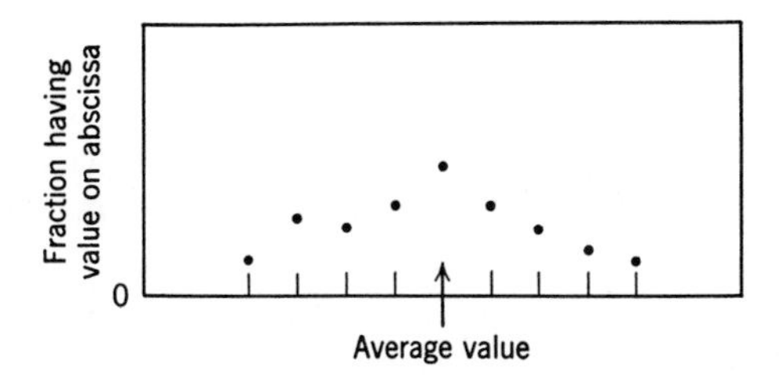

1.11 Is a function of a random variable also a random variable? Explain.

CHAPTER 2

Probability Distributions and Sample Statistics

Probability, according to the frequency theory of probability (see Appendix A), corresponds to the longrun fraction of a specific outcome from among all possible outcomes of an experiment. Other semantic relations between the experiment and the mathematical representation of the experiment are shown below.

Experiment	Mathematical Representation
Random outcome	Random variable
List of experimental outcomes	Sample space
All possible outcomes	Population
Asymptotic relative frequency of an outcome ("in the long run")	Probability of an event
List of asymptotic relative frequencies of each outcome	Probability (density) function
Cumulative sum of relative frequencies	Probability distribution

What the analyst would like to do is replace a large mass of experimental data by a few easily grasped numbers. Under favorable circumstances, he is able to associate the experimental data with a known mathematical function, a probability function, or density, which corresponds reasonably well with the relative frequency of the data. Then he can use the probability function or density to make various predictions about the random variable which is the subject of experimentation. Often, however, only a modest amount of experimental data is available, and it is of such a nature that the experimentalist can at the best make estimates of the ensemble mean and perhaps the ensemble variance of the random variable.

We shall describe a few of the most useful probability density functions in this chapter. In addition we shall describe some of the characteristics of ensemble averages such as the mean, variance, covariance, and correlation coefficient, all of which have applications in process analysis. Then we shall look at the first of the two principal methods of estimating ensemble averages, namely (1) sample averages and (2) time averages. Included in the presentation will be selected sampling distributions which will be of aid in subsequent discussions of interval estimation and hypothesis testing. Time averages will be taken up in Chapter 12.

2.1 PROBABILITY DENSITY FUNCTIONS AND PROBABILITY DISTRIBUTIONS

To simplify the notation we shall denote the *probability distribution function* of $X(t)$ by

$$P\{X(t) \leq x\} \equiv P(x; t) \qquad (2.1\text{-}1)$$

where x is a number. Thus, in Equation 2.1-1 the argument on the left-hand side reads: "all of the values of the random variable $X(t)$ less than or equal to a deterministic variable x." The reason for using the symbol x rather than some constant k is that in many applications the limiting quantity will itself be a deterministic variable. $P(x; t)$ is sometimes termed a first-order probability distribution function because the probability distribution involves only one random variable at a time.

We can give a physical interpretation to $P(x; t)$ from a frequency point of view. Suppose we carry out an experiment by measuring the temperature of a fluid many times. We secure a number of records comprising a family of curves of $X(t)$, some of which are shown in Figure 2.1-1. From each record at time $t = t_1$, we note whether or not $X(t_1) \leq x$. Let the total number of time records at t_1 for which $X(t_1) \leq x$ be n_{t_1} and the total number of records be N. In the limit as $N \rightarrow \infty$,

$$\underset{N\rightarrow\infty}{P(x; t)} = \frac{n_{t_1}}{N}$$

Clearly, $P(x; t)$ ranges between 0 and 1. Figure 2.1-2*a* illustrates the probability distribution function which might be observed if the distribution is a function of

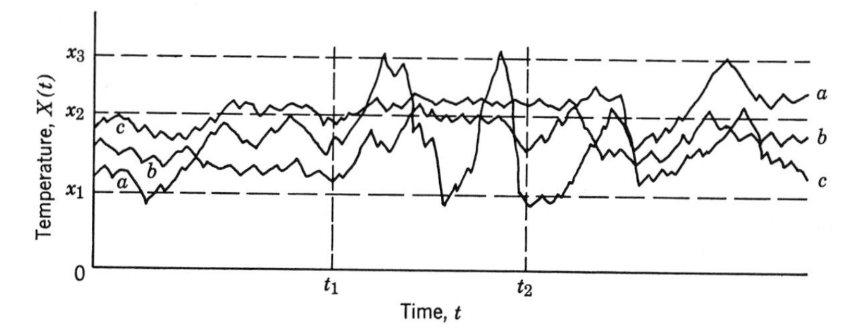

FIGURE 2.1-1 Repeated measurements of fluid temperature at a fixed point; x_1, x_2, and x_3 represent different levels of the random variable $X(t)$.

time. Figure 2.1-2*b* illustrates the case in which the probability distribution is independent of time.

Imagine now that we had examined the experimental records both at $t = t_1$ and at another time, $t = t_2$. Then the joint distribution of the random variables $X(t_1)$ and $X(t_2)$ can be denoted by

$$P(x_1, x_2; t_1, t_2) \equiv P\{X(t_1) \leq x_1; X(t_2) \leq x_2\} \quad (2.1\text{-}2)$$

where $P(x_1, x_2; t_1, t_2)$ is known as the second-order probability distribution of the variable $X(t)$, and x_1 and x_2 are two numbers. The qualification "second order" refers to the joint distribution of the same random variables observed at two different times. From the frequency viewpoint, $P(x_1, x_2; t_1, t_2)$ is the limit as $N \to \infty$ of the joint event $\{X(t_1) \leq x_1\}$ and $\{X(t_2) \leq x_2\}$ in a two-dimensional space. If X does not vary with time, then the functional dependency on t can be omitted.

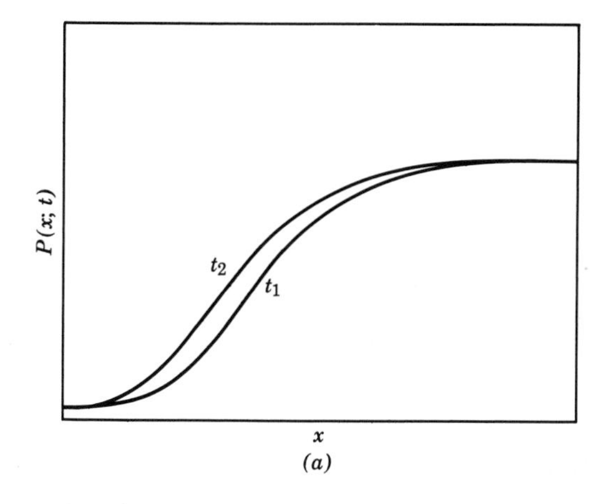

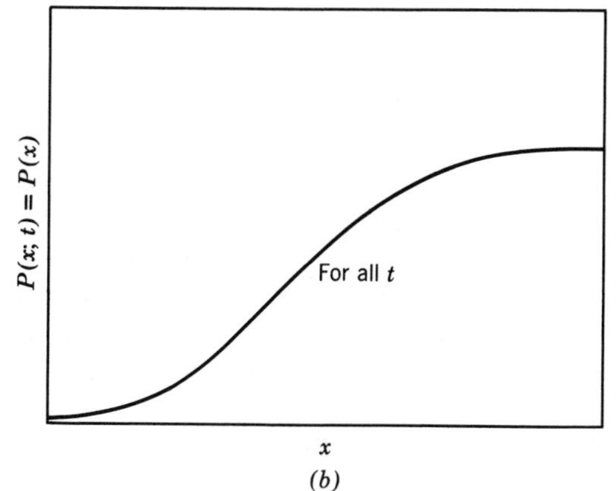

FIGURE 2.1-2 Probability distributions: (*a*) probability distribution as a function of time, and (*b*) probability distribution independent of time.

Corresponding to each probability distribution is a *probability density function* defined as follows:

$$p(x; t) = \frac{\partial P(x; t)}{\partial x} \quad (2.1\text{-}3a)$$

$$p(x_1, x_2; t_1, t_2) = \frac{\partial^2 P(x_1, x_2; t_1, t_2)}{\partial x_1 \, \partial x_2} \quad (2.1\text{-}3b)$$

Note that the lower case p designates the probability density, whereas the capital P designates the probability distribution. Figure 2.1-3 illustrates typical process records and their corresponding first-order probability density functions. The reason for the term "density" becomes meaningful if it is observed that in order for $P(x; t)$ to be dimensionless, the units of $p(x; t)$ must be the reciprocal of the units of x; that is, $p(x; t)$ is the probability per unit value of x. (In some texts the notation $p(x; t)\,dx$ is employed to denote the probability that x lies in the interval x to $x + dx$).

Up to this point we have been concerned with the probability density function and the probability distribution function for *continuous* variables. The *probability function* (not a density) for a *discrete* variable $X(t)$ is $P(x_k; t) \equiv P\{X(t) = x_k\}$, and the probability distribution function is a sum rather than an integral

$$P(x; t) = P\{X(t) \leq x_k\} = \sum_{i=1}^{k} P(x_i; t)$$

The relation between the probability density and the probability distribution for a continuous variable can also be expressed as

$$P(x; t) = \int_{-\infty}^{x} p(x'; t)\, dx'$$

$$P(x_2; t) - P(x_1; t) = \int_{x_1}^{x_2} p(x; t)\, dx$$

$$= P\{x_1 \leq X(t) \leq x_2\}$$

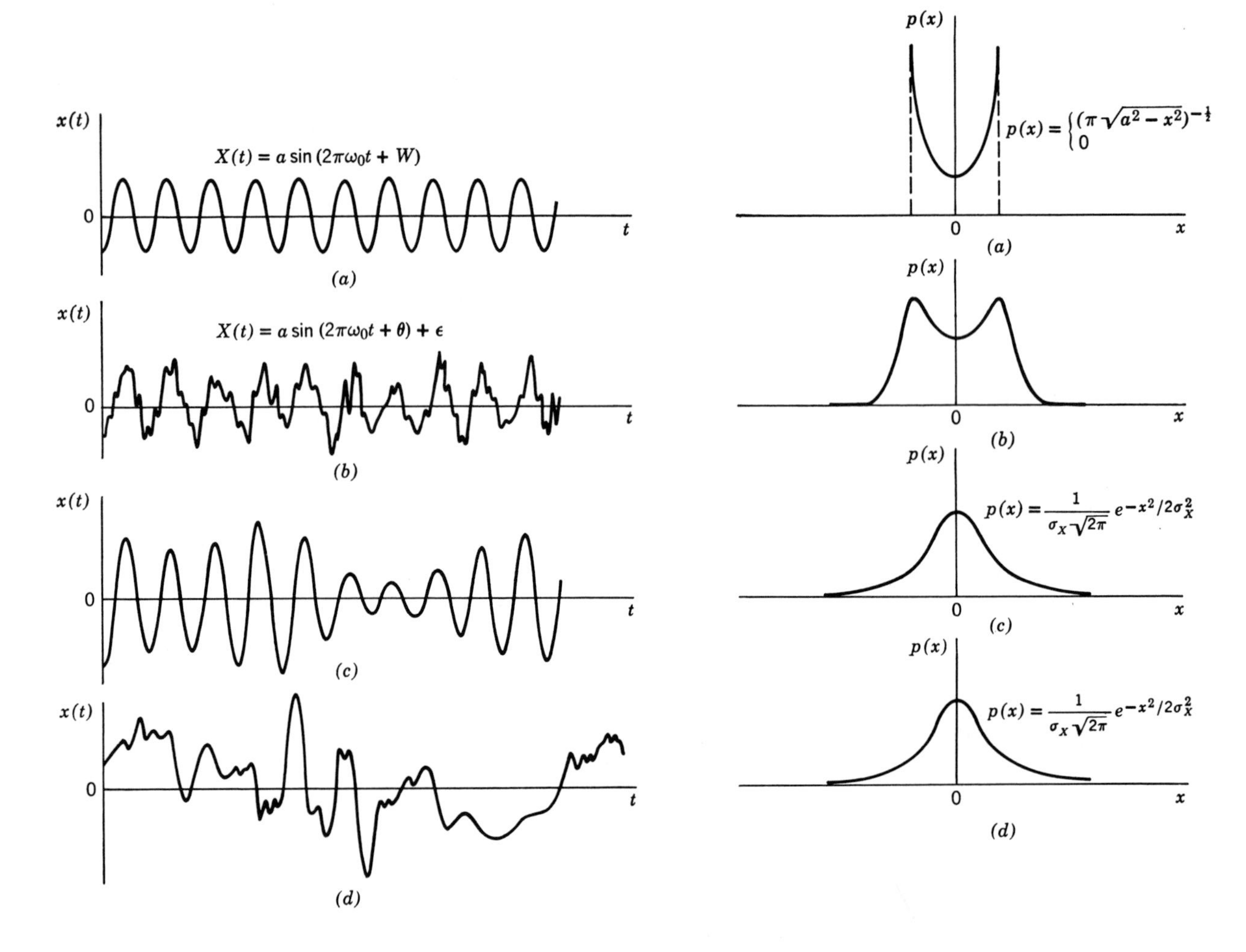

FIGURE 2.1-3 Typical process records (left) and their corresponding (time-independent) probability densities (right): (*a*) sine wave (with random initial phase angle W, (*b*) sine wave plus random noise, (*c*) narrow-band random noise, and (*d*) wide-band random noise. (From J. S. Bendat and A. G. Piersol, *Measurement and Analysis of Random Data*, John Wiley, New York, 1966, pp. 17–18.)

(where the primes are dummy variables). Consequently, $P\{X = x_0\} = 0$ since the interval for integration is zero. In addition, by definition,

$$\int_{-\infty}^{\infty} p(x; t)\, dx \equiv 1$$

Similar relations can be written for the second-order probability distribution. The relation between the first- and second-order densities is

$$p(x_1, t_1) = \int_{-\infty}^{\infty} p(x_1, x_2; t_1, t_2)\, dx_2 \qquad (2.1\text{-}4)$$

$p(x_1, t_1)$ is called the *marginal probability density function* of $X(t_1)$, i.e., the probability density of $X(t_1)$ irrespective of the values assumed by $X(t_2)$.

A joint probability distribution between two different random variables, say $X(t)$ and $Y(t)$, termed a *bivariate distribution*, can be written as

$$P(x, y; t) = P\{X(t) \le x;\ Y(t) \le y\}$$
$$= \int_{-\infty}^{x} \int_{-\infty}^{y} p(x', y'; t)\, dy'\, dx' \qquad (2.1\text{-}5)$$

Figure 2.1-4 illustrates two typical time-independent bivariate probability density functions.

In later chapters we shall make use of the *conditional probability density*. The conditional probability distribution of the random variable Y, assuming the random variable X is equal to the value x, is defined as

$$P(y \mid X = x) = \lim_{\Delta x \to 0} P(y \mid x < X \le x + \Delta x)$$

where the vertical line denotes "given." Then, by making use of the continuous variable analog of Equation A-8 in Appendix A for the upper and lower bounds,

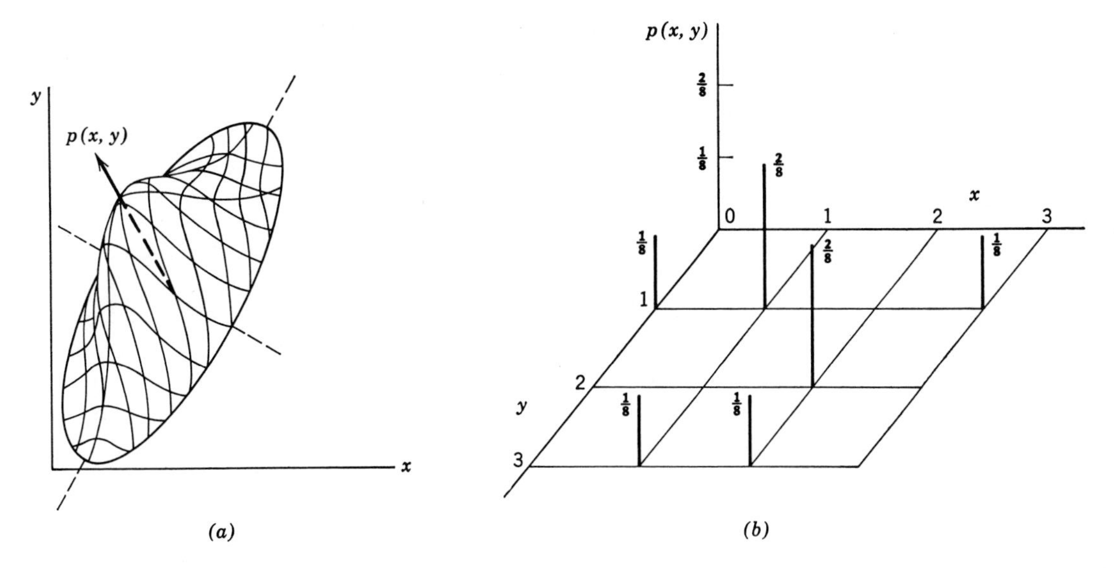

FIGURE 2.1-4 Illustrations of two time-invariant bivariate probability density functions: (*a*) bivariate probability density function for continuous variables, and (*b*) bivariate probability function for discrete variables.

$$P(y \mid X = x) = \lim_{\Delta x \to 0} \frac{P(x + \Delta x, y) - P(x, y)}{P(x + \Delta x) - P(x)}$$

$$= \frac{\partial P(x, y)/\partial x}{\partial P(x)/\partial x}$$

The corresponding probability density is obtained by differentiating $P(y \mid X = x)$ with respect to y:

$$p(y \mid X = x) = \frac{p(x, y)}{p(x)} \tag{2.1-6}$$

To simplify the notation, the conditional probability density is usually written as

$$p(y \mid x) \equiv p(y \mid X = x)$$

Because

$$p(y) = \int_{-\infty}^{\infty} p(x, y)\, dx$$

and the joint density from Equation 2.1-6 is

$$p(x, y) = p(y \mid x)p(x)$$

we can write

$$p(y) = \int_{-\infty}^{\infty} p(y \mid x)p(x)\, dx$$

In other words, to remove the condition $X = x$, we multiply the conditional density by the density of X and integrate over all values of X.

By generalization of Equation 2.1-5, an n-dimensional probability distribution function can be defined. The study of n *different* random variables $X_1, X_2, \ldots, X_n$ is equivalent to the consideration of a single n-dimensional random vector $\mathbf{X} = (X_1, X_2, \ldots, X_n)$. The one-dimensional variables $X_1, X_2, \ldots, X_n$ are said to be stochastically *independent*† if, for all permissible values of the variables and all joint distribution functions,

$$P(x) \equiv P(x_1, \ldots, x_n; t_1, \ldots, t_n) = P(x_1; t_1) \cdots P(x_n; t_n) \tag{2.1-7a}$$

An equivalent relation among the density functions is

$$p(x_1, \ldots, x_n; t_1, \ldots, t_n) = p(x_1; t_1) \cdots p(x_n; t_n) \tag{2.1-7b}$$

The analogous expression for independent discrete variables is an extension of Equation A-6 in Appendix A:

$$P(x_{1k}, \ldots, x_{nk}) \equiv P\{X_1(t) \le x_{1k} \text{ and } \cdots \text{ and } X_n(t) \le x_{nk}\}$$
$$= P(x_{1k}) \cdots P(x_{nk})$$

Example 2.1-1 Bivariate Distribution Function

Let $p(x, y; t_1, t_2) \ge 0$ be a bivariate probability density function for the two random variables $X(t)$ and $Y(t)$. Then $P(a_1 < X \le a_2, b_1 < Y \le b_2)$

$$= \int_{a_1}^{a_2} \int_{b_1}^{b_2} p(x, y; t_1, t_2)\, dy\, dx \tag{a}$$

$P(-\infty < X \le x, -\infty < Y \le y)$

$$= \int_{-\infty}^{x} \int_{-\infty}^{y} p(x', y'; t_1, t_2)\, dy'\, dx' \tag{b}$$

and

$$\int_{-\infty}^{\infty} \int_{-\infty}^{\infty} p(x, y; t_1, t_2)\, dy\, dx = 1 \tag{c}$$

† Stochastic independence of two random variables can be interpreted as follows. If the value of one variable is fixed, the probability of obtaining the value of the other variable is not affected. Continuous processes are notorious for having variables such that previous values do influence later values.

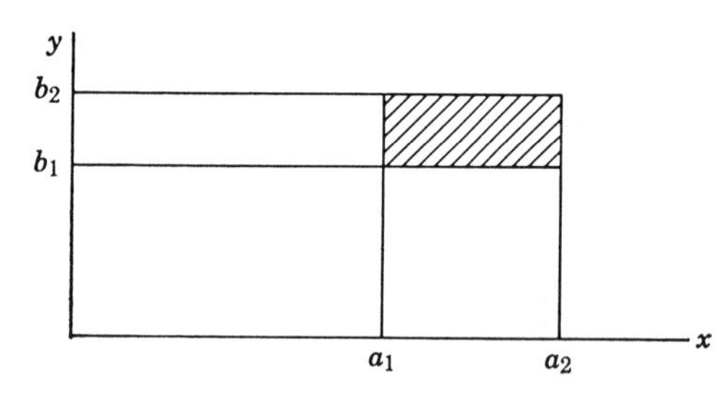

FIGURE E2.1-1A

To provide a simple numerical example, suppose that $p(x, y; t_1, t_2)$ is independent of time and is equal to

$$p(x, y; t_1, t_2) = e^{-(x+y)} \qquad \text{for } X \geq 0,\ Y \geq 0$$

$$p(x, y; t_1, t_2) = 0 \qquad \text{elsewhere}$$

Then

$$P(\tfrac{1}{2} < X < 2; 0 < Y < 4) = \int_0^4 \int_{1/2}^2 e^{-x} e^{-y}\, dx\, dy$$
$$= (e^{-1/2} - e^{-2})(1 - e^{-4}) = 0.462$$

Note also that Equation (c) holds true for

$$\int_0^\infty \int_0^\infty e^{-(x+y)}\, dx\, dy = \int_0^\infty e^{-x}[-e^{-y}]_0^\infty\, dy = 1$$

The lower limit of $-\infty$ can be replaced by zero because of the definition of $p(x, y)$.

The probability distribution $P(x, y)$ can be interpreted geometrically in terms of the rectangle of Figure E2.1-1*a*. Consider the following sets of events (E) in relation to Figure E2.1-1*a*.

$$E_1 = (X \leq a_2,\ Y \leq b_2)$$
$$E_2 = (X \leq a_1,\ Y \leq b_1)$$
$$E_3 = (X \leq a_2,\ Y \leq b_1)$$
$$E_4 = (X \leq a_1,\ Y \leq b_2)$$

The event of interest (denoted by the shaded area) can be written as

$$E = (a_1 < X \leq a_2, b_1 < Y \leq b_2)$$

Now, keeping in mind that the probability $P(E)$ corresponds to the double integral of the density over the designated region, from Figure E2.1-1a we conclude that

$$P(E) = P\{a_1 < X \leq a_2, b_1 < Y \leq b_2\}$$
$$= [P(E_1) - P(E_3)] + [P(E_2) - P(E_4)]. \qquad \text{(d)}$$

An important concept applied to a random variable is the idea of *stationarity*. A stochastic variable is termed *stationary in the strict sense* or *strongly stationary* if the probability density functions of all orders are invariant with respect to a shift in the time origin. In particular, if a is a constant, either positive or negative,

$$p(x; t) = p(x; t + a) = p(x) \qquad (2.1\text{-}8)$$

from which we can conclude that the first order of probability density function of a stationary process is *independent of time*. If we examine the second order of probability density, we can write

$$p(x_1, x_2; t_1, t_2) = p(x_1, x_2; t_1 + a, t_2 + a)$$
$$= p(x_1, x_2; \tau) \qquad (2.1\text{-}9)$$

where $\tau = t_2 - t_1$. Thus, if the variable $X(t)$ is stationary the second order density depends only on the *difference* between the times of observation and not on when the time record was initiated, a very important point.

Stationary random variables are far easier to treat than nonstationary ones. Nonstationary data, such as are illustrated in Figure 2.1-5, are obtained during unsteady-state operating conditions caused by a change in: (1) the input to a process, (2) a process parameter, or (3) the environment surrounding the process. Unfortunately, no general techniques exist which can be substituted for the ones used for stationary processes; each process or class of processes must be treated as a special case. In Section 3.7-5 we shall discuss tests to detect whether or not process variables are stationary.

2.2 ENSEMBLE AVERAGES: THE MEAN, VARIANCE, AND CORRELATION COEFFICIENT

The first type of average we shall consider is the *ensemble expectation* of a function $f[X(t_1), X(t_2), \ldots, X(t_n)]$ of a random variable $X(t)$, which is defined as

$$\mathscr{E}\{f\} = \int_{-\infty}^{\infty} \cdots \int_{-\infty}^{\infty} f[x(t_1) \cdots x(t_n)]$$
$$\times\ p(x_1, \ldots, x_n; t_1, \ldots, t_n)\, dx_1 \cdots dx_n \qquad (2.2\text{-}1)$$

where p is the joint density function and $\mathscr{E}$ stands for expected value. Note that $\mathscr{E}\{f\}$ is *not* a random variable, but may be a function of $t_1, \ldots, t_n$.

Each ensemble average is a function describing certain characteristics about the random variable $X(t)$ such as its central tendency or dispersion or it is a function from which these characteristics can be derived. As is the common practice, we shall not specifically include the word ensemble in the name of the function each time but shall imply ensemble by the symbol for the function.

To carry out operations such as differentiation and integration on the ensemble variable $X(t)$ calls for some special definitions for continuity and convergence which need not be of concern here. However, in order to reduce the number of algebraic manipulations in subsequent sections, we shall list here a few simple rules for linear operators acting on random variables. Mathematical proofs of these rules can be found in most texts on statistics or random processes.

If $\mathscr{H}$ is a linear time invariant operator (described in

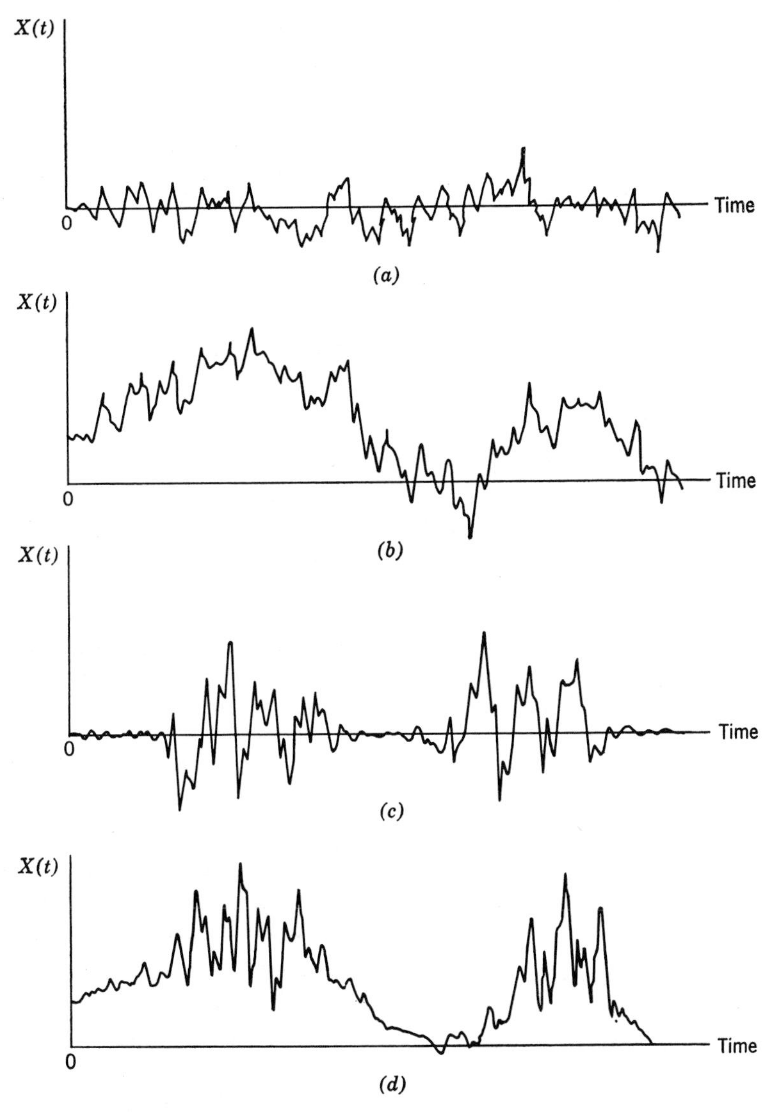

FIGURE 2.1-5 Examples of stationary and nonstationary data: (*a*) Stationary data, (*b*) time-varying mean value, (*c*) time-varying mean square value, and (*d*) time-varying mean and mean square value. (From J. S. Bendat and A. G. Piersol, *Measurement and Analysis of Random Data*, John Wiley, New York, 1966, p. 334.)

more detail in Appendix B), $X(t)$ is a random variable, and

$$Y(t) = \mathscr{H}[X(t)]$$

the procedure of taking an expected value is commutative with the linear operation

$$\mathscr{E}\{Y(t)\} = \mathscr{E}\{\mathscr{H}[X(t)]\} = \mathscr{H}[\mathscr{E}\{X(t)\}] \quad (2.2\text{-}1a)$$

Examples of linear operators are moments when they represent expected values, derivatives of the first degree, definite integrals, and sums. For example:

EXPECTED VALUE OF A DERIVATIVE.

$$\mathscr{E}\left\{\frac{dY}{dt}\right\} = \frac{d\mathscr{E}\{Y\}}{dt}$$

and

$$\mathscr{E}\left\{\frac{dY^{(n)}}{dt}\right\} = \frac{d\mathscr{E}\{Y^{(n)}\}}{dt} \quad (2.2\text{-}1b)$$

EXPECTED VALUE OF AN INTEGRAL. If

$$Y = \int_a^b X(t)\psi(t)\,dt$$

and ψ is a deterministic function

$$\mathscr{E}\{Y\} = \int_a^b \mathscr{E}\{X(t)\}\psi(t)\,dt = \int_a^b \mu_{X^{(t)}}\psi(t)\,dt \quad (2.2\text{-}1c)$$

where $\mu_{X^{(t)}} = \mathscr{E}\{X(t)\}$ as defined in Section 2.2-1.

EXPECTED VALUE OF A SUM. If

$$Y = \sum_{i=1}^{n} a_i X_i$$

then

$$\mathscr{E}\{Y\} = \sum_{i=1}^{n} a_i \mathscr{E}\{X_i\} = \sum_{i=1}^{n} a_i \mu_{X_i} \qquad (2.2\text{-}1d)$$

2.2-1 Mean

The *ensemble mean* of a stochastic variable is the expected value of the variable

$$\mu_X(t) = \mathscr{E}\{X(t)\} = \int_{-\infty}^{\infty} x p(x; t)\, dx \qquad (2.2\text{-}2)$$

If $p(x; t)$ is independent of time (X is stationary), then $\mu_X(t) = \mu_X$ is a constant. The ensemble mean is a measure of the central tendency of a random variable. In effect, it is the deterministic variable used in a process model when error is ignored.

The expected value of the sum of two random variables $X(t)$ and $Y(t)$, namely $W(t) = X(t) + Y(t)$, is

$$\mathscr{E}\{W(t)\} = \mathscr{E}\{X(t)\} + \mathscr{E}\{Y(t)\}$$

or

$$\mu_W(t) = \mu_X(t) + \mu_Y(t) \qquad (2.2\text{-}3)$$

The expected value of the product of two *independent* random variables, $Z(t) = X(t)Y(t)$, is

$$\mathscr{E}\{X(t)Y(t)\} = \mathscr{E}\{X(t)\}\mathscr{E}\{Y(t)\} \qquad (2.2\text{-}4)$$

because $p(x, y; t) = p(x; t)p(y; t)$ by Equation 2.1-7,

$$\mathscr{E}\{Z\} = \int_{-\infty}^{\infty}\int_{-\infty}^{\infty} xyp(x, y; t_1, t_2)\, dx\, dy$$

$$= \left[\int_{-\infty}^{\infty} xp(x; t)\, dx\right]\left[\int_{-\infty}^{\infty} yp(y; t)\, dy\right] = \mu_X(t)\mu_Y(t)$$

Example 2.2-1 Ensemble Mean

A special type of Brownian motion with negligible acceleration describes the movement of a particle hit by a large number of other particles in a fluid. On a molecular scale the motion is quite complicated, but on a macroscopic scale we are interested in determining the expected value of the motion denoted by the random variable $X(t)$. If for a one-dimensional motion the starting location is arbitrarily assigned a zero value, $X(0) = 0$, then the one-dimensional probability density of $X(t)$ is

$$p(x; t) = \frac{1}{\sqrt{2\pi\alpha t}} e^{-x^2/2\alpha t} \qquad (a)$$

where α is a constant. A typical section of path might be as shown in Figure E2.2-1.

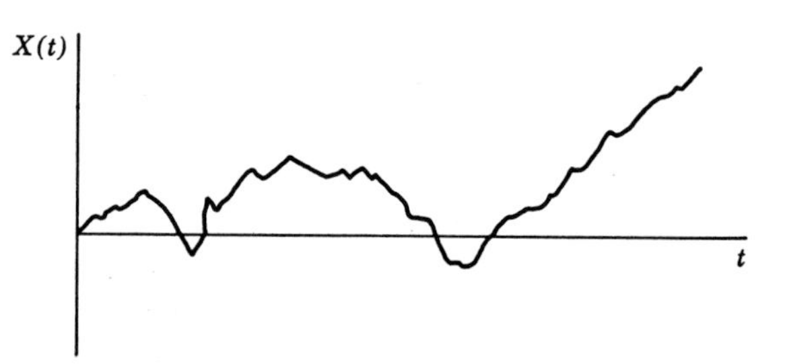

FIGURE E2.2-1

The expected value of $X(t)$ can be calculated using Equation 2.2-2,

$$\mu_X(t) = \int_{-\infty}^{\infty} \frac{x}{\sqrt{2\pi\alpha t}} e^{-x^2/2\alpha t}\, dx \qquad (b)$$

The integral can be split into two parts, one from $-\infty$ to 0 and the other from 0 to ∞, that cancel because the integrand is the product of an odd and an even function. Consequently,

$$\mathscr{E}\{X(t)\} = \mu_X(t) = 0 \qquad (c)$$

If, however, we inquire as to the expected value of the square of $X(t)$, we find that because the integrand is the product of two even functions,

$$\mathscr{E}\{X^2(t)\} = \int_{-\infty}^{\infty} \frac{x^2}{\sqrt{2\pi\alpha t}} e^{-x^2/2\alpha t}\, dx = \alpha t \qquad (d)$$

The expected value of the square of a random variable is used to indicate the intensity of the variable; the positive square root is commonly called the rms value.

Example 2.2-2 Ensemble Mean for a Dynamic Model of a Stochastic Process

Assume that a process is represented by a first-order linear differential equation in which the input $X(t)$ and the output $Y(t)$ are stochastic variables:

$$\frac{dY(t)}{dt} + aY(t) = X(t) \qquad Y(0) = 0$$

What is the $\mathscr{E}\{Y\}$?

Solution:

Take the expected value of both sides of the differential equation and the initial condition, and exchange the operations of the expected value and differentiation as indicated by Equation 2.2-1b:

$$\frac{d\mathscr{E}\{Y(t)\}}{dt} + a\mathscr{E}\{Y(t)\} = \mathscr{E}\{X(t)\} \qquad \mathscr{E}\{Y(0)\} = 0$$

If we let $\mu_Y(t) = \mathscr{E}\{Y(t)\}$ and $\mu_X = \mathscr{E}\{X(t)\}$ = a constant, then we can solve the deterministic ordinary differential equation

$$\frac{d\mu_Y(t)}{dt} + a\mu_Y(t) = \mu_X \qquad \mu_Y(0) = 0 \qquad (a)$$

The solution of Equation (a) is

$$\mu_Y(t) = \frac{\mu_X}{a}[1 - e^{-at}] \qquad (b)$$

Equation (b) is the usual deterministic solution found in texts on differential equations and on process analysis of deterministic processes.

2.2-2 Autocorrelation Function

The ensemble *autocorrelation function* of a random variable $X(t)$, $r_{XX}(t_1, t_2)$, characterizes the dependence of values of $X(t)$ at one time with values at another time:

$$r_{XX}(t_1, t_2) = \mathscr{E}\{X(t_1)X(t_2)\}$$
$$= \int_{-\infty}^{\infty}\int_{-\infty}^{\infty} x_1x_2p(x_1, x_2; t_1, t_2)\,dx_1\,dx_2 \quad (2.2\text{-}5)$$

Note that $r_{XX}(t_1, t_2)$ is not a random variable, and that we use a lower case Roman (rather than Greek) letter to conform to common usage. Figure 2.2-1 illustrates the autocorrelation functions for the process records shown in Figure 2.1-3. The most important uses in model building of the autocorrelation function are in data processing and parameter estimation as described in Chapter 12.

The autocorrelation function of a *stationary variable* is, by making use of Equation 2.1-9, solely a function of τ, the time difference $(t_2 - t_1)$

$$r_{XX}(t_1, t_2) = r_{XX}(\tau) = \mathscr{E}\{X(t + \tau)X(t)\}$$
$$= r_{XX}(-\tau)$$
$$= \int_{-\infty}^{\infty}\int_{-\infty}^{\infty} x_1x_2p(x_1, x_2, \tau)\,dx_1\,dx_2 \quad (2.2\text{-}6)$$

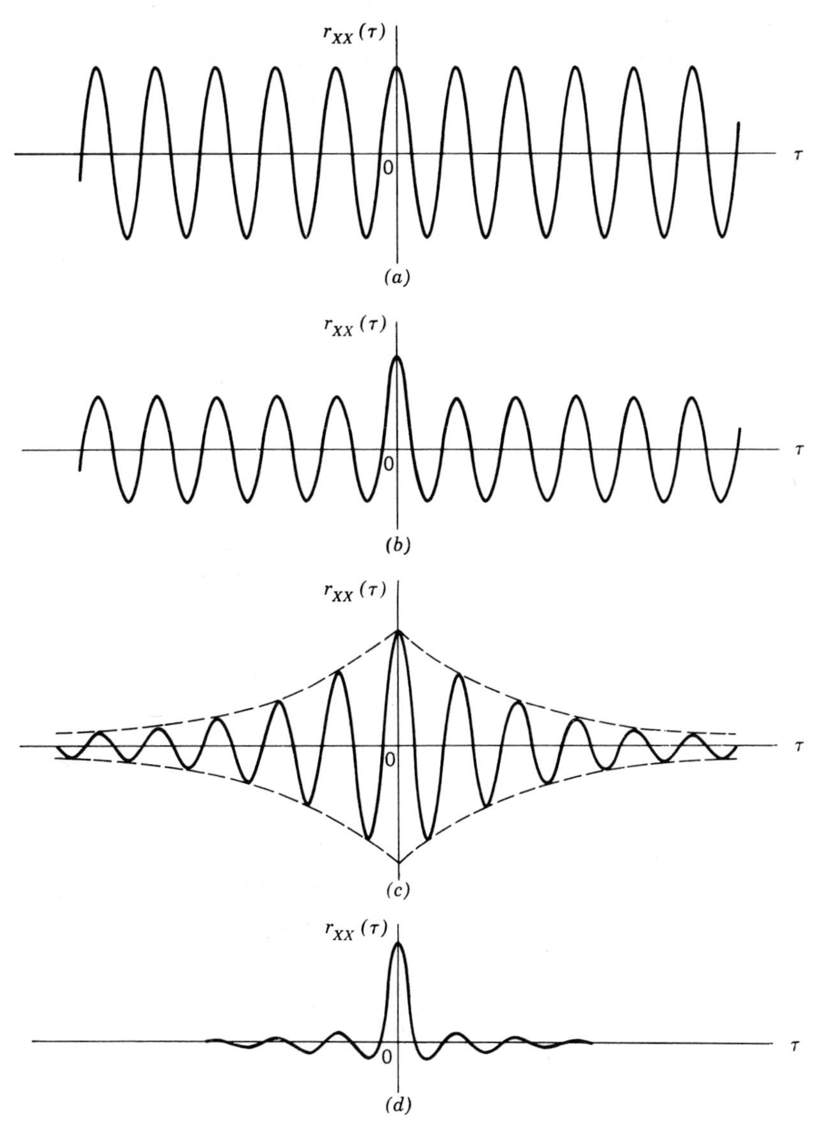

FIGURE 2.2-1 Autocorrelation function plots (autocorrelograms) corresponding to the process records in Figure 2.1-3: (*a*) sine wave, (*b*) sine wave plus random noise, (*c*) narrow-band random noise, and (*d*) wide-band random noise. (From J. S. Bendalt and A. G. Piersol, *Measurement and Analysis of Random Data*, John Wiley, New York, 1966, p. 20.)

The autocorrelation functions $r_{XX}(\tau)$ and $r_{XX}(-\tau)$ are even functions of τ.

A variable is termed *stationary in the wide sense* (or weakly stationary) if it meets just two requirements:

$$\mathscr{E}\{X(t)\} = \mu_X \qquad (2.2\text{-}7a)$$

$$\mathscr{E}\{X(t + \tau)X(t)\} = r_{XX}(\tau) \qquad (2.2\text{-}7b)$$

where μ_X is a constant and r_{XX} depends only on $t_2 - t_1 = \tau$. If a random variable that can be represented by the normal probability distribution is stationary in the wide sense, it is also stationary in the strict sense because the normal distribution, as we shall see, is completely specified by μ_X and r_{XX}, but this conclusion is not generally applicable for other distributions. From a practical viewpoint, if a process is identified as being weakly stationary, the higher order ensemble averages are usually assumed to be stationary also.

Example 2.2-3 Autocorrelation Function

Example 2.2-1 gave the probability density function for a particle in one-dimensional Brownian motion. To compute directly the autocorrelation function for the same particle using Equation 2.2-5, we need the second-order probability density function

$$p(x_1, x_2, t_1, t_2) = \frac{1}{2\pi\alpha^2}\frac{1}{\sqrt{t_1(t_2 - t_1)}}\exp\left[-\frac{x_1^2}{2\alpha^2 t_1} - \frac{(x_2 - x_1)^2}{2\alpha^2(t_2 - t_1)}\right] \qquad (a)$$

Note that because $X(t_1)$ and $X(t_2)$ are not independent, the product of the first-order probability densities is not equal to Equation (a).

However, rather than directly integrate to obtain $r_{XX}(t_1, t_2)$ as indicated by Equation 2.2-5, it is alternately possible to use the property of the Brownian particle that, although $X(t_1)$ and $X(t_2)$ are not independent variables, changes in position over two nonoverlapping intervals are independent. Specifically, $X(t_1)$ and $[X(t_2) - X(t_1)]$ are independent variables. Thus, by Equation 2.2-4,

$$\mathscr{E}\{[X(t_1)][X(t_2) - X(t_1)]\} = \mathscr{E}\{X(t_1)\}\mathscr{E}\{X(t_2) - X(t_1)\} = 0$$

Also, by Equation 2.2-3,

$$\mathscr{E}\{[X(t_1)][X(t_2) - X(t_1)]\} = \mathscr{E}\{X(t_1)X(t_2)\} - \mathscr{E}\{X^2(t_1)\}$$

From Equation (c) in Example 2.2-1, we know that

$$\mathscr{E}\{X^2(t_1)\} = \alpha t_1.$$

Consequently (for $t_2 > t_1$),

$$r_{XX}(t_1, t_2) = \mathscr{E}\{X(t_1)X(t_2)\} = \mathscr{E}\{X^2(t_1)\} = \alpha t_1 \qquad (b)$$

The same result is obtained by direct integration using Equation 2.2-5.

2.2-3 Variance

Just as the mean characterizes the central value of a random variable, a single parameter can be used to characterize its dispersion or scatter about the mean. The classic example of a hunter who shoots all around a duck and misses with each shot illustrates the significance of the dispersion of data. By expectation the duck is dead, but the practical consequence is of little help to the hunter. Figure 2.2-2 illustrates that while two discrete random variables may have the same mean, they may have quite different degrees of dispersion.

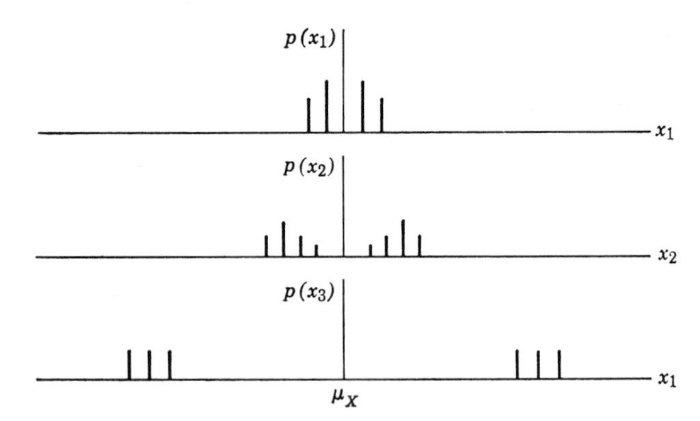

FIGURE 2.2-2 Dispersion of random variables with identical means.

The variance of the stochastic variable $X(t)$ is defined as

$$\sigma_X^2(t) = \mathscr{E}\{[X(t) - \mu_X(t)]^2\} \equiv \text{Var}\{X(t)\}$$
$$= \mathscr{E}\{X^2(t) - 2X(t)\mu_X(t) + \mu_X^2(t)\}$$
$$= \mathscr{E}\{X^2(t)\} - \mu_X^2(t)$$

As an example, because the expected value of the position of the Brownian particle in Example 2.2-1 is zero, the variance can be computed directly from $\mathscr{E}\{X^2(t)\}$.

The variance of a sum of stochastic variables $W = a_1X + a_2Y + \cdots$ can be determined as follows. Subtract the expected value of the sum, namely $\mu_W = a_1\mu_X + a_2\mu_Y + \cdots$, from $W = a_1X + a_2Y + \cdots$ and square the resulting equation to get

$$(W - \mu_W)^2 = [a_1(X - \mu_X) + a_2(Y - \mu_Y) + \cdots]^2$$
$$= a_1^2(X - \mu_X)^2 + a_2^2(Y - \mu_Y)^2 + \cdots$$
$$+ 2a_1a_2(X - \mu_X)(Y - \mu_Y) + \cdots$$

Next, take the variance of both sides

$$\text{Var}\{W\} = \mathscr{E}\{(W - \mu_W)^2\}$$
$$= a_1^2\mathscr{E}\{(X - \mu_X)^2\} + a_2^2\mathscr{E}\{(Y - \mu_Y)^2\} + \cdots$$
$$+ 2a_1a_2\mathscr{E}\{(X - \mu_X)(Y - \mu_Y)\} + \cdots \qquad (2.2\text{-}9)$$

For the special case in which the crossproduct terms vanish in Equation 2.2-9 because *all* the successive pairs of random variables are *independent* (see Section 2.2-5), Equation 2.2-9 reduces to

$$\text{Var}\{W\} = a_1^2\,\text{Var}\{X\} + a_2^2\,\text{Var}\{Y\} + \cdots \qquad (2.2\text{-}9a)$$

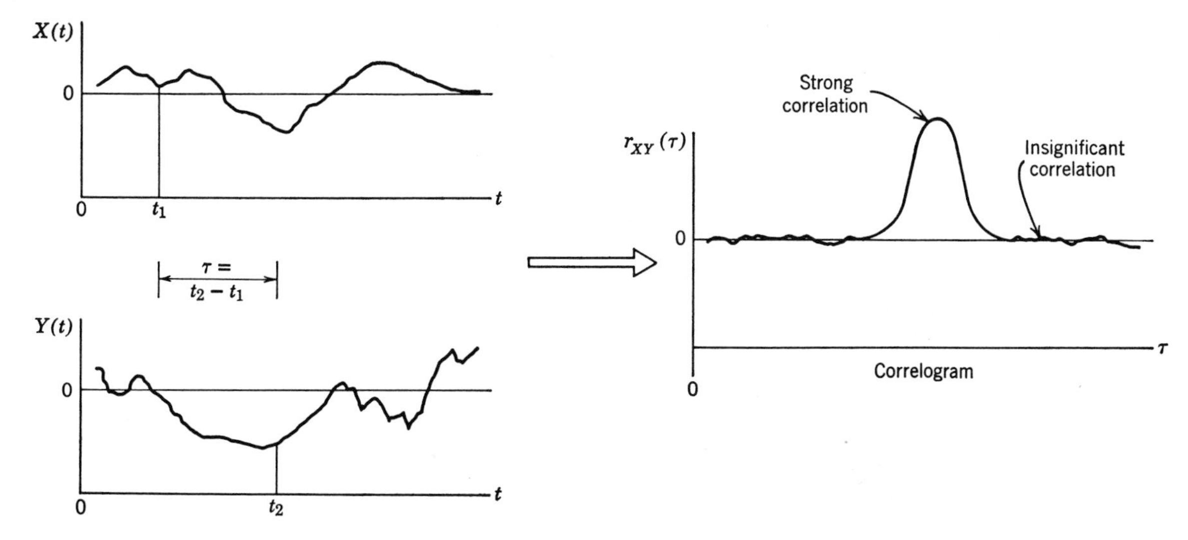

FIGURE 2.2-3 Crosscorrelation function.

The *standard deviation* is the positive square root of the variance and will be denoted by $\sigma_X(t)$. If the variable $X(t)$ is stationary, the functional dependence on t can be deleted. The *coefficient of variation* is a dimensionless form of the standard deviation which provides information on the relative dispersion of $X(t)$:

$$\gamma_X(t) = \frac{\sigma_X(t)}{\mu_X(t)}$$

The *autocovariance* of the stochastic variable $X(t)$ is the covariance of the random variables $X(t_1)$ and $X(t_2)$:

$$\sigma_{XX}(t_1, t_2) = \mathscr{E}\{[X(t_1) - \mu_X(t_1)][X(t_2) - \mu_X(t_2)]\} \quad (2.2\text{-}10)$$

For a stationary ensemble,

$$\sigma_{XX}(\tau) = r_{XX}(\tau) - \mu_X^2 \quad (2.2\text{-}10a)$$

2.2-4 Crosscorrelation Function

The *crosscorrelation function* for two random variables $X(t)$ and $Y(t)$ is used to characterize the dependence of one variable on the other:

$$r_{XY}(t_1, t_2) = \mathscr{E}\{X(t_1)Y(t_2)\} = r_{YX}(t_2, t_1)$$

$$= \int_{-\infty}^{\infty}\int_{-\infty}^{\infty} xyp(x, y; t_1, t_2)\, dx\, dy \quad (2.2\text{-}11)$$

Note that r_{XY} is not a random variable but may be time dependent. Two random variables are *uncorrelated*† if $r_{XY}(t_1, t_2) = \mu_X(t_1)\mu_Y(t_2)$ and are termed *orthogonal* if $r_{XY}(t_1, t_2) = 0$. If the ensembles are stationary, by making use of Equation 2.1-9 we have

$$r_{XY}(t_1, t_2) = r_{XY}(\tau) = r_{YX}(-\tau) \quad (2.2\text{-}12)$$

† Two random variables X and Y are uncorrelated if $\mathscr{E}\{XY\} = \mathscr{E}\{X\}\mathscr{E}\{Y\}$ and are independent if $p(x, y) = p(x)p(y)$. If X and Y are independent, then they are also uncorrelated (see Section 2.2-1). If $\mathscr{E}\{XY\} = 0$, X and Y are orthogonal. Uncorrelatedness is a weaker condition than independence, because if X and Y are uncorrelated, then in general $\mathscr{E}\{f(X)g(Y)\} \neq \mathscr{E}\{f(X)\}\mathscr{E}\{g(Y)\}$. But if X and Y are independent, $\mathscr{E}\{f(X)g(Y)\} = \mathscr{E}\{f(X)\}\mathscr{E}\{g(Y)\}$.

Figure 2.2-3 illustrates figuratively a correlogram for two random variables $X(t)$ and $Y(t)$. $r_{XY}(\tau)$ does not have a maximum at $\tau = 0$ as does $r_{XX}(\tau)$, nor is $r_{XY}(\tau)$ an even function as is $r_{XX}(\tau)$. But to calculate $r_{XY}(\tau)$ and $r_{YX}(\tau)$, it is only necessary to carry out the computations for $\tau \geq 0$ because of the symmetric properties of these two functions.

Crosscorrelation functions can be used in process analysis to:

1. Help check for statistical independence between two random variables.
2. Estimate system impulse and frequency responses without putting a pulse or sinusoidal input into the process (see Chapter 12).
3. Predict delay errors in stationary processes for control studies (the crosscorrelation function for linear processes will peak at a time displacement equal to the time required for a signal to pass through the process).
4. Estimate amplitudes and Fourier components of variables corrupted by uncorrelated noise and/or other signals. (The noise contribution to $r_{XY}(\tau)$ vanishes.)
5. Determine transmission paths for an input to a large linear system. (Separate peaks occur in the crosscorrelogram corresponding to each path.)

Example 2.2-4 Ensemble Mean and Autocorrelation Function of a Stochastic Variable in a Linear Ordinary Differential Equation

Many process models, particularly in control work, are expressed as an nth-order ordinary differential equation:

$$a_n Y^{(n)}(t) + a_{n-1} Y^{(n-1)}(t) + \cdots + a_0 Y(t) = X(t) \qquad t > 0 \quad \text{(a)}$$

where

$X(t)$ = the system input and is a random variable

$Y(t)$ = the system output and is a random variable as a consequence of $X(t)$

$Y^{(n)}(t)$ = nth derivative of $Y(t) = d^n Y(t)/dt^n$

a_i = constants—not random variables

The initial conditions also are random variables:

$$Y^{(n-1)}(0) = Y^{(n-2)}(0) = \cdots = Y(0) = 0 \qquad \text{(b)}$$

Suppose that we are interested in determining the ensemble mean and autocorrelation of $Y(t)$ in terms of known input and output data because the probability density of Y is not known. If we take the expected value of both sides of Equations (a) and (b), by application of Equations 2.2-1 we obtain

$$a_n\mu_{Y^{(n)}} + a_{n-1}\mu_{Y^{(n-1)}} + \cdots + a_0\mu_Y = \mu_X \qquad \text{(c)}$$

$$\mu_{Y^{(n-1)}}(0) = \mu_{Y^{(n-2)}}(0) = \cdots = \mu_Y(0) = 0 \qquad \text{(d)}$$

$$\mu_{Y^{(n)}}(t) = \mathscr{E}\{Y^{(n)}(t)\}$$

$$\mu_X(t) = \mathscr{E}\{X(t)\}$$

Thus, the deterministic model for μ_Y, Equation (c) and (d), will give the solution we seek for μ_Y or

$$\mu_Y(t) = c_1\mu_1(t) + c_2\mu_2(t) + \cdots + c_n\mu_n(t) + \mu_p(t) \qquad \text{(e)}$$

where $\mu_p(t)$ is a particular solution of Equation (c), and the remainder of the right-hand side of Equation (e) is the complementary function. Consequently, given the expected value of $X(t)$ and the values of the coefficients in Equation (c), it is possible to find the deterministic solution to the models represented by Equations (c) and (d).

From the definition of the autocorrelation function

$$r_{XX}(t_1, t_2) = \mathscr{E}\{X(t_1)X(t_2)\}$$

it can be demonstrated that

$$r_{XX'}(t_1, t_2) = \frac{\partial r_{XX}(t_1, t_2)}{\partial t_2}$$

$$r_{X'X'}(t_1, t_2) = \frac{\partial r_{XX'}(t_1, t_2)}{\partial t_1}$$

Hence,

$$r_{X'X'}(t_1, t_2) = \frac{\partial^2 r_{XX}(t_1, t_2)}{\partial t_1\, \partial t_2} = \mathscr{E}\{X'(t_1)X'(t_2)\} \qquad \text{(f)}$$

For a stationary process, $r_{XX}(t_1, t_2) = r_{XX}(\tau)$ and

$$r_{XX'}(\tau) = \frac{dr_{XX}(\tau)}{d\tau}$$

$$r_{X'X'}(\tau) = \frac{dr_{XX'}(\tau)}{d\tau} = \frac{d^2 r(\tau)}{d\tau^2}$$

In general,

$$r_{X^{(n)}Y^{(m)}}(t_1, t_2) = \mathscr{E}\left\{\frac{d^n X(t_1)}{dt_1^n}\frac{d^m Y(t_2)}{dt_2^m}\right\}$$

$$= \frac{\partial^{n+m} r_{XY}(t_1, t_2)}{\partial t_2^n\, \partial t_2^m}$$

The autocorrelation $r_{YY}(t_1, t_2)$ of $Y(t)$ can now be obtained as follows. First, multiply Equations (a) and (b), with $t = t_2$, by $X(t_1)$:

$$X(t_1)[a_n Y^{(n)}(t_2) + \cdots + a_0 Y(t_2)] = X(t_1)X(t_2) \qquad \text{(g)}$$

$$X(t_1)Y^{(n-1)}(0) = \cdots = X(t_1)Y(0) = 0 \qquad \text{(h)}$$

and take the expected value of both sides, term by term, using Equation 2.2-1b to obtain

$$a_n\frac{\partial^n r_{XY}(t_1, t_2)}{\partial t_2^n} + a_{n-1}\frac{\partial^{n-1} r_{XY}(t_1, t_2)}{\partial t_2^{n-1}} + \cdots$$

$$+ a_0 r_{XY}(t_1, t_2) = r_{XX}(t_1, t_2) \qquad \text{(i)}$$

$$\frac{\partial^{n-1} r_{XY}(t_1, 0)}{\partial t_2^{n-1}} = \cdots = r_{XY}(t_1, 0) = 0 \qquad \text{(j)}$$

Equation (i) is actually an *ordinary differential* equation in $r_{XY}(t_1, t_2)$ with t_2 as the independent variable and t_1 as a parameter. Thus, given the autocorrelation function $r_{XX}(t_1, t_2)$, Equations (i) and (j) can be used to compute the crosscorrelation function $r_{XY}(t_1, t_2)$.

Next, Equations (a) and (b) with $t = t_1$ are postmultiplied by $Y(t_2)$:

$$[a_n Y^{(n)}(t_1) + \cdots + a_0 Y(t_1)]Y(t_2) = X(t_1)Y(t_2) \qquad \text{(k)}$$

$$Y^{(n-1)}(0)Y(t_2) = \cdots = Y(0)Y(t_2) = 0 \qquad \text{(1)}$$

and again the expected value of both sides is taken term by term to yield an ordinary differential equation in $r_{YY}(t_1, t_2)$:

$$a_n\frac{\partial^n r_{YY}(t_1, t_2)}{\partial t_1^n} + a_{n-1}\frac{\partial^{n-1} r_{YY}(t_1, t_2)}{\partial t_1^{n-1}} + \cdots$$

$$+ a_0 r_{YY}(t_1, t_2) = r_{XY}(t_1, t_2) \qquad \text{(m)}$$

$$\frac{\partial^{n-1} r_{YY}(0, t_2)}{\partial t_1^{n-1}} = \cdots = r_{YY}(0, t_2) = 0 \qquad \text{(n)}$$

To obtain $r_{YY}(t_1, t_2)$, Equations (i) and (j) first must be solved for $r_{XY}(t_1, t_2)$, assuming $r_{XX}(t_1, t_2)$ is known, and then the result introduced into the right-hand side of Equation (m) which can then be solved for the desired $r_{YY}(t_1, t_2)$ subject to Equation (n).

As an example of an application of the above equations, suppose the input to a well-mixed tank, as illustrated in Figure E2.2-4, is represented by the Brownian random variable of Example 2.2-1. We have previously derived the mean, variance, and autocorrelation functions for the concentration. Hence, we can write

$$\mathscr{E}\{C_0(t)\} = 0 \qquad \text{(o)}$$

$$\mathscr{E}\{[C_0(t_1) - 0][C_0(t_2) - 0]\} = r_{C_0C_0}(t_1, t_2) = \alpha t_1$$

where α is the parameter in the probability density for C_0.

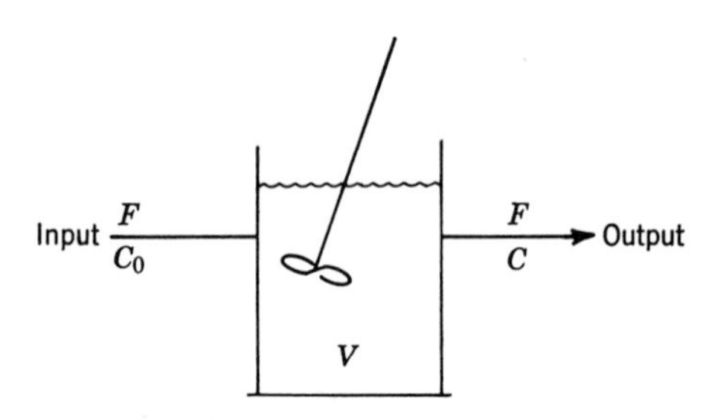

FIGURE E2.2-4 Model of well-mixed tank:

$$V\frac{dC}{dt} = FC_0 - FC; \quad C(0) = 0$$

where

C_0 = input concentration, a random variable
C = concentration, a random variable
F = feed rate
V = volume of fluid

Equations (i) and (j) for the model are

$$t^* \frac{dr_{C_0C}(t_1, t_2)}{dt_2} + r_{C_0C}(t_1, t_2) = r_{C_0C_0}(t_1, t_2);\ r_{C_0C_0}(t_1, t_2) = \alpha t_1$$

$$r_{C_0C}(t_1, 0) = 0$$

and have the solution

$$r_{C_0C}(t_1, t_2) = \alpha t_1[1 - e^{-t_2/t^*}] \tag{p}$$

where $t^* = V/F$.

Equations (m) and (n) are

$$t^* \frac{dr_{CC}(t_1, t_2)}{dt_1} + r_{CC}(t_1, t_2) = r_{C_0C}(t_1, t_2)$$

$$r_{CC}(0, t_2) = 0$$

and have the solution

$$r_{CC}(t_1, t_2) = \alpha[1 - e^{-t_2/t^*}]\left[e^{-t_1/t^*} - 1 - \frac{t_1}{t^*}\right] \tag{q}$$

Thus, the autocorrelation function of the tank output can be determined even if its probability density function is not known.

2.2-5 Crosscovariance and Correlation Coefficient

The analyst is frequently called upon to determine qualitatively and, insofar as possible, quantitatively, whether an association exists between two variables. He might inquire, for example, for a particular reactor whether the increase of pressure increases the yield. If the joint probability distribution for the two variables is known, it is possible to calculate a measure of the *linear* association between the two variables, termed the ensemble *correlation coefficient*. No distinction is made between the variables as to which is independent and which is dependent.

The *crosscovariance function* (sometimes abbreviated Covar) for two random variables $X(t)$ and $Y(t)$ is defined as

$$\sigma_{XY}(t_1, t_2) = \mathscr{E}\{[X(t_1) - \mu_X(t_1)][Y(t_2) - \mu_Y(t_2)]\}$$
$$= \int_{-\infty}^{\infty}\int_{-\infty}^{\infty} (x - \mu_X)(y - \mu_Y) \cdot p(x, y; t_1, t_2)\, dx\, dy \tag{2.2-13}$$

For a stationary ensemble,

$$\sigma_{XY}(\tau) = r_{XY}(\tau) - \mu_X\mu_Y \tag{2.2-13a}$$

Since the magnitude of the covariance depends upon the units of X and Y, two standardized (dimensionless) variables can be formed:

$$\left(\frac{X - \mu_X}{\sigma_X(0)}\right) \quad \text{and} \quad \left(\frac{Y - \mu_Y}{\sigma_Y(0)}\right)$$

where the argument (0) indicates $\tau = 0$. The *correlation coefficient* for a stationary ensemble is the crosscovariance of these two standardized variables:

$$\rho_{XY}(\tau) = \frac{\sigma_{XY}(\tau)}{\sigma_X(0)\sigma_Y(0)} \tag{2.2-14}$$

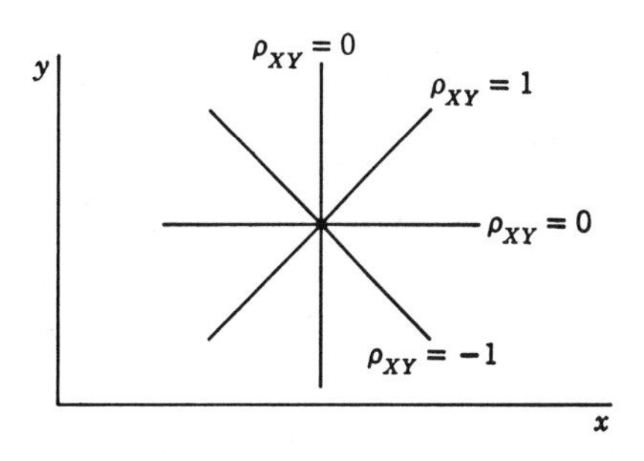

FIGURE 2.2-4 Correlation coefficient at its extreme values and at zero for $\sigma_X^2 = \sigma_Y^2$.

If X and Y are uncorrelated, their ensemble covariance and correlation coefficient are zero. If X and Y are independent, their ensemble covariance and correlation coefficient are also zero; however, the converse is not true. That is, if $\rho_{XY} = 0$, X and Y are not *necessarily* independent (although they may be). For example, two random variables can each be normally distributed and uncorrelated but dependent on each other; they must be distributed by a joint normal distribution to be independent. Pairwise independence among sets of many random variables is not sufficient to indicate independence of the sets.

The correlation coefficient reduces to one number the measure of the *linear relationship* between two variables. A positive correlation means that σ_{XY} is positive (the standard deviation is never negative), while a negative σ_{XY} means that large values of one variable are associated with small values of the other. Figure 2.2-4 illustrates lines of 0, 1, and -1 for the correlation coefficient.

Example 2.2-5 Ensemble Correlation Coefficient

The joint probability density function for two random variables X and Y is given as

$$p(x, y) = x + y \qquad \text{for } 0 \le X \le 1,\quad 0 \le Y \le 1$$
$$p(x, y) = 0 \qquad \text{otherwise}$$

Find the ensemble correlation coefficient.

Solution:

$$\mu_X = \int_0^1\int_0^1 x(x + y)\, dx\, dy = \tfrac{7}{12}$$

$$\mu_Y = \int_0^1\int_0^1 y(x + y)\, dx\, dy = \tfrac{7}{12}$$

$$\mu_{X^2} = \int_0^1\int_0^1 x^2(x + y)\, dx\, dy = \tfrac{5}{12}$$
$$= \mu_{Y^2}$$

$$\sigma_X^2 = \sigma_Y^2 = \tfrac{5}{12} - (\tfrac{7}{12})^2 = \tfrac{11}{144}$$

$$\sigma_{XY} = \mathscr{E}\{(X - \mu_X)(Y - \mu_Y)\} = \mathscr{E}\{XY\} - \mu_X\mu_Y$$
$$= \tfrac{1}{3} - \tfrac{49}{144} = -\tfrac{1}{144}$$

since

$$\mathscr{E}\{XY\} = \int_0^1 \int_0^1 (xy)(x + y)\, dx\, dy = \tfrac{1}{3}$$

Then

$$\rho_{XY} = \frac{\sigma_{XY}}{\sigma_X \sigma_Y} = -\frac{\frac{1}{144}}{\frac{11}{144}} = -\frac{1}{11}$$

The result indicates little correlation between X and Y.

Table 2.2-1 summarizes the ensemble parameters described so far.

TABLE 2.2-1 SUMMARY OF ENSEMBLE PARAMETERS

Parameter	Name of Function	Expected Value
$\mu_X(t)$	Mean	$\mathscr{E}\{X(t)\}$
$\mu_X^2(t)$	Mean square	$\mathscr{E}\{X^2(t)\}$
$\sigma_X^2(t)$	Variance	$\mathscr{E}\{[X(t) - \mu_X(t)]^2\}$
$\sigma_X(t)$	Standard deviation	$+\sqrt{\sigma_X^2(t)}$
$\gamma_X(t)$	Coefficient of variation	$\sigma_X(t)/\mu_X(t)$
$r_{XX}(\tau)$	Autocorrelation*	$\mathscr{E}\{X(t_1)X(t_2)\}$
$r_{XY}(\tau)$	Crosscorrelation*	$\mathscr{E}\{X(t_1)Y(t_2)\}$
$\sigma_{XX}(\tau)$	Autocovariance*	$\mathscr{E}\{[X(t_1) - \mu_X(t_1)] \cdot [X(t_2) - \mu_X(t_2)]\}$
$\sigma_{XY}(\tau)$	Crosscovariance*	$\mathscr{E}\{[X(t_1) - \mu_X(t_1)] \cdot [Y(t_2) - \mu_Y(t_2)]\}$
$\rho_{XY}(\tau)$	Correlation coefficient*	$\sigma_{XY}(\tau)/\sigma_X(0)\sigma_Y(0)$

* For stationary variables.

2.2-6 Moments of a Random Variable

Moments of a random variable have an analogy in mechanics. Recall that the *first moment* of mass is the product of mass and the moment arm, and that the center of mass is the first moment divided by the mass. Both the ensemble mean and the variance are moments in which the probability density function is the weighting function. One is a *raw moment* and the other a *central moment*. Refer to Table 2.2-1 for time-invariant moments for a single variable through order n. The third central moment proves to be a measure of the symmetry of the distribution of a random variable with respect to the mean; the fourth central moment characterizes the sharpness of the peak about the mode.

Moments for a pair of (time-independent) random

TABLE 2.2-2 RAW AND CENTRAL MOMENTS FOR A RANDOM TIME-INDEPENDENT VARIABLE

Continuous	Discrete		Moment
	Raw Moments		
$\int_{-\infty}^{\infty} x^0 p(x)\, dx = 1$	$\sum_{i=1}^{\infty} x_i^0 P(x_i) = 1$	μ_0	Zeroth moment
$\int_{-\infty}^{\infty} x p(x)\, dx = \mu_X$	$\sum_{i=1}^{\infty} x_i P(x_i) = \mu_X$	μ_1	First moment (ensemble mean of X)
$\int_{-\infty}^{\infty} x^2 p(x)\, dx = \mu_{X^2}$	$\sum_{i=1}^{\infty} x_i^2 P(x_i) = \mu_{X^2}$	μ_2	Second moment
$\vdots$	$\vdots$		$\vdots$
$\int_{-\infty}^{\infty} x^n p(x)\, dx = \mu_{X^n}$	$\sum_{i=1}^{\infty} x_i^n P(x_i) = \mu_{X^n}$	μ_n	nth moment
	Central Moments		
$\int_{-\infty}^{\infty} (x - \mu_X)^0 p(x)\, dx = 1$	$\sum_{i=1}^{\infty} (x_i - \mu_X)^0 P(x_i) = 1$	$\mathscr{M}_0$	Zeroth moment
$\int_{-\infty}^{\infty} (x - \mu_X) p(x)\, dx = 0$	$\sum_{i=1}^{\infty} (x_i - \mu_X) P(x_i) = 0$	$\mathscr{M}_1$	First moment
$\int_{-\infty}^{\infty} (x - \mu_X)^2 p(x)\, dx = \sigma_X^2$	$\sum_{i=1}^{\infty} (x_i - \mu_X)^2 P(x_i) = \sigma_X^2$	$\mathscr{M}_2$	Second moment (ensemble variance of X)
$\vdots$	$\vdots$		$\vdots$
$\int_{-\infty}^{\infty} (x - \mu_X)^n p(x)\, dx$	$\sum_{i=1}^{\infty} (x_i - \mu_X)^n P(x_i)$	$\mathscr{M}_n$	nth moment

variables can be defined as follows for continuous variables:

$$\mu_{ij} = \int_{-\infty}^{\infty}\int_{-\infty}^{\infty} x_1^i x_2^j p(x_1, x_2)\, dx_2\, dx_1 \qquad (2.2\text{-}15)$$

Central moments correspond to employing the weighting function of $(x_1 - \mu_1)^i(x_2 - \mu_2)^j$ instead of $x_1^i x_2^j$. For example:

$$\mu_{10} = \int_{-\infty}^{\infty}\int_{-\infty}^{\infty} x_1^1 x_2^0 p(x_1, x_2)\, dx_2\, dx_1 = \mu_{X_1} = \mathscr{E}\{X_1\}$$

$$\mu_{01} = \int_{-\infty}^{\infty}\int_{-\infty}^{\infty} x_1^0 x_2^1 p(x_1, x_2)\, dx_2\, dx_1 = \mu_{X_2} = \mathscr{E}\{X_2\}$$

$$\mu_{11} = \int_{-\infty}^{\infty}\int_{-\infty}^{\infty} x_1^1 x_2^1 p(x_1, x_2)\, dx_2\, dx_1 = \mathscr{E}\{X_1 X_2\}$$

$$\begin{aligned}\mathscr{M}_{11} &= \int_{-\infty}^{\infty}\int_{-\infty}^{\infty} (x_1 - \mu_{X1})^1 (x_2 - \mu_{X_2})^1 p(x_1, x_2)\, dx_2\, dx_1 \\ &= \mathscr{E}\{X_1 X_2\} - \mu_{X_1}\mu_{X_2} = \mu_{11} - \mu_{10}\mu_{01} \\ &= \sigma_{XY}\end{aligned}$$

Example 2.2-6 Moments

Show that the second moment about the value $x = c$ is greater than the second central moment.

Solution:

$$\begin{aligned}\mathscr{E}\{(X - c)^2\} &= \mathscr{E}\{(X - \mu_X + \mu_X - c)^2\} \\ &= \mathscr{E}\{(X - \mu_X)^2\} \\ &\quad + 2\mathscr{E}\{(X - \mu_X)(\mu_X - c)\} + \mathscr{E}\{(\mu_X - c)^2\} \\ &= \mathscr{M}_2 + (\mu_X - c)^2\end{aligned}$$

Note that

$$\mathscr{E}\{(X - \mu_X)(\mu_X - c)\} = (\mu_X - c)\mathscr{E}(X - \mu_X) = (\mu_X - c)0$$

since $\mathscr{M}_1 \equiv 0$.

2.3 THE NORMAL AND χ^2 PROBABILITY DISTRIBUTIONS

We shall next consider briefly two probability distributions which are employed in subsequent chapters. More complete details concerning the characteristics of these probability distributions can be found in the references listed at the end of this chapter. The objective of this section is to delineate the properties of and basic assumptions lying behind the normal and chi-square (χ^2) distributions so that they can be appropriately employed in the analysis of experimental data. Tables 2.3-1 and 2.3-2 list other time-invariant discrete and continuous distributions which are not discussed. Figure 2.3-1 illustrates the probability function and cumulative probability distribution for the discrete binomial random variable. Figure 2.3-2 illustrates the probability densities for several continuous random variables whose characteristics are given in Table 2.3-2.

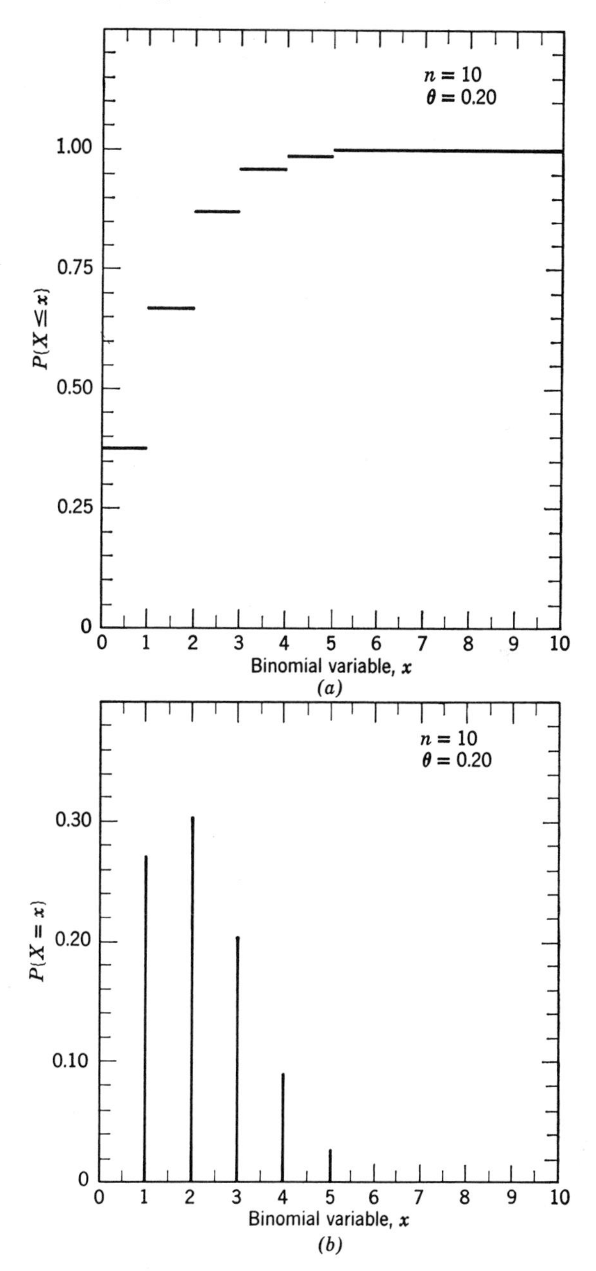

FIGURE 2.3-1 Binomial distribution: (*a*) binomial distribution, and (*b*) binomial probability function.

2.3-1 The Normal Probability Distribution

In the eighteenth and early nineteenth centuries, mathematicians and experimenters in the physical sciences developed a probability density function that represented quite well the errors of observation. Their work yielded the normal (Gaussian) probability density function for the random variable X, i.e., the familiar bell-shaped curve shown in Figure 2.3-3. This is generated by the equation

$$p(x) = \frac{1}{\sigma_X\sqrt{2\pi}} \exp - \left(\frac{(x - \mu_X)^2}{2\sigma_X^2}\right) \qquad -\infty < x < \infty \qquad (2.3\text{-}1)$$

TABLE 2.3-1 DISCRETE PROBABILITY FUNCTIONS*

Name	Probability Function $P(x) = P\{X = x\}$	Applications and Remarks	Mean $\mu_X = \mathscr{E}\{X\}$	Variance $\sigma_X^2 = \mathrm{Var}\{X\}$
Binomial	$P(x) = \binom{n}{x}\theta^x(1-\theta)^{n-x}$ $x = 0, 1, 2, \ldots, n$	Applications in sampling, inspection plans, coin tossing, or any experiment in which: 1. There are a fixed number of outcomes, n. 2. The outcome of each trial must be a dichotomy, i.e., a "success" or a "failure"; x = number of successes. 3. All trials have an identical probability of "success," θ. 4. The trials are independent of each other.	$n\theta$	$n\theta(1-\theta)$
Poisson	$P(x) = \frac{(n\theta)^X}{x!}e^{-n\theta}$ $x = 0, 1, 2, \ldots$	Applications in auto traffic, telephone circuits, computer loading, sampling, and radioactive decay with short half-lives. Events must be independent and rare. It can be used as an approximation to the binomial function when n is large and P is small, since the binomial becomes the Poisson function as $n \to \infty$ with $n\theta$ constant.	$n\theta$	$n\theta$
Multinomial	$P_n(x_1, x_2, \ldots, x_k) = \frac{n!}{x_1!\,x_2!\cdots x_k!}(\theta_1)^{x_1}(\theta_2)^{x_2}\cdots(\theta_k)^{x_k}$ $n = \sum_{i=1}^{k} x_i$	Applications in sampling. A multivariate discrete function can be regarded as a generalization of the binomial function. Up to k possible outcomes exist, each of which is mutually exclusive. The probability of the first event x_1 is θ_1, of x_2 is θ_2, etc., and $\theta_1 + \theta_2 + \cdots + \theta_k = 1$. Each trial must be independent, and the probability of each outcome must be the same from trial to trial. $P_n(x_1, \ldots, x_k)$ is the probability that on n trials a success for variable 1 occurs exactly x_1 times, for variable 2 exactly x_2 times, etc.	Each variable $= n\theta_i$	Each variable $= n\theta_i(1-\theta_i)$
Hypergeometric	$P(x) = \frac{\binom{D}{x}\binom{N-D}{n-x}}{\binom{N}{n}}$ $x = 0, 1, 2, \ldots$ D = total number of defectives in the N total items	Applications in the analysis of sampling without replacement, i.e., sampling by attributes. For a finite number N of items which can be classified either "good" or "bad," "success" or "failure," if samples of a size n are drawn from N one at a time *without replacing* the items withdrawn, then the probability of obtaining exactly x "failures" in a sample n in $P(x)$.	$\frac{nD}{n}$	$\frac{nD(N-D)(N-n)}{N^2(N-1)}$

* The symbol $\binom{n}{x}$ means the number of combinations of x things taken from a total of n things without regard to order;

$$\binom{n}{x} \equiv \frac{n!}{x!\,(n-x)!}$$

By direct integration it can be shown that the two parameters in Equation 2.3-1 are the mean and variance of X:

$$\int_{-\infty}^{\infty} p(x)\,dx = 1 \qquad \text{(zeroth moment)}$$

$$\mathscr{E}\{X\} = \int_{-\infty}^{\infty} xp(x)\,dx = \mu_X \qquad \text{(first moment)}$$

$$\mathscr{E}\{(X-\mu_X)^2\} = \int_{-\infty}^{\infty} (x-\mu_X)^2 p(x)\,dx = \sigma_X^2$$

(second central moment)

Example 2.3-1 Zeroth Moment of the Normal Probability Density Function

Show that

$$\int_{-\infty}^{\infty} p(x)\,dx = 1$$

Solution:

We want to show that

$$\frac{1}{\sigma_X\sqrt{2\pi}}\int_{-\infty}^{\infty} \exp -\left(\frac{(x-\mu_X)^2}{2\sigma_X^2}\right) dx = 1 \qquad \text{(a)}$$

The calculation can be made relatively brief if we make two changes which do not affect the value of the integral. These changes are:

1. Shift the origin on the x axis for integration from $x = 0$ to $x = \mu_{X'}$ so that $\mu_X \equiv 0$.

2. Square both sides of Equation (a); y and z are dummy variables.

Then

$$\left[\int_{-\infty}^{\infty} \frac{1}{\sigma_X\sqrt{2\pi}} \exp\left(-\frac{x^2}{2\sigma_X^2}\right) dx\right]^2$$

$$= \frac{1}{2\pi\sigma_X^2}\left[\int_{-\infty}^{\infty} \exp\left(-\frac{y^2}{2\sigma_X^2}\right) dy\right]\left[\int_{-\infty}^{\infty} \exp\left(-\frac{z^2}{2\sigma_X^2}\right) dz\right]$$

$$= \frac{1}{2\pi\sigma_X^2}\int_{-\infty}^{\infty}\int_{-\infty}^{\infty} \exp\left(-\frac{y^2+z^2}{2\sigma_X^2}\right) dy\,dz$$

TABLE 2.3-2 PROBABILITY DENSITY FUNCTIONS FOR A SINGLE CONTINUOUS VARIABLE

Name	Density Function	Applications or Remarks	Mean $\mu_X = \mathscr{E}\{X\}$	Variance $\sigma_X^2 = \text{Var}\{X\}$
Log-normal	$p(x) = \frac{1}{\sqrt{2\pi}\beta} \exp\left[-\frac{(\ln x - \alpha)^2}{2\beta^2}\right]$ $(0 \leq x < \infty)$ $\alpha = \mathscr{E}\{\ln X\}$ $\beta^2 = \text{Var}\{\ln X\}$	Applies to situations in which several independent factors influence the outcome of an event not additively but according to the magnitude of the factor. Applications are to particle sizes, condensation, aerosols, petrology, economics, and photographic emulsions. A variable X has a log-normal distribution if log X has a normal distribution. The distribution is similar in form to the gamma and Weibull distributions.	$e^{\alpha + \left(\frac{\beta^2}{2}\right)}$	$e^{2\alpha + \beta^2}(e^{\beta^2} - 1)$
Exponential	$p(x) = \left(\frac{1}{\theta}\right) e^{-x/\theta}$ $(0 \leq x < \infty)$	Applies to constant, instantaneous, failure rate, i.e., a first-order ordinary differential equation. x is the random variable and θ is a time constant. The distribution is an excellent model for the failure behavior of many types of complex systems, particularly for those parts and systems which are so complex that many deterioration mechanisms with different failure rates exist.	θ	θ^2
Weibull	$p(x) = \alpha\beta x^{\alpha-1} e^{-\beta x^\alpha}$ $(0 \leq x < \infty)$	Applies to life testing such as first failure among a large number of items (α is related to the failure rate), corrosion resistance, return of goods by week after shipment, and reliability.	$(\beta)^{-1/\alpha}\Gamma\left(\frac{1}{\alpha}+1\right)$	$(\beta)^{-2/\alpha}\left[\Gamma\left(\frac{2}{\alpha}+1\right) - \Gamma^2\left(\frac{1}{\alpha}+1\right)\right]$
Gamma	$p(x) = \frac{\beta^{\alpha+1}}{\Gamma(\alpha+1)} e^{-\beta x} x^\alpha$ $\left(\begin{matrix} 0 \leq x < \infty \\ \alpha > -1 \end{matrix}\right)$	Similar applications to the above.	$\frac{\alpha+1}{\beta}$	$\frac{\alpha+1}{\beta^2}$

Finally, introducing polar coordinates yields

$$\left[\int_{-\infty}^{\infty} \frac{1}{\sigma_X\sqrt{2\pi}} \exp\left(-\frac{x^2}{2\sigma_X^2}\right) dx\right]^2 = \frac{1}{2\pi}\int_0^{2\pi}\int_0^{\infty} \frac{r}{\sigma_X^2} \exp\left(-\frac{r^2}{2\sigma_X^2}\right) dr\, d\theta = 1$$

If the square of any real quantity equals 1, the quantity itself is 1.

By a simple transformation of variables to the standardized (or "unit"—for U) random variable U,

$$U = \frac{(X - \mu_X)}{\sigma_X} \tag{2.3-2}$$

we obtain the probability density function called the *standard normal probability density function.* Note that because $P(u) = P(x)$, where u is the upper limit of integration corresponding to x, $p(u)\, du = p(x)\, dx$; hence,

$$p(u) = \frac{1}{\sqrt{2\pi}} e^{-u^2/2} \tag{2.3-3}$$

which is shown in the upper part of Figure 2.3-4. The moments of the standard normal random variable U are:

$$\int_{-\infty}^{\infty} p(u)\, du = 1 \qquad \text{(zeroth moment)}$$

$$\int_{-\infty}^{\infty} up(u)\, du = 0 \qquad \text{(first moment)}$$

$$\int_{-\infty}^{\infty} u^2 p(u)\, du = 1 \qquad \text{(second moment)}$$

Figure 2.3-4 portrays the relationship between $p(u)$ and $p(x)$.

The *standard normal probability distribution*

$$P(u) = P\{U \leq u\} = \frac{1}{\sqrt{2\pi}} \int_{-\infty}^{u} e^{-u'^2/2}\, du' \tag{2.3-4}$$

is shown in Figure 2.3-5 and tabulated in Appendix C. By taking advantage of the symmetry of $p(u)$, we can compute, for example, from Table C1

$$P\{0 \leq U \leq 1\} = P\{U \leq 1\} - P\{U \leq 0\} = 0.841 - 0.500 = 0.341$$

which is equivalent to the area under the curve $p(u)$ from U equal to 0 to 1. As another example,

$$P\{-3.2 \leq U \leq -0.3\} = 0.999 - 0.618 = 0.381$$

Reference tables of the standard normal probability distribution function are not all based on the function given by Equation 2.3-4. The largest and most comprehensive tables† give

$$F(u) = \frac{1}{\sqrt{2\pi}} \int_{-u}^{u} e^{-t^2/2}\, dt \tag{2.3-5a}$$

† Nat. Bur. of Standards, *Guide to the Tables of the Normal Probability Integral,* Applied Mathematics Series 23, U.S. Government Printing Office, Washington, D.C., 1952.

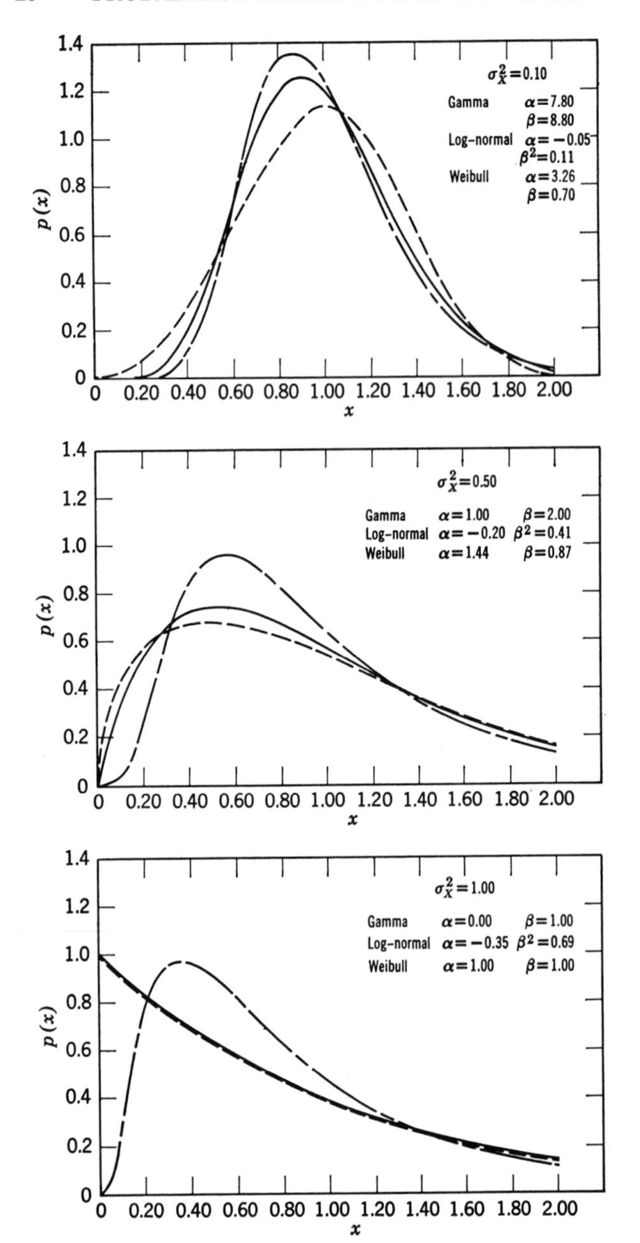

FIGURE 2.3-2 The gamma, log-normal, and Weibull probability densities. (Parameters are identified in Table 2.3-2.)

while other reference books give

$$G(u) = \frac{2}{\sqrt{\pi}}\int_0^u e^{-t^2}\,dt \tag{2.3-5b}$$

One useful relation is

$$P(u) = \int_{-\infty}^{u} \frac{1}{\sqrt{2\pi}} e^{-t^2}\,dt = \frac{1}{2} + \int_0^u \frac{1}{\sqrt{2\pi}} e^{-t^2}\,dt \tag{2.3-6}$$

as can be intuitively seen from the symmetry of Figure 2.3-5.

Example 2.3-2 Mean and Variance of the Standardized Normal Variable

Show that the expected value of U is 0 and that the variance of U is 1.

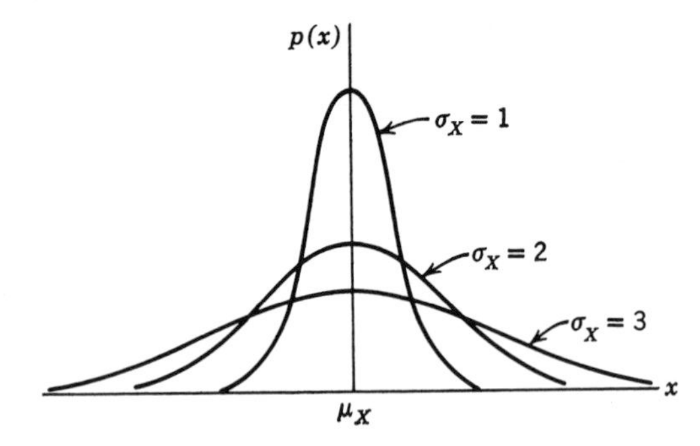

FIGURE 2.3-3 Normal probability density function for various dispersions. Peak is at $(\mu_X, 1/\sigma_X\sqrt{2\pi})$. Points of inflection are at $x = \mu_X + \sigma_X$.

Solution:

$$\mathscr{E}\{U\} = \int_{-\infty}^{\infty} up(u)\,du$$

$$= \frac{1}{\sqrt{2\pi}}\int_{-\infty}^{\infty} u\,e^{-u^2/2}\,du \tag{a}$$

Let $u^2/2 = t$. Then $u\,du = dt$; when $u = -\infty$, $t = \infty$. Consequently,

$$\mathscr{E}\{U\} = \frac{1}{\sqrt{2\pi}}\left[\int_{\infty}^{0} e^{-t}\,dt + \int_0^{\infty} e^{-t}\,dt\right] = 0 \tag{b}$$

The same conclusion could be reached by noting that u is

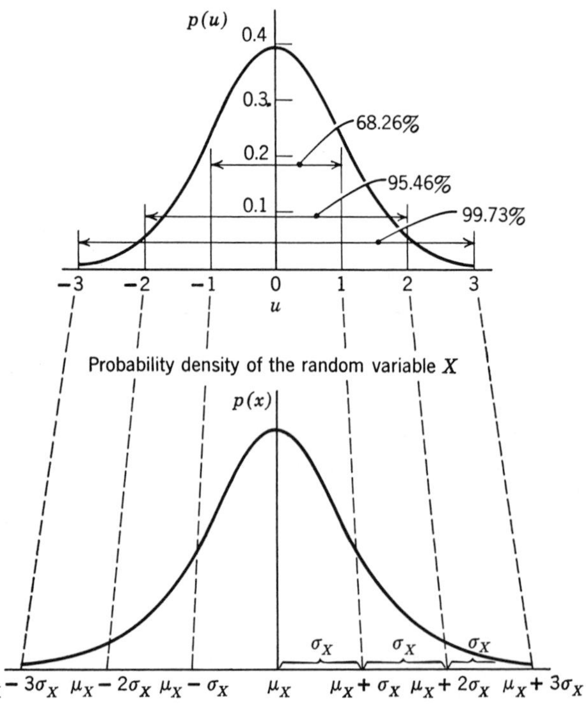

FIGURE 2.3-4 Relation between the standardized normal random variable U and the normal random variable X. The percentages refer to the area under the curve within the indicated bounds on the basis of a total area of 100 percent.

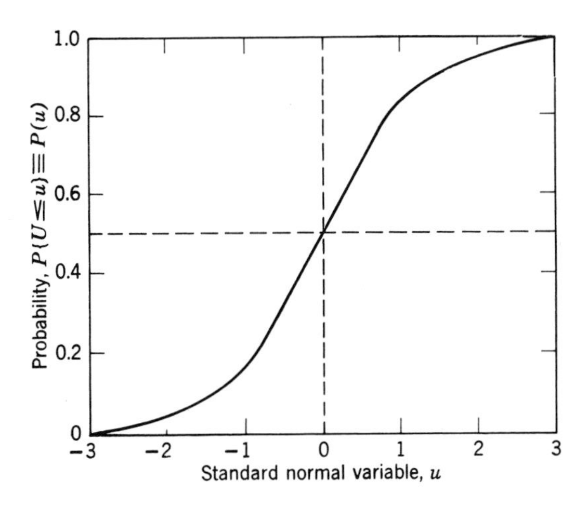

FIGURE 2.3-5 Normal (Gaussian) probability distribution function.

an odd function, $e^{-u^2/2}$ is an even function, and their product integrated over a symmetric interval is zero.

$$\text{Var}\{U\} = \int_{-\infty}^{\infty} (u - 0)^2 p(u)\, du \qquad \text{(c)}$$

$$= \frac{1}{\sqrt{2\pi}} \int_{-\infty}^{\infty} u^2 e^{-u^2/2}\, du = \frac{2}{\sqrt{2\pi}} \int_0^{\infty} u^2 e^{-u^2/2}\, du$$

$$= \frac{2}{\sqrt{\pi}} \int_0^{\infty} t^{1/2} e^{-t}\, dt \qquad \text{(d)}$$

Since the gamma function is

$$\Gamma(n) = \int_0^{\infty} t^{n-1} e^{-t}\, dt = (n - 1)!$$

and

$$\Gamma\left(\frac{3}{2}\right) = \frac{\sqrt{\pi}}{2}$$

the integral in Equation (d) is

$$\Gamma\left(\frac{3}{2}\right) = \frac{\sqrt{\pi}}{2}$$

and

$$\text{Var}\{U\} = \frac{2}{\sqrt{\pi}} \left(\frac{\sqrt{\pi}}{2}\right) = 1$$

Before we assume that experimental data are represented by the normal probability distribution, if sufficient data are available it is desirable to: (1) examine their relative frequency distribution by tests for goodness of fit as described in Section 3.7-7, (2) plot the cumulative frequencies on normal probability paper† which linearizes $P(x)$ by use of a special scale, or (3) carry out other appropriate tests described in Chapter 3. Although the normal probability distribution truly represents many collections of experimental data, it is also often ascribed to data, for convenience, when the variables are continuous but not normally distributed because:

1. The variable can be transformed and the transformed variable will be normally distributed.
2. Sums of random variables, variables not themselves normally distributed, are approximately normally distributed as the sample size $\rightarrow \infty$.
3. The error introduced by using statistical tests based on a normal probability distribution for experimental data of another reasonably symmetric distribution is small.

Example 2.3-3 Graphical Validation of Normality of Experimental Data‡

The data in Table E2.3-3 are the diameters in microns of two hundred particles from a sample of material on an oilfield pipeline filter screen. The number of particles falling within selected cell boundaries gives the frequencies as grouped data. Grouping of data into cells removes the erratic behavior of small batches of data while retaining the predominant characteristics of the data as a whole. Choice of cell range and the number of cells should not cause the loss of too much information relating to the data. Cell bounds are usually chosen so that ten to twenty cells of equal width result. As so often occurs, the data in this example have been classified into cells of unequal size because of the manner in which the particle sizes were collected. It was desired to obtain some idea of the distribution of the particle sizes.

The first step in the preparation of the data for plotting is to arrange the observed classes of the random variable X in ascending order, as shown in the first column of the table. The frequency of each class is listed and the cumulative frequency computed. The values of x are ranked, with the number 1 assigned to the lowest rank. If a value of x has a frequency greater than 1, successive ranks are assigned for each value (i.e., three observations require assignment of three successive ranks). For each value of x, the average rank is calculated by

$$m = \frac{\sum \text{ranks}}{\text{observed frequency}}$$

Finally, the following relation§

$$P = \frac{m}{n + 1}$$

where n = the total sample size, gives the relative dependent variable for plotting.

† The use of special graph paper which linearizes the normal and many other distributions is described in the booklet by J. R. King, "Graphical Data Analysis with Probability Papers," available from Team, 104 Belrose Ave., Lowell, Mass., 1965, and in the article by E. B. Ferrell, *Ind. Qual. Control*, p. 12, July 1958.

‡ The data and graphs in this example have been taken from a paper by C. Lewis, "Applications of Statistics and Computers," *Symposium*, Mar. 1962, edited by R. E. Streets and R. D. Quillan, Southwest Research Institute, San Antonio, Texas, 1962.

§ Due to E. J. Gumbel, *Statistics of Extremes*, Columbia Univ. Press, New York, 1958.

TABLE E2.3-3 PARTICLE SIZE DISTRIBUTION OF TWO HUNDRED PARTICLES OF A SAMPLING FROM AN OILFIELD PIPELINE FILTER SCREEN

Diameter (microns)	Number of Particles	Cumulated Frequency	Ranks	Average Rank (m)	$\frac{m}{n+1}$	Percent
Under 0.30	2	2	1–2	$1\frac{1}{2}$	$1\frac{1}{2}/200$	0.75
0.31–0.40	33	35	3–35	19	19/200	9.50
0.41–0.50	67	102	36–102	69	69/200	34.50
0.51–1.00	5	107	103–107	105	105/200	52.50
1.01–2.00	63	170	108–170	139	139/200	69.50
2.01–4.00	5	175	171–175	173	173/200	86.50
4.01–6.00	11	186	176–186	181	181/200	90.50
6.01–8.00	1	187	187	187	187/200	93.50
8.01–10.00	11	198	188–198	193	193/200	96.50
10.01–20.00	1	199	199	199	199/200	99.50

Because a probability distribution plot reads "equal to or less than," the upper boundary of each cell should be plotted. A plot of the data in Table E2.3-3 gave a badly skewed distribution, as indicated by the extreme curvature in Figure E2.3-3a. This curve, coupled with the physical constraint that the diameter measurements must all be positive and approach zero (there can be no negatively sized particles), suggested that a log-normal plot be made as shown in Figure E2.3-3b.

The line in Figure E2.3-3b is essentially straight down to a diameter of about 0.5 micron, where it makes a sudden break toward zero diameter. Usually, this indicates the presence of some physical condition prohibiting values below (or above) a particular level. In the present instance, however, such an interpretation did not appear reasonable because there was no technical reason why particles smaller than about 0.3 micron could not exist.

After further inquiry the answer was found to be the limit of resolution of the microscope used for measurement. Further investigation showed that approximately 5 percent of the particles picked up by the screen were less than 0.1 micron in diameter, as might be predicted by extending the heavy line in Figure E2.3-3b to 0.1 micron. Very few of the particles (less than one per thousand) could be expected to be larger than about 50 microns.

The multivariate normal probability density, Equation 2.3-7, is nothing more than a generalization of the univariate density. It is written in matrix notation† (refer to Appendix B) for compactness:

$$p(\mathbf{x}) = k\, e^{-q/2} \tag{2.3-7}$$

† The balance of Subsection 2.3-1 can be taken up in conjunction with the matrix notation of Chapter 5, if preferred.

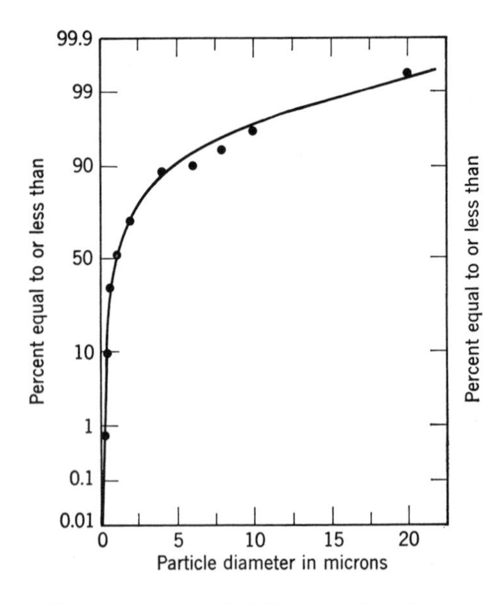

FIGURE E2.3-3A Normal probability plot of particle size data of Table E2.3-3.

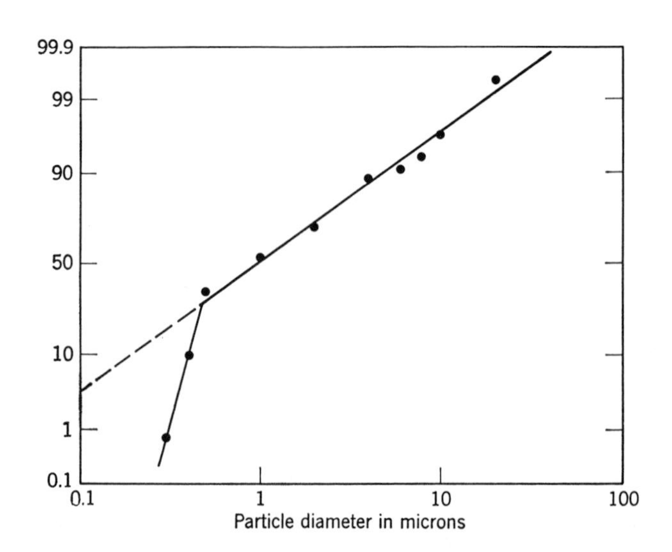

FIGURE E2.3-3B Log-normal plot of particle size data of Table E2.3-3.

where

$$k = \text{positive constant} = \frac{(\det \mathbf{f}^{-1})^{1/2}}{(2\pi)^{n/2}} = \frac{1}{(2\pi)^{n/2}|\mathbf{f}|^{1/2}}$$

termed a normalization factor such that

$$\int\int\cdots\int k \exp\left(-\frac{q}{2}\right) dx_1\, dx_2 \ldots dx_n = 1$$

and

$$q = (\mathbf{x} - \boldsymbol{\mu})^T \mathbf{f}^{-1}(\mathbf{x} - \boldsymbol{\mu})$$

$$\boldsymbol{\mu}^T = [\mu_1, \mu_2, \ldots, \mu_n]$$

$$\mathbf{x}^T = [x_1, x_2, \ldots, x_n]$$

$$\mathbf{f} = \text{Covar}\{\mathbf{X}_i\mathbf{X}_j\}$$

$$= \begin{bmatrix} \sigma_{11} & \cdots & \sigma_{1n} \\ \vdots & & \\ \sigma_{n1} & \cdots & \sigma_{nn} \end{bmatrix} \quad \text{an } n \times n \text{ matrix}$$

$$|\mathbf{f}| = \text{determinant of } \mathbf{f}$$

The μ's and σ's are constants; the μ_i's represent the ensemble means of the respective X's; the σ_{ij}'s represent the variance and covariances of (X_iX_j). (Note that $\sigma_{ii} \equiv \sigma_i^2$.)

Example 2.3-4 Bivariate Normal Density Function

We shall formulate the bivariate normal probability density, the case of $n = 2$ in Equation 2.3-7. The bivariate density has direct applications to turbulent velocity fields, mapping, and targets, as well as in empirical model building.

$$(\mathbf{x} - \boldsymbol{\mu})^T = [(x_1 - \mu_1), (x_2 - \mu_2)]$$

$$\mathbf{f} = \begin{bmatrix} \sigma_{11} & \sigma_{12} \\ \sigma_{21} & \sigma_{22} \end{bmatrix} \qquad \det \mathbf{f} = |\mathbf{f}| = \sigma_{11}\sigma_{12} - \sigma_{12}^2 \quad (\text{since } \sigma_{12} = \sigma_{21})$$

$$= \sigma_1^2\sigma_2^2 - \sigma_{12}^2$$

$$\mathbf{f}^{-1} = \begin{bmatrix} \dfrac{\sigma_{22}}{\det \mathbf{f}} & \dfrac{-\sigma_{12}}{\det \mathbf{f}} \\ \dfrac{-\sigma_{21}}{\det \mathbf{f}} & \dfrac{\sigma_{11}}{\det \mathbf{f}} \end{bmatrix} \qquad \text{with } \sigma_{11} \equiv \sigma_1^2, \quad \sigma_{22} \equiv \sigma_2^2$$

$$q = [(x_1 - \mu_1), (x_2 - \mu_2)] \begin{bmatrix} \dfrac{\sigma_2^2}{|\mathbf{f}|} & \dfrac{-\sigma_{12}}{|\mathbf{f}|} \\ \dfrac{-\sigma_{21}}{|\mathbf{f}|} & \dfrac{\sigma_1^2}{|\mathbf{f}|} \end{bmatrix} \begin{bmatrix} (x_1 - \mu_1) \\ (x_2 - \mu_2) \end{bmatrix}$$

$$= \frac{[(x_1 - \mu_1)^2\sigma_2^2 - 2(x_1 - \mu_1)(x_2 - \mu_2)\sigma_{12} + (x_2 - \mu_2)^2\sigma_1^2]}{\sigma_1^2\sigma_2^2 - \sigma_{12}^2}$$

Now let

$$\rho = \rho_{12} = \frac{\sigma_{12}}{\sigma_1\sigma_2}$$

so that $\sigma_{12} = \sigma_1\sigma_2\rho$:

$$\frac{\sigma_2^2}{\sigma_1^2\sigma_2^2 - \sigma_{12}^2} = \frac{\sigma_2^2}{\sigma_1^2\sigma_2^2 - \sigma_1^2\sigma_2^2\rho^2} = \frac{1}{\sigma_1^2(1 - \rho^2)}$$

$$\frac{\sigma_{12}}{\sigma_1^2\sigma_2^2 - \sigma_{12}^2} = \frac{\sigma_1\sigma_2\rho}{\sigma_1^2\sigma_2^2(1 - \rho^2)} = \frac{\rho}{\sigma_1\sigma_2(1 - \rho^2)}$$

$$\frac{\sigma_1^2}{|\mathbf{f}|} = \frac{1}{\sigma_2^2(1 - \rho^2)}$$

Then

$$q = \frac{\left[\left(\dfrac{x_1 - \mu_1}{\sigma_1}\right)^2 - 2\rho\left(\dfrac{x_1 - \mu_1}{\sigma_1}\right)\left(\dfrac{x_2 - \mu_2}{\sigma_2}\right) + \left(\dfrac{x_2 - \mu_2}{\sigma_2}\right)^2\right]}{(1 - \rho^2)}$$

$$p(x_1, x_2) = \frac{1}{(2\pi)\sigma_1\sigma_2\sqrt{1 - \rho^2}} \exp -\left(\frac{u_1^2 - 2\rho u_1u_2 + u_2^2}{2(1 - \rho^2)}\right)$$

where

$$u_i = \frac{x_i - \mu_i}{\sigma_i} \qquad i = 1, 2, \ldots$$

Tables of the bivariate normal distribution are available.†

For the important but special case in which X_1 and X_2 are stochastically independent, $\rho = 0$ and

$$p(x_1, x_2) = \frac{1}{2\pi\sigma_1\sigma_2} \exp -\left(\frac{u_1^2 + u_2^2}{2}\right)$$

2.3-2 The χ^2 Distribution

The χ^2 distribution has many theoretical and practical applications, some of which will be described in Chapter 3.‡ These include:

1. Testing the goodness of fit of experimental observations to hypothesized probability distributions.
2. Obtaining confidence limits for the variance and the standard deviation.
3. Testing the independence of variables.
4. Deriving the sampling distribution for the standard deviation, the covariance, the coefficient of variation, etc.

Let $X_1, X_2, \ldots, X_\nu$ be a set of ν-independent random variables, each of which is represented by a normal distribution with the respective parameters (μ_1, σ_1^2), $(\mu_2, \sigma_2^2), \ldots, (\mu_\nu, \sigma_\nu^2)$. If we calculate the squares of the standard normal variables, U_i^2,

$$U_i^2 = \left(\frac{X_i - \mu_i}{\sigma_i}\right)^2 \tag{2.3-8}$$

† Nat. Bur. of Standards, *Tables of the Bivariate Normal Distribution Function and Related Functions,* Applied Mathematics Series 50, U.S. Government Printing Office, Washington, D.C., 1959.

‡ A good general reference for the use of χ^2 is: A. E. Maxwell, Analyzing Qualitative Data, John Wiley, New York, 1961.

and sum the U_i^2's, we define a *new* random variable χ^2 ("chi-square") as follows:

$$\chi^2 = U_1^2 + U_2^2 + \cdots + U_\nu^2$$
$$= \sum_{i=1}^{\nu} U_i^2 = \sum_{i=1}^{\nu} \left(\frac{X_i - \mu_i}{\sigma_i}\right)^2 \qquad (2.3\text{-}9)$$

In Equation 2.3-9, ν is called the "number of degrees of freedom" for χ^2. The distribution of χ^2 depends only on ν because the U's are standardized. If the ν observations are *independent*, then the number of degrees of freedom is equal to ν; however, a degree of freedom is lost for each constraint placed on the ν observations.

The *probability density for* χ^2 can be shown to be

$$p(\chi^2) = \frac{1}{(2)^{\nu/2}\Gamma(\nu/2)} (\chi^2)^{\frac{\nu}{2}-1} e^{-\frac{\chi^2}{2}} \qquad (0 < \chi^2 < \infty) \qquad (2.3\text{-}10)$$

and is illustrated in Figure 2.3-6. Some special cases of the χ^2 density of interest are: (1) the square root of χ^2 for $\nu = 2$, called the Rayleigh density function, (2) χ^2 for $\nu = 4$, the Maxwell function for molecular speeds, and (3) $\sqrt{2\chi^2}$ for $\nu > 30$, which is distributed approximately as a normal variable with $\mu = \sqrt{2\nu - 1}$ and $\sigma^2 = 1$. The ensemble mean of χ^2 is equal to the number of degrees of freedom:

$$\mathscr{E}\{\chi^2\} = \mathscr{E}\left\{\sum_{i=1}^{\nu} U_i^2\right\} = \sum_{i=1}^{\nu} \mathscr{E}\{(U_i - 0)^2\} = 1 + 1 + \cdots = \nu$$

because the variance of $U_i = 1$, i.e., $\mathscr{E}\{(U_i - 0)^2\} = 1$. The variance of χ^2 can be shown by direct integration to be

$$\operatorname{Var}\{\chi^2\} = 2\nu$$

The probability distribution for χ^2 is

$$P(\chi_*^2) \equiv P\{\chi^2 \le \chi_*^2\}$$
$$= \frac{1}{(2)^{\nu/2}\Gamma(\nu/2)} \int_0^{\chi_*^2} (\chi^2)^{\frac{\nu}{2}-1} e^{-\frac{\chi^2}{2}} d(\chi^2) \qquad (2.3\text{-}11)$$

Tables of $P(\chi_*^2)$ are available as well as tables of $P(\chi_*^2/\nu)$ and $P\{\chi^2 > \chi_*^2\}$; Table 2.3-3 is a brief extract from Tables C2 in Appendix C. An entry is interpreted for ν degrees of freedom as the upper limit in the integral in Equation 2.3-11. For example, for a $P\{\chi^2 \le \chi_*^2\} = 0.95$ or $P\{\chi^2 > \chi_*^2\} = 0.05$, χ_*^2 can be read for $\nu = 1$ or $\nu = 10$ degrees of freedom as 3.841 or 18.307, respectively.

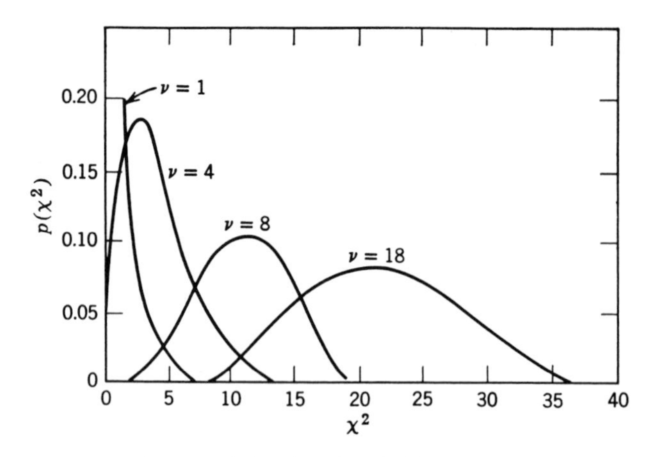

FIGURE 2.3-6 The χ^2 probability density.

2.4 SAMPLE STATISTICS AND THEIR DISTRIBUTIONS

Usually the probability distribution for a process variable is unknown, so the equations in Section 2.2 cannot be directly applied to calculate the ensemble mean, the ensemble variance, and other ensemble averages. While the analyst would like to obtain an estimate of the probability density function for a process variable, this is difficult; in most instances he must settle for merely estimates of the ensemble mean, variance, etc. Two general methods of making an estimate of an ensemble average will be described in this book. One concerns the use of a finite *random sample* of observations or measurements obtained by *repeated* experiments, as discussed in this section. The other method concerns the use of a single time record for *one* experiment, as discussed in Section 12.3.

The term *sample statistic* or just *statistic* refers to a numerical value calculated from a sample of observations or measurements of a random variable. Thus, an estimate of a parameter in a probability density function, probability distribution, or process model or an estimate of an ensemble average obtained from experimental observations is a statistic. A statistic has a dual meaning; it refers to both the rule for calculating the statistic (i.e., a function) and the value of the statistic. The meaning will be clear from the context. Keep in mind that statistics are random variables.

In this section we shall describe the sample mean and sample variance of the random variable X and also their

TABLE 2.3-3 DISTRIBUTION OF χ^2

ν = Degrees Freedom	Probability of a value of χ^2 less than shown in table $P\{\chi^2 \le \chi_*^2\}$						
	0.01	0.05	0.50	0.90	0.95	0.99	0.999
1	1.57×10^{-4}	0.00395	0.455	2.706	3.841	6.635	10.827
10	2.558	3.940	9.342	15.987	18.307	23.209	29.588

probability distributions under specific assumptions about the distribution for the random variable X itself. Sample averages will be denoted by a bar superimposed over the random variable(s) involved, except for the sample variance and correlation coefficient which use other notation for historical reasons. Unless otherwise stated, each finite set of samples is regarded as being statistically independent of any other set if the experiments used to collect the sample are independent and the random variables are statistically independent.

2.4-1 The Sample Mean and Variance

The *sample mean* is generally the most efficient (see Chapter 3) statistic to use in characterizing the central value of experimental data; that is, it requires less data to achieve the same degree of certainty. Let X be a random variable.† A sample of n total observations yields one group of n_1 observations of X denoted by X_1, another of n_2 observations of X denoted by X_2, etc. Then the sample mean is

$$\bar{X} = \frac{1}{n}\sum_i X_i n_i \tag{2.4-1}$$

where $\sum n_i = n$. If $n_i = 1$, then the upper limit of the summation is n. The sample mean is itself a random variable and, being an estimate of μ_X, is often designated $\hat{\mu}_X$.

There are two main reasons why an experimenter makes replicate measurements. One is that the average of the individual results is more representative than any single result. The other is that the dispersion of the individual readings can be evaluated. These objectives may be thwarted unless considerable care is given to the data collection process as described in Chapter 8.

The *sample variance* of the random variable X is a random variable which is the best single estimate of σ_X^2. It is calculated as follows:

$$s_X^2 \equiv \hat{\sigma}_X^2 = \frac{1}{n-1}\sum_i (X_i - \bar{X})^2 n_i \tag{2.4-2}$$

Observe that in the denominator of Equation 2.4-2 the term $(n - 1)$ and not n appears, because the expectation of $\{1/(n-1)\}\sum(X_i - \bar{X})^2 n_i$ is σ_X^2 whereas the expectation of $(1/n)\sum(X_i - \bar{X})^2 n_i$ is $\{(n-1)/n\}\sigma_X^2$. Thus, the latter calculation gives a biased estimate (see Equation 2.4-9 below). (Heuristically, the argument for using $(n - 1)$ instead of n is that one of the n degrees of freedom among the n data values is eliminated when the sample mean is computed. One constraint is placed on the data values; hence the denominator, which represents the degrees of freedom, should be $(n - 1)$.) The sample variance is often more conveniently calculated from Equation 2.4-3a or 2.4-3b.

$$\begin{aligned} s_X^2 &= \frac{1}{n-1}\left[\sum n_i X_i^2 - 2\bar{X}\sum n_i X_i + (\bar{X})^2 \sum n_i\right] \\ &= \frac{1}{n-1}\left[\sum n_i X_i^2 - 2\bar{X}\bar{X}\sum n_i + (\bar{X})^2 \sum n_i\right] \\ &= \frac{1}{n-1}\left[\sum n_i X_i^2 - (\bar{X})^2 \sum n_i\right] \qquad (2.4\text{-}3a) \\ &= \frac{n}{n-1}[(\overline{X_i^2}) - (\bar{X})^2] \qquad (2.4\text{-}3b) \end{aligned}$$

The sample coefficient of variation is

$$c = \frac{s_X}{\bar{X}} \tag{2.4-4}$$

Always be aware that squaring or multiplying first and subtracting afterwards can lead to serious computational round-off error. Thus, for the two equal expressions,

$$\sum_{i=1}^{n}(x_i^2) - \frac{\left(\sum_{i=1}^{n} x_i\right)^2}{n} = \sum_{i=1}^{n}(x_i - \bar{x})^2$$

if $x_1 = 9000$, $x_2 = 9001$, and $x_3 = 9003$, the value of the left-hand relation, using single-precision arithmetic and 8 decimal digits, is 0, and, using single-precision arithmetic and 27 digits in binary arithmetic (which is equivalent to about 8 decimal digits), is 4.0. On the other hand, the value of the right-hand side of the expression by either treatment is correct to 8 decimal digits at 4.6666667

Example 2.4-1 Comparison of Sample Statistics and Their Expected Values

Table 2.3-1 gives the binomial probability density which represents a coin-tossing experiment. Suppose that a coin is tossed 5 times; let X be the number of heads in 5 tosses. If the probability of obtaining a head on each toss is $\theta = \frac{1}{2}$, the second row of Table E2.4-1 shows the probability of obtaining 0, 1, 2, 3, 4, or 5 heads, respectively, in a total of 5 tosses. For this special type of experiment, the sample mean and variance can be compared with the ensemble mean and variance. The experimental data in the third row of the table represent the sum of the results of several experiments of 5 tosses each carried out by different individuals using the same coin.

The calculations are:

$$\mu_X = \sum p(x_i)x_i = 0(\tfrac{1}{32}) + 1(\tfrac{5}{32}) + 2(\tfrac{10}{32}) + 3(\tfrac{10}{32}) + 4(\tfrac{5}{32}) + 5(\tfrac{1}{32}) = 2.5$$

$$\bar{X} = \frac{1}{n}\sum n_i X_i = \tfrac{1}{30}[(0)(1) + 1(6) + 2(10) + 3(7) + 4(5) + 5(1)] = 2.4$$

† The argument of time can be omitted from X since the sampling can be conducted at one time or at different times for a stationary ensemble. The important point is that the data be collected from different experiments and not from one experiment at different times.

TABLE E2.4-1

Values of Random Variable X	0	1	2	3	4	5	Sum
Theoretical probability density							
$p(x) = \dfrac{n!}{n!(n-x)!}\left(\dfrac{1}{2}\right)^x\left(\dfrac{1}{2}\right)^{n-x}$	$\frac{1}{32}$	$\frac{5}{32}$	$\frac{10}{32}$	$\frac{10}{32}$	$\frac{5}{32}$	$\frac{1}{32}$	1
Experimental data (30 tries) $n_i =$	1	6	10	7	5	1	30

$$\sigma_X^2 = \sum p(x_i)(x_i - \mu_X)^2 = \sum p_i(x_i)x_i^2 - \bar{x}^2 \sum p_i$$

$$= [0^2(\tfrac{1}{32}) + 1^2(\tfrac{5}{32}) + 2^2(\tfrac{10}{32}) + 3^2(\tfrac{10}{32}) + 4^2(\tfrac{5}{32}) + 5^2(\tfrac{1}{32}) - (2.5)^2] = 1.25$$

$$s_X^2 = \frac{\sum n_i(X_i - \bar{X})^2}{n-1}$$

$$= \tfrac{1}{29}[(0 - 2.4)^2(1) + (1 - 2.4)^2(6) + (2 - 2.4)^2(10) + (3 - 2.4)^2(7) + (4 - 2.4)^2(5) + (5 - 2.4)^2(1)]$$

$$= 1.42$$

It can easily be shown by interchange of the operators $\mathscr{E}$ and $\sum$ that

$$\mathscr{E}\{\bar{X}\} = \mathscr{E}\left\{\frac{1}{n}\sum_{i=1}^{n} X_i n_i\right\} = \frac{1}{n}\sum_{i=1}^{n}\mathscr{E}\{X_i n_i\}$$

$$= \frac{1}{n}\sum_{i=1}^{n} n_i\mu_X = \mu_X \tag{2.4-5}$$

and by using Equation 2.2-9a for *independent* variables with $\text{Var}\{X_i\} = \sigma_X^2$ that

$$\text{Var}\{\bar{X}\} = \text{Var}\left\{\frac{1}{n}\sum_{i=1}^{n} X_i n_i\right\} = \frac{1}{n^2}\sum_{i=1}^{n} n_i \,\text{Var}\{X_i\}$$

$$= \frac{1}{n^2}(n\sigma_X^2) = \frac{\sigma_X^2}{n} \tag{2.4-6}$$

The positive square root of $\text{Var}\{\bar{X}\}$ is termed the *standard error* or *sample standard deviation.* Thus the sample means themselves are random variables with an expected value the same as that of X and with an ensemble standard deviation of $\sigma_X/\sqrt{n}$. Figure 2.4-1 indicates how the dispersion is reduced for increasing sample sizes as called for by Equation 2.4-6.

One important theorem in statistics, the *central limit theorem*, states that under fairly general conditions the sum of n independent random variables tends to the normal distribution as $n \to \infty$. Thus, the probability density for sample means computed from nonnormal random variables will be more symmetric than the underlying distribution and have less dispersion, as illustrated in Figure 2.4-2.

The sample mean can be transformed to a standard normal variable (which was previously defined by Equation 2.3-2) as follows:

$$U = \frac{\bar{X} - \mu_X}{\sigma_X/\sqrt{n}} \tag{2.4-7}$$

We next show that the expected value of s_X^2, as defined by Equation 2.4-2, is σ_X^2. We split $(X_i - \bar{X})$ into two parts:

$$(X_i - \bar{X}) = (X_i - \mu_X) - (\bar{X} - \mu_X)$$

and replace $(X_i - \bar{X})$ in Equation 2.4-2:

$$(n-1)s_X^2 = \sum [(X_i - \mu_X) - (\bar{X} - \mu_X)]^2 n_i$$

$$= \sum (X_i - \mu_X)^2 n_i - 2\sum (X_i - \mu_X)(\bar{X} - \mu_X)n_i + \sum (\bar{X} - \mu_X)^2 n_i$$

$$= \sum (X_i - \mu_X)^2 n_i - 2n(\bar{X} - \mu_X)^2 + n(\bar{X} - \mu_X)^2$$

$$= \sum (X_i - \mu_X)^2 n_i - n(\bar{X} - \mu_X)^2 \tag{2.4-8}$$

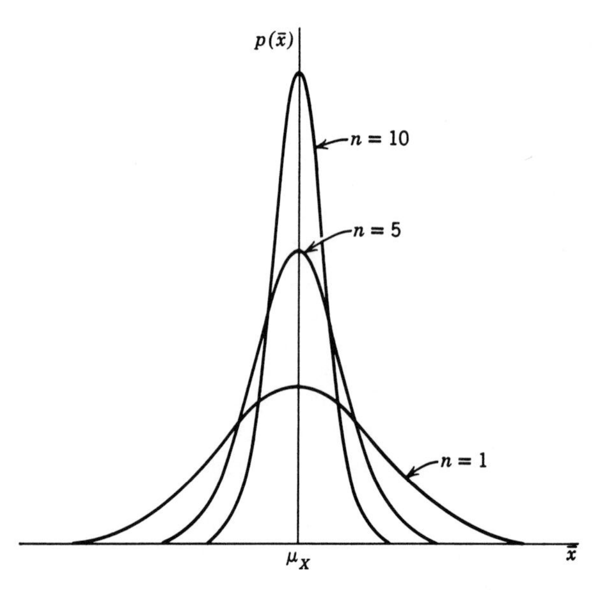

FIGURE 2.4-1 Reduction of dispersion as the sample size increases according to the relation $\text{Var}\{\bar{X}\} = \sigma_X^2/n$.

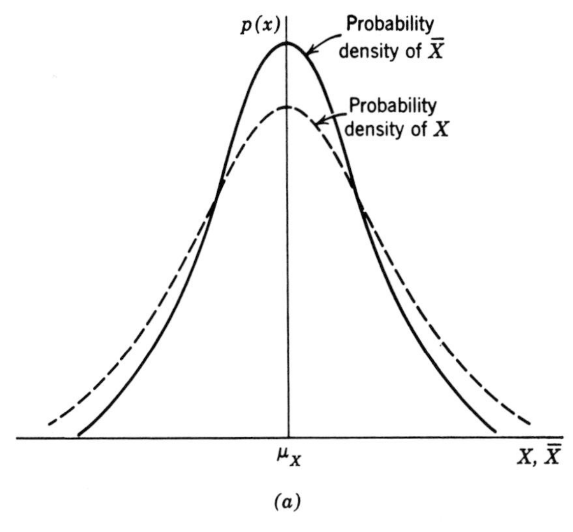

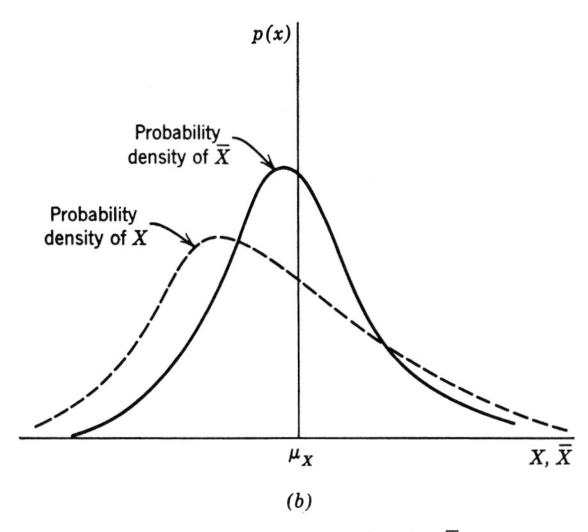

FIGURE 2.4-2 Probability densities of sample means showing the reduced dispersion for the $\bar{X}$ distribution: (a) distribution of X is normal, and (b) distribution of X is not normal.

Next we take the expected value of both sides of Equation 2.4-8:

$$\mathscr{E}\{(n-1)s_X^2\} = \mathscr{E}\left\{\sum (X_i - \mu_X)^2 n_i - n(\bar{X} - \mu_X)^2\right\}$$

$$= n\sigma_X^2 - n \operatorname{Var}(\bar{X}) = n\sigma_X^2 - n\frac{\sigma_X^2}{n}$$

$$= \sigma_X^2(n-1) \qquad (2.4\text{-}9)$$

Consequently, the expected value of the sample variance is the ensemble variance.

To establish the distribution of s_X^2 for n independent observations from a normally distributed population with a mean of μ_X and a variance of σ_X^2 is beyond our scope here, although it can be obtained from the χ^2 partition theorem described in several references at the end of this chapter. All we need here is to note that

$$\sum_{i=1}^{n} \left(\frac{X_i - \bar{X}}{\sigma_X}\right)^2 n_i$$

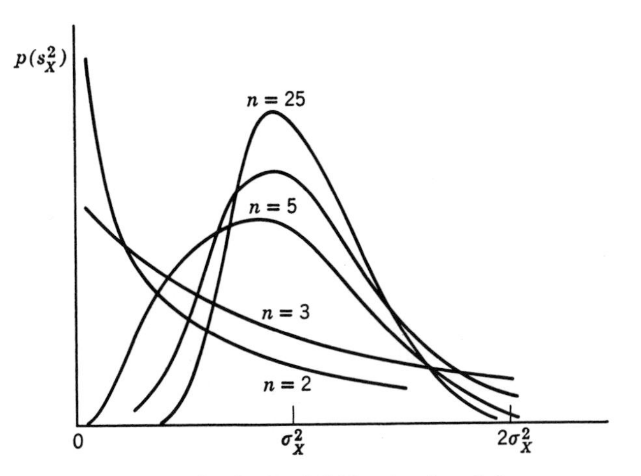

FIGURE 2.4-3 Probability density of s_X^2.

has a χ^2 distribution with $(n-1)$ degrees of freedom; i.e., it is equal χ^2 if the degrees of freedom are $(n-1)$. Consequently, we can write

$$s_X^2 = \sigma_X^2 \frac{\chi^2}{\nu}, \qquad \nu = n - 1 \qquad (2.4\text{-}10)$$

and determine the distribution of s_X^2 (and s_X) from the χ^2 distribution. Figure 2.4-3 shows the probability density of s_X^2.

The ensemble variance of the sample variance itself is defined as

$$\operatorname{Var}\{s_X^2\} = \mathscr{E}\{(s_X^2 - \sigma_X^2)^2\}$$

and can be determined from

$$\operatorname{Var}\{s_X^2\} = \operatorname{Var}\left\{\sigma_X^2 \frac{\chi^2}{\nu}\right\} = (\sigma_X^2)^2 \operatorname{Var}\left\{\frac{\chi^2}{\nu}\right\} = \frac{2\sigma_X^4}{\nu} \qquad (2.4\text{-}11)$$

(Recall that the Var $\{\chi^2\}$ is 2ν.)

For k samples drawn from a normally distributed population, each having the same variance σ_X^2 but not (necessarily) the same mean, a pooled estimate, s_p^2, of σ_X^2 is

$$s_p^2 = \frac{\sum_{i=1}^{k} \nu_i s_i^2}{\sum_{i=1}^{k} \nu_i} \qquad (2.4\text{-}12)$$

where ν_i is the number of degrees of freedom associated with each s_i^2. Thus, by taking a large number of small samples, it is possible to get s_p^2, an estimate of σ_X^2, based on an effectively larger number of degrees of freedom than could be obtained by taking one large sample with the same total number of observations, as indicated in the following example.

Example 2.4-2 Variance Reduction by Pairing of Samples

Suppose a product is formed on two different shifts, A and B, or by two different processes, A and B. The product may be the same or different in various characteristics. But in one specific characteristic, the percentage of a chemical component as determined by titration, the outputs from A and B are supposed to be the same—apart from random normal deviations. In Chapter 3 we shall describe how to determine if the outputs from A and B are the same. However, for the moment, we shall assume that they are the same. Then we can calculate the sample variance using Equation 2.4-2.

Now, on the other hand, suppose we carry out the titrations on pairs of outputs, one of which is selected from A and the other from B. Let X_{i_1} be the result from A and X_{i_2} be the result from B on the ith titration, as shown in Table E2.4-2. If for each pair of titrations, $\bar{X}_i = (X_{i_1} + X_{i_2})/2$, then the sum of the squares of the deviations is, for the ith titration pair,

$$(X_{i_1} - \bar{X}_i)^2 + (X_{i_2} - \bar{X}_i)^2 = \frac{(X_{i_1} - X_{i_2})^2}{2} = \frac{D_i^2}{2} \qquad \text{(a)}$$

where D_i = the difference in measurements. Furthermore, the variance for a pair of measurements is

$$s_i^2 = \frac{1}{2-1}\frac{(X_{i_1} - X_{i_2})^2}{2} = \frac{D_i^2}{2} \qquad \text{(b)}$$

Then the pooled s_p^2 for k sets of measurements is

$$s_p^2 = \frac{\sum_{i=1}^{k} \nu_i s_i^2}{\sum_{i=1}^{k} \nu_i} = \frac{1}{K}\sum_{i=1}^{k} \nu_i \frac{D_i^2}{2} = \frac{1}{2K}\sum D_i^2$$

where K = the total number of degrees of freedom $= \sum \nu_i$.

If the data in the table were to be (incorrectly) treated as individual measurements, $\bar{X} = 70.89$ and the variance calculated from Equation 2.4-2 is

$$s_X^2 = \frac{1}{19}\sum_{i=1}^{20} (X_i - 70.89)^2 = \frac{112.07}{19} = 5.89$$

with 19 degrees of freedom. On the other hand, if the data are (correctly) treated as pairs, from Equation (c)

$$s_p^2 = \frac{3.03}{(2)(10)} = 0.152$$

with 10 degrees of freedom.

TABLE E2.4-2

A	B	D = Difference	D^2
73.2	74.0	0.8	0.64
68.2	68.8	0.6	0.36
70.9	71.2	0.3	0.09
74.3	74.2	−0.1	0.01
70.7	71.8	1.1	1.21
66.6	66.4	−0.2	0.04
69.5	69.8	0.3	0.09
70.8	71.3	0.5	0.25
68.8	69.3	0.5	0.25
73.3	73.6	0.3	0.09
		Total	3.03

We can interpret the results as follows. If for one pair of observations, we note that

$$D_i - \bar{D} = (X_{i_1} - \bar{X}_1) - (X_{i_2} - \bar{X}_2)$$

where $\bar{D} = \sum_{i=1}^{n} D_i/n$, then

$$\frac{\sum (D_i - \bar{D})^2}{n-1} = \frac{\sum (X_{i_1} - \bar{X}_1)^2}{n-1} + \frac{\sum (X_{i_2} - \bar{X}_2)^2}{n-1} - \frac{2\sum (X_{i_1} - \bar{X}_1)(X_{i_2} - \bar{X}_2)}{n-1}$$

or

$$s_D^2 = s_{\bar{X}_1}^2 + s_{\bar{X}_2}^2 - 2s_{X_1X_2} \qquad \text{(c)}$$

Note that the variance of the differences depends on the correlation (covariance in Equation (c)) between pairs of observations. Consequently, it is quite desirable to arrange pairs of observations in the expectation of obtaining high positive correlations, thus reducing the variance. By choosing pairs so that the characteristic of interest in each pair is similar even if the characteristic differs widely from one pair to another, variation between pairs will not affect the variance of the mean difference, because the latter depends only upon differences within pairs.

2.4-2 The t Distribution

The t distribution (or the Student t distribution, so called because of its publication by W. S. Gosset under the pen name of "Student") is employed in making tests and establishing confidence limits for the mean. These tests will be described in Chapter 3. The random variable t represents the ratio of two independent random variables, U, the standardized normal variable, and $\sqrt{\chi^2/\nu}$:

$$t = \frac{U}{\sqrt{\chi^2/\nu}} = \frac{U}{s_X/\sigma_X} = \left(\frac{\bar{X} - \mu_X}{\sigma_X/\sqrt{n}}\right)\frac{1}{s_X/\sigma_X} = \frac{\bar{X} - \mu_X}{s_{\bar{X}}} \qquad (2.4\text{-}13)$$

where $\bar{X}$ is the sample mean and $s_{\bar{X}}$ is the sample standard deviation. The probability density function for t is

$$p(t) = \frac{1}{\sqrt{\pi\nu}}\frac{\Gamma\left(\frac{\nu+1}{2}\right)}{\Gamma\left(\frac{\nu}{2}\right)}\left(1 + \frac{t^2}{\nu}\right)^{-\left(\frac{\nu+1}{2}\right)} \quad (-\infty < t < \infty) \qquad (2.4\text{-}14)$$

where ν is the number of degrees of freedom associated with $s_{\bar{X}}^2$. Figure 2.4-4 illustrates $p(t)$ for various degrees of freedom, ν. Equation 2.4-13 indicates that the sample standard deviation of $\bar{X}$ is used in calculating t whereas to calculate U, the value of σ_X has to be known.

In the limit as $\nu \to \infty$, the t probability density becomes identical with the standard normal probability density,

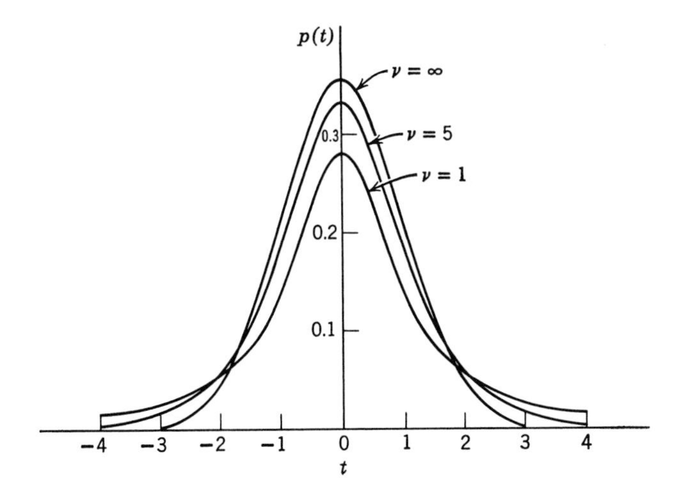

FIGURE 2.4-4 The Student t probability density.

as can be ascertained from Equation 2.4-14 as ν becomes very large.

Figure 2.4-5 illustrates the t probability distribution. Tables of the t probability distribution exist in practically all statistical texts and in Appendix C of this text.† The t distribution gives the probability of t being less than or equal to a selected value of t:

$$P(t) = P\{t \leq t_*\} = \int_{-\infty}^{t_*} p(t)\, dt$$

Some tables record for each degree of freedom, ν, the probability of obtaining a larger absolute value of t than that listed in the table. Other tables use the symmetric property of the t density and record only the probability of obtaining a larger value of t than that listed in the

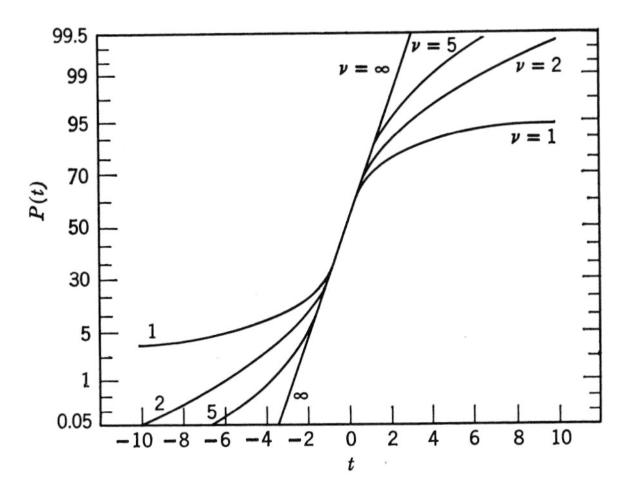

FIGURE 2.4-5 Probability distribution for t.

† More complete tables can be found in: G. U. Yule and M. G. Kendall, *Introduction to the Theory of Statistics*, Griffin, London, 1940, Appendix Table 5; M. Merrington, *Biometrika* **32**, 300, 1941; and R. A. Fisher and F. Yates, *Statistical Tables*, Oliver and Boyd, Edinburgh, Scotland, 1938.

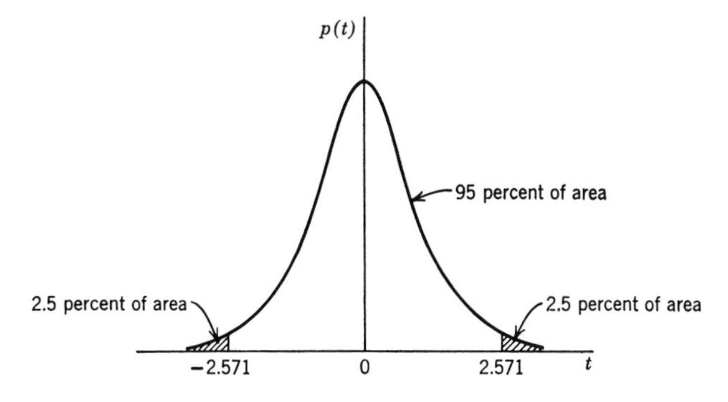

FIGURE 2.4-6 Graphical interpretation of the Student t tables for $\nu = 5$.

table. As an example of the entries in Table C3 in which $P(t) \equiv P\{t \leq t_*\}$, the listing for $\nu = 5$ is

$P(t)$	0.75	0.90	0.95	0.975	0.99
t	0.727	1.476	2.015	2.571	3.365

The listed values can be interpreted to state that 95 percent of the area under the t probability density curve lies within the t values of -2.571 to $+2.571$, and 5 percent of the area (with symmetry) lies outside these values. Examine Figure 2.4-6.

Example 2.4-3 *t* Distribution

If $P\{-2 \leq t \leq t_*\} = 0.25$, what is t_* for $\nu = 10$?

Solution:

From Table C.3 in Appendix C for the t distribution, the $P\{t \leq 2\} \cong 0.96$; hence $P\{t > 2\} \cong 1 - 0.96 = 0.04$. By symmetry, $P\{t \leq -2\} = 0.04$. The total area from $-\infty$ up to t^* is $P = 0.04 + 0.25 = 0.29$, which corresponds to $P\{t \leq t_*) = 0.29$. By use of symmetry again, $P\{t \geq -t_*\} \cong 1 - 0.29 = 0.71$, and from Table C.3, $t_* = -0.56$.

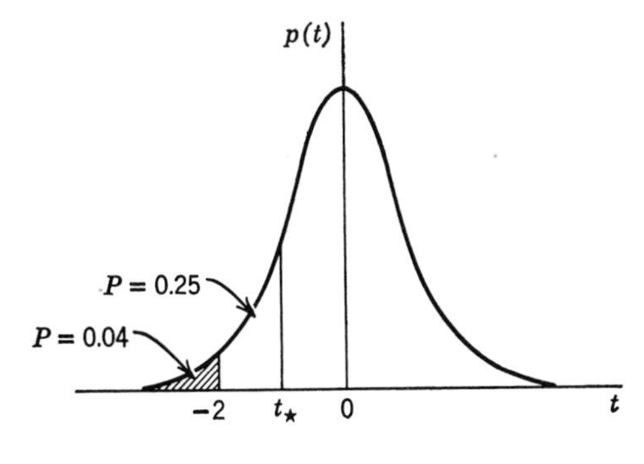

FIGURE E2.4-3

2.4-3 Distribution of the Variance Ratio

A useful distribution developed by R. A. Fisher (the basis of the descriptor F) for the analysis of variance and in model building, topics to be discussed in subsequent chapters, is the distribution of the variance ratio. If two samples are taken, one consisting of n_1 independent measurements of a normal random variable X_1 which has a mean of μ_1 and a variance of σ_1^2, and the other sample consisting of n_2 independent measurements of the normal random variable X_2 which has a mean of μ_2 and a variance of σ_1^2, then the random variable F is defined as

$$F(\nu_1, \nu_2) = \frac{s_1^2/\sigma_1^2}{s_2^2/\sigma_2^2} \qquad (2.4\text{-}15)$$

with $\nu_1 = n_1 - 1$ and $\nu_2 = n_2 - 1$ degrees of freedom. The degrees of freedom associated with the numerator and denominator are those associated with s_1^2 and s_2^2, respectively, and may differ from $(n - 1)$ if the sample variances are calculated by an equation other than Equation 2.4-2. If $\sigma_1^2 = \sigma_2^2 = \sigma^2$ and Equation 2.4-10 is used, F can be related to χ^2:

$$F(\nu_1, \nu_2) = \frac{s_1^2}{s_2^2} = \frac{\chi_1^2/\nu_1}{\chi_2^2/\nu_2} \qquad (2.4\text{-}16)$$

In the argument of F the degrees of freedom for the numerator of Equation 2.4-16 are given as the first number.

Tables of the probability distribution of F, $P(F) = \int_0^{F_*} p(F)\,dF$, are in Appendix C; refer to Table C.4.

The probability density of F is given by

$$p(F) = \frac{\Gamma\left(\frac{\nu_1 + \nu_2}{2}\right)}{\Gamma\left(\frac{\nu_1}{2}\right)\Gamma\left(\frac{\nu_2}{2}\right)} (\nu_1^{\nu_1/2}\,\nu_2^{\nu_2/2}) \frac{F^{\left(\frac{\nu_1}{2}-1\right)}}{(\nu_2 + \nu_1 F)^{\left(\frac{\nu_1+\nu_2}{2}\right)}} \qquad (2.4\text{-}17)$$

and is illustrated in Figure 2.4-7.

The ensemble mean and variance of F are

$$\mathscr{E}\{F\} = \frac{\nu_2}{\nu_2 - 2} \qquad \nu_2 > 2 \qquad (2.4\text{-}18)$$

$$\operatorname{Var}\{F\} = \frac{2\nu_2^2(\nu_1 + \nu_2 - 2)}{\nu_1(\nu_2 - 2)^2(\nu_2 - 4)} \qquad (2.4\text{-}19)$$

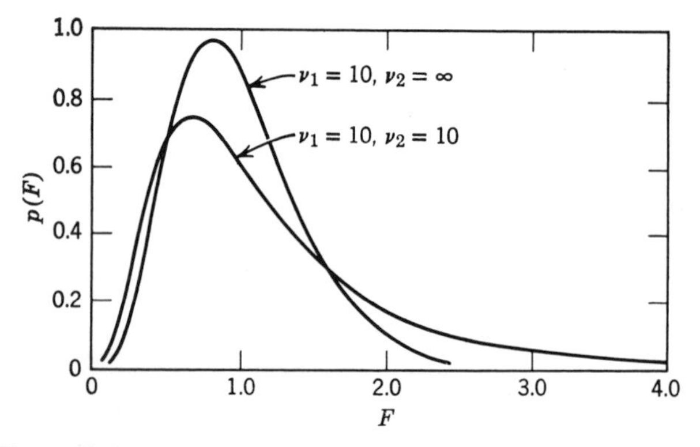

FIGURE 2.4-7 Probability density of F for various values of ν_1, ν_2.

A useful relation is

$F(\nu_2, \nu_1)$ for $P\{F \le F_\alpha\} = k$ is equal to

$$\frac{1}{F(\nu_1, \nu_2) \text{ for } P\{F \le F_\alpha\} = 1 - k}$$

Example 2.4-4 Variance Ratio

Let $\nu_1 = 10$, $\nu_2 = 4$. Then for $P\{0 \le F \le F_*\} = 0.95$, what is F_*?

Solution:

If $P\{0 \le F \le F_*\} = 0.95$, then $P\{F > F_*\} = 0.05$. From Table C.4 in Appendix C, $F_*(10, 4) = 5.96$.

It is also true that $1/5.96 = F_*$ for $P\{F \le F_\alpha\} = 0.05$ with $\nu_1 = 4$ and $\nu_2 = 10$ degrees of freedom.

2.4-4 "Propagation of Error"

A useful feature of experimentation is that experimental measurements can be used to estimate the ensemble mean and variance of a variable which cannot be directly measured. For example, in a material balance, if all the random variables except one are measured, the ensemble mean and variance of the remaining variable can be estimated. We now inquire as to how the engineer can predict the ensemble mean and variance of an unmeasured variable from the ensemble mean and variance of measured variables.

The ensemble mean of a linear function of random variables is equal to the same linear combination of the respective means, as indicated by Equation 2.2-1d. Thus, if $Y = aX + b$,

$$\mathscr{E}\{Y\} = a\mathscr{E}\{X\} + b \qquad (2.4\text{-}20)$$

The ensemble variance of a linear function of random variables is given by Equation 2.2-9 or 2.2-9a. For example, for the single random variable X,

$$\operatorname{Var}\{Y\} = a^2 \operatorname{Var}\{X\} \qquad (2.4\text{-}21)$$

We shall now illustrate the application of Equations 2.2-1d, 2.2-9a, 2.4-20, and 2.4-21.

Example 2.4-5 Controller Error

A process controller as indicated in Figure E2.4-5, senses the values of two streams, and produces an output signal. Each of the sensed streams has error; in addition, the controller introduces error into the output signal. The functional relationship between y and the x's is

$$y = 100 + a_1x_1 + a_2x_2$$

The units of x_i and y are millivolts. The "errors" in the signals shown below as percentages of the value of the ensemble means of the signals represent three standard deviations in the units of x. The expected values of the errors are zero. For the indicated values, calculate the "error" in

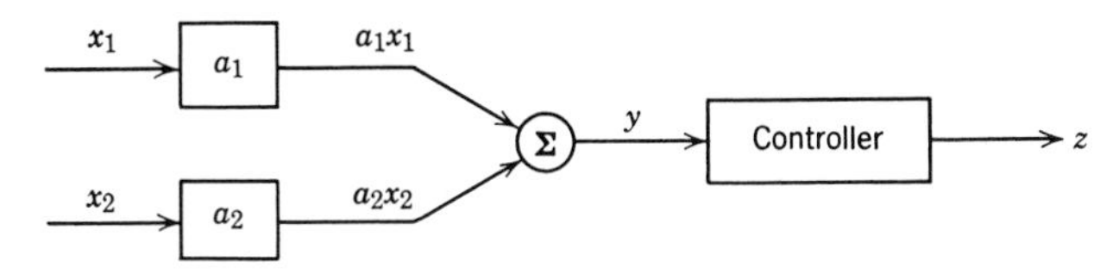

FIGURE E2.4-5

y and z, similarly expressed as three standard deviations. The gain of the controller is unity.

Signal	Constants a_i	μ_X	"Error" $(3\sigma_X)$
x_1	5	100	5%
x_2	2	150	4%
Controller			2% of mean value

Solution:

Let us *assume* that $X_1 = x_1 + \epsilon_1$ and $X_2 = x_2 + \epsilon_2$, where X_1 and X_2 are stochastically *independent* variables. Then $Y = a_1x_1 + a_1\epsilon_1 + a_2x_2 + a_2\epsilon_2 + 100$.

The output of the controller is

$$\mathscr{E}\{Y\} = a_1\mathscr{E}\{x_1\} + a_2\mathscr{E}\{x_2\} + 100$$
$$= 5(100) + 2(150) + 100 = 900 \text{ mv}$$

The variance of the variable Y is

$$\text{Var}\{Y\} = a_1^2 \text{Var}\{X_1\} + a_2^2 \text{Var}\{X_2\}$$
$$= a_1^2 \text{Var}\{\epsilon_1\} + a_2^2 \text{Var}\{\epsilon_2\}$$
$$3\sigma_1 = 0.05(100) \quad \text{or} \quad \sigma_1 = \tfrac{5}{3}; \qquad \sigma_1^2 = (\tfrac{5}{3})^2$$
$$3\sigma_2 = 0.04(150) \quad \text{or} \quad \sigma_2 = \tfrac{6}{3}; \qquad \sigma_2^2 = (\tfrac{6}{3})^2$$
$$\text{Var}\{Y\} = 25(\tfrac{5}{3})^2 + 4(\tfrac{6}{3})^2 = \tfrac{769}{9} (\text{mv})^2$$

The variance of the controller is

$$3\sigma_{\text{contr}} = 0.02(900) \quad \text{or} \quad \sigma_{\text{contr}} = \tfrac{18}{3}; \qquad \sigma_{\text{contr}} = (\tfrac{18}{3})^2$$

Then, assuming that the error introduced by the controller is additive to the error of y

$$\text{Var}\{Z\} = \tfrac{769}{9} + \tfrac{324}{9} = \tfrac{1093}{9} (\text{mv})^2$$
$$\sigma_Z = \sqrt{\tfrac{1093}{9}} = \tfrac{33}{3}$$

so that $3\sigma_Z = 33$ mv. The percent "error" in Z is $(\tfrac{33}{900})(100) = 3.7$ percent.

If the functional relationship between variables is nonlinear, the function must first be linearized in order to apply Equations 2.4-20 and 2.4-21 or 2.2-1d and 2.2-9. The mean and variance computed for the linearized expressions are only approximate and apply only in the vicinity of the state about which the variables in the function have been linearized.

The basic technique underlying linearization is the expansion of the troublesome function in a Taylor series about a mean or reference value of the variable in the domain of interest. A Taylor series for a function of one variable, $f(x)$, about an interval centered on $x = a$ is

$$f(x) = f(a) + \frac{df(a)}{dx}(x-a) + \frac{d^2f(a)}{dx^2}\frac{(x-a)^2}{2!} + \cdots \tag{2.4-22}$$

Linearization is achieved by dropping the second- and higher-order terms.

For example, in the function

$$y = e^{-x}$$

the term e^{-x} is nonlinear; a graph of e^{-x} appears in Figure 2.4-8. Now if we are only interested in small values of x, i.e., values of x only slightly removed from $x = 0$, we may expand e^{-x} about $x = 0$ by using Equation 2.4-22:

$$e^{-x} = 1 - x + \frac{1}{2!}x^2 + \cdots$$
$$\cong 1 - x$$

The general procedure of linearization, then, is to expand any nonlinear functions in a Taylor series about some mean or other constant value of the variables and to retain only the linear terms.

For a function of several variables, the truncated Taylor series can be expressed as

$$f(x_1, x_2, \ldots, x_n) \cong f(x_1^0, x_2^0, \ldots, x_n^0) + \sum_{i=1}^{n} \frac{\partial f(x_1^0, x_2^0, \ldots, x_n^0)}{\partial x_i}(x_i - x_i^0) \tag{2.4-23}$$

where the superscript zeros refer to the reference state for the expansion.

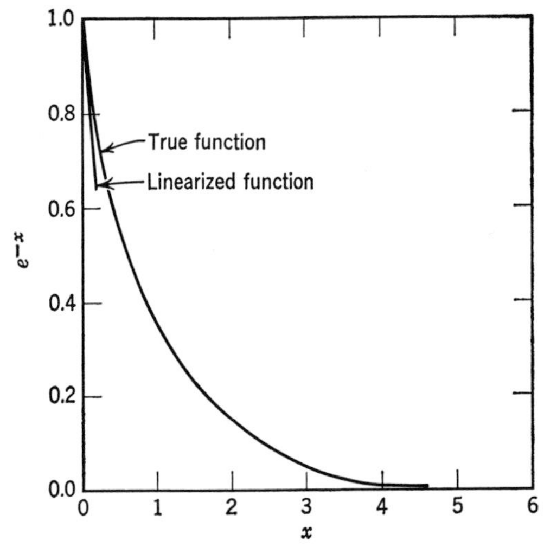

FIGURE 2.4-8 A graph of the function e^{-x} versus x.

For example, for a function of two variables x and y expanded about x_0 and y_0,

$$f(x, y) \cong f(x_0, y_0) + \frac{\partial f(x_0, y_0)}{\partial x}(x - x_0) + \frac{\partial f(x_0, y_0)}{\partial y}(y - y_0)$$

Keep in mind that the partial derivatives are constants that have been evaluated by introducing x_0 and y_0 into the appropriate expression.

Once the function has been linearized by Equation 2.4-22 or 2.4-23, the mean and variance are (assuming the random variables are independent so that Equation 2.2-9a applies)

$$\mathscr{E}\{f(X_1, \ldots, X_n)\} \cong [f(x_1^0, \ldots, x_n^0)] + \sum_{i=1}^{n}\left[\frac{\partial f(x_1^0, \ldots, x_n^0)}{\partial x_i}\right] \times [\mathscr{E}\{(X_i - x_i^0)\}] \qquad (2.4\text{-}24)$$

$$\operatorname{Var}\{f(X_1, \ldots, X_n)\} \cong \sum_{i=1}^{n}\left[\frac{\partial f(x_1^0, \ldots, x_n^0)}{\partial x_i}\right]^2 \operatorname{Var}\{X_i\} \qquad (2.4\text{-}25)$$

where x_i^0 might be the value of the sample mean $\bar{X}_i$, for example.

A special case of Equation 2.4-25 occurs when the original function is of the form

$$Y = cX_1^{a_1}X_2^{a_2}\cdots X_n^{a_n}$$

because then

$$\operatorname{Var}\{Y\} \cong \left(\frac{a_1 y^0}{x_1^0}\right)^2 \operatorname{Var}\{X_1\} + \cdots + \left(\frac{a_n y^0}{x_n^0}\right)^2 \operatorname{Var}\{X_n\}$$

or, as is more commonly encountered,

$$\left(\frac{\sigma_y}{y^0}\right)^2 \cong a_1^2\left(\frac{\sigma_{X_1}}{x_1^0}\right)^2 + \cdots + a_n^2\left(\frac{\sigma_{X_n}}{x_n^0}\right)^2 \qquad (2.4\text{-}26)$$

Example 2.4-6 Mean and Variance of a Nonlinear Function of a Random Variable

Van der Waals' equation can be solved explicitly for P as follows:

$$P = \frac{nRT}{(V - nb)} - \frac{n^2a}{V^2}$$

where

P = pressure (a random variable)
n = number of moles
V = volume (a random variable)
a, b = constants

Assuming $n = 1$ g-mole,

$$a = 1.347 \times 10^6 \text{ atm}\left(\frac{\text{cm}^3}{g\text{-mole}}\right)^2$$

$$b = 38.6\left(\frac{\text{cm}^3}{g\text{-mole}}\right),$$

for air, $T = 300°$K, and that the ensemble mean and variance for V are, respectively, 100 cm^3 and 1 cm^6, find the mean and variance of P in atm. The ideal gas constant

$$R = 82.06\,\frac{(\text{cm}^3)(\text{atm})}{(°\text{K})(g\text{-mole})}$$

Solution:

Since Van der Waals' equation is nonlinear in V, it must first be linearized. Expand the function in a Taylor series, dropping terms of higher order than the first.

$$P \cong \left[\frac{nRT}{V_0 - nb} - \frac{n^2a}{V_0^2}\right] + \left[-\frac{nRT}{(V_0 - nb)^2} + 2\frac{n^2a}{V_0^3}\right](V - V_0)$$

$$\cong \left[\frac{nRT}{V_0 - nb} - \frac{3n^2a}{V_0^3} + \frac{V_0 nRT}{(V_0 - nb)^2}\right] + \left[-\frac{nRT}{(V_0 - nb)^2} + 2\frac{n^2a}{V_0^3}\right]V$$

$$= \alpha + \beta V$$

Then apply Equations 2.4-20 and 2.4-21:

$$\mathscr{E}\{P\} = \alpha + \beta\mathscr{E}\{V\}$$

$$\operatorname{Var}\{P\} = \beta^2 \operatorname{Var}\{V\}$$

$$\alpha = \left[\frac{(82.06)(300)}{(100 - 38.6)} - \frac{(3)(1.347 \times 10^6)}{(100)^2} + \frac{(100)(82.06)(300)}{(100 - 38.6)^2}\right]$$

$$= 648 \text{ atm}$$

$$\beta = \left[-\frac{(82.06)(300)}{(100 - 3816)^2} + \frac{(2)(1.347)10^6}{10^6}\right]\frac{\text{atm}}{\text{cm}^3}$$

$$= -3.84 \text{ atm/cm}^3$$

$$\mathscr{E}\{P\} = 264 \text{ atm}$$

$$\operatorname{Var}\{P\} = 14.75 \text{ atm}^2$$

These results hold for small perturbations about V_0.

Example 2.4-7† Estimate of Error in a Heat Transfer Coefficient

Consider a laboratory experiment dealing with the unsteady-state heating of water in a steam-jacketed open kettle. The apparent overall heat transfer coefficient is given by

$$U_a = \frac{WC_p}{A\,\Delta T_a}\left(\frac{dT}{dt}\right)$$

† Adapted from D. A. Ratkowsky, *J. Chem. Eng. Ed.* **3**, 3, 1965.

where

W = weight of water, lb

C_p = heat capacity of water, Btu/(lb)(°F)

A = area of kettle in contact with water through which heat transfer can take place, ft^2

ΔT_a = apparent temperature difference between steam and water at any instant, $T_s - T_w$, °F

(dT/dt) = slope at any instant of the curve of water temperature versus time

We assume that all the listed variables are random variables. The initial temperature is room temperature.

Find the expected value of U_a and its precision at the condition when $\Delta T_a = 60$°F.

Solution:

From Equation 2.4–26, assuming that the variables are independent,

$$\left(\frac{\sigma_{U_a}}{U_a}\right)^2 = \left(\frac{\sigma_W}{W}\right)^2 + \left(\frac{\sigma_{C_p}}{C_p}\right)^2 + \left(\frac{\sigma_A}{A}\right)^2 + \left(\frac{\sigma_{\Delta T_a}}{\Delta T_a}\right)^2 + \left(\frac{\sigma_{dT/dt}}{dT/dt}\right)^2$$

We shall consider the measurements and estimated variance of each term in sequence.

σ_W: 200 lb of water were measured out in 25-lb batches. If each batch were weighed within maximum error limits of ± 0.30lb, assume that the error represents three sample standard deviations. Therefore $s_{W_i} = 0.30/3 = 0.10$ lb.

$$W = W_1 + W_2 + \cdots + W_8$$

so that by Equation 2.2-9a, if the weighings were independent,

$$\sigma_W^2 \approx s_W^2 = 8s_{W_i}^2 = 0.08 \text{ lb}^2$$

σ_{C_p}: The heat capacity of water is known with sufficiently great precision that we can assume that there is no uncertainty in the value of C_p; i.e., $\sigma_{C_p}^2 = 0$.

σ_A: As heating continued, expansion of the water took place, causing the wetted area to increase. However, the term "apparent heat transfer coefficient" implies that the increase in the area was ignored in favor of using the wetted area at room temperature. From measurements of the liquid depth, and a knowledge of the geometry of the kettle, it was estimated that $A = 8.74$ ft^2 with an uncertainty of $3s_A = 0.45$ ft^2. Therefore,

$$s_A = \frac{0.45}{3} = 0.15 \text{ ft}^2$$

$$\sigma_A^2 \approx s_A^2 = 0.0225 \text{ ft}^2$$

$\sigma_{\Delta T_a}$: The temperature of the steam, assumed to be saturated steam, was determined from the steam pressure which was measured with a mercury manometer. Pressure variations were kept within the error limits of ± 1 inch Hg, i.e., ± 0.5 psi., about a set value of 5 psig. The steam temperature therefore varied between maximum limits of ± 1.5°F. Thus,

$$s_{T_s} = \frac{1.5}{3} = 0.5°\text{F}$$

$$s_{T_s}^2 = 0.25 \text{ (°F)}^2$$

The water temperature was measured by using the average value of two thermocouples, each thermocouple indicating between error limits of ± 0.5°F.

$$T_w = (\tfrac{1}{2})(T_{w_1} + T_{w_2})$$

$$s_{T_w}^2 = (\tfrac{1}{4})(s_{T_{w_1}}^2 + s_{T_{w_2}}^2) = (\tfrac{2}{4})\left(\frac{0.5}{3}\right)^2$$

$= 0.0138$ (°F)^2

Therefore, since $\Delta T_a = T_s - T_w$

$$\sigma_{\Delta T_a}^2 \approx s_{\Delta T_a}^2 = 0.25 + 0.01 = 0.26 \text{ (°F)}^2$$

$\sigma_{dT/dt}$: The derivative of temperature with respect to time at the particular time t where $\Delta T_a = 60$°F was determined from the tangent drawn to a plot of water temperature versus time. After several trials, taking into account the various possibilities for drawing a smooth curve through the points and considering the precision of drawing a tangent to a curve, a reasonable estimate for the derivative dT/dt was 3.0°F/min with variance

$$\sigma_{dT/dt}^2 \approx s_{dT/dt}^2 \cong 0.048 \text{ (°F/min)}^2$$

The average value of the apparent overall heat transfer coefficient, U_a, was calculated as

$$\bar{U}_a = \frac{(200)(1)(3.0)(60)}{(8.74)(60)} = 68.6 \text{ Btu/(hr)(ft}^2\text{)(°F)}$$

and the estimated variance was

$$\hat{\sigma}_{U_a}^2 \approx s_{U_a}^2 = (68.6)^2\left[\frac{0.08}{(200)^2} + \frac{0.0225}{(8.74)^2} + \frac{0.26}{(60)^2} + \frac{0.048}{(3.0)^2}\right]$$

$$= (68.6)^2[2 \times 10^{-6} + 2.95 \times 10^{-4} + 7.22 \times 10^{-5} + 0.00533]$$

$$= (4706)(0.00570) = 26.82$$

The estimated standard deviation of U_a was

$$\hat{\sigma}_{U_a} \approx s_{U_a} = 5.18$$

The estimate of σ_{U_a} is only approximate because the sample variances themselves were only approximate and the ensemble variance for U_a was estimated from a linearized relation. Another way to estimate σ_{U_a} would be to calculate s_{U_a} from repetitive experiments. A reexamination of the error analysis shows that the largest contribution to the experimental error lies in the term involving dT/dt. The error in U_a could best be reduced by reducing the error in the temperature-time curve and the evaluation of its slope.

Keep in mind that the error analysis as outlined in this section encompasses only one phase in the analysis of measurement error. As discussed in the introduction to this chapter, a fixed displacement or bias can contribute to the overall error and cannot be overcome by replication. If the engineer wants to estimate, for example, pressure with a Bourdon gauge, he cannot simply take ten Bourdon gauges and then average the resulting ten readings unless the factory or some other source has recently calibrated each instrument. Without such calibrations the instruments from one lot, for example, might all be biased to read high. Proper calibration, either by adjustment of the gauge or by application of correction constants, would help to ensure that the readings are unbiased. Unfortunately, we do not always work with instruments that have just been calibrated; consequently, we must not forget to consider possible bias.

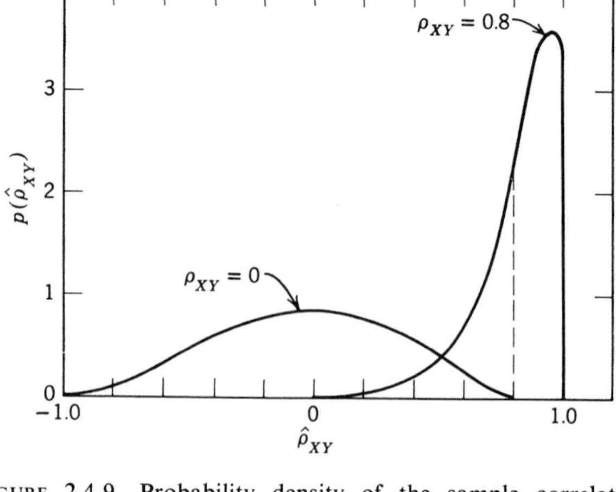

FIGURE 2.4-9 Probability density of the sample correlation coefficient, $\hat{\rho}_{XY}$.

2.4-5 Sample Correlation Coefficient

The *sample correlation coefficient*, $\hat{\rho}_{XY}$, is an estimate of the ensemble correlation coefficient, ρ_{XY}, which is defined in Equation 2.2-14. The sample covariance is a random variable defined as

$$s_{XY} = \frac{1}{(n-1)}\sum_{i=1}^{n}(X_i - \bar{X})(Y_i - \bar{Y})n_i \quad (2.4\text{-}27)$$

and may alternately be computed, noting that

$$\sum_{i=1}^{n}(X_i - \bar{X})(Y_i - \bar{Y})n_i = \sum_{i=1}^{n} n_i X_i Y_i - n\bar{X}\bar{Y}$$

as

$$s_{XY} = \frac{1}{(n-1)}\left[\sum_{i=1}^{n} n_i X_i Y_i - \frac{1}{n}\sum_{i=1}^{n} n_i X_i \sum_{i=1}^{n} n_i Y_i\right] \quad (2.4\text{-}27a)$$

Consequently, the sample correlation coefficient is

$$\hat{\rho}_{XY} = \frac{s_{XY}}{s_X s_Y} \qquad (-1 \le \hat{\rho}_{XY} \le +1)$$

$$= \frac{1}{(n-1)}\sum_{i=1}^{n}\left(\frac{X_i - \bar{X}}{s_X}\right)\left(\frac{Y_i - \bar{Y}}{s_Y}\right)n_i \quad (2.4\text{-}28)$$

If an empirical model is proposed and a series of designed experiments carried out to determine the relationship between two variables, and the observations for one or both variables contain error, then the procedure of regression analysis can be applied as described in Chapter 4 and 5. On the other hand, if one simply measures or observes two variables in a random sample, it is possible to calculate a measure of the *linear* association between the variables, namely the sample correlation coefficient. No distinction is made between the variables as to which is the independent and which the dependent one. If, for some reason, one of the variables, although random in the population, is sampled only in a limited range or picked at preselected values, then the sample correlation coefficient is a distorted estimate of the ensemble correlation coefficient.†

The distribution of the sample correlation coefficient, $\hat{\rho}_{XY}$, is quite complicated. It is symmetrical only for $\rho_{XY} = 0$ and very skewed if $|\rho_{XY}|$ is large, unless n is very large (refer to Figure 2.4-9). Fisher described a transformation of $\hat{\rho}_{XY}$,

$$Z^* = \tanh^{-1}\hat{\rho}_{XY} = \tfrac{1}{2}\ln\frac{1+\hat{\rho}_{XY}}{1-\hat{\rho}_{XY}}$$

where Z^* is approximately normally distributed for any ρ_{XY} and moderate values of n.

† Refer to C. Eisenhart, *Ann. Math. Stat.* **10**, 162, 1939; and M. Ezekiel, *Methods of Correlation Analysis*, John Wiley, New York, 1941, Chapter 20.

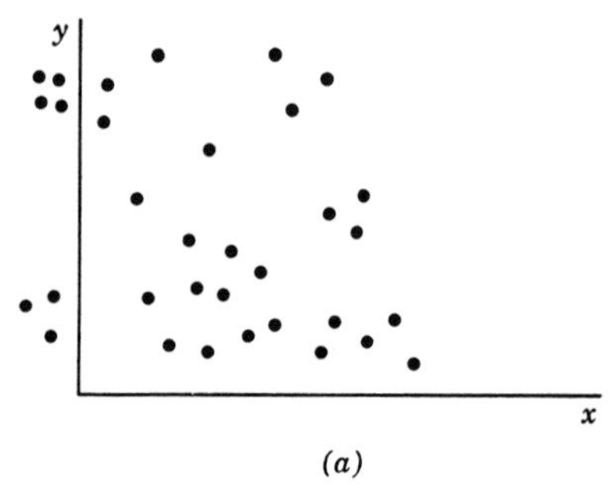

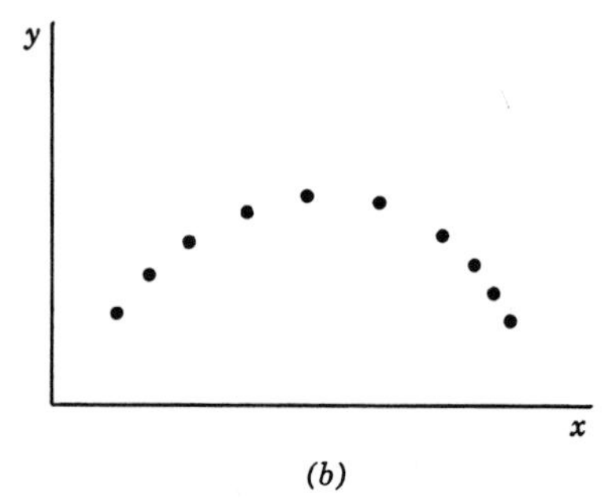

FIGURE 2.4-10 Scatter diagrams of hypothetical data with essentially zero correlation.

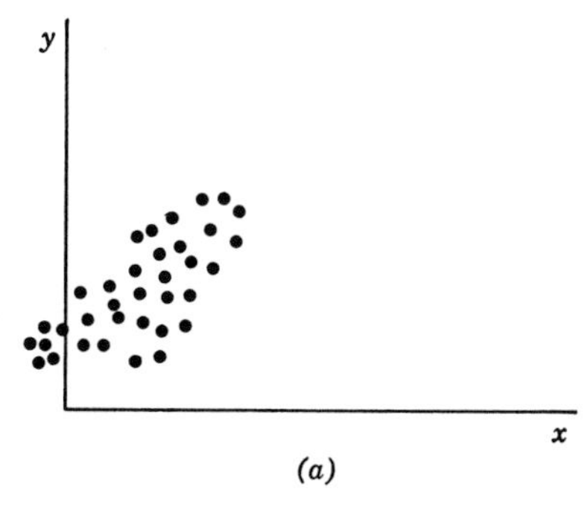

(a)

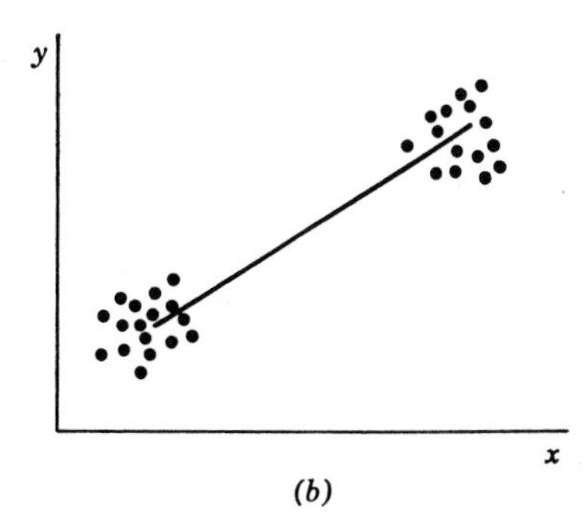

(b)

FIGURE 2.4-11 Scatter diagrams for data with high positive correlation.

The sample correlation coefficient is an estimate of ρ_{XY}; tests which are made† are based on an assumed joint normal distribution for *both* X and Y. Nonnormality can lead to quite biased estimates and hence erroneous conclusions.

When interpreting sample correlation coefficients, it is wise to observe certain precautions. As Figure 2.4-10*b* shows in a qualitative way, the sample correlation coefficient can be quite close to zero and yet the variables X and Y be related rather well by a nonlinear function. If the sample correlation coefficients were to be calculated for the data in Figure 2.4-10*b*, the coefficient would be near zero. We can conclude that a *nonlinear* relation can exist between two variables which will not be detected by the analyst who uses only the sample correlation coefficient as his measure of the relation. Figure 2.4-11 illustrates the necessity of using homogeneous data to avoid a spurious correlation which arises when two nonhomogeneous groups of data are combined in calculating the sample correlation coefficient. Finally, it is essential to keep in mind that a significant correlation does *not* prove that a *causal* relationship exists between two variables.

† For various tests which can be made for ρ_{XY} together with tables and charts, see E. S. Pearson and H. O. Hartley, *Biometrica Tables for Statisticians*, Vol. I (2nd ed.), Cambridge Univ. Press, 1958; R. A. Fisher and F. Yates, *Statistical Tables for Biological, Agricultural, and Medical Research* (3rd ed.), Oliver and Boyd, Edinburgh, 1948; and F. N. David, *Tables of the Correlation Coefficients*, Biometrika Office, University College, London, 1938.

Example 2.4-8 Sample Correlation Coefficient

Eight lots of polymer were taken randomly from a process and two characteristics measured: (1) sedimentation rate and (2) crystallinity. What is the sample correlation coefficient between these two variables?

Sedimentation rate	15	11	8	8	6	4	3	1
Crystallinity	8	8	7	5	4	3	2	1

Solution:

The sample correlation coefficient can be calculated from Equation 2.4-28 as shown in Table E2.4-8.

In general, a value of $\rho_{XY} = 0.937$ is "high"; consult the aforementioned references for appropriate tests.

TABLE E2.4-8

X_i, Crystallinity	Y_i Sedimentation Rate	$(X_i - \bar{X})$	$(Y_i - \bar{Y})$	$(X_i - \bar{X})^2$	$(Y_i - \bar{Y})^2$	$(X_i - \bar{X})(Y_i - \bar{Y})$
1	1	−4	−6	16	36	24
2	3	−3	−4	9	16	12
3	4	−2	−3	4	9	6
4	6	−1	−1	1	1	1
5	8	0	1	0	1	0
7	8	2	1	4	1	2
8	11	3	4	9	16	12
8	15	3	8	9	64	24
40	56	0	0	52	144	81
$\bar{X} = 5$	$\bar{Y} = 7$					

$$\hat{\rho}_{XY} = \frac{\sum (X_i - \bar{X})(Y_i - \bar{Y})}{\sqrt{\sum (X_i - \bar{X})^2 \sum (Y_i - \bar{Y})^2}}$$

$$= \frac{81}{\sqrt{(52)(144)}} = 0.937$$

Supplementary References

Bennett, C. A. and Franklin, N. L., *Statistical Analysis in Chemistry and the Chemical Industry*, John Wiley, New York, 1954.
Brownleee, K. A., *Statistical Theory and Methodology in Science and Engineering*, John Wiley, New York, 1965.
Dixon, W. J. and Massey, F. J., *Introduction to Statistical Analysis* (2nd ed.), McGraw-Hill, New York, 1957.
Dodge, H. F. and Romig, H. G. *Sampling Inspection Tables* (2nd ed.), John Wiley, New York, 1959.
Feller, W., *An Introduction to Probability Theory and its Applications*, Vol. 1 (2nd ed.), John Wiley, New York, 1968.
Hald, A., *Statistical Theory with Engineering Applications*, John Wiley, New York, 1952.
Hoel, P. G., *Introduction to Mathematical Statistics* (3rd ed.), John Wiley, New York, 1962.
Huff, D., *How to Lie with Statistics*, Norton, New York, 1954.
Kenney, J. F. and Keeping, E. S., *Mathematics of Statistics*, D. Van Nostrand, New York, 1951.
Mandel, J., *The Statistical Analysis of Experimental Data*, John Wiley, New York, 1964.
Mosteller, F., Rourke, R. E. K., and Thomas, G. B., *Probability and Statistics*, Addison Wesley, Reading, Mass., 1961.
Wallis, W. A. and Roberts, H. V., *Statistics: A New Approach*, Free Press, Glencoe, Ill., 1956.
Wilks, S. S., *Mathematical Statistics*, John Wiley, New York, 1962.
Wine, R. L., *Statistics for Scientists and Engineers*, Prentice-Hall, Englewood Cliffs, N.J., 1964.

Problems

2.1 Accumulate 20 pennies. Predict what type of probability function you would expect to find for each of the following experiments, and write its equation down or draw a sketch. Then take the pennies and perform the indicated experiments. Compare your experimental relative frequencies with the predicted probabilities.

(a) The distribution of the sizes of the pennies rounded off to the nearest inch.
(b) The distribution of the sizes of the pennies rounded off to the nearest 0.5 mm.
(c) The distribution of the sizes of the pennies rounded off to the nearest $\frac{1}{1000}$ of an inch. Use a micrometer.
(d) The distribution of heads when each coin is tossed once; the distribution of tails; the distribution of standing on edge. Are there are other outcomes?
(e) The distribution of the years of manufacture on the coins.
(f) The distribution of the *ending* (right-hand) digits of the year of manufacture; the distribution of the *first* (left-hand) digits.

2.2 Draw a graph of the probability function and the cumulative probability function of a discrete random variable which has the cumulative probability distribution given by

$$P\{X \leq x\} = \begin{cases} 0 & x < 0 \\ \dfrac{x+1}{n+1} & x = 0, 1, 2, \ldots, n \\ 1 & x > n \end{cases}$$

What is $P\{X = 3\}$?

2.3 Given that the probability distribution function for a continuous random variable is

$$P(x) = P\{X \leq x\} = \begin{cases} 0 & x < 0 \\ \dfrac{x}{n} & 0 \leq x \leq n \\ 1 & x > n \end{cases}$$

plot the probability distribution (versus x) and determine and plot the relation for the probability density.

2.4 The Rayleigh probability density function is

$$p(r) = \frac{r}{\sigma^2} e^{-r^2/2\sigma^2} \qquad r > 0$$

where σ^2 is a constant. Determine the Rayleigh probability distribution function, $P(r)$, and plot both $p(r)$ and $P(r)$ for several values of σ^2. ($P(r)$ corresponds to the probability that a point on a plane, whose coordinates are independent normally distributed random variables, lies within a circle of radius r.)

2.5 By analogy with thermodynamics, the "entropy" for a discrete probability function can be defined as

$$H(n) = -\sum_{k=1}^{n} P(x_k) \ln P(x_k)$$

where $P(x_k) = P\{X = x_k\}$. Under what circumstances is $H = 0$, and what is the interpretation of your answer? Under what circumstances does the greatest uncertainty exist, and what is $P(x_k)$ then? What is $H(n)$ then? Under what circumstances is the entropy a minimum. What is $H(n)$ then?

2.6 The Maxwell probability density is

$$p(x) = \frac{\sqrt{2}}{\alpha^3\sqrt{\pi}} x^2 e^{-x^2/2\alpha^2} U(x)$$

Show that $p(x) \geq 0$ and that $\int_{-\infty}^{\infty} p(x)\,dx = 1$. $U(x)$ is the unit step function and α is a constant.

2.7 The joint probability of X and Y is given in the table below. Show that X and Y are independent.

		x			
		1	2	3	4
	0	$\frac{1}{24}$	$\frac{1}{12}$	$\frac{1}{12}$	$\frac{1}{24}$
y	1	$\frac{1}{12}$	$\frac{1}{6}$	$\frac{1}{6}$	$\frac{1}{12}$
	2	$\frac{1}{24}$	$\frac{1}{12}$	$\frac{1}{12}$	$\frac{1}{24}$

2.8 If the probability density of X is

$$p(x) = \frac{1}{\sqrt{2\pi}\sigma_X} \exp\left[-\frac{(x-\mu_X)^2}{2\sigma_X^2}\right]$$

and the probability density of Y is

$$p(y) = \frac{1}{\sqrt{2\pi}\sigma_Y} \exp\left[-\frac{(y-\mu_Y)^2}{2\sigma_Y^2}\right]$$

what is the probability density of X and Y, $p(x \cap y) = p(x, y)$, if X and Y are independent random variables? ($\cap$ is defined in Appendix A.)

2.9 What is the meaning of "joint probability distribution"? Give an example of both a discrete and a continuous joint distribution. Draw a picture and label axes.

2.10 Is it possible to integrate and differentiate random variables?

2.11 What is the expected value of each of the following quantities? (Y is a random variable; y is a deterministic variable; a is a constant.)

(a) $\dfrac{d^2 Y}{dt^2}$

(b) $\dfrac{d^2 y}{dt^2}$

(c) $f(Y) = \int_0^a (6Y + 5)\, e^{-t}\, dt$

(d) $f(Y) = \sum_{i=1}^{n} i Y_i^2$

2.12 What is the expected value of the dependent variable in each of the following differential equations? (X and Y are random variables; a and b are constants; y is deterministic.)

(a) $\dfrac{d^2 Y}{dt^2} + a\dfrac{dY}{dt} = X(t)$

(b) $\dfrac{\partial Y}{\partial t} + a\dfrac{\partial Y}{\partial x} = b(Y - y)$

2.13 Find the ensemble mean (expected value) of:

(a) The Rayleigh random variable (see Problem 2.4)
(b) The random variable X which is uniformly distributed in the interval $a \le X \le b$, and zero elsewhere.

2.14 For which of the following probability densities is the random variable X stationary in the strict sense?

(a) $p(x) = \dfrac{1}{\sqrt{2\pi}\,\sigma_X} \exp\left[-\left(\dfrac{x-\mu_X}{2\sigma_X^2}\right)\right]$ (normal random variable)

(b) $p(x) = e^{-\lambda t}\dfrac{(\lambda t)^x}{x!}$ (Poisson random variable)

2.15 Find the ensemble mean of the dependent variable in the following process models (capital letters are the random variables).

(a) Heat transfer

$$\frac{d^2 T}{dx^2} = 0 \qquad T(x_1) = T_{10} \quad T(x_2) = T_{20}$$

(b) Mass transfer

$$\frac{\partial C}{\partial t} = a\frac{\partial^2 C}{\partial x^2} \qquad C(0, x) = 0 \quad C(t, 0) = C_0 \quad \lim_{x\to\infty} C(t, x) = 0$$

2.16 Find the ensemble mean and variance of the random variable X which is represented by the rectangular density

$$p(x) = \frac{1}{\alpha} \quad \text{for } \left(-\frac{\alpha}{2} \le X \le \frac{\alpha}{2}\right)$$

$$p(x) = 0 \quad \text{elsewhere}$$

2.17 If the variance of a random variable X is 0.75, what is the variance of the following random variables?

(a) $5X$

(b) $\dfrac{X}{2}$

(c) $(X + 7)$

(d) $\left(\dfrac{X-3}{2}\right)$

2.18 Under what circumstances is $\mathscr{E}\{X\}\mathscr{E}\{Y\} = \mathscr{E}\{XY\}$?

2.19 State in each case whether the random variable X is a stationary random variable (in the weak sense) or not and explain why.

(a) $X(t) = \cos(at + \Gamma)$ (Γ is a random variable)
(b) $X(t) = A\cos wt + B\sin wt$ (A and B are random variables)
(c) $X(t) = aY + bt$ (Y is a random variable)
(d) $X = aY + b$ (Y is a random variable)

2.20 Given the random variable $Y(t)$ below and a corresponding probability density, calculate its autocorrelation function.

Variables:

(a) $Y(t) = A\,e^{i\omega t}$ (A is a random variable; i is the $\sqrt{-1}$)
(b) $Y(t) = A_1\cos\omega t + A_2\sin\omega t$ (A_1, A_2 are independent random variables)
(c) $Y = Ax + b$ (A is a random variable independent of time)

Densities:

(a) $p(a) = \dfrac{1}{b}$ $\quad |A| \le \dfrac{b}{2}$

$p(a) = 0$ $\quad |A| > \dfrac{b}{2}$

(b) $p(a) = c\,e^{-a^2/b}$ where c is a constant to be determined

2.21 Determine the autocovariance for the random variables of Problem 2.20.

2.22 Are X and Y independent variables? The boxes give the values for the joint probability density.

		X=1	X=2	X=3
Y	1	$\frac{1}{4}$	$\frac{1}{8}$	$\frac{1}{8}$
	2	$\frac{1}{8}$	$\frac{1}{16}$	$\frac{1}{16}$
	3	$\frac{1}{8}$	$\frac{1}{16}$	$\frac{1}{16}$

2.23 If X is a random variable with a mean of μ_X, what are the $\mathscr{E}\{X\}$, $\mathscr{E}\{2X\}$, $\mathscr{E}\{X + 1\}$, $\mathscr{E}\{2X + 1\}$, $\mathscr{E}\{X^2\}$, and $\mathscr{E}\{X - \mu_X\}$? Note $\mathscr{E}\{X^2\} \neq \mathscr{E}\{(X)\}^2$. Determine the corresponding variances, i.e., Var $\{X\}$, Var $\{2X\}$, etc.

2.24 The joint probability density function for two random variables X and Y is given by

$$p(x, y) = x + y \qquad \text{for } 0 \leq X \leq 1$$
$$0 \leq Y \leq 1$$
$$p(x, y) = 0 \qquad \text{elsewhere}$$

Find the ensemble correlation coefficient between X and Y.

2.25 Given the indicated joint probability density, calculate the crosscorrelation function $r_{XY}(t_1, t_2)$ of the random variables X and Y.

(a) X and Y are independent random variables uniformly distributed in the intervals $(0, a)$ and $(0, b)$, respectively.

(b) X and Y are jointly normal random variables whose probability density is

$$p(x, y) = A \exp - (ax^2 + bxy + cy^2 + dx + ey)$$

where

$$ax^2 + bxy + cy^2 + dx + ey \geq 0$$

2.26 A lumped stochastic process can be represented by the following model:

$$\frac{dY(t)}{dt} + 2Y(t) = X(t)$$
$$Y(0) = 0$$

where $X(t)$ is a stationary random input to the process with $\mathscr{E}\{X(t)\} = 2$ and $r_{xx}(\tau) = 4 + 2e^{-|\tau|}$ where $\tau = t_2 - t_1$. Find:

(a) $\mathscr{E}\{Y(t)\}$
(b) $r_{XY}(t_2, t_1)$
(c) $r_{YY}(t_2, t_1)$

2.27 The random variables X and Y are independent with the respective probability densities $p(x) = e^{-x}$ and $p(y) = e^{-y}$, with $X \geq 0$ and $Y \geq 0$. Calculate the crosscovariance and the correlation coefficient of X and Y.

2.28 Define correlation in terms of covariance and variance, and briefly discuss the statement that "independent variables are uncorrelated, but not all uncorrelated variables are independent."

(a) Give an example in which zero correlation implies independence.
(b) Give an example in which zero correlation does not imply independence.

Use a bivariate distribution in both (a) and (b).

2.29 If the random variable X is uniformly distributed in the interval $-a$ to a, $p(x) = 1/2a$ in the interval and zero elsewhere. What are the zeroth, first, and second raw and central moments of X?

2.30 What are the zeroth, first, and second raw moments of X if $p(x)$ is the exponential probability density $p(x) = k\,e^{-x}$? What is k? How can k be evaluated?

2.31 Prove that the peak of the standard normal probability density curve is at μ_X and the inflection points at $\mu_X \pm \sigma_X$.

2.32 What are

(a) $P\{U > 0.4\}$
(b) $P\{U > -0.4\}$
(c) $P\{|U| < 0.4\}$

for the standard normal random variable?

2.33 Assume that density of a product is represented by the normal distribution, and it is *known* that the μ of the density is 6.4 g/cc and σ^2 is 1.4 $(\text{g/cc})^2$. What is the lowest value of the density that can be in the upper 15 percent of all the densities?

2.34 If the distribution of the diameters of screw threads can be represented by the normal distribution, and the diameter has an ensemble mean of 0.520 inch and an ensemble standard deviation of 0.008 inch, determine the percentage of threads with diameters: (a) between 0.500 and 0.525 inch, (b) greater than 0.550 inch, and (c) equal to 0.520 inch.

2.35 What is the probability of the standard normal random variable having a value between: (a) 0 and 1, (b) -2 and 0, (c) -3 and 3, and (d) 0.5 and 0.52?

2.36 Prove that the mean of the standardized normal variable is 0 and its variance is 1. What is the probability that a variable (represented by the standard normal distribution) is exactly 1?

2.37 If Y is a standard normal random variable, find:

(a) $P\{Y > 0.2\}$
(b) $P\{0.2 < Y < 0.3\}$
(c) $P\{-0.4 \leq Y \leq 1.0\}$
(d) $P\{Y > 2\}$

2.38 By integration of the probability density, show that the variance of U, the standard normal variable, is 1.

2.39 By use of normal probability paper, determine if the following data can be represented by the normal distribution.

Number	Value of Variable (in)
5	6.00–6.19
18	6.20–6.29
42	6.30–6.39
27	6.40–6.49
8	6.50–6.59

2.40 Select 100 digits at random from the phone book by selecting the next to the last digit in 100 telephone numbers. Can these data be represented by the normal distribution?

2.41 Compute the following probabilities for χ^2 for $\nu = 10$.
(a) $P\{\chi^2 < 10\}$
(b) $P\{\chi^2 > 10\}$
(c) $P\{5 < \chi^2 < 15\}$
(d) $P\{\chi^2 = 3\}$

2.42 Show that the variance of χ^2 is 2ν where ν is the number of degrees of freedom.

2.43 Make a plot of $P\{\chi^2 \leq \chi_\alpha^2\} = P(\chi_\alpha^2)$ as given by Equation 2.3-11 for $\nu = 4$.

2.44 To illustrate the concepts involved in the distribution of the sample mean, carry out the following simple experiment. Use the equation

$$Y = 10\left(1 + \frac{X}{2}\right)^2\left(1 - \frac{X}{3}\right) \qquad 0 \leq X \leq 9$$

to calculate 40 values of the random variable Y from 40 random values of X. (Look at the middle digit in telephone numbers in the telephone book for the X values if a table of random values is not available.) Make a plot of the relative frequency of Y versus Y to illustrate the distribution of Y values. Locate the sample mean, and place lines at $\pm s_Y$, $\pm 2s_Y$, and $\pm 3s_Y$. Then group your data into 10 groups of 4 each and find $\bar{Y}$ for each group. Plot the relative frequency of $\bar{Y}$. Use class limits such as 5.0 to 5.5, etc.; place lines at $\pm 1s_Y/\sqrt{n}$, $\pm 2s_Y/\sqrt{n}$, and $\pm 3s_{\bar{Y}}/\sqrt{n}$ on the second diagram.

2.45 A sample of lightbulbs from Lot 16 fails as follows:

Lifetime (hr)	Number
2000–2999	12
3000–3999	64
4000–4999	35
5000–5999	14

Find the sample average lifetime for this lot.

2.46 Find the sample mean, variance, and standard deviation of the following experimental data.
(a) Background counts prior to detection of a radioactive sample; each count is for two minutes:

12, 15, 10, 18, 14

(b) Counts recorded for the radioactive sample; each count is for two minutes:

95, 92, 103, 89, 88, 95, 90, 93, 89, 102

2.47 Eight pairs of analyses were carried out on batches of acorns to determine their fiber content. The results were as shown below.

Day of Analysis	Tree A (%)	Tree P (%)	Difference (%)
1	37	37	0
2	35	38	3
3	43	36	−7
4	34	47	13
5	36	48	12
6	48	57	9
7	33	28	−5
8	33	42	9

Analyze the data first as if there were 16 unpaired analyses, and determine the variance for the sample of 16, Then take into account the fact that pairs of analyses were carried out, and calculate the sample variance. Which variance is less?

2.48 A new technique has been employed in the manufacture of a solid-solution organic fluor having a high efficiency and short resolving time suitable for application in scintillation counters for particle detection. Although the recipe is fairly straightforward, experience has shown that success is not so much a result of doing the right things as it is a consequence of avoiding the wrong ones. Table P2.48 gives data on the relative sensitivity of various specimens.

TABLE P2.48

			Composition		
Code Number	Relative Sensitivity	Number of Samples	X_1 (grams), Terphenyl	X_2 (grams), TPBD	X_3 (grams), Zinc Stearate
53	29.4	2	207	25	8.3
54	26.9	3	212	25	7.9
55	26.3	5	220	25	7.2
57	21.2	2	210	25	8.0
59	26.3	2	205	25	7.7
60	23.1	3	213	25	8.2
61	26.8	3	200	25	7.8
63	25.4	2	217	25	7.8

(a) Determine the following for each random variable X: (1) sample mean of X_i, (2) sample variance of X_i, and (3) sample standard deviation of X_i.

(b) Determine the following for the variable $Z = X_1 + X_2 + X_3$: (1) Z itself, (2) estimated mean of Z, (3) estimated variance of Z, and (4) estimated standard deviation of Z.

2.49 For $P = 0.99$, determine the values of t for:
(a) A symmetric interval about $t = 0$ (two tailed).
(b) A one-sided interval for $t = -\infty$ to t.

2.50 What are:

(a) $P\{t \leq 3\}$ for $\nu = 4$ degrees of freedom?
(b) $P\{|t| < 2\}$ for $\nu = 30$ degrees of freedom?
(c) $P\{t = 5\}$ for $\nu = 4$ degrees of freedom?
(d) $P\{t > 6.2053\}$ for $\nu = 2$ degrees of freedom?

2.51 For $\nu = 5$ and $\{t \leq t_*\} = 0.10$, what is t_*?

2.52 Given that $P\{F > F_*\} = 0.05$, compute F_* for the variance ratio distribution with $\nu_1 = \nu_2 = 5$ and for $\nu_1 = 3$ and $\nu_2 = 10$.

2.53 If F_* is 7.00 for $\nu_1 = 6$ and $\nu_2 = 5$, what is $P\{F \leq F_*\}$?

2.54 Dalton's law for a binary is

$$Y_A = \frac{P_A}{p_T}$$

At three atmospheres, find the mean and variance of the random variable, the mole fraction, Y_A, in terms of the mean and variance of the random variable, the partial pressure, P_A. p_T is not a random variable.

2.55 The saturation (humidity) curve can be calculated from the relation

$$H_s = \frac{P_s}{p_T - P_s}$$

where:

H_s = molal humidity at saturation
P_s = vapor pressure of water
p_T = total pressure, not a random variable

Find the mean and variance of H_s in terms of the mean and variance of P_s.

2.56 The method of Cox (for vapor pressure charts) was equivalent to representing the vapor pressure of a substance by the following equation:

$$\ln P^* = a - \frac{b}{T - 43}$$

where a and b are constants. Express the mean and variance of P^* in terms of the mean and variance of T. If $\mathscr{E}\{T\} = 100°\text{C}$ and $\text{Var}\{T\} = 1°\text{C}^2$, what are $\mathscr{E}\{P^*\}$ and $\text{Var}\{P^*\}$ if $a = 9.80$ and $b = 2800°\text{K}^{-1}$?

2.57 Find the mean (expected value) and variance of the dependent variable in terms of the mean and variance of the independent variable(s) in the following functions:

(a) $k = k_0(1 + aT)$

where:

k = thermal conductivity, a random variable
T = temperature, a random variable

(b) $k = a_0 + a_1T + a_2T^2$

(c) $k = k_0\left(\frac{492 + b}{T + b}\right)\left(\frac{T}{492}\right)^{3/2}$

(d) $q = UA\,\Delta T$

where A = area and is a constant, and U and ΔT are random variables.

2.58 If the total gas feed to a catalytic cracker of 650 SCF/min is composed of:

μ, SCF/min	Stream	Variance, σ^2
100	Fresh feed	250
350	Recycle	500
170	Inert gas	150
30	Instrumental analysis stream	10
650		

and the ensemble variances are as shown above in appropriate units:

(a) Calculate the ensemble standard deviation of the flow of each stream (show units).
(b) Calculate the upper and lower limits on the stream flow in units of SCF/min for each stream based on 95-percent confidence limits for a normal variable (for $\mu \pm 1.96\sigma$).
(c) Add together all the upper limits; add together the lower limits; compare the difference in totals with the corresponding $\mu_X \pm 1.96\sigma$ limits on the total stream of 650 SCF/min. Which calculation is a better estimate of the dispersion of the stream flow?

2.59 The same reactor as in Problem 2.58 has a bleed stream amounting to 10 percent of the 650 SCF/min. If the ensemble standard deviations were 5 percent of the flow of the in and out streams, and the bleed stream was not measured, what would the 95-percent confidence limits ($\mu \pm 1.96\sigma$) be for the bleed stream in SCF/min? See the diagram.

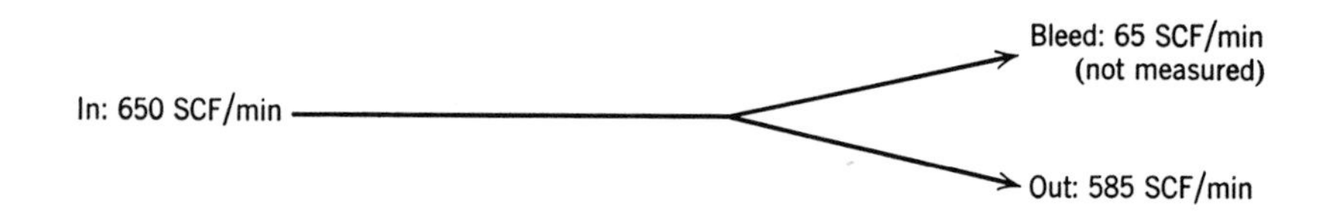

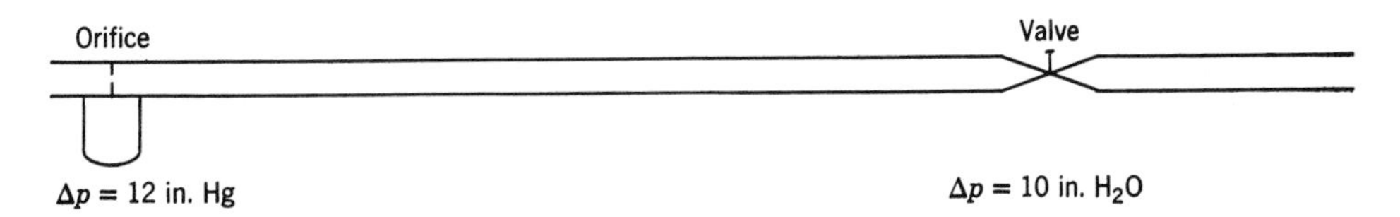

2.60 Pressure drops have been measured on a laboratory experiment as shown in the above diagram. The drop through the piping is estimated to be 2 psia. The standard deviations (calculated by a student) are: orifice = 0.2 in Hg, valve = 0.2 in H_2O, and piping = 0.05 psia. Estimate the precision of the overall (pipe including orifice and valve) pressure drop. Will your answer percentagewise depend on the system of units employed?

2.61 Suppose that the ensemble standard deviation for the mass velocity G, where G is expressed in lb/(hr)(ft²), is 50. What is the corresponding ensemble standard deviation for G when G is expressed in g/(cm)²(sec)?

2.62 The Fanning equation for friction loss in turbulent flow is

$$\Delta P = \frac{2fV^2L\rho}{g_c D}$$

The symbols and their estimated standard deviations are (all variables are random variables):

ΔP = pressure drop due to friction, lb/ft² (1.0)
f = friction factor (not known)
V = average velocity of fluid, ft/sec (0.5)
L = length of pipe, ft (0.1)
D = I.D. (inside diameter) of pipe, ft (0.01)
ρ = density of fluid, lb/ft³ (0.1)
g_c = conversion factor (not a random variable)

For a 100-foot pipe, 2.16 inches I.D., with water (ρ = 62.41 lb/ft³) flowing at 10 ft/sec, the measured Δp is 59.71 psia. Estimate f and the standard deviation of f.

2.63 Calculate the average lb/hr of gas produced by a catalytic cracking unit and the related standard deviation, based on the following data:
F = feed rate, 1000 bbl/hr, with σ_F = 10 bbl/hr
G = gas output, lb/hr, with σ_G to be found
f = feed gravity, 25° API (317 lb/bbl), with σ_f = 0.40° API (0.82 lb/bbl)
L = liquid output, 750 bbl/hr, with σ_L = 10 bbl/hr
p = liquid product gravity, 33° API (30 lb/bbl), with σ_p = 0.30° API (0.56 lb/bbl)
C = coke yield, 16,000 lb/hr, with σ_C = 800 lb/hr

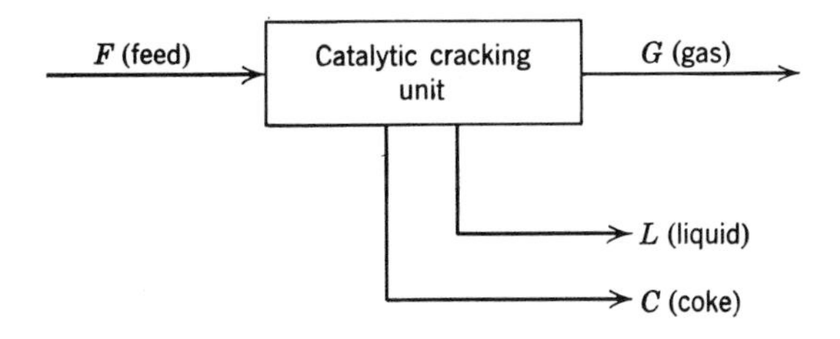

2.64 A plate and frame filter press with 1 ft² of filter area is being operated under conditions such that the filtration equation is

$$\frac{dV}{dt} = \frac{Ap}{\mu\alpha(W/A)}$$

where:

V = ft³ of filtrate
t = time, minutes
p = pressure drop
μ = viscosity of fluid
α = average specific cake resistance ($0 \leq \alpha \leq 1$)
W = weight of dry cake solids
A = filtration area

If p, V, t, and A are measured and μ is known to be 2 cp at 70°F (assumed to be an exact value), find the approximate expression for the mean and variance of α in terms of the measured variables.

2.65 An equation for heat transfer coefficients for steam condensing inside tubes is

$$\frac{dD_\epsilon}{k} = 0.032\left(\frac{D_2}{D_1}\right)\left(\frac{D_\epsilon G}{\eta}\right)^{0.8}\left(\frac{C_p\mu}{k}\right)^{0.4}$$

where:

η = viscosity
h = heat transfer coefficient
G = mass velocity per unit area
D_ϵ = (4)(mean hydraulic radius)
C_p = heat capacity
k = thermal conductivity
D_1, D_2 = diameter of tubes 1 and 2, respectively

Rank the dimensionless ratios according to the relative contributions of each dimensionless ratio to the total variance of (dD_ϵ/k) if the standard deviations σ and mean values (μ) are:

	$\left(\frac{D_2}{D_1}\right)$	$\left(\frac{D_\epsilon G}{\eta}\right)$	$\left(\frac{C_p\mu}{k}\right)$
μ	3	100,000	0.77
σ	0.5%	4%	1%

(σ is expressed as a percent of the mean.) Assume that σ for the constant, 0.032, is zero (not really true).

To reduce the variance contributed by the three dimensionless ratios, which factor (ratio) would you initially try to measure or control more carefully? Which quantity in the factor?

2.66 The following diagrams illustrate typical observations from experiments. For each diagram, estimate roughly the sample correlation coefficient by inspection of the diagram (no calculations are required).

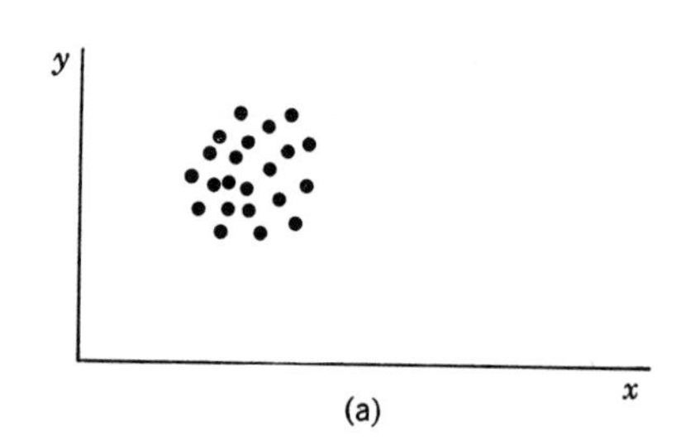

(a)

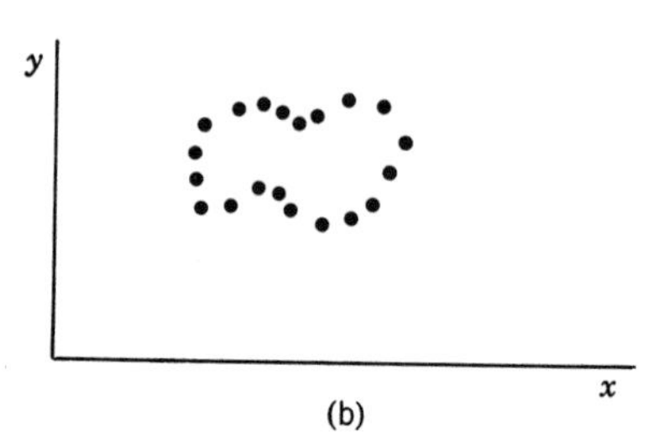

(b)

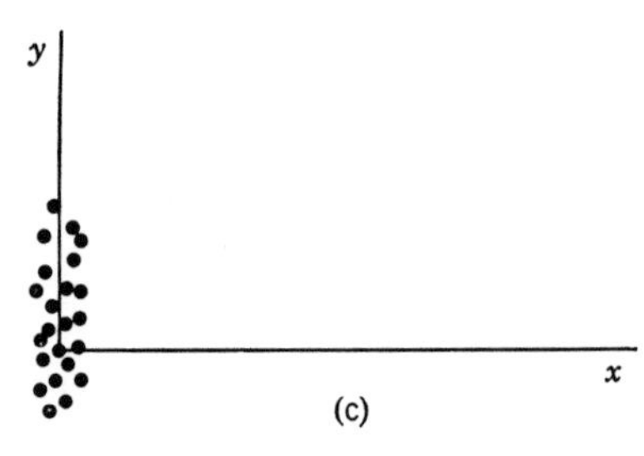

(c)

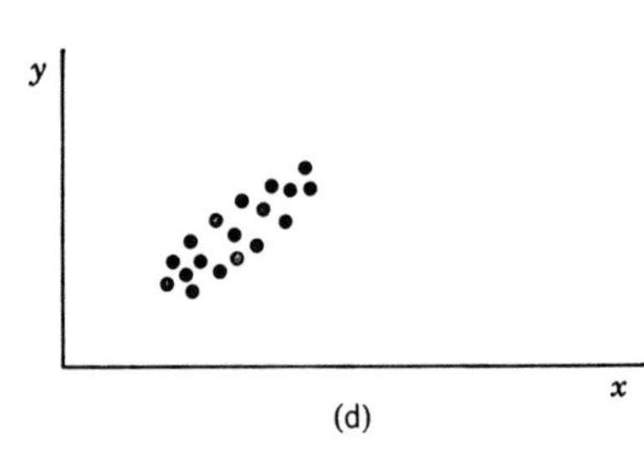

(d)

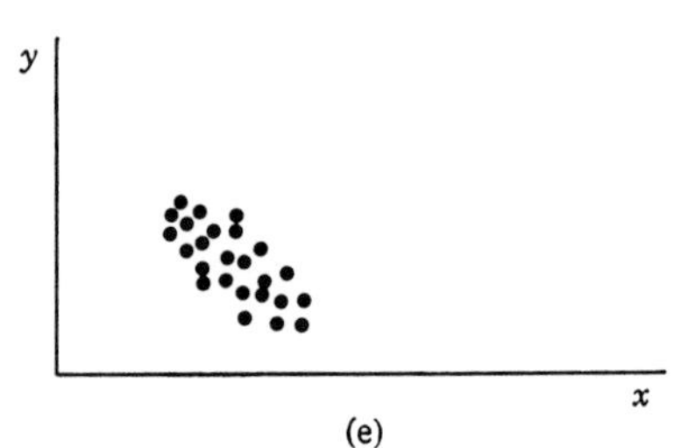

(e)

2.67 A series of experiments was carried out to determine the neutron energy distribution inside a fast reactor. All the experiments involved nuclear plates which served as detectors because their small size enabled them to be placed into the reactor confines without disturbing the reactivity or neutron spectrum.

Two people, E. A, and J. E., counted the number of tracks per 0.1 mev in the same plate with the results shown in the following table. Calculate the sample correlation coefficient between the individual observations.

Incident Proton Energy Interval	Number of Tracks (per mev)	
	E.A.	J.E.
0.3–0.4	12	11
0.4–0.5	32	30
0.5–0.6	26	59
0.6–0.7	21	22
0.7–0.8	3	17
0.8–0.9	9	8
0.9–1.0	9	5
1.0–1.1	6	4
1.1–1.2	5	1
1.2–1.3	4	5

2.68 In a fluidized bed oxidation process, seven runs were carried out at 375°C. The conversion of different naphthalenic feeds to phthalic anhydride (PA) is shown below. What are the sample correlation coefficients between the percent conversion and: (a) the contact time and (b) the air-feed ratio? What is the sample correlation coefficient between the contact time and the air-feed ratio? What interpretation can you give your results:

Run	Contact Time (sec)	Air-Feed Ratio (air/g feed)	Mole percent Conversion to PA
1	0.69	29	50.5
2	0.66	91	30.9
3	0.45	82	37.4
4	0.49	99	37.8
5	0.48	148	19.7
6	0.48	165	15.5
7	0.41	133	49.0

CHAPTER 3

Statistical Inference and Applications

One main purpose of experimentation is to draw inferences about an ensemble from samples of the ensemble. We can identify three different types of inferences which find extensive use in process analysis, namely: (1) parameter estimation, (2) interval estimation, and (3) hypothesis testing. All of these types will be described in this chapter and will be applied here and in subsequent chapters.

3.1 INTRODUCTION

If an engineer wants to make the best estimate he can of one or more parameters of a probability distribution or a proposed process model, the problem is termed one of *parameter estimation.* By parameters we mean those coefficients that identify or describe the probability distribution of a random variable, such as the ensemble mean and variance in the normal probability distribution, or the coefficients in an empirical process model. Estimation of a single value for a parameter is termed *point estimation.* For example, consider a probability density function of known mathematical form of one random variable X, $p(x, \theta)$, which contains one parameter θ which is unknown. A random sample $(x_1, x_2, \ldots, x_n)$ is taken of the random variable. An *estimate* is made of the value of θ, based on the collected experimental data, by calculating a statistic, say the sample mean $\bar{X}$. We say that $\bar{X} = \hat{\mu}_X$, where the superscript caret (^) means estimator of the superscripted variable.

A second type of estimation, *interval estimation,* is concerned with the estimation of the interval that will include the ensemble parameter for a specified probability. Clearly, the parameter estimate is only one useful statistic; the interval estimate is even more informative.

Interval estimation is closely related to *hypothesis testing.* In hypothesis testing, one or more mathematical functions are proposed as representing some feature of experimental data. The functions may be similar in form and differ only in parameter values, or they may differ in form as well. Hypotheses are stated, a criterion of some sort is constructed, data are gathered, the analysis is carried out, and a decision is reached. For example, given $p(x, \theta)$ and some criterion, after collecting a sample $(x_1, x_2, \ldots, x_n)$ containing n observations, we wish to accept or reject the hypothesis that: θ has some value θ_1, or that θ is greater than θ_2, or even that $p(x, \theta)$ has the assumed mathematical form. In Chapters 4 and 5, hypothesis testing in connection with empirical process models will be described.

To obtain "good" estimates, it is necessary that they, in so far as possible, be: (1) unbiased, (2) consistent, (3) efficient, and (4) sufficient.

UNBIASED. An estimate $\hat{\theta}$ of a parameter θ is said to be unbiased if its expected value, $\mathscr{E}\{\hat{\theta}\}$, is equal to the ensemble value θ. For example, the most commonly used estimate of the ensemble mean is $\bar{X}$, the sample average, which is an unbiased estimate of μ_X. On the other hand, it was shown in Section 2.4 that if the sample variance is defined as

$$s_X'^2 = \frac{1}{n} \sum (X_i - \bar{X})^2 n_i$$

instead of

$$s_X^2 = \frac{1}{(n-1)} \sum (X_i - \bar{X})^2 n_i$$

$s_X'^2$ is a biased estimate of σ_X^2.

CONSISTENT. An estimator is said to be consistent if the estimate tends to approach the ensemble value more and more closely as the sample size is increased; that is, the $\mathscr{E}\{(\hat{\theta} - \theta)^2\}$ approaches zero as the sample size n or the record time t_f approaches infinity. More precisely, the probability that the estimates converge to the true value must approach one as the sample size approaches infinity:

$$P\{\hat{\theta} \to \theta\} \to 1 \qquad \text{as } n \to \infty$$

EFFICIENT. In a few unimportant (to us) cases, inconsistent estimates exist, but more often several consistent estimates of a given parameter exist. The question arises as to which estimate should be used. By comparing the variances of the estimates, you can select the most *efficient* estimate in the sense that it has the smallest variance.† For example, the mean and the median of n observations of a normally distributed random variable have an expected value of μ and variances of σ^2/n and $(\sigma^2/n)(\pi/2)$, respectively. Thus, the variance of the mean is less than the variance of the median, and the former is more efficient. The criteria of unbiasedness and minimum variance cannot be considered separately, because a slightly biased estimate with a small variance may be preferable to an unbiased estimate with a large variance.

SUFFICIENT. If $\hat{\theta}$ is a *sufficient* estimate of θ, there is no other estimate of θ that can be made from a sample of a population which will furnish additional information about θ. Fisher‡ showed that the amount of measurable information contained in an estimate is the reciprocal of its variance; hence the definition of sufficient is equivalent to the requirement for minimum variance. A sufficient estimate is necessarily most efficient and also, consequently, consistent. If we assume that a sufficient estimate exists, the method of maximum likelihood, described in the next section, will lead to this estimate. $\bar{X}$ and s_X^2 prove to be sufficient estimates of μ_X and σ_X^2 for a normal distribution.

We now turn to methods of estimating parameters.

3.2 PARAMETER ESTIMATION TECHNIQUES

Quite a number of techniques exist by which parameters can be estimated, not all of which can be applied effectively to any given problem. We shall describe here only three techniques.

1. The method of maximum likelihood (used in Chapters 4, 5, 8, and 9).
2. The method of moments (used in Chapter 9).
3. Bayes' estimates (used in Chapters 8 and 9).

The method of least squares will be discussed in Chapters 4, 5, and subsequent chapters.

3.2-1 Method of Maximum Likelihood

A well-known and desirable estimation procedure (when it can be carried out) is that of maximum likelihood§ introduced by R. A. Fisher which leads asymptotically to estimates with the greatest efficiency but not necessarily unbiased. A desirable feature of the maximum likelihood method is that, under certain conditions (which are not too rigorous), the estimated parameters are normally distributed for large samples. In this section we shall apply the method of maximum likelihood to estimate the parameters in probability density functions. In Chapters 4 and 5 we shall apply the method to estimate coefficients in a linear empirical process model.

Suppose that $p(x; \theta_1, \theta_2, \ldots)$ is a probability density function of *known* form for the random variable X, a function which contains one or more parameters $\theta_1, \theta_2, \ldots$. Also, suppose that we do not know what the values of $\theta_1, \theta_2, \ldots$ are. How can the most suitable values of $\theta_1, \theta_2, \ldots$ be estimated? One way is to collect a random sample of values of the random variable X, $\{x_1, x_2, \ldots, x_n\}$, and select the values of $\theta_1, \theta_2, \ldots$, now regarded as random variables, that maximize the *likelihood function* $L(\theta_1, \theta_2, \ldots \mid x_1, x_2, \ldots, x_n)$, a function described in Appendix A in connection with Bayes' theorem. Such estimators, $\hat{\theta}_1, \hat{\theta}_2, \ldots$, are known as maximum likelihood estimators. In effect, the method selects those values of $\theta_1, \theta_2, \ldots$ that are at least as likely to generate the observed sample as any other set of values of the parameters if the probability density of the random variable X were to be extensively simulated through use of the probability density $p(x \mid \theta_1, \theta_2 \ldots)$.‖ In making a maximum likelihood estimate, we assume that the form of the probability density is connect (only the θ's need be determined) and that all possible values of θ_i are equally likely before experimentation.

The likelihood function for the parameters given one observation is just the probability density in which the observation is regarded as a fixed number and the parameters as the variables:

$$L(\theta_1, \theta_2, \ldots \mid x_1) = p(x_1; \theta_1, \theta_2, \ldots)$$

where the lower case x's and the number subscripts designate the value of the respective observation that is inserted into the probability density function. The likelihood function for the parameters based on several observations is the product of the individual functions if the observations are independent:

$$L(\theta_1, \theta_2, \ldots \mid x_1, x_2, \ldots, x_n) = \prod_{i=1}^{n} L(\theta_1, \theta_2, \ldots \mid x_i) = p(x_1; \theta_1, \theta_2, \ldots)p(x_2; \theta_1, \theta_2, \ldots)\cdots p(x_n; \theta_1, \theta_2, \ldots) \quad (3.2\text{-}1)$$

In choosing as estimates of θ_i the values that maximize L for the given values $(x_1, x_2, \ldots, x_n)$, it turns out that it

† M. G. Kendall and A. Stuart, *The Advanced Theory of Statistics*, Vol. 2, Charles Griffin, London, 1961.

‡ R. A. Fisher, *Proceed. Camb. Phil. Soc.*, **22**, 700, 1925.

§ R. A. Fisher, *Contributions to Mathematical Statistics*, John Wiley, New York, 1950.

‖ The expression $p(x \mid \theta_1, \theta_2, \ldots)$ was termed a conditional probability density function in Section 2.1. The solid vertical line is read as "given." If the values of θ are fixed, then $p(x \mid \theta)$ designates the probability density of the random variable X given the value of θ. On the other hand, $p(\theta \mid x)$ designates that an observation of x is on hand and can be regarded as given (fixed), and that θ is a variable which is a function of the observed value of X.

is more convenient to work with the $\ln L$ than with L itself:

$$\ln L = \ln p(x_1; \theta_1, \theta_2, \ldots) + \ln p(x_2; \theta_1, \theta_2, \ldots) + \cdots$$
$$= \sum_{i=1}^{n} \ln p(x_i; \theta_1, \theta_2, \ldots) \quad (3.2\text{-}2)$$

The $\ln L$ can be maximized with respect to the vector $\boldsymbol{\theta}$ by equating to zero the partial derivatives of $\ln L$ with respect to each of the parameters:

$$\frac{\partial \ln L}{\partial \theta_1} = \frac{\partial \sum_{i=1}^{n} \ln p(x_i; \theta_1, \theta_2, \ldots)}{\partial \theta_1} = 0$$
$$\frac{\partial \ln L}{\partial \theta_2} = \frac{\sum_{i=1}^{n} \ln p(x_i; \theta_1, \theta_2, \ldots)}{\partial \theta_2} = 0 \quad (3.2\text{-}3)$$
$$\vdots$$

Solution of Equations 3.2-3 yields the desired estimates $\hat{\theta}_1, \hat{\theta}_2, \ldots$. (Often iterative rather than analytical methods must be employed to obtain $\hat{\boldsymbol{\theta}}$.) By carrying out this operation, under fairly unrestrictive conditions, it can be shown that as n approaches infinity the maximum likelihood estimates have the desirable asymptotic properties:

(1) $\lim_{n \to \infty} \mathscr{E}\{\hat{\theta}_i\} = \theta_i$

(2) $[\sqrt{n}(\hat{\theta}_i - \theta_i)]$ is normally distributed

and for the case of two parameters:

(3) $\lim_{n \to \infty} [\text{Var}\{\hat{\theta}_i\}] = \frac{1}{n}\left[\mathscr{E}\left\{\left(\frac{\partial \ln p}{\partial \theta_i}\right)^2\right\}\right]^{-1} \frac{1}{(1 - \rho^2_{\hat{\theta}_1\hat{\theta}_2})}$

where $\rho_{\hat{\theta}_1\hat{\theta}_2}$ is the coefficient of correlation of the two $\hat{\theta}$'s. Extension of (3) to more than two parameters requires the use of matrix notation and will not be shown here.

Maximum likelihood estimates are not necessarily unbiased; for example, the maximum likelihood estimate of the variance of a normal random variable is biased as demonstrated in Example 3.2-1. Maximum likelihood estimates, however, are efficient and, hence, consistent estimates. Furthermore, where a sufficient estimate can be obtained, the maximum likelihood method will obtain it. Finally, if $\hat{\theta}$ is a maximum likelihood estimator of θ, then $f(\hat{\theta})$ is a maximum likelihood estimator of $f(\theta)$, a function of θ.

Example 3.2-1 Maximum Likelihood Estimation of the Parameters in the Normal Probability Density Function

Find the maximum likelihood estimates of θ_1 and θ_2 in the normal probability density function

$$p(x; \theta_1, \theta_2) = \frac{1}{\theta_2\sqrt{2\pi}} \exp -\left[\frac{1}{2}\left(\frac{x - \theta_1}{\theta_2}\right)^2\right]$$

Solution:

First, form the likelihood function for the sample $\{x_1, x_2, \ldots, x_n\}$ of measurements of the random variable X:

$$L(\theta_1, \theta_2 \mid x_1, \ldots, x_n) \equiv L = \prod_{i=1}^{n} p(x_i; \theta_1, \theta_2)$$
$$= \frac{1}{(\theta_2\sqrt{2\pi})^n} \exp -\left[\frac{1}{2}\sum_{i=1}^{n}\left(\frac{x_i - \theta_1}{\theta_2}\right)^2\right] \quad (a)$$

and, for convenience, take the logarithm of both sides of Equation (a)

$$\ln L = -n \ln(\theta_2\sqrt{2\pi}) - \frac{1}{2}\sum_{i=1}^{n}\left(\frac{x_i - \theta_1}{\theta_2}\right)^2 \quad (b)$$

Next, obtain the maximum likelihood estimates by equating the partial derivatives of $\ln L$ to zero:

$$\frac{\partial \ln L}{\partial \theta_1} = 0 = \frac{1}{\theta_2^2}\sum_{i=1}^{n}(x_i - \theta_1) \quad (c)$$

$$\frac{\partial \ln L}{\partial \theta_2} = 0 = -n\frac{1}{\theta_2} + \frac{1}{\theta_2^3}\sum_{i=1}^{n}(x_i - \theta_1)^2$$
$$= -\frac{1}{\theta_2}\left[n - \frac{1}{\theta_2^2}\sum_{i=1}^{n}(x_i - \theta_1)^2\right] \quad (d)$$

Consequently, the maximum likelihood estimates of θ_1 and θ_2 are

$$\hat{\theta}_1 = \frac{\sum_{i=1}^{n} x_i}{n} = \bar{X}$$

$$\hat{\theta}_2 = \left[\frac{1}{n}\sum_{i=1}^{n}(x_i - \bar{X})^2\right]^{1/2} = \left[\frac{n-1}{n}s_X^2\right]^{1/2}$$

Thus, $\hat{\theta}_1$ and $\hat{\theta}_2$ are asymptotically (as $n \to \infty$) efficient estimates of μ_X and σ_X; $\hat{\theta}_2$ is a biased estimator because

$$\mathscr{E}\{\hat{\theta}_2\} = \mathscr{E}\left\{\left[\frac{n-1}{n}s_X^2\right]^{1/2}\right\} \neq \sigma_X$$

However, $\hat{\theta}_1$ is an unbiased estimator because $\mathscr{E}\{\hat{\theta}_1\} = \mathscr{E}\{\bar{X}\} = \mu_X$. Note that $\hat{\theta}_1$ and $\hat{\theta}_2$ are independent estimates.

3.2-2 Method of Moments

One of the oldest methods of estimating parameters, that developed by Karl Pearson, is termed the *method of moments*. As applied to a probability density function involving n parameters $(\theta_1, \theta_2, \ldots, \theta_n)$, the technique calls for calculating the first n moments of the random variable X:

$$\mu_i = \sum_{-\infty}^{\infty} x^i P(x_k; \theta_1, \ldots, \theta_n) \quad \text{(discrete)}$$

or

$$\mu_i = \int_{-\infty}^{\infty} x^i p(x; \theta_1, \ldots, \theta_n)\, dx \quad \text{(continuous)}$$

and equating these to the sample moments obtained from experimental data. Then the n values of $\hat{\theta}_i$ can be calculated (perhaps with some difficulty). The method of moments does not always yield efficient estimates as does the maximum likelihood method, but it always yields consistent estimates.

As an example of the application of the method of moments, Table 2.3-1 indicates that the first moment of the binomial random variable is $n\theta$, and we showed in Section 2.3 that the first moment of the normal random variable is μ_X while the second central moment is σ_X^2. On equating the sample moments to the corresponding moments of X, we find as estimates of θ, μ_X, and σ_X^2 the following:

$$n\hat{\theta} = \frac{\sum n_i X_i}{\sum n_i} = \bar{X} \quad \text{or} \quad \hat{\theta} = \frac{\bar{X}}{n} \quad \text{(binomial)}$$

$$\hat{\mu}_X = \frac{\sum n_i X_i}{\sum n_i} = \bar{X} \quad \text{(normal)}$$

$$\hat{\sigma}_X^2 = \frac{\sum (X_i - \bar{X})^2 n_i}{\sum n_i}$$

More often the estimates are obtained with considerably greater difficulty.

Example 3.2-2 Method of Moments

In an experiment, observations come from one of two populations, but it is not known either before or after an observation has been made which population has been sampled. The probability density function for the two populations (A and B) are known to be of the following forms:

$$p_A(y) = \frac{1}{\sqrt{2\pi}} e^{-(y-\alpha)^2/2} \qquad (-\infty < y < \infty)$$

$$p_B(y) = \frac{1}{\sqrt{2\pi}} e^{-(y-\beta)^2/2} \qquad (-\infty < y < \infty)$$

Let ω be the probability that an observation drawn at random comes from population A. Then the probability density function for Y_i $(i = 1, \ldots, n)$ is

$$p(y_i) = \omega p_A(y_i) + (1 - \omega)p_B(y_i), \qquad (-\infty < y_i < \infty)$$

From the sample of n observations, estimates of α, β, and ω are required.

Solution:

The method of maximum likelihood would require that

$$\prod_{i=1}^{n} p(y_i; \alpha, \beta, \omega)$$

be maximized with respect to the three parameters, but the resulting equations are transcendental and virtually intractable. Fortunately, in this problem the method of moments can provide estimates. With the use of the definition of the expected value, it can be shown that

$$\mathscr{E}\{Y\} = \omega\alpha + (1 - \omega)\beta \qquad \text{(a)}$$

$$\mathscr{E}\{Y^2\} = \omega(1 + \alpha^2) + (1 - \omega)(1 + \beta^2) \qquad \text{(b)}$$

$$\mathscr{E}\{Y^3\} = \omega(3\alpha + \alpha^3) + (1 - \omega)(3\beta + \beta^3) \qquad \text{(c)}$$

(Note that with three parameters, three moments are required.)

Let the sample moments be calculated as follows:

$$a_1 = \frac{\sum Y_i}{n} \qquad a_2 = \frac{\sum (Y_i - a_1)^2}{n} \qquad a_3 = \frac{\sum (Y_i - a_1)^3}{n}$$

Then equate the sample moments to their expectations to yield the equations

$$\omega(\alpha - \beta) = a_1 - \beta \qquad \text{(d)}$$

$$\omega(1 - \omega)(\alpha - \beta)^2 = a_2 - 1 \qquad \text{(e)}$$

$$\omega(1 - \omega)(1 - 2\omega)(\alpha - \beta)^3 = a_3 \qquad \text{(f)}$$

Recall that $\mathscr{E}\{(Y) - (\mathscr{E}\{Y\})^2\} = \mathscr{E}\{Y^2\} - (\mathscr{E}\{Y\})^2$. From Equation (d)

$$\omega = \frac{a_1 - \beta}{\alpha - \beta} \qquad \text{(g)}$$

By substituting Equation (d) in Equations (e) and (f) and letting $u = a_1 - \beta$, $v = \alpha - a_1$, Equations (e) and (f) are reduced to

$$uv = a_2 - 1 \qquad \text{(h)}$$

$$uv(v - u) = a_3 \qquad \text{(i)}$$

The solutions of Equations (h) and (i) lead to the estimates

$$\hat{\alpha} = a_1 + \frac{1}{2(a_2 - 1)}[a_3 + \sqrt{a_3^2 + 4(a_2 - 1)^3}] \qquad \text{(j)}$$

$$\hat{\beta} = a_1 + \frac{1}{2(a_2 - 1)}[a_3 - \sqrt{a_3^2 + 4(a_2 - 1)^3}] \qquad \text{(k)}$$

whereupon the estimate of ω, $\hat{\omega}$ may be obtained from Equation (g). Asymptotic variances of the estimates may be calculated if required.

3.2-3 Bayes' Estimates

The modern Bayesian approach to estimation rests on the use of *a priori* information; that is, known or assumed distributions of the parameters to be estimated are employed. While the classical approach to estimation and the Bayesian approach differ to some extent, they do have several common features. Both postulate or involve:

1. Existence of parameter states (values).
2. Possibility of experimentation to shed light on the parameter states.
3. Sampling to yield information on the random variable(s).
4. Optimal decision rules (optimal in terms of the consequences of decisions which result from the rules).

The major difference between the approaches is that the classical statistician makes a decision on the basis of a sample that depends on the parameter state and the type of experiment. In contrast, the Bayesian advocate begins his analysis of the problem by specifying a prior probability density function for the parameter state based on past experience and all other available information. The parameter itself is regarded as a random variable. He then uses a risk (or loss) function related to the worth of the experimental information, together with Bayes' theorem (Equation A-2 in Appendix A), to reach a decision.

If no risk function is known, the posterior distribution itself in Bayes' theorem can be maximized. As sample information becomes available, the analyst again uses his prior probability density function, plus Bayes' theorem, to obtain a posterior probability density function describing his new state of knowledge about the parameter. The posterior probability density function serves as a basis for any decisions and also as a prior probability density function for further analysis.

To be specific, we shall assume that:

1. Several observations are taken of the random variable X, designated by a vector $\mathbf{X}$.
2. Some general functional relationship, $X = f(\boldsymbol{\theta}, \boldsymbol{\epsilon})$, exists between X and the set (vector) of parameters to be estimated which is designated by $\boldsymbol{\theta}$, where $\boldsymbol{\epsilon}$ is a vector of unobservable random errors.
3. The analytical form of the joint probability density function $p(\boldsymbol{\theta}, \boldsymbol{\epsilon})$ is known.

A Bayesian estimate $\hat{\boldsymbol{\theta}}$ of $\boldsymbol{\theta}$ is made as follows. Either Bayes' theorem, Equation A-2 in Appendix A, is used to ascertain the posterior density $p(\boldsymbol{\theta} \mid \mathbf{x})$

$$p(\boldsymbol{\theta} \mid \mathbf{x}) = \frac{p(\mathbf{x} \mid \boldsymbol{\theta})p(\boldsymbol{\theta})}{p(\mathbf{x})} \tag{3.2-4}$$

where $p(\mathbf{x} \mid \boldsymbol{\theta}) = L(\boldsymbol{\theta} \mid \mathbf{x})$, or occasionally it is more convenient to use Equation 2.1-6:

$$p(\boldsymbol{\theta} \mid \mathbf{x}) = \frac{p(\boldsymbol{\theta}, \mathbf{x})}{p(\mathbf{x})} \tag{3.2-5}$$

In either case, the first step is to write the probability density for $p(\mathbf{x})$, which can be done, at least in principle, from the information provided by the known density function $p(\boldsymbol{\theta}, \boldsymbol{\epsilon})$ and the known functional relation between X and $\boldsymbol{\theta}$ and $\boldsymbol{\epsilon}$. If Equation 3.2-4 is used to obtain the posterior density, the second step is to evaluate $p(\mathbf{x} \mid \boldsymbol{\theta})$. This conditional density can also be obtained from the known relations in assumptions (2) and (3) above. The third step is to obtain $p(\boldsymbol{\theta})$ from $p(\boldsymbol{\theta}, \boldsymbol{\epsilon})$ by integration over all values of $\boldsymbol{\epsilon}$. If Equation 3.2-5 is used, the probability density $p(\boldsymbol{\theta}, \mathbf{x})$ must be obtained from the known relations in assumptions (2) and (3), but the analytical treatment in general is difficult or impossible.

The final step in both routes, once the posterior density $p(\boldsymbol{\theta} \mid \mathbf{x})$ containing all the knowledge about $\boldsymbol{\theta}$ from the measurements is written, is to optimize $p(\boldsymbol{\theta} \mid \mathbf{x})$ in some sense. For the special case in which the probability $P\{\hat{\boldsymbol{\theta}} = \boldsymbol{\theta}\}$ is maximized, one maximizes $p(\boldsymbol{\theta} \mid \mathbf{x})$ itself with respect to $\boldsymbol{\theta}$ to obtain the value of $\hat{\boldsymbol{\theta}}$ at the peak of the curve of the density (the mode). When the prior density $p(\boldsymbol{\theta})$ is uniform, this estimate is identical to the maximum likelihood estimate. Many other methods of optimization can be carried out but they are beyond our scope here. Examples of Bayes estimates appear in Chapters 8 and 9.

3.3 INTERVAL ESTIMATION

In the previous two sections, we described certain ways to obtain point estimates of parameters and some of the criteria for assessing the merit of the estimators. An even more meaningful statement than the point estimate can be made in terms of a *confidence interval* estimate. The confidence interval is calculated from the observations in a sample; it includes the fixed value of the ensemble parameter within (or on one of) the interval limits, termed *confidence limits*, for a specified degree of assurance, called the *confidence coefficient*. Johnson and Leone quote a revealing analogy between the confidence interval and horseshoe tossing.†

> A confidence interval and statements concerning it are somewhat like the game of horseshoe tossing. The stake is the parameter in question. (It never moves, regardless of some sportsmen's misconceptions.) The horseshoe is the confidence interval. If out of 100 tosses of the horseshoe one rings the stake 90 times on the average, he has 90 percent assurance (or confidence) of ringing the stake. The confidence interval, just like the horseshoe, is the variable. The parameter, just like the stake, is the constant. At any *one* toss (or one interval estimation) the stake (or parameter) is either enclosed or not. We make a *probability* statement about the variable quantities represented by the positions of the "arms" of the horseshoe.

To make an interval estimate, the general procedure is:

1. Write a probability statement in mathematical symbols involving the ensemble parameter of interest.
2. Rearrange the argument of the statement so that the ensemble parameter is bounded by statistics that can be calculated from a sample.

As an example, let us consider interval estimation for the unknown ensemble mean, μ_X, of the normal random

† N. L. Johnson and F. C. Leone, *Statistical and Experimental Design*, Vol. I, John Wiley, New York, 1964, p. 188.

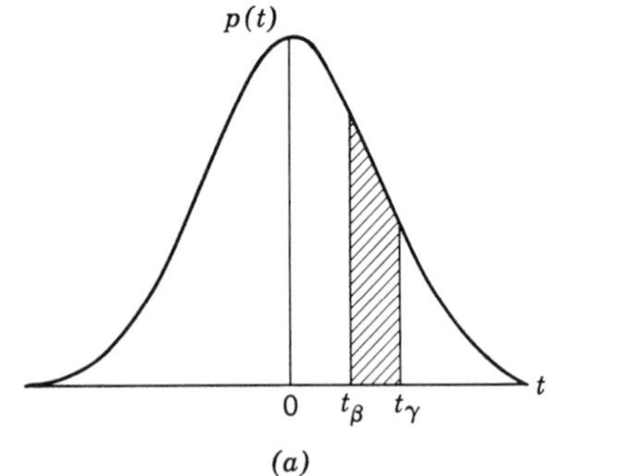

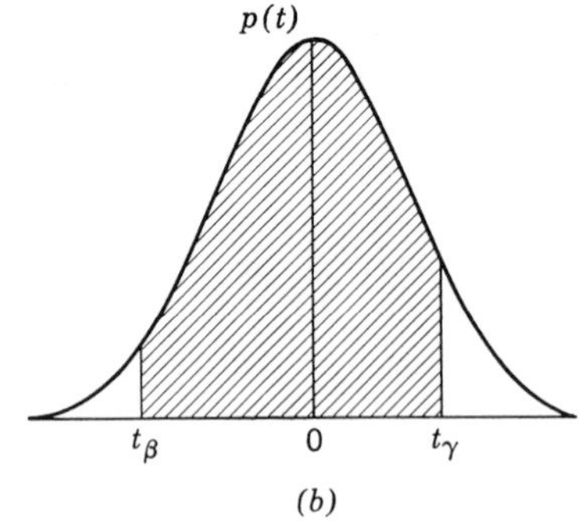

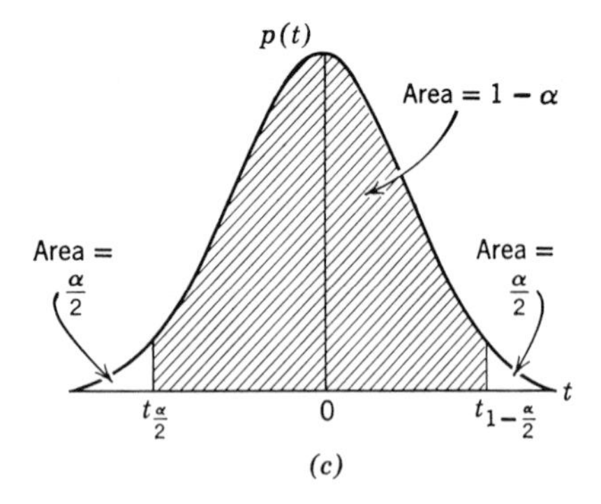

FIGURE 3.3-1 Symmetric and asymmetric bounds about $t = 0$.

variable X through use of the sample mean $\bar{X}$ and the sample variance $s_{\bar{X}}^2$. In Section 2.4-2 which described the t distribution, it was noted that $t = (\bar{X} - \mu_X)/s_{\bar{X}}$ was a random variable with a known probability density as given by Equation 2.4-14. It follows that probability statements can be made concerning the value of t prior to collecting the sample, such as

$$P\{t \le t_\gamma\} = P\left\{\frac{\bar{X} - \mu_X}{s_{\bar{X}}} \le t_\gamma\right\} = \gamma \qquad (3.3\text{-}1)$$

$$P\{t > t_\gamma\} = 1 - \gamma$$

and

$$P\{t_\beta < t \le t_\gamma\} = P\left\{t_\beta < \frac{\bar{X} - \mu_X}{s_{\bar{X}}} \le t_\gamma\right\}$$
$$= P(t_\gamma) - P(t_\beta) = \gamma - \beta \qquad (3.3\text{-}2)$$

where the subscript γ identifies the upper limit and β the lower on the integral $\int_{t_\beta}^{t_\gamma} p(t)\,dt$. If in Equation 3.3-2 the indices γ and β are symmetric about $t = 0$, the interval about t is symmetric—examine Figure 3.3-1. For Figure 3.3-1c, in order to make the area under the probability distribution function outside the interval equal to $(\alpha/2 + \alpha/2) = \alpha$, we let $\gamma = \beta = \alpha/2$. Thus

$$P\left\{t_{\frac{\alpha}{2}} < \frac{\bar{X} - \mu_X}{s_{\bar{X}}} \le t_{1-\frac{\alpha}{2}}\right\} = 1 - \alpha \qquad (3.3\text{-}3)$$

After the sample has been collected, the values $\bar{X}$ and $s_{\bar{X}}$ are regarded as fixed numbers; the probability statements no longer apply inasmuch as $(\bar{X} - \mu_X)/s_{\bar{X}}$ either is in the interval ($P = 1$) or is not ($P = 0$), although which statement is correct is not known. However, the interval itself is a random variable. If the sampling were repeated many times with $\bar{X}$ and $s_{\bar{X}}$ calculated for each sample, one would expect $(\bar{X} - \mu_X)/s_{\bar{X}}$ to fall within the defined interval for about the fraction of samples indicated on the right-hand side of the probability statements. It is in this framework that we speak of an interval, itself a random variable, which includes μ_X the ensemble parameter, for a certain degree of uncertainty. Such a statement is a confidence statement, and the associated interval is the confidence interval, while the degree of trust associated with the confidence statement is the confidence coefficient.

The symmetric confidence interval for the ensemble mean can be identified by rearranging the argument of P in Equation 3.3-3 as follows, using the equality:

$$-(t_{1-\frac{\alpha}{2}}) = t_{\frac{\alpha}{2}}$$

($t_{1-\frac{\alpha}{2}}$ is the positive value of t).

$$(-t_{1-\frac{\alpha}{2}})s_{\bar{X}} < \bar{X} - \mu_X \le (t_{1-\frac{\alpha}{2}})s_{\bar{X}}$$

$$\bar{X} + t_{1-\frac{\alpha}{2}}s_{\bar{X}} > \mu_X \ge \bar{X} - t_{1-\frac{\alpha}{2}}s_{\bar{X}}$$

or

$$\bar{X} - t_{1-\frac{\alpha}{2}}s_{\bar{X}} \le \mu_X < \bar{X} + t_{1-\frac{\alpha}{2}}s_{\bar{X}} \qquad (3.3\text{-}4)$$

The confidence coefficient for the interval given by Equation 3.3-4 is $1 - \alpha$.

A similar interval can be developed for the ensemble mean from the distribution of the random variable

$$U = \frac{\bar{X} - \mu_X}{\sigma_{\bar{X}}}$$

given by Equation 2.4-7 if $\sigma_{\bar{X}}$ is known:

$$\bar{X} - U_{1-\frac{\alpha}{2}}\sigma_{\bar{X}} \le \mu_X < \bar{X} + U_{1-\frac{\alpha}{2}}\sigma_{\bar{X}} \qquad (3.3\text{-}5)$$

To obtain the confidence interval for the ensemble variance, $\sigma_{\bar{X}}^2$, of the random variable X, we make use of the χ^2 probability distribution and write

$$P\{\chi_\beta^2 < \chi^2 \le \chi_\gamma^2\} = P\{\chi_\gamma^2\} - P\{\chi_\beta^2\} \qquad (3.3\text{-}6)$$

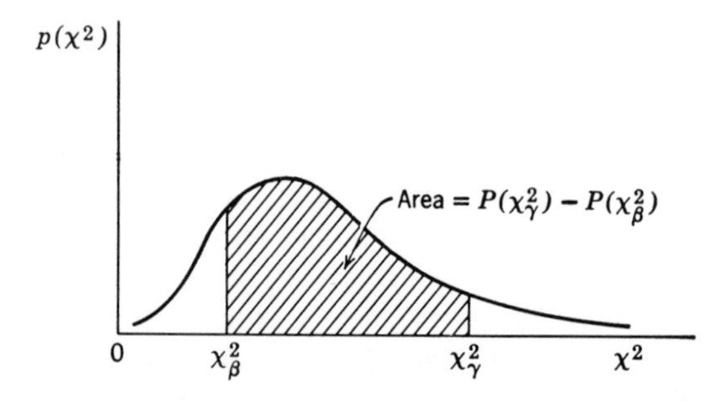

FIGURE 3.3-2 Graphical representation of the probability statement (Equation 3.3-6) for the χ^2 distribution.

Figure 3.3-2 is a graphical representation of how the probabilities in Equation 3.3-6 can be interpreted as areas under the χ^2 probability density.

Substitution of χ^2 from Equation 2.4-10 into the argument of Equation 3.3-6 gives as the argument

$$\chi_\beta^2 < \frac{s_X^2 \nu}{\sigma_X^2} \leq \chi_\gamma^2$$

and rearrangement yields the confidence interval for σ_X^2:

$$\frac{1}{\chi_\beta^2} > \frac{\sigma_X^2}{s_X^2 \nu} \geq \frac{1}{\chi_\gamma^2}$$

$$\frac{s_X^2 \nu}{\chi_\gamma^2} \leq \sigma_X^2 < \frac{s_X^2 \nu}{\chi_\beta^2}$$

If

$$\beta = \frac{\alpha}{2} \quad \text{and} \quad \gamma = 1 - \frac{\alpha}{2}$$

$$\frac{s_X^2 \nu}{\chi_{1-\frac{\alpha}{2}}^2} \leq \sigma_X^2 < \frac{s_X^2 \nu}{\chi_{\frac{\alpha}{2}}^2} \tag{3.3-7}$$

for a confidence coefficient of $(1 - \alpha)$.

Other ensemble averages can be treated similarly if the distribution of their sample estimates is known. Even if the distribution of the sample statistic is unknown, a confidence interval can be specified for any random variable X with a finite variance σ_X^2 through use of the Chebyshev inequality. This states that the probability is at least $[1 - (1/h^2)]$ of obtaining a standardized variable of value equal to or less than a number h.

Let $f(X)$ be a nonnegative function of the random variable X. We show first that if $\mathscr{E}\{f(X)\}$ exists, then for every positive constant c,

$$P\{f(X) \geq c\} \leq \frac{\mathscr{E}\{f(X)\}}{c} \tag{3.3-8}$$

Let ξ be the set of x such that $f X) \geq c$ and ξ^* be the remaining x. Then,

$$\mathscr{E}\{f(X)\} = \int_{-\infty}^{\infty} f(x)p(x)\,dx$$

$$= \int_\xi f(x)p(x)\,dx + \int_{\xi^*} f(x)p(x)\,dx \tag{3.3-9}$$

Because each integral in the sum on the right-hand side of Equation 3.3-9 is nonnegative,

$$\mathscr{E}\{f(X)\} \geq \int_\xi f(x)p(x)\,dx$$

By definition $f(X) \geq c$ for some c; hence

$$\mathscr{E}\{f(X)\} \geq c\int_\xi p(x)\,dx = cP\{f(X) \geq c\} \tag{3.3-10}$$

from which Equation 3.3-8 can be obtained.

Chebyshev's inequality follows if we let

$$f(X) = (X - \mu_X)^2$$

$$c = h^2\sigma_X^2, \qquad h > 1$$

because then

$$P\{(X - \mu_X)^2 \geq h^2\sigma_X^2\} \leq \frac{\mathscr{E}\{(X - \mu_X)^2\}}{h^2\sigma_X^2} \equiv \frac{1}{h^2}$$

and

$$P\{|X - \mu_X| \geq h\sigma_X\} \leq \frac{1}{h^2} \tag{3.3-11}$$

As an example of the application of Equation 3.3-11 for $h = 2$, at least $[1 - (\frac{1}{2})^2] = \frac{3}{4}$ of the occurrences of the random variable $\bar{X}$ should lie within $\pm 2\sigma_{\bar{X}}$ of $\mu_{\bar{X}}$ no matter what the distribution of $\bar{X}$.

We now illustrate the calculation of the confidence interval for the ensemble mean and variance.

Example 3.3-1 Confidence Interval for the Ensemble Mean and Variance

From the following eight volumetric titration analyses, find the confidence interval for the ensemble mean and variance of the normal random variable X.

Values of X in cc	
76.48	76.25
76.43	76.48
77.20	76.48
76.45	76.60

Solution:

$$\bar{X} = \frac{1}{8}\sum X_i = 76.546$$

$$s_X^2 = \frac{\sum (X_i - \bar{X})^2}{n - 1} = \frac{0.5543}{7} = 0.0790\ \text{cc}^2 \qquad \nu = n - 1 = 7$$

$$s_{\bar{X}} = \sqrt{\frac{s_X^2}{n}} = \sqrt{\frac{0.0790}{8}} = 0.099\ \text{cc}$$

Using Table C.3 in Appendix C, for a 95-percent confidence coefficient $(1 - \alpha = 0.95; \alpha/2 = 0.025)$ and for a symmetric interval, we find $t_{0.975} = 2.36$.

The symmetric confidence interval by Equation 3.3-4 is

$$76.55 - 0.099(2.36) \leq \mu_X < 76.55 + 0.099(2.36)$$

or

$$76.31 \leq \mu_X < 76.79$$

The interpretation of the confidence interval is: the probability is 0.95 that the interval between 76.31 and 76.79 contains the ensemble mean.

The confidence interval for $\alpha = 0.05$ for σ_X^2 from Equation 3.3-7 is

$$\frac{s_X^2 \nu}{\chi_{1-\frac{\alpha}{2}}^2} \leq \sigma_X^2 \quad \frac{s_X^2 \nu}{\chi_{\frac{\alpha}{2}}}$$

$$\frac{(0.0790)(7)}{16.013} \leq \sigma_X^2 < \frac{(0.0790)(7)}{1.690}$$

$$0.03452 \leq \sigma_X^2 < 0.3262$$

Example 3.3-2 Process Flow Error

Examine the subsystem illustrated in the block diagram of Figure E3.3-2. The "errors" on the inputs and output,

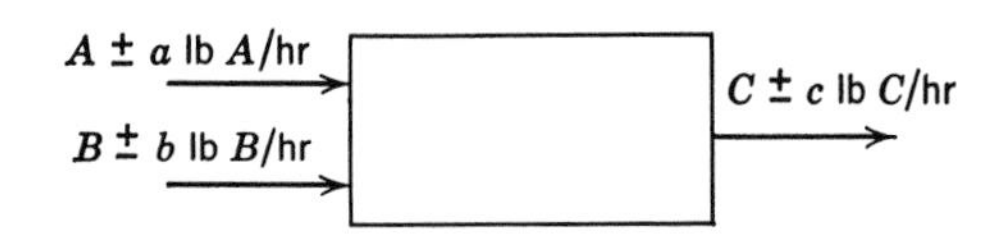

FIGURE E3.3-2

indicated as plus or minus, as commonly encountered are interpreted to mean the limits of the confidence interval for a confidence coefficient $(1 - \alpha)$. μ_A, μ_B, and μ_C are the ensemble values of the flow rates. Unobservable measurement errors in A and B exist, ϵ_A and ϵ_B, which are normal random variables with a mean of zero and variances of $\sigma^2_{\epsilon_A}$ and $\sigma^2_{\epsilon_B}$, respectively. A random sample of A is taken and, independently, a random sample of B, from which the following sample statistics are calculated:

Material	Sample Mean (lb/hr)	Sample Standard Deviation (lb/hr)	Number in Sample
A	10	0.20	5
B	5	0.10	5

The problem is to find the confidence interval for μ_C for a confidence coefficient of $(1 - \alpha) = 0.95$ from the given information about A and B.

Solution:

By a material balance (on the expected values)

$$\mu_A + \mu_B = \mu_C$$

so that the $\bar{C} = \bar{A} + \bar{B} = 10 + 5 = 15$ lb/hr. Also, because the variables are independent, the variance of C is (using Equation 2.2-9a)

$$\text{Var}\{C\} = \text{Var}\{A\} + \text{Var}\{B\}$$

or (using Equation 2.4-10)

$$\frac{s^2_C\nu_C}{\chi^2} = \frac{s^2_A\nu_A}{\chi^2} + \frac{s^2_B\nu_B}{\chi^2}$$

The Var $\{C\}$ can be estimated by

$$s^2_C = \frac{4(0.20)^2 + 4(0.10)^2}{8} = 0.025 \qquad \nu_C = 8$$

The confidence interval for μ_C (using Equation 3.3-4) is (with $t_{1-\frac{\alpha}{2}} = 2.306$)

$$15 - (2.31)(0.025)^{1/2} \le \mu_C < 15 + (2.31)(0.025)^{1/2}$$

$$14.64 \le \mu_C < 15.36$$

3.4 HYPOTHESIS TESTING

Testing is related to interval estimation but has a different viewpoint. In *hypothesis testing*, one tests a hypothesis H_0 against one or more alternate hypotheses $H_1, H_2, \ldots$ that are spelled out or implied. For example, the hypothesis H_0 might be that $\mu = 16$; two alternate hypotheses might be $H_1: \mu > 16$, and $H_2: \mu < 16$. Or the hypothesis to be tested might be that there is no improvement in a process output, with the alternate hypothesis implied that there is an improvement.

Suppose that we know the probability density function $p(\hat{\theta})$ for an estimate $\hat{\theta}$ (which is an unbiased estimate of θ). We assume that the representation of the random variable $\hat{\theta}$ by $p(\hat{\theta})$ is correct and that the ensemble value of θ is, say, θ_0, and we ask the following question: If we presume as true the hypothesis that $\theta = \theta_0$, by how much must $\hat{\theta}$ differ from θ_0 before we reject the hypothesis because it seems to be wrong? Figure 3.4-1 helps to answer the question. If the hypothesis $\theta = \theta_0$ is true, $\mathscr{E}\{\hat{\theta}\} = \theta_0$ as is shown in the figure. The probability that the value of $\hat{\theta}$ would be equal to or less than $\theta_{\frac{\alpha}{2}}$ is

$$P\{\hat{\theta} \le \theta_{\frac{\alpha}{2}}\} = \int_{-\infty}^{\theta_{\frac{\alpha}{2}}} p(\hat{\theta})\, d\hat{\theta} = \frac{\alpha}{2} \qquad (3.4\text{-}1)$$

and because of symmetry

$$P\{\hat{\theta} > \theta_{1-\frac{\alpha}{2}}\} = \int_{\theta_{1-\frac{\alpha}{2}}}^{\infty} p(\hat{\theta})\, d\hat{\theta} = \frac{\alpha}{2} \qquad (3.4\text{-}2)$$

To reach a decision concerning the hypothesis, we select a value of α, which is termed the *level of significance* for the test, before collecting the sample; α is usually arbitrarily selected to be small enough so that the user regards it quite improbable that $\hat{\theta}$ will exceed the selected value of $\theta_{1-\frac{\alpha}{2}}$ or be less than $\theta_{\frac{\alpha}{2}}$. For example, α might be 0.05 or 0.01. Then the sample is collected and $\hat{\theta}$ is calculated. If $\hat{\theta}$ is larger than $\theta_{1-\frac{\alpha}{2}}$ or smaller than $\theta_{\frac{\alpha}{2}}$, the hypothesis is rejected. Otherwise, it is accepted. The range of values of $\hat{\theta}$ for which the hypothesis is rejected is called the *region of rejection*; the range of $\hat{\theta}$ for which the hypothesis is accepted is called the *region of acceptance*.

The test described above is a two-sided test. A one-sided test can be based on either $\hat{\theta}$ being greater than some $\theta_{1-\alpha}$, with the hypothesis $\theta = \theta_0$ being rejected if $\hat{\theta}$ is indeed greater than $\theta_{1-\alpha}$, or on $\hat{\theta}$ being less than θ_α. Rejecting the hypothesis does not mean discarding it offhand, but it instead calls for a careful examination of the experimental procedure and data to ascertain if any-

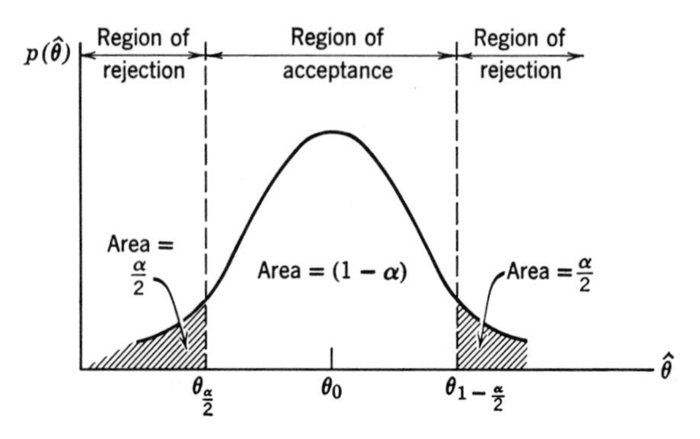

FIGURE 3.4-1 Regions of rejection and acceptance for a symmetric hypothesis test.

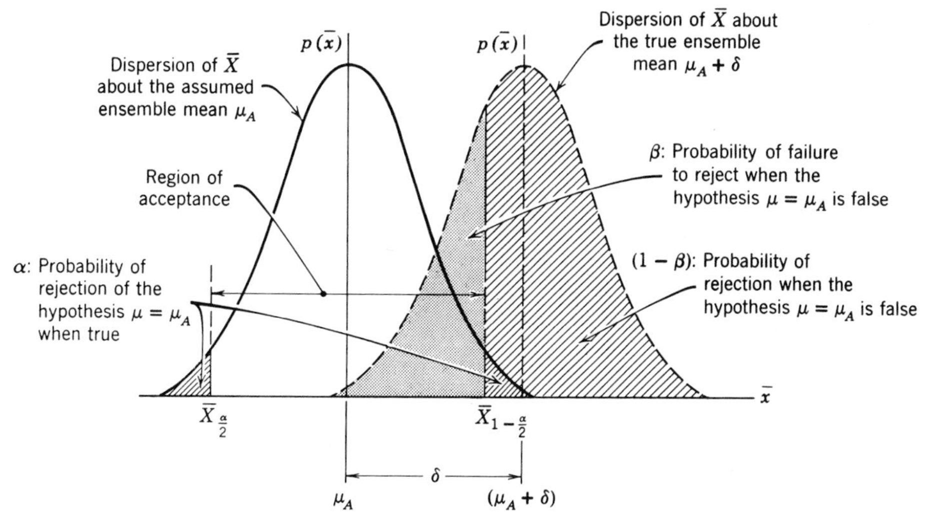

FIGURE 3.4-2 Error of the second kind.

thing went wrong with the experiment. Investigation into the causes of defects in the method of procedure can be most rewarding.

The simplest structure for testing is to imagine that a dichotomy of states exist for the random variable:

1. H_0: x is the true state of the random variable (*the null hypothesis*).

2. H_1: x is not the true state of the variable (*the alternate hypothesis*).

For example, two values of a parameter can represent a probability density. We hypothesize under H_0 that the probability density of a random variable is $p(x; \theta_0)$ and under the alternate hypothesis that the probability density is $p(x; \theta_1)$. Or, as another example, hypothesis H_0 is that the ensemble mean of a process variable has not changed after a process modification, while H_1 is that the process mean has changed. Tests which involve several alternative hypotheses simultaneously are also available, but their description is beyond our scope here.

In hypothesis testing, a decision is made as follows. Based on the assumption that the null hypothesis is true, if the statistic calculated from the random experimental sample falls outside the region of acceptance, the null hypothesis is rejected and H_1 is accepted. Otherwise, H_0 is accepted and H_1 rejected.

Two types of errors can be distinguished in testing a hypothesis:

AN ERROR OF THE FIRST KIND (TYPE 1 ERROR). This error is caused by rejecting the hypothesis when it is true.

AN ERROR OF THE SECOND KIND (TYPE 2 ERROR). This error is caused by not rejecting the hypothesis when it is false.

Clearly, the Type 1 error exists because α is selected to be some nonzero value. When the hypothesis is true and $\alpha = 0.05$, for example, in 5 percent of the tests the hypothesis will be rejected when it is true.

Figure 3.4-2 illustrates the Type 2 error as applied to the ensemble mean. In this illustration we set up the hypothesis that $\mu = \mu_A$. But to demonstrate the Type 2 error, also assume that the true value of μ is really equal to $\mu = \mu_A + \delta$, as shown in Figure 3.4-2. A value of α is selected, which fixes the region of rejection indicated by the heavily shaded areas. In this case the hypothesis $\mu = \mu_A$ is false, yet there is a probability, β, that the sample mean will fall within the region of acceptance. If the hypothesis $\mu = \mu_A$ is true, as assumed, the two-sided test indicated in Figure 3.4-2 will lead to the *correct* decision in $100(1 - \alpha)$ percent of the tests and to the *wrong* decision (rejection) of $100(\alpha)$ percent of the tests, as explained previously. However, if the hypothesis is actually *false*, then the probability of $\bar{X}$ falling in the region of rejection can be calculated if the value of δ is known or assumed.

The probability β is the probability of not detecting a difference when it exists. Figure 3.4-3 shows typical plots of β versus the difference d as a function of the sample size; these curves are termed *operating characteristic curves* (OC curves). The probability $(1 - \beta)$ is termed the *power of the test* to discriminate, and it represents the probability of making a correct decision when the hypothesis is actually wrong. As δ increases, $(1 - \beta)$ increases and β decreases.

From the description of the two kinds of errors, it will be observed that an attempt to decrease one kind of error will result in an increase in the other type of error. The only way to reduce both types of errors simultaneously is to increase the sample size which, in practice, may prove to be expensive. Perhaps one type of error may have less serious consequences than the other, in which case some suitable decision can be reached concerning the selection of values for α and the number of observations

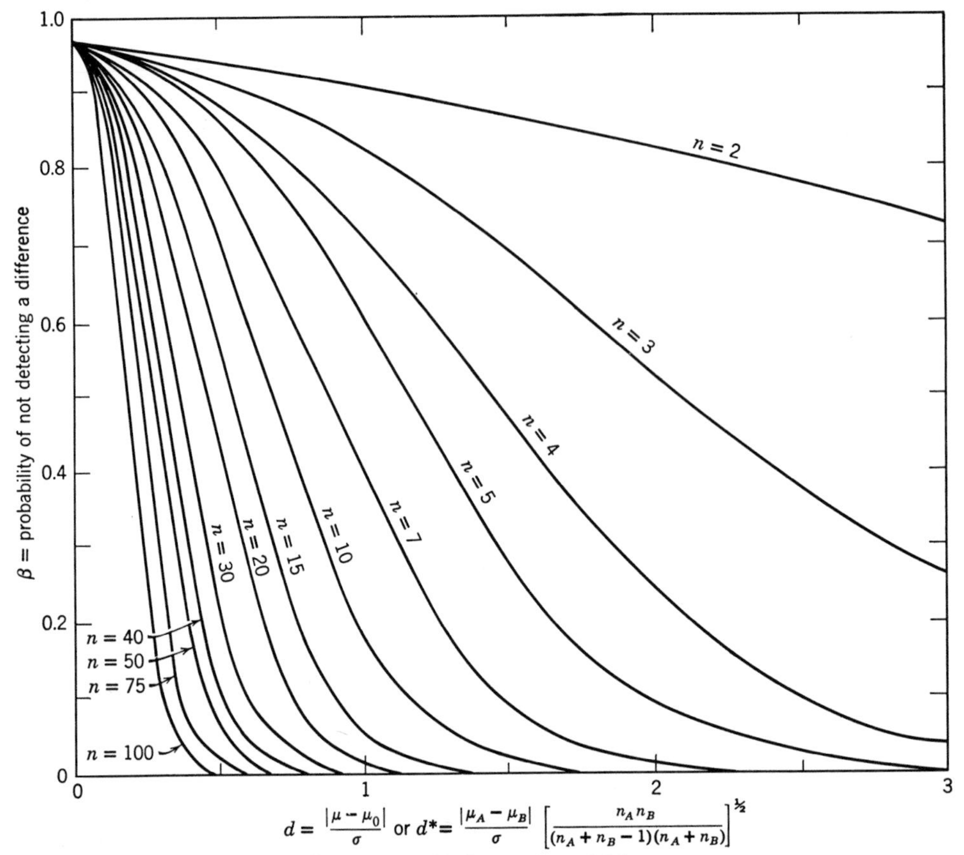

$d = \frac{|\mu - \mu_0|}{\sigma}$ or $d^* = \frac{|\mu_A - \mu_B|}{\sigma}\left[\frac{n_A n_B}{(n_A + n_B - 1)(n_A + n_B)}\right]^{1/2}$

FIGURE 3.4-3*a* OC curves for the two-sided *t*-test ($\alpha = 0.05$):

$\delta = |\mu - \mu_0|$ or $\delta^* = |\mu_A - \mu_B|$ with a sample of size n

σ = ensemble standard deviation—must be estimated if not known

μ_0 = known ensemble mean

μ_A and μ_B = ensemble means of samples A and B, respectively

(Adapted with permission from C. D. Ferris, F. E. Grubbs, and C. L. Weaver, "Operating Characteristics for the Common Statistical Tests of Significance," *Annals Math. Stat.* **17** (2), 178–197, June 1946.)

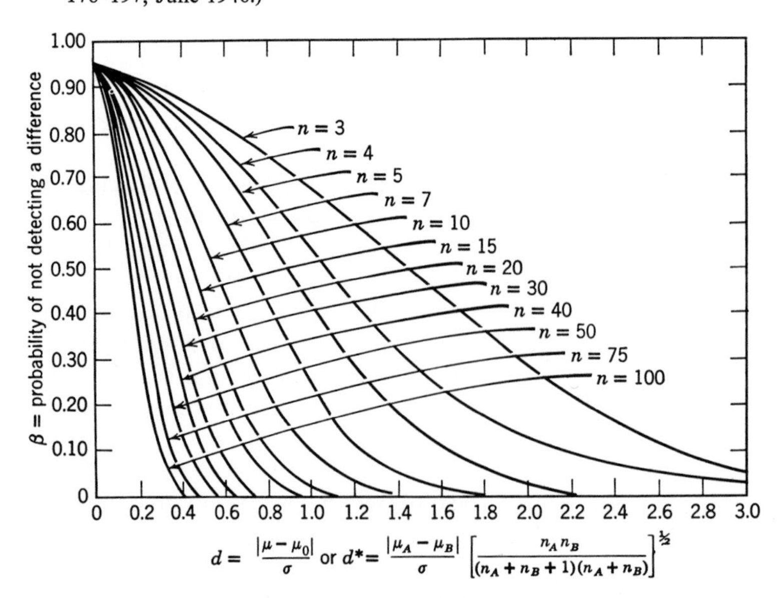

$d = \frac{|\mu - \mu_0|}{\sigma}$ or $d^* = \frac{|\mu_A - \mu_B|}{\sigma}\left[\frac{n_A n_B}{(n_A + n_B + 1)(n_A + n_B)}\right]^{1/2}$

FIGURE 3.4-3*b* OC curves for the one-sided *t*-test ($\alpha = 0.05$):

$\delta = |\mu - \mu_0|$ or $\delta^* = |\mu_A - \mu_B|$ with a sample of size n

σ = ensemble standard deviation—must be estimated if not known

μ_0 = known ensemble mean

μ_A and μ_B = ensemble means of samples A and B, respectively

(Adapted with permission from A. H. Bowker and G. J. Lieberman, *Engineering Statistics*, Prentice-Hall, Englewood Cliffs, N.J.1959.)

to be taken. The best practice takes into account the instruments, process design, and costs so as to make an economic decision for α and β.

The concepts of the power and operating characteristic of a test apply equally well to tests for ensemble variances, as described in Section 3.6, and to other parameters as they do to tests for the ensemble mean.

Example 3.4-1 Hypothesis Test for the Mean

Suppose that a process variable X, which is a random variable, is known to have an ensemble mean of $\mu_X = 6.80$. A sample is taken of the variable, and it is found for a sample size of $n = 9$ that $\bar{X} = 6.50$ and $s_X^2 = 0.25(s_X = 0.50)$. We test the hypothesis H_0 that the random variable has the same ensemble mean as in the past, namely $\mu_X = 6.80$. The alternate hypothesis H_1 is that $\mu \neq 6.80$. If α, the significance level, is selected as 0.05, the region of acceptance for a symmetrical two-sided t-test is as follows:

Accept H_0 if $|\bar{X} - \mu_X| < t_{1-\frac{\alpha}{2}} s_{\bar{X}}$. Otherwise, reject H_0 and accept H_1.

Here

$$s_{\bar{X}} = \frac{s_X}{\sqrt{n}} = \frac{0.50}{3} = 0.167$$

and $t_{1-\frac{\alpha}{2}}$ for $n - 1 = 8$ degrees of freedom and $\alpha = 0.05$ is 2.306.

$$0.30 = |6.50 - 6.80| < (2.306)(0.167) = 0.39$$

Hence the hypothesis H_0 is accepted. Figure E3.4-1 illustrates the regions of acceptance and rejection.

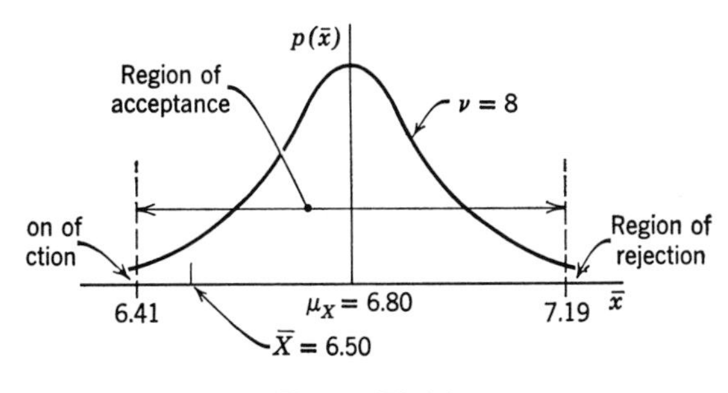

FIGURE E3.4-1

Example 3.4-2 Power of a Test for the Mean

In this example we assume that the hypothesis H_0 (that $\mu = \mu_0 = 6.80$, as described in Example 3.4-1) is correct. Then, if in reality $\mu > \mu_0$ (for a one-sided test) or $\mu \neq \mu_0$ (for a two-sided test), we can calculate the power of the t-test used in Example 3.4-1 to discriminate. The power of the test is

$$1 - \beta = P\left\{\left|\frac{\bar{X} - \mu_0}{s_X/\sqrt{n}}\right| > t_{1-\frac{\alpha}{2}}; \mu = \mu_1\right\}$$

(two-sided symmetric test)

$$1 - \beta = P\left\{\frac{\bar{X} - \mu_0}{s_X/\sqrt{n}} > t_{1-\alpha}; \mu = \mu_1\right\}$$

(one-sided test)

If we write

$$\frac{\bar{X} - \mu_0}{s_X/\sqrt{n}} = \frac{\bar{X} - \mu_1}{s_X/\sqrt{n}} + \frac{\mu_1 - \mu_0}{\sigma_X/\sqrt{n}}\frac{\sigma_X}{s_X}$$
$$= t_1 + \lambda\sqrt{\nu/\chi^2}$$

where $\lambda = (\mu_1 - \mu_0)/(\sigma_X/\sqrt{n})$, we find that the power depends on a combination of the t distribution about μ_1, the χ^2 distribution, the distance between the means, and ν. Approximate relations† for the power in terms of the standard normal random variable are

$$(1 - \beta) \approx P\{U_1 \le u_1\} + P\{U_2 \le u_2\} \quad \text{(two-sided test)}$$

$$(1 - \beta) \approx P\{U \le u\} \quad \text{(one-sided test)}$$

where

$$u_1 = \frac{t_{\frac{\alpha}{2}} - \lambda}{\sqrt{1 + (t^2_{\frac{\alpha}{2}}/2\nu)}}$$

$$u_2 = \frac{t_{\frac{\alpha}{2}} + \lambda}{\sqrt{1 + (t^2_{\frac{\alpha}{2}}/2\nu)}}$$

$$u = \frac{t_\alpha + \lambda}{\sqrt{1 + (t^2_{1-\alpha}/2\nu)}}$$

and where U or U_i is an approximate standard normal random variable.

If $\mu_X = 6.80$, and we assume that $\sigma_X = 0.40$, $\alpha = 0.05$, and $\nu = n - 1 = 8$, the power of the two-sided test against a mean of $\mu_1 = 7.10$ is calculated as follows:

$$\lambda = \frac{7.10 - 6.80}{0.40/\sqrt{9}} = \frac{0.30}{0.133} = 2.25$$

$t_{\frac{\alpha}{2}} = -2.306$ from Table C.3 in Appendix C.

$$\sqrt{1 + (t^2_{\frac{\alpha}{2}}/2\nu)} = \sqrt{1 + [(-2.306)^2/(8)2]} = 1.15$$

$$u_1 = \frac{-2.306 - 2.25}{1.15} = -3.96$$

$$u_2 = \frac{-2.306 + 2.25}{1.15} = -0.0488$$

$$(1 - \beta) \approx (0) + (1.000 - 0.519) = 0.481$$

$$\beta \approx 0.519$$

The same operating characteristic of the test (for $n = 9$), β, can be read from Figure 3.4-3a for

$$d = \frac{|\mu_0 - \mu_1|}{\sigma_X} = \frac{0.30}{0.40} = 0.75$$

but to fewer significant figures. If σ_X were not known but had to be estimated by s_X, then too large a value for σ_X would underestimate $|\mu_0 - \mu_1|/\sigma_X$ and overestimate β, while the contrary would be true if σ_X were underestimated.

Example 3.4-3 Determination of Sample Size

Suppose that the experimenter wants to determine how big a sample to take in order to raise the power of the test used in Example 3.4-1 from 0.481 to, say, $(1 - \beta) = 0.80$.

† A. Hald, *Statistical Theory with Engineering Applications.* John Wiley, New York, 1952, p. 392.

Values of $(1 - \beta)$ could be calculated for a series of values of the sample size n and given values of σ_X and α, and the n selected which gave $(1 - \beta)$ close to 0.80. Figure 3.4-3 can also be used to compute β, and the value of n can be read directly for a calculated d. Based on the data of Example 3.4-2,

$$d = \frac{0.30}{0.40} = 0.75 \qquad \alpha = 0.05$$

and for $(1 - \beta) = 0.80$, $\beta = 0.20$, Figure 3.4-3*a* gives $n \cong 16$.

3.4-1 Sequential Testing

It is quite possible in practice, long before the nth observation calculated in Example 3.4-3 is reached, to ascertain whether or not H_0 should be accepted or rejected by a *sequential testing* plan. In sequential testing, a test is executed after each additional observation is collected, starting after the first, until the hypothesis is accepted or rejected. After each test, one of the following decisions is made:

1. Accept the hypothesis H_0.
2. Reject the hypothesis H_0.
3. Make one more observation.

Thus, instead of having two regions, a region of rejection and a region of acceptance, we have a third region as well, one of no decision except to require further experimentation. (Examine Figure 3.4-4.) Upper and lower limits are determined for a test statistic whose nature depends upon the test being carried out. As soon as the value of the test statistic falls below the lower limit, H_0 is accepted; or, as soon as it exceeds the upper limit, H_0 is rejected. After one of these events occurs, the sampling and testing are terminated. Otherwise, an additional observation is taken.

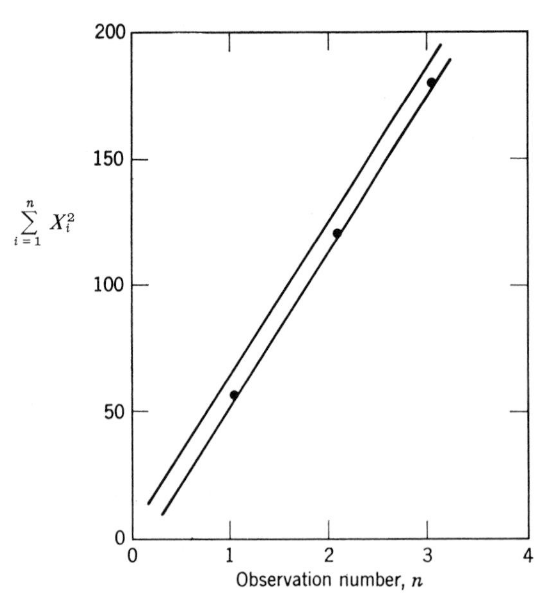

FIGURE 3.4-4 Sequential test chart to detect a difference in ensemble means for gasoline knock rating, $\alpha = \beta = 0.05$, $\mu_1 = 55$, $\mu_2 = 65$, and $\sigma_X^2 \cong 9.5$.

To illustrate one type of test that can be employed, we shall describe the probability ratio test devised by Wald. This test is based upon an assumed sequence of independent observations of the random variable X from a normally distributed population with known variance but unknown mean. The null hypothesis is that $\mu_X = \mu_1$, and the alternate hypothesis is that $\mu_X = \mu_2$. Under these assumptions the likelihood function of the observations defined by Equation 3.2-1 will be one of the following:

$$L_1 = (\sqrt{2\pi}\sigma_X)^{-n} \exp\left[-\frac{1}{2\sigma_X^2}\sum_{i=1}^{n}(X_i - \mu_1)^2\right]$$

or

$$L_2 = (\sqrt{2\pi}\sigma_X)^{-n} \exp\left[-\frac{1}{2\sigma_X^2}\sum_{i=1}^{n}(X_i - \mu_2)^2\right]$$

The test involves the calculation of the ratio (L_2/L_1) after each observation $X_1, \ldots, X_n$. When the ratio exceeds an upper limit, l_l, the hypothesis that $\mu_X = \mu_1$ is accepted. If the ratio falls below a lower limit, l_u, the hypothesis that $\mu_X = \mu_2$ is accepted. If the ratio lies within these bands

$$l_l < \frac{L_2}{L_1} < l_u \tag{3.4-3}$$

one more observation is made. The lower and upper bands are selected so that the power is α when $\mu_X = \mu_1$ and $1 - \beta$ when $\mu_X = \mu_2$. Wald showed that

$$l_u \simeq \frac{1 - \beta}{\alpha}$$

$$l_l \simeq \frac{\beta}{1 - \alpha}$$

and that the probability is 1 that the sequential test will terminate with a choice of one of the hypotheses.

Introduction of L_1, L_2, and the approximations for the upper and lower limits into Equation 3.4-3 yields

$$\frac{\beta}{1 - \alpha} < \exp\left[-\frac{1}{2\sigma_X^2}\sum_{i=1}^{n}\{(X_i - \mu_2)^2 - (X_i - \mu_1)^2\}\right] < \frac{1 - \beta}{\alpha}$$

or

$$\ln\left(\frac{\beta}{1 - \alpha}\right) < -\frac{1}{2\sigma_X^2}\sum_{i=1}^{n}[(X_i - \mu_2)^2 - (X_i - \mu_1)^2] < \ln\left(\frac{1 - \beta}{\alpha}\right)$$

which reduces to

$$\frac{\sigma_X^2}{\mu_2 - \mu_1} \ln\left(\frac{\beta}{1-\alpha}\right) + n\bar{\mu} < \sum_{i=1}^{n} X_i < \frac{\sigma_X^2}{\mu_2 - \mu_1} \ln\left(\frac{1-\beta}{\alpha}\right) + n\bar{\mu} \tag{3.4-4}$$

where $\bar{\mu} = (\mu_1 + \mu_2)/2$. Thus, in a test for one of two ensemble means, the sum of the observations up through the nth observation can be bounded if σ_X^2 is known and values of μ_2 and μ_1 and α and β are chosen. Figure 3.4-4 shows how the bounds on $\sum_{i=1}^{n} X_i$ increase as n increases, as indicated by the terms $n\bar{\mu}$ in Equation 3.4-4. The data in the figure are for knockmeter readings of gasoline with $\mu_1 = 55$ and $\mu_2 = 65$ for two different octane numbers; $\alpha = \beta = 0.05$; and σ_X^2 estimated from earlier tests by $s_X^2 = 9.5$ with 20 degrees of freedom. Inequality 3.4-4 is then approximately

$$60n - 2.80 < \sum_{i=1}^{n} X_i < 60n + 2.80$$

If the ensemble mean, μ_X, is known, and a sequential test is to be carried for two alternate hypotheses with respect to the standard deviation

$$H_1\colon \quad \sigma_X = \sigma_1$$
$$H_2\colon \quad \sigma_X = \sigma_2$$

the likelihood functions can be formed as before by placing $\mu_1 = \mu_2 = \mu_X$ and replacing the standard deviations with σ_1 and σ_2, respectively. The analogous equation to Equation 3.4-4 is

$$\frac{2\ln\left(\dfrac{\beta}{1-\alpha}\right) + n\ln\left(\dfrac{\sigma_2^2}{\sigma_1^2}\right)}{\dfrac{1}{\sigma_1^2} - \dfrac{1}{\sigma_2^2}} < \sum_{i=1}^{n} (X_i - \mu_X)^2 < \frac{2\ln\left(\dfrac{1-\beta}{\alpha}\right) + n\ln\left(\dfrac{\sigma_2^2}{\sigma_1^2}\right)}{\dfrac{1}{\sigma_1^2} - \dfrac{1}{\sigma_2^2}} \tag{3.4-5}$$

If μ_X is unknown, substitute

$$\sum_{i=1}^{n} (X_i - \mu_X)^2 \cong \sum_{i=1}^{n} (X_i - \bar{X})^2$$

and replace n by $(n - 1)$ in the upper and lower bounds of Equation 3.4-5.

Many other sequential tests can be carried out, as described in the references at the end of this chapter.

3.5 HYPOTHESIS TESTS FOR MEANS

Table 3.5-1 summarizes certain tests which enable the analyst to tell if the ensemble mean of a new product or a variable: (1) is different from, (2) exceeds, or (3) is less

TABLE 3.5-1 TESTS FOR COMPARING THE MEAN OF A NEW PRODUCT OR A VARIABLE WITH A STANDARD*

Hypothesis	Knowledge of the Standard Deviation of a New Product or Variable	Test to Be Made Is† (If the inequality is satisfied) the hypothesis is accepted)	Remarks
$\mu \neq \mu_0$	σ unknown; s from sample used	$\lvert\bar{X} - \mu_0\rvert > t_{1-\frac{\alpha}{2}}\left(\dfrac{s_X}{\sqrt{n}}\right)$	Two-sided t-test
	σ known	$\lvert\bar{X} - \mu_0\rvert > U_{1-\frac{\alpha}{2}}\left(\dfrac{\sigma_X}{\sqrt{n}}\right)$	Two-sided U-test
$\mu > \mu_0$	σ unknown; s from sample used	$(\bar{X} - \mu_0) > t_{1-\alpha}\left(\dfrac{s_X}{\sqrt{n}}\right)$	One-sided t-test
	σ known	$(\bar{X} - \mu_0) > U_{1-\alpha}\left(\dfrac{\sigma_X}{\sqrt{n}}\right)$	One-sided U-test
$\mu < \mu_0$	σ unknown; s from sample used	$(\mu_0 - \bar{X}) > t_{1-\alpha}\left(\dfrac{s_X}{\sqrt{n}}\right)$	One-sided t-test
	σ known	$(\mu_0 - \bar{X}) > U_{1-\alpha}\left(\dfrac{\sigma_X}{\sqrt{n}}\right)$	One-sided U-test

* Adapted from M. G. Natrella, *Experimental Statistics*, Nat. Bur. of Standards, Handbook 91, U.S. Dept. of Commerce, Washington, D.C., 1963.

† In each case look up t or U for the selected significance level α; t is for the $n - 1$ degrees of freedom. The tests presume an underlying normal population.

than the ensemble mean of a standard product or variable. The hypothesis selected presumes we *know* the value of the standard ensemble mean, μ_0, from past experience or otherwise. (In the tables which follow, the subscript zero will refer to the standard mean while the absence of a subscript zero will refer to the mean being tested.) After each test is made, as indicated in the third column of the tables, a decision can be reached as follows:

1. If the inequality proves to be true, that is if the calculated difference exceeds the right-hand side of the inequality, the hypothesis *is accepted.*
2. If the inequality does not prove to be true, that is if the calculated difference does not exceed the right-hand side of the inequality, then the hypothesis *is rejected*, and there is little likelihood that the hypothesis is correct.

NBS Handbook 91† provides detailed charts to simplify the calculation of the operating characteristics of each test and also provides tables to establish the sample size required to detect a difference for each test.

The decision rules shown in Table 3.5-1 are now illustrated by an example.

Example 3.5-1 Hypothesis Test for the Mean

Ten different resistance thermometers are calibrated against a standard whose reading is 1000 mv. After receipt by a laboratory, the ten thermometers read:

986	1002
1005	996
991	998
994	1002
983	983

Can these deviations be regarded as being caused by the normal variation of the random variable, the reading in mv, or has some factor (perhaps during shipment or manufacture) affected their performance?

Solution:

We shall test the hypothesis that the ensemble mean of the readings of the ten thermometers, μ, has changed from $\mu_0 = 1000$ by selecting as H_0 the first hypothesis in Table 3.5-1, namely $\mu \neq \mu_0$. The test to be made, since σ_X is unknown, is

$$|\bar{X} - \mu_0| \overset{?}{>} t_{1-\frac{\alpha}{2}}\left(\frac{s_X}{\sqrt{n}}\right)$$

If we choose $\alpha = 0.05$, so that $\alpha/2 = 0.025$, and $t_{1-\frac{\alpha}{2}} = 2.26$, we can calculate

$$\bar{X} = \frac{\sum X_i}{n} = 994.0$$

$$s_X^2 = \frac{\sum (X_i - \bar{X})^2 n_i}{n - 1} = 64.9$$

$$\nu = n - 1 = 9$$

$$|\bar{X} - \mu_0| = 6.0$$

$$t_{1-\frac{\alpha}{2}}\left(\frac{s_X}{\sqrt{n}}\right) = 2.26\left(\frac{64.9}{10}\right)^{1/2} = 2.26(2.55) = 5.76$$

We observe that $6 > 5.76$, and conclude for a significance level of $\alpha = 0.05$ (but not for $\alpha = 0.01$) that the hypothesis H_0 should be accepted.

Table 3.5-2 summarizes tests which can be carried out with respect to the ensemble means of two products (or variables), both of which are sampled. Of interest is to test whether the:

1. Averages of two products (or variables) differ, without caring which is larger.
2. Average of product (or variable) A exceeds that of product (or variable) B.

Again there exist subclasses of the tests, depending upon the extent of the information available about the standard deviation of the variable being measured. Again, too, if the difference calculated is greater than the right-hand side of the inequality, then the hypothesis is accepted; otherwise it is rejected. Operating characteristic curves and tables to determine the sample size can be found in NBS Handbook 91 for each test.

To illustrate the general procedure of developing a hypothesis test to compare two means, we shall outline how the first test listed in Table 3.5-2 is established. The other tests can be developed in a similar manner.

Assume that we have samples of normal random variables A and B as follows, with the ensemble means and variances as indicated.

	A	B
Sample values	$X_{A_1}, X_{A_2}, \ldots, X_{A_{n_A}}$	$X_{B_1}, X_{B_2}, \ldots, X_{B_{n_B}}$
Ensemble mean	μ_A	μ_B
Ensemble variance	σ_A^2	σ_B^2

Sample statistics can be calculated as follows:

$$\bar{X}_A = \frac{1}{n_A}\sum_{i=1}^{n_A} X_{A_i}$$

$$\bar{X}_B = \frac{1}{n_B}\sum_{i=1}^{n_B} X_{B_i}$$

$$s_A^2 = \frac{1}{n_A - 1}\sum_{i=1}^{n_A}(X_{A_i} - \bar{X}_A)^2$$

$$s_B^2 = \frac{1}{n_B - 1}\sum_{i=1}^{n_B}(X_{B_i} - \bar{X}_B)^2$$

$$\nu_A = n_A - 1$$

$$\nu_B = n_B - 1$$

† M. G. Natrella, *Experimental Statistics*, Nat. Bur. of Standards Handbook 91, U.S. Dept. of Commerce, Washington, D.C., 1963.

TABLE 3.5-2 TESTS FOR COMPARING THE MEANS OF TWO PRODUCTS OR VARIABLES*

Hypothesis	Knowledge of the Standard Deviation of A and B	Test to Be Made Is† (If the inequality is satisfied the hypothesis is accepted.)	Remarks
$\mu_A \neq \mu_B$	$\sigma_A \cong \sigma_B$ both unknown	$\lvert \bar{X}_A - \bar{X}_B \rvert > t_{1-\frac{\alpha}{2}} s_p \left(\frac{n_A + n_B}{n_A n_B}\right)^{1/2}$	$s_p = \left[\frac{(n_A - 1)s_A^2 + (n_B - 1)s_B^2}{n_A + n_B - 2}\right]^{1/2}$ $\nu = n_A + n_B - 2$
	$\sigma_A \neq \sigma_B$ both unknown	$\lvert \bar{X}_A - \bar{X}_B \rvert = t'\left(\frac{s_A^2}{n_A} + \frac{s_B^2}{n_B}\right)^{1/2}$	t' = value of $t_{1-\frac{\alpha}{2}}$ for ν degrees of freedom $\nu = \dfrac{(s_A^2/n_A + s_B^2/n_B)^2}{\dfrac{(s_A^2/n_A)^2}{n_A + 1} + \dfrac{(s_B^2/n_B)^2}{n_B + 1}} - 2$
	σ_A and σ_B both known	$\lvert \bar{X}_A - \bar{X}_B \rvert > U_{1-\frac{\alpha}{2}} \left(\frac{\sigma_A^2}{n_A} + \frac{\sigma_B^2}{n_B}\right)^{1/2}$	
$\mu_A > \mu_B$	$\sigma_A \cong \sigma_B$ both unknown	$(\bar{X}_A - \bar{X}_B) > t_{1-\alpha} s_p \left(\frac{n_A + n_B}{n_A n_B}\right)^{1/2}$	$s_p = \left[\frac{(n_A - 1)s_A^2 + (n_B - 1)s_B^2}{n_A + n_B - 2}\right]^{1/2}$ $\nu = n_A + n_B - 2$
	$\sigma_A \neq \sigma_B$ both unknown	$(\bar{X}_A - \bar{X}_B) > t'\left(\frac{s_A^2}{n_A} + \frac{s_B^2}{n_B}\right)$	t' = value of $t_{1-\alpha}$ for ν degrees of freedom $\nu = \dfrac{(s_A^2/n_A + s_B^2/n_A)^2}{\dfrac{(s_A^2/n_A)^2}{n_A + 1} + \dfrac{(s_B^2/n_B)^2}{n_B + 1}} - 2$
	σ_A and σ_B both known	$(\bar{X}_A - \bar{X}_B) = U_{1-\alpha} \left(\frac{\sigma_A^2}{n_A} + \frac{\sigma_B^2}{n_B}\right)^{1/2}$	

* Adapted from NBS Handbook 91.
† n_A and n_B observations are made to obtain samples A and B. t is for $n_A + n_B - 2$ degrees of freedom. s_p and other pooled values of s are discussed in Section 2.4-1. The tests presume an underlying normal population.

The two sample means, $\bar{X}_A$ and $\bar{X}_B$, are normally distributed with parameters $(\mu_A, \sigma_A^2/n_A)$ and $(\mu_B, \sigma_B^2/n_B)$, respectively. Also, the difference between the two means $D = \bar{X}_A - \bar{X}_B$ is normally distributed about $\delta = \mu_A - \mu_B$ with the variance

$$\text{Var}\{D\} = \frac{\sigma_A^2}{n_A} + \frac{\sigma_B^2}{n_B} \qquad (3.5\text{-}1)$$

If s_A^2 does not differ significantly from s_B^2, we set up the test hypotheses

$$\mu_A = \mu_B \qquad \text{and} \qquad \sigma_A^2 \cong \sigma_B^2 = \sigma^2$$

If these hypotheses are true, then we know that $\bar{X}_A - \bar{X}_B = D$ is normally distributed about $\delta = \mu_A - \mu_B = 0$ with the variance

$$\text{Var}\{D\} = \sigma^2\left(\frac{1}{n_A} + \frac{1}{n_B}\right) \qquad (3.5\text{-}2)$$

By using Equation 2.4-12, we can compute the following estimate of σ^2:

$$s_p^2 = \frac{\nu_A s_A^2 + \nu_B s_B^2}{\nu_A + \nu_B} \qquad \text{for } \nu = \nu_A + \nu_B = n_A + n_B - 2$$

Also

$$t = \frac{D}{s_D} = \frac{\bar{X}_A - \bar{X}_B}{s_p\sqrt{\dfrac{1}{n_A} + \dfrac{1}{n_B}}}$$

has a t-distribution with $\nu = n_A + n_B - 2$ degrees freedom. A significant value of t is interpreted to mean that $\mu_A \neq \mu_B$.

Example 3.5-2 Comparison of Two Means

Two different grades of gasolines were used to find the number of miles per gallon obtained under highway travel. Each grade (90 octane and 94 octane) was used in five identical automobiles traveling over the same route with the following results:

	94 Octane	90 Octane
Sample mean (miles/gal)	22.7	21.3
Sample standard deviation (miles/gal)	0.45	0.55

For a significance level of $\alpha = 0.05$, are the grades different? If so, is the 94-octane gasoline significantly better than the 90-octane gasoline?

Solution:

First, we test the hypothesis that $\mu_{94} \neq \mu_{90}$. We assume that $\sigma_{94} \cong \sigma_{90}$; a method of checking this assumption will be given in Section 3.6.

$$|\bar{X}_{94} - \bar{X}_{90}| \overset{?}{>} t_{1-\frac{\alpha}{2}} s_p \left(\frac{5+5}{25}\right)^{1/2}$$

$$\left(\frac{5+5}{25}\right)^{1/2} = 0.632$$

$$s_p = \left[\frac{4(0.45)^2 + 4(0.55)^2}{5+5-2}\right]^{1/2} = (0.252)^{1/2} = 0.50$$

$$t_{1-\frac{\alpha}{2}} = 2.306$$

$$\nu = 5 + 5 - 2 = 8$$

$$|\bar{X}_{94} - \bar{X}_{90}| = 22.7 - 21.3 = 1.4$$

$$(2.306)(0.50)(0.632) = 0.73$$

Since $1.4 > 0.73$, the hypothesis is accepted and $\mu_{94} \neq \mu_{90}$.

Next, we test the hypothesis that $\mu_{94} > \mu_{90}$, assuming still that $\sigma_{94} \cong \sigma_{90}$.

$$(\bar{X}_{94} - \bar{X}_{90}) \overset{?}{>} t_{1-\alpha} s_p \left(\frac{n_{94} + n_{90}}{n_{94} n_{90}}\right)^{1/2}$$

$$(\bar{X}_{94} - \bar{X}_{90}) = 1.4$$

$$t_{1-\alpha} s_p \left(\frac{5+5}{(5)(5)}\right)^{1/2} = (1.860)(0.50)(0.632) = 0.59$$

Again the hypothesis is accepted.

All the tests outlined so far are based on certain assumed characteristics of the random variable involved. In practice, of course, some or all of these assumptions may not hold true. Aberrations are more serious for some tests than for others. Those tests that are relatively insensitive to changes from the assumed characteristics are termed *robust*. Because several assumptions are involved in each test, robustness is interpreted in terms of the separate effects of deviations from normality, independence, equal variance, and randomness.

The underlying assumptions for the t-tests are: (1) the random variables being measured are normally distributed, and (2) the samples are random ones. Decisions made on the basis of the t-test (and other tests) depend, sometimes critically, on the degree of approximation of the experimental conditions to the assumed ones.

The effect of nonnormality on the Student t-test has been studied and illustrated by many investigators. As a rough rule-of-thumb, the classical application of the t-test to a comparison of means is relatively unaffected by aberration of the underlying random variable from normality.

Walsh† examined the influence of nonrandomness of the sample on the Student t-test for large numbers of observations. It was found that even a slight deviation from the supposed randomness led to substantial changes in the significance level and confidence coefficient. Modified tests which were not sensitive to the requirement of a random sample are described in his report. Alternates to the t-test will be discussed in Section 3.7.

3.6 HYPOTHESIS TESTS FOR VARIABILITY

The objective of this section is to summarize certain tests that enable the analyst to reach a decision concerning the variability of a product or variable. Corresponding to the previous section, we can test whether the ensemble variance of a new product or variable: (1) is different from, (2) exceeds, or (3) is less than a standard ensemble variance of a random variable with the aid of the χ^2 distribution originally described in Section 2.3-2. For two products or variables, designated A and B, we can test whether the ensemble variance of A differs from that of B or exceeds that of B with the aid of the variance ratio (F) distribution originally described in Section 2.4-3. In Table 3.6-1, the subscript zero will refer to the standard variance while the absence of a subscript will refer to the variance being tested. The tests are based upon the assumption that the observations are taken randomly of a normal random variable. The decision is based on the test shown in the second column of Table 3.6-1. Refer to NBS Handbook 91 for operating characteristic curves and tables for sample size determination.

To illustrate how the tests are formulated, consider the F-test in the fourth row of Table 3.6-1. We shall hypothesize that $\sigma_1^2 = \sigma_2^2$, i.e., $(\sigma_1^2/\sigma_2^2) = 1$, and use the sample variance ratio to test if σ_1^2/σ_2^2 is greater than or less than unity. If the hypothesis is true, then the region of acceptance for equal tail areas is defined through the probability statement

$$P\left\{F_{\frac{\alpha}{2}}(\nu_1, \nu_2) < \frac{s_1^2}{s_2^2} \leq F_{1-\frac{\alpha}{2}}(\nu_1, \nu_2)\right\} = 1 - \alpha$$

Because $F_{\frac{\alpha}{2}}(\nu_1, \nu_2) = 1/(F_{1-\frac{\alpha}{2}}(\nu_1, \nu_2)) < 1$ always, the left-hand inequality in the probability statement is always satisfied, and we need only test to determine if $s_1^2/s_2^2 \leq F_{1-\frac{\alpha}{2}}$.

† J. E. Walsh, RAND Corp. Rept. P-129, Aug. 8, 1950.

Example 3.6-1 Hypothesis Test for Variability

Twin pilot plant units have been designed and put into operation on a given process. The output for the first ten material balances obtained on each of the two units are listed below (basis is 100 lb):

	Unit A (lb)	Unit B (lb)
	97.8	97.2
	98.9	100.5
	101.2	98.2
	98.8	98.3
	102.0	97.5
	99.0	99.9
	99.1	97.9
	100.8	96.8
	100.9	97.4
	100.5	97.2
$\bar{X}$	99.9	98.1
$s_{\bar{X}}^2$	1.69	1.44

Is the variability (variance) of the material balance significantly different between the two units?

Solution:

The hypothesis H_0 is that $\sigma_A^2 = \sigma_B^2$. The degrees of freedom for each unit are 9. We form the variance ratio

$$\frac{s_A^2}{s_B^2} = \frac{1.69}{1.44} = 1.17$$

to test the hypothesis as indicated in the fourth row of Table 3.6-1. From Table C.4 in Appendix C for $\alpha = 0.05$, $F_{0.95}(9, 9) = 4.03$; hence the hypothesis is accepted and there is no significant difference in variability between the two units.

Example 3.6-2 Combined Tests for the Variance and Mean

In a catalytic reactor the distribution of yields from catalyst A and catalyst B gave the following data:

Catalyst A	Catalyst B
$\bar{X}_A = 1.219$	$\bar{X}_B = 1.179$
$s_A^2 = 0.028$	$s_B^2 = 0.0193$
$s_A = 0.456$	$s_B = 0.439$
$n_A = 16$	$n_B = 15$

As a first hypothesis we shall assume: $\sigma_A^2 = \sigma_B^2$. Based on the test in row four of Table 3.6-1, we can calculate the variance ratio:

$$\frac{s_A^2}{s_B^2} = \frac{0.2080}{0.1930} = 1.08$$

From Appendix C, Table C.4 for $\alpha = 0.05$ and for $\nu_A = (n_A - 1) = 15$ and $\nu_B = (n_B - 1) = 14$, the value of $F_{1-\frac{\alpha}{2}}(15, 14) = 2.95$. Thus the hypothesis is accepted and we decide that σ_A^2 does not differ significantly from σ_B^2.

Once this fact has been established, we can pool the sample variances:

$$s_p^2 = \frac{(n_A - 1)s_A^2 + (n_B - 1)s_B^2}{(n_A - 1) + (n_B - 1)} = \frac{15(0.208) + 14(0.193)}{15 + 14}$$
$$= 0.201$$

TABLE 3.6-1 COMPARISON OF TWO PRODUCTS OR VARIABLES WITH REGARD TO THEIR VARIABILITY*

Hypothesis	Test to Be Made Is		Decision	Remarks
$\sigma^2 = \sigma_0^2$	$s^2 \frac{\nu}{\chi^2_{1-\frac{\alpha}{2}}} < \sigma_0^2 < s^2 \frac{\nu}{\chi^2_{\frac{\alpha}{2}}}$	$\nu = n - 1$	If within range, hypothesis is accepted	Two-sided χ^2 test
$\sigma^2 > \sigma_0^2$	$\sigma^2 > s^2 \frac{\nu}{\chi^2_{\alpha}}$	$\nu = n - 1$	If test inequality is true, hypothesis is accepted	One-sided χ^2 test
$\sigma^2 < \sigma_0^2$	$s^2 \frac{\nu}{\chi^2_{1-\alpha}} > \sigma_0^2$	$\nu = n - 1$	If test inequality is true, hypothesis is accepted	One-sided χ^2 test
$\sigma_A^2 = \sigma_B^2$†	$\frac{1}{F_{1-\frac{\alpha}{2}}[(n_B - 1), (n_A - 1)]} < \frac{s_A^2}{s_B^2} < F_{1-\frac{\alpha}{2}}[(n_A - 1), (n_B - 1)]$	$\nu_1 = n_A - 1$, $\nu_2 = n_B - 1$	If within range, hypothesis is accepted	Two-sided F ratio test. Note that $1/F_{1-\frac{\alpha}{2}}$ is always less than unity; hence only the upper limit need be compared
$\sigma_A^2 > \sigma_B^2$‡	$\frac{s_A^2}{s_B^2} > F_{1-\alpha}[\nu_1, \nu_2]$	$\nu_1 = n_A - 1$, $\nu_2 = n_B - 1$	If inequality is accepted, hypothesis is accepted	One-sided F ratio test

* Adapted from NBS Handbook 91.
† The alternate hypothesis is $\sigma_A^2 \neq \sigma_B^2$.
‡ The alternate hypothesis is $\sigma_A^2 = \sigma_B^2$.

We know that the difference between the sample means is

$$D = 1.219 - 1.179 = 0.040$$

and, from Equation 3.5-2, that

$$\text{Var}\{D\} \approx s_p^2\left(\frac{1}{n_A} + \frac{1}{n_B}\right) = 0.201(\tfrac{1}{16} + \tfrac{1}{15}) = 0.026$$

Also, once we know that $\sigma_A^2 \cong \sigma_B^2$, a test can be made as described in Table 3.5-2 based on the hypothesis $\mu_A \neq \mu_B$.

$$|\bar{X}_A - \bar{X}_B| \overset{?}{>} t_{1-\frac{\alpha}{2}} s_p \left(\frac{n_A + n_B}{n_A n_B}\right)^{1/2} \qquad \nu = n_A + n_B - 2$$

$$0.040 \overset{?}{>} (2.045)(0.201)\left[\frac{31}{(16)(15)}\right]^{1/2} \qquad \nu = 29$$

$$0.040 \overset{?}{>} 0.145$$

Since $0.040 < 0.145$, the hypothesis $\mu_A \neq \mu_B$ is rejected, and we conclude that $\mu_A = \mu_B$.

The F-test is applied to two variances. A commonly used test to detect differences among two or more variances is *Bartlett's test.* M. S. Bartlett devised a test to determine the homogeneity of two or more variances by comparing the logarithm of the average variance with the sum of the logarithms of the separate variances. The formulas necessary for the use of this test are based on the hypothesis $H_0: \sigma_1^2 = \sigma_2^2 = \cdots = \sigma_n^2 = \sigma^2$ and the presumption that the variables measured are normally distributed. The same critical limits hold as in the F-test except that there are n samples. If the test hypothesis is correct, a pooled s^2

$$s^2 = \frac{\sum_{i=1}^{k} \nu_i s_i^2}{\sum_{i=1}^{k} \nu_i} = \frac{1}{\left(\sum_{i=1}^{n} p_i - n\right)} \sum_{i=1}^{n} (p_i - 1)s_i^2 \qquad (3.6\text{-}1)$$

(where p_i = number of replicates in a sample) has an s^2 distribution with a mean of σ^2 and ν degrees of freedom, where $\nu = \sum_{i=1}^{n} \nu_i$. Bartlett showed that

$$\Lambda = -\frac{1}{c} \sum_{i=1}^{n} p_i \ln\left(\frac{s_i^2}{s^2}\right) \qquad (3.6\text{-}2)$$

where

$$c = 1 + \frac{1}{3(n-1)}\left(\sum_{i=1}^{n} \frac{1}{p_i} - \frac{1}{\sum_{i=1}^{k} p_i}\right)$$

has an approximate χ^2 distribution with $(n - 1)$ degrees of freedom. For large values of p_i, $c \cong 1$.

For the special case where all the p_i's are equal so that $\sum p_i = np_i$,

$$\chi^2 \cong \frac{1}{c} np_i\left(\ln s^2 - \frac{1}{n} \sum_{i=1}^{n} \ln s_i^2\right) \qquad (3.6\text{-}3)$$

where $c = \{1 + [(n + 1)/3np_i]\}$. If the value of χ^2 calculated by Equation 3.6-2 or 3.6-3 exceeds the value of $\chi^2_{1-\alpha}$ for $(k - 1)$ degrees of freedom, the test hypothesis that $\sigma_1^2 = \sigma_2^2 \cdots$ is rejected. Certain application restrictions and supplementations to Bartlett's test are described in Hald,† of which the most important and critical is that the observations must be normally distributed.

Example 3.6-3 Test for Nonconstant σ^2

Ten replicate measurements were made for corrosion loss, Y, at four different values of alloy concentration, X. Results are shown in Table E3.6-3 and Figure E3.6-3.

TABLE E3.6-3 RESULTS OF CORROSION EXPERIMENTS

i	X_i	p_i	Y_{i1}	Y_{i2}	Y_{i3}	Y_{i4}	Y_{i5}
1	1.28	10	6.34	6.36	6.41	6.42	6.80
2	1.30	10	5.95	6.04	6.11	6.31	6.36
3	1.40	10	5.23	5.27	5.32	5.39	5.40
4	1.48	10	4.55	4.65	4.68	4.68	4.72

i	Y_{i6}	Y_{i7}	Y_{i8}	Y_{i9}	Y_{i10}	$\bar{Y}_i$	s_i^2
1	6.85	6.91	6.91	7.02	7.12	6.71	0.091
2	6.52	6.60	6.62	6.64	6.71	6.39	0.076
3	5.52	5.52	5.53	5.60	5.78	5.46	0.020
4	4.73	4.78	4.78	4.84	4.86	4.72	0.009

A test can be made to ascertain if the variances at the different values of X_i are the same (homogeneity of variance) or not by using Bartlett's test. If Λ exceeds the value of χ^2 determined from the tables in Appendix C for a given α,

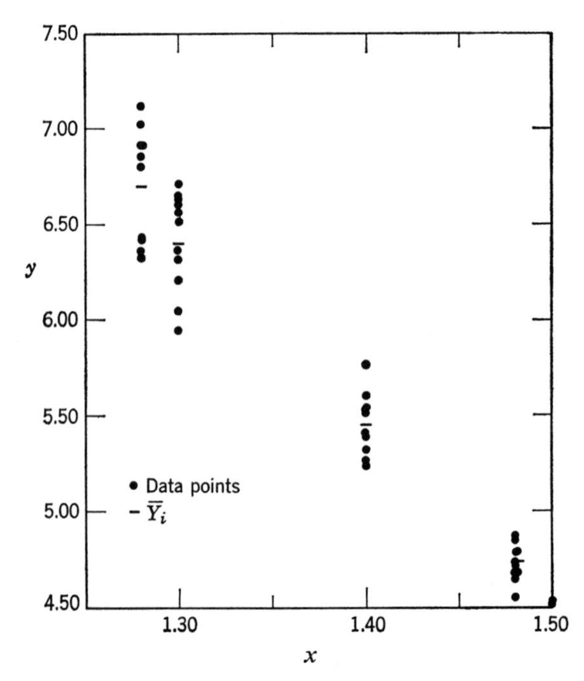

FIGURE E3.6-3 Results of corrosion experiments on alloys.

† A. Hald, *Statistical Theory with Engineering Applications*, John Wiley, New York, 1952, pp. 290–298.

then the hypothesis H_0 that the variances are the same is rejected. Here $n = 4$, $p_i = 10$, $\sum p_i = 40$.

$$c = 1 + \frac{1}{3(n-1)}\left(\sum_{i=1}^{n}\frac{1}{p_i} - \frac{1}{\sum_{i=1}^{n} p_i}\right) = 1 + \frac{1}{3(3)}\left(\frac{4}{10} - \frac{1}{10}\right)$$

$$\cong 1$$

$$s^2 = \frac{1}{\left(\sum_{i=1}^{n} p_i - n\right)}\sum_{i=1}^{n}(p_i - 1)s_i^2 = \frac{1}{(40-4)}\sum (9)s_i^2$$

$$\cong 0.049$$

$$\Lambda = -\frac{1}{c}\sum_{i=1}^{n} p_i \ln\left(\frac{s_i^2}{s^2}\right) = +10\sum_{i=1}^{4}\ln\left(\frac{s_i^2}{0.049}\right)$$

$$\cong 15.3$$

For $\alpha = 0.05$, with $(n - 1) = 3$ degrees of freedom, χ^2 from the appendix tables is 7.81; thus the hypothesis of equal variances for the X_i's is rejected. Figure E3.6-3 illustrates how the dispersion varies as a function of x.

In many experiments the analyst is justified in assuming that the random variables being observed are normally distributed; he can carry out suitable tests, some of which are described in Section 3.7, for normality. But suppose that the random variables being observed are not normally distributed. What can be said then in regard to the application of the F-tests? The F-test and especially Bartlett's test for comparison of variances are quite sensitive to nonnormality; they must be modified or different tests employed if the normality assumption is violated.

Another method of attack on the problem of nonnormality of the measured random variable is to use a transformation of variables to render the data more nearly normal and to reduce the differences between individual variances of groups of data. Transformations may be viewed as scale changes imposed on the original variables to adjust them to a more favorable orientation. For example, a logarithmic transformation changes the very skewed probability density illustrated in Figure 3.6-1*a* to the considerably more normal appearing density in Figure 3.6-1*b*. However, one must be careful not to inadvertently create new difficulties when attempting to resolve existing problems by transformations.

Example 3.6-4 Transformation of a Probability Density

Rose and English† investigated the distribution of the breaking strengths of identical paper sacks containing identical amounts of material dropped under controlled conditions. A representative relative frequency distribution of 200 sacks is illustrated in Figure E3.6-4a versus the drop

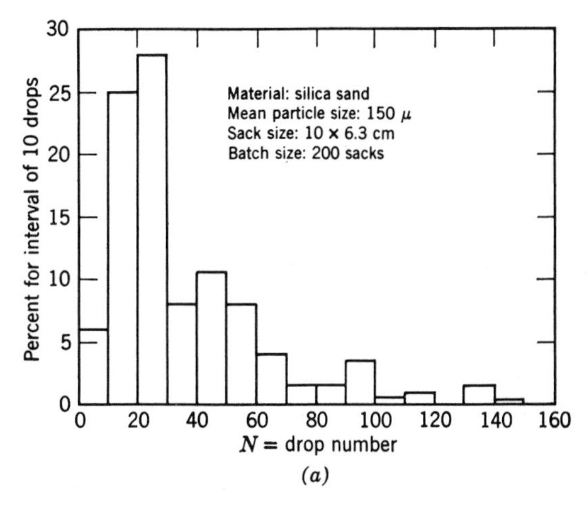

FIGURE E3.6-4A Distribution of the strengths of identical sacks. (From H. E. Rose and J. E. English, *Chem. Eng.*, Sept. 1966, p. 165, with permission.)

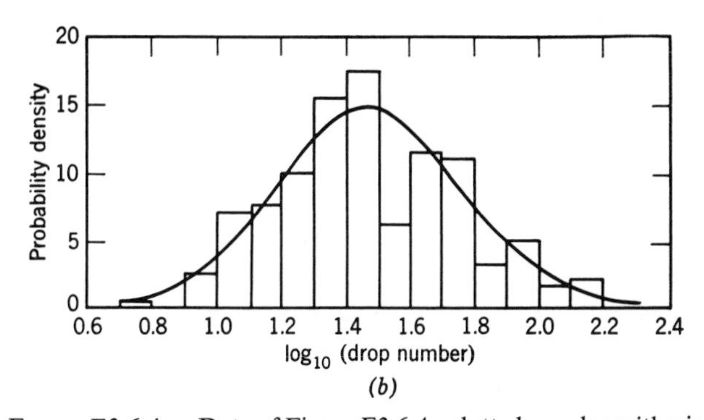

FIGURE E3.6-4B Data of Figure E3.6-4a plotted on a logarithmic-normal basis. (From H. E. Rose and J. E. English, *Chem. Eng.*, Sept. 1966, p. 165, with permission.)

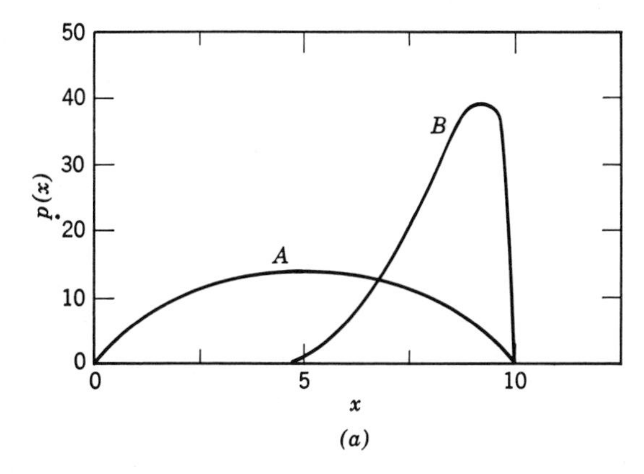

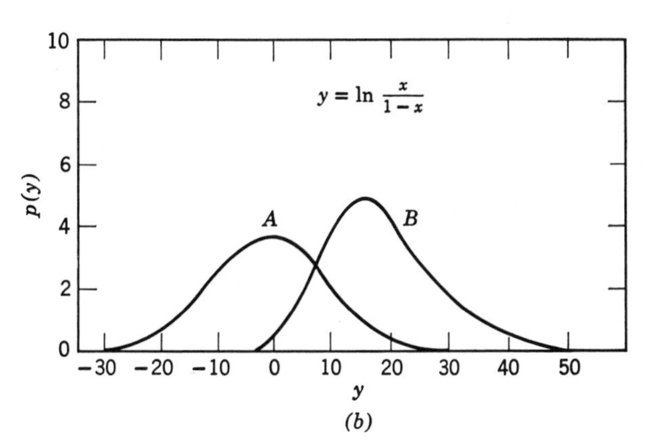

FIGURE 3.6-1 A logarithmic transformation.

number, i.e., the number of drops from a specified height that a sack will withstand before failure. The same data are plotted as the logarithm of the drop number in Figure E3.6-4b.

Rose and English were able to associate, by means of appropriate statistical tests, the logarithmic-normal distribution with the drop number-relative frequency distribution. Then they investigated theoretically why such a probability distribution should be expected. Also, once the underlying distribution was verified, only two parameters, the mean and the standard deviation, were needed to characterize nearly 2000 tests with different sack fillings.

If data are easy and inexpensive to collect, the simplest method of normalization is to average groups of data and make tests on the group averages. The central limit theorem mentioned in Section 2.4-1, which states that the distribution of a sum of n random (not necessarily normal) variables tends to approach the normal distribution as the sample size becomes large, provides the rationale for such treatment.

Sample variances, if obtained from a population with the same σ^2, can be pooled to improve the estimate of σ^2. However, if the sample variances are based on a nonhomogeneous population, the pooled s^2 is not a valid estimate of σ^2; confidence intervals and tests of significance based on the pooled s^2 are then distorted.

One final matter should be mentioned. Since the presentations of information in the form of confidence limits and as hypothesis tests make use of the same basic parameters, we might ask which presentation is more meaningful? One can conclude that if the confidence interval does not include the sample mean, the null hypothesis is rejected—the same conclusion as is obtained from a hypothesis test. However, the use of confidence interval statements can be more meaningful because they give the analyst a picture of the degree of uncertainty in the parameters rather than simply a yes or no answer as is obtained from hypothesis testing.

3.7 NONPARAMETRIC (DISTRIBUTION-FREE) TESTS

All the tests presented up to this point have explicitly involved the assumption that the random variables of interest were represented by a known probability distribution, usually the normal distribution. Such tests are known as parametric tests. Other types of tests exist, including rank correlation and sign tests, which do not require such assumptions and are known as *nonparametric* tests or *distribution-free* tests. (The distribution-free characteristic really applies only to the significance level of the test and only for samples of continuous variables. In many nonparametric tests, probability statements do depend on the probability distribution of the random variable.) Nonparametric methods can be used in tests of hypotheses, to find interval and even point estimates of parameters, and so forth. For example, a nonparametric estimate of the ensemble mean is the *median* of a random sample (the middle value for n odd and the average of the two middle values for n even); a nonparametric estimate of the standard deviation is the *range* (the absolute value of the difference between the highest and lowest values in the sample). Neither of these statistics is particularly efficient as compared with the sample mean and sample standard deviation, respectively, that we described previously.

We shall consider only a few nonparametric tests, mainly those that can be substituted for the parametric tests of means and variances, described in Sections 3.4 and 3.5, and also those that are useful in establishing the stationarity, randomness, and normality of random variables. Most texts on statistics include a chapter describing various types of nonparametric tests; Savage† prepared an excellent bibliography showing applications of the tests.

3.7-1 Sign Test for Median Difference in Paired Observations

The simplest nonparametric test which can be used in lieu of the t-test is the sign test for paired observations. Suppose that n pairs of measurements are taken of a random variable, one of each pair under condition A and the other under condition B. If zero differences are impossible, the differences $A_i - B_i$ can be either positive or negative; the positive outcome is distributed as a binomial variable with $\theta = \frac{1}{2}$. (Zero differences obtained on calculation have been treated in many ways, none of which is completely satisfactory. But if the proportion of zero differences is low, say less than 5 percent, the pairs of zero observations can either be omitted from consideration or divided equally between the plus and minus categories.) Because the sign test is based on the binomial distribution, the binomial events must be independent (refer to Table 2.3-1); that is, the sign difference for one pair of measurements must have no influence on the sign difference for any other pair of measurements, and the sample must be random. Also the outcomes must be continuous.

For every $A_i - B_i$ difference, the $P\{A_i > B_i\} = P\{A_i < B_i\} = \frac{1}{2}$ if $\mathscr{E}\{A_i - B_i\} = 0$. The sign test simply tests the hypothesis that the parameter θ in the binomial density has the value $\frac{1}{2}$, which in terms of the experiment tests the null hypothesis that the population of $A - B$

† H. E. Rose and J. E. English, *The Chemical Engineer*, Sept. 1966, p. 165.

† I. R. Savage, *Bibliography of Nonparametric Statistics*, Harvard Univ. Press, Cambridge, Mass., 1962.

differences has a median of zero. Let r be the number of occurrences of the less frequent sign and $n - r$ be the number of occurrences of the more frequent sign after the zero differences have been divided up. Then the cumulative probability of obtaining r or fewer signs if the null hypothesis (H_0: there is no difference in the effects of A and B) is true is

$$P = \sum_{i=0}^{r} \binom{n}{i} (\tfrac{1}{2})^n \tag{3.7-1}$$

For a two-sided symmetric test, one rejects the null hypothesis if

$$P \leq \frac{\alpha}{2} \quad \text{or} \quad P \geq 1 - \frac{\alpha}{2}$$

If a one-sided test is used and the alternate hypothesis is that the median difference is less than zero, the null hypothesis is rejected if $P \leq \alpha$; the opposite alternative hypothesis calls for rejection of the null hypothesis if $P \geq 1 - \alpha$. As applied to a normally distributed difference, the one-sided sign test has an asymptotic efficiency relative to the t-test of $2/\pi = 0.637$.

Example 3.7-1 Sign Test

Table E3.7-1 lists ten pairs of measurements of the percentage of sulfur dioxide in the exit gas of a smoke stack for two degrees of fuel pulverization, A and B. We shall test the assumption that the two degrees of pulverization produce the same amount of the pollutant sulfur dioxide.

Six pluses and two minuses are found. If the zero differences are distributed equally between plus and minus, we have $r = 3$ and $(n - r) = 7$. Then, making use of a table of probabilities of the binomial variable, we calculate

$$P\{r \leq 3\} = P\{r = 0\} + P\{r = 1\} + P\{r = 2\} + P\{r = 3\}$$
$$= 0.001 + 0.010 + 0.044 + 0.117 = 0.172$$

If α had been selected as 0.05, $\alpha/2 = 0.025$, and because $0.172 > 0.025$, the null hypothesis would be accepted. Table C.5† in Appendix C gives the number of pluses for rejection for various critical regions. In using the table, the zero differences are discarded. Thus, $n = 8$, $r = 2$, and the critical r from the table for $\alpha/2 = 0.025$ is zero. Since $2 > 0$, the hypothesis is accepted.

† Taken from W. J. Dixon and F. J. Massey, *Introduction to Statistical Analysis*, McGraw-Hill, New York, 1951.

TABLE E3.7-1 PERCENT SULFUR DIOXIDE

	Sample Number									
Pulverization	1	2	3	4	5	6	7	8	9	10
A	2.4	2.7	2.0	1.9	2.2	2.3	2.3	2.1	2.4	2.6
B	2.6	2.6	2.0	1.8	2.0	2.0	2.4	2.1	2.1	2.5
$A - B$	−	+	0	+	+	+	−	0	+	+

3.7-2 Mann-Whitney U^*-Test

The Mann-Whitney U^*-test is the most powerful alternate to the t-test among the nonparametric tests. The general procedure, first recommended by Wilcoxon and others, was refined and formalized in tables by Mann and Whitney‡ to test whether or not two populations are identical. Suppose we take a sample of n observations (designated as x's) and a sample of m observations (designated as y's) of the presumably same continuous ensemble. Next the $m + n$ observations are arranged in a list in order of increasing value irrespective of the sample. Each ordered observation is replaced with an x or y, depending upon the sample from which it came. The result is a pattern of n x's and m y's intermixed. If the $m + n$ observations were all different, there would be $(m + n)!$ distinguishable patterns. However, for each truly distinguishable pattern, there are $n!$ permutations of x's with each other which do not change the pattern and, similarly, $m!$ permutations of y's. Therefore, there are

$$\frac{(m+n)!}{m!\,n!} = \binom{m+n}{m}$$

distinguishable patterns.

If two samples are drawn from the same ensemble, each of the patterns is equally likely; but if they come from different ensembles, one would expect to find patterns in which the x's cluster at one end of the list and the y's at the other. Let the test statistic U^* be the number of times a y precedes an x. U^* is the number of y's preceding the smallest x plus the number of y's preceding the next larger x, including all the y's counted in the first batch, and so on until the number of y's preceding the last x in the list is counted and included in the sum. The probability of U^* occurring when the null hypothesis is true is just that fraction of the $\binom{m+n}{m}$ total possible patterns in which the U^*'s are as big as or bigger than that obtained in the experiment. (The null hypothesis, H_0, is that each of the $\binom{m+n}{m}$ patterns is

‡ H. B. Mann and D. R. Whitney, *Ann. Math. Stat.* **18**, 50, 1947.

equally likely; hence, in effect, the two samples were drawn randomly and independently from the same population). The test will be significant at the significance level α when $P\{U^* \leq U^*_\alpha\} = \alpha$. In the case of ties, one recommended procedure is to give each member of the tied group the average of the ranks of the tied members when tallied consecutively. If the rank sum, i.e., sum of the values assigned the ranks, is not an integer, it should be rounded off to the nearest integer; other procedures are described in the references at the end of this chapter.

To carry out the test, we need to let x_i, a member of the smaller sample, be the ith x (in order of increasing value) and also be counted as the rth entry in the list when both the x's and y's are counted. Let u_i be the number of y's preceding x_i. T_x will be the rank sum of the x ranks (T is the Wilcoxon T† critical values for which are given in many statistics books). U^*, the Mann-Whitney statistic, is related to T_x as follows:

$$T_x = \sum_{i=1}^{n} r_i = \sum_{i=1}^{n} (i + u_i)$$

$$= n\left(\frac{n+1}{2}\right) + \sum_{i=1}^{n} u_i = n\left(\frac{n+1}{2}\right) + U^* \qquad (3.7\text{-}2)$$

where n is the number in the smaller sample. T_y, the sum of the y ranks, can also be related to U^*. The sum of all ranks is simply the number of ranks times the average rank or $(m + n)[(m + n + 1)/2]$. T_y is $(m + n)[(m + n + 1)/2] - T_x$ or

$$T_y = mn + \frac{m(m+1)}{2} - U^* \qquad (3.7\text{-}3)$$

Consequently, the statistic U^* does not have to be enumerated by calculating $\sum u_i$ (which can be quite tedious) but can be evaluated from Equation 3.7-2 or 3.7-3.

The number of y's which either precede or follow an x is m, the sample size. Because there are n x's, the number of y's either preceding or following all the x's is equal to mn. Consequently, $mn - U^*$ is the number of times a y follows an x, and is also the number of times an x precedes a y. Most tables, such as Table C.6 in Appendix C, list only the smaller of U^* or $U^{*\prime} = mn - U^*$. In Table C.6, m is the smaller sample and n is the larger sample. For large samples outside the table values, U^* has a mean of $\mathscr{E}\{U^*\} = mn/2$ and a variance of

$$\text{Var}\{U^*\} = \frac{mn(m+n+1)}{12}$$

† F. Wilcoxon, *Some Rapid Approximate Statistical Procedures*, American Cyanamid Co., New York, 1949.

and an (approximate) standard normal variable is

$$Z = \frac{U^* - \dfrac{mn}{2} - \dfrac{1}{2}}{\sqrt{\text{Var}\{U^*\}}} \qquad (3.7\text{-}4)‡$$

The asymptotic efficiency of the Mann-Whitney test is $3/\pi$ or 0.955 relative to the t-test when both tests are applied to a normal population with homogeneous variances. The superiority of the t-test is thus slight; if the data depart from normality, the Mann-Whitney test may be more powerful. Computational details are now illustrated by an example.

Example 3.7-2 Mann-Whitney Test

Let a supplier of a catalyst provide two samples to try out, A and B. The gain in yield for each sample is tabulated in increasing order of gain, and a second list is prepared (not shown) of the merger of both samples in increasing order of gain.

A		*B*	
Gain, %	Rank	Gain, %	Rank
−1.4	1	−0.3	5
−1.2	$2\frac{1}{2}$	0.5	8
−1.2	$2\frac{1}{2}$	0.7	9
−1.0	4	0.8	10
−0.2	6	0.9	11
0.2	7	1.5	12
		2.4	13
Rank sum	23	Rank sum	68

In the first list the rank of each gain, as determined from the second list, is placed in the second column of the table, the ranks going from 1 to 13.

One way to compute U^* is to replace the observations in the second list by A or B, depending upon which sample the observations came from:

$$A\ A\ A\ A\ B\ A\ A\ B\ B\ B\ B\ B\ B \qquad \text{(a)}$$

The number of times a B precedes an A is 2. A value of U^* as small as or smaller than this could be obtained from the following arrangements:

$$\begin{array}{ll} A\ A\ A\ A\ A\ A\ B\ B\ B\ B\ B\ B\ B & U^* = 0 \\ A\ A\ A\ A\ A\ B\ A\ B\ B\ B\ B\ B\ B & U^* = 1 \\ A\ A\ A\ A\ A\ B\ B\ A\ B\ B\ B\ B\ B & U^* = 2 \\ A\ A\ A\ A\ B\ A\ A\ B\ B\ B\ B\ B\ B & U^* = 2 \end{array} \qquad \text{(b)}$$

In total there are $\binom{6+7}{6} = 1716$ possible patterns. Hence the significance level for a one-sided test of the hypothesis that A either equals or exceeds B would be at $\frac{4}{1716}$. In other words, the probability of U^* being equal to

‡ Subtraction of $\frac{1}{2}$ corrects for continuity.

or less than 2 is 0.0023. Consequently, if one has in mind a significance level of 0.05 as being appropriate, the hypothesis that the catalysts have the same effect is rejected. For a two-sided test, there are four mirror images of the above patterns with U^*'s of 40, 40, 41, and 42, respectively. Hence the significance level for a two-sided test would be at $\frac{8}{1716}$.

Rather than count the patterns as above, the value of U^* can be ascertained with much less difficulty from Equations 3.7-2 and 3.7-3. Then the corresponding significance level can be obtained from Table C.6 in Appendix C, or from the normal approximate, Equation 3.7-4.

EQUATION 3.7-2:

$$T_x = 23 \qquad U^* = 23 - \frac{6(6+1)}{2} = 2$$

EQUATION 3.7-3:

$$T_y = 68 \qquad U^* = \frac{7(7+1)}{2} + (7)(6) - 68 = 2$$

From Table C.6 for $m = 7$, $n = 6$, and $U^* = 2$, the significance level α can be read as 0.002. Note that when $n = 8$ and $m = 8$, the normal approximation is quite good.

3.7-3 Siegel-Tukey Test for Dispersion

This test† is a nonparametric test which can be used as an alternate to the F-test to test the null hypothesis that the dispersions of the underlying populations of two independent samples are the same (against the alternate hypothesis that they are not). To carry out the test, list the value of each measurement in ascending order, with the most negative values at the head of the list and the most positive values at the bottom. Identify each value as belonging to sample A or B. Assign rank 1 to the smallest value, rank 2 to the largest value, rank 3 to the next largest value, rank 4 to the second smallest value, rank 5 to the third smallest value, rank 6 to the third largest value, and so forth, assigning ranks after the first in sequential pairs in rotation from the head to the foot of the list. Ties are resolved as was explained in Section 3.7-2.

Finally the ranks of sample A and of sample B are summed, and the approximate standard normal variable Z is calculated (more exact tables can be used in lieu of Z):

$$Z = \frac{\left| R_1 - \frac{n_1(n_1+n_2+1)}{2} \right| - \frac{1}{2}}{\sqrt{\frac{n_1(n_1+n_2+1)n_2}{12}}} \qquad \begin{pmatrix} n_2 > 10 \\ n_1 > 10 \end{pmatrix} \qquad (3.7\text{-}5)$$

where n_1 and n_2 are the sample sizes, $n_1 < n_2$, and R_1 = rank sum of the sample associated with n_1. Equation 3.7-5 is sufficiently accurate for engineering purposes, even for small samples of size less than ten.

† S. Siegel and J. W. Tukey, *J. Amer. Stat. Assn.* **55**, 429, 1960.

Example 3.7-3 Nonparametric Test for Dispersion

We rank the data from Example 3.7-2 as shown below:

Value	Sample	Rank
−1.4	A	1
−1.2	A	$4\frac{1}{2}$
−1.2	A	$4\frac{1}{2}$
−1.0	A	8
−0.3	B	9
−0.2	A	12
0.2	A	13
0.5	B	11
0.7	B	10
0.8	B	7
0.9	B	6
1.5	B	3
2.4	B	2

The sum of ranks of A = 33; the sum of ranks of B = 47. Sample A is smaller so that

$$R_1 = 33$$
$$n_1 = 6$$
$$n_2 = 7$$

$$Z = \frac{\left| 33 - \frac{6(6+7+1)}{2} \right| - \frac{1}{2}}{\sqrt{\frac{6(6+7+1)7}{12}}} = 0.496$$

From Table C.1 of the standard normal variable in Appendix C, for $\alpha = 0.05$, $z = 1.96$; consequently we accept the null hypothesis (by a two-sided test) that the dispersions of A and B are the same.

3.7-4 Tests for Stationarity

Two nonparametric tests are described in this subsection which can be used to ascertain whether or not data from a single time record are stationary. If stationarity can be demonstrated for a single time record, then the ensemble can be assumed stationary for most practical purposes. Furthermore, weak stationarity, as defined in Section 2.2-3, is actually what will be tested. The rationale of extending the umbrella of stationarity to other parameters than the ensemble mean and autocorrelation functions is rigorous for a normally distributed random variable and is observed to be effectively true for most other distributions encountered in practice. The length of the time record to be tested must be long enough, of course, to encompass a trend (nonstationarity) if one exists. A record that is too short will reveal nothing of a long-term trend, for example. Both run tests and trend-inversion tests can be used to test for stationarity.

A run is just a sequence of observations that are preceded and followed by a different observation (or by no observation at all). Thus, if the symbol + is assigned to a value of a variable above the sample median and a symbol − to a value below the sample median in the following sequence

5 1 6 4 2 7 5 9 8 7
− − + − − + − + + +
2 1 2 1 1 3

six runs can be detected. Like events may cluster, as indicated by an unusually small number of runs, patterns in the runs, runs of unexpected length, and other run statistics which can be used to test for randomness of pattern arrangement against the alternate hypothesis of sequential dependency. By judicious definition of the two types of events (designated + and − above), a run test can be employed not only to test for a trend in a sequentially sampled random variable but for many other characteristics which need not concern us here. Run tests are deficient in two respects—most are weak (have low power) and inefficient.

WALD-WOLFOWITZ TOTAL NUMBER OF RUNS TEST. The Wald-Wolfowitz test is not very powerful nor efficient but can be used to determine if observations of a random variable are independent (if they are, there is no trend). A series of observations is taken and their sample median determined. Each observation is assigned a + or − according to whether its value is above or below the median. If the pattern arrangement of +'s and −'s is such that the +'s and −'s are random and independent (the null hypothesis), there is no clustering. A brief table for the test statistic U^+, the number of runs, is in Appendix C (Table C.7). Also, the mean and variance of the random variable U^+ are

$$\mu_{U^+} = \frac{2n_1n_2}{n_1 + n_2} + 1 \tag{3.7-6}$$

$$\sigma^2_{U^+} = \frac{2n_1n_2[2n_1n_2 - (n_1 + n_2)]}{(n_1 + n_2)^2(n_1 + n_2 - 1)} \tag{3.7-7}$$

where n_1 is the number of +'s and n_2 is the number of −'s, and $n_1 + n_2 =$ the total number of observations. Consequently, for large samples the approximate standard normal variable

$$Z = \frac{|U^+ - \mu_{U^+}| - \frac{1}{2}}{\sigma_{U^+}} \tag{3.7-8}$$

can be used. A two-sided test for a given α is usually employed.

SUM OF SQUARED LENGTH TEST. Inasmuch as the Wald-Wolfowitz test does not directly take into account the length of the runs, considerable information is ignored. Ramachandran and Ranganathan† suggested a more powerful test. A run consists of a sequence of like signs; for example, in the arrangement given earlier, there were three runs of length 1, two runs of length 2, and one run of length 3. The test statistic, N, is the sum of the squares of the run lengths, or

$$N = \sum_j j^2 n_j \tag{3.7-9}$$

where j is the length of the run and n_j is the number of runs of length j. For the pattern just given, $N = 3(1^2) + 2(2^2) + 1(3^2) = 20$.

Table C.8 in Appendix C lists $P\{N \geq N_\alpha\} \leq \alpha$ for values of n equal to half the number of values in the time record, $n \leq 15$. For the example, $n = 5$ and $\alpha = 0.05$, so that $N_\alpha = 38$; hence the hypothesis that the sample does not have a trend is accepted.

INVERSIONS AS A TEST FOR LINEAR TREND. If a series of n measurements is arranged in the order taken, and a designated number is followed by a smaller number, an inversion is said to exist. Thus, in the sequence

3 5 1 4 2 6

there are six inversions: 3 is followed by two smaller numbers, 1 and 2; 5 is followed by three smaller numbers, 1, 4, and 2; and 4 is followed by one smaller number, 2. If the order of the numbers in the sequence is random, then each of the $n!$ permutations of the n numbers is equally probable; the *a priori* probability of obtaining a random sequence with exactly I^* inversions is simply the number of permutations containing exactly I^* inversions divided by $n!$, the total number of possible permutations. The number of times a number is followed by a larger number in the sequence is the compliment of I^* and is designated as T^*. A third measure which can be used is $S^* = T^* - I^*$. Mann‡ tabulated exact T^* probabilities for $3 \leq n \leq 10$, and Kendall§ listed probabilities for S^*. I^* has a mean and variance of

$$\mu_{I^*} = \frac{n(n-1)}{4} \tag{3.7-10}$$

$$\sigma^2_{I^*} = \frac{2n^3 + 3n^2 - 5n}{72} \tag{3.7-11}$$

and as n becomes large the approximate standard normal variate can be used

$$Z = \frac{I^* - \mu_{I^*}}{\sqrt{\sigma^2_{I^*}}} \tag{3.7-12}$$

(To correct for continuity, positive numerators should be decreased by $\frac{1}{2}$ and negative numerators increased by $\frac{1}{2}$.) If ties exist and are assigned the midrank, use the S^* tables instead of the T^* or I^* tables.

† G. Ramachandran and J. Ranganathan, *J. Madras Univ. Sect. B8* **23**, 76, 1953.

‡ H. B. Mann, *Econometrica* **13**, 245, 1945.

§ M. G. Kendall, *Rank Correlation Methods* (2nd ed.), Hafner, New York, 1955.

The assumptions behind the S^* or I^* test are that the observations have been taken independently and at random on a continuously distributed variable. When used as a test of randomness and compared with tests for a regression coefficient (discussed in Chapters 4 and 5), the I^* has an asymptotic relative efficiency of $(3/\pi)^{1/3} = 0.98$; hence it is equal to or superior to most other nonparametric tests for trend. The null hypothesis is that the observations are independent observations of a variable X when no trend exists; a two-sided test is used.

To ascertain whether or not a single time record represents stationary data, the time record is divided up so that n representative increments of equal time are obtained. High-frequency data can be in contiguous intervals, but low-frequency data require some interval between the selected portions of the record. The simplest procedure is to compute the mean and mean square for each of the n intervals and to arrange the results in sequence:

$$\langle {}^1X \rangle,\ \langle {}^2X \rangle,\ \ldots,\ \langle {}^nX \rangle$$

$$\langle {}^1X^2 \rangle, \langle {}^2X^2 \rangle, \ldots, \langle {}^nX^2 \rangle$$

where the presuperscript denotes the portion of the time record and $\langle\ \ \rangle$ denotes time average. Each of the two series of values can be tested for trend as described earlier.

One assumes that if the mean square (or variance) of the random variable X is stationary, then the autocorrelation function of X is also stationary. (The ensemble mean square of $X(t)$ is nothing more than the ensemble autocorrelation function at $\tau = 0$, $r_{xx}(0)$.) The basis for this assumption is that it would be most unusual for a nonstationary variable to have a time-varying autocorrelation function for $\tau > 0$ and not have $r_{XX}(0)$ varying also. Use of the mean square saves a tremendous amount of computation. But if the assumption is not valid, then Bendat and Piersol† suggested the following procedure which detects trends in the power spectrum and, hence, in the autocorrelation function:

1. Filter the sample record into c contiguous narrow bandwidth frequency intervals.
2. Divide each interval into n equal time intervals as before.
3. Compute a mean square value for each time interval within each frequency interval, giving a total of cn time averages:

$$\langle {}^{11}X^2 \rangle, \langle {}^{12}X^2 \rangle, \ldots, \langle {}^{1n}X^2 \rangle$$

$$\langle {}^{21}X^2 \rangle, \langle {}^{22}X^2 \rangle, \ldots, \langle {}^{2n}X^2 \rangle$$

$$\cdots\cdots\cdots\cdots$$

$$\langle {}^{c1}X^2 \rangle, \langle {}^{c2}X^2 \rangle, \ldots, \langle {}^{cn}X^2 \rangle$$

4. Test the time sequence in each frequency interval for trends; c tests will be required (plus one for the mean, as before). Rejection by any one test constitutes rejection of the null hypothesis of stationarity for a significance level (Type I error) of $\alpha' = 1 - (1 - \alpha)^{1/c}$, if α is the significance level which would be accepted for a single nonparametric test.

† J. S. Bendat and A. G. Piersol, *Measurement and Analysis of Random Data*, John Wiley, New York, 1966, p. 222.

Example 3.7-4 Tests for Stationarity

A time record of yield has been chopped up into ten segments, and the time average yield (in percent) of each segment is arranged sequentially below:

Period	Time Average
1	36.5
2	43.0
3	44.5
4	38.9
5	38.1
6	32.6
7	38.7
8	41.7
9	41.1
10	36.8

Test at a significance level of $\alpha = 0.05$ for stationarity both by the Wald-Wolfowitz test and the inversion test.

Solution:

WALD-WOLFOWITZ TEST. By inspection of the sequence, the median value of the ten values is $(38.7 + 38.9)/2 = 38.8$. A plus is assigned to a value above 38.8 and a minus to a value below, yielding the following sequence:

$$- \,\vdots\, + \;\; + \;\; + \,\vdots\, - \;\; - \;\; - \,\vdots\, + \;\; + \,\vdots\, -$$

There are five runs in total, and $n_1 = n_2 = 5$. For $\alpha = 0.05$, from Table C.7 in Appendix C, $U^+_{1-\frac{\alpha}{2}} = 2$ and $U^+_{\frac{\alpha}{2}} = 9$; hence the hypothesis that the data do not have a trend is accepted.

INVERSION TEST. We calculate the I^* statistic, the number of times a number is followed by a smaller number.

Value	Number of Inversions
36.5	1
43.0	7
44.5	7
38.9	4
38.1	2
32.6	0
38.7	1
41.7	2
41.1	1
36.8	0
Total	25

From Table C.9 in Appendix C for $\alpha = 0.05$ and $n = 10$, $I^*_{1-\frac{\alpha}{2}} = 11$ and $I^*_{\frac{\alpha}{2}} = 33$; hence the null hypothesis is accepted again.

To determine stationarity, the sequence of mean square values would also have to be formed and tested. For the given time record the null hypothesis was accepted by both tests; hence the mean square values are not tabulated.

3.7-5 Tests for Randomness

The nonparametric tests described above in connection with stationarity also in effect test for randomness, except for possible periodic components. If the segments of the time record pass the stationarity test, then periodic components which are not detected by visual inspection of the time record or by the test for stationarity are best detected by visual inspection of the time average power spectral density or autocorrelation function (defined in Section 12.3-3). Because a sine wave will have an autocorrelation function which will persist over all values of τ, as opposed to random data for which $r(\tau_{XX}) \rightarrow 0$ as $\tau \rightarrow \infty$ (for $\mu_X = 0$), the time average autocorrelation function can be plotted and examined. In this connection, refer back to the autocorrelograms in Figure 2.2-1. A periodic component in the data will show up as a peak in the power spectral density function, especially when the amplitude of the periodic component becomes larger than the associated noise.

3.7-6 Tests for Goodness of Fit and Independence

Tests to ascertain whether or not experimental data are represented by a normal (or other distribution) are of some importance, and the χ^2 test is one of the best known. The test is only approximate and sometimes misleading because of the many discongruities between the theoretical requirements and the actual practice in execution. The test applies to *enumerated* data, i.e., counted outcomes; consequently, continuous records must be converted to digital form before applying the test. We shall now illustrate the application of the χ^2 test to two important problems: (1) testing for goodness of fit, and (2) testing for independence between random variables.

TESTING GOODNESS OF FIT. To represent a random variable by a chosen probability distribution, the analyst must ask: Is the postulated probability density representative of the observed relative frequency distribution?

In Table 2.3-1 the mean and variance of the multinomial distribution for mutually exclusive events are listed as

$$\mathscr{E}\{X_i\} = n\theta_i \qquad \text{Var}\,\{X_i\} = n\theta_i(1 - \theta_i) \qquad (0 \le i \le k)$$

where θ_i is the parameter in the multinomial corresponding to the multinomial variable, X_i; θ_i is the probability that event i will occur x_i times in n trials where the $\sum x_i = n$. An approximate standard normal variable can be formed for each random variable as follows:

$$Z_i = \frac{X_i - n\theta_i}{\sqrt{n\theta_i(1 - \theta_i)}} \tag{3.7-13}$$

which will be approximately normally distributed for large values of $n\theta(1 - \theta)$ with a mean of zero and a variance of 1. Furthermore the variable

$$\sum_{i=1}^{k} Z_i^2 = \sum_{i=1}^{k} \frac{(X_i - n\theta_i)^2}{n\theta_i(1 - \theta_i)}$$

also can be formed and will be *approximately* distributed by the χ^2 distribution with k degrees freedom *if* the X_i's are *independent* of each other. It turns out for certain reasons, too detailed to go into here, that the random variable

$$\tilde{\chi}^2 = \sum_{i=1}^{k} \frac{(X_i - n\theta_i)^2}{n\theta_i} \qquad \nu = k - 1 \tag{3.7-14}$$

is more properly used and is better represented by the χ^2 distribution with ν degrees of freedom.

If the parameters of the probability density of the random variable are not known so that estimates $\hat{\theta}_i$ must be made of θ_i, then

$$\tilde{\chi}^2 = \sum_{i=1}^{k} \frac{(X_i - n\hat{\theta}_i)^2}{n\hat{\theta}_i} \qquad \nu = k - 1 - g \tag{3.7-15}$$

where the number of degrees of freedom is reduced by g linear constraints, one for each estimate. One restriction on Equation 3.7-15 is that $n\hat{\theta}$ must be greater than 5; if not, groups must be combined.

Equation 3.7-15 can be reformulated in slightly different notation to give

$$\tilde{\chi}^2 = \sum_{i=1}^{k} \frac{(n_i - n_i^*)^2}{n_i^*} \tag{3.7-16}$$

where n_i = observed number of occurrences of X_i, and n_i^* = theoretical number of occurrences of X_i calculated on the basis of the postulated probability density.

The goodness of fit is determined by calculating $\tilde{\chi}^2$ in Equation 3.7-16 and comparing this value with the one selected from the tables of χ^2 for a selected significance level, say $\alpha = 0.05$. A one-sided test can be used. If the calculated value of $\tilde{\chi}^2$ exceeds the preselected value of $\chi^2_{1-\alpha}$, one rejects the null hypothesis that the two distributions are the same, i.e., that the experimental relative frequency distribution is represented by the postulated probability density. (Also, if the value of χ^2 is less than χ^2_α, the empirical relative frequency distribution and probability density do not agree.) The χ^2 test for goodness of fit should be used with some caution and supplemented by other tests because it is essentially an approximate test. However, it certainly is a convenient test. A more exact analysis can be made by direct use of the multinomial distribution probabilities, if needed. For a very large number of occurrences, refer to Hodges and Lehmann.†

Example 3.7-5 χ^2 Test

Rubber from a reclaiming plant is classified as grade A, B, C, or D. Previous experience has shown that the distribution of product has been: A, 53.4 percent; B, 26.6 percent; C, 13.3 percent; and D, 6.7 percent. Last week's run was:

Grade	Batches
A	340
B	130
C	100
D	30

Has there been a change in the distribution of products?

Solution:

The procedure is to tabulate the observed frequencies n_i and compute the theoretical frequencies n_i^* based on a total equal to the sum of the observed frequencies.

Grade	Observed n_i	Theoretical n_i^*	$\frac{(n_i - n_i^*)^2}{n_i^*}$
A	340	320	$\frac{400}{320}$
B	130	160	$\frac{900}{160}$
C	100	80	$\frac{400}{80}$
D	30	40	$\frac{100}{40}$
Sum	600	600	14.4

The number of degrees of freedom is $\nu = k - 1 - g = 4 - 1 = 3$ ($g = 0$ since the n_i^*'s are computed from a known probability density). From Table C.2 in Appendix C, χ^2 for $\nu = 3$ and, for example, a probability equal to 0.95 is 7.81. Certainly 14.4 exceeds 7.81; in fact, for $P = 0.99$, $\chi^2_{1-\alpha} = 11.34$. A change in the process is suspected.

Example 3.7-6 Generation of Random Digits

A proposed method of generation of random digits is used 250 times to yield the following data. Does the technique actually produce random digits?

Digit	Number of Occurrences
0	27
1	18
2	23
3	31
4	21
5	23
6	28
7	25
8	22
9	32

† J. L. Hodges and E. L. Lehmann, *J. Stat. Soc.* **B16**, 261, 1954.

Solution:

Presumably, if the observed digits are random, each will occur with a probability of 0.1, or the number of theoretical occurrences (out of 250) would be 25. We compute

$$\sum_{i=1}^{k} \frac{(n_i - n_i^*)^2}{n_i^*} = \frac{(27 - 25)^2}{25} + \frac{(18 - 25)^2}{25} + \cdots + \frac{(32 - 25)^2}{25}$$
$$= 7.2$$

and the number of degrees of freedom ν is $(k - 1) = (10 - 1) = 9$. For $\nu = 9$, from Table C.2 in Appendix C, we find for $\alpha = 0.10$ that $\chi^2_{1-\alpha} = 14.68$, a value clearly greater than 7.2; hence we can accept the hypothesis that the digits are random.

Example 3.7-7 Testing Proposed Distributions

Failure of certain components of a missile were tabulated by Connor,‡ as shown in columns 1 and 2 of Table E3.7-7. Two proposed probability densities were compared to the observed relative frequency distribution in order to: (1) summarize the data by a simple function with one or two coefficients which contained all the known information, and (2) gain insight into the underlying causes of failure. The estimated mean of the experimental data was introduced as the single parameter into the Poisson density (refer to Table 2.3-1), the probability of each event (number of failures) was calculated, and each probability was then multiplied by the total number of failures, 473, to obtain the predicted distribution listed in column 3 of Table E3.7-7.

TABLE E3.7-7

Number of Failures	Observed	Poisson	Negative Binomial
0	331	317	333
1	104	127	100
2	27	25	29
3	8	3	8
4	1	1	2
5	2	0	1
Total	473	473	473
$\sum_{i=1}^{k} \frac{(n_i - n_i^*)^2}{n_i^*}$ Calculated:		17.2	0.31
$\chi^2_{0.99}$ from Table C.2:		9.21	6.63

We compare 17.2 with the value of $\chi^2_{1-\alpha} = 9.21$ from Table C.2 in Appendix C (for $\alpha = 0.01$ and $\nu = k - 1 - g = 4 - 1 - 1 = 2$ degrees of freedom) and do not find a good fit. Note that the classes of size less than 5 must be combined so that $k = 4$. However, the same test for the negative binomial density

$$p(x) = \binom{r + i - 1}{i} \theta^r (1 - \theta)^i \qquad i = 1, 2, \ldots$$
$$r = \text{positive integer}$$
$$0 < \theta < 1$$

‡ W. S. Connor, *Ind. Eng. Chem.* **52** (2), 74A, 1960; **52** (4), 71A, 1960.

indicates a suitable fit. The negative binomial density has two coefficients which must be estimated, r and θ, so that $\nu = 4 - 1 - 2 = 1$. Both the Poisson and negative binomial distributions have the same mean, $n\theta$, but the variance of the Poisson distribution is $n\theta$ and that of the negative binomial distribution $(n\theta/\theta) = n$; that is, the latter is dispersed more extensively, as required here. The rationale and implications of the agreement with the negative binomial density are discussed in the original article by Connor.

A second technique for goodness of fit, which will just be mentioned here, is the Kolmogorov-Simirnov test. This test inquires as to whether the cumulative relative frequency distribution of a variable (obtained by sampling) is represented by a probability distribution. If a random variable is presumed to have a probability distribution $P_0(x)$, and $S(x)$ is the observed empirical cumulative relative frequency distribution, the distribution of $D = \max |P_0(x) - S(x)|$ is known† and can be employed in tests of goodness of fit.

TESTING INDEPENDENCE OF VARIABLES. Suppose n pairs of experimental measurements are taken for two supposedly independent (in a statistical sense) variables. If the n data pairs are classified according to either quantitative or qualitative ranges of the two variables, the χ^2 test can be used to test the hypothesized independence of the two variables. The null hypothesis is that the variables are independent.

TABLE 3.7-1 TWO-WAY CLASSIFICATION*

		Classifications of Variable Y ($j \rightarrow$)				Row Sums: $\sum_{j=1}^{p}$
		y_1	y_2	$\cdots$	y_p	
Classifications of Variable X ($i \downarrow$)	x_1	f_{11}	f_{12}		f_{1p}	$f_{1.}$
	x_2	f_{21}	f_{22}		f_{2p}	$f_{2.}$
	$\vdots$					
	x_m	f_{m1}	f_{m2}		f_{mp}	$f_{m.}$
	Column Sums: $\sum_{i=1}^{m}$	$f_{.1}$	$f_{.2}$		$f_{.p}$	n

* The dot indicates summation over the variable replaced by the dot.

---

† Refer to F. J. Massey, "The Kolmogorov-Simirnov Test for Goodness of Fit," *J. Amer. Stat. Assn.* **46**, 68, 1951; L. H. Miller, "Table of Percentage Points for the Kolmogorov Statistics," *J. Amer. Stat. Assn.* **51**, 111, 1956; and J. Rosenblatt, *Ann. Math. Stat.* **33**, 513, 1962.

Consider the classification made in the Table 3.7-1 in which the number of outcomes is tabulated in each cell; i.e., f_{ij} = number of occurrences for the pair x_i and y_j, a range or class of X and Y. Let the probability of obtaining the count f_{ij} be denoted by θ_{ij}; its estimate is then $\hat{\theta}_{ij}$. We can form

$$\sum_{i=1}^{m}\sum_{j=1}^{p}\frac{(f_{ij} - n\theta_{ij})^2}{n\theta_{ij}} \cong \chi^2 \qquad \nu = mp - 1 \qquad (3.7\text{-}17)$$

The left-hand side of Equation 3.7-17 is approximately distributed as χ^2.

If $p(x_i)p(y_i) = p(x_i, y_i)$, so that the random variables X and Y are independent and, consequently, $\theta_i\theta_j = \theta_{ij}$, then we can estimate θ_i and θ_j by

$$\hat{\theta}_i \cong \frac{f_{i.}}{n}$$

$$\hat{\theta}_j \cong \frac{f_{.j}}{n}$$

so that

$$\hat{\theta}_{ij} \cong \frac{f_{i.}f_{.j}}{n^2} \qquad (3.7\text{-}18)$$

If Equation 3.7-18 is introduced into Equation 3.7-17, we obtain

$$\tilde{\chi}^2 = \sum_{i=1}^{m}\sum_{j=1}^{p}\frac{\left(f_{ij} - \dfrac{f_{i.}f_{.j}}{n}\right)^2}{\dfrac{f_{i.}f_{.j}}{n}} = n\left(\sum_{i=1}^{m}\sum_{j=1}^{p}\frac{f_{ij}^2}{f_{i.}f_{.j}} - 1\right) \qquad (3.7\text{-}19)$$

which is approximately represented by the χ^2 distribution with ν degrees of freedom. In this development there are $(m + p - 1)$ constraints introduced in finding $f_{i.}$ and $f_{.j}$ so that

$$\nu = mp - (m + p - 1) = mp - m - p + 1$$

Another way to look at the degrees of freedom for Equation 3.7-19 is to observe from the table that each marginal total must sum to $n = mp$, so that the degrees of freedom are reduced by one in each case to

$$(m - 1)(p - 1) = mp - m - p + 1$$

A third way to find ν is to note that m parameters have been estimated for the θ_i, but since $\sum_{i=1}^{m}\hat{\theta}_i = 1$, only $(m - 1)$ of these estimates are independent. Similarly, in estimating θ_j, only $(p - 1)$ degrees of freedom remain. Hence

$$(mp - 1) - (m - 1) - (p - 1) = (m - 1)(p - 1)$$

degrees of freedom remain, as above.

If $\tilde{\chi}^2$ calculated by Equation 3.7-19 proves to be greater than that found from the table of χ^2 for a preselected significant level, then the variables are not independent.

At least five predicted counts are needed per cell; otherwise the cells must be combined.

Example 3.7-8 Test of Independence

Eighty-seven rockets yield data on range and deflection as shown in the following table. For a confidence coefficient of 0.95, we test the hypothesis that the two measurements of range and deflection are independent.

	Deflection (mils)			
Range (yd)	−250 to −50	−50 to +50	50 to 200	Total
0–1200	5	9	7	21
1200–1800	7	5	9	21
1800–2700	8	21	16	45
Total	20	35	32	87

Solution:

The minimum predicted frequency is greater than 5. The degrees of freedom are 4.

$$\chi^2 = 87\left[\left(\frac{5^2}{(21)(20)} + \frac{9^2}{(21)(35)} + \frac{7^2}{(21)(32)} + \frac{7^2}{(21)(20)} + \frac{5^2}{(21)(35)} + \cdots\right) - 1\right]$$

$$= 87(0.232) = 20.2$$

We find that $\chi^2_{0.95} = 9.488$ from Table C.2 in Appendix C for $\alpha = 0.05$. The variables are not independent.

3.8 DETECTION AND ELIMINATION OF OUTLIERS

Even carefully planned and executed experiments can yield inhomogeneous data. Changed conditions during an experiment may remain undetected so that anomalous measurements, often termed "blunders," "wild" values, or outliers, are made. Or, aberrant measurements may be due to errors in the operation of recording devices, which if known would cause the recorded values to be rejected. Or, key-punch errors, inverted digits, or misplaced decimal points may contaminate otherwise valid data. On the other hand, the outlier may be simply one of the extreme values in a probability distribution for a random variable which occur quite naturally but infrequently and should not be rejected.

When the analyst knows that an abnormal error or blunder has been made, he does not hesitate to discard such an observation. When he does not have enough practical grounds to either accept or reject an extreme observation, he must resort to some kind of statistical judgment. He would like to answer the question: "What is the probability that the observed differences are due solely to random sampling errors?" in such a way that there is little doubt that certain observations will be rejected.

The approach to the problem of analyzing outlying observations depends upon the objective at hand. If the analyst is solely interested in determining whether an observation is an outlier in order, perhaps, to investigate the condition or conditions that may have led to this extreme observation, then the test for such an outlying observation is an end in itself. If, on the other hand, he is interested in deleting the outliers in order to obtain a more accurate estimate of some population parameter, say the population mean, then he is interested not only in a test for an outlying observation but also in the estimation of the parameter subsequent to the outlier test. Thus, he would also consider the possible bias of the estimate and its variance, taking proper account of the use of the outlier test. If the sample data, subsequent to an outlier test, are to be used to test hypotheses about a population parameter, then he is interested not only in a criterion for an outlier but also in the power of subsequent tests of hypotheses.

Tests for outliers generally have one of the following objectives:

1. To prune the observations prior to analysis (rejection of outliers).
2. To ascertain that outliers are present, indicating a need for reexamination of the data generation.
3. To pinpoint observations that may be of special interest just because they are extremes.

We shall be concerned with the first type of test.

The classical method of handling the problem of detecting a point outlier is to assume that the sample observations are of a normally distributed random variable, to devise an appropriate outlier test statistic sensitive to the kind of wildness envisioned, to derive the distribution of this test statistic under the null hypothesis that all observations come from the same normal population, and then to reject the hypothesis if the calculated test statistic for it is unlikely to have occurred in random sampling. The usual test statistic is based on the idea that the analyst can look at the sample results of an experiment and note that he has a discordant observation. The test statistic, referred to as the extreme deviate statistic, involves the difference between the extreme value and the sample mean value and either the ensemble standard deviation or an estimate of it obtained from the sample at hand and/or from an independent sample. The theory and practice of the rejection of outliers are not firmly resolved and, to quote Gumbel:†

> The rejection of outliers on a purely statistical basis is and remains a dangerous procedure. Its very existence may be proof that the underlying population is, in reality, not what is was assumed to be.

† E. J. Gumbel, *Technometrics* **2**, 165, 1960.

We shall follow Anscombe† in the use of test rules. Given a sample of observations $X_1, X_2, \ldots, X_n$ $(n \geq 3)$, which is assumed to be a random sample of a normal random variable X with the parameters μ_X and σ_X^2, we compute

$$Y_i = X_i - \bar{X} \qquad i = 1, 2, \ldots, n$$

where $\bar{X} = \sum_{i=1}^{n} X_i/n$. If a single X_i is omitted, the sample average of the remaining observations is

$$\sum_{\substack{j=1 \\ j \neq i}}^{n} \frac{X_j}{\nu} = \bar{X} - \frac{Y_i}{\nu} \qquad \nu = n - 1 \tag{3.8-1}$$

If several observations are omitted, $X_1, X_2, \ldots, X_r$, the sample average is

$$\bar{X} - \frac{(Y_1 + Y_2 + \cdots + Y_r)}{n - r} \tag{3.8-2}$$

If the subscript M is used to designate the observation which has the greatest residual, $Y_M = X_M - \bar{X}$, Anscombe suggests the following rule (for the case in which σ_X^2 is unknown). For a given c, reject X_M if $|Y_M| > cs_X$. Otherwise, do not reject X_M. For large samples, if X_M is rejected, the reduced sample is treated as a new sample that can be subjected to further analysis. Each time, μ_X is estimated from the observations retained after deleting X_M. The c can change with sample size, and Anscombe gives c implicitly in terms of t:

$$\left[\frac{nc^2(\nu + \nu_0 - 1)}{\nu(\nu + \nu_0 - nc^2/\nu)}\right]^{1/2} \cong t_{1-\frac{\alpha}{2}}^{(\nu_0 + \nu - 1)} \tag{3.8-3}$$

and explicitly by the following approximate relation in terms of the F distribution:

$$c \approx \left(\frac{\nu}{n}\right)^{1/2}\left(\frac{3F_{1-q}}{1 + [(3F_{1-q} - 1)/(\nu + \nu_0)]}\right)^{1/2} \tag{3.8-4}$$

where $\nu = n - 1$ and ν_0 is any other additional degrees of freedom which accompany the estimate of σ_X^2 other than from the sample of size n. (The positive square root is taken for c.)

The test is carried out, using Equation 3.8-4, as follows. Multiply the allowable fractional increase in σ_X^2 if no rejection is to take place (the "premium") by ν/n. Denote this product by q, and find the corresponding upper percentage point of the variance ratio, F_{1-q}, for 3 and $\nu + \nu_0 - 1$ degrees of freedom. Calculate c from Equation 3.8-4 and carry out the test for X_M. The "premium" depends on how much one fears spurious observations, but some small fractional increase in σ_X^2 should be acceptable, say 0.02. As an example, if $n = 4$, $\nu = 3$, and $(\nu/n) = 0.75$, for a "premium" of 0.02, $q = (0.02)(0.75) = 0.05$. We look up $F_{1-0.05}$ for 3 and 3 degrees of freedom, respectively. $F_{1-q} = 9.28$. Then

$$c = (0.75)^{1/2}\left(\frac{3F_{0.95}}{1 + [(3F_{0.95} - 1)/(3)]}\right)^{1/2} = 0.831$$

X_M would be rejected if $|Y_M| > 0.831s_X$.

Outliers in regression analysis will be treated in Chapters 4 and 5.

† F. J. Anscombe, *Technometrics* **2**, 123, 1960.

Example 3.8-1 Test of an Outlier

In the series

x_1	x_2	x_3	x_4	x_5
23.2	23.4	23.5	24.1	25.5

ascertain whether or not x_5 is an outlier to be deleted from the sample.

Solution:

Compute $\bar{X} = 23.9$ and then $Y_i = X_i - \bar{X} = 25.5 - 23.9 = 1.6$; $s_X = 0.77$. For $\alpha = 0.05$, $\nu = 4$, and $n = 5$, from Equation 3.8-3 we compute by trial and error that

$$\left[\frac{5c^2(3)}{3\left(3 - \frac{4c^2}{3}\right)}\right]^{1/2} = 2.776^3$$

and $c = 1.49$. The test is

$$|1.6| > (1.49)(0.77) = 1.05$$

and observation x_5 is retained.

3.9 PROCESS CONTROL CHARTS

Hypothesis testing can be applied in a quite simple and yet practical way to assist in process quality control. Control charts are a graphical means of analysis which have proved easy to maintain and use under plant operating conditions. Figure 3.9-1 illustrates a typical process control chart based on the sample mean. The general procedure in preparing a control chart is: (1) to collect a sample, (2) to compute an appropriate statistic such as the sample mean, the range, or the cumulative sum, and (3) to plot the statistic on a chart as a function of sample sequence or time.

Superimposed on the chart in some manner are the

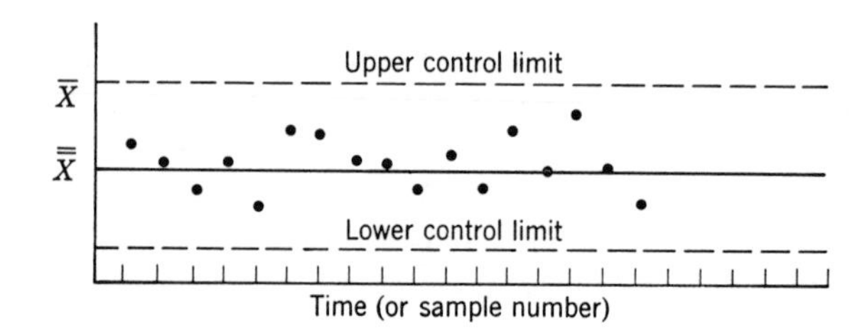

FIGURE 3.9-1 A typical process quality control chart.

rules for making a decision as to whether the process variable is "in control" or not. Figure 3.9-1 illustrates an upper control limit and a lower control limit. As long as the statistic being plotted falls within these two bounds, the process is deemed to be in control. The decision rules used to fix the lines can be based on an assumed distribution for the observed random variable, usually the normal distribution, or they can be based on a nonparametric analysis as discussed in Section 3.7.

If the statistic being plotted exceeds a control limit, the decision is reached that the process is "not in (statistical) control"; the breaking out indicates abnormal performance. Even the accumulation of an undue number of points on one side of the central line can be interpreted as a shift of some type in the process. Control charts can be used to provide.

1. Both a signal that a change has occurred in the process and an estimate of the amount of change required for corrective action.
2. Solely a signal that a change has occurred in the process so that the operator can be made aware that the process needs his attention.
3. Estimates of the times in the past during which changes in the process have occurred and thus assist in assigning causes for the changes.
4. Measures of the quality of output for classification by period.

Because of the way control limits are placed in practice and because of the lack of information about the probability distribution of the random variable being measured, exact probability statements of the type used in Sections 3.3 and 3.4 are usually avoided. Control charts are especially valuable when used as simple graphical aids to let the process operator, who is untrained in statistical techniques, get a mental picture of the process history and interpret whether or not the quality of the product is at a satisfactory level.

The use of process control charts in the process industries has not been as fruitful as, for example, in the automotive parts industry because the goal in the former is often not to control an output variable, such as the yield, to within a given range but to maximize the variable. Improvement is synonymous with optimization. In traditional applications the goal is to produce a product with increased uniformity. A second difficulty in applying process control charts in the process industries is that the assignment of the causes of an "out of control" process is never obvious except for aberrant values which may be caused by improper feed of raw materials, improper setting of control variables, equipment failure, failure to follow the proper operating instructions, etc. In any case, such causes are usually corrected before their effects are detected on control charts. But a shift in level and/or cyclical fluctuations in a process are difficult to ascribe to assignable causes, some of which may be unobservable variables or in the surrounding environmental conditions.

Several kinds of control charts will be briefly described in this section, each of them based on the plotting of a different statistic:

1. Shewhart control charts ($\bar{X}$, R, and s charts).
2. Geometric moving average (exponentially weighted moving average) charts.
3. Cumulative sum charts.
4. Multivariate control charts.

Many other types of charts are equally useful, and these are described in the references at the end of this chapter. Table 3.9-1 characterizes the relative effectiveness of four types of control charts in detecting various changes in the process.

Designing a process control chart, that is establishing the central line and control limits, requires some thought and investigation into the process itself. We shall assume that the process and sampling points are clearly defined, lag and dead times have been taken into account, and a suitable sampling method and sampling interval have

TABLE 3.9-1 RELATIVE EFFECTIVENESS OF CONTROL CHARTS IN DETECTING A CHANGE IN A PROCESS*

	Control Chart			
Cause of Change	Mean ($\bar{X}$)	Range (R)	Standard Deviation (s)	Cumulative Sum (CS)
Gross error (blunder)	1	2	—	3
Shift in average	2	—	3	1
Shift in variability	—	1	—	—
Slow fluctuation (trend)	2	—	—	1
Rapid fluctuation (cycle)	—	1	2	—

* 1 = most useful, 2 = next best, 3 = least useful, and — = not appropriate.

been selected. Then the sampling procedure itself must be investigated so that the precision of the data to be used is known (and is at an acceptably low level). Large samples make for more sensitive tests, but the time element may be such that the sample can consist only of a single reading, say from a gas chromatograph. The economics of sampling, cost of off-specification material, cost of taking a corrective action, etc., are vital considerations in designing a control chart but factors beyond our scope here.

3.9-1 Shewhart Control Charts

The Shewhart control chart for $\bar{X}$ was one of the initial tools of statistical quality control.† A sample of a presumed normal random variable (Burr‡ indicates that the effect of nonnormality is slight and provides tables of compensating coefficients, if required) with a mean of μ_X and a variance of $\sigma_{\bar{X}}^2$ is taken, $\bar{X}$ is computed, and then $\bar{X}$ is plotted as illustrated in Figure 3.9-1. For a selected value of α, often set at 0.0027 so that $1 - \alpha = 0.9973$, upper and lower control limits are calculated, using $\sigma_{\bar{X}}$ or its estimated value, and placed on the chart about the known or estimated value of μ_X. When a sample average falls outside the control limits, one concludes that the process is "out of control." One important decision which must be made is what value of α to choose; the narrower the band of control limits, the more frequent the "out of control" signals will be when unwarranted. Another important decision is what sample size, n, to use. A common value of n is 5. A second statistic which usually accompanies the $\bar{X}$ plot is R, the range of X in the sample. The arithmetic mean of the ranges, $\bar{R}$, can be used as an estimate of the sample dispersion, and the arithmetic mean of the $\bar{X}$'s, $\bar{\bar{X}}$, can be used to estimate μ_X.

† W. A. Shewhart, *Economic Control of Quality of Manufactured Product*, D. Van Nostrand, Princeton, N.J., 1931.
‡ I. W. Burr, *Ind. Qual. Control* **23**, 563, 1967.

The range is a somewhat more convenient measure of dispersion to calculate than the standard deviation. The advantage of plotting the range as well as $\bar{X}$ on control charts is that abnormal variations are more easily detected. The range is a rough measure of the "rate of change" of the variable being observed. A point out of control on the range chart, when the mean is still within the control limits, sounds the alarm well in advance of a change in the mean.

If there is no damage or cost when one of the two control limits is exceeded, but if the opposite is true if the other control limit is exceeded, the mean of the process can be shifted so that the important limit is farther from the mean and the other limit is ignored. If one of the limits turns out to be greater than the physical limit of the process, for example, a value in percent less than 0 or greater than 100, the control limit is usually made to coincide with the physical limit.

Figure 3.9-2 illustrates a process control chart in which $\bar{X}$ and R are plotted together. The 7:30 a.m. range is out of control, indicating that the decrease in yield is at a too rapid rate and leading to an out of control condition on $\bar{X}$ by 8:30 a.m. If the reason for the change is known, such as a previous control valve or temperature adjustment, then no action is required. But if the cause is unknown, deciding from the chart what

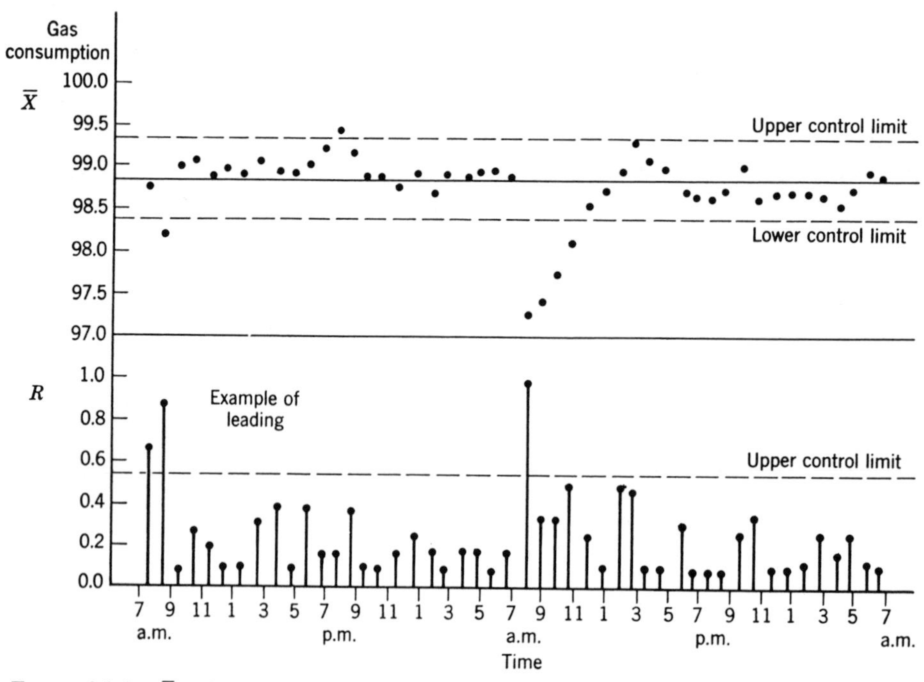

FIGURE 3.9-2 $\bar{X}$ and R control charts (lower control limit on range is not shown).

variable to adjust and how much action is required is not always easy.

If the average range $\bar{R}$ is employed to estimate the variance of the statistic being plotted, which in turn is used to establish the control limits, special tables have been prepared, such as Table 3.9-2 (for $\alpha = 0.0027$), that tabulate the proper constant A_2 by which to multiply $\bar{R}$ in order to calculate the upper and lower (symmetric) control limits. A_2 is an appropriate constant based on the distribution of $(\bar{X} - \bar{\bar{X}})/\bar{R}$. When the subgroup size n is 5, $A_2 = 0.577$; the control limits are then set at $\bar{\bar{X}} \pm A_2\bar{R}$. The null hypothesis for the test being applied on the control chart is that the expected value of X is a specified value, μ_0.

Unfortunately, $\bar{\bar{X}}$ and $A_2\bar{R}$ are not very accurate estimates of μ_X and $3\sigma_{\bar{X}}$ unless the number of successive samples used to obtain these estimates is quite large, at least 25. Thus, when only a small number of subgroups has been collected, $\bar{\bar{X}} \pm A_2\bar{R}$ may differ greatly from $\mu_X \pm 3\sigma_{\bar{X}}$. One consequence of this difference is that more than 0.27 percent of the future values of $\bar{X}$ may fall outside of $\bar{\bar{X}} \pm A_2\bar{R}$, even when the process is in control. Table 3.9-3 indicates the probability, assuming the process is in control, that a randomly selected $\bar{X}$ will fall outside the limits $\bar{\bar{X}} \pm 0.577\bar{R}$ based on m samples, each of size $n = 5$.

Table 3.9-4 gives values of A_2 for a sample of $n = 5$ for various significance levels, α, for an increasing

TABLE 3.9-2 FACTORS FOR COMPUTING CONTROL CHART LINES* ($\alpha = 0.0027$)

Sample Size	Chart for Averages				Chart for Standard Deviations						Chart for Ranges					
	Factors for Control Limits				Factors for Central Line		Factors for Control Limits				Factors for Central Line		Factors for Control Limits			
n	A	A_0	A_1	A_2	c_2	$1/c_2$	B_1	B_2	B_3	B_4	d_2	$1/d_2$	D_1	D_2	D_3	D_4
2	2.121	3.760	3.760	1.880	0.5642	1.7725	0	1.843	0	3.267	1.128	0.8862	0	3.686	0	3.267
3	1.732	3.070	2.394	1.023	0.7236	1.3820	0	1.858	0	2.568	1.693	0.5908	0	4.358	0	2.575
4	1.500	2.914	1.880	0.729	0.7979	1.2533	0	1.808	0	2.266	2.059	0.4857	0	4.698	0	2.282
5	1.342	2.884	1.596	0.577	0.8407	1.1894	0	1.756	0	2.089	2.326	0.4299	0	4.918	0	2.115
6	1.225	2.899	1.410	0.483	0.8686	1.1512	0.026	1.711	0.030	1.970	2.534	0.3946	0	5.078	0	2.004
7	1.134	2.935	1.277	0.419	0.8882	1.1259	0.105	1.672	0.118	1.882	2.704	0.3698	0.205	5.203	0.076	1.924
8	1.061	2.980	1.175	0.373	0.9027	1.1078	0.167	1.638	0.185	1.815	2.847	0.3512	0.387	5.307	0.136	1.864
9	1.000	3.030	1.094	0.337	0.9139	1.0942	0.219	1.609	0.239	1.761	2.970	0.3367	0.546	5.394	0.184	1.816
10	0.949	3.085	1.028	0.308	0.9227	1.0837	0.262	1.584	0.284	1.716	3.078	0.3249	0.687	5.469	0.223	1.777
11	0.905	3.136	0.973	0.285	0.9300	1.0753	0.299	1.561	0.321	1.679	3.173	0.3152	0.812	5.534	0.256	1.744
12	0.866	3.189	0.925	0.266	0.9359	1.0684	0.331	1.541	0.354	1.646	3.258	0.3069	0.924	5.592	0.284	1.716
13	0.832	3.242	0.884	0.249	0.9410	1.0627	0.359	1.523	0.382	1.618	3.336	0.2998	1.026	5.646	0.308	1.692
14	0.802	3.295	0.848	0.235	0.9453	1.0579	0.384	1.507	0.406	1.594	3.407	0.2935	1.121	5.693	0.329	1.671
15	0.775	3.347	0.816	0.223	0.9490	1.0537	0.406	1.492	0.428	1.572	3.472	0.2880	1.207	5.737	0.348	1.652
16	0.750	3.398	0.788	0.212	0.9523	1.0501	0.427	1.478	0.448	1.552	3.532	0.2831	1.285	5.779	0.364	1.636
17	0.723	3.448	0.762	0.203	0.9551	1.0470	0.445	1.465	0.466	1.534	3.588	0.2787	1.359	5.817	0.379	1.621
18	0.707	3.497	0.738	0.194	0.9576	1.0442	0.461	1.454	0.482	1.518	3.640	0.2747	1.426	5.854	0.392	1.608
19	0.688	3.545	0.717	0.187	0.9599	1.0418	0.477	1.443	0.497	1.503	3.689	0.2711	1.490	5.888	0.404	1.596
20	0.671	3.592	0.697	0.180	0.9619	1.0396	0.491	1.433	0.510	1.490	3.735	0.2677	1.548	5.922	0.414	1.586
21	0.655	3.639	0.679	0.173	0.9638	1.0376	0.504	1.424	1.523	1.477	3.778	0.2647	1.606	5.950	0.425	1.575
22	0.640	3.684	0.662	0.167	0.9655	1.0358	0.516	1.415	0.534	1.466	3.819	0.2618	1.659	5.979	0.434	1.566
23	0.626	3.729	0.647	0.162	0.9670	1.0342	0.527	1.407	0.545	1.455	3.858	0.2592	1.710	6.006	0.443	1.557
24	0.612	3.773	0.632	0.157	0.9684	1.0327	0.538	1.399	0.555	1.445	3.895	0.2567	1.759	6.031	0.452	1.548
25	0.600	3.816	0.619	0.153	0.9696	1.0313	0.548	1.392	0.565	1.435	3.931	0.2544	1.804	6.058	0.459	1.541
>25	$\frac{3}{\sqrt{n}}$		$\frac{3}{\sqrt{n}}$		1.0000	1.0000	†	‡	‡	‡						

* Use explained in Table 3.9-5. The relation $A_0 = 3\sqrt{n}\, d_2$ holds. Adapted with permission of the American Society for Testing Materials from *ASTM Manual on Quality Control of Materials*, Philadelphia, Jan. 1951, p. 115.

† $1 - \frac{3}{\sqrt{2n}}$

‡ $1 + \frac{3}{\sqrt{2n}}$

TABLE 3.9-3 EFFECT OF NUMBER OF SAMPLES ON FRACTION OF X VALUES FALLING OUTSIDE CONTROL LIMITS (SAMPLE SIZE = 5)

m	Probability of an $\bar{X}$ Falling Outside Control Limits
5	0.0120
10	0.0067
15	0.0051
20	0.0044
25	0.0040
50	0.0033
100	0.0030
∞	0.0027

number of samples; methods of calculating these values are described by Hillier.†

Since the value of A_2 decreases as the number of subgroups used for calculating $\bar{\bar{X}}$ and $\bar{R}$ increases, one can

† F. S. Hillier, Stanford Univ. Tech. Rept. 63, Sept. 4, 1962; Stanford Univ. Tech. Rept. 83, Oct. 4, 1965.

TABLE 3.9-4 VALUE OF A_2 FOR VARIOUS SIGNIFICANCE LEVELS (SAMPLE SIZE = 5)

m \ α	0.001	0.0027	0.01	0.025	0.05
1	2.27	1.74	1.21	0.911	0.720
2	1.19	1.00	0.781	0.637	0.532
3	0.960	0.834	0.673	0.562	0.477
4	0.864	0.760	0.624	0.527	0.451
5	0.811	0.720	0.596	0.507	0.436
6	0.779	0.695	0.579	0.495	0.426
7	0.756	0.677	0.564	0.485	0.418
8	0.738	0.662	0.556	0.477	0.412
9	0.729	0.655	0.551	0.474	0.410
10	0.719	0.647	0.545	0.470	0.407
15	0.687	0.621	0.527	0.455	0.396
20	0.672	0.609	0.518	0.449	0.391
25	0.663	0.602	0.513	0.445	0.387
50	0.649	0.590	0.505	0.439	0.383
100	0.640	0.583	0.500	0.434	0.379
∞	0.633	0.577	0.495	0.431	0.377

TABLE 3.9-5 CALCULATION OF UPPER AND LOWER CONTROL LIMITS

Sample Statistic Controlled	Central Line	Lower Control* Limit (LCL)	Upper Control Limit (UCL)	Sample size
		μ_X and σ_X Specified		
Mean $\bar{X}$	μ_X	$\mu_X - A$	$\mu_X + A$	n
Range R	$d_2\sigma_X$	$D_1\sigma_X$	$D_2\sigma_X$	Small, preferably 10 or fewer
Standard deviation s_X	$\sigma_X c_2\sqrt{\frac{n}{n-1}}$	$\sigma_X B_1\sqrt{\frac{n}{n-1}}$	$\sigma_X B_2\sqrt{\frac{n}{n-1}}$	n
		μ_X and σ_X Unknown		
Mean $\bar{X}$	$\bar{\bar{X}}$	$\bar{\bar{X}} - A_2\bar{R}$	$\bar{\bar{X}} + A_2\bar{R}$	Small, preferably 10 or fewer
Mean $\bar{X}$	$\bar{\bar{X}}$	$\bar{\bar{X}} - \bar{s}_X A_1\sqrt{\frac{n-1}{n}}$	$\bar{\bar{X}} + s_X A_1\sqrt{\frac{n-1}{n}}$	25 or fewer, constant size
Mean $\bar{X}$	$\bar{\bar{X}}$	$\bar{\bar{X}} - \frac{3\bar{s}_X}{\sqrt{\bar{n}}}$	$\bar{\bar{X}} + \frac{3\bar{s}_X}{\sqrt{\bar{n}}}$	Sample sizes ≥ 25, may vary slightly
Sum S	$\bar{S}$	$\bar{S} - A_0\bar{R}$	$\bar{S} + A_0\bar{R}$	Small, preferably 10 or fewer
Range R	$\bar{R}$	$D_3\bar{R}$	$D_4\bar{R}$	Small, preferably 10 or fewer
Standard deviation s_X	$\bar{s}_X$	$B_3\bar{s}_X$	$B_4\bar{s}_X$	25 or fewer, constant size
Standard deviation s_X	$\bar{s}_X$	$\bar{s}_X - \frac{3\bar{s}_X}{\sqrt{2\bar{n}}}$	$\bar{s}_X + \frac{3\bar{s}_X}{\sqrt{2\bar{n}}}$	Sample sizes ≥ 25, may vary slightly

* $\bar{n}$ is the average sample size for variable sample sizes.

have tighter control limits, and certainly more up-to-date ones, by computing new control limits after additional samples are obtained. But it is undesirable, both from the standpoint of the effort involved and the psychological impact on the workers affected by the control charts, to revise the control limits too frequently. When control limits are revised, an $\bar{X}$ may at one time lie inside the control limits and later on outside them. Because the control limits are established for a process in control, any values of $\bar{X}$ and its related R should be deleted from the calculations if they prove subsequently to be outside the revised control limits; one should use the number of subgroups "in control," not the total number of subgroups, for the value of m when finding the value of A_2 from Table 3.9-4.

Other types of control charts besides the $\bar{X}$ and R charts can be prepared. Table 3.9-5 summarizes some of these and their related control limits. The various constants are tabulated in Table 3.9-2 (for $\alpha = 0.0027$). In Table 3.9-5 the sum of the kth sample is $S_k = \sum_{i=1}^{n_k} X_{ik}$, and the notation $\bar{s}_X$ stands for the arithmetic mean of the standard deviations of the samples:

$$\bar{s}_X = \frac{n_1 s_1 + n_2 s_2 + \cdots + n_k s_k}{n_1 + n_2 + \cdots + n_k}$$

Control charts also can be based on runs of specific length above or below the central line or on the number of runs in a given series of samples. Nonparametric tests, such as were described in Section 3.7, are the bases of the decisions to be made.

Example 3.9-1 Establishing $\bar{X}$ and R Chart Control Limits

Table E3.9-1 lists the sample means for the percentage yields from a continuous reactor. Determine the central line and upper and lower control limits for an $\bar{X}$ and an R chart from Table 3.9-2. Each value of $\bar{X}$ and R in the table has been prepared from three analyses.

TABLE E3.9-1

Sample	Sample Mean $\bar{X}$	Range R	Sample	Sample Mean $\bar{X}$	Range R
1	64.97	9.8	14	66.60	0.6
2	64.60	9.8	15	66.12	6.3
3	64.12	8.4	16	63.22	7.5
4	68.52	3.9	17	62.85	6.7
5	68.35	7.6	18	62.37	4.9
6	67.87	8.7	19	61.97	6.7
7	64.97	0.1	20	61.60	9.9
8	64.60	9.7	21	61.12	6.9
9	64.12	7.7	22	65.72	0.1
10	63.22	7.5	23	65.35	8.3
11	62.85	1.2	24	64.87	5.2
12	62.37	9.8	25	61.97	3.2
13	66.97	6.4			

Solution:

$$\sum \bar{X} = 1611.29 \qquad \sum R_i = 156.9$$

$$\bar{\bar{X}} = 64.452 \qquad \bar{R} = 6.28$$

The control limits for the mean are: upper, 64.452 + 1.023(6.28); and lower, 64.452 − 1.023(6.28). The value of $A_2 = 1.023$ was taken from Table 3.9-2 for $n = 3$. The control limits for the range are:

$$D_3\bar{R} = (0)(6.28) = 0$$
$$D_4\bar{R} = 2.575(6.28) = 16.17$$

Example 3.9-2 Initiation of Control Charts

This example illustrates the initiation of $\bar{X}$ and R control charts for the data of Example 3.9-1, except that we assume for the purposes of illustration that the sample size is 5 in

TABLE E3.9-2 DATA FOR THE $\bar{X}$ CHART

Sample Number	$\bar{X}$	R	$\bar{\bar{X}}$	$\bar{R}$	m	A_2*	LCL, $\bar{\bar{X}} - A_2\bar{R}$	UCL, $\bar{\bar{X}} + A_2\bar{R}$
1	68.2	7						
2	66.2	3						
3	72.4	6						
4	67.8	2						
5	67.0	8	68.32	5.20	5	0.720	64.6	72.1
Revised			67.30	5.00	4	0.760	63.5	71.1
6	66.8	4					63.5	71.1
7	67.0	4					63.5	71.1
8	65.8	7					63.5	71.1
9	62.6	8					63.5	71.1
10	69.0	4					63.5	71.1
11	67.6	8					63.5	71.1
12	66.0	9	67.14	5.60	10	0.647	63.8	70.8

* Values taken from Table 3.9-4; the value from Table 3.9-2 would be 0.577.

order to make use of Table 3.9-4. To quickly obtain information for the $\bar{X}$ chart, it might be decided to initiate use of the chart as soon as possible by setting control limits based on the first five initial samples. It was decided to use the usual value of $\alpha = 0.0027$. Table E3.9-2 lists the values of A_2 and the control limits after five samples.

Notice that the value of $\bar{X}$ for the third sample exceeded the upper control limit after the limit had been established based on the first five samples; consequently the sample was excluded from all subsequent calculations of $\bar{\bar{X}}$ and of the control limits. This exclusion resulted in a revised $\bar{\bar{X}}$ and control limits based on four samples. Similarly, when $\bar{X}$ for the ninth sample fell below the lower control limit, the sample was excluded from subsequent calculations. Investigation of the process conditions for the third and ninth subgroup revealed that the process was susceptible to temporary shifts in the true process average. Corrective action removed the causes of these shifts. At the end of ten valid samples, the control limits were recalculated as indicated on the last line of Table E3.9-2.

Example 3.9-3 Corrective Action Based on Control Charts

This example is taken from Breunig† and illustrates the usefulness of control charts in pointing out undesirable process conditions with subsequent detection of the proper corrective action. For some time management had suspected that an unsatisfactory situation existed in the production of Vitamin A in certain multiple vitamin products. But until the dramatic evidence of control charts was available, little corrective action had been taken. Because the determination of Vitamin A in the control laboratory was one of the few instances where duplicate determinations were routinely made, an estimate of laboratory variability was available. Although there may be some question as to the propriety of judging lot-to-lot variation against laboratory precision alone, this was the best available estimate of within-lot variation. When $\bar{X}$ charts based upon laboratory variability were prepared, the sample means were found to fluctuate so widely that very few were in control as evidenced by Figure E3.9-3*a* for Product H and Figure E3.9-3*e* for Product M. An investigation was made into the raw material-handling procedures as a first step.

Vitamin A, supplied in the form of acetate or palmitate salts, is quite susceptible to atmospheric oxidation so that an excess amount is normally included in the formulation. Geometric configuration of the side chain of the Vitamin A molecule engenders cis and trans isomers. Originally the trans isomer predominates, but in time it is partially converted by oxidation to the cis isomer until an equilibrium mixture is reached consisting of approximately 66 percent trans and 33 percent cis. Although there appears to be some recent evidence to the contrary, the assumption has long been made that there is relatively little physiological difference in response to the two isomers. The analytical determination of Vitamin A was based upon an ultra-violet extinction procedure which assumed that only the trans isomer was present.

† H. L. Breunig, *Ind. Qual. Control* **21**, (2), 79, 1964.

The investigation revealed that Vitamin A palmitate, for instance, was purchased in bulk form in 50-kilogram containers as a semisolid; an assay was made upon arrival. It was purchased from several sources, including brokers who may have pooled batches from still other sources. The potency specification was that it contain "70 percent activity" by the U.V. assay. Based upon the original determined activity, Vitamin A was introduced into the manufacturing process at the appropriate potency level.

The 50-kilogram containers were stored in a cold room. Requisitions were filled by removing the drum from the chill room, warming, dipping out the proper amount to fill the order, and returning the container to the cold room. Obviously, as this procedure was repeated several times, oxidation was taking place. Furthermore, those drums which contained an insufficient amount of material for filling requisitions were stored until several had accumulated and the contents were then mixed together. Thus, constantly shifting isomer ratios were experienced which showed up in the high variability of the finished product assay values on the control charts. There also appeared to be evidence of additional potency loss that was not explained by the cis-trans isomerization.

The first attack upon the problem included consultation by the purchasing department with certain reliable suppliers who agreed to provide Vitamin A as "prepackaged" material at no extra cost. A scheme was worked out whereby all manufacturing requisitions could be filled by combinations of three standard package sizes which were sealed under nitrogen in the plant of the supplier and not opened until needed, although one package of each shipment was checked for identity. This procedure led to a strong dependence upon "vendor certification" of potency. At first the revised packaging appeared to give satisfactory results, but, as is noted in Figures E3.9-3*b* and *c* for Product H, variation of the sources of material was not the sole difficulty. Purchase of the material was being made on a "100 percent trans" level, but obviously some shifting to cis was still taking place.

The next step in the attempt to attain quality control was directed to the problem of the cis-trans isomer ratio. The suppliers agreed to provide an equilibrium mixture of 66 percent trans- 33 percent cis in their prepackaged material. The arrow in Figure E3.9-3*c* for Product H indicates when the first lots containing "preisomerized" Vitamin A were received for analysis. At this time a significant upward shift in Vitamin A content occurred, so new control had to be established as shown in Figure E3.9-3*d* and *f* for Product H and Product M, respectively. The process control charts, after the corrective action described above was taken, indicated far more uniform control of the Vitamin A content of two typical vitamin products. Not only were nearly all the lot means in control but so also were the within-lot and between-lot ranges.

3.9-2 Acceptance Control Charts

A discussion of sampling plans is beyond our scope here. In acceptance sampling some characteristic of the

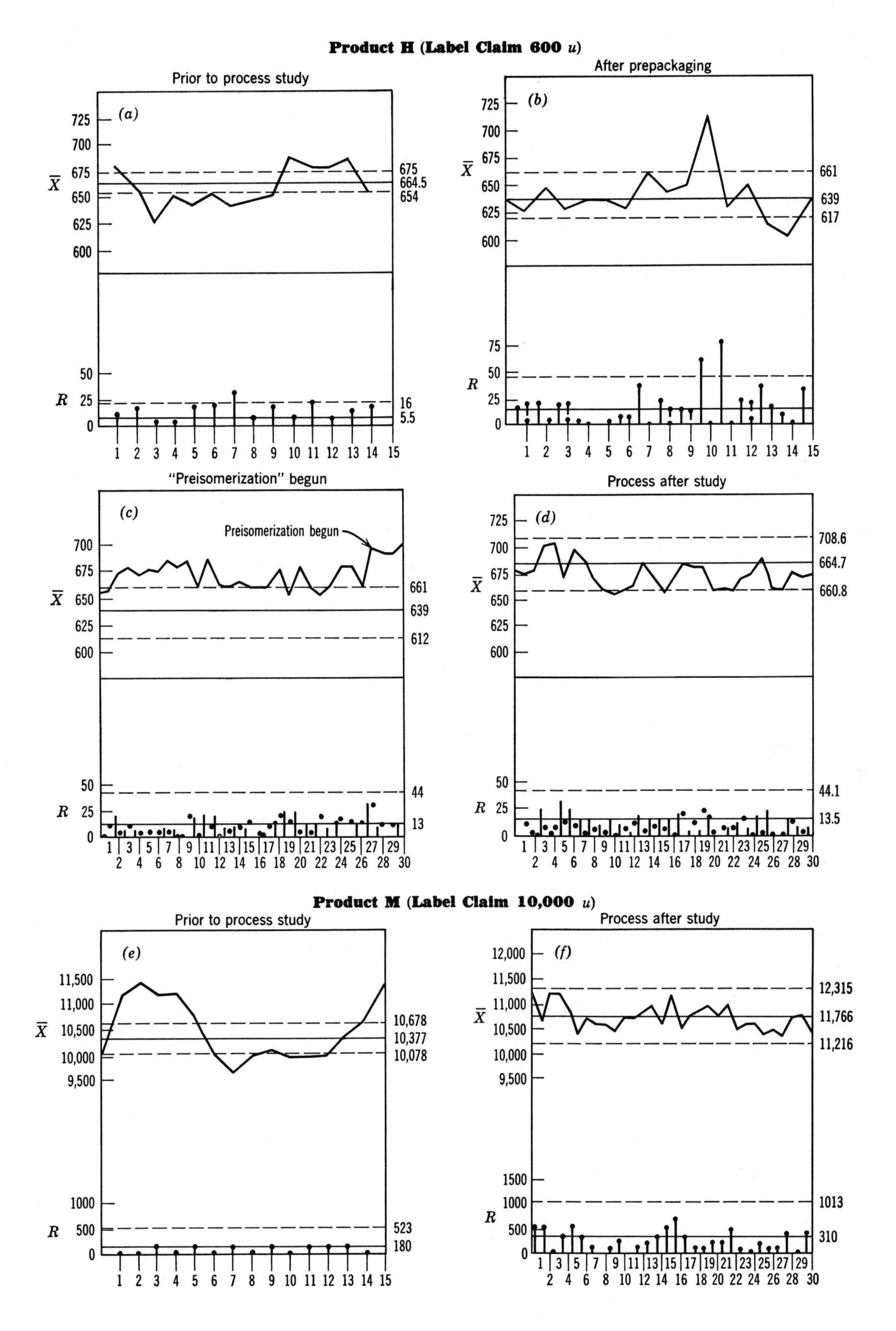

FIGURE E3.9-3 Charts at several stages of investigation into two products.

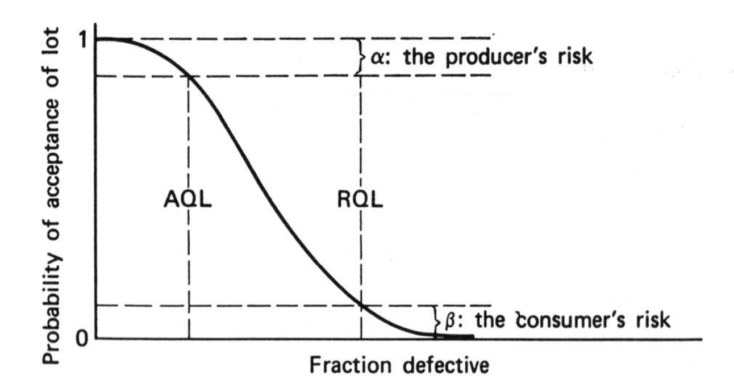

FIGURE 3.9-3 Operating characteristic curve for a sampling plan.

product is measured, a random sample of size n is selected according to some sampling plan, the sample mean and standard deviation are computed, and a significance test is carried out for a null hypothesis. Associated with the null hypothesis are alternate hypotheses and errors of the first and second kind. The consumer who buys the product sets a limit below which the product is unsatisfactory for his use, and he is the one who determines the alternate hypothesis. Figure 3.9-3 illustrates the operating characteristic curve of a typical sampling plan. For the producer, the null hypothesis H_0 is that the product is acceptable, and even if he makes an acceptable product, 100α percent of it will be deemed unacceptable because the process sample statistic is stochastic in nature. In this sense, α is termed the *producer's risk*, and the related level of the process fraction defective is termed the *acceptable quality level* (AQL). If the producer makes some 100β percent defective product, which is not detected as being defective because of the stochastic nature of the sample statistic, the probability β is called the *consumer's risk*, and the alternate hypothesis, H_1, establishes a *rejectable quality level* (RQL).

In designing an acceptance control chart, the acceptable process level (APL) is based on α and the rejectable process level (RPL) is based on β. As long as the product statistic being monitored lies between the APL and the RPL, the process is deemed to be in control. Note that the control limits depend on α, β, *and* n. Freund† gave some examples of acceptance control charts.

3.9-3 Geometric Moving Average Control Chart

The geometric moving average control chart,‡ also known as the exponentially weighted moving average control chart, and the cumulative sum chart (to be described in Section 3.9-4), proves to be of most use where the specifications must be tight so that a sensitive control scheme is needed. These techniques combine information from past samples with that from the current one and, in effect, make use of more information than do the Shewhart charts with the result that they have the ability to detect a smaller shift in the process level. Of course, the disadvantage is that the old information submerges possible small shifts in the process level signalled only by the new information. The geometric (exponentially) weighted moving average chart gives more weight to recent measurements than to old ones by computing a weighted linear combination of a sample statistic such as $\bar{X}$. The most recent value is assigned a weight of w, with $0 \le w \le 1$, and the older weighted statistic is assigned a weight of $1 - w$. Thus, if:

Z_k^* = weighted average of the sample statistic after sample k
Z_k = value of the kth sample statistic
k = current measurement; $(k - 1)$ = next most recent measurement, etc.; $0 \le i \le k$
$\bar{Z}$ = central line on the control chart

then

$$Z_0^* = \bar{Z}$$
$$Z_1^* = wZ_1 + (1 - w)Z_0^*$$
$$Z_2^* = wZ_2 + (1 - w)Z_1^*$$
$$\vdots$$
$$Z_k^* = w\left[\sum_{i=0}^{k-1} (1 - w)^i Z_{k-i}\right] + (1 - w)^k Z_0^* \qquad (3.9\text{-}1)$$

If $w = 1$, all the weight is placed on the current data, and a Shewhart type chart is obtained. If $w = 0$, no weight is given to the current data, so in effect no current sample need be taken!

It can be shown that the expected value of Z_k^*, if $Z_i = \bar{X}_i$, is

$$\mathscr{E}\{Z_k^*\} = \mu_0$$

and the variance of Z_k^* is

$$\mathrm{Var}\,\{Z_k^*\} = \sigma_{\bar{X}}^2[1 - (1 - w)^{2k}]\frac{w}{2 - w}$$

As k becomes large,

$$\mathrm{Var}\,\{Z_k^*\} = \sigma_{\bar{X}}^2 \frac{w}{2 - w}$$

Control limits can be drawn on a typical chart at appropriate distances from $\bar{Z}$. The exponentially weighted moving average chart is compared with other control charts in Figure 3.9-7.

3.9-4 Cumulative Sum Control Charts

Cumulative sum control charts, as the name indicates make use of cumulative sums of a random variable or a

† R. A. Freund, *Ind. Qual. Control* **14**, 13, Oct. 1957.
‡ S. W. Roberts, *Technometrics* **1**, 239, 1959.

function of a random variable starting from a given reference time. For example, the statistic summed may be:

1. The variable itself.
2. The difference between the measured value of the variable and its expected value.
3. The difference between the measured value of the variable and a target value.
4. Successive differences between values of a variable or absolute differences.
5. The sample mean.
6. The range.

Table 3.9-6 lists relations for calculating the statistic for typical cumulative sum charts; each sum is based on a sample of size n.

TABLE 3.9-6 COMPUTATIONAL RELATIONS FOR CUMULATIVE SUM CHARTS

Type of Chart	Cumulative Sum
Deviation from reference (target) value, h	$\sum_{i=1}^{n} (\bar{X}_i - h)$
Absolute value of a deviation from its expected absolute value	$\sum_{i=1}^{n} [\lvert X_i - \bar{X}\rvert - \mathscr{E}\{\lvert X_i - \bar{X}\rvert\}]$
Successive differences	$\sum_{i=1}^{n} (D_i); \quad D_i = (X_j - X_{j-1})$
Absolute value of successive differences from the expected absolute value	$\sum_{i=1}^{n} [\lvert D_i\rvert - \mathscr{E}\{\lvert D_i\rvert\}]$
Range of two successive pairs of observations from the expected value	$\sum_{i=1}^{n} [R_i - \mathscr{E}\{R_i\}]$

The major advantage of the cumulative sum charts, as contrasted with the Shewhart charts, is that they are more sensitive to moderate deviations in the process statistic of interest from its expected value; they "damp out" random noise while "amplifying" true process changes. True, the Shewhart charts can be made more sensitive by using, in addition to the control limits given in Table 3.9-5 (for one statistic), one or more of the following criteria:†

1. "Warning" lines within the control limits and "action" lines at the usual control limits.
2. Runs of the statistic, such as three consecutive points outside control lines placed at $\pm\sigma$ or seven consecutive points on one side of the central line.

These alternate decision rules make use of some of the extra information retained in a control chart beyond that provided by the current sample. Cumulative sum charts also take into account more than the current sample; consequently, substantially more information can be tied into the decision rules.

Not only is there a difference in the type of visual record which appears on a cumulative sum chart, but the criteria for taking action are different. Control limits for a cumulative sum chart evolve from the distribution of the statistic being plotted; however, the control limits are not drawn on the chart but are provided through use of a special template or overlay mask. What is of interest in the cumulative sum chart is not the absolute value of the sum but the *slope* of the curve comprised of successive (recent) points. Each type of chart requires a different template to indicate the degree of slope.

Figure 3.9-4 illustrates a typical template together with the rules for its construction and use based on the distribution of the statistics being plotted, assuming the random variable was normal. After each point is plotted, the reference point P on the mask is placed over this most recent point. The observer then looks to see whether a previously plotted point appears (or disappears if the mask is not transparent) beneath the mask when the mask is correctly positioned for a given decision rule. (Note the analogy with the Wald sequential test of Section 3.4-1.) When such an event occurs, the process is said to be "out of control." For V-shaped masks, it is suggested that the visual impact of a change is optimal when one horizontal step is about equal to a 2σ vertical step.‡

Since a V-shaped mask can be designed by establishing just two parameters, θ, the half-angle of the V, and d, the lead length of the V, as indicated in Figure 3.9-4, the question naturally arises as to what interpretation can be given to suitable combinations of θ and d in terms of the power of the decision rules. To answer this question, we must first discuss the topic of average run length. The *average run length* (ARL) refers to the number of *samples* collected before an action signal occurs. ARL is a measure of how frequently one must interfere with a process if one follows the appropriate decision rules based on a selected α. Because the average run length is a random variable whose distribution depends on the criteria selected for "in control," in a rough way it is a measure of the relative efficiency of a control scheme.

To examine in a fair fashion the relative performance of a Shewhart chart and a cumulative sum chart, suppose we choose the decision rules for each such that they have

† G. P. Moore, *Biometrica* **45**, 89, 1958; E. S. Page, *Biometrica* **42**, 243, 1955; and H. Weiler, *J. Amer. Stat. Assn.* **48**, 816, 1953 and **49**, 298, 1954.

‡ E. S. Page, *Technometrics* **3**, 1, 1961.

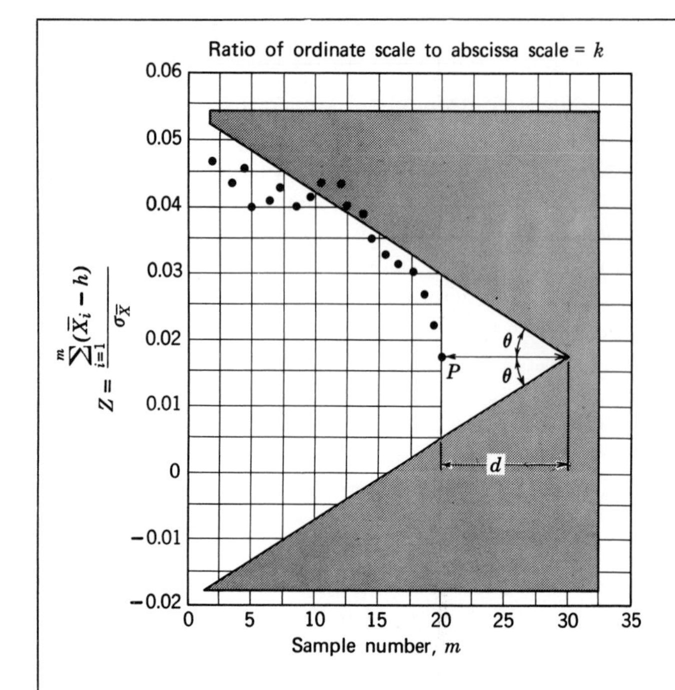

Application Rules

1. Place point P on the most recently plotted point on control chart.
2. A change has occurred if any plotted point is covered by the mask.

Sample Mean Template Design (deviation from target h)

Plot $Z = \frac{\sum_{i=1}^{m} (\bar{X}_i - h)}{\sigma_{\bar{X}}}$ versus m

$$\theta = \text{arc tan}\left(\frac{\delta}{2k}\right)$$

$$d = \frac{-2 \ln \alpha}{\delta^2}$$

α = significance level (for a two-sided test the fraction falling outside the region of acceptance will be 2α)

$$\delta = \frac{D}{\sigma_{\bar{X}}}$$

D = minimum shift in the process mean which is to be detected

Note: $\sigma_{\bar{X}}$ can be approximated by a pooled sample variance

$$s_p = \left(\sum_{i=1}^{m} \frac{s_i^2}{n}\right)^{1/2}$$

where m is the number of the last sample

Sample Range Template Design

Plot (for $n < 10$) $\xi = \frac{\sum_{i=1}^{m} R_i^2}{(\sigma_X c)^2}$ versus $m\nu_1$

$$\theta = \text{arc tan}\left[\frac{2 \ln (\sigma_1/\sigma_0)}{1 - (\sigma_0/\sigma_1)^2}\right]$$

$$d = -\frac{\ln \alpha}{\ln (\sigma_1/\sigma_0)}$$

σ_0^2 = standard variance
σ_1^2 = hypothesized variance to be tested
n = sample size

Values of c and ν_1

Sample size n	c	ν_1
3	1.378	1.93
4	1.302	2.95
5	1.268	2.83
6	1.237	4.69
7	1.207	5.50
8	1.184	6.26
9	1.164	6.99
10	1.146	7.69

FIGURE 3.9-4 Template for cumulative sum control chart. (Rules taken from N. L. Johnson and F. C. Leone, *Ind. Qual.* Control **18** (12), 15; **19** (1), 29; **19** (2), 22, 1962.)

exactly the same average run lengths, ARL_I, when the process is in control. A certain step change is next made in the process level, anywhere from say 0 to 3 standard deviations from the original level, and the average run lengths, ARL_{II}, are subsequently calculated between the initiation of the change in process level and the detection of the change.

For the Shewhart chart, the average run length can be calculated from the following relations:†

$$ARL_I = \frac{1}{\alpha}$$

$$ARL_{II} = \frac{1}{1 - \beta}$$

† P. L. Goldsmith and H. Whitfield, *Technometrics* **3**, 1, 1961; and W. D. Ewan, *Technometrics* **5**, 1, 1963.

for a process change of $k\sigma$ from the target, where β = probability of a point falling inside the control limits when the process level is off the target ($\mu \neq \mu_0$). Unfortunately, there is no analytical way to calculate ARL_I and ARL_{II} for cumulative sum charts so that the comparisons here have been taken from the results of Goldsmith and Whitfield who evaluated the two ARL's by Monte Carlo simulation on a digital computer.† Figure 3.9-5 compares the ARL_{II}'s for four different ARL_I's as a function of k. One observes that the Shewhart charts in general are less efficient than the cumulative sum charts, especially for large values of the parameter ARL_I, i.e., small values of α.

† P .L. Goldsmith and H. Whitfield, *Technometrics* **3**, 1, 1961.

To return now to the question of the design of V-shaped masks, one can specify how long an ARL_I to have while in control and how short an ARL_{II} to have to detect a given size process change. One would like to specify ARL_I to be as long as feasible and ARL_{II} to be as short as feasible. (The method of design described by Johnson and Leone and outlined in Figure 3.9-4 is based solely on the distribution of the statistic Z, the cumulative deviation from a target value, and assumes that the measured variable is a normally distributed random variable.) The charts in Figure 3.9-6 show the average run length ARL_{II} after a process change was introduced of k units of standard deviation. Using the charts to obtain d and θ (defined in Figure 3.9-4), it is

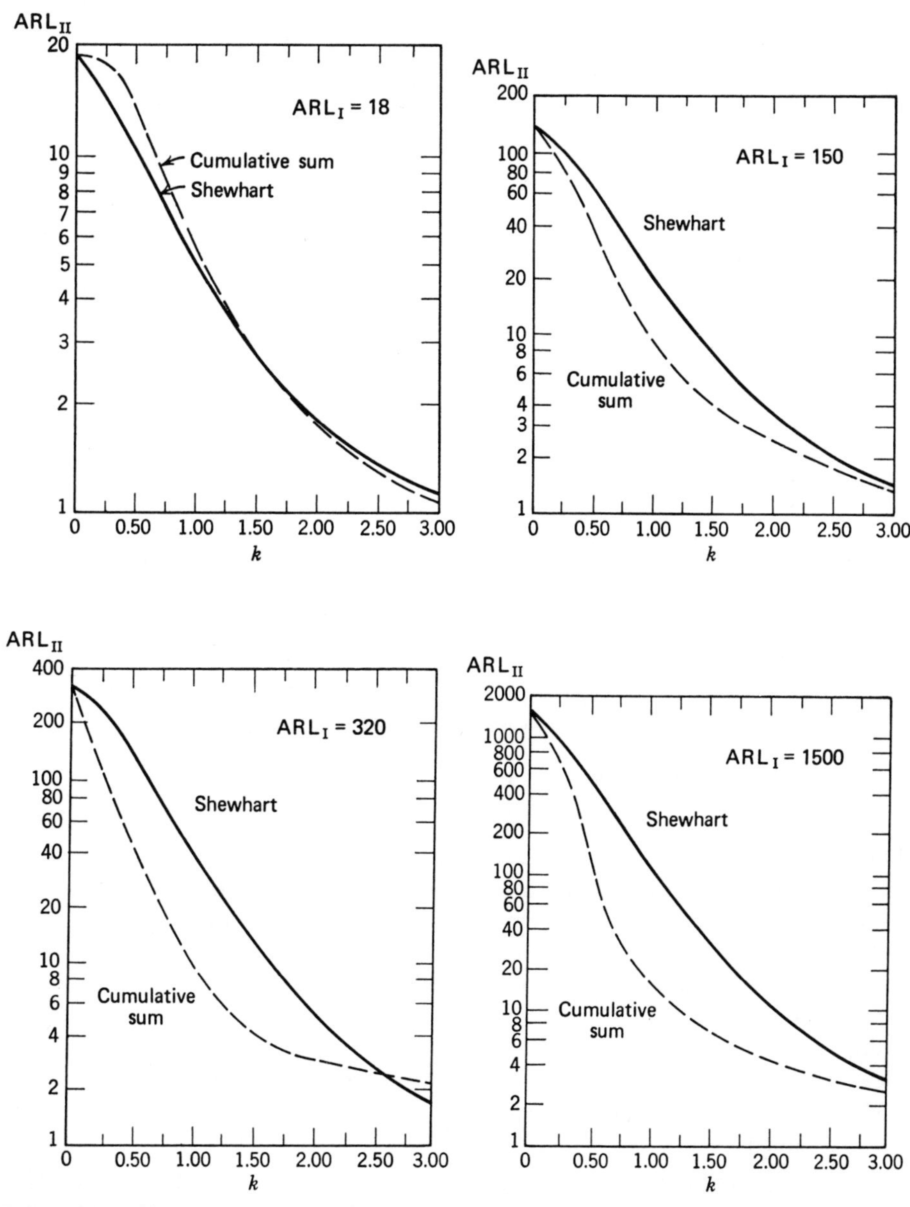

FIGURE 3.9-5 Average run lengths (ARL_{II}) after a process change was introduced of k units of the standard deviation until the process change was detected for four different significance levels ($ARL_I = 1/\alpha$), based on a normal independent variable with a variance of σ^2.

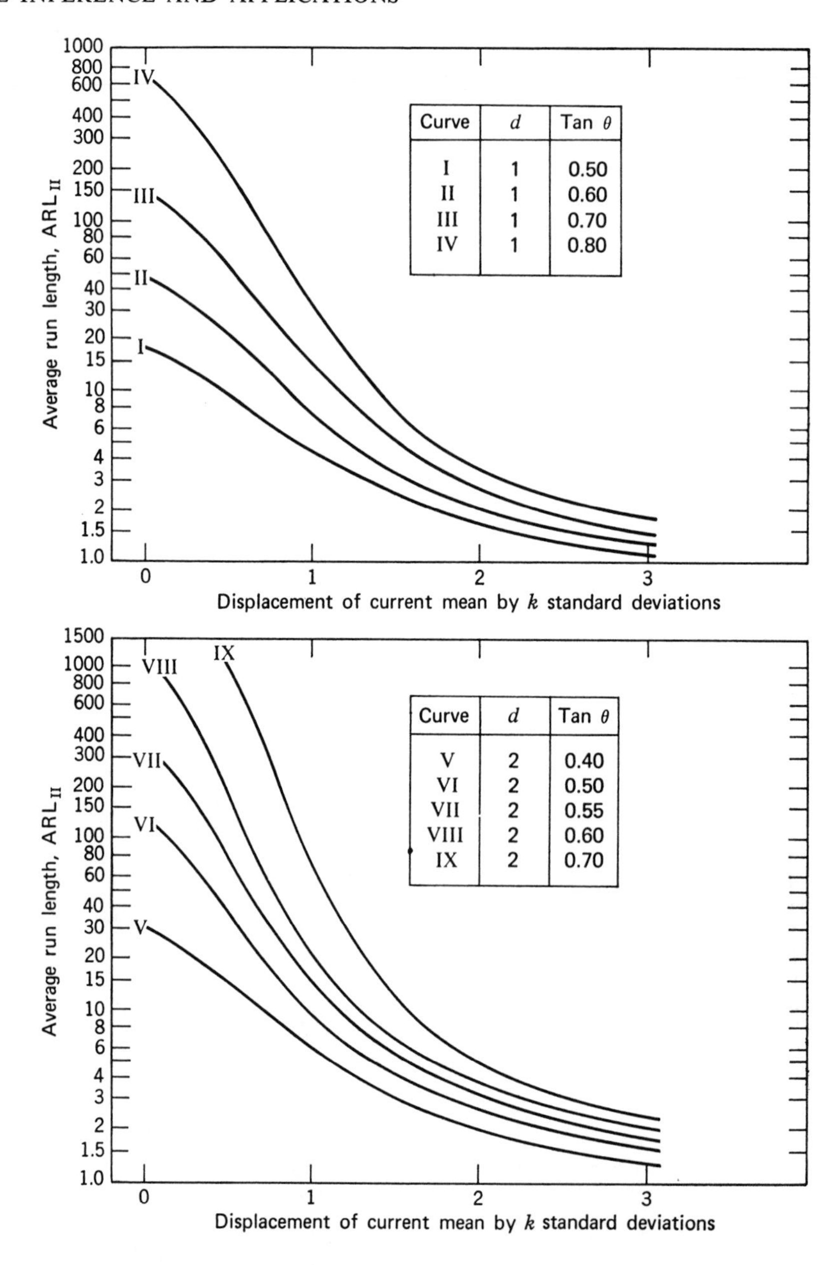

FIGURE 3.9-6 Design of V masks using average run lengths.

assumed that the plotting interval on the horizontal axis for the process statistic is equal to 2σ on the vertical axis, resulting in a 45° angle for the mean path of the process statistic if the process mean shifts 2σ. If the plotting interval on the horizontal axis is some other multiple, $q\sigma$, on the vertical axis, then the values of $\tan\theta$ given for the chart must be multiplied by $2/q$.

One can either pick a d and θ, assume a k, and evaluate ARL_{II} from Figure 3.9-6 and ARL_{I} from the following empirical relation for ARL_{I} in terms of d and θ

$$\log_{10}(\log_{10}\text{ARL}_{\text{I}}) = -0.5244 + 0.0398d + 1.1687\tan\theta + 1.2641(\tan\theta)(\log_{10}d)$$

or proceed in the reverse order. Suppose, for example we want ARL_{I} to be 200 and ARL_{II} to be 8 for a shift of one standard deviation in the process mean ($k = 1$). From the equation we obtain

$$0.886 = 0.0398d + 1.1687\tan\theta + 1.2641\tan(\theta)(\log_{10}d) = \psi$$

From Figure 3.9-6 we find for $\text{ARL}_{\text{II}} = 8$ and $k = 1$

d	$\tan\theta$	ψ
1	0.61	0.751
2	0.47	0.807
5	0.30	0.813
8	0.24	0.872

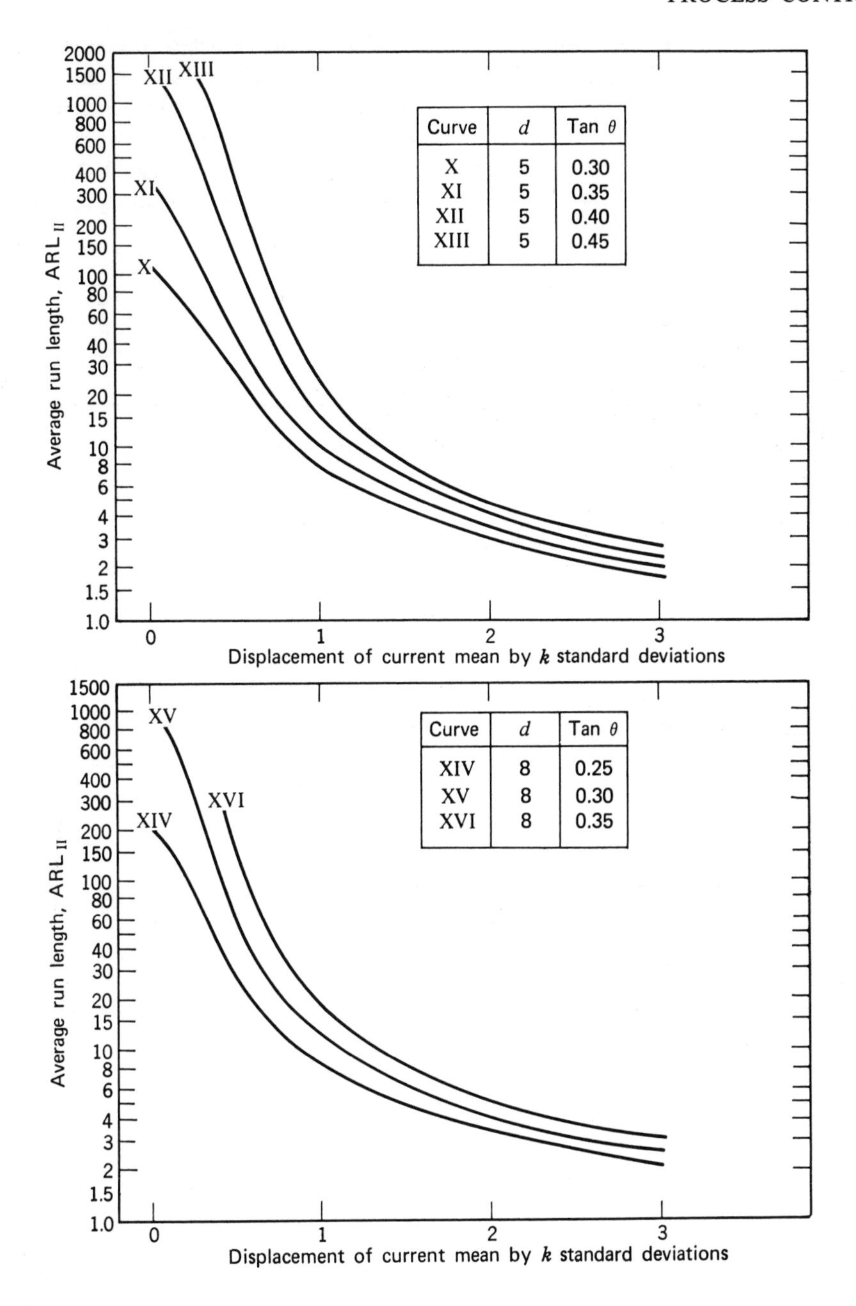

FIGURE 3.9-6 (*continued*)

so that $d \cong 8$ and $\tan \theta \cong 0.24$ is approximately the desired design.

Roberts† compared several types of control charts, using one or more tests in connection with each chart. The simulated observations were taken from a table of normal random variates having an expected value of zero and a variance of 1. After 100 numbers had been selected, a constant of 1 was added to all the numbers commencing with the 101st so as to represent a 1σ shift in the process mean between the 100th and 101st observation. Table 3.9-7 lists the equations used in plotting the charts and the tests employed. Figures 3.9-7*a–d* show that the number of subsequent samples until corrective action was called for was 19 for most of the tests.

Below Figure 3.9-7*a* are tabulated the data for a run sum test based on where $\bar{X}$ falls. A run is a sequence of values of $\bar{X}$ falling within specified limits or being above or below a limit. A run terminates when a value of $\bar{X}$ falls on the opposite side of the central line. The run sum is the sum of the scores assigned the plotted values of $\bar{X}$. A two-sided runs test is illustrated underneath Figure 3.9-7*a* in which the following values are assigned to each point above μ_0

† S. W. Roberts, *Technometrics* **8**, 411, 1966.

Band	Value Assigned
$\mu_0 < \bar{X}_i \le \mu_0 + \sigma_{\bar{X}}$	0
$\mu_0 + \sigma_{\bar{X}} \le \bar{X}_i < \mu_0 + 2\sigma_{\bar{X}}$	1
$\mu_0 + 2\sigma_{\bar{X}} \le \bar{X}_i < \mu_0 + 3\sigma_{\bar{X}}$	2
$\mu_0 + 3\sigma_{\bar{X}} \le \bar{X}_i < \infty$	3

and a similar series of values are assigned for values of $\bar{X}$ below μ_0. The process is deemed out of control when the cumulated score reaches a selected value.

To compare the relative effectiveness of each of the methods listed in Table 3.9-7, Table 3.9-8 lists the expected value of the number of samples required after a

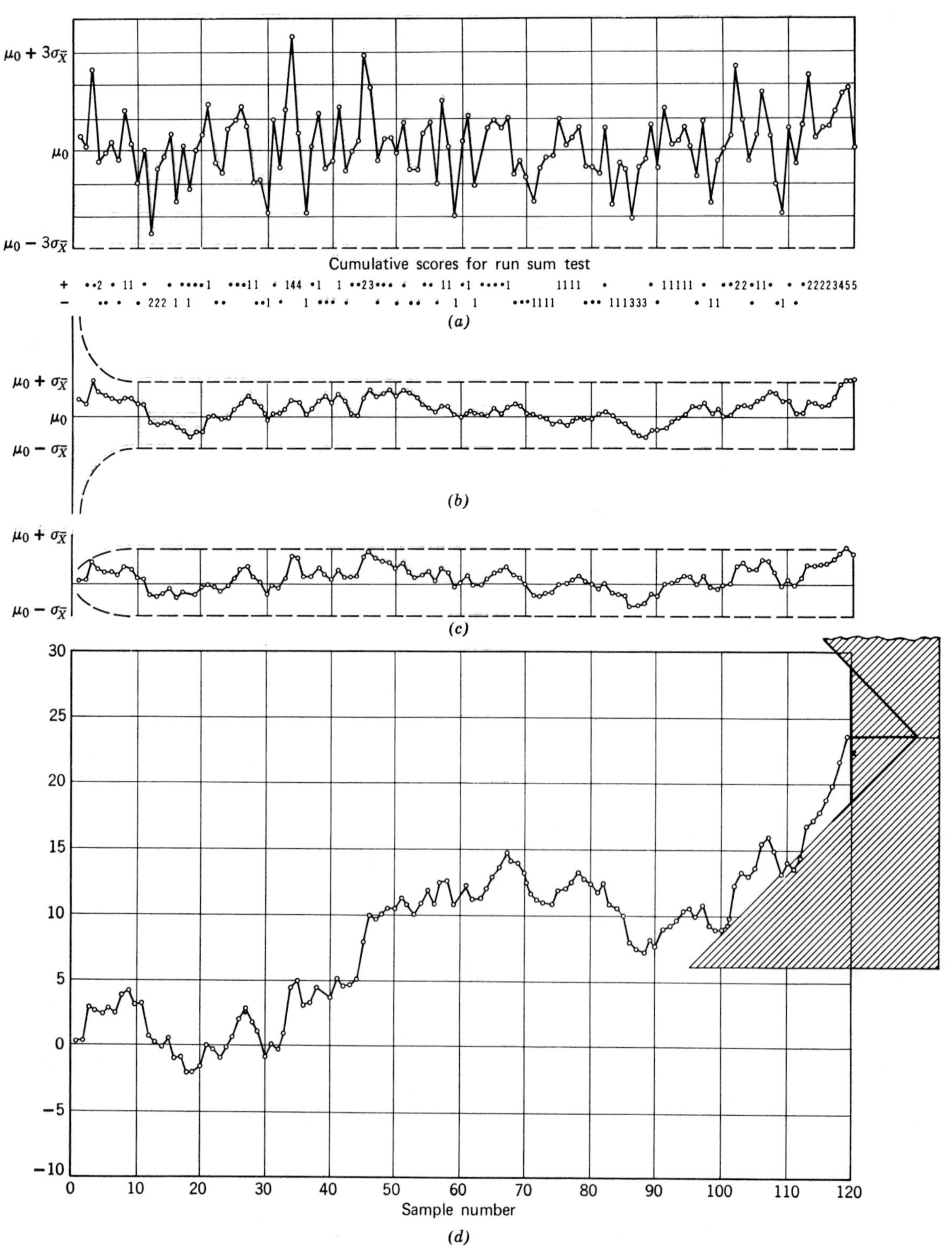

FIGURE E3.9-7 (*a*) Standard $\bar{X}$ control chart, (*b*) moving average chart, (*c*) geometric moving average chart, and (*d*) cumulative sum chart. (From S. W. Roberts, *Technometrics* **8**, 412, 1966.)

TABLE 3.9-7

Type of Chart	Calculation of Plotted Point	Test Used for Corrective Action
Shewhart $\bar{X}$	$\bar{X}_i = \frac{1}{n}\sum_{j=1}^{n} X_{ij}$	$\bar{X}_i > \mu_0 \pm 3\sigma_{\bar{X}}$
Shewhart $\bar{X}$ plus runs test	$\bar{X}_i$ plus run count above or below central line	$\bar{X}_i > \mu_0 \pm 3\sigma_{\bar{X}}$ plus two-sided runs test*
Moving average	$\bar{X}(i) = \frac{\bar{X}_{i-k+1} + \cdots + \bar{X}_{i-1} + \bar{X}_i}{k}$	$\bar{X}(i) > \mu_0 \pm \frac{A_k\sigma_{\bar{X}}}{\sqrt{k}}$ for $k = 9$
	$k = \min\{i, 9\}$ $A_k = 3.0$	$\bar{X}(i) > \mu_0 + \frac{A_k\sigma_{\bar{X}}}{\sqrt{i}}$ for $k < 9$
Geometric moving average	$Z_i^* = 0.2\bar{X}_i + 0.8Z_{i-1}^*$	$Z_i^* > \mu_0 \pm A_r\sigma_{\bar{X}}\left(\frac{w}{2-w}\right)^{1/2}$
	$Z_0^* = \mu_0 \quad w = \frac{2}{k+1} = \frac{2}{10}$ $A_r = 3.0$	for i large
Cumulative sum	$S_i = (\bar{X}_i - \mu_0) + S_{i-1}$ $S_0 = 9, i = 0, \ldots, n$	Point falls outside arms of V-shaped mask

* Process deemed out of control if one or more of the following occurs: (1) $\bar{X}_i > \mu_0 \pm 3\sigma_{\bar{X}}$; (2) $\bar{X}_i$ and either $\bar{X}_{i-1}$ or $\bar{X}_{i-2}$ fall between $2\sigma_{\bar{X}}$ and related $3\sigma_{\bar{X}}$ control levels; (3) $\bar{X}_{i-7}, \bar{X}_{i-6}, \ldots, \bar{X}_i$ all fall on the same side of μ_0.

change in the process variable value has occurred from μ_0 to $\mu_0 + k\sigma_{\bar{X}}$ (where k is a constant) until the shift in μ would be detected. The entries in the table are based on essentially the same sensitivity for each test in calling for corrective action in the absence of a shift in the process mean. Except for small values of k the tests roughly prove to be equally effective.

TABLE 3.9-8

	k				
Test or Chart	0	0.5	1.0	2.0	3.0
Shewhart $\bar{X}$ supplemented by the two-sided runs test	740	79	19	4.4	1.9
Runs sum test	740	50	12	3.8	2.4
Moving average for $k = 8$	740	40	10	4.6	3.3
Shewhart $\bar{X}$ supplemented by moving average for $k = 8$	740	50	11	3.7	1.9
Geometric moving average with $w = 0.25$	740	40	10	3.5	2.2
Shewhart $\bar{X}$ supplemented by geometric moving average with $w = 0.25$	740	50	12	3.3	1.7
Cumulative sum of 5	740	34	10	4.3	2.9

3.9-5 Control Charts for Several Variables

If two or more variables are observed and a sample statistic for each variable is plotted on individual control charts, the process may be termed as being out of control as soon as one chart shows an out of control condition. But such a decision rule is improper if the variables have a joint distribution. Suppose that two variables have a joint normal distribution and that α is selected as 0.05. If charts plotting the variable itself are separately maintained, the probability that both of the variables will fall within the control limits at the same time would be $(0.95)(0.95) = 0.9025$; hence the true Type I error is more nearly 0.10 instead of 0.05. The true control region is an ellipse, with all points on the perimeter having equal probability of occurring, rather than being a square or a rectangle. If the variables are correlated, the region is an ellipse rotated so that the major axes are no longer aligned with the coordinates $x_1 - x_2$. Figure 4.3-3 illustrates such a region.

To obtain one common statistic calculated from values of many variables that can be plotted on a control chart, Jackson† suggested that the statistic T^2 be used, where T^2 is Hotelling's T^2.‡ T^2 is simply the locus of the ellipse of the confidence region and, for two jointly distributed normal random variables X and Y, is given in terms of the sample size n, the sample means, and the sample variances as follows:

$$T_i^2 = \frac{ns_X^2 s_Y^2}{s_X^2 s_Y^2 - s_{XY}^2}\left[\frac{(X_i - \bar{X})^2}{s_X^2} + \frac{(Y_i - \bar{Y})^2}{s_Y^2} - \frac{2s_{XY}(X_i - \bar{X})(Y_i - \bar{Y})}{s_X^2 s_Y^2}\right] \quad (3.9\text{-}2)$$

† J. E. Jackson, *Technometrics* **1**, 359, 1959.
‡ H. Hotelling, "Multivariate Quality Control" in *Techniques of Statistical Analysis*, ed. by C. Eisenhart, M. W. Hastay, and W. A. Wallis, McGraw-Hill, New York, 1947, pp. 11–84.

All values of T_i^2 greater than that given by Equation 3.9-2 represent an out of control condition. T^2 can be related to the F distribution

$$T_\alpha^2 = \frac{2(n-1)F_\alpha}{n-2} \tag{3.9-3}$$

where F_α has 2 and $n - 2$ degrees of freedom.

For p variables, T^2 is best expressed in matrix notation (refer to Appendix B)

$$T^2 = \mathbf{X}\mathbf{s}^{-1}\mathbf{X}^T \tag{3.9-4}$$

where

$$\mathbf{X} = [X_1 - \bar{X}_1, X_2 - \bar{X}_2, \ldots, X_p - \bar{X}_p]$$

and the sample covariance matrix is

$$\mathbf{s} = \begin{bmatrix} s_{X_1}^2 & s_{X_1X_2} & \cdots & \\ s_{X_1X_2} & s_{X_2}^2 & \cdots & \\ & & \ddots & \\ & & & s_{Xp}^2 \end{bmatrix}$$

T_2^α is distributed as $p\nu F_\alpha/(\nu - p + 1)$ where F_α has p and $(\nu - p + 1)$ degrees of freedom with ν being the number of degrees of freedom in estimating the sample variances and usually equal to $n - 1$.

Supplementary References

General

Anscombe, F. J., "Rejection of Outliers," *Technometrics* **2**, 123, 1960.

Birnbaum, A., "Another View on the Foundation of Statistics," *Amer. Stat.*, p. 17, Feb. 1962.

Brownlee, K. A., *Statistical Theory and Methodology in Science and Engineering*, John Wiley, New York, 1960.

Cochran, W. G., *Sampling Techniques*, John Wiley, New York, 1950.

David, H. A. and Paulson, A. S., "The Performance of Several Tests for Outliers," *Biometrika* **52**, 2129, 1965.

Dixon, W. J., "Rejection of Observations" in *Contribution to Order Statistics*, ed. by A. E. Sarhan and B. G. Greenberg, John Wiley, New York, 1962, pp. 299–342.

Fraser, D. A. S., *Nonparametric Methods in Statistics*, John Wiley, New York, 1957.

Gumbel, E. J., *Statistics of Extremes*, Columbia Univ. Press, New York, 1958.

Hald, A., *Statistical Theory with Engineering Applications*, John Wiley, New York, 1952.

Johnson, N. L. and Leone, F. C., *Statistics and Experimental Design in Engineering and the Physical Sciences*, John Wiley, New York, 1964.

Kendall, M. G., *Rank Correlation Methods* (2nd ed.), Griffin, London, 1955.

Lehmann, E. L., *Testing Statistical Hypotheses*, John Wiley, New York, 1959.

Mendenhal, W., "A Bibliography of Life Testing and Related Topics," *Biometrika* **45**, 521, 1958.

Thrall, R. M., Coombs, C. H., and Davis, R. L., *Decision Processes*, John Wiley, New York, 1954.

Wald A., *Statistical Decision Functions*, John Wiley, New York, 1950.

Wilks, S. S., *Mathematical Statistics*, John Wiley, New York, 1963.

Woodward, R. H. and P. L. Goldsmith, *Cumulative Sum Techniques*, Oliver and Boyd, Edinburgh, 1964.

Bayes Estimation

Gupta, S. S. and Soebel, M., *Biometrika* **49**, 495, 1962.

Guttman, I., *Ann. Inst. Stat. Math.* **13**, 9, 1961.

Jeffreys, H., *Theory of Probability* (3rd ed.), Clarendon Press, Oxford, 1961.

Raiffa, H. and Schlaifer, R., *Applied Decision Theory*, Harvard Univ. Press, 1961.

Savage, L. J., *Foundations of Statistics*, John Wiley, New York, 1954.

Tiao, G. C. and A. Zellner, *Biometrika* **51**, 219, 1964.

Rejection of Outliers

Bross, I. D. J., "Outliers in Patterned Experiments—A Strategic Appraisal," *Technometrics* **3**, 91, 1961.

David, H. A. and Paulson, A. S., "The Performance of Several Tests for Outliers," *Biometrika* **52**, 42a, 1965.

Ferguson, T. S., "Rules for the Rejection of Outliers," *Rev. Inst. Instl. Statisque* **29**, 23, 1961.

Ferguson, T. S., "On the Rejection of Outliers" in *Proceed. Fourth Berkeley Symp. Math. Statis. & Prob.*, ed. by J. Neyman, Vol. 1, p. 253, Univ. of Calif. Press, Berkeley, 1961.

Grubbs, F. G., "Sample Criteria for Testing Outlying Observations," *Ann. Math. Stat.* **21**, 27, 1950.

Proscham, F., "Testing Suspected Observations," *Ind. Qual. Control* **13**, 14, 1957.

Tukey, J. W., "The Future of Data Analysis," *Ann. Math. Stat.* **33**, 1, 1962.

Process Control Charts

Aroian, L. A. and Levene, H., "The Effectiveness of Quality Control Charts," *J. Amer. Stat. Assn.* **45**, 550, 1950.

Barnard, G. A., "Cumulative Charts and Stochastic Process," *J. Royal Stat. Soc.* **B21**, 1959.

Duncan, A. J., "The Economic Design of $\bar{X}$ Charts to Maintain Current Control of a Process," *J. Amer. Stat. Assn.* **51**, 228, 1956.

Ewan, W. D., "When and How to Use Cu-Sum Chart," *Technometrics* **5**, 1, 1963.

Goldsmith, P. L. and Whitfield, H., "Average Run Lengths in Cumulative Chart Quality Control Schemes," *Technometrics* **3**, 11, 1961.

Ireson, W. G., Resnikoff, G. J., and Smith, B. E., "Statistical Tolerance Limits for Determining Process Capability," *J. Ind. Eng.* **12**, 126, 1961.

Johnson, N. L. and Leone, F. C., "Cumulative Sum Control Charts," *Ind. Qual. Control* **18** (12), 15, 1961; **19** (1), 29, 1962; **19** (2), 22, 1962.

Kemp, K. W., "The Average Run Length of the Cumulative Sum Charts when a V-mask is Used," *J. Royal Stat. Soc.* **B23**, 149 1961.

Mitra, S. K., "Tables for Tolerance Limits for a Normal Population Based on Sample Mean and Range or Mean Range," *J. Amer. Stat. Assn.* **52**, 88, 1957.

Page, E. S., "Cumulative Sum Control Charts," *Technometrics* **3**, 1, 1961.

Page, E. S., "Controlling the Standard Deviation by Cusums and Warning Lines," *Technometrics* **5**, 307, 1963.

Patnaik, P. B., "The Use of Mean Range as an Estimator of Variance in Statistical Tests," *Biometrika* **37**, 78, 1950.

Woodward, R. H. and Goldsmith, P. L., *Cumulative Sum Techniques*, Oliver and Boyd, Edinburgh, 1964.

Problems

3.1 The logarithmic series probability function for a random variable X is

$$P(x_k, \theta) \equiv P\{X = x_k\} = \frac{-\theta^x}{x \ln(1-\theta)}$$

$$x = 1, 2, \ldots, \infty$$
$$0 \leq \theta \leq 1$$

Given that a sample of experimental observations of size n has been collected, find the maximum likelihood estimate of θ.

3.2 Consider the joint normal probability density for the random variables X and Y with the common parameter μ:

$$p(x, y) = \frac{1}{2\pi\sigma_X\sigma_Y} \exp\left[-\frac{1}{2}\left\{\left(\frac{x-\mu}{\sigma_X}\right)^2 + \left(\frac{y-\mu}{\sigma_Y}\right)^2\right\}\right]$$

Find the maximum likelihood estimates of μ, σ_X^2 and σ_Y^2 for n *independent* observations of which n_X are made on the X variable (only) and n_Y are made on the Y variable (only).

3.3 Find the maximum likelihood estimate of the parameter λ in the Poisson probability function

$$P(x, \lambda) = \frac{e^{-\lambda}\lambda^x}{x!}$$

3.4 Compute the first and second moments of the exponential distribution (listed in Table 2.3-2) and equate them to the first and second sample moments obtained from an experiment to estimate θ. Do both moments give the same estimate of θ? Which would be the best estimate to use?

3.5 Can a confidence interval for a random variable which is not a normal random variable be estimated? Explain.

3.6 Based on the following grades:

Student Number	Grade
1	95
2	92
3	90
4	86
5	86
6	80
7	75
8	72
9	64
10	60

find the values of $\bar{X}$ and s^2. If we assume that the grades have been taken from a normally distributed population, determine the cut-off points for the grades of A, B, C, D, and F based on the following rules:

(a) The A's should equal the D's plus the F's.
(b) The B's should equal the C's.

3.7 Measurement of the density of 20 samples of fertilizer gave a mean CaO content of 8.24 percent and a standard deviation of 0.42 percent (percent of 100 percent). What are the two-sided symmetric confidence limits for a confidence coefficient of $(1 - \alpha)$ equal to (a) 0.95 and (b) 0.99 for (1) the ensemble mean and (2) the ensemble variance? From (2) calculate the confidence limits on the ensemble standard deviation.

3.8 Given that the sample standard deviation for the total pressure in a vapor–liquid equilibria experiment is 2.50 atm, find (a) the 95-percent and (b) the 99-percent confidence limits for (1) the ensemble standard deviation and (2) the ensemble variance. Eight individual values of the pressure were measured.

3.9 The velocity of a space missile after its fuel is gone is

$$v = v_g \ln\left(\frac{m_b + m_p}{m_b}\right)$$

where:

v_g = exhaust velocity of the gases, a random variable
m_b = rocket weight after burning, a random variable
m_p = propellant weight, a random variable

Find the sample variance of v in terms of the sample variances of v_g, m_p, and m_b.

3.10 Assume that a considerable number of fluid velocity measurements made in the laboratory give a sample mean of 4.60 ft/sec and a sample variance of 0.6 ft^2/sec^2. Suppose the next velocity measurement made has a value of:

(a) 7.60 ft/sec.
(b) 5.60 ft/sec.

What conclusions would you draw in each case?

3.11 The Deil Co., a cleaning-agent manufacturer, has a slogan "Dial for Deil." It claims that Deil is at least 90 percent effective in cleaning boiler scale or "your money back." The government has taken the company to court citing false advertising. As proof, the company cites a sample of 10 random applications in which an average of 81 percent of the boiler scale was removed. The government says 81 percent does not equal 90 percent. The company says that the test is only a statistical sample, and the true effectiveness may easily be 90 percent. Who is correct and why? The data were as follows.

Number	Removed
1	93
2	60
3	77
4	92
5	100
6	90
7	91
8	82
9	75
10	50

3.12 Data for the cooling of superheated steam without condensation have been correlated by

$$\frac{hD}{k} = 0.021\left(\frac{DG}{\mu}\right)^{0.8}$$

TABLE P3.12

Symbol	Physical Quantity	Value of Sample Mean	One Sample Standard Deviation*
h	Heat transfer coefficient, Btu/(hr)(ft^2)(°F)		
D	Tube diameter, ft	0.20	$\frac{1}{2}$
k	Thermal conductivity Btu/(hr)(ft)(°F)	0.0441	2
G	Mass velocity, lb/(hr)(ft^2)	20,000	5
μ	Viscosity, lb/(hr)(ft)	0.109	1

* Expressed as a percent of sample mean.

Find the sample mean of the heat transfer coefficient based on the calculated values given in Table P3.12. Find the sample standard deviation for h, and express it as a percent of the sample mean for h. Assume that the variables are random normal variables and that 10 measurements were made in obtaining each standard deviation. Estimate the confidence interval for the ensemble mean heat transfer coefficient and for the ensemble standard deviation of the heat transfer coefficient.

3.13 Five thermocouples are calibrated against a standard whose reading is 250°C. It was found that

$$\bar{X} = 248.5°\text{C}$$
$$s_{\bar{X}}^2 = 70(°\text{C})^2$$

Assume that the hypothesis is $\mu_X = \mu_0 = 250.0°\text{C}$ and estimate the power of the t-test to discriminate for $\alpha = 0.05$ if $\mu_X = 248.5°\text{C}$.

3.14 Pressure gauges are being manufactured to sell for $1.25 wholesale. A sample of 20 gauges out of 200 is characterized as follows when connected to a standard gauge at 30 psia:

$$\bar{X} = 29.1$$
$$s_X = 1.2$$

Using a symmetric two-tailed test with $\alpha = 0.05$, answer the following questions:

(a) What is the region of rejection?
(b) What is the region of acceptance?
(c) What is the power of the test toward an ensemble mean 90 percent of the standard of 30 psia (i.e., toward an ensemble mean of 27 psia)?
(d) Would you pass or reject this sample?

3.15 Prepare an operating characteristic (OC) curve and a power curve based on the following information about the random variable X:

$$\mu_X = 30.0$$
$$\sigma_X = 2.4$$
$$n = 64.0$$

Plot β and $(1 - \beta)$ versus selected values of possible μ's above and below 30.0 for $\alpha = 0.01$.

3.16 Can the power of a test ever be bigger than the fraction α, the significance level?

3.17 Classify the following results of hypothesis testing as to: (1) error of the "first kind," (2) error of the "second kind," (3) neither, and (4) both. The hypothesis being tested is designated as H_0.

(a) H_0 is true, and the test indicates H_0 should be accepted.
(b) H_0 is true, and the test indicates H_0 should be rejected.
(c) H_0 is false, and the test indicates H_0 should be accepted.
(d) H_0 is false, and the test indicates H_0 should be rejected.

3.18 A manufacturer claimed his mixer could mix more viscous materials than any rival's mixer. In a test of stalling speed on nine viscous materials, the sample mean viscosity for stall was 1600 poise with a sample standard deviation of 400 poise. Determine whether or not to reject the following hypotheses H_1 based on $\alpha = 0.05$.

(a) The true ensemble stalling speed of the mixer is at 1700 poise ($H_{1_A}: \mu = 1700$ poise versus the alternate hypothesis $H_{1_B}: \mu \neq 1700$ poise).
(b) The true stalling speed is 1900 poise (H_{2_A}: $\mu = 1900$ poise versus the alternate hypothesis $H_{2_B}: \mu \neq 1900$).
(c) The true stalling speed is greater than 1400 poise (H_{1_A}: $\mu > 1400$ poise versus the alternate hypothesis $H_{2_A}: \mu \leq 1400$ poise).

3.19 A new design has been devised to improve the length of time a circular-type aspirin pill die can be used before it has to be replaced. The old die in 10 trials gave an average life of 4.4 months with a standard deviation of 0.05 month. The proposed die in 6 trials had an average life of 5.5 months with a standard deviation of 0.9 months. Has the die been improved? (Use $\alpha = 0.05$ as the significance level.)

3.20 Two chemical solutions are measured for their index of refraction with the following results:

	A	B
Index of refraction	1.104	1.154
Standard deviation	0.011	0.017
Number in sample	5	4

Do they have the same index of refraction? (Use $\alpha = 0.05$.)

3.21 In a test of six samples of oil, the sample mean for the specific viscosity was 0.7750 cp, with a sample standard deviation of 1.45×10^{-2} cp. The specifications call for a mean (ensemble mean) of 0.8000 cp. Is the oil on specification or not? What confidence coefficient should be selected? If the oil was supposed to be at least 0.8000 cp., would this change your answer?

3.22 A liquid-liquid batch extractor removes component A from a solution of A and B by use of a solvent. For a long time the mass fraction A in the extract has been $\omega_A = 0.30$ with a standard deviation of 0.02. By rearrangement of the baffles in the extractor, it is believed that the value of ω_A can be increased. Formulate a test which tells after nine samples have been run whether or not the baffling is effective at the 0.05 level of significance. (Hint: Determine the hypothesis first.)

Suppose that indeed the rearrangement of baffles does improve the value of ω_A to 0.45. Under the decision rule formulated, what is the probability of deciding that no change has taken place even though the new set-up differs from the old one? (Assume for simplicity that the standard deviation remains the same.) If $\omega = 0.35$, would your answer change?

3.23 One hundred crude oil samples are taken from a pipeline and found to have a mean sulfur content of 1.60 wt. percent with standard deviation of 0.12 wt. percent. If μ is the ensemble mean sulfur content of the crude oil (based on past experience), test the hypothesis that $\mu = 1.70$ wt. percent against the hypothesis that $\mu \neq 1.70$ wt. percent for two significance levels: (a) $\alpha = 0.05$ and (b) $\alpha = 0.01$.

Also test the hypothesis that $\mu = 1.70$ wt. percent against the alternate hypothesis that $\mu < 1.70$ wt. percent, using the same significance levels.

3.24 A sample of 16 centrifugal pumps purchased from manufacturer A lasted an average of 150 days before breaking down; the standard deviation for breakdown was 25 days. Another batch of 10 centrifugal pumps purchased from manufacturer B lasted an average of 120 days before breaking down; their standard deviation was 12 days. For $\alpha = 0.05$:

(a) Find the confidence interval for the ensemble mean lifetime of the pumps from manufacturer A. What assumption have you made about the distribution of the random variable, the lifetime?

(b) State whether or not the pumps from manufacturer A are better than, the same as, or poorer than those from manufacturer B. Show the calculations for your choice.

3.25 Gas from two different sources is analyzed and yields the following methane content (mole percent):

Source 1	Source 2
64.0	69.0
65.0	69.0
75.0	61.5
67.0	67.5
64.5	64.0
74.0	
75.0	

Is there a significant difference in the methane content from the two sources?

3.26 From the following data, determine if there is a significant difference in pressure gauge performance (readings are in mm as a deviation from 760).

Trial	Gauge 1	Gauge 2
1	4.4	3.2
2	−1.4	7.7
3	3.2	6.4
4	0.2	2.7
5	−5.0	3.1
6	0.3	0.6
7	1.2	2.6
8	2.2	2.2
9	1.3	2.2

What is the 95-percent confidence interval for the ensemble standard deviation for gauge 1 and gauge 2, respectively?

3.27 A gas chromatographic apparatus has analyzed the concentration of methane gas with a variance of 0.24 during the last two months. Another person uses it for 25 samples and the variance is 0.326. Is 0.326 significantly larger than 0.24, i.e., is the new operator doing something wrong? Use $\alpha = 0.05$. Use an F-test. Also compute the confidence limits for σ^2.

3.28 Four temperature controllers are monitoring the temperature in a stream. Each of them is from a different manufacturer. Past experience over the last four years has shown the following number of maintenance jobs on each instrument:

Manufacturer identification	1	2	3	4
Number of repairs	46	33	38	49

Your company is now about to purchase six more temperature controllers. Your assistant says that obviously the ones from manufacturer No. 2 are the ones to get. Do you believe that the instrument from manufacturer No. 2 is clearly the best?

3.29 Apply Bartlett's test to the data of Dorsey and Rosa below who measured the ratio of the electromagnetic to the electrostatic unit of electricity. During the observations they assembled, disassembled, and cleaned their apparatus many times.

Group of Data	Number of Observations	Variances $\times 10^8$
1	11	1.5636
2	8	1.1250
3	6	3.7666
4	24	4.1721
5	15	4.2666

Are the variances homogeneous?

3.30 Seven samples of two different solid-state circuits have been tested for maximum current output and the following data observed (in amps):

Circuit A	Circuit B
0.18	0.21
0.24	0.29
0.18	0.14
0.17	0.19
1.03	0.46
0.14	0.08
0.09	0.14

Based on the sign test, do circuits A and B differ in average performance? Is the current output of circuit A higher than that of B?

3.31 Apply the sign test to the data of Problem 2.47.

3.32 Two shifts have submitted plant data for the yields of wax:

I	II
40	47
27	42
39	41
46	34
32	45
46	52
40	49
44	35
48	43
	44

Do the shifts differ in performance as determined by the Mann-Whitney test?

3.33 Apply the Mann-Whitney test to the data of Problem 3.25.

3.34 The following data represent precipitation for various months and runoff at a gauging station.

Month	Precipitation (mm)	Runoff (mm)
1	350	0.0
2	370	29.6
3	461	15.2
4	306	66.5
5	313	2.3
6	455	0.5
7	477	102
8	250	12
9	546	6.1
10	274	6.2

(a) Determine whether or not these records represent stationary random variables.
(b) Would you reject the runoff value in the seventh month as being an extreme value?
(c) Is there a linear trend in either time record (with runoff 102 deleted)?

3.35 Vibration tests were carried out in the laboratory, and the output of the accelerometer was recorded on a portable tape recorder. The tape output was processed in an analog computer to yield: (1) the mean square values of the accelerometer signal, and (2) the power spectral density of the signal. There was some question as to whether or not the data were stationary. Consequently, the continuous mean square signal was sampled at one-half second intervals, and the amplitude of the signal (from the bottom of the chart) at each sampling time was measured on the chart paper. The following table lists the results of one such sampling sequence.

Time (sec)	Amplitude (chart divisions)
6.5	7
7.0	6
7.5	10
8.0	3
8.5	15
9.0	8
9.5	5
10.0	7
10.5	13
11.0	3
11.5	26
12.0	9
12.5	5
13.0	12
13.5	10
14.0	4
14.5	12
15.0	2
15.5	4
16.0	5
16.5	11
17.0	7
17.5	7
18.0	8
18.5	12
19.0	4
19.5	6
20.0	3
20.5	11
21.0	10

The sampling time was chosen so as to provide at least five smoothing time values (t_f values used in calculating the mean sequence) in the interval between samples. Hence the listed data are believed to be statistically independent.

Is the random variable being measured stationary? Does it contain a periodic component and, if so, how will this affect your answer?

3.36 A digital record of the output of an instrument monitoring hydrocarbons downtown in a large city shows an unusually high series of values from 4:30 to 5:30 p.m. Are these possibly extreme values in the data record caused by some defect in the instrument so that they should be rejected? Explain.

Time (p.m.)	Hydrocarbon (ppm)
1:30	123
2:00	179
2:30	146
3:00	190
3:30	141
4:00	206
4:30	407
5:00	564
5:30	530
6:00	273
6:30	199
7:00	142
7:30	171

3.37 Biological oxygen demand, BOD (in mg/liter), in a river has been measured at a junction of two small rivers. Various domestic and industrial discharges upstream on each river affect the measured values of BOD, but sampling errors have also been noted in the past. Should Sample No. 63-4 be discarded? How about Sample No. 63-9? Explain. The samples were taken in sequence at two-hour intervals.

Sample Number	BOD
63.2	6.5
63.3	5.8
63-4	16.7
63-5	6.4
63-6	7.0
63-7	6.3
63-8	7.0
63-9	9.2
63-10	6.7
63-11	6.7

3.38 Nitrogen removal by a biological process from a waste water stream is calculated at periodic intervals at station 16-A. At the end of each reporting period, a report must be made of the average nitrogen removed. The concentrations of NH_3, nitrates, and nitrites, in total expressed as moles of N_2/liter, recorded for the last period were:

0.0127	0.0176
0.0163	0.0170
0.0159	0.0147
0.0243	0.0168

Should the value of 0.243 be discarded as an extreme value? Explain.

3.39 Can the following data be represented by the normal distribution?

Number of accidents	3	19	16	10	11
Time of day	7–8	8–9	9–10	10–11	11–12

3.40 Do the following repeated pressure measurements under the same conditions indicate that measurements from the apparatus can be represented by the normal distribution?

Frequency	Range of Measurement (deviation, psia)
2	−0.4 to −0.5
4	−0.3 to −0.4
9	−0.2 to −0.3
22	−0.1 to −0.2
27	0.0 to −0.1
11	0.0 to 0.1
7	0.1 to 0.2
3	0.2 to 0.3
1	0.3 to 0.4

3.41 The following data were determined from screening galena. The mesh categories given are for successively smaller wire spacings. Apply the χ^2 test to establish whether or not the data are normally distributed.

Mesh	Number of Particles Retained on Screen
3–4	10
4–6	40
6–8	81
8–10	115
10–14	160
14–20	148
20–28	132
28–35	81
35–48	62
48–65	41
65–100	36
100–150	22
150–200	19
200	53

3.42 From the observed data for 120 tosses of a die, decide whether or not it is a fair die.

Face which fell up	1	2	3	4	5	6
Observed frequency	25	17	15	23	24	16

3.43 A cylindrical mixing vessel was packed in three segregated layers with red, white, and blue balls, about 20 mesh in size, to ascertain the completeness of mixing. The initial quantities of balls were: blue, 50 percent; red, 30 percent; and white, 20 percent. After 23 revolutions of the mixer, the following distribution prevailed at the 12 sampling points in the mixer (12 samples were taken to provide uniform sampling). Was the set of balls well mixed at that time?

Each sample had only 2 degrees of freedom because each sample was fixed at 30 balls and the white balls were determined by difference. The total degrees of freedom were 24.

Sample Position	Number of Particles by Color Distribution		
	Red	White	Blue
1	3	9	18
2	11	1	18
3	10	5	15
4	10	5	15
5	11	5	14
6	6	4	20
7	17	2	11
8	16	4	10
9	13	6	11
10	8	10	12
11	7	7	16
12	8	2	20

For the same run, the following information is of interest:

Revolutions of mixer	2	5	11	23	35	55
χ^2	330	300	90	46	21	30

3.44 A recent article used various equations from literature to predict theoretical vapor compositions for carbon tetrachloride vapor–liquid equilibria. The theoretical predicted mole fractions, y_t, were compared with the experimental mole fractions, y_e, by means of the χ^2 test as follows:

$$\chi^2 = \frac{\sum (y_e - y_t)^2}{y_t}$$

In one series of runs at 760 mm, χ^2 at the 1-percent probability level, i.e., $P\{\chi^2 \geq \chi^2_{1-\alpha}\} = 0.01$, was:

Equation Number	χ^2
1	1.04
2	0.57
3	39.21
4	57.03

What can you conclude from these experiments?

3.45 The following data on the strength of test bars of a new polymer have been collected and placed into intervals for convenience in calculation.

Yield (lb)	Molecular Weight		
	0–10^4	1×10^4 to 5×10^4	5×10^4 to 1×10^6
0–100	3	5	6
100–150	8	7	9
150–200	6	6	5
>200	8	7	10

At the 5-percent level of significance, determine if the two variables, yield and molecular weight, are independent.

3.46 In a series of leaching tests on uranium ores (U.S.A.E. Document MITG-A63, Mar. 1949), a balance of radioactivity was calculated from data on flow rates and activities and is given in Table P3.46. Measurements of flow rates were subject to large errors because of fluctuations and the short periods covered, but a reasonably good overall balance was reached. Are the "in" and "out" measurements independent of where the sample was taken?

TABLE P3.46 BALANCE OF RADIOACTIVITY IN CYCLIC LEACHING, PLANT SOLUTIONS, NOVEMBER 22

Sample From	Beta-Activity Product, Counts		
	In	Out	pH
Agitator 1	12,780	1,100	3.8
Thickener *A*	1,100	1,440	4.0
Agitator 2	130	23,780	1.3
Agitator 21	23,780	22,980	1.4
Thickener *B*	27,520	15,510	1.9
Thickener *C*	3,610	5,200	2.5
Thickener *D*	750	930	2.8
Thickener *E*	60	130	3.4

3.47 In another series of tests, the counts for a uranium solution have been tabulated by shift and date. Are the shift and date data independent as was anticipated in advance? Data are counts in 10 minutes.

Date	Shift *A*	Shift *B*	Shift *C*
21	64	37	90
22	191	320	330
23	154	240	250
24	105	220	180
25	94	72	66
26	57	85	140

3.48 Data have been collected every hour for calcium gluconate to substantiate a label claim of 1.000 gram. Prepare an $\bar{X}$ chart and an R chart for the process. Indicate the upper and lower control limits on both charts. Initially, use samples of five and adjust the control limits as additional samples are taken into account.

Sample Number	Assay (X)
1	0.968
2	0.952
3	0.945
4	0.958
5	0.965
6	0.955
7	0.956
8	0.958
9	0.965
10	0.954
11	0.968
12	0.979
13	0.971
14	0.947
15	0.968
Sum	14.409

3.49 A series of 30 individual sequential measurements of the random variable X was taken as tabulated in Table P3.49. Prepare a Shewhart chart for X, based on confidence limits at $\bar{\bar{X}} \pm 3\sigma$, and a cumulative sum chart for X.

Establish when the first of the 30 points, if any, goes out of control. Indicate which of the remaining points are out of control. Estimate the standard deviation from only those points in control; the first 12 points may be used initially. Estimate $\bar{\bar{X}}$ for the Shewhart chart from the same points. Delete any points out of control and recompute the parameters.

TABLE P3.49 SET OF CONSECUTIVE MEASUREMENTS

(1) Point Number	(2) Individual Result, X	(3) Target or Mean, h	(4) Deviation from Target, D	(5) Cumulative Deviation $\sum D$
1	16	10	6	6
2	7	10	−3	3
3	6	10	−4	−1
4	14	10	4	3
5	1	10	−9	−6
6	18	10	8	2
7	10	10	0	2
8	10	10	0	2
9	6	10	−4	−2
10	15	10	5	3
11	13	10	3	6
12	8	10	−2	4
13	20	10	10	14
14	12	10	2	16
15	9	10	−1	15
16	12	10	2	17
17	6	10	−4	13
18	18	10	8	21
19	14	10	4	25
20	15	10	5	30
21	16	10	6	36
22	9	10	−1	35
23	6	10	−4	31
24	12	10	2	33
25	10	10	0	33
26	17	10	7	40
27	13	10	3	43
28	9	10	−1	42
29	19	10	9	51
30	12	10	2	53

3.50 A portion of an ammonia plant consists of a gas purification unit, an NH_3 synthesis unit, and an air oxidation unit as shown in Figure P3.50. On Friday you are assigned the job of setting up statistical controls on the N_2 stream concentration from the gas purifier. The last 60 analyses are listed in Table P3.50a. One analysis is made each four hours (twice a shift). On Monday morning you are to report if a process control chart can be successfully employed to control the N_2 concentration and, if so, to recommend an appropriate type of Chart(s). Saturday's, Sunday's, and Monday's production data are shown in Table 3.50b. Is the N_2 stream in or out of control on Monday morning?

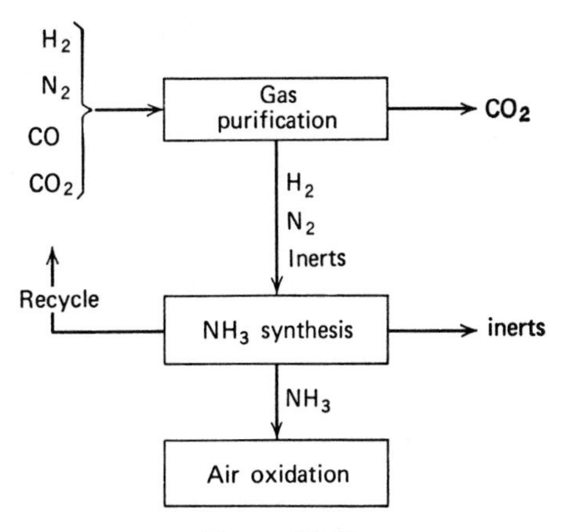

FIGURE P3.50

TABLE P3.50a

Sample Number	Percent N_2	Sample Number	Percent N_2
1	24.5	31	28.3
2	24.2	32	27.3
3	28.3	33	25.8
4	29.8	34	26.0
5	26.4	35	27.5
6	29.0	36	25.2
7	27.0	37	25.8
8	27.0	38	25.5
9	22.4	39	22.8
10	25.3	40	21.7
11	30.9	41	24.7
12	28.6	42	25.6
13	28.0	43	26.5
14	28.2	44	24.6
15	26.4	45	22.0
16	23.4	46	22.7
17	25.1	47	22.0
18	25.0	48	21.0
19	23.3	49	20.7
20	23.0	50	19.6
21	23.2	51	20.6
22	24.9	52	20.0
23	25.2	53	21.2
24	24.4	54	21.4
25	24.1	55	29.6
26	24.0	56	29.4
27	26.6	57	29.0
28	22.1	58	29.0
29	23.2	59	28.5
30	23.1	60 Saturday 8 a.m.	28.7

TABLE P3.50b N_2 STREAM ANALYSIS

Day	Shift	Time	Percent N_2
Saturday	1	12 noon	27.6
Saturday	2	4 p.m.	25.6
Saturday	2	8 p.m.	29.6
Sunday	3	12 midnight	30.7
Sunday	3	4 a.m.	30.0
Sunday	1	8 a.m.	30.6
Sunday	1	12 noon	31.7
Sunday	2	4 p.m.	29.6
Sunday	2	8 p.m.	30.6
Monday	3	12 midnight	28.1
Monday	3	4 a.m.	26.5
Monday	1	8 a.m.	27.5

PART II

Development and Analysis of Empirical Models

Three general categories of models were listed in Chapter 1: (1) models based on transport phenomena principles, (2) models based on balances on entities, and (3) empirical models. Because many processes cannot be satisfactorily represented by the first two types of models due to a lack of understanding about the process or because of the complexity of the process, empirical models act as appropriate substitutes. A typical example is the fitting of a polynomial or similar function to experimental data in order to predict the process response as a function of one or more independent variables. Of course, empirical models used to represent a process have limited value when the engineer wants to verify a theory or to make predictions beyond the range of the variables for which empirical data were collected during the model evolution.

How empirical models can be developed and interpreted is the subject of Part II of this text. Statistical techniques provide a guide to data collection and model building, the two interrelated activities underlying the description, explanation, and prediction of process phenomena.

CHAPTER 4

Linear Models With One Independent Variable

If the important variables for a process are known or sought but the process model is unknown, an empirical approach to model building is required. The development of empirical models to represent a continuous process involves postulation of a model, experimentation to collect empirical data, "fitting" of the model, i.e., estimation of the model coefficients, and evaluation of results. The strategy of empirical model building is described in detail in Chapter 8. In this chapter we shall be concerned with just one phase of model building, namely the estimation of the coefficients in a linear model, and certain related matters such as the estimation of confidence regions and the application of hypothesis tests. By starting the discussion of coefficient estimation with a linear model which incorporates two coefficients and just one independent variable, $\eta = \alpha + \beta x$, it is possible to defer the use of matrix notation until Chapter 5 and to provide illustrations in two dimensions of certain significant points which cannot easily be illustrated graphically for a more complex model. It also is feasible for the reader to follow through the examples with hand calculations; for models with many variables, computations on a digital computer are almost essential.

In discussing linear models, the word linear has meaning as applied to the independent variables of the model and also as applied to the parameters (coefficients) in the model. We shall be concerned with the latter connotation; that is, a linear model in this chapter is one having a linear combination of the parameters. By independent variables we mean those variables which are under the control of the experimenter. It is not necessary for the independent variables to be functionally independent in a mathematical sense nor that they be orthogonal (as described in Chapter 8). For example, a typical linear model is

$$\eta = \beta_0 + \beta_1 x_1 + \beta_2 x_2 + \cdots + \beta_p x_p$$

where η is the dependent variable (the *response*), and x_i's are the independent variables, and the β's are the model parameters (coefficients). The independent variables themselves may be nonlinear as in the following model,

$$\eta = \beta_0 + \beta_1 x_1 + \beta_2 x_1^2 + \cdots + \beta_p x_1^p$$

which is an equation linear in the parameters.

Additional examples are:

1. Linear in β, nonlinear in x:

$$\eta = \beta_0 + \beta_1 x_1 + \beta_2 x_2 + \beta_3 x_1 x_2 + \beta_4 \ln x_1$$

2. Linear in x, nonlinear in β:

$$\eta = \beta_0 + \beta_1 \beta_2 x_1 + \beta_1 x_2 + \beta_2 x_3$$

3. Nonlinear in x and β:

$$\eta = e^{\beta_1 x_1} + e^{\beta_2 x_2}$$
$$\eta = \sqrt{\beta_0 + \beta_1 x_1 + \beta_2 x_2}$$

A general form for a linear model in the sense used here is

$$\eta = \beta_0 f_0(x_1, x_2, \ldots) + \beta_1 f_1(x_1, x_2, \ldots) + \cdots$$

in which the functions f are of known form. Clearly, the response is linear with respect to the parameters if the first partial derivatives of the response with respect to each of the parameters are not functions of the parameters.

Because one of the most difficult aspects of model building is to select an appropriate initial form for the model, we shall first consider ways to select a suitable model of one independent variable. Then we shall describe how to estimate the parameters in the (linear) model $\eta = \alpha + \beta x$.

4.1 HOW TO SELECT A FUNCTIONAL RELATIONSHIP

Given a set of data involving one independent and one dependent variable, how can a functional relationship between them be identified? Certainly no universal guides exist. If the experimental data, when plotted on arithmetic coordinate paper, do not approximate a straight line, $y = a + bx$ where y is the measured dependent variable, but do seem to approximate a smooth curve, the shape of the curve and/or an understanding of the nature of the experiment may suggest the equation of a curve that will fit the points most closely. A helpful way of verifying the appropriateness of a particular equation for a set of data is to transform the

TABLE 4.1-1 TRANSFORMATIONS TO LINEAR FORM FOR A FUNCTION OF ONE VARIABLE

	Equation	Coordinates for Straight Line		Straight-Line Equation	Remarks
		x-axis	y-axis		
Use of reciprocals of logarithms:					
(1)	$\frac{1}{y} = \alpha + \beta x$	x	$\frac{1}{y}$	$\frac{1}{y} = \alpha + \beta x$	Asymptotes: $x = \frac{-\alpha}{\beta}$, $y = 0$
(2)	$y = \alpha + \frac{\beta}{x}$	$\frac{1}{x}$	y	$y = \alpha + \frac{\beta}{x}$	Asymptotes: $x = 0$, $y = \alpha$
(3)	$\frac{x}{y} = \alpha + \beta x$ $\left(\text{or } y = \frac{x}{\alpha + \beta x}\right)$	x	$\frac{x}{y}$	$\frac{x}{y} = \alpha + \beta x$	Asymptotes: $x = \frac{-\alpha}{\beta}$, $y = \frac{1}{\beta}$
	or $\frac{1}{y} = \frac{\alpha}{x} + \beta$	$\frac{1}{x}$	$\frac{1}{y}$	$\frac{1}{y} = \beta + \frac{\alpha}{x}$	
(3a)	$y = \frac{x}{\alpha + \beta x} + \gamma$	x	$\frac{x - x_1}{y - y_1}$	$\frac{x - x_1}{y - y_1} = (\alpha + \beta x_1) + \frac{\beta}{\alpha}(\alpha + \beta x_1)x$	Asymptotes: $x = -\alpha/\beta$, $y = (1/\beta) + \gamma$. Same curve as (3) shifted up or down by a distance of γ
		where (x_1, y_1) is any point on the experimental curve			
(4)	$y = \alpha x^{\beta}$	$\log x$	$\log y$	$\log y = \log \alpha + \beta \log x$	If β is +, curve has parabolic shape and passes through origin and $(1, \alpha)$. If β is −, curve has hyperbolic shape, passes through $(1, \alpha)$, and is asymptotic to x- and y-axes
(4a)	$y = \alpha x^{\beta} + \gamma$	$\log x$	$\log (y - \gamma)$	$\log (y - \gamma) = \log \alpha + \beta \log x$	First approximate γ by the equation $\gamma = (y_1 y_2 - y_3^2)/(y_1 + y_2 - 2y_3)$, where $y_3 = \alpha x_3^{\beta} + \gamma$, $x_3 = \sqrt{x_1 x_2}$, and (x_1, y_1) and (x_2, y_2) are experimental points
(4b)	$y = \gamma 10^{\alpha x^{\beta}}$	$\log x$	$\log (\log y - \log \gamma)$	$\log (\log y - \log \gamma) = \log \alpha + \beta \log x$	After taking logarithms of the original equation, follow method (4a)
(5)	$y = \alpha \beta^{x}$ (equivalent forms $y = \alpha \gamma^{\beta_2 x}$, $y = 10^{\alpha_1} + \beta_1 x$, $y = \alpha(10)^{\beta_1 x}$)	x	$\log y$	$\log y = \log \alpha + x \log \beta$	Passes through the point $(0, \alpha)$

data to a linear form so that a straight-line plot, $y' = a' + b'x'$, is obtained. Table 4.1-1 summarizes a few suggested methods of straight-line transformation. A collection of graphs (Figure 4.1-1) illustrates the effect of changing the coefficients for many of the equations presented in the table. For numerical examples and a more detailed explanation of the methods of transformation, refer to the references at the end of this chapter. (Special graph paper is available to facilitate logarithmic, reciprocal, square root, etc., transformations from the Codex Book Co., Norwood, Mass.)

If a straight line is achieved through some such type of transformation, the parameters of the nonlinear model can be estimated from the modified linear model. However, a problem arises if the unobservable error, ϵ, is added to the dependent variable as

$$Y = y + \epsilon$$

because after the transformation, some complex function of the error results rather than ϵ being added to the transformed variable. For example, if the model is Equation (4) in Table 4.1-1, the observed dependent variable is

$$Y = \alpha x^{\beta} + \epsilon \tag{4.1-1}$$

Clearly Equation 4.1-1 is not the same as

$$\log Y = \log \alpha + \beta \log x + \epsilon \quad (4.1\text{-}2)$$

because the logarithm of the right-hand side of Equation 4.1-1 is not equal to the right-hand side of Equation 4.1-2. In some instances, Equation 4.1-1 may represent the actual situation, in which case nonlinear estimation of α and β is required. In other instances, Equation 4.1-2 is more correct, depending upon the details of experimentation. This matter is discussed again in Section 6.5 after we treat nonlinear models. For the purposes of this chapter, we shall assume that function of interest is linear in the parameters.

Example 4.1-1 Determination of Functional Form

The data in the first two columns below represent a series of dependent variable measurements (Y) for corresponding (coded) values of the independent variable (x). Find a suitable functional form to represent the dependence of Y on x.

Solution:

Several differences and ratios can be computed, some of which are shown in Table E4.1-1.

TABLE E4.1-1

x	Y	$\log Y$	ΔY	$\Delta^2 Y$	x/Y	$\Delta(x/Y)$
1	62.1	1.79246			0.01610	
2	87.2	1.93962	25.1		0.02293	0.00683
3	109.5	2.03941	22.3	−2.8	0.02739	0.00446
4	127.3	2.10483	17.8	−4.5	0.03142	0.00403
5	134.7	2.12937	7.4	−10.4	0.03712	0.00570
6	136.2	2.13386	1.5	−5.9	0.04405	0.00693
7	134.9	2.13001	−1.3	−2.8	0.05189	0.00784

Next we test several possible models; η is the expected value of Y at the given x.

MODEL $\eta = \alpha + \beta x$: Not satisfactory because $\Delta Y/\Delta x$ is not constant.

MODEL $\eta = \alpha\beta^x$: Transform to $\log \eta = \log \alpha + (\log \beta)x$. Not satisfactory because $\Delta \log Y/\Delta x$ is not constant.

MODEL $\eta = \alpha x^\beta$: Transform to $\log \eta = \log \alpha + \beta \log x$. Not satisfactory because $\Delta \log Y/\Delta \log x$ is not constant.

MODEL $\eta = \alpha + \beta x + \gamma x^2$: Not satisfactory because $\Delta^2 Y/\Delta x^2$ is not constant.

MODEL $\eta = x/(\alpha + \beta x)$: The model is the same as $(x/\eta) = \alpha + \beta x$. Since $\Delta(x/Y)$ is roughly constant, this model would fit the data better than any of the previous models but it is not necessarily the best possible model.

4.2 LEAST SQUARES PARAMETER ESTIMATION

Once the functional form for an empirical model has been chosen, process data can be collected by a suitably designed experiment (discussed in Chapter 8) and the parameters in the model can be estimated. The procedure of estimation to be used in this chapter is called linear estimation or regression analysis.† The analyst wants to obtain the "best" estimates in some sense of the parameters in the model, and often the criterion of "best" depends upon the character of the model. We shall first mention estimation of the parameters of a linear (in the parameters) model when the probability density function and its parameters are known, and then we shall describe estimation when the probability density function is unknown.

Optimal estimation is difficult to carry out except in certain special cases. If we want to calculate the parameters in a model in a given relation $Y = f(X)$ where both X and Y are random variables and the joint probability density $p(x, y)$ is known, the most acceptable criterion of best is the mean square estimate:

Minimize

$$\mathscr{E}\{[Y - f(X)]^2\} = \int_{-\infty}^{\infty} \int_{-\infty}^{\infty} [y - f(x)]^2 p(x, y)\, dx\, dy \quad (4.2\text{-}1)$$

The function $f(X)$ that minimizes the expectation in Equation 4.2-1 is the conditional expected value of Y, given X, i.e., $f(X) = \mathscr{E}\{Y \mid X\}$, as can be shown by introducing the relation $p(x, y) = p(y \mid x)p(x)$ into the right-hand side of Equation 4.2-1 so that

$$\text{Min}\, \mathscr{E}\{[Y - f(X)]^2\}$$

$$= \text{Min}\left[\int_{-\infty}^{\infty} [y - f(x)]^2 p(y \mid x)\right] \int_{-\infty}^{\infty} p(x) dx$$

The first integral is the second moment of the conditional density $p(y \mid x)$ which will be a minimum for every x if $f(x) = \int_{-\infty}^{\infty} y p(y \mid x)\, dy = \mathscr{E}\{Y \mid x\}$. The function $f(x) = \mathscr{E}\{Y \mid x\}$ is the *regression curve*.

For the special case in which $f(X) = \beta_0 + \beta_1 X$, we can calculate the two constants β_0 and β_1, which minimize the function in Equation 4.2-1, if we know $p(x, y)$:

$$\text{Min}\, \mathscr{E}\{[Y - (\beta_0 + \beta_1 X)]^2\}$$

$$= \text{Min} \int_{-\infty}^{\infty} \int_{-\infty}^{\infty} (y - \beta_0 - \beta_1 x)^2 p(x, y)\, dx\, dy \quad (4.2\text{-}2)$$

By differentiation of $\mathscr{E}\{[Y - (\beta_0 + \beta_1 X)]^2\}$ with respect to β_0 and equation of the result to zero, we get

$$\mathscr{E}\{-2[Y - \beta_0 - \beta_1 X]\} = 0$$

or

$$\mathscr{E}\{Y\} = \beta_0 + \beta_1 \mathscr{E}\{X\}$$

$$\mu_Y = \beta_0 + \beta_1 \mu_X \quad (4.2\text{-}3)$$

† The latter is used because the first published investigations dealt with the regression of inherited factors.

$$\frac{1}{y} = \alpha + \beta x$$

A. $\frac{1}{y} = -0.1 + 0.3x$

B. $\frac{1}{y} = 0.1 + 0.3x$

C. $\frac{1}{y} = -0.5 + 0.3x$

D. $\frac{1}{y} = 0.5 + 0.3x$

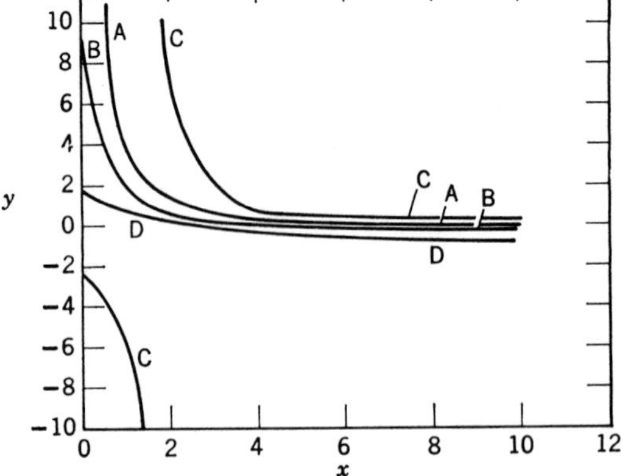

Equation (1)

$$y = \alpha + \frac{\beta}{x}$$

A. $y = -0.1 + \frac{0.3}{x}$

B. $y = 2 + \frac{0.3}{x}$

C. $y = 4 + \frac{0.3}{x}$

D. $y = 6 + \frac{0.3}{x}$

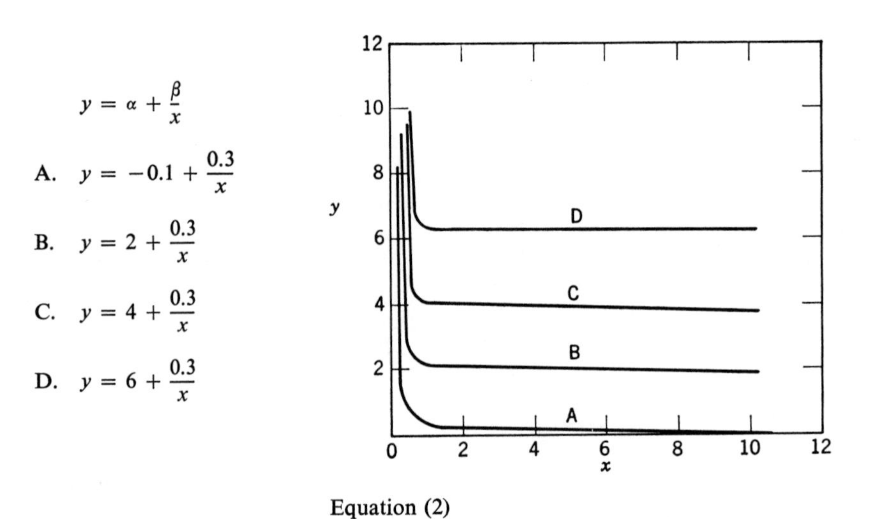

Equation (2)

$$\frac{x}{y} = \alpha + \beta x$$

A. $\frac{x}{y} = -0.1 + 0.3x$

B. $\frac{x}{y} = 0.1 + 0.3x$

C. $\frac{x}{y} = -0.4 + 0.3x$

D. $\frac{x}{y} = 4 + 0.3x$

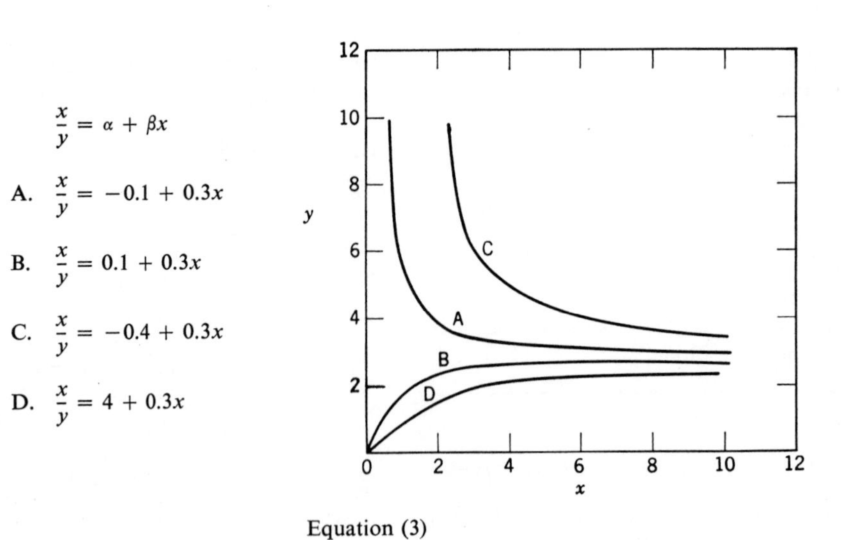

Equation (3)

FIGURE 4.1-1 Graphs of Equations (1) through (5), Table 4.1-1.

$y = \alpha x^{\beta}$

A. $y = 4x^{0.5}$

B. $y = 4x^{0.3}$

C. $y = 4x^{-0.3}$

D. $y = 4x^{-0.5}$

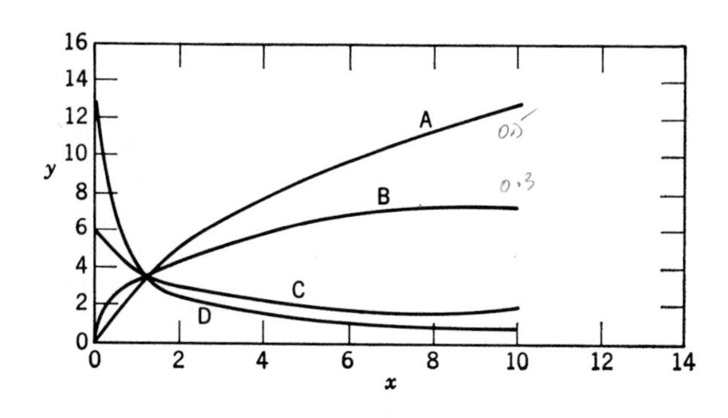

Equation (4)

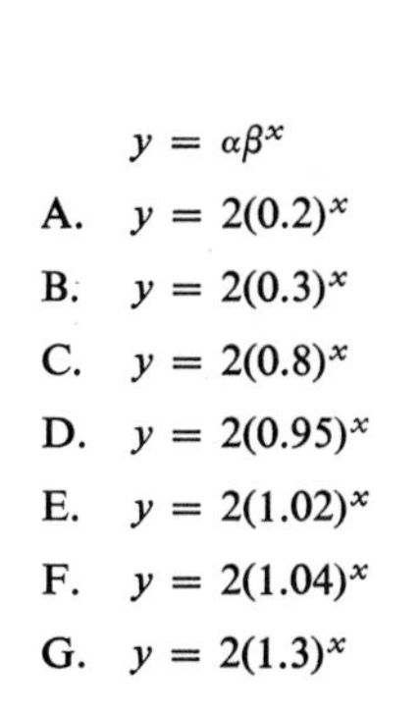

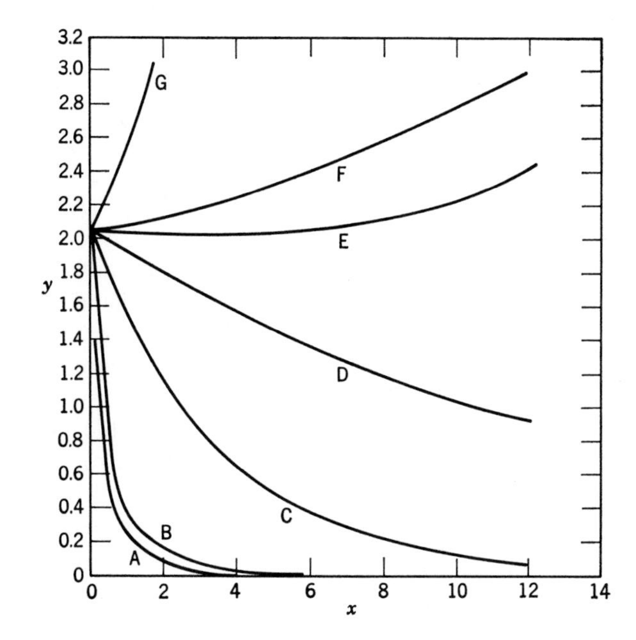

Equation (5)

FIGURE 4.1-1 (cont.)

Introduction of β_0 from Equation 4.2-3 into Equation 4.2-2, differentiation with respect to β_1, and equation of the result to zero yield

$$\frac{\partial \mathscr{E}[(Y - \mu_Y) - \beta_1(X - \mu_X)]^2\}}{\partial \beta_1} = 0$$

$$\mathscr{E}\{(Y - \mu_Y)(X - \mu_X)\} = \beta_1 \mathscr{E}\{(X - \mu_X)^2\}$$

or

$$\beta_1 = \frac{\sigma_{XY}}{\sigma_X^2} \tag{4.2-4}$$

Once β_1 is determined, β_0 can be determined from Equation 4.2-3.

Now suppose that the probability density $p(x, y)$ is not known. Instead we plan to collect some experimental data and, on the basis of the data, obtain the best estimates for the parameters in a linear model. Whether only the dependent variable in the model or whether both the independent and dependent variables are random variables makes a vast difference in the computational details and degree of difficulty of obtaining the parameter estimates. Section 4.5 discusses estimation when both the independent and dependent variables are random variables, as illustrated in Figure 4.2-1*b*.

We shall start with the easiest case, namely that only the dependent variable is a random variable, as shown in Figure 4.2-1*a*. Specifically the model is of the form

$$\bar{Y}_i = \beta_0 + \beta_1(x_i - \bar{x}) + \epsilon_i \tag{4.2-5}$$

where $\bar{Y}_i$ is the sample average of the measured replicate values of the dependent variable Y obtained at a given value of x, x_i; and ϵ_i is the unobservable random error representing the difference $(\bar{Y}_i - \mathscr{E}\{\bar{Y}_i \mid x\}) = \epsilon_i$ which has a known distribution (usually the normal distribution)

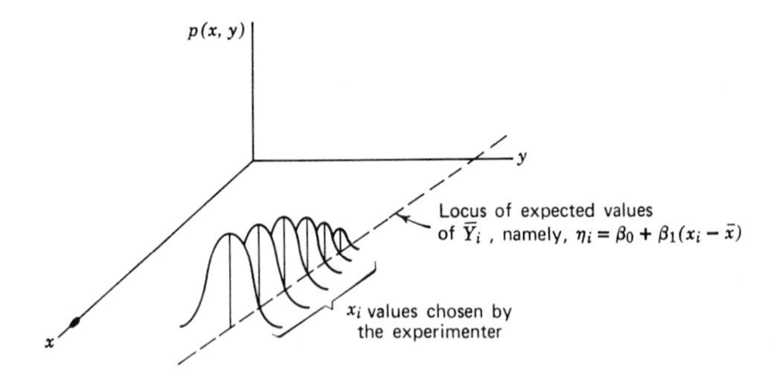

FIGURE 4.2-1*a* Representation of a linear model for the case in which the dependent variable alone is a random variable.

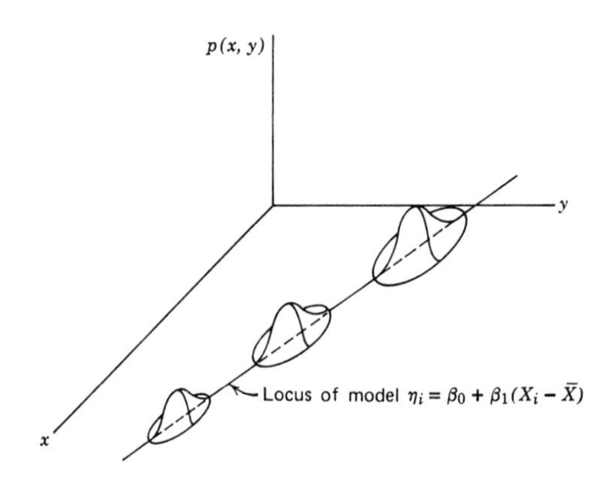

FIGURE 4.2-1*b* Representation of a linear model when both variables are random variables.

with an expected value of zero and a variance of $\sigma^2_{\epsilon_i}$. The unobservable errors in the observations used to compute $\bar{Y}_i$ are presumed to have an expected value of zero and to be independent of x, ϵ, or previous errors. We shall assume initially that $\sigma^2_{\epsilon_i}$ is a constant, independent of x (in Section 4.3), and subsequently that $\sigma^2_{\epsilon_i}$ varies with x (in Section 4.4). Another way to describe the same model is to say that

$$\mathscr{E}\{\bar{Y}_i \mid x_i\} = \eta_i = \beta_0 + \beta_1(x_i - \bar{x}) \qquad (4.2\text{-}6)$$

which states that the expected value of $\bar{Y}_i$, at a specified x_i, is equal to $\beta_0 + \beta_1(x_i - \bar{x})$.

Thus, the four basic assumptions underlying the estimation procedure are:

1. The expected value of $\bar{Y}_i$, given x_i, is a linear (in the parameters) function.
2. The values of x selected for experimentation are not random variables.
3. The variance of ϵ_i, $\sigma^2_{\epsilon_i}$, equals the variance of $\bar{Y}_i$, $\sigma^2_{\bar{Y}_i}$, and may be a constant or vary with x.
4. The observations of Y are mutually independent, which is the same as saying that the errors ϵ_i are statistically independent.

Based on these assumptions only, the method of least squares yields unbiased estimators, b_0 and b_1,† of β_0 and β_1 which have, according to Markov's theorem, the smallest variances among the group of all possible unbiased linear estimators. Least squares is the descriptive term for the procedure which obtains the estimates b_0 and b_1 by minimizing the sum of the squares of the deviations between the observed values, $\bar{Y}_i$, and the expected values of $\bar{Y}_i$, η_i:

$$\text{Minimize} \sum_{i=1}^{n} (\bar{Y}_i - \eta_i)^2$$

From one viewpoint, least squares is nothing more than a method of solving an overdetermined set of equations in the parameter space of β_0 and β_1, if each data pair is regarded as being an equation. For example,

$$16.08 - \beta_0 + 1.80\beta_1 = 0$$
$$16.32 - \beta_0 + 2.10\beta_1 = 0$$
$$16.77 - \beta_0 + 2.40\beta_1 = 0$$

The difference between the sum of the squares of the left-hand sides of the equations and zero is minimized to get the best estimates of the β's.‡

Figure 4.2-2 illustrates the estimated regression line $\hat{Y} = b_0 + b_1(x - \bar{x})$, the true model $\eta = \beta_0 + \beta_1(x - \bar{x})$, and the notation employed so far. Y_{ij} designates one (the jth) observation or measurement, $1 \le j \le p_i$, of the dependent variable Y at x_i, and $\bar{Y}_i$ is the sample mean of the observations at x_i, $1 \le i \le n$.

To estimate a confidence region for the variables β_0 and β_1 and to apply statistical tests, a fifth assumption is required, namely:

5. The conditional distribution of $\bar{Y}_i$, given x_i, is normal about $\eta_i = \mathscr{E}\{\bar{Y}_i \mid x_i\}$.

In practice, experimental data may not fulfill the five requirements. Some of the common departures from the assumptions are:

1. The range of variation of the independent variable x is so small that the variation in the dependent variable is of the order of magnitude of the error in measurement of the dependent variable. As an extreme example, repetition of the same value of x 100 times will provide only one value of $\bar{Y}_i$ for estimation, not 100 values. Because the number of data points must be at the very least equal to the number of model parameters to be estimated, variation of the independent variable within a narrow range represents ineffective experimentation.

† Although the estimates of β_0 and β_1, $\hat{\beta}_0 = b_0$ and $\hat{\beta}_1 = b_1$, are themselves **random variables**, we shall use lower case Roman letters to designate the estimates because of custom.
‡ W. E. Smith, *Technometrics* **8**, 675, 1966.

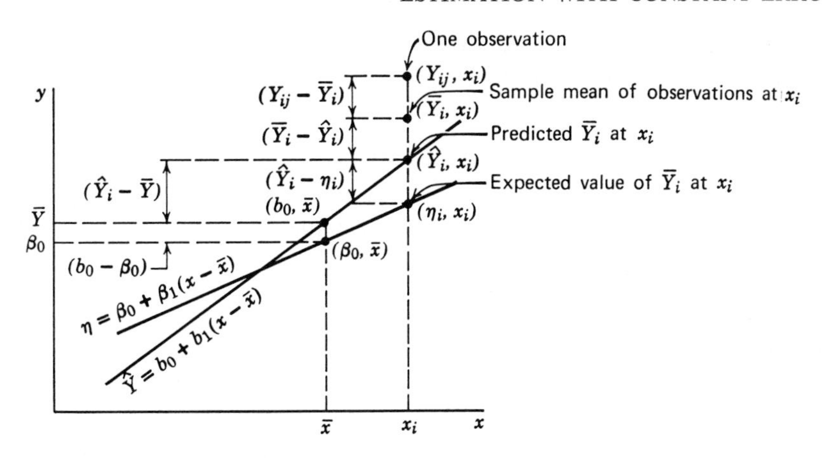

FIGURE 4.2-2 Relationships among the experimental observations, mean of observations, and estimated and theoretical linear models.

For a model containing several independent variables, the investigator will find that if he holds certain of the important process independent variables essentially constant, regression analysis will lead to the conclusion that they are not significant variables. Furthermore, independent variables that are not controlled but are simply observed are likely to behave as random variables. The essence of experimentation is, insofar as possible, to make definite changes in the experimental conditions (the independent variables) and to let the dependent variable be the random variable.

2. The errors in the observations of the dependent variable are not independent. Process measurements taken in time can incorporate serial correlation of errors, a correlation which perhaps changes with time. Because every production process is affected by independent variables not subject to control by the experimenter, such as aging of the plant, scaling in a heat exchanger, uncontrollable change in raw materials, meteorological changes, and personnel changes, assumption four above may prove unrealistic. Sometimes these noncontrollable variables are termed *latent variables.*

Consequently, a passive collection of unplanned data from a process must be analyzed with considerable judgment. The best experimental technique is to make deliberate changes in all the controllable variables, as described in Chapter 8. The investigator, before carrying out the least squares estimation, should make sure insofar as possible that he has information on the interval of variation of x relative to the possible overall range of variation, the magnitude of the errors in the independent and dependent variable(s), and the details of possible extraneous factors. Techniques to assist in overcoming deficiencies in the underlying assumptions in the estimation procedure are discussed in Sections 4.5, 4.6, and 5.4. Proper methods of experimental design to avoid the defects in the first place are described in Chapter 8.

4.3 ESTIMATION WITH CONSTANT ERROR VARIANCE

We shall write the empirical model whose coefficients are to be estimated as

$$\eta = \beta_0 + \beta_1(x - \bar{x}) \tag{4.3-1}$$

instead of as

$$\eta = \beta_0' + \beta_1 x \tag{4.3-2}$$

because, first, the estimates b_0 and b_1 of β_0 and β_1 can be obtained without solving coupled simultaneous sets of equations, as is the case if the linear model is expressed in the form of Equation 4.3-2, and, second, the estimates of β_0 and β_1 are stochastically independent whereas the estimates of β_0' and β_1 are not. Models with several independent variables yield better conditioned matrices if the form of Equation 4.3-1 is used (refer to Chapter 5). We seek estimates for β_0 and β_1 which are unbiased and have minimum variance. We assume that $\sigma^2_{\bar{Y}_i} = \sigma^2_{\epsilon_i}$ = constant.

4.3-1 Least Squares Estimates

Legendre's method of least squares was to minimize the sum of the squares of the deviations in the y direction in Figure 4.2-2. Gauss's and Laplace's development, on the other hand, minimized the sum of the squares of the weighted deviations (described in Section 4.4). In this section we shall minimize

$$\phi = \sum_{i=1}^{n} (\bar{Y}_i - \eta_i)^2 p_i$$

$$= \sum_{i=1}^{n} p_i[\bar{Y}_i - \beta_0 - \beta_1(x - \bar{x})]^2 \tag{4.3-3}$$

where p_i is the number of replicate measurement of the dependent variable for a given x_i, by equating the partial derivatives of ϕ with respect to β_0 and with respect to β_1 equal to zero. (It is not difficult to show that this procedure yields a minimum rather than a maximum for ϕ by examining the second partial derivatives of ϕ.)

$$\frac{\partial\phi}{\partial\beta_0} = 0 = \frac{\partial\left\{\sum_{i=1}^{n} p_i[\bar{Y}_i - \beta_0 - \beta_1(x_i - \bar{x})]^2\right\}}{\partial\beta_0}$$

$$= -2\sum_{i=1}^{n} p_i[\bar{Y}_i - \beta_0 - \beta_1(x_i - \bar{x})] \tag{4.3-4}$$

$$\frac{\partial\phi}{\partial\beta_1} = 0 = \frac{\partial\left\{\sum_{i=1}^{n} p_i[\bar{Y}_i - \beta_0 - \beta_1(x_i - \bar{x})]^2\right\}}{\partial\beta_1}$$

$$= -2\sum_{i=1}^{n} p_i[\bar{Y}_i - \beta_0 - \beta_1(x_i - \bar{x})](x_i - \bar{x})$$

Collecting terms, we obtain the *normal equations* in which the model parameters β_0 and β_1 have been replaced by their estimates:

$$\sum_{i=1}^{n} p_i\bar{Y}_i = b_0\sum_{i=1}^{n} p_i + b_1\sum_{i=1}^{n} p_i(x_i - \bar{x}) \tag{4.3-5a}$$

$$\sum_{i=1}^{n} p_i(\bar{Y}_i)(x_i - \bar{x}) = b_0\sum_{i=1}^{n} p_i(x_i - \bar{x}) + b_1\sum_{i=1}^{n} p_i(x_i - \bar{x})^2 \tag{4.3-5b}$$

Note that

$$\sum_{i=1}^{n} p_i(x_i - \bar{x}) \equiv 0$$

Hence, as mentioned earlier, Equation 4.3-5a can be solved for b_0 separately from Equation 4.3-5b, and Equation 4.3-5b can be solved separately for b_1:

$$\hat{\beta}_0 \equiv b_0 = \frac{\sum_{i=1}^{n} p_i\bar{Y}_i}{\sum_{i=1}^{n} p_i} = \bar{Y} \tag{4.3-6}$$

$$\hat{\beta}_1 \equiv b_1 = \frac{\sum_{i=1}^{n} p_i\bar{Y}_i(x_i - \bar{x})}{\sum_{i=1}^{n} p_i(x_i - \bar{x})^2} \tag{4.3-7}$$

4.3-2 Maximum Likelihood Estimates

Exactly the same estimates of β_0 and β_1 can be obtained by the method of maximum likelihood, if assumption five of Section 4.2 is added to the other assumptions at the beginning. We form the likelihood function described in Section 3.2-1 based on the probability density function

$$p(y \mid x; \beta_0, \beta_1, \sigma^2_{\bar{Y}_i}) = \frac{1}{\sqrt{2\pi}\,\sigma_{\bar{Y}_i}} \exp\left[-\frac{1}{2\sigma^2_{\bar{Y}_i}}(Y - \eta)^2 p_i\right]$$

$$L(\beta_0, \beta_1, \sigma^2_{\bar{Y}_i} \mid y, x) \equiv L$$

$$= \prod_{i=1}^{n} \frac{1}{\sqrt{2\pi}\,\sigma_{\bar{Y}_i}} \exp\left[-\frac{1}{2\sigma^2_{\bar{Y}_i}}(\bar{Y}_i - \eta_i)^2 p_i\right] \tag{4.3-8}$$

In Equation 4.3-8 the parameters are the variables, and the values of Y and x are given. Then

$$\ln L = -n \ln \sqrt{2\pi} - \frac{n}{2}\ln \sigma^2_{\bar{Y}_i} - \frac{\sum_{i=1}^{n} p_i\{\bar{Y}_i - [\beta_0 + \beta_1(x_i - \bar{x})]^2\}}{2\sigma^2_{\bar{Y}_i}}$$

To obtain the maximum likelihood estimates, we place

$$\frac{\partial \ln L}{\partial\beta_0} = \frac{\partial \ln L}{\partial\beta_1} = \frac{\partial \ln L}{\partial(\sigma^2_{\bar{Y}_i})} = 0$$

and obtain three equations:

$$\sum_{i=1}^{n} [\{\bar{Y}_i - [\hat{\beta}_0 + \hat{\beta}_1(x_i - \bar{x})]\}\{p_i\}] = 0 \tag{4.3-9a}$$

$$\sum_{i=1}^{n} [\{\bar{Y}_i - [\hat{\beta}_0 + \hat{\beta}_1(x_i - \bar{x})]\}\{p_i(x_i - \bar{x})\}] = 0 \tag{4.3-9b}$$

$$\sum_{i=1}^{n} p_i\{\bar{Y}_i - [\hat{\beta}_0 + \hat{\beta}_1(x_i - \bar{x})]\}^2 - n\hat{\sigma}^2_{\bar{Y}_i} = 0 \tag{4.2-9c}$$

of which the first two are the same as Equations 4.3-5a and 4.3-5b, respectively, and yield Equations 4.3-6 and 4.3-7 for b_0 and b_1. Equation 4.3-9c yields a biased estimate of $\sigma^2_{\bar{Y}_i}$

$$\hat{\sigma}^2_{\bar{Y}_i} = \frac{1}{n}\sum_{i=1}^{n} p_i\{\bar{Y}_i - [b_0 + b_1(x_i - \bar{x})]\}^2$$

$$= \frac{1}{n}\sum_{i=1}^{n} p_i(\bar{Y}_i - \hat{Y}_i)^2$$

as we shall see.

4.3-3 Expected Values and Variances of Estimators and Analysis of Variance

The probability distribution functions for b_0 and b_1 can be obtained either from the addition theorem for the normal distribution or from the partition theorem for the χ^2 distribution. However, we shall omit the details, which can be found in books on statistics. Because b_0 and b_1 are linear combinations of $\bar{Y}_i$, we can conclude that they will each have approximately normal distributions. We are interested first in finding the expected value and variance of b_0 and b_1, because these will be needed to carry out appropriate analyses used in model building.

The expected values of b_0 and b_1 are, respectively (each sum is from $i = 1$ to n and $\sum p_i(x_i - \bar{x}) \equiv 0$),

$$\mathscr{E}\{b_0\} = \mathscr{E}\left\{\frac{\sum p_i \bar{Y}_i}{\sum p_i}\right\} = \frac{\sum p_i \mathscr{E}\{\bar{Y}_i\}}{\sum p_i}$$

$$= \frac{\sum p_i \mathscr{E}\{\beta_0 + \beta_1(x_i - \bar{x}) + \epsilon_i\}}{\sum p_i}$$

$$= \frac{\sum p_i \beta_0}{\sum p_i} = \beta_0$$

$$\mathscr{E}\{b_1\} = \mathscr{E}\left\{\frac{\sum p_i \bar{Y}_i(x_i - \bar{x})}{\sum p_i(x_i - \bar{x})^2}\right\} = \frac{\sum p_i(x_i - \bar{x})\mathscr{E}\{\bar{Y}_i\}}{\sum p_i(x_i - \bar{x})^2}$$

$$= \frac{\sum p_i(x_i - \bar{x})[\beta_0 + \beta_1(x_i - \bar{x})]}{\sum p_i(x_i - \bar{x})^2} = \beta_1$$

Consequently, the estimates b_0 and b_1 are unbiased. A similar analysis gives the variances of b_0 and b_1, respectively ($\sigma^2_{\bar{Y}_i}$ is a constant here), as

$$\text{Var}\{b_0\} \equiv \mathscr{E}\{(b_0 - \beta_0)^2\}$$

$$\text{Var}\left\{\frac{\sum p_i \bar{Y}_i}{\sum p_i}\right\} = \frac{(\sum p_i)\,\text{Var}\{\bar{Y}_i\}}{(\sum p_i)^2}$$

$$= \frac{(\sum p_i)\sigma^2_{\bar{Y}_i}}{(\sum p_i)^2} = \frac{\sigma^2_{\bar{Y}_i}}{\sum p_i} \qquad (4.3\text{-}10)$$

$$\text{Var}\{b_1\} \equiv \mathscr{E}\{(b_1 - \beta_1)^2\} = \text{Var}\left\{\frac{\sum p_i \bar{Y}_i(x_i - \bar{x})}{\sum p_i(x_i - \bar{x})^2}\right\}$$

$$= \frac{\sigma^2_{\bar{Y}_i}}{\sum p_i(x_i - \bar{x})^2} \qquad (4.3\text{-}11)$$

A model which is a line through the origin, $\eta = \beta x$, can be treated as a special case of the general development outlined above. The estimate of the slope can be shown to be

$$\hat{\beta} = b = \frac{\sum_{i=1}^{n} p_i \bar{Y}_i x_i}{\sum_{i=1}^{n} p_i (x_i)^2} \qquad (4.3\text{-}7a)$$

and the variance of the estimated slope can be shown to be

$$\text{Var}\{b\} = \frac{\sigma^2_{\bar{Y}_i}}{\sum_{i=1}^{n} p_i(x_i)^2} \qquad (4.3\text{-}11a)$$

All that remains is to find an unbiased estimate of $\sigma^2_{\bar{Y}_i}$, which can be obtained with the aid of the following theorem (the χ^2 partition theorem):

If the sum of squares of n variables, $W_1, W_2, \ldots, W_n$, is partitioned into k sums of squares, $S_1, S_2, \ldots, S_k$, with $\nu_1, \nu_2, \ldots, \nu_k$ degrees of freedom, respectively, then

$$\chi^2 = \sum_{i=1}^{n} W_i^2 = S_1 + S_2 + \cdots + S_k$$

Also, the necessary and sufficient conditions that $S_1, S_2, \ldots, S_k$ are stochastically independent and each distributed as χ^2 with $\nu_1, \nu_2, \ldots, \nu_k$ degrees of freedom, respectively, are that

$$\nu_1 + \nu_2 + \cdots + \nu_k = n$$

The proof of the theorem can be found in several references at the end of this chapter.

The partitioning of interest is carried out as follows. If both sides of the identity

$$(Y_{ij} - \eta_i) = (Y_{ij} - \bar{Y}_i) + (\bar{Y}_i - \hat{Y}_i) + (\hat{Y}_i - \eta_i)$$

$$= (Y_{ij} - \bar{Y}_i) + (\bar{Y}_i - \hat{Y}_i) + (b_0 - \beta_0) + (b_1 - \beta_1)(x_i - \bar{x})$$

are squared and summed over i and j, the crossproduct terms are easily shown to be zero, either because of the constraints imposed by the least square minimization, Equations 4.3-4, or because the sum on j vanishes. For example, the crossproduct term

$$\sum_{i=1}^{n} \sum_{j=1}^{p_i} (Y_{ij} - \bar{Y}_i)(\bar{Y}_i - \hat{Y}_i) = 0$$

because of the initial summation on j,

$$\sum_{j=1}^{p_i} (Y_{ij} - \bar{Y}_i) = 0$$

Crossproduct terms such as

$$\sum_{i=1}^{n} (\bar{Y}_i - \hat{Y}_i)(b_1 - \beta_1)(x_i - \bar{x})p_i = 0$$

because of the second of Equations 4.3-4. After dropping the crossproduct terms, the following is obtained for the sum of squares:

$$\sum_{i=1}^{n} \sum_{j=1}^{p_i} (Y_{ij} - \eta_i)^2$$

Total sum of squares between the experimental data points and the expected value of Y given x

$$= \sum_{i=1}^{n} \sum_{j=1}^{p_i} (Y_{ij} - \bar{Y}_i)^2 + \sum_{i=1}^{n} p_i(\bar{Y}_i - \hat{Y}_i)^2$$

Sum of squares of deviations within data sets; "error sum of squares" — Sum of squares of deviations about empirical regression line; "residual sum of squares"

$$+ (b_0 - \beta_0)^2 \sum_{i=1}^{n} p_i + (b_1 - \beta_1)^2 \sum_{i=1}^{n} p_i(x_i - \bar{x})^2$$

Sum of squares for deviation between b_0 and β_0 — Sum of squares for deviation between b_1 and β_1

(4.3-12)

The interpretation of each of the sums of squares in Equation 4.3-12 can be carried out best by examining Figure 4.2-2. The first term on the right-hand side of the equality sign is a measure of the experimental error obtained in each of the separate experiments conducted at the various values of x; the second term is a measure of the success of the linear model in fitting the experimental data. The left-hand side of Equation 4.3-12 is a sum of squares analogous to Equation 2.3-9 with $\sum_{i=1}^{n} p_i$ degrees of freedom and is distributed as $\sigma_{Y_i}^2\chi^2$. Each term on the right-hand side of Equation 4.3-12 can be shown to be distributed as $\sigma_{Y_i}^2\chi^2$ with $(\sum_{i=1}^{n} p_i - n)$, $(n-2)$, 1, and 1 degree of freedom, respectively.

The error sum of squares has n constraints imposed, one for each $\bar{Y}_i$ that is calculated. The residual sum of squares has two constraints imposed on the n data points, one for each of Equations 4.3-3, leaving two degrees of freedom to be divided among the remaining two sums of squares, or one each since each has a single variable b_0 and b_1, respectively. It can also be concluded that b_0 is a random variable distributed normally about β_0, b_1 is a random variable distributed normally about β_1, and b_0 is stochastically independent of b_1.

If we estimate $\sigma_{Y_i}^2$ from the second term on the right-hand side of Equation 4.3-12, which represents the sum of the squares of the residuals

$$s_r^2 = \frac{1}{n-2}\sum_{i=1}^{n} p_i(\bar{Y}_i - \hat{Y}_i)^2 \qquad (4.3\text{-}13)$$

it is easy to show that $\mathscr{E}\{s_r^2\}$ is an unbiased estimate of $\sigma_{Y_i}^2$, *if the model is correct*, when we recall from Section 2.3-2 that $\mathscr{E}\{\chi^2 \text{ (for d.f.} = n)\} = n$:

$$\mathscr{E}\left\{\frac{1}{n-2}\sum_{i=1}^{n} p_i(\bar{Y}_i - \hat{Y}_i)^2\right\}$$

$$= \frac{1}{n-2}\mathscr{E}\{\sigma_{Y_i}^2\chi^2 \text{ (for d.f.} = n-2)\}$$

$$= \frac{\sigma_{Y_i}^2}{n-2}\mathscr{E}\{\chi^2 \text{ (for d.f.} = n-2)\} = \sigma_{Y_i}^2$$

(Note that the maximum likelihood estimate of $\sigma_{Y_i}^2$ proved to be a biased estimate of $\sigma_{Y_i}^2$.) If the linear model is *not* correct, then the expected value of s_r^2 is not $\sigma_{Y_i}^2$, i.e., s_r^2 is a biased estimate of $\sigma_{Y_i}^2$.

The expected value of

$$s_e^2 = \frac{\sum_{i=1}^{n}\sum_{j=1}^{p_i}(Y_{ij} - \bar{Y}_i)^2}{\sum_{i=1}^{n} p_i - n} \qquad (4.3\text{-}14)$$

is also an unbiased estimate of $\sigma_{Y_i}^2$; hence s_e^2 can also be used to estimate $\sigma_{Y_i}^2$; s_e^2 is a measure of the dispersion caused by experimental error, in contrast with the lack of fit represented by Equation 4.3-13. Consequently, before reaching any decisions about the model, the analyst should test the hypothesis that the linear model $\eta = \beta_0 + \beta_1(x - \bar{x})$ represents the experimental data satisfactorily by forming the variance ratio (s_r^2/s_e^2). (Refer to Section 3.6.) If

$$\frac{s_r^2}{s_e^2} > F_{1-\alpha}$$

where $F_{1-\alpha}$ is taken from the appropriate table for a selected value of α, the hypothesis that the linear model is adequate is rejected. Another model should be selected.

If the calculated variance ratio is less than $F_{1-\alpha}$, the hypothesis that the linear model is an adequate fit is accepted (is a plausible model but not necessarily the correct one). In this case the variances s_r^2 and s_e^2, since they both estimate $\sigma_{Y_i}^2$, can be pooled as follows to get a better estimate of $\sigma_{Y_i}^2$ with $(\sum_{i=1}^{n} p_i - 2)$ degrees of freedom. In the pooling, each variance is weighted by its respective number of degrees of freedom, as indicated earlier by Equation 2.4-12:

$$s_{\bar{Y}_i}^2 = \frac{\sum_{i=1}^{n}\sum_{j=1}^{p_i}(Y_{ij} - \bar{Y}_i)^2 + \sum_{i=1}^{n} p_i(\bar{Y}_i - \hat{Y}_i)^2}{\left(\sum_{i=1}^{n} p_i - n\right) + (n-2)}$$

$$= \frac{\sum_{i=1}^{n}\sum_{j=1}^{p_i}(Y_{ij} - \hat{Y}_i)^2}{\sum_{i=1}^{n} p_i - 2} \qquad (4.3\text{-}15)$$

Of course, if replicate values of Y_{ij} at values of x_i are not available, then $\sigma_{Y_i}^2$ must be estimated solely from s_r^2, with the result that $s_{\bar{Y}_i}^2$ is inflated if the model is an improper one. Without replicate data the F test for the hypothesis of linearity cannot be carried out, but the data can be plotted and examined visually. A test for the hypothesis that $\beta_1 = 0$ can be carried out, as will be described shortly.

Table 4.3-1 summarizes the various sums of squares and their respective degrees of freedom which are used in what is termed the *analysis of variance*, an analysis based on the partition theorem for χ^2 and the variance ratio (F) test. The sums of squares divided by their respective degrees of freedom are termed the *mean squares*. Each variance in Table 4.3-1 can be used as an estimator of $\sigma_{Y_i}^2$, but because Var $\{b_0\}$ and Var $\{b_1\}$ are usually not known, the pooled $s_{\bar{Y}_i}^2$ is used as the estimator of $\sigma_{Y_i}^2$. Estimated variances of b_0 and b_1 can, in turn, be obtained from $s_{\bar{Y}_i}^2$ if the latter is substituted for $\sigma_{Y_i}^2$ in Equations 4.3-10 and 4.3-11.

When carrying out calculations by hand, the following identities may be useful:

TABLE 4.3-1 PARTITION OF VARIATION ABOUT THE MODEL $\eta = \beta_0 + \beta_1(x - \bar{x})$

Source of Variation	Sum of Squares	Degrees of Freedom	Mean Squares
1. Deviation of b_0 from β_0	$(b_0 - \beta_0)^2 \sum_{i=1}^{n} p_i$	1	$s_1^2 = \text{Var}\{b_0\} \sum_{i=1}^{n} p_i$
2. Deviation of b_1 from β_1	$(b_1 - \beta_1)^2 \sum_{i=1}^{n} p_i(x_i - \bar{x})^2$	1	$s_2^2 = \text{Var}\{b_1\} \sum_{i=1}^{n} p_i(x_i - \bar{x})^2$
3. About the regression line	$\sum_{i=1}^{n} p_i(\bar{Y}_i - \hat{Y}_i)^2$	$n - 2$	$s_r^2 = \frac{1}{n-2} \sum_{i=1}^{n} p_i(\bar{Y}_i - \hat{Y}_i)^2$
4. Within sets (error of experiment)	$\sum_{i=1}^{n} \sum_{j=1}^{p_i} (Y_{ij} - \bar{Y}_i)^2$	$\sum_{i=1}^{n} p_i - n$	$s_e^2 = \frac{\sum_{i=1}^{n} \sum_{j=1}^{p_i} (Y_{ij} - \bar{Y}_i)^2}{\sum_{i=1}^{n} p_i - n}$
Total about the expected values of $\hat{Y}_i$	$\sum_{i=1}^{n} \sum_{j=1}^{p_i} (Y_{ij} - \eta_i)^2$	$\sum_{i=1}^{n} p_i$	

$$\sum_{i=1}^{n} p_i(\hat{Y}_i - \bar{Y})^2 = b_1^2 \sum_{i=1}^{n} p_i(x_i - \bar{x})^2$$

$$\sum_{i=1}^{n} p_i(\bar{Y}_i - \hat{Y}_i)^2 = \left[\sum_{i=1}^{n} p_i \bar{Y}_i^2 - \frac{\left(\sum_{i=1}^{n} p_i \bar{Y}_i\right)^2}{\sum_{i=1}^{n} p_i} \right] - \frac{[\sum p_i(x_i - \bar{x})(\bar{Y}_i - \bar{Y})]^2}{\sum p_i(x_i - \bar{x})^2}$$

$$\sum_{i=1}^{n} p_i(x_i - \bar{x})(\bar{Y}_i - \bar{Y}) = \sum_{i=1}^{n} p_i(x_i - \bar{x})\bar{Y}_i$$

Another analysis of variance can be carried out, somewhat different from the previous one, by expanding $(Y_{ij} - \bar{Y})$ instead of $(Y_{ij} - \eta_i)$, where $\bar{Y} = \sum Y_{ij}/\sum p_i$. The term $(Y_{ij} - \bar{Y})$ can be split up as follows:

$$(Y_{ij} - \bar{Y}) = (Y_{ij} - \bar{Y}_i) + (\bar{Y}_i - \hat{Y}_i) + (\hat{Y}_i - \bar{Y})$$

As before, both sides of this expression can be summed over i and j, and the partitioning and distribution of the sums of the squares are analogous to that described earlier. Table 4.3-2 summarizes the results. The sum of squares in the second and third rows of the table are exactly the same as listed in Table 4.3-1. The sum of squares in the fourth row has associated with it the total degrees of freedom $(\sum p_i)$ less 1, the 1 being for the constraint imposed in calculating $\bar{Y}$. As a consequence, the sum of squares in the first row can have only one degree of freedom associated with it.

TABLE 4.3-2 PARTITION OF VARIATION ABOUT THE MEAN $\bar{Y}$

Source of Variation	Sum of Squares	Degrees of Freedom	Mean Squares
1. Deviation between values on the regression line and the mean (due to regression)	$\sum_{i=1}^{n} p_i(\hat{Y}_i - \bar{Y})^2$	1	$s_3^2 = \sum_{i=1}^{n} p_i(\hat{Y}_i - \bar{Y})^2$
2. Deviation about the regression line (deviation from regression)	$\sum_{i=1}^{n} p_i(\bar{Y}_i - \hat{Y}_i)^2$	$n - 2$	$s_r^2 = \frac{\sum_{i=1}^{n} p_i(\bar{Y}_i - \hat{Y}_i)^2}{n-2}$
3. Deviation within sets (residual error)	$\sum_{i=1}^{n} \sum_{j=1}^{p_i} (Y_{ij} - \bar{Y}_i)^2$	$\sum_{i=1}^{n} p_i - n$	$s_e^2 = \frac{\sum_{i=1}^{n} \sum_{i=1}^{p_i} (Y_{ij} - \bar{Y}_i)^2}{\sum_{i=1}^{n} p_i - n}$
4. Total	$\sum_{i=1}^{n} \sum_{j=1}^{p_i} (Y_{ij} - \bar{Y})^2$	$\sum_{i=1}^{n} p_i - 1$	

We can first test the hypothesis concerning linearity of the model by forming the variance ratio s_r^2/s_e^2 and employing the F-test, as explained before. If the variance ratio is not significant, the hypothesis concerning the linear form of the model is accepted. Next, we can test the hypothesis that $\beta_1 = 0$ by forming the variance ratio $s_3^2/s_{\bar{Y}_i}^2$. If the value of $s_3^2/s_{\bar{Y}_i}^2$ is greater than the value of $F_{1-\alpha}$ from the tables for a selected α, the hypothesis that $\beta_1 = 0$ is rejected. Figure 4.3-1 illustrates the situation (a) in which the experimental data are fitted by the estimated regression line significantly better than by a line of zero slope, as opposed to the situation (b) where a line of zero slope fits as well. The hypothesis that $\beta_1 = 0$ or β_1 is any other value could also be tested through use of a t-test based on Equation 4.3-20 below.

Another form for the analysis of variance that is quite useful is to split $(Y_{ij} - 0)$ as follows:

$$(Y_{ij} - 0) = (Y_{ij} - \bar{Y}_i) + (\bar{Y}_i - \hat{Y}_i) + (\hat{Y}_i - 0)$$

Again, each side of the equation is summed over i and j. The following partition for the sums of squares results:

$$\sum_{i=1}^{n}\sum_{j=1}^{p_i}(Y_{ij} - 0)^2 = \sum_{i=1}^{n}\sum_{j=1}^{p_i}(Y_{ij} - \bar{Y}_i)^2 + \sum_{i=1}^{n} p_i(\bar{Y}_i - \hat{Y}_i)^2 + \sum_{i=1}^{n} p_i(\hat{Y}_i - 0)^2$$

The first two terms on the right-hand side of the equality sign are the same as those in rows 3 and 2 of Table 4.3-2, respectively. The last term, which represents the deviations of the predicted values of Y about the axis at zero, can itself be partitioned as follows if the estimate regression equation is introduced for $\hat{Y}_i$:

$$\sum_{i=1}^{n} p_i(\hat{Y}_i - 0)^2 = b_0^2 \sum_{i=1}^{n} p_i + b_1^2 \sum_{i=1}^{n} p_i(x_i - \bar{x})^2$$

$$= \sum_{i=1}^{n} (\bar{Y} - 0)^2 p_i + \sum_{i=1}^{n} (\hat{Y}_i - \bar{Y})^2 p_i$$

Each of the two terms on the right-hand side of the last equality can be interpreted as sums of squares related to whether only β_0 or both β_0 and β_1 are included in the model. Suppose that only β_0 were included in the model and β_1 were deleted so that the model was $Y_i = \beta_0 + \epsilon_i$. Then the least squares estimate of β_0 would be $b_0 = \bar{Y}$, and

$$\sum_{i=1}^{n} p_i \hat{Y}_i^2 = \sum_{i=1}^{n} p_i b_0^2 = \bar{Y}^2 \sum_{i=1}^{n} p_i$$

Note the correspondence of this sum of squares with the first term in the partitioned sum of squares, $\sum (\bar{Y} - 0)^2 p_i$. Consequently, we conclude that the second term

$$b_1^2 \sum_{i=1}^{n} p_i(x_i - \bar{x})^2 = b_1 \sum_{i=1}^{n} p_i \bar{Y}_i(x_i - \bar{x})$$

$$= \sum_{i=1}^{n} (\hat{Y}_i - \bar{Y})^2 p_i$$

represents the contribution to the sum of squares effected by adding b_1 to a model $Y_i = \beta_0 + \epsilon_i$, one that already contains an intercept. Table 4.3-3 summarizes this third partitioning of the sum of squares. Variance ratio tests can be carried out to determine if β_1 and β_0 can be deleted from the model by forming the variance ratios $s_4^2/s_{\bar{Y}_i}^2$ and $s_5^2/s_{\bar{Y}_i}^2$. If a ratio exceeds the value of $F_{1-\alpha}$, the corresponding parameter makes a significant contribution to the model.

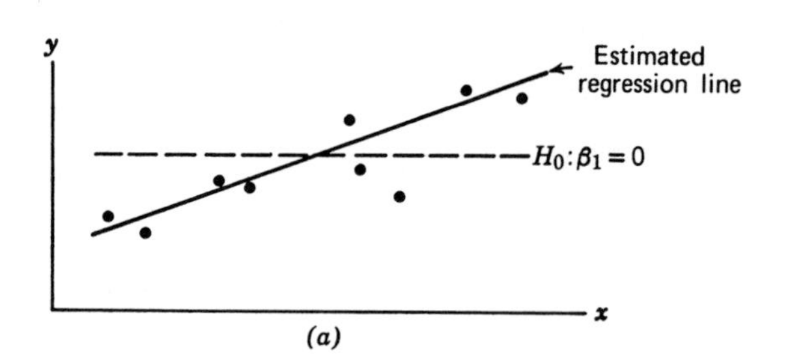

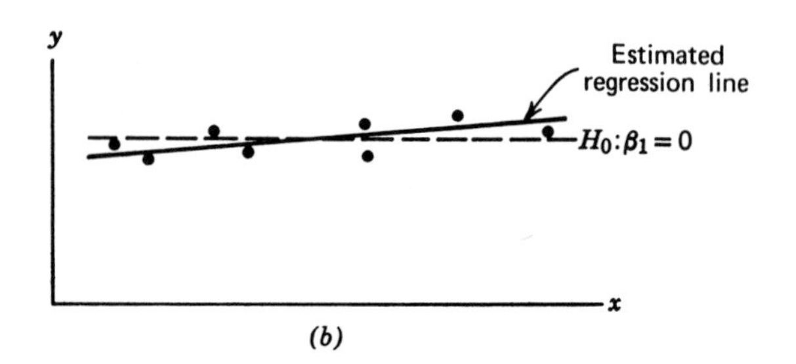

FIGURE 4.3-1 Experimental data for the test of the hypothesis $\beta_1 = 0$: (a) hypothesis rejected, and (b) hypothesis accepted.

Example 4.3-1 Estimation and Analysis of Simulated Data for a Linear Model

As an example in which the model is known, we assume that $\eta = \beta_0' + \beta_1 x$ with $\beta_0' = 10$ and $\beta_1 = 0.2$. "Observed" values or Y are simulated by adding to η errors from a table of normal random deviates with a mean of zero and a variance of 1,† as shown in Table E4.3-1a.

We want: (1) to compute the estimates b_0' and b_1 of β_0' and β_1, respectively, from the "observed" values of Y; (2) to determine the confidence intervals for β_0, β_1, and η (the latter as a function of x), respectively; (3) to plot the estimated regression line, the confidence limits about the

† Taken from M. G. Natrella, *Experimental Statistics*, NBS Handbook 91, Supt. Documents, Washington, D.C., 1963.

TABLE 4.3-3 PARTITION OF VARIATION ABOUT ZERO

Source of Variation	Sum of Squares	Degrees of Freedom	Mean Squares
1. Due to regression: b_0	$\bar{Y}^2 \sum_{i=1}^{n} p_i$	1	$s_5^2 = \bar{Y}^2 \sum_{i=1}^{n} p_i$
b_1, after allowance for b_0	$b_1 \sum_{i=1}^{n} p_i \bar{Y}_i(x_i - \bar{x})$	1	$s_4^2 = b_1 \sum_{i=1}^{n} p_i \bar{Y}_i(x_i - \bar{x})$
2. Deviation from regression	$\sum_{i=1}^{n} p_i(\bar{Y}_i - \hat{Y}_i)^2$	$n - 2$	$s_r^2 = \dfrac{\sum_{i=1}^{n} p_i(\bar{Y}_i - \hat{Y}_i)^2}{n-2}$
3. Deviation within sets (residual error)	$\sum_{i=1}^{n}\sum_{j=1}^{p_i} (Y_{ij} - \bar{Y}_i)^2$	$\sum_{i=1}^{n} p_i - n$	$s_e^2 = \dfrac{\sum_{i=1}^{n}\sum_{j=1}^{p_i} (Y_{ij} - \bar{Y}_i)^2}{\sum_{i=1}^{n} p_i - n}$
4. Total	$\sum_{i=1}^{n}\sum_{j=1}^{p_i} (Y_{ij} - 0)^2$	$\sum_{i=1}^{n} p_i$	

TABLE E4.3-1a

Data Set	x	Error ϵ	η	"Observed" Y
1	10.00	0.05	12.00	12.05
2	10.00	−0.52	12.00	11.48
3	20.00	−1.41	14.00	12.59
4	20.00	1.82	14.00	15.82
5	30.00	1.35	16.00	17.35
6	30.00	0.42	16.00	16.42
7	40.00	−1.76	18.00	16.24
8	40.00	−0.96	18.00	17.04
9	50.00	0.56	20.00	20.56
10	50.00	−0.72	20.00	19.28

line, and the sample means, $\bar{Y}_i$, at each x_i; and (4) to prepare an analysis of variance.

Solution:

Let the significance level be $\alpha = 0.05$. The calculations will be carried out in detail so that the separate steps can be followed. (See Table E4.3-16.)

TABLE E4.3-16

x_i	p_i	$\bar{Y}_i = \dfrac{\sum_{j=1}^{p_i} Y_{ij}}{p_i}$	$(x_i - \bar{x})$	$\bar{Y}_i(x_i - \bar{x})$	$(x_i - \bar{x})^2$	$\bar{Y}_i^2$
10	2	11.765	−20	−235.3	400	138.41
20	2	14.215	−10	−142.2	100	201.78
30	2	16.885	0	0	0	285.10
40	2	16.640	10	166.4	100	276.89
50	2	19.920	20	398.4	400	396.80
Total	10	79.425	—	187.3	1000	1298.98

$$b_0 = \bar{Y} = \frac{\sum \bar{Y}_i p_i}{\sum p_i} = \frac{(79.425)(2)}{10} = 15.89$$

$$b_1 = \frac{\sum p_i \bar{Y}_i(x_i - \bar{x})}{\sum p_i(x_i - \bar{x})^2} = \frac{2(187.3)}{2(1000)} = 0.1873$$

$$\hat{Y} = 15.89 + 0.1873(x - 30) = 10.26 + 0.1873x$$

$$s_r^2 = \frac{1}{n-2} \sum_{i=1}^{n} p_i(\bar{Y}_i - \hat{Y}_i)^2$$

$$= \left(\frac{1}{n-2}\right)\left\{\left[\sum_{i=1}^{n} p_i \bar{Y}_i^2 - \frac{\left(\sum_{i=1}^{n} p_i \bar{Y}_i\right)^2}{\sum_{i=1}^{n} p_i}\right] - \frac{\left[\sum_{i=1}^{n} p_i(x_i - \bar{x})(\bar{Y}_i)\right]^2}{\sum_{i=1}^{n} p_i(x_i - \bar{x})^2}\right\}$$

$$= \tfrac{1}{3}\left\{\left[2(1298.98) - \frac{[2(79.425)]^2}{10} - \frac{[2(187.3)]^2}{2(1000)}\right]\right\}$$

$$= \tfrac{1}{3}(5.02) = 1.67$$

$$s_e^2 = \frac{\sum_{i=1}^{n}\sum_{j=1}^{p_i} (Y_{ij} - \bar{Y}_i)^2}{\sum_{i=1}^{n} p_i - n} = \tfrac{1}{5}(6.95) = 1.39$$

Note that excessive roundoff error can seriously distort the numerical results unless all significant figures are retained in the calculations. For example, if the values of $\bar{Y}_i$ are rounded to the fourth significant figure instead of to the fifth, the sum of the squares of the deviations is affected in the third significant figure.

The variance ratio $s_r^2/s_e^2 = 1.20$ can be formed and the F-test utilized to see if the two mean squares are significantly different. Since for $\alpha = 0.05$, from Table C.4 in Appendix C, $F(3, 5) = 5.41$, we conclude that the mean squares are not significantly different and that the linear model $\eta = \beta_0 \ldots \beta_1 x$ adequately represents the simulated data. Next, the mean squares are pooled by using Equation 4.3-15:

$$s_{\bar{Y}_i}^2 = \frac{5.02 + 6.95}{8} = 1.50$$

From Equation 4.3-10 with the pooled $s_{\bar{Y}_i}^2$ used as the estimate of $\sigma_{\bar{Y}_i}^2$, the *estimated* variance of b_0 is

$$\widehat{\text{Var}}\{b_0\} = \frac{s_{\bar{Y}_i}^2}{\sum_{i=1}^{5} p_i} = \frac{1.50}{10} = 0.15$$

From Equation 4.3-11

$$\widehat{\text{Var}}\{b_1\} = \frac{s_{\bar{Y}_i}^2}{\sum_{i=1}^{5} p_i(x_i - \bar{x})^2} = \frac{1.50}{2000} = 7.5 \times 10^{-4}$$

Also, if $\hat{Y} = b_0' + b_1 x$, from Equation 4.3-18 we calculate

$$\widehat{\text{Var}}\{b_0'\} = 1.50\left[\frac{1}{10} + \frac{(30)^2}{2000}\right] = 0.825$$

4.3-4 Confidence Interval and Confidence Region

Because the estimated regression line $\hat{Y} = b_0 + b_1(x - \bar{x})$ is linear in the estimated coefficients, and because b_0 and b_1 are *independent random variables* so that the covariance terms vanish,

$$\text{Var}\{\hat{Y}_i\} = \text{Var}\{b_0\} + (x_i - \bar{x})^2 \text{Var}\{b_1\}$$

$$= \sigma_{\bar{Y}_i}^2\left[\frac{1}{\sum_{i=1}^{n} p_i} + \frac{(x_i - \bar{x})^2}{\sum_{i=1}^{n} p_i(x_i - \bar{x})^2}\right] \quad (4.3\text{-}16)$$

Note that the minimum variance is at $\bar{x}$.

A single new observation of Y, Y_{ij}^*, at x_i will be distributed about η_i with a variance of $\sigma_{\bar{Y}_i}^2$ independently of $\hat{Y}_i$, so that if the deviation of Y_{ij}^* from the predicted regression line $\hat{Y}_i$ is $(Y_{ij}^* - \hat{Y}_i)$, the variance of the deviation is

$$\sigma_D^2 = \text{Var}\{Y_{ij}^* - \hat{Y}_i\} = \text{Var}\{Y_{ij}^*\} + \text{Var}\{\hat{Y}_i\}$$

$$= \sigma_{\bar{Y}_i}^2\left[1 + \frac{1}{\sum_{i=1}^{n} p_i} + \frac{(x_i - \bar{x})^2}{\sum_{i=1}^{n} p_i(x_i - \bar{x})^2}\right] \quad (4.3\text{-}17)$$

If the model had been originally formulated simply as $\eta = \beta_0' + \beta_1 x$, then the estimated regression line would have been $\hat{Y} = b_0' + b_1 x = (b_0 - b_1\bar{x}) + b_1 x$, so that $b_0' = b_0 - b_1\bar{x}$. The variance of b_0' is equal to Var $\{b_0\}$ + $\bar{x}^2$ Var $\{b_1\}$ or

$$\text{Var}\{b_0'\} = \sigma_{\bar{Y}_i}^2\left[\frac{1}{\sum_{i=1}^{n} p_i} + \frac{\bar{x}^2}{\sum_{i=1}^{n} p_i(x_i - \bar{x})^2}\right] \quad (4.3\text{-}18)$$

and the Var $\{b_1\}$ remains the same as given by Equation 4.3-11.

To obtain the *confidence interval* for β_0, since b_0 is distributed normally about β_0, we can form the dimensionless Student t:

$$t = \frac{b_0 - \beta_0}{s_{b_0}} = \frac{b_0 - \beta_0}{s_{\bar{Y}_i}\Big/\left(\sum_{i=1}^{n} p_i\right)^{1/2}}; \quad \nu = \sum_{i=1}^{n} p_i - 2 \quad (4.3\text{-}19)$$

which has a t-distribution with $(\sum p_i - 2)$ degrees of freedom. The pooled estimate, $s_{\bar{Y}_i}^2$, from Equation 4.3-15 would be presumably used since the pooled estimate is a better estimate of $\sigma_{\bar{Y}_i}^2$ than is either s_e^2 or s_r^2 alone. If replicate values of Y are not available to calculate s_e^2, then s_r replaces $s_{\bar{Y}_i}^2$ in the relations below and the degrees of freedom are those associated with s_r^2, namely $(n - 2)$. $s_{\bar{Y}_i}$ is called the *standard error of estimate*. The confidence interval for β_0 is

$$b_0 - t_{1-\frac{\alpha}{2}} s_{b_0} \leq \beta_0 < b_0 + t_{1-\frac{\alpha}{2}} s_{b_0} \quad (4.3\text{-}20)$$

Similarly for β_1:

$$t = \frac{b_1 - \beta_1}{s_{b_1}} = \frac{b_1 - \beta_1}{s_{\bar{Y}_i}\Big/\left[\sum_{i=1}^{n} p_i(x_i - \bar{x})^2\right]^{1/2}}; \quad \nu = \sum p_i - 2 \quad (4.3\text{-}21)$$

$$b_1 - t_{1-\frac{\alpha}{2}} s_{b_1} \leq \beta_1 < b_1 + t_{1-\frac{\alpha}{2}} s_{b_1} \quad (4.3\text{-}22)$$

The $100(1 - \alpha)$ percent confidence interval for the expected value of $\bar{Y}_i$ given x_i, η_i, is determined similarly:

$$t = \frac{\hat{Y} - \eta}{s_{\hat{Y}}}; \quad \nu = \sum_{i=1}^{n} p_i - 2 \quad (4.3\text{-}23)$$

$$(\hat{Y} - t_{1-\frac{\alpha}{2}} s_{\hat{Y}}) \leq \eta < (\hat{Y} + t_{1-\frac{\alpha}{2}} s_{\hat{Y}}) \quad (4.3\text{-}24)$$

where $s_{\hat{Y}}^2$ is obtained by using Equation 4.3-16 with $s_{\bar{Y}_i}^2$ replacing $\sigma_{\bar{Y}_i}^2$. Finally, if one additional value of x_i were selected, say x_i^*, the confidence interval for the expected value of the additional observation Y_{ij}^* would be (using Equation 4.3-17)

$$\hat{Y}_i - t_{1-\frac{\alpha}{2}} s_D \leq \eta_i^* < \hat{Y}_i + t_{1-\frac{\alpha}{2}} s_D \quad (4.3\text{-}25)$$

To obtain the confidence interval for $\bar{Y}_m$, the mean of m observations at an additional x, replace the first number, 1, in the square brackets in Equation 4.3-17 with $1/m$ because the variance of $\bar{Y}_m$ is $\sigma_{\bar{Y}_i}^2/m$.

The interpretation of all these confidence intervals is the same as that given in Section 3.3, namely that with $100(1 - \alpha)$ percent confidence, the interval calculated includes β_0, β_1, or η, as the case may be, if the assumptions of Section 4.2 are met. The confidence limits for η given by Equation 4.3-24 can be plotted on a chart together with the experimental data, as shown in Figure 4.3-2. Note that while the estimated regression line is

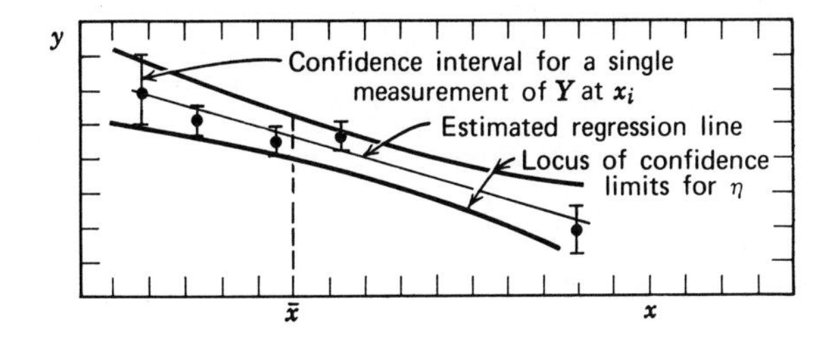

FIGURE 4.3-2 Estimated regression line $\hat{Y} = b_0 + b_1(x - \bar{x})$ with confidence interval for η.

straight, the loci of the confidence limits are curved with a minimum separation occurring at $\bar{x}$.

Many additional confidence questions can be asked that we do not have the space to discuss, such as what is the confidence interval expected if another experimenter were to repeat the experiment at the same values of x_i or, if the experiment were to be repeated at another set of values of x_i, how would the results differ. Further details can be found in the references at the end of the chapter.

So far in discussing the confidence interval for β_0, β_1, or η, we have been concerned with a single parameter. Thus, the confidence interval for β_0 is concerned with the interval that includes the intercept for models with the same slope; the confidence interval for β_1 is concerned with the interval that includes the slope for models with the same intercept. However, if we inquire as to what Model (line) could have been the source of the experimental data, taking into account the slope and intercept simultaneously, it is necessary to make an estimate of a *joint confidence region* for β_0 and β_1. The rectangular region outlined by the two estimated individual confidence intervals and the ellipse defining the jointly estimated confidence region may contain quite different values of the β's. Refer to Figure 4.3-3.

We can estimate a joint confidence region for β_0 and β_1 in the linear model $\eta = \beta_0 + \beta_1(x - \bar{x})$ as follows. We have already said that $(b_0 - \beta_0)^2 \sum_{i=1}^n p_i$ is distributed as $\sigma_{\bar{Y}_i}^2\chi^2$ with 1 degree of freedom and that $(b_1 - \beta_1)^2 \sum_{i=1}^n p_i(x_i - \bar{x})^2$ is also distributed as $\sigma_{\bar{Y}_i}^2\chi^2$ with 1 degree of freedom. Because these terms are independent, their sum is likewise distributed as $\sigma_{\bar{Y}_i}^2\chi^2$ but with 2 degrees of freedom:

$$(b_0 - \beta_0)^2 \sum_{i=1}^n p_i + (b_1 - \beta_1)^2 \sum_{i=1}^n p_i(x_i - \bar{x})^2 = \sigma_{\bar{Y}_i}^2\chi^2 \qquad \nu = 2 \tag{4.3-26}$$

Equation 4.3-26 could have included the crossproduct term $2(b_0 - \beta_0)(b_1 - \beta_1) \sum_{i=1}^n (x_i - \bar{x})$ but inasmuch as $\sum_{i=1}^n (x_i - \bar{x}) = 0$ for Model 4.3-1, this term has been omitted in Equations 4.3-26 through 4.3-28.

The expected value of the left-hand side of Equation 4.3-26 is $\sigma_{\bar{Y}_i}^2\mathscr{E}\{\chi^2(\text{d.f.} = 2)\} = 2\sigma_{\bar{Y}_i}^2$, so the expected value of one-half of the left-hand side is equal to $\sigma_{\bar{Y}_i}^2$. As before, we can form a variance ratio which has an F-distribution

$$\frac{\frac{1}{2}\left[(b_0 - \beta_0)^2 \sum_{i=1}^n p_i + (b_1 - \beta_1)^2 \sum_{i=1}^n p_i(x_i - \bar{x})^2\right]}{s_{\bar{Y}_i}^2} = F \tag{4.3-27}$$

in which the numerator has 2 and the denominator has $(\sum_{i=1}^n p_i - 2)$ degrees of freedom, respectively, if a pooled estimate of $\sigma_{\bar{Y}_i}^2$ is used. Because $P\{F \le F_{1-\alpha}\} = 1 - \alpha$ designates a critical level, we can rewrite Equation 4.3-27 as

$$(b_0 - \beta_0)^2 \sum_{i=1}^n p_i + (b_1 - \beta_1)^2 \sum_{i=1}^n p_i(x_1 - \bar{x})^2 = 2s_{\bar{Y}_i}^2F_{1-\alpha}$$
$$\nu = \left(2, \sum p_i - 2\right) \tag{4.3-28}$$

which represents an ellipse in parameter space, i.e., in the coordinates (β_0, β_1), for a given $100(1 - \alpha)$ percent joint confidence region. Equation 4.3-28 delineates the locus of the boundary of an area (that is itself a random variable) that includes the parameters β_0 and β_1 with $100(1 - \alpha)$ percent confidence. Note that the contour has been given for Model 4.3-1. If Model 4.3-2 were the one of interest instead, we know that $\beta_0' = \beta_0 - \beta_1\bar{x}$, and the critical confidence contour in (β_0', β_1) space could be computed from the critical contour in (β_0, β_1) space as given by Equation 4.3-28.

Figure 4.3-3 illustrates a confidence region for the model $\eta = \beta_0' + \beta_1 x$ in which both points A and B are within the individual confidence limits but B lies outside of the joint confidence region for $\alpha = 0.05$. The principal axes of the ellipse are at an angle to the coordinate axes β_0' and β_1 because the estimates b_0' and b_1 are correlated. Figure E4.3-1b in Example 4.3-1 illustrates an elliptical contour in (β_0, β_1) space that is not rotated.

Figure 4.3-3 illustrates only one contour, that for $\alpha = 0.05$. We can break ϕ, the sum of squares, into two

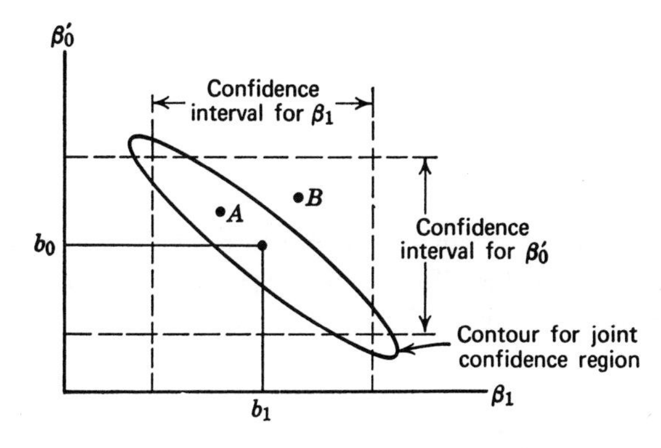

FIGURE 4.3-3 Individual confidence intervals versus joint confidence region for the model $\eta = \beta_0' + \beta_1 x$.

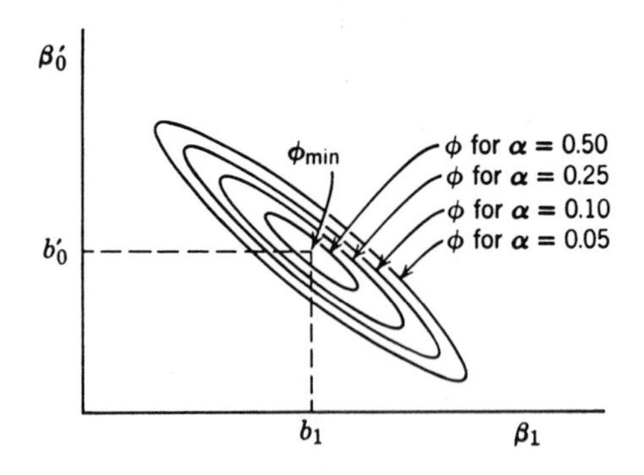

FIGURE 4.3-4 Sums of squares contours for various significance levels for the model $\eta = \beta_0' + \beta_1 x$.

parts that provide information about the character of the *sum of squares surface* (all sums are from $i = 1$ to n):

$$\phi = \sum (\bar{Y}_i - \eta_i)^2 p_i = \sum (\bar{Y}_i - \hat{Y}_i)^2 p_i + \sum (\hat{Y}_i - \eta_i)^2 p_i$$

$$= \phi_{\min} + (b_0 - \beta_0)^2 \sum p_i + (b_1 - \beta_1)^2 \sum p_i (x_i - \bar{x})^2$$

$$= \phi_{\min} + 2 s_{\bar{Y}_i}^2 F_{1-\alpha} = \phi_{\min} + 2\frac{\phi_{\min}}{n-2} F_{1-\alpha}$$

$$= \phi_{\min}\left[1 + \frac{2}{n-2} F_{1-\alpha}\right] \tag{4.3-28a}$$

In the above we have assumed that the model is correct so that $s_{\bar{Y}_i}^2$ can be replaced by s_r^2. Figure 4.3-4 illustrates several contours for various values of α; $\phi_{\min}$ is at the center and the values of ϕ increase as α becomes smaller. The contours are just projections onto the $\beta_0 - \beta_1$ plane of the quadratic surface designated ϕ.

4.3-5 The Estimated Regression Equation in Reverse

Before illustrating with examples the calculations described up to this point, we shall turn to one final topic. Once an estimated regression equation has been determined, how can a value of x, a nonstochastic variable, be predicted from an observed value of Y? This is the so-called inverse estimation problem. If we introduce into the estimated regression equation $\hat{Y} = b_0 + b_1(x - \bar{x})$ a new observed value Y^* (or the mean of several observations at the same x_i) and solve for x, we obtain

$$X(Y^*) = \frac{Y^* - b_0}{b_1} + \bar{x} \tag{4.3-29}$$

where X is a random variable because Y^*, b_0, and b_1 are random variables. Taking the expected value of both sides of Equation 4.3-29, we get

$$\mu_X^*(Y^*) = \frac{\eta - \beta_0}{\beta_1} + \bar{x} \tag{4.3-30}$$

or

$$\eta - \beta_0 - \beta_1(\mu_X^* - \bar{x}) = 0$$

Finally, if we form

$$Z = Y^* - b_0 - b_1(\mu_X^* - \bar{x})$$

the expected value of Z is

$$\mathscr{E}\{Z\} = \eta - \beta_0 - \beta_1(\mu_X^* - \bar{x})$$

and the variance of Z is

$$\operatorname{Var}\{Z\} = \operatorname{Var}\{Y^*\} + \operatorname{Var}\{b_0\} + (\mu_X^* - \bar{x})^2 \operatorname{Var}\{b_1\}$$

Hald† or Brownlee‡ shows that a t-variate can be formed:

$$t = \frac{Z - 0}{s_Z} = \frac{Y^* - b_0 - b_1(\mu_X^* - \bar{x})}{s\left[\dfrac{1}{m} + \dfrac{1}{\sum_{i=1}^{n} p_i} + \dfrac{(\mu_X^* - \bar{x})^2}{\sum_{i=1}^{n} p_i (x_i - \bar{x})^2}\right]^{1/2}}$$

$$\nu = \sum p_i - 2 \tag{4.3-31}$$

where s is an estimate of $\sqrt{\sigma_{\bar{Y}_i}^2}$. Consequently, the confidence interval for μ_X^* proves to be

$$\bar{x} + \frac{Y^* - b_0}{b_3} - t_{1-\frac{\alpha}{2}} \frac{s}{b_3}\left[\left(\frac{1}{m} + \frac{1}{\sum p_i}\right)\frac{b_3}{b_1} + \frac{[X(Y^*) - \bar{x}]^2}{\sum p_i (x_i - \bar{x})^2}\right]^{1/2} \le \mu_X^* \le \bar{x} + \frac{Y^* - b_0}{b_2} - t_{1-\frac{\alpha}{2}} \frac{s}{b_2}$$

$$\times \left[\left(\frac{1}{m} + \frac{1}{\sum p_i}\right)\frac{b_2}{b_1} + \frac{[X(Y^*) - \bar{x}]^2}{\sum p_i (x_i - \bar{x})^2}\right]^{1/2} \tag{4.3-32}$$

where

$$b_j = b_1 - \frac{t_{1-\frac{\alpha}{2}}^2 s^2}{b_1 \sum p_i (x_i - \bar{x})^2} \qquad j = 2, 3$$

Krutchkoff § called attention to Eisenhart's suggestion‖ to write the inverse of model 4.3-2 as:

$$x = \frac{\eta}{\beta_1} - \frac{\beta_0'}{\beta_1}$$

or, with $\eta = Y + \epsilon$,

$$X = \gamma + \delta Y + \epsilon' \tag{4.3-33}$$

where

$$\gamma = -\frac{\beta_0'}{\beta_1}$$

$$\delta = \frac{1}{\beta_1}$$

$$\epsilon' = -\frac{\epsilon}{\beta_1}$$

† A. Hald, *Statistical Theory with Engineering Applications*, John Wiley, New York, 1952, p. 550.

‡ K. A. Brownlee, *Statistical Theory and Methodology in Science and Engineering*, John Wiley, New York, 1960, Chapter 11.

§ R. G. Krutchkoff, *Technometrics* **9**, 425, 1967.

‖ C. Eisenhart, *Ann. Math. Stat.* **10**, 162, 1939.

The least squares estimates of γ and δ are

$$d = \hat{\delta} = \frac{\sum_{i=1}^{n} p_i(x_i - \bar{x})(\bar{Y}_i - \bar{Y})}{\sum_{i=1}^{n} p_i(\bar{Y}_i - \bar{Y})^2}$$

$$c = \hat{\gamma} = \bar{x} - d\bar{Y}$$

and thus

$$\hat{x} = c + dY^* \tag{4.3-34}$$

where $\hat{x}$ is the estimate of x given a measurement of Y, Y^*. Krutchkoff concluded that Eisenhart's inverse approach is a more satisfactory method of estimating x given a Y^*.

Example 4.3-1 (*continued*)

The confidence interval for β_0' is calculated from Equation 4.3-18 and the equation for b_0' analogous to Inequality 4.3-20 ($t = 2.306$ for the pooled variance with $3 + 5 = 8$ degrees of freedom) is

$$10.26 - (2.306)(0.825)^{1/2} \leq \beta_0' < 10.26 + (2.306)(0.825)^{1/2}$$
$$8.16 \leq \beta_0' < 12.36$$

Similarly, from Inequality 4.3-20 the confidence interval for β_0 is

$$14.99 \leq \beta_0 < 16.77$$

The confidence interval for β_1 from Inequality 4.3-22 is

$$0.1874 - 2.306(7.5 \times 10^{-4})^{1/2} \leq \beta_1 < 0.1874 + 2.306(7.5 \times 10^{-4})^{1/2}$$
$$0.124 \leq \beta_1 < 0.251$$

Note that the true values of β_0' and β_1 (which we know here) fall within the confidence interval. Finally, the confidence interval for η is, from Inequality 4.3-24

$$\hat{Y} - 2.306 s_{\hat{Y}} \leq \eta < \hat{Y} + 2.306 s_{\hat{Y}}$$

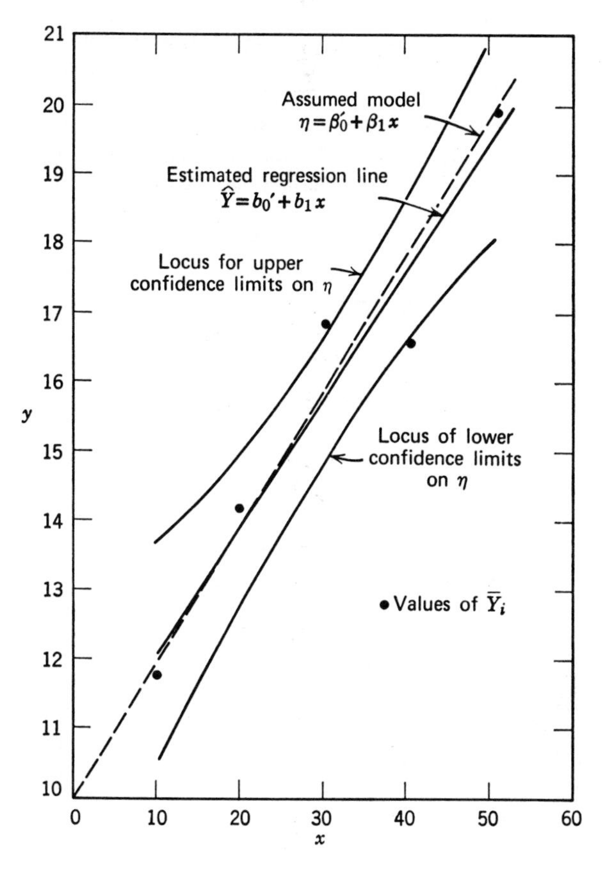

FIGURE E4.3-1A The model, the estimated regression line, and the values of $\bar{Y}_i$.

We can calculate selected values of the confidence limits to be used in plotting. At:

$$x = 10: \quad 12.13 - 1.55 \leq \eta < 12.13 + 1.55$$
$$x = 20: \quad 14.01 - 1.09 \leq \eta < 14.01 + 1.09$$
$$x = 30: \quad 15.88 - 0.89 \leq \eta < 15.88 + 0.89$$
$$x = 40: \quad 17.76 - 1.09 \leq \eta < 17.76 + 1.09$$
$$x = 50: \quad 19.63 - 1.55 \leq \eta < 19.63 + 1.55$$

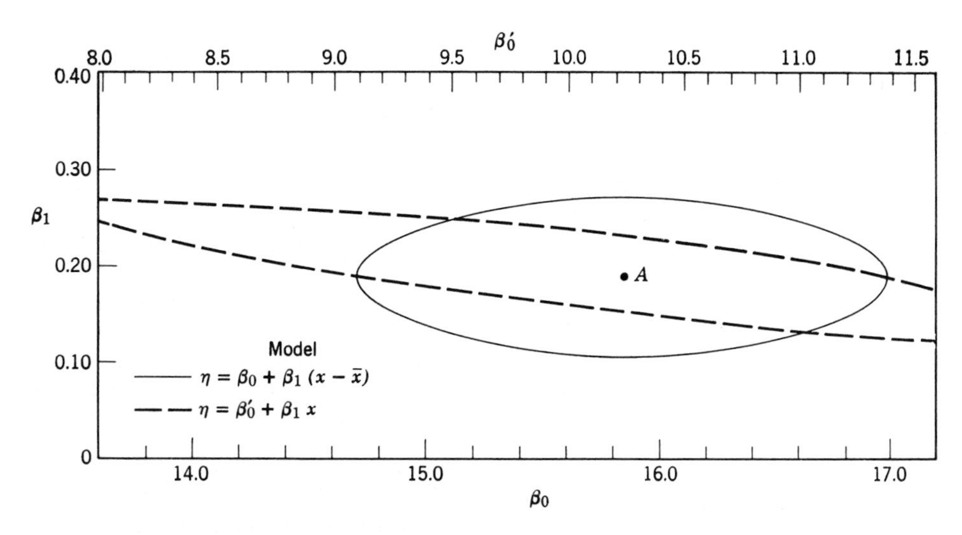

FIGURE E4.3-1B Contours for 95-percent joint confidence region.

Figure E4.3-1a shows the values of $\bar{Y}_i$ and compares the assumed model and the regression line which estimates the model.

Figure E4.3-1b shows a joint confidence region for β_0 and β_1 given by Equation 4.3-28. Note that the ellipse is aligned with the major axes and that the estimated b_0 and b_1 fall within the ellipse. Contrast the joint confidence region for β_0 and β_1 with that for β_0' and β_1, the latter ellipse being rotated, and also with the individual confidence regions computed separately for β_0, β_1, and β_0'.

The analysis of variance corresponding to Table 4.3-3 is shown in Table E4.3-1c.

TABLE E4.3-1c

Source of Variation	Sum of Squares	ν = d.f.	Mean Squares
Due to regression:			
b_0: $\bar{Y}^2 \sum p_i$	252.81	1	252.81
b_1, after allowing for b_0: $\sum p_i(\hat{Y}_i - \bar{Y})^2$	70.28	1	70.28
Deviation of residuals: $\sum p_i(\bar{Y}_i - \hat{Y}_i)^2$	5.02	3	1.67 } pooled 1.50
Deviation within sets (error): $\sum\sum(Y_{ij} - \bar{Y}_i)^2$	6.95	5	1.39 }
Total	335.06	10	

The hypothesis $\beta_1 = 0$ can be tested by forming the variance ratio $s_3^2/s_{\bar{Y}_i}^2$

$$\frac{s_3^2}{s_{\bar{Y}_i}^2} = \frac{70.28}{1.50} = 36.8$$

which is significantly greater than $F_{0.05}(1, 5) = 6.61$ from Table C.4 in Appendix C; hence the hypothesis that $\beta_1 = 0$ is rejected. Clearly, the ratio $s_5^2/s_{\bar{Y}_i}^2 = 252.81/1.50$ indicates that β_0 is a significant parameter in the model. A two-sided t-test with H_0 being that $\beta = 0$ for 8 degrees of freedom (from Table C.3 in Appendix C, for $\alpha = 0.05$, $t = 2.306$):

$$t = \frac{b_1 - 0}{s_{b_1}} = \frac{0.1874}{\sqrt{7.5 \times 10^{-4}}} = 6.85$$

also indicates that the hypothesis that $\beta = 0$ should be rejected because $6.85 > 2.306$.

As a matter of interest, the F-test for the hypothesis that $\beta_1 = 0$ is exactly the same as the t-test for $\beta_1 = 0$ if $s_r^2 = s_{\bar{Y}}^2$ because

$$\frac{s_3^2}{s_r^2} = \frac{\sum p_i(\hat{Y}_i - \bar{Y})^2/1}{s_r^2}$$

and, from the estimated regression equation $\hat{Y}_i = b_0 + b_1(x_i - \bar{x})$ with $b_0 = \bar{Y}$,

$$\sum p_i(\hat{Y}_i - \bar{Y})^2 = b_1^2 \sum (x_i - \bar{x})^2$$

so that

$$\frac{s_3^2}{s_r^2} = \frac{b_1^2 \sum (x_i - \bar{x})^2}{s_r^2} = \left[\frac{b_1[\sum (x_i - \bar{x})^2]^{1/2}}{s_{\bar{Y}_i}}\right]^2$$

By introducing the estimate of the variance of b_1 based on Equation 4.3-11,

$$\frac{s_3^2}{s_r^2} = \left(\frac{b_1 - 0}{s_{b_1}}\right)^2 = t^2$$

Because the variable $F[1, (n - 2)]$ is the square of $t[n - 2]$ exactly the same test is executed by using either the F- or the t-test.

By way of illustration, we might also test the hypothesis that $\beta_1 = 0.150$ through use of a t-test:

$$t = \frac{0.1874 - 0.150}{\sqrt{7.5 \times 10^{-4}}} = \frac{0.0374}{2.74 \times 10^{-2}} = 1.37$$

This hypothesis would have to be accepted because $1.37 < 2.306$. Other hypotheses could be formed and tested for both β_0' and β_1 or for the two jointly.

Example 4.3-2 Simple Linear Regression

This example illustrates the analysis of actual experimental data (listed in Table 4.3-2a) collected to determine the relationship between a rotameter reading (x) and the flow rate (Y). The rotameter reading could be measured with great precision and was stable at a fixed level, so x could essentially be considered a deterministic variable.

TABLE E4.3-2a

Data Set	Y(cc/min)	x(in)
1	112	1.14
2	115	1.37
3	152	1.89
4	199	2.09
5	161	2.45
6	209	2.04
7	237	2.73
8	231	3.04
9	233	3.19
10	259	3.09
11	287	3.05
12	240	3.10
13	281	3.34
14	311	3.75
15	392	4.19
16	357	4.59

We want: (1) to estimate the values of β_0' and β_1 in the proposed model $\eta = \beta_0' + \beta_1 x$; (2) to determine the confidence intervals (for a confidence level of $1 - \alpha = 0.95$) for β_0', β_1, and η; (3) to make a plot illustrating the estimated regression equation, the confidence limits for η, and the data points; and (4) to test to see if the slope of the regression line differs from zero.

Solution:

This example differs from the previous one in that real experimental data are involved and no replicate measurements have been obtained. Consequently, it is *not* possible to test to determine if a straight line is a suitably fitting model for the data by means of an F-test. However, a graph can be prepared of the data and inspected for the appropriateness of the linear model, as illustrated by Figure E4.3-2.

The preliminary calculations (not all are needed for any given calculational scheme; all sums are from $i = 1$ to 16 since $n = 16$ and $p_i = 1$) are:

$$\bar{x} = \frac{\sum x_i}{\sum p_i} = \frac{45.05}{16} = 2.8156$$

$$\sum x_i^2 = 140.454$$

$$\bar{Y} = \frac{\sum Y_i}{\sum p_i} = \frac{3776}{16} = 236$$

$$\frac{(\sum x_i)^2}{n} = 126.844$$

$$\sum x_i Y_i = 1.707 \times 10^4$$

$$\sum (x_i - \bar{x})^2 = \sum x_i^2 - \frac{(\sum x_i)^2}{n}$$

$$= 13.610$$

$$\sum Y_i^2 = 985{,}740$$

$$\sum Y_i(x_i - \bar{x}) = \sum Y_i x_i - \bar{x} \sum Y_i$$

$$= 11{,}707 - 2.8156(3776) = 1075$$

$$\frac{(\sum Y_i)^2}{n} = 891{,}136$$

$$\sum (Y_i - \bar{Y})^2 = \sum Y_i^2 - \frac{(\sum Y_i)^2}{n} = 94{,}604$$

Next the estimated parameters can be calculated:

$$b_0 = \frac{\sum Y_i}{n} = \bar{Y} = 236$$

$$b_0' = 13.506$$

$$b_1 = \frac{\sum Y_i(x_i - \bar{x})}{\sum (x_i - \bar{x})^2} = \frac{1075}{13.61} = 79.02$$

$$s_{\bar{Y}_i}^2 = \frac{1}{n-2} \sum (\bar{Y}_i - \hat{Y}_i)^2$$

$$= \frac{1}{n-2}\left\{\sum (\bar{Y}_i - \bar{Y})^2 - \frac{[\sum \bar{Y}_i(x_i - \bar{x})]^2}{\sum (x_i - \bar{x})^2}\right\}$$

$$= \tfrac{1}{14}\left\{94{,}604 - \frac{(1075)^2}{13.61}\right\}$$

$$= 687$$

$$s_{\bar{Y}_i} = 26.2$$

$$s_{b_0}^2 = \frac{s_{\bar{Y}_i}^2}{n} = \frac{687}{16} = 42.8$$

$$s_{b'_0}^2 = 443.0$$

$$s_{b_0} = 6.41$$

$$s_{b'_0} = 21.1$$

$$s_{b_1}^2 = \frac{s_{\bar{Y}_i}^2}{\sum (x_i - \bar{x})^2} = \frac{687}{13.61} = 50.4$$

The estimated regression line proves to be

$$\hat{Y} = 236 + 79.02(x - 2.816)$$

$$= 13.51 + 79.02x$$

while the confidence intervals for β_0', β_0, β_1, and η, based on $t_{1-\frac{\alpha}{2}} = t_{0.975} = 2.145$ for 14 degrees of freedom, are:

$$[13.51 - 2.145(21.1)] \leq \beta_0' < [13.51 + 2.145(21.1)]$$

$$-31.63 \leq \beta_0' < 58.65$$

$$\left[236 - 2.145\left(\frac{26.2}{\sqrt{16}}\right)\right] \leq \beta_0 < \left[236 + 2.145\left(\frac{26.2}{\sqrt{16}}\right)\right]$$

$$222 \leq \beta_0 < 250$$

$$\left[79.02 - 2.145\left(\frac{26.2}{\sqrt{13.61}}\right)\right] \leq \beta_1 < \left[79.02 + 2.145\left(\frac{26.2}{\sqrt{13.61}}\right)\right]$$

$$63.78 \leq \beta_1 < 94.26$$

$$[\hat{Y} - 2.145(s_{\hat{Y}})] \leq \eta < [\hat{Y} + 2.145(s_{\hat{Y}})]$$

where

$$s_{\hat{Y}} = s_{\bar{Y}_i}\left[\frac{1}{16} + \frac{(x - \bar{x})^2}{13.61}\right]^{1/2}$$

Figure E4.3-2 shows the estimated regression equation, the locus of the confidence limits for a significance level of $\alpha = 0.05$, and the experimental data. Although many of the individual experimental data points fall outside the 95-percent confidence limits, remember that the confidence limits are for the mean of a sample of Y at a given value of x, and in this example we do not have any replicate measurements. From Equation 4.3-25, we can calculate the con-

TABLE E4.3-2b

Data Point	$\hat{Y}_i$	$ts_{\hat{Y}_i}$	$(\bar{Y}_i - \hat{Y}_i)$
1	103.59 +/−	29.14	8.40
2	121.76 +/−	26.12	−6.76
3	162.85 +/−	19.90	−10.85
4	178.66 +/−	17.88	20.33
5	207.10 +/−	15.11	−46.10
6	174.70 +/−	18.36	34.29
7	229.23 +/−	14.11	7.76
8	253.73 +/−	14.46	−22.73
9	265.58 +/−	15.16	−32.58
10	257.68 +/−	14.66	1.31
11	254.52 +/−	14.49	32.47
12	258.47 +/−	14.70	−18.47
13	277.43 +/−	16.16	3.56
14	309.83 +/−	20.00	1.16
15	344.60 +/−	25.21	47.39
16	376.21 +/−	30.46	−19.21

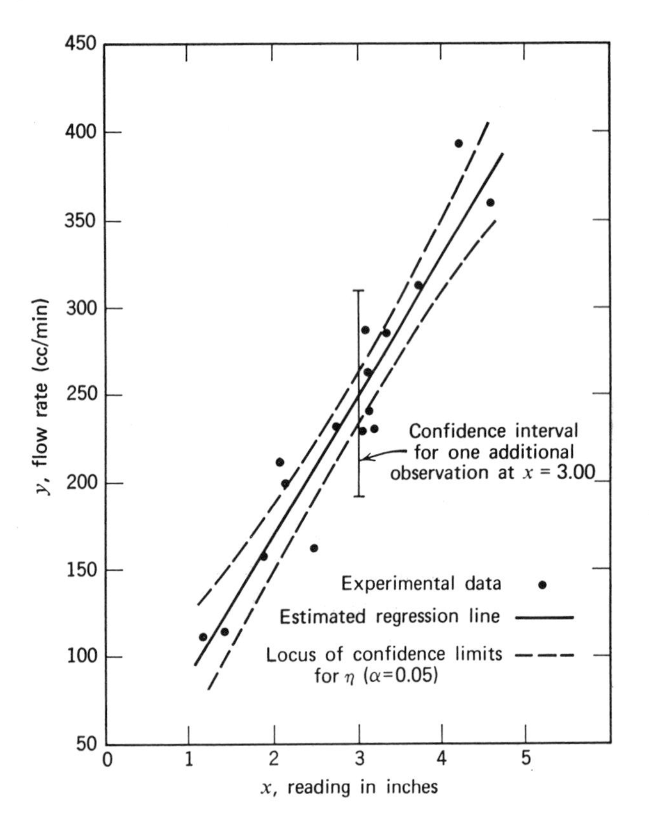

FIGURE E4.3-2 Estimated regression line, confidence limits, and data.

fidence limits about $\hat{Y}_i$ for *one* additional measurement of Y at, say, $x_i = 3.00$:

$$s_D = 26.2\left[1 + \frac{1}{16} + \frac{(3.00 - 2.8156)^2}{13.61}\right]^{1/2} = 27.03$$

$$\hat{Y}_i \pm t_{0.975}s_D = 250.6 \pm (2.145)(27.03)$$
$$= 250.6 \pm 58.0$$

As shown in the figure, this interval encompasses a much larger span than do the dashed lines.

Although no replicate values are available so that s_e^2 can be calculated and used in a test of the appropriateness of the model, the linear form actually used appears to be suitable from visual examination of the figure. A test of the hypothesis that $\beta_1 = 0$ can be made based on the analysis of variance in Table E4.3-2c. The variance ratio is

TABLE E4.3-2c

Source of Variation	Sum of Squares	Degree of Freedom	Mean Square
Due to regression:			
b_0: $\bar{Y}^2 \sum p_i$	891,136	1	891,136
b_1, after allowing for b_0: $\sum p_i(\hat{Y}_i - \bar{Y}_i)^2$	84,988	1	84,988
Deviation about the empirical regression line: $\sum(\bar{Y}_i - \hat{Y}_i)^2$	9,616	14	687
Total	985,740	16	

84,988/687 = 123; clearly the hypothesis that $\beta_1 = 0$ is rejected.

Example 4.3-3 Correlation of Engineering Data by Dimensional Analysis

Rowe† presented an interesting example of the dangers of the blind usage of the results of least squares estimation in the correlation of engineering data. Fluidized beds, in general, cannot be represented by transport phenomena models; consequently, empirical models are a natural approach to obtain functional relationships between the dimensionless groups of variables (and coefficients) which are involved in heat transfer, mass transfer, and momentum transfer in the bed. Rowe simulated 45 sets of experimental data by selecting from a table of random variables values of d, the particle diameter; v, the air velocity; ΔT, the temperature difference between the wall and the bed; and h, the interphase heat-transfer coefficient for a hypothetical bed 12 inches in diameter. These variables, together with certain physical properties: μ (air viscosity), ρ (air density), k (thermal conductivity of air), c_p (air heat capacity), σ (air bubble surface tension), β (coefficient of expansion), and g (acceleration of gravity), were combined in suitable dimensionless groups.

The following are typical comments on the treatment of the simulated data feigning the report of an engineer working with real experimental data.

NUSSELT NUMBER-REYNOLDS NUMBER CORRELATION: The first attempt at a correlation was to plot the Nusselt number, $\mathrm{Nu} = hd/k$, against the Reynolds number, $\mathrm{Re} = vd\rho/\mu$, as shown in Figure E4.3-3a where the line log (Nu) = log (0.13) + 0.79 log (Re) has been drawn through the data. The correlation is not very good, although the index found for Re is near to the value 0.8 which often occurs in empirical heat transfer/fluid-flow correlations. The dashed lines represent the locus of the confidence limits for $\alpha = 0.05$. There seem to be some "rogue" points suggesting that the apparatus was not always working properly.

STANTON NUMBER-REYNOLDS NUMBER CORRELATION: It has been argued that in a fluidized bed there is considerable dis-

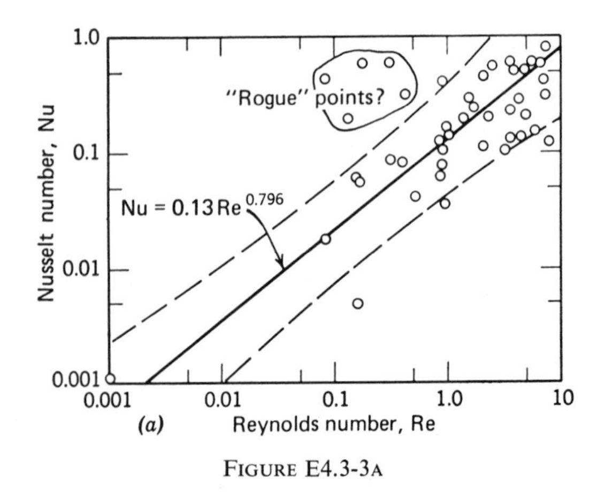

FIGURE E4.3-3A

† P. M. Rowe, *Trans. Inst. Chem. Eng.* (London) **41**, CE 70, Mar. 1963.

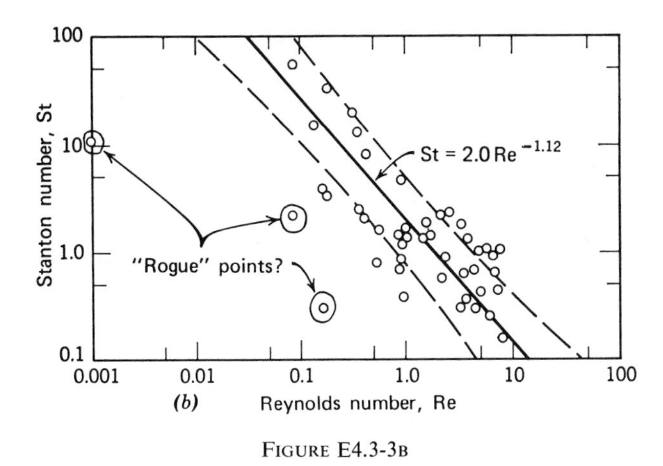

FIGURE E4.3-3B

sipation of momentum, and it has been suggested that there may be a relationship between energy loss and heat transfer. Consequently, by analogy with the relationship between friction and heat transfer in pipes (the well-known j-factors), the Stanton number, $\mathrm{St} = h/c_p v \rho$, was calculated and is plotted against the Reynolds number in Figure E4.3-3b. This correlation is an improvement on the previous one; the rough rule, $\mathrm{St} = 2.0\ \mathrm{Re}^{-1.12}$ is a guide in design but subject to a rather large error.

WEBER NUMBER-STANTON NUMBER CORRELATION: The Reynolds number may not be the best parameter to describe a fluidized bed from a heat transfer point of view, and alternative groups were considered. Bubbles are highly characteristic of gas-fluidized beds and are known to affect heat transfer. The bubbles have surface energy associated with them; consequently, it was reasoned that a Weber number, $\mathrm{We} = v^2\, d\rho/\sigma$, might characterize the bed. Fluidized beds do not have a surface tension in the usual sense, but an arbitrary value was used simply to examine the concept. Figure E4.3-3c is a plot of the Stanton number against the Weber number. The correlation is still not good but appears promising, especially as the law may be written $\mathrm{St} = 0.2\ \mathrm{We}^{-1/2}$, which is simple and suggests a theoretical basis for the law.

NUSSELT NUMBER-GRASHOF NUMBER CORRELATION: Most fluidized particles exhibit an "up the middle and down the sides" pattern of movement strongly suggestive of convective circulation; on this basis a Grashof number was calculated as $\mathrm{Gr} = d^3\rho^2 g\beta\,\Delta T/\mu^2$. Figure E4.3-3d is a plot of the Nusselt number against the Grashof number. This plot was modified slightly by multiplying the Nusselt number by the ratio of particle to bed diameter, $\mathrm{Nu}\,(d/D)$, as in Figure E4.3-3e, which seems to yield a moderately good correlation. The relationship $\mathrm{Nu}\,(d/D) = 0.26\ \mathrm{Gr}^{0.82}$ is proposed as a basis for design. It is seen to hold approximately over five orders of magnitude of each parameter, and precise relations of such wide application are hard to obtain.

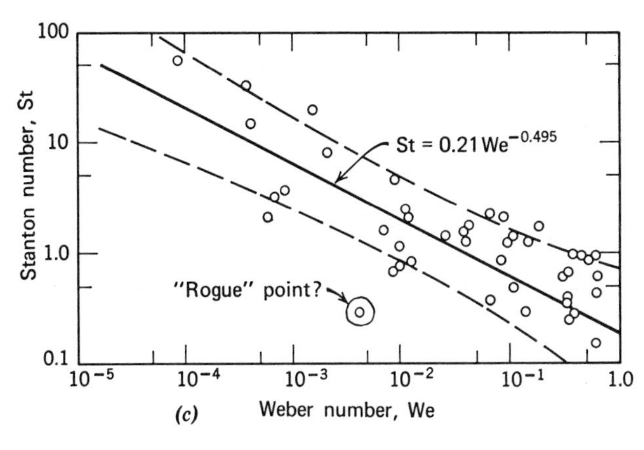

FIGURE E4.3-3C

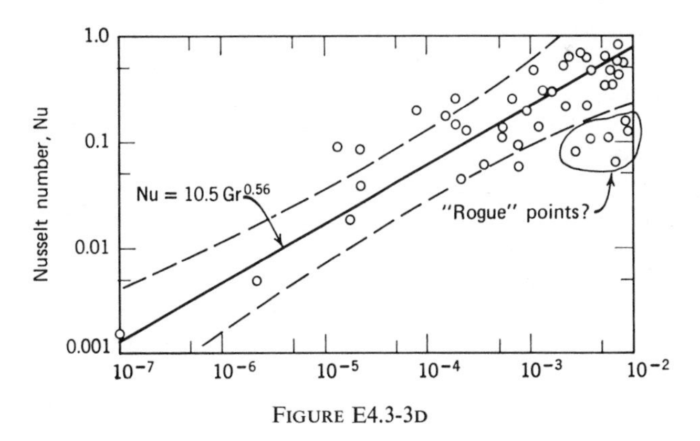

FIGURE E4.3-3D

Of course, all the ascribed relationships are fictitious. There are two principal reasons for the apparent reasonableness of the linear relations (on log-log paper) developed. The first is the use of log-log paper for plotting data; the second is the inclusion of the same variable on each axis. A logarithmic plot distorts the data because the more or less uniform distribution of random points found on an arithmetic basis appears as a concentration of data in the upper right-hand corner, as illustrated in Fig. E4.3-3a.

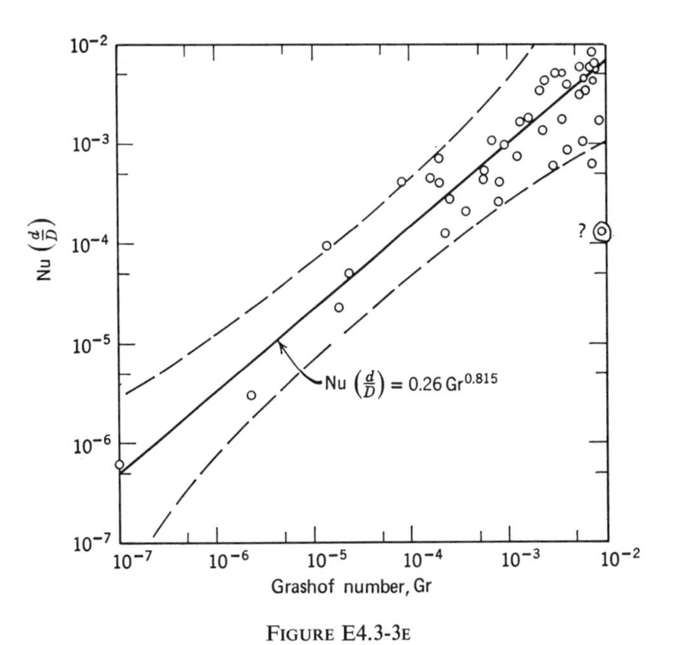

FIGURE E4.3-3E

When the axes are labeled as dimensionless groups, it may not be immediately apparent that the same variable occurs in both groups being plotted. For example, Figure E4.3-3a is a plot not of Nu against Re but of hd against vd because in Nu $= hd/k$, k is a constant (only one value was used, that for air), and in Re $= dv\rho/\mu$, ρ and μ are constants. When d is by chance large, Nu and Re also are large so that an artificial association is produced. Similarly, Figure E4.3-3b is really a plot of h/v against vd. Figure E4.3-3c is a plot of h/u against u^2d. Figure E4.3-3d is a plot of hd against d^3. The convincing Figure E4.3-3e is a plot of hd^2 against d^3. In addition to forming a false association, raising the same quantity to some power has the effect of apparently increasing the range of the observations. The original data span only two orders of magnitude, but Figure E4.3-3e appears to cover five orders.

To avoid spurious relations such as illustrated above, the experiments should be designed as described in Chapter 8. Apparent "rogue" points should never be ignored until it can be shown that there are reasonable experimental grounds for excluding them. If, in retrospect, it can be seen that the apparatus may have been faulty during a certain period, all the observations made during that period should be discarded. It is particularly unsound to use dimensionless groups when, in fact, only one component of the group was varied unless there are special reasons for believing that the group is more meaningful than the single variable that it includes. It should always be made clear when only one component of the group has been varied. When dimensionless groups are combined, it is essential to examine the resulting combination and to avoid including the same variable in two groups that are to be related in a model.

In most experimental work a variable is calculated from the primary experimental observations. For example, a heat-transfer coefficient is calculated from a measured heat flux, surface area, and temperature drop. A Reynolds number may be calculated from a measured length and velocity together with properties of the fluid read from tables. It is essential to calculate the error of the derived quantity from the known error of the measured components and to make sure that the assumptions underlying the least squares estimation are fulfilled. Most dimensionless groups include physical properties whose values are read from standard handbooks. The errors in the quoted values should be taken into account when assessing the error of the resulting dimensionless group because they are sometimes considerable. There is no easy way of judging the error of published data of this kind when it is not specifically quoted, but some idea can be formed by comparing values from different sources and by considering the method of measurement used.

In conclusion, it has been shown that the estimation of linear relations using dimensionless groups as variables is fraught with danger.

4.4 ESTIMATION WITH ERROR VARIANCE A FUNCTION OF THE INDEPENDENT VARIABLE

In this section we briefly consider for Model 4.3-1 the case in which the variance of Y is a function of x.

Typical data from the dissolution of surface active agents, absorption of gases, diffusion in solids, and chemical reactors yield dependent variables that are a declining function of distance or time. If the error in measurement is at a fixed level, the relative error in the dependent variable increases with the increase of the independent variable. Figure 4.4-1 illustrates the increasing relative error for the counting of radioactive tracer as a function of time.† The counting rate R was the ratio of the counting rate at any time to the counting rate at infinity. The model used to represent the data was

$$-\frac{dR}{dt} = k(R-1) \qquad R(0) = 0$$

and could be integrated to yield

$$\ln(1-R) = \beta_0 + \beta_1 t$$

Figure 4.4-1 shows the estimated regression equations and the locus of the confidence limits obtained if, in the least squares method, each data point was weighted inversely proportional to the variance of the data point with the variance estimated from replicate measurements.

To take into account varying error as a function of the independent variable, exactly the same analysis as was carried out in Section 4.3 is employed except that we assume

$$\text{Var}\{\bar{Y}_i \mid x_i\} = \sigma^2_{\bar{Y}_i}[f(x)]^2$$

where the functional relation to x, $f(x)$, is known and $\sigma^2_{\bar{Y}_i}$ is unknown. Because the general procedure was spelled out in detail in Section 4.3, we omit much of the intermediate detail in the following discussion. The

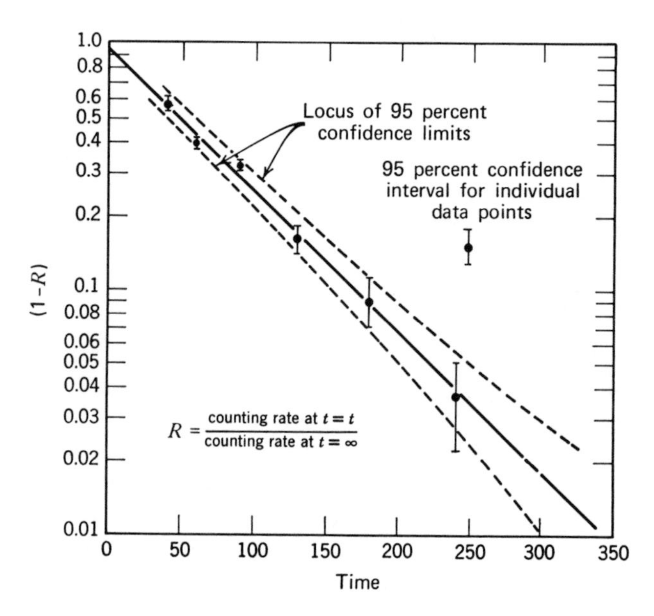

FIGURE 4.4-1 Radioactive tracer counting illustrates the increasing error with time of measurement.

† J. C. Wang and D. M. Himmelblau, *AIChE J.* **4**, 574, 1964.

observed Y_{ij}'s are still assumed to be normally distributed about $\eta_i = \beta_0 + \beta_1(x_i - \bar{x})$ but with variance $\sigma^2_{\bar{Y}_i}[f(x)]^2$, so that

$$U_{ij} = \frac{Y_{ij} - \eta_i}{\sigma_{\bar{Y}_i} f(x_i)}$$

is normally distributed with parameters (0, 1). Recall that

$$\chi^2 = \sum_{i=1}^{n} \sum_{j=1}^{p_i} U_{ij}^2$$

If we call $w_i = (1/f(x_i))^2$ the weight, and multiply χ^2 by $\sigma^2_{\bar{Y}_i}$, we obtain

$$\sigma^2_{\bar{Y}_i}\chi^2 = \sum_{i=1}^{n} \sum_{j=1}^{p_i} w_i(Y_{ij} - \eta_i)^2 \tag{4.4-1}$$

The sum of squares on the right-hand side of Equation 4.4-1 can be partitioned as in Section 4.3 by an identical procedure except that w_i must be included in each sum. The least squares procedure gives:

1. Estimated regression equation $\hat{Y} = b_0 + b_1(x - \bar{x})$.

2. $$\bar{x} = \frac{\sum_{i=1}^{n} w_i p_i x_i}{\sum_{i=1}^{n} w_i p_i}$$

3. $$b_0 = \bar{Y} = \frac{\sum_{i=1}^{n} w_i p_i \bar{Y}_i}{\sum_{i=1}^{n} w_i p_i}$$

4. $$b_1 = \frac{\sum_{i=1}^{n} w_i p_i (x_i - \bar{x})\bar{Y}_i}{\sum_{i=1}^{n} w_i p_i (x_i - \bar{x})^2}$$

5. $$\operatorname{Var}\{b_0\} = \frac{\sigma^2_{\bar{Y}_i}}{\sum_{i=1}^{n} w_i p_i}$$

6. $$\operatorname{Var}\{b_1\} = \frac{\sigma^2_{\bar{Y}_i}}{\sum_{i=1}^{n} w_i p_i (x_i - \bar{x})^2}$$

An analysis of variance can be carried out which corresponds to Tables 4.3-1 and 4.3-2. The variance ratio s_r^2/s_e^2 can be formed exactly as in Section 4.3 and an F-test carried out. If the ratio s_r^2/s_e^2 is not significant, the pooled variance is

$$s^2_{\bar{Y}_i} = \frac{\sum_{i=1}^{n} \sum_{j=1}^{p_i} w_i(Y_{ij} - \bar{Y}_i)^2 + \sum_{i=1}^{n} w_i p_i(\bar{Y}_i - \hat{Y}_i)^2}{\left(\sum_{i=1}^{n} p_i - n\right) + (n - 2)} \tag{4.4-2}$$

which corresponds to Equation 4.3-15.

The estimated response $\hat{Y}$ is normally distributed about $\eta = \beta_0 + \beta_1(x - \bar{x})$ and, as before,

$$\operatorname{Var}\{\hat{Y}\} = \operatorname{Var}\{b_0\} + (x - \bar{x})^2 \operatorname{Var}\{b_1\}$$

Let us now consider two interesting and practical special cases which contrast the use of weights with the assumption of unity (no) weights. Let the linear model be a line through the origin with slope of β:

$$\mathscr{E}\{\bar{Y}_i \mid x\} = \beta x$$

We examine the estimate of β, b, for three cases:

(I) $\operatorname{Var}\{\bar{Y}_i \mid x\} = \sigma^2_{\bar{Y}_i} x$; a weight of $\frac{1}{x}$

(II) $\operatorname{Var}\{\bar{Y}_i \mid x\} = \sigma^2_{\bar{Y}_i} x^2$; a weight of $\frac{1}{x^2}$

(III) $\operatorname{Var}\{\bar{Y}_i \mid x\} = \sigma^2_{\bar{Y}_i}$; a weight of 1

Then the corresponding estimates of β and the variances of b become (all sums are from $i = 1$ to n):

$$b_{\text{I}} = \frac{\sum p_i \bar{Y}_i}{\sum p_i x_i} \qquad \operatorname{Var}\{b_{\text{I}}\} = \frac{\sigma^2_{\bar{Y}_i}}{\sum p_i x_i}$$

$$b_{\text{II}} = \frac{\sum \frac{p_i \bar{Y}_i}{x_i}}{\sum p_i} \qquad \operatorname{Var}\{b_{\text{II}}\} = \frac{\sigma^2_{\bar{Y}_i}}{\sum p_i}$$

$$b_{\text{III}} = \frac{\sum p_i x_i \bar{Y}_i}{\sum p_i x_i^2} \qquad \operatorname{Var}\{b_{\text{III}}\} = \frac{\sigma^2_{\bar{Y}_i}}{\sum p_i x_i^2}$$

In each case a different slope will be obtained. Use of weights should be based on physical grounds, i.e., information about the variability of $\sigma^2_{\bar{Y}_i}$ obtained from the experiment or elsewhere.

For example, if replicate values of Y are taken at each x_i as in Example 3.6-3, then estimates of $\sigma^2_{\bar{Y}_i}$ can be obtained at each x_i by using Equation 2.4-2. Least squares can be applied to estimate the required functional dependence of $\operatorname{Var}\{Y\}$ on x. Or, estimation of the functional dependence can be made from analysis of known instrumental errors in the measuring instruments.

Example 4.4-1 Weighted Linear Regression

Example 4.3-1 is repeated in part with the revised premise that three types of weighting are to be compared:

1. Weight = 1 for each $\bar{Y}_i$.
2. Weight = $1/x_i$ for each $\bar{Y}_i$.
3. Weight = $1/x_i^2$ for each $\bar{Y}_i$.

Solution:

TABLE E4.4-1

Quantity	Results for Weights of 1	$\frac{1}{x}$	$\frac{1}{x^2}$
$\sum_{i=1}^{n} w_i p_i x_i$	300	10	0.456
$\sum_{i=1}^{n} w_i p_i$	10	0.456	2.928×10^{-2}
$\bar{x} = \frac{\sum_{i=1}^{n} w_i p_i x_i}{\sum_{i=1}^{n} w_i p_i}$	30.0	21.9	15.6
$\sum w_i p_i \bar{Y}_i$	158.8	6.526	0.3806
$\bar{Y} = \frac{\sum_{i=1}^{n} w_i p_i \bar{Y}_i}{\sum_{i=1}^{n} w_i p_i}$	15.88	14.31	12.99
$b_0 = \bar{Y} = \frac{\sum_{i=1}^{n} w_i p_i \bar{Y}_i}{\sum_{i=1}^{n} w_i p_i}$	15.88	14.31	12.99
x_i		$(x_i - \bar{x})$	
10	−20	−11.9	−5.6
20	−10	−1.9	4.4
30	0	8.1	14.4
40	10	18.1	24.4
50	20	28.1	34.4
$\sum_{i=1}^{n} w_i p_i (x_i - \bar{x}) \bar{Y}_i$	374.62	15.868	0.592
$\sum_{i=1}^{n} w_i p_i (x_i - \bar{x})^2$	2000	81.022	2.874
$b_1 = \frac{\sum_{i=1}^{n} w_i p_i (x_i - \bar{x}) \bar{Y}_i}{\sum_{i=1}^{n} w_i p_i (x_i - \bar{x})^2}$	0.1874	0.1959	0.2057
$\widehat{\mathrm{Var}}\{b_0\}$	0.1674	0.1132	0.0548
$\widehat{\mathrm{Var}}\{b_1\}$	8.3×10^{-4}	6.4×10^{-4}	5.6×10^{-4}
$\widehat{\mathrm{Var}}\{b_0'\}$	0.921	0.421	0.193

Figure E4.4-1 illustrates the locus of the three estimated regression equations and the respective confidence intervals for η.

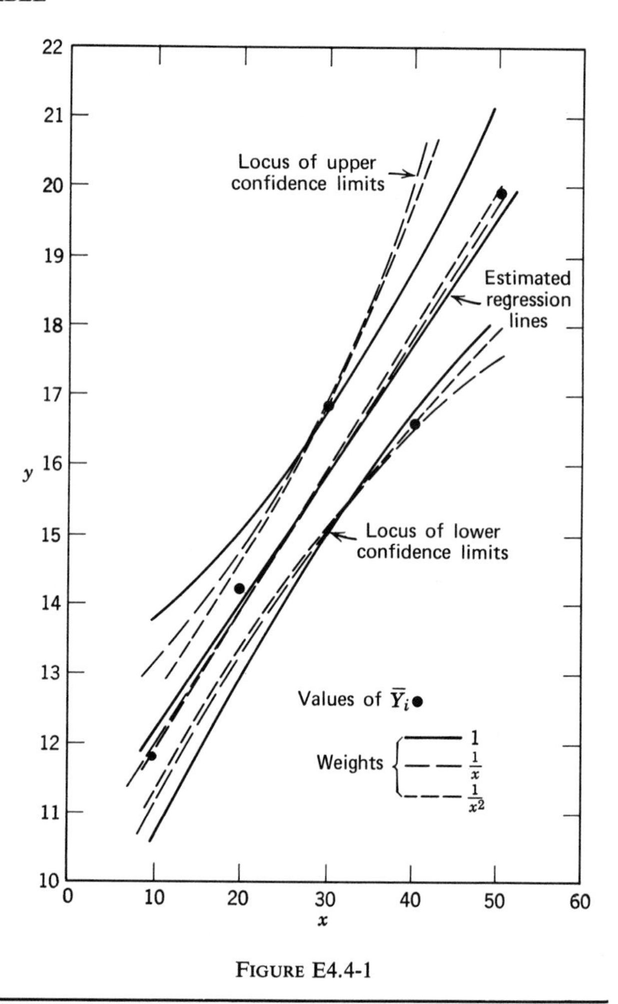

FIGURE E4.4-1

4.5 ESTIMATION WITH BOTH INDEPENDENT AND DEPENDENT VARIABLES STOCHASTIC

If both the independent and dependent variables are random and normally distributed, the estimation of coefficients in a linear (in the coefficients) model and the designation of suitable tests and confidence intervals become quite difficult. In spite of the attention that this important problem has attracted, completely satisfactory techniques are yet to be devised. Several different methods of attack, beyond the scope of this text, have been proposed, the references for which can be found at the end of the chapter.

To provide a contrast with the empirical model in which the dependent variable only is stochastic and also to illustrate some of the difficulties involved in estimation, one of the many methods of estimation, that of maximum likelihood, is described here. The technique is not presented because it is more widely applicable or better than other techniques—the best treatment when both variables are stochastic is far from being resolved.

We can only treat the case of a simple linear model with one dependent variable, Y, and one independent variable, X, with X and Y jointly distributed by a normal distribution according to the density given in Example 2.3-4. Nevertheless, this simple model will indicate one approach to estimation without the introduction of obscuring mathematical details. We assume that the model is

$$\eta_i = \beta_0 + \beta_1(\mu_i - \bar{\mu}) \qquad (4.5\text{-}1)$$

where

$\eta_i = \mathscr{E}\{Y_i\}$
$\mu_i = \mathscr{E}\{X_i\}$
$\bar{\mu}$ = some appropriate average of the μ_i's
β_0, β_1 = intercept and slope, respectively, of the model graph

Both η_i and μ_i are nonrandom variables, and we define the errors U_i and V_i and their variances as

$$(Y_i - \eta_i) = V_i \qquad \text{Var}\{V_i\} = \sigma_v^2$$
$$(X_i - \mu_i) = U_i \qquad \text{Var}\{U_i\} = \sigma_u^2$$
$$\text{Covar}\{U_i V_j\} = \rho_{uv}\sigma_u\sigma_v$$

We want to estimate β_0 and β_1 (and carry out hypothesis tests) through use of the method of maximum likelihood.

The likelihood function is formed for n sets of observations, each containing p_i replicates exactly as described in Section 4.3. As usual, errors from one pair of observations to the next are assumed independent.

We write

$$L = \left[\frac{1}{2\pi\sigma_u\sigma_v\sqrt{1-\rho_{uv}^2}}\right]^{\left(\sum_{i=1}^n p_i\right)}$$
$$\times \exp\left[-\frac{1}{2(1-\rho_{uv}^2)}\sum_{i=1}^n\sum_{j=1}^{p_i}\left\{\left(\frac{Y_{ij}-\eta_i}{\sigma_v}\right)^2 - 2\rho_{uv}\left(\frac{Y_{ij}-\eta_i}{\sigma_v}\right)\left(\frac{X_{ij}-\mu_i}{\sigma_u}\right) + \left(\frac{X_{ij}-\mu_i}{\sigma_u}\right)^2\right\}\right] \qquad (4.5\text{-}2)$$

To save space we do not show the logarithm of the likelihood function but only list the results after summation over the index j of the minimization of $\ln L$ with Equation 4.5-1 substituted for η_i in Equation 4.5-2:

Partial differentiation with respect to β_0 yields

$$\sum_{i=1}^n p_i\left[(\bar{Y}_i - \eta_i) - \frac{\rho_{uv}\sigma_v}{\sigma_u}(\bar{X}_i - \mu_i)\right] = 0 \qquad (4.5\text{-}3)$$

Partial differentiation with respect to β_1 yields

$$\sum_{i=1}^n p_i\left\{\left[(\bar{Y}_i - \eta_i) - \frac{\rho_{uv}\sigma_v}{\sigma_u}(\bar{X}_i - \mu_i)\right][\mu_i - \bar{\mu}]\right\} = 0 \qquad (4.5\text{-}4)$$

Partial differentiation with respect to μ_i yields

$$(\bar{Y}_i - \eta_i) - \frac{\rho_{uv}\sigma_v}{\sigma_u}(\bar{X}_i - \mu_i)\left[\frac{\beta_1 - \left(\dfrac{\sigma_v}{\sigma_u\rho_{uv}}\right)}{\beta_1 - \left(\dfrac{\rho_{uv}\sigma_v}{\sigma_u}\right)}\right] = 0 \qquad (4.5\text{-}5)$$

The combination of Equations 4.5-3 and 4.5-4 yields, after some manipulation,

$$\hat{\mu} = \frac{\sum_{i=1}^n p_i\mu_i}{\sum_{i=1}^n p_i} = \frac{\sum_{i=1}^n p_i\bar{X}_i}{\sum_{i=1}^n p_i} = \bar{X} \qquad (4.5\text{-}6)$$

$$b_0 = \frac{\sum_{i=1}^n p_i\eta_i}{\sum_{i=1}^n p_i} = \frac{\sum_{i=1}^n\sum_{j=1}^{p_i} Y_{ij}}{\sum_{i=1}^n p_i} = \bar{Y} \qquad (4.5\text{-}7)$$

However, the equation for $\hat{\beta}_1$ proves to be of quadratic form, indicating a more difficult computation than required for simple estimation with error in only one variable:

$$\sum_{i=1}^n p_i[(\bar{Y}_i - \bar{Y}) - \hat{\beta}_1(\bar{X}_i - \bar{X})(\bar{Y}_i - \bar{Y})]$$

$$= \left[\frac{\rho_{uv}\sigma_v}{\sigma_u}\right]\left[\frac{\hat{\beta}_1 - \left(\dfrac{\sigma_v}{\sigma_u\rho_{uv}}\right)}{\hat{\beta}_1 - \left(\dfrac{\rho_{uv}\sigma_v}{\sigma_u}\right)}\right](\bar{X}_i - \bar{X}) \qquad (4.5\text{-}8)$$

To solve Equation 4.5-8 for $\hat{\beta}_1$, we need to have the values for or estimates of σ_v, σ_u, and ρ_{uv}. As long as the experiment can be designed to collect replicate data with identified (Y_i, X_i) pairs, estimates of σ_v, σ_u, and ρ_{uv} can be made as follows for each pair:

$$\hat{\sigma}_{u_i}^2 = s_{u_i}^2 = \frac{\sum_{j=1}^{p_i}(X_{ij} - \bar{X}_i)^2}{p_i - 1}$$

$$\hat{\sigma}_{v_i}^2 = s_{v_i}^2 = \frac{\sum_{j=1}^{p_i}(Y_{ij} - \bar{Y}_i)^2}{p_i - 1}$$

$$(\hat{\rho}_{uv}s_u s_v)_i = \frac{\sum_{j=1}^{p_i}(X_{ij} - \bar{X}_i)(Y_{ij} - \bar{Y}_i)}{p_i - 1}$$

Homogeneity of variance can be tested as described in Chapter 3, and pooled estimates can be formed by summing over the index i, if warranted. If estimates of σ_u, σ_v, and ρ_{uv} cannot be made from the experimental data, certain assumptions can be presumed concerning these values or their ratios, and β_1 can again be estimated.

Discussion of the confidence intervals for β_0 and β_1 is beyond our scope here; the interested reader is referred to the references at the end of the chapter. Extension of the method of maximum likelihood to multivariate problems is possible in principle but results in sets of nonlinear equations, which are often difficult to solve for the desired parameter estimates. Satisfactory, generally applicable techniques of estimation for Model 4.3-1 when both variables are stochastic have yet to be devised.

4.6 ESTIMATION WHEN MEASUREMENT ERRORS ARE NOT INDEPENDENT

As mentioned in Section 4.2, in many instances the errors ϵ_i in the model

$$\bar{Y}_i = \beta_0 + \beta_1(x_i - \bar{x}) + \epsilon_i$$

are not independent as was assumed in Sections 4.3, 4.4, and 4.5. Over a period of time, earlier process yields, temperatures, or flow rates may affect later observations; hence the $\bar{Y}_i$ and, in effect, the ϵ_i are not statistically independent. A typical example is the sampling at intervals of the concentration of a reaction product from a well-mixed tank. We examine two common examples of the lack of independence in this section.

4.6-1 Cumulative Data

One characteristic feature of certain special experiments is the use of the same batch of materials for the entire series of measurements, for example, a series of measurements of the volumetric displacement of the same fluid made as a function of pressure or a component repeatedly sampled in time in a reaction vessel. If the unobservable error in the first observation is designated as ϵ_1, the error in the second observation ϵ_2' includes ϵ_1 plus a random component introduced aside from ϵ_1, or $\epsilon_2' = \epsilon_1 + \epsilon_2$. The error in the third observation is $\epsilon_3' = \epsilon_1 + \epsilon_2 + \epsilon_3$, and so forth. Mandel, whose analysis we follow,† distinguished between the usually assumed type of independent measuring error in the dependent variable and a "cumulative" or interval error in which each new observation includes the errors of the previous observations. Cumulative errors, arising because of fluctuations as a function of time in the process itself due to small changes in operating conditions (temperature, pressure, humidity, etc.), are not independent—only the differences in measurement from one period to the next are independent. Thus, if we consider the simplest case, a model without an intercept, two models are:

1. Model *A*—independent error:

$$Y_i = \beta x_i + \epsilon_i \qquad i = 1, 2, \ldots, n \tag{4.6-1}$$

2. Model *B*—cumulative data:

$$Y_j - Y_{j-1} = \beta(x_j - x_{j-1}) + \epsilon_j$$
$$j = 1, 2, \ldots, n; x_0 = 0$$

or

$$Y_i = \sum_{j=1}^{i} (Y_j - Y_{j-1}) = \beta \sum_{j=1}^{i} (x_j - x_{j-1}) + \sum_{j=1}^{i} \epsilon_j$$

$$= \beta x_i + \sum_{j=1}^{i} \epsilon_j \tag{4.6-2}$$

† J. Mandel, *J. Amer. Stat. Assn.* **52**, 552, 1957. Also refer to J. L. Jaech, *J. Amer. Stat. Assn.* **59**, 863, 1964.

in which ϵ_i are independent random variables with $\mathscr{E}\{\epsilon_i\} = 0$ and $\mathrm{Var}\{\epsilon_i\} = \sigma_1^2$, a constant. The ϵ_j are also random independent variables because they represent differences such as $\epsilon_3' - \epsilon_2' = (\epsilon_1 + \epsilon_2 + \epsilon_3) - (\epsilon_1 + \epsilon_2) = \epsilon_3$. We assume that $\mathscr{E}\{\epsilon_j\} = 0$ and $\mathrm{Var}\{\epsilon_j\} = \sigma_1^2(x_j - x_{j-1})$; i.e., the variance of ϵ_j can be a function of the test interval. Figure 4.6-1 illustrates the simulation of Equations 4.6-1 and 4.6-2 for $\beta = 3$, with ϵ_i and ϵ_j being normal random deviates and $\sigma_{\epsilon_i}^2 = \sigma_{\epsilon_j}^2 = 1$. Note how the experimental points tend to stay on one side of the regression line of best fit for Model *B*. Although the proposed Model *B* may be applied only in certain types of experiments, it has been found to resolve the observed trends about the regression line in these cases quite well.

The line of best fit for Model *A* from Equation 4.3-7a as applied to Model *A* is

$$b_A = \hat{\beta}_A = \frac{\sum_{i=1}^{n} x_i Y_i}{\sum_{i=1}^{n} x_i^2} \tag{4.6-3}$$

and the variance of b_A is given by Equation 4.3-11a as

$$\mathrm{Var}\{b_A\} = \frac{\sigma_1^2}{\sum_{i=1}^{n} x_i^2} \tag{4.6-4}$$

The best unbiased linear estimate of the slope of Model *B* is obtained by minimizing the weighted sum of squares:

$$\sum_{j=1}^{n} \left(\frac{1}{x_j - x_{j-1}}\right)[(Y_j - Y_{j-1}) - \beta(x_j - x_{j-1})]^2$$

as described in Section 4.4, which gives

$$b_B = \hat{\beta}_B = \frac{\sum_{j=1}^{n} (Y_j - Y_{j-1})}{\sum_{j=1}^{n} (x_j - x_{j-1})} \tag{4.6-5}$$

If the test interval is uninterrupted by a gap, Equation 4.6-5 in terms of the cumulative data is nothing more than

$$b_B = \frac{Y_n}{x_n} \tag{4.6-6}$$

The interpretation of Equation 4.6-6 is that the best estimate of the slope of Model *B* is made by taking the last value of the dependent variable and dividing it by the last value of the independent variable! Although the intermediate results might seem useless, they are not because: (1) they help decide if the model is really linear, and (2) they are needed to reduce the variance of the estimator of the slope which is

$$\mathrm{Var}\{b_B\} = \frac{\sigma_1^2}{\sum_{j=1}^{n} (x_j - x_{j-1})} \tag{4.6-7}$$

If σ_1^2 is not known, the Var $\{b_B\}$ can be estimated by

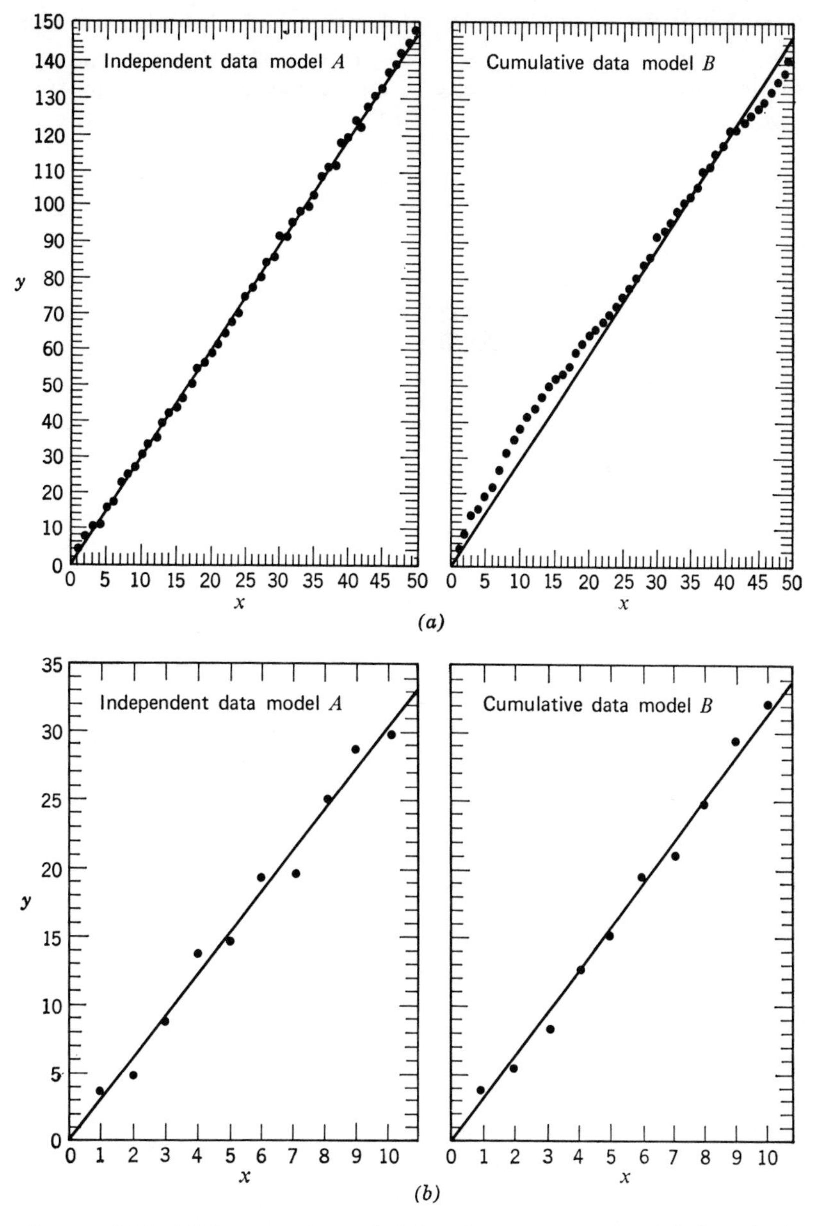

FIGURE 4.6-1 Fitting experimental data when errors are not independent: (*a*) comparison of independent and cumulative data; $n = 50$; and (*b*) comparison of independent and cumulative data; $\eta = 10$. (From J. Mandel, *J. Amer. Stat. Assn.* **52**, 552, 1957.)

$$\widehat{\mathrm{Var}}\,\{b_B\} = \left(\frac{1}{n-1}\right)\left(\frac{1}{\sum_{j=1}^{n}(x_j - x_{j-1})}\right)\sum_{j=1}^{n}\frac{1}{(x_j - x_{j-1})} \times [(Y_j - Y_{j-1})b_B(x_j - x_{j-1})^2] \qquad (4.6\text{-}8)$$

For the special case in which each increment $(x_j - x_{j-1}) = 1$, Equation 4.6-6 becomes

$$b_B = \frac{Y_n}{n}$$

and Equation 4.6-7 becomes

$$\mathrm{Var}\,\{b_B\} = \frac{\sigma_1^2}{n}$$

If the experiment is best represented by Model B but, through ignorance or otherwise, is treated as being represented by Model A, Mandel showed that for the case in which $x_1 = 1, x_2 = 2, \ldots, x_n = n$, i.e., no interval gap and unit changes in x,

$$\tilde{b} = \frac{\sum_{i=1}^{n} x_i Y_i}{\sum x_1^2} \qquad (4.6\text{-}9)$$

$$\mathrm{Var}\,\{\tilde{b}\} = \left(\frac{6}{5}\right)\frac{2n^2 + 2n + 1}{n(n+1)(2n+1)}\sigma_1^2 \qquad (4.6\text{-}10)$$

where the overlay tilde ($\sim$) indicates an estimate incorrectly calculated.

Consider the ratio of the variance of the incorrect estimate to the variance of the correct estimate:

$$\frac{\text{Var}\{\tilde{b}\}}{\text{Var}\{b_B\}} = \left(\frac{6}{5}\right)\frac{2n^2 + 2n + 1}{(n + 1)(2n + 1)}$$

As $n \to \infty$, the ratio $\to$ 1.2; hence the estimate $\tilde{b}$ is only slightly less efficient than the correct estimate b_B. However, if the variance of $\tilde{b}$ also is calculated as if the data were represented by Model A when in fact Model B applies, it can be shown that the expected value of the incorrectly calculated variance is

$$\mathscr{E}\{\widetilde{\text{Var}\{\tilde{b}\}}\} = \frac{3(n + 2)\sigma_1^2}{5n(n + 1)(2n + 1)}$$

Then we can form the ratio

$$\frac{\mathscr{E}\{\widetilde{\text{Var}\{\tilde{b}\}}\}}{\text{Var}\{b_B\}} = \frac{3(n + 2)}{5(n + 1)(2n + 1)} \tag{4.6-11}$$

As n becomes large the ratio becomes quite small, indicating that the standard error of the slope is drastically underestimated. For example, for $n = 10$, the square root of the right-hand side of Equation 4.6-11 is only 0.175. Hence, application of the equations in section 4.3 to Model B data will result in considerable overestimation of the precision of the estimate of the slope, although the estimate of the slope itself will be quite good, because

$$\frac{\tilde{b}}{b_B} = \left(\frac{\sum_{i=1}^{n} x_i Y_i}{\sum_{i=1}^{n} x_i^2}\right)\left(\frac{x_n}{Y_n}\right) \tag{4.6-12}$$

Note that the ratio of the expected values of $\tilde{b}$ and b_B, respectively, is unity (for $\Delta x_j \equiv 1$):

$$\mathscr{E}\{\tilde{b}\} = \mathscr{E}\left\{\frac{\sum_{i=1}^{n} x_i\left(\beta_i x_i + \sum_{j=1}^{n} \epsilon_j\right)}{\sum_{i=1}^{n} x_i^2}\right\} = \beta$$

$$\mathscr{E}\{b_B\} = \mathscr{E}\left\{\frac{\beta x_n + \sum_{j=1}^{n} \epsilon_j}{n}\right\} = \beta$$

Mandel also examined the behavior of the residuals in Models A and B. He demonstrated, as indicated in Figure 4.6-1, that for large n the data represented by Model A tend to be scattered at random above and below the regression line, whereas data represented by Model B tend to remain on one side of the line for long sequences. This trait will assist in discriminating between the models (if a large number of observations can be made at different x's).

Example 4.6-1 Estimation for Cumulative Data

This example illustrates the analysis of cumulative data for a chemical reaction. Samples were periodically removed and analyzed during an experiment, yielding the following data:†

Time (min)	Log (fraction of sucrose remaining × 10)
0	1.000
10	0.954
20	0.895
30	0.843
40	0.791
50	0.735
60	0.685
70	0.628
80	0.581

Analysis of the data by the conventional relations given in Section 4.3 gave, for the model $\eta = \beta_0' + \beta_1 x$,

	Estimate	Standard Error
$\tilde{b}_0'$	1.0024	0.00170
$\tilde{b}_1$	−0.005303	0.0000357
Square root of sum of squares of residuals	0.00276	

Consequently, the estimated confidence interval for β_1 was $-0.005387 \leq \beta_1 < -0.005219$ for a significance level of $\alpha = 0.05$ and $t = 2.365$. However, the correct analysis using a cumulative error model should have been:

	Estimate	Standard Error
b_0'	1.00	0
b_1	−0.005238	0.000166
Square root of sum of squares of residuals	0.00469	

The intercept has zero error since it is simply the first measured value. The correctly estimated confidence interval is $-0.00563 \leq \beta_1 < -0.00485$, illustrating how the precision determined by the wrong method of analysis appears to be much greater than it should be.

4.6-2 Correlated Residuals

It is well known that data collected at a sequential series of values of time are liable to have correlated error residuals. It is then natural to ask how good the least squares estimation procedure of Section 4.3 or 4.4 is. This problem was investigated by Grenander‡ and Rosenblatt§ for time series; they concluded that if significant correlations actually exist, the estimates of the variances of the parameters in the least square solution will be biased and inefficient.

† J. Mandel, *Technometrics* **6**, 225, 1964.
‡ U. Grenander, *Ann. Math. Stat.* **25**, 253, 1954.
§ M. Rosenblatt, *Probability and Statistics*, John Wiley, New York, 1960, p. 246.

We have the space here only to outline the suggestions of Wold† concerning one appropriate estimation procedure for the coefficients and their variances in simple linear models with correlated residual errors. We shall defer to Section 5.4 the consideration of models with several independent variables. For methods of identification and estimation of the parameters in time series, that is, empirical models that are explicitly functions of time, consult other references.‡

To ascertain whether or not a sequential series of values are indeed correlated, a test for serial correlation should be carried out. The Durbin-Watson test§ for serial correlation of ϵ's was designed to apply to independent variables which are *exogenous*, that is the ϵ's are statistically independent of the x's. Hence the test is not strictly applicable if, as in a time series, some of the x's are lagged. The test is quite straightforward—all that need be done is to compute the statistic D for a series of n observations.

$$D = \frac{\sum_{t=2}^{n} (E_t - E_{t-1})^2}{\sum_{t=1}^{n} E_t^2} \qquad (4.6\text{-}13)$$

where E_t denotes the residual $(Y_t - \hat{Y}_t)$ at t, and $(E_t - E_{t-1})$ is the successive first difference.

Figure 4.6-2 illustrates the distribution of D and the regions of acceptance and rejection for serial correlation. Table C.10 in Appendix C lists the values of the upper, D_u, and lower, D_l, bounds for the test. If D calculated in Equation 4.6-13 is less than D_l or exceeds $(4 - D_l)$, then serial correlation is presumed to exist. If D falls within D_u and $(4 - D_u)$, the opposite is true. In the regions marked by a questionmark the test is inconclusive.

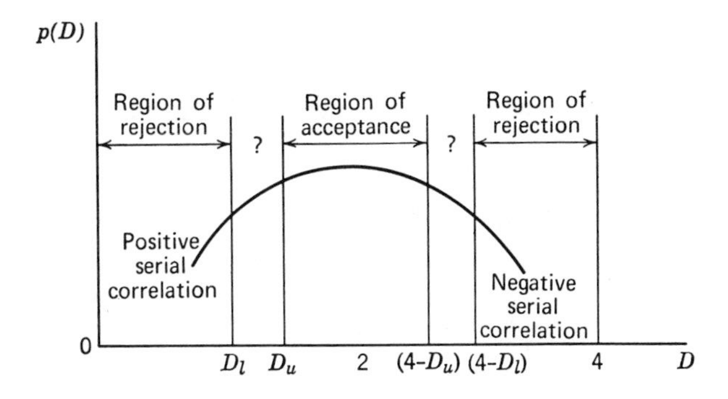

FIGURE 4.6-2 Distribution of D used in testing for serial correlation (the null hypothesis is that there is no serial correlation).

† H. Wold, *Bull. Inst. Int. Stat.* **32** (2), 1960; H. Wold, in *Proceed. 4th Berkeley Symp. Math. Stat. Prob.* **1**, 719, 1961.

‡ G. E. P. Box, G. M. Jenkins, and D. W. Bacon, "Models for Forecasting Seasonal and Nonseasonal Time Series" in *Spectral Analysis of Time Series*, ed. by D. P. Harris, John Wiley, New York, 1967; G. E. P. Box and G. M. Jenkins, *Time Series Analysis, Forecasting, and Control*, Holden-Day, San Francisco, 1969.

§ J. Durbin and G. S. Watson, *Biometrika* **38**, 159, 1951.

Example 4.6-2 Durbin-Watson Test for Serial Correlation

The following data represent the flow rates through a water-driven turbine as a function of the gate opening in inches:

Gate Opening (in)	Flow Rate (ft^3/sec)
1.1	8.92
2.3	15.51
2.9	20.08
2.5	16.38
3.5	19.53
4.0	22.12
4.7	24.60
5.0	25.35
5.1	25.01
4.5	23.03
5.5	29.47
6.0	32.97
6.3	35.05
6.5	36.58
6.7	38.30
6.9	40.06

1. Assume that a linear model with uncorrelated residuals, $\eta = \beta_0' + \beta_1 x$, represents the data. Find b_0' and b_1, find the estimates of β_0' and β_1, and calculate the 16 residuals. Examine the residuals.

2. Apply the Durbin-Watson test for serial correlation. Are the residuals correlated?

Solution:

The estimated regression equation from least squares is

$$\hat{Y} = 2.792 + 5.0101x$$

An F-test of the significance of β_1 indicates that it is a significant component of the model.

The residuals are:

0.616
1.193
2.756
1.061
−0.799
−0.714
−1.742
−2.495
−3.336
−2.310
−0.880
0.113
0.690
1.218
1.936
2.694

Clearly the residuals are not randomly distributed about zero as they should be. (The analysis of residuals is discussed in Section 7.1.)

From Equation 4.6-13, the statistic D is

$$D = \frac{\sum_{i=2}^{16} (E_t - E_{t-1})^2}{\sum_{i=1}^{16} E_t^2} = \frac{17.221}{51.149} = 0.336$$

From Table C.10 in Appendix C, for a two-sided test with $\alpha = 0.05$, $\alpha/2 = 0.025$, and $K = 1$, we find $D_l = 0.97$. Consequently, the hypothesis that the unobservable errors are uncorrelated is rejected. The residuals are positively correlated.

The first model to be examined is

$$Y_t = \beta x_t + \epsilon_t \tag{4.6-14}$$

where t represents the index for time. We let $x_1, x_2, \ldots, x_n$ be n consecutive values of the independent variable and $Y_1, Y_2, \ldots, Y_n$ be the observed values of the dependent variable. Although the residuals may be correlated with each other, the residuals each have an expected value of zero and are assumed to be not correlated with x_t:

$$\mathscr{E}\{\epsilon_t\} = 0$$

$$\mathscr{E}\{x_t \epsilon_t\} = 0$$

The least square estimator b of the coefficient β is obtained as described in Section 4.3:

$$b = \frac{\sum_{t=1}^{n} x_t Y_t}{\sum_{t=1}^{n} x_t^2} \tag{4.6-15}$$

The deviation of b from β can be written by introducing Equation 4.6-14 for Y_t in Equation 4.6-15:

$$b - \beta = \frac{\sum_{t=1}^{n} x_t \epsilon_t}{\sum_{t=1}^{n} x_t^2}$$

If we take terms with the same lag in time, the square of the deviation can be written as

$$(b - \beta)^2 = \frac{\sum_{t-1}^{n} x_t^2 \epsilon_t^2 + 2 \sum_{t=1}^{n-1} x_t x_{t+1} \epsilon_t \epsilon_{t+1} + \cdots + 2x_1 x_n \epsilon_1 \epsilon_n}{\left(\sum_{t=1}^{n} x_t^2\right)^2}$$

We are interested in obtaining the expected value of $(b - \beta)^2$, but since the values of $\epsilon_1, \epsilon_2, \ldots$ are unknown, we must replace terms such as $\mathscr{E}\{\sum_{t=1}^{n-1} x_t x_{t+1} \epsilon_t \epsilon_{t+1}\}$ with an estimate $\sum_{t=1}^{n-1} x_t x_{t+1} E_t E_{t+1}$ where

$$E_t = Y_t - b x_t$$

and thus obtain an estimate for the variance of b. Also, because the correlation $x_t x_{t+k} \epsilon_t \epsilon_{t+k}$ drops off as the lag in time between two terms increases, in the approximation for the variance of b we shall delete all terms after $t + k$ as being negligible. Consequently,

$$\widehat{\text{Var}}\{b\} \cong \frac{\sum_{t=1}^{n} x_t^2 E_t^2 + 2 \sum_{t=1}^{n-1} x_t x_{t+1} E_t E_{t+1} + \cdots + 2 \sum_{t=1}^{n-k} x_t x_{t+k} E_t E_{t+k}}{\left(\sum_{t=1}^{n} x_t^2\right)^2} \tag{4.6-16}$$

A similar but more complicated treatment of the model

$$Y_t = \beta_0' + \beta_1 x_t + \epsilon_t \tag{4.6-17}$$

can be carried out with the results listed below:

$$b_1 = \frac{\sum_{t=1}^{n} (x_t - \bar{x}) Y_t}{\sum_{t=1}^{n} (x_t - \bar{x})^2} \tag{4.6-18}$$

$$b_0' = \bar{Y} - b\bar{x} \tag{4.6-19}$$

$$\bar{x} = \frac{1}{n} \sum_{t=1}^{n} x_t$$

$$\bar{Y} = \frac{1}{n} \sum_{t=1}^{n} Y_t$$

$$\widehat{\text{Var}}\{b_0'\} \cong \frac{C_0}{n} - \frac{2\bar{x}C_1}{n \sum_{t=1}^{n} (x_t - \bar{x})^2} + \bar{x}^2 \text{ Var}\{b_1\} \tag{4.6-20}$$

$$\widehat{\text{Var}}\{b_1\} \cong \frac{C_2}{\left[\sum_{t=1}^{n} (x_t - \bar{x})^2\right]^2} \tag{4.6-21}$$

where

$$C_0 = \sum_{t=1}^{n} E_t^2 + 2 \sum_{t=1}^{n-1} E_t E_{t+1} + \cdots + 2 \sum_{t=1}^{n-k} E_t E_{t+k}$$

$$C_1 = \sum_{t=1}^{n} (x_t - \bar{x}) E_t^2 + \sum_{j=1}^{k} \sum_{t=1}^{n-j} (x_t - \bar{x}) E_t E_{t+j} + \sum_{j=1}^{k} \sum_{t=1}^{n-j} (x_{t+j} - \bar{x}) E_t E_{t+j}$$

$$C_2 = \sum_{t=1}^{n} (x_t - \bar{x})^2 E_t^2 + 2 \sum_{j=1}^{k} \sum_{t=1}^{n-j} (x_t - \bar{x})(x_{t+j} - \bar{x}) E_t E_{t+j}$$

Also

$$\widehat{\text{Covar}}\{b_0', b_1\} \cong \frac{C_1}{n \sum_{t=1}^{n} (x_t - \bar{x})^2} - \bar{x} \text{ Var}\{b_1\} \tag{4.6-22}$$

In addition to Equations 4.6-14 and 4.6-17, many other models have been proposed in which the error residuals are not independent, but we do not have the space to describe them here. Methods of detecting and treating various other types of models with correlated residuals and for estimating the degree of bias introduced by ignoring the correlation when it exists can be found in the references at the end of the chapter.

4.7 DETECTION AND REMOVAL OF OUTLIERS

In Section 3.8 we examined the problem of outliers or extreme points and described some tests for detecting outliers in a sample. Even carefully prepared experiments may yield inhomogeneous data because uncontrolled experimental conditions may change, the experimenter may make a mistake when taking or recording an observation, and so forth. If the experimenter does not detect those "wildshots," "blunders," outliers, or other anomalous observations, they will be incorporated with the valid observations used to estimate the regression line; they may cause substantial displacements in the estimated parameters and especially in the estimates of the variances of the estimated parameters which are strongly influenced by extreme values. In such circumstances the experimentalist would like to delete the outliers. On the other hand, he does not want to suppress any observations that deviate considerably from a preconceived (linear here) trend because the deviations may contain valid information, such as that the linear model is wrong.

One difficulty in rejecting outliers in connection with linear regression analysis (in contrast with the methods of Section 3.8) is that a pattern has been imposed on the data; that is, a functional relationship has been assumed. Consequently, we are forced to examine over again what an outlier means. In Section 3.8 the outlier conceptually was an observation isolated from the others which could be tested for through its numerical value. However, in regression analysis, the numerical value, the location of the value, and the character of the model have to be taken into consideration. Although the pattern of observations is characterized by a linear model, this is a man-made restriction and the process variable is under no obligation to meet such a constraint. If an observation were out of line and the adjacent observations showed a similar tendency, though perhaps to a lesser degree, we would be much more likely to regard the observation as representing real departure from the *assumed model* than as an outlier. On the other hand, an observation standing out from its nearest neighbors would much more likely be regarded as an outlier. To qualify as an outlier, then, an observation should significantly disrupt the *correctly* assumed trend in the (linear) model.

Although statistical techniques might appear to offer objective guides for rejection of outliers, personal opinions and prejudices do enter into the choice of methods used in data analysis. Probably no criterion is better than the judgment exercised by an experienced experimenter who is thoroughly familiar with his measurement techniques.

One way to reach a decision on a supposed outlier through use of statistical tools is to examine the residuals. If residuals $(Y_i - \hat{Y}_i)$ are calculated including the outlier, then every residual is affected. Also, the residuals may be correlated even without outliers. Hence, the use of residuals as a basis for rejection of outliers has some disadvantages. Nevertheless, suppose we wish to test whether a single observation deviates far enough from the value predicted by the estimated regression line to cause us to classify the observation as an outlier. We assume that the model has been shown to be an appropriate one. The residuals are supposed to be normally distributed with zero ensemble mean and an ensemble variance of $\sigma^2_{\bar{Y}_i}$. If we calculate‡

$$\frac{|E^\dagger - \bar{E}^\dagger|}{s_{Y_i}} = V \tag{4.7-1}$$

where $E^\dagger$ is the residual for the suspected outlier, $\bar{E}^\dagger$ is the mean of all the other residuals, and s_{Y_i} is based on Equation 4.3-15 with the suspected observation deleted, the values tabulated in Table 4.7-1 based on the distribution of V can be used as critical values to accept or reject the observation. If V exceeds the value listed in the table, the hypothesis H_0: $E^\dagger$ does not correspond to an outlier, is rejected.

TABLE 4.7-1 CRITICAL VALUES OF V USED TO DETERMINE WHETHER OR NOT TO REJECT AN OUTLIER

Sample Size	Significance Level α for One-Sided Test	
n	0.05	0.01
3	123	31.4
4	7.17	16.27
5	5.05	9.00
6	4.34	6.85
7	3.98	5.88
8	3.77	5.33
9	3.63	4.98
10	3.54	4.75
15	3.34	4.22
20	3.28	4.02
25	3.26	3.94

This table is abridged, with permission, from F. S. Acton, *Analysis of Straight Line Data*, John Wiley, New York, 1959, p. 261.

Unfortunately, if there is more than one anomalous observation, the V criterion will not in general detect the anomalous values, especially when there are less than 30 residuals to be analyzed. If more than one extreme value is suspect, one of the Dixon criteria can be employed.§ These statistics characterize the deviation of one or

‡ W. R. Thompson, *Ann. Math. Stat.* **6**, 214, 1962.

§ W. J. Dixon, *Biometrics* **9**, 74, 1953.

TABLE 4.7-2

Number of Residuals	Compute r_{ij}	
	If E_n is Suspect	If E_1 is Suspect
$3 \leq n \leq 7$	r_{10}: $(E_n - E_{n-1})/(E_n - E_1)$	$(E_2 - E_1)/(E_n - E_1)$
$8 \leq n \leq 10$	r_{11}: $(E_n - E_{n-1})/(E_n - E_2)$	$(E_2 - E_1)/(E_{n-1} - E_1)$
$11 \leq n \leq 13$	r_{21}: $(E_n - E_{n-2})/(E_n - E_2)$	$(E_3 - E_1)/(E_{n-1} - E_1)$
$14 \leq n \leq 25$	r_{22}: $(E_n - E_{n-2})/(E_n - E_3)$	$(E_3 - E_1)/(E_{n-2} - E_1)$

several elements from neighboring terms in the series. We assume that all the residuals except the outliers are from a normal population with unknown mean and variance; we arrange them in order from highest to lowest so that $E_1 \leq E_2 \leq \cdots \leq E_n$. The null hypothesis is that E_n (or E_1) is not an outlier. Choose α, the significance level, and compute the statistic indicated as shown in Table 4.7-2.

For a two-sided test (the outlier can be too large as well as too little), look up $r_{1-\frac{\alpha}{2}}$ from Table C.11 in Appendix C; if r_{ij} is greater than $r_{1-\frac{\alpha}{2}}$, reject the suspect observation; otherwise retain it. A one-sided test can be made by ascertaining whether $r_{ij} > r_{1-\alpha}$.

The Dixon criteria are optimal for small displacements and are independent of the number of errors, whereas the other criterion, that given by Equation 4.7-1, is optimal when there is only one anomalous value and is independent of the size of the residual. Additional references for analogous tests can be found at the end of Chapter 3.

Supplementary References

General

Acton, F. S., *Analysis of Straight Line Data*, John Wiley, New York, 1959.

Brownlee, K. A., *Statistical Theory and Methodology in Science and Engineering* (2nd ed.), John Wiley, New York, 1965.

Cramer, H., *Mathematical Methods of Statistics*, Princeton Univ. Press, Princeton N.J., 1954.

Davies, O. L., *Statistical Methods in Research and Production* (3rd ed.), Oliver and Boyd, London, 1958.

Feller, W., *An Introduction to Probability Theory and Its Applications*, John Wiley, New York, 1957.

Fisher, R., *Statistical Methods and Scientific Inference*, Oliver and Boyd, London, 1956.

Guest, P. G., *Numerical Methods of Curve Fitting*, Cambridge Univ. Press, Cambridge, England, 1961.

Natrella, M. G., "Experimental Statistics," NBS Handbook 91, U.S. Government Printing Office, Washington, D.C., 1963.

Plackett, R., *Principles of Regression Analysis*, Oxford Univ. Press, Oxford, England, 1960.

Williams, E. J., *Regression Analysis*, John Wiley, New York, 1959.

Both Variables Stochastic

Acton, F. S., *Analysis of Straight Line Data*, John Wiley, New York, 1959, Chapter 5.

Bartlett, M. S., "Fitting a Straight Line When Both Variables Are Subject to Error," *Biometrics* **5**, 207, 1949.

Basman, R. L., "An Expository Note on the Estimation of Simultaneous Structural Equations," *Biometrics* **16**, 464, 1960.

Berkson, J., "Are There Two Regressions," *J. Amer. Stat. Assn.* **45**, 164, 1950.

Clutton-Brock, M., "Likelihood Distributions for Estimating Functions When Both Variables Are Subject to Error," *Technometrics* **9**, 261, 1967.

Cochran, W. G., "Some Consequences When the Assumptions for the Analysis of Variance are not Satisfied," *Biometrics* **3**, 22, 1948.

Deming, W. E., "The Application of Least Squares," *London, Edinburgh, and Dublin Phil. Mag.* **11**, Series 7, 1930.

Eisenhart, C., "The Assumptions Underlying Analysis of Variance," *Biometrics* **3**, 1, 1947.

Halperin, M., "Fitting of Straight Lines and Prediction When Both Variables are Subject to Error," *J. Amer. Stat. Assn.* **56**, 657, 1961.

Keeping, E. S., "Note on Walds' Method of Fitting a Straight Line When Both Variables are Subject to Error," *Biometrics* **12**, 445, 1956.

Kerridge, D., "Errors of Prediction in Multiple Regression With Stochastic Regression Variables," *Technometrics* **9**, 309, 1967.

Madansky, A., "The Fitting of Straight Lines When Both Variables Are Subject to Error," *J. Amer. Stat. Assn.* **54**, 173, 1959.

Wald, A., "The Fitting of Straight Lines When Both Variables are Subject to Error," *Ann. Math. Stat.* **11**, 284, 1940.

Determination of Functional Form

Davis, D. S., *Nomography and Empirical Equations*, Reinhold, New York, 1955, pp. 3–79.

Dolby, J. L., "A Quick Method for Choosing a Transformation," *Technometrics* **5**, 317, 1963.

Hoerl, A. E., "Fitting Curves to Data" in *Chemical Business Handbook*, McGraw-Hill, New York, 1954.

Johnson, L. H., *Nomography and Empirical Equations*, John Wiley, New York, 1952, pp. 95–146.

Lipka, J. *Graphical and Mechanical Computation, Part II*, John Wiley, New York, 1918, pp. 120–169.

Running, T. R., *Empirical Formulas*, John Wiley, New York, 1917.

Scarborough, J. B., *Numerical Mathematical Analysis*, Johns Hopkins Press, Baltimore, 1958, pp. 455–489.

Estimation with Correlated Residuals

Duncan, D. B. and Jones, R. H., "Multiple Regression with Stationary Errors," *J. Amer. Stat. Assn.* **62**, 917, 1967.

Durbin, J., "Estimation of Parameters in Time-series Regression Models," *J. Royal Stat. Soc.* **B22**, 139, 1960.

Goldberger, A. S., *Economic Theory*, John Wiley, New York, 1964.
Johnson, J. *Econometric Methods*, McGraw-Hill, New York, 1963.
Sargan, J. D., "The Maximum Likelihood Estimation of Economic Relationships with Auto Regressive Residuals," *Econometrica* **29**, 414, 1961.
Zellner, A., "Econometric Estimation with Temporally Dependent Disturbance Terms," *Int. Econ. Rev.* **2** (2), 164, 1961.

Problems

4.1 State for each of the equations below whether it is linear or nonlinear in the parameters (x is the independent variable):

(a) $y = \beta_0 + \beta_1 x_1 + \beta_2 x_2$

(b) $y = \dfrac{1}{\beta_0 + \beta_1 x_1}$

(c) $y = e^{-\beta_0 x + \beta_1}$

(d) $\ln y = \beta_0 + \beta_1 x$

(e) $\dfrac{1}{y} = \beta_0 + \dfrac{\beta_1}{x}$

(f) $y = \beta_0 x_1^{\beta_1} x_1^{\beta_2}$

4.2 Transform each of the nonlinear models in Problem 4.1 to one linear in the parameters.

4.3 In which of the following models can the parameters be estimated by linear estimation techniques?

(a) $y = \beta_0 + \beta_1 x + \beta_2 x^2$
(b) $y = \beta_1 x_1 + \beta_2 x_2$
(c) $xy = \beta_1 x + \beta_2$
(d) $y = (\beta_1) \ln x + \beta_2$
(e) $y = e^{\beta_1 x}$

4.4 Under what circumstances can equations, nonlinear in the parameters, by fit by linear regression?

4.5 Determine the best functional relation to fit the following data:

(a)

x	Y
1	5
2	7
3	9
4	11

(b)

x	Y
2	94.8
5	87.9
8	81.3
11	74.9
14	68.7
17	64.0

(c)

x	Y
2	0.0245
4	0.0370
8	0.0570
16	0.0855
32	0.1295
64	0.2000
128	0.3035

(d)

x	Y
0	8290
20	8253
40	8215
60	8176
80	8136
100	8093

4.6 Can the parameter α in the model $y = (x_1 + x_2)/(\alpha + x_3)$ be estimated by a linear regression computer routine. Explain. Will fitting the transformed model:

(a) $y = k(x_1 + x_2 - x_3 y)$, where $k = \dfrac{1}{\alpha}$

or

(b) $\dfrac{1}{y} = \alpha\left(\dfrac{1}{x_1 + x_2}\right) + \dfrac{x_3}{x_1 + x_2}$

accomplish your objective? Explain.

4.7 From the values of x and Y given, determine the functional form of a suitable linear relation between Y and x. Do not evaluate the coefficients; just ascertain the form of the function $Y = f(x)$.

x	Y
0	0
0.1	1.333
0.2	1.143
0.3	0.923
0.4	0.762
0.5	0.645
0.6	0.558
0.7	0.491
0.8	0.438
0.9	0.396
1.0	0.360

4.8 A rate model for a batch reactor is

$$r = \frac{kK_A \dfrac{(1-x)}{(1+x)} p}{1 + K_A\left(\dfrac{1-x}{1+x}\right)p + K_w\left(\dfrac{x}{1+x}\right)p}$$

where k, K_A, and K_w are coefficients, x is the independent variable, and p is another independent variable. It is proposed to write the model as

$$r = \frac{1 - xp}{\beta_1 + \beta_2 x}$$

where

$$\beta_1 = \frac{1}{kK_A} + \frac{p}{k}$$

$$\beta_2 = \frac{1}{kK_A} + \frac{K_w - K_A}{kK_A} p$$

and to estimate β_1 for data when $x = 0$ and to estimate β_2 from data using fixed values of x but a function of p. Comment on this proposal.

4.9 A model is proposed:

$$y = \frac{\beta_1 \beta_2 x}{(1 + \beta_2 x)^2}$$

Can it be made linear in the parameters as follows:

$$\left(\frac{x}{y}\right)^{1/2} = \frac{1}{(\beta_1\beta_2)^{1/2}} + \frac{\beta_2}{(\beta_1\beta_2)^{1/2}} x$$

and the parameters estimated by linear least squares?

4.10 Below are listed several experiments. For each, tell in one or two sentences whether the method of least squares is appropriate for: (1) estimating the parameters in a linear empirical model, and (2) estimating the confidence interval for the parameters.

(a) A graduate student, in order to calibrate a refractometer, analyzes several samples of solutions containing known concentrations of benzene. He collects data in terms of grams of benzene/gram of solutions versus the refractive index.

(b) A biologist measures the length of the forearm of sons and fathers to see if there is any relationship.

(c) A chemist weights the displacement of water by solid spheres of known radii with weights that are not accurate.

(d) Two proportional counters (for beta radiation) measures the same radioactive source for a series of sources. (Both have a number of types of error involved in their output.)

(e) A chemist attempts to pour x grams of sodium carbonate on a scale. The sodium carbonate is added to an exactly known volume of water. The pH of the solution is then measured.

(f) In a more careful trial, he tries to weigh x grams but fails. He reads the scale and records x plus an error. Then the refractive index is measured.

(g) A technician measures the hardness of epoxy paint on steel sheets at an oceanshore laboratory. Then he goes to a laboratory in the desert and makes similar measurements on other samples. He would like to relate hardness to location.

4.11 The following rocks were measured for threshold contact pressure and shear strength in psia.

In reply to the question as to whether or not linear regression analysis can be applied to the data, the following answers were received. For each answer, tell whether or not the reasoning and answer are correct and, if not, wherein the fallacy lies.

	Threshold Contact Pressure (10^{-3})	Shear Strength (10^{-3})
Basalt	634	4.5
Sandstone, A	570	6.5
Granite	494	8.5
Polomite	364	9.0
Marble	102	4.6
Sandstone, B	86	3.0
Limestone	50	3.0
Shale	3	1.2

(a) The data when plotted are not represented by a straight line but do seem to fall along a smooth curve. If the data can be transformed to a linear form so that a straight line is obtained, then linear regression analysis can be applied.

(b) There is a definite relation between the pressure and the shear strength. Regression analysis can be used to determine the relationship between these two variables.

(c) Linear regression analysis cannot be used on the data. A plot of the experimental data points does not yield a straight line, and there are no constant intervals in either of the measurements by which differences could be taken in order to establish a suitable polynomial expression.

(d) Linear regression analysis cannot be applied to the above data because the data are so scattered and inconsistent that it is not possible to find differences that are constant for constant changes.

(e) One cannot apply regression analysis to the above data because they consist of eight different sets of data, each of which is unrelated to the others.

(f) Linear regression analysis cannot be applied to the experimental data because the value of the correlation coefficient indicates that there is practically no correlation between the two measured quantities. The basic assumption of linear regression analysis is that one quantity is a linear function of the other.

4.12 You are given data for Y versus x and asked to fit an empirical model of the form:

$$\eta = \alpha + \beta x$$

where β is a *known* value. Given an equation to calculate the best estimate of α.

4.13 Show that Equation 4.3-7a, page 113, is correct for a line through the origin by application of the maximum likelihood method.

4.14 In fitting a linear empirical model to some experimental data, discuss briefly the following points:

(a) Can an equation of the form $\eta = \alpha + \beta x$ be used or must you use $\eta = \alpha + \beta(x - \bar{x})$?

(b) Is it necessary to replicate points?

(c) How can you tell if a line through the origin fits better than a line of the form $\eta = \alpha + \beta x$?

(d) In the analysis of variance, of what use is the

sum of the squares of the deviations between the grand mean $\bar{Y}$ and the predicted values of Y, $\hat{Y}$?

4.15 You are asked to fit a line $\eta = 6.2 + \beta x$ to some experimental data. Derive the equations which will enable you to estimate:
(a) β, the slope.
(b) $\sigma_{Y_i}^2$, the variance.
(c) The confidence interval for η.

4.16 Given a model

$$\eta = \beta_1 \cos \omega x + \beta_2 \sin \omega x + \beta_3$$

with a period

$$T = \frac{2\pi}{\omega}$$

given the fact that the graph of the model extends over an integral number of periods, and given pairs of (Y, x) data points:
(a) Derive the normal equations.
(b) Set up the simplest expressions possible to estimate β_1, β_2, and β_3.
(c) Find the sum of the squares of the residuals

$$\sum (\hat{Y}_i - Y_i)^2$$

(d) State the number of degrees of freedom for (c).
(e) Find the variances of β_1 and β_2.

In the case of a periodic function such as the above, the amplitude of the wave is $(\beta_1^2 + \beta_2^2)^{1/2}$; the "intensity" is $(\beta_1^2 + \beta_2^2)$. Explain how you might test whether $(\beta_1^2 + \beta_2^2)$ is significantly different from zero. In other words, is the wave a reality? List all the assumptions required.

4.17 The following data represent bursting strengths of aluminum foil:

Disk Thickness (in)	Bursting Pressure (psia)
0.001	1
0.002	5
0.003	15
0.0045	21
0.005	22
0.008	47
0.010	57

(a) Estimate β_0 and β_1 in the linear model $\eta = \beta_0 + \beta_1 x$.
(b) Estimate the variance of β_0; β_1; $\hat{Y}$.
(c) Find if the parameters β_0 and β_1 are significantly different from zero.

Note: Use the $\alpha = 0.05$ significance level if needed in any of the above.

4.18 Derive an equation to estimate β in terms of the observed data pairs (Y, x) for a model of the form $\eta = \alpha + \beta x$, where α is a known constant. Does it make any difference if the x values are calculated about the mean of the independent variable, or about some arbitrary origin or about $x = 0$?

4.19 The following data have been collected:

x	Y
10	1.0
20	1.26
30	1.86
40	3.31
50	7.08

Which of the following three models best represents the relationship between Y and x?

$$\eta = e^{\alpha + \beta x}$$
$$\eta = e^{\alpha + \beta_1 x + \beta_2 x}$$
$$\eta = \alpha x^{\beta}$$

(Do not find the values of the estimated parameters.)

4.20 Take the following series of values and fit a model of the form $\eta = \beta x$; repeat with $\eta = \alpha + \beta x$. Determine the confidence intervals on η and β for the first model, and for η, α, and β for the second. Prepare an analysis of variance. Which estimated regression equation gives the best fit?

x:	9	8	7	7	6	4	3	3	1	2
Y:	7	9	7	8	7	3	6	1	2	2

Make a graph of each estimated regression equation, plot the given points, and put on each side of $\hat{Y}$ lines showing the locus of the confidence limits for η at a 5-percent significance level. Plot the joint confidence region for $P = 0.95$ in parameter space of α versus β.

4.21 Given the following equilibrium data for the distribution of SO_3 in hexane, determine a suitable linear (in the parameters) empirical model to represent the data.

x, Pressure (psia)	Y, Weight Fraction Hexane
200	0.846
400	0.573
600	0.401
800	0.288
1000	0.209
1200	0.153
1400	0.111
1600	0.078

4.22 To date there have been no really successful generalized correlations of pressure drop for finned tubes with gas flows. The standard errors of estimate for most friction factor correlations are of the order of ± 40 percent of the predicted value, a quantity excessive for engineering use. Therefore, the pressure drop data below are presented for individual tubes without attempting to correlate all the data into a single generalized equation. Estimate α_0 and α_1 in the model

$$\sigma \frac{\Delta p}{n} = \alpha_0 v^{\alpha_1}$$

where $\sigma \, \Delta p/n$ is regarded as the random variable, v (ft/min) is regarded as a deterministic variable,

Δp = pressure drop, σ = relative density (to that of air), and n = the number of tubes, a constant. What precautions are needed to ensure that $\sigma \Delta p/n$ is indeed a random variable and v is a deterministic one?

v	$\sigma \frac{\Delta p}{n}$ (in. H_2O)
400	0.0125
470	0.0165
590	0.0215
610	0.0225
620	0.0235
840	0.0420
950	0.0530
1200	0.0750
1400	0.0970
1550	0.120

(a) Prepare an analysis of variance and determine if the model represents the data better than a horizontal line.

(b) Calculate the confidence limits on α_0 and α_1.

(c) Calculate the confidence limits on the expected value of $\sigma \Delta p/n$ for the first, fourth, seventh, and tenth points of the data. Prepare two lines on each side of the predicted response to indicate the locus of the confidence limits for a significance level of 0.05. Also, make a plot of the confidence region in parameter space.

4.23 A number of experiments have been carried out in order to build an empirical model of a chemical reactor. The following data represent product yields, Y_i, in lb moles/lb feed.

Reaction Time (hr)				
1	2	3	7	28
2.11	2.34	2.47	2.51	2.62
2.12	2.38	2.44	2.48	2.62
2.07	2.39	2.38	2.53	2.60
		2.38	2.52	2.55
		2.41	2.55	2.57

Fit a linear model, $Y = \beta_0 + \beta_1(x - \bar{x}) + \epsilon$, by the data. Use Bartlett's test to ascertain whether or not the variances are the same at each time period; if not, use weighted least squares in which the variance is a function of time.

4.24 If an independent variable in a model is found to have a significant effect on the dependent variable by a statistical test, does this also mean the independent variable should be considered as one cause of the value of the dependent variable in a physical sense?

4.25 Suppose that several values of x are selected and Y is measured for a model $\eta = \alpha + \beta x$. It is possible to use the same data to fit a model $x = \alpha' + \beta' Y$?

4.26 For the liquid flow rate of 4000 lb/(hr)(ft^2), Hudson† measured the foam height as a function of the gas rate in a distillation column tray.

† University of Texas Ph.D. Dissertation, 1968.

Foam Height (in)	Gas Rate (lb/(hr)(ft^2))
7.56	200
6.53	250
5.09	300
4.56	350
3.51	450
2.56	600
2.28	800

Find a linear relation for the data between foam height and gas rate.

4.27 The following data have been obtained for two-phase flow:‡

λ_{mix}/λ_0	Re
1.75	800
1.68	900
1.44	1200
1.30	1600
1.31	2100
1.29	2750
1.15	3750
1.00	6000

where:

Re = Reynolds number calculated on the reduced velocity of the liquid phase

λ_0 = coefficient of hydraulic resistance for single-phase flow

λ_{mix} = coefficient of hydraulic resistance for a vapor-liquid mixture

Find a linear relation between λ_{mix}/λ_0 and Re.

4.28 By dimensional analysis for a problem in heat transfer, it was shown that

$$\text{Nu} = \alpha\, \text{Re}^{\beta}$$

where Nu = Nusselt number, Re = Reynolds number, and α and β are constants. Assuming that the Nu only is a random variable, obtain the best estimates of α and β based on the following data:

Re:	100	100	200	200	300	300	400	400	500	500
Nu:	31	36	39	40	40	42	43	45	46	49

4.29 The barometric pressure is related to the height by the following equation:

$$\frac{p}{p_0} = e^{-kz/T}$$

where p is the barometric pressure, p_0 is the pressure at $z = 0$, z is the height, k is a constant, and T is the temperature. At approximately constant T, six measurements of pressure were made by a group climbing a mountain:

Level (ft)	Pressure (in Hg)
3290	27.2
3610	23.4
3940	19.8
4600	14.3
4930	12.2
5260	10.5

‡ *Int. Chem. Eng.* **6**, 43, 1966.

What are the best estimates of k and p_0? Does your p_0 agree with 29.9 in Hg at sea level ($z = 0$)? What is the joint confidence region for k and p_0?

4.30 The precision of measurement of the data for $\sigma\ \Delta p/n$ given in Problem 4.22 is roughly proportional to the value of v. Estimate α_0 and α_1 in the model of Problem 4.22 by using an appropriate weighted least squares technique. What difference does the revised procedure have on the confidence interval for $\sigma\ \Delta p/n$ at $v = 1200$?

4.31 An experimenter attempts to fit a model of the form $Y = \beta x + \epsilon$ in which the ensemble standard deviation of Y is proportional to Y itself. Obtain a weighted least squares estimate of β by minimizing Equation 4.3-3 in the chapter.

4.32 The following data are to be treated in three different ways:

(a) Y is the random variable and x is a deterministic variable. Obtain the estimated regression equation for the model $Y = \beta_0 + \beta_1 x + \epsilon$, and estimate the confidence interval for η at $x = 62$.

(b) Y is the random variable and x is a deterministic variable. Predict X for a measured Y of 150 for the model listed in (a). Can you estimate a confidence interval for X?

(c) Both X and Y are random variables. Estimate the model parameters in $Y = \beta_0 + \beta_1 X$. Can you estimate the confidence interval for Y at a given value of X; for X at a given value of Y?

X or x	Y
60	110
60	135
60	120
62	120
62	140
62	130
62	135
64	150
64	145
70	170
70	185
70	160

4.33 The following extract is from the *J. Chem. Educ.* **42**, 609, 1965. Comment on the extract. Is the author correct or not?

> Currently, a common linear regression analysis in physical organic chemistry is that of the regression of $\log k$ on σ to obtain a Hammett equation, $\log k = \log k^\circ + \rho\sigma$, where $\log k^\circ$ is the regression intercept and ρ is the regression slope. This equation is useful in several ways; one of these is the estimation of new k-values from the corresponding σ-values, if available. It has been stated in the literature that not the above equation but, instead, the equation for the regression of σ on $\log k$ should be used for the estimation of new σ-values from experimental k-values. This proposal is questionable for two reasons. First, for many reaction series of *m*- and *p*-substituted benzene derivatives, it is obvious that there is a linear functional relationship between $\log k$ as the dependent variable and σ as the independent variable. Second, even though there are uncertainties in σ-values, they are generally less than the uncertainties in experimental k-values. For these reasons, the equation for the regression of $\log k$ on σ is equally applicable to the estimation of new k-values from known σ-values and of new σ-values from experimental k-values.

4.34 A first-order homogeneous reaction of N_2O_5 in a lumped tank is to be modelled by

$$\ln \frac{p}{p_0} = k(t - t_0)$$

where p = partial pressure of the N_2O_5, p_0 = partial pressure at $t = t_0$, t_0 is the reference time, and k is a constant whose value is to be estimated. Use the data of Daniels and Johnson† to estimate k, taking into account that the p values were obtained continuously from the same tank ($\hat{k}$ was reported to be in the range of 0.0096 to 0.0078 min^{-1}).

Estimate the variance of $\hat{k}$.

Time (min)	$p_{N_2O_5}$ (mm Hg)
0	308.2
20	254.4
30	235.5
40	218.2
50	202.2
60	186.8
100	137.2
140	101.4
200	63.6

4.35 The following data have been obtained from a batch reaction used to make ethylene glycol (E.G.) from ethylene oxide (E.O.) with a 9-percent ethylene oxide feed:

	Effluent Composition	
Holding Time	E.O.	E.G.
31.2	1.27	9.53
25.0	1.52	9.23
18.7	1.93	8.85
12.5	2.62	8.03
6.25	4.07	6.38
3.12	5.62	4.50

Explain how you would determine the best linear relationship between the E.O. and E.G. compositions, the independent and dependent variables, respectively. State all assumptions.

4.36 Apply the Dubin-Watson test for serial correlation to the data of Examples 4.3-1 and 4.3-2. Would you change the estimation procedure in these examples? If so, in what way?

† *J. Amer. Chem. Soc.* **43**, 53, 1921.

4.37 In the following data it is suspected that the unobservable errors in the model $Y = \beta_0' + \beta_1 x + \epsilon$ are correlated. Estimate β_0' and β_1, first assuming the errors are not correlated and then assuming the errors are correlated.

Time (hr)	Response (mv)
51	70
52	64
53	60
54	49
55	47
56	31
57	44
58	38
59	34

Apply the Durbin-Watson test for serial correlation. What conclusion can you reach?

4.38 In calibrating a flow-measuring device for which the deterministic relation between velocity v and pressure drop Δp is

$$v = c\sqrt{\Delta p} \qquad c = \text{constant}$$

the following data were obtained. Would you reject any of the observations as outliers?

v (ft/sec)	Δp (in. Hg)
0.927	0.942
0.907	0.903
0.877	0.823
0.719	0.780
0.761	0.757
0.644	0.684
0.508	0.603

4.39 The information given below is intended to confirm that the investment advisory service has merit. Does the TPO performance line actually demonstrate this?

1. Using the November 20 issue, 77 stocks on the New York Stock Exchange had "long" TPO's. This means that these 77 stocks had short-term technical strength as well as high long-term profit potential.
2. The price appreciation of the 77 stocks during the next 10 weeks was calculated and plotted on the diagram. Each dot represents the price appreciation of one of the 77 stocks. For example, a \$10 stock with a market value of \$12 at the end of the 10-week holding period would be plotted as a 20-percent price appreciation.

Conclusions:

1. The higher the 12-month profit potential of a TPO stock, the better the chance of a short-term profit. This is illustrated by the TPO average performance line.
2. The average appreciation of all stocks was 20.5–6.5 percent better than the 14-percent rise in the Dow Jones Industrial Average during the same time period.
3. Although individual stocks may or may not live up to their rating, on the *average* the computer ratings are accurate. Equal dollar investment in 10 stocks is required to give a 9 out of 10 chance that the average price appreciation will be within 5 percent of the average performance line.
4. Since the market went up 14 percent during the same period, the stocks (points) under the dashed line would have shown a loss in a sideways market.

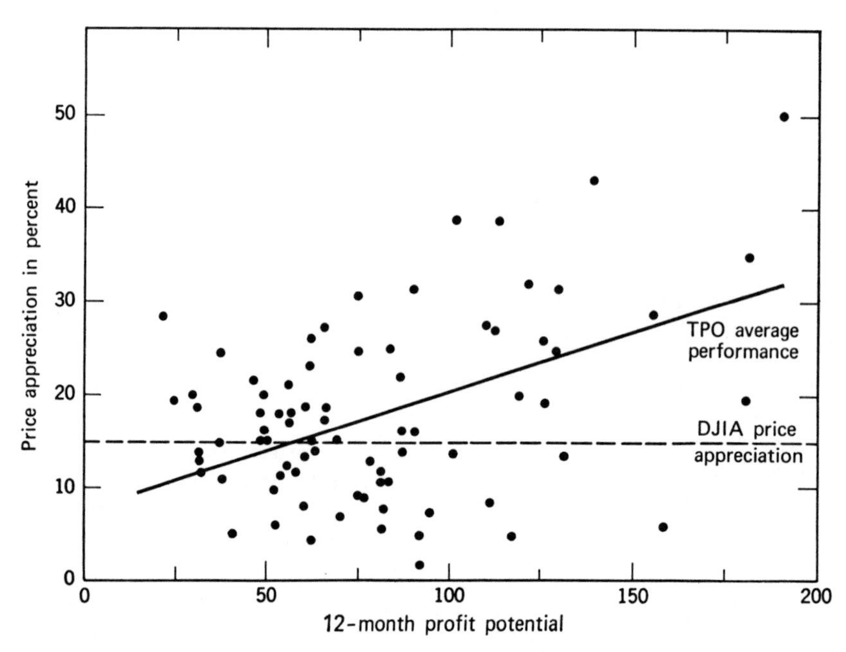

FIGURE P4.39 Ten-week price appreciation of New York Stock Exchange stocks published in the November 20, 1962, issue of an investment service.

CHAPTER 5

Linear Models With Several Independent Variables

In this chapter we are interested in exactly the same estimation problem as in Chapter 4, but the problem is complicated by the use of a model with several independent variables. Given that n sets of experimental data have been collected, hopefully based on a plan such as described in Chapter 8, how can the best estimates of the parameters in a proposed process model be obtained? How can the confidence intervals of the parameters be formed, and how can hypothesis tests similar to those described in Chapter 4 be executed? One additional factor of interest, arising because of the large number of computations involved, is how to carry out the computations on a digital computer. The generic term often applied to the estimation procedure described here is *multiple regression*.

5.1 ESTIMATION OF PARAMETERS

We are interested in estimating the parameters in the model

$$\mathscr{E}\{Y \mid x\} = \eta = \beta_0 + \beta_1(x_1 - \bar{x}_1) + \beta_2(x_2 - \bar{x}_2) + \cdots + \beta_q(x_q - \bar{x}_q) \qquad (5.1\text{-}1)$$

or its equivalent

$$Y = \beta_0 + \beta_1(x_1 - \bar{x}_1) + \beta_2(x_2 - \bar{x}_2) + \cdots + \beta_q(x_q - \bar{x}_q) + \epsilon \qquad (5.1\text{-}1a)$$

where ϵ is the unobservable error which causes Y to differ from η. An alternate form of the same model is

$$\eta = \beta_0' + \beta_1 x_1 + \beta_2 x_2 + \cdots + \beta_q x_q \qquad (5.1\text{-}2)$$

All the assumptions listed in Section 4.2 still are in effect. The estimated regression equation corresponding to Equation 5.1-1 will be written as an extension of the corresponding equation of Section 4.3:

$$\hat{Y} = b_0 + b_1(x_1 - \bar{x}_1) + b_2(x_2 - \bar{x}_2) + \cdots + b_q(x_q - \bar{x}_q)$$

The x's now may be different variables such as flow rate, pressure (p) and concentration (c), or they may be products of variables such as p^2, pc, and pc^2.

5.1-1 Least Squares Estimation

Exactly the same least squares procedures as described in Sections 4.3 and 4.4 can be carried out to obtain estimates of the parameters in Equation 5.1-1. We minimize the weighted sum of the squares (in "ordinary" least squares the weights are all unity) of the deviations between the observations of Y, $\bar{Y}_i$, and the corresponding expected values of Y_i, η_i:

$$\text{Minimize } \phi = \sum_{i=1}^{n} w_i(\bar{Y}_i - \eta_i)^2 = \sum_{i=1}^{n} w_i \epsilon_i^2 \qquad (5.1\text{-}3)$$

with respect to the coefficients $\beta_0, \beta_1, \beta_2, \ldots, \beta_q$. In Equation 5.1-3 the weights may be proportional to $(\sigma^2_{\bar{Y}_i})$, which ensures that the points with the largest variances will have the least influence in determining the best fitting line through the data, or some other scheme of weighting may be employed. The subscripts to be employed are:

Index of data sets (matrix rows)	$1 \leq i \leq n$
Index of coefficients (matrix columns)	$0 \leq k \leq q$

To minimize ϕ, we take the partial derivatives of ϕ with respect to each β_i and equate the resulting expressions to zero (the extremum can be shown to be a minimum):

$$\frac{\partial \phi}{\partial \beta_0} = 0 = -2 \sum_{i=1}^{n} w_i[\bar{Y}_i - \beta_0 - \beta_1(x_{i1} - \bar{x}_1) - \beta_2(x_{i2} - \bar{x}_2) - \cdots - \beta_q(x_{iq} - \bar{x}_q)]$$

$$\frac{\partial \phi}{\partial \beta_1} = 0 = -2 \sum_{i=1}^{n} w_i[\bar{Y}_i - \beta_0 - \beta_1(x_{i1} - \bar{x}_1) - \beta_2(x_{i2} - \bar{x}_2) - \cdots - \beta_q(x_{iq} - \bar{x}_q)](x_{i1} - \bar{x}_1)$$

$$\cdots\cdots$$

$$\frac{\partial \phi}{\partial \beta_q} = 0 = -2 \sum_{i=1}^{n} w_i[\bar{Y}_i - \beta_0 - \beta_1(x_{i1} - \bar{x}_1) - \beta_2(x_{i2} - \bar{x}_2) - \cdots - \beta_q(x_{iq} - \bar{x}_q)](x_{iq} - \bar{x}_q) \qquad (5.1\text{-}4)$$

By rewriting Equations 5.1-4 we obtain the set of simultaneous normal equations, Equations 5.1-5, equivalent to Equations 4.3-5:

$$b_0\left(\sum_{i=1}^{n} w_i\right) + b_1(0) + b_2(0) + \cdots + b_q(0) = \sum_{i=1}^{n} w_i \bar{Y}_i$$

$$b_0(0) + b_1\left[\sum_{i=1}^{n} w_i(x_{i1} - \bar{x}_1)(x_{i1} - \bar{x}_1)\right] + b_2\left[\sum_{i=1}^{n} w_i(x_{i2} - \bar{x}_2)(x_{i1} - \bar{x}_1)\right] + \cdots$$
$$+ b_q\left[\sum_{i=1}^{n} w_i(x_{iq} - \bar{x}_q)(x_{i1} - \bar{x}_1)\right] = \sum_{i=1}^{n} w_i \bar{Y}_i(x_{i1} - \bar{x}_1$$

$$b_0(0) + b_1\left[\sum_{i=1}^{n} w_i(x_{i1} - \bar{x}_1)(x_{i2} - \bar{x}_2)\right] + b_2\left[\sum_{i=1}^{n} w_i(x_{i2} - \bar{x}_2)(x_{i2} - \bar{x}_2)\right] + \cdots$$
$$+ b_q\left[\sum_{i=1}^{n} w_i(x_{iq} - \bar{x}_q)(x_{i2} - \bar{x}_2)\right] = \sum_{i=1}^{n} w_i \bar{Y}_i(x_{i2} - \bar{x}_2)$$

$$\cdots\cdots\cdots$$

$$b_0(0) + b_1\left[\sum_{i=1}^{n} w_i(x_{i1} - \bar{x}_1)(x_{iq} - \bar{x}_q)\right] + b_2\left[\sum_{i=1}^{n} w_i(x_{i2} - \bar{x}_2)(x_{iq} - \bar{x}_q)\right] + \cdots$$
$$+ b_q\left[\sum_{i=1}^{n} w_i(x_{iq} - \bar{x}_q)(x_{iq} - \bar{x}_q)\right] = \sum_{i=1}^{n} w_i \bar{Y}_i(x_{iq} - \bar{x}_q) \tag{5.1-5}$$

Note that inasmuch as

$$\sum_{i=1}^{n} w_i(x_{ik} - \bar{x}_k) \equiv 0$$

the first equation in 5.1-5 yields

$$\hat{\beta}_0 = b_0 = \frac{\sum_{i=1}^{n} w_i \bar{Y}_i}{\sum_{i=1}^{n} w_i} = \bar{Y} \tag{5.1-6}$$

The term *normal equations* has a geometric interpretation as follows. If the observations $\bar{Y}_i$ are interpreted as the components of a vector of observations $\mathbf{Y}$ with a base at the origin in observation space, as illustrated in Figure 5.1-1*a*, we want to select the values of b_k (which exist in parameter space) that yield values of $\hat{Y}_i$ that minimizes ϕ. The components of $\hat{Y}_i$ can similarly be interpreted as forming a vector in observation space, $\hat{\mathbf{Y}}$, and the various choices of b's form a plane of estimates of η. We want the estimate of η, $\hat{\mathbf{Y}}$, which yields the shortest distance between the head of the vector $\mathbf{Y}$ and the surface of estimates of η. The normal equations are those that determine the b's such that the vector $(\mathbf{Y} - \hat{\mathbf{Y}})$ passes through the head of $\mathbf{Y}$ and is perpendicular (normal) to the surface comprised of all the possible values of the estimates of η. This perpendicular vector is the one that ensures that ϕ is a minimum.

Figure 5.1-1*d* illustrates geometrically the unobservable error vector $\boldsymbol{\epsilon}$ which is the difference between the vector $\boldsymbol{\eta}$ and the vector of observations $\mathbf{Y}$. There is a given probability that the head of the vector $\boldsymbol{\eta}$ will be found in some sphere of radius ϵ about the head of $\mathbf{Y}$. The vector $\hat{\mathbf{Y}}$ has components b_1 and b_2 in the directions x_1 and x_2, respectively. Because a normal projection is used to project $\mathbf{Y}$ onto $\hat{\mathbf{Y}}$ through least squares, the vector $\boldsymbol{\epsilon}$ projects as the smallest possible circle on the $x_1 - x_2$ plane. Further projection on the x_1 and x_2 axes results in confidence intervals for β_1 and β_2, respectively. When the correlation between x_1 and x_2 is large, that is the cosine of the angle between x_1 and x_2 approaches 1, the projection of the circle on the x axes is longer than if x_1 and x_2 are orthogonal, i.e., perpendicular, when the projections have shortest length. *Orthogonality* of two variables, x_k and x_j, means that

$$\sum_{i=1}^{n} x_{ij}x_{ik} = 0 \qquad \text{if } k \neq j$$

We now outline the estimation procedure in matrix notation† inasmuch as the display of Equations 5.1-5 clearly calls for some compact way of presenting the same results. Let the n experimental data sets and the $q + 1$ parameters be represented as follows:

$$\mathbf{Y} = \begin{bmatrix} \bar{Y}_1 \\ \bar{Y}_2 \\ \vdots \\ \bar{Y}_n \end{bmatrix} \text{ an } n \times 1 \text{ matrix} \qquad \hat{\mathbf{Y}} = \begin{bmatrix} \hat{Y}_1 \\ \hat{Y}_2 \\ \vdots \\ \hat{Y}_n \end{bmatrix} \text{ an } n \times 1 \text{ matrix}$$

† Readers unfamiliar with matrix notation should first refer to Appendix B.

$$\boldsymbol{\beta} = \begin{bmatrix} \beta_0 \\ \beta_1 \\ \vdots \\ \beta_q \end{bmatrix} \quad \text{a } q \times 1 \text{ matrix}$$

$$\mathbf{x} = \begin{bmatrix} 1 & (x_{11} - \bar{x}_1) & (x_{12} - \bar{x}_2) & \cdots & (x_{1q} - \bar{x}_q) \\ 1 & (x_{21} - \bar{x}_1) & (x_{22} - \bar{x}_2) & \cdots & (x_{2q} - \bar{x}_q) \\ \vdots & \vdots & \vdots & & \vdots \\ 1 & (x_{n1} - \bar{x}_1) & (x_{n2} - \bar{x}_2) & \cdots & (x_{nq} - \bar{x}_q) \end{bmatrix}$$

an $n \times (q + 1)$ matrix

In the first column the 1's are dummy variables which are needed only if Model 5.1-1 is to have an intercept and yet be represented in matrix form as $\boldsymbol{\eta} = \mathbf{x}\boldsymbol{\beta}$ with the corresponding estimated regression equation $\hat{\mathbf{Y}} = \mathbf{xb}$. Each $\mathbf{x}_i$ vector is

$$\mathbf{x}_i = [1 \; (x_{i1} - \bar{x}_1) \cdots (x_{iq} - \bar{x}_q)]$$

It would be equally possible to represent the model as $\boldsymbol{\eta} = \boldsymbol{\beta}_0 + \acute{\mathbf{x}}\boldsymbol{\beta}$ with the corresponding regression equation $\hat{\mathbf{Y}} = \bar{\mathbf{Y}} + \acute{\mathbf{x}}\mathbf{b}$ where $\acute{\mathbf{x}}$ is identical to $\mathbf{x}$ except that the first column of dummy 1's is deleted.

The weights have the same interpretation as in Section 4.4 and are

$$\mathbf{w} = \begin{bmatrix} w_1 & 0 & \cdots & 0 \\ 0 & w_2 & \cdots & 0 \\ \vdots & \vdots & & \vdots \\ 0 & 0 & \cdots & w_n \end{bmatrix} \quad \text{an } n \times n \text{ matrix}$$

while the residuals are represented by $E_i = (\bar{Y}_i - \hat{Y}_i)$ or

$$\mathbf{E} = \begin{bmatrix} E_1 \\ \vdots \\ E_n \end{bmatrix} \quad \text{an } n \times 1 \text{ matrix}$$

The sum of the squares of the unobservable errors is

$$\phi = \sum_{i=1}^{n} w_i \epsilon_i^2 = \boldsymbol{\epsilon}^T \mathbf{w} \boldsymbol{\epsilon} \tag{5.1-7}$$

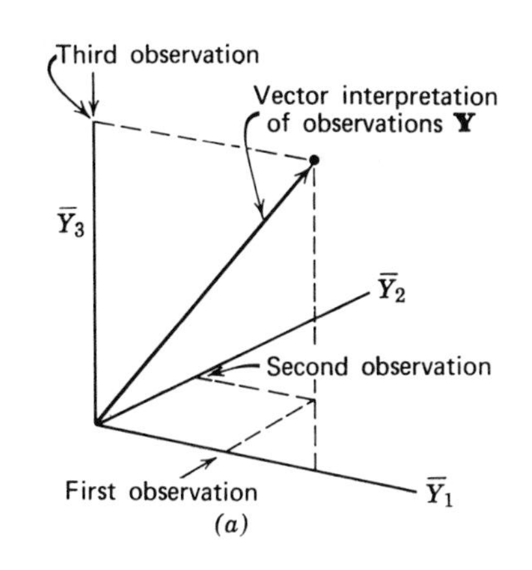

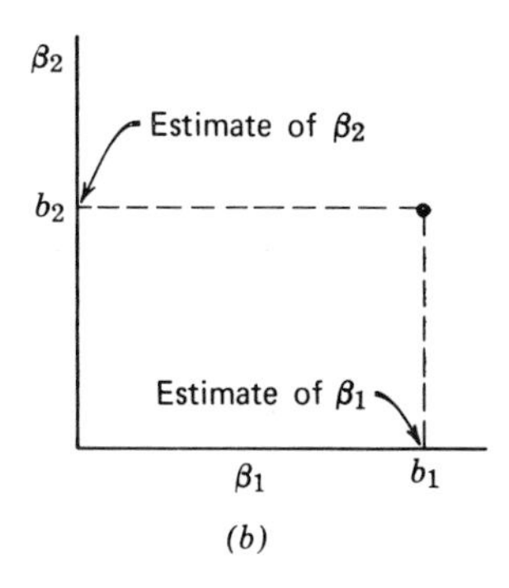

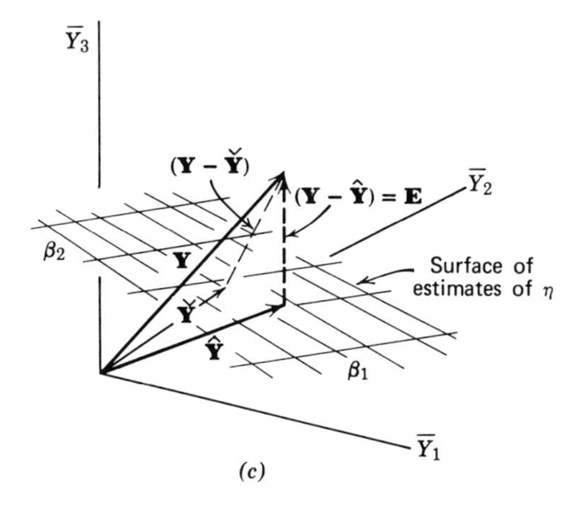

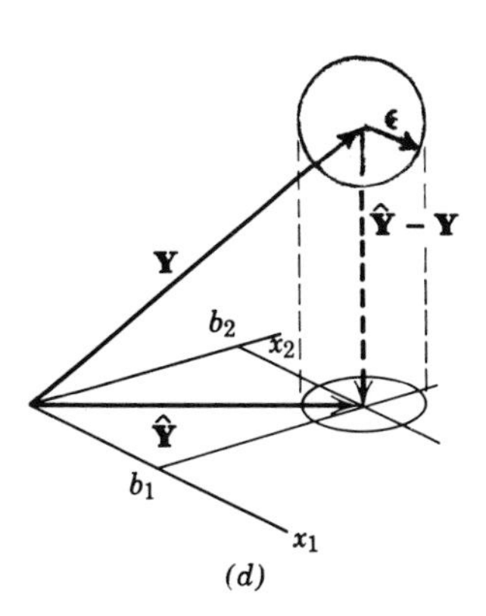

FIGURE 5.1-1 Interpretation of the vector of residuals as a normal to the surface of estimates of η: (*a*) observation space (three observations), (*b*) parameter space (two parameters), (*c*) normal vector $\mathbf{E}$ in observation space ($\hat{\mathbf{Y}}$ is the best estimate of η; $\check{\mathbf{Y}}$ is a poorer estimate of η), and (*d*) experimental observation space (two independent variables).

Because $\boldsymbol{\epsilon} = \mathbf{Y} - \boldsymbol{\eta} = \mathbf{Y} - \mathbf{x}\boldsymbol{\beta}$,

$$\phi = (\mathbf{Y} - \mathbf{x}\boldsymbol{\beta})^T\mathbf{w}(\mathbf{Y} - \mathbf{x}\boldsymbol{\beta}) \tag{5.1-7a}$$

which can be minimized with respect to all the β_k's by taking $\partial\phi/\partial\boldsymbol{\beta}$ and equating the resulting matrix to a zero matrix:†

$$\frac{\partial\phi}{\partial\boldsymbol{\beta}} = \begin{bmatrix} \frac{\partial\phi}{\partial\beta_0} \\ \vdots \\ \frac{\partial\phi}{\partial\beta_q} \end{bmatrix} = \frac{2[\partial(\mathbf{Y} - \mathbf{x}\boldsymbol{\beta})^T]\mathbf{w}(\mathbf{Y} - \mathbf{x}\boldsymbol{\beta})}{\partial\boldsymbol{\beta}}$$

$$= -2\mathbf{x}^T\mathbf{w}(\mathbf{Y} - \mathbf{x}\boldsymbol{\beta}) = \mathbf{0} \tag{5.1-8}$$

(Recall that **Y** and **x** act as constants in the differentiation.)

Equation 5.1-8 with the estimate **b** replacing $\boldsymbol{\beta}$ becomes

$$\mathbf{x}^T\mathbf{w}\mathbf{Y} = \mathbf{x}^T\mathbf{w}\mathbf{x}\mathbf{b} \tag{5.1-9}$$

which is exactly the same as Equations 5.1-5. This can be demonstrated by expansion of Equation 5.1-9 into elements and subsequent multiplication. The solution of matrix Equation 5.1-9 for **b** is

$$\mathbf{b} = (\mathbf{x}^T\mathbf{w}\mathbf{x})^{-1}(\mathbf{x}^T\mathbf{w}\mathbf{Y}) = \mathbf{c}\mathbf{G}, \qquad (\mathbf{x}^T\mathbf{w}\mathbf{x}) \neq 0 \tag{5.1-10}$$

where, to simplify the notation, we shall let

$$(\mathbf{x}^T\mathbf{w}\mathbf{x})^{-1} \equiv \mathbf{a}^{-1} \equiv \mathbf{c} \quad \text{and} \quad (\mathbf{x}^T\mathbf{w}\mathbf{Y}) \equiv \mathbf{G}.$$

The matrix $\mathbf{x}^T\mathbf{w}\mathbf{x}$ is symmetric, as can be seen from the terms in Equation 5.1-5; hence the matrix **c** is also symmetric so that $\mathbf{c}^T = \mathbf{c}$.

5.1-2 Maximum Likelihood Estimation

Minimizing the sum of the squares of ϕ gives rise to the same estimate of $\boldsymbol{\beta}$ as does minimizing the variance of an arbitrary linear function of the elements of $\boldsymbol{\beta}$. It should be noted that the procedure above is independent of any severe restrictions on the distribution of the unobservable errors ϵ, but it can be demonstrated that *if* the errors are assumed to be normally distributed, the maximum likelihood estimate also yields Equation 5.1-9. If the *multivariate* normal probability density function, Equation 2.3-6, represents the distribution of the set of unobservable errors in Model 5.1-1a,

$$p(\epsilon) = k \exp\left(-\tfrac{1}{2}\boldsymbol{\epsilon}^T\mathbf{f}^{-1}\boldsymbol{\epsilon}\right) \tag{5.1-11}$$

where $\epsilon_i = (\bar{Y}_i - \eta_i)$ is the error matrix‡

$$\boldsymbol{\epsilon} = \begin{bmatrix} \epsilon_1 \\ \vdots \\ \epsilon_n \end{bmatrix}$$

and k is a normalization factor defined in Section 2.3, then a likelihood function can be written after the observations are made similar to Equation 4.3-8 in which the **Y** and **x** matrices are regarded as given and the $\boldsymbol{\beta}$ matrix (and perhaps **f**) is the variable. The natural logarithm of the likelihood function is

$$\ln L(\boldsymbol{\beta}, \mathbf{f} \mid \mathbf{y}, \mathbf{x}) \equiv \ln L = n \ln k - \tfrac{1}{2}[(\mathbf{Y} - \mathbf{x}\boldsymbol{\beta})^T\mathbf{f}^{-1}(\mathbf{Y} - \mathbf{x}\boldsymbol{\beta})] \tag{5.1-11a}$$

where

$$\mathbf{f} = \text{Covar}\{\boldsymbol{\epsilon}\} \equiv \mathscr{E}\{\boldsymbol{\epsilon}\boldsymbol{\epsilon}^T\} = \begin{bmatrix} \sigma_1^2 & \sigma_{12} & \cdots & \sigma_{1n} \\ \sigma_{21} & \sigma_2^2 & \cdots & \sigma_{2n} \\ \cdot & \cdot & \cdot & \cdot \\ \sigma_{n1} & \sigma_{n2} & \cdots & \sigma_n^2 \end{bmatrix}$$

Assuming that $\mathscr{E}\{\epsilon_i\} = 0$, that $\text{Var}\{\epsilon_i\} = \mathscr{E}\{(\epsilon_i - 0)(\epsilon_i - 0)\} = \mathscr{E}\{\epsilon_i^2\} = \sigma_i^2$, and also that $\text{Var}\{\epsilon_i\epsilon_j\} = 0$, we can reduce the covariance matrix **f** to

$$\mathbf{f} = \begin{bmatrix} \sigma_1^2 & 0 & \cdots & 0 \\ 0 & \sigma_2^2 & \cdots & 0 \\ \cdot & \cdot & \cdot & \cdot \\ 0 & 0 & \cdots & \sigma_n^2 \end{bmatrix}$$

We can then identify

$$\mathbf{f}^{-1} = \frac{1}{\sigma_i^2}\mathbf{I} = \mathbf{w}$$

Finally, if all the elements on the main diagonal of **f** are equal and equal to $\sigma_{\bar{Y}_i}^2$, a constant, **f** reduces to

$$\mathbf{f} = \sigma_{\bar{Y}_i}^2\mathbf{I} \tag{5.1-12}$$

To save space we shall use **w** in many of the following equations instead of $(1/\sigma_{\bar{Y}_i}^2)\mathbf{I}$. For later use, note that $\boldsymbol{\epsilon}^T\mathbf{w}\boldsymbol{\epsilon} = \boldsymbol{\epsilon}\boldsymbol{\epsilon}^T\mathbf{w}^T$. If the elements in the matrix **f** are not simplified as shown but retained, the type of estimate obtained is known as a Markov estimate.

Maximization of $\ln L$ in Equation 5.1-11a with $\mathbf{f} = \sigma_{\bar{Y}_i}^2\mathbf{I}$ can be accomplished by differentiation of $\ln L$ with respect to $\boldsymbol{\beta}$ and $\sigma_{\bar{Y}_i}^2$, yielding the normal Equations 5.1-9 and a biased estimate for $\sigma_{\bar{Y}_i}^2$, just as in Section 4.3. Minimization of $(\mathbf{Y} - \mathbf{x}\boldsymbol{\beta})^T(\sigma_{\bar{Y}_i}^2)^{-1}(\mathbf{Y} - \mathbf{x}\boldsymbol{\beta})$ maximizes $\ln L$; compare with Equation 5.1-7a. As a matter of interest, if all the elements in the matrix **f** are retained, minimization of $(\mathbf{Y} - \mathbf{x}\boldsymbol{\beta})^T\mathbf{f}^{-1}(\mathbf{Y} - \mathbf{x}\boldsymbol{\beta})$ leads to the Markov estimate

$$\mathbf{b} = (\mathbf{x}^T\mathbf{f}^{-1}\mathbf{x})^{-1}(\mathbf{x}^T\mathbf{f}^{-1}\mathbf{Y})$$

with

$$\text{Var}\{\mathbf{b}\} = (\mathbf{x}^T\mathbf{f}^{-1}\mathbf{x})^{-1}$$

but the elements of **f** must be evaluated by replicate experimentation or some other method to carry out the calculations. A discussion of estimation schemes in which the elements of **f** are estimated along with the

† We make use of the following property in differentiating a matrix. If

$$\mathbf{h} = \mathbf{q}^T\mathbf{a}\mathbf{q}$$

then

$$d\mathbf{h} = 2(d\mathbf{q}^T)(\mathbf{a}\mathbf{q})$$

which can be verified by decomposition into matrix elements and appropriate manipulation of the elements.

‡ Note that $\boldsymbol{\epsilon}$ is *not* the same as the matrix $\mathbf{E} = (\mathbf{Y} - \hat{\mathbf{Y}})$, identified previously.

parameters is beyond our scope here. Unless specifically stated otherwise, we shall assume that Equation 5.1-12 holds in all cases.

5.1-3 Correlation and Bias in the Estimates

Because the unobservable errors ϵ_i are literally that, unobservable, they can only be estimated by using $\hat{\boldsymbol{\epsilon}} = \mathbf{Y} - \mathbf{x}\hat{\boldsymbol{\beta}}$ so that

$$\hat{\boldsymbol{\epsilon}} = \mathbf{Y} - \mathbf{x}(\mathbf{x}^T\mathbf{w}\mathbf{x})^{-1}\mathbf{x}^T\mathbf{w}\mathbf{Y} = \mathbf{M}\mathbf{Y}$$

where

$$\mathbf{M} = \mathbf{I} - \mathbf{x}(\mathbf{x}^T\mathbf{w}\mathbf{x})^{-1}\mathbf{x}^T\mathbf{w}$$

The matrix $\mathbf{M}$ can be interpreted as the matrix that transforms the "true" errors $\boldsymbol{\epsilon}$ linearly into the estimates of the errors $\hat{\boldsymbol{\epsilon}}$ for, (with $\mathbf{M}\mathbf{x} = 0$)

$$\hat{\boldsymbol{\epsilon}} = \mathbf{M}\mathbf{Y} = \mathbf{M}(\mathbf{x}\boldsymbol{\beta} + \boldsymbol{\epsilon}) = \mathbf{M}\boldsymbol{\epsilon}$$

Thus we find that the estimated errors are functions not only of $\boldsymbol{\epsilon}$ but also of $\mathbf{M}$ and hence of the values of the independent variables in $\mathbf{x}$. Consequently, even if the assumptions about $\mathscr{E}\{\boldsymbol{\epsilon}\} = 0$ and $\mathscr{E}\{\boldsymbol{\epsilon}\boldsymbol{\epsilon}^T\} = 0$ hold, the estimated errors are in general neither uncorrelated nor do they have constant variance:

$$\mathscr{E}\{\hat{\boldsymbol{\epsilon}}\hat{\boldsymbol{\epsilon}}^T\} = \mathscr{E}\{\mathbf{M}\boldsymbol{\epsilon}\boldsymbol{\epsilon}^T\mathbf{M}^T\} = \sigma^2\mathbf{M}\mathbf{M}^T$$

If we use Equation 5.1-10, we can show that the expected value of $\mathbf{b}$ is $\boldsymbol{\beta}$:

$$\begin{aligned}\mathscr{E}\{\mathbf{b}\} &= \mathscr{E}\{(\mathbf{x}^T\mathbf{w}\mathbf{x})^{-1}(\mathbf{x}^T\mathbf{w}\mathbf{Y})\} = \mathscr{E}\{(\mathbf{x}^T\mathbf{x})^{-1}[(\mathbf{x}^T(\mathbf{x}\boldsymbol{\beta} + \boldsymbol{\epsilon})]\} \\ &= \boldsymbol{\beta} \qquad (5.1\text{-}13)\end{aligned}$$

and the covariance of $\mathbf{b}$ is†

$$\begin{aligned}\text{Covar}\,\{\mathbf{b}\} &= \mathscr{E}\{(\mathbf{b} - \boldsymbol{\beta})(\mathbf{b} - \boldsymbol{\beta})^T\} \\ &= \mathscr{E}\{[(\mathbf{x}^T\mathbf{w}\mathbf{x})^{-1}(\mathbf{x}^T\mathbf{w}\mathbf{Y}) - \boldsymbol{\beta}] \\ &\qquad \times [(\mathbf{x}^T\mathbf{w}\mathbf{x})^{-1}(\mathbf{x}^T\mathbf{w}\mathbf{Y}) - \boldsymbol{\beta}]^T\}\end{aligned}$$

If we substitute $\mathbf{Y} = \mathbf{x}\boldsymbol{\beta} + \boldsymbol{\epsilon}$,

$$\begin{aligned}\text{Covar}\,\{\mathbf{b}\} &= \mathscr{E}\{[\mathbf{c}\mathbf{x}^T\mathbf{w}(\mathbf{x}\boldsymbol{\beta} + \boldsymbol{\epsilon}) - \boldsymbol{\beta}][\mathbf{c}\mathbf{x}^T\mathbf{w}(\mathbf{x}\boldsymbol{\beta} + \boldsymbol{\epsilon}) - \boldsymbol{\beta}]^T\} \\ &= \mathscr{E}\{(\mathbf{c}\mathbf{x}^T\mathbf{w}\boldsymbol{\epsilon})(\mathbf{c}\mathbf{x}^T\mathbf{w}\boldsymbol{\epsilon})^T\} = \mathscr{E}\{\mathbf{c}\mathbf{x}^T\mathbf{w}\boldsymbol{\epsilon}\boldsymbol{\epsilon}^T\mathbf{w}^T\mathbf{x}\mathbf{c}\} \\ &= \mathbf{c}\mathbf{x}^T\mathbf{w}\mathscr{E}\{\boldsymbol{\epsilon}\boldsymbol{\epsilon}^T\mathbf{w}^T\}\mathbf{x}\mathbf{c} = \sigma^2_{\bar{Y}_i}\mathbf{c} \qquad (5.1\text{-}14)\end{aligned}$$

Since $\sigma^2_{\bar{Y}_i}$ is not usually known, it is necessary to obtain an unbiased estimate of $\sigma^2_{\bar{Y}_i}$, $s^2_{\bar{Y}_i}$, presumably a pooled estimate, of or s_r^2 if replicate measurements are not available but the model is an adequate one. Because the assumptions of Section 4.2 apply here as they did in Sections 4.3 and 4.4, s_r^2 is simply the sum of the squares of the residuals divided by the number of degrees of freedom (the number of data sets, n, less the number of constraints imposed while minimizing ϕ, namely the number of parameters being estimated):

$$s_r^2 = \frac{\phi_{\min}}{n - (q + 1)} = \frac{(\mathbf{E}^T\mathbf{w}\mathbf{E})}{n - q - 1} \qquad (5.1\text{-}15)$$

Thus, in the absence of an error estimate s_e^2,

$$\widehat{\text{Covar}}\,\{\mathbf{b}\} = \frac{(\mathbf{E}^T\mathbf{w}\mathbf{E})}{n - q - 1}\,\mathbf{c} \qquad (5.1\text{-}16)$$

whereas if s_e^2 is available, $s^2_{\bar{Y}_i}$ can be calculated by Equation 4.3-15.

Keep in mind that $\mathbf{b}$ is an unbiased estimate of $\boldsymbol{\beta}$ in Model 5.1-1 *only if* the model is correct. Suppose that the model is not

$$\eta_{\text{I}} = \mathbf{x}\boldsymbol{\beta}$$

but instead actually contains additional terms as follows

$$\eta_{\text{II}} = \mathbf{x}\boldsymbol{\beta} + \mathbf{x}^*\boldsymbol{\beta}^*$$

Then, the expected value of b according to Model I is

$$\mathscr{E}\{\mathbf{b}\} = \boldsymbol{\beta}$$

from Equation 5.1-13. But the expected value of $\mathbf{b}$, assuming Model II applies, is

$$\begin{aligned}\mathscr{E}\{\mathbf{b}\} &= (\mathbf{x}^T\mathbf{w}\mathbf{x})^{-1}\mathbf{x}^T\mathbf{w}\mathscr{E}\{\mathbf{Y}\} \\ &= (\mathbf{x}^T\mathbf{w}\mathbf{x})^{-1}\mathbf{x}^T\mathbf{w}(\mathbf{x}\boldsymbol{\beta} + \mathbf{x}^*\boldsymbol{\beta}^*) = \boldsymbol{\beta} + \mathbf{x}^\dagger\boldsymbol{\beta}^* \qquad (5.1\text{-}17)\end{aligned}$$

where $\mathbf{x}^\dagger$ is termed an *alias matrix*, i.e., the estimate of $\boldsymbol{\beta}$ is biased. However, Equation 5.1-14 for the variance of $\mathbf{b}$ can be shown to be true if Model II instead of Model I is correct, but, of course, the estimate of $\sigma^2_{\bar{Y}_i}$ will be incorrect.

5.1-4 Sensitivity of the Estimates

The *sensitivity* of the residual sum of the squares, $\phi_{\min}$, and the sensitivity of the elements of the matrix $\mathbf{b}$ can be calculated for a change in $\bar{Y}_i$ (the observation(s) of the ith data set). By sensitivity is meant the fractional change in $\phi_{\min}$ (or b_k) produced by a fractional change in $\bar{Y}_i$. Calculation of the sensitivity can assist in the interpretation of predictions based on the regression equation and in the design of the experiments. Because

$$\phi_{\min} = \sum_{i=1}^{n} w_i(\bar{Y}_i - \hat{Y}_i)^2$$

the sensitivity of $\phi_{\min}$ with respect to $\bar{Y}_i$ is

$$\begin{aligned}\frac{\partial\phi_{\min}/\phi_{\min}}{\partial\bar{Y}_i/\bar{Y}_i} &= \frac{\partial\ln\phi_{\min}}{\partial\ln\bar{Y}_i} = \frac{\partial\phi_{\min}}{\partial\bar{Y}_i}\left(\frac{\bar{Y}_i}{\phi_{\min}}\right) \\ &= \frac{2\bar{Y}_i\sum_{i=1}^{n} w_i(\bar{Y}_i - \hat{Y}_i)}{\sum_{i=1}^{n} w_i(\bar{Y}_i - \hat{Y}_i)^2} \qquad (5.1\text{-}18)\end{aligned}$$

† Zeros make up the elements in $(\mathbf{b} - \boldsymbol{\beta})$ for conformability with $(\mathbf{b} - \boldsymbol{\beta})^T$.

In matrix notation using Equation 5.1-10,

$$\begin{bmatrix} b_0 \\ b_1 \\ \vdots \\ b_q \end{bmatrix} = [(\mathbf{x}^T\mathbf{w}\mathbf{x})^{-1}\mathbf{x}^T\mathbf{w}] \begin{bmatrix} \bar{Y}_1 \\ \bar{Y}_2 \\ \vdots \\ \bar{Y}_n \end{bmatrix}$$

and the sensitivity of b_k for $\bar{Y}_i$ is

$$\frac{\partial b_k}{\partial \bar{Y}_i}\left(\frac{\bar{Y}_i}{b_k}\right) = [(\mathbf{x}^T\mathbf{w}\mathbf{x})^{-1}\mathbf{x}^T\mathbf{w}]_{ki}\left(\frac{\bar{Y}_i}{b_k}\right) \qquad (5.1\text{-}19)$$

where $[(\mathbf{x}^T\mathbf{w}\mathbf{x})^{-1}\mathbf{x}^T\mathbf{w}]_{ki}$ is the element in the kth row and ith column of $(\mathbf{x}^T\mathbf{w}\mathbf{x})^{-1}\mathbf{x}^T\mathbf{w}$.

To illustrate a sensitivity calculation, we use the data of Example 4.3-2 for the tenth set of data and b_1:

$$\frac{\partial \phi_{\min}}{\partial \bar{Y}_{10}}\left(\frac{\bar{Y}_{10}}{\phi_{\min}}\right) = \frac{2(259)(0.03)}{(14)(687)} = 6.1 \times 10^{-3}$$

$$\frac{\partial b_1}{\partial \bar{Y}_{10}}\left(\frac{\bar{Y}_{10}}{b_1}\right) = \frac{(x_{10} - \bar{x})}{\sum (x_i - \bar{x})^2}\left(\frac{259}{79.02}\right)$$

$$= \frac{(3.09 - 2.816)}{(13.61)}\left(\frac{259}{79.02}\right)$$

$$= 6.58 \times 10^{-2}$$

In other words, a 10-percent change in $\bar{Y}_i$ will produce a 6.1×10^{-2}-percent change in $\phi_{\min}$ and a 0.658-percent change in b_1. Both of these sensitivities are quite low, which is all to the good in model building.

5.1-5 Computational Problems

We shall now briefly mention some of the practical problems which arise in the machine computation of parameter estimates and allied calculations discussed above. The most elusive difficulties are the following:

1. Loss of significant digits in subtracting approximately equal numbers. As has been observed in the numerical examples in Chapter 4, many of the terms which are subtracted from each other are nearly equal. Two numbers with five significant digits, each of which agrees in the first two digits, retain only three significant digits on subtraction. One partial aid in overcoming loss of significant digits is use of double-precision arithmetic.

2. Roundoff error. Freund† and Smiley‡ demonstrated the magnitude of rounding error in computations by floating-point arithmetic. The use of double-precision arithmetic and more careful attention to the significance of individual variables at the intermediate calculation stages are prescribed as antidotes for rounding error.§

3. Matrix **a** becomes ill conditioned. The least squares solution may be very sensitive to small perturbations in the elements of **a**. For example, as an extreme case, consider Equation 5.1-9, **ab** = **G**. Suppose that

$$\mathbf{a} = \begin{bmatrix} 1 & 1 \\ 1 & 1 \end{bmatrix} \quad \text{and} \quad \mathbf{G} = \begin{bmatrix} 1 \\ 0 \end{bmatrix},$$

then the det **a** = 0, **a** is singular and a plot of the two equations represented by **ab** = **G**, Figure 5.1-2*a*, shows that they are two parallel lines with a slope of −1. Now suppose because of numerical or experimental error the matrix **a** is

$$\mathbf{a} = \begin{bmatrix} 1 & 1 \\ 1 & 1+\epsilon \end{bmatrix}$$

where ϵ is a small perturbation. **a** is no longer singular, although close to it, and is termed an *ill-conditioned matrix*. The two corresponding equations, illustrated in Figure 5.1-2*b*, now intersect at a point whose value becomes more uncertain the smaller the value of ϵ. As $\epsilon \to 0$, the lines again become parallel.

Matrix **a** can become ill conditioned by improper selection of experimental values of the independent

† R. J. Freund, *Amer. Stat.*, **17** Dec. 1963, p. 13.
‡ K. W. Smiley, *Amer. Stat.*, **18** Oct. 1964, p. 26.
§ Also see: M. J. Gaber, *Comm. ACM* **7**, 721, 1964; and R. H. Wampler, *J. Res Nat. Bur. Standards* **73B** (in press), who evaluated twenty different computer programs.

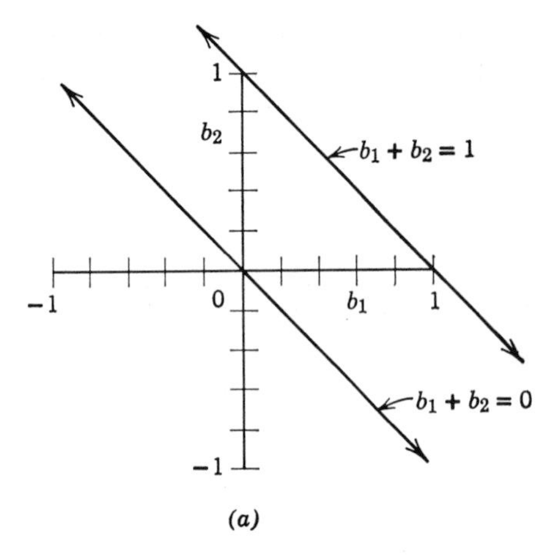

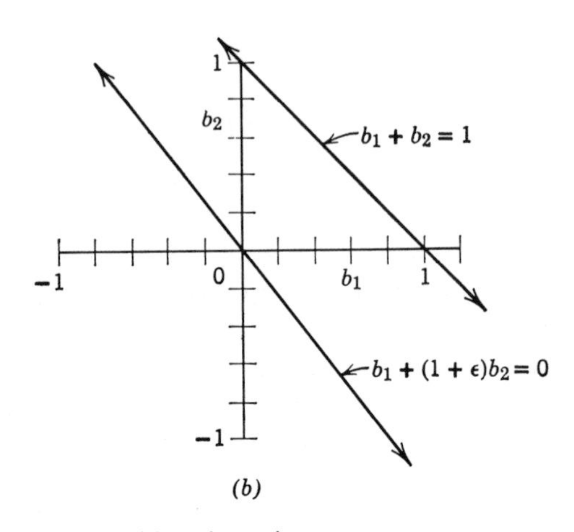

FIGURE 5.1-2 Graph of the equations yielding ill-conditioned matrices.

variable. For example, suppose that the model is $\eta = \beta_0' + \beta_1 x$, and three observations are taken at $x = 19.9$, 20.0, and 20.1 units. Then

$$\mathbf{x} = \begin{bmatrix} 1 & 19.9 \\ 1 & 20.0 \\ 1 & 20.1 \end{bmatrix} \qquad \mathbf{a} = [\mathbf{x}^T\mathbf{x}] = \begin{bmatrix} 3 & 60.0 \\ 60.0 & 1200.02 \end{bmatrix}$$

and the det $\mathbf{a} = (3)(1200.02) - (60)(60) = 3600.06 - 3600.00 = 0.06$. Suppose that the numbers to the right of the decimal points in the matrix $\mathbf{x}$ are the last significant figures. Then it is clear that rounding of the elements to be subtracted at four digits will give det $\mathbf{a} = 0$, that rounding at five digits will also give det $\mathbf{a} = 0$, and that six digits are needed to obtain 0.06. However, if the model is written as $\eta = \beta_0 + \beta_1(x - \bar{x})$:

$$\mathbf{x} = \begin{bmatrix} 1 & -0.1 \\ 1 & 0.0 \\ 0 & 0.1 \end{bmatrix} \qquad \mathbf{a} = [\mathbf{x}^T\mathbf{x}] = \begin{bmatrix} 3 & 0 \\ 0 & 0.02 \end{bmatrix}$$

the det $\mathbf{a}$ still is equal to 0.06 but the calculation $(3)(0.02) - (0)(0) = 0.06$ indicates $\mathbf{a}$ is well conditioned. Also, the contours of the sum of the squares surface are much more circular. Use of orthogonal experimental designs, as described in Chapter 8, and use of Model 5.1-1 rather than Model 5.1-2 are recommended methods of avoiding having to work with an ill-conditioned $\mathbf{a}$ matrix.

Example 5.1-1 Estimation with Orthogonal Variables

Data obtained from experiments based on the experimental set-up shown in Figure 8.1-1 (known as a two-level factorial experiment) are:

Temperature, T (°F)	Pressure, p (atm)	Yield, Y (%)
160	1	4
160	1	5
160	7	10
160	7	11
200	1	24
200	1	26
200	7	35
200	7	38

(The yield is the dependent variable.) Estimate the coefficients in a linear model of the form

$$\eta = \beta_0 + \beta_1 x_1 + \beta_2 x_2$$

Solution:

The values of the independent variables can be coded so that the calculations are easier to follow. Let

$$x_1 = \tilde{T} = \frac{T - 180}{20}, \qquad x_2 = \tilde{p} = \frac{p - 4}{3}$$

The coded data are as shown in Table E5.1-1.

Notice that $\bar{x}_1 = 0$, $\bar{x}_2 = 0$, and that the independent variables are orthogonal because $\sum x_0 x_1 = \sum x_0 x_2 = \sum x_1 x_2 = 0$. The column under x_0 contains the dummy variable 1 in order for the model to include an intercept. All the weights will be unity.

TABLE E5.1-1

x_0	x_1	x_2	Y_{ij} = yield	Calculation of Error within Sets: $\bar{Y}_i$	$\Delta = (Y_{ij} - \bar{Y}_i)$	Δ^2
1	−1	−1	4	4.5	−0.5	0.25
1	−1	−1	5		+0.5	0.25
1	−1	1	10	10.5	−0.5	0.25
1	−1	1	11		+0.5	0.25
1	1	−1	24	25	−1.0	1.00
1	1	−1	26		1.0	1.00
1	1	1	35	36.5	−1.5	2.25
1	1	1	38		1.5	2.25
						Sum = 7.50

Based on the coded variables, the matrices used in the estimation were (the number of digits retained has been truncated at four from the eight actually used):

$$\mathbf{a} = (\mathbf{x}^T\mathbf{x}) = \begin{bmatrix} 8.000 & 0.000 & 0.000 \\ 0.000 & 8.000 & 0.000 \\ 0.000 & 0.000 & 8.000 \end{bmatrix}$$

$$\mathbf{c} = (\mathbf{x}^T\mathbf{x})^{-1} = \begin{bmatrix} 0.125 & 0.000 & 0.000 \\ 0.000 & 0.125 & 0.000 \\ 0.000 & 0.000 & 0.125 \end{bmatrix}$$

$$\mathbf{G} = (\mathbf{x}^T\mathbf{Y}) = \begin{bmatrix} 153.0 \\ 92.99 \\ 35.00 \end{bmatrix}$$

The estimated regression coefficients were computed to be

$$\mathbf{b} = \begin{bmatrix} b_0 \\ b_1 \\ b_2 \end{bmatrix} = (\mathbf{x}^T\mathbf{x})^{-1}(\mathbf{x}^T\mathbf{Y}) = \begin{bmatrix} 19.125 \\ 11.625 \\ 4.375 \end{bmatrix}$$

Consequently, the estimated regression equations were:

$$\text{Coded:} \quad \hat{Y} = 19.125 + 11.625x_1 + 4.375x_2$$

$$\text{Uncoded:} \quad \hat{Y} = -91.333 + 0.58125T + 1.4588p$$

Orthogonal designs for the independent variables simplify the detailed calculations and are more efficient than nonorthogonal designs because they obtain more information for a given amount of experimentation. This matter will be discussed in more detail in Chapter 8.

Figure E5.1-1 illustrates the contours of the estimated regression equation in observation space. In this example $s_e^2 = (7.50/4) = 1.875$ and $s_r^2 = (15.12/1) = 15.12$. The variance ratio $(15.12/1.875) = 8.06$ is greater than $F_{0.95}(1,4) = 7.71$; hence the model can be improved by one of the methods discussed in Chapters 7 and 8.

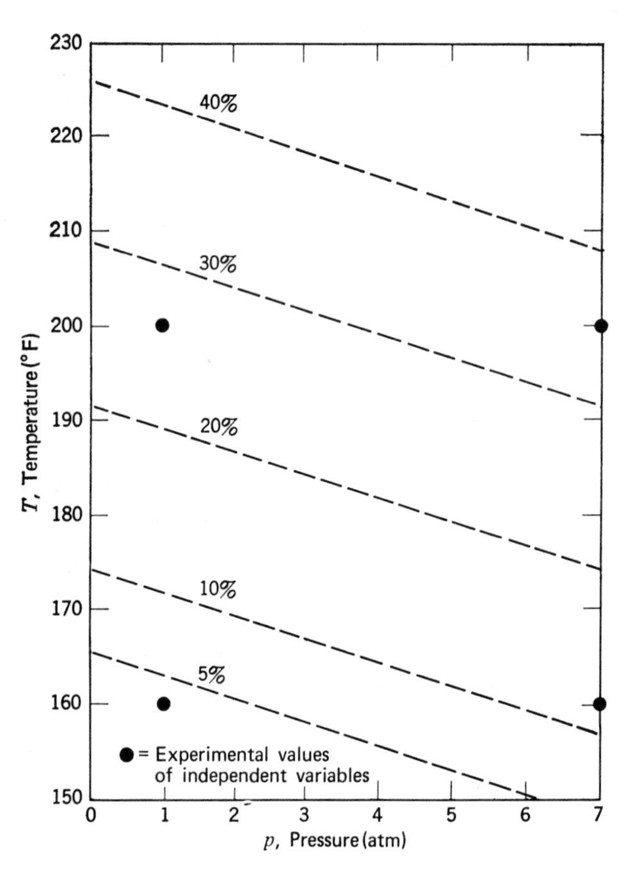

FIGURE E5.1-1

Example 5.1-2 Harmonic Analysis

The solution of certain types of differential equations and the approximation of most periodic responses can be expressed by an empirical model which is linear in the coefficients but not in the independent variables:

$$\eta = \alpha_0 + \alpha_1 \cos x + \beta_1 \sin x + \alpha_2 \cos 2x + \beta_2 \sin 2x + \cdots + \alpha_m \cos mx + \beta_m \sin mx \quad \text{(a)}$$

The scale of x should be chosen so that the fundamental period is 2π on the x-scale, in which case the parameters α_j and β_j for $j = 1, 2, \ldots, m$ depend on the choice of origin on the x-scale. However, the amplitude, $(\alpha_j^2 + \beta_j^2)^{1/2}$, of the jth harmonic is invariant under translation of axis. If the terms corresponding to the jth harmonic are written as

$$\alpha_j \cos jx + \beta_j \sin jx = \rho_j \sin (jx + \theta_j) \quad \text{(b)}$$

where

$$\rho_j = (\alpha_j^2 + \beta_j^2)^{1/2}, \qquad \theta_j = \tan^{-1}\frac{\alpha_j}{\beta_j}$$

it is evident that the amplitude, ρ_j, is not altered by an arbitrary choice of origin. However, the phase angle, θ_j, does depend on the location of the origin. In harmonic analysis it is customary to estimate and/or test hypotheses on the amplitudes of the various harmonics rather than on the parameters α_j and β_j. With this exception, the development of harmonic analysis follows the usual linear regression analysis.

Suppose we consider a special type of harmonic analysis in which the n observations are taken at values of x which are equally spaced over one cycle of the periodic function. It will be seen that the calculations in this important case are particularly simple because of the orthogonality of the data for all parameters.

The values of the independent variable may, without loss of generality, be taken as $x_t = tr$, with the data sets at $t = 0, 1, 2, \ldots, n - 1$, and $r = 2\pi/n$. (The letter t is used here because of the predominance of applications in which time is the independent variable. The value of n might be 24, for example, if the period was one day.) Thus for each observation:

$$Y_t = \alpha_0 + \sum_{j=1}^{m} (\alpha_j \cos jtr + \beta_j \sin jtr) + \epsilon_t, \qquad t = 0, 1, 2, \ldots, n - 1 \quad \text{(c)}$$

(Note that the number of observations must be $n \geq 2m + 1$ for determinacy.)

In the normal equations the following typical sums vanish because of orthogonality:

$$\sum_{t=0}^{n-1} \cos jtr \sin jtr = 0$$

$$\sum_{t=0}^{n-1} \cos jtr \cos ktr = \sum_{t=0}^{n-1} \sin jtr \sin ktr = 0, \qquad j, k = 1, 2, \ldots, m; j \neq k$$

but the squared functions do not:

$$\sum_{t=0}^{n-1} \cos^2 jtr = \sum_{t=0}^{n-1} \sin^2 jtr = \frac{n}{2}$$

Consequently, the normal equations are

$$n\alpha_0 = \sum_{t=0}^{n-1} Y_t$$

$$\frac{n}{2}\alpha_j = \sum_{t=0}^{n-1} Y_t \cos jtr \quad \text{(d)}$$

$$\frac{n}{2}\beta_j = \sum_{t=0}^{n-1} Y_t \sin jtr \qquad j = 1, 2, \ldots, m$$

and the least squares estimates of the model parameters are

$$a_0 = \hat{\alpha}_0 = \frac{1}{n}\sum Y_t = \bar{Y}$$

$$a_j = \hat{\alpha}_j = \frac{2}{n}\sum Y_t \cos jtr \quad \text{(e)}$$

$$b_j = \hat{\beta}_j = \frac{2}{n}\sum Y_t \sin jtr$$

The usual assumptions on the ϵ_t's give the variances of

$$\sigma_{\hat{\alpha}_0}^2 = \frac{\sigma_{Y_t}^2}{n} \quad (\text{f}_1)$$

$$\sigma_{\hat{\alpha}_j}^2 = \sigma_{\hat{\beta}_j}^2 = \frac{2\sigma_{Y_t}^2}{n} \quad (\text{f}_2)$$

where $\sigma^2_{\bar{Y}_i}$ is the variance of ϵ_t. An unbiased estimate of $\sigma^2_{\bar{Y}_i}$ is given by $s^2_{r_i}$ which can be calculated by using Equation 5.1-15 or

$$s^2_{r_i} = \frac{\sum_{t=0}^{n-1} Y_t^2 - na_0^2 - \frac{n}{2}\sum_{j=1}^{m}(a_j^2 + b_j^2)}{(n - 2m - 1)} \tag{g}$$

The variances of the estimated parameters can themselves be estimated by $s^2_{a_0}$, $s^2_{a_j}$, and $s^2_{b_j}$ when $s^2_{\bar{Y}_i}$ is substituted for $\sigma^2_{\bar{Y}_i}$ in the right-hand sides of Equations (f).

Under the assumption that the ϵ's are normal, Table E5.1-2a gives an analysis of variance for testing that the amplitudes of the harmonics differ from zero. Under the normality assumptions on the ϵ_t's, the jth harmonic amplitude may be tested by an F-test, using the variance ratio $\frac{1}{4}n(a_j^2 + b_j^2)/s^2_{\bar{Y}_i}$. For example, if the variance ratio for the first harmonic alone is significant, then the empirical data represent a sine wave. The hypothesis that each of the amplitudes of the harmonics is zero can be tested in turn. Also, once the estimated coefficients have been evaluated, $\hat{\rho}_j$ and $\hat{\theta}_j$ can be calculated.

TABLE E5.1-2a ANALYSIS OF VARIANCE FOR HARMONIC ANALYSIS

Source of Variation	ν = Degree of Freedom	Sum of Squares	Mean Square
First harmonic	2	$\frac{1}{2}n(a_1^2+b_1^2)$	$\frac{1}{4}n(a_1^2+b_1^2)$
Second harmonic	2	$\frac{1}{2}n(a_2^2+b_2^2)$	$\frac{1}{4}n(a_2^2+b_2^2)$
⋮	⋮		
mth harmonic	2	$\frac{1}{2}n(a_m^2+b_m^2)$	$\frac{1}{4}n(a_m^2+b_m^2)$
Residual	$n-2m-1$	(Difference)	$s^2_{\bar{Y}_i}$
Total	$n-1$	$\sum Y_t^2 - na_0^2$	

As an example of estimation of the parameters in harmonic analysis, the following data, taken from the periodic output for a steady-state process, were fitted by Equation (c) with $m=4$:

x (time)	Y (volts)
0	0.972
$\pi/6$	−0.653
$\pi/3$	−0.353
$\pi/2$	2.063
$2\pi/3$	3.803
$5\pi/6$	2.798
π	−0.977
$7\pi/6$	−4.391
$4\pi/3$	−4.709
$3\pi/2$	−2.165
$5\pi/3$	2.324
$11\pi/6$	1.048
2π	0.814

Here $n = 12$ (2π initiates a new cycle).

From Equations (e) the nine estimated parameters were:

$$\begin{aligned} a_0 &= -0.0153 & b_1 &= 2.0768 \\ a_1 &= 0.9334 & b_2 &= -2.8978 \\ a_2 &= 0.0391 & b_3 &= 0.0027 \\ a_3 &= 0.0625 & b_4 &= -0.0377 \\ a_4 &= 0.0030 \end{aligned}$$

and the estimated variance from Equation 5.1-15 was $s_r^2 = 3.249 \times 10^{-3}$. Replicate data from earlier runs indicated that $s_e^2 = 1.12 \times 10^{-3}$ with 4 degrees of freedom; $F_{1-\alpha}(3, 4) = 6.59$ from Table C.4 in Appendix C; hence the model was deemed adequate. The pooled variance was $s^2_{\bar{Y}_i} = 2.03 \times 10^{-3}$.

The mean squares corresponding to Table E5.1-2a are shown in Table E5.1-2b. Additional harmonics could be added to the model, and possibly some of them would prove to be significant.

TABLE E5.1-2b

	Mean Square	Variance Ratio $\frac{n}{4}(a_j^2 + b_j^2)/s^2_{\bar{Y}_i}$
First harmonic	15.552	Significant*
Second harmonic	25.197	Significant*
Third harmonic	11.73	Significant*
Fourth harmonic	4.29×10^{-3}	Not significant

* $F_{0.95}(2, 7) = 4.74$

5.1-6 Estimation Using Orthogonal Polynomials

In using polynomials as empirical models, the **a** matrix can become quite ill conditioned. For example, when the number of coefficients reaches nine, a computer program using about eight significant figures will not give meaningful results. Consequently, in lieu of fitting fairly high-order polynomials to data, it is more effective to fit orthogonal polynomials (or to use a computer program involving orthogonal transformations). The resulting orthogonal polynomial can, if desired, be transformed into an ordinary polynomial after the curve-fitting process is completed.

If the experimental data are equally spaced with respect to the independent variable x† and arranged as a series of pairs which can be arbitrarily numbered $(0, Y_0)$, $(1, Y_1)$, $(2, Y_2), \ldots, (n, Y_n)$, a model can be formed from a combination of orthogonal functions:

$$Y_q(x) = \beta_0 P_{0,n}(x) + \beta_1 P_{1,n}(x) + \cdots + \beta_q P_{q,n}(x) + \epsilon \tag{5.1-20}$$

† W. E. Milne, *Numerical Calculus*, Princeton Univ. Press, 1954.

and the sum of the squares of the unobservable errors in Equation 5.1-3 minimized to obtain the estimated parameters b_k. The orthogonal functions are themselves polynomials $P_{m,n}(x)$, such as

$$P_{0,n}(x) = 1$$

$$P_{1,n}(x) = 1 - 2\frac{x}{n}$$

$$P_{2,n}(x) = 1 - 6\frac{x}{n} + 6\frac{x(x-1)}{n(n-1)}$$

$$P_{3,n}(x) = 1 - 12\frac{x}{n} + 30\frac{x(x-1)}{n(n-1)} - 20\frac{x(x-1)(x-2)}{n(n-1)(n-2)}$$

and in general

$$P_{m,n}(x) = \sum_{k=0}^{m} (-1)^k \frac{(m+k)!}{(m-k)!\,(k!)^2}\frac{x^{(k)}}{n^{(k)}} \qquad (5.1\text{-}21)$$

where the notation $x^{(k)}$ (or $n^{(k)}$) means $x(x-1)(x-2)\cdots(x-k+1)$, m is the degree of the polynomial, and x takes on integer values from 0 to n. These polynomials have the very useful orthogonal property that

$$\sum_{x=0}^{n} P_{m,n}(x)P_{q,n}(x) = 0, \qquad \text{if } q \neq m$$

$$\sum_{x=0}^{n} P_{m,n}^2(x) = \frac{(n+m+1)(m+n)^{(m)}}{(2m+1)n^{(m)}} \qquad \text{if } q = m$$

Because of the orthogonality property, all the off-diagonal terms in the equation equivalent to Equation 5.1-5 vanish, and each coefficient can be determined independently from the others by

$$b_m = \frac{\sum_{x=0}^{n} Y(x)P_{m,n}(x)}{\sum_{x=0}^{n} P_{m,n}^2(x)} \qquad m = 0, 1, 2, \ldots, q \qquad (5.1\text{-}22)$$

In addition to the advantage of not having to solve a system of equations for the parameters, the use of orthogonal polynomials has another advantage. If one has already obtained the mth degree polynomial, a fit to an $(m+1)$st degree polynomial requires only one new coefficient, b_{m+1}, be determined; all other coefficients remain the same.

If the experimental data are not equally spaced with respect to the independent variable, the simple polynomials $P_{m,n}(x)$ are no longer applicable. Suitable polynomials do exist, but in addition to depending on m and n, they also depend on the particular spacing of the unequally spaced points. Thus, every individual, unequally spaced, curve-fitting problem will lead to a regression equation that is a linear combination of its own special orthogonal polynomials. Let us fit a polynomial by the weighted least squares method to the data represented by the nonequally spaced points $(x_1, Y_1), (x_2, Y_2), \ldots, (x_n, Y_n)$ using nonzero (positive) weights $w(x_k)$, $k = 1, 2, \ldots, n$. The approximating function is to be

$$\hat{Y}_q(x) = b_0P_0(x) + b_1P_1(x) + \cdots + b_qP_q(x) \qquad (5.1\text{-}23)$$

The general orthogonality condition will be

$$\sum_{k=1}^{n} w(x_k)P_j(x_k)P_i(x_k) = 0 \qquad \text{if } i \neq j \text{ for } i,j = 0, 1, 2, \ldots, q$$

These polynomials may be found recursively as follows. Let:

$$P_{-1}(x) = 0$$
$$P_0(x) = 1$$
$$P_1(x) = (x - \alpha_1)P_0(x)$$
$$P_2(x) = (x - \alpha_2)P_1(x) - \beta_1P_0(x)$$
$$P_3(x) = (x - \alpha_3)P_2(x) - \beta_2P_1(x)$$
$$\cdots$$
$$P_{j+1}(x) = (x - \alpha_{j+1})P_j(x) - \beta_jP_{j-1}(x)$$
$$\cdots$$

The α's and β's are constants to be determined so that the general orthogonality relationships are satisfied. It can be shown† that if α_{j+1} and β_j are calculated as follows:

$$\alpha_{j+1} = \frac{\sum_{k=1}^{n} w(x_k)x_k[P_j(x_k)]^2}{\sum_{k=1}^{n} w(x_k)[P_j(x_k)]^2}$$

$$\beta_j = \frac{\sum_{k=1}^{n} w(x_k)[P_j(x_k)]^2}{\sum_{k=1}^{n} w(x_k)[P_{j-1}(x_k)]^2}$$

$P_{j+1}(x)$ will be orthogonal in the sense desired to both $P_j(x)$ and $P_{j-1}(x)$.

Example 5.1-3 Orthogonal Polynomials

A waste-treatment pond was not reducing organic compounds adequately to meet the existing standards. A new bacterial culture has been introduced into the pond. The data in Table E5.1-3a have been taken as a function of x, the time in hours; Y is the voltage reading in the pond effluent stream monitoring device. It is desired to fit the 41 data points to an orthogonal polynomial and to terminate the fitting when the sum of the squares for the last term becomes insignificant at the 5-percent significance level.

† G. E. Forsythe, *J. Soc. Ind. Appld. Math.* **5**, 74, 1957.

TABLE E5.1-3a

x	Y	x	Y
0	14.534	210	15.386
10	15.144	220	14.716
20	15.831	230	14.029
30	16.435	240	13.293
40	17.034	250	12.590
50	17.567	260	11.871
60	18.050	270	11.168
70	18.440	280	10.393
80	18.764	290	9.640
90	19.028	300	8.998
100	19.193	310	8.311
110	19.248	320	7.625
120	19.226	330	6.949
130	19.100	340	6.301
140	18.880	350	5.619
150	18.578	360	5.021
160	18.187	370	4.389
170	17.748	380	3.823
180	17.243	390	3.109
190	16.644	400	2.603
200	16.072		

Solution ·

Because the intervals for x are equally spaced, we can use Equation 5.1-22 to estimate the coefficients in Equation 5.1-20. The results for the first few polynomials and for several coefficients are:

Polynomials

$$P_0 = 1$$

$$P_1 = 1 - \frac{2x}{41}$$

$$P_2 = 1 - \frac{6x}{41} + \frac{6x}{41}\left(\frac{x-1}{40}\right)$$

etc.

$$b_0 = \frac{\sum_{x=0}^{n} Y_x(1)}{\sum_{x=0}^{n} (1)^2} = \bar{Y}$$

$$b_1 = \frac{\sum_{x=0}^{n} Y_x\left(1 - \frac{2x}{41}\right)}{\sum_{x=0}^{n}\left(1 - \frac{2x}{41}\right)^2}$$

etc.

Coefficients

$b_0 = 13.337$
$b_1 = -0.391$
$b_2 = -0.019$
$b_3 = 0.892 \times 10^{-3}$
$b_4 = 0.801 \times 10^{-5}$
$b_5 = -0.999 \times 10^{-5}$
$b_6 = 0.365 \times 10^{-6}$
$b_7 = 0.343 \times 10^{-7}$
$b_8 = -0.122 \times 10^{-7}$
$b_9 = -0.519 \times 10^{-9}$

Table E5.1-3b lists the sum of the squares removed from $\sum_{i=1}^{41} (Y_i - 0)^2$ as each additional term is added to the model. The interpretation of each sum of squares in terms of the F-test is the same as that described in Section 4.3, and it is discussed again in Section 5.3.

As can be seen in Table E5.1-3b, significance is established for each of the zeroth through seventh degree terms added.

TABLE E5.1-3b

Term Added		Degree of Freedom, ν	Sum of Squares	Mean Square	Variance Ratio*
	Total $\sum (Y_i - 0)^2$	41	8449.79		
0	Removed	1	7293.49	7293.49	
	Residual	40	1156.30	28.907	
1	Removed	1	880.6600	880.6600	124.60
	Residual	39	275.6453	7.678	
2	Removed	1	236.2725	236.2725	228.04
	Residual	38	39.3728	1.0361	
3	Removed	1	38.1157	38.1157	1121.71
	Residual	37	1.2571	0.03398	
4	Removed	1	0.1545	0.01595	5.23
	Residual	36	1.0976	0.03049	
5	Removed	1	1.0356	1.0356	60.21
	Residual	35	0.0620	0.01720	
6	Removed	1	0.0394	0.0394	59.27
	Residual	34	0.0226	0.0006647	
7	Removed	1	0.0077	0.0077	17.05
	Residual	33	0.0149	0.0004515	
8	Removed	1	0.0013	0.0013	3.06
	Residual	32	0.0136	0.000425	
9	Removed	1	0.00001	0.00001	0.20
	Residual	31	0.0136	0.000438	

* $F_{0.95}(1, \nu)$ ranges from 4.09 to 4.16.

Thus we start the estimated regression equation with an intercept and terminate with $P_8(x)$ to get Equation 5.1-23.

To express $\hat{Y}$ as a polynomial in x, we need to introduce each $P_m(x)$ into the estimated regression equation. After some considerable algebraic manipulations, we find:

$$\hat{Y} = 14.521 + 0.06587x - 0.3311 \times 10^{-4}x^2 - 0.6112 \times 10^{-6}x^3 - 0.2283 \times 10^{-7}x^4 + 0.1758 \times 10^{-9}x^5 - 0.5225 \times 10^{-12}x^6 + 0.7245 \times 10^{-15}x^7 - 0.3920 \times 10^{-18}x^8$$

5.2 CONFIDENCE INTERVALS AND HYPOTHESIS TESTS

We now turn to consideration of: (1) estimating confidence intervals for the model parameters, (2) estimating a joint confidence region for the parameters, and (3) executing hypothesis tests correspond to those in Section 4.3.

5.2-1 Confidence Intervals and Region

A confidence interval can be estimated for each individual parameter β_k in the vector $\boldsymbol{\beta}$ through use of the t distribution. The standard error of the estimate for b_k comes from the estimates of elements on the main diagonal of Equation 5.1-14:

$$s_{b_k} = \hat{\sigma}_{b_k} = \sqrt{s_{Y_i}^2 c_{kk}}$$

The confidence interval for β_k for a significance level α is formed exactly as in Section 4.3:

$$(b_k - t_{1-\frac{\alpha}{2}} s_{Y_i}\sqrt{c_{kk}}) \le \beta_k < (b_k + t_{1-\frac{\alpha}{2}} s_{Y_i}\sqrt{c_{kk}}); \quad \nu = n - q - 1 \quad (5.2\text{-}1)$$

Because $b_0 = \bar{Y}$,

$$s_{b_0} = \sqrt{s_{Y_i}^2 c_{00}} = \sqrt{\frac{s_{Y_i}^2}{\sum w_i}}$$

(recall that c_{00} is just the inverse of the number of data sets if the weights are unity) and the confidence interval for the intercept is

$$(b_0 + t_{1-\frac{\alpha}{2}} s_{b_0}) \le \beta_0 < (b_0 + t_{1-\frac{\alpha}{2}} s_{b_0}) \quad (5.2\text{-}2)$$

For Model 5.1-2 where

$$\beta_0' = \beta_0 - \sum_{k=1}^{q} \beta_k \bar{x}_k$$

the variance of b_0' is

$$\text{Var}\{b_0'\} = \text{Var}\{b_0\} + \sum_{k=1}^{q} \bar{x}_k^2 \,\text{Var}\{b_k\}$$

The confidence limits for β_0' for a selected significance level α are given by Equation 5.2-2 with s_{b_0}' substituted for s_{b_0} and b_0' substituted for b_0.

We saw in Section 4.3 that the variance of $\hat{Y}_i$ in the regression equation $\hat{Y}_i = b_0 + b_1(x_i - \bar{x})$ was

$$\text{Var}\{\hat{Y}_i\} = \text{Var}\{b_0\} + (x_i - \bar{x})^2 \,\text{Var}\{b_1\}$$

Similarly, the variance of the matrix $\hat{\mathbf{Y}}$ for Model 5.1-1 is, using Equation 5.1-14,

$$\text{Var}\{\hat{\mathbf{Y}}\} = \text{Var}\{\mathbf{xb}\} = \mathbf{x}\,\text{Var}\{\mathbf{b}\}\mathbf{x}^T = \sigma_{Y_i}^2 \mathbf{xcx}^T \quad (5.2\text{-}3)$$

For a single data set

$$\text{Var}\{\hat{\mathbf{Y}}_i\} = [1\ (x_{i1} - \bar{x}_1) \cdots (x_{iq} - \bar{x}_q)] \times \begin{bmatrix} c_{00} & \cdots & c_{0q} \\ \vdots & & \vdots \\ c_{q0} & & c_{qq} \end{bmatrix} \begin{bmatrix} 1 \\ (x_{i1} - \bar{x}_1) \\ \vdots \\ (x_{iq} - \bar{x}_q) \end{bmatrix} \sigma_{Y_i}^2 = \mathbf{x}_i \mathbf{c} \mathbf{x}_i^T \sigma_{Y_i}^2$$

(The elements in the first row and column of $\mathbf{c}$ are all zero except for c_{00}.)

The confidence interval for η_i for a given significance level α employs the estimated standard error $s_{\hat{Y}_i} = s_{Y_i}\sqrt{\mathbf{x}_i \mathbf{c} \mathbf{x}_i^T}$.

$$(\hat{Y}_i - t_{1-\frac{\alpha}{2}} s_{\hat{Y}_i}) \le \eta_i < (\hat{Y}_i + t_{1-\frac{\alpha}{2}} s_{\hat{Y}_i}) \quad \nu = (n - q - 1) \quad (5.2\text{-}4)$$

If we want to use the empirical model to predict, two types of predictions can be made: *point predictions* and *interval predictions*. The acid test of the predictive ability of a model, of course, is to compare the prediction with the corresponding experimental data. In predicting, we presume that the assumptions underlying the random variable being predicted do not change (or, if they change in some fashion, take the change into account). The (point) predicted value, Y_{n+1}^*, for one additional observation or one additional time period is based on the relation

$$Y_{n+1}^* = \eta_{n+1} + \epsilon_{n+1}$$

so that $\mathscr{E}\{Y_{n+1}^*\} = \eta_{n+1}$ as long as $\mathscr{E}\{\epsilon_{n+1}\} = 0$. If we use the best estimate of η_{n+1}, $\hat{Y}_{n+1}$, the variance of Y_{n+1}^* is (with $\sigma_{\epsilon_i}^2 = \sigma_{\epsilon_{n+1}}^2 = \sigma_{Y_i}^2$)

$$\sigma_{Y_{n+1}^*}^2 = \text{Var}\{Y_{n+1}^*\} = \text{Var}\{\hat{Y}_{n+1}\} + \text{Var}\{\epsilon_{n+1}\} = \sigma_{Y_i}^2 \mathbf{x}_{n+1}\mathbf{c}\mathbf{x}_{n+1}^T + \sigma_{Y_i}^2 = \sigma_{Y_i}^2(\mathbf{x}_{n+1}\mathbf{c}\mathbf{x}_{n+1}^T + 1)$$

(If m replicate observations are taken for x_{n+1}, $\text{Var}\{\epsilon_{n+1}\} = \sigma_{Y_i}^2/m$.) The confidence interval for η_{n+1} can be formed by using Equation 5.2-4 but replacing $s_{\hat{Y}_i}$ with $s_{Y_{n+1}^*} = s_{Y_i}\sqrt{\mathbf{x}_{n+1}\mathbf{c}\mathbf{x}_{n+1}^T + 1}$ as in Section 4.3.

A joint confidence region for the β's for a given significance level α can be formed exactly as described in

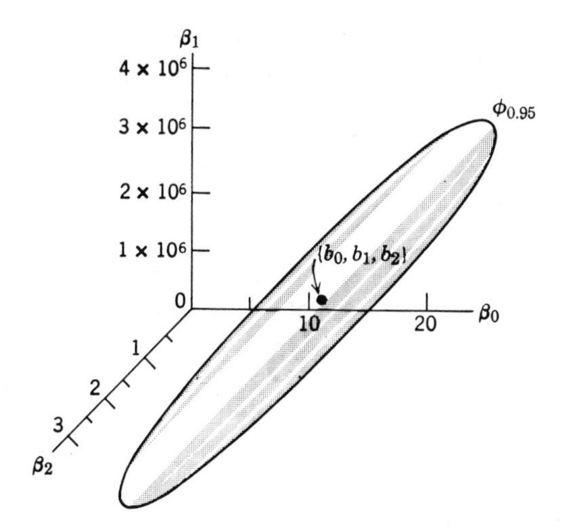

FIGURE 5.2-1 Approximate 95-percent confidence region in parameter space for a linear model.

Section 4.3; in matrix notation the equivalent of Equation 4.3-28 is

$$(\boldsymbol{\beta} - \mathbf{b})^T\mathbf{x}^T\mathbf{w}\mathbf{x}(\boldsymbol{\beta} - \mathbf{b}) = s_{\bar{Y}_i}^2(q + 1)F_{1-\alpha} \quad (5.2\text{-}5)$$

where $F_{1-\alpha}$ is the upper F value for $(q + 1)$ and $n - (q + 1)$ degrees of freedom, respectively. Figure 5.2-1 illustrates the confidence region for a linear (in the parameters) model of the effect of a magnetic field on the vaporization of water. A confidence region that is long and attenuated, such as that of Figure 5.2-1, implies that the parameter values have been poorly estimated; a small, spherical-shaped region is more desirable. The long, narrow shape of the region results primarily from a high degree of correlation among the various parameter estimates. One practical implication of this high correlation is that if a wrong value of one parameter is inadvertently estimated, this value will be balanced in the fitting procedure by a compensating wrong value of another parameter to give an overall fit for the model which will be nearly as good as that obtained using the

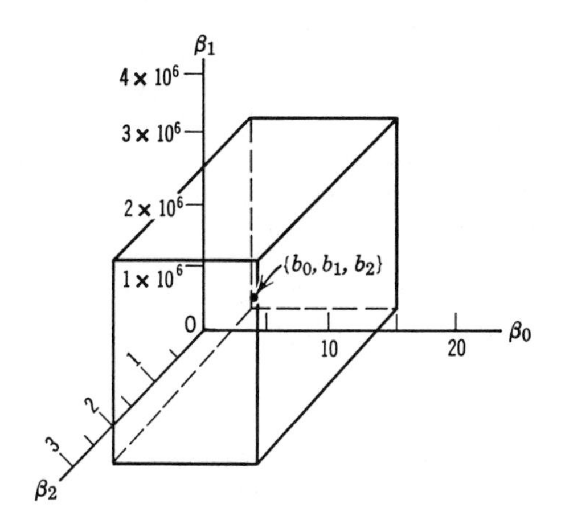

FIGURE 5.2-2 Erroneous confidence region based on individual confidence limits.

best estimates of the parameters. This matter is discussed in Chapter 8 as related to the design of experiments. Figure 5.2-2 portrays the volume in parameter space blocked out by the individual confidence intervals. Compare Figure 5.2-2 with the correct joint confidence region of Figure 5.2-1. The sum of the squares surface analogous to Equation 4.3-28a is

$$\phi_{1-\alpha} = \phi_{\min}\left[1 + \frac{m}{n - m}F_{1-\alpha}\right]$$

5.2-2 Hypothesis Tests

The hypothesis tests summarized here are the analogs of those developed in Section 4.3.

1. To test the hypothesis that all the $\beta_1 = \beta_2 = \cdots = \beta_q = 0$, form the variance ratio

$$\frac{s_3^2}{s_{\bar{Y}_i}^2} = \frac{\mathbf{b}^T\mathbf{G}/q}{s_{\bar{Y}_i}^2} \quad (5.2\text{-}6)$$

If the variance ratio is greater than $F_{1-\alpha}(q, n - q - 1)$ for a significance level α, then the hypothesis is rejected.

2. Another hypothesis that can be tested is that certain of the β's are zero. Split the β's into two groups, labeled I and II, and test the hypothesis that all the β's in group II are zero (without assuming anything about those in group I). A variance ratio is formed in which the numerator represents the mean square for group II:

$$\frac{\dfrac{(\mathbf{b}^T\mathbf{G})_{(\mathrm{I}+\mathrm{II})} - (\mathbf{b}^T\mathbf{G})_{\mathrm{I}}}{\nu_{(\mathrm{I}+\mathrm{II})} - \nu_{\mathrm{I}}}}{s_{\bar{Y}_i}^2} \quad (5.2\text{-}7)$$

If the variance ratio is greater than

$$F_{1-\alpha}[(\nu_{(\mathrm{I}+\mathrm{II})} - \nu_{\mathrm{I}}), (n - q - 1)],$$

the hypothesis is rejected. This test helps to ascertain if certain variables should be included or excluded from a model. Note, however, that if the hypothesis is accepted that $\boldsymbol{\beta}_{\mathrm{II}} = 0$ and the coefficients are deleted from the model, bias will exist in the estimate of the nonzero coefficients as explained in connection with Equation 5.1-17.

3. To test the hypothesis that β_k has a specified value, β_k^*, compute

$$t = \frac{b_k - \beta_k^*}{s_{\bar{Y}_i}\sqrt{c_{kk}}} \quad (5.2\text{-}8)$$

If t is greater than $t_{1-\frac{\alpha}{2}}$ for $(n - q - 1)$ degrees of freedom, the hypothesis that $\beta_k = \beta_k^*$ is rejected.

4. To test the hypothesis that $\boldsymbol{\beta} = \boldsymbol{\beta}^*$, where $\boldsymbol{\beta}^*$ is a matrix of specified values of $\boldsymbol{\beta}$, form the variance ratio

$$\frac{(\mathbf{b} - \boldsymbol{\beta}^*)^T\mathbf{x}^T\mathbf{x}(\mathbf{b} - \boldsymbol{\beta}^*)/(q + 1)}{s_{\bar{Y}_i}^2} \quad (5.2\text{-}9)$$

If the variance ratio is greater than $F_{1-\alpha}(q + 1, n - q - 1)$ for a confidence level of α, the hypothesis is rejected.

We can test that the underlying structure and assumptions of the model are the same in the prediction period as in the sample period using H_0, the null hypothesis (no difference). If the hypothesis is accepted, the estimated equation satisfactorily applies both to the original data collected and to the new data. If H_0 is rejected, either the characteristics of the random variable Y changed *or* the model was not adequate to encompass the new data. Either (or both) explanations are plausible. To carry out the test, the proper statistic to use is the t statistic computed for Y^*_{n+1}.

In multiple regression the F-test for the hypothesis that $\boldsymbol{\beta} = 0$ is not exactly the same as the t-test (i.e., t^2) for $\boldsymbol{\beta} = 0$, as pointed out in Example 4.3-1. Consequently, different conclusions can sometimes be reached, depending upon which test is used. Consider a model similar to Model 5.1-1a:

$$Y_i - \bar{Y} = \sum_{j=1}^{q} \beta_j(x_{ij} - \bar{x}_j) + \epsilon_i \qquad \begin{matrix} i = 1, \ldots, n \\ \beta_j > 0 \end{matrix} \tag{5.2-10}$$

The independent variables x_{ij} can each be scaled so that

$$\sum_{i=1}^{n} (x_{ij} - \bar{x}_j)^2 = n \qquad j = 1, \ldots, q$$

From Equation 2.4-32, the sample correlation coefficient among the x's is

$$\hat{\rho}_{jk} = \frac{1}{n-1} \frac{\sum_{i=1}^{n} (x_{ij} - \bar{x}_j)(x_{ik} - \bar{x}_k)}{s_{x_j} s_{x_k}}$$

so that

$$\sum_{i=1}^{n} (x_{ij} - \bar{x}_j)(x_{ik} - \bar{x}_k) = n\hat{\rho}_{jk} \tag{5.2-11}$$

Equation 5.2-6 gives the variance ratio to be used in the F-test with q replacing $q + 1$ degrees of freedom, because in Equation 5.2-10 $\bar{Y}$ is not counted as a parameter:

$$\frac{s_3^2}{s_{\bar{Y}_i}^2} = \frac{\mathbf{b}^T\mathbf{G}/q}{s_{\bar{Y}_i}^2}$$

We shall assume next that $\mathbf{w} \equiv \mathbf{I}$; hence $\mathbf{G}$, defined in connection with Equation 5.1-10, is $\mathbf{G} \equiv \mathbf{x}^T\mathbf{Y}$. Then

$$\frac{s_3^2}{s_{\bar{Y}_i}^2} = \frac{\mathbf{b}^T(\mathbf{x}^T\mathbf{Y})}{q s_{\bar{Y}_i}^2} = \frac{\mathbf{b}^T(\mathbf{x}^T\mathbf{x}\mathbf{b})}{q s_{\bar{Y}_i}^2} = \frac{\mathbf{b}^T\mathbf{a}\mathbf{b}}{q s_{\bar{Y}_i}^2}$$

The elements in Equation 5.2-11 can be written as $a_{jk} = n\hat{\rho}_{jk}$ so that

$$\frac{s_3^2}{s_{\bar{Y}_i}^2} = n \frac{\sum_{j=1}^{q} b_j^2 + \sum_{j=1}^{q} \sum_{\substack{k=1 \\ j \neq k}}^{q} b_j b_k \hat{\rho}_{jk}}{q s_{\bar{Y}_i}^2} \tag{5.2-12}$$

Introduction of Equation 5.2-8 for the b's in Equation 5.2-12 yields a relation between $(s_3^2/s_{\bar{Y}_i}^2)$ and t_j:

$$\frac{s_3^2}{s_{\bar{Y}_i}^2} = \frac{n}{q}\left[\sum_{j=1}^{q} c_{jj}t_j^2 + \sum_{j=1}^{q} \sum_{\substack{k=1 \\ j \neq k}}^{q} \hat{\rho}_{jk} t_j t_k \sqrt{c_{jj}c_{kk}}\right] \tag{5.2-13}$$

If the independent variables are all uncorrelated, $\hat{\rho}_{jk} = 0$, $c_{jj} \equiv a_{jj}^{-1} = 1/n$, and Equation 5.2-13 reduces to

$$\frac{s_3^2}{s_{\bar{Y}_i}^2} = \frac{1}{q} \sum_{j=1}^{q} t_j^2 \tag{5.2-14}$$

(If $q = 1$, Equation 5.2-14 is the same as the relation given in Example 4.3-1.) With three or more residual degrees of freedom, the significance level of $F(q, n-q-1)$ is lower than the significance level of $F(1, n - q - 1)$ which corresponds to the significance level of t. Thus, the possibility exists that some or all of the coefficients may prove to be nonsignificant by t-tests whereas the variance ratio is significant by one of the F-tests. The explanation is that a significant variance ratio does not indicate the significance of any given coefficient but merely the existence of at least one linear combination of coefficients that is significantly different from zero. If the independent variables are highly correlated and $\hat{\rho}_{jk} > 0$, the variance ratio can become quite large relative to the t_j^2's.

Example 5.2-1 Estimation Without Replication or Proper Experimental Design

A major problem which constantly faces engineers is that of corrosion. By use of electrical resistance probes, the corrosion rate in a suction header of two furnace feed pumps in a thermal cracking plant was measured. The probes themselves were made from 5 percent Cr–$\frac{1}{2}$ percent Mo 40-mil diameter wire. Along with the corrosion rate, readings were taken of the: (1) sulfur content of the oil, (2) temperature at the probe, (3) temperatures in the two cracking coils, and (4) rate of flow of the charge. See Table E5.2-1a. It was estimated that the wire diameter could be measured to within 4 microinches.

Based on the corrosion data provided, estimate the parameters in a linear model including all five variables listed in the table. Determine whether or not each variable might be deleted from the model by testing whether or not its associated coefficient might be equal to zero. Do the data meet the assumptions for estimation as described in Section 4.2?

Solution:

Because no replicate data are available, it is impossible to obtain an estimate of the experimental error with which to test the hypothesis that a proper model is a linear one. The range of temperatures is quite narrow. Hence the temperatures may prove to have little influence on the corrosion

TABLE E5.2-1a

Day	Y 5 percent Cr–½ percent Mo Corrosion Rate (in./yr)	x_1 Total Sulfur in Feed Stock	x_2 Flow Rate at Probe (bbl/day)	x_3 Temperature of Probe (F°)	x_4 Temperature in Cracking Coil 1 (°F)	x_5 Temperature in Cracking Coil 2 (°F)
1	0.117	0.041	16.9	753	922	885
2	0.107	0.041	17.0	748	925	885
3	0.088	0.040	17.1	749	925	886
4	0.077	0.041	16.6	747	925	887
5	0.091	0.042	17.0	745	934	895
6	0.040	0.008	17.5	743	940	905
7	0.048	0.007	35.0	762	936	904
8	0.022	0.008	34.5	760	935	895
9	0.077	0.041	33.8	752	928	887
10	0.121	0.041	33.6	752	928	887
11	0.143	0.044	33.2	749	930	887

rate; if included they may cause the matrix $\mathbf{a} = \mathbf{x}^T\mathbf{x}$ to be ill conditioned. Another defect in the data provided is that only two levels of sulfur and flow rate are available. Suppose, nevertheless, that a model $\eta = \beta_0 + \beta_1 x_1 + \beta_2 x_2 + \beta_3 x_3 + \beta_4 x_4 + \beta_5 x_5$ is proposed as the model to fit the data, with the x's designated as in Table E5.2-1a. A computer routine for regression analysis was used with the results indicated in Table E5.2-1b (the numbers have been truncated to save space). The point estimates of the β's, the individual confidence intervals for the β's, and the confidence intervals for the η's (for $w_i \equiv 1$) which are tabulated in Table E5.2-1c indicate the unsatisfactory nature of the experiment with respect to reaching a decision on the terms to include in the suggested model.

If we successively form the variance ratio described in Section 5.2-2, page 155, to see if each one of the β's could be equal to zero, we find the results shown in Table E5.2-1d. (The notation SS refers to the difference in the sum of the squares with $\beta_k = 0$ and with $\beta_k \neq 0$; "x_k removed" means that the hypothesis being tested is that $\beta_k = 0$.) Because for $\alpha = 0.05$, $F_{1-\alpha}[1, (n - q - 1)] = F_{0.95}[1, 5] = 6.61$, each hypothesis that $\beta_k = 0$ in turn can be accepted. We see that no matter how sophisticated the treatment of the data, one cannot make a silk purse out of a sow's ear!

TABLE E5.2-1b

$$\mathbf{a} = \begin{bmatrix} 1.10 \times 10^{1} & 3.54 \times 10^{-1} & 2.72 \times 10^{2} & 8.26 \times 10^{3} & 1.02 \times 10^{4} & 9.80 \times 10^{3} \\ 3.54 \times 10^{-1} & 1.38 \times 10^{-2} & 8.35 \times 10^{1} & 2.65 \times 10^{2} & 3.28 \times 10^{2} & 3.14 \times 10^{2} \\ 2.72 \times 10^{2} & 8.35 \times 10^{0} & 7.52 \times 10^{3} & 2.04 \times 10^{5} & 2.53 \times 10^{5} & 2.42 \times 10^{5} \\ 8.26 \times 10^{3} & 2.65 \times 10^{2} & 2.04 \times 10^{5} & 6.20 \times 10^{6} & 7.68 \times 10^{6} & 7.36 \times 10^{6} \\ 1.02 \times 10^{4} & 3.28 \times 10^{2} & 2.53 \times 10^{5} & 7.68 \times 10^{6} & 9.51 \times 10^{6} & 9.11 \times 10^{6} \\ 9.80 \times 10^{3} & 3.14 \times 10^{2} & 2.42 \times 10^{5} & 7.36 \times 10^{6} & 9.11 \times 10^{6} & 8.73 \times 10^{6} \end{bmatrix}$$

$$\mathbf{c} = \begin{bmatrix} 5.41 \times 10^{4} & -9.88 \times 10^{3} & 1.69 \times 10^{1} & -3.10 \times 10^{1} & -4.65 \times 10^{1} & 1.38 \times 10^{1} \\ -9.88 \times 10^{3} & 3.08 \times 10^{3} & -1.89 \times 10^{0} & 4.87 \times 10^{0} & 4.49 \times 10^{0} & 2.22 \times 10^{0} \\ 1.69 \times 10^{1} & -1.89 \times 10^{0} & 8.00 \times 10^{-3} & -1.12 \times 10^{-2} & -1.82 \times 10^{-2} & 9.35 \times 10^{-3} \\ -3.10 \times 10^{1} & 4.87 \times 10^{0} & -1.12 \times 10^{-2} & 2.06 \times 10^{-2} & 2.70 \times 10^{-2} & -1.06 \times 10^{-2} \\ -4.65 \times 10^{1} & 4.49 \times 10^{0} & -1.82 \times 10^{-2} & 2.70 \times 10^{-2} & 6.38 \times 10^{-2} & -3.68 \times 10^{-2} \\ 1.38 \times 10^{1} & 2.22 \times 10^{0} & 9.35 \times 10^{-3} & -1.06 \times 10^{-2} & -3.68 \times 10^{-2} & 3.14 \times 10^{2} \end{bmatrix}$$

$$\mathbf{G} = \begin{bmatrix} 9.31 \times 10^{-1} \\ 3.49 \times 10^{-2} \\ 2.27 \times 10^{1} \\ 6.99 \times 10^{2} \\ 8.64 \times 10^{2} \\ 8.28 \times 10^{2} \end{bmatrix} \qquad \mathbf{b} = \begin{bmatrix} 0.6751 \\ 2.3064 \\ 0.0012 \\ -0.0007 \\ -0.0021 \\ 0.0020 \end{bmatrix}$$

TABLE E5.2-1c

Confidence Intervals on the β's; Pairs of Confidence Limits Are:

$\beta_0 = 0.675 +/- 15.562$
$\beta_1 = 2.306 +/- 3.714$
$\beta_2 = 0.001 +/- 0.005$
$\beta_3 = -0.000 +/- 0.009$
$\beta_4 = -0.002 +/- 0.016$
$\beta_5 = 0.002 +/- 0.011$

Confidence Intervals on the η_i's; Pairs of Confidence Limits Are:

	Percent Deviation Between Predicted and Experimental Values
$\eta_1 = 0.098 +/- 0.040$	15.9
$\eta_2 = 0.095 +/- 0.027$	10.8
$\eta_3 = 0.094 +/- 0.024$	− 7.4
$\eta_4 = 0.099 +/- 0.029$	−29.4
$\eta_5 = 0.100 +/- 0.062$	−10.7
$\eta_6 = 0.031 +/- 0.060$	20.3
$\eta_7 = 0.044 +/- 0.059$	7.3
$\eta_8 = 0.031 +/- 0.061$	−42.2
$\eta_9 = 0.110 +/- 0.035$	−43.8
$\eta_{10} = 0.109 +/- 0.028$	9.7
$\eta_{11} = 0.114 +/- 0.039$	19.8

Whether or not the experimental data meet the assumptions of Section 4.2 is difficult to tell without additional information. For example, the temperatures may represent random variables instead of fixed levels of temperature. If the temperatures are omitted from the role of variables, and if small variations in the sulfur content and flow rate are ignored as well, then in effect replicate values of the corrosion rate exist and can be used as a measure of the experimental error, which is evidently quite large. Finally, because measurements are made serially in time on the same wire, the data may fall into the category discussed in Section 4.6 for nonindependent errors. To sum up, this example illustrates the difficulty of extracting information from an experiment which has been completed without prior attention to setting up an efficient statistical design. Similar problems arise in the analysis of historical data.

5.3 ANALYSIS OF VARIANCE

An analysis of variance can be carried out on the model with several independent variables that is a direct extension of the analysis previously described in Section 4.3. Table 5.3-1, based on Model 5.1-1, corresponds to Table 4.3-2 of simple linear regression. The sum of the squares of the residuals between $\bar{Y}_i$ and $\hat{Y}_i$, as well as the sum of the squares of the deviations between the $\bar{Y}_i$ and the grand mean $\bar{Y}$, is computed. The table corresponding to Table 4.3-1 is not shown, although it can easily be written as an extension of the two-parameter case. In matrix notation the reduction in the sum of the "squares due to $\boldsymbol{\beta}$", or $\sum_{i=1}^{n} w_i(\hat{Y}_i - 0)^2$, is

$$\sum_{i=1}^{n} w_i \hat{Y}_i^2 = \hat{\mathbf{Y}}^T\mathbf{w}\hat{\mathbf{Y}} = (\mathbf{xb})^T\mathbf{w}(\mathbf{xb})$$
$$= \mathbf{b}^T(\mathbf{x}^T\mathbf{wx})\mathbf{b} = \mathbf{b}^T(\mathbf{x}^T\mathbf{wY}) = \mathbf{b}^T\mathbf{G}$$

We can give a geometric interpretation to $\sum (Y_i - 0)^2$ in terms of Figure 5.1-1. The square of the length $\hat{\mathbf{Y}}$ is just split up into the sum of squares of its components in the x-plane.

The number of degrees of freedom $(n - q - 1)$ shown in Table 5.3-1 in row two equals the number of *independent* measurements that are available for estimating the parameters; it consists of the total number of sets of data less the number of constraints which are established by the least squares method. For example, with eight data points (values of $\bar{Y}_i$ and x_i), we have eight total degrees of freedom. Fitting an equation with three parameters, including the intercept as one parameter, introduces three constraints (three degrees of freedom "absorbed") and leaves five degrees of freedom as the "residual" degrees of freedom.

Additional valuable information about a model can be obtained by computing the sum of squares (SS) corresponding to removing one or more variable from a model initially containing all the variables. The SS between the predicted values from the regression equation and the mean, $\sum (\hat{Y}_i - \bar{Y})^2$, is thereby reduced; the SS removed from $\sum (\hat{Y}_i - \bar{Y})^2$ can be tested by an F-test, as will be explained shortly, to evaluate the significance of one or a group of the independent variables in the model.

TABLE E5.2-1d

Source of Variation	ν = Degree of Freedom	SS	Mean Square	Variance Ratio
x_0 removed (intercept)	1	8.416×10^{-6}	8.416×10^{-6}	1.243×10^{-2}
x_1 removed	1	1.724×10^{-3}	1.724×10^{-3}	2.547×10^{0}
x_2 removed	1	1.850×10^{-4}	1.850×10^{-4}	2.733×10^{-1}
x_3 removed	1	2.246×10^{-5}	2.246×10^{-5}	3.319×10^{-2}
x_4 removed	1	7.281×10^{-5}	7.281×10^{-5}	1.075×10^{-1}
x_5 removed	1	1.328×10^{-4}	1.328×10^{-4}	1.962×10^{-1}

TABLE 5.3-1 ANALYSIS OF VARIANCE

Source of Variation	Degrees of Freedom (ν)	Sum of Squares (SS)	Mean Square
Due to regression	q	$\sum_{i=1}^{n} w_i(\hat{Y}_i - \bar{Y})^2$	$s_2^2 = \dfrac{\sum w_i(\hat{Y}_i - \bar{Y})^2}{q}$
Deviation about the empirical regression line (deviations from regression)	$n - q - 1$	$\sum_{i=1}^{n} w_i(\bar{Y}_i - \hat{Y}_i)^2$	$s_r^2 = \dfrac{\sum w_i(\bar{Y}_i - \hat{Y}_i)^2}{n - q - 1}$
Subtotal	$n - 1$	$\sum_{i=1}^{n} w_i(\bar{Y}_i - \bar{Y})^2$	$s_t^2 = \dfrac{\sum w_i(\bar{Y}_i - \bar{Y})^2}{n - 1}$
Deviations within sets (residual error)	$\sum_{i=1}^{n} p_i - n$	$\sum_{i=1}^{n}\sum_{j=1}^{p_i} (Y_{ij} - \bar{Y}_i)^2$	$s_e^2 = \dfrac{\sum_{i=1}^{n}\sum_{j=1}^{p_i} (Y_{ij} - \bar{Y}_i)^2}{\sum_i p_i - n}$

Suppose we assign to one parameter in Model 5.1-1—the last term for convenience in notation but the results are valid for any term—a value ξ, perhaps zero. The model can then be written

$$\eta - \xi(x_q - \bar{x}_q) = \beta_0 + \sum_{k=1}^{q-1} \beta_k(x_k - \bar{x}_k) \quad (5.3\text{-}1)$$

In the following discussion the weights will be suppressed to save space. To obtain the estimated coefficients in the regression equation, one could minimize the sum of the squares:

$$\phi_{q-1} = \sum_{i=1}^{n} [Y_i - (\eta_i - \xi(x_{iq} - \bar{x}_q))]^2$$

to obtain the estimates $\mathbf{b}^*$ of $\boldsymbol{\beta}$. The estimates, of course, would not be the same as the $\mathbf{b}$ obtained without assignment of the value for β_q. The sum of the squares of the residuals and the normal equations for the case in which $\beta_q = \xi$ can be expressed as follows. Let us partition Model 5.1-1, $\boldsymbol{\eta} = \mathbf{x}\boldsymbol{\beta}$, into two parts:

$$\boldsymbol{\eta} = [\mathbf{x}^*\mathbf{x}_q]\begin{bmatrix}\boldsymbol{\beta}^* \\ \boldsymbol{\beta}_q\end{bmatrix} = \mathbf{x}^*\boldsymbol{\beta}^* + \mathbf{x}_q\boldsymbol{\beta}_q$$

where the last term represents the term $(x_q - \bar{x}_q)\xi$. Then the normal equations and the reduction in the sum of the squares "due to $\boldsymbol{\beta}$" would be:

	One Parameter Assigned	Full Model
Normal equations	$(\mathbf{x}^{*T}\mathbf{x}^*)\mathbf{b}^* = \mathbf{x}^{*T}\mathbf{Y}$	$(\mathbf{x}^T\mathbf{x})\mathbf{b} = \mathbf{x}^T\mathbf{Y}$
Reduction in sum of squares "due to β" $\sum(\hat{Y}_i - 0)^2$	$\mathbf{b}^{*T}\mathbf{x}^{*T}\mathbf{Y}$	$\mathbf{b}^T\mathbf{x}^T\mathbf{Y}$

We now want to relate the parameters $\mathbf{b}^*$ and $\mathbf{b}$ to each other, find how to evaluate the components of $(\mathbf{x}^{*T}\mathbf{x}^*)^{-1}$ in terms of the components of $(\mathbf{x}^T\mathbf{x})^{-1}$, and find the difference in the sum of the squares $(\mathbf{b}^T\mathbf{x}^T\mathbf{Y} - \mathbf{b}^{*T}\mathbf{x}^{*T}\mathbf{Y})$. We first state the important relationships and then indicate how they can be obtained.

RELATION BETWEEN $\mathbf{b}^*$ AND $\mathbf{b}$. If b_i is a regression coefficient in Model 5.1-1 and if b_i^* is the corresponding regression coefficient in the reduced model in which the parameter β_k is assigned the value ξ, the b's are related by

$$b_i^* = b_i - \frac{c_{ik}}{c_{kk}}(b_k - \xi) \quad (5.3\text{-}2)$$

RELATION BETWEEN THE COMPONENTS OF $(\mathbf{x}^T\mathbf{x})^{-1}$ AND $(\mathbf{x}^{*T}\mathbf{x}^*)^{-1}$. The relation between the components of the $(q - 1)$ by $(q - 1)$ matrix $\mathbf{c}^* \equiv (\mathbf{x}^{*T}\mathbf{x}^*)^{-1}$ in terms of the original q by q matrix $\mathbf{c} \equiv (\mathbf{x}^T\mathbf{x})^{-1}$ when the parameter $\beta_k = \xi$ is

$$c_{ij}^* = c_{ij} - \frac{c_{ki}c_{kj}}{c_{kk}} \quad (5.3\text{-}3)$$

REDUCTION IN SUM OF SQUARES. If ϕ_q is the sum of squares for the original model and ϕ_{q-1}^* is the sum of squares for the model with $\beta_k = \xi$, then the difference in the sum of squares $\sum(\bar{Y}_i - \bar{Y})^2$ is obtained from

$$\phi_{q-1}^* = \phi_q + \frac{(\xi - b_k)^2}{c_{kk}} \quad (5.3\text{-}4)$$

We now describe how Equations 5.3-2 through 5.3-4 can be derived. First, we write the normal equations in summation notation (omitting the equation for the intercept $b_0 = \bar{Y}$ which is the same for either model) for the full model:

$$\sum_{i=1}^{n} (x_{ik} - \bar{x}_k)\bar{Y}_i = \sum_{i=1}^{n}\sum_{j=1}^{q} (x_{ik} - \bar{x}_k)(x_{ij} - \bar{x}_j)b_j \qquad k = 1, \ldots, q \quad (5.3\text{-}5)$$

and for the model with β_q assigned as ξ:

$$\sum_{i=1}^{n}(x_{ik}-\bar{x}_k)[Y_i-(x_{iq}-\bar{x}_q)\xi]$$
$$=\sum_{i=1}^{n}\sum_{j=1}^{q-1}(x_{ik}-\bar{x}_k)(x_{ij}-\bar{x}_j)b_j^* \qquad k=1,\ldots,q-1 \tag{5.3-6}$$

From the first $(q-1)$ equations of Equation 5.3-5, we subtract the respective equation in Equation 5.3-6 to get

$$\xi\sum_{i=1}^{n}(x_{ik}-\bar{x}_k)(x_{iq}-\bar{x}_q)$$
$$=\sum_{i=1}^{n}\left[\sum_{j=1}^{q-1}(x_{ik}-\bar{x}_k)(x_{ij}-\bar{x}_j)(b_j-b_j^*)+(x_{ik}-\bar{x}_k)(x_{iq}-\bar{x}_q)b_q\right] \tag{5.3-6a}$$

or, if we denote $\sum_{i=1}^{n}(x_{ip}-\bar{x}_p)(x_{iq}-\bar{x}_q)$ by $\{\mathbf{x}^T\mathbf{x}\}_{pq}$, Equation 5.3-6a becomes

$$(\xi-b_q)\{\mathbf{x}^T\mathbf{x}\}_{kq}=\sum_{j=1}^{q-1}\{\mathbf{x}^T\mathbf{x}\}_{kj}(b_j-b_j^*)$$
$$k=1,\ldots,q-1 \tag{5.3-7}$$

Because $(\mathbf{x}^T\mathbf{x})^{-1}\equiv\mathbf{c}$ is the inverse of $(\mathbf{x}^T\mathbf{x})$, we have

$$\sum_{j=1}^{n}\{\mathbf{x}^T\mathbf{x}\}_{kj}\{(\mathbf{x}^T\mathbf{x})^{-1}\}_{jl}=\delta_{kl} \qquad k,l=1,\ldots,q$$

where $\delta_{ki}=0$ if $k\neq l$ and $\delta_{kl}=1$ if $k=l$. Next, Equation 5.3-7 is multiplied from the right by $\{(\mathbf{x}^T\mathbf{x})^{-1}\}_{kl}\equiv c_{kl}$; the result is summed over k from $k=1$ to $k=q-1$ to obtain

$$(\xi-b_q)\sum_{k=1}^{q-1}\{\mathbf{x}^T\mathbf{x}\}_{qk}c_{kl}=\sum_{j=1}^{q-1}(b_j-b_j^*)\sum_{k=1}^{q-1}\{\mathbf{x}^T\mathbf{x}\}_{jk}c_{kl}$$
$$l=1,\ldots,q$$

or

$$(\xi-b_q)[\delta_{ql}-\{\mathbf{x}^T\mathbf{x}\}_{qq}c_{ql}]=\sum_{j=1}^{q-1}(b_j-b_j^*)[\delta_{jl}-\{\mathbf{x}^T\mathbf{x}\}_{jq}c_{ql}]$$
$$l=1,\ldots,q \tag{5.3-8}$$

For $l=q$, Equation 5.3-8 becomes

$$(\xi-b_q)[1-\{\mathbf{x}^T\mathbf{x}\}_{qq}c_{qq}]=-c_{qq}\sum_{i=1}^{q-1}(b_j-b_j^*)\{\mathbf{x}^T\mathbf{x}\}_{jq} \tag{5.3-9a}$$

For $l\neq q$, Equation 5.3-8 becomes

$$(\xi-b_q)[-\{\mathbf{x}^T\mathbf{x}\}_{qq}c_{ql}]=(b_l-b_l^*)-c_{ql}\sum_{j=1}^{q-1}(b_j-b_j^*)\{\mathbf{x}^T\mathbf{x}\}_{jq}$$
$$l=1,\ldots,q-1 \tag{5.3-9b}$$

Equation 5.3-9a can be used to eliminate the summation over j from Equation 5.3-9b to yield the desired relation between $\mathbf{b}^*$ and $\mathbf{b}$:

$$b_l^*=b_l-(b_q-\xi)\frac{c_{ql}}{c_{qq}} \qquad l=1,\ldots,q-1 \tag{5.3-2a}$$

We now turn to evaluation of the components of $(\mathbf{x}^{*T}\mathbf{x}^*)^{-1}$ in terms of the components of $(\mathbf{x}^T\mathbf{x})^{-1}$. If the set of Equations 5.3-6 are formally solved for the coefficients b_k^*, we find:

$$b_k^*=\sum_{j=1}^{q-1}c_{kj}^*\sum_{i=1}^{n}(x_{ij}-\bar{x}_j)[Y_i-(x_{iq}-\bar{x}_q)]$$
$$k=1,\ldots,q-1 \tag{5.3-10}$$

where $\mathbf{c}^*$ is the inverse matrix of the reduced $(q-1)$ by $(q-1)$ matrix $(\mathbf{x}^{*T}\mathbf{x}^*)$. A similar solution of Equations 5.3-5 for b_k gives

$$b_k=\sum_{j=1}^{q}c_{kj}\sum_{i=1}^{n}(x_{ij}-\bar{x}_j)Y_i \qquad k=1,\ldots,q \tag{5.3-11}$$

We subtract the first $(q-1)$ equation of the set of Equations 5.3-11 from the set of Equations 5.3-10 and get

$$b_k^*-b_k=\sum_{j=1}^{q-1}(c_{kj}^*-c_{kj})\sum_{i=1}^{n}(x_{ij}-\bar{x}_j)Y_i$$
$$-\xi\sum_{j=1}^{q-1}c_{kj}^*\sum_{i=1}^{n}(x_{ij}-\bar{x}_j)(x_{iq}-\bar{x}_q)$$
$$-c_{kq}\sum_{i=1}^{n}(x_{iq}-\bar{x}_q)Y_i \qquad k=1,\ldots,q-1 \tag{5.3-12}$$

By use of Equation 5.3-11 with $k=q$, the set of Equations 5.3-2a can be written as

$$b_k^*-b_k=\xi\frac{c_{qk}}{c_{qq}}-\frac{c_{qk}}{c_{qq}^*}\sum_{j=1}^{q}c_{qj}\sum_{i=1}^{n}(x_{ij}-\bar{x}_j)Y_i$$
$$k=1,\ldots,q-1 \tag{5.3-13}$$

By equating the right-hand sides of Equations 5.3-12 and 5.3-13, we get

$$\sum_{j=1}^{q-1}\left[c_{kj}^*-c_{kj}+\frac{c_{qk}c_{qj}}{c_{qq}}\right]\sum_{j=1}^{n}(x_{ij}-\bar{x}_j)Y_i$$
$$=\xi\left[\sum_{j=1}^{q-1}c_{kj}^*c_{jk}+\frac{c_{qk}}{c_{qq}}\right] \qquad k=1,\ldots,q-1 \tag{5.3-14}$$

The set of Equations 5.3-14 must be independent of the value of ξ; hence the quantity in the square brackets on the right-hand side of Equation 5.3-14 must vanish. The summation over i in Equation 5.3-14 can be eliminated by use of the normal Equations 5.3-5 to reduce Equation 5.3-14 to

$$\sum_{l=1}^{q}b_l\sum_{j=1}^{q-1}\left[c_{kj}^*-c_{kj}+\frac{c_{qj}c_{qk}}{c_{qq}}\right]\{\mathbf{x}^T\mathbf{x}\}_{jl}=0$$
$$k=1,\ldots,q-1 \tag{5.3-15}$$

Because the coefficients of the b_l in Equation 5.3-15 do not depend on the Y_i, these equations are independent of the b_l and each coefficient vanishes. Using the orthogonality property of $c_{kj}c_{jl}=\delta_{kl}$ to evaluate the sums in the resulting expressions, we get

$$\sum_{j=1}^{q-1}c_{kj}^*\{\mathbf{x}^T\mathbf{x}\}_{jl}=\delta_{kl}-\frac{c_{qk}}{c_{qq}}\delta_{ql} \qquad \begin{matrix}l=1,\ldots,q\\k=1,\ldots,q-1\end{matrix} \tag{5.3-16}$$

Multiplication of Equation 5.3-16 from the right by c_{lm} with $m = 1, \ldots, q-1$ in turn and summation over all values of l give the desired expression:

$$c_{km}^{*} = c_{km} - \frac{c_{qm}c_{qk}}{c_{qq}} \qquad k, m = 1, \ldots, q-1 \qquad (5.3\text{-}3a)$$

Finally, in order to evaluate the sum of squares, we write, using Equations 5.3-5,

$$\phi_q = \sum_{i=1}^{n} Y_i^2 - \sum_{k=1}^{q} b_k \sum_{j=1}^{q} \{\mathbf{x}^T\mathbf{x}\}_{kj} b_j \qquad (5.3\text{-}17)$$

Similarly, for the model with $\beta_q \equiv \xi$, using Equations 5.3-6,

$$\phi_{q-1}^{*} = \sum_{i=1}^{n} Y_i^2 - 2\xi\{\mathbf{x}^T\mathbf{Y}\}_q - \sum_{k=1}^{q-1} b_k^{*}$$

$$\times \left[\sum_{j=1}^{q} \{\mathbf{x}^T\mathbf{x}\}_{kj} b_j - \{\mathbf{x}^T\mathbf{x}\}_{kq}\xi\right] + \xi^2\{\mathbf{x}^T\mathbf{x}\}_{qq} \qquad (5.3\text{-}18)$$

After the coefficients b_k^{*} are eliminated with the aid of Equations 5.3-2a,

$$\phi_{q=1}^{*} - \phi_q = \frac{(\xi - b_q)^2}{c_{qq}} \qquad (5.3\text{-}4a)$$

Omitting one term from the model amounts to letting $\xi = 0$. Thus, from Equations 5.3-4 if β_k is omitted from the model

$$\Delta SS = \frac{b_k^2}{c_{kk}} \qquad (5.3\text{-}19)$$

The quantity ΔSS is often termed the sum of the squares for x_k *adjusted for all the other variables* or the sum of the squares given the other variables. The ΔSS for any *group* of p variables adjusted for all the others is computed by

$$\Delta SS = \mathbf{b}_p^T \mathbf{c}_p^{-1} \mathbf{b}_p \qquad (5.3\text{-}20)$$

where $\mathbf{b}_p$ is a single-column vector (matrix) composed of the selected groups of b_k's only, and $\mathbf{c}_p$ is the matrix of the related (c_{jk}) elements.

For example, suppose we want to measure the combined effect of x_1, x_2, and x_4 removed from a model based on $x_1, x_2, x_3, x_4, \ldots, x_q$. Each x is associated with a corresponding b. Then

$$\mathbf{b}_p^T \mathbf{c}_p^{-1} \mathbf{b}_p = [b_1 b_2 b_4] \begin{bmatrix} c_{11} & c_{12} & c_{14} \\ c_{21} & c_{22} & c_{24} \\ c_{41} & c_{42} & c_{44} \end{bmatrix}^{-1} \begin{bmatrix} b_1 \\ b_2 \\ b_4 \end{bmatrix}$$

In general, removing a term from a model by letting $\beta_k = 0$ as computed by Equation 5.3-19 will *not* yield the same ΔSS as is computed by removing the corresponding term from the model after several other parameters have been removed first. It is only when the $(x_k - \bar{x}_k)$'s form an orthogonal set that the ΔSS will agree. Consequently, the sum of the ΔSS computed from applying successively Equation 5.3-19 to each parameter cannot be expected to equal the total $\sum_{i=1}^{n} w_i(\hat{Y}_i - 0)^2$ for the full model unless the independent variables are orthogonal.

To sum up the discussion so far, an F-test can be applied to test the significance of a variable (or a group of variables) in the full model by computing the variance ratio $(s^2/s_{\hat{Y}_i}^2)$, where s^2 is computed as follows:

	ΔSS	ν	s^2
A single x_k	(b_k^2/c_{kk})	1	b_k^2/c_{kk}
A group of p x's	$\mathbf{b}_p^T\mathbf{c}_p^{-1}\mathbf{b}_p$	p	$\dfrac{\mathbf{b}_p^T\mathbf{c}_p^{-1}\mathbf{b}_p}{p}$

If the variance ratio exceeds the value of $F_{1-\alpha}$ from Table C.4 in Appendix C for the selected significance level, then the variable (or group of variables) makes a significant contribution to the full model.

Because the independent variables x_i may be correlated, the results of the t-test and the F-test may be misleading if interpreted in terms of a physical variable affecting the dependent variable in the model. The apparent significant contribution to a model of a single variable x_k may be really due to the facts that Y is influenced by x_n and that x_k and x_n are highly correlated; x_n may not even be measured.

To prepare a table for the analysis of variance in which several variables are successively removed from the model, we can proceed as follows:

1. Remove the first variable and calculate ΔSS by using Equation 5.3-19.
2. Remove the first *and* second variables and calculate the combined ΔSS by using Equation 5.3-20. Subtract from this combined ΔSS the ΔSS calculated in step 1 to give the *net* ΔSS of removing the second variable.
3. Remove the first, second, and third variables, calculate ΔSS by using Equation 5.3-20, subtract the ΔSS for the first two variables, and so forth.

Each additional variable removed will provide a new combined ΔSS from which the previous ΔSS can be subtracted to yield the residual effect of removing the additional variable. The total of all the ΔSS for each stage calculated in this way will equal the sum of the squares $\sum_{i=1}^{n} w_i(\hat{Y}_i - \bar{Y})^2$ listed in the first row of Table 5.3-1. Keep in mind that if the x's are not orthogonal, the order of removal of the variables is important inasmuch as different ΔSS_i will be obtained depending upon the sequence of removal.

Table 5.3-2 summarizes the splitting of the $\sum_{i=1}^{n} w_i(\hat{Y}_i - \bar{Y})^2$ into parts, the associated degrees of freedom, and the variances which can each be used in an F-test of the variance ratio $s^2/s_{\hat{Y}_i}^2$. The fourth, fifth, and

TABLE 5.3-2 ANALYSIS OF VARIANCE IN WHICH SUCCESSIVE VARIABLES ARE REMOVED

Source of Variation	Degrees of Freedom (ν)	Sum of Squares (SS)	Mean Square
1. Due to removing x_1 (e.g., $b_1 = 0$)	1	$\Delta SS_1 = \dfrac{b_1^2}{c_{11}}$ $\Delta SS_1 = \Delta SS_1 - 0 = \Delta SS_1$	$\dfrac{b_1^2}{c_{11}} = s^2$
2. Due to removing x_2 (e.g., $b_2 = 0$) following the removal of x_1	1	$\Delta SS_{1+2} = \mathbf{b}_2^T \mathbf{c}^{-1} \mathbf{b}_2$ $\Delta SS_2 = \Delta SS_{1+2} - \Delta SS_1$	$\dfrac{\Delta SS_2}{1}$
3. Due to removing x_2 (e.g., $b_3 = 0$) following the removal of x_1 and x_2	1	$\Delta SS_{1+2+3} = \mathbf{b}_3^T \mathbf{c}^{-1} \mathbf{b}_3$ $\Delta SS_3 = \Delta SS_{1+2+3} - \Delta SS_{1+2}$	$\dfrac{\Delta SS_3}{1}$
etc.			
4. Subtotal	q	$\sum_{i=1}^{n} w_i(\hat{Y}_i - \bar{Y})^2$	
5. Due to removing as the last step b_0 (the intercept)	1	$\sum_{i=1}^{n} w_i(\bar{Y} - 0)^2 = (\bar{Y})^2 \sum_{i=1}^{n} w_i$	$\dfrac{(\bar{Y})^2 \sum_{i=1}^{n} w_i}{1}$
6. Deviations about regression line	$n - q - 1$	$\sum_{i=1}^{n} w_i(\bar{Y}_i - \hat{Y}_i)^2 = (\mathbf{Y} - \mathbf{xb})^T \mathbf{w} (\mathbf{Y} - \mathbf{xb})$	$s_r^2 = \dfrac{\sum_{i=1}^{n} w_i(\bar{Y}_i - \hat{Y}_i)^2}{n - q - 1}$
7. Total	n	$\sum_{i=1}^{n} w_i \bar{Y}_i^2$	
8. Deviations within sets		See Table 5.3-1	

sixth rows of the table represent the partition of the sum of squares $\sum_{i=0}^{n} w_i(\bar{Y}_i - 0)^2$ into three parts:

$$\sum_{i=1}^{n} w_i(\bar{Y}_i - 0)^2 = (\bar{Y})^2 \sum_{i=1}^{n} w_i + \sum_{i=1}^{n} w_i(\bar{Y}_i - \hat{Y}_i)^2 + \sum_{i=1}^{n} w_i(\hat{Y}_i - \bar{Y})^2$$

The first, second, and third rows represent the partitioning of $\sum (\hat{Y}_i - \bar{Y})^2$ into the sum of squares associated with each parameter. In Section 7.2 a method of building up a model, term by term, is described based on those ideas.

The squared multiple correlation coefficient of Y is a measure of the overall fit of an empirical model. The *multiple correlation coefficient* is a measure of the overall degree of linear association between Y, the dependent variable, and the set of x's, the independent variables. We can give an informative geometric interpretation to the estimated multiple correlation coefficient, namely that the cosine of the angle between the vectors $\hat{\mathbf{Y}}$ and $\mathbf{Y}$ is the multiple correlation coefficient. From a related viewpoint, it is the simple correlation between the observed values, $\bar{Y}_i$, and those predicted by the regression equation, $\hat{Y}_i$.

The square of the estimated multiple correlation coefficient can be calculated by

$$\hat{\rho}_\pi^2 = \frac{\sum_{i=1}^{n} w_i(\bar{Y}_i - \bar{Y})^2 - \sum_{i=1}^{n} w_i(\bar{Y}_i - \hat{Y}_i)^2}{\sum_{i=1}^{n} w_i(\bar{Y}_i - \bar{Y})^2} \qquad 0 \le \hat{\rho}_\pi^2 \le 1$$

$$= \frac{\sum_{i=1}^{n} w_i(\hat{Y}_i - \bar{Y})^2}{\sum_{i=1}^{n} w_i(\bar{Y}_i - \bar{Y})^2} = \frac{(\hat{\mathbf{Y}} - \bar{\mathbf{Y}})^T \mathbf{w} (\hat{\mathbf{Y}} - \bar{\mathbf{Y}})}{(\mathbf{Y} - \bar{\mathbf{Y}})^T \mathbf{w} (\mathbf{Y} - \bar{\mathbf{Y}})} \tag{5.3-21}$$

Because a regression equation with q parameters will fit *exactly* q observations, one must be careful to interpret $\hat{\rho}_\pi^2$ in light of the residual degrees of freedom of the model. Nevertheless, $\hat{\rho}_\pi^2$ finds use as a single number which is a measure of the overall fit of a regression equation. Kramer† gives tables that assist in the estimation of the confidence interval for ρ_π.

In addition to using $\hat{\rho}_\pi^2$ directly as an aid in model building, suppose we calculate

$$\hat{\rho}_{\pi,i}^2 = 1 - \frac{1}{a_{ii} c_{ii}} \tag{5.3-22}$$

where a_{ii} is the ith diagonal element of the matrix $\mathbf{a} = \mathbf{x}^T \mathbf{w} \mathbf{x}$ and c_{ii} is the ith element of $\mathbf{c} = (\mathbf{x}^T \mathbf{w} \mathbf{x})^{-1}$. If $\hat{\rho}_{\pi,i}^2$ is high (approaching the value of one), then the variable x_i is virtually, if not completely, useless as a significant component of the model.

† K. H. Kramer, *J. Amer. Stat. Assn.* **58**, 1082, 1963.

Contrast $\hat{\rho}^2_{\pi,i}$ for the variables in Example 5.1-1 with those in Example 5.2-1 (such as the variable x_2):

Example 5.1-1

$$1 - \frac{1}{(8.00)(0.125)} = 0$$

Example 5.2-1

$$1 - \frac{1}{(7.52 \times 10^3)(8.00 \times 10^{-3})} = 0.98$$

The results substantiate the conclusion that x_2 does contribute to the model in Example 5.1-1 but not to the model of Example 5.2-1.

Example 5.3-1 Analysis of Variance

To illustrate the difficulty of interpreting the analysis of variance when the independent variables are not orthogonal, we use the data of Gorman and Toman† who simulated experimental data with the model

$$Y_i = 1 + x_{i1} + x_{i2} + \epsilon_i$$

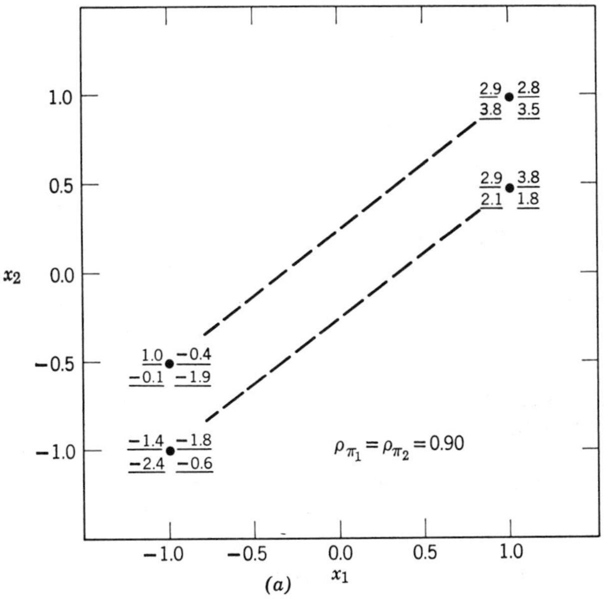

FIGURE E5.3-1a Simulated data for the model $Y_i = 1 + x_{i1} + x_{i2} + \epsilon_j$ (values of the response are underlined). (From Gorman and Toman, *Technometrics* **8**(1), 598, Feb. 1966.)

† J. W. Gorman and R. J. Toman, *Technometrics* **8**, 27, 1966.

where ϵ was a normal random deviate with $\mathscr{E}\{\epsilon_i\} = 0$ and $\text{Var}\{\epsilon_i\} = 1$. Four sets of simulated observations were generated at the points

$$\begin{array}{lrrrr} x_1: & -1 & -1 & 1 & 1 \\ x_2: & -1 & -\frac{1}{2} & \frac{1}{2} & 1 \end{array}$$

The responses, Y_i, are shown in Figure E5.3-1a. Note that $\bar{x}_1 = \bar{x}_2 = 0$.

Four models were selected and the coefficients estimated to give the corresponding estimated regression equations:

$$\begin{aligned} \hat{Y}_{\text{III}} &= 0.94 + 0.85x_1 + 1.55x_2 \\ \hat{Y}_{\text{II}} &= 0.94 + 3.01x_1 \\ \hat{Y}_{\text{I}} &= 0.94 + 2.57x_2 \\ \hat{Y}_{\text{IV}} &= 0.94 \end{aligned}$$

The coefficients, except $b_0 = \bar{Y}$, differ among the first three equations because the b's in the equations with only one independent variable are biased estimates of the respective β's.

Table E5.3-1a is an analysis of variance which indicates that the first three models are better than Model IV, but it is impossible to discriminate among Models I, II, and III. The error sum of squares was

$$s_e^2 = \frac{4.29 + 1.71 + 0.69 + 2.43}{4(3)} = 0.76$$

which is less than $\sigma_\epsilon^2 = 1$. However, for $\alpha = 0.05$, the values of $F_{1-\alpha}$ for (13, 12), (14, 12), and (15, 12) degrees of freedom from Table C.4 in Appendix C are: 2.66, 2.63, and 2.62, respectively. Consequently, each model is deemed appropriate except Model IV (since 5.15 > 2.62). An F-test or t-test for the hypothesis that each coefficient in Models I, II, and III is zero is rejected. Removing x_1 or x_2 from the full model yields the following variance ratios:

	s^2	Estimated $s^2_{\bar{Y}_i}$	$\dfrac{s^2}{s^2_{\bar{Y}_i}}$	$F_{0.95}(1, 12)$
Remove x_1	$\left(\dfrac{12.48 - 10.07}{1}\right)$	0.76	3.17	4.67
Remove x_2	$\left(\dfrac{11.23 - 10.07}{1}\right)$	0.76	1.53	4.67

The F-tests indicate that a significant reduction in the sum of the squares of the residuals is not obtained when x_1 or x_2 is removed from the full model.

If the effect of both x_1 and x_2 is desired in the regression equation, then the full model should be retained. If only $\hat{Y}$ is of interest, then an abbreviated model, I or II, will be

TABLE E5.3-1a

Model	Degrees of Freedom (ν)	Sum of Squares of the Residuals ($\phi_{\min}$)	Estimated Mean Square	Estimated Multiple Correlation Coefficient ($\hat{\rho}^2_\pi$)
Model III (x_1 and x_2)	13	10.07	0.775	0.87
Model II (x_1)	14	12.48	0.89	0.84
Model I (x_2)	14	11.23	0.80	0.85
Model IV (intercept only)	15	77.28	5.15	—

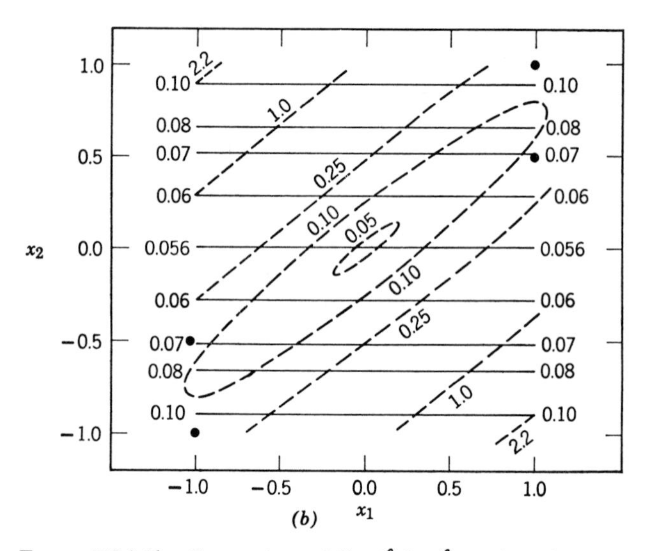

FIGURE E5.3-1b Comparison of Var $\hat{Y}$ for $\hat{Y}_{\text{II}} = b_0 + b_2x_2$ and $\hat{Y}_{\text{III}} = b_0 + b_1x_1 + b_2x_2$. (From Gorman and Toman, *Technometrics* **8**(1), 598, Feb. 1966.)

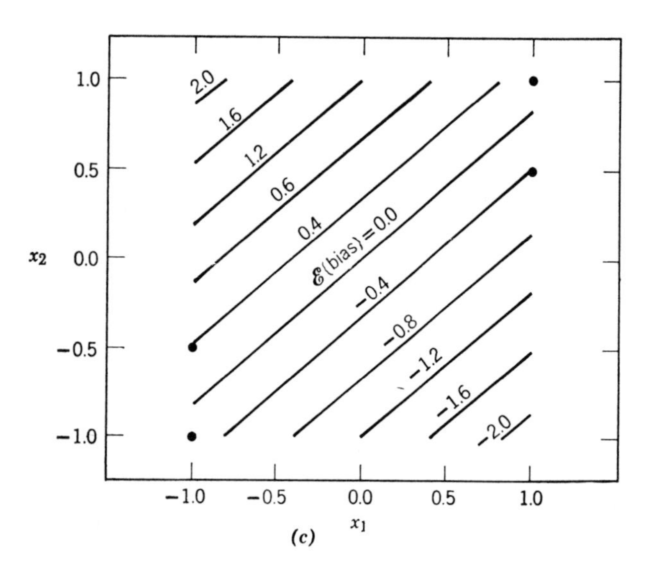

FIGURE E5.3-1c Expected value of bias for $\hat{Y} = b_0 + b_2x_2$ when the true equation is $\eta = 1 + x_1 + x_2$. (From Gorman and Toman, *Technometrics* **8**(1), 598, Feb. 1966.)

satisfactory. However, some danger exists in such simplification, as demonstrated by Figure E5.3-1b, because the variance of $\hat{Y}$ as predicted by Model II, for example, as contrasted with that predicted by Model III is misleading. For example, if $\hat{Y}$ and its variance are to be predicted for $x_1 = -1$, $x_2 = 1$, Figure E5.3-1b shows that Var $\{\hat{Y}_{\text{II}}\} \cong 0.12$ whereas Var $\{\hat{Y}_{\text{III}}\} = 2.3$. Hence the variance of $\hat{Y}$ is substantially underestimated by Model II. Model II also yields biased estimates of $\hat{Y}$. Figure E5.3-1c shows that the bias at $x_1 = -1$, $x_2 = 1$ is about 2.2, and even in the vicinity of the experimental data the expected value of the bias is from 0 to 0.4.

Of course, the example "data" above have been simulated. What is important is that the variables in the experimental space be as orthogonal to each other as possible.

Example 5.3-2 Analysis of Variance

Further analysis of Example 5.2-1 is carried out in this example. The analysis of variance corresponding to Table 5.3-1 is shown in Table E5.3-2a. Because $F_{1-\alpha}(5, 5) = 5.05$, the hypothesis that $\boldsymbol{\beta} = 0$ would have to be accepted, as concluded in connection with Example 5.2-1.

An analysis of variance corresponding to Table 5.3-2 consists of removing successive groups of variables. As an example, two different groups are selected for removal from the full model: (1) x_3, x_4, and x_5, and (2) x_5, x_4, x_3, x_2, and x_1. When x_3 is removed first followed by x_4, the sum of the squares for x_4, $\Delta SS = 7.281 \times 10^{-5}$, obtained in Example 5.2-1 does not agree with 5.615×10^{-5} listed in Table E5.3-2b. A similar lack of agreement is observed for x_5(1.38×10^{-4} versus 6.051×10^{-5}). The sum of squares for removing x_3, x_4, and x_5 all at once, $\Delta SS = 1.391 \times 10^{-4}$, does not prove to be significant by an F-test. The analysis does indicate that x_1 may be a significant variable.

5.4 ESTIMATION WHEN ERRORS ARE NOT INDEPENDENT

Ordinary least squares estimation will fail to yield satisfactory (in the sense of Section 3.1) point and interval estimates for the model parameters if the unobservable but hypothesized errors are not independent. This section continues the presentation of Section 4.6, but it is directed toward consideration of a

TABLE E5.3-2a ANALYSIS OF VARIANCE

Source of Variation	Degrees of Freedom (ν)	Sum of Squares (SS)	Mean Square	Variance Ratio
Due to regression				
$\sum w_i(\bar{Y}_i - \bar{\bar{Y}})^2$	5	1.031×10^{-2}	2.062×10^{-3}	3.05
Deviation from regression				
$\sum w_i(\bar{Y}_i - \hat{Y}_i)^2$	5	3.383×10^{-3}	6.767×10^{-4}	
Total	10	1.370×10^{-2}		

TABLE E5.3-2b

Source of Variation	Degrees of Freedom (ν)	SS (10^5)	Mean Square	Variance Ratio
x_3 removed	1	2.246	2.246	3.319
x_4 removed after removing x_3	1	5.615	5.615	8.297
x_5 removed after removing x_3 and x_4	1	6.051	6.051	8.941
Total removed: x_3, x_4, and x_5	3	13.91	4.637	6.852

Source of Variation	Degrees of Freedom (ν)	SS (10^4)	Mean Square (10^4)	Variance Ratio
x_5 removed	1	1.32	1.32	0.196
x_4 removed after removing x_5	1	0.026	0.026	0.003
x_3 removed after removing x_4 and x_5	1	0.036	0.036	0.005
x_2 removed after removing x_3, x_4, and x_5	1	2.83	2.83	0.418
x_1 removed after removing x_2, x_3, x_4, and x_5	1	98.6	98.6	14.6
Intercept removed after removing all the x's	1	787	787	116

model with several independent variables. Major assumption No. 4 of Section 4.2 is now voided; we assume instead that the ϵ's in Model 5.1-2 are serially correlated, that is, $\mathscr{E}\{\epsilon_t\epsilon_{t+1}\} \neq 0$, but still that

$$\mathscr{E}\{\epsilon_t\} = 0$$
$$\mathscr{E}\{x_{t1}\epsilon_t\} = \mathscr{E}\{x_{t2}\epsilon_t\} = \cdots = \mathscr{E}\{x_{tq}\epsilon_t\} = 0 \quad (5.4\text{-}1)$$

Here, as in Section 4.6, the subscript t designates a sequence of sets of data in time, $t = 1, 2, \ldots, n$ from a stationary process.

To determine whether or not the estimation procedures of this section are required, a test for serial correlation should be executed such as the Durbin-Watson test described in Section 4.6. If the test shows little or no serial correlation, the usual least squares procedure should be satisfactory. In matrix notation, the residuals are $\mathbf{E} = (\mathbf{Y} - \hat{\mathbf{Y}})$.

If serial correlation is established (with values of $\hat{Y}$ being calculated by the best regression equation), we can extend the technique described for the model of Section 4.6 which contained a single independent variable to the case of several independent variables. The estimates of the β's in Model 5.1-2 can be written as follows in a different form than in Section 5.1:

$$b_p = \frac{\sum_{k=1}^{q} \Delta_{pk} \sum_{t=1}^{n} (x_{tk} - \bar{x}_k) Y_t}{\Delta} \quad (5.4\text{-}2)$$

where Δ is the determinant of the $q \times q$ matrix having a typical element $\sum_{t=1}^{n} (x_{ti} - \bar{x}_i)(x_{tj} - \bar{x}_j)$, i.e., an element of the matrix $(\dot{\mathbf{x}}^T\dot{\mathbf{x}})$ in which the column of 1's is omitted from the design matrix $\mathbf{x}$, and Δ_{ij} is a minor of Δ. Also, $b_0' = \bar{Y} - b_1\bar{x}_1 - \cdots - b_q\bar{x}_q$.

As in Section 4.6, we can form the differences by introducing Y_t into Equation 5.4-2:

$$(b_p - \beta_p) = \frac{\sum_{k=1}^{q} \Delta_{pd} \sum_{t=1}^{n} (x_{tk} - \bar{x}_k)\epsilon_t}{\Delta} \quad (5.4\text{-}3)$$

$$(b_0' - \beta_0') = \bar{\epsilon} - \sum_{k=1}^{q} (b_k - \beta_k)\bar{x}_k \quad (5.4\text{-}4)$$

where

$$\bar{\epsilon} = \frac{\sum_{t=1}^{n} \epsilon_t}{n}$$

If we take terms with the same time lag, we can write expressions for the expected values of $(b_p - \beta_p)^2$, $(b_r - \beta_r)(b_s - \beta_s)$, $(b_0 - \beta_0)^2$, and so forth, the details of which can be found in Lyttkens†. We shall write here

† E. Lyttkens, "Standard Errors of Regression Coefficients in the Case of Autocorrelated Residuals," in *Proceed. Symposium Time Series Analysis*, ed. by M. Rosenblatt, John Wiley, New York, 1963, p. 49.

just the final expression which gives an approximate estimate of the covariance of (b_r, b_s):

$$\text{Covar}\,\{b_r, b_s\} \sim \frac{\sum_{i=1}^{q}\sum_{j=1}^{q} \Delta_{ri}\,\Delta_{sj}\acute{c}_{ij}}{\Delta^2} \tag{5.4-5}$$

where.

$$\acute{c}_{ij} = g_{ij}(0) + g_{ij}(1) + g_{ij}(-1) + g_{ij}(2) + g_{ij}(-2) + \cdots + g_{ij}(k_0) + g_{ij}(-k_0)$$

$$g_{ij}(k) = \sum_{t=1}^{n} (x_{ti} - \bar{x}_i)(x_{t+k,j} - \bar{x}_j)E_t E_{t+k}$$

$$g_{ij}(-k) = g_{ji}(k)$$

$$k_0 = \text{truncation index for the data}$$

$$E_t = \bar{Y}_t - b_0 - b_1 x_{t1} - \cdots - b_q x_{tq} = \bar{Y}_t - \hat{Y}_t$$

Equation 5.4-5 provides an estimate to use in establishing a joint confidence region and in hypothesis testing.

Many other models exist in which the ϵ's are not independent but we do not have the space to describe them; references are given at the end of Chapter 4.

5.5 ESTIMATION FOR MULTIRESPONSE MODELS

Often a process will have more than one dependent variable or response, as illustrated in Figure 5.5-1. Each output can be represented as a linear combination of the inputs. For example, the model of Figure 5.5-1 might be represented as:

$$Y_1 = \beta_{11}x_1x_2 + \beta_{12}x_2x_3 + \beta_{12}x_2x_3 + \epsilon_1$$

$$Y_2 = \beta_{21}\,e^{x_1} + \beta_{22}x_2 + \beta_{23}x_3 + \epsilon_2$$

Certain features of estimation for multiresponse models are different than those for the single response models. The remarks in this section will apply to both linear and nonlinear models, so the general form for the model can be written as

$$\eta_1 = \mathscr{E}\{Y_1 \mid \mathbf{x}\} = f_1(\boldsymbol{\beta}, \mathbf{x})$$

$$\eta_2 = \mathscr{E}\{Y_2 \mid \mathbf{x}\} = f_2(\boldsymbol{\beta}, \mathbf{x})$$

$$\cdots\cdots$$

$$\eta_v = \mathscr{E}\{Y_v \mid \mathbf{x}\} = f_v(\boldsymbol{\beta}, \mathbf{x})$$

where the index v designates the last equation. In general, there is no universal choice of a "best" criterion to use in estimating $\boldsymbol{\beta}$.

Inasmuch as maximum likelihood estimates of parameters have been demonstrated to have desirable properties in several earlier sections, it is natural to investigate the maximum likelihood estimates for multiresponse process models.

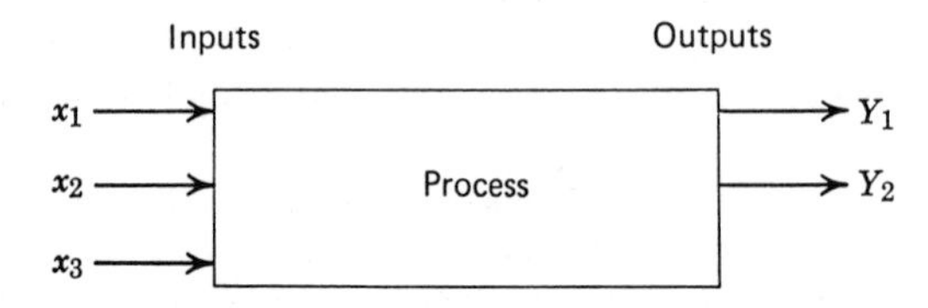

FIGURE 5.5-1 A multiresponse process.

Consider a model in which the observations Y are related to the responses as follows:

$$\begin{aligned} Y_{1i} &= \eta_{1i} + \epsilon_{1i} \\ Y_{2i} &= \eta_{2i} + \epsilon_{2i} \\ &\ \vdots \\ Y_{vi} &= \eta_{vi} + \epsilon_{vi} \end{aligned} \qquad i = 1, 2, \ldots, n \tag{5.5-1}$$

(The first index refers to the model number and the index i refers to the experimental data set number.) If we assume that the errors ϵ_{ri} are each normally distributed and independent with zero expected value and fixed variances σ_{rr}, which may be different for each model, and the covariances between models are σ_{rs}, a probability density function can be written identical to Equation 5.1-11 for the observations for one response:

$$p(\mathbf{y}_r \mid \mathbf{x}, \boldsymbol{\beta}, \sigma_{rr}) = k^n \exp\left[-\tfrac{1}{2}\boldsymbol{\epsilon}_r^T \frac{1}{\sigma_{rr}} \boldsymbol{\epsilon}_r\right] \tag{5.5-2}$$

where k has been given in connection with Equation 2.3-6 and $\boldsymbol{\epsilon}_r$ has been earlier defined as

$$\boldsymbol{\epsilon}_r = \begin{bmatrix} Y_{r1} - \eta_{r1} \\ \vdots \\ Y_{rn} - \eta_{rn} \end{bmatrix}$$

If for all the responses the observations Y_{ri} and Y_{sj} are statistically independent, then the joint probability density function of all the $n \times v$ observations is analogous to Equation 5.1-11 (keep in mind that the covariances σ_{rs} between models are not necessarily zero):

$$p(\mathbf{y}_1, \ldots, \mathbf{y}_v \mid \mathbf{x}, \boldsymbol{\beta}, \boldsymbol{\Gamma}) = k^* \exp\left\{-\tfrac{1}{2}[\boldsymbol{\epsilon}_1^T \cdots \boldsymbol{\epsilon}_v^T]\boldsymbol{\Gamma}^{-1}\begin{bmatrix}\boldsymbol{\epsilon}_1 \\ \vdots \\ \boldsymbol{\epsilon}_v\end{bmatrix}\right\} \tag{5.5-3}$$

where k^* is the normalizing constant not needed for what follows, and $\boldsymbol{\Gamma}$ is the covariance matrix *between models*

$$\boldsymbol{\Gamma} = \begin{bmatrix} \sigma_{11} & \sigma_{12} & \cdots & \sigma_{1v} \\ \sigma_{12} & \sigma_{22} & & \sigma_{2v} \\ \vdots & \vdots & & \vdots \\ \sigma_{1v} & \sigma_{2v} & & \sigma_{vv} \end{bmatrix}$$

After taking the data, we consider the observations as fixed and the parameters $\boldsymbol{\beta}$ as variables. Thus the likelihood function is

$$L(\boldsymbol{\beta} \mid \mathbf{y}_1, \ldots, \mathbf{y}_v; \mathbf{x}, \boldsymbol{\Gamma}) = k^* \exp\left\{-\frac{1}{2}\sum_{r=1}^{v}\sum_{s=1}^{v} \sigma^{rs}\left[\sum_{i=1}^{n} (Y_{ri} - \eta_{ri})(Y_{si} - \eta_{si})\right]\right\} \tag{5.5-4}$$

where σ^{rs} is the element of $\mathbf{\Gamma}^{-1}$. We have replaced the matrix notation in the exponent of Equation 5.5-3 with the equivalent summations. Because the double summation over r and s gives a positive quantity, minimization of it with respect to $\boldsymbol{\beta}$ is desired.

To minimize the double sum over r and s, estimates of the elements of $\mathbf{\Gamma}$ are needed, say, from replicate experiments for each model:

$$s_1^2 = \hat{\sigma}_{11} = \frac{\sum_{j=1}^{p} (Y_{1j} - \bar{Y}_1)^2}{p - 1}$$

$$s_r^2 = \hat{\sigma}_{rr} = \frac{\sum_{j=1}^{p} (Y_{rj} - \bar{Y}_r)^2}{p - 1}$$

$$s_{12} = \hat{\sigma}_{12} = \frac{\sum_{j=1}^{p} (Y_{1j} - \bar{Y}_1)(Y_{2j} - \bar{Y}_2)}{p - 1}$$

$$s_{rs} = \hat{\sigma}_{rs} = \frac{\sum_{j=1}^{n} (Y_{rj} - \bar{Y}_r)(Y_{sj} - \bar{Y}_s)}{p - 1}$$

where $\bar{Y}_r = \sum_{j=1}^{p} Y_{rj}/p$.

When the errors associated with the observations of the different dependent variables do not have equal variances, but σ_{rs} can be assumed to be zero for all $r \neq s$, we need only minimize

$$\psi_1 = \sum_{r=1}^{v} \sum_{i=1}^{n} w_i(Y_{ri} - \eta_{ri})^2 \qquad (5.5\text{-}5)$$

where

$$w_i = \frac{1}{\sigma_{rr}} \equiv \frac{1}{\sigma_r^2}$$

This is equivalent to minimizing the weighted trace of

$$\boldsymbol{\psi} = \begin{bmatrix} \boldsymbol{\epsilon}_1^T\boldsymbol{\epsilon}_1 & \cdots & \boldsymbol{\epsilon}_1^T\boldsymbol{\epsilon}_v \\ \vdots & & \vdots \\ \boldsymbol{\epsilon}_1^T\boldsymbol{\epsilon}_v & \cdots & \boldsymbol{\epsilon}_v^T\boldsymbol{\epsilon}_v \end{bmatrix} \qquad (5.5\text{-}6)$$

where $\boldsymbol{\epsilon}_r^T\boldsymbol{\epsilon}_s = \sum_{i=1}^{n} (Y_{ri} - \eta_{ri})(Y_{si} - \eta_{si})$. When $\sigma_{rs} = 0$, there is no correlation among the various observations Y_{1i} to Y_{vi} on one experiment. Each σ_r^2 is the variance for the fit of model r and is estimated by s_r^2. Certain useful information results from minimizing Equation 5.5-5 in addition to the parameter estimates. One obtains:

1. Values of $s_{b_j}^2$, the estimated parameter variances.
2. Values of the residuals between the observed and estimated values of the dependent variable.

Finally, if the variances for each σ_r^2 are all the same, minimization of ψ_1 is the same as minimization of the trace of the matrix 5.5-6. It has been suggested that, when the elements of $\mathbf{\Gamma}$ are known to be nonzero but cannot be estimated, the determinant of $\boldsymbol{\psi}$ be minimized.†

† G. E. P. Box and N. R. Draper, *Biometrika* **52**, 355, 1965.

In general, one of the above criteria should be employed, depending upon the experimenter's knowledge or lack of knowledge about the unobservable errors among the model responses.

Other criteria which have been used to secure estimates of the parameters in multiresponse models are:

1. Maximize the square of the smallest correlation coefficient.
2. Maximize the square of the largest correlation coefficients.
3. Maximize the sum of the squared correlation coefficients.
4. Maximize the square of the product of the correlation coefficients.

Criterion No. 3 gives the best average multiple correlation coefficient. The equation for the square of the estimated multiple correlation coefficient is

$$\hat{\rho}_{\pi}^2 = 1 - \frac{\sum_{i=1}^{n} w_i(\bar{Y}_i - \hat{Y}_i)^2}{\sum_{i=1}^{n} w_i(\bar{Y}_i - \bar{Y})^2}$$

and the sum of the squared estimated multiple correlation coefficients is just

$$\psi_2 = \sum_{r=1}^{v} \hat{\rho}_{\pi_r}^2 \qquad (5.5\text{-}7)$$

To obtain estimates of the coefficients β_k in linear models, ψ_2 can be differentiated with respect to each of the coefficients, the resulting expressions equated to zero and solved simultaneously. Although the algebra is tedious, the procedure is straightforward. Coefficients in nonlinear models can be estimated by the iterative optimization techniques described in Chapter 6.

To determine whether a multiresponse model adequately represents the experimental data, the variance ratio

$$\frac{\psi_{\min}/(n - m)}{s_e^2}$$

can be formed and tested, where m is the total number of coefficients determined. Also, the residuals should be randomly distributed without correlation and outliers as explained in Chapter 7.

5.6 ESTIMATION WHEN BOTH INDEPENDENT AND DEPENDENT VARIABLES ARE STOCHASTIC

How to estimate the parameters in an empirical model when some independent variables as well as dependent variables are random variables is a problem of common occurrence, and has been considered briefly for a model with one independent variable in Section 4.5.

One approach to estimation when more than one of the variables are random variables is termed the *recursive method.*† We ask the question: If there are many random variables, in which direction should the sum of the squares be minimized? Recall from Figure 4.2-2 that with a single dependent random variable, only one direction can be selected, namely that of Y itself. No completely satisfactory answer can be given to the question for the multivariate model. However, in the recursive structure the covariance matrix of the errors for any one set of data is diagonal, and the matrix of the coefficients of the jointly random variables has nothing but zeros to one side of the diagonal.

Examine the following structure:

$$Y_1 = \sum_{k=1}^{q} \gamma_{1k} x_k + \epsilon_1$$

$$\beta_{21} y_1 + Y_2 = \sum_{k=1}^{q} \gamma_{2k} x_k + \epsilon_2$$

$$\beta_{31} y_1 + \beta_{32} y_2 + Y_3 = \sum_{k=1}^{q} \gamma_{3k} x_k + \epsilon_3$$

$$\vdots \qquad\qquad \vdots$$

$$\beta_{l1} y_1 + \beta_{l2} y_2 + \cdots + Y_l = \sum_{k=1}^{q} \gamma_{lk} x_k + \epsilon_l \qquad (5.6\text{-}1)$$

where the Y's are observable random variables, the x's are predetermined deterministic variables, the β's and γ's are coefficients to be estimated and the ϵ's are independent unobservable random variables with $\mathscr{E}\{\epsilon_i\} = 0$ and with constant finite variances and covariances.

The parameters β and γ can be consistently estimated in recurrent steps by least squares as follows. In the first equation, Y_1 is the only random variable, and $\gamma_{11}, \gamma_{12}, \ldots, \gamma_{1q}$ are estimated by minimizing the sum of the squares in the Y_1 direction. Then, with Y_1 in the second equation predetermined by the first equation, i.e., $y_1 = \hat{Y}_1$, and Y_2 acting as the random dependent variable, the parameters $\beta_{21}, \gamma_{21}, \gamma_{22}, \ldots, \gamma_{2q}$ can be estimated by minimizing the sum of the squares in the Y_2 direction. Repetition for each of the l random variables yields estimates in the last equation of all the parameters. How satisfactory the usual statistical inferences are based on the sum of the squares from the last equation is not well known.

The references at the end of this chapter discuss some techniques which have been used in economics to treat cases where certain of the independent variables are stochastic variables.

It is also of interest to discuss the consequences of estimating model parameters for a model such as Equation 5.1-2 by the method of least squares, even though direct application of the method gives estimates with undesirable statistical characteristics. Kerridge‡ considered the model

$$Y_i = \beta_0' + \mathbf{X}_i\boldsymbol{\beta} + \epsilon_i \qquad (5.6\text{-}2)$$

in which the $Y, X_1, X_2, \ldots, X_q$ are jointly normally distributed. If the coefficients in Model 5.6-2 were estimated by the usual least squares procedure and if one more observation were made, Kerridge developed an expression for the error of prediction

$$\Delta_{n+1} = (\hat{Y}_{n+1} - \tilde{\hat{Y}}_{n+1}),$$

where $\hat{Y}_{n+1}$ is the correct predicted value of Y for the $(n + 1)$th set of X's and $\tilde{\hat{Y}}_{n+1}$ is the predicted value of Y obtained by assuming the X's are not random variables. We omit here the details of the development of the probability distribution of Δ_{n+1} and simply write the final result:

$$\Delta_{n+1} = U\sigma\left(1 + \frac{1}{n}\right)^{1/2}\left(\frac{\chi^2_{n-q} + \chi^2_q}{\chi^2_{n-q}}\right)^{1/2} \qquad (5.6\text{-}3)$$

where U is the standardized normal random variable and the subscripts on χ^2 indicate the associated number of degrees of freedom.

For practical purposes the mean square error may be of greater interest than the value of Δ_{n+1} itself:

$$\sigma^2_{\Delta_{n+1}} = \sigma^2_\epsilon\left(1 + \frac{1}{n}\right)\left(\frac{n-2}{n-q-2}\right)$$

Thus, if n is large and the number of independent variables q is small, the variance of $\hat{Y}_{n+1}$ is

$$\text{Var}\{\hat{Y}_{n+1}\} \cong \sigma^2_\epsilon + \text{Var}\{\tilde{\hat{Y}}_{n+1}\}$$

where the $\text{Var}\{\tilde{\hat{Y}}_{n+1}\}$ can be computed from the equations in Section 5.2. However, if n is small, say 10, and q is large, say 5, then the variance of Δ_{n+1} becomes large:

$$\sigma^2_{\Delta_{n+1}} = \sigma^2_\epsilon(1.1)(\tfrac{8}{3}) = 2.93\sigma^2_\epsilon$$

In fact, when $(n - q) = 2$, the variance $\sigma^2_{\Delta_{n+1}}$ is infinite.

Supplementary References

General

Action, F. S., *Analysis of Straight Line Data*, John Wiley, New York, 1959.

Cramer, H., *Mathematical Methods of Statistics*, Princeton Univ. Press, Princeton, N.J., 1954.

Davies, O. L., *Statistical Methods in Research and Production* (3rd. ed.), Oliver and Boyd, London, 1958.

Draper, N. R. and Smith, H., *Applied Regression Analysis*, John Wiley, New York, 1966.

Feller, W., *An Introduction to Probability Theory and Its Applications*, John Wiley, New York, 1957.

† R. H. Strotz and H. O. A. Wold, *Econometrica* **28**, 417, 1960.

‡ D. Kerridge, *Technometrics* **9**, 309, 1967.

Fisher, R., *Statistical Methods and Scientific Inference*, Oliver and Boyd, London, 1956.

Guest, P. G., *Numerical Methods of Curve Fitting*, Cambridge Univ. Press, Cambridge, England, 1961.

Mood, A. and Graybill, F., *Introduction to the Theory of Statistics*, McGraw-Hill, New York, 1963.

Plackett, R., *Principles of Regression Analysis*, Oxford Univ. Press, Oxford, England, 1960.

Smile, K. W., *An Introduction to Regression and Correlation*, Academic Press, New York, 1966.

Williams, E. J., *Regression Analysis*, John Wiley, New York, 1959.

Sources of Digital Computer Programs

ALSQ—A FORTRAN IV subroutine to solve the linear least squares problem, written by G. W. Stewart, III, Union Carbide Corp., Oak Ridge, Tenn. This program uses a modification of the algorithm by:

Businger, P. and Golub, G. H. "Linear Least Squares Solutions by Householder Transformations," *Num. Math.* **7**, 269–276, 1965.

BÖJRCK-GOLUB—A FORTRAN V program to solve the linear least squares problem, written by Roy H. Wampler, National Bureau of Standards, using the Björck-Golub algorithm described in:

Björck, A., "Solving Linear Least Squares Problems by Gram-Schmidt Orthogonalization," *BIT* **7**, 1–21, 1967.

Björck, A., "Iterative Refinement of Linear Least Squares Solutions, I," *BIT* **7**, 257–278, 1967.

Björck, A. and G. H. Golub, "ALGOL Programming, Contribution No. 22: Iterative Refinement of Linear Least Square Solutions by Householder Transformation," *BIT* **7**, 322–337, 1967.

Björck, A., "Iterative Refinement of Linear Least Squares Solutions, II," *BIT* **8**, 8–30, 1968.

BMDO2R, Stepwise Regression—One of the Biomedical Computer programs, written in FORTRAN and listed in:

Dixon, W. J. (ed.), *BMD Biomedical Computer Programs*, Health Sciences Computing Facility, Univ. of Calif., Los Angeles, 1964. Revised 1965 and 1967.

BMDO3R, Multiple Regression with Case Combinations—One of the Biomedical Computer Programs, written in FORTRAN.

BMDO5R, Polynomial Regression—One of the Biomedical Computer Programs, written in FORTRAN.

LINFIT—A program which fits a linear function to collected data via least squares. Optional constraints may be applied to the fitting coefficients to make them nonnegative, add to a constant, etc. This is one of eighteen statistical routines written by J. R. Miller. This library of routines exists in the Project MAC 7094. See:

Miller, J. R., *On-Line Analysis for Social Scientists*, MAC-TR-40, Project MAC, Mass. Inst. of Tech., Cambridge, Mass., 1967.

LINFIT-A—Another program written in BASIC for linear least squares fitting and computing correlations developed at Dartmouth College, Hanover, N. H., and available in the C-E-I-R Multi-Access Computer Services Library programs documentation, MAC 71-7-1, 1967: Addendum, MAC 71-7-1, A 12-368, 1968.

LSCF—A least squares polynomial curve-fitting subroutine written in BASIC developed at Dartmouth College, Hanover, N.H. and available in the C-E-I-R Multi-Access Computer Services Library.

LSFITW—A least squares curve-fitting program written in BASIC. Adapted by J. B. Shumaker, National Bureau of Standards, from the ORTHO algorithm by P. J. Walsh. This is available in the C-E-I-R Multi-Access Computer Services Library. See:

Walsh, P. J., "Algorithm 127, ORTHO," *Communications of the ACM* **5**, 511–513, 1962.

LSTSQ—A FORTRAN IV subroutine which solves for X the overdetermined system $AX = B$ of m linear equations in n unknowns for p right-hand sides. It was written by P. Businger, Computation Center, University of Texas, using the Businger-Golub algorithm.

MATH-PACK, ORTHLS, Orthogonal Polynomial Least-Squares Curve Fitting—Written in FORTRAN V, one of the Univac 1108 MATH-PACK programs.

Univac 1108 Multi-Processor System, MATH-PACK Programmers Reference, UP-7542, Univac Division of Sperry Rand Corporation, 1967.

MPR3, Stepwise Multiple Regression with Variable Transformations—A FORTRAN II program written by M. A. Efroymson, Esso Research and Engineering Co., Madison, N.J., using the Efroymson algorithm. This is available in the SHARE library: 7090-G2 3145MPR3. See:

Efroymson, M. A., "Multiple Regression Analysis" in *Mathematical Methods for Digital Computers*, Vol. 1, ed. by A. Ralstmand and H. S. Wilf, John Wiley, New York, 1960.

OMNITAB—A general-purpose computer program for statistical and numerical analysis. OMNITAB allows the user to communicate with a computer in an efficient manner by means of simple English sentences. It was developed at the National Bureau of Standards and is available in an ASA FORTRAN version. See:

Hilsenrath, J., Ziegler, G., Messina, C. G., Walsh, P. J., and Herbold, R., *OMNITAB, A Computer Program for Statistical and Numerical Analysis*, Nat. Bur. of Standards Handbook 101, U.S. Government Printing Office, Washington, D.C., 1966. Reissued Jan. 1968, with corrections.

ORTHO—A program written by P. J. Walsh, formerly with the National Bureau of Standards, which uses a Gram-Schmidt orthonormalization process for least squares curve fitting. ORTHO has been written as an ALGOL procedure and a FORTRAN program (see OMNITAB), and a BASIC program (see LSFITWA).

POLRG, Polynomial Regression—One of the programs of the IBM System 360 Scientific Subroutine Package written in FORTRAN IV. See:

IBM Application Program, System/360 Scientific Subroutine Package (360A-CM-03X) Version III, Application Description, H20-0166-5, 1968.

IBM Application Program, System/360 Scientific Subroutine Package (360A-CM-03X) Version III, Programmer's Manual, H20-0205-3, 1968.

STAT-PACK, GLH, General Linear Hypotheses—One of the Univac 1108 STAT-PACK programs, written in FORTRAN V. See:

Univac 1108 Multi-Processor System, STAT-PACK Programmers Reference, UP-7502, Univac Division of Sperry Rand Corporation, 1967.

STAT-PACK, REBSOM, Back Solution Multiple Regression—One of the Univac 1108 STAT-PACK programs, written in FORTRAN V.

STAT-PACK, RESTEM, Stepwise Multiple Regression—One of the Univac 1108 STAT-PACK programs, written in FORTRAN V.

STAT20***—A program written in BASIC for stepwise multiple linear regression, developed at Dartmouth College, Hanover, N.H., and available in the C-E-I-R Multi-Access Computer Services Library. (See LINFIT above.)

STAT21***—A program written in BASIC for multiple linear regression with detailed output, developed at Dartmouth College, Hanover, N.H., and available in the C-E-I-R Multi-Access Computer Services Library. (See LINFIT above.)

Problems

5.1 The data in Table P5.1 were collected from different wells. Can you apply the least square methods of Section 5.1 to the data to estimate the parameters in a linear model? To estimate confidence limits of the parameters?

TABLE P5.1

Well Number	Depth (ft)	Specific Gravity	Dissolved Solids (ppm)	Total Hardness as $CaCO_3$ (ppm)
E-71	225	43.1	2320	1030
E-73	220	55.5	2320	1140
I-2	203	34.8	2660	1280
I-22	150	78.8	3060	1140
I-24	136	77.4	4460	1640
I-29	140	20.6	2160	673
I-46	210	31.5	2540	868

5.2 Obtain the least squares estimates of β_1 and β_2 in the model $Y = \beta_1 + e^{-\beta_2 x} + \epsilon$. Point out some of the difficulties. Let the weights be unity.

5.3 Given the model

$$Y = \beta_0 + \beta_1 x_1 + \beta_2 x_2 + \epsilon$$

and using expanded notation, obtain the matrix **a**, invert the **a** matrix, and calculate the **c** matrix. Find the Var $\{b_1\}$, Var $\{b_2\}$, and Var $\{b_0\}$ in the expanded notation.

5.4 Given that a matrix **Y** is distributed by a multivariate normal distribution with the parameters $(\mathbf{x}\boldsymbol{\beta}, \mathbf{I}\sigma^2)$, show that the least squares estimates of $\boldsymbol{\beta}$ are equivalent to the maximum likelihood estimates of $\boldsymbol{\beta}$, and show that $\hat{\sigma}^2 = \mathbf{E}^2/n$ is the maximum likelihood estimate of σ^2. Note:

$$\mathbf{E}^2 = \sum (\bar{Y}_i - \hat{Y}_i)^2 = (\mathbf{Y} - \mathbf{x}\mathbf{b})^T(\mathbf{Y} - \mathbf{x}\mathbf{b})$$

5.5 Show that the expected value of ϕ (the sum of the squares of the residuals given by Equation 5.1-15) is equal to $(n - q - 1)\sigma^2_{\bar{Y}_i}$. One useful relation is

$$\mathscr{E}\{Q\} = \sigma^2 \text{ Trace } [\mathbf{M}] + \boldsymbol{\eta}^T\mathbf{M}\boldsymbol{\eta}$$

where **Q** is the quadratic form $\mathbf{Y}^T\mathbf{M}\mathbf{Y}$, **M** is an $n \times n$ matrix, $\boldsymbol{\eta} = \mathscr{E}\{\mathbf{Y}\}$, and Var $\{\mathbf{Y}\} = \sigma^2\mathbf{I}$.

5.6 Show that the least squares estimates of β_0, β_1, and β_2 in the model $\eta = \beta_0 + \beta_1(x_1 - \bar{x}_1) + \beta_2(x_1 - \bar{x}_2)$ are the same as would be obtained from the model $\eta = \beta_0' + \beta_1 x_1 + \beta_2 x_2$, in which $\beta_0' = \beta_0 - \beta_1\bar{x}_1 - \beta_2\bar{x}_2$.

5.7 To test whether a line passes through a given point (Y_0, x_0), show that the deviation $\Delta = b_1 + b_2 x_0 - Y_0$ has a variance

$$\text{Var }\{\Delta\} = \sigma^2_{\bar{Y}_i}\left[\frac{1}{n} + \frac{(x_0 - \bar{x})^2}{\sum (x_n - \bar{x})^2}\right]$$

The model for the line is $Y = \beta_1 + \beta_2 x + \epsilon$; $n =$ the number of data sets; $\sigma^2_{\bar{Y}_i} = \text{Var }\{\epsilon_\cdot\}$.

5.8 A student is asked to estimate the coefficient in a model

$$Y = \beta_1 x_1 + \beta_2 x_2 + \epsilon \tag{a}$$

from given experimental data. He suggests that first he should use the model $Y = \beta_1' x_1 + \epsilon$ to estimate β_1' and then use the model $X_2 = \beta_2' x_1 + \epsilon$ to estimate β_2'. Then either model (a) $(Y - b_1' x_1) = \beta_3 x_2 + \epsilon$ or (b) $Y = \beta_4(x_2 - b_2' x_1) + \epsilon$ will yield the correct estimate of β_2 in model (a); i.e., either b_3 or b_4 will be b_2. Is the student correct? Explain.

5.9 Consider the following model:

$$\eta = \beta_0 + \beta_1 x_1 + \beta_2 x_2 + \beta_3 x_3$$

If the design matrix (matrix of **x**'s) is arranged as follows $(x_0 \equiv 1)$:

Y	x_1	x_2	x_3
Y_1	−1	−1	−2
Y_2	1	−1	−1
Y_3	−1	0	−2
Y_4	1	0	−1
Y_5	1	1	−2
Y_6	1	1	−1
Y_7	−1	1	1
Y_8	1	1	2

the least square estimates of the β's given by the normal equations can be expressed as linear functions of the eight observations, Y_i.

Show, for example, that

$$\begin{aligned} b_0 = {} & 0.2193(Y_1 + Y_2 + Y_3 + \cdots + Y_8) \\ & - 0.0288(-Y_1 + Y_2 - Y_3 + \cdots + Y_8) \\ & - 0.1042(-Y_1 - Y_2 + \cdots + Y_8) \\ & + 0.0814(-2Y_1 - Y_2 - 2Y_2 - \cdots + 2Y_8) \end{aligned}$$

Consolidate the equations for b_0, b_1, b_2, and b_3 in terms of each Y_i and confirm the entries in Table P5.9. Each column represents the multipliers of the observations Y_j for the expression $b_j = \sum_{i=1}^{8} a_{ij} Y_i$.

TABLE P5.9

	b_0	b_1	b_2	b_3
Y_1	0.1895	−0.1275	−0.1792	−0.0163
Y_2	0.2133	0.1514	−0.3192	0.0618
Y_3	0.0853	−0.1633	0.0684	−0.0847
Y_4	0.1091	0.1156	−0.0716	−0.0066
Y_5	−0.0765	0.0831	0.2444	−0.1597
Y_6	0.0049	0.0798	0.1760	−0.0750
Y_7	0.2253	−0.2090	0.1108	0.1010
Y_8	0.2491	0.0699	−0.0292	0.1791

Show that the Var $\{b_0\}$ is simply

$$\text{Var}\,\{b_j\} = \sigma_{Y_i}^2\left(\sum_{i=1}^{8} a_{ij}^2\right)$$

$$\text{Covar}\,\{b_j b_k\} = \sigma_{Y_i}^2\left(\sum_{i=1}^{8} a_{ij}a_{ik}\right)$$

Suppose that the observation of Y_3 is biased by 10 percent. How much of a change (in percent) will there be in the estimate of β_2? Determine the two most important observations insofar as their contributions to the variance of b_3. Which observation contributes the most to the variance of $\hat{Y}$?

5.10 Temperature-yield data for a batch chemical reaction have been collected as follows:

Temperature (°C)	Yield (percent)
200	6
210	7
220	8
230	11
240	18

Execute a regression analysis and an analysis of variance as follows:

(a) Code the x variables as

$$x = \frac{T - 190}{10}$$

(b) Find $\mathbf{x}^T\mathbf{x}$ and $\mathbf{x}^T\mathbf{Y}$.

(c) From these, compute $\mathbf{b}$, the estimate of $\boldsymbol{\beta}$, in the linear model

$$\eta = \beta_0 + \beta_1 x_1$$

(d) Write the estimated regression equation, both in coded (x) and uncoded form; i.e.,

$$\hat{Y} = \tilde{b}_0 + \tilde{b}_1 T$$

$$\hat{Y} = b_0 + b_1 x_1$$

(e) Tabulate the analysis of variance for $\alpha = 0.05$ for both the coded and uncoded models.

(f) Plot the experimental data on a graph of Y versus T together with the regression line.

(g) Plot the joint confidence region for β_0 and β_1.

5.11 The data in Table P5.11 have been taken from T. Kunungi, T. Tamura, and T. Naito, *Chem. Eng. Prog.* **57**, 43, 1961. Assume a linear model

$$Y = \alpha + \beta_1 x_1 + \beta_2 x_2 + \beta_3 x_3 + \epsilon$$

and determine the best estimates of α and the β's, in other words, compute a and the b_i's; the confidence limits on α and β_i; find the multiple correlation coefficient; carry out an analysis of variance; determine if any of the variables can be dropped from the model; and, finally, give the confidence interval at selected values of the independent variables for $\eta = \mathscr{E}\{Y \mid \mathbf{x}\}$.

TABLE P5.11

Conversion of n-heptane to Acetylene (percent) Y_i	Reactor Temperature (°C) x_1	Ratio of H_2 to n-heptane (mole ratio) x_2	Contact Time (sec) x_3
49.0	1300	7.5	0.012
50.2	1300	9.0	0.012
50.5	1300	11.0	0.0115
48.5	1300	13.5	0.013
47.5	1300	17.0	0.0135
44.5	1300	23.0	0.012
28.0	1200	5.3	0.040
31.5	1200	7.5	0.038
34.5	1200	11.0	0.032
35.0	1200	13.5	0.026
38.0	1200	17.0	0.034
38.5	1200	23.0	0.041
15.0	1100	5.3	0.084
17.0	1100	7.5	0.098
20.5	1100	11.0	0.092
29.5	1100	17.0	0.086

5.12 The analysis of labor costs involved in the fabrication of heat exchangers can be used to predict the costs of a new exchanger of the same class. Let the cost be expressed as a linear equation

$$C = \beta_0 + \beta_1 N + \beta_2 A$$

where β_0, β_1, and β_2 are constants, N = number of tubes, and A = shell-surface area. Estimate the constants β_0, β_1, and β_2 from the data below.

Eliminate β_0 from the model by calculating the mean of all the C's $\bar{C}$, and estimate β_1' and β_2' in the model

$$C = \bar{C} + \beta_1'(N - \bar{N}) + \beta_2'(A - \bar{A})$$

where $\bar{N}$ and $\bar{A}$ are the means of N and A, respectively.

Predict the cost of a 350 psia working pressure exchanger with 240 16-foot long tubes and a shell I.D. of 1 foot, 6 inches. Can you estimate what are reasonable cost limits, in dollars, at this pressure? What assumptions must be made about the variables C, A, and N?

TABLE P5.12 DATA FOR MILD-STEEL FLOATING-HEAD EXCHANGERS (0–500 WORKING PRESSURE)

Labor Cost ($)	Area (A)	Number of Tubes (N)
310	120	550
300	130	600
275	108	520
250	110	420
220	84	400
200	90	300
190	80	230
150	55	120
140	64	190
100	50	100

5.13 The phase-transition boundaries, in general, and liquidus and solidus curves, in particular, of the alloy phase (temperature versus composition) diagram are generally determined by various metallographic methods. The liquidus and solidus curves define the temperatures at which a cooling melt (of given solute concentration) begins to freeze and completes freezing, respectively. Because a pure metal contains no solute, most of the sources of the unobservable errors in the temperatures are avoided, and it is possible to say that the melting temperature (T_m) at $c_s = c_l = 0$ is a known value. Similarly, the temperature at the eutectic point (T_w) where the concentration is c_{s_w} or c_{l_w} is known.

With these two boundary conditions specified, determine the best polynomial models for the liquidus curve ($c_s = f_s(T)$) and solidus curve ($c_l = f_l(T)$) from the following data.

T (°C)	c_s, y_0	c_l, y_0
900	0	0
850	1.0	6.25
800	2.0	13.00
750	3.0	20.25
700	4.0	28.00
650	5.0	36.25
600	6.0	45.00

5.14 Will the inclusion of an additional independent variable in a model, even though it is shown not to be significant by an F-test, always improve the model (in the sense that the sum of the squares of the residuals $= \sum (\bar{Y}_i - \hat{Y}_i)^2$ will be reduced) or at least not make it worse? Explain.

5.15 A total of 142 datum points was collected for the purpose of determining a standard octane curve.† In addition to a standard curve, it was necessary to determine the precision of the derived curve in terms of the single sets of data used and to estimate the limits of predictability of the derived curve for evaluation of yields of a new catalyst preparation.

The results of the statistical calculations are:

Model: $f = \alpha + \beta_1 t_1 + \beta_2 t_2^2$
f = octane number
t = temperature, °C
Estimated coefficients: $\hat{\alpha} = 3.13$
$\hat{\beta}_1 = 0.258$
$\hat{\beta}_2 = 0.001$
Sum of squares of residuals: 9.439
Degrees of freedom: 8

Are $\hat{\beta}_1$, $\hat{\beta}_2$, and $\hat{\alpha}$ significant (i.e., differ from zero)?

5.16 In an empirical correlation, the friction factor f is to be made a function of the Re $= Dv\rho/\mu$ and the tube roughness ξ:

$$\ln f = \beta_0 + \beta_1 \ln \text{Re} + \beta_2 \ln \xi \qquad \text{(a)}$$

The friction factor itself is computed from the mechanical energy balance in which the "lost energy" is related to the friction factor or, in terms of measured quantities,

$$\Delta p = \frac{2f\rho L v^2}{D}$$

where

Δp = pressure drop
ρ = fluid density
L = tube length
v = fluid velocity
D = tube diameter
μ = fluid viscosity

† M. Greyson and J. Cheasley, *Petrol. Ref.* **38** (8), 135, 1959.

Is it possible to estimate a joint confidence region for the three parameters, β_0, β_1, and β_2, in Equation (a)? Note that both f (the dependent variable) and Re (an independent variable) contain some of the same measured quantities.

5.17 Levenspiel, Weinstein, and Li† employed the method of least squares to estimate the parameters in a dimensionless correlation using the data of Sieder and Tate.‡ Their results were

$$\text{Nu} = 0.973\ \text{Re}^{0.288}\ \text{Pr}^{0.243}\left(\frac{\mu_a}{\mu_w}\right)^{0.142}$$

where

Nu = Nusselt number for heat transfer

Pr = Prandtl number

Re = Reynolds number

μ = viscosity of fluids a and w

Sixty-seven data sets were used and the variance of the residuals $s_r^2 = \sum (Y_i - \hat{Y}_i)^2/(67 - 4) = 0.0026$; $\sum Y_i^2 = 108.9004$. The elements of the **c** matrix for the model $\log \text{Nu} = \beta_0 + \beta_1 \log \text{Re} + \beta_2 \log \text{Pr} + \beta_3 \log (\mu_a/\mu_w)$ were, omitting the first row and column,

$$\begin{bmatrix} 0.1455 & 0.1749 & 0.0134 \\ 0.1749 & 0.2627 & 0.0179 \\ 0.0134 & 0.0179 & 0.0126 \end{bmatrix}$$

What are the confidence limits on the estimated coefficients?

By a theoretical analysis of heat transfer, the expression for the heat transfer coefficient in the form of a dimensionless group is

$$\text{Nu} = 0.402\ \text{Re}^{1/3}\ \text{Pr}^{1/3}\left(\frac{\mu_1}{\mu_2}\right)^{0.142}$$

Can the experimental equation be correct? Explain.

5.18 The following expression for the efficiency of a fractionating column was given as

$$E = 10.84A^{-0.28}h^{0.241}\left(\frac{L}{V}\right)^{0.024} G^{-0.013} \cdot \left(\frac{\sigma}{\mu_L V_G}\right)^{0.044}\left(\frac{\mu_L}{\rho_L D_L}\right)^{0.137} \alpha^{-0.028}$$

Comment on the appropriateness of each variable. Can any of the variables or groups of variables be eliminated? The data used to obtain the above expression were based on seven different articles in the literature.

Notation	Range
μ_L = liquid viscosity, poise	
ρ_L = liquid density, g/cc	
D_L = liquid diffusivity, cm²/sec	
σ = interfacial tension, dynes/cm	
α = relative volatility, dimensionless	1.6–20.8
A = fraction free area, dimensionless	0.041–0.125
$\frac{L}{V}$ = reflux ratio, dimensionless	0.83–70
h = height, in.	0.25–5
G = gas velocity based on column crossection, lb/(hr)(ft²)	100–2000
$\left(\frac{\sigma}{\mu_L V_G}\right)$	70–609
$\left(\frac{\mu_L}{\rho_L D_L}\right)$	27.7–520

5.19 An analysis of variance has been prepared based on some unreplicated experimental data for two models:

Model I: $Y = \beta_0 + \beta_1 x + \beta_2 x^2 + \epsilon$

Model II: $Y = \beta_1 x + \beta_2 x^2 + \epsilon$

Model I	d.f.	SS
Due to regression	2	99,354
Departure from origin	1	103
Deviation about regression line	33	863
Total	36	100,390

Model II	d.f.	S
Due to regression	2	21,621
Deviation about regression line	33	863
Total	35	22,484

Is Model II (a line through the origin) as good as model as Model I (the one with an intercept)?

5.20 Based on the listed experimental data:

(a) For the model $\eta = \beta_0 x_0 + \beta_1 x_1$, find b_0 and b_1 and prepare an analysis of variance.

(b) For the model $\eta = \beta_0 x_0 + \beta_1 x_1 + \beta_2 x_1^2$, find b_0, b_1, and b_2 and prepare an analysis of variance. Do you find a difference in removing b_0, b_1, b_2 versus b_0, b_2, b_1 versus b_1, b_2, b_0?

$$\mathbf{Y} = \begin{bmatrix} 6.4 \\ 5.6 \\ 6.0 \\ 7.5 \\ 6.5 \\ 8.3 \\ 7.7 \\ 11.7 \\ 10.3 \\ 17.6 \\ 18.0 \\ 18.4 \end{bmatrix} \qquad \mathbf{x} = \begin{bmatrix} 1 & 1 \\ 1 & 1 \\ 1 & 1 \\ 1 & 2 \\ 1 & 2 \\ 1 & 3 \\ 1 & 3 \\ 1 & 4 \\ 1 & 4 \\ 1 & 5 \\ 1 & 5 \\ 1 & 5 \end{bmatrix}$$

(columns of $\mathbf{x}$: x_0, x_1)

† O. Levenspiel, N. J. Weinstein, and J. C. R. Li, *Ind. Eng. Chem.* **48**, 324, 1956.
‡ E. Sieder and G. E. Tate, *Ind. Eng. Chem.* **28**, 1429, 1936.

5.21 Hydrocyclones are used extensively in the mineral industry and the pulp and paper industry for operations such as classification, thickening, and dewatering. This wide acceptance has been achieved because the hydrocyclone is efficient and has no moving parts.

Preliminary experiments were done with a small glass hydrocyclone, using a water medium to which various quantities of sugar had been added to increase the medium density. These experiments indicated that the variables influencing the throughput of the hydrocyclone, Q, were: the pressure drop across it, P, and the density, ρ, and viscosity, μ, of the medium. In addition, it seemed reasonable to believe that the geometry of the hydrocyclone, as represented by a characteristic diameter, D, influenced the throughput:

$$Q = f(P, \rho, \mu, D)$$

With the assumption of a simple power relationship between the variables, the following equation was obtained:

$$Q = KP^a\rho^b\mu^c D^d \qquad \text{(a)}$$

When the appropriate mass, length, and time units were substituted in Equation (a), the following relationship was obtained:

$$(L^3T^{-1}) \simeq (ML^{-1}T^{-2})^a(ML^{-3})^b(ML^{-1}T^{-1})^c L^d$$

Equating the indicates of the three basic dimensions, M, L, and T, on both sides of this equation yielded:

$$a = \frac{d-1}{2}, \qquad b = \frac{d-3}{2}, \qquad c = 2 - d$$

or

$$Q = KP^{(d-1)/2}\rho^{(d-3)/2}\mu^{(2-3)}D^d \qquad \text{(b)}$$

which give

$$Q = \frac{K\mu^2}{(P\rho^3)^{0.5}}\left(\frac{(P\rho)^{0.5}D}{\mu}\right)^d$$

When KD^d was equated to a new constant K_1 (for a given hydrocyclone) and Equation (b) was rearranged:

$$\frac{Q\rho}{\mu} = K_1\left(\frac{(P\rho)^{0.5}}{\mu}\right)^{d-1} \qquad \text{(c)}$$

The coefficients K_1 and d in Equation (c) were estimated from the data in Table P5.21. K proved to be 3970 and d proved to be 1.904. A straight line was obtained on a log-log plot which matched the data well. Comment on this experiment and the subsequent statistical analysis. Would the model be an improvement over $Q = K_2P^m$?

5.22 If the two variables X_1 and X_2 are distributed by a bivariate normal distribution with means μ_1 and μ_2 and the variance-covariance matrix

$$\begin{bmatrix} \sigma_{11} & \sigma_{12} \\ \sigma_{21} & \sigma_{22} \end{bmatrix}$$

TABLE P5.21 EXPERIMENTAL RESULTS USED TO CALCULATE THE PARAMETERS $Q\rho/\mu$ AND $(P\rho)^{0.5}/\mu$

ρ (g/ml)	μ (poise)	P (psig)	Q (ml/min)	$\frac{(P\rho)^{0.5}}{\mu}$	$\frac{Q\rho}{\mu} \times 10^{-4}$
1.199	0.0846	2	3550	18.4	5.03
		4	5030	25.9	7.13
		6	6150	31.8	8.72
		8	7110	36.6	10.1
		10	7910	41.0	11.2
		11	8230	42.9	11.7
1.164	0.0498	4	5070	43.4	11.9
		6	6130	53.2	14.3
		8	6940	61.2	16.2
		10	7680	68.5	18.0
1.122	0.0288	4	5060	73.5	19.7
		6	6060	90.1	23.6
		8	6970	104	27.2
		10	7620	116	29.7
		11	8040	122	31.3
1.000	0.0127	4	5000	157	39.4
		6	6060	193	47.7
		8	6760	223	53.2
		10	7470	249	58.8
		12	8180	273	64.4
0.989	0.0054	4	4540	369	83.1
		6	5480	451	100
		8	6240	520	114
		10	6850	583	125
		12	7540	638	138

find the maximum likelihood estimates for μ_1, μ_2, σ_{11}, σ_{22}, and $\sigma_{12} = \sigma_{21}$.

5.23 Fit a second-degree polynomial to the following data in which the dependent variable is known to be correlated in time.

Time, t	Adhesiveness, Y
1	21
2	9
3	14
4	16
5	10
6	1
7	14
8	14
9	26
10	40
11	41
12	59
13	74
14	91
15	105

Discuss the data itself, carry out appropriate tests, and interpret your results.

5.24 An experiment was performed in 1959 in a first course in the mechanics of materials. A TV section of 125 students was subdivided into five sections of 25 each in four separate rooms (two sections of 25 each in one larger room with two TV sets). Some 225 other students were taught in the conventional way by different experienced instructors who used conventional methods. A common final examination was given.

The results of this experiment are given as a graph in Figure P5.24. The plotted points represent the means for the respective sections identified by the numeral. Since the point for the TV section falls on the regression line for the entire group, the achievement of this group is equal to the average of all students taking the course, relative abilities being considered.

Comment on the use of least squares to obtain the estimated regression curve.

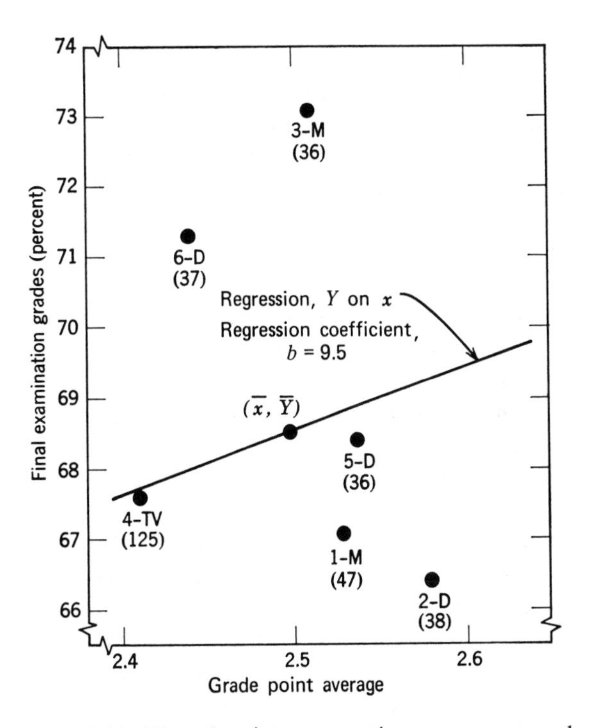

FIGURE P5.24 Plotted points are section averages; numbers identify sections; letters identify instructors; numbers of students are in parentheses. (From *J. Eng. Ed.* **52**, 316, 1962.)

CHAPTER 6

Nonlinear Models

As explained in the introduction to Chapter 4, the term "nonlinear" as applied to models in this part of the text means the model is nonlinear in the parameters (coefficients) to be estimated (and more than likely is also nonlinear in the independent variables). Not only is the estimation of the parameters in nonlinear models more difficult than in linear models, for reasons to be explained shortly, but confidence intervals for the parameters, hypothesis tests, and all the matters described in Chapters 4 and 5 are considerably more difficult to calculate and interpret. Often we shall rely on approximate rather than exact methods.

Before discussing the details of fitting empirical data by nonlinear models, we need to outline the notation and assumptions which form the basis of the nonlinear estimation methods. Then we shall describe several techniques which have been effectively employed to estimate model parameters. At the same time, we shall indicate what the difficulties are in carrying out these techniques. Next will come a discussion of the error involved in the estimated parameters and, finally, a summary of methods of estimation when the variables or parameters in the model are constrained in some way.

6.1 INTRODUCTION

Suppose we have a random observable dependent variable, the response, either $\bar{Y}_i$ or Y_i, $i = 1, \ldots, n$, depending on whether or not several replicates are taken, and several nonrandom independent (controllable) variables x_k, $k = 1, 2, \ldots, q$. (We shall use Y_i rather than $\bar{Y}_i$ in this chapter because very often, in experimentation with nonlinear models, replicate experiments are not made.) Both Y_i and x_k are presumed to be continuous variables, i.e., real numbers in some finite or perhaps infinite range. Let β_j, $j = 1, \ldots, m$, be the parameters in the model

$$\eta = \eta(x_1, x_2, \ldots, x_q; \beta_1, \beta_2, \ldots, \beta_m) \qquad (6.1\text{-}1)$$

or, in matrix notation,

$$\boldsymbol{\eta} = \boldsymbol{\eta}(\mathbf{x}; \boldsymbol{\beta}) \qquad (6.1\text{-}2)$$

where

$$\mathbf{x} = \begin{bmatrix} x_{11} & x_{12} & \cdots & x_{1q} \\ x_{21} & x_{22} & \cdots & x_{2q} \\ \vdots & & & \vdots \\ x_{n1} & x_{n2} & \cdots & x_{nq} \end{bmatrix} \qquad \boldsymbol{\beta} = \begin{bmatrix} \beta_1 \\ \beta_2 \\ \vdots \\ \beta_m \end{bmatrix}$$

and $n > m$. Each observed Y_i for a given set of x's denoted by $\mathbf{x}_i = [x_{i1}, x_{i2}, \ldots, x_{iq}]$, is related to the expected value of Y_i, $\mathscr{E}\{Y_i \mid \mathbf{x}_i\} = \eta_i$, by

$$Y_i = \eta_i + \epsilon_i \qquad i = 1, 2, \ldots, n \qquad (6.1\text{-}3)$$

where ϵ represents some type of unobservable "error."

Two general types of errors can be considered. One is error in the *measurement* of the experimental dependent variable; the other is the error in the *form* of the model. If model error and measurement error are both present in an experiment, ϵ must represent the combination of both effects. We assume that $\boldsymbol{\epsilon}$, the vector of errors, is represented by a probability density function of known form involving a set of unknown parameters $\boldsymbol{\theta}$. As in Sections 4.3 and 5.1, for a given $\mathbf{Y}$ and $\mathbf{x}$, the estimates $\hat{\boldsymbol{\beta}}$ and $\hat{\boldsymbol{\theta}}$ can be regarded as the variables and we can write a likelihood function L:

$$L(\hat{\boldsymbol{\beta}}, \hat{\boldsymbol{\theta}} \mid \mathbf{Y}, \mathbf{x}) \qquad (6.1\text{-}4)$$

If the estimation problem is set up in this way, we can see that the information to be obtained about $\hat{\boldsymbol{\beta}}$ and $\hat{\boldsymbol{\theta}}$ depends on $\mathbf{x}$, that is, on how the experiment is designed. Certainly, if the variables selected for $\mathbf{x}$ are not chosen according to some effective scheme, the estimates of $\boldsymbol{\beta}$ and $\boldsymbol{\theta}$ may not be very precise.

To obtain the estimates of $\boldsymbol{\beta}$ and $\boldsymbol{\theta}$, we shall assume that the values of Y_i in the vector $\mathbf{Y}$ are random observations from the distribution of Y_i about η_i, and that Y_i can be interpreted as indicated in Equation 6.1-3. The expected value of ϵ_i for a particular set of x's will be zero and the variance of ϵ_i will be $\sigma^2_{\epsilon_i} = \sigma^2_{Y_i}$. When Equation 6.1-3 is stated explicitly as

$$\mathbf{Y} = \boldsymbol{\eta}(x_1, \ldots, x_q; \beta_1, \ldots, \beta_m) + \boldsymbol{\epsilon} \qquad (6.1\text{-}5)$$

Equation 6.1-5 is called the *regression equation.*

A different nonlinear model can be proposed which is of great practical significance, although we shall not make use of it because of its complexity, namely a model in which **Y** and **X** are *jointly distributed* random variables:

$$\mathbf{Y} = \boldsymbol{\eta}(X_1, \ldots, X_q; \beta_1, \ldots, \beta_m) + \boldsymbol{\epsilon} \qquad (6.1\text{-}6)$$

In Equation 6.1-6 the X_k are observed values of random variables, while in Equation 6.1-5 the x_k are fixed numbers.

As in linear estimation, we would like to obtain both the estimates **b** of the parameters $\boldsymbol{\beta}$ in the nonlinear model of Equation 6.1-2 as well as the estimates $\hat{\boldsymbol{\theta}}$ of the parameters $\boldsymbol{\theta}$ in the probability density for $\boldsymbol{\epsilon}$, because the values of $\hat{\boldsymbol{\theta}}$ can provide estimates of the dispersion of the values **b** about the true values $\boldsymbol{\beta}$. However, it proves far easier to obtain **b** than to obtain $\hat{\boldsymbol{\theta}}$, and we shall restrict our attention to the former. A wide choice of numerical estimation techniques are available, some of which are "better" in certain senses than others.

The maximum likelihood technique to estimate $\boldsymbol{\beta}$ and $\boldsymbol{\theta}$, that is the procedure to obtain the values of **b** and $\hat{\boldsymbol{\theta}}$ which make Equation 6.1-4 a maximum, has been described in Sections 4.3 and 5.1, but it has two handicaps, First, it depends upon knowing some functional form for the likelihood Equation 6.1-4. Second, in general, the procedure cannot be carried out analytically for nonlinear models. While it is always wise, if possible, to explore the nature of the likelihood function in the vicinity of the maximum, as a practical matter one most often assumes that the following basic premises of Section 4.2 hold true (whether they do or not in a real experiment):

1. The error ϵ_i is normally distributed.
2. The variance of Y_i given $\mathbf{x}_i$ is constant (or possibly some function of $\mathbf{x}_i$).

Because the least squares estimation technique is the easiest to execute, we shall use it here in preference to the maximum likelihood method. Also, many of the desirable properties of the least squares estimates (consistent, efficient, unbiased, and minimum variance) are independent of the normality assumption for the linear model and approximately so for the nonlinear model. If the probability density of $\boldsymbol{\epsilon}$ has a single parameter σ_Y^2, the likelihood function has the same contours as the function expressing the sum of the squares of the deviations between the observed values **Y** and the predicted ones. Thus, the estimation technique to be used will be the same as that used in Chapters 4 and 5, namely the method of least squares.

6.2 NONLINEAR ESTIMATION BY LEAST SQUARES

Recall that in Chapters 4 and 5 we minimized the sum of squares function, ϕ, to obtain the desired parameter estimates. Exactly the same technique will be used here. We want to

$$\text{Minimize } \phi = \sum_{i=1}^{n} w_i[Y_i - \eta_i(\mathbf{x}_i, \boldsymbol{\beta})]^2 \qquad (6.2\text{-}1)$$

where w_i represents appropriate weights, perhaps unity, and Y_i is the single observation made at $\mathbf{x}_i$. Figure 6.2-1 illustrates the geometric interpretation of the method of least squares as applied to a nonlinear model. As in Figure 5.1-1, we look for the shortest vector from the point P in observation space to the *curved* surface, which is the locus of predictions of **Y** for a given set of estimated parameters, $\tilde{b}_1$ and $\tilde{b}_2$. In parameter space the contours of the sum of squares, ϕ, will not be elliptical but might appear as shown in the right-hand side of Figure 6.2-1.

Posed in the form of Equation 6.2-1, the nonlinear estimation problem appears as simply an *optimization problem* in parameter space in which the Y's and x's are given numbers and the β's are the variables. Many of

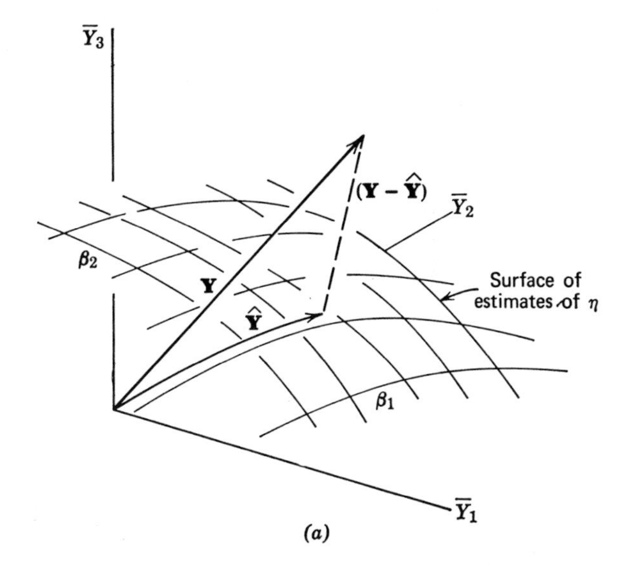

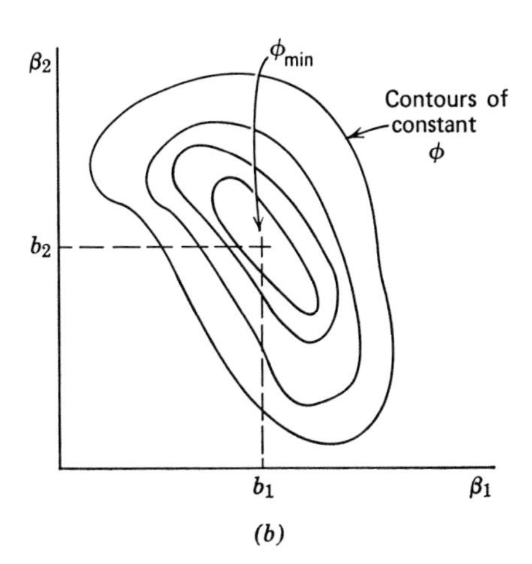

FIGURE 6.2-1 Geometric interpretation of least squares for a nonlinear model: (*a*) observation space (three observations), and (*b*) parameter space (two parameters).

the tools of deterministic nonlinear optimization can be successfully brought to bear on this problem. From the user's viewpoint, these optimization techniques fall into two broad classes: (1) derivative-free methods and (2) derivative methods. If in the search for a minimum of ϕ, the partial derivatives of ϕ (with respect to β_k) must be calculated, then we shall classify the method as a derivative-type method; otherwise, the method will be termed a derivative-free method of estimation.

While the advantages of not having to compute analytical derivatives of the function ϕ can be overrated, inasmuch as numerical derivatives can be substituted for the analytical ones, the calculation and evaluation of numerical derivatives take a substantial amount of computer time. Moreover, near the minimum of ϕ the error in the derivatives rises; hence, termination of the iterative procedure leads to oscillation.

We shall describe five of the more effective optimization techniques, namely:

1. Derivative-free methods:
   (a) Simplex method.
   (b) Direct search method.
2. Derivative methods:
   (a) Gauss-Seidel.
   (b) Gradient methods.
   (c) Marquardt's method.

All of the effective procedures are iterative ones which are best executed on a digital or hybrid computer. Wilde and Beightler† and Beveridge and Schechter‡ describe a number of other nonlinear optimization methods.

The reason why iterative methods of optimization are required and why the direct application of classical calculus fails to yield estimates of the parameters in a nonlinear model can be demonstrated by the following model:

$$\eta = \frac{\beta_1 x}{x + \beta_2} \tag{6.2-2}$$

or

$$Y = \frac{\beta_1 x}{x + \beta_2} + \epsilon \tag{6.2-3}$$

The unweighted sum of the squares of the unobservable errors, ϵ, is

$$\phi = \sum_{i=1}^{n} (Y_i - \eta_i)^2$$

$$= \sum_{i=1}^{n} Y_i^2 - 2\beta_1 \sum_{i=1}^{n} \frac{x_i Y_i}{x_i + \beta_2} + \beta_1^2 \sum_{i=1}^{n} \frac{x_i^2}{(x_i + \beta_2)^2} \tag{6.2-4}$$

† D. J. Wilde and C. S. Beightler, *Foundations of Optimization*, Prentice-Hall, Englewood Cliffs, N.J., 1967.

‡ G. Beveridge and R. S. Schechter, *Optimization—Theory and Practice*, McGraw-Hill, New York (1970).

By partially differentiating Equation 6.2-4, first with respect to β_1 and then with respect to β_2, and by equating each of the partial derivatives to zero, we obtain a pair of normal equations incorporating the estimates b_1 and b_2:

$$b_1 \sum_{i=1}^{n} \frac{x_i^2}{(x_i + b_2)^2} - \sum_{i=1}^{n} \frac{x_i Y_i}{x_i + b_2} = 0 \tag{6.2-5}$$

$$b_1 \sum_{i=1}^{n} \frac{x_i^2}{(x_i + b_2)^3} - \sum_{i=1}^{n} \frac{x_i Y_i}{(x_i + b_2)^2} = 0 \tag{6.2-6}$$

Note that Equations 6.2-5 and 6.2-6 are themselves nonlinear equations, so that we have converted the original optimization problem into a root-finding problem of a degree of difficulty equal to or greater than the optimization problem. Because finding the roots of a set of nonlinear equations involves some type of iterative technique, it seems to be equally (and perhaps more) feasible to minimize ϕ by directly using an iterative procedure to minimize the original objective function, Equation 6.2-4. We shall now describe two derivative-free methods of minimization which have proved to be remarkably flexible, easy to use, and relatively trouble free.

6.2-1 Direct Search Methods

The direct search method proposed by Hooke and Jeeves† has some distinct advantages in nonlinear estimation from the viewpoint of the user. No derivatives need be calculated and an acceleration phase is built directly into the logical scheme. The disadvantage of direct search methods is that they are slow in comparison with the derivative or simplex methods, especially as the number of parameters becomes large.

The direct search algorithm operates in the following manner. Initial values (guesses), $\mathbf{b}^{(0)}$, for all the β's‡ must be provided, as well as an initial incremental change for each parameter, $\Delta b_j^{(0)}$. ϕ is first evaluated at the initial point, $\mathbf{b}^{(0)}$. Each $b_j^{(0)}$ of the set $\mathbf{b}^{(0)}$ is changed in turn by $+\Delta b_j^{(0)}$ and, if ϕ is improved, $b_j^{(0)} + \Delta b_j^{(0)}$ is adopted as a new estimate of β_j, $b_j^{(1)}$. If ϕ is not improved, $b_j^{(0)} - \Delta b_j^{(0)}$ is tested. If no improvement is experienced for either $\pm\Delta b_j^{(0)}$, $b_j^{(1)} = b_j^{(0)}$. This process is continued for all the β_j's to complete an "exploratory move." The new estimated parameters define a vector in parameter space that represents a successful direction to reduce ϕ. A series of accelerating steps, or "pattern moves," is made along this vector as long as ϕ is reduced. The magnitude of the pattern move in each coordinate direction is proportional to the number of prior successful moves in that direction. If ϕ is not improved by one

† R. Hooke and T. A. Jeeves, *J. Assn. Compt. Mach.* **8**, 212, 1961.

‡ All the intermediate estimates of the model parameters will be designated as $\mathbf{b}^{(\ )}$; it is the terminal vector of estimated parameters which is the best estimate of $\boldsymbol{\beta}$.

of these pattern moves, a new exploratory move is made in order to define a new successful direction. If an exploratory move fails to give a new successful direction, the Δb_j's are reduced gradually until either a new successful direction can be defined or each Δb_j becomes smaller than some predetermined tolerance. Failure to improve ϕ for a very small Δb_j indicates that a local optimum has been reached.

Two basic tests have been employed to determine when the search should terminate. One test is made on the fractional change in the individual estimated parameters, Δb_j, i.e., on the step sizes. Minimum desirable values of the fractional change in the variables are read into the computer program, and the test is conducted after each exploratory search failure. Another test occurs after each exploratory search or pattern move; the change in the value of ϕ is compared to a specified fraction read into the computer program. If the value of ϕ has not decreased from the value on the previous move by an amount greater than the specified fraction, an exploratory search or pattern move is considered a failure. The calculations terminate when both tests are satisfied on a specified number of cycles.

It is very easy to add simple constraints to the search routine. For example, if one wishes to restrict the b_j's to positive numbers only, as required in certain categories of engineering problems in which the b_j's represent physical quantities which cannot be negative, one can readily build into the computer program the constraint

$$l_j \leq b_j \leq u_j$$

where

l_j = lower bound of the search for b_j

u_j = upper bound of the search for b_j

and choose $l_j = 0$ and u_j to be some very large number.

Example 6.2-1 Direct Search Technique in Estimation

A comparison can be made between estimation by direct search and linear estimation for nonlinear models if the models can be linearized by suitable transformations and then treated by linear analysis. Such a comparison is made in this example; some of the typical problems encountered in iterative nonlinear estimation are pointed out. Eleven sets of simulated data were prepared for the model

$$\eta = \alpha x_1^{\beta_1} x_2^{\beta_2}$$

or

$$\log \eta = \log \alpha + \beta_1 \log x_1 + \beta_2 \log x_2$$

by arbitrarily selecting $\alpha = 1.0$, $\beta_1 = 3.0$, and $\beta_2 = 0.5$. The values of η were then perturbed by normal random deviates with variances of 10 and 100, and also by uniform deviations of ± 0.1 percent, ± 1 percent, and ± 10 percent with random sign allocation.

The simulated data were fit using three criteria, Equations (a), (b), and (c) below, respectively. The results appear in Table E6.2-1 for two initial starting vectors.

$$\text{Minimize} \sum_{i=1}^{n} (Y_i - \eta_i)^2 \tag{a}$$

$$\text{Minimize} \sum_{i=1}^{n} \left(\frac{Y_i - \eta_i}{\eta_i}\right)^2 \tag{b}$$

$$\text{Minimize} \sum_{i=1}^{n} (\log Y_i - \log \eta_i)^2 \tag{c}$$

The third criterion, Equation (c), is the criterion used in least squares linear estimation. In all of the searches the initial fractional step size for each Δb_j was arbitrarily set at 0.30, Past experience with the direct search technique indicated that $\Delta b_j = 0.30$ was a reasonable compromise between a large initial step size, which might have to be reduced substantially before ϕ would be reduced, and too small a step size, which would cause innumerable, time-consuming, small steps. The criterion for stopping the search was a change of less than 0.01 percent in the value of each of the estimated parameters. In using a variance of 100 to calculate the simulated Y_i values, a negative value was obtained for one Y_i which precluded calculation by criterion (c) for this one case.

The estimated parameters by direct search using criterion (c), can be seen to give essentially the same values as the linearized technique for this simple problem, although they require, in general, more time to compute. Since the three criteria are not identical, the comparisons made in the table also demonstrate the effect of changing the criterion itself. If the unobservable error, ϵ, is added to $\ln \eta$, then linear least squares gives the desired estimates of the parameters. However, if the proper model is $Y = \eta + \epsilon$, then the estimated parameters can be quite different from those obtained from the model $\ln Y = \ln \eta + \epsilon$, particularly as the error increases.

Figure E6.2-1 illustrates the progress of one search. Note that although the sum of the squares of the residuals steadily decreases, the values of the estimated parameters do not change monotonically. Other initial guesses for the parameters than those listed in Table E6.2-1 were tested and yielded essentially the same final answers for both the estimated parameters and the sum of the squares of residuals. However, the choice of the initial estimates of the parameters is by no means as simple as it seems. Unsuitable choices for the initial guesses can introduce scaling difficulties. By selecting starting guesses for α, β_1, and β_2 which led to the extremely small initial predicted values of Y, it was observed that the search program would not operate. For example, if the initial guesses were chosen to be -5, -5, and -5, respectively, then a comparison between the first few initial simulated and predicted values of Y revealed the following (for the data without error):

Y Simulated Data	Y Calculated from the Model using $a^{(0)} = b_1^{(0)} = b_2^{(0)} = -5$	Percent Deviation
0.48000×10^2	-0.2584×10^{-8}	-0.1857×10^{13}
0.61094×10^3	-0.19622×10^{-7}	-0.3113×10^{13}
0.12626×10^4	-0.34845×10^{-6}	-0.3623×10^{12}

TABLE E6.2-1 RESULTS OF NONLINEAR ESTIMATION BY DIRECT SEARCH USING SIMULATED DATA FOR THE MODEL $\eta = \alpha x_1^{\beta_1} x_2^{\beta_2}$

Direct Search—Initial Guesses I: $a = 0.05$, $b_1 = 4.0$, $b_2 = 0.7$

	Degree of Error Introduced into Y					
	None	±0.1%	±1%	±10%	Var {Y}=10	Var {Y}=100
a	0.999	0.988	0.886	.273	1.232	5.543
b_1	3.000	3.004	3.047	3.516	2.907	2.229
b_2	0.500	0.502	0.521	0.714	0.485	0.397
Minimum of Equation (a)	4.8×10^{-3}	2.15	214.3	2.1×10^{5}	760.9	7.8×10^{5}
Number of exploratory searches	418	370	310	161	468	489
a	0.999	0.996	0.930	0.628	0.716	19.704
b_1	3.000	3.001	3.026	3.159	3.107	1.701
b_2	0.500	0.501	0.518	0.629	0.608	0.244
Minimum of Equation (b)	1.7×10^{-7}	7.7×10^{-6}	9.4×10^{-4}	7.2×10^{-2}	4.5×10^{-2}	1.99
Number of exploratory searches	143	150	167	215	280	231
a	1.002	0.996	0.931	0.651	0.699	
b_1	2.999	3.001	3.025	3.142	3.115	
b_2	0.500	0.501	0.517	0.617	0.610	
Minimum of Equation (c)	3.2×10^{-7}	7.7×10^{-6}	9.4×10^{-4}	7.7×10^{-2}	4.4×10^{-2}	
Number of exploratory searches	142	170	157	120	105	
Linear regression analysis						
a	1.000	0.996	0.931	0.652	0.695	
b_1	3.000	3.001	3.026	3.143	3.118	
b_2	0.500	0.501	0.618	0.618	0.612	

Direct Search—Initial Guesses II: $a = 5.0$, $b_1 = 5.0$, $b_2 = 5.0$

	Degree of Error Introduced into Y					
	None	±0.1%	±1%	±10%	Var {Y}=10	Var {Y}=100
a	0.998	0.973	0.889	0.276	1.231	5.640
b_1	3.000	3.011	3.046	3.513	2.908	2.222
b_2	0.500	0.503	0.521	0.712	0.485	0.394
Minimum of Equation (a)	3.5×10^{-2}	4.42	214	2.1×10^{5}	761	7.78×10^{5}
Number of exploratory searches	149	123	221	183	169	73
a	1.001	0.997	0.930	0.625	0.717	19.55
b_1	2.999	3.000	3.026	3.162	3.106	1.704
b_2	0.499	0.500	0.518	0.630	0.607	0.247
Minimum of Equation (b)	4.1×10^{-6}	7.8×10^{-6}	9.4×10^{-4}	7.2×10^{-2}	4.5×10^{-2}	1.99
Number of exploratory searches	105	70	127	114	80	94
a	0.999	0.995	0.931	0.651	0.694	
b_1	3.000	3.001	3.026	3.143	3.118	
b_2	0.500	0.501	0.517	0.618	0.612	
Minimum of Equation (c)	9.2×10^{-8}	7.7×10^{-6}	9.5×10^{-4}	7.7×10^{-2}	4.4×10^{-2}	
Number of exploratory searches	125	127	96	100	109	

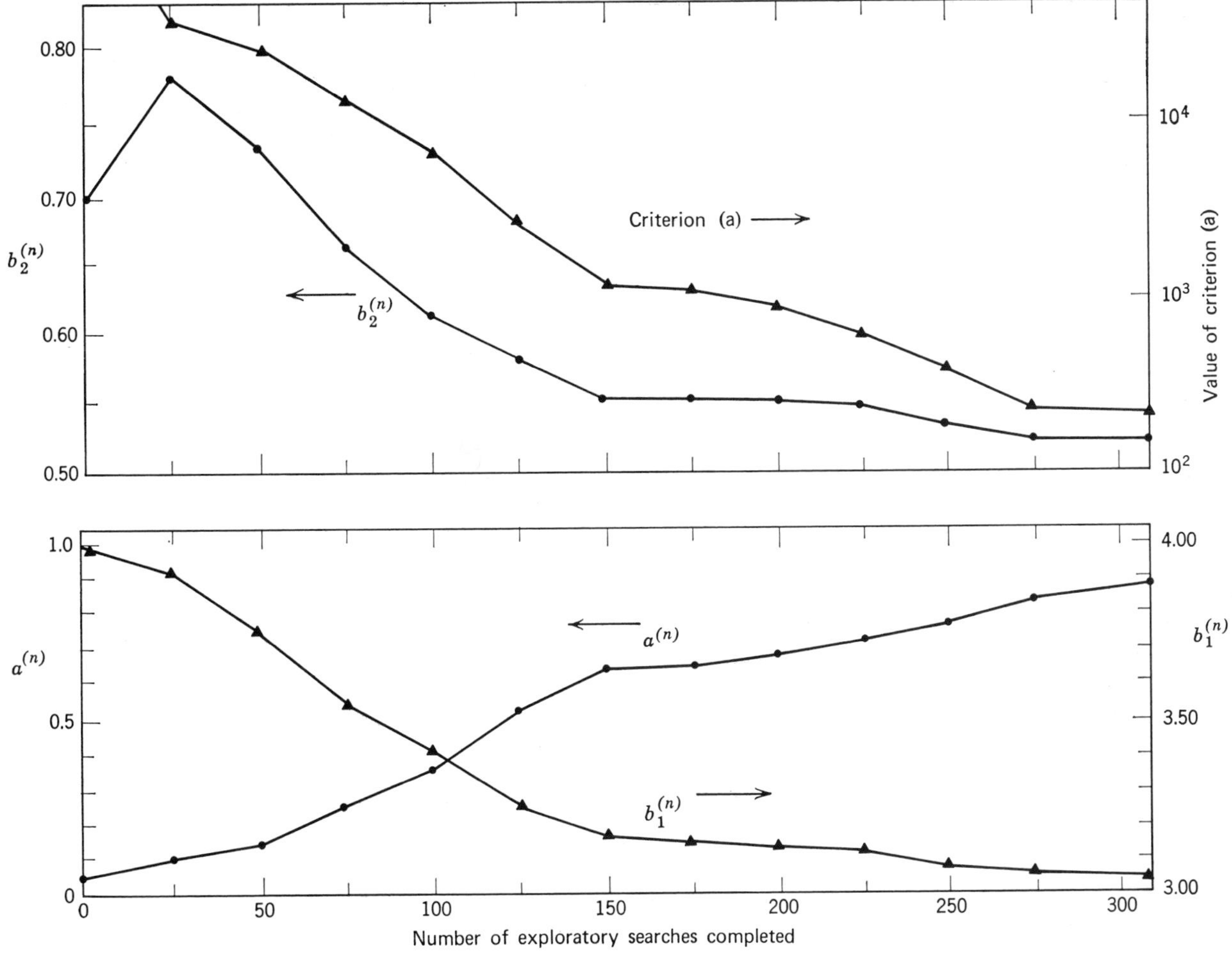

FIGURE E6.2-1 Progress of direct search for initial guesses $a^{(0)} = 0.050$, $b_1{}^{(0)} = 4.000$, and $b_2^{(0)} = 0.700$ for criterion (a). Error is uniform ± 1 percent.

No successful exploratory or pattern search moves were possible in such circumstances, because the effect of any change in the parameters was well beyond the significant figures in the simulated values of Y.

Although this example is somewhat specialized, it does bring out many of the typical problems encountered in nonlinear estimation, including:

1. How to choose initial guesses for the parameters.
2. How to choose a suitable step size for minimization.
3. How the model, and thus the criterion for optimization, should be written.

6.2-2 Flexible Geometric Simplex Method

A second derivative-free method of minimization of a nonlinear objective function is by use of regular patterns of search involving *simplexes*. These techniques have proved very successful in finding an extremum of an unconstrained objective function, as well as a constrained extremum, and are especially effective as the number of model parameters increases. For two parameters, a regular simplex is an equilateral triangle (three points); for three parameters, the design is a regular tetrahedron (four points). See Figure 6.2-2.

In the search for a minimum of the sum of the squares of the deviations, ϕ, trial values of the model parameters can be selected at points in *parameter space* located at the vertices of the simplex, as originally suggested by Spendley, Hext, and Himsworth† in connection with experimental designs. The sum of the squares of the deviations is evaluated at each of the vertices of the simplex; a projection is made from the point yielding the highest value of the objective function, point A in Figure 6.2-2, through the centroid of the simplex. Point A is deleted and a new simplex, termed a *reflection*, is formed composed of the remaining old points and one new point, B, located along the projected line at the proper distance from the centroid. Continuation of this procedure, always deleting the vertex that yields the highest value of the objective function, plus rules for reducing the size of the simplex and rules to prevent cycling in the vicinity of the

† N. Spendley, G. R. Hext, and F. R. Himsworth, *Technometrics* **4**, 441, 1962.

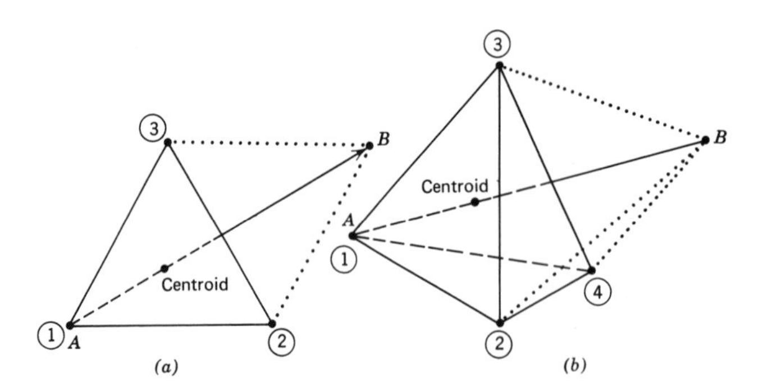

FIGURE 6.2-2 Regular simplexes for two and three independent parameters. The (1) represents the lowest response. The arrows point in the direction of greatest improvement. (*a*) Two variable simplex. (*b*) Three variable simplex.

extremum, permits a derivative-free search in which the step size is, in essence, fixed at successively reduced levels but the direction of search is permitted to change. Figure 6.2-3*a* illustrates the successive simplexes formed for a two-parameter model with a well-behaved sum of squares objective function.

Certain practical difficulties in the original procedure, namely that it did not provide for acceleration of the search, and encountered difficulty in carrying on the search in curving valleys or on curving ridges led to several improvements (for example the complex method of M. J. Box, *Compt. J.* **7**, 42, 1965, using nonregular polyhedrons). We describe here a straightforward technique, in which the simplex is permitted to alter in shape,† that has proved to be very effective and easily implemented on a digital computer.‡ In general it is recommended as better than the previously described direct search method because it takes less computer time even though the convergence to termination is slow.

† J. A. Nelder and R. Mead, *Compt. J.* **7**, 308, 1965.

‡ Available from U.S. Naval Ordnance Test Station, China Lake, Calif., as Publication 2698, LSQ2, Jan. 1967. Also available as NOLSQ4.

As before we want to minimize ϕ, where ϕ is defined by Equation 6.2-1. To simplify the notation, we show only the functional dependence of ϕ on the estimated parameters, suppressing in the notation the dependence on the independent variables, $\mathbf{x}$:

$$\phi = \phi(\mathbf{b}) \qquad \mathbf{b} = [b_1, b_2, \ldots, b_m]^T \tag{6.2-7}$$

Let $\mathbf{b}_i = [b_{1,i}, b_{2,i}, \ldots, b_{m,i}]$ be the vector specifying the m coordinates in the parameter space $b_1, b_2, \ldots, b_m$ of the vertex i. There will be $m + 1$ vertices comprising the simplex, and each will be specified by a vector, $\mathbf{b}_i$. Let ϕ_i denote the corresponding values of the objective function. The initial simplex will be a regular simplex (it does not have to be) with vertex 1 as the origin. From texts on analytical geometry, it can be shown that the coordinates of the vertices of the regular simplex are designated as shown in Table 6.2-1.

For example, for $m = 2$ and $a = 1$, the triangle given

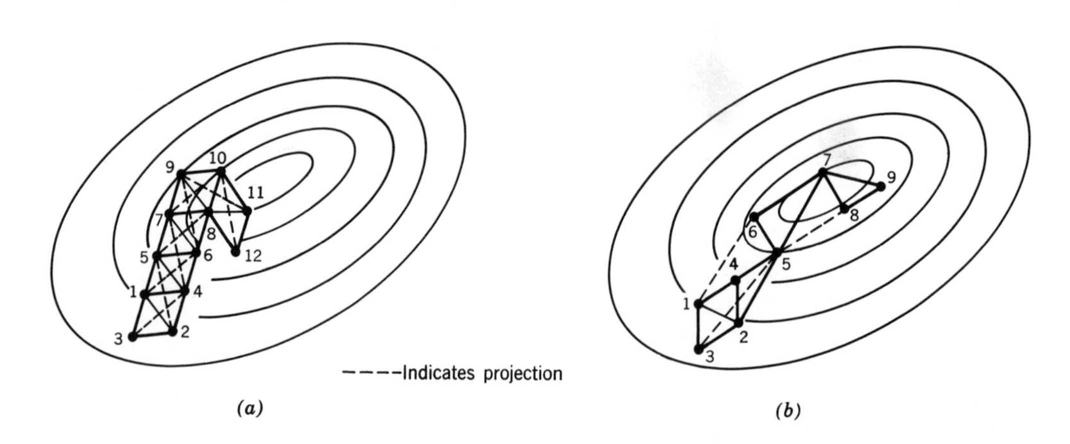

FIGURE 6.2-3 Sequence of simplexes obtained in minimizing the sum of the squares of the deviations: (*a*) regular simplex, and (*b*) variable size simplex.

TABLE 6.2-1

Vertex	Coordinates				
	$b_{1,i}$	$b_{2,i}$	$\cdots$	$b_{m-1,i}$	$b_{m,i}$
1	0	0	$\cdots$	0	0
2	π_1	π	$\cdots$	π	π
3	π	π_1	$\cdots$	π	π
$\vdots$	$\vdots$	$\vdots$		$\vdots$	$\vdots$
m	π	π	$\cdots$	π_1	π
$m + 1$	π	π	$\cdots$	π	π_1

where

$$\pi_1 = \frac{a}{m\sqrt{2}}[\sqrt{m+1} + n - 1]$$

$$\pi = \frac{a}{m\sqrt{2}}[\sqrt{m+1} - 1]$$

a = length of the path between two vertices

in Figure 6.2-2 has the following coordinates for the three vertices:

Vertex	$b_{1,i}$	$b_{2,i}$
1	0	0
2	0.965	0.259
3	0.259	0.965

We shall let

$$\phi_u = \max_i \{\phi_i\} \text{ with the corresponding } \mathbf{b}_{i=u}$$

$$\phi_l = \min_i \{\phi_i\} \text{ with the corresponding } \mathbf{b}_{i=l}$$

and let $\mathbf{c}$ be the centroid of all the points of the simplex with $i \neq u$, i.e., omitting the worst point. The procedure consists of sequentially replacing vertex $\mathbf{b}_u$ with a new vertex according to the following scheme.

1. First, *reflection* of the simplex is carried out to obtain a vertex designated by $\mathbf{b}^*$ with coordinates given by

$$\mathbf{b}^* = (1 + \gamma_r)\mathbf{c} - \gamma_r\mathbf{b}_u \qquad (6.2\text{-}8)$$

where γ_r is the so-called reflection coefficient, a positive constant determined by the user that may be unity. Let $\phi(\mathbf{b}^*) = \phi^*$. After $\mathbf{b}_u$ is reflected, one of three outcomes can exist:

(a) If $\phi_1 < \phi^* < \phi_u$, replace $\mathbf{b}_u$ by $\mathbf{b}^*$. The resulting simplex is used as a new starting simplex in step 1.

(b) If $\phi^* < \phi_u$, *expand* $\mathbf{b}^*$ to $\mathbf{b}^{**}$ by the relation

$$\mathbf{b}^{**} = \gamma_e\mathbf{b}^* + (1 - \gamma_e)\mathbf{c} \qquad (6.2\text{-}9)$$

where γ_e is the expansion coefficient (say a value of 2). If $\phi^{**} < \phi_1$, replace $\mathbf{b}_u$ by ϕ^{**} and start step 1 again. If $\phi^{**} > \phi_1$, the expansion has failed; replace $\mathbf{b}_u$ by $\mathbf{b}^*$ and start step 1 again.

(c) If $\phi^* > \phi_i$ for all $i \neq u$, that is, replacing $\mathbf{b}_u$ by $\mathbf{b}^*$ leaves $\mathbf{b}^*$ as the point that gives the maximum ϕ, then define a new $\mathbf{b}_u$ which is either the old $\mathbf{b}_u$ or is $\mathbf{b}^*$, whichever gives the lower value of ϕ. Afterwards a *contraction* is carried out, denoted by $\mathbf{b}^{**}$, and computed as follows:

$$\mathbf{b}^{**} = \gamma_c\mathbf{b}_u + (1 - \gamma_c)\mathbf{c} \qquad (6.2\text{-}10)$$

where γ_c is the contraction coefficient, $0 \leq \gamma_c \leq 1$ (usually $\frac{1}{2}$). Replace $\mathbf{b}_u$ by $\mathbf{b}^{**}$ and start step 1 again, unless the vertex obtained by contraction is worse than the max of $\{\phi(\mathbf{b}_u), \phi(\mathbf{b}^*)\}$, that is, $\phi^{**} > \min\{\phi_u, \phi^*\}$, in which case replace *all* the $\mathbf{b}_i$ by $\frac{1}{2}(\mathbf{b}_i + \mathbf{b}_l)$ and go back to step 1.

2. The search is *terminated* when

$$\sqrt{\frac{\sum_i (\phi_i - \bar{\phi})^2}{m}} \leq \epsilon$$

where ϵ is an arbitrarily chosen small number and $\bar{\phi}$ is the average value of ϕ.

Figure 6.2–4 is a flow chart of the program logic.

Example 6.2-2 Flexible Simplex Method

To illustrate the flexible simplex method, the data of Example 4.3-2 were fitted by the linear model $\eta = \beta_0' + \beta_1 x$. The starting vector was $b_0'^{(0)} = 1$, $b_1^{(0)} = 1$, at which the sum of the square of the deviations, $\phi^{(0)}$, was 9.55×10^5. Figure E6.2-2 illustrates the progress of the search for 85 successive reflections, expansions, and contractions (which took 1.76 seconds on a CDC 6600 computer), at which stage the search terminated giving $b_0' = 13.506$, $b_1 = 79.021$, and $\phi_{\min} = 9.616 \times 10^3$ compared with 13.51, 79.02, and 9.617×10^3, respectively, from Example 4.3-2. Starting at other starting vectors yielded identical results.

Example 6.2-3 Nonlinear Estimation of a Stream Flow Model

A model proposed to predict excess stream flow above a normal level of flow was

$$\frac{Q}{Q^*} = \frac{f\beta^n\sqrt{\dfrac{n}{2\pi}}\left(\dfrac{t}{\bar{t}}\right)^{n-1} exp\left[n\left(1 - \dfrac{\beta t}{\bar{t}}\right)\right]}{\left(1 + \dfrac{1}{12n}\right)} + \frac{(1-f)\left(\dfrac{\beta}{\alpha}\right)^m\sqrt{\dfrac{m}{2\pi}}\left(\dfrac{t}{\bar{t}}\right)^{m-1} exp\left[m\left(1 - \dfrac{\beta t}{\alpha\bar{t}}\right)\right]}{\left(1 + \dfrac{1}{12m}\right)}$$

where

$\beta = f + (1 - f)\alpha$

Q = predicted excess channel flow rate above the normal channel flow rate, the dependent variable

Q^* = ioput flow rate, a known value

t = time, an independent variable

$\left.\begin{matrix} n \\ m \\ f \\ \alpha \end{matrix}\right\}$ = model parameters characterizing hold up, by passing, and stagnancy, to be estimated

$\bar{t}$ = mean residence time for the channel, a known value

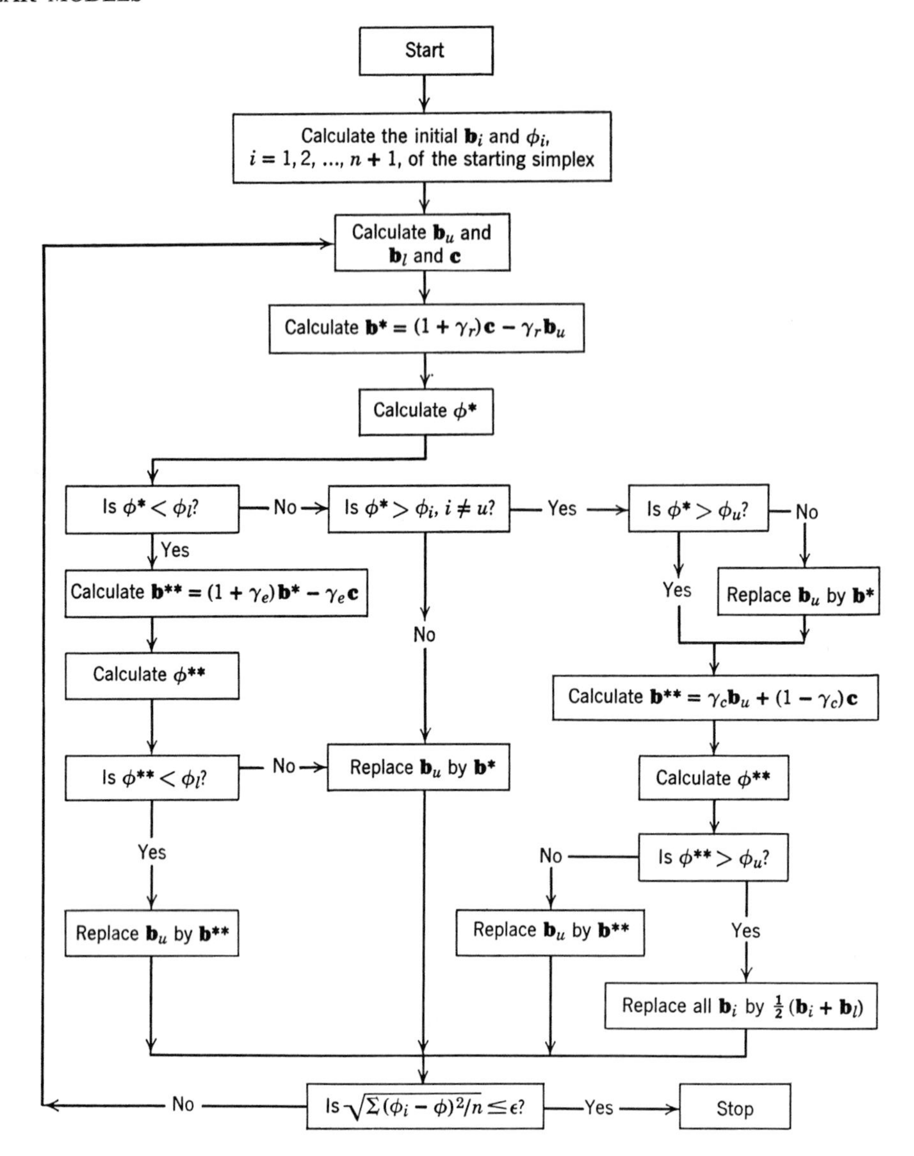

FIGURE 6.2-4 Information flow chart for flexible simplex method.

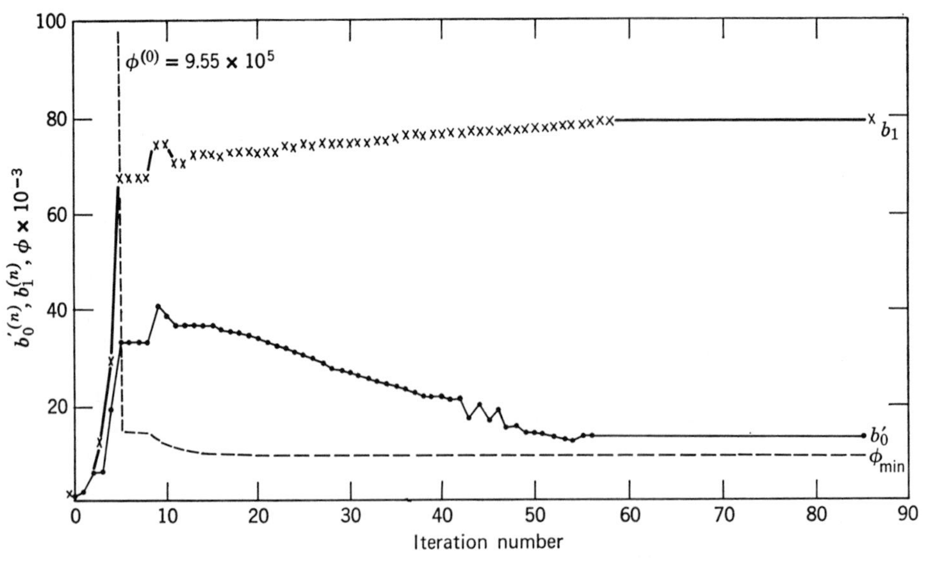

FIGURE E6.2-2

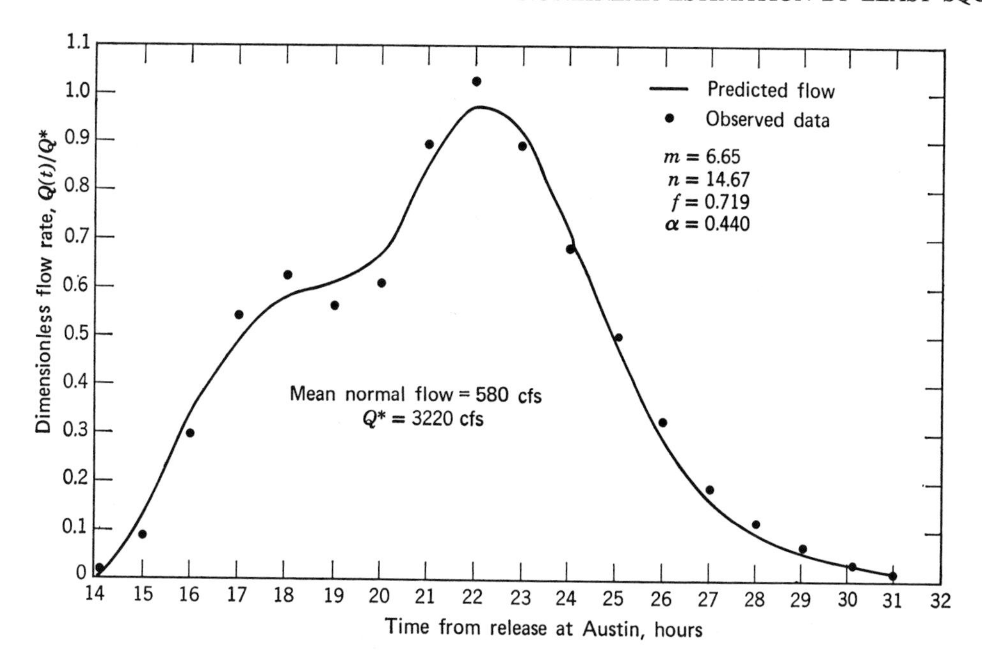

FIGURE E6.2-3 Predicted and experimental excess flow rates at Smithville, Texas.

TABLE E6.2-3a

Stage Number*		ϕ	m	n	f	α
1	*A*	2.443	2.035	1.077	0.898	1.013
	B	0.470	1.077	4.035	0.898	1.013
	C	2.460	1.131	1.119	0.155	1.080
	D	2.768	1.152	1.143	0.730	1.979
	E	0.107	2.000	4.000	0.040	2.000
5	*A*	0.675	1.888	3.895	0.577	1.203
	B	0.471	1.077	4.035	0.898	1.013
	C	0.496	2.410	5.423	0.895	1.220
	D	0.977	1.969	3.973	0.266	0.574
	E	0.489	1.814	4.460	0.652	0.613
10	*A*	0.480	1.555	5.53	0.970	0.966
	B	0.470	1.077	4.403	0.898	1.013
	C	0.432	1.875	4.924	0.832	1.027
	D	0.470	1.489	4.630	0.751	0.466
	E	0.413	1.675	4.647	0.731	0.853
20	*A*	0.241	2.354	6.467	0.867	0.366
	B	0.250	2.202	6.177	0.579	0.385
	C	0.225	2.546	6.625	0.836	0.359
	D	0.249	2.393	6.243	0.778	0.489
	E	0.314	1.953	5.972	0.916	0.391
50	*A*	0.143	3.731	8.587	0.727	0.472
	B	0.169	3.470	8.047	0.730	0.439
	C	0.147	3.670	8.481	0.729	0.472
	D	0.167	3.228	7.822	0.788	0.384
	E	0.126	4.167	9.458	0.713	0.500
184	*A*	0.0219	6.651	14.672	0.719	0.440
	B	0.0219	6.653	14.676	0.719	0.440
	C	0.0219	6.653	14.675	0.719	0.440
	D	0.0219	6.651	14.672	0.719	0.440
	E	0.0219	6.651	14.671	0.719	0.440

* *A*, *B*, *C*, *D*, and *E*, refer to the simplex vertices.

The model was fitted, using the flexible simplex method, to data provided by the Lower Colorado River Authority for the Colorado River below Austin, Texas. The data consisted of stream flow measurements at Smithville, Texas, a town about 45 miles downstream from Austin. The data were for water released at Austin on August 4, 1966, which gave a crest at Smithville about one day later. Figure E6.2-3 compares the data and estimated Q/Q^*. Table E6.2-3a illustrates the path of the search in parameter space. The search was terminated when the "volume" of the simplex was reduced below 10^{-7} (after approximately 8 seconds of central processing time on a CDC 6600 computer).

Several starting vectors were used, all of which yielded essentially the same values of the estimated parameters. The lowest value of ϕ obtained was 0.0216 for which $m = 6.13, n = 15.08, f = 0.705$ and $\alpha = 0.448$. Interaction among the parameters (refer to Section 6.3) accounts for these differences.

The parameters in the same model were estimated, using the direct search technique, in about the same time. The results are given in Table E6.2-3b. The estimates for m and n are somewhat different from those in Table E6.2-3a because the direct search program, using the same percentage change in the coefficients, terminated earlier than did the

TABLE E6.2-3b

Exploratory Search Number	ϕ	m	n	α	f
0	1.067	2.000	4.000	0.040	2.000
5	0.434	4.400	7.600	0.028	0.800
10	0.048	3.200	36.40	0.280	0.800
20	0.026	3.136	23.83	0.399	0.716
50	0.026	3.133	23.92	0.398	0.717

simplex search (at a ϕ of 0.026 versus a ϕ of 0.0219 for the simplex method).

6.2-3 Linearization of the Model

We now consider the first of the derivative-type methods of minimizing the sum of the squares of the deviations, that is, methods that require the numerical or analytical computation of first (and for some methods second) derivatives. Among the many varieties of derivative-type methods, we have the space to describe just those two most widely used:

1. Linearization of the process model itself.
2. Linearization of the criterion, that is, linearization of the function ϕ.

Wilde and Beightler† described a number of additional methods.

The first technique to be described here has been called by many names including the Newton-Raphson method, the Gauss-Newton method, and the Gauss-Seidel method, though Gauss deserves the lion's share of the credit. The method is very simple in concept: linearize the model in a truncated Taylor series in order to make use of linear analysis, and attain the desired minimum of the sum of the squares of the deviations by an iterative sequence of calculations. Initial guesses are made for the parameters; cyclically, new estimates are obtained by a method which has its foundations in the Newton-Raphson algorithm. The calculations are repeated until a criterion for convergence is met.

We begin by expanding η in a truncated Taylor series (refer to Section 2.4-4) about $\mathbf{b}^{(0)}$, the initial guess for $\boldsymbol{\beta}$. Weights are included as in Chapter 5. The initial guess for β_j is designated $b_j^{(0)}$. If

$$\phi = \sum_{i=1}^{n} w_i(Y_i - \eta_i)^2 \tag{6.2-11}$$

where η_i refers to the model with the vector for the ith data set introduced, $\mathbf{x}_i$, by minimizing ϕ we can find an improved estimate of β_j. (If η were truly a linear function, only one step would be needed to reach the minimum of ϕ.) We expand η as follows:

$$\begin{aligned}\eta &= \eta_0 + \left(\frac{\partial \eta}{\partial \beta_1}\right)_0 (\beta_1 - b_1^{(0)}) + \cdots + \left(\frac{\partial \eta}{\partial \beta_m}\right)_0 (\beta_m - b_m^{(0)}) \\ &= \eta_0 + \sum_{j=1}^{m} \left(\frac{\partial \eta}{\partial \beta_j}\right)_0 \Delta b_j^{(0)}\end{aligned} \tag{6.2-12}$$

where $\Delta b_j^{(0)} = \beta_j - b_j^{(0)}$, the subscript 0 on η means η evaluated using $b_1^{(0)}, \ldots, b_m^{(0)}$, and the subscript 0 on the partial derivatives means the same thing. To relieve the user of one of the most burdensome features of employing derivative methods, computer programs have been written which approximate the partial derivatives by partial difference quotients:

$$\left(\frac{\partial \eta_i}{\partial \beta_j}\right) \cong \left(\frac{\delta \eta_i}{\delta b_j}\right)_0 = \frac{\eta_i(\mathbf{x}_i; b_1^{(0)}, \ldots, b_j^{(0)} + \delta b_j^{(0)}, \ldots, b_m^{(0)}) - \eta_i(\mathbf{x}_i; b_1^{(0)}, \ldots, b_j^{(0)}, \ldots, b_m^{(0)})}{\delta b_j^{(0)}}$$

where δ represents a small perturbation. However, it may not always be possible to compute a derivative numerically with the required accuracy. If the regression curve for the model is flat, the quantity in the denominator grows small and the relative error in the approximate derivative can increase drastically. (The same feature is true in the next section in connection with the numerical approximation of the derivatives of ϕ. As ϕ approaches its minimum, the relative errors in the numerically computed derivatives become larger. Consequently, the search for the minimum of ϕ can oscillate and/or become very inefficient.)

After the linear approximation for η, Equation 6.2-12, is introduced into Equation 6.2-11, the partial derivatives of ϕ with respect to each of the $\Delta b_j^{(0)}$ can be equated to zero, as explained in Section 5.1

$$\frac{\partial \sum_{i=1}^{n} w_i \left[Y_i - (\eta_i)_0 - \sum_{j=1}^{m} \left(\frac{\partial \eta_i}{\partial \beta_j}\right)_0 \Delta b_j^{(0)}\right]^2}{\partial(\Delta b_j^{(0)})} = 0 \tag{6.2-13}$$

Equation 6.2-13 yields a set of m linear equations corresponding to the normal equations of Chapter 5:

$$2\sum_{i=1}^{n} w_i \left[Y_i - (\eta_i)_0 - \left(\frac{\partial \eta_i}{\partial \beta_1}\right)_0 \Delta b_1^{(0)} - \left(\frac{\partial \eta_i}{\partial \beta_2}\right)_0 \Delta b_2^{(0)} + \cdots\right]\left(\frac{\partial \eta_i}{\partial \beta_1}\right)_0 = 0$$

$$\cdots\cdots$$

$$2\sum_{i=1}^{n} w_i \left[Y_i - (\eta_i)_0 - \left(\frac{\partial \eta_i}{\partial \beta_1}\right)_0 \Delta b_1^{(0)} - \left(\frac{\partial \eta_i}{\partial \beta_2}\right)_0 \Delta b_2^{(0)} + \cdots\right]\left(\frac{\partial \eta_i}{\partial \beta_m}\right)_0 = 0$$

Let $E_i^{(0)} = Y_i - (\eta_i)_0$. Then these linear equations can be written as follows:

$$\begin{aligned}&\Delta b_1^{(0)} \sum_{i=1}^{n} w_i \left(\frac{\partial \eta_i}{\partial \beta_1}\right)_0 \left(\frac{\partial \eta_i}{\partial \beta_1}\right)_0 + \Delta b_2^{(0)} \sum_{i=1}^{n} w_i \left(\frac{\partial \eta_i}{\partial \beta_2}\right)_0 \left(\frac{\partial \eta_i}{\partial \beta_1}\right)_0 + \cdots \\ &\qquad + \Delta b_m^{(0)} \sum_{i=1}^{n} w_i \left(\frac{\partial \eta_i}{\partial \beta_m}\right)_0 \left(\frac{\partial \eta_i}{\partial \beta_1}\right)_0 = \sum_{i=1}^{n} w_i E_i^{(0)} \left(\frac{\partial \eta_i}{\partial \beta_1}\right)_0 \\ &\cdots\cdots \\ &\Delta b_1^{(0)} \sum_{i=1}^{n} w_i \left(\frac{\partial \eta_i}{\partial \beta_1}\right)_0 \left(\frac{\partial \eta_i}{\partial \beta_m}\right)_0 + \Delta b_2^{(0)} \sum_{i=1}^{n} w_i \left(\frac{\partial \eta_i}{\partial \beta_2}\right)_0 \left(\frac{\partial \eta_i}{\partial \beta_m}\right)_0 + \cdots \\ &\qquad + \Delta b_m^{(0)} \sum_{i=1}^{n} w_i \left(\frac{\partial \eta_i}{\partial \beta_m}\right)_0 \left(\frac{\partial \eta_i}{\partial \beta_m}\right)_0 = \sum_{i=1}^{n} w_i E_i^{(0)} \left(\frac{\partial \eta_i}{\partial \beta_m}\right)_0\end{aligned} \tag{6.2-14}$$

† D. J. Wilde and C. S. Beightler, *Foundations of Optimization*, Prentice-Hall, Englewood Cliffs, N.J., 1967.

We now want to solve for the Δb_j's.

The array of equations can be made much more compact by introducing the following matrix notation:

$$[X_{ij}] = \frac{\partial \eta_i(\mathbf{x}_i; \mathbf{b})}{\partial \beta_j} \qquad \begin{matrix} i = 1, 2, \ldots, n \\ j = 1, 2, \ldots, m \end{matrix}$$

$$\mathbf{X}^{(0)} = \begin{bmatrix} \left(\frac{\partial \eta_1}{\partial \beta_1}\right)_0 & \cdots & \left(\frac{\partial \eta_1}{\partial \beta_m}\right)_0 \\ \vdots & & \vdots \\ \left(\frac{\partial \eta_n}{\partial \beta_1}\right)_0 & \cdots & \left(\frac{\partial \eta_n}{\partial \beta_m}\right)_0 \end{bmatrix} \text{an } n \times m \text{ matrix}$$

$$\mathbf{B}^{(0)} = \begin{bmatrix} \Delta b_1^{(0)} \\ \Delta b_2^{(0)} \\ \vdots \\ \Delta b_m^{(0)} \end{bmatrix}$$

Then:

$$\mathbf{A}^{(0)} = (\mathbf{X}^T\mathbf{w}\mathbf{X})^{(0)}$$

$$= \begin{bmatrix} \sum_{i=1}^{n} w_i \left(\frac{\partial \eta_i}{\partial \beta_1}\right)_0 \left(\frac{\partial \eta_i}{\partial \beta_1}\right)_0 & \cdots & \sum_{i=1}^{n} w_i \left(\frac{\partial \eta_i}{\partial \beta_m}\right)_0 \left(\frac{\partial \eta_i}{\partial \beta_1}\right)_0 \\ \vdots & & \vdots \\ \sum_{i=1}^{n} w_i \left(\frac{\partial \eta_i}{\partial \beta_1}\right)_0 \left(\frac{\partial \eta_i}{\partial \beta_m}\right)_0 & \cdots & \sum_{i=1}^{n} w_i \left(\frac{\partial \eta_i}{\partial \beta_m}\right)_0 \left(\frac{\partial \eta_i}{\partial \beta_m}\right)_0 \end{bmatrix}$$

$$\mathbf{Z}^{(0)} = (\mathbf{X}^T\mathbf{w}\mathbf{E})^{(0)} = \begin{bmatrix} \sum_{i=1}^{n} w_i E_i^{(0)} \left(\frac{\partial \eta_i}{\partial \beta_1}\right)_0 \\ \vdots \\ \sum_{i=1}^{n} w_i E_i^{(0)} \left(\frac{\partial \eta_i}{\partial \beta_m}\right)_0 \end{bmatrix}$$

and Equations 6.2-14 can be written in matrix notation as

$$(\mathbf{X}^T\mathbf{w}\mathbf{X})^{(0)}\mathbf{B}^{(0)} = (\mathbf{X}^T\mathbf{w}\mathbf{E})^{(0)} \tag{6.2-15}$$

or

$$\mathbf{A}^{(0)}\mathbf{B}^{(0)} = \mathbf{Z}^{(0)}$$

so that

$$\mathbf{B}^{(0)} = \mathbf{C}^{(0)}\mathbf{Z}^{(0)} \tag{6.2-16}$$

where $\mathbf{C}^{(0)} = (\mathbf{A}^{(0)})^{-1}$. Observe the close analogy between the development in Section 5.1 and that above.

Once the vector $\mathbf{B}^{(0)}$ is calculated by Equation 6.2-16, a new estimate of each β_j is obtained by repeating the calculation with $b_j^{(1)}$, the improved estimate of β_j, replacing $b_j^{(0)}$ in Equation 6.2-16 and in the matrix elements $[X_{ij}]$. The recursion relation

$$b_j^{(n+1)} = b_j^{(n)} + h_j^{(n)} \Delta b_j^{(n)} \tag{6.2-17}$$

is used to calculate $b_j^{(1)}$; h_j is an "acceleration factor," i.e., a factor supplied by the user to speed up the progress of the search for the minimum of ϕ. In effect, Δb_j determines the direction of the search for the minimum ϕ in parameter space, and h_j determines the step length. In the Gauss-Seidel method, $h_j \equiv 1$. Other techniques to evaluate h_j, which have been used with greater effectiveness, are:

1. Select the length of each step along the vector $\mathbf{B}$ proportional to the slope of the approximating plane for the objective function ϕ.
2. Select equal size steps until an increase in ϕ is experienced.
3. Use a Fibonacci search along the direction of decrease in ϕ to locate its lowest point.
4. Use a multiple bigger or smaller than unity of some initially chosen step size, a multiple proportional to the number of previous successful moves in the given coordinate direction.

The motivation underlying the adjustment of h_j is that the minimization can be accelerated or decelerated automatically to: (1) speed up the initial approach to the minimum of ϕ and also (2) slow down the final approach to avoid excessive oscillation. Successive vectors $\mathbf{B}$ are calculated until each Δb_j is small enough and/or the absolute or relative change in ϕ drops below a predetermined number, in which case the search for the minimum ϕ is terminated.

Certain practical difficulties that arise in the procedure will now be mentioned.

1. How can suitable initial guesses for the $b^{(0)}$'s be obtained? Because the function ϕ is nonlinear, more than one minimum may exist in ϕ—a feature absent from the linear analysis of Chapter 5. Consequently, if the initial guesses for the parameters are too far away from the estimates that minimize ϕ, the search may not terminate at the global (lowest) minimum for ϕ but at some other minimum. Figure 6.2-5 displays in two dimensions what might happen with a poor choice for the initial vector $\mathbf{b}^{(0)}$. One suggestion to obtain suitable initial guesses for the β's is to plot the response as a function of a single variable, holding all other variables constant, and to take some asymptotic value or other clearly indicated value for $b_j^{(0)}$. Then the initial values for other parameters can be based on the(se) initially selected values. Often, approximate values of the β's will be known from earlier studies or from physical reasoning. The ultimate resort is just to try several starting vectors $\mathbf{b}^{(0)}$ in the feasible range and ascertain whether or not they all yield the same value for the minimum of ϕ. Kittrell, Mezaki, and Watson† described other techniques to obtain initial parameter estimates.

2. The objective function may become unbounded in the range of the search for the minimum of ϕ, or the first partial derivatives of the model may become

† J. R. Kittrell, R. Mezaki, and C. C. Watson, *Ind. Eng. Chem.* **57**, 19, 1965.

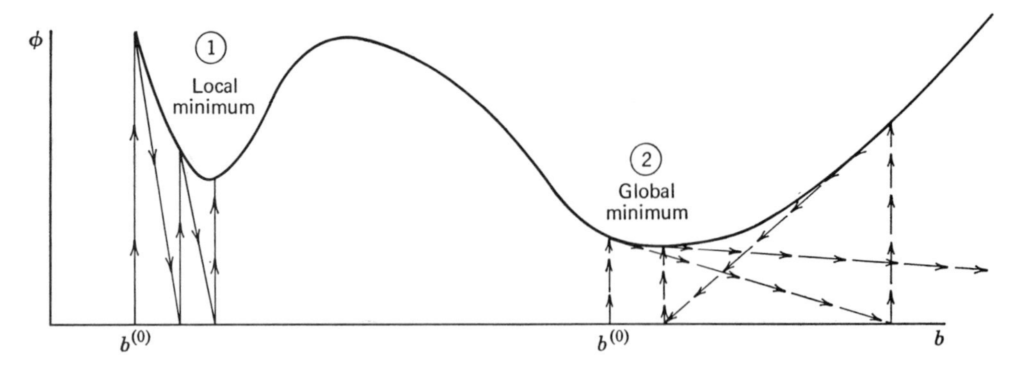

FIGURE 6.2-5 Effect of the initial guess for β on: (1) convergence to a local minimum in the sum of the squares of the deviations (solid line), and (2) oscillation in ϕ in the search sequence (dashed line) and subsequent divergence.

unbounded. Models with polynomials in the denominator are particularly subject to this problem, as for example

$$\eta = \frac{\beta_0 + \beta_1 x_1}{\beta_2 x_1 + \beta_3 x_2}$$

in which both η and the partial derivative of η with respect to β_2

$$\frac{\partial \eta}{\partial \beta_2} = -\frac{(\beta_0 + \beta_1 x_1)x_2}{(\beta_2 x_1 + \beta_3 x_2)^2}$$

become unbounded when $b_2^{(n)} x_1 = -b_3^{(n)} x_2$. The only ways to overcome this difficulty are to restrict the region of search for the β's and/or to be quite careful in the original construction of the process model.

3. The matrix $\mathbf{A}^{(0)}$ may become singular because of redundancy among the data, or almost singular at one or more values of $\boldsymbol{\beta}$ in the search. A proper experimental design for the original collection of data can overcome this difficulty.

4. The iterative technique at some stage may increase rather than decrease ϕ. Refer to the dashed line in Figure 6.2-5. Suitable logical steps introduced into the computer code will avoid this outcome, such as testing to see if $\phi^{(n+1)} < \phi^{(n)}$ at each cycle and, if not, cutting the value of h_j by a preselected factor.

We now give an example of the Gauss-Seidel method. With initial estimates of the parameters far from the final estimates, it is a characteristically slow method but converges rapidly near termination (in contrast to the method of steepest descent, described in the next section, which converges very slowly).

Example 6.2-4 Application of the Gauss-Seidel Method

Simulated data have been prepared to demonstrate estimation by the Gauss-Seidel method. The simulated data in Table E6.2-4a were generated by adding to the function $\eta = 3x_1 + 3e^{-x_2/2}$ random errors with a mean of 0 and a variance of 0.01.

TABLE E6.2-4a SIMULATED DATA

x_1	x_2	y (Exact)	Y (Simulated)
0.0	0.0	3.00	2.93
0.0	1.0	1.82	1.95
0.0	2.0	1.10	0.81
0.0	3.0	0.67	0.58
1.0	0.0	6.00	5.90
1.0	1.0	4.82	4.74
1.0	2.0	4.10	4.18
1.0	2.0	4.10	4.05
2.0	0.0	9.00	9.03
2.0	1.0	7.82	7.85
2.0	2.0	7.10	7.22
2.5	2.0	8.60	8.50
2.9	1.8	9.92	9.81

We shall now assume that we know nothing at all about the generation of the data in Table E6.2-4a, but merely that we have the data and want to estimate the parameters in the known (or assumed) model

$$\eta = \beta_1 x_1 + \beta_2 e^{\beta_3 x_2} \qquad \text{(a)}$$

by minimizing

$$\phi = \sum_{i=1}^{n} [Y_i - \beta_1 x_1 - \beta_2 e^{\beta_3 x_2}]^2 \qquad \text{(b)}$$

First, we have to determine the initial estimates $b_1^{(0)}$, $b_2^{(0)}$, and $b_3^{(0)}$. We might just select, out of thin air,

$$b_1^{(0)} = b_2^{(0)} = b_3^{(0)} = 1$$

or

$$b_1^{(0)} = b_2^{(0)} = b_3^{(0)} = 0$$

Instead, for illustrative purposes, we shall obtain estimates near the true minimum by some preliminary graphical work. We know from Equation (a) that if x_2 is held constant, we obtain a straight line whose slope $(\partial\eta/\partial x_1)_{x_2}$ is β_1. Figure

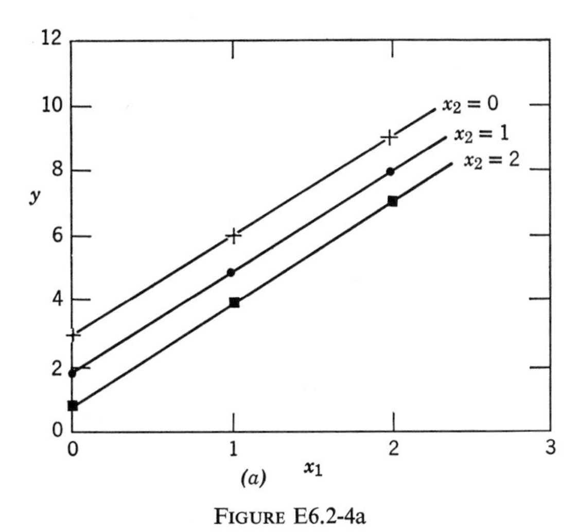

FIGURE E6.2-4a

E6.2-4a illustrates a plot of Y versus x_1 with x_2 fixed at several different values. The slopes approximating β_1 are:

x_2	$b_2^{(0)}$
0	3.1
1	2.9
2	3.2

so that an average value of 3.1 can be used for $b_1^{(0)}$. Note that if the x_2 values had not been replicated, the above procedure would require considerable interpolation among the data points.

To get $b_2^{(0)}$ and $b_3^{(0)}$, we can form

$$(\eta - 3.1x_1) = \beta_2\, e^{\beta_3 x_2}$$

$$\ln(\eta - 3.1x_1) = \ln \beta_2 + \beta_3 x_2 \tag{b}$$

Figure E6.2-4b is a plot of the values of $\ln(Y - 3.1x_1)$ for some of the data sets versus x_2; the slope of the line is an estimate of $b_3^{(0)}$ while the intercept is approximately $\ln(b_2^{(0)})$. Figure E6.2-4b gives $b_2^{(0)} \sim 2.9$ and

$$b_3^{(0)} \sim [(\ln 2.88 - \ln 1.02)/(0 - 2)] = -0.52.$$

These estimates are close to the true parameters because of the small error variance chosen for the simulated data.

Next we must decide on a weighting scheme. In this example all the data sets will be weighted equally; i.e.,

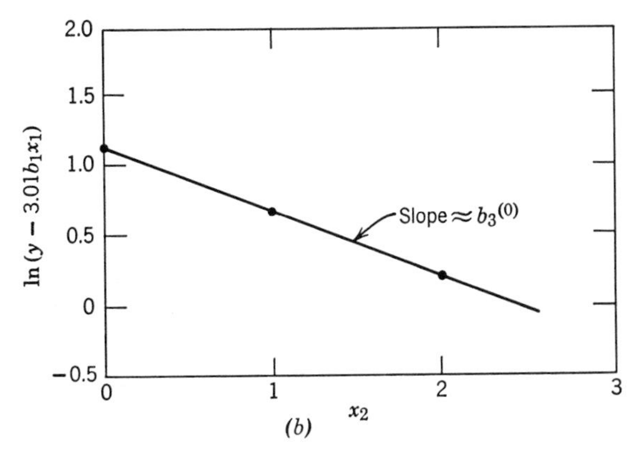

FIGURE E6.2-4b

$w_i = 1$. Finally, a termination criterion must be selected to let the computer know when to stop, say for each b_j when

$$\frac{b_j^{(n)} - b_j^{(n-1)}}{b_j^{(n-1)}} < 10^{-6} \tag{c}$$

The partial derivatives of η are

$$\frac{\partial \eta}{\partial \beta_1} = x_1$$

$$\frac{\partial \eta}{\partial \beta_2} = e^{\beta_3 x_2}$$

$$\frac{\partial \eta}{\partial \beta_3} = \beta_2 x_2\, e^{\beta_3 x_2}$$

Consequently,

$$\left(\frac{\partial \eta_1}{\partial \beta_1}\right)_0 = x_{11} = 0.0$$

$$\left(\frac{\partial \eta_2}{\partial \beta_1}\right)_0 = x_{21} = 0.0$$

$$\vdots$$

$$\left(\frac{\partial \eta_1}{\partial \beta_2}\right)_0 = e^{-0.52(0)} = 1$$

$$\left(\frac{\partial \eta_2}{\partial \beta_2}\right)_0 = e^{-0.52(1)} = 0.594$$

$$\vdots$$

$$\left(\frac{\partial \eta_1}{\partial \beta_3}\right)_0 = (2.9)(0)e^{-0.52(0)} = 0$$

$$\left(\frac{\partial \eta_2}{\partial \beta_3}\right)_0 = (2.9)(1)\, e^{-0.52(1)} = 1.723$$

$$\vdots$$

These elements comprise the matrix $\mathbf{X}^{(0)}$, from which the matrix $(\mathbf{X}^{(0)})^T\mathbf{X}^{(0)} \equiv \mathbf{A}^{(0)}$ can be computed. The elements of $\mathbf{E}^{(0)}$ are

$$E_1^{(0)} = Y_1 - (\eta_1)_0 = Y_1 - (3.1x_{11} + 2.9\, e^{-0.52x_{12}})$$

$$= 2.93 - [3.1(0) + 2.9\, e^{-0.52(0)}] = 0.03$$

$$E_2^{(0)} = Y_2 - (\eta_2)_0 = 1.95 - [3.1(0) + 2.9\, e^{-0.52(1)}] = 0.23$$

etc.

from which the matrix $(\mathbf{X}^{(0)T}\mathbf{E}^{(0)})$ can be computed. Then $\mathbf{B}^{(0)}$ can be calculated from Equation 6.2-16, and the vector $\mathbf{b}^{(1)}$ can be computed from Equation 6.2-17.

Table E6.2-4b lists the progress of the Gauss-Seidel method by cycles (only four significant figures are shown in

TABLE E6.2-4b PROGRESS OF THE GAUSS-SEIDEL METHOD

Cycle Number	$b_1^{(n)}$	$b_2^{(n)}$	$b_3^{(n)}$	$\phi^{(n)}$
0 (initial guesses)	3.1	2.9	−0.52	0.1981
1	3.017	2.958	−0.5222	0.1573
2	3.017	2.958	−0.5220	0.1574
3	3.017	2.958	−0.5220	0.1574

the table). Note that with the initial guesses close to the final estimates of $\boldsymbol{\beta}$ and with a well-behaved objective function, only a very few cycles are needed to meet the termination criterion.

To make the iterative procedure converge faster and to avoid oscillation with less favorable objective functions than used in Example 6.2-4, several authors have suggested that the behavior of the sum of the squares of the deviations, ϕ, be automatically explored during each cycle of iteration in the directions given by the elements of $\boldsymbol{\beta}$.† Hartley, after solving Equation 6.2-16 but prior to applying Equation 6.2-17, calculated several values of ϕ for different values of $v_j \Delta b_j^{(n)}$ in the range $0 \leq v_j \leq 1$. Other ranges can be used, of course. The value of v_j for which $\phi^{(n)}(v)$ was a minimum, say v_j^*, was used to obtain the $(n + 1)$st vector of b's by placing $v_j^* = h_j$ in Equation 6.2-17. If a parabola is used to fit the values of $\phi^{(n)}(v)$ as illustrated in Figure 6.2-6:

$$\phi^{(n)}(v_j) = a_0 + a_1 v_j + a_2 v_j^2$$

only three values of ϕ are required to obtain a_0, a_1, and a_2, and $v_j^* = -a_1/2a_2$. Hartley demonstrated that this modification of the Gauss-Seidel method converged to a minimum ϕ under certain specific conditions. However, because the given requirements cannot in general be established prior to finding the minimum of ϕ, the practical merit of Hartley's method lies in the evaluation of h_j in Equation 6.2-17 by a flexible objective rather than a subjective criterion. The suggested exploration technique has an advantage over the method of steepest descent, which will be discussed shortly, in that the scaling (selection of the magnitude) of the moves to improve ϕ is controlled. Any method that continuously adjusts h_j so that ϕ can only decrease and never increase will avoid some of the difficulties of the standard Gauss-Seidel method in which h_j is unity.

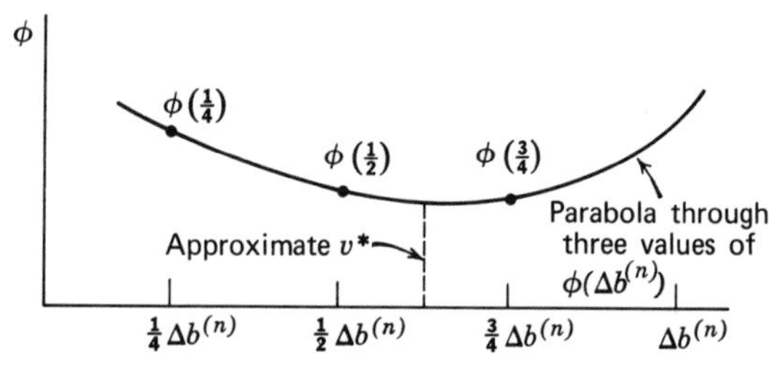

FIGURE 6.2-6 Exploration of the surface ϕ.

† G. E. P. Box, *Bull. Inst. Int. Stat.* **36**, 215, 1958; G. W. Booth and T. I. Peterson, IBM Share Program Paper No. 687 WL NLI, 1958; H. O. Hartley, *Technometrics* **3**, 269, 1961.

Example 6.2-5 Modified Gauss-Seidel Techniques

Strand, Kohl, and Bonham‡ used a version of Hartley's modification of the Gauss-Seidel method to fit values for the Thomas-Fermi-Dirac potential for atoms as tabulated by Thomas. The model was

$$\eta(x, \beta_1, \ldots, \beta_6) = \beta_1 e^{-\beta_2 x} + \beta_3 e^{-\beta_4 x} + \beta_5 e^{-\beta_6 x}$$

‡ T. G. Strand, D. A. Kohl, and R. A. Bonham, *J. Chem. Physics* **39**, 1307, 1963.

TABLE E6.2-5 LEAST SQUARE FIT OF THE THOMAS-FERMI-DIRAC POTENTIAL FUNCTION

	n Cycle Number	$b_1^{(n)}$	$b_2^{(n)}$	$b_3^{(n)}$	$b_4^{(n)}$	$b_5^{(n)}$	$b_6^{(n)}$	$\hat{\sigma}_r 10^3$ *
				Modified Gauss-Seidel Method				
A:	0	0.4660	1.1420	0.5410	6.4470	0.1000	9.9990	83.7
	1	0.4607	1.3458	0.4526	5.9621	0.0820	8.9045	7.415
	2	0.5261	1.6027	0.4302	6.9199	0.0414	9.0207	4.048
	5	0.5696	1.6775	0.3845	7.6433	0.0438	10.993	3.533
	10	0.4786	1.5047	0.4063	5.4189	0.1125	12.708	3.016
B:	0	5.180	1.5770	0.3910	6.2190	0.0890	25.000	6.624
	1	0.5625	1.6825	0.3674	6.5011	0.0699	22.126	2.390
	2	0.5313	1.6276	0.3892	5.9050	0.0792	21.504	1.923
	4	0.4578	1.4974	0.4758	5.2726	0.0651	28.376	0.786
	9	0.4256	1.4431	0.4918	4.8668	0.0823	24.116	0.572
				Unmodified Method				
A0†	10	0.5897	1.7177	0.2792	8.2340	0.1292	8.2392	3.574
A2	9	0.5899	1.7182	−199.47	8.2426	199.88	8.1085	402.7
B1	10	1.2226	−0.7341	1.9944	2.0019	0.2384	−7.8010	10^{12}

$$* \; \hat{\sigma}_r = \left(\frac{\phi(\mathbf{b}^{(n)})}{6}\right)^{1/2}$$

† Values at cycle number n with starting vector *A*0, etc, as designated in the upper portion of the table.

The parameter v was allowed to vary between -1.5 and 1.75 in intervals of 0.25. Fourteen values of $\phi^{(n)}(v)$ were calculated on each cycle for each $b_j^{(n)}$; the smallest value of $\phi^{(n)}(v)$ was ascertained at v_i. The minimum v^* was found as the minimum of a parabola through the points $[v_{i-1}, \phi(v_{i-1})]$, $[v_i, \phi(v_i)]$, and $[v_{i+1}, \phi(v_{i+1})]$ (unless v_i was -1.5 or 1.75).

Iteration was terminated when

$$\frac{\phi^{(n+1)} - \phi^{(n)}}{\phi^{(n)}} < 10^{-4}$$

or if $\phi^{(n+1)}$ increased. Table E6.2-5 lists selected results comparing the modified Gauss-Seidel method and the unmodified method. The modified procedure clearly exhibits a superior performance.

6.2-4 Linearization of the Objective Function

Another way to minimize the sum of the squares of the deviations, ϕ, is to linearize the objective function itself. Such methods include the well-known method of steepest descent, the conjugate gradient method,† and Marquardt's method.‡ The gradient of ϕ, i.e., grad ϕ or $\nabla\phi$, is a vector perpendicular to the surface ϕ in parameter space which extends in the direction of the maximum increase in ϕ at a given point. The negative of the gradient extends in the direction of steepest descent. Figure 6.2-7 illustrates the geometric interpretation of ϕ, $\nabla\phi$, and $-\nabla\phi$ in a space of two parameters, β_1 and β_2. The closed curves represent contours of constant ϕ which are of increasing value proceeding from the minimum ϕ.

Suppose we expand ϕ in a truncated Taylor series about $\mathbf{b}^{(0)}$:

$$\phi \simeq (\phi)_0 + \sum_{j=1}^{m} \left(\frac{\partial \phi}{\partial \beta_j}\right)_0 (\beta_j - b_j^{(0)})$$

(the notation was defined in Section 6.2-3). The magnitudes of the components of $-\nabla\phi$:§

$$-\nabla\phi|_{\mathbf{b}^{(0)}} = -\left(\frac{\partial \phi}{\partial \beta_1}\right) \boldsymbol{\delta}_{\beta_1} - \left(\frac{\partial \phi}{\partial \beta_2}\right)_0 \boldsymbol{\delta}_{\beta_2} - \cdots - \left(\frac{\partial \phi}{\partial \beta_m}\right) \boldsymbol{\delta}_{\beta_m}$$

evaluated at $\mathbf{b}^{(0)}$ are identical to the respective terms in the first-order expansion of ϕ in parameter space, and the components are used to establish the direction of search in the *method of steepest descent*.

We assume that ϕ, given by Equation 6.2-1, is single valued, is continuous, and has a single minimum in the region of search. By finding the components of the vector $-\nabla\phi$, it is possible to carry out an iterative sequence of calculations and reduce the value of ϕ to at least a local minimum. The general procedure is to:

1. Compute analytically (or numerically) the components of $-\nabla\phi$ and evaluate them at $\mathbf{b}^{(0)}$.

† R. Fletcher and C. M. Reeves, *Compt. J.* **7**, 149, 1964.

‡ D. W. Marquardt, *J. Soc. Ind. Appld. Math.* **11**, 431, 1963.

§ $\boldsymbol{\delta}\beta_1$ is a unit vector in the β_1 direction; $\boldsymbol{\delta}\beta_j$ is a unit vector in the β_j direction.

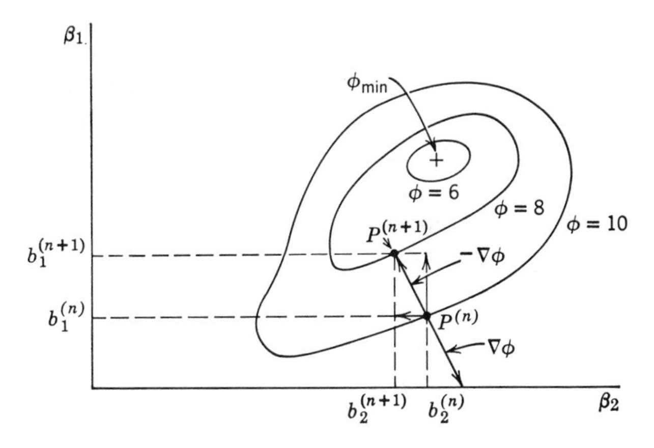

FIGURE 6.2-7 Geometric representation of the sum of the squares of the deviations, ϕ, $\nabla\phi$, and the direction of steepest descent, $-\nabla\phi$, in parameter space at point P.

The unit vector $-\nabla\phi/\|-\nabla\phi\|$ is computed to ascertain the components of the direction of search:

$$\frac{-\nabla\phi}{\|-\nabla\phi\|} = \frac{-\dfrac{\partial \phi}{\partial \beta_1}\boldsymbol{\delta}_{\beta_1} - \dfrac{\partial \phi}{\partial \beta_2}\boldsymbol{\delta}_{\beta_2} - \cdots}{\sqrt{\left(-\dfrac{\partial \phi}{\partial \beta_1}\right)^2 + \left(-\dfrac{\partial \phi}{\partial \beta_2}\right)^2 + \cdots}} \tag{6.2-18}$$

As an example, for the linear equation $z = 2\beta_1 - \beta_2$:

$$\nabla z = 2\boldsymbol{\delta}_{\beta_1} - \boldsymbol{\delta}_{\beta_2}$$

$$\frac{-\nabla z}{\|-\nabla z\|} = \frac{-2}{\sqrt{5}}\boldsymbol{\delta}_{\beta_1} + \frac{1}{\sqrt{5}}\boldsymbol{\delta}_{\beta_2}$$

2. The components of $-\nabla\phi/\|-\nabla\phi\|$ evaluated at $\mathbf{b}^{(0)}$ establish the direction of search for the minimum of ϕ. (In the above example, if z were ϕ, the intial Δb's are $\Delta b_1^{(0)} = -2/\sqrt{5}$ and $\Delta b_2^{(0)} = 1/\sqrt{5}$; the components of the gradient above are not functions of $\boldsymbol{\beta}$ because the example is linear in $\boldsymbol{\beta}$.) Each new cycle of $b_j^{(n)}$'s is computed from the previous cycle (starting with $\mathbf{b}^{(0)}$) by means of Equation 6.2-17:

$$b_j^{(n+1)} = b_j^{(n)} + h_j^{(n)} \Delta b_j^{(n)}$$

3. The sequence of iterative moves continues until the process terminates because ϕ is less than a specified criterion or because the process diverges or oscillates and no further reduction in ϕ can be achieved.

Improper scaling (i.e., the relative magnitudes of the components of $-\nabla\phi$) can cause difficulty in minimizing ϕ. If the hyperspace is badly elongated, as illustrated for two dimensions in Figure 6.2-8, the method of steepest descent may take an excessively long time to converge because the direction of steepest descent proves to be nearly perpendicular to the direction that will minimize ϕ. The negative of the gradient of ϕ points in the direction that minimizes ϕ only in a *local* region and not in the direction of the *global* minimum of ϕ, the minimum desired, unless the contours are arcs of circles with ϕ_{min} as a center.

Marquardt† observed in practice that for elongated ridges the method of steepest descent and the Gauss-Seidel method gave directions of search nearly orthogonal to each other. He suggested a compromise between the two methods. Marquardt's method improves the conditioning of the matrix of partial derivatives, $(\mathbf{X}^T\mathbf{wX}) \equiv \mathbf{A}$. Suppose that in Equation 6.2-15 a diagonal matrix is added to $\mathbf{A}$:

$$(\mathbf{A} + \lambda\mathbf{I})\mathbf{B} = \mathbf{Z} \tag{6.2-19}$$

where $\lambda \geq 0$. When $\lambda = 0$, Equation 6.2-19 is identical to Equation 6.2-15 and $\mathbf{B}^{(n)}$ is computed from Equation 6.2-16 as in the Gauss-Seidel method. When $\lambda \to \infty$, $\lambda\mathbf{I} \gg \mathbf{A}$ in some sense, and $\mathbf{B}$ is computed essentially as

$$\mathbf{B} = \frac{1}{\lambda}\mathbf{Z}$$

In the method of steepest descent, the components of the *unit* vector in the optimal direction can be multiplied by the step size $h^{(n)}$ to give

$$\mathbf{B}^{(n)} = \frac{h^{(n)}(-\nabla\phi(\mathbf{b}^{(n)}))}{\|-\nabla\phi(\mathbf{b}^{(n)})\|} = \begin{bmatrix} -\dfrac{\partial\phi(\mathbf{b}^{(n)})}{\partial\beta_1} \\ \vdots \\ -\dfrac{\partial\phi(\mathbf{b}^{(n)})}{\partial\beta_m} \end{bmatrix} \frac{h^{(n)}}{\|-\nabla\phi(\mathbf{b}^{(n)})\|}$$

Because

$$\frac{\partial\phi(\mathbf{b}^{(n)})}{\partial\beta_j} = -2\sum_{i=1}^{n} w_i[Y_i - \eta_i(\mathbf{b}^{(n)})]\frac{\partial\eta_i(\mathbf{b}^{(n)})}{\partial\beta_j} = -Z_j^{(n)}$$

is the negative of the typical element in the matrix $\mathbf{Z}$, we find that

$$\begin{bmatrix} -\dfrac{\partial\phi(\mathbf{b}^{(n)})}{\partial\beta_1} \\ \vdots \\ -\dfrac{\partial\phi(\mathbf{b}^{(n)})}{\partial\beta_m} \end{bmatrix} \equiv \mathbf{Z}^{(n)}$$

† D. W. Marquardt, *J. Soc. Ind. Appld. Math.* **11**, 431, 1963.

Consequently, for the case in which $\lambda \to \infty$, the identification is made

$$\lambda \Leftrightarrow \|-\nabla\phi(\mathbf{b}^{(n)})\|/h^{(n)}$$

Thus, we find that Equation 6.2-19 encompasses both the method of steepest ascent and the Gauss-Seidel method as limiting cases. Intermediate values of λ represent a composite of the two directions of search. In general, λ decreases as the computations proceed.

We seek a small value of λ where conditions are such that the unmodified Gauss-Seidel method (which has quadratic convergence) would converge satisfactorily. Large values of λ should be used *only* where necessary to satisfy the condition that ϕ on the $(r + 1)$st cycle should be less than ϕ on the rth cycle:

$$\phi^{(r+1)} < \phi^{(r)}$$

Specifically, λ can be chosen as follows. Let $\nu > 1$ and let $\lambda^{(r-1)}$ denote the value of λ from the previous iteration (the initial $\lambda^{(0)} \approx 10^{-2}$). Compute $\phi(\lambda^{(r-1)})$ and $\phi(\lambda^{(r-1)}/\nu)$. Three conditions exist which govern the choice of $\lambda^{(r)}$:

1. If $\phi(\lambda^{(r-1)}/\nu) \leq \phi^{(r)}$, then let $\lambda^{(r)} = \lambda^{(r-1)}/\nu$.
2. If $\phi(\lambda^{(r-1)}/\nu) > \phi^{(r)}$ and $\phi(\lambda^{(r-1)}) \leq \phi^{(r)}$, then let $\lambda^{(r)} = \lambda^{(r-1)}$.
3. If $\phi(\lambda^{(r-1)}/\nu) > \phi^{(r)}$ and $\phi(\lambda^{(r-1)}) > \phi^{(r)}$, increase λ by successive multiplication by ν until, for some small w, $\phi(\lambda^{(r-1)}\cdot\nu^w) \leq \phi^{(r)}$. Then let $\lambda^{(r)} = \lambda^{(r-1)}\cdot\nu^w$.

Case No. 3 is met only rarely, such as when large correlations between parameter estimates exist that cause unreasonably large values of λ. In this case, Case No. 3, certain special additional refinements exist in Marquardt's method. We shall not describe these refinements but they can be found in the original reference.

Marquardt recommended that the elements of $\mathbf{A}$ and $\mathbf{Z}$ be scaled as follows to make the objective function less elongated:

$$A_{ij}^* = \xi_i\xi_j A_{ij}$$
$$Z_j^* = \xi_j Z_j$$

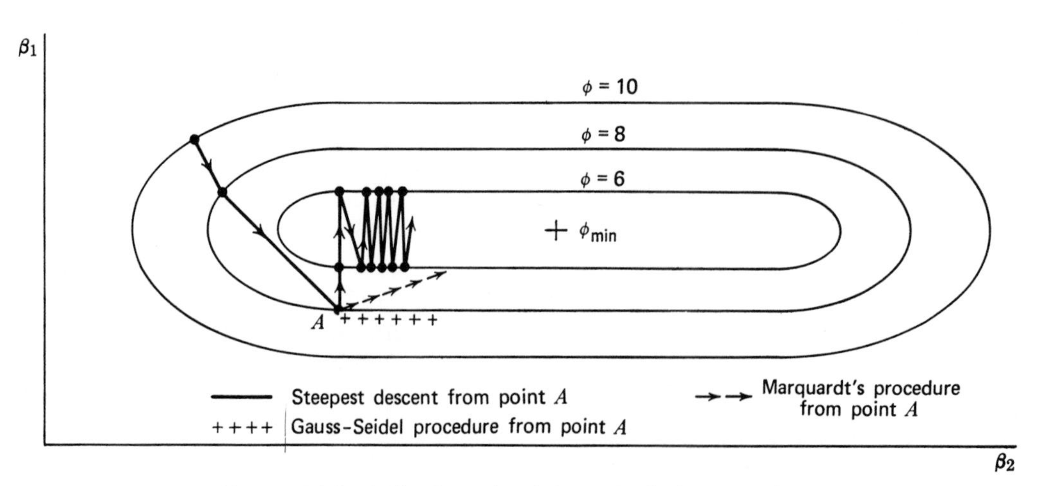

FIGURE 6.2-8 A disadvantage of the method of steepest descent.

where the (*) designates the scaled element and the scale factor is $\xi_i = (A_{ii})^{-1/2}$. The scaled elements of $\mathbf{B}^*$ are converted back to the elements of $\mathbf{B}$ by

$$\Delta b_j = \frac{\Delta b_j^*}{\xi_j}$$

However, Ball† showed that the recommended scaling is exactly equivalent to replacing the matrix $\mathbf{I}$ in Equation 6.2-19 by the diagonal matrix $\mathbf{D}$

$$(\mathbf{A} + \lambda\mathbf{D})\mathbf{B} = \mathbf{Z} \qquad (6.2\text{-}20)$$

where the elements of $\mathbf{D}$ are made up of the elements from the main diagonal of $\mathbf{A} = \mathbf{X}^T\mathbf{X}$. Equations 6.2-19 and 6.2-20 were equally effective in estimation in a number of tests on kinetic rate models.

Marquardt's method has been implemented by the IBM Share Library Program No. 3094, dated March 1964, and is recommended as being quite effective. It is definitely superior to either the Gauss-Seidel method or the method of steepest descent. Because either analytical or numerical derivatives at the minimum of ϕ are available, it is superior to the flexible simplex method in that subsequent estimates of the precision of the parameters are easy to make. On the other hand, the flexible simplex method has the advantage that the partial derivatives of ϕ need not be calculated at all, thus saving considerable computer time in estimation. For very complex models, the flexible simplex method has proved the more effective in estimating the parameters in simulation studies.

Example 6.2-6 Nonlinear Estimation by Marquardt's Method

The same model as was used in Example 6.2-3 was fit by Marquardt's method as executed by the IBM Share Library Program No. 3094. Some initial difficulty was encountered in minimization starting with various initial parameter vectors, because the routine tended to make some of the parameters unbounded or zero. After parameter limits were added to the computer routine, the same minimum was obtained as in the direct search technique but a higher minimum was obtained than with the flexible simplex method (refer to Example 6.2-3). Table E6.2-6 is a summary of the progress of the search; λ is the parameter in Equation 6.2-19; γ is the angle, in degrees, between the direction of search indicated by the linearization of the model and that indicated by the gradient method.

Numerical partial derivatives generated by the computer routine were employed in the estimation which caused the computer time to be about twice that for the simplex or direct search methods.

The matrix of correlation coefficients between the elements of the $(\mathbf{X}^T\mathbf{X})^{-1}$ matrix (which roughly shows the correlation among the estimated parameters—see Section 6.4) was

$$\begin{bmatrix} 1.000 & 0.047 & 0.853 & -0.711 \\ 0.047 & 1.000 & 0.008 & 0.285 \\ 0.853 & 0.008 & 1.000 & -0.851 \\ -0.711 & 0.285 & -0.851 & 1.000 \end{bmatrix}$$

Based on a value of $t_{1-\frac{\alpha}{2}}$ of 2.00, the individual confidence limits on the parameters (see Section 6.4) were

	Lower	Upper
m	2.65	3.61
n	16.1	31.7
f	0.29	0.49
α	0.64	0.78

A joint confidence region could be estimated as described in Section 6.4.

---

† W. E. Ball, *Ind. Eng. Chem. Fundamentals* **6**, 475, 1967.

TABLE E6.2-6

Cycle Number	ϕ	m	n	f	α	λ	γ
0	1.284	6.00	25.0	0.500	1.000	10^{-2}	27.3
1	0.831	1.00	19.2	0.432	0.760	10^{-3}	29.9
5	0.0284	3.22	25.1	0.371	0.722	10^{-7}	52.3
10	0.0267	3.13	23.9	0.398	0.717	10^{-8}	50.5
15	0.0267	3.13	23.9	0.398	0.717	10^{-8}	50.9

6.3 RESOLUTION OF CERTAIN PRACTICAL DIFFICULTIES IN NONLINEAR ESTIMATION

Any of the procedures to minimize the sum of the squares of the deviations described in Section 6.2 can fail to reach a global minimum because: (1) of improper initial guesses for the parameters, and/or (2) the objective function becomes unbounded, as described in Section 6.2.

Additional difficulties discussed in this section, which may be encountered together or separately, are *improper scaling*, *parameter interaction*, and *null effect*.

IMPROPER SCALING. Scaling difficulties can occur when the value of one of the terms in the objective function is of a much different order of magnitude than another relative to the significant figures in each term. Then the objective function is insensitive to changes in the values of the parameters in the small term. For example, the value of an objective function

$$\phi = 100\beta_1^2 - 0.010\beta_2^2$$

would be unaffected by changes in β_2 unless the values of β_2, because of its physical units, are much greater than β_1. If the values of β_2 are of the same magnitude as β_1, one or both variables should be multiplied by scaling factors which convert the two terms on the right-hand side of the equation to roughly equal magnitude. Let

$$\tilde{\beta}_1 = 10\beta_1 \qquad \beta_1^2 = 10^{-2}\tilde{\beta}_1^2$$

$$\tilde{\beta}_2 = 10^{-1}\beta_2 \qquad \beta_2^2 = 10^2\tilde{\beta}_2^2$$

Then the terms in the objective function become the same order of magnitude. After the minimum is found for

$$\phi = \tilde{\beta}_1^2 - \tilde{\beta}_2^2$$

the values of the estimates of the β's, namely b_1 and b_2, could be determined from the estimates $\tilde{b}_1$ and $\tilde{b}_2$.

It is clear from this example that spending some time in proper scaling before attempting a minimization is a sound practice. Poor scaling can lead to poor estimates of the model parameters. However, scaling for nonlinear models cannot usually be effected in advance for all ranges of the independent variables.

PARAMETER INTERACTION. This term is used to describe the adverse mutual influence of the estimate of one parameter on that of another. If one parameter is incorrectly estimated, the other is also incorrectly estimated (biased), but the combined effect of the estimated parameters when introduced into the model may yield quite reasonable predictions. Parameter interaction may be illustrated by examining an extremely simple objective function in which two parameters are multiplied by each other:

$$\phi = 2\beta_1\beta_2 + 10$$

The individual estimates of β_1 and β_2 can range over any series of values for a given estimate of the product $\beta_1\beta_2$. Thus, once a parameter has been assigned a given value, the other parameter will compensate to make the product satisfactory, even though both estimates are badly biased. Scaling is more difficult if interaction exists. Quadratic functions, as explained in Appendix Section B.5, can be transformed to canonical form so that the interaction term is removed. New coordinate axes are defined, as shown in Figure 8.2-2 by the dashed lines, about which the quadratic surface is symmetric. For example, the surface

$$\phi = 7\beta_1^2 + 6\beta_2^2 + 5\beta_3^2 - 4\beta_1\beta_2 - 4\beta_2\beta_3 - 6\beta_1 - 24\beta_2 + 18\beta_3 + 18$$

can be transformed to

$$\phi - 18 = 3\tilde{\beta}_1^2 + 6\tilde{\beta}_2^2 + 9\tilde{\beta}_3^2$$

by a translation of origin and rotation of axes (refer to Example 8.2-1). In the new coordinate system, the scaling of each term is decidedly clearer than in the original coordinate system. Nonlinear objective functions (in the parameters) become quadratic functions only if the model is linearized by some suitable transformation or by expansion in a truncated Taylor series, and ϕ is defined by Equation 6.2-1.

A more subtle example, but one just as vulnerable to interaction among the parameters, involves a model such as $\eta = \beta_1 e^{\beta_2 x}$ in which β_1 in effect multiplies β_2, as becomes clear if we expand the exponential $e^{\beta_2 x} \cong 1 + \beta_2 x + (\beta_2 x)^2/2 + \cdots$. The method of steepest ascent is particularly inhibited by parameter interaction and poor scaling.

It is worthwhile examining the elements of the **A** matrix in order to obtain information about the interaction of variables. The smaller the off-diagonal elements are in relation to the main-diagonal elements, the less likely **A** is to be singular and the less interaction will exist between parameters.

Example 6.3-1 Reduction of Interaction of Parameters by Transformation of Variables

A difficult nonlinear expression to fit because of the interaction between k and E is the Arrhenius rate equation, $r = k\,e^{-E/T}$, and similar equations where the preexponential factor k and the energy of activation E are constants to be estimated and r and T are the measured dependent and independent variables, respectively. If we form the elements of the matrix **A**:

$$\frac{\partial r}{\partial k} = e^{-E/T} = \frac{r}{k}$$

$$\frac{\partial r}{\partial E} = -\frac{k}{T}e^{-E/T} = -\frac{r}{T}$$

$$\mathbf{A} = \begin{bmatrix} \frac{1}{k^2}\sum r_i^2 & -\frac{1}{k}\sum \frac{r_i^2}{T_i} \\ -\frac{1}{k}\sum \frac{r_i^2}{T_i} & \sum\left(\frac{r_i^2}{T_i}\right)^2 \end{bmatrix}$$

and calculate the determinant of **A**, we obtain

$$\det(\mathbf{A}) = \frac{1}{k^2}\left[\sum r_i^2 \sum\left(\frac{r_i^2}{T_i}\right)^2 - \left(\sum\frac{r_i^2}{T_i}\right)^2\right] \qquad \text{(a)}$$

Because r_i^2, $(r_i^2/T_i)^2$, and (r_i^2/T_i) are all positive for any range of r_i and T_i, the det (**A**) can be quite small if T_i takes on only a small range of values, and the matrix **A** becomes singular as the values of the two terms in the brackets approach each other. On the other hand, if a transformation of variable is carried out so that

$$T^* = \frac{T - \bar{T}}{T} = 1 - \frac{\bar{T}}{T} \qquad \text{(b)}$$

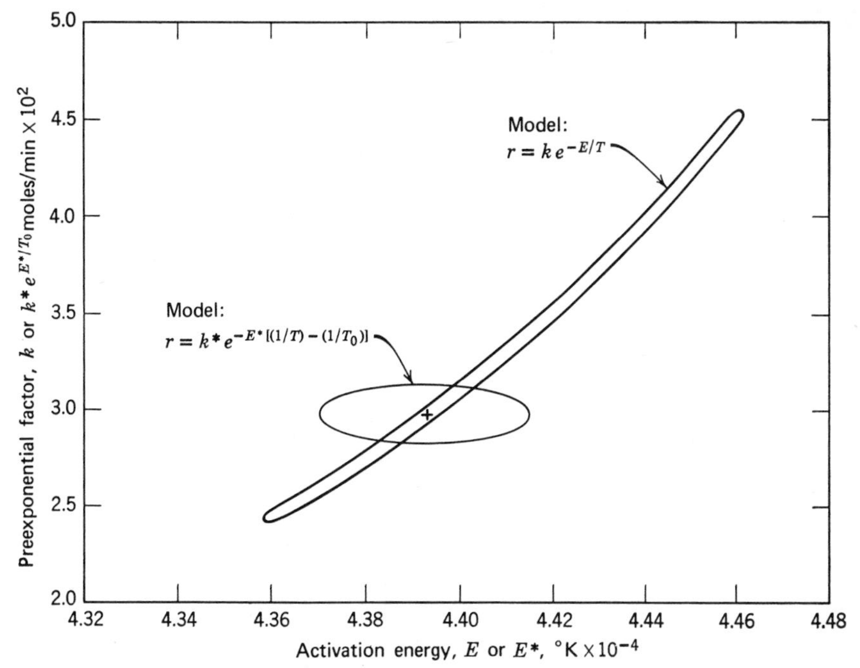

FIGURE E6.3-1 Comparison of the approximate 95-percent confidence regions obtained by fitting two related models.

where $\bar{T}$ is the average value of the absolute temperature, the variable T^* can assume both positive and negative values. Reformation of the Arrhenius rate expression as

$$r = \tilde{k}\, e^{\tilde{E}T^*} = \tilde{k}\, e^{\tilde{E}\left(1-\frac{\bar{T}}{T}\right)}$$
$$= (\tilde{k}\, e^{\tilde{E}})(e^{-\tilde{E}\bar{T}/T}) \tag{c}$$

makes it possible to identify

$$k = \tilde{k}\, e^{\tilde{E}} \quad \text{and} \quad E = \tilde{E}\bar{T}$$

In terms of the transformed temperature, the partial derivatives of r are

$$\frac{\partial r}{\partial \tilde{k}} = e^{\tilde{E}T^*} = \frac{r}{\tilde{k}}$$

$$\frac{\partial r}{\partial \tilde{E}} = \tilde{k}T^*\, e^{\tilde{E}T^*} = rT^*$$

and the equation for det (**A**) corresponding to Equation (a) is

$$\det(\mathbf{A}) = \frac{1}{\tilde{k}^2}\left[\sum r_i^2 \sum (r_i T_i^*)^2 - \left(\sum r_i^2 T_i^*\right)^2\right] \tag{d}$$

Here, the second term in the brackets, with T_i^* taking on both positive and negative values, will be relatively small. Hence, the det (**A**) will not approach zero, and **A** will not be singular.

A related, commonly used transformation is to let

$$r = k^*\, e^{-E^*\left(\frac{1}{T}-\frac{1}{T_0}\right)} \tag{e}$$

where T_0 may be $\bar{T}$ or some other arbitrary temperature. Figure E6.3-1 contrasts the approximate 95 percent confidence region obtained for the model $r = k\, e^{-E/T}$ with that obtained for Model (e).

Example 6.3-2 Scale Factors and Transformations

Fariss and Law† calculated the best fitting coefficients k_1, A_1, k_2, and A_2 for the following nonlinear objective function

$$r = k_1 u_1\, e^{a_1/(t+273)} + k_2 u_2\, e^{a_2/(t+273)} \tag{a}$$

where t is in degrees Centigrade. One hundred experimental data points were simulated by using the following constants:

$$k_1 = 20e^{11.82033} \qquad a_1 = 5{,}000$$
$$k_2 = 2e^{47.28132} \qquad a_2 = 20{,}000$$

and by using random values of u_1 and u_2 in the range 0 to 1 and random values of t in the range 100 to 200. A normally distributed error with a mean of zero and a variance of

$$\sigma_\epsilon^2 = 0.01 + (0.05r)^2 \tag{b}$$

was added to each deterministic data point:

$$R_i = r_i + \epsilon_i \tag{c}$$

The sum of the squares of the deviations, ϕ, given by Equation 6.2-1 with the weights defined as $w_i = \sigma_i^{-1}$, was minimized to estimate k_1, k_2, a_1, and a_2. The initial guesses for the parameters and the final results by a derivative-type estimation technique appear in Table E6.3-2.

The derivatives were calculated analytically. The following scale factors, used in the calculations,

Parameter	k_1	k_2	a_1	a_2
Factor used	k_1	k_2	1000	1000

were divided into k_1, k_2, a_1, and a_2, respectively, at the beginning of each iteration cycle.

The transformation for the absolute temperature given in Example 6.3-1 was also applied to Equation (a); $\bar{T}$ was set

† R. H. Fariss and V. J. Law, Paper presented at the Houston AICE meeting, Feb. 18, 1967.

TABLE E6.3-2

	k_1	k_2	a_1	a_2	ϕ
Initial guesses	$10e^{20}$	$5e^{20}$	8,460	8,460	7,671
Estimated parameters at the minimum ϕ	$20.085e^{11.7358}$	$1.9220e^{47.7613}$	4,964	20,203	90.7
Model parameters in Equation (a)	$20e^{11.82033}$	$2e^{47.28132}$	5,000	20,000	100

equal to 423 ($\bar{t} = 150$). Minimization of the transformed objective function gave the following results:

	$\tilde{k}_1$	$\tilde{k}_2$	$\tilde{a}_1$	$\tilde{a}_2$
Starting guesses	10	5	20	20
Scale factors	10	5	20	20
Estimated parameters at the minimum ϕ	20.085	1.9220	11.7358	47.7613

Several different minimization techniques were used to minimize the sum of the squares of the deviations. The authors' experiences can be summarized as follows. (The functional evaluation count below was based on $\frac{1}{100}$ of the sum of: (1) the number of calculations of $(R_i - r_i)$, plus (2) the number of evaluations of a partial derivative of r with respect to one parameter.)

1. Unweighted steepest descent with untransformed temperature but with scaling and an adjustable acceleration factor h: After 579 functional evaluations, ϕ was reduced to only 3615; after 1019 evaluations, ϕ was still 3523. Conclusion: not an effective procedure.

2. Steepest descent with transformed temperature, an adjustable acceleration factor h, and with scale factors ξ_j based on

$$\ln \xi_j = (\ln q)\left\{1 - \exp\left[-\left(\frac{0.5}{\ln q}\right)\ln\left(\frac{\partial^2\phi/\partial\beta_1^2}{\partial^2\phi/\partial\beta_j^2}\right)\right]\right\}$$

where q = maximum scale factor = 100: After 206 functional evaluations, ϕ was 100; after 471 evaluations, ϕ was 90.95. Conclusion: effective but characteristically (for steepest descent) slow.

3. Gauss-Seidel with untransformed temperature: Obtained a ϕ of 2904 after 1265 functional evaluations, and a ϕ of 90.7 after 1317 functional evaluations. Conclusion: slow except for last 10 evaluations.

4. Gauss-Seidel with transformed temperature: Obtained a ϕ of 90.7 with 28 functional evaluations. Conclusion: effective and quick.

5. Marquardt's method with the initial $\lambda = 1$ (a large-sized selection) and λ adjusted each 10 cycles: For the ultransformed problem, ϕ was 90.7 after 1501 functional evaluations; for the transformed problem, ϕ was 90.7 after 21 functional evaluations. Conclusions: essentially as effective and rapid as Gauss-Seidel.

A second transformation of Equation (a) to

$$r' = u_1\, e^{[k_1' - a_1/(t+c)]} + u_2\, e^{[k_2' - a_2/(t+c)]} \tag{d}$$

by letting

$$k' = \ln k$$

was carried out. The starting values for k' were the logarithms of the original starting values; the scale factors were taken to be 1 for both k_1' and k_2'. The results were essentially the same as with the transformation of Example 6.3-1, although the number of functional evaluations was slightly greater.

Estimation was also carried out by using fine-mesh forward difference schemes to approximate the derivatives numerically. With a parameter increment of 0.001 times the scale factor, little difference was experienced between the two methods. However, larger meshes indicated that additional functional evaluations were required.

NULL EFFECT. This can be illustrated by using as an example the following objective function:

$$\phi = \beta_1^2 + 2\beta_1\beta_2 + \beta_2^2 + 2$$
$$= (\beta_1 + \beta_2)^2 + 2$$

After the transformation $\beta_1 + \beta_2 = \tilde{\beta}_1$ is made, we find

$$\phi = \tilde{\beta}_1^2 + 2$$

Observe that only one variable is left, $\tilde{\beta}_1$. The geometric interpretation of $\phi = (\beta_1 + \beta_2)^2 + 2$ is shown by the slanted lines in Figure 8.2-2e; in the new coordinates the values of ϕ are all parallel to the dashed axis, $\tilde{\beta}_2$ (which corresponds to $\tilde{x}_2$ in the figure). Although both β_1 and β_2 appear to be parameters, in truth there is only one parameter which must be varied to minimize ϕ, namely $\tilde{\beta}_1$ (corresponding to $\tilde{x}_1$ in the figure). The Gauss-Seidel method is particularly vulnerable to the null effect of a parameter because the matrix **A** tends to be singular when such an effect exists. On the other hand, the method of steepest ascent continues to operate in the presence of unrecognized null effects with the penalty of a greater series of zig-zag steps. As to the procedure of Marquardt, the influence of the null effect depends on the value of λ. For small λ the Marquardt method is similar to the Gauss-Seidel method and is vulnerable to the null effect; for large λ the Marquardt method corresponds more closely to steepest descent. Direct search methods for problems in which the null effect exists encounter difficulty mainly in improper scaling and parameter interaction.

6.4 HYPOTHESIS TESTS AND THE CONFIDENCE REGION

In addition to estimating the parameters in a nonlinear model, the analyst would like to obtain some measure of the dispersion of the parameter estimates and also some measure of dispersion of the predicted dependent variable $\hat{Y}$. We shall describe below three methods to accomplish these objectives.

6.4-1 Linearization of the Model in the Region About the Minimum Sum of the Squares

An *approximate* confidence region for the parameters can be constructed by linearizing the nonlinear model (as described in Section 6.2-3) about the least square estimate **b** in parameter space. The variances and co-variances of **b** are then given approximately by the analog of Equation 5.1-14, that is, by

$$\widehat{\text{Covar}}\{\mathbf{b}\} \cong (\mathbf{X}^T\mathbf{w}\mathbf{X})^{-1}\sigma_{\bar{Y}_i}^2 = \mathbf{C}\sigma_{\bar{Y}_i}^2 \qquad (6.4\text{-}1)$$

Each element of the matrix **X** is evaluated at **b**. If the derivatives of η cannot be calculated analytically, they can still be evaluated numerically. To estimate $\sigma_{\bar{Y}_i}^2$, *if the model is correct*, one can compute

$$s_r^2 = \frac{\mathbf{E}^T\mathbf{w}\mathbf{E}}{n - m} \qquad (6.4\text{-}2)$$

at the minimum ϕ and then

$$\widehat{\text{Covar}}\{\mathbf{b}\} \cong (\mathbf{X}^T\mathbf{w}\mathbf{X})^{-1}\frac{\mathbf{E}^T\mathbf{w}\mathbf{E}}{n - m} \qquad (6.4\text{-}3)$$

where n is the number of data sets and m is the number of parameters estimated. Of course, if the model used is incorrect, Equation 6.4-2 will give a biased estimate of $\sigma_{\bar{Y}_i}^2$; hence, as usual it is desirable to obtain s_e^2 from replicate data in order to determine how well the model represents the data.

From Equation 6.4-3 the *approximate* confidence intervals for the individual β's can be calculated as described in Section 5.2 for the linear models; the confidence interval for η can be approximated by using $s_{\hat{Y}_i}^2$ calculated as follows. First, obtain

$$s_{b_j}^2 = \widehat{\text{Var}}\{b_j\} \cong s_{\bar{Y}_i}^2 C_{jj}$$

where C_{jj} is a diagonal element of $\mathbf{C} = (\mathbf{X}^T\mathbf{w}\mathbf{X})^{-1}$. Then use the linearized (about **b**) predicted response $\hat{Y}$ to obtain

$$s_{\hat{Y}_i}^2 \equiv \widehat{\text{Var}}\{\hat{Y}_i\} \cong \sum_{j=1}^{m}\left(\frac{\partial \hat{Y}}{\partial b_j}\right)^2 \widehat{\text{Var}}\{b_j\}$$

$$+ \sum_{i=1}^{m}\sum_{\substack{j=1\\ i\neq j}}^{m}\left(\frac{\partial \hat{Y}}{\partial b_i}\right)\left(\frac{\partial \hat{Y}}{\partial b_j}\right)\widehat{\text{Covar}}\{b_i b_j\}$$

$$\cong s_{\bar{Y}_i}^2 \sum_{i=1}^{m}\sum_{j=1}^{m}\left(\frac{\partial \hat{Y}}{\partial b_i}\right)\left(\frac{\partial \hat{Y}}{\partial b_j}\right)C_{ij} \qquad (6.4\text{-}4)$$

An approximate joint confidence region, ellipsoidal in shape, can also be formed from the quadratic form corresponding to that used in Section 5.2 for the linear models

$$(\boldsymbol{\beta} - \mathbf{b})^T(\mathbf{X}^T\mathbf{w}\mathbf{X})(\boldsymbol{\beta} - \mathbf{b}) = s_{\bar{Y}_i}^2 m F_{1-\alpha}[m, n - m] \qquad (6.4\text{-}5)$$

where $F_{1-\alpha}$ is the upper limit of the F-distribution for m and $(n - m)$ degrees of freedom. The graph of Equation 6.4-5 can be drawn in two or three dimensions, as illustrated in Figure 6.4-1 which compares a true sum of squares surface with the contours determined from Equation 6.4-5 for simulated data from the model

$$\eta = \frac{\beta_1}{\beta_1 - \beta_2}(e^{-\beta_2 x} - e^{-\beta_1 x})$$

Note that β_1 is estimated more precisely than β_2.

The approximate contours for the sum of squares surface can be written as in Section 5.2:

$$\phi_{1-\alpha} = \phi_{\min} + s_{\bar{Y}_i}^2 m F_{1-\alpha}[m, (n - m)]$$

$$= \phi_{\min}\left[1 + \frac{m}{n - m}F_{1-\alpha}(m, n - m)\right] \qquad (6.4\text{-}6)$$

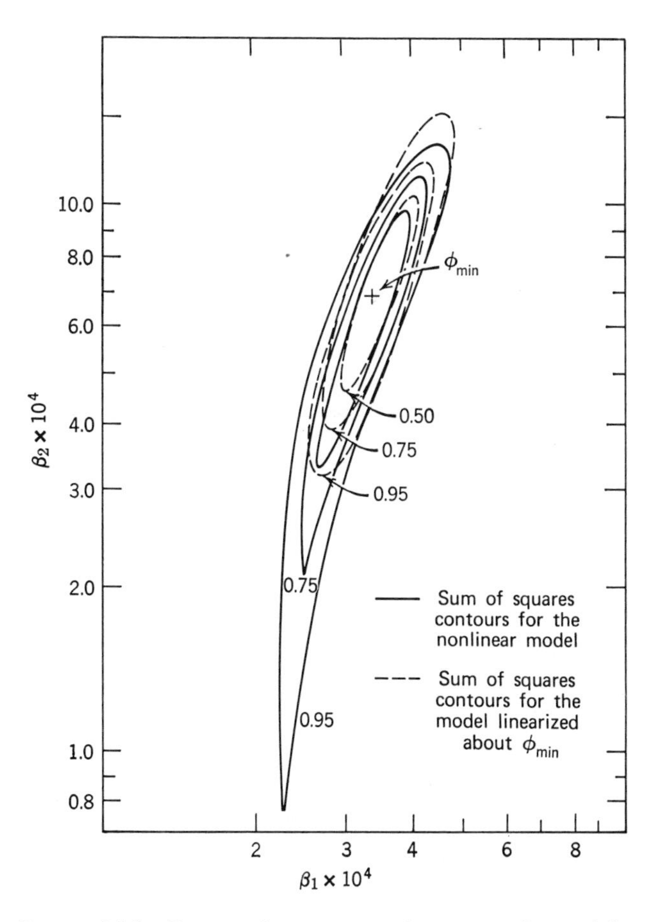

FIGURE 6.4-1 Contours for a true sum of squares surface and the corresponding contours based on Equation 6.4-5. Numbers next to contours indicate probability for indicated confidence region. (From G. E. P. Box and W. G. Hunter, *Technometrics* **4**, 301, 1962.)

where $\phi_{1-\alpha}$ is the approximate value of the sum of squares contour for the confidence level $(1-\alpha)$ and $\phi_{\min} = \sum_{i=1}^{n} w_i(Y_i - \hat{Y}_i)^2$.

Guttman and Meeter† discussed measures of nonlinearity that indicate when the degree of nonlinearity is small enough to justify using linear theory as an approximating theory for nonlinear models. Beale‡ also provided some additional insight into estimating confidence limits for nonlinear models.

Example 6.4-1 Approximate Joint Confidence Region for a Nonlinear Model

To illustrate the estimated individual confidence intervals for estimated parameters and the joint estimate of the confidence region, a two-parameter model used in a chemical kinetic study will be described. Data were collected for a hydrogenation reaction in a tubular flow reactor represented by the empirical model (at constant temperature)

$$r = \frac{\beta_0 p}{1 + \beta_1 p}$$

where r = initial reaction rate and p = total pressure. It was assumed that $R = r + \epsilon$. The data at 164°C were:

R, g-moles/(hr)(g catalyst)	p, psia
0.0680	20
0.0858	30
0.0939	35
0.0999	40
0.1130	50
0.1162	55
0.1190	60

A portion of the results calculated by a modified version of Marquardt's method (IBM Share Program SD No. 3094) employing analytical partial derivations were

p	$(R - \hat{R}) \times 10^3$
20	0.433
30	−0.656
35	−0.0619
40	−0.605
50	1.636
55	0.282
60	−1.006

$$\mathbf{X}^T\mathbf{X} = \begin{bmatrix} 0.0191 & 0.178 \\ 0.178 & 1.703 \end{bmatrix}$$

$$\mathbf{c} = (\mathbf{X}^T\mathbf{X})^{-1} = \begin{bmatrix} 2020 & -211.1 \\ -211.1 & 26.65 \end{bmatrix}$$

$$\mathbf{b} = \begin{bmatrix} 5.154 \times 10^{-3} \\ 2.628 \times 10^{-2} \end{bmatrix}$$

† I. Guttman and D. A. Meeter, *Technometrics* **7**, 623, 1965.
‡ E. M. L. Beale, *J. Royal Stat. Soc.* **B22**, 41, 1960.

The sum of the squares of the residuals was $\phi_{\min} = 4.76 \times 10^{-6}$, λ ranged during the search from 10^{-2} to 10^{-6}, and the estimated parameter correlation matrix was

$$\begin{bmatrix} 1.000 & 0.9902 \\ 0.9902 & 1.000 \end{bmatrix}$$

Note that the parameters are highly correlated.

For $\alpha = 0.05$ and $\nu = 7 - 2 = 5$ degrees of freedom, $t_{1-\frac{\alpha}{2}} = t_{0.975} = 2.571$; hence the individual parameter confidence intervals, calculated as in Section 5.2, for the linearized model are

$$b_0 - t_{1-\frac{\alpha}{2}} s_{\bar{Y}_i} \sqrt{C_{11}} \le \beta_0 < b_0 + t_{1-\frac{\alpha}{2}} s_{\bar{Y}_i} \sqrt{C_{11}}$$

$$b - t_{1-\frac{\alpha}{2}} s_{\bar{Y}_i} \sqrt{C_{22}} \le \beta_1 < b_1 + t_{1-\frac{\alpha}{2}} s_{\bar{Y}_i} \sqrt{C_{22}}$$

As an estimate of $s_{\bar{Y}_i}$,

$$s_r = \sqrt{s_r^2} = \sqrt{\frac{4.7604 \times 10^{-6}}{7-2}}$$

$$= \sqrt{0.952 \times 10^{-6}} = 0.975 \times 10^{-3}$$

is used since no replicate data are available. If the model is a poor one, s_r is a poor estimate of $s_{\bar{Y}_i}$. The respective confidence intervals are then

$$-0.107 \le \beta_0 < 0.117$$

$$0.0119 \le \beta_1 < 0.0263$$

An *approximate* joint confidence interval is defined by Equation 6.4-5:

$$[(\beta_0 - 5.154 \times 10^{-3})(\beta_1 - 2.628 \times 10^{-2})] \begin{bmatrix} 0.0191 & 0.178 \\ 0.178 & 1.703 \end{bmatrix}$$

$$\begin{bmatrix} (\beta_0 - 5.154 \times 10^{-3}) \\ (\beta_1 - 2.628 \times 10^{-2}) \end{bmatrix} = (0.952 \times 10^{-6})(2)(5.79)$$

or

$$\beta_0^2 + 18.64\beta_1\beta_0 + 89.13\beta_1^2 - 0.499\beta_0 - 4.875\beta_1 + 0.00698 = 0 \quad \text{(a)}$$

and is illustrated in Figure E6.4-1. Note the typical long attenuated region which is characteristic of cases in which the parameter estimates are correlated. The sum of squares contour written in the form of Equation 6.4-6 is

$$\phi_{1-\alpha} = 4.76 \times 10^{-6}\left[1 + \frac{2}{7-2} F_{1-\alpha}(2, 5)\right] \quad \text{(b)}$$

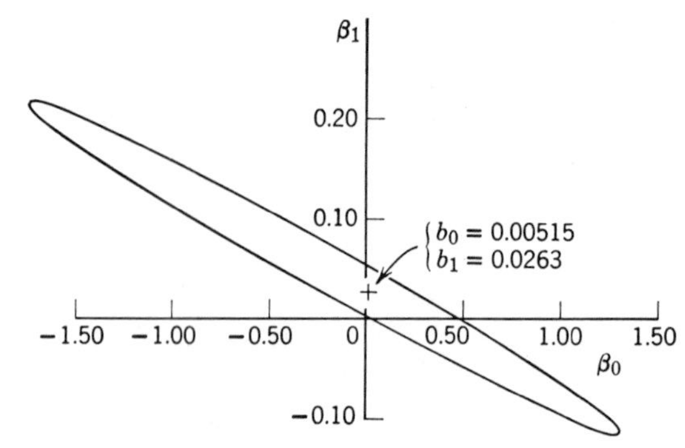

FIGURE E6.4-1 Approximate 95-percent confidence interval contour.

6.4-2 The Williams Method

Williams† reviewed some of the alternative methods of establishing parameter confidence intervals and also suggested a method for models containing one nonlinear coefficient. To clarify the technique, the model

$$\eta = \beta_0 + \beta_1 e^{-\kappa x} \tag{6.4-7}$$

will be employed for which the estimated regression equation is

$$\hat{Y} = b_0 + b_1 e^{-kx}$$

where k is the estimate of κ. Equation 6.4-7 in general can be represented as

$$\eta = \beta_0 + \beta_1 f(\kappa, x) \tag{6.4-8}$$

in which $f(\kappa, x)$ contains the nonlinearity. The usual assumption (that Y is distributed normally with constant variance σ^2) is presumed to hold, and the criterion used in estimation is the minimization of the sum of the squares of the deviations, ϕ, in Equation 6.2-1.

First we linearize Equation 6.4-7 by means of a truncated Taylor series expansion about an assumed $k^{(0)}$

$$\eta \cong \beta_0 + \beta_1\left[f(x, k^{(0)}) + \frac{df(x, k^{(0)})}{d\kappa}(\kappa - k^{(0)})\right]$$

$$= \beta_0 + \beta_1 f(x, k^{(0)}) + b_2 \frac{df(x, k^{(0)})}{d\kappa} \tag{6.4-9}$$

where $b_2 = \beta_1(\kappa - k^{(0)})$. The parameters β_0 and β_1, as well as b_2 and their variances, can be estimated by linear analysis. After the first iteration, the second approximation for k starts with an assumed

$$k^{(1)} = k^{(0)} + \frac{b_2}{b_1} \tag{6.4-10}$$

and the iteration continues. At such time as

$$b_2^{(n+1)} = b_1^{(n+1)}(k^{(n+1)} - k^{(n)}) \to 0$$

or

$$\frac{b_2^{(n+1)}}{b_1^{(n+1)}} \to 0$$

the iteration can be terminated. Thus the linear coefficient b_2 vanishes at the minimum least squares estimate.

To test the null hypothesis $\kappa = k^{(n)}$, where $k^{(n)}$ is any in the sequence of values of k, we check to see if b_2 is significantly different from zero. If it is, the null hypothesis is rejected. Williams established the confidence limits for κ as the values of k for which b_2 was not significantly different from zero at an assumed probability level. In the iterative process a number of values of $k^{(n)}$ are established, but it may be necessary to start with different values of $k^{(0)}$ in order to encompass the necessary range of values of b_2.

† E. J. Williams, *J. Royal Stat. Soc.* **B24**, 125, 1962.

From linear analysis we know that the sum of the squares of the deviations for any variable adjusted for all the others is

$$\Delta SS_{b_i} = \frac{b_i^2}{c_{ii}}$$

where b_i is the linear regression coefficient and c_{ii} is an element on the main diagonal of the inverse matrix **c**. By selecting an initial set of values of the coefficient $k^{(0)}$ and/or using values of k developed during the iteration process, it is possible to plot ΔSS_{b_2} versus k and use the significance level associated with ΔSS_{b_2} to ascertain the confidence limits for κ as shown in the example below. Halperin‡ extended Williams's procedure to a broader class of regression functions with more than one nonlinear parameter. He also pointed out that the Williams method will not yield exact confidence regions for the linear parameters independent of the nonlinear ones.

Example 6.4-2 Confidence Limits for a Parameter in a Nonlinear Model

Williams fit the following data (x = independent variable and Y = dependent variable, with 4 degrees of freedom for each entry):

$\bar{Y}_i$	x_i
51.6	0.4
53.4	1.4
20.0	5.4
−4.2	19.5
−3.0	48.2
−4.8	95.9

to the model $\eta = \beta_0 + \beta_1 e^{-\kappa x}$. An initial estimate for κ was $k^{(0)} = 0.165$. The regression results were

First Iteration	Second Iteration
$b_1^{(1)} = 65.276$	$b_1^{(2)} = 65.262$
$b_2^{(1)} = 0.0518$	$b_2^{(2)} = -0.0269$
$(k^{(1)} - k^{(0)}) = 0.0008$	$(k^{(2)} - k^{(1)}) = -0.0004$
$k^{(1)} = 0.166$	$k^{(2)} = 0.166$
	$b_0^{(2)} = -4.85$

at which stage the analysis was terminated.

Additional values of b_2 were determined for $0.05 \leq k^{(0)} \leq 0.40$, and ΔSS for b_2 was plotted versus the values of k. See Figure E6.4-2. The sum of the squares, $\sum_{j=1}^{4}(Y_{ij} - \bar{Y}_i)^2$, for the six data sets was 1108.80. Consequently, the error variance was

$$s_e^2 = \frac{1108.80}{24} = 46.20$$

The sum of the squares of the residuals, ϕ, for $k = 0.166$, divided by the number of degrees of freedom, 3, gave $s_r^2 = 24.66$.

Suppose an F-test were to be carried out at, for example, a significance level of 0.01 for which, from Table C.4 in

‡ M. Halperin, *J. Royal Stat. Soc.* **B24**, 330, 1963.

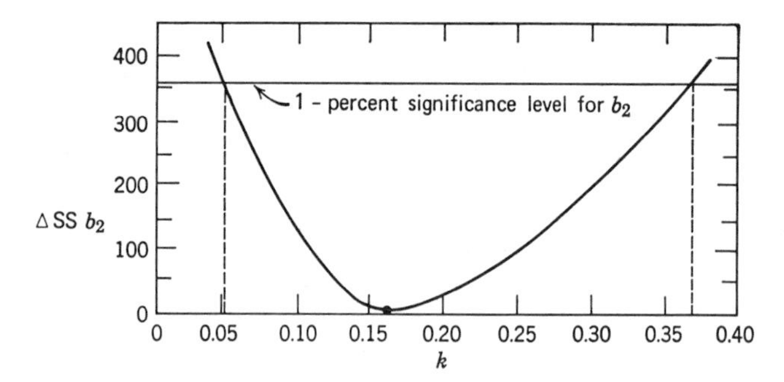

FIGURE E6.4-2 (From E. J. Williams, *J. Royal Stat. Soc.* **B24**, 125, 1962, with permission.)

Appendix C, we find $F_{0.99}[1, 24] = 7.28$. If we equate $(s_{b_2}^2/s_e^2)$ to 7.28, we can compute $s_{b_2}^2 = (7.28)(46.20) = 361.4$ at the 1 percent level. Because the degrees of freedom associated with b_2 are just 1, $s_{b_2}^2 = (\Delta SS_{b_2}/1)$ and

$$\Delta SS_{b_2} = s_{b_2}^2 = 361.4$$

At the intersections of $\Delta SS_{b_2} = 361.4$ with the curve for ΔSS_{b_2} versus k, we can read $k = 0.057$ and $k = 0.372$. These values are the 99 percent confidence limits for κ. They are exact in the sense that any values outside these limits are rejected at the 1-percent level of significance by an exact test on the associated value of b_2.

6.4-3 The Method of Hartley and Booker

Hartley and Booker† suggested an alternate method to the least squares technique which can provide both the parameters and their confidence limits. It yields estimates $\tilde{\mathbf{b}}$ which are asymptotically (as $n \to \infty$) 100-percent efficient under fairly general assumptions (the details of which can be found in their article). Consider the case of n sets of observations with m parameters. Let $n = 6$ and $m = 3$ and the model be $\eta = \beta_0 + \beta_1 e^{-\beta_2 x}$. We can describe the data by a set of six nonlinear equations:

$$\left.\begin{aligned} Y_{11} &= \tilde{\beta}_0 + \tilde{\beta}_1 e^{\tilde{\beta}_2 x_{11}} \\ Y_{12} &= \tilde{\beta}_0 + \tilde{\beta}_1 e^{\tilde{\beta}_2 x_{12}} \end{aligned}\right\} \text{Group 1; } k = 2$$

$$\left.\begin{aligned} Y_{21} &= \tilde{\beta}_0 + \tilde{\beta}_1 e^{\tilde{\beta}_2 x_{21}} \\ Y_{22} &= \tilde{\beta}_0 + \tilde{\beta}_1 e^{\tilde{\beta}_2 x_{22}} \end{aligned}\right\} \text{Group 2; } k = 2$$

$$\left.\begin{aligned} Y_{31} &= \tilde{\beta}_0 + \tilde{\beta}_1 e^{\tilde{\beta}_2 x_{31}} \\ Y_{32} &= \tilde{\beta}_0 + \tilde{\beta}_1 e^{\tilde{\beta}_2 x_{32}} \end{aligned}\right\} \text{Group 3; } k = 2$$

Although this is an overdetermined set of equations and thus calls for statistical treatment, we could average the equations into h sets of two equations each ($k = 2$) so that $h = m$ (= 3 here). We would then obtain a completely determined system of nonlinear equations in which $\tilde{\beta}_j$ is a consistent estimator of β_j and

$$\bar{Y}_h = \bar{f}(h, \tilde{\boldsymbol{\beta}}) \tag{6.4-11}$$

† H. O. Hartley and A. Booker, *Ann. Math. Stat.* **36**, 638, 1965.

where

$$\bar{Y}_h = \frac{1}{k}\sum_{i=1}^{k} Y_{hi}$$

$$\bar{f}(h, \tilde{\boldsymbol{\beta}}) = \frac{1}{k}\sum_{i=1}^{k} f(x_{hi}, \tilde{\boldsymbol{\beta}})$$

For example, for group 1:

$$\bar{Y}_1 = \tfrac{1}{2}\sum_{i=1}^{2} Y_{1i} = \tfrac{1}{2}(Y_{11} + Y_{12})$$

$$\begin{aligned} \bar{f}(1, \tilde{\boldsymbol{\beta}}) &= \tfrac{1}{2}\sum_{i=1}^{2} f(x_{1i}, \tilde{\boldsymbol{\beta}}) \\ &= \tfrac{1}{2}[\tilde{\beta}_0 + \tilde{\beta}_1 e^{\tilde{\beta}_2 x_{11}} + \tilde{\beta}_0 + \tilde{\beta}_1 e^{\tilde{\beta}_2 x_{12}}] \\ &= \tilde{\beta}_0 + \frac{\tilde{\beta}_1}{2}(e^{\tilde{\beta}_2 x_{11}} + e^{\tilde{\beta}_2 x_{12}}) \end{aligned}$$

The solution of the nonlinear Equation 6.4-11 for $\tilde{\boldsymbol{\beta}}$ presumably can be carried out by the Newton-Raphson method, by one of the search methods, or by one of the optimization methods described in the references at the end of this chapter.

After the values of the elements of $\tilde{\boldsymbol{\beta}}$ have been established by solving Equation 6.4-11, the $\tilde{\boldsymbol{\beta}}$ are used as the starting values to carry out a one-step iteration by the Gauss-Seidel method. At the termination of the first iteration, Hartley and Booker showed that one obtains asymptotically 100-percent efficient estimates of $\boldsymbol{\beta}$, $\tilde{\boldsymbol{\beta}}^*$. When the regression equations are linear, $\tilde{\boldsymbol{\beta}}^*$ agrees with the standard least squares estimators, $\mathbf{b}$.

If the experimental values of x are repeated for each of the k trials in the hth group of data, Hartley and Booker described how to obtain the confidence interval for each of the m functions $\bar{f}(h, \boldsymbol{\beta})$ based on values of $\bar{Y}_h$.

6.5 TRANSFORMATIONS TO LINEAR FORM

Certain classes of nonlinear models can be easily transformed to linear form, and the linear model can be treated by linear analysis. For example, taking logarithms of both sides of

$$\eta = \beta_0 x_1^{\beta_1} x_2^{\beta_2} \tag{6.5-1}$$

yields

$$\log \eta = \log \beta_0 + \beta_1 \log x_1 + \beta_2 \log x_2 \tag{6.5-2}$$

a model linear in the coefficients. However, note that minimization of $\sum_{i=1}^{n} (\log Y_i - \log \eta_i)^2$ is not the same as minimization of $\sum_{i=1}^{n} (Y_i - \eta_i)^2$.

If a linearizing transform $\mathscr{T}$ exists that will transform the nonlinear model into linear form, in order that the least squares estimates $b_1, \ldots, b_m$ of the related model parameters $\beta_1, \ldots, \beta_m$ possess optimal properties (i.e., unbiasedness, minimum variance, etc.) when estimated in the transformed regression equation, it is necessary that the assumptions concerning the additive unobservable random error be applicable to the transformed model rather than to the original model. Thus, for the transformed model and for an observed Y_i corresponding to a set of independent variables $\mathbf{x}_i$, we assume that

$$\mathscr{T}(Y_i) = \mathscr{T}[\eta_i(x_1, \ldots, x_q; \beta_1, \ldots, \beta_m)] + \epsilon_i \quad (6.5\text{-}3)$$

where the random variable ϵ_i is independently distributed with zero mean and constant variance. For example, for Equation 6.5-1,

$$\log Y_i = \log \beta_0 + \beta_1 \log x_{1i} + \beta_2 \log x_{2i} + \epsilon_i \qquad i = 1, 2, \ldots, n \quad (6.5\text{-}4)$$

The effect of the additive error in Equation 6.5-3 can be related back to the untransformed model only by examining each model as a separate case. The usual assumptions of linear analysis described in Section 4.2 lead to the best linear unbiased estimates of the parameters $\beta_1, \ldots, \beta_m$ in Equation 6.5-4. However, the estimation procedure produces the best linear unbiased estimate of $\log \beta_0$, not of β_0 itself. In terms of the nonlinear model of Equation 6.5-1, the additive error in Equation 6.5-4 corresponds to a multiplicative type of error in the untransformed model

$$Y_i = \beta_0 x_{1i}^{\beta_1} x_{2i}^{\beta_2} \cdots x_{qi}^{\beta_q} \phi_i \qquad i = 1, 2, \ldots, n \quad (6.5\text{-}5)$$

where $\phi_i = \log^{-1} \epsilon_i$ is a positive error. The usual tests of hypotheses and confidence intervals require that ϕ_i be lognormally distributed. We can conclude that the logarithmic transform and subsequent least squares analysis are justified if the error ϕ is proportional to Y rather than being a fixed value independent of the value of Y. For example, rulers have a fixed error whatever the value of the measured distance, whereas many observed process variables come from measuring devices in which the error indeed is proportional to the value of the variable.

As another example, suppose the model is

$$x = \beta_1 e^{\beta_2 Y}$$

and that Y is the dependent random variable available for fixed values of x. Since $\log x = \log \beta_1 + \beta_2 Y$,

$$Y_i = \frac{\log x_i}{\beta_2} - \frac{\log \beta_1}{\beta_2} + \epsilon_i$$

and linear regression yields the best estimates of the new parameters:

$$\tilde{\beta}_1 = \log\left(\frac{1}{\beta_1}\right)^{1/\beta_2}$$

$$\tilde{\beta}_2 \cong \frac{1}{\beta_2}$$

The calculation of the approximate confidence limits of the old parameters can be from those of the new parameters.

As a final example, replacement of the nonlinear Hougen-Watson-Langmuir-Hinshelwood type of rate equation, such as the "dual site" model

$$R_0 = \frac{kKc}{(1 + Kc)^2} + \epsilon$$

where

R_0 = rate of reaction, dependent random variable
c = concentration, independent deterministic variable
k, K = constants

by the linear model

$$\left(\frac{c}{R_0}\right)^{1/2} = \frac{1}{\sqrt{kK}} + \frac{K}{\sqrt{kK}}c + \epsilon'$$

in order to use a linear (in the parameters) estimation routine ignores the fact that the additive error ϵ' in the linear model is not the same as ϵ in the original model. The relation evolved from the linear model would be

$$R_0 = \frac{kKc}{(1 + Kc)^2} + \frac{kK}{(1 + Kc)^2}(R_0 \epsilon'^2 - 2\sqrt{R_0 c}\,\epsilon')$$

Additional information pertaining to transformation of the independent variables can be found in Box and Tidwell† and of the dependent variable in Box and Cox.‡

6.6 ESTIMATION WITH THE PARAMETERS AND/OR VARIABLES SUBJECT TO CONSTRAINTS

The idea of imposing constraints on the parameters and/or the variables in a process model comes about quite naturally. For example, in certain essentially empirical models for chemical kinetics such as

$$r = \frac{kK_A p_A p_B}{1 + K_A p_A + K_B p_B}$$

where r is a rate of reaction, k and K are constants, and p is the pressure, arguments on physical grounds lead to the conclusion that k, K_A, and K_B must be nonnegative. Consequently, fitting the model without restricting the region of search for the parameters estimates to $k \geq 0$, $K_A \geq 0$, and $K_B \geq 0$ will lead to unreasonable, often negative, estimates. An example of constraints on the independent variables occurs when the independent variables x_i represent mass fractions, in which case $\sum x_i \equiv 1$.

In general, constraints can be classified into two types:

1. Equality constraints.
2. Inequality constraints:

The strategy of optimization (minimization for least squares) of an objective function subject to inequality

† G. E. P. Box and P. W. Tidwell, *Technometrics* **4**, 531, 1962.
‡ G. E. P. Box and D. R. Cox, Dept. of Stats. Tech. Rept. No. 26, Univ. of Wis., Madison, Mar. 1964.

and/or equality constraints has come to be known as mathematical programming. *Quadratic programming* refers to optimization of a quadratic objective function subject to linear (in the parameters) inequality constraints. *Nonlinear programming* is the generic term applied to optimization of a nonlinear objective function subject to both nonlinear equality and inequality constraints.

Certain commonly encountered simple constraints can be accommodated without the use of nonlinear programming. For example, the "trivial" constraints that

$$\beta_j > k_j$$

where k_j is a positive constant can be handled by letting

$$\beta_j = k_j + e^{\beta_j^*}$$

where β_j^* is the parameter to be estimated.

Although β_j^* may range from $-\infty$ to $+\infty$, β_j will always be greater than k_j.

The more general constraints:

$$g_i(\mathbf{Y}, \mathbf{x}, \boldsymbol{\beta}) > 0 \qquad i = 1, 2, \ldots, q$$

can be arranged as penalty functions added to ϕ and the following sum minimized

$$\text{Minimize } \tilde{\phi} = \phi + \sum_{i=1}^{q} \lambda_i (g_i)^r \qquad (6.6\text{-}1)$$

where r is an even power to assure that the added terms are in fact all added to ϕ, and λ is a scaling factor. For the constraint $b_j > k_j$ or $b_j - k_j > 0$, the function g_i for $r = 2$ and $\lambda = 1$ would be

$$g_i = 0 \qquad \text{if } b_j > k_j$$

$$g_i = (b_j - k_j)^2 \qquad \text{if } b_j \leq k_j$$

Equality constraints:

$$g_j(\mathbf{Y}, \mathbf{x}, \boldsymbol{\beta}) = 0 \qquad j = 1, 2, \ldots, r$$

can be treated in the same fashion with each $\lambda_j g_j^2(\mathbf{Y}, \mathbf{x}, \boldsymbol{\beta})$ added to ϕ as a penalty function. When the constraint is violated, a dominant penalty is added; when the constraint is almost satisfied, a negligible penalty is added.

However, the penalty function approach has its disadvantages, particularly in regard to the selection of the scaling factors, λ. Moreover, since several computer codes are available to carry out nonlinear programming directly, it usually proves simpler to employ one of the available codes rather than to develop a new code for a special problem. We do not have the space here to describe the various algorithms which have been suggested to accommodate both inequality and equality constraints into the optimization schemes discussed in Section 6.2, since they are all quite involved. However, Table 6.6-1 lists references to iterative types of codes. It is also possible to seek extrema by analytical methods, if the constraints are solely equality constraints, by means of Lagrangian multipliers.

TABLE 6.6-1 COMPUTER CODES FOR OPTIMIZATION SUBJECT TO CONSTRAINTS

Name	Technique	Reference
POP/360	Iterative linear programming, numerical derivatives	1
Simplex Search	Simplex method	2
SUMT	Penalty function, analytic derivatives	3

[1] IBM Corp. Share Library.
[2] Shell Development Corp., Emeryville, Calif.
[3] G. P. McCormick, *Mang. Sci.* **10**, 360, 1964; Research Analysis Corp., McLean, Va.

Constraints on the independent variables of the model, either recognized or unrecognized, cause difficulty in the estimation of the model parameters. To give a simple example, if the model is

$$\eta = \beta_0 + \beta_1 x_1 + \beta_2 x_2$$

and the constraint is

$$x_1 + x_2 = 1$$

clearly the model is equivalent to

$$\eta = \beta_0^* + \beta_1^* x_1$$

in which x_2 plays no role. Consequently, the remarks of Section 6.3 about the null effect apply. Each such equality constraint removes one degree of freedom among the independent variables in the model.

Supplementary References

Bartlett, M. S., "The Use of Transformations," *Biometrics* **3**, 39, 1947.

Beale, E. M. L., "Confidence Regions in Nonlinear Estimation," *J. Royal Stat. Soc.* **B22**, 41, 1960.

Box, G. E. P. and Coutie, G. A., "Application of Digital Computers in the Exploration of Functional Relationships," *Proceed. Inst. Elec. Eng.* **103**, Pt. B, Supplement 1, 100, 1956.

Box, G. E. P. and Cox, D. R., "An Analysis of Transformations," Dept of Stats. Tech. Rept. No. 26, Univ. of Wis., Madison, 1964.

Box, G. E. P. and Tidwell, P., "Transformation of the Independent Variables," *Technometrics* **4**, 531, 1962.

Curry, H. B., "The Method of Steepest Descent for Nonlinear Minimization Problems," *Quart. Appld. Math.* **2**, 258, 1944.

Dickinson, A. W., in *Chemical Division Transactions of the American Society for Quality Control*, 7959, p. 181.

Dolby, J. L., "A Quick Method for Choosing a Transformation," *Technometrics* **5**, 317, 1963.

Goldberger, A. S., *Econometric Theory*, John Wiley, New York, 1964.

Hartley, H. O., "The Modified Gauss-Newton Method for the Fitting of Nonlinear Regression Functions by Least Squares," *Technometrics* **3**, 269, 1961.

Kale, B. K., "On the Solution of the Likelihood Equation by Iteration Processes," *Biometrika* **48**, 452, 1951.

Kale, B. K., "On the Solution of the Likelihood Equation by Iteration Processes. The Multiparametric Case," *Biometrika* **49**, 479, 1962.

Levenberg, K., "A Method for the Solution of Certain Nonlinear Problems in Least Squares," *Quart. Appld. Math.* **2**, 1964, 1944.

Marquardt, D. L., "An Algorithm for Least Squares Estimation of Nonlinear Parameters," *J. Soc. Ind. Appld. Math.* **2**, 431, 1963.

Pereyra, V., "Iterative Methods for Solving Nonlinear Least Squares Problems," *SIAM J. Num. Anal.* **4**, 27, 1967.

Tukey, J. W., "On the Comparative Anatomy of Transformations," *Ann. Math. Stat.* **28**, 602, 1957.

Turner, M. E., Monroe, R. J., and Lucas, H. L., "Generalized Asymptotic Regression and Nonlinear Path Analysis," *Biometrics* **17**, 120, 1961.

Williams, E. J., *Regression Analysis*, John Wiley, New York, 1959.

Surveys of Optimization Techniques for Nonlinear Objective Functions

Dorn, W. S., "Nonlinear Programming—A Survey," *Mang. Sci.* **9**, 171, 1963.

Spang, H. A., "A Review of Minimization Techniques for Nonlinear Functions," *SIAM Rev.* **4**, 343, 1962.

Wilde, D. J. and Beightler, C. F., *Foundations of Optimization*, Prentice-Hall, Englewood Cliffs, N.J., 1967.

Wolfe, P., "Methods of Nonlinear Programming," in *Nonlinear Programming*, ed. by J. Abadie, North-Holland, New York, 1967, Chapter 6.

Zoutendijk, G., "Nonlinear Programming: A Numerical Survey," *SIAM J. on Control* **4**, 194, 1966.

Computer Routines for Nonlinear Estimation

Derivative-type methods:

Booth, G. W. and Peterson, T. I., *AIChE Computer Program Manual No. 3*, Dec. 1960, AICE, 345 E, 47th St., New York (FORTRAN).

Efroymson, M. A. and Mathew, D., "Nonlinear Regression Program with Nonlinear Equations," Esso Res. and Develop. Co. (FORTRAN).

Marquardt, D. W., *et al.*, "NLIN, Least Squares Estimation of Non-linear Parameters," IBM Share Program SD 3094, 1964 (FORTRAN).

Moore, R. H., Zeigler, R. K., and McWilliams, P., "PAKAG," Los Alamos Scientific Laboratory, Albuquerque, N.M. (FORTRAN and FAP).

Derivative-free methods:

Beisinger, Z. E. and Bell, S., "H2 SAND MIN," Sandia Corp. (FORTRAN). Direct Search.

Lindamood, G. E., "AP MINS," John Hopkins Univ. Baltimore, Md., SD 1259 (FAP). Direct Search.

Kaupe, A. F., "Collected Algorithms from the Association of Computing Machinery," *Algorithm 178*, New York, annually (Algol). Direct Search.

Rosen, J. B., "GP90 Gradient Projection Method for Non-linear Programming" (FAP).

Problems

6.1 Using the following data (x_i = independent variable and Y_i = dependent variable):

x_i	Y_i
0.4	51.6
1.4	53.4
5.4	20.0
19.5	−4.2
48.2	−3.0
95.9	−4.8

find the best fitting parameters for the equation

$$Y = b_0 + b_1 e^{b_2 x} + \epsilon$$

6.2 Fit the Antoine equation

$$\log_{10} p = A - \frac{B}{T}$$

where p is the vapor pressure in millimeters of mercury, T is the temperature in degrees Kelvin, and A and B are constants to be determined.

TABLE P6.2 VAPOR PRESSURE-TEMPERATURE DATA FOR SULFURIC ACID-WATER SYSTEM (95 PERCENT WEIGHT H_2SO_4 IN H_2O)*

Vapor Pressure, p (mm Hg)		Absolute Temperature, T (°K)	
a	b	a'	b'
0.00150	8.39000	308.16	438.16
0.00235	10.30000	313.16	443.16
0.00370	12.90000	318.16	448.16
0.00580	15.90000	323.16	453.16
0.00877	20.20000	328.16	458.16
0.01330	24.80000	333.16	463.16
0.01960	30.70000	338.16	468.16
0.02880	36.70000	343.16	473.16
0.04150	45.30000	348.16	478.16
0.06060	55.00000	353.16	483.16
0.08790	66.90000	358.16	488.16
0.12300	79.80000	363.16	493.16
0.17200	95.50000	368.16	498.16
0.23700	115.00000	373.16	503.16
0.32100	137.00000	378.16	508.16
0.43700	164.00000	383.16	513.16
0.59000	193.00000	388.16	518.16
0.78800	229.00000	393.16	523.16
1.07000	268.00000	398.16	528.16
1.42000	314.00000	403.16	533.16
1.87000	363.00000	408.16	538.16
2.40000	430.00000	413.16	543.16
3.11000	500.00000	418.16	548.16
4.02000	580.00000	423.16	553.16
5.13000	682.00000	428.16	558.16
6.47000	790.00000	433.16	563.16

* Data for sulfuric acid are from J. H. Perry, *Chemical Engineers Handbook*, McGraw-Hill, New York, 1963.

6.3 Fit the data below to the model

$$y = \frac{A_1 x_1 \log_e \left(\frac{A_2}{x_2}\right)}{e^{A_3 x_3} + A_4}$$

Find A_1, A_2, A_3, and A_4 by least squares.

Y	x_1	x_2	x_3
0.81028	1.0000	0.1000	0.1000
8.1028	10.000	0.1000	0.1000
12.154	15.000	0.1000	0.1000
5.0514	5.0000	0.1000	0.1000
60.771	75.000	0.1000	0.1000
0.68833	1.0000	0.1000	1.0000
6.8833	10.000	0.1000	1.0000
10.325	15.000	0.1000	1.0000
3.4417	5.0000	0.1000	1.0000
51.625	75.000	0.1000	1.0000
0.30451	1.0000	1.0000	0.1000
3.0451	10.000	1.0000	0.1000
4.5676	15.000	1.0000	0.1000
1.5225	5.000	1.0000	0.1000
22.838	75.000	1.0000	0.1000

6.4 Fit the data below to the model

$$y = \frac{e^{-x}}{x}\left[\frac{a_1 + a_2 x + a_3 x^2 + x^3}{a_4 + a_5 x + a_6 x^2 + x^3}\right]$$

by least squares.

Y	x
1.9697	0.1000
3.867×10^{-1}	0.6700
1.226×10^{-1}	1.3400
4.611×10^{-2}	2.0100
1.877×10^{-2}	2.6800
1.5805×10^{-2}	4.6900
7.2126×10^{-3}	5.3600
3.3327×10^{-3}	6.0300
1.5553×10^{-3}	6.7000
7.3177×10^{-4}	4.73700
3.4665×10^{-4}	8.0400
1.6516×10^{-4}	8.7100
7.9076×10^{-5}	9.3800
3.8024×10^{-5}	10.050

6.5 Given the data below, estimate a_1, a_2, a_3, and a_4 in the model

$$y = (a_1 + a_2 x_1^2 + \cos a_3 x_2)^{\sin(a_4 x_3)}$$

Y	x_1	x_2	x_3
7.7385×10^{-4}	75	33	75
4.2372×10^{-4}	68	15	68
8.8133×10	39	9	39
4.5851×10	16	25	16
9.4883	58	48	58
1.1336×10^{-3}	53	5	53
1.2052×10^{2}	61	63	61
1.0767×10^{-1}	47	72	47
4.3098×10^{-1}	99	29	99
8.0050×10^{-1}	33	17	33
6.1111×10^{-4}	97	80	97
3.1792×10^{3}	29	61	29
4.40359	1.6	23	16
1.4448×10^{2}	13	32	13
7.5917×10^{-3}	72	77	72
2.6723×10^{-4}	43	67	43
3.6466×10^{-5}	84	34	84
9.5717×10^{-5}	100	15	100
4.7435×10^{-5}	81	13	81
2.4336×10	63	11	63

6.6 Select a series of (simulated) data points from the Steam Tables for water vapor, or for another gas from data available in the literature, and fit the (simulated) data to the following equations of state for 1 mole.

(a) Van der Waals:

$$\left(p + \frac{a}{V^2}\right)(V - b) = RT$$

(b) Macleod:

$$\pi(V - b') = RT$$

$$\pi = p + \frac{a}{V^2}$$

$$b' = b_0(1 - B\pi + C\pi^2)$$

(c) Clausius:

$$p = \frac{RT}{(V - b)} - \frac{a}{T(V + c)^2}$$

(d) Lorentz:

$$p = \frac{RT}{V^2}(V + b) - \frac{a}{V^2}$$

(e) Dieterici:

$$p = \frac{RT}{(V - b)} e^{-a/VRT}$$

(f) Berthelot:

$$p = \frac{RT}{(V - b)} - \frac{a}{TV^2}$$

(g) Wohl:

$$p = \frac{RT}{(V - b)} - \frac{a}{V(V - b)} + \frac{c}{TV^3}$$

(h) Keyes:

$$p = \frac{RT}{(V - \delta)} - \frac{A}{(V - l)^2}$$

$$\delta = \beta e^{-a/V}$$

(i) Kammerlingth-Onnes:

$$pV = RT\left[1 + \frac{B}{V} + \frac{C}{V^2} + \cdots\right]$$

(j) Holborn:

$$pV = RT[1 + B'p + C'p^2 + \cdots]$$

(k) Beattie-Bridgeman:

$$pV = RT + \frac{\beta}{V} + \frac{\gamma}{V^2} + \frac{\delta}{V^3}$$

$$\beta = RTB_0 - A_0 - \frac{Rc}{T^2}$$

$$\gamma = -RTB_0 b + aA_0 + \frac{RB_0 c}{T^2}$$

$$\delta = \frac{RB_0 bc}{T^2}$$

(l) Benedict-Webb-Rubin:

$$pV = RT + \frac{\beta}{V} + \frac{\sigma}{V^2} + \frac{\eta}{V^4} + \frac{w}{V^5}$$

$$\beta = RTB_0 - A_0 - \frac{C_0}{T^2}$$

$$\sigma = bRT - a + \frac{c}{T^2} \exp -\frac{\gamma}{V^2}$$

$$\eta = c\gamma \exp -\frac{\gamma}{V^2}$$

$$w = a\alpha$$

The notations (use consistent units) are:

p = pressure
V = volume/mole
T = absolute temperature
R = gas constant

All other symbols are coefficients to be determined. Discuss what should be done if the independent variables as well as the dependent variables are stochastic.

6.7 A compartment-type experimental dryer was built to simulate the drying conditions of a commercial leather drier.† After a certain drying time (t), the thickness of the leather (L) and the mass velocity of the drying air (G) were measured. The drying coefficient (β) was then calculated for that run from the equation

$$\frac{dW}{dt} = -\beta W \Delta H$$

where

L = thickness of leather
W = free moisture content of leather
ΔH = unsaturation of air
β = drying coefficient
t = time elapsed
G = mass velocity of drying air

The purpose of the experiment was to find the value of B as a function of the parameters G and L in the nonlinear equation

$$B = \frac{mG^n}{L}$$

where m and n are the constants to be determined.

† O. A. Hougen, "Rate of Drying Chrome Leather," *Ind. Eng Chem.*, pp. 333–339, Mar. 1934.

TABLE P6.7 ORIGINAL DATA

Runs	$B\left(\frac{\text{kg evap.}}{(\text{meter})^2(\text{min})}\right)$	L (min)	$G\left(\frac{\text{kg}}{(\text{meter})^2(\text{min})}\right)$
1	1.305	1.05	14.1
2	1.90	1.17	42.7
3	2.71	1.06	42.0
4	2.61	1.00	42.5
5	2.48	1.04	42.3
6	3.61	1.13	71.4
7	3.48	1.02	70.0
8	4.95	1.02	117.0
9	4.38	1.00	115.0
10	4.63	1.06	112.0
11	4.65	1.04	114.0
12	3.18	1.32	99.6
13	3.55	1.43	99.6

Find the best values of m and n for the following data. The values in the article were $m = 0.282$ and $n = 0.2$, obtained by graphical methods. Comment on the possible bias in the parameters m and n. What are their respective estimated variances?

6.8 Nitrogen oxide is absorbed in a reacting solution to produce a product; the data are given below. Estimate the coefficients β_1, β_2, and β_3 in the model

$$y = \beta_1 e^{\beta_2 x} x^{\beta_3}$$

Y Nitrogen Oxide Absorbed (g/liter)	x Concentration of Product (g/liter)
0.09	15.1
0.32	57.3
0.69	103.3
1.51	174.6
2.29	191.5
3.06	193.2
3.39	178.7
3.63	172.3
3.77	167.5

6.9 Diffusion data have been collected in the laboratory to fit an equation:

$$N = c\sqrt{\frac{D}{\pi t}}$$

where

N = moles absorbed/min
c = concentration, moles/cc
D = diffusion coefficient, cm^2/min
t = time, min.

The variables N, c, and t are measured. You are asked to calculate D from 12 sets of data. How would you go about this? Give the equations you would use in terms of the notations above. Give a means of calculating the confidence limits on a predicted value of D. How should the magnitude of the experimental error in the work be determined?

6.10 In Example 3.6-3, Bartlett's test demonstrated that the variances at four values of x_i were not homogeneous, i.e., were not the same. Since there is no physical reason to presume any particular functional relation between Var $\{\bar{Y}_i \mid x\}$ and x, the form of the relation can be established by regression analysis in fitting the best curve through the $s_i^2(x)$ data.

However, a large number of functional forms for Var $\{\bar{Y}_i \mid x\}$ can be proposed, many of which will prove to be essentially equivalent. For example, some possible regression equations are

$$s^2(x) = e^{-x^a}$$

$$s^2(x) = \frac{1}{a(x+b)^2}$$

$$s^2(x) = \frac{1}{ax+b}$$

$$s^2(x) = \frac{1}{x^a}$$

and so forth. From the data in Example 3.6-3, determine a suitable functional form for $s^2(x)$ for use in equations incorporating nonunity weights. Also test the resulting predicted values of s^2 for each of the four x's to see if they meet the modified Bartlett's test based on weighted variances:

$$s_i'^2 = \frac{s_i^2}{s^2(x_i)}$$

(the pooled variance is weighted also). Make a plot of the predicted values of $s^2(x)$ versus x, and show the experimental points for comparison.

6.11 The density and viscosity of anhydrous hydrazine were measured at elevated temperatures ranging from 288.16 to 449.83 degrees Kelvin.† Samples of 99.6-percent purity were prepared by stirring commercial-grade (about 97 percent) hydrazine over barium oxide for several hours, vacuum distilling, and passing the condensate twice through a packed column of Linde Molecular Sieves. The hydrazine content of each sample was determined by an iodine titration method. Density determinations were made with sealed borosilicate glass pycnometers.

The following equation has been suggested to express the relationship between the absolute viscosity, η, the density, ρ, and the temperature, T:

$$\eta = \rho^k \exp\left(a + \frac{b}{T} + \frac{c}{T^3}\right)$$

Determine the constants a, b, c, and k; the confidence limits on the constants; and the confidence limits on the predicted values of η. What basic assumptions have you made? List them.

† R. C. Ahlert, G. L. Bauerle, and J. V. Leece, "Density and Viscosity of Anhydrous Hydrazine at Elevated Temperatures," *J. Chem. Eng. Data*, Jan. 1962.

TABLE P6.11 DATA

Temperature (°K)	η (centipoise)	ρ (g/cc)
288.16	1.0275	1.0114
310.94	0.7268	0.9934
338.72	0.5363	0.9672
366.49	0.4028	0.9392
394.27	0.3266	0.9124
422.05	0.2728	0.8862
449.83	0.2344	0.8575

6.12 Fit the Berthelot equation of state (for 1 mole):

$$p = \frac{RT}{V-b} - \frac{a}{TV^2}$$

for SO_2 to the following data:‡

V (cm³/g)	P (atm)	Mass of Gas (g)	t (°C)
67.810	5.651	0.2948	50
50.882	7.338		
45.280	8.118		
72.946	5.767	0.2948	75
49.603	8.237		
23.331	15.710		
80.170	5.699	0.2948	100
45.664	9.676		
25.284	16.345		
15.285	24.401		
84.581	5.812	0.2948	125
42.675	11.120		
23.480	19.017		
14.735	27.921		
23.913	20.314	1.9533	150
18.241	25.695		
7.2937	51.022		
4.6577	63.730		
20.685	26.617	1.9533	200
10.595	47.498		
5.8481	74.190		

6.13 A model for a surface reaction mechanism:

$$r = \frac{kK_2\left(p_2 - \frac{p_3}{K}\right)}{1 + K_1p_1 + K_2p_2 + K_3p_3} \tag{a}$$

where

k, K = constants
r = dependent variable, stochastic
p = independent variables, deterministic

has been linearized as follows for regression analysis:

$$\frac{p_2 - (p_3/K)}{r} = \frac{1}{kK_2} + \frac{K_1}{kK_2}p_1 + \frac{1}{k}p_2 + \frac{K_3}{kK_2}p_3 \tag{b}$$

‡ T. L. Kang, Ph.D. Thesis, Univ. of Texas, 1960.

How satisfactory would the values of k and K be if obtained from the coefficients of p_i in Equation (b)? What could be said about the confidence intervals on k and the K's?

6.14 Linearize the model

$$y = \beta_1 e^{-\beta_2 x} + \beta_3 e^{-\beta_4 x}$$

by a truncated Taylor series and form the set of equations corresponding to Equation 6.2-12. Indicate what the elements of the **P** and **Z** matrices are.

6.15 Find the gradient of the objective function

$$\phi_1 = \sum_{i=1}^{n} (y_i - \eta_i)^2$$

where

$$\eta = \alpha + \beta_1 e^{\beta_2 x_1} + \beta_3 e^{\beta_4 x_2}$$

At the point where the respective estimates of the parameters are: $a = 10$, $b_1 = 1$, $b_2 = 2$, $b_3 = 2$, and $b_4 = 1$, give the components of the unit vector in the direction of steepest descent from the point (10, 1, 2, 2, 1).

6.16 The power function $y = \alpha x_1^{\beta_1} x_2^{\beta_2}$ can be linearized by taking logarithms of both sides of the equation. However, if it is desired to use as a criterion $\sum (Y_i - y_i)^2$ rather than $\sum (\ln Y_i - \ln y_i)^2$, what weights could be used in conjunction with the latter criterion to give approximately the same results as minimizing the former?

CHAPTER 7

Identification of the Best Models

Chapters 4, 5, and 6 have been primarily concerned with the mechanics of parameter point and interval estimation and hypothesis tests for a given empirical model. In this Chapter we inquire as to which among several models is best, inasmuch as the main interest of the analyst is to select an appropriate model from among many feasible models. For example, if the objective is just to fit a polynomial to some experimental data in order to use the data in the form of a function (rather than using tabulated data) in a computer program, a natural question to ask is: How many terms should be employed in the polynomial? In other instances, too many independent variables exist in an empirical model; then the analyst is interested in finding a smaller set which contains the important variables, deleting those of slight importance from the model. Chapter 8 describes the intermeshing of experimentation with estimation and model building, but we are interested now in the analysis of process data, taken in perhaps undesigned experiments, for which the estimation and analysis may occur as afterthoughts. A discussion of the methods of experimentation to discriminate effectively among models is deferred until Section 8.5.

What we would like to do is select the "best" model when more than one is possible or proposed. For the linear models described in Chapter 5, it is impractical to employ computer procedures to evaluate *all* the possible regression equations obtainable by deleting and adding successive sets of variables, even if we ignore possible transformations. Instead, the analyst works with a limited number of different models and discriminates among them by one or more systematic procedures. As to the criterion of "best," one or a combination of the following is commonly used:

1. Fewest coefficients consistent with reasonable error.
2. Simplest form consistent with reasonable error.
3. Rationale based on physical grounds ("seems to follow . . .'s law").
4. Minimum sum of squares of deviations between predicted and empirical values.
5. Minimum variance, $s^2_{\bar{Y}_i}$ ($s^2_{\bar{Y}_i}$ is not an unbiased variance estimate since it contains a systematic component that exists because of the difference between the estimated function and the true function; refer to Section 5.2-2).

To help identify suitable models, we shall discuss first the analysis of residuals, which should be one of the first phases in any model-evaluation program. Then we shall describe stepwise regression for linear models, which is a systematic estimation procedure that simultaneously isolates the important independent variables. Finally, we shall discuss certain tests that have been proposed to discriminate among two and among more than two models.

7.1 ANALYSIS OF RESIDUALS

The analysis of variance has been used in Sections 4.3 and 5.3 to establish whether or not a linear model represents the data adequately. But the variance ratio test to determine if (s_r^2/s_e^2) is greater than $F_{1-\alpha}$ only demonstrates that the overall fit of the model is satisfactory. Important discrepancies can still exist, even though the model passes the F-test. These discrepancies often can be detected through the *analysis of residuals*, that is, by examining the set of deviations between the experimental and predicted values of the dependent variable, $(\bar{Y}_i - \hat{Y}_i) = E_i$.

Certain underlying assumptions have been outlined for regression analysis, such as independence of the unobservable errors ϵ, constant variance, and normal distribution for ϵ. If the model represents the data adequately, the residuals should possess characteristics that agree with, or at least do not refute, the basic assumptions. The analysis of residuals is thus a way of checking that one or more of the assumptions underlying regression analysis is not violated. For example, if the model fits well, the residuals should be randomly distributed about the $\hat{Y}$ predicted by the regression equation. Systematic departures from randomness indicate that the model is not satisfactory; examination of the patterns formed by the residuals can provide clues as to how the model can be improved.

Examinations of plots of the residuals versus $\hat{Y}_i$, x_i, or time, or a plot of the frequency of the residuals versus the magnitude of the residuals, all have been suggested as numerical and/or graphical aids to assist in the analysis of residuals.† A study of the signs of the residuals (+ or −) and sums of signs can be used.‡ Many of the nonparametric tests described in Section 3.7 can also be employed. In the following example (adapted from Nelson), five features of residual analysis are presented:

1. Detection of an outlier (an extreme observation).
2. Detection of a trend in the residuals.
3. Detection of an abrupt shift in level of the experiment.
4. Detection of changes in the error variance (usually assumed to be constant).
5. Examination of residuals to ascertain if they are represented by a normal distribution.

When residuals are used to elucidate the adequacy of a model, keep in mind that as more and more independent variables are added to the model, the residuals become less and less informative. Each residual is in effect a weighted average of the ϵ's; as more unnecessary x's are added to a model, the residuals become more alike, each one reflecting an indiscriminate average of all the ϵ's instead of reflecting primarily one ϵ.

† F. J. Anscombe, *Proceed. Fourth Berkeley Symposium on Math. Stat. and Probability* **1**, 19, 1963; F. J. Anscombe and J. W. Tukey, *Technometrics* **5**, 141, 1963; G. E. P. Box, *Annals. N.Y. Acad. Sci.* **86**, 792, 1960; R. J. Freund, R. W. Vail, and C. W. Clunies-Ross, *J. Amer. Stat. Assn.* **56**, 98, 1961; L. S. Nelson, *Chem. Div., Amer. Soc. Qual. Control Trans.*, Houston, Texas, 1955, p. 111.

‡ N. R. Draper and H. Smith, *Applied Regression Analysis*, John Wiley, New York, 1966, Chapter 3.

In carrying out the analysis of residuals, the analyst will quickly discover that graphical presentation of the residuals materially assists in the diagnosis because one aberration, such a single extreme value, can simultaneously affect several of the numerical tests.

Example 7.1-1 Analysis of Residuals

Simulated data were generated by the following model:

$$Y_i = 10.8 + 0.40x_1 - 0.20x_2 + \epsilon_i$$

in which x_1 and x_2 took on integer values 1, 2, 3, 4, 5, and 6, and the errors, ϵ_i, were random, normally distributed with an expected value of 0 and a variance of 1. Note that the model is linear and no replicate measurements were generated. The simulated data are listed in Table E7.1-1a; the numbers in parentheses above each Y_i indicate the order in which the Y_i's were calculated. Each cell will have a corresponding residual.

Application of the linear estimation procedure described in Section 5.1 yielded the following estimated regression equation:

$$\hat{Y} = 10.52 + 0.49x_1 - 0.19x_2 \qquad \text{(a)}$$

and the analysis of variance given in Table E7.1-1b.

The linear term for x_1 was significant at $P = 0.001$, while the linear term for x_2 was significant at $P = 0.025$. Note that s_r^2 was 0.73 as compared with $\sigma^2 = 1$.

DETECTION OF AN OUTLIER. If the value 9.6 in the cell ($x_1 = 1, x_2 = 6$) is replaced by 13.9 (a value 3.9 standard deviations higher than 10), a new analysis of variance indicates that the variable x_2 was not significant (see Table E7.1-1c). The mean square for the residuals, s_r^2, was still close to unity and the estimated regression equation was $\hat{Y} = 10.32 + 0.39x_1$. Figure E7.1-1a is a plot of the residuals calculated by using this equation. The circled point is an outlier. If the data had been taken in a real experiment, the adequacy of the experimental conditions for which the encircled point was collected should be reexamined.

TABLE E7.1-1a

		x_1							
		1	2	3	4	5	6	Sum	Mean
x_2	1	(28) 10.3	(3) 11.3	(1) 13.0	(31) 12.0	(36) 10.6	(9) 13.2	70.4	11.73
	2	(6) 11.6	(13) 10.4	(4) 11.5	(25) 11.9	(24) 13.3	(19) 14.2	72.9	12.15
	3	(8) 9.7	(30) 12.2	(15) 10.9	(5) 11.8	(34) 12.1	(16) 12.8	69.5	11.58
	4	(27) 11.4	(17) 11.5	(29) 11.0	(12) 11.5	(7) 11.3	(18) 14.5	71.2	11.87
	5	(32) 10.3	(33) 10.6	(22) 10.0	(35) 11.3	(11) 11.4	(10) 12.8	66.4	11.07
	6	(20) 9.6	(2) 9.4	(14) 11.8	(23) 10.2	(21) 12.7	(26) 12.4	66.1	11.02
	Sum	62.9	65.4	68.2	68.7	71.4	79.9	416.5	
	Mean	10.48	10.90	11.37	11.45	11.90	13.32		11.57

TABLE E7.1-1b ANALYSIS OF VARIANCE

Source of Variation	ν d.f.	Sum of Squares	Mean Square	Variance Ratio
x_1	1	25.51	25.51	37.5
x_2	1	3.68	3.68	5.4
Deviation from regression:	33	24.05	0.73	
Total	35			

TABLE E7.1-1c

Source of Variation	ν d.f.	Sum of Squares	Mean Square	Variance Ratio
x_1	1	16.01	16.01	14.0
x_2	1	0.75	0.75	—
Deviation from regression	33	37.51	1.14	
Total	35			

DETECTION OF A TREND. Because the underlying process may shift with time, the experimenter must always check for a trend in the residuals. To simulate this characteristic of a real process, the values 0, 0.1, 0.2, 0.3, . . . , 3.5 were added to each of the entries in Table E7.1-1a in the sequence given by the precedence order numbers (those in parentheses). The new simulated data are listed in Table E7.1-1d.

The analysis of variance for the data in Table E7.1-1d indicated that x_1 was a significant variable but that x_2 was not. See Table E7.1-1e. Note that s_r^2 became somewhat larger than 1. The regression equation used in calculating the residuals, which are plotted in Figure E7.1-1b, was

$$\hat{Y} = 11.53 + 0.51x_1 \qquad \text{(b)}$$

Figure E7.1-1b clearly brings out the trend of the residuals; a least square line which best fit the residuals indicated that the trend could be removed if 0, 0.09, 0.18, . . . , 3.15 were taken from the simulated data in order of their initial sequence.

Other trends can be expected to occur in experimental data, such as a diverging or converging spread of values or a nonlinear trend. A graph such as Figure E7.1-1b can help in the detection of such a trend.

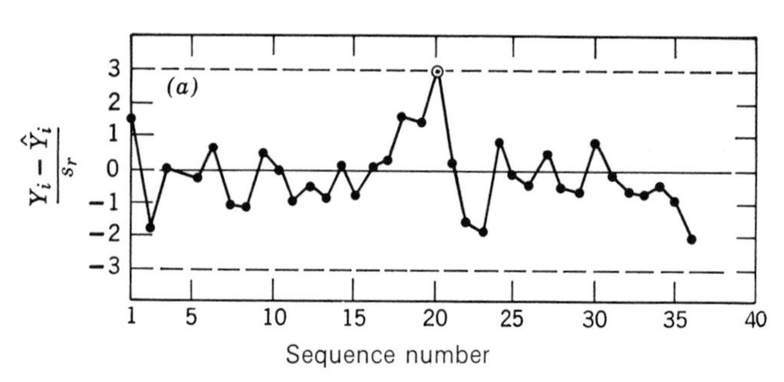

FIGURE E7.1-1a

TABLE E7.1-1e

Source of Variation	ν d.f.	Sum of Squares	Mean Square	Variance Ratio
x_1	1	27.31	27.31	15.7
x_2	1	1.43	1.43	—
Deviation from regression	33	57.44	1.74	
Total	35			

DETECTION OF AN ABRUPT SHIFT IN LEVEL. To simulate an abrupt shift in process level, each value in the sequence of data in Table E7.1-1a numbered 18 through 36 was increased by 3. The analysis of variance for the new data indicated again that x_1 was a significant variable ($P = 0.01$)

TABLE E7.1-1d

		x_1: 1	2	3	4	5	6	Sum	Mean
x_2	1	13.0	11.5	13.0	15.0	14.1	14.0	80.6	13.43
	2	12.1	11.6	11.8	14.3	14.6	16.0	81.4	13.57
	3	10.4	15.1	12.3	12.2	15.4	14.3	97.7	13.28
	4	14.0	13.1	13.8	12.6	11.9	16.2	81.6	13.60
	5	13.4	13.8	12.1	14.7	12.4	13.7	80.1	13.35
	6	11.5	9.5	13.1	12.4	14.7	14.9	76.1	12.68
Sum		74.4	74.6	76.1	81.2	84.1	89.1	479.5	
Mean		12.40	12.43	12.68	13.53	14.02	14.85		13.32

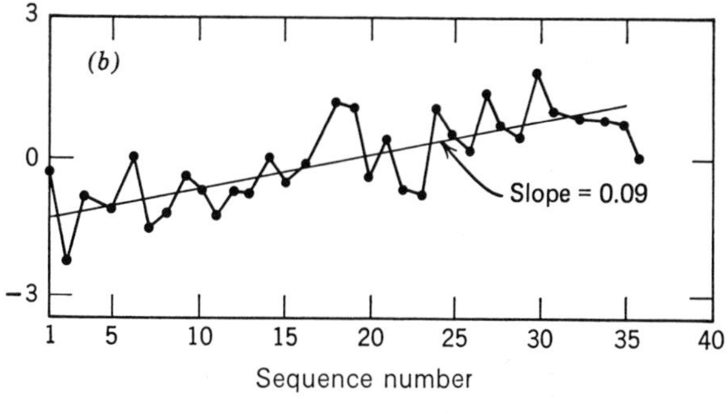

FIGURE E7.1-1b

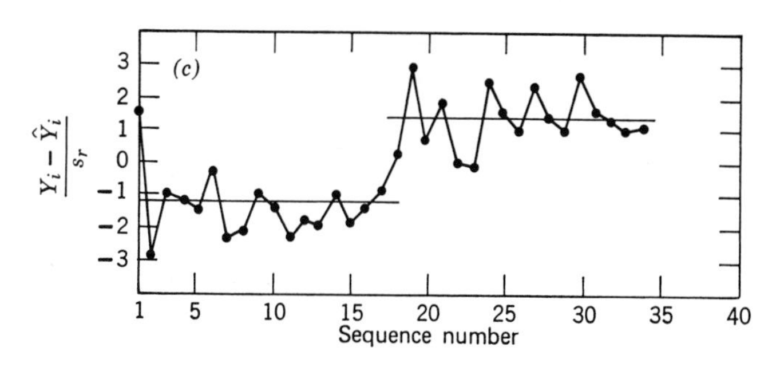

FIGURE E7.1-1c

and x_2 was not. See Table E7.1-1f. (Observe that the mean square of the residuals is now significantly different from 1.) A plot of the residuals based on predicted values of Y from the equation of best fit

$$\hat{Y} = 11.44 + 0.465x_1 \tag{c}$$

clearly shows what has happened; examine Figure E7.1-1c. The difference between the sample mean values for the two groups of residuals was computed as 2.8.

TABLE E7.1-1f

Source of Variation	ν d.f.	Sum of Squares	Mean Square	Variance Ratio
x_1	1	22.63	22.63	7.6
x_2	1	0.46	0.46	—
Deviation from regression	33	98.24	2.98	
Total	35			

DETECTION OF CHANGES IN THE ERROR VARIANCE. A major assumption in the analysis of variance usually is that the error variance is constant. It is of some importance to ascertain if this assumption is not fulfilled. The analysis in Chapter 5 gives an average value for the error variance, and if the error variance is not homogeneous, this average value may not correctly represent any part of the experiment. To simulate nonhomogeneity of variance, each error added to all the values at experimental levels 5 and 6 for x_1 in Table E7.1-1a was multiplied by five to yield the simulated data listed in Table E7.1-1g.

An analysis of variance of the data in Table E7.1-1g yielded the results shown in Table E7.1-1h. The effect of variable x_1 showed up strongly ($P = 0.001$), but the effect of factor x_2 was at the 10 percent level of significance. The residual variance s_r^2 was very much greater than one. If variable x_2 was ignored, the estimated regression equation was

$$\hat{Y} = 8.39 + 1.19x_1 \tag{d}$$

Table E7.1-1i lists residuals calculated with Equation (d). Analysis of the ranges of the residuals in Table E7.1-1i suggested that the two highest levels of variable x_1 had a significantly different variance from the other levels.

TABLE E7.1-1h

Source of Variation	ν d.f.	Sum of Squares	Mean Square	Variance Ratio
x_1	1	148.93	148.93	20.6
x_2	1	20.81	20.81	2.9
Deviation from regression	33	236.27	7.16	
Total	35			

TABLE E7.1-1g

		x_1 1	2	3	4	5	6	Sum	Mean
x_2	1	10.3	11.3	13.0	12.0	22.4	14.2	83.2	13.87
	2	11.6	10.4	11.5	11.9	16.7	19.8	81.9	13.65
	3	9.7	12.2	10.9	11.8	12.8	13.4	70.8	11.80
	4	11.4	11.5	11.0	11.5	8.6	23.1	77.1	12.85
	5	10.3	10.6	10.0	11.3	10.0	15.1	67.3	11.22
	6	9.6	9.4	11.8	10.2	17.1	13.0	72.0	12.00
	Sum	62.9	65.4	68.2	68.7	87.6	98.6	452.3	
	Mean	10.48	10.90	11.37	11.45	14.60	16.43		12.56

TABLE E7.1-1i

		x_1						
		1	2	3	4	5	6	Range
	1	0.72	0.53	1.04	−1.15	8.06	−1.33	9.39
	2	2.02	−0.37	−0.46	−1.25	2.36	4.27	5.52
	3	0.12	1.43	−1.06	−1.35	−1.54	−2.13	3.56
x_2	4	1.82	0.73	−0.96	−1.65	−5.74	7.57	13.31
	5	0.72	−0.17	−1.96	−1.85	−4.34	−0.43	5.06
	6	0.02	−1.37	−0.16	−2.95	2.76	−1.63	5.71
Range		2.00	2.80	3.00	1.80	13.80	9.70	

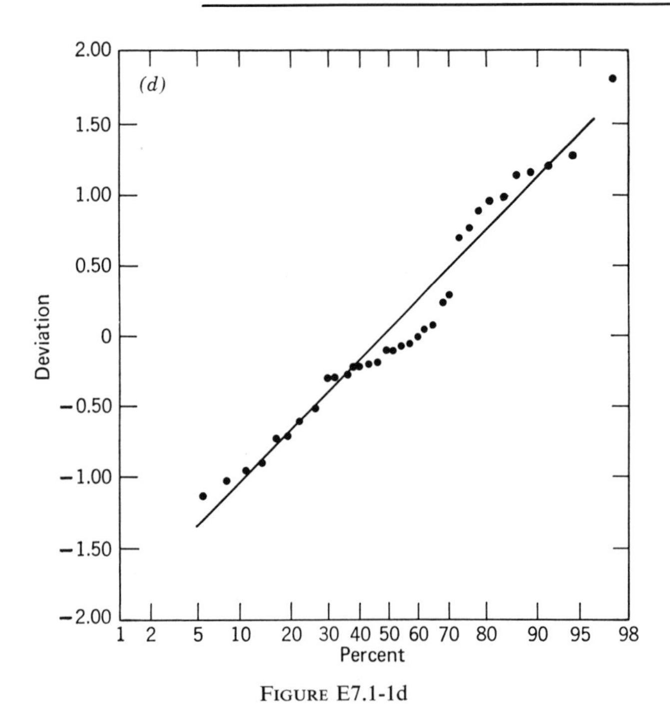

FIGURE E7.1-1d

EXAMINATION OF RESIDUALS TO ASCERTAIN IF THEY ARE NORMALLY DISTRIBUTED. Figure E7.1-1d is a plot (according to the method outlined in Example 2.3-3) on normal probability paper of the 36 residuals based on the data of Table E7.1-1a with $\hat{Y}$ given by Equation (a). The residuals appear to be normally distributed. A plot of relative frequency versus the value of the residual will give the familiar bell-shaped curve about a mean of zero; it also can be used to check for outliers.

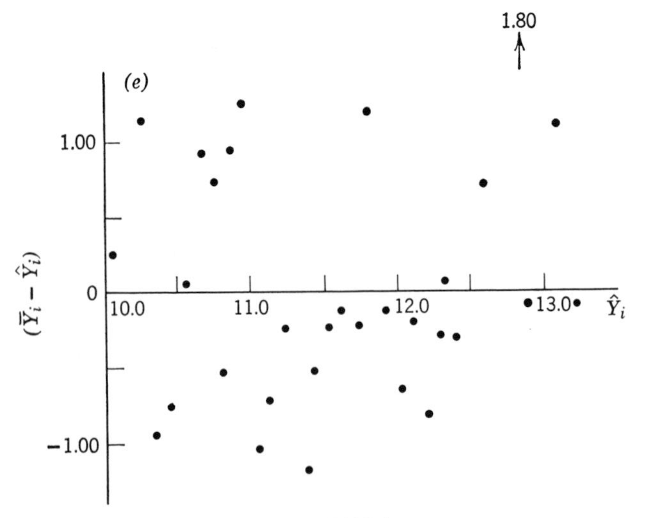

FIGURE E7.1-1e

Figure E7.1-1e is a plot of the residuals based on Equation (a) versus the predicted response $\hat{Y}$. No anomalous data are noted, except perhaps the one residual equal to 1.80. Factors such as: (1) systematic departures from the estimated regression equation (because the model is not adequate), or (2) nonconstant variance can be detected in figures such as E7.1-1e.

7.2 STEPWISE REGRESSION

Stepwise regression consists of sequentially adding (and/or deleting) a variable to an initial linear model and testing at each stage to see if the added variable is significant or not. For the reasons given in Section 5.3, the procedure is most effective when the independent variables are orthogonal because the calculational sequence is then unimportant. (Orthogonal experimental designs to implement this requirement can be found in Chapter 8.) At each stage in the computation, a decision is made as to whether to terminate the computer run with the current regression equation or to move to the next stage by replacing the current regression equation by a new one. By starting to build the model from scratch and by using orthogonal independent variables, a unique model is obtained for a predesignated criterion of best. For nonorthogonal variables, using a different starting equation can lead to a different terminal equation; hence, there may be several adequate models. However, each terminating equation is locally optimal in the sense that it is "better" than the others tested.

Efroymson† described a procedure (the "forward procedure") in which each of the independent variables (and corresponding parameters) is introduced into the model one at a time. At each stage that independent variable is added which causes the greatest reduction in

† M. A. Efroymson, "Multiple Regression Analysis," in *Mathematical Methods for Digital Computers*, ed. by A. Ralston and H. S. Wilf, John Wiley, New York, 1960, p. 191.

the sum of the squares of the residuals, provided the reduction is significant. Essentially the same algorithm can be used to delete independent variables one by one from a full model (the "backward procedure") if the subsequent introduction of another variable renders a model variable insignificant. Several variations of Efroymson's algorithm are available as computer routines.†

In determining at each stage which independent variable to add to a model, use is made of the partial correlation coefficient which we have not yet defined. Suppose that in a model with three variables, defined as X_1, X_2, and X_3, we treat the variables X_1 and X_3 as the dependent and independent random variables, respectively, in a model

$$\mathscr{E}\{X_1 \mid X_3\} = \mu_{x_1} + \beta_{13}(X_3 - \mu_{x_3}) \qquad (7.2\text{-}1)$$

Then, in turn, we treat X_2 and X_3 as the dependent and independent random variables, respectively, in a model

$$\mathscr{E}\{X_2 \mid X_3\} = \mu_{x_2} + \beta_{23}(X_3 - \mu_{x_3}) \qquad (7.2\text{-}2)$$

From the bivariate normal distribution developed in Example 2.3-4, after integration over X_3 we can obtain the conditional probability density:

$$p(X_1 \mid X_3) = \frac{1}{\sqrt{2\pi}\sigma_{X_1}\sqrt{1-\rho_{X_1X_3}^2}} \cdot \exp\left[-\frac{1}{2}\left\{\frac{x_1 - \left[\mu_{X_1} + \rho_{X_1X_3}\left(\frac{\sigma_{X_1}}{\sigma_{X_3}}\right)(x_3 - \mu_{X_3})\right]}{\sigma_{X_1}\sqrt{1-\rho_{X_1X_3}^2}}\right\}^2\right]$$

The expected value of $(X_1 \mid X_3)$ is

$$\mathscr{E}\{X_1 \mid X_3\} = \mu_{X_1} + \rho_{X_1X_3}\left(\frac{\sigma_{X_1}}{\sigma_{X_2}}\right)(X_3 - \mu_{X_3}) \quad (7.2\text{-}3)$$

and the variance is

$$\operatorname{Var}\{X_1 \mid X_3\} = \sigma_{X_1}^2(1 - \rho_{X_1X_3}^2) \qquad (7.2\text{-}4)$$

Similar relations can be obtained for $(X_2 \mid X_3)$, and they will be used shortly.

Next we form the deviations between X_1 and its expected value and between X_2 and its expected value, using Equations 7.2-1 and 7.2-2 as follows:

$$X_{1\cdot 3} \equiv X_1 - \mathscr{E}\{X_1 \mid X_3\} = X_1 - \mu_{X_1} - \beta_{13}(X_3 - \mu_{X_3}) \qquad (7.2\text{-}5)$$

$$X_{2\cdot 3} \equiv X_2 - \mathscr{E}\{X_2 \mid X_3\} = X_2 - \mu_{X_2} - \beta_{23}(X_3 - \mu_{X_3}) \qquad (7.2\text{-}6)$$

† W. J. Dixon, ed., *BMD: Biomedical Computer Programs*, Health Sciences Computing Facility, UCLA, Los Angeles, Calif., 1964; H. Thornber, *A Manual for B34T—A Stepwise Regression Program*, Center for Mathematical Studies in Business and Economics, Univ. of Chicago, Chicago, Ill., 1966.

The symbolism $X_{1\cdot 3}$ denotes a variable X_1 after "eliminating" the effect of variable X_3; a similar connotation applies for X_2 to the symbol $X_{2\cdot 3}$. The partial correlation coefficient between X_1 and X_2, eliminating X_3, is defined as the correlation coefficient of $X_{1\cdot 3}$ and $X_{2\cdot 3}$:

$$\rho_{12\cdot 3} = \frac{\operatorname{Covar}\{X_{1\cdot 3}, X_{2\cdot 3}\}}{\sqrt{\operatorname{Var}\{X_{1\cdot 3}\}\operatorname{Var}\{X_{2\cdot 3}\}}} \qquad (7.2\text{-}7)$$

We now proceed to express $\rho_{12\cdot 3}$ in terms of the usual correlation coefficients $\rho_{X_1X_2}$, $\rho_{X_1X_3}$, and $\rho_{X_2X_3}$, coefficients in which the effect of X_3 is not eliminated; the correlation between X_1 and X_2, for example, may exist solely because both are related to X_3 in some fashion. From Equation 7.2-5, employing Equation 2.2-9,

$$\operatorname{Var}\{X_{1\cdot 3}\} = \operatorname{Var}\{X_1\} + \beta_{13}^2 \operatorname{Var}\{X_3\} - 2\beta_{13}\operatorname{Covar}\{X_1, X_3\}$$

and by comparing Equation 7.2-1 with 7.2-3, we see that $\beta_{13} = \rho_{X_1X_2}(\sigma_{X_1}/\sigma_{X_3})$. Consequently,

$$\begin{aligned}\operatorname{Var}\{X_{1\cdot 3}\} &= \sigma_{X_1}^2 + \rho_{X_1X_2}^2\left(\frac{\sigma_{X_1}}{\sigma_{X_3}}\right)^2\sigma_{X_3}^2 - 2\rho_{X_1X_3}\left(\frac{\sigma_{X_1}}{\sigma_{X_3}}\right)(\rho_{X_1X_3}\sigma_{X_1}\sigma_{X_3}) \\ &= \sigma_{X_1}^2(1 - \rho_{X_1X_3}^2) \qquad (7.2\text{-}8)\end{aligned}$$

Similarly, $\operatorname{Var}\{X_{2\cdot 3}\} = \sigma_{X_2}^2(1 - \rho_{X_2X_3}^2)$. Finally,

$$\begin{aligned}\operatorname{Covar}\{X_{1\cdot 3}, X_{2\cdot 3}\} &= \mathscr{E}\{X_{1\cdot 3}, X_{2\cdot 3}\} - \mathscr{E}\{X_{1\cdot 3}\}\mathscr{E}\{X_{2\cdot 3}\} \\ &= \sigma_{X_1}\sigma_{X_2}(\rho_{X_1X_2} - \rho_{X_1X_3}\rho_{X_2X_3}) \qquad (7.2\text{-}9)\end{aligned}$$

We are now able to express $\rho_{12\cdot 3}$ as

$$\rho_{12\cdot 3} = \frac{\rho_{X_1X_2} - \rho_{X_1X_3}\rho_{X_2X_3}}{\sqrt{(1 - \rho_{X_1X_3}^2)(1 - \rho_{X_2X_3}^2)}} \qquad (7.2\text{-}10)$$

The estimated correlation coefficients from a sample can be used in lieu of the respective ensemble correlation coefficient to give an estimate of $\rho_{12\cdot 3}$, $\hat{\rho}_{12\cdot 3}$. Although the calculation given by Equation 7.2-10 can be extended to obtain the partial correlation coefficient between two variables, eliminating the effect of several others, in general the quickest way to calculate the estimated partial correlation coefficient between Y and a single x among many x's, say x_j, eliminating the effect of all the other x's, is to include x_j in the model and then compute‡

$$\hat{\rho}_{Yx_j\cdot x_k}(k \neq j) = \frac{\dfrac{b_j}{s_{b_j}}}{\left(\dfrac{b_j}{s_{b_j}} + n - q - 1\right)^{1/2}} \qquad (7.2\text{-}11)$$

We return now to the construction of a model by stepwise regression. A typical computer program will

‡ R. L. Gustafson, *J. Amer. Stat. Assn.* **56**, 363, 1961.

accept a list of a suggested sequence of variables to include in the model and/or it will select from the total list the one with the largest partial correlation coefficients $\hat{\rho}_{Yx_j \cdot x_k}$. After one variable is included, a second can be added which has the largest partial correlation coefficient among the remaining variables, and so forth. Various types of tests of significance can be applied to determine whether or not the added variable is a significant one. But, in general, either the variance ratio test described in Sections 5.2 and 5.3 or a test on the statistic $Z = b_j/s_{bj}$ is employed.

In the example below a t-test of significance for $\hat{\rho}_{Yx_j \cdot x_k}$ is made. If the ensemble parameter $\rho_{Yx_j \cdot x_k}$ is zero, i.e., the expected value of $\hat{\rho}_{Yx_j \cdot x_k}$ is zero, the estimated variance of $\hat{\rho}_{Yx_j \cdot x_k}$ can be shown to be equal to $(1 - \hat{\rho}^2_{Yx_j \cdot x_k})/(\nu)$, where $\nu =$ the number of data sets less the number of terms absorbed in the correlation. If the correlation is between Y and one x_j, $\nu = n - 2$; if it is between Y and one x_j after incorporating one x_k in the model, $\nu = n - 3$, and so forth. Thus, the statistic t is

$$t = \frac{\hat{\rho}_{Yx_j \cdot x_k}}{[(1 - \hat{\rho}^2_{Yx_j \cdot x_k})/\nu]^{1/2}} \tag{7.2-12}$$

A variable once included in the model can be later deleted if it no longer makes a significant contribution to the sum of the squares.

Example 7.2-1 Stepwise regression

Stepwise regression was used to identify the terms in a model of an extruder that related the shear stress to the temperature, viscosity, and pressure:

$$V = \beta_0 + \beta_1(x_1 - \bar{x}_1) + \beta_2(x_2 - \bar{x}_2) + \beta_3(x_3 - \bar{x}_3) + \epsilon$$

where, in coded units,

$$V = \text{shear stress}$$
$$x_1 = \text{viscosity}$$
$$x_2 = \text{temperature}$$
$$x_3 = \text{pressure}$$

The matrix of observations was (with $Y = V$)

Y	x_1	x_2	x_3
3.98	1	0	0
−5.10	0	2	−1
−1.03	−1	3	2
9.00	4	10	1
32.0	2	0	8

In the first stage of the analysis the reductions in the sum of the squares, using in turn each of the individual terms to initiate the model, were:

Adding the Term	Reduction in Sum of Squares	Mean Square	Square of the Partial Correlation Coefficient $\hat{\rho}^2_{Yx_j \cdot x_k}$
x_1	28.9	28.9	0.271
x_2	23.5	23.5	0.028
x_3	722.0	722.0	0.857

Inasmuch as the partial correlation coefficient for x_3 was the largest coefficient, x_3 became the key variable. The significance of the term involving x_3 was tested by a t-test of significance, using Equation 7.2-2:

$$t = \frac{(0.857)^{1/2}(5 - 2)^{1/2}}{(1 - 0.857)^{1/2}} = 4.24$$

From Table C.3 in Appendix C, $t_{0.95} = 2.35$ for 3 degrees of freedom, indicating that x_3 was a significant variable and should be incorporated in the model.

Next, the sum of squares and the partial correlation coefficients for the remaining two terms were reevaluated by assuming that x_3 was already incorporated in the model (as described in Section 5.3 by suitable manipulation of matrix elements) to give

Adding the Term	Reduction in Sum of Squares	Mean Square	Square of the Partial Correlation Coefficient $\hat{\rho}^2_{Yx_j \cdot x_k \mid x_3}$
x_1	91.7	91.7	0.758
x_2	68.7	68.7	0.057

The key variable was now x_1; the t-test based on

$$t = \left[\frac{(0.758)(5 - 3)}{(1 - 0.758)}\right]^{1/2} = 2.10$$

indicated that variable x_1 did not contribute (nor would x_2) significantly to the model.

As a consequence of the forward stepwise regression procedure, the estimated regression equation would be

$$\hat{Y} = \bar{Y} + b_3(x_3 - \bar{x}_3)$$

or

$$\hat{Y} = 0.201 + 0.38(x_3 - \bar{x}_3)$$

(The intercept of the model was a significant term.) The model was so simple that no tests for stepwise removal of a variable had to be carried out as might be the case for a more complicated model.

7.3 A GRAPHICAL PROCEDURE FOR SCREENING MODELS

If there are q independent variables proposed for incorporation in a linear (in the parameters) model, there are 2^q possible models. Clearly, as q becomes large, it is not possible, even if the analyst wishes, to evaluate all the possible models. Also, if the experiment that provides the data has not been designed properly, inadvertently or otherwise, stepwise regression may lead to confusing results, especially when the independent variables are highly correlated. Furthermore, instead of one best model existing, several equally good models may exist. In view of these rather common handicaps, Gorman and Toman suggested a simple graphical method of screening models which has considerable merit.† The

† J. W. Gorman and R. J. Toman, *Technometrics* **8**, 27, 1966.

development and the example which follow have been taken from their work.

They used a graphical method for isolating the best among several regression equations based on a statistic originally proposed by Mallows.‡ For n data points and a linear model containing q parameters, an estimate of the sum of the squares of the bias error plus the residual error, both divided by $\sigma^2_{\bar{Y}_i}$, is given by the statistic C_q:

$$C_q = \frac{\sum_{i=1}^{n} [\bar{Y}_i - \hat{Y}_i(q)]^2}{s^2_{\bar{Y}_i}} - (n - 2q) \tag{7.3-1}$$

When a q-term model has negligible bias,

$$\sum_{i=1}^{n} [\bar{Y}_i - \hat{Y}_i(q)]^2$$

is equal to $s_r^2(n - q)$ (refer to Table 5.3-1 where there are $(q + 1)$ parameters involved), so that

$$C_q \approx \frac{(n - q)s_r^2}{s^2_{\bar{Y}_i}} - (n - 2q) \tag{7.3-2}$$

However, $s_r^2 \approx s^2_{\bar{Y}_i}$, so that

$$C_q \approx q \tag{7.3-3}$$

Equation 7.3-3 is interpreted as follows. When C_q calculated by Equation 7.3-1 is plotted against q for different values of q, models leading to unbiased estimated regression equations will yield C_q's that tend to cluster about the line $C_q = q$. A point well away from the line $C_q = q$ represents a biased equation. An unbiased estimate of $\sigma^2_{\bar{Y}_i}$ must be used for $s^2_{\bar{Y}_i}$ in Equation 7.3-1, because if only s_r^2 is available, the graphical approach causes the values of C_q to be near the line given by Equation 3.3-3.

Figure 7.3-1 illustrates the graphical procedure of Gorman and Toman, using the data of Example 5.3-1 as shown in Table 7.3-1. Model I has the least bias and also the least sum of the squares. A point such as (A) in Figure 7.3-1 would have a smaller sum of squares but more bias. If an equation is needed just for interpolation in the region in which the data are collected, it may be useful to accept the equation corresponding to point (A)

‡ C. Mallows, Paper presented at the Central Regional Meeting, I.M.S., Manhattan, Kansas, May 7-9, 1964.

TABLE 7.3-1

Model	Variables	q	$\sum(\bar{Y}_i - \hat{Y}_i)^2$	C_q
IV	None	1	77.28	87.6
III	x_1, x_2	3	10.07	3.3
II	x_1	2	12.48	4.4
I	x_2	2	11.23	2.8

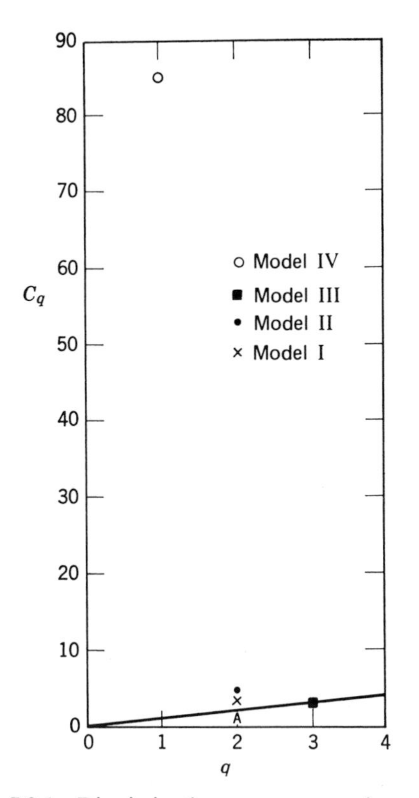

FIGURE 7.3-1 Discrimination among regression equations.

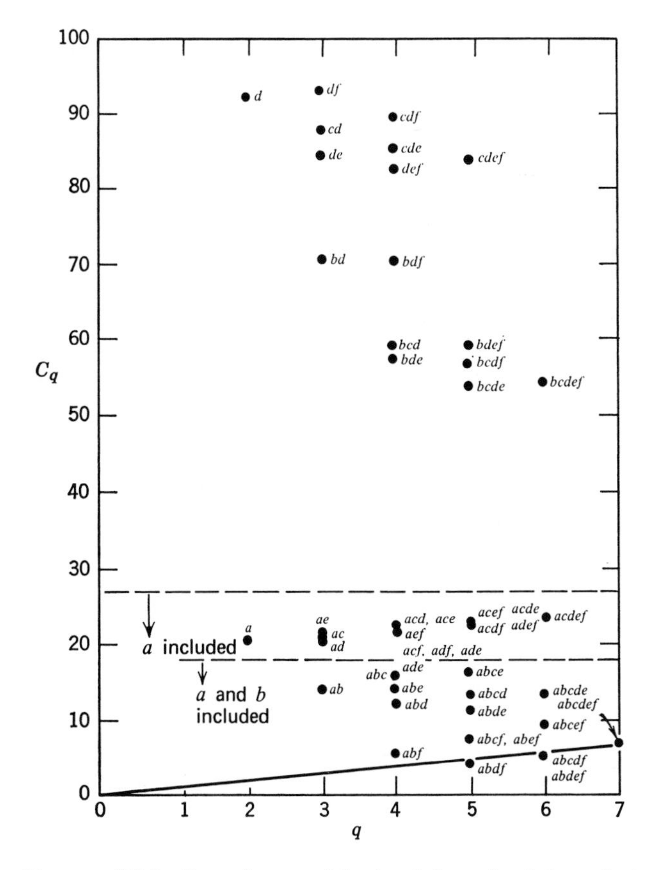

FIGURE 7.3-2 Screening models involving six independent variables, x_a through x_f.

in order to use a simpler equation with a smaller variance, s_r^2, and to accept some bias. If the number of variables is large, Gorman and Toman suggested fractional factorial designs (described in Chapter 8) to reduce the extent of the calculations. Figure 7.3-2, based on the work of Gorman and Toman, is a plot corresponding to Figure 7.3-1, except that six variables are involved in the possible models, x_a through x_f. Forty-eight models were considered (only those with estimated C_q's less than 100 are plotted) of which the models containing the variables (the x's are suppressed) *abf*, *abdf*, *abdef*, and *abcdf* are candidates for further examination.

7.4 COMPARISON OF TWO REGRESSION EQUATIONS

We now turn to methods of discrimination among models that apply equally well to linear *or* nonlinear models. We shall first examine tests that can be applied to two estimated regression equations to determine which is the best. In the next section we shall take up the important case of discrimination among many estimated regression equations.

Many different comparison tests can be executed depending upon the specific hypothesis selected. In the unsymmetrical test suggest by Hoel,† an estimated regression equation $\hat{Y}_1$ is presumed to represent correctly the data and a test is carried out to see whether $\hat{Y}_1$ should be abandoned in favor of $\hat{Y}_2$, another regression equation. The test is made by finding the slope λ of the relation

$$Z \equiv (Y - \hat{Y}_1) = \lambda(\hat{Y}_2 - \hat{Y}_1) \qquad (7.4\text{-}1)$$

where Y represents the empirical measurement, $\bar{Y}_i$, and Z is the dependent variable. If λ is significantly positive, then $\hat{Y}_1$ is rejected in favor of $\hat{Y}_2$.

To select among two estimated regression equations which, initially at least, are equally feasible, one can use the symmetric test of Williams and Kloot.‡ The null hypothesis is that the two (perhaps nonlinear) regression equations are of equal ability in predicting the values of Y. The test is carried out by estimating the slope λ (obtained by linear regression) of the equation which passes through the origin:

$$Z \equiv [Y - \tfrac{1}{2}(\hat{Y}_1 + \hat{Y}_2)] = \lambda(\hat{Y}_2 - \hat{Y}_1) \qquad (7.4\text{-}2)$$

The argument underlying Equation 7.4-2 is as follows. Suppose two models exist:

$$\eta_1 = \eta_1(\mathbf{x}, \boldsymbol{\beta})$$

$$\eta_2 = \eta_2(\mathbf{x}, \boldsymbol{\beta})$$

in which the $\mathbf{x}$ and $\boldsymbol{\beta}$ vectors do not have to be the same. Furthermore, suppose that Model 1 is correct so that

$$Y = \eta_1 + \epsilon$$

† P. G. Hoel, *J. Amer. Stat. Assn.* **42**, 605, 1947.
‡ E. J. Williams and N. H. Kloot, *Aust. J. Appld. Sci* **4**, 1, 1953.

If we define the variable

$$Z \equiv [Y - \tfrac{1}{2}(\hat{Y}_1 + \hat{Y}_2)]$$

and substitute in for Y, we obtain

$$Z = \eta_1 - \tfrac{1}{2}(\hat{Y}_1 + \hat{Y}_2) + \epsilon$$

Let us now assume that $\mathbf{b}$ is very close to $\boldsymbol{\beta}$ and, consequently, $\hat{Y}_1$ is very close to η_1. Then, if we replace η_1 with $\hat{Y}_1$:

$$Z \cong -\tfrac{1}{2}(\hat{Y}_2 - \hat{Y}_1) + \epsilon \qquad (7.4\text{-}3)$$

and a plot of $[Y - \frac{1}{2}(\hat{Y}_1 + \hat{Y}_2)]$ versus $(\hat{Y}_2 - \hat{Y}_1)$ should have a slope of approximately $-\frac{1}{2}$ if the hypothesis about Model 1 being correct is true. We can infer that a significant negative λ indicates that $\hat{Y}_1$ is a better estimated regression equation than $\hat{Y}_2$; hence Model 1 is better than Model 2. A similar analysis with the supposition that $\hat{Y}_2$ is the correct equation leads to the conclusion that $\lambda = \frac{1}{2}$; i.e., a significant positive slope should be found. If λ is not significantly different from zero, no choice can be made between $\hat{Y}_1$ and $\hat{Y}_2$.

Example 7.4-1 The Williams and Kloot Test

The following data are for flood damage as a function of the discharge in Little Lehigh Creek in the Lehigh River Basin. These data represent an undesigned experiment, but they fulfill reasonably well the estimation assumption of independent error since they were collected at different time periods (the dollar values have been adjusted to a 1956 price level).

x Discharge, cfs $\times 10^{-3}$	Y Damage, \$ $\times 10^{-3}$ (1956 price levels)
61	0
64	50
70	100
75	150
83	180
88	210
94	250
100	290
105	340
112	420
120	520
127	670
134	810
142	1200
150	1600
160	2100
170	2500
180	2900
190	3300
200	3700

A plot (see Figure E7.4-1a) of the data indicated that a power series might fit the data well. To keep the model as

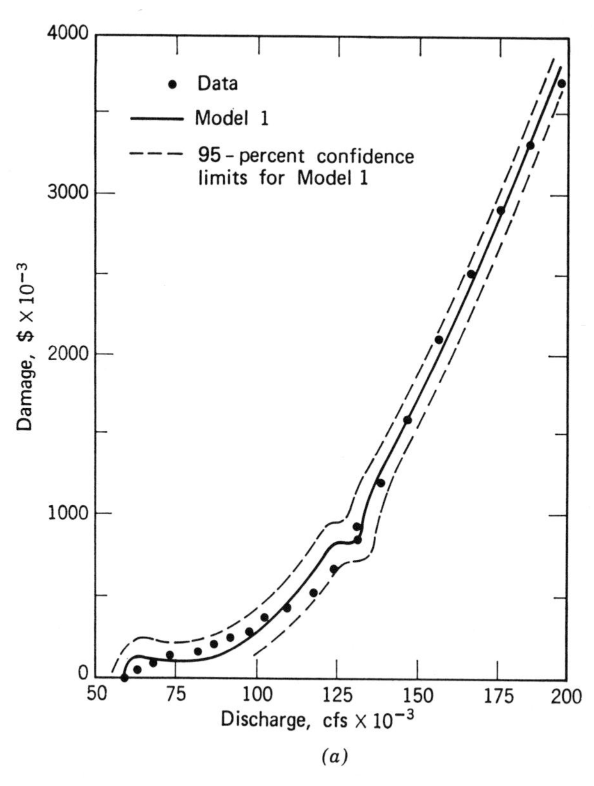

(a)

FIGURE E7.4-1a

TABLE E7.4-1a ESTIMATED REGRESSION EQUATIONS FOR MODELS*

Model 1: $\hat{Y}_1 = b_0 + b_1x + b_2x^2 + b_3x^3 + \dfrac{b_4}{x - 60}$

$b_0 = 2840 \pm 1490$
$b_1 = -74.1 \pm 37.3$
$b_2 = 0.572 \pm 0.298$
$b_3 = -8.92 \times 10^{-4} \pm 7.58 \times 10^{-4}$
$b_4 = 267 \pm 289$

Model 2: $\hat{Y}_2 = b_0 + b_1x + b_2x^2 + b_3x^3$

$b_0 = 1990 \pm 1280$
$b_1 = -55.3 \pm 33.7$
$b_2 = 0.437 \pm 0.280$
$b_3 = -580 \times 10^{-4} \pm 7.33 \times 10^{-4}$

Model 3: $\hat{Y}_3 = b_0 + b_1x + b_2x^2$

$b_0 = 1050 \pm 452$
$b_1 = -292 \pm 7.6$
$b_2 = 0.217 \pm 0.030$

Model 4: $\hat{Y}_4 = b_0 + b_1x + b_2x^2 + \dfrac{b_4}{x - 60}$

$b_0 = 1190 \pm 589$
$b_1 = -31.3 \pm 9.36$
$b_2 = 0.224 \pm 0.035$
$b_3 = -115 \pm 297$

* $\pm$ indicates value to be added to b_k for 95 percent confidence interval for β_k.

TABLE E7.4-1b RESIDUAL SUM OF SQUARES

Model	ν d.f.	$\sum_{i=1}^{n} (\bar{Y}_i - \hat{Y}_i)^2$, (SSR × 10^{-3})	Mean Square × 10^{-3} (s_r^2)
1	15	120	8.0
2	16	151	9.4
3	17	178	10.5
4	16	171	10.7

simple as possible, linear regression equations containing x, x^2, and x^3 were fit. To improve the fit in the vicinity of $x = 61$, a variable $1/(x - 60)$ was added to the polynomial. A term incorporating such a variable will have a large value near its pole but a negligible value for large x.

The independent variables were scaled as follows to make the terms in the regression equation of roughly the same order of magnitude:

$$x_1 = x(10^{-2})$$
$$x_2 = x^2(10^{-4})$$
$$x_3 = x^3(10^{-6})$$
$$x_4 = \frac{1}{x - 60}(10)$$
$$Y^* = Y(10^{-3})$$

Table E7.4-1a lists the regression coefficients for four linear models as estimated by a least squares procedure with $w_i = 1$.

Table E7.4-1b lists the sum of the squares of the residuals for each model and the respective mean squares, s_r^2. There

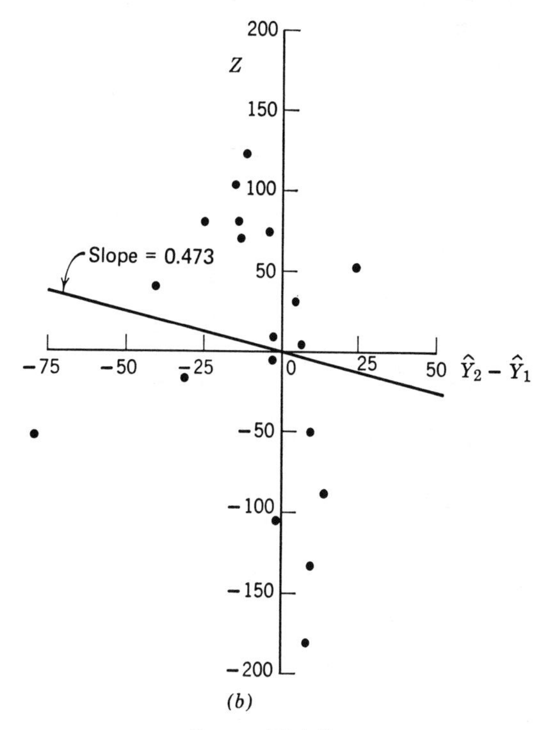

(b)

FIGURE E7.4-1b

TABLE E7.4-1c CALCULATIONS FOR THE WILLIAMS AND KLOOT TEST

Y	$\hat{Y}_1$	$\hat{Y}_2$	$\hat{Y}_2 - \hat{Y}_1$	$(\frac{1}{2})(\hat{Y}_1 + \hat{Y}_2)$	$Z = Y - (\frac{1}{2})(\hat{Y}_1 + \hat{Y}_2)$
0	−19	118	137	50	−50
50	143	64	−79	103	−53
100	126	96	−30	111	−11
150	110	69	−41	90	40
180	113	88	−25	100	80
210	136	122	−14	129	81
250	186	182	−4	184	66
290	257	261	4	259	31
340	334	341	7	337	3
420	465	475	10	470	−50
520	647	657	10	652	−132
670	833	841	8	837	−167
910	835	888	48	859	51
1200	1305	1304	−1	1305	−105
1600	1593	1587	−6	1590	10
2100	1984	1971	−13	1978	122
2500	2404	2389	−15	2397	103
2900	2847	2834	−13	2840	60
3300	3308	3304	−4	3306	−6
3700	3782	3796	14	3789	−89

appear to be no significant differences in the fits of the models. By arbitrarily eliminating the two models with the largest variances, the choice of models is reduced to a choice between Model 1 and Model 2. The only difference between them is that Model 1 contains the extra term $\beta_4/(x - 60)$. We shall use the test of Williams and Kloot to ascertain which of the two models is the best.

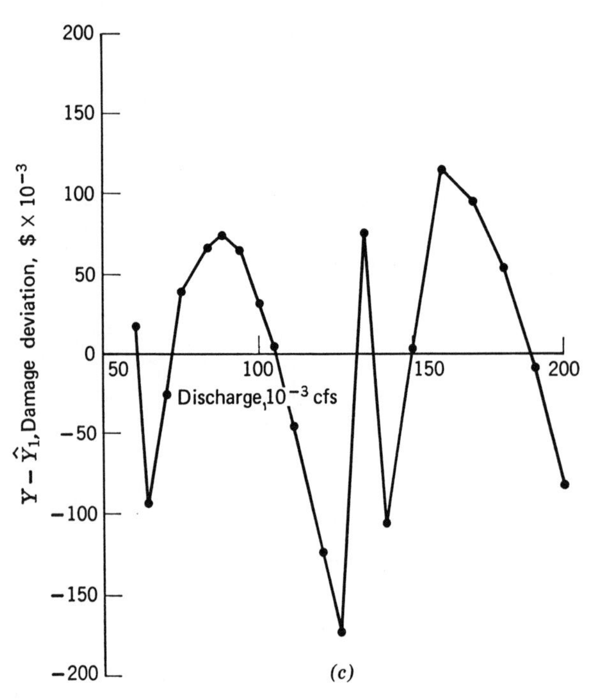

FIGURE E7.4-1c Residuals for Model 1.

Table E7.4-1c lists the data and calculations needed for Equation 7.4-2. Figure E7.4-1b is a plot of $[Y-(\frac{1}{2})(\hat{Y}_1+\hat{Y}_2)]$ versus $(\hat{Y}_2 - \hat{Y}_1)$ for all the values of Z except the first row in Table E7.4-1c in which a negative $\hat{Y}_1$ appears. The slope of the best fitting line through the original, computed from Equation 4.3-7a, was -0.473. However, the Var $\{b\} \sim 8000/13{,}104 = 0.61$, and the confidence interval for β for a significance level of $\alpha = 0.05$ ($t_{1-\frac{\alpha}{2}} = 2.13$ for 15 degrees of freedom), $-2.13 \leq \beta < 1.19$, does not lead to the conclusion that Model 1 is any better than Model 2.

Figure E7.4-1c is a plot of the residuals for Model 1. Although no longrange trends are visible, the residuals are not randomly distributed. Several shortrange trends are visible in the discharge regions of 90 to 125 × 10^{-3} cfs and 160 to 200 × 10^{-3} cfs. The existence of shortrange trends in the residuals does not invalidate the model, but it does demonstrate that the model can be improved somewhat.

7.5 COMPARISON AMONG SEVERAL REGRESSION EQUATIONS

To compare several linear or nonlinear (in the coefficients) estimated regression equations simultaneously, Wilks† developed a test in which all the regression equations are considered to be on an equal footing. The test is posed in terms of the homogeneity of the residual sums of the squares for different regression equations. Williams‡ gave a lucid description of the Wilks test and

† S. S. Wilks, *Ann. Math. Stat.* **17**, 257, 1946.
‡ E. J. Williams, *Regression Analysis*, John Wiley, New York, 1959.

TABLE 7.5-1 ANALYSIS OF VARIANCE FOR MODEL SELECTION

Source of Variation	Degrees of Freedom	Sum of Squares	Mean Square
Improvement of $\hat{Y}^*$ over $\bar{\hat{Y}}$	$p - 1$	$\sum_{i=1}^{n} (Y_i - \bar{\hat{Y}})^2 - \sum_{i=1}^{n} (Y_i - \hat{Y}^*)^2$	$s_3^2 = \dfrac{\sum_{i=1}^{n} (Y_i - \bar{\hat{Y}})^2 - \sum_{i=1}^{n} (Y_i - \hat{Y}^*)^2}{p - 1}$
Deviation from $\hat{Y}^*$	$n - p + 1$	$\sum_{i=1}^{n} (Y_i - \hat{Y}^*)^2$	$s_2^2 = \dfrac{\sum_{i=1}^{n} (Y_i - \hat{Y}^*)^2}{n - p + 1}$
Deviation from $\bar{\hat{Y}}$	n	$\sum_{i=1}^{n} (Y_i - \bar{\hat{Y}})^2$	$s_1^2 = \dfrac{\sum_{i=1}^{n} (Y_i - \bar{\hat{Y}})^2}{n}$

also pointed out that it is essentially an approximate method since the sums of the squares for the equations which are not the "true" ones contain an extra systematic component not present for the "true" equation.

Recall that for a single estimated regression equation an analysis of variance can be made (Tables 5.3-1 and 5.3-2) which leads to the F-test as an overall test of significance for regression. An F-test can be used to discriminate among different estimated regression equations if they are assembled in a linear combination as follows. Let the various regression equations to be compared be designated $\hat{Y}_1, \hat{Y}_2, \ldots, \hat{Y}_p$, and let $\hat{Y}^*$ be a linear combination of the regression equations

$$\hat{Y}^* = b_1^* \hat{Y}_1 + b_2^* \hat{Y}_2 + \cdots + b_p^* \hat{Y}_p \tag{7.5-1}$$

with the b^*'s chosen so that each regression equation contributes toward $\hat{Y}^*$ according to its fitness as an estimator of $\hat{Y}^*$. For convenience we shall adjust the b^*'s so that $\sum_{k=1}^{p} b_k^* = 1$. It would also seem wise to restrict the β_k^*'s to be $0 \leq \beta_k^* \leq 1$. Define

$$\bar{\hat{Y}} = \frac{1}{p} \sum_{k=1}^{p} \hat{Y}_k \tag{7.5-2}$$

Suppose we consider $\hat{Y}_{ij}$, where the index i indicates the predicted value of Y for the ith set of data by the jth regression equation, as the independent variable with one set of $\hat{Y}_{ij}$'s existing for each observed dependent variable Y_i ($\bar{Y}_i$ if replicate observations are made). The test involves determining if the compound variable $\hat{Y}^*$ is a significant improvement over the average predictor $\bar{\hat{Y}}$. Table 7.5-1 summarizes the calculations needed for the analysis of variance. If the variance ratio s_3^2/s_2^2 with $(p - 1)$ and $(n - p + 1)$ degrees of freedom for the numerator and denominator, respectively, is greater than $F_{1-\alpha}$, then the null hypothesis that the compound function makes no significant improvement over the average $\bar{\hat{Y}}$ is rejected.

Once the b^*'s are computed, as described below, their order of rank is a rough measure of the effectiveness of each regression equation in fitting the experimental data. Furthermore, any two b^*'s can be tested as described in Section 5.2 to determine if there is a significant difference between them and, thus, to conclude whether one estimated regression equation is better than another.

The scope of the calculations can be simplified by computing the following quantities. To make the notation clear, we shall define a $p \times p$ matrix $\mathbf{V}$ whose element $[V_{jk}]$ is

$$[V_{jk}] = \sum_{i=1}^{n} (Y_i - \hat{Y}_{ij})(Y_i - \hat{Y}_{ik}) \qquad \begin{Bmatrix} 1 \leq i \leq n \\ 1 \leq j \leq p \\ 1 \leq k \leq p \end{Bmatrix}$$

where Y_i is the observed experimental dependent variable in the ith data set and $\hat{Y}_{ij}$ is the predicted response. As examples,

$$V_{11} = \sum_{i=1}^{n} (Y_i - \hat{Y}_{i1})^2$$

$$V_{12} = \sum_{i=1}^{n} (Y_i - \hat{Y}_{i1})(Y_i - \hat{Y}_{i2})$$

The elements of the inverse matrix $\mathbf{V}^{-1}$ are $[V_{jk}]^{-1}$. In terms of this notation,

$$\sum_{i=1}^{n} (Y_i - \bar{\hat{Y}})^2 = \frac{1}{p^2} \sum_{j=1}^{p} \sum_{k=1}^{p} [V_{jk}] \tag{7.5-3}$$

$$\sum_{i=1}^{n} (Y_i - \hat{Y}^*)^2 = \frac{1}{\sum_{j=1}^{p} \sum_{k=1}^{p} [V_{jk}]^{-1}} \tag{7.5-4}$$

$$b_k^* = \frac{\sum_{j=1}^{p} [V_{jk}]^{-1}}{\sum_{j=1}^{p} \sum_{k=1}^{p} [V_{jk}]^{-1}} \tag{7.5-5}$$

$$\operatorname{Var}\{b_k^*\} = \sigma_{\hat{Y}\cdot}^2\left\{[V_{kk}]^{-1} - \frac{\left(\sum_{j=1}^{p}[V_{jk}]^{-1}\right)^2}{\sum_{j=1}^{p}\sum_{k=1}^{p}[V_{jk}]^{-1}}\right\}$$

$$= \sigma_{\hat{Y}\cdot}^2\left\{[V_{kk}]^{-1} - (b_k^{*2})\sum_{j=1}^{p}\sum_{k=1}^{p}[V_{jk}]^{-1}\right\} \qquad (7.5\text{-}6)$$

$$\operatorname{Covar}\{b_r^*, b_s^*\} = \sigma_{\hat{Y}\cdot}^2\left\{[V_{rs}]^{-1} - \frac{\sum_{j=1}^{p}[V_{jr}]^{-1}\sum_{j=1}^{p}[V_{js}]^{-1}}{\sum_{j=1}^{p}\sum_{k=1}^{p}[V_{jk}]^{-1}}\right\}$$

$$= \sigma_{\hat{Y}\cdot}^2\left\{[V_{rs}]^{-1} - b_r^* b_s^*\sum_{j=1}^{p}\sum_{k=1}^{p}[V_{jk}]^{-1}\right\} \qquad (7.5\text{-}7)$$

The variance $\sigma_{\hat{Y}\cdot}^2$ can be estimated from

$$s_{\hat{Y}\cdot}^3 = \frac{1}{(n-p+1)}\left(\frac{1}{\sum_{j=1}^{p}\sum_{k=1}^{p}[V_{jk}]^{-1}}\right) \qquad (7.5\text{-}8)$$

To test the hypothesis that b_j^* is different from some constant γ, we can compute

$$t = \frac{b_j^* - \gamma}{s_{b_j^*}}$$

and for a significance level α, we can determine if t is greater than $t_{1-\frac{\alpha}{2}}$ obtained from the tables for $(n-p)$ degrees of freedom. To test for a difference between two parameters, b_j^* and b_k^*, and some constant γ (perhaps 0), a value of t is computed as follows:

$$t = \frac{(b_j^* - b_k^*) - \gamma}{[\operatorname{Var}\{b_j^*\} + \operatorname{Var}\{b_k^*\} - 2\operatorname{Covar}\{b_j^*, b_k^*\}]^{1/2}}$$

If the null hypothesis is rejected, then b_j^* and b_k^* are different, and $\hat{Y}_j$ is a better fitting equation than $\hat{Y}_k$. Another essentially equivalent test would be to follow the scheme outlined in Table 5.3-2 and remove in sequence a $\hat{Y}_j$ from Equation 7.5-1, testing each time to see if the deletion is significant.

Example 7.5-1 The Wilks Test

To illustrate the numerical computations in the Wilks test, the following data were generated using the model $Y = 2 + x^2 + \epsilon$:

x	Y
1	3.1
2	5.9
3	11.1
4	17.8
5	27.2

Two linear models were fit to the data:

$$\eta_1 = b_{01} + b_{11}x_1 \qquad \text{with } x_1 = x$$
$$\eta_2 = b_{02} + b_{21}x_2 \qquad \text{with } x_2 = x^2$$

The respective estimated regression equations were determined by the methods of Section 4.3 as

$$\hat{Y}_1 = -4.83 + 6.01x$$
$$\hat{Y}_2 = 2.20 + 1.00x^2$$

The elements of the **V** matrix were next calculated:

$$V_{11} = \sum_{i=1}^{5}(Y_i - \hat{Y}_{i1})(Y_i - \hat{Y}_{i1}) = 1.1883 \times 10^3$$

$$V_{12} = V_{21} = 1.1795 \times 10^3$$
$$V_{22} = 1.1807 \times 10^3$$

and thereafter the inverse matrix $\mathbf{V}^{-1}$:

$$\mathbf{V}^{-1} = \begin{bmatrix} 0.10001 & -0.09991 \\ 0.09991 & 0.10065 \end{bmatrix}$$

The estimates of $\mathbf{b}^*$ were

$$\mathbf{b}^* = \begin{bmatrix} 0.120 \\ 0.880 \end{bmatrix}$$

We conclude, as expected, that Model 2 is the best.

Example 7.5-2 Selection Among Nonlinear Models

An example based on simulated data illustrates the difficulty of selecting the best model among several nonlinear models.† Fifteen data points (shown in Figure E7.5-2) were generated from the model

$$\eta = 10 + 100(1 - e^{-0.115x}) \qquad \text{(a)}$$

Random deviations with a mean of zero and a standard deviation of σ were selected from a truncated (at 2σ) normal distribution and added to each value of η.

Five different nonlinear models were fitted to the simulated data

(1) $\eta = \beta_0 + \beta_1(1 - e^{-\beta_2 x})$

(2) $\eta = \beta_0 + \beta_1\left(\frac{\beta_2 x}{1 + \beta_2 x}\right)$

(3) $\eta = \beta_0 + \beta_1[\tan^{-1}(\beta_2 x)]$

(4) $\eta = \beta_0 + \beta_1[\tanh(\beta_2 x)]$

(5) $\eta = \beta_0 + \beta_1 e^{-\beta_2/x}$

with the corresponding estimated regression equations shown in Figure E7.5-2.

Each equation fitted the data well and closely resembled the shape of the other four curves. As increasing error (from $\sigma^2 = 0$ to $\sigma^2 = 100$) was introduced, the estimates of the parameters in each model remained fairly stable, and $s_{\hat{Y}_i}^2$ increased in proportion to σ^2. The covariances, i.e., elements off the main diagonal of Covar $\{\mathbf{b}\}$, remained very much less than the variances.

† W. L. Wilcoxson, U.S. Naval Civil Engineering Laboratory Tech. Rept. R419, Port Hueneme, Calif., Dec. 1965.

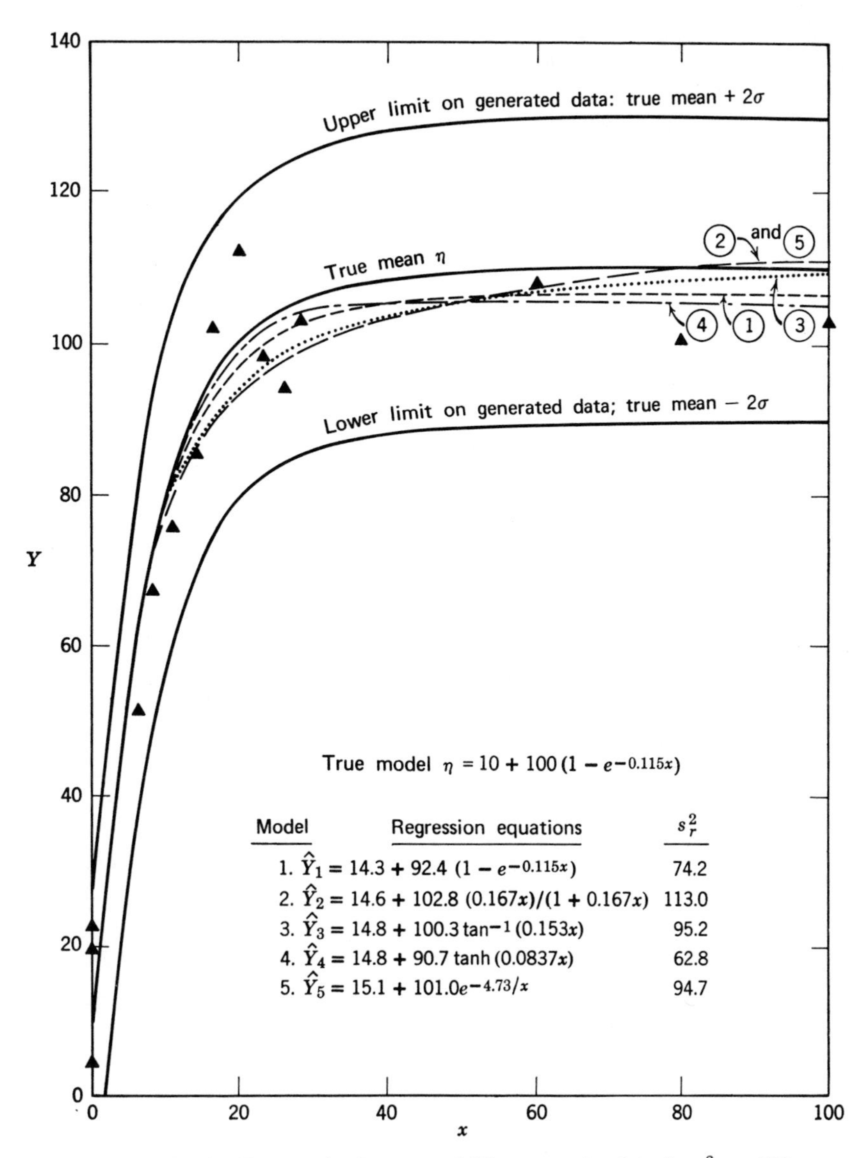

FIGURE E7.5-2 Test results for a set of 15 generated points for $\sigma^2_{Y_i} = 100$.

TABLE E7.5-2 TEST RESULTS OF THE MODEL SELECTION PROGRAM

Degree of Random Error Introduced Into True Model, σ	Fraction of Cases in which Correct Model was Selected by: Coefficients **b***			Minimum $s^2_{Y_i}$		
	C*	A	N	C	A	N
0	1.0	—	—	1.0	—	—
0.5	—	0.8	0.2	0.8	0.2	—
1.0	—	0.5	0.5	0.8	0.2	—
2.5	—	0.4	0.6	0.6	0.4	—
5.0	—	0.2	0.8	0.4	0.6	—
10.0	—	—	1.0	—	1.0	—

* C = correct model selected; A = another model selected, N = no model selected.

Wilcoxson carried out the Wilks test by using the five regression equations; he concluded that the coefficients **b*** were not arranged in descending order according to increasing values of $s^2_{Y_i}$. He also showed that selection by using the minimum $s^2_{Y_i}$ was in general a better selection tool for identifying the true model than using the b*'s. Examine Table E7.5-2.

Supplementary References

Gorman, J. W. and Roman, R. J., "Selection of Variables for Fitting Equations to Data," *Technometrics* **8**, 27, 1966.

Hoel, P. G., "On the Choice of Forecasting Formulas," *J. Amer. Stat. Assn.* **42**, 605, 1947.

Hotelling, H., "The Selection of Variates for Use in Prediction with Some Comments on the General Problem of Nuisance Parameters," *Ann. Math. Stat.* **11**, 271, 1940.

Larson, H. J. and Bancroft, T. A., "Sequential Model Building for Prediction in Regression Analysis," *Ann. Math. Stat.* **34**, 462, 1963.

Mezaki, R. and Kittrell, J. R., "Discrimination Between Two Rival Models through Nonintrinsic Parameters," *Can. J. Chem. Eng.* **44**, 285, 1966.

Newton, R. G. and Spurrell, D. J., "A Development of Multiple Regression for the Analysis of Routine Data," *Appl. Stat.* **16**, 52, 1967.

Problems

7.1 Carry out an analysis of residuals based on the estimated regression equation determined in Example 4.3-2. What interpretation can you give to the experiment and model of that example in addition to that already provided in the example?

7.2 The data of Example 7.4-1 were fitted by the linear model

$$Y = \alpha + \beta_1 x + \beta_2 x^2 + \beta_3 x^3 + \frac{\beta_4}{x - 60} + \epsilon$$

and the following estimates of the coefficients obtained

$$a = 2.8389$$
$$b_1 = -7.4057$$
$$b_2 = 5.7223$$
$$b_3 = -0.8916$$
$$b_4 = -0.02669$$

In addition, the computer printout indicated the following residuals (truncated):

Data Set Number	Residual
1	0.018
2	−0.093
3	−0.026
4	0.040
5	0.067
6	0.074
7	0.064
8	0.033
9	0.006
10	−0.045
11	−0.127
12	−0.163
13	0.075
14	−0.105
15	0.006
16	0.116
17	0.096
18	0.053
19	−0.008
20	−0.082

Carry out an analysis of residuals as indicated in Section 7.1.

7.3 The following model was fitted to the experimental data designated "actual value" in the IBM printout by stepwise regression.

$$\text{VISI} = 14.637262 - 11.204441\ \text{PULSE} + 1.4795947\ (\text{PULSE})^2$$

where

$$\text{VISI} = \text{visibility factor}$$
$$\text{PULSE} = \log_{10}(\text{number of pulses})$$

MATRICES FOR P7.3

R MATRIX =						
1.0000	0.9755	0.9313	0.8880	0.8515	0.8224	−0.9410
0.9755	1.0000	0.9878	0.9647	0.9407	0.9195	−0.8509
0.9313	0.9878	1.0000	0.9937	0.9812	0.9678	−0.7673
0.8880	0.9647	0.9937	1.0000	0.9966	0.9897	−0.7007
0.8515	0.9407	0.9812	0.9966	1.0000	0.9981	−0.6501
0.8224	0.9195	0.9678	0.9897	0.9981	1.0000	−0.6121
−0.9410	−0.8509	−0.7673	−0.7007	−0.6501	−0.6121	1.0000

SIGMA = 0.8681 F1 (ENTER) = 2.500 F2 (REMOVE) = 2.500

$B(0) = 0.1483 \times 10^2$	$SB(0) = 0.3069 \times 10^0$
$B(1) = -0.1120 \times 10^2$	$SB(1) = 0.1135 \times 10^1$
$B(2) = 0.1479 \times 10^1$	$SB(2) = 0.2477 \times 10^0$
B(3) = 0.	SB(3) = 0.
B(4) = 0.	SB(4) = 0.
B(5) = 0.	SB(5) = 0.
B(6) = 0.	SB(6) = 0.

VISIBILITY

NUMBER PULSES	ACTUAL VALUE	PREDICTED VALUE	DIFFERENCE	PERCENT ERROR
6000	-0.7100×10^1	-0.6574×10^1	0.5257×10^0	−7.4
6000	-0.6300×10^1	-0.5420×10^1	0.8796×10^0	−14.0
900	-0.3300×10^1	-0.3310×10^1	-0.1042×10^{-1}	0.3
900	-0.2800×10^1	-0.3310×10^1	-0.5104×10^0	18.2
200	-0.3000×10^0	-0.1278×10^0	0.1721×10^0	−57.4
200	0.3200×10^1	0.4268×10^1	0.1068×10^1	33.4
50	0.1040×10^2	0.9628×10^1	-0.7717×10^0	−7.4
50	-0.6000×10^1	-0.5574×10^1	-0.5742×10^0	9.6
12	-0.5100×10^1	-0.5420×10^1	-0.3203×10^0	6.3
12	-0.1200×10^1	-0.1278×10^0	-0.1027×10^1	−110.7
3	0.3400×10^1	0.4268×10^1	0.8688×10^0	25.6

Key: R is the normalized covariance matrix. It consists of the sums of products and crossproducts normalized with respect to the diagonal elements.
SIGMA is the sum of the squares of the residuals.
B( ) is the vector of estimated parameters.
SB( ) is the vector of estimated standard deviations of B( ).
F1 (ENTER) and F2 (REMOVE) are the Fisher F-values for adding and removing variables from the model.

(a) Has the model been properly constructed?
(b) Analyze and interpret the residuals.

7.4 Refer to Problem 5.9. If a residual is defined as $D_i = Y_i - \hat{Y}_i$, express each residual as a function of all the observations (with the aid of the table in Problem 5.9). For example:

$$D_1 = 0.471\,Y_1 - 0.257\,Y_2 - 0.350\,Y_3 - \cdots + 0.150\,Y_8$$

Prepare a table of coefficients for each residual in the form of Y_i versus D_j.

(a) Show that $\mathscr{E}\{D_j\} = 0$.
(b) Determine the $\mathscr{E}\{D_j^2\}$ for each D_j, and sum them to obtain

$$\sum \mathscr{E}\{D_j^2\} = 4.000\sigma_{Y_i}^2$$

Note that the sum is the trace of the main diagonal in the table, and that 4.000 represents the degrees of freedom, namely 8 observations less 4 parameters. Also note that:

1. Var $\{D_j\} = \mathscr{E}\{(D_j - 0)^2\}$ is the main diagonal of the table.
2. Covar $\{D_j D_k\}$ is the off-diagonal element in the table.
3. Calculate the correlation coefficient between D_7 and D_8, $\rho_{D_7 D_8}$.

7.5 Obtain a stepwise regression program and ascertain the best model to fit the data of: (a) Problem 5.11 and (b) Problem 5.21. Many stepwise regression programs also plot the residuals and carry out an analysis of residuals which aid in the interpretation of the model's adequacy at various stages of its construction.

7.6 The data in the Table P7.6a represent the specific fuel consumption of a jet engine at 25,000 feet and a Mach number of 0.4. Several different functions were used as models to fit the data; refer to Table P7.6b.

(a) By stepwise regression, determine if a fourth-order polynomial is the polynomial of best fit.
(b) Which model in Table P7.6b is the best and which is the worst?

TABLE P7.6a

Thrust t	Specific Fuel Consumption Y
2,000	1.295
3,000	1.088
4,000	1.010
5,000	0.963
6,000	0.935
7,000	0.920
8,000	0.912
9,000	0.910
10,000	0.912
11,000	0.918
12,000	0.929
13,000	0.940
14,000	0.952
15,000	0.966
16,000	0.980
17,000	0.994
18,000	1.010

TABLE P7.6b

		Fourth-Degree Polynomial		$\hat{Y}=\frac{a_0}{x+a_1+a_2x}$		$\hat{Y}=\frac{a_0}{x+a_1+a_2x+a_3x^2}$		$\hat{Y}=\frac{a_0}{x+0.65}+a_1+a_2x+a_3x^2$	
x	Y	$\hat{Y}$	Error	$\hat{Y}$	Error	$\hat{Y}$	Error	$\hat{Y}$	Error
1	1.295	1.270	0.025	1.314	−0.019	1.299	−0.004	1.294	0.001
2	1.088	1.123	−0.035	1.060	0.028	1.080	0.008	1.092	−0.004
3	1.010	1.022	−0.012	0.982	0.028	1.003	0.007	1.005	0.005
4	0.963	0.959	0.004	0.949	0.014	0.964	−0.001	0.960	0.003
5	0.935	0.922	0.013	0.933	0.002	0.940	−0.005	0.935	+0.000
6	0.920	0.906	0.014	0.927	−0.007	0.926	−0.006	0.921	−0.001
7	0.912	0.903	0.007	0.925	−0.013	0.918	−0.006	0.914	−0.002
8	0.910	0.909	0.001	0.926	−0.016	0.914	−0.004	0.912	−0.002
9	0.912	0.918	−0.006	0.930	−0.018	0.915	−0.003	0.915	−0.003
10	0.918	0.928	−0.010	0.935	−0.017	0.918	−0.000	0.920	−0.002
11	0.929	0.938	−0.009	0.941	−0.012	0.925	0.004	0.928	0.001
12	0.940	0.945	−0.005	0.948	−0.008	0.934	0.006	0.938	0.002
13	0.952	0.952	−0.000	0.956	−0.004	0.946	0.006	0.950	0.002
14	0.966	0.959	0.007	0.964	0.002	0.960	0.006	0.963	0.003
15	0.980	0.970	0.010	0.972	0.008	0.977	0.003	0.978	0.002
16	0.994	0.988	0.006	0.981	0.013	0.996	−0.002	0.995	−0.001
17	1.010	1.019	−0.009	0.990	0.020	1.018	−0.008	1.013	−0.003
Note: $x=(t/1000)-1$		Coefficients of fitted polynomial in this form were not calculated		$a_0 = 0.53069$ $a_1 = 0.77242$ $a_2 = 0.010964$		$a_0 = 0.42106$ $a_1 = 0.88852$ $a_2 = -0.011786$ $a_3 = 0.00105556$		$a_0 = 0.91148$ $a_1 = 0.73665$ $a_2 = 0.004888$ $a_3 = 0.004897$	

7.7 The following experimental data

x	Y	$s_{\bar{Y}_i}$
0.12	3.85	0.09
0.56	9.42	0.15
0.83	12.90	0.42
1.36	17.36	0.42
1.48	19.31	0.23
1.73	22.73	0.27
2.20	32.89	0.36
2.57	44.51	0.83
2.83	53.01	0.52
3.01	62.09	0.61
3.32	81.00	0.93
3.62	102.11	0.86
3.90	124.00	0.71

were used to estimate the coefficients in a model of the form

$$\eta = \sum_{k=0}^{q} \alpha_k x^k$$

for q = 1, 2, 3, 4, and 5. The results obtained by stepwise regression were

$m = n - q - 1$
n = 13 = number of data sets
q = number of parameters
ϕ = sum of the squares of the residuals

n	1	2	3	4	5
m	11	10	9	8	7
a_0	−0.5567	4.8593	1.9142	1.8899	1.9802
a_1	18.9928	−1.5784	17.0535	17.2574	16.2962
a_2	—	7.3924	−8.0616	−8.3539	−6.2119
a_3	—	—	3.0132	3.1472	1.4636
a_4	—	—	—	−0.01877	0.5136
a_5	—	—	—	—	−0.0580
ϕ_{min}	1004.01	147.1809	1.4002	1.6371	1.8340

Which polynomial is the most suitable to represent the data? Explain why.

7.8 Spouting is a technique for contacting gases (or liquids) with solid particles, usually one-eighth inch in diameter or larger; see Figure P7.8. The technique has been used to dry wheat and wood chips, for low temperature carbonization, etc. A gas is passed through a conical opening in the bottom of the contacting apparatus. After a certain gas velocity is reached, the solid particles rise rapidly through the center of the bed and move down at the sides.

The central stream of particles which forms the "spout" may oscillate and may vary in diameter as it rises through the apparatus. For design purposes, it is important to know the diameter of the spout in the apparatus. To obtain a correlation for spout diameter as a function of operating conditions, data on nine materials were listed in Table P7.8a; the

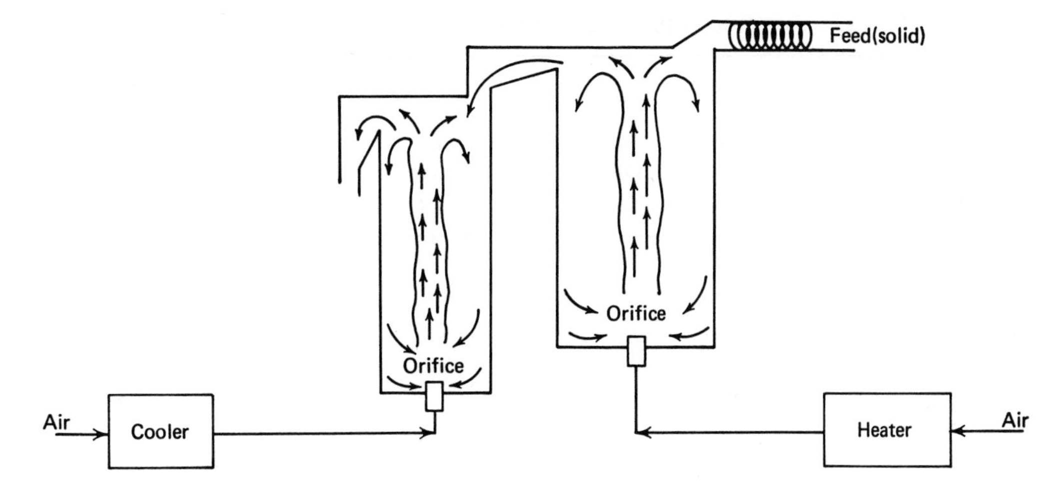

FIGURE P7.8

known physical properties of the materials were listed in Table P7.8b. Although the spout diameters did vary along the bed height, variations were small, and average values of the spout diameter were tabulated.

Make a list of the possible models which can represent the relationship between the spout diameter and the independent variables. Screen these models by the method of Gorman and Toman. Then repeat the screening by using the method of Wilks. Compare the two techniques from the viewpoint of their ability to discriminate between models and the time required to reach a decision concerning any particular model relative to the others.

TABLE P7.8a DATA ON SPOUT DIAMETER

Material	Column Diameter, In, d_c	Orifice Diameter, In, d_o	Bed Height, In, L	Mass Velocity, Lb/(Hr) (Sq Ft), G	Average Spout Diameter, In, d_s
Polystyrene 1	6	0.50	12	326.0	1.13
				356.2	1.20
				416.3	1.27
				476.1	1.33
			8	259.5	1.07
				296.4	1.13
				355.8	1.16
	4	0.50	6	320.2	0.72
				373.6	0.80
				453.6	0.85
		0.25	12	598.9	1.06
				667.2	1.12
Polystyrene 2	6	0.50	12	595.0	1.47
				654.0	1.52
				713.8	1.58
			8	506.0	1.38
				536.2	1.40
				594.8	1.50

(continued)

TABLE P7.8a DATA ON SPOUT DIAMETER (*Continued*)

Material	Column Diameter, In, d_c	Orifice Diameter, In, d_o	Bed Height, In, L	Mass Velocity, Lb/(Hr) (Sq Ft) G	Average Spout Diameter, In, d_s
	4	0.50	6	762.7	1.10
				890.0	1.20
Barley	6	0.50	12	635.0	1.35
				705.0	1.47
				775.4	1.49
			8	528.4	1.31
				563.6	1.35
				634.0	1.44
Polyethylene	6	0.50	12	512.0	1.37
				530.0	1.43
				619.0	1.50
				706.0	1.62
				795.0	1.68
				883.6	1.75
			8	442.0	1.23
				486.0	1.34
				530.0	1.37
				573.6	1.50
				618.2	1.50
Millet	6	0.50	12	386.2	1.02
				416.0	1.10
				575.6	1.16
				535.2	1.25
				594.0	1.28
			8	326.0	1.00
				356.4	1.05
				416.0	1.10
				475.6	1.14
	4	0.50	6	455.0	0.80
				567.0	0.85
				624.3	0.90
Wheat	6	0.50	12	704.0	1.39
				776.0	1.45
				845.4	1.50
			8	563.6	1.35
				634.2	1.42
				704.0	1.45
		0.25	12	795.9	1.47
				884.3	1.53
			8	663.4	1.34
				707.5	1.44
				795.9	1.53
	4	0.625	10	954.8	1.12
				1028.2	1.25
				1175.1	1.28

(*continued*)

TABLE P7.8a DATA ON SPOUT DIAMETER (*Continued*)

Material	Column Diameter, In, d_c	Orifice Diameter In, d_o	Bed Height, In, L	Mass Velocity, Lb/(Hr) (Sq Ft) G	Average Spout Diameter, In, d_s
	4	0.50	6	889.0	1.00
				1017.0	1.12
				1193.5	1.20
				1273.3	1.20
		0.25	8	881.3	1.19
				1028.2	1.22
				1175.1	1.31
			10	999.2	1.24
				1175.1	1.31
	6*	0.75	24	1054.0	1.63
				1160.0	1.70
				1265.0	1.75
	24*	4.00	72	736.3	3.25
		3.00	72	638.9	3.13
		2.00	72	593.1	3.13
			60	604.6	2.95
				664.6	3.18
				724.8	3.53
			36	428.0	2.53
			48	462.0	2.69
			60	541.4	2.80
Rice	6	0.50	12	535.2	1.13
				594.0	1.27
				654.0	1.33
			8	416.1	1.06
				476.0	1.11
				535.8	1.25
	4	0.50	6	663.0	0.94
				928.6	1.10
	4	0.25	10	800.6	1.14
				934.1	1.28
				1067.5	1.33
			8	694.1	1.04
				934.1	1.17
				1067.5	1.28
Corn	6	0.50	12	1195.2	2.12
				1326.6	2.28
			8	840.0	1.75
				972.0	1.98
Flaxseeds	6	0.50	12	356.4	1.08
				416.0	1.17
	4	0.50	12	426.9	0.73
				453.9	0.80
				624.4	0.90

* B. Thonley, J. B. Saunby, K. B. Mathur, and G. L. Osberg, *Can. J. Chem. Eng.* **37**, 184, 1959.

TABLE P7.8b PHYSICAL PROPERTIES OF SPOUTED PARTICLES

Material	Absolute Density, Lb/Cu Ft	Bulk Density, Lb/Cu Ft	Percent Void	Particle Diameter, In	Shape Factor
Polystyrene 1	66.02	40.00	39.41	0.0616	1.141
Polystyrene 2	66.02	37.00	43.94	0.1254	1.176
Polyethylene	57.58	36.97	35.80	0.1350	1.020
Millet	73.69	45.37	38.44	0.0783	1.070
Rice	90.95	56.47	37.92	0.1071	1.041
Barley	79.87	45.24	43.36	0.1458	1.141
Corn	83.75	46.39	44.60	0.2857	1.500
Flaxseeds	70.51	43.88	37.77	0.0824	1.050
Wheat	87.40	53.98	39.24	0.1420	1.073

7.9 Kabel† obtained the following data from undesigned experiments for the dehydration of alcohol (A) to form ether (E) and water (W). The sum of the partial pressures, $\sum p_i$, was held constant at 1 atmosphere. Kittrell‡ proposed 47 models to represent the data; the models could be reduced to the following five general models (the variables and coefficients are deterministic):

	Model	Mechanism
(a)	$r = a_1$	Desorption
(b)	$r = \dfrac{b_1 + b_2 p_A}{p_A}$	Adsorption
(c)	$r = \dfrac{(c_1 + c_2 p_A)^2}{p_A}$	Adsorption
(d)	$r = \left(\dfrac{d_1 + d_2 p_A}{p_A}\right)^2$	Surface reaction
(e)	$r = \dfrac{e_1 + e_2 p_A}{p_A^2}$	Surface reaction

Fit the models, assuming $R = r + \epsilon$, and apply the screening method of Gorman and Toman to discriminate among them. All the a's, b's, c's, d's, and e's are nonnegative constant parameters. List the models in two categories—"keep" for further analysis and "reject." How can $s^2_{\bar{Y}_i}$ be estimated?

Repeat, using the Wilks procedure.

TABLE P7.9 RATE OF REACTION DATA

R, Reaction Rate, g-moles/(g-cat) (Min)	p_A Alcohol Partial Pressure, Atm	p_E Ether Partial Pressure, Atm	p_W Water Partial Pressure, Atm
0.00000385	0.27308	0.05391	0.67301
0.00000600	0.33130	0.02480	0.64390
0.00000900	0.38090	0.00000	0.61910
0.00006890	0.73027	0.13487	0.13487
0.00009366	0.87864	0.06068	0.06068
0.00013291	1.00000	0.00000	0.00000
0.00005780	0.65488	0.19911	0.14601
0.00010180	0.83945	0.10683	0.05373
0.00013647	0.94690	0.05310	0.00000
0.00007235	0.65848	0.23221	0.10931
0.00010550	0.80092	0.16099	0.03809
0.00013035	0.87710	0.12290	0.00000
0.00010674	0.73596	0.24142	0.02262
0.00006986	0.66312	0.27784	0.05904
0.00013766	0.78120	0.21880	0.00000
0.00006541	0.43042	0.54909	0.02049
0.00004646	0.32314	0.60273	0.07413
0.00008877	0.47140	0.52860	0.00000
0.00004815	0.44593	0.49093	0.06313
0.00007427	0.51953	0.45414	0.02634
0.00010270	0.57220	0.42780	0.00000
0.00005019	0.54041	0.37769	0.08189
0.00007652	0.60424	0.34578	0.04998
0.00010354	0.70420	0.29580	0.00000
0.00007309	0.58620	0.38655	0.02725
0.00005695	0.55396	0.40267	0.04337
0.00010984	0.64070	0.35930	0.00000
0.00006421	0.71662	0.14169	0.14169
0.00009257	0.82606	0.08697	0.08697
0.00012128	1.00000	0.00000	0.00000
0.00003912	0.59116	0.08192	0.32692
0.0005768	0.75500	0.00000	0.24500
0.0001957	0.52736	0.01222	0.46042
0.0001608	0.30069	0.02556	0.47376
0.0002430	0.55180	0.00000	0.44820
0.0004715	0.52422	0.22389	0.25189
0.00005209	0.62260	0.17470	0.20270

7.10 In developing empirical mathematical models, instead of using a polynomial it is suggested that a product form of a model will be better:

$$Y = cf_1(x_1)f_2(x_2)\cdots f_n(x_n) + \epsilon$$

Specifically take the model

$$Y = \alpha(1 + \beta_1 x_1)(1 + \beta_2 x_2) + \epsilon$$

† R. L. Kabel, Ph.D. Thesis, Univ. of Washington, Seattle, 1961; R. L. Kabel and L. N. Johnson, *AICE J.* **8**, 621, 1962.
‡ J. R. Kittrell, Ph.D. Thesis, Univ. of Wisconsin, Madison, 1966.

The following simulated observations were generated with a normal random error with the parameters (0, 1) for $\alpha = 2$, $\beta_1 = \frac{1}{4}$, and $\beta_2 = \frac{1}{2}$:

x_1	x_2	η	Y
0	0	2	1.344
1	1	3.75	2.972
2	1	4.5	4.852
2	2	6	7.352
3	2	7	7.017

Would it be possible to use a linear estimation procedure to estimate the coefficients α, β_1, and β_2? Explain.

An estimated regression equation was obtained using the above data:

$$\hat{Y} = 1.862(1 + 0.024x_1)(1 + 1.176x_2)$$

with $\phi_{\min} = 0.388$. Is this equation the best one to represent the simulated data? Test it against a linear model with three parameters: $\eta = \beta_0 + \beta_1 x_1 + \beta_2 x_2$. Propose several additional models and determine which is the best.

CHAPTER 8

Strategy for Efficient Experimentation

Up to this point we have paid only minimal attention to one of the most important aspects of empirical model building, namely how to obtain an appropriate model with a minimum of experimentation. A trial-and-error approach to the design and execution of experiments cannot only be time consuming and expensive but even self-defeating. Whatever the objectives of an experiment, something more than an analysis of the experimental results is required if the experimenter wants to obtain estimates of the parameters in his model with small confidence regions. No amount of analysis will ever overcome the handicap of poorly designed experiments. On the other hand, if the experimenter plans his experiments so as to provide a maximum amount of information, he can achieve his goals most effectively.

This chapter outlines some basic methods for efficient experimentation which have proved useful in the planning of experiments and their sequential execution. These methods are designed to assist in:

1. The selection of the best model among the set of plausible models.
2. The efficient estimation of the parameters in the selected model.

Both objectives are sought simultaneously and usually sequentially, because the experimenter generally does not know what variables to measure nor their range nor what series of experiments to run until his experimental program is at least partially completed. In practice, he plans one or more experiments, carries them out, analyzes the results, and modifies his experimental plan accordingly. This strategy of experimentation is outlined in Figure 8.0-1. The designs we seek are the values of the independent variables. The methods to be described cannot replace imagination or judgment, but they can save time and money and can provide some objective data to substantiate any decisions made on the basis of the partial or complete experiments.

Before proceeding, we should remark that every series of experiments should have carefully defined objectives or criteria expressed in mathematical terms insofar as possible. The designated objectives underlie the choice of: (1) the controllable variables involved in the experiment, (2) the procedure, and (3) the methods of analysis of the results of the experiment. Table 8.0-1 is a catalog of certain practical aspects of experimentation which must always be taken into account.

8.1 RESPONSE SURFACE METHODS

One major contributor to the practice of effective experimentation has been G. E. P. Box and his co-workers (see references at the end of this chapter) who developed statistical techniques in the design and analysis of experiments, termed "response surface methodology,"

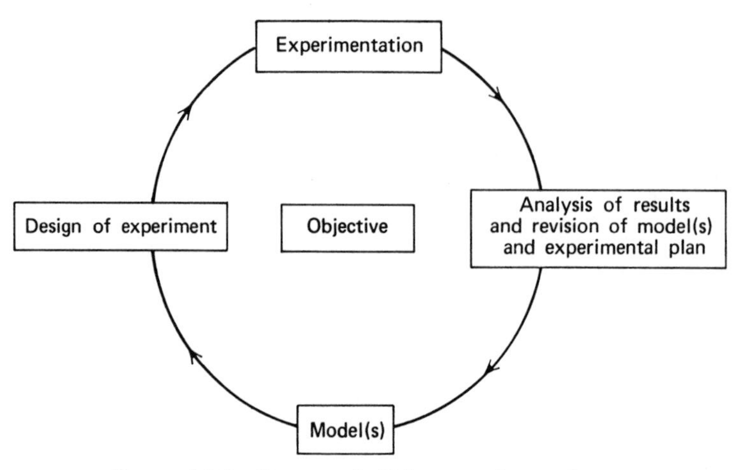

FIGURE 8.0-1 Strategy of efficient experimentation.

TABLE 8.0-1 CHECKLIST FOR EXPERIMENTATION

Statement of Objectives

1. Why is the work to be done? What questions will be answered by the experiment?
2. What are the consequences of a failure to find an effect or to claim one when it does not really exist?
3. What is the experimental space to be covered?
4. What is the time schedule?
5. What is the allowable cost?
6. What previous information is there about the experiment or its results?
7. Is an optimum among the variables sought or only the effect of the variables?

Type of Model(s) to be Used

1. Will empirical or transport phenomena models be used?
2. Is the form of the model correct or is the form to be determined?
3. What will be the independent and dependent variables?

Experimental Program

1. What are the variables to be measured? How will they be measured and in what sequence?
2. Which variables are initially considered most important? Which least important? Can the desired effect be detected?
3. What extraneous or disturbing factors must be controlled, balanced, or minimized?
4. What kind of control of the variables is desirable?
5. Are the variables independent or functions of other variables?
6. How much dispersion can be expected in the test results? Will the dispersion be different at different levels of the variables?

Replication and Analysis

1. What is the experimental unit and how are the experiments to be replicated—all at once, sequentially, or in a group?
2. What are the number and type of tests to be carried out?
3. How are the data to be analyzed and interpreted?

which can be applied to solve practical process problems. The strategy of response surface methodology, as indicated in Figure 8.0-1, is to decide on a model in which the response is expressed as a function of the independent variables presumed to be involved in the process.

This model provides the basis for new experimentation, which in turn leads to a new model, and the entire cycle is repeated. Compared with the method of holding all the variables constant except one that is varied, thus testing one variable at a time, the response surface method is much more efficient. Even compared with orthodox experimental design methods in which all combinations of the factors, or suitable fractions thereof, are tested, the response surface method proves to be more effective for continuous processes.

A second attractive feature of response surface methods is that conclusions can be drawn from the very first experiments. The experimentation can be terminated any time further experimentation appears uneconomic. Finally, response surface methods are a prelude to the determination of optimum operating conditions for a process (discussed in Section 8.3).

Before discussing response surface methods, it should be pointed out that the models to be described have only *one* dependent variable. If more than one response is being observed, surfaces can be constructed for each and they can be studied in juxtaposition to one another. This can often best be done by means of superimposing the various contour plots; refer also to Example 8.1-5 and Section 8.3-1.

To answer the question as to what is a proper response, a brief scrutiny by the analyst of any continuous process will reveal a number of independent and dependent variables and some of intermediate character whose classification is by no means clear. Dependent variables are considered to be *responses* to independent (preferably controllable) variables. Typical examples of the former are product purity, yield, and weight. Examples of the latter are temperature, pressure, flow rates, concentration, and time of reaction. Before starting any program of experimentation, it is essential to be able to measure quantitatively both the response variable and the independent variables and to control the independent variables if meaningful results are to be achieved.

8.1-1 One-Dimensional Models

By first examining the detailed calculations for one dependent variable and one independent variable (the "one" dimension in the section heading refers to the number of coordinates needed to represent geometrically the independent variables), the response surface method can be illustrated both analytically and graphically. Then, higher dimensional examples can be examined and a general procedure outlined.

Example 8.1-1 One-Dimensional Experiment

We start by assuming we know nothing about the relation between the shear stress of a non-Newtonian fluid (the response) and the temperature (the independent variable). We carry out an experiment, collecting five sets of data:

t, Temperature (°F)	V, Shear Stress (dynes/cm^2)
60	15
80	17
100	20
120	28
140	45

To fit a straight line to the data, a trial model is used of the form

$$V = \beta_0 + \beta_1 t_1 + \epsilon$$

where ϵ represents the unobservable random error and V is a random variable. To simplify the calculations (if done by hand rather than on a computer), the temperature can be coded as follows:

$$x = \frac{t - 50}{10}$$

and the shear stress can be coded as

$$Y = V - 15$$

These transformations lead to the following matrices (the column of dummy variables, 1, appears in order to have an intercept in the model):

Uncoded

$$\mathbf{t} = \begin{bmatrix} 1 & 60 \\ 1 & 80 \\ 1 & 100 \\ 1 & 120 \\ 1 & 140 \end{bmatrix} \qquad \mathbf{V} = \begin{bmatrix} 15 \\ 17 \\ 20 \\ 28 \\ 45 \end{bmatrix}$$

$$\mathbf{t}^T\mathbf{t} = \begin{bmatrix} 5 & 500 \\ 500 & 54{,}000 \end{bmatrix}$$

$$\mathbf{t}^T\mathbf{V} = \begin{bmatrix} 125 \\ 13{,}920 \end{bmatrix}$$

Coded

$$\mathbf{x} = \begin{bmatrix} 1 & 1 \\ 1 & 3 \\ 1 & 5 \\ 1 & 7 \\ 1 & 9 \end{bmatrix} \qquad \mathbf{Y} = \begin{bmatrix} 0 \\ 2 \\ 5 \\ 13 \\ 30 \end{bmatrix}$$

$$\mathbf{x}^T\mathbf{x} = \begin{bmatrix} 5 & 25 \\ 25 & 165 \end{bmatrix}$$

$$\mathbf{x}^T\mathbf{Y} = \begin{bmatrix} 50 \\ 392 \end{bmatrix}$$

From the least squares analysis, the estimate of $\boldsymbol{\beta}$ is:

Uncoded

$$\mathbf{b} = (\mathbf{t}^T\mathbf{t})^{-1}(\mathbf{t}^T\mathbf{V}) = \begin{bmatrix} -10.5 \\ 0.355 \end{bmatrix}$$

Coded

$$\tilde{\mathbf{b}} = (\mathbf{x}^T\mathbf{x})^{-1}(\mathbf{x}^T\mathbf{Y}) = \begin{bmatrix} -7.75 \\ 3.55 \end{bmatrix}$$

Consequently, the estimated first-order regression equations are, respectively,

Uncoded: $\hat{V} = -10.5 + 0.355t$

Coded: $\hat{Y} = -7.75 + 3.55x$

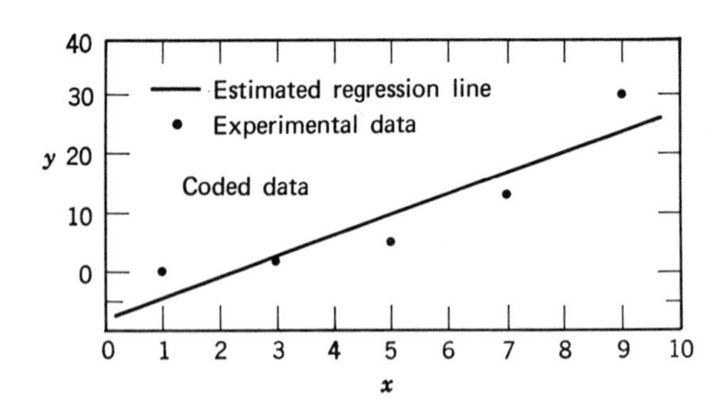

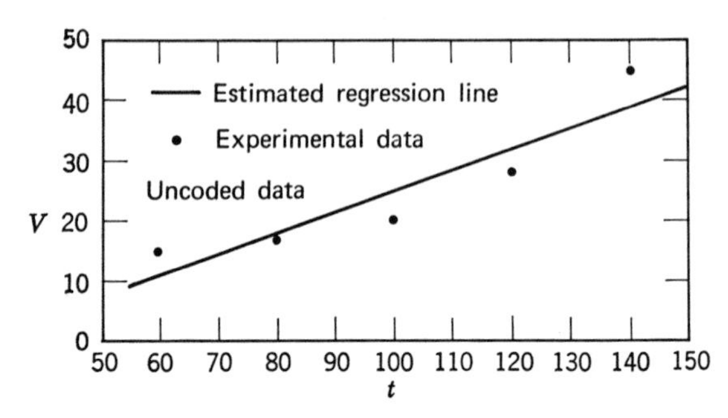

FIGURE E8.1-1

These equations are plotted in Figure E8.1-1 together with the experimental data points. Next we carry out an analysis of variance as shown in Table E8.1-1a.

Both constants are meaningful. But a glance at the plot of the original data indicates that a linear first-order model does not fit the data very well! This paradox brings out an important point. The sum of the squares of the residuals about the empirical regression line, $\phi_{min} = 93.9$, will

TABLE E8.1-1a ANALYSIS OF VARIANCE OF CODED DATA

Source of Variation	SS	d.f.	Mean Square	Variance Ratio $\frac{s^2}{31.3}$
Effect of removing b_1 $\sum_{i=1}^{n} p_i(\hat{Y}_i - \bar{Y})^2$	504.1	1	504.1	16.1
Effect of removing b_0 (with b_1 already removed) $\sum_{i=1}^{n} p_i(\bar{Y} - 0)^2$	500	1	500	16.0
Deviation about the empirical regression line $\sum_{i=1}^{n} p_i(Y_i - \hat{Y}_i)^2$	93.9	3	31.3	
Total $\left(\sum Y_i^2\right)$	1098.0	5		

Note: $F_{0.95}(1, 3) = 10.13$

TABLE E8.1-1b ANALYSIS OF VARIANCE OF UNCODED DATA

Source of Variation	SS	d.f.	Mean Square	Variance Ratio $\left\}\frac{s^2}{31.3}\right.$
Effect of removing b_1	504.1	1	504.1	16.1
Effect of removing b_0 (with b_1 already removed)	3125	1	3125	100
Deviation about the empirical regression line	93.9	3	31.3	
Total	3723.0	5		

Note: $F_{0.95}(1, 3) = 10.13$

include the effect of the fit of the empirical regression line, and it will **also include any effect of selecting an inappropriate mathematical model.** Visual observation points out that some type of curve will fit the data better than a straight line.

To decide if a model is appropriate, one might try all sorts of models and see which give the lowest variances for the variation about the empirical regression line. A better way, if at all feasible, **is to make replicate experiments** and thus obtain an estimate of the experimental error which is independent of the mathematical model. As explained in Chapter 4, the F-test can be employed to ascertain if the fit about the estimated regression line is significantly different from the experimental error. As a matter of policy, it is wise to have sufficient replication to yield several degrees of freedom for the error estimation, thereby reducing the error mean square to as low a value as is economically feasible.

Example 8.1-2 One-Dimensional Experiment with Replication

Replicate experiments (the series of experiments done completely over) have been carried out to extend the experiment of Example 8.1-1 as follows:

Temperature (°F)	Shear Stress (dynes/cm²)
60	14, 15*, 16
80	16, 17*, 18
100	20, 20*, 21
120	26, 26, 28*
140	44, 45*, 46

* = old values.

A first-order model, $Y = \beta_0 + \beta_1 x_1 + \epsilon$, can be fitted to the coded data (coded as in the previous example), yielding the following matrices:

a Matrix

1.500×10^1	7.500×10^1
7.500×10^1	4.950×10^2

Inverted **a** Matrix (**c** Matrix)

2.270×10^{-1}	-4.166×10^2
-4.166×10^{-2}	8.333×10^{-3}

g Matrix

1.470×10^2	1.153×10^3

Regression Coefficients

$b_0 = -7.616$
$b_1 = 3.483$

The corresponding analysis of variance is shown in Table E8.1-2a.

TABLE E8.1-2a

Source of Variation	SS	d.f.	Mean Square	Variance Ratio $\left\}\frac{s^2}{0.933}\right.$
Effect of removing b_1	1456	1	1456	1560*
Effect of removing b_0 (with b_1 already removed)	1440	1	1440	1500*
Deviation about the empirical regression line	309.0	3	103.0	110*
Deviation within sets	9.33	10	0.933	
Total	3215	15		

Note: $F_{0.95}(1, 10) = 4.96$
$F_{0.95}(3, 10) = 3.71$

* Significant.

A quick computation provides a result similar to Example 8.1-1, namely the residual sum of squares,

$$\begin{array}{r} 309.0 \\ 9.33 \\ \hline 318.33 \end{array}$$

which leads to

$$\frac{318.33}{13} = 24.4$$

compared with 31.3 in Example 8.1-1.

The F-test points out that the deviation about the first-order regression line is significantly higher than the experimental error, and that the model can be materially improved.

Once the first-order linear model is known to be unsatisfactory, another first-order model can be tried, one perhaps obtained from some type of transformation. Of course, if such a transformation is carried out, the additive unobservable error will be added to the transformed response rather than to the original response. If no better success is achieved with these other models, a second-order model can be tried. It is best to proceed in as simple a fashion as possible rather than indiscriminately picking a complicated model.

With the model

$$Y = \beta_0 + \beta_1 x_1 + \beta_2 x_1^2 + \epsilon$$

the matrices are:

a Matrix

1.500×10^1	7.500×10^1	4.950×10^2
7.500×10^1	4.950×10^2	3.675×10^3
4.950×10^2	3.675×10^3	2.900×10^4

Inverted **a** Matrix (**c** Matrix)

7.050×10^{-1}	-2.946×10^{-1}	2.529×10^{-2}
-2.946×10^{-1}	1.571×10^{-1}	-1.488×10^{-2}
2.529×10^{-2}	-1.488×10^{-2}	1.488×10^{-3}

g Matrix

1.470×10^2	1.153×10^3	9.458×10^3

Regression Coefficients

$$b_0 = 3.210$$
$$b_1 = -2.885$$
$$b_2 = 0.636$$

The analysis of variance is shown in Table E8.1-2b.

The second-order model also proves to be inadequate. By removing b_2 and then b_1, the first two sums of the squares in the analysis of variance would become 1676 and 53, respectively. These differences exist because the independent variables x_1 and x_1^2 are not orthogonal, but the conclusion that the model is inadequate is not changed. A third-degree polynomial can be tested as a model

$$Y = \beta_0 + B_1 x_1 + \beta_2 x_1^2 + \beta_3 x_1^3 + \epsilon$$

Or perhaps a linearized form of a nonlinear (in the coefficients) model such as

$$\log Y = \alpha + \beta x + \epsilon$$

might better fit the data and give a nonsignificant *F*-test for the deviations about the empirical regression line. These calculations will not be shown here in order to save space, but the procedure should now be clear.

TABLE E8.1-2b

Source of Variation	SS	d.f.	Mean Square	Variance Ratio $\frac{s^2}{0.933}$
Effect of removing b_1	1456	1	1456	1560*
Effect of removing b_2 (with b_1 already removed)	273	1	273	292*
Effect of removing b_0 (with b_1 and b_2 already removed)	1440	1	1440	1500*
Deviation about the empirical regression line	36	2	18.2	19.5*
Deviation within sets	9.33	10	0.933	
Total	3215	15		

Note: $F_{0.95}(1, 10) = 4.96$; $F_{0.95}(2, 10) = 4.10$

* Significant.

8.1-2 Two-Dimensional Models and Experimental Designs

Next let us consider models involving two geometric coordinates and one dependent variable. By employing *orthogonal designs*, the response surface method can be made most efficient. When the estimates of the coefficients are orthogonal, the sum of the squares of the deviations associated with the coefficients are orthogonal also, and the maximum amount of information can be obtained for a given amount of experimentation.

Orthogonal experimental designs are arrangements of the independent variables such that for all pairs *j*, *k*, the sum over the data sets $i = 1, 2, \ldots, n$, vanishes, i.e., $\sum_i x_{ij}x_{ik} = 0$, for $j \neq k$. In the following discussion the term *dimension* will refer to the number of geometric coordinates required to represent the response excluding the coordinate for the response itself. The term *order* of the model will refer to the degree of the polynomial forming the model (first degree $\Leftrightarrow$ first order, second degree $\Leftrightarrow$ second order, etc.). The order of the design is related to the order of the model; that is, a first-order design can be used with linear first-order models of one, two, three, or more dimensions but not with second-order models. While second-order designs can be used for first-order models, more experimentation would have to be carried out with such designs than is actually needed.

A very simple orthogonal two-level factorial† experi-

† A factor denotes any of the experimental variables which are deliberately varied from experiment to experiment, such as temperature, pressure, time, and concentration. A factor may be qualitative, such as "high" or "low," "present" or "absent," when quantitative values cannot be assigned to the experimental variables. A *factorial* experiment is one in which all possible combinations of the factors are used. For example, if two variables are to be controlled at two levels each, then there are four possible combinations of experimental conditions; the experiment is termed a two-level factorial experiment, often abbreviated 2^2 design.

mental design (2^2 design) suitable for the first-order two-dimensional model

$$Y = \beta_0 x_0 + \beta_1 x_1 + \beta_2 x_2 + \epsilon \qquad (8.1\text{-}1)$$

is (in coded form)

Levels ($x_0 \equiv 1$ at all levels)	
x_1	x_2
−1	−1
1	−1
−1	1
1	1

To provide the replication needed for an estimate of the experimental error, the entire series of values might be run two or three times. An alternate means of providing replication with less experimentation is to replicate the center point of the design, (0, 0). We assume in the regression analysis that the error is the same at each point. The estimated coefficients are easy to calculate by Equation 5.1-10 because of the orthogonality of x_0, x_1, and x_2. Note that $\sum x_j = 0$, $\sum x_j x_k = 0$, and $\sum x_j x_j \neq 0$. The estimated coefficients are

$$b_0 = \frac{\sum_i \bar{Y}_i x_{i0}}{n} \qquad (8.1\text{-}2)$$

$$b_1 = \frac{\sum_i \bar{Y}_i x_{i1}}{n} \qquad (8.1\text{-}3)$$

$$b_2 = \frac{\sum_i \bar{Y}_i x_{i2}}{n} \qquad (8.1\text{-}4)$$

where $n = \sum_i x_{i0} x_{i0} = \sum_i x_{i1} x_{i1} = \sum_i x_{i2} x_{i2}$.

A word might now be said about coding. In Example 8.1-1 we coded the temperatures to set up a series of integers. In the 2^2 factorial design, each value can be portrayed geometrically as the corners of a square in two dimensions (see Figure 8.1-1). If x_1 represents the coded temperature and x_2 represents the coded pressure, the analyst sets up a series of four experiments at pairs of (t, p) values such that when properly transformed the factorial design is obtained. For example, the temperature might be held at 200 and 240°F, and the pressure at 1 and 5 atmosphere. Then the coding would be

$$x_1 = \frac{t\,(°\text{F}) - 220}{20}$$

$$x_2 = \frac{p\,(\text{atm}) - 3}{2}$$

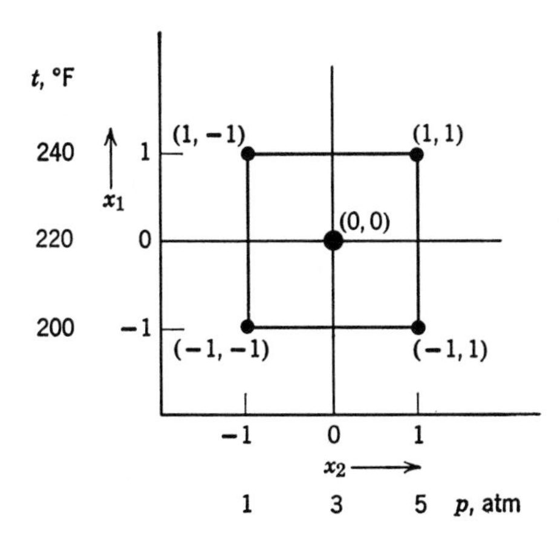

FIGURE 8.1-1 Two-level factorial design for the factors temperature and pressure.

The design in Figure 8.1-1 also will provide an estimate of the lack of fit due to:

1. Interaction between x_1 and x_2; i.e., a b_{12} coefficient can be obtained in an estimated regression equation such as Equation 8.1-5.
2. Effects of quadratic terms (by taking the difference between the average of the four peripheral values and the average of the center values).

Another often employed simple design for first-order two-dimensional models is the so-called "equilateral triangle" design shown in Figure 8.1-2. The matrix of the x elements for the equilateral triangle design is

$$\mathbf{x} = \begin{bmatrix} 1 & -\sqrt{\frac{3}{2}} & -\sqrt{\frac{1}{2}} \\ 1 & \sqrt{\frac{3}{2}} & -\sqrt{\frac{1}{2}} \\ 1 & 0 & \sqrt{2} \\ 1 & 0 & 0 \end{bmatrix} \quad (\text{columns } x_0, x_1, x_2)$$

Note that $\sum x_{i1} = 0$, $\sum x_{i2} = 0$, and $\sum x_{i1} x_{i2} = 0$. To get an estimate of the error, two or more sets of data can be taken at the point (0, 0). The difference between the average of the three peripheral values and the average of the center values provides a measure of lack of fit. Because the design in Figure 8.1-2 contains only four sets of the independent variables, at the very maximum only a four-parameter model can be used. Neither it nor the

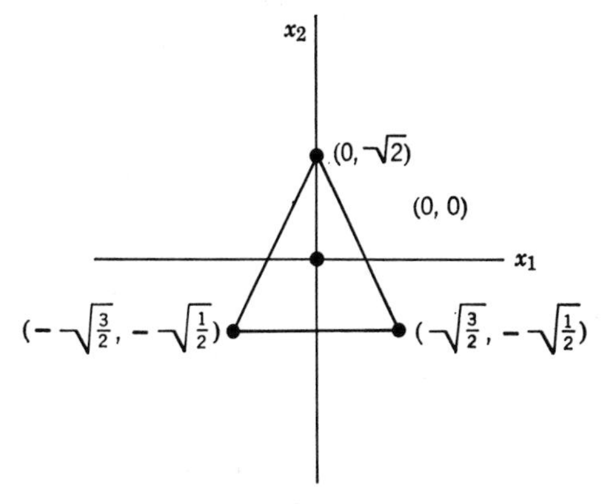

FIGURE 8.1-2 Equilateral triangle design.

design in Figure 8.1-1 are suitable for second-order models (of two dimensions).

If a first-order model proves inadequate to represent the experimental data, the experimenter usually next considers a second-order model. Because a complete second-order two-dimensional model,

$$Y = \beta_0 + \beta_1 x_1 + \beta_2 x_2 + \beta_{11} x_1^2 + \beta_{22} x_2^2 + \beta_{12} x_1 x_2 + \epsilon \quad (8.1\text{-}5)$$

incorporates six coefficients, to fit Equation 8.1-5 three additional estimated coefficients have to be determined beyond those in a linear model, namely b_{11}, b_{22}, and b_{12}, the latter being associated with the so-called *interaction* term, the $x_1 x_2$ term. Experimental designs of the type shown in Figures 8.1-1 and 8.1-2 are not adequate for estimating six constants.

Examination of the matrix of the variables in Model 8.1-5 in terms of the previously mentioned two-level factorial design:

x_0	x_1	x_2	x_1^2	x_2^2	x_{12}
1	−1	−1	1	1	1
1	1	−1	1	1	−1
1	−1	1	1	1	−1
1	1	1	1	1	1

reveals that the columns headed x_0, x_1^2, and x_2^2 are identical. Hence there is no way to distinguish between the coefficients b_0, b_{11}, and b_{22} with the type of designs described so far (this characteristic is called *confounding*), although b_{12} can be determined. Since all the lack of fit can be explained by calculating a b_{12} term, it becomes apparent that the mere addition of the interaction term to a first-order model is not sufficient justification to use the model

$$Y = \beta_0 x_0 + \beta_1 x_1 + \beta_2 x_2 + \beta_{12} x_1 x_2 + \epsilon \quad (8.1\text{-}6)$$

in lieu of a fullscale second-order model, Model 8.1-5, without trying Model 8.1-5 first.

A geometric interpretation of the 2^2 design as shown in Figure 8.1-1 is that it can be rotated 45° to yield the design shown in Figure 8.1-3. This latter design is a perfectly good design with a matrix of

x_0	x_1	x_2	$x_1 x_2$
1	1	0	0
1	0	1	0
1	−1	0	0
1	0	−1	0

but it will not provide any information about the interaction between x_1 and x_2. Since one does not usually

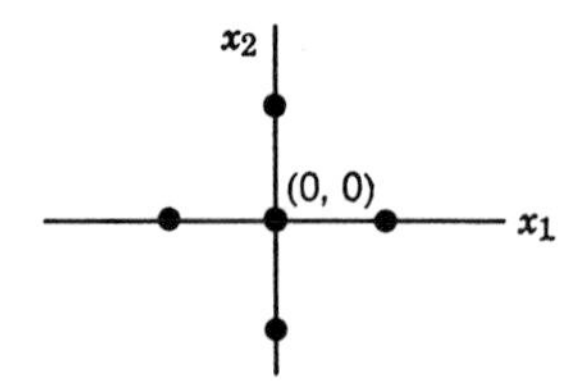

FIGURE 8.1-3 Design of Figure 8.1-1 rotated 45°.

know the orientation of the response surface, at least at the start of an experimental program, the use of Model 8.1-6 without first trying the fullscale second-order model can be very misleading.

Rotatable designs are those in which the variance of $\hat{Y}$ is the same for all the peripheral points in the design. For the first-order designs, using the estimated regression equation

$$\hat{Y} = b_0 + b_1 x_1 + b_2 x_2$$

the variance of $\hat{Y}$ at $\mathbf{x}_i$ is

$$\text{Var}\{\hat{Y}_i\} = \text{Var}\{b_0\} + x_{i1}^2 \text{Var}\{b_1\} + x_{i2}^2 \text{Var}\{b_2\}$$

Because the Covar $\{\mathbf{b}\} = \sigma_{Y_i}^2 \mathbf{c}$ and the elements of $\mathbf{c}$ on the main diagonal are all the same except for the leading element

$$\mathbf{c} = \begin{bmatrix} (n+k)^{-1} & 0 & 0 \\ 0 & n^{-1} & 0 \\ 0 & 0 & n^{-1} \end{bmatrix}$$

the variance of $\hat{Y}_i$ is

$$\text{Var}\{\hat{Y}_i\} = \sigma_{Y_i}^2 \left(\frac{1}{n+k} + \frac{x_{i1}^2}{n} + \frac{x_{i2}^2}{n} \right)$$

where k is the number of replicates. In as much as $x_{i1}^2 = x_{i2}^2 = 1$ except at the center of the design, the Var $\{\hat{Y}_i\}$ is invariant at the peripheral points.

Typical examples of two-dimensional, rotatable, second-order designs are the vertices and at least one center point of any $(n - 1)$ dimensional regular polygon which can be inscribed in a circle. A few such polygons are illustrated in Figure 8.1-4. As an example, the variance of $\hat{Y}_i$ using the octagon design, the $\mathbf{c}$ matrix for which is given in Example 8.1-3, is

$$\begin{aligned} \text{Var}\ \hat{Y}_i &= \text{Var}\{b_0\} + (x_{i1}^2 + x_{i2}^2)(0.1250) \\ &\quad + (x_{i1}^4 + x_{i2}^4)(0.15625) + (x_{i1}x_{i2})^2(0.250) \\ &\quad + 2[(x_{i1}^2 + x_{i2}^2)(-0.1250) + (x_{i1}x_{i2})^2(0.03125)] \\ &= \text{Var}\{b_0\} - 0.1250(x_{i1}^2 + x_{i2}^2) \\ &\quad + 0.15625(x_{i1}^4 + x_{i2}^4) + 0.3125(x_{i1}x_{i2})^2 \end{aligned}$$

The design points in sequence following Figure 8.1-4 yield

$x_{i1}^2 + x_{i2}^2$	$x_{i1}^4 + x_{i2}^4$	$(x_{i1}x_{i2})^2$
2	2	1
2	2	1
2	2	1
2	2	1
2	4	0
2	4	0
2	4	0
2	4	0

and also yield the same Var $\{\hat{Y}_i\}$ for each point:

$$\text{Var}\{\hat{Y}_i\} = \text{Var}\{b_0\} + 0.375$$

An example of a nonrotatable design in which each independent variable is fixed at the $+1$, 0, and -1 levels, and all combinations of all levels are used, is shown in

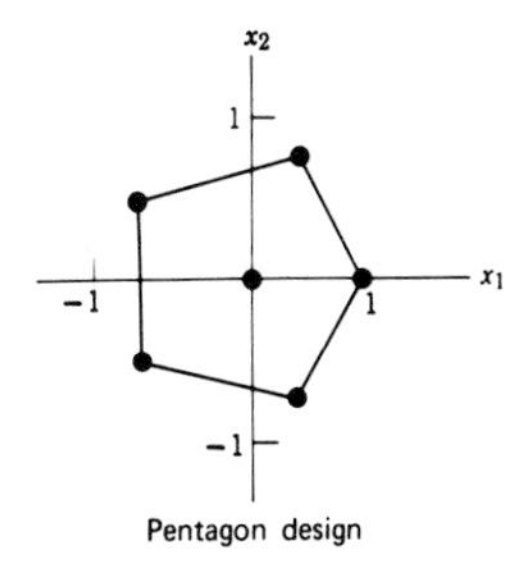

Experimental Levels

x_1	x_2
1.000	0
0.309	0.951
−0.809	0.588
−0.809	−0.588
0.309	−0.951
0	0

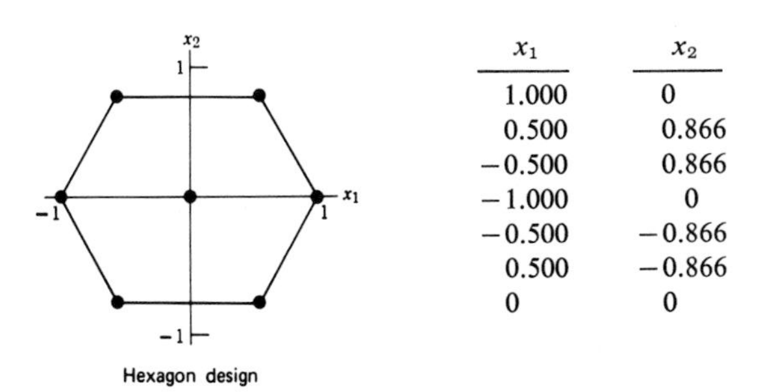

x_1	x_2
1.000	0
0.500	0.866
−0.500	0.866
−1.000	0
−0.500	−0.866
0.500	−0.866
0	0

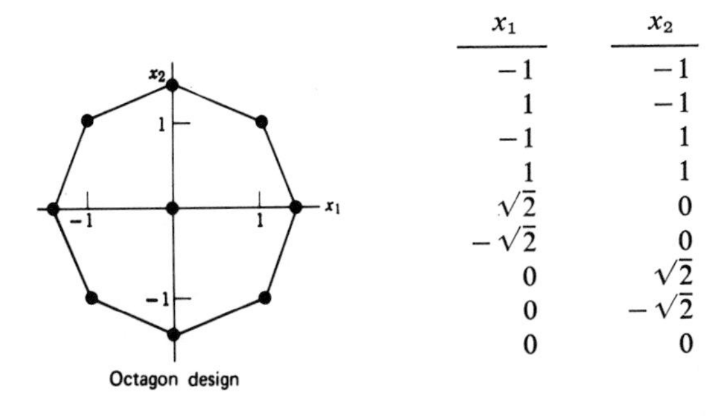

x_1	x_2
−1	−1
1	−1
−1	1
1	1
$\sqrt{2}$	0
$-\sqrt{2}$	0
0	$\sqrt{2}$
0	$-\sqrt{2}$
0	0

Figure 8.1-4 Simple two-dimensional rotatable designs.

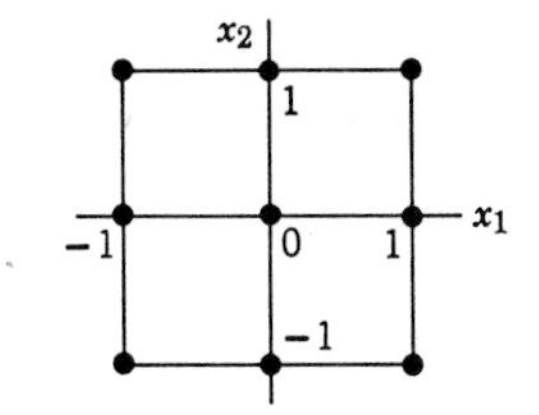

Experimental Levels

x_1	x_2
−1	−1
0	−1
1	−1
−1	0
0	0
1	0
−1	1
0	1
1	1

FIGURE 8.1-5 3^2 design (nonrotatable).

Figure 8.1-5. For this design the number of experiments can be generalized as 3^k, where k is the number of variables. Figure 8.1-5 is for $k = 2$.

Since the experimenter does not know in advance how the response surface will be oriented with respect to his x's, use of a rotatable design is sound policy.†

In initiating an experimental program, assuming no prior information on the nature of the response surface is known, the experimentalist would probably start with a first-order model and see how well it fitted. If it did not fit the data, then a second-order model would be tested for fit. To do this efficiently, the initial first-order design should preferably be one that can be expanded into a second-order design merely by the addition of extra data points. A simple first-order design which can be augmented to form a rotatable second-order design, specifically the octagonal design, is illustrated in Figure 8.1-4. In using the designs, the data collection should be *randomized*; that is, the different combinations of the experimental variables should be randomized.

We now turn to some examples to illustrate the principles and techniques presented up to this point.

Example 8.1-3 Sequential Experimentation to Obtain a Suitable Model

The following data were obtained for a first-order design (Figure 8.1-1), which can be augmented to an octagon design if needed.

† For pertinent references on the construction of rotatable designs, see: J. S. Hunter, *Amer. Soc. Qual. Control Trans.* **15** (7), 1958; G. E. P. Box and J. S. Hunter, *Ann. Math. Stat.* **28**, 1957; W. G. Cochran and G. M. Cox, *Experimental Designs* (2nd. ed.), John Wiley, New York, 1957.

Y	x_1	x_2
24.500	−1	−1
60.141	1	−1
54.890	−1	1
67.712	1	1
77.870	0	0
78.933	0	0
70.100	0	0

A linear model $Y = \beta_0 + \beta_1 x_1 + \beta_2 x_2 + \epsilon$ was fitted to the observations, including the replicated point, with the following results (here $x_0 \equiv 1$ and the replicated data were assigned a weight of 3):

$$\mathbf{a} = \begin{bmatrix} 7 & 0 & 0 \\ 0 & 4 & 0 \\ 0 & 0 & 4 \end{bmatrix} \qquad \mathbf{a}^{-1} = \begin{bmatrix} 0.1428 & 0 & 0 \\ 0 & 0.250 & 0 \\ 0 & 0 & 0.250 \end{bmatrix}$$

$$\mathbf{G} = \begin{bmatrix} 434.1 \\ 48.46 \\ 37.96 \end{bmatrix}$$

$$\hat{Y} = 62.020 + 12.116x_1 + 9.490x_2$$

The analysis of variance is shown in Table E8.1-3a.

The degrees of freedom for the deviations about the empirical regression line are equal to the number of different sets of x's less the number of coefficients, $(5 - 3) = 2$, and the degrees of freedom for the deviations of the replicated point are equal to the total number of replicated data values less the number of constraints imposed, $(3 - 1) = 2$. The constraint comes from the calculation of $\bar{Y}$ at the center. The analysis of variance shows that the first-order model is not an adequate fit.

Additional experimental values to complete an octagon design were next collected:

Y	x_1	x_2
79.162	1.414	0
53.095	−1.414	0
71.328	0	1.414
38.609	0	−1.414
80.131	0	0

TABLE E8.1-3a

Source of Variation	SS	d.f.	Mean Square	Variance Ratio
Effect of removing b_2	360.3	1	360.3	Significant
Effect of removing b_1	587.2	1	587.2	Significant
Effect of removing b_0	26,926.1	1	29,926.1	Significant
Deviation about the empirical regression line	1,103.2	2	551.6	Significant
Deviation at replicated point	46.5	2	23.3	
Total		7		

These values were processed with the following results:

$$\mathbf{a} = \begin{bmatrix} 12.00 & 0 & 0 & 8.000 & 8.000 & 0 \\ 0 & 8.000 & 0 & 0 & 0 & 0 \\ 0 & 0 & 8.000 & 0 & 0 & 0 \\ 8.000 & 0 & 0 & 12.00 & 4.000 & 0 \\ 8.000 & 0 & 0 & 4.000 & 12.00 & 0 \\ 0 & 0 & 0 & 0 & 0 & 4.000 \end{bmatrix}$$

$$\mathbf{c} = \begin{bmatrix} 0.2500 & 0 & 0 & -0.1250 & -0.1250 & 0 \\ 0 & 0.1250 & 0 & 0 & 0 & 0 \\ 0 & 0 & 0.1250 & 0 & 0 & 0 \\ -0.1250 & 0 & 0 & 0.15625 & 0.03125 & 0 \\ -0.1250 & 0 & 0 & 0.03125 & 0.15625 & 0 \\ 0 & 0 & 0 & 0 & 0 & 0.2500 \end{bmatrix}$$

$$\hat{Y} = 76.75 + 10.66x_1 + 10.53x_2 - 7.50x_1^2 - 13.08x_2^2 - 5.70x_1x_2$$

The analysis of variance is shown in Table E8.1-3b.

TABLE E8.1-3b

Source of Variation	SS	d.f.	Mean Square	Variance Ratio
Effect of removing b_{12}	130.17	1	130.17	Not significant
Effect of removing b_{22}	1,094.86	1	1,094.86	Significant
Effect of removing b_{11}	359.94	1	359.94	Significant
Effect of removing b_2	886.88	1	886.88	Significant
Effect of removing b_1	910.11	1	910.11	Significant
Effect of removing b_0	23,567.5	1	23,567.5	Significant
Deviation about empirical regression line	178.14	3	59.38	Not significant
Deviation at replicated point	61.67	3	20.56	
Total		12		

Note: $F_{0.95}(1, 3) = 10.12$

The number of degrees of freedom for the deviations about the estimated regression line are equal to 9 sets of x's less 6 constants, or 3, and the degrees of freedom for the replicated point are equal to the total number of replicated data points, 4, less the number of constraints, or 1, for a net of 3. The analysis of variance shows that the second-order model is a good fit as well as the fact that the crossproduct term might be deleted from the model.

Once a suitable model has been obtained, it can be used for optimization and further analysis.

Example 8.1-4 The Problem of Insufficient Replicated Values†

Consider the following matrix based on a 2^2 factorial design with a repeated center point:

Y	x_1	x_2
80.8	−1	−1
85.1	1	−1
82.9	−1	1
71.9	1	1
82.9	0	0
81.1	0	0

The estimated regression equation based on a first-order model is

$$\hat{Y} = 80.8 - 1.7x_1 - 2.8x_2$$

The analysis of variance is as shown in Table E8.1-4a.

TABLE E8.1-4a

Source of Variation	SS	d.f.	Mean Square	Variance Ratio
Effect of removing b_0	39,155.7	1	39,155.7	24,400
Effect of removing b_1	11.2	1	11.2	7.0
Effect of removing b_2	30.8	1	30.8	19.2
Deviation about empirical regression line	63.0	2	31.5	19.7
Deviation at replicated point	1.6	1	1.6	
Total	39,262.3	6		

Note: $F_{0.95}(2, 1) = 200$; $F_{0.95}(1, 1) = 161$

None of the mean squares is significantly different from 1.6 except for 39,155.7. This paradox develops because only 1 degree of freedom is associated with the variance of the error. A brief examination of the tables for F will show that a few replications at (0, 0) will materially improve the experimenter's ability to evaluate the model ($\alpha = 0.05$):

$$F_{1,1} = 161.00 \qquad F_{2,1} = 200.00$$
$$F_{1,2} = 18.51 \qquad F_{2,2} = 19.00$$
$$F_{1,3} = 10.13 \qquad F_{2,3} = 9.55$$
$$F_{1,4} = 7.71 \qquad F_{2,4} = 6.94$$
$$F_{1,5} = 6.61 \qquad F_{2,5} = 5.79$$

We shall assume that these runs have been carried out and that the F-test proves significant, so additional data are collected to fit a second-order model:

Y	x_1	x_2
81.7	$\sqrt{2}$	0
82.9	$-\sqrt{2}$	0
57.7	0	$\sqrt{2}$
84.7	0	$-\sqrt{2}$
83.8	0	0
80.9	0	0

The estimated regression equation is:

$$\hat{Y} = 82.18 - 1.05x_1 - 6.11x_2 + 0.92x_1^2 - 4.63x_2^2 - 1.19x_1x_2$$

The new analysis of variance is shown in Table E8.1-4b.

TABLE E8.1-4b

Source of Variation	SS	d.f.	Mean Square	Variance Ratio
Effect of removing b_0	76,225.0	1	76,225.0	Significant
First-order terms	305.6	2	152.8	Significant
Second-order terms	184.5	3	61.5	Significant
Deviation about regression line (lack of fit)	157.1	3	52.4	Significant
Deviation at replicated point	6.0	3	2.0	
Total	76,880.2	12		

Note: $F_{0.95}(3, 3) = 9.28$

It appears as if the model is still not satisfactory. Upon reexamining the model and data, the value of Y of 57.7 looks suspicious and should be checked, or perhaps the point $(0, \sqrt{2})$ should be redone.

† J. S. Hunter, *Ind. Qual. Control* **15** (8), 6, 1959.

8.1-3 Three- (and Higher) Dimensional Models and Experimental Designs

The *first-order model* for three controllable variables is

$$Y = \beta_0 + \beta_1x_1 + \beta_2x_2 + \beta_3x_3 + \epsilon \qquad (8.1\text{-}7)$$

A two-level factorial design for three variables, 2^3, is satisfactory for estimating the coefficients. Replication will provide an estimate of the error. In experimental space the design is a cube with corners at the circles, as indicated in Figure 8.1-6.

To cut down the amount of experimentation needed, a so-called *half-replicate* can be used. The first set of points (solid circles) in Figure 8.1-6 form a tetrahedron as do the second set (open circles). The half-replicate is an orthogonal design itself which provides unconfounded estimates of the β's. In cutting down the experimental work to one-half, it is necessary to give up some of the advantages of the full design. First, it becomes impossible to test lack of fit unless center points are taken. Second,

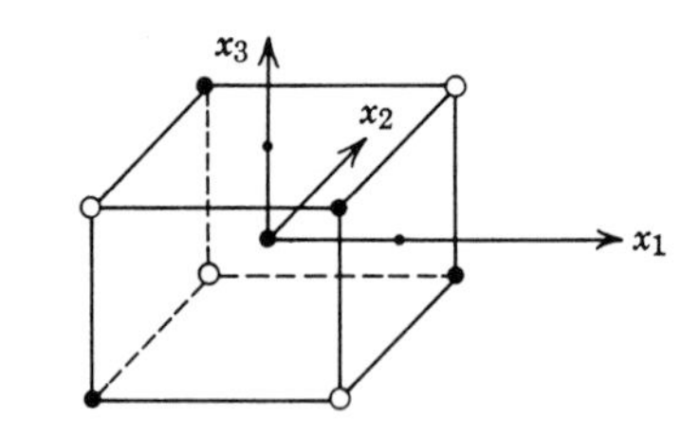

2^3 Design

x_1	x_2	x_3	Points used for Half-Replicates	
−1	−1	−1	√	
1	−1	−1		√
−1	1	−1		√
1	1	−1	√	
−1	−1	1		√
1	−1	1	√	
−1	1	1	√	
1	1	1		√

FIGURE 8.1-6 Two-level, three-dimensional, factorial design (replicate points not listed).

if an interaction term does exist, $x_i x_j$, it will bias estimates of the coefficient of the succeeding variable x_k. For example, if an $x_1 x_2$ interaction exists, it will bias the estimate of β_3. By running the full factorial, not only can one get unbiased estimates of β_i but also of β_{12}, β_{13}, and β_{23}.

Higher dimensional first-order models are of the form

$$Y = \sum_{i=0}^{k} \beta_i x_i + \epsilon \tag{8.1-8}$$

Full factorial designs require $2k$ experimental points. All that has been said about unbiased estimates and the use of center-point replication and half-replicates applies to Equation 8.1-8. Table 8.1-1 gives the four- and five-dimensional full factorial and half-replicate first-order factorial designs. One can see that with five variables a half-replicate can be quite a saving in work. Additional details concerning these designs can be found in the references listed at the end of this chapter.

The linear *second-order model* with three independent variables is

$$Y = \beta_0 x_0 + \beta_1 x_1 + \beta_2 x_2 + \beta_3 x_3 + \beta_{11} x_1^2 + \beta_{22} x_2^2 + \beta_{33} x_3^2 + \beta_{12} x_1 x_2 + \beta_{13} x_1 x_3 + \beta_{23} x_2 x_3 + \epsilon \tag{8.1-9}$$

Again, rotatable designs exist which can be used in conjunction with Equation 8.1-9. They can be described geometrically as the vertices of a icosahedron (12 points), a dodecahedron (20 points), and a so-called central composite figure (14 points), the latter being formed from the vertices of a cube (8 points) plus the vertices of an octahedron (6 extra points). In addition, each

TABLE 8.1-1 FULL FACTORIAL AND HALF-REPLICATE EXPERIMENTAL DESIGNS FOR FOUR AND FIVE VARIABLES

$k = 4$

x_1	x_2	x_3	x_4	x_1	x_2	x_3	x_4
−1	−1	−1	−1	1	−1	−1	−1
1	1	−1	−1	−1	1	−1	−1
1	−1	1	−1	−1	−1	1	−1
−1	1	1	−1	1	1	1	−1
1	−1	−1	1	−1	−1	−1	1
−1	1	−1	1	1	1	−1	1
−1	−1	1	1	1	−1	1	1
1	1	1	1	−1	1	1	1
0	0	0	0 *	0	0	0	0 *
0	0	0	0 *	0	0	0	0 *
.	.	.	. *	.	.	.	. *

$k = 5$

x_1	x_2	x_3	x_4	x_5	x_1	x_2	x_3	x_4	x_5
−1	−1	−1	−1	−1	−1	−1	−1	−1	1
1	1	−1	−1	−1	1	1	−1	−1	1
1	−1	1	−1	−1	1	−1	1	−1	1
−1	1	1	−1	−1	−1	1	1	−1	1
1	−1	−1	1	−1	1	−1	−1	1	1
−1	1	−1	1	−1	−1	1	−1	1	1
−1	−1	1	1	−1	−1	−1	1	1	1
1	1	1	1	−1	1	1	1	1	1
1	−1	−1	−1	1	1	−1	−1	−1	−1
−1	1	−1	−1	1	−1	1	−1	−1	−1
−1	−1	1	−1	1	−1	−1	1	−1	−1
1	1	1	−1	1	1	1	1	−1	−1
−1	−1	−1	1	1	−1	−1	−1	1	−1
1	1	−1	1	1	1	1	−1	1	−1
1	−1	1	1	1	1	−1	1	1	−1
−1	1	1	1	1	−1	1	1	1	−1
0	0	0	0	0 *	0	0	0	0	0 *
0	0	0	0	0 *	0	0	0	0	0 *
.	.	.	.	. *	.	.	.	.	. *

* Replicated center points.

design includes a center point. Figure 8.1-7 illustrates the central composite design. As usual, replicate points at the origin are needed to estimate the experimental error and obtain the estimate of β_0.

Second-order models with more than three variables can be expressed mathematically as

$$Y = \sum_{i=0}^{k} \beta_i x_i + \sum_{\substack{i=1 \\ i \le j}}^{k} \sum_{j=1}^{k} \beta_{ij} x_i x_j + \epsilon \tag{8.1-10}$$

A few rotatable designs (omitting central points) that can be used for such models are listed in Table 8.1-2.

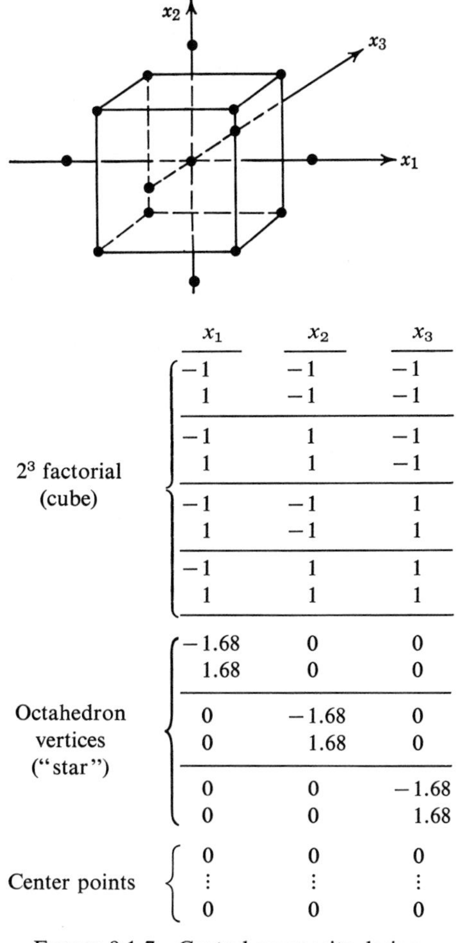

	x_1	x_2	x_3
2^3 factorial (cube)	−1	−1	−1
	1	−1	−1
	−1	1	−1
	1	1	−1
	−1	−1	1
	1	−1	1
	−1	1	1
	1	1	1
Octahedron vertices ("star")	−1.68	0	0
	1.68	0	0
	0	−1.68	0
	0	1.68	0
	0	0	−1.68
	0	0	1.68
Center points	0	0	0
	⋮	⋮	⋮
	0	0	0

FIGURE 8.1-7 Central composite design.

TABLE 8.1-2 ROTATABLE CENTRAL COMPOSITE DESIGNS FOR $k \geq 4$*

Factorial	Star				
	x_1	x_2	x_3	x_4	
$k=4$: 2^4 factorial +	±2	0	0	0	
	0	±2	0	0	
	0	0	±2	0	
	0	0	0	±2	
	8 points				
	x_1	x_2	x_3	x_4	x_5
$k=5$: 2^5 factorial +	±2.378	0	0	0	0
	0	±2.378	0	0	0
	0	0	±2.378	0	0
	0	0	0	±2.378	0
	0	0	0	0	±2.378
	10 points				
$k=k$: 2^k factorial +	k by k matrix in which the main diagonal has the numbers $\pm 2^{k/4}$				
	$2k$ points				

* Each design must be augmented by replicate points at the center of the design.

Example 8.1-5 Development of Empirical Models for Water-Soluble Films†

Commercial water-soluble packaging films must have a number of desirable properties including: (1) good water solubility and (2) toughness and stretchability. Bentz and Roth investigated these properties for a low molecular weight polyvinyl alcohol (PVA-L) on which was grafted ethyl acrylate (EA). After grafting, but before casting, a high molecular weight PVA (PVA-H) was blended with the grafted material to increase the tensile strength of the film.

The independent variables and the coded and actual levels for a central composite design for experimentation are listed in Table E8.1-5a.

TABLE E8.1-5a

	Coded Levels				
Variable	−1.68	−1	0	1	1.68
Ceric ion conc., meq/g PVA	0.029	0.112	0.233	0.255	0.437
EA in graft, wt percent	5	15.1	30	44.9	55
PVA-H in blend, wt percent	43	40.5	50	59.5	66

Measurements of the desirable properties which represented the dependent variables in category (2) were made according to ASTM standards. These properties were:

1. Tensile strength.
2. Tear strength (tab tear).
3. Tear strength (Elmendorf tear).
4. Elongation.
5. Initial modulus.

Solubility was measured in terms of rupture time and total dissolution. Table E8.1-5b shows the results for four films at the center point of the design (0, 0, 0). Measurements in italics were rejected on the basis of t-tests.

The large error in the measurements of tab tear strength, rupture time, and solution time prohibited development of first- or second-order models that had meaning. Figures E8.1-5a, E8.1-5b, and E8.1-5c illustrate the response surfaces within cubes with sides at distances of 2.5 coded units from the center of the design for three responses: (1) tensile strength, (2) elongation at break, and (3) initial modulus. Figure E8.1-5a represents a first-order model and shows the planes of constant tensile strength in the cube. The intersections in any face are a series of parallel straight lines.

Figure E8.1-5b corresponds to a series of cylindrical annuli lying at an angle to the cube. At lower ceric concentrations, the elongation is never more than 215 percent. For packaging, a high elongation is desirable. Elongation does increase with ceric acid concentration as evidenced by the 275-percent contour. Without a suitably designed

† A. P. Bentz and R. W. Roth, *J. Appld. Polymer Sci.* **9**, 1095, 1965.

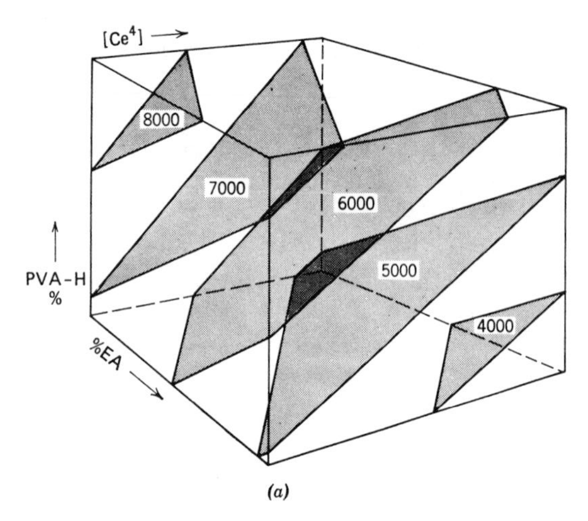

FIGURE E8.1-5a Tensile strength. (Reproduced with permission of John Wiley, publishers of *J. Polymer Sci.*)

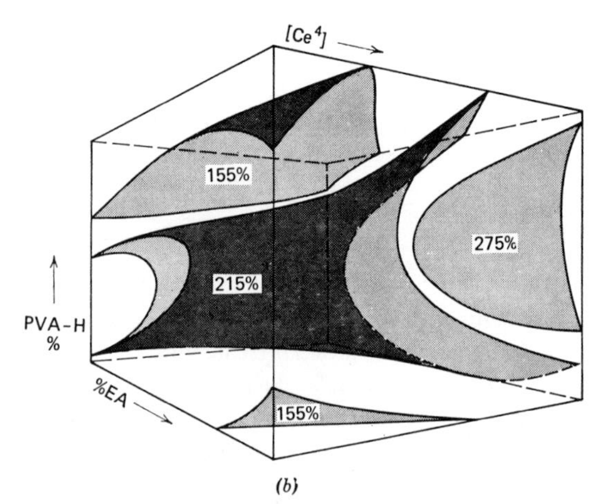

FIGURE E8.1.5b Elongation at break. (Reproduced with permission of John Wiley, publishers of *J. Polymer Sci.*)

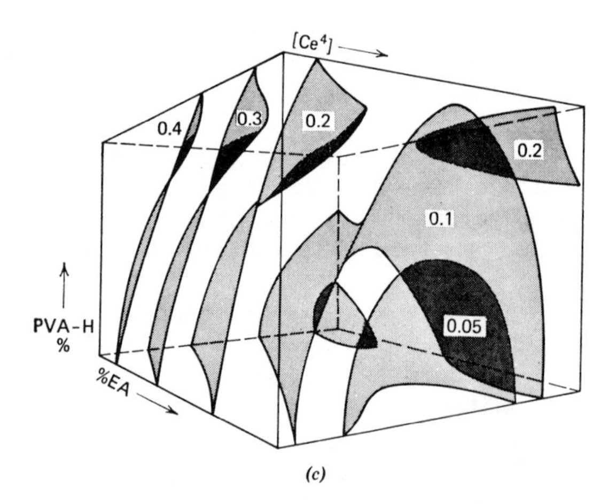

FIGURE E8.1-5c Modulus. (Reproduced with permission of John Wiley, publishers of *J. Polymer Sci.*)

experiment, one might have concluded that, for medium levels of EA and PVA-H, ceric ion concentration had little effect on elongation.

Figure E8-1.5c illustrates saddle-shaped surfaces. At low ceric ion concentration, the surfaces are closer together; at high ceric ion concentrations there is only a slight change with high molecular weight PVA. To minimize the modulus at high ceric concentrations, a decrease in PVA-H and EA is necessary. At low, low ceric ion concentration, the PVA-H would have to be reduced but the EA would have to be increased.

These results emphasize the merit of a sound statistical approach. The authors used the regression equations obtained from their preliminary experiments to predict the factors needed to improve films. They were then able to develop films successfully with these properties.

8.1-4 Blocking

In any sequential series of experiments extending over a period of time or using several batches of material or carried out on different shifts, a change may take place in the experimental environment. To avoid the bias in the average level of response caused by such changes, normally the levels in the experimental design should be run in random order. However, in certain cases the effect of variables can be segregated by suitable *blocking* of the raw materials or sequence of trials.

Schemes to block the second-order rotatable designs so that linear, cross-product, and square coefficients are not influenced by block-to-block differences can be found in G. E. P. Box and J. S. Hunter, *Ann. Math. Stat.* **28**, 195, 1957. Fortunately, the 2^k factorial, the half-replicate, and the "star" designs are by themselves rotatable first-order designs. To make the variables in the second-order model independent of any block effects, the number of center points must be weighted properly.

As long as the points comprising a rotatable design can be broken into blocks, each with an equal number of points and each by itself forming a first-order rotatable design, the number of points required at the center of each block remains equal, and the full second-order design remains rotatable. For example, the octagon design can be broken into two blocks, each block having four peripheral points plus two center points for replication. However, it is not always possible to split a second-order design into blocks, each with an equal number of points and each being a rotatable first-order design. For example, the central composite design for $k = 3$ variables can be split into three rotatable first-order designs: two provided by the two one-half replicates of the 2^3 factorial, each with four peripheral points, and the third block provided by the star design with six peripheral points. It is now impossible to guarantee rotatability of the first-order designs for each block *and* orthogonality among the estimated coefficients in the second-order model. However, if the axis arm of the star design is changed from 1.682 units (the distance for

TABLE E8.1-5b FILM TEST RESULTS OF REPLICATED CENTER POINT*

Film	Tensile Strength (psi)	Tear Strength (g)		Solubility		Elongation at Break (Percent)	Modulus $\times 10^{-6}$ (psi)
		Tab	Elmendorf	Rupture Time (sec)	Solution Time (min)		
1	4,920	21	3.5, 3.0	20	3	219	0.12
	5,690	*33*	4.0, 3.0				
	4,920	24	3.0, 3.0				
	4,920	25	3.0, 3.5				
	4,770	23	3.5, *4.5*				
2	5,500	38	3.0, 2.5	15	2	234	0.12
	4,920	47	2.5, 2.5				
	7,230	49	3.0, 3.0				
	6,620	55	3.0, 3.0				
	5,830	56	3.5, 3.5				
3	6,920	18	2.5, *4.0*	20	2	243.5	0.15
	6,620	19	3.0, 3.5				
	6,770	19	3.0, 3.0				
	10,170	18	3.0, 3.0				
	6,670	18	3.5, 3.0				
4	5,690	24	3.5, 4.0	20	3.5	253	0.09
	5,330	24	4.5, 3.0				
	7,540	20	3.5, 4.0				
	5,080	25	3.5, 3.0				
	5,080	23	3.0, 3.5				

* Italic responses are to be questioned.

rotatability) to 1.633 units, and two center points are placed in each block, then orthogonality is obtained among the coefficient estimates. The design is only slightly nonrotatable, the change from 1.682 to 1.633 being barely noticeable.

The central composite design for $k = 4$ variables can be arranged into rotatable first-order designs which give orthogonality among the coefficient estimates if two blocks are formed from the half-replicates of the 2^4 factorial (each has eight peripheral points) and from the star design (also with eight peripheral points). The axis arm for rotatability as well as orthogonality is 2.000. Each block should contain at least two center points.

Tables giving the design coordinates for rotatable and orthogonally blocked second-order designs can be found in Box and Hunter.

The following example on the use of blocking has been taken from J. S. Hunter, *Ind. Qual. Control* **15** (6) 16, (7) 7, and (8) 6, 1959.

Example 8.1-6 Blocking

It will be observed that the first, third, and fifth entries in the hexagon design, Figure 8.1-4, form an equilateral triangle with the length of the sides equal to $\sqrt{3}$. Similarly, the second, fourth, and sixth entries form an opposite triangle. Suppose that the first triangular design is used to fit a first-order model:

$$Y = \beta_0 x_0 + \beta_1 x_1 + \beta_2 x_2 + \epsilon$$

Vector of Observations Y	Matrix of Independent Variables x_0	x_1	x_2
96.0	1	1.000	0
76.7	1	−0.500	0.866
64.8	1	−0.500	−0.866
97.4	1	0	0
93.0	1	0	0

The estimated regression equation proves to be $\hat{Y} = 85.58 + 16.83x_1 + 6.90x_2$ while, the analysis of variance is as shown in Table E8.1-6a.

Because only one degree of freedom is involved in the error variance, the adequacy of the fit proves to be marginal. Suppose that the experimenter wishes to augment the equilaterial design to provide adequate information for estimating all the coefficients in a second-order model in this experimental region. Imagine that a second equilateral triangle (the even-number runs) with two center points is added to the first; the two blocks of experiments together

TABLE E8.1-6a

Source of Variation	d.f.	SS	Variance Ratio
Due to removing b_0	1	36,551	
Due to removing b_1	1	431.7	
Due to removing b_2	1	70.8	
Deviation about empirical regression line	1	370.2	38.24
Deviation at replicated point (error)	1	9.68	
Total ($\sum Y_i^2$)	5	37,433	

Note: $F_{0.90}(1, 1) = 39.86$

form a hexagon design. The matrix of all the independent variables and the vector of observations are

Vector of Observations Y; Matrix of Independent Variables x_0, x_1, x_2

$$
\begin{bmatrix} 96.0 \\ 78.7 \\ 76.7 \\ 54.6 \\ 64.8 \\ 78.9 \\ 97.4 \\ 90.5 \\ 93.0 \\ 86.3 \end{bmatrix}
\qquad
\begin{bmatrix} 1 & 1.000 & 0 \\ 1 & 0.500 & 0.866 \\ 1 & -0.500 & 0.866 \\ 1 & -1.000 & 0 \\ 1 & -0.500 & -0.866 \\ 1 & 0.500 & -0.866 \\ 1 & 0 & 0 \\ 1 & 0 & 0 \\ 1 & 0 & 0 \\ 1 & 0 & 0 \end{bmatrix}
$$

The estimated regression equation is

$$\hat{Y} = 91.80 + 16.48x_1 + 3.38x_2 - 16.50x_1^2 - 17.20x_2^2 - 6.98x_1x_2$$

Now let us suppose that the second block of experiments gives responses uniformly depressed by 10; that is, the second observation would be 78.7 − 10 = 68.7 and so forth. The new vector of observations would then be

Vector of Observations Y; Matrix of Independent Variables x_0, x_1, x_2

$$
\begin{bmatrix} 96.0 \\ 68.7 \\ 76.7 \\ 44.6 \\ 64.8 \\ 68.9 \\ 97.4 \\ 80.5 \\ 93.0 \\ 76.3 \end{bmatrix}
\qquad
\begin{bmatrix} 1 & 1.0 & 0 \\ 1 & 0.5 & 0.866 \\ 1 & -0.5 & 0.866 \\ 1 & -1.0 & 0 \\ 1 & -0.5 & -0.866 \\ 1 & 0.5 & -0.866 \\ 1 & 0 & 0 \\ 1 & 0 & 0 \\ 1 & 0 & 0 \\ 1 & 0 & 0 \end{bmatrix}
$$

We have in effect blocked separately two different groups of data. The estimated regression equation would be

$$\hat{Y} = 76.48 + 16.48x_1 + 3.38x_2 - 16.50x_1^2 - 17.20x_2^2 - 6.98x_1x_2$$

With the expection of the b_0 coefficient, the fitted model is identical to the first one. The shape of the fitted contour surface thus remains unchanged, despite the fact that in the second block all yields were depressed by 10.

An analysis of variance can be prepared to show the difference, if any, between the data in the two blocks. The SS for the effect "due to blocks" is calculated by

$$\sum_{i=1}^{2} \frac{(B_i)^2}{n_i} - \frac{\left(\sum_{i=1}^{10} Y_i\right)^2}{10}$$

where B_i = total sum of the squares for block i and n_i = number of observations in block i:

$$\frac{(427.9)^2}{5} + \frac{(339.0)^2}{5} - \frac{(766.9)^2}{10} = 790.3; \qquad \text{d.f.} = 1$$

The expanded analysis of variance is shown in Table E8.1-6b.

TABLE E8.1-6b

Source of Variation	d.f.	SS	Mean Square	Variance Ratio
Due to blocks	1	790.3		Significant
Due to removing b_0	1	58,813		Significant
Due to removing linear terms	2	849.3		Significant
Due to removing second-order terms	3	718.4		Significant
Deviation about empirical regression line	1	1.60	1.60	Not significant
Deviation at replicated point (error)	2	18.5	9.25	
Total	10	61,191		

The number of different data sets less the number of coefficients is 7 − 6 = 1 degree of freedom for the deviations about the empirical regression line. The error sum of squares is obtained from within each block and then pooled. Since there are two repeated observations in each block, the error sum of squares is given by $\sum d^2/2$ or

$$\frac{(4.4)^2 + (4.2)^2}{2} = 18.5$$

with 2 degrees of freedom. We observe that the lack of fit term is not significant, that the second-order terms are significant, and that the contribution of the block effect is significant. However, the differences between the blocks have not altered the shape of the fitted contours, since all the estimated coefficients associated with x_1, x_2, x_1^2, x_2^2, and x_1x_2 have remained unchanged.

8.1-5 Practical Considerations

In a continuous process which is operating and must continue to operate, a number of practical difficulties in the precise use of response surface methods, as described above, are encountered:

1. Use of orthogonal designs can be impractical. Preselection of the levels of the independent variables may yield uneconomic or dangerous operating conditions so that orthogonal designs may be incompatible with production requirements. Adjustment of one variable can throw others out of adjustment. Replicate points at the center may be impractical to obtain.

2. Randomization may be impractical; the selection of operating points cannot be run in random order. Randomization removes the possibility of confusing the effect of an uncontrolled independent variable with the observed effect of a controlled variable.

Two other problems associated with response surface methods are related to the assumptions underlying the linear regression techniques employed. These are yet unresolved and hence will only be briefly mentioned here.

3. *Effect of errors in the levels of the design.* A basic assumption in regression analysis, as discussed in Chapter 4, is that all the unobservable error occurs in the dependent variable, Y, and none in the independent variable. Obviously, errors will exist in the x's which are of unknown character, and the experimental data will be analyzed as if no errors exist. Box† demonstrated that the standard statistical analysis is justified even if the experimental levels are in error as long as the response is linear in x. Each case for a quadratic (in x) function and other nonlinear functions has to be considered separately because the size of the error transmitted from the x's to the response depends upon the slope of the response surface at the design point. Two-level factorial and some fractional factorial designs appear to be quite robust with respect to errors in the x's.

4. *Designs to use for correlated data.* A second basic assumption of the regression analysis technique used in response surface methods is that the experimental responses are uncorrelated, but it is well known that data collected from continuous processes are quite likely to be subject to trends or drifts in the mean level (low-frequency changes) and to serial or time correlation in spite of deliberate attempts to avoid them. Section 5.4 described some of the difficulties in the analysis of correlated data.

Finally, we should mention that the orthogonal designs illustrated in this section are just one among the many useful classes of designs. A vast literature exists concerning other experimental designs that have been proposed and applied, examples of which can be found in the general references at the end of this chapter.

† G. E. P. Box, *Technometrics* **5**, 247, 1963.

8.2 CANONICAL ANALYSIS

In this section we are interested in the graphical and analytical interpretation of the response surface (i.e., the estimated regression equation) itself. Surfaces that are a function of two experimental variables can easily be represented in a two-dimensional plane in which the response appears as contours of constant value. Surfaces that are a function of three experimental variables also can be illustrated, although less readily. Lind and Young‡ described a three-dimensional device to display response surfaces, and computer programs can be prepared to plot response surfaces and illustrate them on cathode ray display tubes. If more than three experimental variables are involved in the response surface, graphical portrayal is difficult.

To gain an understanding of the nature of the response surface, canonical analysis, as described by Box and Wilson,§ can be effectively used. Canonical analysis transforms the estimated regression equation into a simpler form and interprets the resulting expression in terms of geometric concepts. We shall initially look at two- and three-dimensional surfaces in order to compare the geometric concepts with the algebraic terms in the regression equation. We shall then explain how the canonical transformation is carried out. And finally, through an example, we shall demonstrate how to interpret a response surface with many variables through use of the canonical equation.

A canonical equation is a second- (or higher) order equation in an original coordinate system which is transformed to a new coordinate system by translation of the center of the old coordinates to the extremum of the response surface with subsequent rotation of the axes to achieve symmetry. Figure 8.2-1 illustrates figuratively the general procedure.

The old coordinates are designated by x_1 and x_2; the new coordinates, *termed principal axes*, by $\tilde{x}_1$ and $\tilde{x}_2$. The two transformations yield a new expression for the response surface, termed the *canonical equation*, expressed in terms of the principal axes, an equation that is much simpler than the original regression equation in as much as all the first-order terms and crossproduct terms have been eliminated. For example, the estimated regression equation for Model 8.1-5 transforms to

$$(\hat{Y} - \hat{Y}_e) = \tilde{b}_{11}\tilde{x}_1^2 + \tilde{b}_{22}\tilde{x}_2^2 \tag{8.2-1}$$

where $\hat{Y}_e$ is the predicted response at the center of the response surface, $\tilde{b}_{11}$ and $\tilde{b}_{22}$ are the transformed estimated parameters, and the overlay tilde designates "in canonical form." The translation indicated in Figure

‡ E. E. Lind and W. R. Young, *Ind. Qual. Control* **23**, 436, 1967.

§ G. E. P. Box and K. B. Wilson, *J. Royal Stat. Soc.* **B13**, 1, 1951.

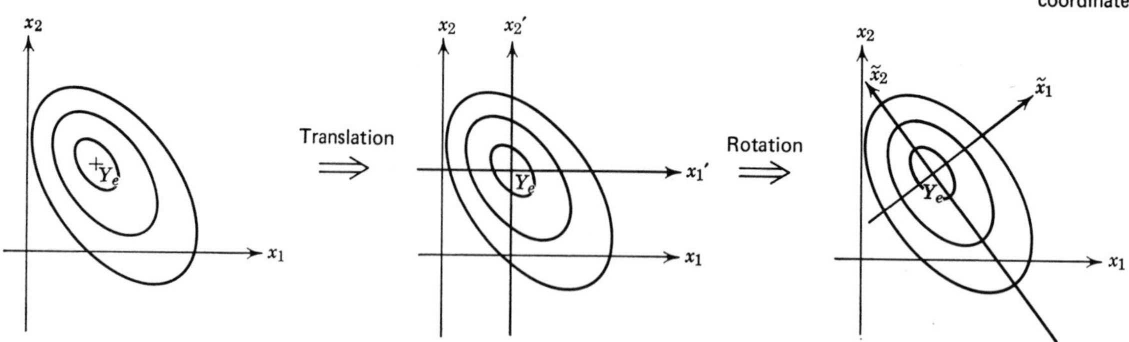

FIGURE 8.2-1 Transformation of coordinates into the principal axes. (Canonical coordinates are the heavy lines.)

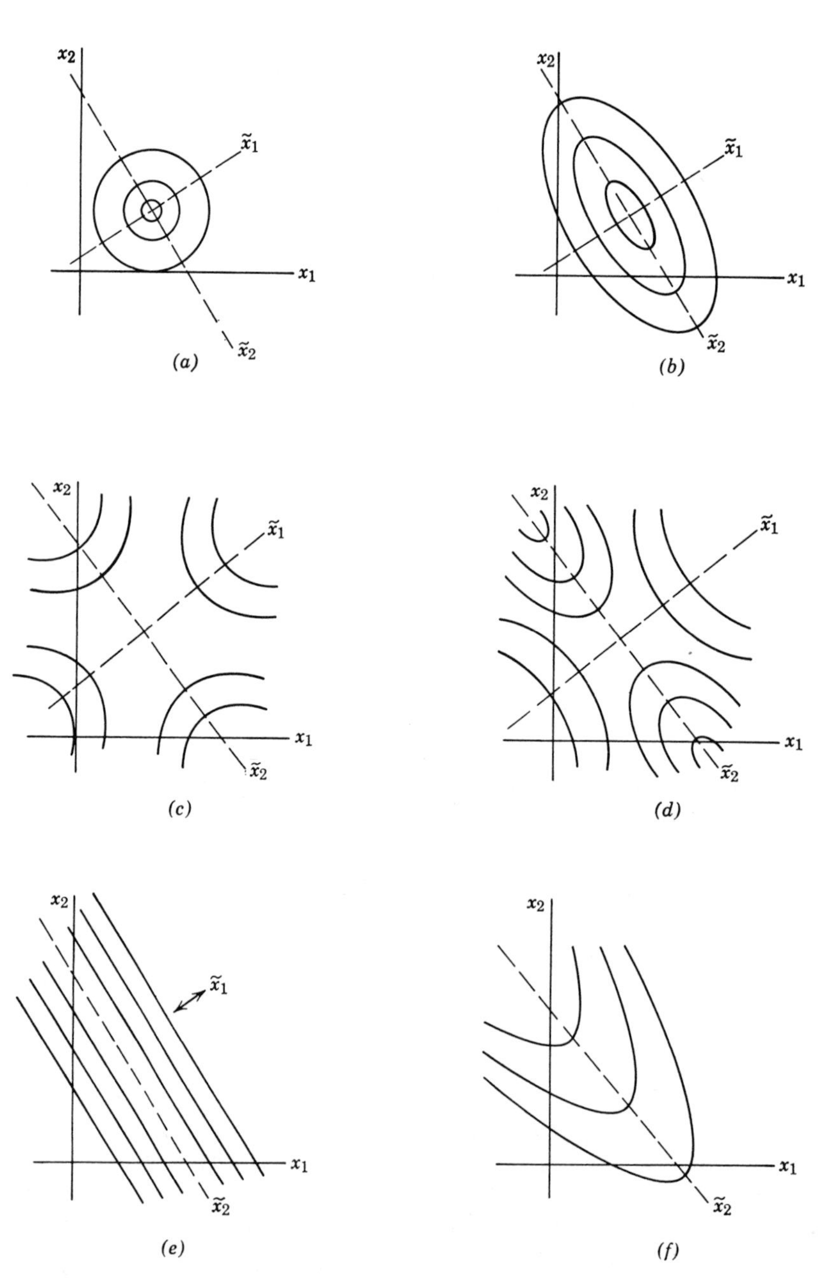

FIGURE 8.2-2 Contours for second-order models with two independent variables. (Adapted from G. E. P. Box, *Biometrics* **10**, 16, 1954.)

TABLE 8.2-1 INTERPRETATION OF THE CANONICAL EQUATION

$$\hat{Y} - \hat{Y}_e = \tilde{b}_{11}\tilde{x}_1^2 + \tilde{b}_{22}\tilde{x}_2^2$$

Case	Coefficient Relations	Signs $\tilde{b}_{11}$	Signs $\tilde{b}_{12}$	Type of Curves	Geometric Interpretation	Center	Figure 8.2-2
1	$\tilde{b}_{11} = \tilde{b}_{22}$	$-$	$-$	Circles	Circular hill	Maximum	(*a*)
2	$\tilde{b}_{11} = \tilde{b}_{22}$	$+$	$+$	Circles	Circular valley	Minimum	(*a*)
3	$\tilde{b}_{11} > \tilde{b}_{22}$	$-$	$-$	Ellipses	Elliptical hill	Maximum	(*b*)
4	$\tilde{b}_{11} > \tilde{b}_{22}$	$+$	$+$	Ellipses	Elliptical valley	Minimum	(*b*)
5	$\tilde{b}_{11} = \tilde{b}_{22}$	$+$	$-$	Hyperbolas	Symmetrical saddle	Saddle point	(*c*)
6	$\tilde{b}_{11} = \tilde{b}_{22}$	$-$	$+$	Hyperbolas	Symmetrical saddle	Saddle point	(*c*)
7	$\tilde{b}_{11} > \tilde{b}_{22}$	$+$	$-$	Hyperbolas	Elongated saddle	Saddle point	(*d*)
8	$\tilde{b}_{22} = 0$	$-$		Straight lines	Stationary ridge	None	(*e*)
9	$\tilde{b}_{22} = 0$	$-$		Parabolas	Rising ridge	At infinity	(*f*)

8.2-1 corresponds to the deletion of the linear terms; the rotation corresponds to the deletion of the crossproduct term appearing in Equation 8.1-5. Figure 8.2-2 illustrates typical examples of two-dimensional response surfaces related to their original and principal axes.

Table 8.2-1 interprets the information provided by the canonical equation in terms of the shape of the response surface. If $|\tilde{b}_{11}| > |\tilde{b}_{22}|$, the contours are elongated along the $\tilde{x}_2$ (smaller coefficient) axis, and vice-versa. If the center on the $\tilde{x}_2$ axis is at infinity and $\tilde{b}_{11}$ is negative, the fitted contours are parabolas as illustrated in Figure 8.2-2*f*. Either of the surfaces shown in Figure 8.2-2*e* or 8.2-2*f* is known as a ridge and appears when one of the coefficients is very small in magnitude compared with the other. Figure 8.2-2 is only an idealization of what is encountered in practice, but it can be of material aid in evaluating the nature of response surfaces.

All that has been said about response surfaces involving two experimental variables can be extended to those involving more variables. Figure 8.2-3 illustrates the three-dimensional canonical equation

$$(\hat{Y} - \hat{Y}_e) = \tilde{b}_{11}\tilde{x}_1^2 + \tilde{b}_{22}\tilde{x}_2^2 + \tilde{b}_{33}\tilde{x}_3^2 \qquad (8.2\text{-}2)$$

When one or more of the b_{ii} becomes small, a ridge is present. The other features described in Table 8.2-1 can also be applied to three- and higher dimensional surfaces. Consequently, one can obtain a mental picture of the surface even though it cannot be represented graphically.

To effect the transformation to canonical form for a second-order model, all that is required is to (1) find $\hat{Y}_e$, the center of the new system in terms of the old coordinates, (2) translate the origin to $\hat{Y}_e$, and (3) rotate the axes about the new origin to obtain the new principal axes $\tilde{x}_1, \tilde{x}_2, \ldots$. We first derive the equations which locate the center of the surface in experimental space. Matrix notation is used for compactness in much of the following discussion.

The complete second-order response surface with three controllable variables associated with Equation 8.1-9 can be written as

$$\hat{Y} = b_0 + b_1x_1 + b_2x_2 + b_3x_3 + b_{11}x_1^2 + b_{22}x_2^2 + b_{33}x_3^2$$
$$+ b_{12}x_1x_2 + b_{21}x_2x_1 + b_{13}x_1x_3 + b_{31}x_3x_1$$
$$+ b_{32}x_3x_2 + b_{23}x_2x_3 \qquad (8.2\text{-}3)$$

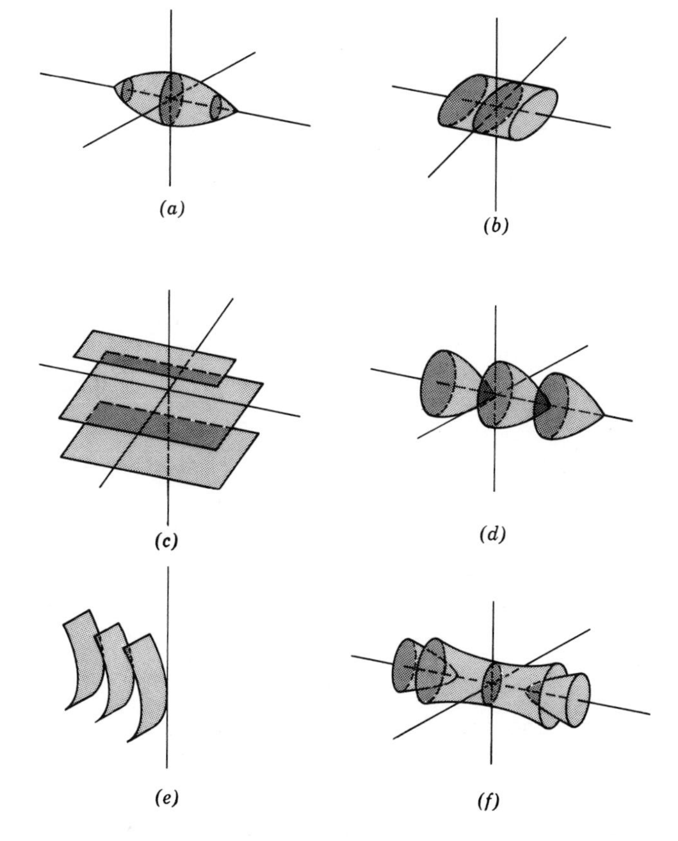

FIGURE 8.2-3 Contours for second-order models with three independent variables. (Adapted from G. E. P. Box, *Biometrics* **10**, 16, 1954.)

or

$$\hat{Y} = b_0 + \mathbf{b}_1\mathbf{x} + \mathbf{x}^T\mathbf{b}_{11}\mathbf{x} \qquad (8.2\text{-}4)$$

where

$$\mathbf{x} = \begin{bmatrix} x_1 \\ x_2 \\ x_3 \end{bmatrix} \qquad \mathbf{b}_1 = [b_1 \quad b_2 \quad b_3]$$

$$\mathbf{b}_{11} = \begin{bmatrix} b_{11} & b_{12} & b_{13} \\ b_{12} & b_{22} & b_{23} \\ b_{13} & b_{23} & b_{33} \end{bmatrix}$$

The crossproduct estimated coefficients are split into two coefficients of equal value to fill the matrix. By definition, $b_{ij} = b_{ji}$. In the off-diagonal element, b_{ij} is one-half of the related coefficient in the usual response surface equation.

The center of the new system in experimental space is to be located at the extremum (minimum or maximum) of the old system; hence, to locate the coordinates of the center equate the partial derivatives of the response function to zero, check for a minimum or maximum by using the second derivatives, and solve the resulting simultaneous equations for the extremum coordinates.

For the three-dimensional second-order Equation 8.2-3, the general term obtained upon differentiation is

$$\frac{\partial \hat{Y}}{\partial x_i} = b_i + 2b_{ii}x_i + 2\sum_{\substack{i \neq j \\ i<j}} b_{ij}x_j = 0 \qquad (8.2\text{-}5)$$

or, in matrix notation,

$$\mathbf{b}_1^T + 2\mathbf{b}_{11}\mathbf{x} = 0 \qquad (8.2\text{-}6)$$

Solution of Equation 8.2-6 for $\mathbf{x}$ gives the set of x's at the extremum:

$$\mathbf{x}_{\text{extremum}} = -(\tfrac{1}{2})\mathbf{b}_{11}^{-1}\mathbf{b}_1^T \equiv \mathbf{x}_e \qquad (8.2\text{-}7)$$

Introduction of $\mathbf{x}_e$ into Equation 8.2-4 enables $\hat{Y}_e$ to be evaluated (one should be sure that Equation 8.2-6 is consistent, i.e., the det $(\mathbf{b}_{11}) \neq 0$, if the surface is to have a center):

$$\hat{Y}_{\text{extremum}} \equiv \hat{Y}_e = b_0 + \mathbf{b}_1\mathbf{x}_e + \mathbf{x}_e^T\mathbf{b}_{11}\mathbf{x}_e \qquad (8.2\text{-}8)$$

Now, to eliminate the first-order terms, define new independent variables which are to be measured from the new center:

$$\acute{x}_i = x_i - x_e$$

or

$$\acute{\mathbf{x}} = \mathbf{x} - \mathbf{x}_e$$

and introduce the relation for $\mathbf{x}$ into Equation 8.2-4:

$$\hat{Y} = b_0 + \mathbf{b}_1(\acute{\mathbf{x}} + \mathbf{x}_e) + (\acute{\mathbf{x}} + \mathbf{x}_e)^T\mathbf{b}_{11}(\acute{\mathbf{x}} + \mathbf{x}_e) \qquad (8.2\text{-}9)$$

Subtract Equation 8.2-8 from Equation 8.2-9 to get

$$\hat{Y} - \hat{Y}_e = [\mathbf{b}_1\acute{\mathbf{x}} + \acute{\mathbf{x}}^T\mathbf{b}_{11}\mathbf{x}_e + \mathbf{x}_e^T\mathbf{b}_{11}\acute{\mathbf{x}}] + \acute{\mathbf{x}}^T\mathbf{b}_{11}\acute{\mathbf{x}} \qquad (8.2\text{-}10)$$

The expression in the square brackets is identical to zero because: (1) premultiplication of Equation 8.2-7 successively by $\mathbf{b}_{11}$ and $\acute{\mathbf{x}}^T$ yields

$$\acute{\mathbf{x}}^T\mathbf{b}_{11}\mathbf{x}_e = -\tfrac{1}{2}(\acute{\mathbf{x}}^T\mathbf{b}_1^T)$$

while (2) the equality of

$$\mathbf{x}_e^T\mathbf{b}_{11}\acute{\mathbf{x}} = \acute{\mathbf{x}}^T\mathbf{b}_{11}\mathbf{x}_e = -\tfrac{1}{2}(\acute{\mathbf{x}}^T\mathbf{b}_1^T)$$

can be demonstrated by multiplication of the elements of the respective matrices, and (3) $(\mathbf{b}_1\acute{\mathbf{x}} - \acute{\mathbf{x}}^T\mathbf{b}_1^T) = 0$. Consequently,

$$\hat{Y} - \hat{Y}_e = \acute{\mathbf{x}}^T\mathbf{b}_{11}\acute{\mathbf{x}} \qquad (8.2\text{-}11)$$

which is an equation with the first-order terms deleted as desired.

The final step is to carry out a rotation of axes for Equation 8.2-11. The technique, described in detail in Appendix B.5, results in an equation without cross-product terms. The essential step is to find a real unitary matrix $\mathbf{U}$ such that, upon introduction of the transformation $\acute{\mathbf{x}} = \mathbf{U}\tilde{\mathbf{x}}$ into Equation 8.2-11, one obtains

$$\hat{Y} - \hat{Y}_e = \tilde{\mathbf{x}}^T\mathbf{U}^T\mathbf{b}_{11}\mathbf{U}\tilde{\mathbf{x}} = \lambda_1\tilde{x}_1^2 + \lambda_2\tilde{x}_2^2 + \cdots \qquad (8.2\text{-}12)$$

where the λ_i are the eigenvalues of the det $(\mathbf{b}_{11} - \lambda\mathbf{I}) = 0$. $\tilde{\mathbf{x}}$ represents the final set of coordinates, shown in Figure 8.2-2, which we have termed the principal axes. If you are not familiar with orthogonal transformations, refer to Appendix B.5 before reading the following example.

Example 8.2-1 Transformation to Canonical Form

Reduce the following equation to canonical form:

$$\hat{Y} = 7x_1^2 + 6x_2^2 + 5x_3^2 - 4x_1x_2 - 4x_2x_3 - 6x_1 - 24x_2 + 18x_3 + 18$$

(Note that the integers are used in the example equation to enable you to follow the calculations by hand—in any model developed by regression analysis, it would be quite unlikely that integers would represent the model coefficients.)

Solution:

1. Check the determinant of the quadratic terms to make sure it is not zero:

$$\det \mathbf{b}_{11} = \begin{vmatrix} 7 & -2 & 0 \\ -2 & 6 & -2 \\ 0 & -2 & 5 \end{vmatrix} = 162 \neq 0$$

2. Locate the center of the surface by using Equation 8.2-7:

$$\mathbf{x}_e = -\tfrac{1}{2}\mathbf{b}_{11}^{-1}\mathbf{b}_1^T$$

$$\text{Adjoint of } \mathbf{b}_{11} = \begin{bmatrix} 26 & 10 & 4 \\ 10 & 35 & 14 \\ 4 & 14 & 38 \end{bmatrix}$$

$$\text{Inverse of } \mathbf{b}_{11} = \mathbf{b}_{11}^{-1} = \frac{1}{162}\begin{bmatrix} 26 & 10 & 4 \\ 10 & 35 & 14 \\ 4 & 14 & 38 \end{bmatrix}$$

$$\mathbf{x}_e = -\frac{1}{2}\left(\frac{1}{162}\right)\begin{bmatrix} 26 & 10 & 4 \\ 10 & 35 & 14 \\ 4 & 14 & 38 \end{bmatrix}\begin{bmatrix} -6 \\ -24 \\ 18 \end{bmatrix} = \begin{bmatrix} 1 \\ 2 \\ -1 \end{bmatrix}$$

The translation is then

$$\acute{x}_1 = x_1 - 1$$
$$\acute{x}_2 = x_2 - 2$$
$$\acute{x}_3 = x_3 + 1$$

and $\hat{Y}_e$ is

$$\hat{Y}_e = 18 + [-6 \quad -24 \quad 18]\begin{bmatrix} 1 \\ 2 \\ -1 \end{bmatrix}$$

$$+ [1 \quad 2 \quad -1]\begin{bmatrix} 7 & -2 & 0 \\ -2 & 6 & -2 \\ 0 & -2 & 5 \end{bmatrix}\begin{bmatrix} 1 \\ 2 \\ -1 \end{bmatrix}$$

$$= -18$$

3. The last step is the rotation of axes for Equation 8.2-11:

$$\hat{Y} - 18 = \acute{\mathbf{x}}^T \mathbf{b}_{11} \acute{\mathbf{x}}$$

as described in Appendix B.5.

(a) First, find the eigenvalues of $\mathbf{b}_{11}$:

$$\det(\mathbf{b}_{11} - \lambda \mathbf{I}) = 0$$

$$\begin{vmatrix} (7-\lambda) & -2 & 0 \\ -2 & (6-\lambda) & -2 \\ 0 & -2 & (5-\lambda) \end{vmatrix} = 0$$

$$\lambda_1 = 3, \qquad \lambda_2 = 6, \qquad \lambda_3 = 9$$

Equation 8.2-12 shows that the canonical form of the original response surface is

$$\hat{Y} - 18 = 3\tilde{x}_1^2 + 6\tilde{x}_2^2 + 9\tilde{x}_3^2 \tag{a}$$

where the $\tilde{x}_i$ are the principal axes. The initial and final stages of the translation and rotation are illustrated graphically in Figure E8.2-1.

(b) The new coordinates $\tilde{x}_i$ can be related to the old coordinates x_i through the unitary matrix $\mathbf{U}$ and the transformation

$$\acute{\mathbf{x}} = \mathbf{U}\tilde{\mathbf{x}}$$

or

$$\tilde{\mathbf{x}} = \mathbf{U}^{-1}\acute{\mathbf{x}} = \mathbf{U}^T(\mathbf{x} - \mathbf{x}_e)$$

Recall that $\mathbf{U}^{-1} = \mathbf{U}^T$.

For $\lambda = 3$, $(\mathbf{b}_{11} - \lambda\mathbf{I}) = 0$ is

$$\begin{bmatrix} (7-3) & -2 & 0 \\ -2 & (6-3) & -2 \\ 0 & -2 & (5-3) \end{bmatrix}\begin{bmatrix} u_1 \\ u_2 \\ u_3 \end{bmatrix} = 0$$

or

$$4u_1 - 2u_2 = 0$$
$$-2u_1 + 3u_2 - 2u_3 = 0$$
$$-2u_2 + 2u_3 = 0$$

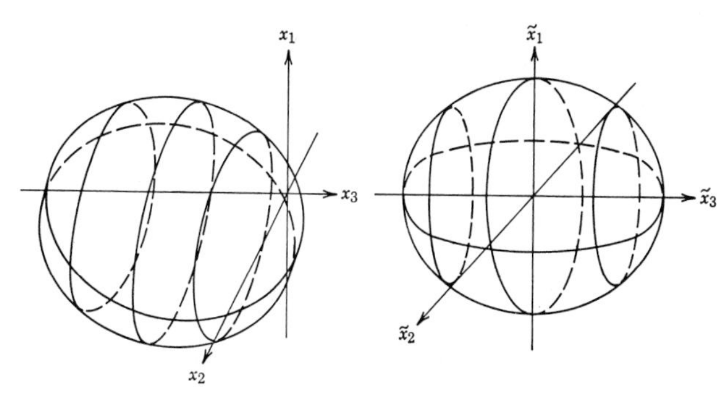

FIGURE E8.2-1

A nontrivial solution of these equations is $u_1 = 1$, $u_2 = 2$, and $u_3 = 2$ which form the orthonormal vector $\mathbf{U}_1$ (refer to Appendix B4-4):

$$\mathbf{U}_1 = \begin{bmatrix} \frac{1}{3} \\ \frac{2}{3} \\ \frac{2}{3} \end{bmatrix}$$

For $\lambda = 6$

$$\begin{bmatrix} 1 & -2 & 0 \\ -2 & 0 & -2 \\ 0 & -2 & -1 \end{bmatrix}\begin{bmatrix} u_1 \\ u_2 \\ u_3 \end{bmatrix} = 0$$

which has a nontrivial solution: $u_1 = 2$, $u_2 = 1$, and $u_3 = -2$, and

$$\mathbf{U}_2 = \begin{bmatrix} \frac{2}{3} \\ \frac{1}{3} \\ -\frac{2}{3} \end{bmatrix}$$

Finally, for $\lambda = 9$

$$\mathbf{U}_3 = \begin{bmatrix} -\frac{2}{3} \\ \frac{2}{3} \\ -\frac{1}{3} \end{bmatrix}$$

The unitary matrix $\mathbf{U}$ is

$$\mathbf{U} = \begin{bmatrix} \frac{1}{3} & \frac{2}{3} & -\frac{2}{3} \\ \frac{2}{3} & \frac{1}{3} & \frac{2}{3} \\ \frac{2}{3} & -\frac{2}{3} & -\frac{1}{3} \end{bmatrix}$$

Consequently,

$$\begin{bmatrix} \tilde{x}_1 \\ \tilde{x}_2 \\ \tilde{x}_3 \end{bmatrix} = \begin{bmatrix} \frac{1}{3} & \frac{2}{3} & \frac{2}{3} \\ \frac{2}{3} & \frac{1}{3} & -\frac{2}{3} \\ -\frac{2}{3} & \frac{2}{3} & -\frac{1}{3} \end{bmatrix}\begin{bmatrix} (x_1 - 1) \\ (x_2 - 2) \\ (x_3 + 1) \end{bmatrix} \tag{b}$$

If the det $\mathbf{b}_{11} = 0$, the response surface degenerates to an infinite cylinder, cone, or paraboloid, and one of the eigenvalues is zero. Equation 8.2-7 can no longer be applied; instead the procedure is as indicated in the next example.

Example 8.2-2 Degenerate Response Surface

Transform the following equation to canonical form:

$$\hat{Y} = 2x_1^2 + 2x_2^2 + 3x_3^2 + 4x_1x_2 + 2x_1x_3 + 2x_2x_3 - 4x_1 + 6x_2 - 2x_3 + 3 \qquad \text{(a)}$$

Solution:

In matrix notation, Equation (a) is equivalent to Equation 8.2-4 with

$$\mathbf{x} = \begin{bmatrix} x_1 \\ x_2 \\ x_3 \end{bmatrix} \qquad \mathbf{b}_{11} = \begin{bmatrix} 2 & 2 & 1 \\ 2 & 2 & 1 \\ 1 & 1 & 3 \end{bmatrix}$$

It is also easy to show that

$$\det \mathbf{b}_{11} = 0$$

so that Equation 8.2-7 cannot be applied. Still, the rotation of axes can be carried out by the methods described in the previous example:

$$\det (\mathbf{b}_{11} - \lambda \mathbf{I}) = 0$$

or

$$\begin{vmatrix} (2-\lambda) & 2 & 1 \\ 2 & (2-\lambda) & 1 \\ 1 & 1 & (3-\lambda) \end{vmatrix} = 0$$

The eigenvalues are obtained from the characteristic equation

$$\lambda^3 - 7\lambda^2 + 10\lambda = 0 \qquad \text{(b)}$$

and are $\lambda = 0$, $\lambda = 2$, and $\lambda = 5$.

For each λ an orthonormal eigenvector can be computed as in the previous example:

$$\lambda = 0 \begin{cases} 2u_1 + 2u_2 + u_3 = 0 \\ 2u_1 + 2u_2 + u_3 = 0 \\ u_1 + u_2 + 3u_3 = 0 \end{cases} \qquad \begin{aligned} u_1 &= 1 \\ u_2 &= -1 \\ u_3 &= 0 \end{aligned}$$

$$\text{norm} = \sqrt{1+1} = \sqrt{2}$$

$$\lambda = 2 \begin{cases} 2u_2 + u_3 = 0 \\ 2u_1 + u_3 = 0 \\ u_1 + u_2 + u_3 = 0 \end{cases} \qquad \begin{aligned} u_1 &= 1 \\ u_2 &= 1 \\ u_3 &= -2 \end{aligned}$$

$$\text{norm} = \sqrt{1+1+4} = \sqrt{6}$$

$$\lambda = 5 \begin{cases} -3u_1 + 2u_2 + u_3 = 0 \\ 2u_1 - 3u_2 + u_3 = 0 \\ u_1 + u_2 - 2u_3 = 0 \end{cases} \qquad \begin{aligned} u_1 &= 1 \\ u_2 &= 1 \\ u_3 &= 1 \end{aligned}$$

$$\text{norm} = \sqrt{1+1+1} = \sqrt{3}$$

The matrix $\mathbf{U}$ is thus

$$\mathbf{U} = \begin{bmatrix} 1/\sqrt{2} & 1/\sqrt{6} & 1/\sqrt{3} \\ -1/\sqrt{2} & 1/\sqrt{6} & 1/\sqrt{3} \\ 0 & -2/\sqrt{6} & 1/\sqrt{3} \end{bmatrix} \qquad \text{(c)}$$

The new coordinates in terms of the original ones are

$$\tilde{\mathbf{x}} = \mathbf{U}^T\mathbf{x} = \begin{bmatrix} 1/\sqrt{2} & -1/\sqrt{2} & 0 \\ 1/\sqrt{6} & 1/\sqrt{6} & -2/\sqrt{6} \\ 1/\sqrt{3} & 1/\sqrt{3} & 1/\sqrt{3} \end{bmatrix} \mathbf{x}$$

The canonical form of Equation (a) is

$$\hat{Y} = 2\tilde{x}_2^2 + 5\tilde{x}_3^2 + \text{linear terms} \qquad \text{(d)}$$

The first-order terms are obtained by introducing x_1, x_2, and x_3 into each linear term in Equation (a), using the transformation $\mathbf{x} = U\tilde{\mathbf{x}}$ or

$$\begin{aligned} x_1 &= \frac{1}{\sqrt{2}}\tilde{x}_1 + \frac{1}{\sqrt{6}}\tilde{x}_2 + \frac{1}{\sqrt{3}}\tilde{x}_3 \\ x_2 &= -\frac{1}{\sqrt{2}}\tilde{x}_1 + \frac{1}{\sqrt{6}}\tilde{x}_2 + \frac{1}{\sqrt{3}}\tilde{x}_3 \\ x_3 &= -\frac{2}{\sqrt{6}}\tilde{x}_2 + \frac{1}{\sqrt{3}}\tilde{x}_3 \end{aligned} \qquad \text{(e)}$$

Example 8.2-3 Canonical Analysis for a Plastics Extruder

The following example is taken from the work of Klein.† In the study of plastic extrusion, data for a composite design with six independent variables were collected to fit a second-order regression equation having as the dependent variable the outlet stock temperature, T. See Table E8.2-3. The calculations were performed for a 2.5-inch diameter extruder using a plasticized polyvinyl chloride at 228 combinations of levels of the six independent variables. The following estimated regression equation was obtained:

TABLE E8.2-3

Variable	Coded:	Design Values −1	0	+1
x_1, Screw speed (rpm)		30	45	60
x_2, Channel depth (in)		0.100	0.120	0.140
x_3, Barrel temperature (°F)		330	340	350
x_4, Flow rate (lb/hr)		50	100	150
x_5, Input temperature (°F)		280	310	340
x_6, Metering zone length (in)		10	19	28

$$\begin{aligned} \hat{T} = {} & 349.392 - 2.369x_1 - 3.78x_1 - 3.78x_2 + 6.077x_3 \\ & + 7.064x_4 + 1.6795x_5 + 3.956x_6 + 3.957x_1^2 \\ & + 0.655x_2^2 + 10.019x_3^2 + 7.602x_4^2 - 0.0428x_5^2 \\ & + 2.33x_6^2 + 0.564x_1x_2 + 0.564x_1x_3 - 10.793x_1x_4 \\ & + 0.473x_1x_5 + 0.003x_1x_6 + 0.011x_2x_3 \\ & - 3.852x_2x_4 + 0.201x_2x_5 - 0.005x_2x_6 \\ & - 1.376x_3x_4 - 0.010x_3x_5 + 0.012x_3x_6 \\ & + 0.942x_4x_5 + 2.739x_4x_6 - 1.847x_5x_6 \end{aligned} \qquad \text{(a)}$$

† I. Klein, Paper presented at the 53rd National AIChE Meeting, Pittsburg, Pa., May 17-20, 1964.

The square of the multiple correlation coefficient of this regression was 0.998, but the equation was rather complicated and the effect of each variable was not easily visualized. Consequently, it was processed by a computer program designed especially for the analysis of response surfaces. Equation (b) is the canonical equation obtained and immediately below are listed the coordinates of the center of system:

$$\hat{T} - 349.96 = -1.388\tilde{x}_1^2 - 0.064\tilde{x}_2^2 + 0.604\tilde{x}_3^2 + 1.201\tilde{x}_4^2 + 9.776\tilde{x}_5^2 + 12.049\tilde{x}_6^2 \qquad \text{(b)}$$

Center of System: (Coded Variables)

$x_{1e} = 2.07$	$x_{4e} = 1.84$
$x_{2e} = 6.57$	$x_{5e} = 4.73$
$x_{3e} = -0.23$	$x_{6e} = 2.87$
	$T_e = 349.96$

Note that two of the coefficients in Equation (b) are negative whereas four are positive. Inspection of Table 8.2-1 leads to the conclusion that Equation (b) is a hyper-saddle surface. Examination of the coordinates of the center of the system, however, indicates that the center is quite outside the range of the design. Whenever this situation arises, the surface in the region of the experiments represents the multidimensional equivalent of either an inclined ridge or an inclined trough. Furthermore, the surface along certain axes is either always increasing or always decreasing; it is likely to be well represented by only linear terms in some of the variables.

Because of the remoteness of the center of the system, it was believed that some crossproduct terms and second-degree terms in Equation (a) did not significantly contribute to the response. This indeed proved to be the case; a simplified equation obtained from a new regression analysis was

$$\hat{T} = 356.57 - 4.0975x_1^2 - 10.5077x_1x_4 - 2.445x_1 + 0.6509x_2^2 - 3.8418x_2x_4 - 3.7421x_2 + 6.1038x_3 + 7.5876x_4^2 + 7.1016x_4 + 1.706x_5 + 3.945x_6 \qquad \text{(c)}$$

which had a multiple correlation coefficient of 0.923. Equation (c) indicates that the dependence of temperature on the variables x_3, x_5, and x_6 is purely linear.

At this stage of the analysis, these latter three variables were temporarily fixed at the zero level to simplify the analysis somewhat. $\hat{T}$ was analyzed as a function of the three remaining variables x_1, x_2, and x_4. The new center of the system was

$$x_{1e} = -0.643$$

$$x_{2e} = 0.707$$

$$x_{4e} = -0.734$$

The new canonical equation obtained was:

$$\hat{T} - 353.426 = -0.7477\tilde{x}_1^2 + 11.6025\tilde{x}_2^2 + 1.4812\tilde{x}_4^2 \qquad \text{(d)}$$

with the corresponding transformation equations:

$$\begin{aligned} \tilde{x}_1 &= 0.538(x_1 + 0.643) + 0.681(x_2 - 0.707) + 0.496(x_4 + 0.734) \\ \tilde{x}_2 &= 0.568(x_1 + 0.643) + 0.142(x_2 - 0.707) - 0.811(x_4 + 0.734) \\ \tilde{x}_4 &= 0.623(x_1 + 0.643) - 0.718(x_2 - 0.707) + 0.310(x_4 + 0.734) \end{aligned} \qquad \text{(e)}$$

Inspection of Equation (d) reveals that one term has a negative coefficient and the other two are positive—again the equation of a saddle surface. But notice that now the center of the fitted system is within the experimental range and the maximum values of the canonical variables $\tilde{x}_1$, $\tilde{x}_2$, and $\tilde{x}_4$ will be small numbers, all of the same order of magnitude. Consequently, the relative smallness of coefficients of the first and third terms of Equation (d) made it possible to drop these terms without introducing an appreciable error in the calculated response, $\hat{T}$. The simplified equation therefore was:

$$\hat{T} - 353.4 = 11.6025\tilde{x}_2^2 \qquad \text{(f)}$$

or, by rearranging,

$$\tilde{x}_2 = \sqrt{\frac{\hat{T} - 353.4}{11.60}} \qquad \text{(g)}$$

By substituting for $\tilde{x}_2$ from Equation (e), Equation (g) becomes

$$0.568x_1 + 0.142x_2 - 0.811x_4 - 0.331 = \sqrt{\frac{\hat{T} - 353.4}{11.60}} \qquad \text{(h)}$$

which is the equation of a family of planes for different values of $\hat{T}$ having a minimum temperature of 353.4°F. Geometrically the equation represents a stationary ridge with parallel planes distributed around the one representing the minimum response as shown in Figure E8.2-3.

Keep in mind that Equation (h) and Figure E8.2-3 are only simplified representations of the original response surface which was a function of six independent variables. To include the effect of the variables x_3, x_5, and x_6, which had been held at zero level for Equation (f), the linear terms deleted from Equation (c) were added to Equation (f) to yield

$$\hat{T} - 353.4 = 11.6025\tilde{x}_2^2 + 6.1038x_3 + 1.706x_5 + 3.945x_6 \qquad \text{(i)}$$

Equations (i) and (e) lead to:

$$0.568x_1 + 0.142x_2 - 0.811x_4 - 0.331 = \sqrt{\frac{\hat{T} - (353.4 + 6.1038x_3 + 1.706x_5 + 3.945x_6)}{11.60}} \qquad \text{(j)}$$

The deleted terms x_3, x_5, and x_6 were previously shown to be linear and to represent an inclined trough or ridge on the response surface. Since all three terms have positive coefficients, for minimization of the response they must be selected at their lowest level, namely -1. This will reduce the term in parentheses, the modified $\hat{T}_e$, from 353.4 to 341.6 and represents an additional reduction of 11.8°F in the response. Equation (j), however, still represents a family of planes with parameter $\hat{T}$, and Figure E8.2-3 is therefore

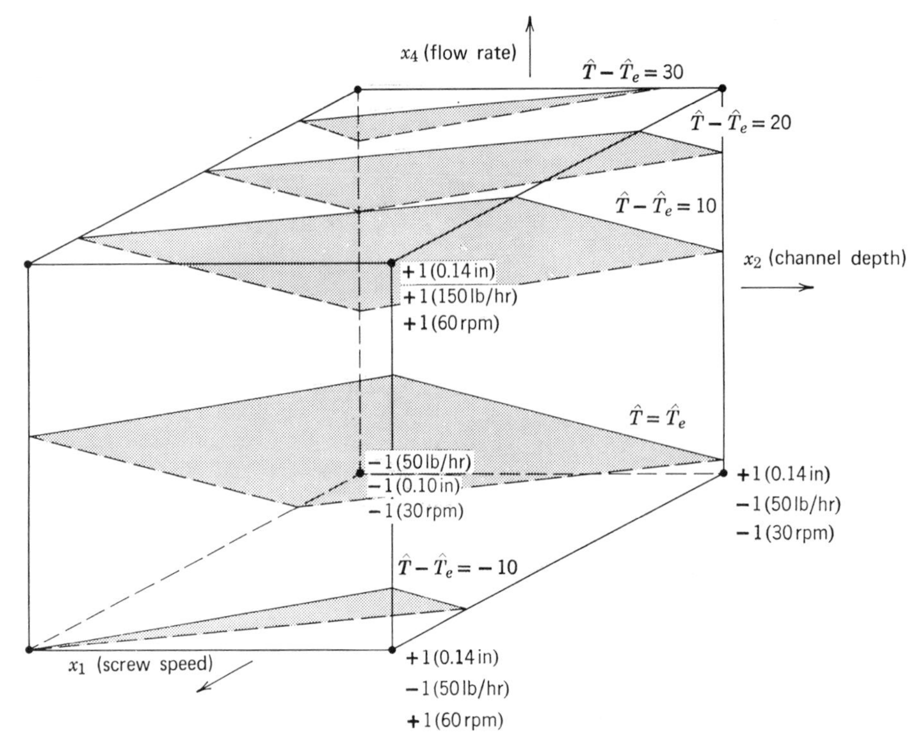

FIGURE E8.2-3 Representation of the system by the simplified Equation (h).

still an accurate representation of the full response surface with the minimum $\hat{T}_e$ now at 341.6°F.

It should be pointed out that the analysis presented applies only to the experimental range investigated and does not necessarily bear any resemblance to the response surface outside the range.

8.3 STRATEGIES FOR PROCESS OPTIMIZATION BY EXPERIMENTATION

This section is concerned with optimization of steady-state processes by a sequence of designed experiments having the objective of improving process performance. Conceptually, optimization can be visualized in three dimensions as an analog of climbing a hill. While objective functions containing two independent variables rarely have the exact shape of the hill shown in Figure 8.3-1, nevertheless the general procedure is to start at some point (set of values of the independent variables) on the hill and climb to the top (or descend to the bottom of a valley in minimization). One has to decide which way is "up" and, subject usually to certain constraints, how to proceed to the top most efficiently. The significant difference between deterministic optimization and experimental optimization is that in the latter the shape of the hill is unknown; consequently, its representation may vary from experiment to experiment.

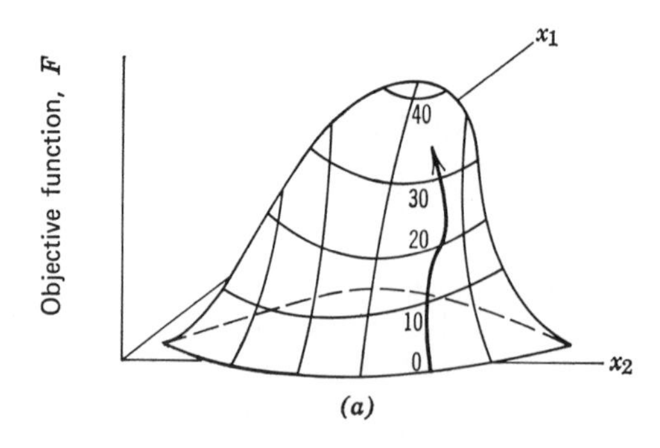

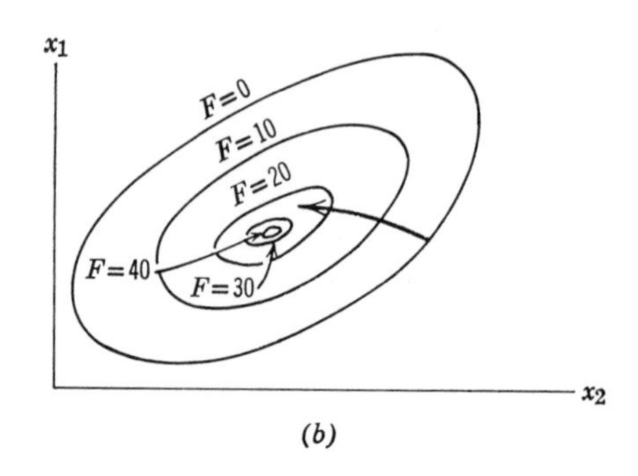

FIGURE 8.3-1 Hill climbing as an analog of optimization by sequential experimentation: (*a*) Three-dimensional representation of the objective function of two independent variables, x_1 and x_2. (*b*) Contour representation of the objective function F of two independent variables, x_1 and x_2.

In a very general sense the analyst would like to optimize the monetary profit of the entire process, but so many qualitative factors become involved with the quantitative ones (i.e., how safe is it, will it last, what about the impurities, can we sell it) that in practice it is not easy to establish a universal objective function. He usually selects one or more of the principal process responses, weights them by some cost values or subjective weights to form the objective function, and decides what constraints need to be satisfied on the responses and controllable variables. As a simple example, an objective function might be just the yield from a chemical reactor and the constraints the fact that the process controllable variables must be positive.

8.3-1 The Box-Wilson (Gradient) Method

In 1951, Box and Wilson popularized a new concept in experimentation by recognizing that for many processes the engineer is not so much interested in testing the significance of the process variables as in simply determining the best operating conditions for the process. Although earlier work had been reported by Hotelling,† the method to be described is that evolved by Box and his coworkers‡ who made use of the experimental designs and response surface analysis described previously in Sections 8.1 and 8.2.

Suppose we wanted to find the extremum (say the maximum) of the response Y, the dependent variable of quadratic Equation 8.1-10. In principle, because the true functional form of the response surface is unknown, any functional form could be used in lieu of Equation 8.1-10, but the latter is easy to fit and interpret, hence its use in the Box-Wilson method. Our objective is to find the combination of values of x_i which optimizes the response within the region of the q-dimensional observation space where experimentation is feasible with the least cost or with as few experimental observations as possible. The number of observations required will, of course, depend upon the desired accuracy and precision of the estimation as well as their cost.

Because the estimated response surface represents the true functional relationship only in a local region, the simplest possible models are used. Usually the experimenter starts working with first-order models until the vicinity of the optimum is reached. From the initial estimated response surface, he calculates the direction of steepest ascent by evaluating the components of the gradient of $\hat{Y}$. The gradient of $\hat{Y}$ can be normalized by dividing by its norm, $\|\nabla \hat{Y}\|$, to give a unit vector

$$\frac{\nabla \hat{Y}}{\|\nabla \hat{Y}\|} = \frac{\dfrac{\partial \hat{Y}}{\partial x_1}\boldsymbol{\delta}_1 + \dfrac{\partial \hat{Y}}{\partial x_2}\boldsymbol{\delta}_2 + \cdots}{\left[\sum_i \left(\dfrac{\partial \hat{Y}}{\partial x_1}\right)^2\right]^{1/2}} \tag{8.3-1}$$

whose components indicate the relative step sizes for steepest ascent in each of the x_i coordinate directions. The unit gradient of the estimated first-order regression equation $\hat{Y} = b_0 + b_1x_1 + b_2x_2$ is

$$\frac{\nabla \hat{Y}}{\|\nabla \hat{Y}\|} = \frac{b_1\boldsymbol{\delta}_1 + b_2\boldsymbol{\delta}_2}{\sqrt{b_1^2 + b_2^2}} \tag{8.3-2}$$

which is a constant vector (does not depend on x).

The unit gradient of the estimated regression equation for the second-order model

$$\hat{Y} = b_0 + b_1x_1 + b_2x_2 + b_{11}x_1^2 + b_{22}x_2^2 + b_{12}x_1x_2 \tag{8.3-3}$$

namely,

$$\frac{\nabla \hat{Y}}{\|\nabla \hat{Y}\|} = \frac{(b_1 + 2b_{11}x_1 + b_{12}x_2)\boldsymbol{\delta}_1 + (b_2 + 2b_{22}x_2 + b_{12}x_1)\boldsymbol{\delta}_2}{[(b_1 + 2b_{11}x_1 + b_{12}x_2)^2 + (b_2 + 2b_{22}x_2 + b_{12}x_1)^2]^{1/2}} \tag{8.3-4}$$

varies with its position in observation space as shown, in Figure 8.3-2, because it is a function of both x_1 and x_2.

Once the direction of steepest ascent is ascertained, the experimenter selects a new local region for experimentation at some distance along the components of the vector of steepest ascent. How far to go before additional experimentation is carried out depends on judgment, the model, costs, etc. In the new region, data are collected to fit a first-order model, and the cycle is repeated. By this step-by-step procedure, points of higher and higher responses are reached.

First-order relations alone cannot, however, be used directly to reach the maximum response if it lies within the feasible region in observation space—and not on a boundary—because, as the experimenter approaches the extremum, the coordinate components of the gradient become smaller and thus more difficult to estimate. After arrival in the vicinity of the optimum, location of the optimum point itself usually requires a series of coordinated experiments and the use of a second-order model.

As a practical matter, in moving along the path of steepest ascent, the analyst may find that one or more of the coefficients in the response surface equation will

† H. Hotelling, "The Experimental Determination of the Maximum of a Function," *Ann. Math. Stat.* **12**, 20, 1941.

‡ G. E. P. Box and K. B. Wilson, *J. Royal Stat. Soc.* **B13**, 1, 1951; G. E. P. Box, "The Exploration and Exploitation of Response Surfaces: Some General Considerations and Examples," *Biometrics* **10**, 16, 1954; G. E. P. Box and P. V. Youle, "The Exploration and Exploitation of Response Surfaces: An Example of the Link Between the Fitted Surface and the Basic Mechanism of the System," *Biometrics* **11**, 287, 1955; G. E. P. Box and J. S. Hunter, "Multi-Factor Experimental Designs for Exploring Response Surfaces," *Ann. Math. Stat.* **28**, 195, 1957; G. E. P. Box and N. R. Draper, "A Basis for the Selection of a Response Surface Design," *J. Amer. Stat. Assn.* **54**, 622, 1959.

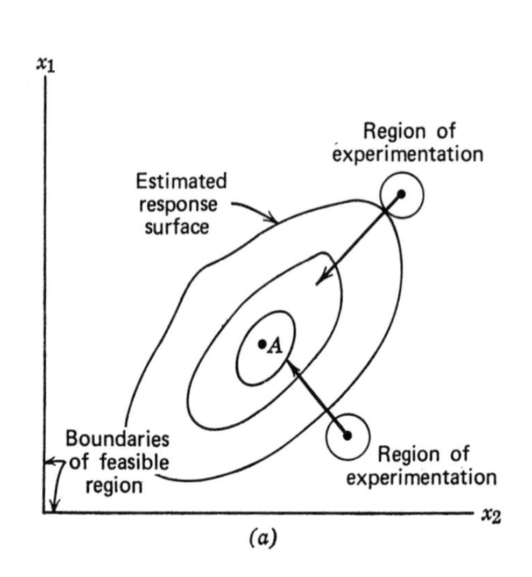

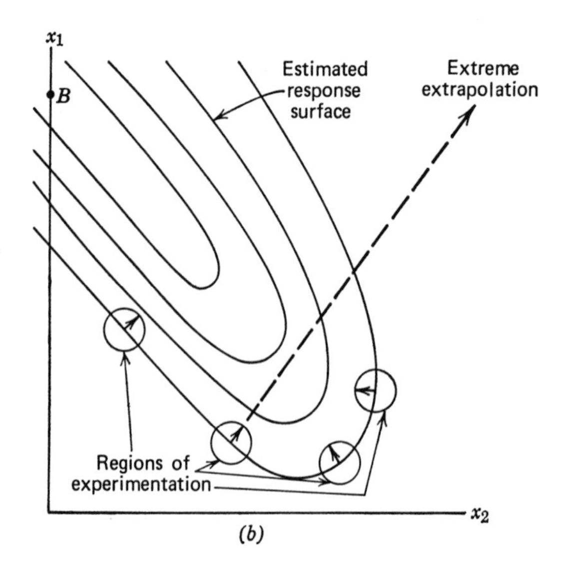

FIGURE 8.3-2 Vectors indicating direction of steepest ascent for a second-order model.

become very small before the others do. Three possible explanations must be considered:

1. The variable is near its optimum level.
2. The coding scheme has produced a coded value too small to have much influence on the fitting process (a scaling problem).
3. The variable really is not a significant one.

It is easy to avoid problem 2 by initially selecting an appropriate coding scheme; moving off the path of steepest ascent will help resolve problem 1. The tests described in Section 5.3 and Chapter 7 can be used to recognize problem 3.

As soon as the region of the extremum is reached and a suitable order model is fitted to the experimental data, the location of the optimum point can be estimated. If the extremum of $\hat{Y}$ is located on a single boundary, as is point B in Figure 8.3-2b, then a one-dimensional search (sequence of experiments with only x_1 changing) along the boundary will yield the maximum value of $\hat{Y}$. In multidimensional observation space, several variables may be at their physical constraining limits; hence, several values may be held constant in the model while at the same time others are permitted to vary. If the extremum lies within the feasible region, as illustrated by point A in Figure 8.3-2a, the extremum can be located reasonably well by analytic maximization (or minimization) of the response function. For two independent variables, according to Lagrange's criterion, a function φ must in some bounded region satisfy the necessary conditions (to give a stationary point):

$$\frac{\partial \varphi}{\partial x_1} = 0 \quad \text{and} \quad \frac{\partial \varphi}{\partial x_2} = 0 \qquad (8.3\text{-}5)$$

and the sufficient conditions

$$\text{for a maximum: } B^2 - AC < 0 \quad \text{and} \quad A + C < 0$$
$$\text{for a minimum: } B^2 - AC < 0 \quad \text{and} \quad A + C > 0$$

where

$$A = \frac{\partial^2 \varphi}{\partial x_1^2}, \qquad B = \frac{\partial^2 \varphi}{\partial x_1\, \partial x_2}, \qquad C = \frac{\partial^2 \varphi}{\partial x_2^2}$$

If $B^2 - AC > 0$, a saddle point exists; if $B^2 - AC = 0$, the nature of the stationary point is undetermined (one examines higher derivatives).

For response functions involving more than two independent variables, it is convenient to form the *Hessian matrix* to check for the sufficient conditions

$$\mathbf{h} = \begin{bmatrix} h_{11} & \cdots & h_{1q} \\ \vdots & & \\ h_{q1} & \cdots & h_{qq} \end{bmatrix}$$

where

$$h_{jk} = \frac{1}{2} \frac{\partial^2 \varphi}{\partial x_j\, \partial x_k}$$

and h_{jk} is evaluated at the stationary point in observation space. If each of the principal minors of h is negative-definite (< 0) at the stationary point, then the response is a maximum; if each of the principal minors of h is positive-definite (> 0), then the response is a minimum.

The experimenter can perform an analysis of variance as described in Section 8.1 by using the experimental data in the vicinity of the optimum to establish the significance of the respective variables and the adequacy of the fit of the model itself. If a second- or higher order equation adequately fits the data, after the optimum is located the response surface can be mapped (for two- or three-dimensional models), and a canonical reduction of the response equation can be carried out as described in Section 8.2. From the signs and relative magnitudes of the canonical coefficients, it is easy to gain insight into the nature of the response surface.

Example 8.3-1 Optimization by Sequential Experimentation

In one application of process optimization, the dependent variable Y, the quality, was related to the following three independent variables:

Variable	Units	Possible Range of Variable
Viscosity (x_1)	poise	1 to 100
Pressure (x_2)	atm	1 to 100
Flow rate (x_3)	lb/min	0 to 100

The objective was to maximize the quality. Each independent variable was coded as follows:

$$z_i = \frac{x_i - 50}{25}$$

To reduce the amount of initial experimentation, a half-replicate of a 2^3 factorial design was carried out in the center of experimental space, with four replicates at the center yielding the following results (in uncoded numbers):

Y	x_1	x_2	x_3
786.8	50.0	50.0	50.0
744.1	50.0	50.0	50.0
642.6	50.0	50.0	50.0
684.9	50.0	50.0	50.0
142.7	25.0	25.0	25.0
955.9	75.0	75.0	25.0
1123.5	75.0	25.0	75.0
1075.6	25.0	75.0	75.0

From the first four data sets, the estimate of the error variance σ_e^2 can be calculated as

$$s_e^2 = \frac{\sum Y_i^2 - \dfrac{(\sum Y_i)^2}{n}}{n - 1} = \frac{(2.0551 - 2.0430) \times 10^6}{3} = 4030$$

$s_e = 63$

The coefficients in the estimated regression equation

$$\hat{Y} = b_0 + b_1x_1 + b_2x_2 + b_3x_3 \qquad \text{(a)}$$

were

$$b_0 = 823$$
$$b_1 = 215$$
$$b_2 = 191$$
$$b_3 = 245$$

Now, an experimenter may be shrewd enough or lucky enough to start the experimentation close to optimum conditions—near the top of the hill so to speak. If so, then the equation of the plane will not be a very good approximation to the response surface because the response surface will have considerable curvature. Or the plane may not be a good approximation in any case. However, in this study, only one degree of freedom was left for the residual sum of the squares ($n - m = 5 - 4$). Instead of carrying out an analysis of variance at this stage of the optimization, in the interests of reducing the experimentation, one move toward the optimum response was made and then a full factorial design was planned. One could, of course, instead make a few check runs in the direction of the gradient to see how well Model (a) held up and continue as far as the equation could be extrapolated with confidence.

The normalized gradient of the response, Equation (a), was

$$\frac{\nabla \hat{Y}}{\|\nabla \hat{Y}\|} = \frac{\dfrac{\partial \hat{Y}}{\partial x_1}\boldsymbol{\delta}_1 + \dfrac{\partial \hat{Y}}{\partial x_2}\boldsymbol{\delta}_2 + \dfrac{\partial \hat{Y}}{\partial x_3}\boldsymbol{\delta}_3}{\left[\sum_{i=1}^{3}\left(\dfrac{\partial \hat{Y}}{\partial x_i}\right)\right]^{1/2}}$$

$$= 0.54\,\boldsymbol{\delta}_1 + 0.48\,\boldsymbol{\delta}_2 + 0.61\,\boldsymbol{\delta}_3 \qquad \text{(b)}$$

On the basis of Equation (b) but in view of the approximate nature of Equation (a), each value of x was changed from a 2^3 design centered about (50, 50, 50) to a 2^3 factorial (part of a central composite design) centered about (85, 85, 85).

To avoid exceeding their given upper limits, the variables were recoded so that

$$z_i = \frac{x_i - 85}{4}$$

and the (uncoded) results from the 2^3 factorial experiment were

Y	x_1	x_2	x_3
2011.0	81.0	81.0	81.0
2432.0	89.0	81.0	81.0
2345.6	81.0	89.0	81.0
2391.7	89.0	89.0	81.0
2449.6	81.0	81.0	89.0
2833.7	89.0	81.0	89.0
2494.3	81.0	89.0	89.0
2629.0	89.0	89.0	89.0
2458.0	85.0	85.0	85.0
2129.9	85.0	85.0	85.0
2121.4	85.0	85.0	85.0
2389.5	85.0	85.0	85.0

After the 2^3 factorial design was completed, the first-order estimated regression equation was determined to be

$$\hat{Y} = 2448 + 123.2z_1 + 16.78z_2 + 153.3z_3 \qquad \text{(c)}$$

An analysis of variance (Table E8.3-1a) was then carried out, using the value of s_e^2 calculated from the four replicate experiments at (85, 85, 85). Clearly the error variance was not constant (compare 30,440 with 4030), and the first-order model was an adequate fit.

Also, the sum of squares associated with each variable as it was removed from the first-order model was as shown in Table E8.3-1b. From this analysis of variance, the variable x_2 might well be deleted from the model.

It is quite common, particularly by the time one is ready to fit a second-degree equation, that some of the variables can be deleted from the model. This, of course, reduces the number of runs required and simplifies experimentation. To protect from being misled by the coding of the variables resulting in too small a change in the variable, one should

TABLE E8.3-1a

Source of Variation	ν = d.f.	SS	Mean Square
Due to regression			
$\sum_{i=1}^{n} (Y_i - \bar{Y})^2$	3	311,684	103,894
Deviation of residuals			
$\sum_{i=1}^{n} (Y_i - \hat{Y}_i)^2$	4	76,788	19,197
Experimental error	3	91,326	30,440

make a run in which the x_i in question is changed by a larger amount. This could be done in the next set of runs needed for the steepest ascent. If the variable(s) is(are) not significant, it will again be without effect (very small b_i) and can be dropped from the design.

The first-order model was not fit again with x_2 deleted, but instead a new direction of steepest ascent was obtained from Equation (c) with x_2 deleted:

$$\frac{\nabla \hat{Y}}{\|\nabla \hat{Y}\|} = \frac{123.2\,\boldsymbol{\delta}_1 + 153.3\,\boldsymbol{\delta}_3}{(123^2 + 154^2)^{1/2}} = 0.62\,\boldsymbol{\delta}_1 + 0.78\,\boldsymbol{\delta}_3 \quad \text{(d)}$$

No particular scale factor has been mentioned for the move in the direction of steepest ascent, but obviously extreme extrapolation can lead away from the optimum rather than toward it. Examine Figure 8.3-2*b* again. One procedure is to make a run at one of the extrapolated points. If the observed response is close to the calculated response, one can extrapolate further. After one or more extrapolations, Model (c) will no longer correctly predict the response. Then a new set of coefficients must be determined by making additional runs in the same way that Model (c) was obtained. However, at this stage in the experiment being described, it appeared as if the optimum operating conditions lay near the upper limits on x_1 and x_3 at least. Consequently, in anticipation of this eventuality, rather than moving strictly in the direction of steepest ascent as dictated by Equation

TABLE E8.3-1b

Source of Variation	ν = d.f.	SS	Variance* Ratio
x_0 removed (intercept)	1	4.7957×10^7	Significant
x_1 removed	1	1.2142×10^5	Significant
x_2 removed	1	2.2512×10^2	Not significant
x_3 removed	1	1.8801×10^5	Significant

* Using a pooled variance $s^2_{\bar{Y}_i} = \dfrac{76,778 + 91,326}{4 + 3} = 24,015$ and $F_{0.95}(1, 7) = 5.59$.

(d), a 2^3 half-replicate design centered about $x_1 = x_2 = x_3 = 97.5$ was selected. The following experimental results were obtained:

Y	x_1	x_2	x_3
3861.5	100.0	100.0	100.0
3284.9	95.0	100.0	100.0
3349.2	100.0	95.0	95.0
3982.2	100.0	95.0	100.0
3483.9	100.0	100.0	95.0

After these experiments, it became clear that the process variability was quite large at these high values of quality, camouflaging the effect of changes in the three variables. A series of three replicate experiments was carried out at $x_1 = 98.0$, $x_2 = 99.0$, and $x_3 = 99.0$:

	x_1	x_2	x_3
3821.0	98.0	99.0	99.0
3471.7	98.0	99.0	99.0
3961.5	98.0	99.0	99.0

from which $s_e^2 = 6.55 \times 10^4$ and $s_e = 254$. On the basis of this rather large mean square, rather than replicating further experiments extensively, which was costly, an investigation was initiated into the reasons for the variability in Y and the experimentation was terminated. However, substantial improvement had been achieved in the operating conditions, and the high dispersion at the extreme operating conditions, the main barrier to possible further improvement, which of course would be much smaller and perhaps not worth the expense of investigation, was clearly isolated.

So far we have considered an objective function comprised of a single variable, the process response. However, in Section 5.5 we discussed ways in which the parameters in a multiresponse model can be estimated. How to optimize by experimentation when multiresponse models are involved is not well defined. One approach, that of using a monovariate response model, is to weight the individual responses suitably so as to form a single response. If a unified objective function cannot be evolved, each response must be individually carried along in the analysis. After each stage in the experimentation, a judgment must be made for each response as to its relative influence in establishing the region for the next series of experiments. For example, Lind, Goldin, and Hickman† ran a series of experiments to ascertain the effect of complexing agents on the yield of an antibiotic. They used second-order models for two responses: (1) cost and (2) yield. Figure 8.3-3 illustrates the two responses at their optima in the form of contour diagrams in two-dimensional observation space. They were able to find new operating conditions that increased the yield by 5 percent and reduced costs by $5 per kilo-

† E. E. Lind, J. Goldin, and J. B. Hickman, *Chem. Eng. Progress* **56** (11), 62, 1960.

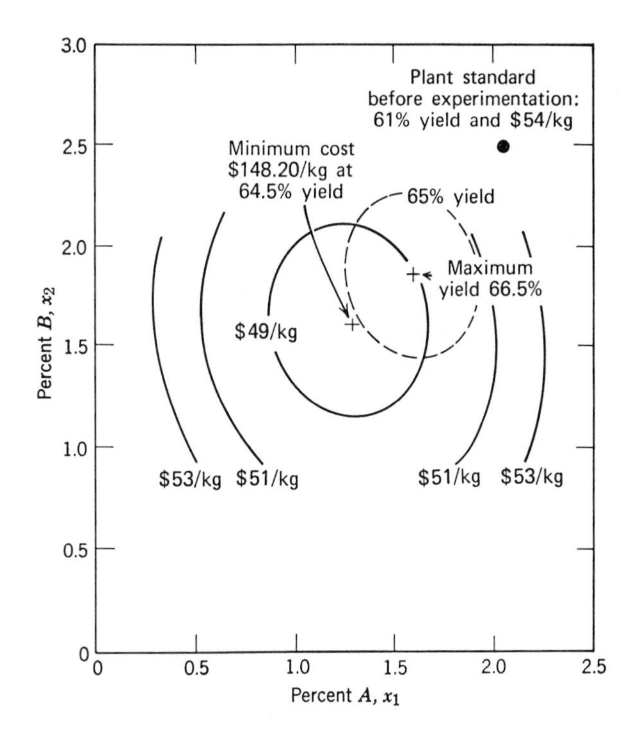

FIGURE 8.3-3 Cost contours and yield contours as a function of two complexing agents, A and B, for a certain antibiotic. (Adapted from E. E. Lind, J. Goldin, and J. B. Hickman, *Chem. Eng. Progr.* **56** (11), 62, 1960, with permission of the publisher, The American Institute of Chemical Engineers.)

gram of product, because the optima both fell within the feasible region for experimentation and the matrix of independent variables for both responses was roughly in the same region. Clearly, such favorable conditions will not always exist. Hill and Hunter† listed a number of other examples of optimization by experimentation with multiple responses.

8.3-2 Evolutionary Operation (EVOP)

Hunter described evolutionary operation as "a method which compels a production process to provide information about itself of immediate applicability without upsetting either production quality or throughput."‡ The technique is a deliberately simplified sequential approach to experimental optimization, one that is oriented toward a direct application by production personnel. A good summary of the theory and computational aspects of EVOP can be found in Box and Hunter.§ Because continuous processes which have been in operation for some time usually have been well studied by engineers, EVOP may have little to offer and has not been as widely adopted as might be expected.

EVOP makes use of small changes in operating conditions, whose effect on the output of the process is likewise small but which are repeated many times until the general effect of the many small changes becomes noticeable. Since the result of each small change wallows in a sea of "noise," it is necessary to detect small differences in performance between the normal process and the perturbed process; this can only be accomplished by extensive replication. EVOP perturbs the process independent variables according to a repeated series of experiments, each of which comprises one *cycle* or experimental design. The cycle is repeated again and again until the *phase* is ended and a new set of operating conditions is established. In practice it has been found convenient to study the effect of only two or three variables at a time on the process response since the results can be portrayed graphically in two or three dimensions.

Past plant records can give information concerning variables to consider, the nature of the random changes in the response, the time for a change in a continuous process to become effective, etc. But the reasons plant data and "natural" variations in operating variables are not too effective in determining the nature of the response surface are that:

1. Many important variables do not change at all.
2. Variables that do change, change over uncontrolled ranges.
3. Variables are not orthogonal; hence it is hard to discriminate the effect of one from the other.
4. Spurious correlation develops because of the time factor; a change caused by time may appear to be due to another variable changing.

Hence the use of EVOP.

As a simple example of EVOP applied to a batch process, assume that a process involves three dependent variables (yield, percent solvent recovered, and particle size) and two independent variables (time and temperature). Current operating procedure calls for 300°C temperature and 3 hours reaction time. Past experience shows that changes of ± 3°C and ± 5 minutes are not injurious to the process; in fact, variations of this magnitude are normally encountered in the process.

A 2×2 factorial design with a center point is set up as follows with the coded temperature equal to $(T-300)/3$ and the coded time equal to $(t - 180)/5$; the circled variables in Figure 8.3-4 designate the responses from the experiments. One batch of material is run at each pair of operating conditions including the center. The points can be run randomly or in some particular order until

† W. J. Hill and W. G. Hunter, "A Review of Response Surface Methodology: A Literature Survey," *Technometrics* **8**, 571, 1966.
‡ J. S. Hunter, *Chem. Eng.*, 193, Sept. 19, 1960.
§ G. E. P. Box and J. S. Hunter, *Technometrics* **1**, 77, 1959. Additional references on experiences with EVOP are listed in W. G. Hunter and J. F. Kittrell, "Evolutionary Operation: A Review," *Technometrics* **8**, 389, 1966.

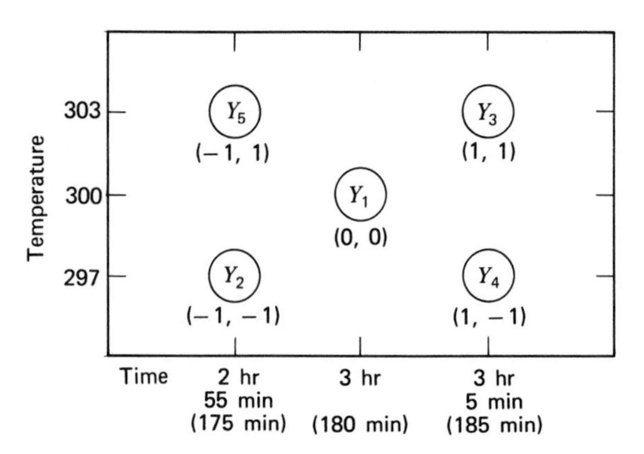

FIGURE 8.3-4 Experimental design for first cycle of EVOP.

all five points have been completed, thereby ending one cycle. Assume the responses are as follows:

Yield

76.4		77.2
	75.6	
75.4		75.8

Percent Solvent Recovered

83.8		83.2
	82.8	
82.2		82.6

Particle Size

4.0		5.2
	4.4	
4.6		5.4

If the process is continuous rather than a batch process, a suitable time should be allowed for the process to settle down to its *steady-state* values after making a change in the setting of the independent variables.

Since the design is orthogonal, it can easily be shown that the *contrasts* (differences in the sample means), such as $(\bar{Y}_1 - \bar{Y}_2)$, are mutually orthogonal. The notations to be used are as follows:

Y_{ij} = measured response for the jth pair of points in the experimental design on the ith cycle; $1 < i < n$; $i < j < k$

$$\bar{Y}_j = \frac{\sum_{i=2}^{n=1} Y_{ij}}{n-2}$$ = average of the responses for point j starting with the second cycle and continuing up to the most recent cycle but one

$D_j = \bar{Y}_j - Y_{ij}$ = difference between average value of Y and the most recent value of Y for the jth point

Two major types of calculations must be made: One is to calculate the contrast $(\bar{Y}_p - \bar{Y}_q)$ to measure the effect of changing a variable, and the other is to calculate the standard deviation to be used as a measure of the error.

To find the effect of time, we form the contrast

$$\frac{\bar{Y}_3 + \bar{Y}_4}{2} - \frac{\bar{Y}_2 + \bar{Y}_5}{2}$$

or

$$\text{time effect} = \tfrac{1}{2}(\bar{Y}_3 + \bar{Y}_4 - \bar{Y}_2 - \bar{Y}_5) \quad (8.3\text{-}6a)$$

Also:

$$\text{temperature effect} = \tfrac{1}{2}(\bar{Y}_3 + \bar{Y}_5 - \bar{Y}_2 - \bar{Y}_4) \quad (8.3\text{-}6b)$$

interaction (time × temperature) effect

$$= \tfrac{1}{2}(\bar{Y}_2 + \bar{Y}_3 - \bar{Y}_4 - \bar{Y}_5) \quad (8.3\text{-}6c)$$

$$\text{change in mean effect} = \tfrac{1}{5}(\bar{Y}_2 + \bar{Y}_3 + \bar{Y}_4 + \bar{Y}_5 - 4\bar{Y}_1) \quad (8.3\text{-}6d)$$

The center point responses enter only in Equation 8.3-6d, which measures the difference between the grand mean of the response for all the exterior points for all the experiments in the cycle less the mean for the center of the design:

$$\tfrac{1}{5}(\bar{Y}_1 + \bar{Y}_2 + \bar{Y}_3 + \bar{Y}_4 + \bar{Y}_5) - \bar{Y}_1 = \Delta$$

If the response surface is convex ("hill-like") as in Figure 8.3-1, the average of the sum of the responses for the points surrounding the center point will always be less than the center point so that Δ will be negative. Conversely, for a concave topography, Δ will be positive. For a response surface represented by parallel lines, Δ will be zero. Equations 8.3-6a and 8.3-6b can be used to move in a direction for improving the response by taking steps in the direction indicated for each effect. If the response is a cost function, Equation 8.3-6d is a measure of the cost of obtaining information.

In addition to the contrasts given by Equations 8.3-6, an estimate of the error variance is required. Box and Hunter recommended that the range be used as an estimate of $\sigma^2_{\bar{Y}_i}$ because of its simplicity; otherwise, use

$$s_e^2 = \sum_{i=2}^{n} s_i^2(n-1)$$

where

$$s_i^2 = \frac{1}{k-1}\left[\frac{i-1}{i}\right]\left[\sum_{j=1}^{k} D_j^2 - \frac{\left(\sum_{j=1}^{k} D_j\right)^2}{k}\right]$$

At the end of a few cycles, one can get a feeling for the effect on the response of each variable and decide to either accumulate more information or perhaps make a specific change in one of the controlled variables so as to:

1. Start a new phase about a new center point chosen in the optimal direction.
2. Explore in the optimal direction.
3. Select levels of variable more widely spaced.
4. Substitute new variables for one or more of the previous ones.

When several simultaneous responses are tallied, a compromise is needed. For example, if at the end of several cycles the effect of the design is to: (1) reduce the cost per batch by 0.3 ± 0.4 units, (2) reduce the yield by 0.5 ± 0.06 percent, and (3) increase the particle size by 1.5 ± 1.1 units, it will be necessary to assess the deleterious effects of (2) and (3) compared to the advantage of (1) through use of an objective cost evaluation or subjective judgment. For a given significance level α and operating characteristic β, the number of cycles required to detect an effect of a variable depends on the allowable increase in variability in the process. For a 30-percent increase in σ, four or five cycles are needed for a 2^2 factorial design, and two to three cycles are needed for a 2^3 factorial design if α and β are about 0.05 to 0.10 each.†

8.3-3 Simplex Designs

An EVOP program does not have clearcut rules as to when or where to move to a new region of experimentation. These decisions are left to the experimenter. Spendley, Hext, and Himsworth,‡ as mentioned in Section 6.2-2, introduced the concept of automatic evolutionary operation through the use of regular simplex designs. They developed a number of rules which can be applied relatively objectively to decide (1) when and (2) where to move the experimental program. These rules have been supplemented by successive investigators to improve the performance of the experimental program from cycle to cycle.

The essence of the rules is to use only the most recent observations and to use frequent but small changes in the independent variables. The strategy which will be described applies to a regular simplex. (The flexible simplex of Section 6.2-1 has not yet been adequately tested for experimental, as opposed to deterministic, optimization. With error present in the response, the possibility appears that the system of simplexes will become hung up on a spuriously high (low) response.) A regular simplex (refer to Figure 6.2-1) is a first-order design with the minimum number of experimental points. If q independent variables are used, the regular simplex contains $q + 1$ points. Given one vertex of the simplex at $[0, 0, \ldots, 0]$ to define the origin, a regular simplex with an edge length a was specified in Section 6.2-2. It is desirable to scale each independent variable so that a unit change in each scaled variable will produce approximately the same change in the response.

Experimental optimization proceeds by constructing a new simplex on the face of the old one opposite the point that yields the poorest response, as shown in Figure 6.2-1. The new point added to the simplex is the location of the new experiment; hence the design incorporates n old experiments and one new one on each move. To find the coordinates of the added vertex, $\mathbf{x}_{\text{new}}$, which is the mirror image in hyperspace of the deleted point, take twice the average of all the coordinates of the common points (those retained in the simplex) and subtract the coordinates of the deleted vertex, $\mathbf{x}_j$:

$$\mathbf{x}_{\text{new}} = \frac{2}{q}(\mathbf{x}_1 + \mathbf{x}_2 + \cdots + \mathbf{x}_{j-1} + \mathbf{x}_{j+1} + \cdots + \mathbf{x}_{q+1}) - \mathbf{x}_j$$

$$= \frac{2}{q}\sum_{i=1}^{q+1} \mathbf{x}_i - \left[\frac{2}{q} + 1\right]\mathbf{x}_j \tag{8.3-7}$$

where $\mathbf{x}_i$ is the vector of the coordinates of vertex i. The coordinates of the centroid of the new simplex are

$$\mathbf{c}_{\text{new}} = \frac{1}{q+1}\left[\sum_{i=1}^{q+1} \mathbf{x}_i - \mathbf{x}_j + \mathbf{x}_{\text{new}}\right] \tag{8.3-8}$$

and the predicted response at $\mathbf{x}_{\text{new}}$ is

$$\hat{Y}_{\text{new}} = \frac{2}{q}\sum_{i=1}^{q+1} \hat{Y}_i - \left[\frac{2}{q} + 1\right]\hat{Y}_j \tag{8.3-9}$$

The consensus of the effective decision rules§ for optimization, using a regular simplex, is as follows:

1. At each stage, ascertain the lowest response and exclude the corresponding vertex from the new simplex. Replace the excluded vertex by the vertex calculated by Equation 8.3-7.
2. To reduce the risk of being hung up on some spuriously high vertex, if a response occurs in $(q + 1)$ successive simplexes and is not eliminated by rule 1, discard the response and replace it with the results of a new experiment.
3. To cycle about an optimum and also be able to follow a ridge, if the new response, Y_{new}, is the lowest response in the new simplex, do not apply rule 1 but return to the previous simplex; instead of x_j, delete the vertex giving the second lowest response. Such a move advances in the second most favorable direction.
4. If the experimental error is too large relative to the expected changes in the response, to avoid error-bias buildup, replace all the old observations with new ones after every $2(q + 1)$ experiment.
5. Replace a new vertex which exceeds a constraint by the alternate vertex selected in rule 3.
6. To accelerate the search for the optimum, apply one of the rules of Section 6.2.

† G. E. P. Box and N. R. Draper, Dept. of Stat. Tech. Rept. 106, Univ. of Wisconsin, Madison, 1967.

‡ W. Spendley, G. R. Hext, and F. R. Himsworth, *Technometrics* **4**, 441, 1962.

§ Spendley *et al.*, *op. cit.*; B. H. Carpenter and H. C. Sweeny, *Chem. Eng.* **72**, 117, July 5, 1965.

7. To close in on the optimum (if it is not changing with time), successively reduce the size of the simplex by a factor such as one-fourth.

Spendley, Hext, and Himsworth compared the regular simplex method by using some simulation data with univariate search, factorial experiments, steepest ascent, and random search; their general conclusion appeared to be that the simplex method is at least as satisfactory as any of the other techniques. The regular simplex is both efficient and rotatable, and it yields an optimal estimate of the slope of a response surface in the presence of error. The best features of the regular simplex method appear to be: (1) the direction of the advance depends solely on the ranking of the responses and not on their absolute values, (2) the technique can be extended to higher dimensions with relative ease, and (3) only one new experiment must be conducted at each stage. The path of ascent followed by the regular simplex method has been observed from simulation studies to be roughly that of steepest ascent, as might be expected from the nature of the local linearization of the response surface related to the simplex.

8.3-4 Optimization by Direct Search

A strategy in which each independent variable is changed in turn cyclically also can be employed to optimize by experimentation. Given that Y is an unknown function of several variables, $x_1, x_2, \ldots, x_q$, in any real experiment the value of x_k will be bounded between l_k, the lower limit, and u_k, the upper limit. Or, in matrix notation with $\mathbf{x} = [x_1, x_2, \ldots, x_q]^T$,

$$\mathbf{l} \leq \mathbf{x} \leq \mathbf{u}$$

gives the admissible region. During experimentation, changes are made in the values of x_k. If the change called for is less than l_k or exceeds u_k, the constraining value is substituted for the designated value.

The search for the optimum set of operating conditions, $\mathbf{x}^*$, takes place in cycles from an initial set of the x's designated $x_k^{(1)}$. To start the nth cycle, changes are made in each x_k from the previous cycle, and the Y's are measured. A single variable x_k may be changed at a time, or all the x's can be changed simultaneously Or, as described in Chapter 6 for the Hooke and Jeeves method,† sequences of individual changes followed by group changes can be carried out Decisions of how to change the x_k's on the $(n + 1)$st cycle are based on the changes in Y observed on the nth cycle.

We now describe one direct search method which does not require the functional form of the model to be written down. Starting with the nth cycle, changes in the set of x's can be executed as follows (the subscript on the vector $\mathbf{x}^{(n)}$ indicates the precedence order of the change, the superscript in parentheses is the cycle index, and $\delta_k^{(n)}$ is the change in $x_n^{(k)}$):

Order, j		Vector Designation	Response
0 (Start)	$[x_1^{(n)} - \delta_1^{(n)}][x_2^{(n)} - \delta_2^{(n)}]\cdots[x_q^{(n)} - \delta_q^{(n)}]$	$\mathbf{x}_0^{(n)}$	$Y_0^{(n)}$
1	$[x_1^{(n)} + \delta_1^{(n)}][x_2^{(n)} - \delta_2^{(n)}]\cdots[x_q^{(n)} - \delta_q^{(n)}]$	$\mathbf{x}_1^{(n)}$	$Y_1^{(n)}$
2	$[x_1^{(n)} - \delta_1^{(n)}][x_2^{(n)} + \delta_2^{(n)}]\cdots[x_q^{(n)} - \delta_q^{(n)}]$	$\mathbf{x}_2^{(n)}$	$Y_2^{(n)}$
$\vdots$	. . . .	$\vdots$	$\vdots$
q	$[x_1^{(n)} - \delta_1^{(n)}][x_2^{(n)} - \delta_2^{(n)}]\cdots[x_q^{(n)} + \delta_q^{(n)}]$	$\mathbf{x}_q^{(n)}$	$Y_q^{(n)}$

(8.3-10)

The δ's can be evaluated in almost any way and may constantly be reduced from cycle to cycle, or they can grow and recede depending upon the accumulated successes or failures in improving Y, as in the Hooke and Jeeves technique. Here the δ's are reduced as follows:

$$\delta_k^{(n)} = \frac{(-1)^{n+1}\delta_k^{(1)}}{n^{1/4}} \tag{8.3-11}$$

with $\delta_k^{(1)} = 0.1(u_k - l_k)$. As an initial condition for the x's, we let

$$x_k^{(1)} = \tfrac{1}{2}(u_k + l_k) \tag{8.3-12}$$

A series of $Y_j^{(n)}$'s is obtained from the series of experiments indicated by the list 8.3-10. From these $Y_j^{(n)}$'s, a decision is reached as to how to modify the x's for the next cycle according to the following rule:

$$x_k^{(n+1)} = x_k^{(n)} + \Delta_k^{(n)} p a_k^{(n)} \tag{8.3-13}$$

The symbol $\Delta_k^{(n)}$ is determined on the nth cycle according to the following rules:

$$\Delta_k^{(n)} = \begin{Bmatrix} +1 & \text{if } Y_j^{(n)} - Y_0^{(n)} > 0 \\ 0 & \text{if } Y_j^{(n)} - Y_0^{(n)} = 0 \\ -1 & \text{if } Y_j^{(n)} - Y_0^{(n)} < 0 \end{Bmatrix} \tag{8.3-14}$$

The constant $a_k^{(n)}$ is an arbitrary value related to $\delta_k^{(n)}$ and can be, for example, calculated by

$$a_k^{(n)} = \frac{(-1)^{n+1}a_k^{(1)}}{n^{3/4}} \tag{8.3-15}$$

where $a_k^{(1)} = 0.2(u_k - l_k) = 2\delta_k^{(1)}$. The constant p can be unity, or it may be equal to any acceleration factor chosen to make the optimization process converge more rapidly than $p = 1$.

After the sequential series of experiments indicated by the list 8.3-10 has been completed, the sign of $\Delta_k^{(n)}$ is known. It may then prove desirable to continue with the same cycle but adjust *all* the x's simultaneously to accelerate convergence as follows:

$$x_k^{(n;h)} = x_k^{(n)} + h\,\Delta_k^{(n)} a_k^{(n)} \tag{8.3-16}$$

where h is a secondary index equal to 1 for the first simultaneous adjustment, to 2 for the next adjustment,

† R. Hooke and T. A. Jeeves, *J. Assn. Compt. Mach.* **8**, 212, 1961.

and so forth. The adjustment indicated by Equation 8.3-16 is continued until the $Y^{(n)}$'s first fail to show an ascending trend in the cycle, i.e., until

$$Y^{(n;1)} < Y^{(n;2)} < \cdots < Y^{(n;h-1)} < Y^{(n;h)}$$

no longer holds (because the last $Y^{(n;h)}$ is smaller than $Y^{(n,h-1)}$. In this event, x_k to start the $(n+1)$st cycle is equated to

$$x_k^{(n+1)} = x_k^{(n;h-1)}$$
$$= x_k^{(n)} + (h-1)\,\Delta_k^{(n)} a_k^{(n)} \qquad (8.3\text{-}17)$$

In general, the sequence described above converges to the local optimum $\mathbf{x}^*$; in practice, it can converge fairly rapidly so that too many cycles are not required. Certain easily satisfied requirements for the process to converge were specified by Fabian.† If the hypothetical process response has only one extremum, the sequence converges to it; if several extrema exist, there is no guarantee, of course, that the global extremum will be selected, but as a practical matter this handicap is not significant. A radical change in starting point can help the experimenter investigate the nature of the objective function but at the expense of additional experimentation. A major advantage of the direct search technique is that, as the number of variables increases beyond three or four, the number of experimental runs for each additional variable is much less than for factorial-type experiments. The experimenter can also attempt to use his subjective knowledge about the experiment to achieve more rapid convergence to the optimum by adjusting p.

Example 8.3-2 Optimization Without a Model‡

An experiment was carried out to find the maximum extinction coefficient of a mixture of citric acid, sodium dihydrogen phosphate, and sodium chloride by varying three variables:

x_1 = wave length
x_2 = citric acid concentration
x_3 = sodium chloride concentration

The extinction coefficient was measured with a photometer. The admissible values for each variable, the necessary constants, and the starting conditions obtained through use of Equations 8.3-11, 8.3-12, and 8.3-15 were:

	u_k	l_k	$x_k^{(1)}$	$\delta_k^{(1)}$	$a_k^{(1)}$
x_1	5000	4000	4500	100	200
x_2	2	0	1	0.2	0.4
x_3	7.5	2.5	5	0.5	1.0

Table E8.3-2a lists some useful values (which have been rounded off) for δ_k and a_k required in the calculations based on $\delta_k^{(n)} = (-1)^{n+1}\,\delta_k^{(1)}/(n)^{1/4}$ and $a_k^{(n)} = (-1)^{n+1} a_k^{(1)}/(n)^{3/4}$.

† V. Fabian, *Czech. Math. J.* **10**, 123, 1960.
‡ Adapted from V. Fabian, *Aplikace Matematiky* **6**, 162, 1961.

TABLE E8.3-2a

n	$\delta_1^{(n)}$	$\delta_2^{(n)}$	$\delta_3^{(n)}$	$a_1^{(n)}$	$a_2^{(n)}$	$a_3^{(n)}$
1	100	0.20	0.5	200	0.4	1
2	−80	−0.20	−0.4	−120	−0.25	−0.6
3	80	0.15	0.4	80	0.20	0.4
4	−70	−0.15	−0.35	−70	−0.14	−0.4

Table E8.3-2b lists a sequence of five cycles of calculation. In the cycle $n = 1$ (and subsequent cycles), the initial conditions are placed on the line with the *. The second line, corresponding to the trial $j = 0$, indicates the execution of the experiment corresponding to the change made in the initial conditions described by the first row of list 8.3-10, namely the subtraction of $\delta_k^{(1)}$ from each variable:

$$(4500 - 100) = 4400; \quad (1 - 0.2) = 0.8; \quad (5 - 0.5) = 4.5$$

The next line, corresponding to $j = 1$, initiates the change of each variable in sequence by $(+\delta_k^{(1)})$. The decision to retain the new value of $x_k^{(1)}$ or the old one, after carrying out the changes indicated by the rows of list 8.3-10, depends upon whether $Y_j^{(1)}$ is increased or decreased. If $Y_j^{(1)}$ improves, the new value of $x_k^{(1)}$ is adopted; if it does not, the old value of $x_k^{(1)}$ is retained. For $j = 1$, $46 < 313$, so $x_1^{(1)}$ remains at 4400; for $j = 2$, $266 < 313$, so $x_2^{(1)}$ remains at 0.8; and so forth. During the first three steps, $j = 1$, $j = 2$, and $j = 3$, the sign of $\Delta_k^{(1)}$ is established for further calculations in cycle 1.

After the experiments outlined by the list 8.3-10 were executed, and the $\Delta_k^{(1)}$'s determined, Equation 8.3-11 was used to accelerate the change in $Y^{(1)}$. For $h = 1$ (row $j = 4$):

$$x_1^{(1;1)} = 4500 + 1(-200) = 4300$$
$$x_2^{(1;1)} = 1 + 1(-0.4) = 0.6$$
$$x_3^{(1;1)} = 5 + 1(-1.0) = 4$$

For $h = 3$ ($j = 6$), the value of $x_2^{(1;3)} = 1 + 3(-0.4) = -0.2$ exceeds the value $l_2 = 0$, hence $x_2^{(1;3)}$ is placed equal to the lower limit 0. The accelerating sequence was continued according to Equation 8.3-11 until $j = 7$ when $Y^{(1;4)} < Y^{(1;3)}$, i.e., $3200 < 4140$. At this step, according to Equation 8.3-17,

$$x_1^{(2)} = 4500 + (4 - 3)(-200) = 3900$$
$$x_2^{(2)} = 1 + (4 - 3)(-0.4) = -0.2; \qquad 0 \text{ is used instead}$$
$$x_3^{(2)} = 5 + (4 - 3)(-1) = 2$$
$$Y^{(2)} = 4140$$

New values of δ_k were computed as in Table E8.3-2a, line 2; new values of a_k were obtained from Equation 8.3-15; and the entire procedure was repeated starting with a new cycle. Figure E8.3-2a illustrates the values of the independent and dependent variables obtained as the optimization process continued. By the end of the fourth cycle, Y became 4800, $x_1 = 3800$, $x_2 = 0$, and $x_3 = 0.6$, at which time 26 experiments had been carried out.

For more accurate location of the optimum near the end of the optimization sequence (or throughout the

TABLE 8.3-2b

Cycle, n	Trial, j	Variables, $x_1^{(n)}$	$x_2^{(n)}$	$x_3^{(n)}$	Extinction, $Y_j^{(n)}$	$\Delta_k^{(n)} a_k^{(n)}$ $k = 1$	2	3
1	*	4500	1	5		200	0.4	1
	0	4400	0.8	4.5	313			
	1	4600	0.8	4.5	46	−200		
	2	4400	1.2	4.5	266		−0.4	
	3	4400	0.8	5.5	288			−1
	4	4300	0.6	4	819			
	5	4100	0.2	3	1696			
	6	3900	0	2	4140			
	7	3700	0	1	3200			
2	*	3900	0	2	4140	−120	−0.25	−0.60
	0	3980	0.20	2.4	2850			
	1	3720	0.20	2.4	4210	−120		
	2	3980	0.00	2.4	3240		−0.25	
	3	3980	0.20	1.6	3120			−0.60
	4	3780	0	1.4	4080			
	5	3660	0	0.8	3100			
3	*	3780	0	1.4	4080	90	0.20	0.40
	0	3700	0	1.0	3700			
	1	3860	0	1.0	4350	90		
	2	3700	0.15	1.0	3400		−0.20	
	3	3700	0	1.8	3600			−0.40
	4	3870	0	1	4400			
	5	3960	0	0.6	3980			
4	*	3870	0	1	4540	−70	−0.14	−0.40
	0	3940	0.15	1.35	4010			
	1	3800	0.15	1.35	4700	−70		
	2	3940	0	1.35	4150		−0.14	
	3	3940	0.15	0.65	4120			−0.40
	4	3800	0	0.6	4800			
	5	3720	0	0.2	4300			
5	*	3800	0	0.6	4800			

process in the absence of interactions), it may be helpful to plot estimates for the derivatives of the hypothetical response surface.

The quotient $[Y_j^{(n)} - Y_0^{(n)}]/[2\delta_k^{(n)}]$ may be regarded as an estimate of the partial derivative of the response surface dependent variable (insofar as such a derivative exists) at the point $\mathbf{x}^{(n)}$. The variables are said not to be interacting if this approximation to the partial derivative depends solely on the kth coordinate $x_k^{(n)}$. If the no-interaction condition is not satisfied, it generally becomes evident that the derivative estimates obtained cannot be connected by a smooth curve.

For the example under consideration, we obtain the following derivatives with respect to the wavelength, x_1, for values of wavelength equal to 4500, 3900, 3780, and 3870 for the first four cycles, respectively:

$$\frac{46 - 313}{200} = -1.34 \qquad \frac{4210 - 2850}{-160} = -8.50$$

$$\frac{4350 - 3400}{160} = 5.94 \qquad \frac{4700 - 4010}{-140} = -4.92$$

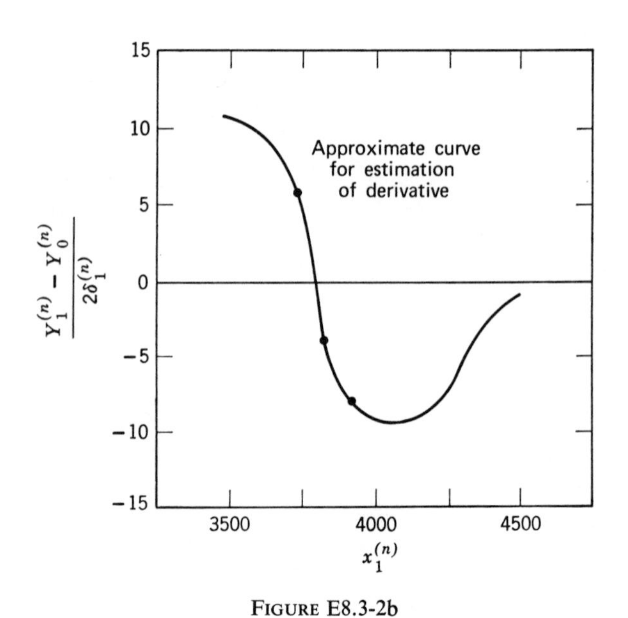

FIGURE E8.3-2b

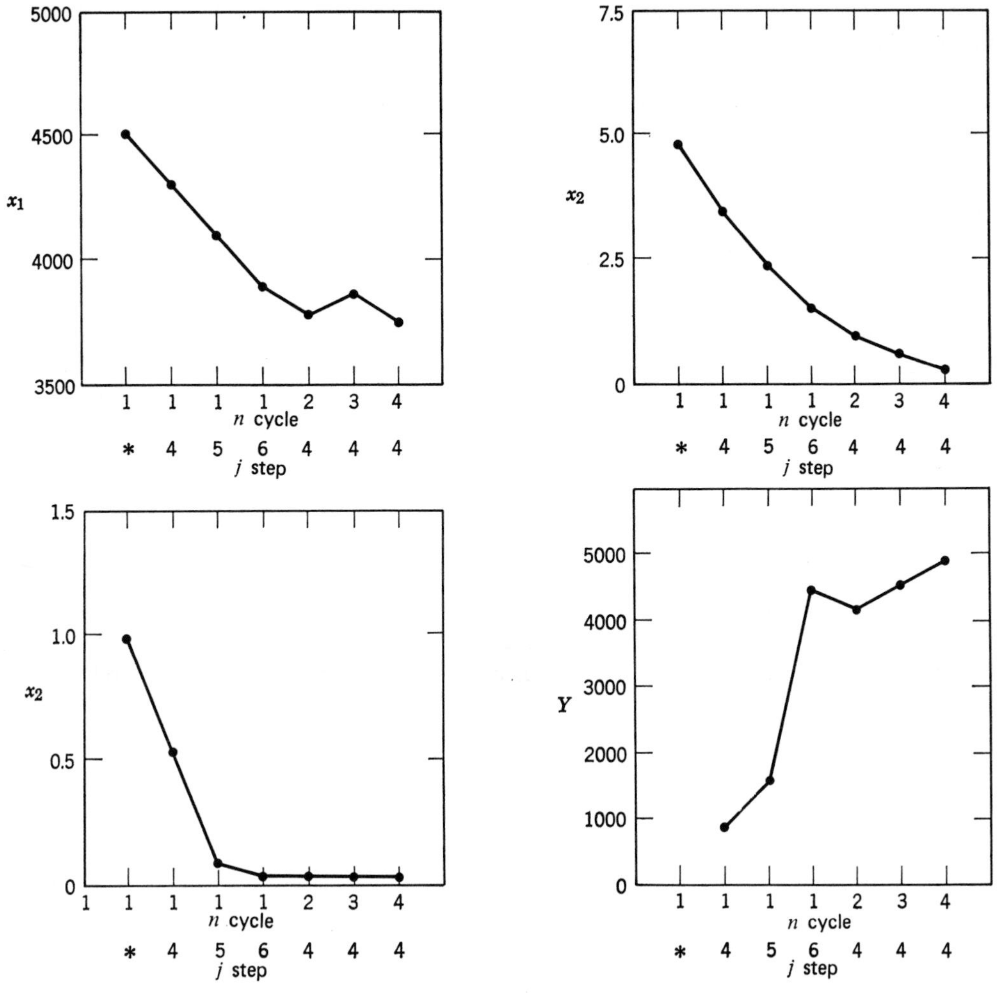

FIGURE E8.3-2a

Refer to Figure E8.3-2b. The approximation to a zero derivative occurs at approximately $x_1 = 3800$.

8.4 SEQUENTIAL DESIGNS TO REDUCE UNCERTAINTY IN PARAMETER ESTIMATES

We shall suppose in the first part of this section that a single adequate model has been selected, perhaps from among a group of several potential models, and that we would like to obtain better estimates of the model coefficients. How should additional experiments be carried out to accomplish this objective most effectively? As we have previously noted, if the experiments are not properly planned, the estimates of the model coefficients may not only be imprecise but also highly correlated. Box, W. G. Hunter, Draper, and their coworkers have made important contributions in developing sequential design procedures to decrease the uncertainty in parameter estimates, especially in connection with nonlinear models. References can be found at the end of this chapter.

We shall use the notation of Chapter 6 here in which

$$\eta = \eta(\boldsymbol{\beta}, \mathbf{x}) \tag{8.4-1}$$

is the model containing m β's and q independent variables x. The observations are

$$Y_i = \eta_i + \epsilon_i \qquad i = 1, 2, \ldots, n$$

where the ϵ_i's are independently normally distributed with zero mean and a variance of σ_Y^2. Also

$$[X_{ij}] = \frac{\partial \eta_i(\mathbf{b}, \mathbf{x}_i)}{\partial \beta_j} \qquad j = 1, 2, \ldots, m$$

is the typical element in the matrix $\mathbf{X}$ defined in Section 6.2-3.

8.4-1 Processes with a Single Response

We shall assume at the beginning that some observations have already been taken and shall inquire as to what values of x should be chosen for the next observation(s) to estimate the β's most effectively in some sense. We shall consider later how to decide on the initial vector $\mathbf{x}$ when no initial observations are available. The procedure rests on the use of the Bayes theorem and an assumed multivariate normal distribution for $\boldsymbol{\beta}$ centered at a vector of estimates, **b**.

Bayes' theorem, given as Equation A-2 in Appendix A, written in terms of the notation of this section is

$$p_{n+1}(\boldsymbol{\beta} \mid \mathbf{y}_{n+1}) = \frac{L(\boldsymbol{\beta} \mid \mathbf{y}_{n+1})p_n(\boldsymbol{\beta})}{\int_{-\infty}^{\infty} L(\boldsymbol{\beta} \mid \mathbf{y}_{n+1})p_n(\boldsymbol{\beta})\, d\boldsymbol{\beta}} \tag{8.4-2}$$

where

$p_{n+1}(\boldsymbol{\beta} \mid \mathbf{y}_{n+1})$ = the posterior probability density for $\boldsymbol{\beta}$ after $(n + 1)$ observations of Y have been obtained

$p_n(\boldsymbol{\beta})$ = the prior probability density function for $\boldsymbol{\beta}$ (before the last observation or group of observations was obtained)

$L(\boldsymbol{\beta} \mid \mathbf{y}_{n+1})$ = the probability density function which is the likelihood (function) of $\boldsymbol{\beta}$ given $\mathbf{y}_{n+1}$ (described in Section 3.2)

and where the subscripts on p and y designate the experimental run numbers. The integral in the denominator of Equation 8.4-2 represents the normalization factor. The dependence of the posterior density function on the independent variables **x** as well as on **Y** is implied, although not specifically stated, in the arguments of the probability density functions in Equation 8.4-2. It is the future values of the x's that are of major interest in establishing the experimental design. From the Bayesian point of view, the posterior probability density function contains all the available information about the β's. A natural criterion to use in fixing the values of an experimental design for a new experiment would be to select those values of x which bring about the most desirable posterior distribution in some sense and, in particular, bring about the biggest desired change from the distribution that existed at the end of the nth experiment.

Once the first n observations have been made, a likelihood function can be written similar to Equation 5.1-11a, except that here the model itself may be nonlinear:

$$L(\boldsymbol{\beta} \mid \mathbf{y}_n, \mathbf{x}, \sigma_Y^2) = \frac{1}{(2\pi)^{n/2}\sigma_Y^n} \exp\left[-\frac{\sum_{i=1}^{n} [Y_i - \eta_i(\boldsymbol{\beta}, \mathbf{x}_i)^2]}{2\sigma_Y^2}\right] \tag{8.4-3}$$

To obtain $p_n(\boldsymbol{\beta})$ for use in Bayes' theorem, we need to have the initial prior probability density for $\boldsymbol{\beta}$ before the n observations were taken, say $p_0(\boldsymbol{\beta})$, and $L(\boldsymbol{\beta} \mid \mathbf{y}_n)$. Then application of Equation 8.4-2 will give the desired posterior density $p_n(\boldsymbol{\beta} \mid \mathbf{y}_n)$, which in turn can be used as the prior density for the succeeding observations. Instead of assuming a uniform probability density for $p_0(\boldsymbol{\beta})$, we follow Draper and Hunter† and assume that prior information is available on the β's from some source so that the initial density for the β's can be given by

$$p_0(\boldsymbol{\beta}) = \frac{1}{(2\pi)^{m/2}|\boldsymbol{\Omega}|^{1/2}} \exp\left[-\tfrac{1}{2}(\boldsymbol{\beta} - \mathbf{b}^{(0)})^T\boldsymbol{\Omega}^{-1}(\boldsymbol{\beta} - \mathbf{b}^{(0)})\right] \tag{8.4-4}$$

where $\mathbf{b}^{(0)}$ is the vector of initial estimates for the β's and

$$\boldsymbol{\Omega} = \begin{bmatrix} \omega_{11} & \cdots & \omega_{1m} \\ \vdots & & \\ \omega_{m1} & \cdots & \omega_{mm} \end{bmatrix}$$

is a covariance matrix for the β's in which each element is specified.

Introduction of Equations 8.4-4 and 8.4-3 into Bayes' theorem gives the desired probability density function after n observations have been made:

$$p_n(\boldsymbol{\beta} \mid \mathbf{y}_n) = k_n \exp\left[\frac{-\sum_{i=1}^{n} [Y_i - \eta_i(\boldsymbol{\beta}, \mathbf{x}_i)]^2}{2\sigma_Y^2}\right] \times \exp\left[-\tfrac{1}{2}(\boldsymbol{\beta} - \mathbf{b}^{(0)})^T\boldsymbol{\Omega}^{-1}(\boldsymbol{\beta} - \mathbf{b}^{(0)})\right] \tag{8.4-5}$$

where k_n represents the appropriate normalizing factor which need not be specifically specified as yet.

Now, let us consider what should be done in the way of setting up the values of $x_{n+1}, x_{n+2}, \ldots$, the design matrix, for n^* additional observations of Y. Equation 8.4-5 now becomes the prior distribution in Bayes' theorem, Equation 8.4-2, and the likelihood function $L(\boldsymbol{\beta} \mid \mathbf{y}_{n+n^*})$ is analogous to Equation 8.4-3 except that n^* products are involved. Thus, the summation in the exponent is from $i = n + 1$ to $i = n + n^*$; i.e., it is over the new observations

$$L(\boldsymbol{\beta} \mid \mathbf{y}_{n+n^*}) = \frac{1}{(2\pi)^{n^*/2}\sigma_Y^{n^*}} \exp\left[-\frac{\sum_{i=n+1}^{n+n^*} [Y_i - \eta_i(\boldsymbol{\beta}, \mathbf{x}_i)]^2}{2\sigma_Y^2}\right] \tag{8.4-6}$$

Introduction of Equations 8.4-5 and 8.4-6 into Equation 8.4-2 yields the desired posterior probability density for the β's:

$$p_{n+n^*}(\boldsymbol{\beta} \mid \mathbf{y}_{n+n^*}) = k_{n+n^*} \exp\left[\frac{\sum_{i=1}^{n+n^*} [Y_i - \eta_i(\boldsymbol{\beta}, \mathbf{x}_i)]^2}{2\sigma_Y^2}\right] \times \exp\left[-\tfrac{1}{2}(\boldsymbol{\beta} - \mathbf{b}^{(0)})^T\boldsymbol{\Omega}^{-1}(\boldsymbol{\beta} - \mathbf{b}^{(0)})\right] \tag{8.4-7}$$

where k_{n+n^*} is the new normalizing factor.

How can Equation 8.4-7 be used for deciding on the values of the x's in the design matrix? The references at the end of this chapter interpret the meaning of "best" as applied to the probability density, Equation 8.4-7, in terms of the objectives of the experimenter. Draper, Hunter, and others, have maximized the posterior density

† N. R. Draper and W. G. Hunter, *Biometrika* **54**, 147, 1967.

given by Equation 8.4-7 both with respect to the β's and with respect to the settings of the new x's $(x_{n+1},\ldots,x_{n+n^*})$. Once σ_Y^2 and $\boldsymbol{\Omega}$ are specified, it would in principle be possible to maximize $p_{n+n^*}(\boldsymbol{\beta} \mid \mathbf{y}_{n+n^*})$ by one of the iterative methods of optimization described in Chapter 6. However, by making a few reasonable assumptions, the computations required can be substantially reduced.

In Chapter 6, the model was expanded as in Equation 6.2-12 in a truncated Taylor series about an assumed $\mathbf{b}^{(0)}$. Here we shall assume that a similar expansion in $\boldsymbol{\beta}$ space is valid in any local restricted region about some selected vector $\mathbf{b}^*$:

$$\eta_i(\boldsymbol{\beta}, \mathbf{x}_i) \simeq \eta_i(\mathbf{b}^*, \mathbf{x}_i) + \sum_{j=1}^{m} (\beta_j - b_j^*)X_{ij} \quad (8.4\text{-}8)$$

Equation 8.4-8 introduces the assumption of *local linearity*, which is not the same as assuming that the model is linear for all values of $\boldsymbol{\beta}$.

The summation term in Equation 8.4-7 is $(\mathbf{Y} - \boldsymbol{\eta})^T (\mathbf{Y} - \boldsymbol{\eta})$; after Equation 8.4-8 is introduced for η_i, the summation becomes

$$\begin{aligned}\sum_{i=1}^{n+n^*} (Y_i - \eta_i)^2 &= [\mathbf{Y} - \boldsymbol{\eta}(\mathbf{b}^*, \mathbf{x}) - \mathbf{X}(\boldsymbol{\beta} - \mathbf{b}^*)]^T \\ &\quad \times [\mathbf{Y} - \boldsymbol{\eta}(\mathbf{b}^*, \mathbf{x}) - \mathbf{X}(\boldsymbol{\beta} - \mathbf{b}^*)] \\ &= [\mathbf{Y} - \boldsymbol{\eta}(\mathbf{b}^*, \mathbf{x})]^T[\mathbf{Y} - \boldsymbol{\eta}(\mathbf{b}^*, \mathbf{x})] \\ &\quad - [\mathbf{Y} - \boldsymbol{\eta}(\mathbf{b}^*, \mathbf{x})]^T[\mathbf{X}(\boldsymbol{\beta} - \mathbf{b}^*)^T] \\ &\quad - [\mathbf{X}(\boldsymbol{\beta} - \mathbf{b}^*)]^T[\mathbf{Y} - \boldsymbol{\eta}(\mathbf{b}^*, \mathbf{x})] \\ &\quad + (\boldsymbol{\beta} - \mathbf{b}^*)^T\mathbf{X}^T\mathbf{X}(\boldsymbol{\beta} - \mathbf{b}^*)]\end{aligned}$$

If for $\mathbf{b}^*$ we use the maximum likelihood estimator of $\boldsymbol{\beta}$, $\mathbf{b}$, then we know from Equation 5.1-8 that the two cross-product terms in the summation vanish. Consequently, the posterior density given by Equation 8.4-7 becomes

$$\begin{aligned}p_{n+n^*}&(\boldsymbol{\beta} \mid \mathbf{y}_{n+n^*}) \\ &= k_{n+n^*}^* \exp\left(-\frac{[\mathbf{Y} - \boldsymbol{\eta}(\mathbf{b}, \mathbf{x})]^T[\mathbf{Y} - \boldsymbol{\eta}(\mathbf{b}, \mathbf{x})]}{2\sigma_Y^2}\right) \\ &\quad \times \exp\left\{-\frac{1}{2\sigma_Y^2}[(\boldsymbol{\beta} - \mathbf{b})^T\mathbf{X}^T\mathbf{X}(\boldsymbol{\beta} - \mathbf{b}) \right. \\ &\quad \left. + (\boldsymbol{\beta} - \mathbf{b}^{(0)})^T\sigma_Y^2\boldsymbol{\Omega}^{-1}(\boldsymbol{\beta} - \mathbf{b}^{(0)})]\right\} \quad (8.4\text{-}9)\end{aligned}$$

The normalizing factor $k_{n+n^*}^*$ now becomes of interest. Equation 8.4-9 is the combination of two multinormal probability densities; hence $(k_{n+n^*}^*)^2$ by analogy with Equation 2.3-7 is proportional to

$$\Delta^{(n+n^*)} \equiv |\mathbf{X}^T\mathbf{X} + \sigma_Y^2\boldsymbol{\Omega}^{-1}| \quad (8.4\text{-}10)$$

where $|\ \ |$ signifies the determinant. If the posterior probability density Equation 8.4-9 is maximized with respect to both $\boldsymbol{\beta}$ and the vector of new observations $(x_{n+1}, \ldots, x_{n+n^*})$, $\boldsymbol{\beta}$ depends on $\mathbf{b}^{(n+n^*)}$ and $\mathbf{b}^{(0)}$ and cannot be determined in advance of taking the observations. But whatever the observations and $\boldsymbol{\beta}$ prove to be, at the extremum the exponent of the last exponential in Equation 8.4-9 becomes zero; hence we conclude that $k_{n+n^*}^*$ and thus Δ given by Equation 8.4-10 is the quantity that determines the value of p_{n+n^*}. Consequently, we should maximize Δ as the criterion for determining the values of $(x_{n+1}, \ldots, x_{n+n^*})$. One difficulty is that the elements of $(\mathbf{X}^T\mathbf{X})$ contain $\boldsymbol{\beta}$, the unknown vector of parameters. Therefore, we substitute the most recent estimate of $\boldsymbol{\beta}$, such as $\mathbf{b}^{(n)}$, for $\boldsymbol{\beta}$ in the elements of $\mathbf{X}$, in which case, to avoid confusion, the matrix $\mathbf{X}$ may be designated by $\mathbf{X}^{(n)}$.

In the determinant in Equation 8.4-10, the matrix $\mathbf{X}$ (for n observations at hand and n^* observations to be taken) is

$$\mathbf{X} = \begin{bmatrix} X_{11} & X_{12} & \cdots & X_{1m} \\ X_{21} & X_{22} & \cdots & X_{2m} \\ \vdots & \vdots & & \vdots \\ X_{n1} & X_{n2} & \cdots & X_{nm} \\ \vdots & \vdots & & \vdots \\ X_{n+n^*,1} & X_{n+n^*,2} & \cdots & X_{n+n^*,m} \end{bmatrix}$$

and the individual elements in $\mathbf{X}$ are

$$X_{ij} = \frac{\partial\eta_i(\mathbf{b}^{(n)}, \mathbf{x}_i)}{\partial\beta_j}$$

The elements of the matrix $\mathbf{X}^T\mathbf{X}$ will be denoted by A_{ij} and, as an example, with three parameters after four experiments with one new experiment being contemplated,

$$\mathbf{X}^T\mathbf{X} = \begin{bmatrix} \left(\sum_{i=1}^{4} X_{i1}^2 + X_{51}^2\right) & \left(\sum_{i=1}^{4} X_{i1}X_{i2} + X_{51}X_{52}\right) & \left(\sum_{i=1}^{4} X_{i1}X_{i3} + X_{51}X_{53}\right) \\ & \left(\sum_{i=1}^{4} X_{i2}^2 + X_{52}^2\right) & \left(\sum_{i=1}^{4} X_{i2}X_{i3} + X_{52}X_{53}\right) \\ \text{symmetric} & & \left(\sum_{i=1}^{4} X_{i3}^2 + X_{53}^2\right) \end{bmatrix}$$

In the above matrix the summed quantities are known, and the terms with the index 5 are sought in terms of the new values of the independent variables for the fifth experiment.

Examination of a model with two parameters provides some insight into the character of the criterion Δ:

$$\Delta = \det\left\{\begin{bmatrix} A_{11} & A_{12} \\ A_{21} & A_{22} \end{bmatrix} + \sigma_Y^2\begin{bmatrix} \omega_{11} & \omega_{12} \\ \omega_{21} & \omega_{22} \end{bmatrix}^{-1}\right\}$$

We shall let ρ be the correlation coefficient between β_1 and β_2, $\omega_{11} = \omega_1^2$, and $\omega_{22} = \omega_2^2$, as in Example 2.3-4. Then

$$\Delta = \det \begin{bmatrix} A_{11} + \dfrac{\sigma_Y^2}{\omega_1^2(1-\rho^2)} & A_{12} - \dfrac{\sigma_Y^2\rho}{\omega_1\omega_2(1-\rho^2)} \\ A_{21} - \dfrac{\sigma_Y^2\rho}{\omega_1\omega_2(1-\rho^2)} & A_{22} + \dfrac{\sigma_Y^2}{\omega_2^2(1-\rho^2)} \end{bmatrix}$$

$$= \frac{(A_{11}A_{22} - A_{12}^2) + \left(A_{11}\dfrac{\sigma_Y^2}{\omega_2^2} + A_{22}\dfrac{\sigma_Y^2}{\omega_1^2} + 2A_{12}\dfrac{\sigma_Y^2\rho}{\omega_1\omega_2}\right)}{(1-\rho^2) + \left(\dfrac{\sigma_Y^2}{\omega_1\omega_2}\right)^2}$$

Suppose $\omega_1 = \omega_2 = 0$. This means that the initial estimates $b_1^{(0)}$ and $b_2^{(0)}$ have zero variance and therefore are certainties. Thus, Δ is a maximum no matter what the A_{ij}'s are, and no additional experiments need be run. Another extreme case is when $\omega_1 = \omega_2 \to \infty$, because then

$$\Delta_1 = \det(\mathbf{X}^T\mathbf{X}) = A_{11}A_{22} - A_{12}^2$$

which is the criterion developed by Box and Lucas† and later investigators. They suggested that the settings of the independent variables be those that minimize the confidence region associated with the parameters $\boldsymbol{\beta}$. If the joint confidence region is given approximately by Equation 6.4-5, which is based on the linearized model, the size of the confidence region is inversely proportional to $\det(\mathbf{X}^T\mathbf{X}) = \Delta_1$, all other factors remaining the same. By maximizing Δ_1, the joint confidence region can be minimized. Exactly the same conclusion could be drawn from Equation 5.1-14, even though the equation is for a linear model, because to make Covar $\{\mathbf{b}\}$ a minimum, with $\mathbf{c}$ equal to the adjoint matrix of $(\mathbf{x}^T\mathbf{w}\mathbf{x})$ divided by the det $(\mathbf{x}^T\mathbf{w}\mathbf{x})$, we see that maximization of Δ_1 minimizes the Covar $\{\mathbf{b}\}$.

The interpretation of infinite variances for $b_1^{(0)}$ and $b_2^{(0)}$ is that joint prior information is completely absent at the initiation of experimentation. Various other intermediate inferences can be drawn given the relative values of ρ, the ω's, and σ_Y^2. In general the better the prior information on a parameter, the greater is the emphasis placed on the other parameter in selecting the x's.

To sum up, the general procedure for parameter estimation is:

1. Select initial estimates of the parameters, their variances, and covariances.
2. Evaluate the elements of $\mathbf{X}$ by using the estimated parameters.
3. Decide on an initial design by maximizing Δ.
4. Carry out the experiments according to the design.
5. Reestimate the values of the parameters.
6. Evaluate the elements of $\mathbf{X}$ by using the new estimates.
7. Select the next design by maximizing Δ.
8. Revert to step 4 and continue until the desired precision in the parameters is obtained or the experimentation must be terminated.

We now illustrate by examples how Equation 8.4-10 can be used to determine more effective experimental designs. Example 8.4-1 describes the initial design; Example 8.4-2 illustrates how additional values of the independent variables can be selected based on prior knowledge gained from earlier observations.

† G. E. P. Box and H. L. Lucas, *Biometrika* **46**, 77, 1959.

Example 8.4-1 Optimal Sequential Designs for a Nonlinear Model

This example illustrates how an initial experimental design is developed, assuming a complete absence of prior information as to the values of the parameters in a model and the covariance elements for the model parameters. Subsequent designs are also obtained after the initial experiments are carried out.

Shaft-grinding mills are used for the combined pulverization and drying of lignite, shale, coal, etc. Inasmuch as no valid theoretically based process models for these devices have been developed to relate the heat transfer to the process independent variables, the following empirical model was developed:

$$\frac{Q}{Q_0} = (1 - e^{\beta_1 x_1})(1 - \beta_2 x_2) + \epsilon \qquad \text{(a)}$$

where

$\dfrac{Q}{Q_0}$ = fraction of heat transferred to the pulverized material (a random variable)
x_1 = concentration of coal in the gases, kg/kg dry gas
x_2 = square of the mass flow rate of gases in the shaft, [kg dry gas/(meter)^2(sec)]^2
β_1, β_2 = coefficients to be estimated

In what follows we shall let Q/Q_0 be denoted by the random variable Y. Several tests were planned on industrial shaft grinders; to reduce the costs of the tests, it was desired to carry out a sequential series of experiments in which Equation 8.4-10 was used as the criterion for establishing the values of x_1 and x_2. The range of coal concentrations was 0.17 to 1.1 and the range of mass flow rates was 0 to 12 (0 to 144 for x_2).

Since no initial estimates of the three parameters in Model (a) were available, by inspection of the equation it could be seen that β_2 had to be small anough so that $(1 - \beta_2 x_2)$ was not negative; hence, $b_2^{(0)} = 0.001$ was a reasonable guess. Because β_1 could be either positive or negative, it was decided to let $b_1^{(0)} = 1$. Furthermore, it was assumed that all elements in the matrix $\boldsymbol{\Omega}^{-1}$ were zero.

The minimum number of observations required to estimate two parameters is two. Consequently, the criterion Δ in Equation 8.4-10 reduced to

$$\Delta_1^{(0)} = |\mathbf{X}^T\mathbf{X}|$$

$$= \begin{vmatrix} \sum_{i=1}^{2} X_{i1}^2 & \sum_{i=1}^{2} X_{i1}X_{i2} \\ \sum_{i=1}^{2} X_{i1}X_{i2} & \sum_{i=1}^{2} X_{i2}^2 \end{vmatrix}$$

$$= \left(\sum_{i=1}^{2} X_{i1}^2\right)\left(\sum_{i=1}^{2} X_{i2}^2\right) - \left(\sum_{i=1}^{2} X_{i1}X_{i2}\right)^2 \qquad \text{(b)}$$

Here, the elements X_{ij} can be obtained analytically:

$$\begin{aligned} X_{i1} &= -x_{i1}[\exp(b_1x_{i1})](1 - b_2x_{i2}) \\ X_{i2} &= -x_{i2}[1 - \exp(b_1x_{i1})] \end{aligned} \qquad \text{(c)}$$

If the model functions are so complex that analytical derivatives cannot be taken, numerical derivatives can be used in their place, as pointed out in Section 6.2-3.

The two pairs of initial values of x_{i1} and x_{i2} can be obtained by introducing the $b^{(0)}$'s into Equations (c) and by maximizing Δ_1 given by Equation (b) with respect to all four of the x's. As might be expected, when functions (c) are introduced into (b), the expression for Δ_1 becomes quite nonlinear in x_{i1} and x_{i2}; hence a numerical optimization scheme is required to find the two pairs of x's. Quite possibly more than one local optimum may exist for such nonlinear functions in a mathematical sense, but because of the restricted range on the x's and the correspondence of the model to a real process, this difficulty is not encountered too often. In such cases, by starting at different initial $\mathbf{x}$ vectors, the difficulty can be disclosed and the largest Δ_1 discovered.

The location of the values of x_{i1} and x_{i2} for the special case in which the *number of experiments is equal to the number of parameters* can also be obtained by maximizing the determinant of $\mathbf{X}$ itself if we make use of two properties of determinants: (a) the determinant of the product of two square matrices is equal to the product of the determinants of the respective matrices, $|A \quad B| = |A|\,|B|$, and (2) the determinant of a square matrix is equal to the determinant of its transpose, $|A| = |A^T|$. Then

$$\Delta_1^{(0)} = |\mathbf{X}^{(0)T}\mathbf{X}^{(0)}| = |\mathbf{X}^{(0)T}|\,|\mathbf{X}^{(0)}| = |\mathbf{X}^{(0)}|^2 \qquad \text{(d)}$$

and it is only necessary to maximize $|\mathbf{X}^{(0)}|$, subject to $0.17 \le x_{i1} \le 1.1$ and $0 \le x_{i2} \le 144$ to find the initial design. Consequently, the following expression was maximized by the flexible simplex method of Chapter 6, in lieu of Equation (b):

$$|\mathbf{X}^{(0)}| = \begin{vmatrix} X_{11} & X_{12} \\ X_{21} & X_{22} \end{vmatrix} = X_{11}X_{22} - X_{21}X_{12}$$

$$\begin{aligned} |\mathbf{X}^{(0)}| &= x_{11}x_{22}(1 - b_2^{(0)}x_{12})\exp(b_1^{(0)}x_{11})[1 - \exp(b_1^{(0)}x_{21})] \\ &\quad - x_{12}x_{21}(1 - b_2^{(0)}x_{22})\exp(b_1^{(0)}x_{21})[1 - \exp(b_1^{(0)}x_{11})] \end{aligned} \qquad \text{(e)}$$

The effect of initial guesses for the x's on the final design is shown in Table E8.4-1a. As might be expected, the design falls at the extremes of the range of the x's—at the upper limit of x_1 in both runs and in one run at the upper limit and in one row at the lower limit of x_2.

TABLE E8.4-1a MAXIMIZATION OF $|\mathbf{X}^{(0)}|$ SUBJECT TO $0.17 \le x_{i1} \le 1.1$ AND $0 \le x_{i2} \le 144$ WITH $b_1^{(0)} = 1$ AND $b_2^{(0)} = 10^{-3}$

Initial Guesses for x				Final Results for x				Value of
x_{11}	x_{21}	x_{12}	x_{22}	x_{11}	x_{21}	x_{12}	x_{22}	$\lvert\mathbf{X}^{(0)}\rvert$
0.2	0.6	40	80	1.10	1.10	144	4.38	927
0.17	1.1	0	144	1.10	1.10	143	0	949
0.9	1.0	120	130	1.10	1.10	144	0	954

The final results listed in Table E8.4-1a can be compared with those obtained by maximization of $\Delta_1^{(0)}$ given by Equation (b) as listed in Table E8.4-1b. Exactly the same values for x_1, namely 1.10 and 1.10, are obtained for both experiments, but the values for x_2 could just as well be 144 and 0 as the 0 and 144 selected for the first two experiments. It is not possible to illustrate the contours of $\Delta_1^{(0)}$ versus x_1 and x_2 because of the four-dimensional experimental space.

As a result of the first two experiments, the following observations were made:

$$Y_1 = 0.646 \quad \text{at} \quad x_1 = 1.10,\ x_2 = 0$$

$$Y_2 = 0.194 \quad \text{at} \quad x_1 = 1.10,\ x_2 = 144$$

From these observations the first two parameters were estimated by using the flexible simplex method of Section 6.2:

$$b_1^{(0)} = 0.944$$

$$b_2^{(0)} = 4.86 \times 10^{-3}$$

To determine the next experimental point, $\Delta_1^{(3)}$ was maximized by using Equations b and c with $b_1^{(0)}$ and $b_2^{(0)}$ introduced for the estimated parameters. Table E8.4-1c lists the results; Figure E8.4-1 shows the contours of $\Delta_1^{(3)}$ versus x_1 and x_2. It was clear that the new run should either be taken at $x_1 = 1.06$ and $x_2 = 0$ or at $x_1 = 1.10$ and $x_2 = 144$. The former was chosen, the Y_3 observed was 0.627, and the resulting parameter estimates were

$$b_1^{(3)} = -0.937$$

$$b_2^{(3)} = 4.85 \times 10^{-3}$$

TABLE E8.4-1b MAXIMIZATION OF $\Delta_1^{(0)} = |\mathbf{X}^{(0)T}\mathbf{X}^{(0)}|$ SUBJECT TO $0.17 \le x_{i1} \le 1.1$ AND $0 \le x_{i2} \le 144$ WITH $b_1^{(0)} = 1$ AND $b_2^{(0)} = 10^{-3}$

Initial Guesses for x				Final Results for x				$\Delta_1^{(0)}$
x_{11}	x_{21}	x_{12}	x_{22}	x_{11}	x_{21}	x_{12}	x_{22}	$\times 10^{-5}$
0.2	0.6	40	80	1.10	1.10	41.9	82.7	0.725
0.17	1.1	0	144	1.10	1.10	2.35	143	8.71
0.9	1.0	120	130	1.10	1.10	144	5.98	8.36
0.17	0.17	0	0	1.10	1.10	0.33	144	9.05
1.1	1.1	144	144	1.10	1.10	144	0	9.08
0.5	0.5	50	50	1.10	1.10	144	0	9.096
0.2	1.0	10	130	1.10	1.10	0	144	9.095

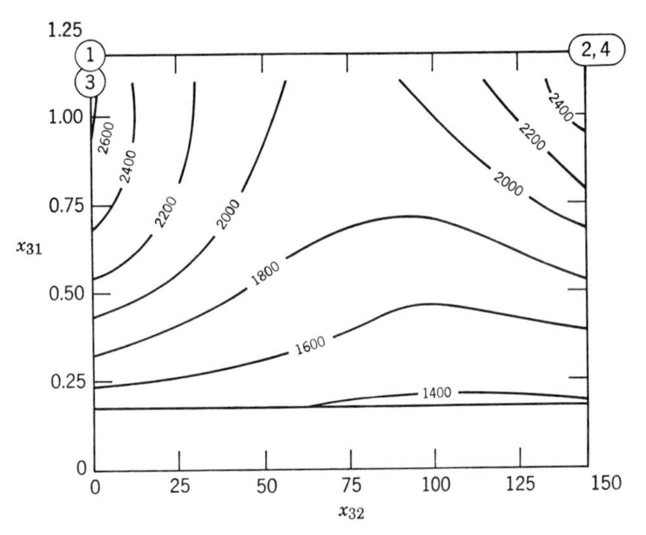

FIGURE E8.4-1 Contours of $\Delta_1^{(3)}$; and location of experiments (designated by circled numbers).

TABLE E8.4-1c RESULTS OF MAXIMIZATION OF $\Delta_1^{(3)}$ TO DETERMINE THE THIRD EXPERIMENTAL POINT

Initial Guesses		Final Results		Value of
x_{31}	x_{32}	x_{31}	x_{32}	$\Delta_1^{(3)} \times 10^{-3}$
0.17	0	1.06	0	2.62
0.20	20	1.06	0	2.62
0.80	100	1.10	144	2.26
1.10	144	1.10	137	2.50
0.17	144	1.10	144	2.62
1.10	0	1.06	0	2.62

Although it was now clear that the experimentation could be terminated, it was decided to carry out one more experiment. Table E8.4-1d lists the results of the maximization of $\Delta_1^{(4)}$ (in which $b_1^{(3)}$ and $b_2^{(3)}$ were employed). The contours of $\Delta_1^{(4)}$ are similar to those illustrated in Figure E8.4-1. The final run was carried out at $x_1 = 1.10$ and $x_2 = 144$ to yield a $Y_4 = 0.173$. The final estimated parameters, using all four runs, were

$$b_1 = -0.937$$

$$b_2 = 4.96 \times 10^{-3}$$

TABLE E8.4-1d RESULTS OF MAXIMIZATION OF $\Delta_1^{(4)}$ TO DETERMINE THE FOURTH EXPERIMENTAL POINT

Initial Guesses		Final Results		
x_{41}	x_{42}	x_{41}	x_{42}	$\Delta_1^{(4)} \times 10^{-3}$
0.17	0	1.07	0	3.96
0.20	20	1.07	0	3.96
0.40	50	1.10	144	5.28
0.80	100	1.10	100	4.04
1.10	144	1.10	144	5.28
0.17	144	1.10	144	5.28
1.10	0	1.07	0	3.96

The next experiment would have been at $x_1 = 1.07$ and $x_2 = 0$. Figure E8.4-1 illustrates the location of the experiments in the $x_1 - x_2$ plane.

Although the variance s_Y^2 was known, because $F_{1-\alpha}[m_1(n - m)]$ was so large—$F_{0.95}(2, 1) = 199$ and $F_{0.94}(2, 2) = 19$, the joint confidence regions for β_1 and β_2 after experiments three and four were run are not shown. The next example does illustrate how the area enclosed by the confidence region shrinks as the sequential experiments continue. From the parameter correlation matrix, it was observed that the correlation coefficient between $b_1^{(3)}$ and $b_2^{(3)}$ was only -0.194.

TABLE E8.4-1e EFFECT OF THE ASSIGNMENT OF VARIOUS WEIGHTS ON THE SELECTION OF THE EXPERIMENTAL DESIGN OBTAINED BY MAXIMIZING $\Delta^{(0)} = |\mathbf{X}^{(0)T}\mathbf{X}^{(0)} + \sigma_Y^2\mathbf{\Omega}^{-1}|$ (FOR INITIAL VALUES OF x_{ij} GIVEN IN TABLE E8.4-1A)

Elements of $\mathbf{\Omega}$ with $\sigma_Y^2 = 1$				Final Design			
ω_{11}	ω_{12}	ω_{21}	ω_{22}	x_{11}	x_{12}	x_{21}	x_{22}
0.5	5	50	500	1.10	144	1.10	0
5	50	500	0.5	1.06	144	1.10	0
50	500	0.5	5	1.04	144	1.10	0.2
500	0.5	5	50	1.10	142	1.10	144
10	20	30	40	1.04	144	1.10	0

To ascertain the effect of prior information on the design of the experiment, several combinations of assumed elements for $\mathbf{\Omega}$ were selected for use with Equation 8.4-10. Then Δ itself was maximized with the results indicated in Table E8.4-1e. Not only did the x_{ij}'s still fall essentially on the boundaries, but the inclusion of prior information clearly made no significant change in the experimental design. Thus letting $\mathbf{\Omega}^{-1} = \mathbf{0}$ was quite satisfactory.

Example 8.4-2 Experimental Design to Reduce Parameter Uncertainty

This example, which is based on the work of Kittrell, Hunter, and Watson,† illustrates a sequential series of designs for a model with three coefficients. It also brings out some additional features of model building not encountered in the previous example. The reaction studied was the catalytic reduction of nitric oxide:

$$NO + H_2 \rightleftharpoons H_2O + \tfrac{1}{2}N_2$$

The reaction mechanism was assumed to be the surface reaction between an adsorbed nitric oxide molecule and one adjacently adsorbed hydrogen molecule, a mechanism that can be represented by the following model:

$$r = \frac{kK_{NO}K_{H_2}p_{NO}p_{H_2}}{(1 + K_{NO}p_{NO} + K_{H_2}p_{H_2})^2} \qquad \text{(a)}$$

† J. R. Kittrell, W. G. Hunter, and C. C. Watson, *AIChE J.* **12**, 5, 1966.

where

r = reaction rate, g-moles/(min)(g catalyst)
k = forward rate constant, g-moles/(min)(g catalyst)
K = equilibrium absorption coefficient, atm^{-1}
p = partial pressure, atm

It will be assumed that the random variable is $R = r + \epsilon$, as usual.

The "experimental observations" given below were in fact a simulated set of deterministic values for r generated using $k = 4.94 \times 10^{-4}$ g-moles/(g catalyst)(min), $K_{NO} = 14.64\ atm^{-1}$, and $K_{H_2} = 19.00\ atm^{-1}$. Added to the deterministic rate was an independent, normal random error with zero mean and $\sigma^2 = 9.508 \times 10^{-12}$. These simulated observations will be treated as real observations, and the simulation has the advantage that it will be possible to compare the estimated model parameters with the assumed ones.

UNPLANNED EXPERIMENT. The one-variable-at-a-time procedure, as used by Ayen and Peters,† was simulated first so that it would be contrasted with planned experiments. The values of the partial pressures and reaction rates are listed in Table E8.4-2a.

TABLE E8.4-2a ONE-VARIABLE-AT-A-TIME DESIGN

Run Number	p NO	p H_2	Simulated Observed Rates, R
1	0.00922	0.0500	2.01
2	0.0136	0.0500	2.52
3	0.0197	0.0500	3.10
4	0.0280	0.0500	3.65
5	0.0291	0.0500	3.82
6	0.0389	0.0500	3.82
7	0.0485	0.0500	4.90
8	0.0500	0.00918	2.02
9	0.0500	0.0184	2.83
10	0.0500	0.0298	3.75
11	0.0500	0.0378	4.32
12	0.0500	0.0491	4.53

† R. J. Ayen and M. S. Peters, *Ind. Eng. Chem. Process Design Develop.* **1**, 204, 1962.

The first row of Table E8.4-2b lists the results of a nonlinear least squares analysis as well as the values of the square root of the determinant of $(\mathbf{X}^T\mathbf{X})$, which is inversely proportional to the volume of the joint confidence region. Figure E8.4-2a illustrates the approximate 95-percent confidence region surface calculated using all 12 runs for the sum of the squares based on Equation 6.4-5. Although no replicate runs were available, the model was assumed to be the correct one, and s_Y^2 was assumed to be equal to $\phi_{min}/(n - m)$. Note that the confidence region does not contain the true value of the forward rate constant.

The region for Model (a) is characteristically large and attenuated, indicating that the coefficients are not adequate estimates, primarily because of a high degree of correlation among the coefficient estimates. What happens is that the least squares fitting procedure compensates for improper values of one coefficient by the choice of the values of the other coefficients so as to yield an overall fit (in the sense of predicting the dependent variable) nearly as good as that obtained with the best estimates of the parameters. For example, values of the predicted rate $\hat{R}$ calculated by using three sets of reasonably spaced estimated parameters, namely those at points A, B, and C in Figure E8.4-2a, are compared in Figure E8.4-2b. Little difference in the predicted rates is observed, and each set of estimated parameters appears to be equally able to represent the experimental data.

PLANNED EXPERIMENTS. The use of planned sequential experimentation is now demonstrated to obtain estimates of the parameters in Model (a) using the same number of experimental runs, namely 12. A two-level factorial design about an arbitrary point in the parameter space was used to get the initial parameter estimates; refer to the first four runs of Table E8.4-2c.

The fifth experimental point was chosen to maximize Δ_1 with the estimated parameters from the fourth run introduced into the elements X_{ij}. A grid with intervals of 0.005 atm between 0 and 0.10 atm was superimposed on the independent variable space, and a search for the settings of the partial pressures which maximized Δ_1 resulted in row 5 of Table E8.4-2c. The fifth experiment was run, the reaction rate observed, and again the parameters k, K_{NO}, and K_{H_2} were estimated by least squares. Afterwards, Δ was maximized to get the experimental settings for the sixth run, and so on until 12 runs had been completed.

TABLE E8.4-2b RESULTS OF THE UNPLANNED DESIGN COMPARED WITH THOSE FROM THE SEQUENTIAL DESIGN

$k \times 10^4$ g-moles / (min)(g catalyst)	K_{NO}, atm^{-1}	K_{H_2}, atm^{-1}	$\phi_{min} \times 10^{11}\left[\frac{\text{g-moles}}{\text{(min)(g catalyst)}}\right]^2$	$\lvert\mathbf{X}^T\mathbf{X}\rvert^{1/2}$
*2.9 ± 0.92	38.0 ± 26.1	38.0 ± 24.9	4.62	5.02×10^{-14}
†4.7 ± 0.53	16:9 ± 4.0	20.2 ± 4.2	6.88	9.02×10^{-13}

* The numbers ± designate the bounds of the 95-percent individual confidence limits.
† Sequential design.

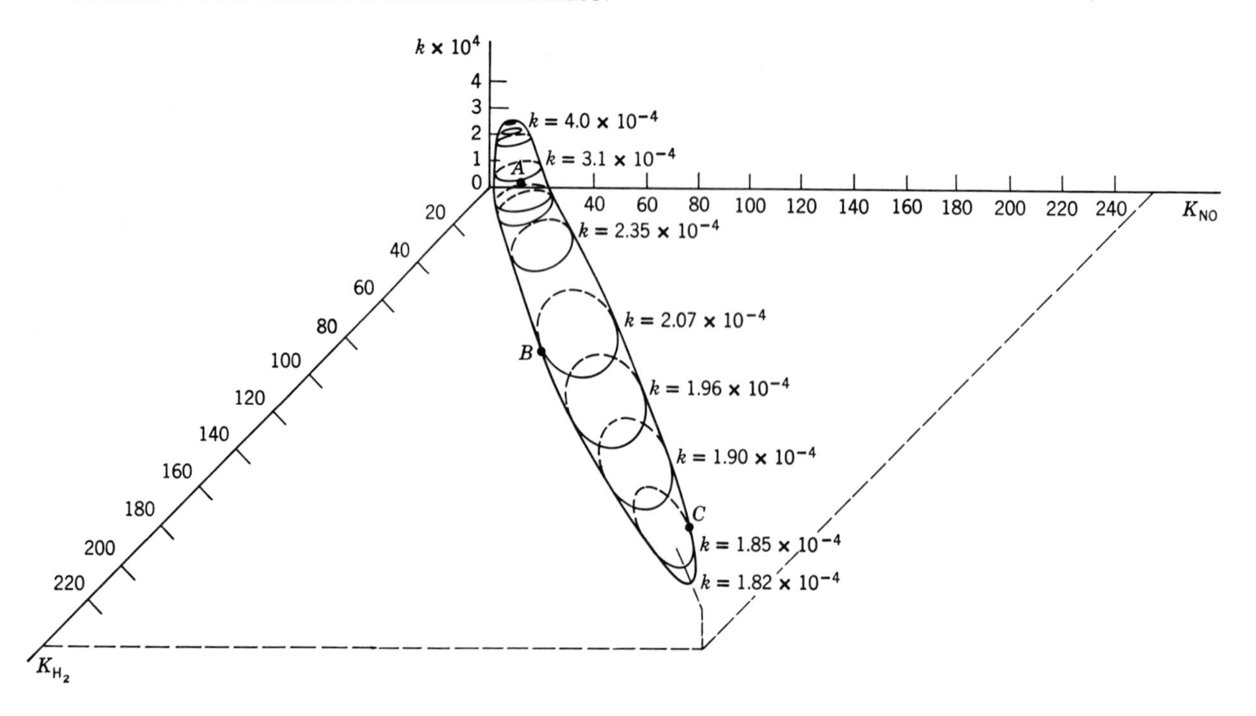

FIGURE E8.4-2a Approximate 95-percent contour for the joint confidence region derived from unplanned experiments. Contours shown on the surface are the loci of lines of constant k in the surface. (Reproduced from J. R. Kittrell, W. G. Hunter, and C. C. Watson, *AIChE J.* **12**, 5, 1966, with permission of the publisher, The American Institute of Chemical Engineers.)

The square root of the determinant Δ_1 is inversely proportional to the size of the confidence region. Figure E8.4-2c contrasts the change in the relative volume of the confidence region by the sequential design procedure with the constant volume of the unplanned design. It can be seen that the estimates from the sequential design procedure after the fifth point (i.e., the first point chosen by the minimum volume design) are as good as those obtained after 12 unplanned experiments. Figure E8.4-2d is a visual representation of the confidence region at the end of 12 experiments and can be compared with Figure E8.4-2a.

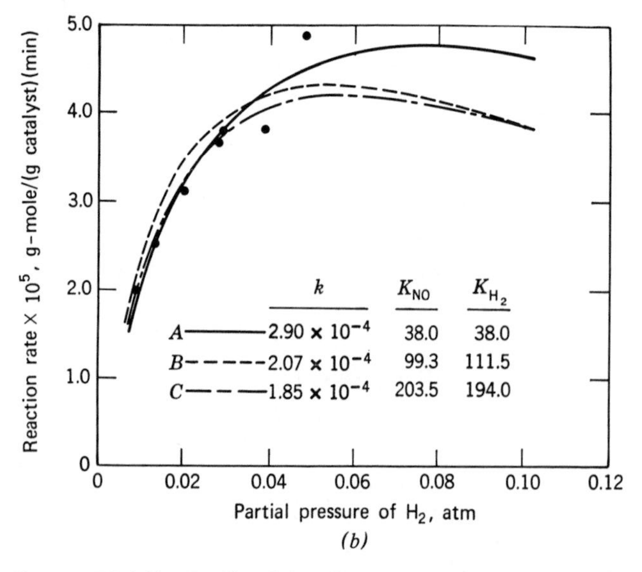

FIGURE E8.4-2b Predicted (the lines) and observed (the solid dots) rates for three different parameter estimates. (Reproduced from J. R. Kittrell, W. G. Hunter, and C. C. Watson, *AIChE J.* **12**, 5, 1966, with permission of the publisher, The American Institute of Chemical Engineers.)

TABLE E8.4-2c EXPERIMENTAL CONDITIONS USING SEQUENTIAL DESIGN

Run Number	p_{NO}, atm	p_{H_2}, atm		$\Delta_1 \times 10^{25}$
1	0.01	0.01	Initial design	—
2	0.01	0.03		—
3	0.03	0.01		—
4	0.03	0.03		0.0003
5	0.10	0.06		0.022
6	0.06	0.10		0.498
7	0.10	0.03		1.14
8	0.10	0.10		1.96
9	0.04	0.10		3.15
10	0.10	0.03		4.60
11	0.10	0.10		6.93
12	0.03	0.03		8.15

Figure E8.4-2e brings out another feature of the planned experiments, namely that the experimental runs, after the initial 2^2 design, fell along the periphery of experimental region. Figure E8.4-2b shows that it was only in the region of high partial pressures that the downward trend of the rate curve could be detected.

CONCLUSIONS. This example demonstrates for a model with three coefficients how planned experiments can be more

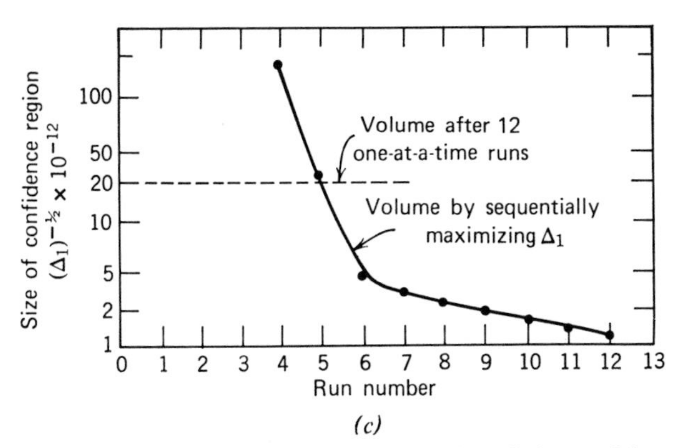

FIGURE E8.4-2c Change in the relative size of the confidence region.

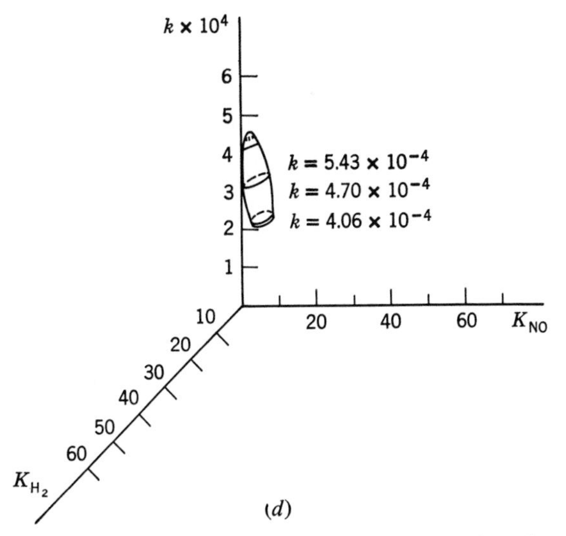

FIGURE E8.4-2d Confidence region after 12 runs using planned experiments. (Reproduced from J. R. Kittrell, W. G. Hunter, and C. C. Watson, *AIChE J.* **12**, 5, 1966, with permission of the publisher, The American Institute of Chemical Engineers.)

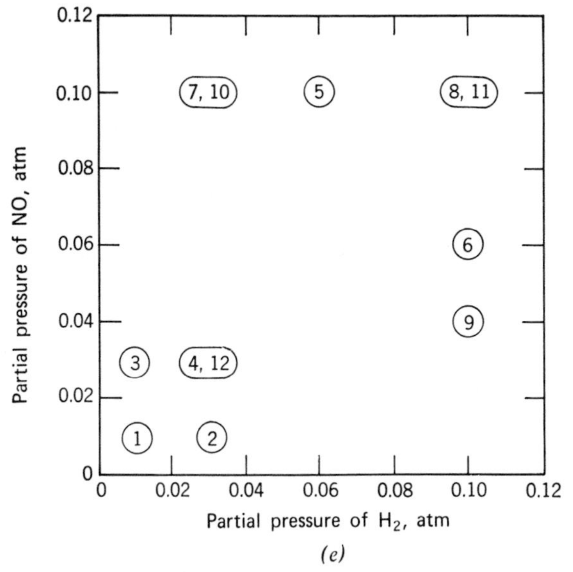

FIGURE E8.4-2e Regions of experimentation for planned experiments; circled numbers correspond to run numbers.

effective than unplanned ones. It also shows that the planned experimental program delineates the important regions in the experimental space for experimentation. One question not answered is whether the added cost and time of estimating the model coefficients and maximizing Δ_1 outweigh the benefits arising from effective experimentation. One can obtain an approximate idea of the magnitude of the expected decrease in the confidence region beyond that which could be obtained, say, with orthogonal experimental designs by practical experience or by simulation studies. One would then be more capable of deciding by which route to proceed, and how far.

8.4-2 Multiresponse Processes

In this section we continue with the development of sequential experimental designs to reduce uncertainty in parameter estimates. The new feature of this section is that multiple responses are observed for a model comprises of several equations of known form, such as

$$\eta_1 = \beta_1 e^{-\beta_2 x}$$

$$\eta_2 = \frac{\beta_1 x_1}{\beta_2 x_2 + \beta_3}$$

We shall designate each response by

$$Y_{ri} = \eta_{ri} + \epsilon_{ri} \qquad \begin{array}{c} 1 \le r \le v \\ 1 \le i \le n \end{array}$$

where the index r refers to the number of the model and the index i to the data set number. The model η_r will be a function of the m parameters β, as before, and the independent variables $x_{r,ik}$, where the superscript k, $1 \le k \le q$, designates the variable number. The errors ϵ_{ri} have $\mathscr{E}\{\epsilon_{ri}\} = 0$, $\mathscr{E}\{\epsilon_{ri}\epsilon_{sj}\} = 0$ for $i \ne j$, $\mathscr{E}\{\epsilon_{ri}\epsilon_{rj}\} = 0$, $\mathscr{E}\{\epsilon_{ri}\epsilon_{ri}\} = \sigma_{rr} \equiv \sigma_r^2$, and $\mathscr{E}\{\epsilon_{ri}\epsilon_{si}\} = \sigma_{rs} = \sigma_{sr}$ for $r \ne s$. Thus the observations on one experiment for one model r and another s may be correlated, and the covariance matrix among models whose elements are σ_{rs} will be denoted by $\mathbf{\Gamma}$; the elements of the inverse $\mathbf{\Gamma}^{-1}$ will be designated by σ^{rs}. Note that for different experiments the errors are independent.

As before we follow the arguments of Draper and Hunter.† We assume that the observations of Y are represented by a multivariate normal distribution analogous to Equation 8.3-3 and that after n experiments the likelihood function for $\boldsymbol{\beta}$ is

$$L(\boldsymbol{\beta} \mid \mathbf{y}_n, \mathbf{x}, \sigma^{rs}) = \frac{|\mathbf{\Gamma}^{-1}|^{n/2}}{(2\pi)^{nq/2}} \exp\left[-\frac{1}{2}\sum_{r=1}^{v}\sum_{s=1}^{v}\sigma^{rs}V_{rs}\right] \tag{8.4-11}$$

where

$$V_{rs}^{(n)} = \sum_{i=1}^{n} [Y_{ri} - \eta_{ri}(\boldsymbol{\beta}, \mathbf{x}_i)]$$

denotes the sum of the sums of squares of deviations of the observed Y_{ri}'s for n experiments.

† N. R. Draper and W. G. Hunter, *Biometrika* **55**, 662, 1968.

Also, as before, we assume that the prior information available after n experiments, but before the n^* additional experiments, leads to a density exactly the same as Equation 8.4-4. Introducing the prior density and the likelihood of Equation 8.4-11 into Bayes' theorem, we obtain the posterior density analogous to Equation 8.4-5. To select n^* additional values of the independent variables, as in Section 8.4-1, the posterior density analogous to Equation 8.4-5 can be used as a prior density together with a likelihood function corresponding to Equation 8.4-11 (except that as before the summation on i is over $i = n + 1$ to $i = n + n^*$) in Bayes' theorem to give the posterior density after $(n + n^*)$ experiments:

$$p_{n+n^*}(\boldsymbol{\beta} \mid \mathbf{y}_{n+n^*}) = \frac{|\boldsymbol{\Omega}|^{-1/2}|\boldsymbol{\Gamma}^{-1}|^{-(n+n^*)/2}}{(2\pi)^{[m+(n+n^*)v]/2}} \cdot \exp\left[-\frac{1}{2}\sum_{r=1}^{v}\sum_{s=1}^{v}\sigma^{rs}V_{rs}^{(n+n^*)}\right] \cdot \exp\left[-\tfrac{1}{2}(\boldsymbol{\beta} - \mathbf{b}^{(0)})^T\boldsymbol{\Omega}^{-1}(\boldsymbol{\beta} - \mathbf{b}^{(0)})\right] \tag{8.4-12}$$

Note the analogy to Equation 8.4-7.

The remainder of the development is exactly the same as in Section 8.4-1. To maximize p_{n+n^*} with respect to $\boldsymbol{\beta}$ and the vector of new values of $\mathbf{x}_{ri} \equiv \mathbf{x}_{r,i}$ for $i = n + 1, \ldots, n + n^*$, we introduce the equivalent of Equation 8.4-8 into Equation 8.4-12, except that now we must expand each model as

$$\eta_r(\boldsymbol{\beta}, \mathbf{x}_{r,i}) \cong \eta_r(\mathbf{b}^*, \mathbf{x}_{r,i}) + \sum_{j=1}^{m}(\beta_j - b_j^*)X_{r,ij}$$

where

$$X_{r,ij} = \frac{\partial\eta_r(\mathbf{b}^*, \mathbf{x}_{r,i})}{\partial\beta_j} \qquad \mathbf{x}_{r,i} = \begin{bmatrix} x_{r,i1} \\ \vdots \\ x_{r,iq} \end{bmatrix}$$

If we replace $\mathbf{b}^*$ with $\mathbf{b}$, the maximum likelihood estimate of $\boldsymbol{\beta}$ after $(n + n^*)$ experiments, the crossproduct terms in the double summation in Equation 8.4-12 vanish for the reasons presented in Section 8.4-1. The following then can be written for the double summation:

$$\begin{aligned}\sum_{r=1}^{v}\sum_{s=1}^{v}\sigma^{rs}V_{rs}^{(n+n^*)} &= \sum_{r=1}^{v}\sum_{s=1}^{v}\sigma^{rs}\sum_{i=1}^{n+n^*}\left[d_{ri} - \sum_{j=1}^{m}(\beta_j - b_j^*)X_{r,ij}\right] \cdot \left[d_{si} - \sum_{j=1}^{m}(\beta_j - b_j^*)X_{s,ij}\right] \\ &= \sum_{r=1}^{v}\sum_{s=1}^{v}\sigma^{rs}\sum_{i=1}^{n+n^*}d_{ri}\,d_{si} \\ &\quad + \sum_{r=1}^{v}\sum_{s=1}^{v}(\boldsymbol{\beta} - \mathbf{b})^T(\sigma^{rs}\mathbf{X}_r^T\mathbf{X}_s)(\boldsymbol{\beta} - \mathbf{b})\end{aligned} \tag{8.4-13}$$

where

$$\mathbf{X}_r = \begin{bmatrix} X_{r,11} & X_{r,12} & \cdots & X_{r,1m} \\ X_{r,21} & X_{r,22} & \cdots & X_{r,2m} \\ \vdots & \vdots & \vdots & \vdots \\ X_{r,(n+n^*)1} & X_{r,(n+n^*)2} & \cdots & X_{r,(n+n^*)m} \end{bmatrix}$$

$$d_{ri} = Y_{ri} - \eta_{ri}(\mathbf{b}, \mathbf{x}_{r,i})$$

To obtain the design for the next n^* experiments, we should maximize the determinant:

$$\Delta = \left|\sum_{r=1}^{v}\sum_{s=1}^{v}\sigma^{rs}\mathbf{X}_r^T\mathbf{X}_s + \boldsymbol{\Omega}^{-1}\right| \tag{8.4-14}$$

which is proportional to the square of the normalization factor that would be obtained after Equation 8.4-13 is introduced into Equation 8.4-12. All the derivatives in $\mathbf{X}_r$ must be evaluated at $\mathbf{b}^{(n)}$, and the elements σ_{rs} are presumed known. Criterion 8.4-14 reduces to the criterion developed by Draper and Hunter† and others if the elements of $\boldsymbol{\Omega}^{-1}$ are zero. Equation 8.4-14 also appeals to common sense because it weights the $\mathbf{X}$ matrices of each model inversely proportional to the error associated with the model.

To give a specific example of Δ for the multiresponse situation, let us take the case of two models ($v = 2$) with three coefficients each ($m = 3$), $n = 0$ (no experiments yet completed), all the ω's zero, and $n^* = 4$ (four experiments to be planned). Then:

$$\Delta = |\sigma^{11}(\mathbf{X}_{1,n+n^*})^T(\mathbf{X}_{1,n+n^*}) + \sigma^{12}(\mathbf{X}_{1,n+n^*})^T(\mathbf{X}_{2,n+n^*}) + \sigma^{12}(\mathbf{X}_{2,n+n^*})^T(\mathbf{X}_{1,n+n^*}) + \sigma^{22}(\mathbf{X}_{2,n+n^*})^T(\mathbf{X}_{2,n+n^*})|$$

where

$$\mathbf{X}_{r,n+n^*} = \begin{bmatrix} X_{r,11} & X_{r,12} & X_{r,13} \\ X_{r,21} & X_{r,22} & X_{r,23} \\ X_{r,31} & X_{r,32} & X_{r,33} \\ X_{r,41} & X_{r,42} & X_{r,43} \end{bmatrix} \quad \text{for model } r$$

and $X_{r,ij}$ has been defined earlier. Estimates of σ_{11}, σ_{22}, and σ_{12} can be obtained from replicate data (or otherwise) as described in Section 5.5. Then $\hat{\sigma}^{11}$, $\hat{\sigma}^{22}$, and $\hat{\sigma}^{12}$ can be used to replace the respective ensemble parameters.

Example 8.4-3 Experimental Designs for Parameter Estimation in Multiresponse Models

Constant pressure, ternary component, vapor-liquid mixtures, which are ideal in the liquid phase and follow the ideal gas law in the vapor phase, can be represented by the following equations:

$$y_A = \frac{\beta_{AC}x_A}{1 + (\beta_{AC} - 1)x_A + (\beta_{BC} - 1)x_B} \tag{a}$$

$$y_B = \frac{\beta_{BC}x_B}{1 + (\beta_{AC} - 1)x_A + (\beta_{BC} - 1)x_B} \tag{b}$$

† N. R. Draper and W. G. Hunter, *Biometrika* **53**, 525, 1966.

where

A, B, C = the three components, respectively
y = the mole fraction in the vapor phase
x = the mole fraction in the liquid phase
β_{ij} = relative volatilities, essentially empirical parameters, to be estimated

Although the assumption that x is a deterministic variable and only $Y = y + \epsilon$ is a random variable stretches the truth for any real series of experiments, for this example, we shall assume that Y is the random dependent variable.

Because we have two parameters to fit, we need to make a minimum of two initial runs. At what values of x_A and x_B should the first two experiments be carried out? Isobaric ternary component experiments for the acetone (A)–benzene (B)–carbon tetrachloride (C) system were planned at 760 mm of Hg pressure. Mixtures of A, B, and C could be prepared at selected values of x_i as measured by refractive index, and Y_i was determined by gas chromatographic analysis.

A study of the literature indicated that although the parameters in Equations (a) and (b) had not been specifically reported, values of Y versus x had been reported† from which initial estimates of the β's could be computed from two Y versus x measurements. For example, at a temperature of 62.4°C and a pressure of 1 atm,

$$x_A = 0.389 \qquad Y_A = 0.572$$
$$x_B = 0.332 \qquad Y_B = 0.200$$
$$x_C = 0.279 \qquad Y_C = 0.228$$

Only two pairs of Y versus x measurements are independent because, by definition, $\sum x_i = 1$ and $\sum Y_i = 1$. Then

$$0.572[1 + (b_{AC} - 1)0.389 + (b_{BC} - 1)0.332] = b_{AC}(0.389)$$
$$0.200[1 + (b_{AC} - 1)0.389 + (b_{BC} - 1)0.332] = b_{BC}(0.332)$$

and

$$b_{BC}^{(0)} = 0.74$$
$$b_{AC}^{(0)} = 1.21$$

for the initial estimates.

The three additional data points chosen and initial estimates of β_{ij} calculated were as follows:

Point Number	x_A	x_B	x_C	Y_A
1	0.234	0.268	0.498	0.441
2	0.288	0.476	0.236	0.480
3	0.624	0.206	0.170	0.723

Point Number	Y_B	Y_C	b_{AC}	b_{BC}
1	0.169	0.390	2.41	0.80
2	0.310	0.210	1.88	0.73
3	0.121	0.156	1.26	0.64

† B. V. Subbarao and C. V. Rao, *J. Chem. Eng. Data* **11**, 158, 1966.

It is apparent that the b's change significantly with composition. Thus, the proposed model may not prove satisfactory over a wide range of compositions. As initial guesses for the parameter, it was decided to use

$$b_{AC}^{(0)} = 1.85$$
$$b_{BC}^{(0)} = 0.72$$

Equation 8.4-11 in this instance becomes

$$\Delta = \left| \sum_{r=1}^{2} \sum_{s=1}^{2} \sigma^{rs} \mathbf{X}_r^T \mathbf{X}_s + \mathbf{\Omega}^{-1} \right| \tag{c}$$

and Δ was maximized to obtain the first two experimental points: (x_{1A}, x_{1B}) and (x_{2A}, x_{2B}). The two matrices $\mathbf{X}_r$ are

$$\mathbf{X}_1 = \begin{bmatrix} X_{1,11} & X_{1,12} \\ X_{1,21} & X_{1,22} \end{bmatrix} \qquad \mathbf{X}_2 = \begin{bmatrix} X_{2,11} & X_{2,12} \\ X_{2,21} & X_{2,22} \end{bmatrix}$$

Specifically,

$$X_{1,i1} = \frac{[1 + (\beta_{AC} - 1)x_{iA} + (\beta_{BC} - 1)x_{iB}]x_{iA} - \beta_{AC}x_{iA}^2}{[1 - (\beta_{AC} - 1)x_{iA} + (\beta_{BC} - 1)x_{iB}]^2}$$

$$X_{1,i2} = \frac{-\beta_{AC}x_{iA}x_{iB}}{[1 + (\beta_{AC} - 1)x_{iA} + (\beta_{BC} - 1)x_{iB}]^2}$$

$$X_{2,i1} = \frac{-\beta_{BC}x_{iB}x_{iA}}{[1 + (\beta_{AC} - 1)x_{iA} + (\beta_{BC} - 1)x_{iB}]^2}$$

$$X_{2,12} = \frac{[1 + (\beta_{AC} - 1)x_{iA} + (\beta_{BC} - 1)x_{iB}]x_{iB} - \beta_{BC}x_{iB}^2}{[1 + (\beta_{AC} - 1)x_{iA} + (\beta_{BC} - 1)x_{iB}]^2}$$

The elements of $\mathbf{\Omega}$ are

$$\mathbf{\Omega} = \begin{bmatrix} \omega_{11} & \omega_{12} \\ \omega_{21} & \omega_{22} \end{bmatrix}$$

and when $\omega_{12} = \omega_{21} \rightarrow 0$, we can write

$$\mathbf{\Omega}^{-1} = \begin{bmatrix} \frac{1}{\omega_{11}} & 0 \\ 0 & \frac{1}{\omega_{22}} \end{bmatrix}$$

Inasmuch as no estimates of the elements of $\mathbf{\Gamma}$ were available nor of the elements of $\mathbf{\Omega}$, a number of different assumptions were made concerning the values of those elements, and Δ in Equation (c) was maximized by the flexible simplex method of Section 6.2 with the results listed in Table E8.4-3. The relation

$$\rho_{12} = \frac{\sigma_{12}}{(\sigma_1^2\sigma_2^2)^{1/2}}$$

was used to calculate σ_{12}.

With a few exceptions, practically all the pairs of experimental points fall at zero mole fraction of component C, an unexpected result. Clearly defined points emerge repeatedly, such as $x_{1A} = 0.280$ and $x_{1B} = 0.720$, no matter what the assumed elements of $\mathbf{\Gamma}$ and $\mathbf{\Omega}$ are. Recalling that $\omega_{11} = \omega_{22} = \infty$ and $\omega_{12} = \omega_{21} = 0$ correspond to the case

TABLE E8.4-3 MAXIMIZATION OF Δ FOR VARIOUS ASSUMED VALUES OF THE ELEMENTS OF $\boldsymbol{\Gamma}$ AND $\boldsymbol{\Omega}$ TO OBTAIN THE INITIAL EXPERIMENTAL DESIGN

(a) $\omega_{11} = \infty, \omega_{22} = \infty, \omega_{12} = \omega_{21} = 0$

ρ_{12}	σ_1^2	σ_2^2	x_{1A}	x_{1B}	x_{2A}	x_{2B}	Objective Function, Δ
0	1	1	0	0.573	0.280	0.720	4.40×10^{-3}
0	0.1	1	0.280	0.720	0	0.580	2.42×10^{-3}
0	1	0.1	0.351	0	0.280	0.720	2.42×10^{-3}
0	10^3	10^{-3}	0.351	0	0.281	0.719	$2.20 \times 10^{+3}$
0	10^{-3}	10^3	0	0.582	0.280	0.720	$2.20 \times 10^{+3}$
0.5	1	1	0	0.582	0.269	0.428	2.11×10^{-3}
0.5	10^3	10^{-3}	0.351	0	0.280	0.720	$2.20 \times 10^{+3}$
0.5	10^{-3}	10^3	0	0.585	0.281	0.719	$2.20 \times 10^{+3}$

(b) $\omega_{11} = 1, \omega_{22} = 1, \omega_{12} = \omega_{21} = 0$

ρ_{12}	σ_1^2	σ_2^2	x_{1A}	x_{1B}	x_{2A}	x_{2B}	Objective Function, Δ
0	1	1	0.280	0.720	0.280	7.720	1.555
0	0.1	1	0.280	0.720	0.280	0.720	1.305
0	1	0.1	0.280	0.720	0.280	0.720	1.305
0	10^3	10^{-3}	0.279	0.721	0.347	0.001	2.36×10^3
0	10^{-3}	10^3	0.282	0.718	0	0.593	2.46×10^3
0.5	1	1	0.281	0.719	0.284	0.716	1.278
0.5	10^3	10^{-3}	0.291	0.709	0.360	0.005	2.33×10^3
0.5	10^{-3}	10^3	0.282	0.718	0	0.593	2.46×10^3

(c) $\omega_{11} = 10^2, \omega_{22} = 10^{-2}, \omega_{12} = \omega_{21} = 0$

ρ_{12}	σ_1^2	σ_2^2	x_{1A}	x_{1B}	x_{2A}	x_{2B}	Objective Function, Δ
0	1	1	0.274	0.726	0.277	0.723	8.307
0	0.1	1	0.280	0.720	0.280	0.720	5.020
0	1	0.1	0.279	0.721	0.279	0.721	5.020
0	10^3	10^{-3}	0.341	0	0.277	0.723	5.85×10^3
0	10^{-3}	10^3	0.280	0.720	0.280	0.720	3.66×10^3
0.5	1	1	0.351	0	0.351	0	4.652
0.5	10^3	10^{-3}	0.351	0	0.280	0.720	5.85×10^3
0.5	10^{-3}	10^3	0.280	0.720	0.280	0.720	3.65×10^3

(d) $\omega_{11} = 10^{-2}, \omega_{22} = 10^2, \omega_{12} = \omega_{21} = 0$

ρ_{12}	σ_1^2	σ_2^2	x_{1A}	x_{1B}	x_{2A}	x_{2B}	Objective Function, Δ
0	1	1	0.280	0.720	0.280	0.720	49.23
0	0.1	1	0.279	0.721	0.281	0.719	27.52
0	1	0.1	0.280	0.720	0.280	0.720	27.52
0	10^3	10^{-3}	0.280	0.720	0.280	0.720	2.41×10^4
0	10^{-3}	10^3	0	0.577	0.280	0.720	2.63×10^4
0.5	1	1	0.280	0.720	0.280	0.720	25.11
0.5	10^3	10^{-3}	0.280	0.720	0.280	0.720	2.41×10^4
0.5	10^{-3}	10^3	0	0.581	0.281	0.719	2.63×10^4

in which nothing initially is known about the precision of the β's but it is known that they are uncorrelated, we can observe the effect of the assumed correlation between models, or lack of it, among the observations on one experiment. If there is no correlation, the effect of high precision ($\sigma^2 = 10^{-3}$) versus low precision ($\sigma^2 = 10^3$) can be seen in the fourth and fifth rows of part (a) of Table E8.4-3. The direction of experimentation is to gain information concerning the parameters with the least precision. Other prior information can be examined analogously in others parts of the table.

8.5 SEQUENTIAL DESIGNS TO DISCRIMINATE AMONG MODELS

Often an experimenter sets up more than one model to represent a process, and it is natural to inquire what would be the best experimental design to use from the viewpoint of discriminating among the proposed models. In addition, what series of experiments will provide the most information with the least effort so that a quantitative decision can be reached as to the better model(s)? It is clear that only in certain critical areas of data collection can a distinction between two (or more) models be established. As an example, Figure 8.5-1 illustrates the hypothetical responses for two multicoefficient models that are a function of a single independent variable. Any data collected in the lightly shaded region will not provide useful information to enable the analyst to discriminate between models A and B; the data taken in the unshaded region are what matter.

We shall look at discrimination between two models first before proceeding to discrimination between several models. Suppose an investigator has carried out n experiments and has in mind two possible models that might represent the process. After the n experimental runs, he is unable to conclude which is the best model; he would like on the $(n + 1)$st run to collect observations that will help him to discriminate as much as possible among the proposed models. How should he select the new set of experimental conditions, i.e., the values of the independent variables, to achieve this objective?

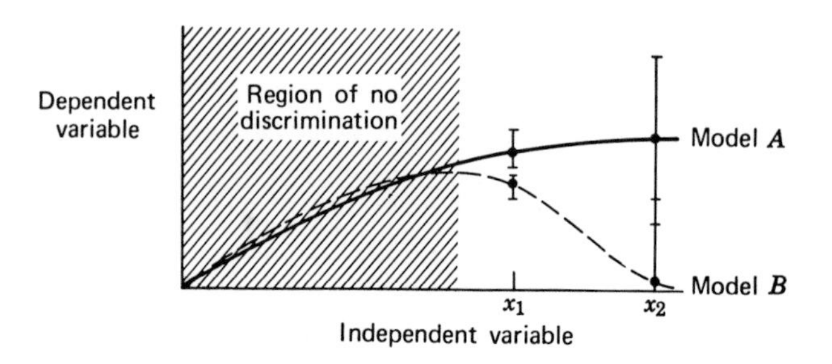

FIGURE 8.5-1 Data collection for discrimination among models.

8.5-1 Model Discrimination Using One Response

Hunter and Reiner† proposed that the $(n + 1)$st experimental point be selected by using the maximum likelihood parameter estimates obtained from the n completed runs so as to strain the incorrect model the most in its attempt to jointly explain the previous data and the new observation. Let the two models under consideration be y_1 and y_2. Then if y_1 is correct, $\mathbf{x}^{(n+1)}$ is chosen to max S_2, and if y_2 is correct, $\mathbf{x}^{(n+1)}$ is chosen to max S_1, where:

$\mathbf{x}^{(n+1)}$ = a matrix composed of elements of the independent variables on the $(n + 1)$st run

$$S_r = \sum_r (Y^{(n)} - \hat{Y}_r^{(n)})^2, \quad r = 1, 2$$

$Y^{(n)}$ = value of the nth observed response

$\hat{Y}_r^{(n)}$ = predicted response for model r, $r = 1, 2$, after n data points have been collected

(Note that the index for the experimental run number is now a superscript in parentheses in order subsequently to avoid excessive subscripts.) The sequential design is essentially insensitive to which model is correct; hence a practical solution was proposed. That is, choose $\mathbf{x}^{(n+1)}$ which maximizes

$$S = [\hat{Y}_1^{(n+1)}(\mathbf{b}^{(n)}) - \hat{Y}_2^{(n+1)}(\mathbf{b}^{(n)})]^2 \tag{8.5-1}$$

where $\hat{Y}_r^{(n+1)}(\mathbf{b}^{(n)})$ is the predicted value of the response for the $(n + 1)$st observation in the rth model, using estimates of the parameters obtained from the previous n runs.

Kullback‡ suggested that a *discriminant function*, e.g., a function used in deciding which of two states of nature is true, could be used to distinguish between two models. Let Y be a random variable that is distributed with a probability density $p_1(y)$ when hypothesis H_1 is true and distributed with the probability density $p_2(y)$ when H_2 is true. These hypotheses will be that Model 1 or Model 2 is the correct model, respectively. Then in some sense the quantity

$$\ln \frac{p_1(y)}{p_2(y)}$$

can be said to be a measure of the odds in favor of choosing H_1 over H_2 or, from the information theory viewpoint, of the information in favor of hypothesis H_1 as opposed to hypothesis H_2.

The "weight of evidence" or expected information in favor of chosing H_1 is defined as

$$I(1:2) = \int_{-\infty}^{\infty} p_1(y) \ln \frac{p_1(y)}{p_2(y)}\, dy \tag{8.5-2a}$$

† W. G. Hunter and A. M. Reiner, *Technometrics* **7**, 307, 1965.

‡ S. Kullback, *Information Theory and Statistics*, John Wiley, New York, 1959.

Similarly, the expected information for discrimination in favor of choosing H_2 is

$$I(2:1) = \int_{-\infty}^{\infty} p_2(y) \ln \frac{p_2(y)}{p_1(y)}\, dy \tag{8.5-2b}$$

Kullback suggested that

$$J(1, 2) = I(1:2) + I(2:1)$$
$$= \int_{-\infty}^{\infty} [p_1(y) - p_2(y)] \ln \frac{p_1(y)}{p_2(y)}\, dy \tag{8.5-2c}$$

be maximized to distinguish between two states of nature.

From the linearized analysis in Chapters 5 and 6, we can further assume that the $(n + 1)$st observation is normally distributed about the expected value for model r, $\mathscr{E}\{Y_r^{(n+1)}\} = y_r^{(n+1)}$, with a variance of σ_Y^2. Furthermore, $y_r^{(n+1)}$ is distributed in a local (linearized) region about its predicted value, $\hat{Y}_r^{(n+1)}$, with a variance of σ_r^2. Consequently, $Y^{(n+1)}$ is distributed about $\hat{Y}_r^{(n+1)}$ with a variance of $\sigma_Y^2 + \sigma_r^2$. We conclude that the probability density of $Y^{(n+1)}$ for the rth model is

$$p_r(y^{(n+1)}) = \frac{1}{\sqrt{2\pi(\sigma_Y^2 + \sigma_r^2)}} \exp\left[-\frac{1}{2}\frac{(Y^{(n+1)} - \hat{Y}^{(n+1)})^2}{\sigma_Y^2 + \sigma_r^2}\right]$$
$$r = 1, 2 \tag{8.5-3}$$

The quantities y_r and σ_r^2 represent the mean and variance of Y when $H = H_1$ or alternately $H = H_2$. Kullback showed that when Equation 8.5-3 is substituted into Equations 8.5-2, it follows after completing the integrations that

$$I(1:2) = \frac{1}{2}\ln\frac{\sigma_Y^2 + \sigma_2^2}{\sigma_Y^2 + \sigma_1^2} + \frac{1}{2}\frac{\sigma_Y^2 + \sigma_1^2}{\sigma_Y^2 + \sigma_2^2} - \frac{1}{2} + \frac{1}{2}\frac{(\hat{Y}_1^{(n+1)} - \hat{Y}_2^{(n+1)})^2}{\sigma_Y^2 + \sigma_2^2}$$

$$I(2:1) = \frac{1}{2}\ln\frac{\sigma_Y^2 + \sigma_1^2}{\sigma_Y^2 + \sigma_2^2} + \frac{1}{2}\frac{\sigma_Y^2 + \sigma_2^2}{\sigma_Y^2 + \sigma_1^2} - \frac{1}{2} + \frac{1}{2}\frac{(\hat{Y}_1^{(n+1)} - \hat{Y}_2^{(n+1)})^2}{\sigma_Y^2 + \sigma_1^2}$$

$$J(1, 2) = \tfrac{1}{2}(\sigma_1^2 - \sigma_2^2)\left(\frac{1}{\sigma_Y^2 + \sigma_2^2} - \frac{1}{\sigma_Y^2 + \sigma_1^2}\right) + \frac{1}{2}\left(\frac{1}{\sigma_Y^2 + \sigma_2^2} + \frac{1}{\sigma_Y^2 + \sigma_1^2}\right)(\hat{Y}_1^{(n+1)} - \hat{Y}_2^{(n+1)})^2$$

Thus, after n observations become available, the appropriate I or J could be maximized with respect to the independent variables in the model to obtain the next $\mathbf{x}$ vector for the new experimental run.

Box and Hill§ described an improved version of Kullback's discriminant functions in which the prior probabilities are included. We do not have the space to

§ G. E. P. Box and W. J. Hill, *Technometrics* **9**, 57, 1967.

go through the development of Hill† but instead employ a heuristic argument to obtain the same result. To make full use of the prior information available about the validity of each model, in addition to computing the estimated coefficients used in $\hat{Y}_r^{(n+1)}$, it would seem reasonable to weight $I(1:2)$ and $I(2:1)$ by the respective prior probabilities of Model 1 or Model 2 being the correct model, $P_1^{(n)}$ and $P_2^{(n)}$, respectively.

Let us extend the previous problem and assume that several competing models exist, among which the $(n + 1)$st experiment is to discriminate. A scalar discriminant function can be formed using the matrix of relative likelihoods and a vector of prior probabilities as follows

$$K_v = [P_1^{(n)} \quad P_2^{(n)} \quad \cdots \quad P_v^{(n)}] \begin{bmatrix} I(1:1) & I(1:2) & \cdots & I(1:v) \\ I(2:1) & I(2:2) & \cdots & I(2:v) \\ \vdots & \vdots & & \vdots \\ I(v:1) & I(v:2) & \cdots & I(v:v) \end{bmatrix} \begin{bmatrix} P_1^{(n)} \\ P_2^{(n)} \\ \vdots \\ P_v^{(n)} \end{bmatrix} \tag{8.5-4}$$

Each element on the main diagonal of the $I(r:s)$ matrix is zero so the discriminant function becomes

$$\begin{aligned} K_v &= P_1P_2J(1, 2) + P_1P_3J(1, 3) + \cdots + P_1P_vJ(1, v) \\ &\quad + P_2P_vJ(2, v) + \cdots + P_{v-1}P_vJ(v-1, v) \\ &= \frac{1}{2}\sum_{r=1}^{v}\sum_{s=r+1}^{v} P_r^{(n)}P_s^{(n)} \left[\frac{(\sigma_r^2 - \sigma_s^2)^2}{(\sigma_Y^2 + \sigma_r^2)(\sigma_Y^2 + \sigma_s^2)} \right. \\ &\quad \left. + (\hat{Y}_r^{(n+1)} - \hat{Y}_s^{(n+1)})^2 \left(\frac{1}{\sigma_Y^2 + \sigma_r^2} + \frac{1}{\sigma_Y^2 + \sigma_s^2}\right)\right] \end{aligned} \tag{8.5-5}$$

(Note that $J(r, r) = 0$.) Equation 8.5-5 is the same equation as that developed by Box and Hill.

One way to obtain the posterior probability that model r is correct after taking n observations is to apply successively for each model Bayes' theorem in the following form:

$$P_r^{(n)} = \frac{P_r^{(n-1)}p_r(y(n))}{\sum_{r=1}^{v} P_r^{(n-1)}p_r(y(n))} \tag{8.5-6}$$

where $P_r^{(n-1)}$ is the prior probability associated with the rth model. The initial probabilities $P_r^{(0)}$ can be set equal to $1/v$ if not known.

In general, neither σ_Y^2 nor σ_r^2, $r = 1, \ldots, v$, will be known; hence the values of these variances must be estimated. To obtain the variance of $\hat{Y}_r$ for a model, we can use the intermediate calculations developed in the least squares estimates of the coefficients $\boldsymbol{\beta}$. If $\hat{Y}_r$ is linearized by expansion in a truncated Taylor series, the variance of $\hat{Y}_r$, σ_r^2, is given by Equation 6.4-4 and the elements of matrix $\mathbf{C}$ are given by Equation 6.4-1. But the estimate of σ_Y^2 given by Equation 6.4-2 cannot be used for each model inasmuch as Equation 6.4-2 is based on the concept that the model is the correct one. Hence, σ_Y^2 should be taken as σ_e^2 and the latter estimated from replicate experiments (as s_e^2).

† W. J. Hill, Ph.D. Dissertation, Univ. of Wisconsin, Madison, 1966.

The sequential procedure to discriminate among models can be summarized as follows:

1. Based on an experimental design selected in some arbitrary or suboptimal way, collect n data points.
2. Estimate the parameters in the v models by linear or nonlinear regression; estimate σ_Y^2 and calculate each σ_r^2, using Equation 6.4-4.
3. Calculate the prior probabilities for the $(n + 1)$st run which are equal to the posterior probabilities for the nth run by using Equations 8.5-6 and 8.5-3 with n substituted for $(n + 1)$ in the latter. The initial P's can all be equal to $1/v$ if no better choice is available.
4. Select the vector of experimental conditions for the $(n + 1)$st run (the vector $\mathbf{x}^{(n+1)}$) by maximizing K_v using a numerical optimization routine.
5. Run an experiment at $\mathbf{x}^{(n+1)}$ and repeat starting with step 2. Figure 8.5-2 illustrates figuratively the probability of each of several competitive models representing the experimental data after a few cycles of the suggested procedure.

The sequential procedure continues until one (or more) $P_r^{(n)}$ reaches a value that causes acceptance of the model by some criterion. Or the experimenter can just observe the trend of the changes in the P_r as the number of experiments increases, drop models with low values of P_r, and add models, if he wishes, terminating the experiments when he feels satisfied with the discrimination actually achieved.

Because in practice the models will usually be nonlinear, the assumed probability distribution underlying the development of Equation 8.5-3 will be only approximately correct. Also, because the variances σ_Y^2 and σ_r^2 must be estimated from experimental data, as must be the coefficients used to predict $\hat{Y}_r$, the vector $\mathbf{x}^{(n+1)}$ which maximizes K_v may be only approximate. However, the

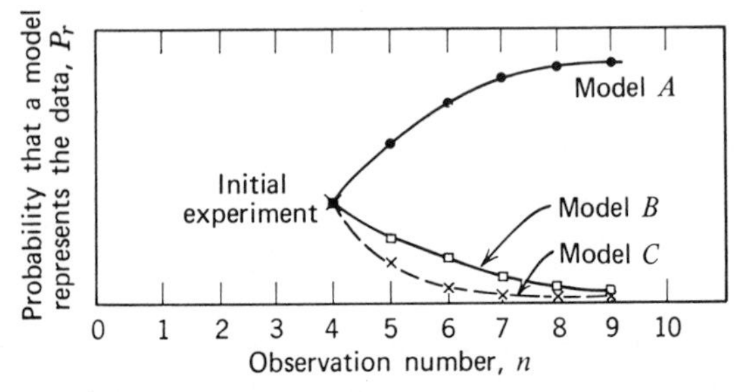

FIGURE 8.5-2 Model discrimination by sequential designs.

sequential nature of the design procedure overcomes these handicaps in analysis.

One desirable feature of the Box and Hill discrimination technique is that it can discriminate between a model and its augmented relative. For example, two models might be

$$\text{I:}\quad y_1 = \frac{\beta_1 x_1^2 + \beta_2 x_2 x_3}{1 + \beta_3 x_1 + \beta_4 x_2 + \beta_5 x_3 + \beta_6 x_4}$$

$$\text{II:}\quad y_2 = \frac{\beta_1 x_1^2 + \beta_2 x_2 x_3}{1 + \beta_3 x_1 + \beta_4 x_2 + \beta_5 x_3}$$

If $\beta_6 \equiv 0$ in Model I, the models are identical. It appears at first glance that model discrimination is impossible because if Model II is correct, Model I also appears to be correct. However, the model discrimination procedure described above works effectively because the variances of the dependent variables, σ_1^2 and σ_2^2, are a function of the number of coefficients in the model. For an equal sum of squares, the model with the smallest number of coefficients will prove to be best.

We now illustrate the detailed calculations of the Box and Hill method by an example.

Example 8.5-1 Sequential Discrimination Between Models

Cattle feedlots rank high on the list of sources of water pollution. To model the runoff from such lots for evaluating pollution control measures, three models were developed of increasing complexity:

1. Stirred tank model:

$$c = \beta_0 \exp(-\beta_1 x) + \epsilon$$

2. Stirred tank model with injection:

$$c = \beta_2 \exp(-\beta_3 x) + \beta_4$$

3. Two stirred tanks in series with injection in first tank:

$$c = \beta_9 \exp(-\beta_5 x) + \beta_6 \exp(-\beta_7 x) + \beta_8$$

The notations are

C = concentration of COD, chemical oxygen demand $= c + \epsilon$
x = water quantity, mm/hr

The ranges of the dependent and independent variables were: C(3000 to 10,000) and x: (0 to 12). It was necessary to estimate the following parameters:

Model 1: β_0 and β_1.
Model 2: β_2, β_3, and β_4.
Model 3: β_5, β_6, β_7, β_8, and β_9.

At least five experiments were required to estimate the parameters in Model 3; hence six experiments plus two replicates were carried out by sprinkling, reasonably uniformly, a small concrete-surfaced feedlot to obtain the following results:

C, mg/liter	x, mm/hr
8140	0.1
7430	1.0
6310	2.0
5510	4.0
5390	5.0
5250	6.0
6140	2.0
6490	2.0

The replicate runs for essentially the same conditions provided the information to estimate the residual variance.

It was not clear from any of the tests in Chapter 7 as to which model represented the data the best. Therefore, it was decided to carry out a sequence of experiments for model discrimination. From the initial eight experiments the estimated coefficients were

$b_0 = 7919$	$b_5 = -2.490 \times 10^{-2}$
$b_1 = 0.07937$	$b_6 = 8.649 \times 10^4$
$b_2 = 3431$	$b_7 = 3.730 \times 10^{-2}$
$b_3 = 0.3630$	$b_8 = -1.908 \times 10^5$
$b_4 = 4797$	$b_9 = 1.102 \times 10^5$

The estimated parameters b_5 through b_9 were for the transformed model

$$c = \beta_9 \exp[-\beta_5(x - \bar{x})] + \beta_6 \exp[-\beta_7(x - \bar{x})] + \beta_8$$

where $\bar{x} = 3.000$. The transformation improved the speed of convergence of the nonlinear estimation routine (Marquardt's method) substantially over that obtained for the untransformed model, as mentioned in Chapter 6.

The sums of the squares of the residuals for each model were:

	Model	$\phi_{\min}$	ν = d.f.
	1	4.416×10^5	4
	2	1.039×10^5	3
Transformed	3	7.217×10^4	1

and from the replicate data $s_e^2 = 3.11 \times 10^4$.

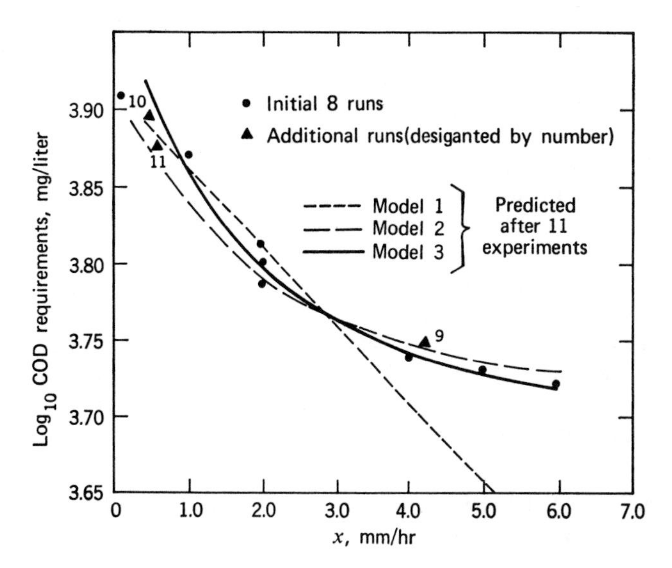

FIGURE E8.5-1a COD concentration in the runoff for a concrete lot.

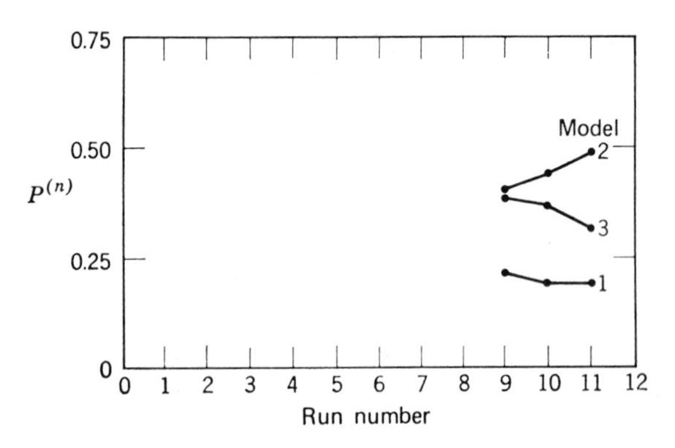

FIGURE E8.5-1b

Two different assumptions were made concerning the $P_r^{(8)}$, one that $P_1^{(8)} = P_2^{(8)} = P_3^{(8)} = \frac{1}{3}$, and the other that $P_r^{(8)}$ was inversely proportional to $\phi_{\min}/\nu$, or

$$P_1^{(8)} = 0.17 \qquad P_2^{(8)} = 0.53 \qquad P_3^{(8)} = 0.30$$

Introduction of these probabilities into Equation 8.5-5 for $P^{(n)}$, together with s_e^2 substituted for σ_Y^2 and values of $\hat{Y}^{(9)}$ calculated using the coefficients estimated after the initial eight runs, yielded the following values of $x^{(9)}$ on maximization of K_v:

	x	P_1	P_2	P_3
Assumed		0.333	0.333	0.333
$x^{(9)}$	4.2			
Posterior		0.22	0.40	0.38
Assumed		0.17	0.53	0.30
$x^{(9)}$	4.2			
Posterior		0.20	0.42	0.38

The posterior values of P were computed from Equations 8.5-6 and 8.5-3. Clearly, the initial values of P have little influence in this problem on the value of x for the ninth run.

Because of the high experimental error at low flow rates, caused mainly by nonuniform coverage of the feedlot by waste material, no experimental runs were carried out at an x of less than 0.1 mm/hr. At the end of 11 runs the experiments had to be terminated for economic reasons, but it was clearly shown that Model 1 was considerably poorer than Models 2 and 3, and that Model 3 was preferred over Model 2 because its $P^{(11)}$ was higher, although not decisively so, and it was a simpler model. Both Models 2 and 3 represented the data satisfactorily.

Figure E8.5-1a illustrates the experimental data after 11 runs, 3 of which were replicate runs, together with the predicted values of the COD based on the coefficients estimated at the end of the eleventh run. Figure E8.5-1b shows how the probabilities for the models changed as the sequential experimentation was carried out.

8.5-2 Model Discrimination in the Case of Multiple Responses

In this section we consider the problem of discriminating among process models that include more than one response. Such models occur quite naturally in multicomponent processes such as those involving physical equilibria or chemical reactions. For example, the reaction

$$A \xrightarrow{k_1} B \xrightarrow{k_2} C$$

might be best represented by one of two models:

Model I	Model II
$r_A = -k_1 c_A$	$r_A = -k_1' c_A^2$
$r_B = k_1 c_A - k_2 c_B$	$r_B = k_1' c_A^2 - k_2' c_B^2$
$r_C = k_2 c_B$	$r_C = k_2' c_B^2$

(Only two of the equations in each model are independent inasmuch as the third equation can be formed by a linear combination of the other two.)

A criterion for discrimination among multiresponse models can be obtained in exactly the same fashion as indicated in Section 8.5-1 for single response models. Although the concepts applied are no different than in Section 8.5-1, the bookkeeping for the variables, parameters, and equations is more complex. Since each model can consist of several equations, a double index is needed to designate the specific equation in a given model. Let y_{rj} stand for the rth model, $r = 1, \ldots, v$, and the jth equation, $j = 1, \ldots, u$, in the model. Corresponding to the jth equation is the jth observation. In vector notation, then:

Observations	Expected Values	Models
$\mathbf{Y} = \begin{bmatrix} Y_1 \\ \vdots \\ Y_j \\ \vdots \\ Y_u \end{bmatrix}$	$\mathbf{y}_r = \begin{bmatrix} y_{r1} \\ \vdots \\ y_{rj} \\ \vdots \\ y_{ru} \end{bmatrix}$	$\mathbf{y} = \begin{bmatrix} \mathbf{y}_1 \\ \vdots \\ \mathbf{y}_r \\ \vdots \\ \mathbf{y}_v \end{bmatrix}$

The additive error for the jth response in the nth run is

$$Y_j^{(n)} = y_{rj} + \epsilon_{rj}^{(n)} \qquad j = 1, \ldots, u \qquad (8.5\text{-}7)$$

We assume that the observations $\mathbf{Y}$ are normally distributed about their expected values, $\mathbf{y}_r$, for a given model with a covariance matrix $\mathbf{\Sigma_Y}$, where

$$\mathbf{\Sigma_Y} = \begin{bmatrix} \sigma_1^2 & \sigma_{12} & \cdots & \sigma_{1u} \\ \sigma_{12} & \sigma_2^2 & \cdots & \sigma_{2u} \\ \vdots & \vdots & & \vdots \\ \sigma_{1u} & \sigma_{2u} & \cdots & \sigma_u^2 \end{bmatrix}$$

Then the probability density for $\mathbf{Y}^{(n+1)}$, given $\mathbf{y}_r$ and $\mathbf{\Sigma_Y}$, is

$$p(\mathbf{y}^{(n+1)} \mid \mathbf{y}_r, \mathbf{\Sigma_Y}) = \frac{|\mathbf{\Sigma_Y}|^{-1/2}}{(2\pi)^{u/2}} \exp\left[-\tfrac{1}{2}(\mathbf{Y}^{(n+1)} - \mathbf{y}_r)^T \mathbf{\Sigma_Y}^{-1} (\hat{Y}^{(n+1)} - \mathbf{y}_r)\right] \qquad (8.5\text{-}8)$$

As in Section 8.5-1, we assume that each model can be locally linearized in parameter space about the estimated parameters $\hat{\boldsymbol{\beta}}$:

$$y_{rj}^{(n)} = y_{rj}(\hat{\boldsymbol{\beta}}_r^{(n)}, \mathbf{x}_{rj}^{(n)}) + \sum_{k=1}^{m} (\beta_{rk} - \hat{\beta}_{rk}^{(n)}) \left[\frac{\partial y_{rj}(\hat{\boldsymbol{\beta}}_r^{(n)}, \mathbf{x}_{rj}^{(n)})}{\partial \beta_{rk}}\right]_{\boldsymbol{\beta}_r = \hat{\boldsymbol{\beta}}_r^{(n)}}$$

To simplify the notation in the subsequent development, we let

$$X_{rj,k}^{(n)} = \left[\frac{\partial y_{rj}(\hat{\boldsymbol{\beta}}_r^{(n)}, \mathbf{x}_{rj}^{(n)})}{\partial \beta_{rk}}\right]_{\boldsymbol{\beta}_r = \hat{\boldsymbol{\beta}}_r^{(n)}}$$

and the matrix $\mathbf{X}_{rj}$ be

$$\mathbf{X}_{rj} = \begin{bmatrix} X_{rj,1}^{(1)} & X_{rj,2}^{(1)} & \cdots & X_{rj,m}^{(1)} \\ \vdots & \vdots & & \vdots \\ X_{rj,1}^{(n)} & X_{rj,2}^{(n)} & \cdots & X_{rj,m}^{(n)} \end{bmatrix}$$

The posterior probability density function for $\boldsymbol{\beta}$ (which is a column vector of all the parameters) after n runs is

$$p(\boldsymbol{\beta} \mid \boldsymbol{\Sigma}_{\mathbf{Y}}, \mathbf{Y}) = \frac{|\mathbf{M}|^{1/2}}{(2\pi)^{m/2}} \exp\left[-\tfrac{1}{2}(\boldsymbol{\beta} - \hat{\boldsymbol{\beta}})^T \mathbf{M}(\boldsymbol{\beta} - \hat{\boldsymbol{\beta}})\right] \quad (8.5\text{-}9)$$

where $\mathbf{M} = \sum_{j=1}^{u} \sum_{l=1}^{u} \sigma^{jl} \mathbf{X}_{rj}^T \mathbf{X}_{rl}$ and σ^{jl} is an element from $\boldsymbol{\Sigma}_{\mathbf{Y}}^{-1}$. In other words, $(\boldsymbol{\beta} - \hat{\boldsymbol{\beta}})$ is normally distributed about $\mathbf{O}$ with a covariance matrix $\mathbf{M}^{-1}$. After n runs the matrix of partial derivatives evaluated at $\hat{\boldsymbol{\beta}}^{(n)}$, but using the $(n + 1)$st matrix of independent variables, is

$$\mathbf{X}_r^{(n+1)} = \begin{bmatrix} X_{r1,1}^{(n+1)} & X_{r1,2}^{(n+1)} & \cdots & X_{r1,m}^{(n+1)} \\ \vdots & \vdots & \vdots & \vdots \\ X_{ru,1}^{(n+1)} & X_{ru,2}^{(n+1)} & \cdots & X_{ru,m}^{(n+1)} \end{bmatrix}$$

The matrix $\mathbf{X}_r^{(n+1)}(\boldsymbol{\beta} - \hat{\boldsymbol{\beta}}^{(n)})$ is normally distributed about $\mathbf{O}$ with a covariance matrix of

$$\mathbf{W}_r^{(n+1)} = \mathbf{X}_r^{(n+1)} \mathbf{M}^{-1} (\mathbf{X}_r^{(n+1)})^T$$

Next, because of the linearization of the models, $\mathbf{y}_r$ is normally distributed about the predicted response, $\hat{\mathbf{Y}}_r^{(n+1)}$, with a covariance matrix $\mathbf{W}_r^{(n+1)}$ and a probability density

$$p(\mathbf{y}_r \mid \boldsymbol{\Sigma}_{\mathbf{Y}}) = \frac{|\mathbf{W}_r^{(n+1)}|^{-1/2}}{(2\pi)^{u/2}} \cdot \exp\left[-\tfrac{1}{2}(\mathbf{y}_r - \hat{\mathbf{Y}}_r^{(n+1)})^T (\mathbf{W}_r^{(n+1)})^{-1} (\mathbf{y}_r - \hat{\mathbf{Y}}_r^{(n+1)})\right] \quad (8.5\text{-}10)$$

Finally, after combining Equations 8.5-8 through 8.5-10 and after some extensive manipulations, the probability density function for $\hat{\mathbf{Y}}_r^{(n+1)}$, given $\boldsymbol{\Sigma}_{\mathbf{Y}}$, is

$$p(\mathbf{Y}_r^{(n+1)} \mid \boldsymbol{\Sigma}_{\mathbf{Y}}) = \frac{|\boldsymbol{\Sigma}_r^{(n+1)}|^{-1/2}}{(2\pi)^u} \exp\left[-\tfrac{1}{2}(\mathbf{Y}^{(n+1)} - \hat{\mathbf{Y}}_r^{(n+1)})^T \cdot (\boldsymbol{\Sigma}_r^{(n+1)})^{-1} (\mathbf{Y}^{(n+1)} - \hat{\mathbf{Y}}_r^{(n+1)})\right] \quad (8.5\text{-}11)$$

where $\boldsymbol{\Sigma}_r^{(n+1)} = \boldsymbol{\Sigma}_{\mathbf{Y}} + \mathbf{W}_r^{(n+1)}$. Equation 8.5-11 corresponds to Equation 8.5-3 of Section 8.5-1.

Introduction of Equation 8.5-11 into Equations 8.5-2 and use of the following two relations for the expected value of quadratic forms:

$$\mathscr{E}\{(\mathbf{Y}^{(n)} - \hat{\mathbf{Y}}_r^{(n)})^T (\boldsymbol{\Sigma}_r^{(n)})^{-1} (\mathbf{Y}^{(n)} - \hat{\mathbf{Y}}_r^{(n)})\} = [\mathscr{E}\{(\mathbf{Y}^{(n)} - \hat{\mathbf{Y}}_r^{(n)})^T\}](\boldsymbol{\Sigma}_r^{(n)})^{-1}[\mathscr{E}\{(\mathbf{Y}^{(n)} - \hat{\mathbf{Y}}_r^{(n)})\}] + \operatorname{trace} \boldsymbol{\Sigma}_r^{(n)} (\boldsymbol{\Sigma}_r^{(n)})^{-1} = \operatorname{trace} \mathbf{I}_u$$

where $\mathbf{I}_u$ is a $u \times u$ identity matrix, and

$$\mathscr{E}\{(\mathbf{Y}^{(n)} - \hat{\mathbf{Y}}_r^{(n)})^T (\boldsymbol{\Sigma}_s^{(n)})^{-1} (\mathbf{Y}^{(n)} - \hat{\mathbf{Y}}_r^{(n)})\} = \operatorname{trace} \boldsymbol{\Sigma}_r^{(n)} (\boldsymbol{\Sigma}_s^{(n)})^{-1}$$

lead to the general Kullback criterion

$$I^{(n+1)}(r:s) = \int_{-\infty}^{\infty} p_r(y) \ln \frac{p_r(y)}{p_s(y)} \, dy^{(n+1)} = \frac{1}{2}\left[\ln \frac{|\boldsymbol{\Sigma}_s^{(n+1)}|}{|\boldsymbol{\Sigma}_r^{(n+1)}|} - \operatorname{trace} \mathbf{I}_u + \operatorname{trace} \boldsymbol{\Sigma}_r^{(n+1)} (\boldsymbol{\Sigma}_s^{(n+1)})^{-1} + (\hat{\mathbf{Y}}_r^{(n+1)} - \hat{\mathbf{Y}}_s^{(n+1)})^T \cdot (\boldsymbol{\Sigma}_s^{(n+1)})^{-1} (\hat{\mathbf{Y}}_r^{(n+1)} - \hat{\mathbf{Y}}_s^{(n+1)})\right] \quad (8.5\text{-}12)$$

A similar expression is obtained for $I^{(n+1)}(s:r)$ by interchanging indexes. Introduction of the quantities given in Equation 8.5-12 into Equation 8.5-4 results in the multiresponse analog of Equation 8.5-5:

$$K_v = \tfrac{1}{2} \sum_{r=1}^{v} \sum_{s=r+1}^{v} P_r^{(n)} P_s^{(n)} \cdot \{\operatorname{trace}[\boldsymbol{\Sigma}_r^{(n+1)} (\boldsymbol{\Sigma}_s^{(n+1)})^{-1} + \boldsymbol{\Sigma}_s^{(n+1)} (\boldsymbol{\Sigma}_r^{(n+1)})^{-1} - 2\mathbf{I}_u] + (\hat{\mathbf{Y}}_r^{(n+1)} - \hat{\mathbf{Y}}_s^{(n+1)})^T \cdot [(\boldsymbol{\Sigma}_s^{(n+1)})^{-1} + (\boldsymbol{\Sigma}_r^{(n+1)})^{-1}](\hat{\mathbf{Y}}_r^{(n+1)} - \hat{\mathbf{Y}}_s^{(n+1)})\} \quad (8.5\text{-}13)$$

Equation 8.5-13 can be used as a design criterion for selecting the values of the independent variables on the $(n + 1)$st run after n runs have been completed.

The discrimination procedure is essentially the same as that presented in Section 8.5-1. An initial set of experimental runs is carried out, the parameters in each model are estimated, the posterior probabilities are calculated by Equation 8.5-6, the next experiment(s) is(are) designed, and the next run is carried out. These steps are repeated until the desired degree of discrimination is achieved. Because of the presence of the prior probabilities in the function for K_v, less emphasis is placed on the poorly fitting models and more emphasis on the better models. Thus, maximizing K_v leads to experimental designs at conditions at which the maximum discrimination takes place between the best models.

8.5-3 Sequential Designs for Simultaneous Discrimination and Parameter Estimation

Hunter and Wichern† simulated the discrimination among three chemical kinetic models, each having two responses:

$$\text{Model I:}\quad r_1 = \frac{k_1 K_A K_B p_A p_B}{(1 + K_A p_A + K_B p_B)^2}$$

$$r_2 = \frac{k_2 K_A K_B p_A p_B}{(1 + K_A p_A + K_B p_B)^2}$$

$$\text{Model II:}\quad r_1 = \frac{k_1 K_A K_B p_A p_B}{(1 + K_A p_A + K_B p_B)^2}$$

$$r_2 = \frac{k_2 K_A p_A p_B}{(1 + K_A p_A + K_B p_B)}$$

$$\text{Model III:}\quad r_1 = \frac{k_1 K_A p_A p_B}{(1 + K_A p_A)}$$

$$r_2 = \frac{k_2 K_A p_A p_B}{(1 + K_A p_A)^2}$$

where A and B are the components, p is the partial pressure, K is the absorption equilibrium constant, k is the rate constant, and r_1 and r_2 are the responses (reaction rates) to be observed. For the simulation, Model III was chosen as being the "true" model, with $k_1 = 0.0005$, $k_2 = 0.16$, and $K_A = 15$. Simulated observations, Y_1 and Y_2, of the reaction rates were prepared by adding to the deterministic r_1 and r_2 a normal error with zero mean and standard deviation of $\sigma_1 = 3.162 \times 10^{-6}$ and $\sigma_2 = 1.0 \times 10^{-4}$, respectively.

It was first assumed that only a single response, Y_2, would be observed. The initial values of P were $P_{\text{I}}^{(0)} = P_{\text{II}}^{(0)} = P_{\text{III}}^{(0)} = \frac{1}{3}$, and the first experiment was a 2^2 factorial design. Figure E8.5-2 illustrates the change in the posterior probabilities as successive simulated experiments were carried out by the procedure discussed in Section 8.5-2. Progress in discrimination between Models I and III was slow, although (the correct) Model III was emerging as the preferred one.

If both responses Y_1 and Y_2 were observed and used to plan the experiments, the posterior probabilities were as listed in Table E8.5-2. After the four initial experiments in a 2^2 factorial design, Model III was fairly well identified as

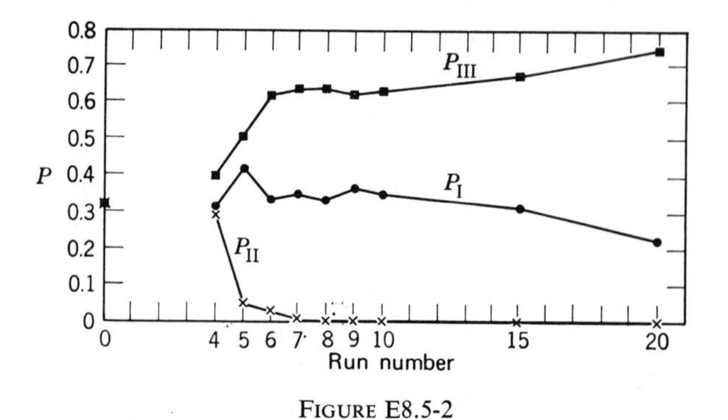

FIGURE E8.5-2

TABLE E8.5-2 POSTERIOR PROBABILITIES WITH BOTH Y_1 AND Y_2 MEASURED

Run	P_{I}	P_{II}	P_{III}
0	0.333	0.333	0.333
4	0.039	0.753×10^{-6}	0.961
5	0.001	0.000	0.999

being the best model. Clearly, measuring both responses, if feasible, in an experiment is more effective in discriminating among tentative models.

8.5-3 Sequential Designs for Simultaneous Discrimination and Parameter Estimation

So far we have discussed designs for parameter estimation separately from those for model discrimination. But suppose, as is often the case, the experimenter wants to obtain designs which will fulfill both objectives simultaneously. What should he do then? Rather than discriminate first and resolve the coefficients subsequently, Hill, Hunter, and Wichern‡ suggested that a weighted criterion be used which involves the criterion K_v given by Equation 8.5-5, or 8.5-13 for multiple responses, and the criterion Δ given by Equation 8.4-10, or 8.4-14 for multiple responses.

Specifically, they proposed that the following criterion be used:

$$C = w_1 D + w_2 E \tag{8.5-14}$$

where

$$D = K_v / K_{v,\max}$$

$$E = \sum_{r=1}^{v} P_r^{(n)} \Delta_r / \Delta_{r,\max}$$

$$w_1 = [v(1 - P_b^{(n)})/(v - 1)]^{\lambda} \qquad 0 < \lambda < \infty$$

$$w_2 = 1 - w_1$$

In Equation 8.5-14, $K_{v,\max}$ and $\Delta_{r,\max}$ are the maximum values of K_v and Δ_r in the experimental region, respectively; Δ_r represents the criterion Δ for a single model, the rth; $P_b^{(n)}$ is the prior probability associated with the best model, b, that is the largest $P^{(n)}$ for a model after the nth observation is completed; and r is the index for the models, $r = 1, 2, \ldots, v$. Under maximum uncertainty with $P = 1/v$, $w_1 = 1$ and $w_2 = 0$, all the weight will be allocated to the discrimination criterion. On the other hand, if $P_b^{(n)} = 1$, $w_1 = 0$ and $w_2 = 1$, the parameter estimation criterion dominates. The choice of λ is up to the experimenter. High values of $\lambda\,(\lambda \gg 1)$ emphasize the

† W. G. Hunter and D. W. Wichern, Depts. of Chem. Eng. and Stat. Tech. Rept. 33, Univ. of Wisconsin, Madison, Oct. 1966.

‡ W. J. Hill, W. G. Hunter, and D. W. Wichern, *Technometrics* **10**, 145, 1968.

parameter estimation criterion whereas low values of λ ($\lambda \ll 1$) emphasize the model discrimination phase. As $\lambda \rightarrow 0$, $w_1 \rightarrow 1$ and $C \rightarrow D$; as $\lambda \rightarrow \infty$, $w_1 \rightarrow 0$ and $C \rightarrow E$. In effect, λ acts as a parameter governing the transition from one criterion to the other. If there is considerable variability in the observations, then initially the probabilities are likely to fluctuate from one run to the next. Hence, a small value of λ should be chosen so that parameter estimation is not prematurely emphasized if P_r increases suddenly solely because of experimental error; λ can be subsequently increased.

Example 8.5-3 Sequential Designs by a Combined Criterion

Hunter, Hill, and Wichern also described a simulated experiment in which the sequential experiments were selected by the combined criterion of Equation 8.5-14. Four models were proposed to represent the reaction $A \rightarrow B$:

$$y_1 = \exp[-x_1 \exp(\beta_{11} - \beta_{12}x_2)] \quad \text{(a)}$$

$$y_2 = [1 + x_1 \exp(\beta_{21} - \beta_{22}x_2)]^{-1} \quad \text{(b)}$$

$$y_3 = [1 + 2x_1 \exp(\beta_{31} - \beta_{32}x_2)]^{-\frac{1}{2}} \quad \text{(c)}$$

$$y_4 = [1 + 3x_1 \exp(\beta_{41} - \beta_{42}x_2)]^{-\frac{1}{3}} \quad \text{(d)}$$

where y is the dependent variable, the concentration; x_1 is the time; and $x_2 = [(1/T) - (1/525)]$, a scaled inverse absolute temperature. In the simulation study, Model 2 was selected as the "true" model, and the data $Y_r = y_r + \epsilon_r$ generated by using $\sigma_\epsilon = 0.05$, $\beta_{21} = 3.53235$, and $\beta_{22} = 5000$. The range for "experimentation" was limited to

$$0 \le x_1 \le 150 \text{ min}$$

$$450 \le T \le 600\ °\text{K}$$

TABLE E8.5-3a DESIGNS USING CRITERION C

Run	$x_1 = t$	T	Y	P_1	P_2	P_3	P_4
1	25	575	0.3961	—	—	—	—
2	25	475	0.7232	—	—	—	—
3	125	475	0.4215	—	—	—	—
4	125	575	0.1297	0.0060	0.4335	0.4087	0.1518
5	150	550	0.1504	0.0004	0.5580	0.3740	0.0676
6	25	525	0.5565	0.0001	0.6278	0.3446	0.0275
7	150	450	0.5542	0.0001	0.5011	0.4541	0.0447
8	150	550	0.0671	0.0002	0.9031	0.0953	0.0014
9	25	600	0.3356	0.0002	0.9208	0.0780	0.0009
10	150	450	0.4842	0.0002	0.9312	0.0680	0.0006
11	25	600	0.3140	0.0002	0.9322	0.0671	0.0005
12	150	450	0.5133	0.0002	0.9307	0.0685	0.0006
13	25	600	0.3500	0.0002	0.9386	0.0607	0.0005
14	150	450	0.4936	0.0002	0.9402	0.0592	0.0004
15	25	600	0.3058	0.0002	0.9381	0.0613	0.0004

Reproduced from *Technometrics* **10**, 152–159, 1968, with permission of the authors and the American Statistical Association.

TABLE E8.5-3b DESIGNS USING CRITERION K_v

Run	$x_1 = t$	T	Y	P_1	P_2	P_3	P_4
1	25	575	0.3961	—	—	—	—
2	25	475	0.7232	—	—	—	—
3	125	475	0.4215	—	—	—	—
4	125	575	0.1297	0.0060	0.4335	0.4087	0.1518
5	150	550	0.1504	0.0004	0.5580	0.3740	0.0676
6	25	525	0.5565	0.0001	0.6278	0.3446	0.0275
7	150	550	0.1558	0.0000	0.6865	0.3020	0.0115
8	150	550	0.0671	0.0000	0.9424	0.0570	0.0005
9	25	525	0.6196	0.0000	0.9895	0.0105	0.0000
10	150	525	0.1427	0.0000	0.9978	0.0022	0.0000
11	25	525	0.5979	0.0000	0.9993	0.0007	0.0000
12	150	525	0.1717	0.0000	0.9997	0.0003	0.0000
13	150	525	0.2419	0.0000	0.9995	0.0005	0.0000
14	25	525	0.5441	0.0000	0.9994	0.0006	0.0000
15	150	525	0.1977	0.0000	0.9996	0.0004	0.0000

Reproduced from *Technometrics* **10**, 152–159, 1968, with permission of the authors and the American Statistical Association.

A preliminary 2^2 factorial design was chosen to generate four initial values of Y at the following levels:

x_1 (min)	T (°K)
25	475
25	575
125	475
125	575

The initial values of the P_r's, presumably in view of the lack of other information, were all set equal to 0.25. After the fourth value of Y was generated, the posterior probabilities were calculated by using Equation 8.5-6 (see row

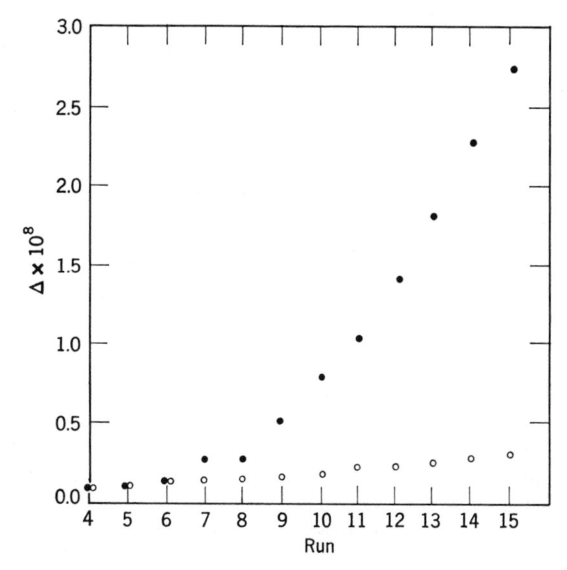

FIGURE E8.5-3a The value of Δ for sequential designs using two different criteria. (Reproduced from *Technometrics* **10**, 152–159, 1968, with permission of the authors and the American Statistical Association.)

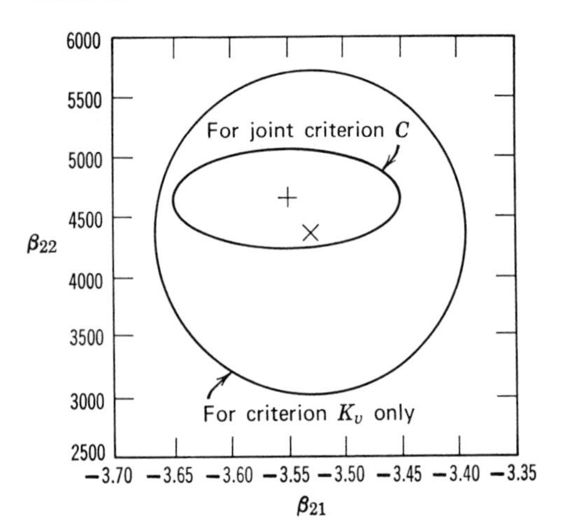

FIGURE E8.5-3b Approximate 95-percent confidence contour for the combined criterion C and the single criterion K- after 15 runs; + and × indicate the respective parameter estimates. (Reproduced from *Technometrics* **10**, 152–159, 1968, with permission of the authors and the American Statistical Association.)

$n = 4$ in Table E8.5-3a. From the initial experiments it appeared as if Models 2 and 3 would be the best.

To determine the settings of x_1 and x_2 for the fifth run, C in Equation 8.5-14 was maximized by search on a grid of mesh size 25. The weights, based on $\lambda = 2$, were

$$w_1 = \left[\frac{4(1 - 0.4335)}{4 - 1}\right]^2 = 0.57$$

$$w_2 = 1 - w_1 = 0.43$$

and the optimal conditions proved to be $x_1 = 150$ min and $T = 550°$K. At this stage the experimental designs began to diverge somewhat from those which would be obtained if K_v alone were the criterion. Tables E8.5-3a and E8.5-3b show the results after each series of hypothetical experiments. Figure E8.5-3a indicates that for the early series of runs, up to say number 8, Δ did not change substantially while P_2 went from 0.4335 to 0.9031. In other words, the early runs were mainly discriminatory and the later runs mainly reduced the confidence region for the model selected. Figure E8.5-3b shows the 95-percent joint confidence region for β_{21} and β_{22} of Model 2 after 15 runs using C and K_v alone as the criteria. The confidence region obtained by using C is clearly smaller than that obtained by using K_v.

Supplementary References

General

Cochran, W. G. and Cox, G. M., *Experimental Designs* (2nd. ed.), John Wiley, New York, 1957.

Cox, D. R., *Planning of Experiments*, John Wiley, New York, 1958.

Davies, O. L., *Design and Analysis of Industrial Experiments*, Hafner, New York, 1956.

Davies, O. L., *Statistical Methods in Research and Production*, Hafner, New York, 1961.

Hill, W. J. and Hunter, W. G., "A Review of Response Surface Methodology: A Literature Survey," *Technometrics* **8**, 571, 1966.

Response Surface Designs for Models Subject to Constraints

Box, G. E. P. and Gardner, C. J., "Constrained Designs," Dept. of Stat. Tech. Rept. 89, Univ. of Wisconsin, Madison, 1966.

Gorman, J. W. and Hinman, J. E., "Simplex Lattice Designs for Multicomponent Systems," *Technometrics* **4**, 463, 1962.

Scheffé, H., "Experiments with Mixtures," *J. Royal Stat. Soc.* **B20**, 344, 1958.

Scheffé, H., "The Simplex Centroid Design for Experiments with Mixtures," *J. Royal Stat. Soc.* **B25**, 235, 1963.

Designs to Adjust for Time Trends

Hill, H. H., "Experimental Designs to Adjust for Time Trends," *Technometrics* **2**, 67, 1960.

Lisenkov, A. N., *et al.*, "Experimental Design Under Conditions of Time Drift," *Industrial Laboratory* (English Translation) **33**, 706, 1967.

Optimization Using Response Surface Methods

Box, G. E. P., "The Exploration and Exploitation of Response Surfaces," *Biometrics* **10**, 16, 1954.

Brooks, S. H., "A Comparison of Maximum Seeking Methods," *Operations Res.*, **7**, 430, 1959.

Brooks, S. H. and Mickey, M. R., "Optimum Estimation of Gradient Direction in Steepest Ascent Experiments," *Biometrics* **17**, 48, 1961.

Hill, W. J. and Hunter, W. G., "A Review of Response Surface Methodology; A Literature Survey," *Technometrics* **8**, 571, 1966.

Evolutionary Operation

Box, G. E. P., "A Simple System of Evolutionary Operation Subject to Empirical Feedback," *Technometrics* **8**, 19, 1966.

Box, G. E. P. and Hunter, J. S., "Condensed Calculations for Evolutionary Operation Programs," *Technometrics* **1**, 77, 1959.

Delver, A. and Organon, N. V., "Gradual Optimization of Production Processes," *Sigma* **7** (3), 45, 1961.

Hunter, W. G. and Kittrell, J. R., "Evolutionary Operation: A Review," *Technometrics* **8**, 389, 1966.

Simplex Methods

Carpenter, B. H. and Sweeny, H. C., "Process Improvement with Simplex Self-Directing Evolutionary Operation," *Chem. Eng.* 117, July 5, 1965.

Lowe, C. W., *Trans. Inst. Chem. Eng.* (London) **42**, T334, 1964.

Spendley, W., Next, G. R., and Himsworth, F. R., "Sequential Application of Simplex Designs in Optimization and Evolutionary Operation," *Technometrics* **4**, 441, 1962.

Adaptive Optimization

Rees, N. W., "Self-adaptive Control Systems," British Chem. Eng. **5**, 106, Feb. 1960.

Designs to Increase Precision of Parameter Estimation

Atkinson, A. C. and Hunter, W. G., "The Design of Experiments for Parameter Estimation," *Technometrics* **10**, 271, 1968.

Box, G. E. P. and Lucas, H. L., "Design of Experiments in Nonlinear Situations," *Biometrics* **46**, 77, 1959.

Draper, N. R. and Hunter, W. G., "The Use of Prior Distribution in the Design of Experiments for Parameter Estimation in Nonlinear Situations," *Biometrica* **54**, 147, 1967.

Elfving, G., "Optimum Allocation in Linear Regression Theory," *Ann. Math. Stat.* **23**, 255, 1952.
Kiefer, J. C., "Optimum Experimental Designs," *J. Royal Stat. Soc.* **B21**, 272, 1959.
Kiefer, J. C., "Optimum Experimental Designs V, with Applications to Systematic and Rotatable Designs," *Proceed. Fourth Berkeley Symposium* **1**, 381, 1961.
Kittrell, J. R., Hunter, W. G., and Watson, C. C., "Obtaining Precise Parameter Estimates for Nonlinear Catalytic Rate Models," *AIChE J.* **12**, 5, 1966.
Stone, M., "Application of a Measure of Information to the Design and Comparison of Regression Experiments," *Ann. Math. Stat.* **30**, 55, 1959.
Wald, A., "On the Efficient Design of Statistical Investigations," *Ann. Math. Stat.* **14**, 134, 1943.

Sequential Designs for Discrimination

Behnken, D. W., "Estimation of Copolymer Reactivity Ratios," *J. Polymer Sci.* **A2**, 645, 1964.
Box, G. E. P. and Hill, J. W.," Discrimination Among Mechanistic Models," *Technometrics* **9**, 57, 1967.
Box, G. E. P. and Lucas, H. L., "Design of Experiments in Nonlinear Situations," *Biometrica* **46**, 77, 1959.
Hunter, W. G. and Mezaki, R., "An Experimental Design Strategy for Distinguishing Among Rival Mechanistic Models," *Can. J. Chem. Eng.* **45**, 247, 1967.
Hunter, W. G. and Reiner, A. M., "Designs for Discriminating Between Two Rival Models," *Technometrics* **1**, 307, 1965.
Keifer, J. and Wolfowitz, J., "Optimal Designs in Regression Problems," *Ann. Math. Stat.* **30**, 271, 1959.
Nelson, A. C., "Some Optimum Regression Designs for Discriminating Between Two Models," Research Triangle Institute, Tech. Rept. 4, Durham, N.C., Jan. 27, 1967.
Peterson, D. W., "Discriminant Functions" (AD465281), Stanford Electronics Lab. Stanford, Calif., Apr. 1965.
Roth, P. M., "Design of Experiments for Discrimination Among Rival Models," Ph.D. Thesis, Princeton Univ., 1966.

Problems

8.1 A replicated two-level factorial experiment is carried out as follows (the dependent variables are yields):

Time (hr)	Temperature (°C)	Yield (%)
1	240	24
5	240	42
1	280	3
5	280	19
1	240	24
5	240	46
1	280	5
5	280	21

Find the coefficients in a first-order model, $Y = \beta_0 + \beta_1 x_1 + \beta_2 x_2 + \epsilon$, and determine by an analysis of variance whether or not the model successfully fits the data.

8.2 Tell whether or not the variables in the following designs comprise an orthogonal set of independent variables; show calculations.

(a)

x_1	x_2	x_3
−1	−1	−1
1	−1	1
−1	1	1
1	1	−1

(b)

x_1	x_2	x_3	x_4
$-\frac{1}{3}$	$-\frac{1}{3}$	−4.84	−2.48
$\frac{1}{3}$	$\frac{1}{3}$	−2.47	−2.01
$-\frac{1}{3}$	$\frac{1}{3}$	−2.46	−2.47
$\frac{1}{3}$	$\frac{1}{3}$	−0.08	−1.99
$-\frac{2}{3}$	0	−4.84	−2.72
$\frac{2}{3}$	0	−0.08	−1.76
0	$-\frac{2}{3}$	−4.85	−2.25
0	$\frac{2}{3}$	−0.08	−2.22
0	0	−2.46	−2.24

8.3 Develop equations to code the following variables to accommodate a:

(a) Two-level factorial design.
(b) Equilateral triangular design.

Range of Variables for Fuel Codes 943F to 977F

Ignition Delay, Degrees, Dependent Variable	Independent Variables	
	Cetane Number	Volume, Percent Olefins
16.2–8.0	22.3–49.6	1.3–60.7

8.4 Determine the suitability of a first-order model in fitting to the data in Table P8.4. Note that the x's are not orthogonal.

TABLE P8.4

Response, Yield	Controlled Variables		Coded Controlled Variables	
	Temperature	Catalyst Nitrite	x_1	x_2
23	110	0.54	−0.55	−1.075
27	117	0.63	0.85	0.725
24	113	0.71	0.05	2.325
22	107	0.48	−1.15	−2.275
32	118	0.57	1.05	−0.475
29	117	0.63	0.85	0.725
18	102	0.59	−2.15	−0.075
31	118	0.60	1.05	0.125

$$x_1 = \frac{\text{temperature} - 112.75}{5.00} \qquad x_2 = \frac{\text{C/N} - 0.59375}{0.05}$$

8.5 Demonstrate, by calculations, that the following second-order designs are rotatable:

(a) Pentagon.
(b) Hexagon.
(c) Central composite.

Demonstrate that the 3^2 design of Figure 8.1-5 is not rotatable.

8.6 An experiment based on the hexagon design was carried out with four replications at the origin, giving the data in Table P8.6.

TABLE P8.6

	Factor Levels		Design Levels	
Yield (%)	Temperature (°C)	Time (hr)	x_1	x_2
96.0	75	2.0	1.000	0
78.7	60	2.866	0.500	0.866
76.7	30	2.866	−0.500	0.866
54.6	15	2.0	−1.000	0
64.8	30	1.134	−0.500	−0.866
78.9	60	1.134	0.500	−0.866
97.4	45	2.0	0	0
90.5	45	2.0	0	0
93.0	45	2.0	0	0
86.3	45	2.0	0	0

Coding: $x_1 = \dfrac{\text{temperature} - 45}{30}$ $\quad x_2 = \dfrac{\text{time} - 2}{1.000}$

Fit a full second-order model to the data, determine if the model fits the data, and, if it does, establish which parameters can be removed from the full model.

8.7 Comstock, Jurnack, and Mooney† investigated the effect of temperature of precipitation, concentration of reactants, and rate of addition of diammonium hydrogen phosphate on the yield of precipitate and on four physical properties of powders:

1. B.E.T. surface area, by gas adsorption.
2. Particle size, by optimal measurement of sedimentation rates.
3. Fisher subsieve size (SSS).
4. Bulk density, by a conventional pyconometer.

A central composite design for four independent variables required 24 points plus replicates. The design was run in three blocks, as indicated in Table P8.7, and arranged so that a first-order response could be estimated after running the first block.

Obtain both a first-order model fit to the first block of data and a second-order fit to all three blocks. Prepare an analysis of variance which indicates the effect of each coefficient, the first and second-order effects for the second-order model, the experimental error, and the lack of fit of the model. The coding of the variables was

$$x_1 = \frac{\text{temperature} - 80}{7.5}$$

$$x_2 = \frac{CaCl_2 \text{ concentration} - 1.25}{0.375}$$

$$x_3 = \frac{(NH_4)_2HPO_4 \text{ concentration} - 1.25}{0.375}$$

$$x_4 = \frac{\text{addition rate} - 180}{85}$$

8.8 Are the following experimental data fitted well or poorly by a first-order model? The matrices are:

Observed Yields; columns of X: x_0, x_1, x_2

$$Y = \begin{bmatrix} 1.0 \\ 1.7 \\ 6.0 \\ 5.2 \\ 7.0 \\ 7.9 \\ 18.0 \\ 19.2 \end{bmatrix} \qquad X = \begin{bmatrix} 1 & -1 & -1 \\ 1 & -1 & -1 \\ 1 & 1 & -1 \\ 1 & 1 & -1 \\ 1 & -1 & 1 \\ 1 & -1 & 1 \\ 1 & 1 & 1 \\ 1 & 1 & 1 \end{bmatrix}$$

8.9 An experiment was designed to determine if the ratio of fresh to recycle monomer in monomer blends affected the polymerization conversion or any of the product qualities.‡ A second factor studied was the level of catalyst concentration in the monomer at the start of the polymerization. Both factors affect the conversion and the molecular weight of the product. The two responses measured were:

1. Conversion.
2. ZST (a measure of the apparent molecular weight of the polymer).

Seven polymerizations were run at each monomer concentration according to the scheme below:

Blend Ratio	Catalyst Concentration 1	2	3
1	7	7	7
2	7	7	7

The results are given in Table P8.9 on page 286.

† A. J. Comstock, S. J. Jurnack, and R. W. Mooney, *Ind. Eng. Chem.* **51**, 325, 1959.

‡ L. A. Pasteelnick and W. B. Leder, *Chem. Eng. Progress* **53**, 392, 1957.

TABLE P8.7 BOX-WILSON DESIGN AND EXPERIMENTAL DATA

	Coded Values of Independent Variables				Particle Size Measurements				
Run Number	x_1 (temperature)	x_2 ($CaCl_2$)	x_3 [$(NH_4)_2HPO_4$]	x_4 (addition rate)	B.E.T. surface area (sq meters/g)	Fisher SSS (microns)	Sedimentation, d_{op} (microns)	Bulk density (g/cu in)	Yield (%)
				Block I					
1	−1	−1	−1	−1	11.0	2.2	4.2	4.7	97.0
2	0	0	0	0	9.4	3.0	4.5	5.6	95.9
3	−1	+1	+1	−1	3.5	3.2	4.7	5.3	98.5
4	+1	−1	−1	+1	21.1	2.2	3.6	6.7	90.5
5	+1	+1	−1	−1	8.9	5.0	4.5	9.0	94.4
6	−1	−1	+1	+1	13.6	1.3	4.4	2.8	95.2
7	+1	+1	+1	+1	9.2	2.9	3.5	5.3	94.6
8	0	0	0	0	11.1	3.4	4.0	5.3	93.4
9	+1	−1	+1	−1	12.3	5.6	4.9	8.8	94.0
10	−1	+1	−1	+1	13.0	2.1	3.9	3.7	92.8
				Block II					
1	+1	+1	−1	+1	10.9	3.8	3.9	6.4	93.0
2	−1	−1	+1	−1	13.1	2.6	4.4	5.6	94.2
3	0	0	0	0	10.5	2.7	5.1	5.5	95.6
4	−1	+1	+1	+1	10.8	1.3	3.4	2.4	98.3
5	+1	+1	+1	−1	5.4	4.5	5.0	6.4	96.3
6	0	0	0	0	11.1	2.2	3.9	5.9	95.1
7	−1	+1	−1	−1	11.2	3.3	4.1	6.4	91.8
8	+1	−1	+1	+1	18.0	2.5	3.6	6.2	92.0
9	−1	−1	−1	+1	15.2	1.2	3.2	2.9	94.8
10	+1	−1	−1	−1	18.4	4.0	3.5	8.5	91.6
				Block III					
1	0	0	0	0	10.6	2.8	3.2	5.6	93.3
2	−2	0	0	0	11.6	1.4	3.5	2.9	95.7
3	0	0	−2	0	20.9	3.1	3.8	7.8	90.0
4	0	0	0	0	10.1	3.3	4.0	5.7	94.5
5	0	0	0	+2	13.0	1.6	3.2	3.6	94.9
6	0	0	0	−2	3.8	11.5	5.8	12.3	94.5
7	0	−2	0	0	27.9	1.3	2.7	5.9	90.0
8	0	+2	0	0	7.6	2.9	3.2	5.0	96.4
9	+2	0	0	0	9.4	4.2	3.4	7.5	95.2
10	0	0	+2	0	15.4	1.6	3.1	3.9	95.3

(a) Develop an orthogonal design for the experiment. Assume both catalyst concentration and blend ratios are coded as follows:

$$\text{blend ratio} = \frac{\text{true blend ratio} - a}{b}$$

$$\text{catalyst concentration} = \frac{\text{true concentration} - c}{d}$$

(b) Determine the best fitting model for the problem; estimate the coefficients.

(c) Predict the location of the optimum catalyst concentration and blend ratio (in coded values) for the first experiment, as reported.

(d) Set up a new design and new location for experimentation. Give the coordinates for the next series of experiments.

8.10 Transform the following response surface into canonical form and find the relations to express the canonical vector $\tilde{\mathbf{x}}$ in terms of the original vector $\mathbf{x}$.

$$\hat{Y} = 95.0 + 0.05x_1 - 1.58x_2 - 8.13x_1^2 - 5.87x_2^2 - 6.25x_1x_2$$

8.11 Transform the following estimated regression equation to the canonical form and interpret the nature of the response surface.

$$\hat{Y} = 60.64 - 3.672x_1 + 11.661x_2 - 3.514x_1^2 - 0.924x_2^2 + 2.220x_1x_2$$

TABLE P8.9

Catalyst Concentration		Blend Ratio 1 Conversion	Blend Ratio 1 ZST	Blend Ratio 2 Conversion	Blend Ratio 2 ZST
		−4.2	−15	1.7	2
		0.5	2	−3.6	−8
		−2.3	1	0.4	8
1		−2.4	−6	0.7	−4
		−0.9	1	−0.9	−2
		−1.1	−2	−3.3	−11
		−1.1	1	3.2	0
	$\sum x_2$	−11.5	−18	−1.8	−15
	$\sum x$	32.17	272	38.44	273
		−0.9	0	−0.7	−4
		0.4	0	−0.7	0
		0.0	1	−0.3	−2
2		0.3	6	0.4	−1
		−2.9	−4	0.3	−3
		1.1	2	1.7	2
		2.1	3	−0.1	−7
	$\sum x_2$	−0.1	8	0.6	−15
	$\sum x$	15.09	66	4.22	−15
		1.4	7	0.1	2
		0.7	1	0.6	0
		1.6	6	−2.3	6
3		2.7	6	1.6	−2
		1.2	0	3.2	−1
		1.3	3	−3.5	−9
		0.3	−2	3.9	10
	$\sum x_2$	9.2	21	3.6	6
	$\sum x$	15.2	135	45.92	226

8.12 Transform the following estimated regression equations to canonical form and give the equations for the coordinate transformations. Find the center of the system in terms of the original coordinates. What type of a surface does the equation represent (maximum, minimum, or saddle point)?

(a) $\hat{Y} = 95.00 + 0.05x_1 - 1.58x_2 - 8.13x_1^2 - 5.87x_2^2 - 6.25x_1x_2$

(b) $\hat{Y} = 19.43 + 8.86x_1 - 0.145x_2 - 2.302x_1^2 + 0.000293x_2^2 + 0.04777x_1x_2$

(c) $\hat{Y} = 349.392 + 3.957x_1^2 + 0.564x_1x_2 + 0.536x_1x_3 - 10.793x_1x_4 + 0.473x_1x_5 - 2.369x_1 + 0.665x_2^2 - 3.852x_2x_4 + 0.201x_2x_5 - 3.78x_2 + 10.019x_3^2 - 1.376x_3x_4 + 6.077x_3 + 7.602x_4^2 + 0.942x_4x_5 + 2.739x_4x_6 + 7.064x_4 - 0.0428x_5^2 - 1.847x_5x_6 + 1.6795x_5 + 2.33x_6^2 + 3.956x_6$

(d) $\hat{Y} = 57.71 + 1.94x_1 + 0.91x_2 + 1.07x_3 - 1.54x_1^2 - 0.26x_2^2 - 0.68x_3^2 - 3.09x_1x_2 - 2.19x_1x_3 - 1.21x_2x_3$

8.13 For the following design matrix and responses, find: (a) the second degree canonical equation of best fit and (b) the coordinates (y, x_1, x_2) of the center of the surface.

x_1	x_2	Y
0	1	93.7
0.866	0.5	98.5
0.866	−0.5	88.8
0	−1	85.8
−0.866	−0.5	92.4
−0.866	0.5	87.8
0	0	97.8
0	0	99.0

8.14 In pressure sintering the important variables are time, temperature, and pressure. The objectives of a study were to obtain the maximum density for sintered alumina (which has a theoretical pressed density of 3.98 g/cc) and to learn as much as possible about the effect of the independent variables on the density. Preliminary work indicated the following range of variables in which further experimentation was to be carried out.

Pressure (psi)	Temperature (°C)	Time (min)
1000	1205	10.2
2200	1695	29.8

A second-order model was chosen to represent the influence of each variable on the pressed density:

$$\rho = \beta_0 + \beta_1x_1 + \beta_2x_2 + \beta_3x_3 + \beta_{11}x_1^2 + \beta_{22}x_2^2 + \beta_{33}x_3^2 + \beta_{12}x_1x_2 + \beta_{13}x_1x_3 + \beta_{23}x_2x_3$$

where:

$$x_1 = (\text{pressure} - 1600)/370$$
$$x_2 = (\text{temperature} - 1450)/150$$
$$x_3 = (\text{time} - 20.0)/6$$
$$\rho = \text{pressured density (g/cc)}$$

The scaling was designed to produce equal changes in density for a 1 unit change in each x.

Experiments were carried out according to the experimental design shown in Table P8.14. Each "group" number represents a successive period of experimentation. Such separation permitted a separate assessment of the influence of each variable independently. Also, each group had a built-in measure of experimental error for control. From the overall design, a test could be made to judge whether the model chosen was adequate.

(a) Obtain the estimated regression equation.
(b) Find the best fitting second-order model.
(c) Carry out an analysis of variance to show the effect of the intercept, the first-order terms, the second-order terms, the groups, the residual error, and the experimental error.

TABLE P8.14 TEST SEQUENCE AND RESULTS

Group Number	Run Number	Experimental Conditions: Coded Variables x_1	x_2	x_3	Pressure (psi)	Temperature (°C)	Time (min)	Results, Density (g/cc)
I	1	+1	−1	+1	1970	1300	26	2.17
	2	+1	+1	−1	1970	1600	14	3.69
	3	0	0	0	1600	1450	20	2.92
	4	−1	+1	+1	1230	1600	26	3.53
	5	−1	−1	−1	1230	1300	14	1.77
	6	0	0	0	1600	1450	20	2.92
II	7	−1	−1	+1	1230	1300	26	1.77
	8	0	0	0	1600	1450	20	2.88
	9	−1	+1	−1	1230	1600	14	3.37
	10	0	0	0	1600	1450	20	2.91
	11	+1	+1	+1	1970	1600	26	3.83
	12	+1	−1	−1	1970	1300	14	2.10
III	13	1.63	0	0	2200	1450	20	3.05
	14	0	0	0	1600	1450	20	2.92
	15	−1.63	0	0	1000	1450	20	2.35
	16	0	0	1.63	1600	1450	29.8	3.04
	17	0	−1.63	0	1600	1205	20	1.66
	18	0	0	0	1600	1450	20	2.87
	19	0	1.63	0	1600	1695	20	3.86
	20	0	0	−1.63	1600	1450	10.2	2.61

(d) Carry out a canonical analysis and indicate the transformed coordinates and value of the maximum density in terms of pressure, temperature, and time.

8.15 The objectives in this problem were: (1) to find the optimum operating conditions for an experiment involving three variables by using the least possible amount of experimentation and (2) to ascertain the nature of the response surface in the vicinity of the optimum. Three independent variables were involved with limiting values as shown:

Temperature, T	460 to 1000°R
Pressure, p	1 to 100 atm
Flow rate, F	0 to 100 lb/min

The dependent variable was the yield (lb/min).

For the first cycle of experimentation, a two-level factorial design (2^3) was chosen. The center of the design was arbitrarily taken to be at values of the independent variables of $T = 600°\text{R}$, $p = 50$ atm, and $F = 50$ lb/min. The independent variables were coded to provide an orthogonal design:

$$x_1 = \frac{T - 600}{10}$$

$$x_2 = \frac{p - 50}{2}$$

$$x_3 = \frac{F - 50}{2}$$

To minimize the amount of experimentation, a half-replicate was used with the center point replicated three times to provide a measure of the experimental error. This resulted in a total of seven experimental points for the first cycle. See Table P8.15.

The resulting equation for the response surface was

$$\hat{Y} = -457.9 + 7.48T - 6.08p + 24.4F$$

$$(\mathbf{x}^T\mathbf{x})^{-1} = \begin{bmatrix} \frac{1}{7} & 0 & 0 & 0 \\ 0 & \frac{1}{4} & 0 & 0 \\ 0 & 0 & \frac{1}{4} & 0 \\ 0 & 0 & 0 & \frac{1}{4} \end{bmatrix}$$

$$\mathbf{x}^T\mathbf{y} = \begin{bmatrix} 3142.31 \\ 29.92 \\ -24.34 \\ 97.62 \end{bmatrix}$$

$$\mathbf{x}^T\mathbf{x} = \begin{bmatrix} 7 & 0 & 0 & 0 \\ 0 & 4 & 0 & 0 \\ 0 & 0 & 4 & 0 \\ 0 & 0 & 0 & 4 \end{bmatrix}$$

Show by an analysis of variance that the first-order model was a poor fit to the experimental data

TABLE P8.15 EXPERIMENTAL DATA

Response	Temperature (°R)	Pressure (atm)	Flow Rate (lb/min)
422.356	590	48	48
425.146	610	52	48
486.126	610	48	52
458.998	590	52	52
449.447	600	50	50
449.962	600	50	50
450.256	600	50	50

in the vicinity of the center of the design. Indicate the direction of steepest ascent; suggest a center point for the next sequence of experimentation.

8.16 As in Problem 8.15, a half-replicate of a 2^3 design with the center point replicated three times was used. The design was centered about:

$$T = 800°\text{R}$$
$$p = 30 \text{ atm}$$
$$F = 70 \text{ lb/min}$$

By using a linear model, the response surface was found to be

$$\hat{Y} = -1012.05 + 12.48T - 8.08p + 32.4F$$

See Table P8.16.

$$\mathbf{x}^T\mathbf{y} = \begin{bmatrix} 6989.18 \\ 49.93 \\ -32.33 \\ 129.53 \end{bmatrix}$$

Carry out the analysis indicated in Problem 8.15.

TABLE P8.16 EXPERIMENTAL DATA

Response	Temperature (°R)	Pressure (atm)	Flow Rate (lb/min)
960.93	790	28	68
969.73	810	32	68
1050.66	810	28	72
1009.53	790	32	72
999.00	800	30	70
999.52	800	30	70
999.81	800	30	70

8.17 After several cycles of experiments, the optimum operating conditions for the problem posed in 8.15 were believed to be the maximum allowable temperature and flow rate and the minimum allowable pressure:

$$T = 1000°\text{R}$$
$$p = 1 \text{ atm}$$
$$F = 100 \text{ lb/min}$$

A series of experiments was then made in the vicinity of the optimum to determine the nature of the response surface in this region.

To make use of both first- and second-order linear models, a full central composite design was run with the center point replicated four times. See Table P8.17a.

TABLE P8.17a EXPERIMENTAL DATA

Point	Response	Temperature (°R)	Pressure (atm)	Flow Rate (lb/min)
1	1915.01	993	1.5	97.0
2	1923.26	997	1.5	97.0
3	1910.08	993	2.5	97.0
4	1918.33	997	2.5	97.0
5	1954.72	993	1.5	99.0
6	1963.18	997	1.5	99.0
7	1951.01	993	2.5	99.0
8	1959.48	997	2.5	99.0
9	1931.61	991.6	2.0	98.0
10	1945.25	998.3	2.0	98.0
11	1943.16	995	1.159	98.0
12	1934.62	995	2.840	98.0
13	1905.17	995	2.0	96.318
14	1971.80	995	2.0	99.681
15	1937.72	995	2.0	98.0
16	1938.16	995	2.0	98.0
17	1938.62	995	2.0	98.0
18	1938.98	995	2.0	98.0

The coded variables were

$$x_1 = \frac{T - 995}{2}$$

$$x_2 = \frac{p - 2.0}{0.5}$$

$$x_3 = \frac{F - 98}{1}$$

and the first-order response surface was

$$\hat{Y} = -2113.61 + 4.18T - 2.16p + 20.21F$$

$$(\mathbf{x}^T\mathbf{x})^{-1} = \begin{bmatrix} \frac{1}{12} & 0 & 0 & 0 \\ 0 & \frac{1}{8} & 0 & 0 \\ 0 & 0 & \frac{1}{8} & 0 \\ 0 & 0 & 0 & \frac{1}{8} \end{bmatrix}; \quad \mathbf{x}^T\mathbf{y} = \begin{bmatrix} 23{,}248{,}55 \\ 33.43 \\ -17.27 \\ 161.71 \end{bmatrix}$$

The analysis of variance is shown in Table P8.17b.

Can the first-order model successfully represent the response surface in the vicinity of the optimum? If so, what are the optimum response and optimum values of T, p, and F?

TABLE P8.17b ANALYSIS OF VARIANCE

Source of Variation	SS	d.f.	Mean Square
Due to b_1	4.50×10^7	1	4.50×10^7
Due to b_2	139.8	1	139.8
Due to b_3	37.2	1	37.2
Due to b_4	3270	1	3270
Deviation about regression line	7.5653	5	1.5131
Error	0.9012	3	0.3004
Total		12	

8.18 The result of fitting experimental data to a full second-order model in the vicinity of the optimum (Problem 8.15 completed) was the response surface:

$$\hat{Y} = -91{,}898 + 173.58T + 70.84p + 73.66F$$
$$- 0.0865T^2 - 0.3668p^2 - 0.3266F^2$$
$$- 0.1424Tp + 0.0091TF + 0.684pF$$

Find the location of the optimum. Sketch the response surface in the vicinity of the optimum in a three-dimensional representation.

8.19 A square design (points 1, 2, 3, and 4) with a center point (point 5) has been used to obtain the response for a process with two presumed independent variables. After four cycles of data taking (in random sequence in each cycle), the responses have been tabulated as follows:

Cycle Number	Date Points 1	2	3	4	5
1	16	10	22	13	16
2	14	12	18	13	14
3	12	11	17	16	15
4	18	11	27	10	15

Work up the EVOP calculations, including the experimental error, and suggest the direction to move for a new sequence of experimentation.

8.20 In a three-variable EVOP program, a 2^3 factorial design was used. One of the half-replicates of the cube was labeled 2, 3, 4, and 5 while the other half-replicate was labeled 7, 8, 9, and 10. The center point of the first half-replicate was labeled 1 while the center point accompanying the second half-replicate was labeled 6. Based on the data in the Table P8.20, estimate the main effects of the independent variables and the interactions between the blocks (the two half-replicates) separately. Include the error limits.

TABLE P8.20 PROCESS RESPONSES

Cycle	First Half-Replicate 1	2	3	4	5	Second Half-Replicate 6	7	8	9	10
1	68	72	53	71	78	75	69	65	68	57
2	72	62	69	85	66	59	67	71	60	73
3	55	72	58	75	69	79	86	56	73	62

8.21 After the third cycle for the design and data shown in Table P8.21 and Figure P8.21, is it possible to determine an optimal direction?

TABLE P8.21

	Operating Conditions 1	2	3	4	5
Sum from previous cycle	17.30	15.06	19.87	17.29	19.44
Average from previous cycle	8.65	7.53	9.94	8.64	9.72

Previous sum = 1.185
Previous average = 1.185

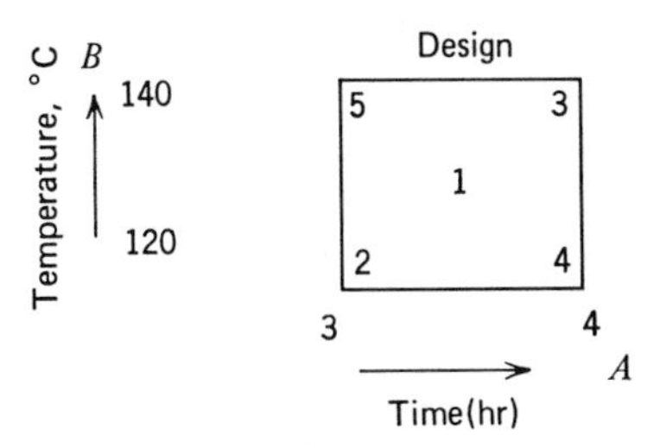

FIGURE P8.21

8.22 It was desired to find the optimal operating conditions for chromatographic separation of iso-octane-heptane. The independent variables and the experimental design were as shown in Table P8.22 where

x_1 = column temperature
x_2 = column length
x_3 = carrier gas flow rate at outlet
x_4 = weight of liquid phase per weight of solid phase

TABLE P8.22

	Variables x_1 (°C)	x_2 (cm)	x_3 (ml/min)	x_4 (%)
Zero level 0	45	230	20	25
Low level −1	35	190	35	20
Upper level +1	55	270	35	30
Starred points: −2	25	150	20	15
+2	65	310	80	35
Variation interval	10	40	15	5
Number of starred points	20	80	30	10

The other column parameters—diameter, particle diameter, type of carrier, etc.—were all held constant.

Sixteen experiments were carried out about an arbitrarily chosen point in experimental space according to a central composite design. The estimated regression equations obtained described the dependence of the separation coefficient, K, and the duration of the analysis, t, on the x's:

$$\begin{aligned} K = 0.87 &- 0.0408x_1 + 0.0672x_2 - 0.0675x_3 + 0.102x_4 \\ &- 0.002x_1^2 - 0.003x_2^2 + 0.006x_3^2 - 0.027x_4^2 \\ &+ 0.0128x_1x_2 - 0.00406x_1x_3 - 0.000313x_1x_4 \\ &+ 0.00094x_2x_3 + 0.0303x_2x_4 - 0.00531x_3x_4 \quad \text{(a)} \end{aligned}$$

$$\begin{aligned} t = 8.8 &- 2.195x_1 + 1.52x_2 - 2.095x_3 + 1.11x_4 \\ &+ 0.37x_1^2 - 0.14x_2^2 + 0.92x_3^2 - 0.06x_4^2 \\ &- 0.495x_1x_2 + 0.673x_1x_3 - 0.371x_1x_4 \\ &- 0.336x_2x_3 + 0.283x_2x_4 - 0.349x_3x_4 \quad \text{(b)} \end{aligned}$$

Illustrate the nature of the response surfaces by plotting jointly contours of K and t versus:

(a) x_1 and x_3 for constant x_2 and x_4.
(b) x_2 and x_4 for constant x_1 and x_3.

At what values of $\mathbf{x}$ should the next series of experiments be conducted?

Estimate $\phi_{\min}$ and the corresponding vector $\mathbf{x}$ for each response. What criterion should be used to obtain the best compromise between K and t?

8.23 Three models have been selected to represent a process. They all yield roughly the same form of the response in the range of interest of the independent variable, x.

Model	
1	$\eta = \alpha_1(1 - e^{-\alpha_2 x})$
2	$\eta = \dfrac{\beta_1\beta_2 x}{1 + \beta_2 x}$
3	$\eta = \gamma_1 x^{\gamma_2}$

The range of x is: $0 \le x \le 10$, and the experimental increments are 0.1. Four initial data points have been determined as follows:

x	Y
0.5	2.95
5.0	9.78
0.5	3.66
5.0	7.59

At what value of x should the fifth experiment be carried out? What are the estimated coefficients in each model after the fourth experiment? What are the prior and posterior probabilities for each model?

8.24 For the same models as in Problem 8.23, the following values exist ($\hat{}$ = estimated value) after the eighth run:

$\hat{\alpha}_1 = 9.31$	$\hat{\beta}_1 = 10.83$	$\hat{\gamma}_1 = 4.61$
$\hat{\alpha}_2 = 0.76$	$\hat{\beta}_2 = 0.87$	$\hat{\gamma}_2 = 0.35$
$P_1 = 0.019$	$P_2 = 0.847$	$P_3 = 0.132$

The coefficient of variation is estimated as 0.153. At what level of x should the next experiment be run? Are additional data needed?

8.25 Three kinetic models are proposed as possible mechanisms for a chemical reaction ($A \leftrightharpoons D + H$):

1. Single site surface reaction controlling, D adsorbed:

$$r = \frac{k_0[p_A - (p_D p_H/K_E)]}{1 + K_A p_A + K_D p_D}$$

2. Adsorption of A controlling, single site, D adsorbed

$$r = \frac{k_0[p_A - (p_D p_H/K_E)]}{1 + K_D p_D + K_{HD} p_D p_H}$$

3. Adsorption of A controlling, single site, H adsorbed

$$r = \frac{k_0[p_A - (p_D p_H/K_E)]}{1 + K_H p_H + K_{HD} p_D p_H}$$

The notations are as follows:

r = rate of reaction for a steady-state process measured in a reactor such as was proposed by Perkins and Rase.†
p = partial pressure of component (as indicated by subscripts)
K_E = reaction equilibrium constant
k_0 = reaction rate constant
K = absorption constants (as indicated by subscripts)

The purpose of the experimental investigation is to determine which model is most appropriate. Six initial runs were conducted with the following results:

p_A	p_H	p_D	$R = r + \epsilon$
1.0	2.0	2.0	−4.39
1.0	0.1	0.1	3.24
5.0	0.1	2.0	6.54
5.0	2.0	0.1	14.80
1.0	2.0	2.0	−2.32
5.0	0.1	2.0	7.19

The ranges for the variables are

$$0 \le p_A \le 5$$
$$0 \le p_H \le 2$$
$$0 \le p_D \le 3$$

What should be the next values of the partial pressures in order to achieve maximum discrimination?

8.26 A 2^2 factorial experiment has been used to collect four data points to help determine the coefficients in the estimated regression equation

$$\hat{Y} = \frac{b_1 x_1}{1 + b_2 x_1 + b_3 x_2}$$

† T. K. Perkins and H. F. Rase, *AIChE J.* **4**, 351, 1958.

The data are as follows:

x_1	x_2	Y
1	1	1.26
2	1	2.19
1	2	0.76
2	2	1.25

Determine the best values of x_1 and x_2 at which to run the fifth experiment. Does the 2^2 factorial design appear to have been a good or a poor design from the viewpoint of its effectiveness in reducing the confidence region for β_1, β_2, and β_3? How can you represent $\Delta^{(5)}$ graphically?

8.27 Table P8.27 shows the results of a model discrimination sequence of designs (starting with three models).

TABLE P8.27

Observa-tion	Partial Pressures (atm) Alcohol (A)	Ether (E)	Water (W)	P_1	P_2	P_3
1	1.00	0.00	0.00			
2	0.80	0.00	0.20			
3	0.60	0.00	0.40			
4	0.70	0.30	0.00			
5	0.50	0.70	0.00			
6	0.25	0.75	0.00	0.713	0.075	0.212
7	3.00	0.00	0.00	0.910	10^{-5}	0.090

Model

1 $$r = \frac{k_A L[p_A - (p_E p_W/K)^{1/2}]}{1 + [(K_A/K)p_E p_W]^{1/2} + K_E p_E + K_W p_W}$$

2 $$r = \frac{k_1 K_A^2[p_A^2 - (p_E p_W/K)]}{[1 + K_A p_A + K_E p_E + K_W p_W]^2}$$

3 $$r = \frac{k_1 K_A L[p_A^2 - (p_E p_W/K)]}{1 + K_A p_A + K_E p_E + K_W p_W}$$

The notations are as follows:

r = reaction rate
k = reaction rate constant (with subscript)
K = equilibrium constant
K = absorption constant (with subscript)
p = partial pressure
L = concentration of active sites

Contrast the experimental design at the end of the seventh experiment for discrimination with that which would be used to obtain the best estimates of the coefficients in Model 1 at the end of the seventh experiment.

8.28 Consider the first-order irreversible reaction

$$A \rightarrow B$$

The differential equations are

$$\frac{dc_A}{dt} = -kc_A \qquad \frac{dc_B}{dt} = kc_A$$

$$c_A(0) = 1 \qquad c_B(0) = 1$$

(a) At what time should the first experiment be taken in order to get the smallest confidence region for k?
(b) Is it necessary to measure both c_A and c_B, or is the measurement of one component sufficient?
(c) At what values of c_A and c_B should the two observations be made? Hint: Let k_0 be the prior assumed value for k.
(d) Suppose three initial experiments are to be made instead of just one. Repeat questions (a) through (c).

8.29 The reaction

$$A \xrightarrow{k_1} B \xrightarrow{k_2} C$$

can be represented by a model with two responses:

$$\frac{dc_A}{dt} = -k_1 c_A \qquad c_A(0) = 1$$

$$\frac{dc_B}{dt} = k_1 c_A - k_2 c_B \qquad c_B(0) = 0$$

where c = concentration and the k's are the reaction rate coefficients to be estimated. After integration of the coupled equations, we obtain

$$\text{Response 1:} \quad c_A = e^{-k_1 t} \qquad (a_1)$$

$$\text{Response 2:} \quad c_B = \frac{k_1}{k_1 - k_2}(e^{-k_2 t} - e^{-k_1 t}) \qquad (a_2)$$

(a) At what values of t should the first two experimental observations be taken if only two observations can be made?
(b) How would your choice of σ_{11}, σ_{12}, $\sigma_{12} = \sigma_{21}$, and ρ_{12} affect your answer to (a)? Try several variations of σ_{rs} and ρ_{rs}.

8.30 As a result of an assumed kinetic scheme, the following nonlinear model was proposed as suitable for a model:

$$\eta = \frac{\beta_1}{\beta_1 - \beta_2}(e^{-\beta_2 x} - e^{-\beta_1 x}) \qquad \beta_i > 0$$

Develop an initial design for this equation and examine the results graphically. What are the two best values of x at which to observe Y?

8.31 This problem has been adapted from the article by Hunter and Atkinson† which used data representative of the isomerization of bicyclohexane.‡ For a first-order irreversible reaction of $A \rightarrow B$,

$$\frac{dc_A}{dt} = -kc_A \qquad c_{A,0}(0) = 1 \qquad (a)$$

† W. G. Hunter and A. C. Atkinson, *Chem. Eng.*, 159, June 6, 1966.
‡ R. Srinivasan and A. A. Levi, *J. Amer. Chem. Soc. 85*, 3363, 1963.

the integrated model is

$$c_A = e^{-kt}$$

or

$$c_A = \exp\left[-k_0 t\, e^{-\Delta E/RT}\right] \qquad \text{(b)}$$

where the rate coefficient k has been replaced by the Arrhenius expression $k = k_0\, e^{-\Delta E/RT}$, R is the gas constant, T is the absolute temperature in °K, k_0 is the frequency factor, and ΔE is the activation energy.

A sequential series of experiments is to be carried out to estimate the β's to within a certain degree of precision.

(a) At what values of the independent variables should the first run(s) be conducted? The feasible values of T range from 600 to 640°K and the times from 15 to 150 minutes.

(b) Two sets of observations were next taken using the above experimental design, yielding the response $Y_1 = 0.912$ and $Y_2 = 0.382$, respectively. Determine the designs for the next run. Plot the ellipse in parameter space for the approximate 95-percent confidence contour. Plot the contours of Δ_1 in the experimental space.

(c) Carry out the next cycle of design specification, assuming that the response from run 3 was $Y_3 = 0.397$. Repeat the plots of question (b). Hint: To reduce parameter interaction, transform the temperature so that Equation (b) may be written as

$$c_A = \exp\left[-k_0 t\, e^{-E\left(\frac{1}{T} - \frac{1}{T_0}\right)}\right]$$

where $E = \Delta E/R$. Let T_0 be 620°K.

Part III

Estimation Using Models Based on Transport Phenomena Principles

In Section 1.1 it was pointed out that process models can be based on empirical relations, population balances, and transport phenomena principles, or combinations thereof. Part III describes how the parameters in models based on transport phenomena principles, including ordinary differential equations, partial differential equations, and transfer functions (frequency response) can be estimated. Four different categories of information may be available to the analyst: (1) the form of the model itself, (2) the model parameters (coefficients), (3) the model inputs, and (4) the model outputs. The type of problem which evolves can be classified by the following arrangement (√ denotes the known factors).

Topic	Model Including Boundary Conditions	Model Parameters	Model Input	Model Output
Model identification			√	√
Parameter estimation	√		√	√
Prediction	√	√	√	
Inverse problem	√	√		√

A quite comprehensive and exceedingly difficult problem is that of characterizing the process by an appropriate model, otherwise known as the *identification* problem. Given a class of process models and a process, the identification problem is to determine the best model in some sense through observations of the output of and input to the process. Specific classes of models might be impulse responses, transfer functions, differential equations, integral equations, difference equations, and so forth. To ascertain the best model among *all* the possible classes of models is probably generalizing the identification problem excessively. Furthermore, we cannot ask that the "true" model of the process be obtained as a result of the identification, because practically all the process models are simplified to such an extent that they can describe the process only approximately even if their parameters are known exactly.

Chapter 7 described various techniques that can be used to discriminate among empirical models; these same techniques can be applied to transport phenomena-based models. In addition to the criteria of representativeness (the differences between the responses of the process and those predicted from the model satisfy some criterion) and simplicity, the criterion of prior knowledge of the form(s) of the model is(are) involved. Certain physical constraints noted in Chapter 1 govern the selection of the class of models.

Once the form of the model is tentatively chosen, and the process inputs and outputs are known or can be obtained, the objective of *parameter* and *state estimation* is to determine the "best" values of the parameters and dependent variables in the model. Part III describes for several different classes of transport phenomena models appropriate techniques of estimation, employing both discrete and continuous experimental data.

CHAPTER 9

Estimation in Process Models Represented by Ordinary Differential Equations

After the algebraic equations, the simplest types of models based on transport phenomena (see Figure 1.1-2) are models containing ordinary differential equations. In this chapter we shall be concerned with estimation of the parameters in models involving single or multiple ordinary differential equations, primarily of the first but also of higher order. Such models represent unsteady-state lumped systems or steady-state distributed parameter systems as encountered, for example, in:

1. Missile trajectory analysis.
2. Chemical reactor analysis.
3. Signal processing.
4. Cybernetics.
5. Transport in human beings.

We shall describe parameter and, to some minor extent, state estimation. Estimation of the "state" of a process means estimation of the process dependent variables. Three types of state estimation can be distinguished. Given the observations over an interval t_0 to t_f, estimation of the state vector at time t can be classified as:

1. Interpolation or smoothing if t is less than t_f.
2. Filtering if $t = t_f$.
3. Prediction or extrapolation if t is greater than t_f.

The instrumentation used to carry out the numerical computations for number 2, which is the topic of interest here, is termed a filter; state estimation often is spoken of as filtering.

We cannot go into some of the very practical computational aspects of estimation, such as:

1. The time requirements for measurement and evaluation.
2. The choice of test inputs to be used, if any.
3. The efficiency of the data collection set-up.
4. The speed of convergence in the estimation procedure.
5. The type and arrangement of the data processing instrumentation, if any.

These factors are not readily susceptible to generalization, and they are best elucidated by specific examples in the current literature. The requirements outlined in previous chapters for experimentation and model evaluation still pertain to transport phenomena-based models. Experimental data should be taken by using a good experimental design (Chapter 8), the designs should be such that the errors in measurement are independent (Chapter 5), and the model in which the coefficients are to be estimated should be an *adequate one* (Chapter 7).

Some of the methods to be described yield unbiased estimates of the parameters while others do not. Estimates of the precision of the parameter estimates are always approximate in cases in which they can be made, because all the models solutions are nonlinear in the parameters and must be linearized (in the parameters) to obtain the precision estimates. In some instances the model differential equations can be solved explicitly for the dependent variable(s); in others this step is not feasible. Frequently the number of experimental observations which can be made and the time available in which to make them are restricted by economical factors or physical obstacles, and often the assumed stationarity and independence among the observational errors are not valid.

We shall commence by reviewing the type of models to be treated and the manner in which the unobservable errors are assumed to be introduced into the deterministic process model. Next, the least squares and maximum likelihood estimation techniques will be applied, after which three somewhat different methods of estimation will be described and illustrated: sequential estimation, quasilinearization, and the "equation error." Both discrete and continuous observations will be treated. The latter are important for estimation in real time, i.e., when the time of estimation is less than the time for the observed process variable to change significantly.

9.1 PROCESS MODELS AND INTRODUCTION OF ERROR

In this section we shall briefly characterize the types of models to be treated and certain of their significant features.

9.1-1 Process Models and Response Error

Table 1.1-2 indicates the relation between the models of this chapter and those of Chapter 10. A model in the sense used here is comprised of (1) the differential equation(s), plus (2) the boundary and/or initial conditions. Both (1) and (2) are required if the model is to have a unique solution. We must distinguish between linear (in the dependent variable(s)) and nonlinear models because analytical solutions for the latter are generally nonexistent and numerical or approximate solutions are needed, leading to estimation problems of a higher degree of complexity. We must also distinguish between initial value models and boundary value models.

In an *initial value model*, for a single differential equation, initial conditions (values of the dependent variable or its derivatives at the coordinate, or time, origin) must be given equal in number to the order of the highest order derivative. For a set of first-order differential equations, one initial condition on the dependent variable is usually given for each equation. If the general solution for the model is known, the arbitrary constants in the general solution arising from integration can be evaluated by substituting the given initial conditions into the general solution and solving the resulting set of equations for the constants. If the general solution is not known and a numerical solution of the differential equation(s) is carried out, the initial conditions give the starting point(s) for the integration which is usually executed by a stepwise or "marching" technique. A simple example of an initial value model in which the differential equation is linear in the dependent variable is a model of an isothermal well-stirred tank in which a first-order reaction takes place (see Figure 9.1-1).

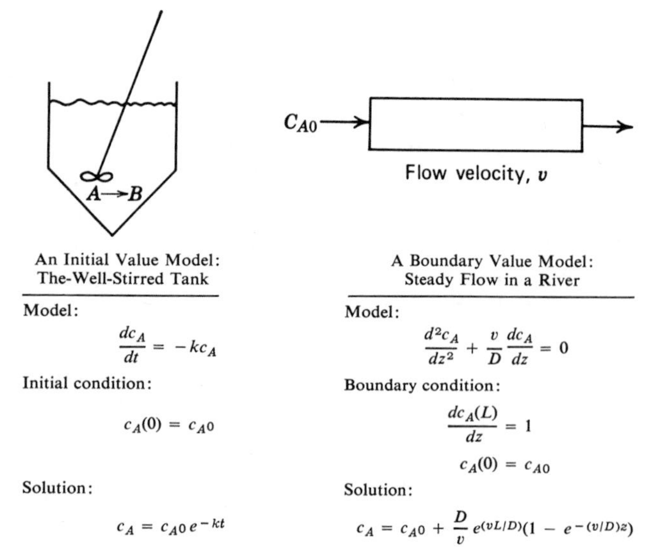

FIGURE 9.1-1 Comparison of an initial value model and a boundary value model; c_A = concentration of A; k is a reaction rate coefficient, t = time; D is a constant.

On the other hand, in a *boundary value model* the proper number of values of the dependent variable or its derivatives is given at various values of the independent variable, some values not being at the origin but (usually) at the end of the range for the independent variable. If the general solution for the model is known, the given values can be substituted into the general solution and the arbitrary constants evaluated. But if a numerical solution of the differential equations is required, the starting values for the numerical integration scheme are missing; hence the solution of a boundary value model is more complex and requires more time to execute than that of an initial value model. For example, we might use an initial value integration scheme but assume the initial values and compare the calculated values of the dependent variable with the given values at the other boundary. By iteration, these are made to agree. Refer to Collatz† and other texts on numerical analysis for specific calculation schemes. One particular method is described in Section 9.5. Figure 9.1-1 shows a typical boundary value model, that for idealized dispersion in a turbulent stream.

It is quite possible that the initial and/or boundary conditions must be estimated along with the model parameter, and we shall subsequently indicate how this may be accomplished.

The simplest model to be considered is the scalar, single, first-order, linear (in the dependent variable), ordinary differential equation with a constant coefficient:

$$\frac{dy(t)}{dt} = \alpha y(t) + x(t) \qquad y(0) = y_0 \tag{9.1-1}$$

which has the well-known solution:‡

$$y(t) = y_0\, e^{\alpha t} + \int_0^t x(\tau)\, e^{\alpha(t-\tau)}\, d\tau \tag{9.1-2}$$

where τ is a dummy variable. In Equation 9.1-1, α is the coefficient, y is the dependent variable often termed the system "state," t is the independent variable (usually but not always time), y_0 is the initial condition independent of time, and $x(t)$ is the deterministic input (or "forcing function"). Conceptually the unobservable error $\epsilon(t)$ is added to $y(t)$ to give the observable dependent variable $Y(t)$; refer to Figure 9.1-2. For discrete observations

$$Y(t_i) = y(t_i) + \epsilon(t_i) \tag{9.1-3a}$$

and for continuous variables

$$Y(t) = y(t) + \epsilon(t) \tag{9.1-3b}$$

If the estimated parameter $\hat{\alpha}$ replaces the model parameter α, the residual error is $E(t) = Y(t) - \hat{Y}(t)$. The

† L. Collatz, *The Numerical Treatment of Differential Equations*, Springer-Verlag, Berlin, 1960.

‡ W. Kaplan, *Ordinary Differential Equations*, Addison-Wesley, Reading, Mass., 1958.

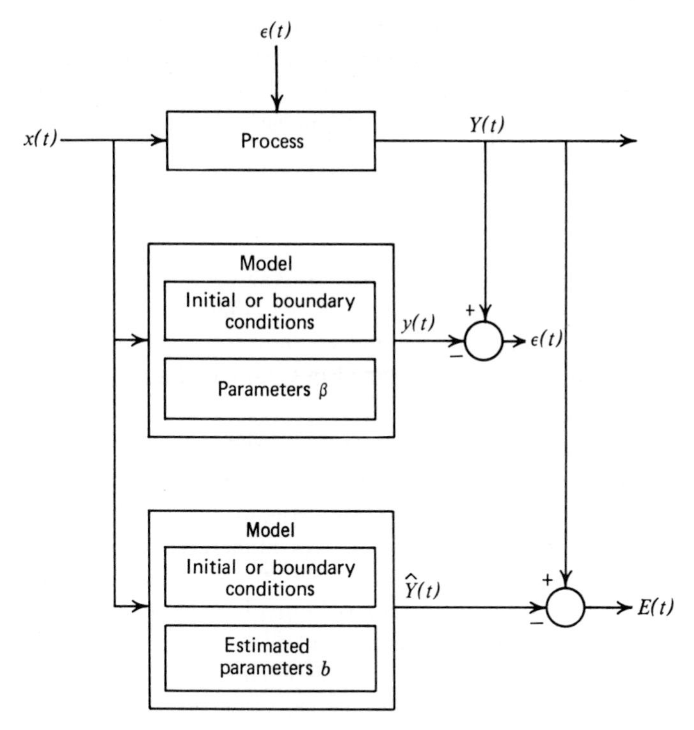

FIGURE 9.1-2 Information flow for the process and the model; $y(t)$ is the deterministic model output; $Y(t)$ is the experimental process output; $\hat{Y}(t)$ is the predicted output.

objective in parameter estimation is to obtain the "best," in the sense described in the next sections, estimate of α based on the observations $Y(t_i)$ or $Y(t)$. To do this we shall need to prescribe what $x(t)$ is and have some information about the nature of $\epsilon(t)$.

A more general model than Equation 9.1-1 is that comprised of a set of simultaneous, first-order, linear (in the dependent variables), ordinary differential equations with constant coefficients, of which the typical equation is

$$\frac{dy_r}{dt} = \sum_{s=1}^{v} \alpha_{rs} y_s + x_r(t) \qquad y_r(0) = y_{r0}$$

$$r = 1, 2, \ldots, v \quad (9.1\text{-}4)$$

Figure 9.1-3 illustrates the model. Equation 9.1-4 is usually written in matrix notation as

$$\frac{d\mathbf{y}}{dt} = \boldsymbol{\alpha}\mathbf{y} + \mathbf{x}(t) \qquad \mathbf{y}(0) = \mathbf{y}_0 \qquad (9.1\text{-}5)$$

where

$$\mathbf{y} = \begin{bmatrix} y_1 \\ y_2 \\ \vdots \\ y_v \end{bmatrix} \begin{matrix} \text{a} \\ v \times 1 \\ \text{matrix} \end{matrix} \qquad \mathbf{x}(t) = \begin{bmatrix} x_1(t) \\ x_2(t) \\ \vdots \\ x_v(t) \end{bmatrix} \begin{matrix} \text{a} \\ v \times 1 \\ \text{matrix} \end{matrix}$$

$$\boldsymbol{\alpha} = \begin{bmatrix} \alpha_{11} & \alpha_{12} & \cdots & \alpha_{1v} \\ \alpha_{21} & \alpha_{22} & \cdots & \alpha_{2v} \\ \vdots & \vdots & & \vdots \\ \alpha_{v1} & \alpha_{v2} & \cdots & \alpha_{vv} \end{bmatrix} \begin{matrix} \text{a} \\ v \times v \\ \text{matrix} \end{matrix}$$

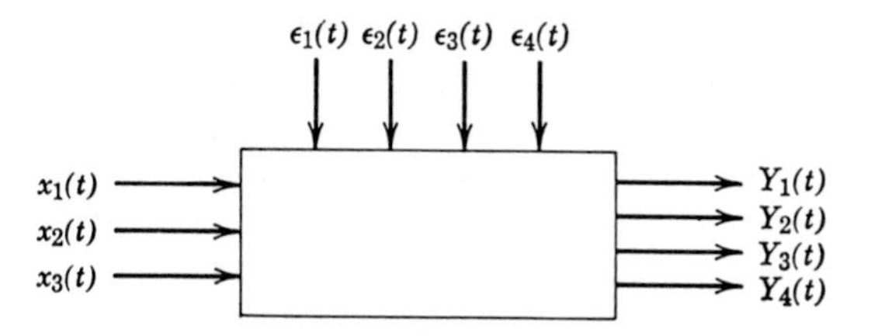

FIGURE 9.1-3 Multivariate process with multiple inputs.

Suppose that the tank in Figure 9.1-1 contains three components that react according to the scheme

$$A \xrightarrow{k_1} B \xrightarrow{k_2} C$$

where the k's represent reaction rate coefficients. Then the individual equations and specified initial conditions in Equation 9.1-5 would be

$$\frac{dc_A}{dt} = -k_1 c_A \qquad c_A(0) = 1$$

$$\frac{dc_B}{dt} = k_1 c_A - k_2 c_B \qquad c_B(0) = 0$$

$$\frac{dc_C}{dt} = k_2 c_B \qquad c_C(0) = 0$$

We assume that $\mathbf{x}(t)$ is given *a priori* but that $\mathbf{y}_0$ and $\boldsymbol{\alpha}$ are to be estimated over the time interval $0 \le t \le t_n$ from discrete observations described by the following relation:

$$\mathbf{Y}(t_i) = \mathbf{h}(t_i)\mathbf{y}(t_i) + \boldsymbol{\epsilon}(t_i) \qquad 1 \le i \le n \quad (9.1\text{-}6)$$

where $\mathbf{Y}(t_i)$ is an $n \times 1$ column vector, $\mathbf{h}(t_i)$ is an $n \times v$ matrix given *a priori*, and $\boldsymbol{\epsilon}(t_i)$ is an $n \times 1$ column vector ("noise" vector) whose elements are the unobservable errors.

The solution to Model 9.1-5 can be written in a form analogous to Equation 9.1-2 as

$$\mathbf{y}(t) = \exp(\boldsymbol{\alpha}t)\mathbf{y}_0 + \int_0^t \exp[\boldsymbol{\alpha}(t-\tau)]\mathbf{x}(\tau)\,d\tau \quad (9.1\text{-}7)$$

For example, the solution to the model of three chemical components in the well-stirred tank is

$$c_A = e^{-k_1 t}$$

$$c_B = \frac{k_1}{k_2 - k_1}(e^{-k_1 t} - e^{-k_2 t})$$

$$c_C = 1 - \frac{1}{k_2 - k_1}(k_2 e^{-k_1 t} - k_1 e^{-k_2 t})$$

Introduction of Equation 9.1-7 into Equation 9.1-6 yields

$$\mathbf{Y}(t_i) = \mathbf{h}(t_i)\left[\exp(\boldsymbol{\alpha}t_i)\mathbf{y}_0 + \int_0^{t_i} \exp[\boldsymbol{\alpha}(t_i-\tau)]\mathbf{x}(\tau)\,d\tau\right] + \boldsymbol{\epsilon}(t_i)$$

$$(9.1\text{-}8)$$

which can be written in general form as

$$\mathbf{Y}(t_i) = \mathbf{\Psi}(\boldsymbol{\alpha}, \mathbf{y}_0, t_i) + \boldsymbol{\epsilon}(t_i) \qquad (9.1\text{-}9)$$

A similar expression can be written for continuous observations by dropping the subscript index i on t. Although Equation 9.1-9 is somewhat formal, it can be seen from the example solution used to illustrate Equation 9.1-7 that the coefficients and the initial conditions are involved quite nonlinearly in the model solution.

It should be pointed out that a model containing one or more higher order linear (in the dependent variable) differential equations with constant coefficients, such as

$$\frac{d^2y}{dt^2} + \alpha_1 \frac{dy}{dt} + \alpha_2 y = x(t)$$

can be transformed to a model containing a set of first-order ordinary differential equations as follows. Let

$$w = \frac{dy}{dt}$$

and

$$\frac{dw}{dt} = \frac{d^2y}{dt^2}$$

Consequently, the second-order differential equation becomes two first-order differential equations:

$$\frac{dy}{dt} = w$$

$$\frac{dw}{dt} = -\alpha_1 w - \alpha_2 y + x(t)$$

However, w is a derivative. For some of the criteria discussed in subsequent sections, the experimental values of the derivative must be observed as functions of time in order to use the above scheme.

Finally, the general nonlinear (in the dependent variables) first-order model is

$$\frac{d\mathbf{y}}{dt} = \mathbf{f}(\boldsymbol{\alpha}, \mathbf{y}, t) \qquad \mathbf{y}(0) = \mathbf{y}_0 \qquad (9.1\text{-}10)$$

where $\mathbf{f}(\boldsymbol{\alpha}, \mathbf{y}, t)$ represents a very general nonlinear function. Equation 9.1-10 will not have an analytical solution except in rare instances, and it must be solved by numerical methods. However, Equation 9.1-3 still applies if the solution to Equation 9.1-10 can be obtained.

9.1-2 Unobservable Error Added to Derivatives

Because of the difficulty of obtaining analytical solutions to the deterministic process model as represented by Equation 9.1-10, experiments have been arranged whereby the vector of derivatives $d\mathbf{Y}/dt$ is measured rather than $\mathbf{Y}$ itself. Carberry† compared chemical reactor configurations which enable the experimenter to observe either c_A or dc_A/dt. Also, in some types of optical instruments the rate of change of the dependent variable is measured rather than the dependent variable itself. In such cases it is assumed that the unobservable error is added to the deterministic derivative dy/dt as follows:

$$\frac{d\mathbf{Y}}{dt} = \frac{d\mathbf{y}}{dt} + \boldsymbol{\epsilon}$$

When the derivative is the observed variable, the estimation procedure does not involve a differential equation at all; the parameters and initial conditions can be estimated by the techniques described in Chapters 5 and 6.

† J. J. Carberry, *Ind. Eng. Chem.* **56** (11), 39, 1964.

9.1-3 Differentiation of Process Data

Another, less satisfactory, approach to eliminating working with derivatives in estimation is to observe Y and to evaluate the derivatives numerically. Two main types of error must be considered in the numerical evaluation of derivatives. One type involves error introduced by the numerical scheme employed, and the other type involves the stochastic error associated with the observations. We examine the numerical error first.

Numerical differentiation of deterministic variables involves evaluation of dy/dt, or higher derivatives, at some arbitrary value of the independent variable t, say t_0, given a series of values of y in the interval about t_0. Most texts treating numerical analysis describe relations that can be used to calculate derivatives from values of y taken at equal or unequal intervals. However, even the use of polynomials, $y = g(t)$, to approximate the values of y requires careful treatment because, as Figure 9.1-4 illustrates, a suitable fitting polynomial at each y may have the wrong slope at each base point as well as elsewhere.

As one would expect, the deterministic error in the derivative is smaller when the data are centered about a t_0 in the middle of the range of t than when the t_0 is placed at one or the other end of the interval for the y's. Any of the interpolation polynomials (divided, forward, central, and backward difference, Lagrange, Gram, etc.) can be used to replace the continuous derivative, depending on the circumstances. Table 9.1-1 lists the numerical errors for several difference formulas used to evaluate the temperature gradient at a wall, $(dT/dz)_{z_0}$,

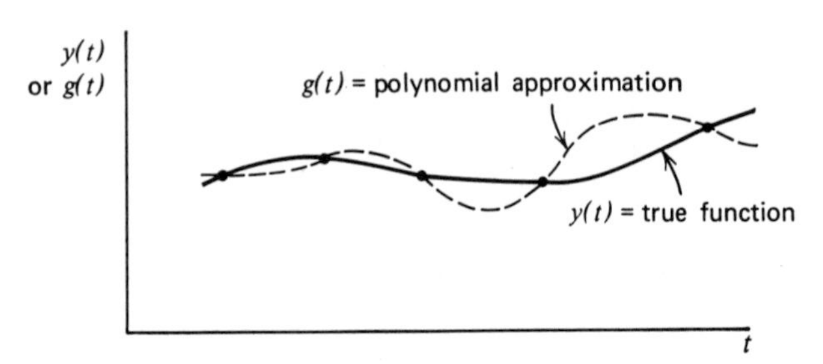

FIGURE 9.1-4 Polynomial approximation of a function.

TABLE 9.1-1 CALCULATION OF THE TEMPERATURE GRADIENT dT/dz AT $z = z_0^*$

Order of Polynomial	Approximation of dT/dz at $z = z_0$	Order of Error Term	Numerical Value of $(dT/dz)_{z=0}$
1	$\dfrac{T_1 - T_0}{h}$	h	4.12
2	$\dfrac{-T_2 + 4T_1 - 3T_0}{2h}$	h^2	4.71
3	$\dfrac{2T_3 - 9T_2 + 18T_1 - 11T_0}{6h}$	h^3	5.25
4	$\dfrac{-3T_4 + 16T_3 - 36T_2 + 48T_1 - 25T_0}{12h}$	h^4	3.47

Measured temperatures were $T_0 = -1.000$, $T_1 = -0.588$, $T_2 = -0.295$, $T \;\; = -0.259$, $T_5 = -0.305$, and $h = 0.1$.

from measured temperatures if the temperatures were treated as deterministic variables.†

The approximations of dT/dz in Table 9.1-1 and the "order of error" obscure a far more important source of error in the evaluation of dT/dz, namely the stochastic error engendered by measuring T. Most approximation schemes for derivatives can be written in the general form:

$$D_{m+1}^k(y) = \frac{c}{h^k}(a_0 y_0 + a_1 y_1 + \cdots + a_m y_m)$$

where a_i is a constant, D is the differential operator, k is the order of the derivative, h is the interval between the y's, and $(m + 1)$ designates the number of data points used. Thus, the variance of the derivative can be estimated by using the propagation of error formulas (assuming that the Y's are stochastically independent—which may be unlikely in practice):

$$\text{Var}\{D_{m+1}^k(Y)\} = \frac{c^2}{h^{2k}}[a_0^2 \text{Var}\{Y_0\} + \cdots + a_m^2 \text{Var}\{Y_m\}] \quad (9.1\text{-}11)$$

If we assume the variances of all the Y's are equal,

$$\text{Var}\{D_{m+1}^k(Y)\} = \frac{c^2}{h^{2k}} \text{Var}\{Y\} \sum_{i=0}^{m} a_i^2 \quad (9.1\text{-}12)$$

From Equation 9.1-12, we observe that the smaller the interval h and the more terms in the formula, the larger the error in the derivative and that the error goes up with the order of the derivative. To take a specific example, suppose all the T's in Table 9.1-1 have a common standard deviation of 0.01 (1 percent of T_0) or a variance of 10^{-4}. Then, for the fourth-order polynomial,

$$\text{Var}\left\{\left(\frac{dT}{dz}\right)_{z=z_0}\right\} \cong \frac{4490}{[12(0.1)]^2}(10^{-4}) = 0.311$$

and

$$\sigma_{(dT/dz)_{z=0}} \cong 0.56$$

or 16 percent of $(dT/dz)_{z=z_0}$.

In other words, if the numerically evaluated derivative is to be used as the response in estimation, the error in the stochastic dependent variable is amplified tremendously. Consequently, we conclude that the general admonition to avoid numerical differentiation of experimental data, if possible, in the estimation of parameters has a sound foundation. In the "equation error" estimation technique, described in Section 9.6, we shall assume that observations of the derivatives themselves are used rather than derivatives calculated from the observations.

† J. O. Wilkes, "The Finite Difference Computation of Natural Convection in an Enclosed Rectangular Cavity," Ph.D. Thesis, Univ. of Michigan, Ann Arbor, 1963.

9.2 LEAST SQUARES ESTIMATION

Inasmuch as least squares parameter estimation does not require prior knowledge of the distribution of unobservable errors, yields unbiased estimates, that is, $\mathscr{E}\{\hat{Y}(t)\} = y(t)$, and results in the minimum variance among all linear unbiased estimators, the least squares technique is used as extensively for transport phenomena models as it is for the empirical models described in Part II of this text. If the observations $\mathbf{Y}$ for the model responses are continuous functions of time from $t = 0$ to $t = t_f$, the Markov (or "rigorous least squares") criterion is to minimize

$$\phi = \tfrac{1}{2}\int_0^{t_f} [\mathbf{Y} - \mathbf{\Psi}]^T \mathbf{\Gamma}^{-1} [\mathbf{Y} - \mathbf{\Psi}]\, dt \quad (9.2\text{-}1)$$

where $\mathbf{\Gamma}$ is the covariance matrix (or perhaps a matrix of appropriate weights) described in Section 5.5, and ϕ is the time integrated value of the error squared ("integral squared error"). If the observations are made at discrete instants of time, t_i, $i = 1, 2, \ldots, n$, the Markov criterion is to minimize

$$\phi = \tfrac{1}{2} \sum_{i=1}^{n} [\mathbf{Y}(t_i) - \mathbf{\Psi}(t_i)]^T \mathbf{\Gamma}^{-1} [\mathbf{Y}(t_1) - \mathbf{\Psi}(t_i)] \quad (9.2\text{-}2)$$

(Sometimes the number $\frac{1}{2}$ is omitted in Equations 9.2-1 and 9.2-2.) If, as in Chapter 5, $\mathbf{\Gamma}$ is a diagonal matrix (all the off-diagonal elements are zero), ϕ becomes a "weighted least squares" criterion; if $\mathbf{\Gamma} = \sigma_\epsilon^2 \mathbf{I}$, ϕ is the "ordinary least squares" criterion.

9.2-1 Discrete Observations

To minimize ϕ for discrete observations, we can formally differentiate Equation 9.2-2 with respect to $\mathbf{y}_0$ and with respect to $\boldsymbol{\alpha}$ and then equate the resulting expressions to a null vector $\mathbf{0}$ and a null matrix $\mathbf{0}$, respectively. We obtain the following set of nonlinear (in the estimators) equations:†

$$\frac{\partial \phi}{\partial \mathbf{y}_0} = \mathbf{0}^T = -\sum_{i=1}^{n} [\mathbf{Y} - \mathbf{\Psi}(\hat{\boldsymbol{\alpha}}, \hat{\mathbf{y}}_0, t_i)]^T \mathbf{\Gamma}^{-1}(t_i) \frac{\partial}{\partial \mathbf{y}_0} \mathbf{\Psi}(\hat{\boldsymbol{\alpha}}, \hat{\mathbf{y}}_0, t_i)$$

$$(9.2\text{-}3)$$

$$\frac{\partial \phi}{\partial \boldsymbol{\alpha}} = \mathbf{0} = -\sum_{i=1}^{n} [\mathbf{Y} - \mathbf{\Psi}(\hat{\boldsymbol{\alpha}}, \hat{\mathbf{y}}_0, t_i)]^T \mathbf{\Gamma}^{-1}(t_i) \frac{\partial}{\partial \boldsymbol{\alpha}} \mathbf{\Psi}(\hat{\boldsymbol{\alpha}}, \hat{\mathbf{y}}_0, t_i)$$

A similar set of equations can be obtained for continuous data except that time integrals replace the discrete summations. Equations 9.2-3 represent the $n + (n \times n) = n(1 + n)$ nonlinear equations used to estimate the $n(1 + n)$ elements in $\boldsymbol{\alpha}$ and $\mathbf{y}_0$.

To obtain estimates of the precision of the estimates $\hat{\boldsymbol{\alpha}}$ and $\hat{\mathbf{y}}_0$, some assumption must be made about the distribution of the unobservable errors, such as that of a joint normal distribution. To obtain estimates of the precision of $\hat{\mathbf{Y}}$, the solution to the model must be expressed approximately as a linear function of the parameters by expanding the solution about the estimated parameters as described in Section 6.4. In circumstances in which an analytical solution for the model is not known, an approximate analytical solution can be substituted and linearized. If it is assumed that $\mathbf{\Gamma} = \sigma_\epsilon^2 \mathbf{I}$, Equation 6.4-3 gives the approximate covariance matrix for the elements of

$$\mathbf{b} = \begin{bmatrix} \hat{\boldsymbol{\alpha}} \\ \hat{\mathbf{y}} \end{bmatrix}$$

and Equation 6.4-5 gives the approximate joint confidence region. If $\mathbf{\Gamma} \neq \sigma_\epsilon^2 \mathbf{I}$, then the elements of $\mathbf{\Gamma}$ must be estimated as described in Section 5.5 and the estimated covariance matrix of $\mathbf{b}$, as described in Section 5.1, is

$$\widehat{\text{Covar}\,\{\mathbf{b}\}} \cong (\hat{\mathbf{X}}^T \hat{\mathbf{\Gamma}}^{-1} \hat{\mathbf{X}})^{-1}$$

where the overlays ^ denote that estimated parameters are used to evaluate the matrix elements.

As an example of Equations 9.2-3 applied to a specific model, we use the scalar Equations 9.1-1 through 9.1-3. By differentiating ϕ:

$$\phi = \tfrac{1}{2} \sum_{i=1}^{n} \left[Y(t_i) - y_0\, e^{\alpha t_i} + \frac{x_0}{\alpha}(1 - e^{\alpha t_i}) \right]^2$$

with respect to y_0 and next with respect to α and then replacing y_0 and α in the resulting expressions with their estimates, we get

$$\sum_{i=1}^{n} \left[Y(t_i) - \hat{y}_0\, e^{\hat{\alpha} t_i} + \frac{x_0}{\hat{\alpha}}(1 - e^{\hat{\alpha} t_i}) \right] e^{\hat{\alpha} t_i} = 0$$

$$\sum_{i=1}^{n} \left[Y(t_i) - \hat{y}_0\, e^{\hat{\alpha} t_i} + \frac{x_0}{\hat{\alpha}}(1 - e^{\hat{\alpha} t_i}) \right] \cdot \left\{ t_i \hat{y}_0\, e^{\hat{\alpha} t_i} - \frac{x_0}{\hat{\alpha}^2} [e^{\hat{\alpha} t_i}(1 - \hat{\alpha} t_i) - 1] \right\} = 0$$

The important points to keep in mind when using Equations 9.2-3 are that: (1) the unobservable error is added in a special way to the deterministic response, (2) all the n responses are used simultaneously in the estimates, and (3) no prior statistical information enters into the criterion except perhaps as introduced through the matrix $\mathbf{\Gamma}$. Because Equations 9.2-3 are nonlinear, some type of numerical solution such as the Newton-Raphson technique is required to ascertain the values of $\hat{\boldsymbol{\alpha}}$ and $\hat{\mathbf{y}}_0$. As explained in Section 6.2, this requirement often leads one to select one of the optimization procedures described in Section 6.2 rather than work with Equations 9.2-3 directly.

9.2-2 Computational Problems

The main difficulty involved in the least squares estimation procedure for transport phenomena models remains the same as that recited in Sections 5.5, 6.2, and 6.3, namely how to carry out successfully the optimiza-

† The symbol $\partial \mathbf{\Psi} / \partial \boldsymbol{\beta}$ means that each element in the matrix

$$\mathbf{\Psi} = \begin{bmatrix} \Psi_1 \\ \vdots \\ \Psi_v \end{bmatrix}$$

is differentiated with respect to β_1 to give $\partial \mathbf{\Psi} / \partial \beta_1$, next with respect to β_2 to give the matrix $\partial \mathbf{\Psi} / \partial \beta_2$, and so on. The elements $\partial \mathbf{\Psi} / \partial \beta_j$ assembled as a column vector comprise

$$\frac{\partial \mathbf{\Psi}}{\partial \boldsymbol{\beta}} = \begin{bmatrix} \dfrac{\partial \mathbf{\Psi}}{\partial \beta_1} \\ \vdots \\ \dfrac{\partial \mathbf{\Psi}}{\partial \beta_m} \end{bmatrix}$$

tion of ϕ for a nonlinear model. Because of the complex nature of the solutions to the model, local optima can be expected to be encountered; hence the choice of initial guesses for the parameters is important. If the initial guesses can be chosen so that the dominant time constants† of the model are of the same order of magnitude as the dominant time constants of the measured responses, the difficulty can be ameliorated.

In addition, one new problem arises related to the experimental design, that of the stability of the deterministic model. If the model is stable, the dependent variable reaches some asymptotic value in time (or distance). To estimate successfully the model parameters, the bulk of the experimental data must be collected in the early portion of the run. See Figure 9.2-1. On the other hand, if the model is unstable so that the deterministic solution increases without bound, the data also have to be properly collected or weighted if the last, and largest, responses are not to overwhelm the earlier ones.

Examples of estimation using nonlinear (in the parameters) analytical solutions to the deterministic process model corresponding to Equation 9.1-7 have been previously given in Chapter 6. Consequently, we show below the application of least squares to a process model which is nonlinear in the dependent variables—analogous to Equation 9.1-10—and hence requires a numerical solution.

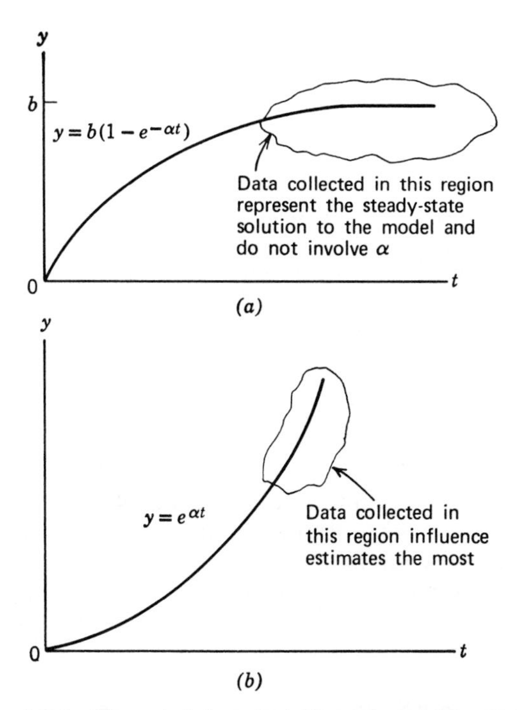

FIGURE 9.2-1 Characteristics of stable and unstable solutions for process models: (*a*) stable solution to deterministic model and (*b*) unstable solution to deterministic model.

† That is, exponential exponents associated with t in the solution to the model, which are the same as eigenvalues of the matrix $\boldsymbol{\alpha}$.

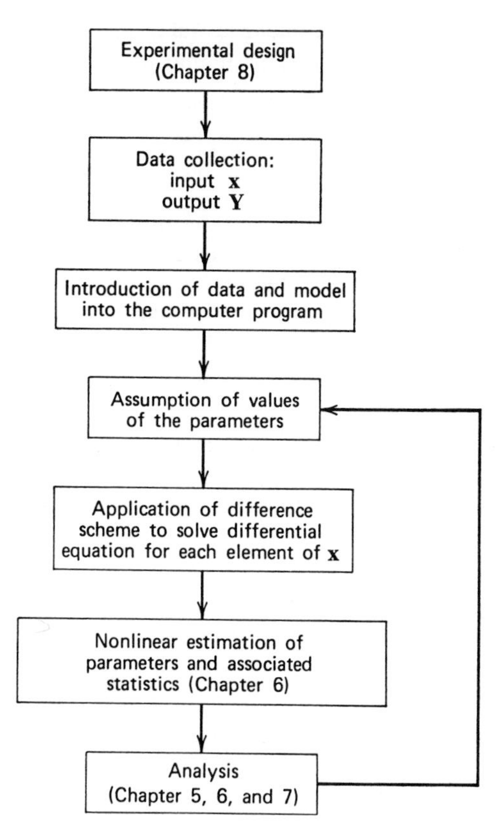

FIGURE 9.2-2 Information flow for estimation of initial value model parameters with the aid of numerical difference schemes.

If difference equations are used to solve the differential equations, the addition of a suitable computer routine to a nonlinear estimation routine is reasonably straightforward. The Adams-Moulton multistep method and the Runge-Kutta single-step method for the initial computations are well known as standard techniques for initial value problems;‡ most computer centers have computer codes to execute the solutions. Analog and hybrid computers, or digital simulators of analog computers (such as MIMIC), provide easy execution of standard codes for those unskilled in programming. It is necessary to provide initial estimates of the model coefficients, which are then improved by some iterative scheme as outlined in the information flow diagram of Figure 9.2-2. Of course, difficulties can be encountered with the numerical schemes and with roundoff. In the Runge-Kutta and other numerical difference schemes, numerical instability can arise if certain stability criteria are violated.§ In some models, large terms of

‡ P. Henrici, *Discrete Variable Methods in Ordinary Differential Equations*, John Wiley, New York, 1962; L. Lapidus, *Digital Computation for Chemical Engineers*, McGraw-Hill, New York, 1962.

§ J. Certaine, "The Solution of Ordinary Differential Equations with Large Time Constants" in *Mathematical Methods for Digital Computers*, ed. by A. Ralston and H. S. Wilf, John Wiley, New York, 1960.

approximately equal value are subtracted from each other, with a resulting loss of significant figures.

Example 9.2.1 Estimation of Kinetic Rate Coefficients using Discrete Observations

Ball and Groenweghe † applied the procedure outlined in Figure 9.2-2 to estimate the kinetic coefficients for the second-order kinetics of certain organometallic compounds represented by the following reaction scheme:

$$1 + 3 \underset{}{\overset{k_1}{\rightleftarrows}} 2 + 2$$

$$2 + 4 \underset{}{\overset{k_2}{\rightleftarrows}} 3 + 3$$

$$3 + 5 \underset{}{\overset{k_3}{\rightleftarrows}} 4 + 4$$

$$1 + 4 \underset{}{\overset{k_4}{\rightleftarrows}} 2 + 3$$

$$2 + 5 \underset{}{\overset{k_5}{\rightleftarrows}} 3 + 4$$

$$1 + 5 \underset{}{\overset{k_6}{\rightleftarrows}} 2 + 4$$

in which the k's are the forward rate constants and the numbers designate the chemical species. The rate constants were first assumed and later verified to be independent of the composition of the mixture. The six equilibrium constants (K_i) for the reactions were evaluated separately so that only half of the twelve rate constants had to be estimated (the reverse rate constant could be derived from the forward one).

The reaction mechanism indicated that five species existed, 1, 2, 3, 4, and 5, but they were constrained by two total material balance equations based on the original compositions at $t = 0$, with the consequence that only three independent differential equations were required to make up the model. The remaining components then could be obtained from the total material balances. The three differential equations were

$$\frac{dc_2}{dt} = 2k_1\left(c_1c_3 - \frac{c_2c_2}{K_1}\right) - k_2\left(c_2c_4 - \frac{c_3c_3}{K_2}\right) + k_4\left(c_1c_4 - \frac{c_2c_3}{K_4}\right) - k_5\left(c_2c_5 - \frac{c_3c_4}{K_5}\right) + k_6\left(c_1c_5 - \frac{c_2c_4}{K_6}\right)$$

$$\frac{dc_3}{dt} = -k_1\left(c_1c_3 - \frac{c_2c_2}{K_1}\right) + 2k_2\left(c_2c_4 - \frac{c_3c_3}{K_2}\right) - k_3\left(c_3c_5 - \frac{c_4c_4}{K_3}\right) + k_4\left(c_1c_4 - \frac{c_2c_3}{K_4}\right) + k_5\left(c_2c_5 - \frac{c_3c_4}{K_5}\right)$$

$$\frac{dc_4}{dt} = -k_2\left(c_2c_4 - \frac{c_3c_3}{K_2}\right) + 2k_3\left(c_3c_5 - \frac{c_4c_4}{K_3}\right) - k_4\left(c_1c_4 - \frac{c_2c_3}{K_4}\right) + k_5\left(c_2c_5 - \frac{c_3c_4}{K_5}\right) + k_6\left(c_1c_5 - \frac{c_2c_4}{K_6}\right)$$

The initial conditions were known deterministic values: $c_1(0) = c_{10}$; $c_2(0) = 0$; $c_3(0) = 0$; $c_4(0) = 0$; and $c_5(0) = c_{50}$. The integration scheme was taken from Fehlberg,‡ who presented a number of fast and accurate numerical integration formulas of the predictor-corrector type. The particular method was to continue the solution $d\mathbf{C}/dt = \mathbf{C}' = \mathbf{f}(t, \mathbf{C})$ from values already found for previous times $t_0, t_1, \ldots, t_n$. Let $\mathbf{C}_n$ be a vector of composition values at t_n. Then the procedure would be to ($h = t_{n+1} - t_n$):

1. Predict (estimate) the next value of the dependent variable vector:

$$\mathbf{C}_{n+1}^* = \mathbf{C}_{n-3} + \frac{4h}{3}(2\mathbf{C}_n' - \mathbf{C}_{n-1}' + 2\mathbf{C}_{n-2}')$$

2. Evaluate the derivative function:

$$\mathbf{C}_{n+1}' = \mathbf{f}(t_{n+1}, \mathbf{C}_{n+1}^*)$$

3. Correct the estimated value:

$$\mathbf{C}_{n+1} = \tfrac{9}{17}\mathbf{C}_n + \tfrac{9}{17}\mathbf{C}_{n-1} - \tfrac{1}{17}\mathbf{C}_{n-2} + h(\tfrac{6}{17}\mathbf{C}_{n+1}' + \tfrac{18}{17}\mathbf{C}_n')$$

4. Evaluate the derivative function:

$$\mathbf{C}_{n+1}' = \mathbf{f}(t_{n+1}, \mathbf{C}_{n+1})$$

5. Check the local truncation error using the value of $\frac{27}{503}(\mathbf{C}_{n+1}^* - \mathbf{C}_{n+1})$. Bal and Groenweghe indicated that the Fehlberg equations required less machine time than Hamming's integration formula.§

Marquardt's method as described in Chapter 6, particularly Equation 6.2-20, was used to carry out the least squares estimation. The required partial derivatives were obtained numerically by integrating the rate equations with slightly perturbed rate constants and then calculating the appropriate differences:

$$\frac{\partial \hat{C}_{ij}}{\partial k_l} = \frac{\hat{C}_{ij}(k_1, \ldots, k_l + \Delta k_l, \ldots, k_L) - \hat{C}_{ij}(k_1, \ldots, k_l, \ldots, k_L)}{\Delta k_l} \qquad l = 1, 2, \ldots, 6$$

Hence, one Marquardt iteration required the integration of the rate equations at least seven times to obtain the necessary number of derivative values.

In one application of the above program, compositions were measured (in moles/liter) for each of the five components for the times (in minutes): 0, 10, 40, 70, 202, 490,

† W. E. Ball and L. C. D. Groenweghe, *Ind. Eng. Chem. Fundamentals* **5**, 181, 1966.

‡ E. Fehlberg, "Numerically Stable Interpolation Formulas with Favorable Error Propagation for First and Second Order Differential Equations," *Tech. Note* **D-599**, National Aeronautics and Space Administration, Mar. 1961.

§ R. W. Hamming, "Stable Predictor-Corrector Methods for Ordinary Differential Equations," *J. ACM* **6**, 37, 1959.

1190, 1453, 2410, 2795, and 3765. Thus a total of (5)(10) = 50 measured composition values were used to estimate the six rate coefficients. Starting estimates of the rate coefficients were obtained by hand calculation of the slopes of the concentration-*versus*-time curves at several points. The decrease in the function ϕ for each iteration was as follows:

Iteration	ϕ	IBM 7040 Machine Time (sec)
0		0
1	0.0278	100
2	0.0254	186
3	0.0215	273
4	0.0206	360

The changes in rate constants were:

	Initial Estimated k Values	Final Estimated k Values
k_1:	0.053	0.0418
k_2:	0.025	0.0272
k_3:	0.0048	0.0218
k_4:	0.0085	0.0208
k_5:	0.0069	0.000532
k_6:	0.0141	0.0108

The estimates tended to converge satisfactorily in very few iterations when the proposed kinetic model adequately represented the chemical system, i.e., when ϕ was small. However, as might be expected, when the model was a poor one, many more iterations were performed and the estimated parameters changed in a seemingly random way.

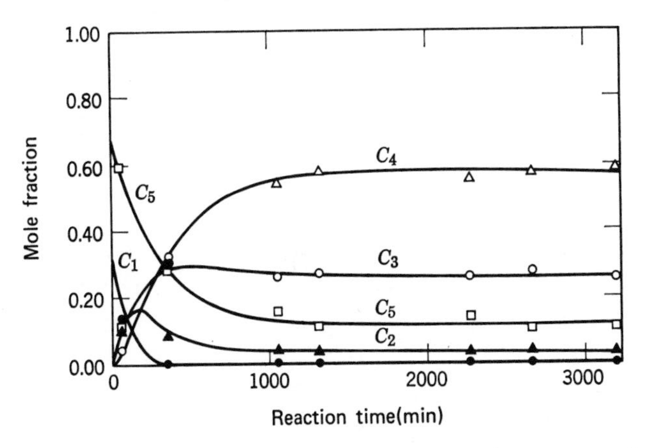

FIGURE E9.2-1 Calculated response curves and experimental data:

Experimental points:
● = C_1 = (*tert*-BuO)$_4$Ti, ▲ = C_2 = (*tert*-BuO)$_3$(Me$_2$N)Ti,
○ = C_3 = *tert*-BuO)$_2$(Me$_2$N)$_2$Ti,
△ = C_4 = (*tert*-BuO)(Me$_2$N)$_3$Ti, □ = C_5 = (Me$_2$N)$_4$Ti

Values of kinetic constants:
$k_1 = 3.74 \times 10^{-3}$, $k_2 = 8.33 \times 10^{-3}$, $k_3 = 5.33 \times 10^{-4}$,
$k_4 = 2.47 \times 10^{-4}$, $k_5 = 1.58 \times 10^{-4}$, $k_6 = 2.58 \times 10^{-5}$

Equilibrium constants:
$K_1 = 7.40 \times 10^{-2}$, $K_2 = 2.97 \times 10^{-1}$, $K_3 = 9.50 \times 10^{-2}$

(Adapted with permission from *Ind. Eng. Chem. Fundamentals* **5** (2), 183, 1966.)

Estimates of the precision of the estimates of the estimators were not given nor were estimates of the precision of $\hat{C}_i$.

Figure E9.2-1 compares the computer output curves for the predicted concentrations (the "state") for a slightly different run with the experimental points. Note that the bulk of the observations of concentration were taken in the steady state ($t > 1000$ min); such an experimental design makes the estimation of a response with a maximum, such as $\hat{C}_2$, or a response that rapidly reduces to zero, such as $\hat{C}_1$, quite difficult. The fact that the initial conditions were known was of material assistance in the estimation.

9.2-3 Continuous Observations

Example 9.2-1 describes least squares estimation oriented toward data processing by an off-line digital computer. In the fields of engineering, biomedicine, geophysics, and oceanography, large volumes of data must be processed in real time on-line, with the result that the model parameters are estimated and adjusted *continuously* by continuously reducing ϕ in Equation 9.2-1.

Several techniques have been applied to minimize ϕ continuously, but we shall restrict our attention to the most common procedure, namely that of steepest descent. In steepest descent, the time rate of change of a model parameter β_j is given by

$$\frac{\partial \beta_j}{dt} = -k \frac{\partial \phi}{\partial \beta_j} \tag{9.2-4}$$

Equation 9.2-4 evolves from the following analysis.

Consider a function of several parameters, $\phi(\beta_1, \beta_2, \ldots, \beta_n)$. To move toward the minimum of ϕ, we need to go a distance ds in parameter space where ds is defined in the Euclidean sense:

$$ds^2 = \sum_{j=1}^{m} d\beta_j^2 \tag{9.2-5}$$

The change of ϕ with respect to s is

$$\frac{d\phi}{ds} = \sum_{j=1}^{m} \frac{\partial \phi}{\partial \beta_j} \frac{d\beta_j}{ds} \tag{9.2-6}$$

and the direction of steepest descent is the biggest negative value of $(d\phi/ds)$ which makes Equation 9.2-6, subject to Equation 9.2-5, stationary.

A Lagrangian function can be formed (see Appendix B)

$$g = \sum_{j=1}^{m} \frac{\partial \phi}{\partial \beta_j} \frac{d\beta_j}{ds} + \lambda \left[1 - \sum_{j=1}^{m} \left(\frac{d\beta_j}{ds}\right)^2\right]$$

After equating the partial derivatives of g with respect to $(d\beta_j/ds)$ to zero, we have

$$\frac{\partial \phi}{\partial \beta_j} + \lambda\left(-2 \frac{d\beta_j}{ds}\right) = 0$$

or

$$\frac{d\beta_j}{ds} = \frac{1}{2\lambda}\frac{\partial \phi}{\partial x_i} \qquad j = 1, \ldots, m \tag{9.2-7}$$

Introduction of Equation 9.2-7 into

$$1 = \sum_{j=1}^{m} \left(\frac{d\beta_j}{ds}\right)^2$$

gives the function for λ:

$$\lambda = \pm\frac{1}{2}\left[\sum_{j=1}^{m}\left(\frac{\partial \phi}{\partial \beta_j}\right)^2\right]^{1/2} \tag{9.2-8}$$

Introduction of Equation 9.2-8 into 9.2-7 gives

$$\frac{d\beta_j}{ds} = \pm\left[\sum_{i=1}^{n}\left(\frac{\partial \phi}{\partial \beta_j}\right)^2\right]^{-1/2}\left(\frac{\partial \phi}{\partial \beta_j}\right) \tag{9.2-9}$$

where the positive sign indicates the rate of change for steepest ascent and the negative sign for steepest descent.

If (ds/dt) is now identified as v, a "velocity," because

$$\frac{d\beta_j}{dt} = \frac{d\beta_j}{ds}\frac{ds}{dt} = \frac{d\beta_j}{ds}v$$

all we have to do is choose v proportional to the magnitude of the gradient:

$$v = k\left[\sum_{j=1}^{m}\left(\frac{\partial \phi}{\partial \beta_j}\right)^2\right]^{1/2} \qquad k > 0$$

to find one expression for the rate of change β_i in the direction of steepest descent:

$$\frac{d\beta_j}{dt} = -k\frac{\partial \phi}{\partial \beta_j} \qquad j = 1, \ldots, m \tag{9.2-4}$$

as previously indicated. Equation 9.2-4 can be compared with the stepwise version of steepest descent given in Section 6.2-4 by letting

$$\frac{d\beta_j}{dt} \cong \frac{\beta_j^{(n+1)} - \beta_j^{(n)}}{\Delta t} = -k\frac{\partial \phi}{\partial \beta_j}$$

or

$$\beta_j^{(n+1)} = \beta_j^{(n)} - k(\Delta t)\frac{\partial \phi}{\partial \beta_i}$$

If $k(\Delta t)$ is denoted by the constant h, the step size, the relationship with Equation 6.2-17a becomes evident. The use of Equation 9.2-4 will be illustrated by the next example.

Example 9.2-2 Continuous Linear Estimation Using Steepest Descent

For a simple illustration of continuous parameter estimation by the method of least squares using steepest descent, consider the determination of the parameters in the linear model $y = \beta_0 + \beta_1 x$. The criterion for estimation is to minimize ϕ:

$$\phi = \tfrac{1}{2}\int_0^t \epsilon^2\, dt' = \tfrac{1}{2}\int_0^t (Y - \beta_0 - \beta_1 x)^2\, dt' \tag{a}$$

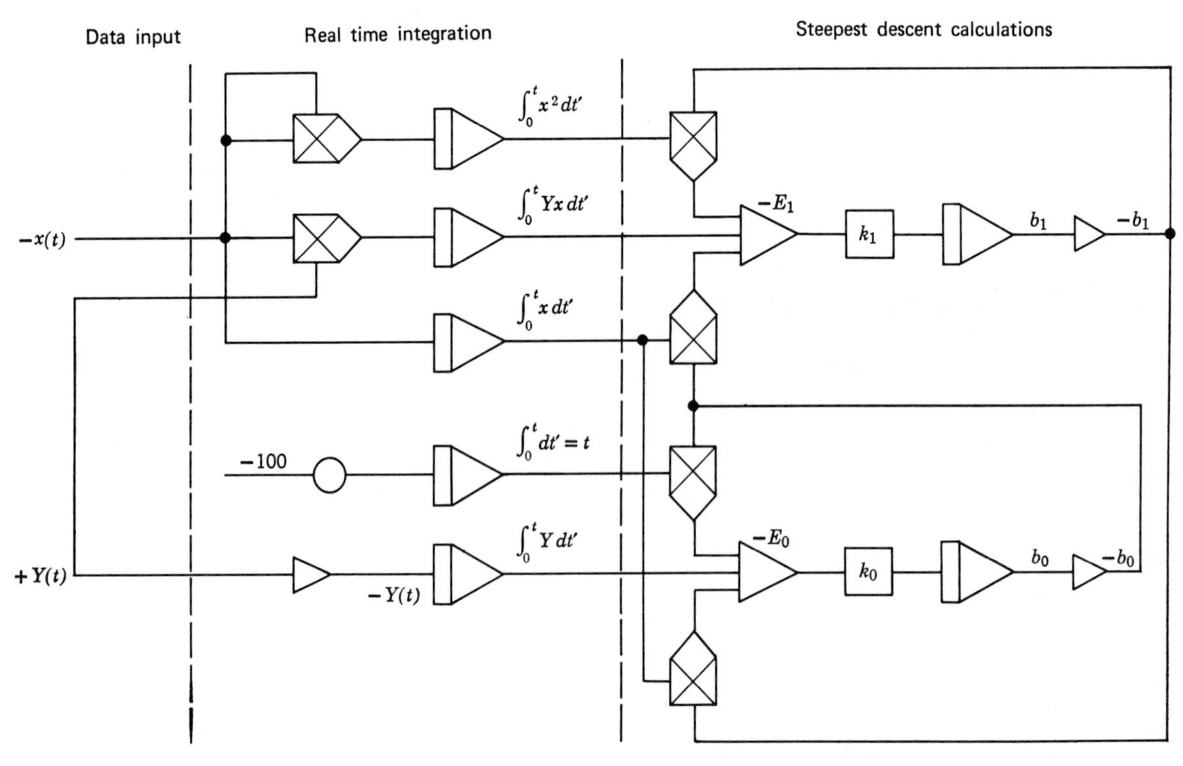

FIGURE E9.2-2

where Y represents the stochastic dependent variable, t' is a dummy variable, and t is real time. For steepest descent, we let

$$\frac{d\beta_0}{d\tau} = -k_0 \frac{\partial \phi}{\partial \beta_0} \qquad (b_1)$$

$$\frac{d\beta_1}{d\tau} = -k_1 \frac{\partial \phi}{\partial \beta_1} \qquad (b_2)$$

where τ is the time scale of operation on the analog or hybrid computer.

The derivatives on the right-hand side of Equations (b) can be evaluated as follows:

$$\frac{\partial \phi}{\partial \beta_0} = -\int_0^t (Y - \beta_0 - \beta_1 x)\, dt' \qquad (c_1)$$

$$\frac{\partial \phi}{\partial \beta_1} = -\int_0^t (Y - \beta_0 - \beta_1 x) x\, dt' \qquad (c_2)$$

Substituting Equations (c) into (b) and denoting the estimates of $\boldsymbol{\beta}$ by $\mathbf{b}$ give

$$\frac{db_0}{d\tau} = k_0\left[\int_0^t Y\, dt' - b_0 t - b_1 \int_0^t x\, dt'\right] \qquad (d_1)$$

$$\frac{db_1}{d\tau} = k_1\left[\int_0^t Yx\, dt' - b_0 \int_0^t x\, dt' - b_1 \int_0^t x^2\, dt'\right] \qquad (d_2)$$

Figure E9.2-2 portrays an analog circuit to carry out the calculations. Note that the values of the estimated parameters do not feed back into the real time calculations (integrations), so the data processing circuits which operate in real time are decoupled from the steepest descent calculations which operate in the scaled time τ.

The model used in Example 9.2-2 was selected to be linear in the parameters for illustrative purposes only. Least squares estimation by steepest descent, as given by Equation 9.2-4, can be applied equally well to process models stated in the form of ordinary differential equations whose solution is given by the continuous analog of Equation 9.1-8 and in which the error is added as in Equation 9.1-9. In general,

$$\frac{d\boldsymbol{\beta}}{dt} = -\mathbf{k}\frac{\partial \phi}{\partial \boldsymbol{\beta}} = \mathbf{k}\int_0^t [\mathbf{Y} - \boldsymbol{\Psi}(\boldsymbol{\beta}, t)] \frac{\partial \boldsymbol{\Psi}(\boldsymbol{\beta}, t')}{\partial \boldsymbol{\beta}}\, dt' \qquad (9.2\text{-}10)$$

The elements of the matrix $(\partial \boldsymbol{\Psi}/\partial \boldsymbol{\beta})$ are termed the *sensitivity coefficients* or parameter influence coefficients; they can be evaluated as follows. Differentiate both sides of Equation 9.1-10 with respect to $\boldsymbol{\beta}$†, assuming $\boldsymbol{\beta}$ is independent of t:

$$\frac{\partial}{\partial \boldsymbol{\beta}}\left(\frac{d\mathbf{y}}{dt}\right) = \left(\frac{\partial \mathbf{f}(\boldsymbol{\beta}, \mathbf{y}, t)}{\partial \mathbf{y}}\right)\left(\frac{\partial \mathbf{y}}{\partial \boldsymbol{\beta}}\right) + \frac{\partial \mathbf{f}(\boldsymbol{\beta}, \mathbf{y}, t)}{\partial \boldsymbol{\beta}}$$

† We use $\boldsymbol{\beta}$ rather than $\boldsymbol{\alpha}$ because the initial conditions can be put in Model 9.1-10 as differential equations by noting that $d\mathbf{y}_0/dt = \mathbf{0}$ and adding these equations to $d\mathbf{y}/dt = \mathbf{f}$.

If $\mathbf{y}$ is continuous and differentiable, the order of differentiation can be exchanged to yield

$$\frac{d}{dt}\left(\frac{\partial \mathbf{y}}{\partial \boldsymbol{\beta}}\right) = \frac{\partial \mathbf{f}(\boldsymbol{\beta}, \mathbf{y}, t)}{\partial \mathbf{y}}\left(\frac{\partial \mathbf{y}}{\partial \boldsymbol{\beta}}\right) + \frac{\partial \mathbf{f}(\boldsymbol{\beta}, \mathbf{y}, t)}{\partial \boldsymbol{\beta}} \qquad (9.2\text{-}11)$$

Equation 9.2-11 is an ordinary differential equation

$$\dot{\mathbf{u}} = \frac{\partial \mathbf{f}}{\partial \mathbf{y}}\mathbf{u} + \frac{\partial \mathbf{f}}{\partial \boldsymbol{\beta}} \qquad \mathbf{u}(0) = \mathbf{0} \qquad (9.2\text{-}12)$$

in the sensitivity coefficients $\mathbf{u} \equiv \partial \mathbf{y}/\partial \boldsymbol{\beta}$ (here we have assumed $\mathbf{h} = \mathbf{I}$, the identity matrix, in Equation 9.1-6). Even if $\hat{\boldsymbol{\beta}}$ changes, as $\phi \to \phi_{\min}$, the estimated parameters approach constant values so that in the end the adjustment of $\hat{\boldsymbol{\beta}}$ meets the assumption that $\hat{\boldsymbol{\beta}}$ is constant.

Thus we see that continuous time estimation by least squares using steepest descent requires only that the input and the output of the process be measured. It has the disadvantage that the partial derivatives in Equations 9.2-4 and 9.2-12 are functionals, that is, they are functions of the coefficients to be estimated. Therefore the gradient of ϕ does not exist unless the coefficients are constant, but this contradicts the original objectives of allowing the coefficients to change with time. Consequently, the steepest descent procedure is satisfactory only when k is small, because the coefficients used to evaluate the response must be those from earlier estimates. How to calculate the precision of the estimates as a function of time is not clear, hence the advantage of sequential estimation to be described in Section 9.4, but at t_f there is no reason why estimates of the precision of the estimated parameters cannot be obtained from samples of the observations as described in Section 9.2-1.

Example 9.2-3 Continuous Estimation

A set of differential equations was proposed to represent a waste-disposal plant containing several units. Only the first two equations are considered here:

$$\frac{dy_1}{dt} = \alpha_{11} y_1 + \alpha_{12} y_2 + x_1(t) \qquad y_1(0) = 0$$

$$\frac{dy_2}{dt} = \alpha_{21} y_1 + \alpha_{22} y_2 + x_2(t) \qquad y_2(0) = 0$$

where y_1 and y_2 are the dependent variables and x_1 and x_2 represent the external inputs to the process units.

A disturbance was introduced at zero time in each process unit by making $x_1(t)$ and $x_2(t)$ each a small step up for a fixed time and then returning to the original reference input. In Figure E9.2-3a the solid line shows the responses recorded for y_1 and y_2 (as deviations from their initial values) as a function of time. The signals from the detectors were fed to a tape recorder, and the information from the tape was processed off-line using the schematic information flow sketched in Figure E9.2-3b. The gains used were each 0.05. The circuits for the hypothetical analog computer are not shown inasmuch as the processing was actually carried out on a CDC 6600 computer using the analog simulator MIMIC. All the integrations, solution of the differential equations, and calculation of the sensitivity coefficients could be executed in a few seconds.

The broken lines in Figure E9.2-3a indicate the predicted responses as a function of time for the estimated coefficients shown in Figure E9.2-3c. Only small step changes could be introduced into the process units if reasonable deviations between the predicted and measured responses for the linear model were to result, no doubt because the actual performance of the equipment was nonlinear for large step changes. Hence the linear model was inadequate.

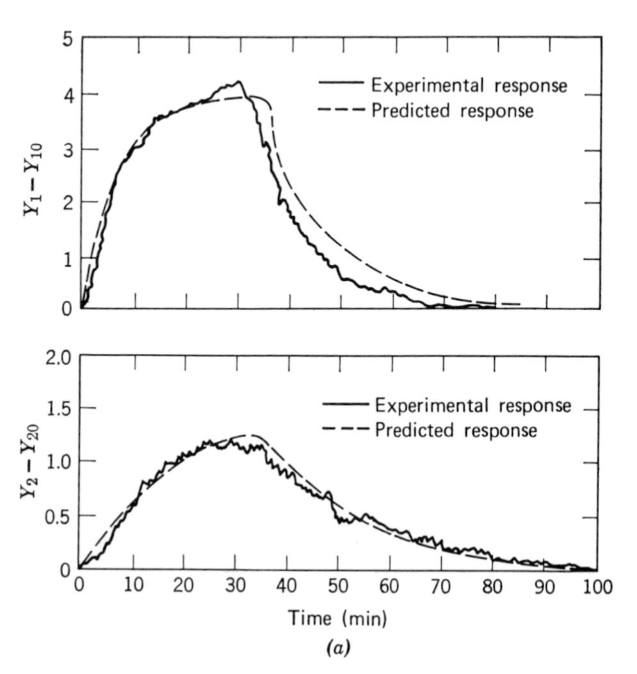

FIGURE E9.2-3a

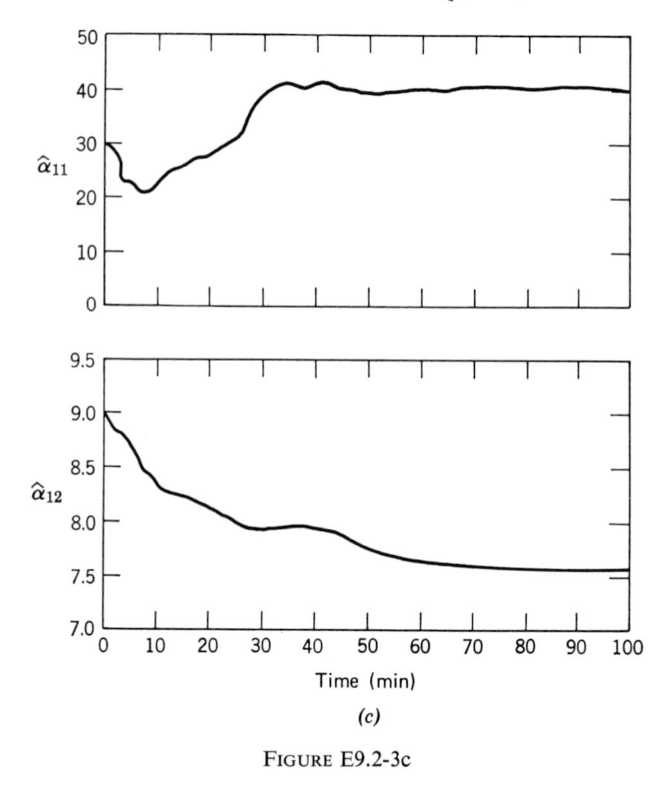

FIGURE E9.2-3c

The parameter estimates will be asymptotically unbiased only if the assumptions hold true concerning the way in which the unobservable errors are added. In practice, many immeasurable and uncontrolled external factors influence the observations so that the estimates may be biased to some extent, as demonstrated by simulation studies.†

† P. C. Young, *Simulation*, 125, Mar. 1968.

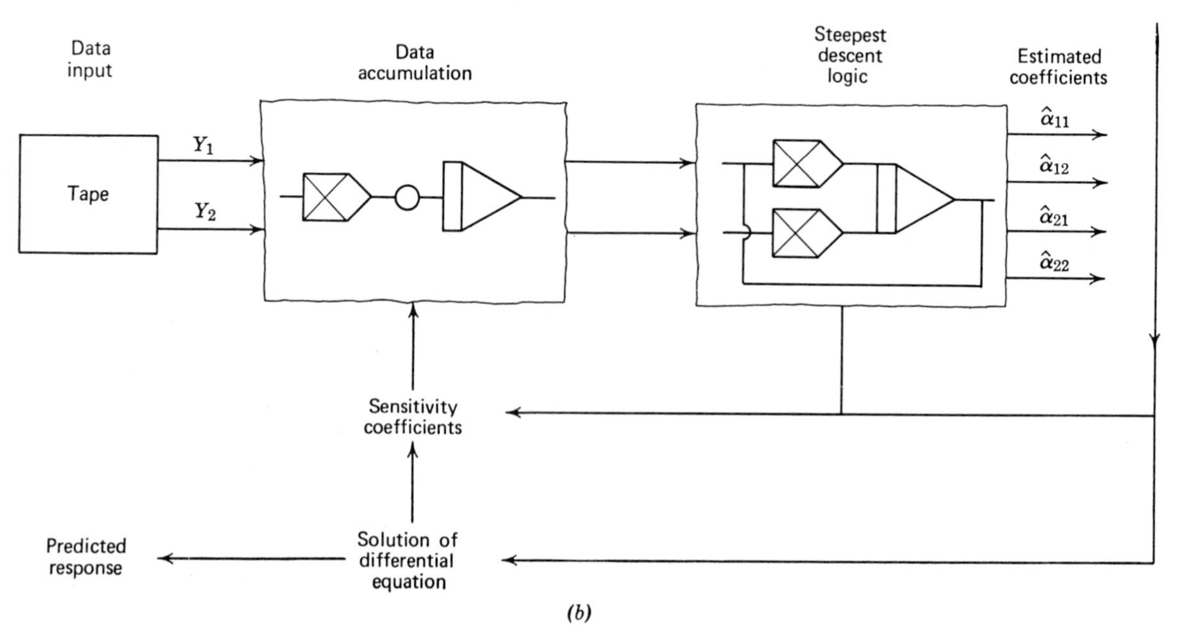

FIGURE E9.2-3b

9.2-4 Repetitive Integration of Experimental Data

Another technique that has been used effectively for both continuous and discrete observations is to integrate repeatedly the experimental data, using a numerical quadrature scheme.† Suppose, for example, that the deterministic process model is

$$\frac{d^2y}{dt^2} + \alpha_0 \frac{dy}{dt} = \alpha_1 y + \alpha_2 y^2 + \alpha_3 e^{\alpha_4 t} \qquad y(0) = y_0$$

If the model is integrated over the interval 0 to t, we get (t' and t'' are dummy variables)

$$\frac{dy}{dt} - \left(\frac{dy}{dt}\right)_{t=0} + \alpha_0(y - y_0)$$

$$= \alpha_1 \int_0^t y\, dt' + \alpha_2 \int_0^t y^2\, dt' + \alpha_3 \int_0^t e^{\alpha_4 t}\, dt'$$

and if integrated again,

$$y - y_0 + \left[\alpha_0 y_0 - \left(\frac{dy}{dt}\right)_0\right] \int_0^t dt' + \alpha_0 \int_0^t y\, dt'$$

$$= \alpha_1 \int_0^t \left[\int_0^t y\, dt'\right] dt'' + \alpha_2 \int_0^t \left[\int_0^t y^2\, dt'\right] dt''$$

$$+ \alpha_3 \int_0^t \left[\int_0^t e^{\alpha_4 t'}\, dt'\right] dt'' \qquad (9.2\text{-}13)$$

The resulting equation is expressed solely in terms of y and integrals of functions of y.

Suppose now that the deterministic variable y is replaced by the stochastic observations Y in the integrals, and that a series of different upper limits are selected for t, such as $t_1, t_2, \ldots, t_n$, with n greater than the number of parameters to be estimated. Then an overdetermined set of equations is obtained that might be solved for the estimated parameters by the method of least squares. Of course, the unobservable error added to y now is immeshed in the integrals themselves; hence the "independent variables" are random variables. Also, because the observed Y's are taken as a sequence in time, the integrals are not statistically independent. Nevertheless, the computations are easy to carry out continuously or off-line for data taken at equal or unequal time increments; simulation studies indicate that the procedure has some merit.‡

Repeated integration of second- and higher order differential equations does require knowing the initial values of the next lower order derivative of y (and x if derivatives of the latter appear in the model). It may be possible to estimate the quantity $[\alpha_0 y_0 - (dy/dt)_0]$ as a whole in Equation 9.2-13, but this procedure has not been tested as yet. If the initial values of y and its derivatives are all zero, there is no problem.

Loeb and Cahen § circumvented the requirement that all the initial conditions for y, x, and their derivatives be known by multiplying each term in the differential equation by a so-called modulus function, a function chosen so that it and its first $(n - 1)$ derivatives vanish at the ends of the interval of integration.

9.3 MAXIMUM LIKELIHOOD ESTIMATION

Maximum likelihood estimates have been described in several earlier sections. They have the desirable characteristics of asymptotic efficiency and normality. Each time they have been associated with the (joint) normal distribution because of mathematical convenience. Consider the joint probability density function (the likelihood function) $p(\boldsymbol{\alpha}, \mathbf{y}_0 \mid \mathbf{y}(t_1), \mathbf{y}(t_2), \ldots, \mathbf{y}(t_n))$ for $\boldsymbol{\alpha}$ and $\mathbf{y}_0$. If a maximum of this function over all choices of $\mathbf{y}_0$ and $\boldsymbol{\alpha}$ can be found, the estimates so obtained are maximum likelihood estimates. The conditions at the maximum can be evolved incorporating prior information as follows.

The posterior probability density $p(\boldsymbol{\alpha}, \mathbf{y}_0 \mid \mathbf{y}(t_1), \mathbf{y}(t_2), \ldots, \mathbf{y}(t_n))$ can be expressed as the ratio of two probability densities if we make use of the analog for continuous variables of Equation A-8 in Appendix A:

$$p(\boldsymbol{\alpha}, \mathbf{y}_0 \mid \mathbf{y}(t_n), \ldots, \mathbf{y}(t_1)) = \frac{p(\boldsymbol{\alpha}, \mathbf{y}_0, \mathbf{y}(t_n), \ldots, \mathbf{y}(t_1))}{p(\mathbf{y}(t_n), \ldots, \mathbf{y}(t_1))} \qquad (9.3\text{-}1)$$

The numerator of the right-hand side of Equation 9.3-1, using Equation A-8a in Appendix A, becomes

$$p(\boldsymbol{\alpha}, \mathbf{y}_0, \mathbf{y}(t_n), \ldots, \mathbf{y}(t_1))$$

$$= p(\mathbf{y}(t_n) \mid \boldsymbol{\alpha}, \mathbf{y}_0, \mathbf{y}(t_{n-1}), \ldots, \mathbf{y}(t_1))$$

$$\cdot p(\boldsymbol{\alpha}, \mathbf{y}_0, \mathbf{y}(t_{n-1}), \ldots, \mathbf{y}(t_1)) \qquad (9.3\text{-}2)$$

These operations can be continued repetitively until we get

$$p(\boldsymbol{\alpha}, \mathbf{y}_0, \mathbf{y}(t_n), \ldots, \mathbf{y}(t_1))$$

$$= p(\boldsymbol{\alpha}, \mathbf{y}_0) \prod_{i=1}^{n} p(\mathbf{y}(t_i) \mid \boldsymbol{\alpha}, \mathbf{y}_0, \mathbf{y}(t_{i-1}), \ldots, \mathbf{y}(t_1)) \qquad (9.3\text{-}3)$$

Examination of Equations 9.1-7 and 9.1-9 shows that $\mathbf{Y}(t_i)$ depends only on t_i, $\mathbf{y}_0$, $\boldsymbol{\alpha}$, and $\boldsymbol{\epsilon}(t_i)$ and is not conditioned by any previous measurements. Consequently, we can write

$$p(\mathbf{y}(t_i) \mid \boldsymbol{\alpha}, \mathbf{y}_0, \mathbf{y}(t_{i-1}), \ldots, \mathbf{y}(t_1)) = p(\mathbf{y}(t_i) \mid \boldsymbol{\alpha}, \mathbf{y}_0) \qquad (9.3\text{-}4)$$

provided Equation 9.1-9 is observed as a constraint. The desired joint conditional probability density function is thus

$$p(\boldsymbol{\alpha}, \mathbf{y}_0 \mid \mathbf{y}(t_n), \ldots, \mathbf{y}(t_1)) = \frac{p(\boldsymbol{\alpha}, \mathbf{y}_0) \prod_{i=1}^{n} p(\mathbf{y}(t_i) \mid \boldsymbol{\alpha}, \mathbf{y}_0)}{p(\mathbf{y}(t_n), \ldots, \mathbf{y}(t_1))} \qquad (9.3\text{-}5)$$

† E. Mishkin and L. Braun, *Adaptive Control Systems*, McGraw-Hill, New York, 1961.

‡ D. M. Himmelblau, C. R. Jones, and K. B. Bischoff, *Ind. Eng. Chem. Fundamentals* **6**, 539, 1967.

§ J. Loeb and G. Cahen, *Automatisme* **8**, 479, 1963.

The function to be maximized will be the logarithm of the likelihood function $L = p(\boldsymbol{\alpha}, \mathbf{y}_0 \mid \mathbf{y}(t_n), \ldots, \mathbf{y}(t_1))$ constrained by Equation 9.1-9, which can be written as $\ln L$ plus the Lagrangian multipliers $\boldsymbol{\lambda}(t_k)$ (refer to Appendix B.6) times the constraint function, or

$$L^* \equiv \ln p(\boldsymbol{\alpha}, \mathbf{y}_0) + \sum_{i=1}^{n} \{\ln p(\mathbf{y}(t_i) \mid \boldsymbol{\alpha}, \mathbf{y}_0) + \boldsymbol{\lambda}^T(t_i) \cdot [\mathbf{y}(t_i) - \boldsymbol{\Psi}(\boldsymbol{\alpha}, \mathbf{y}_0, t_i) - \boldsymbol{\epsilon}(t_i)]\} - \ln [p(\mathbf{y}(t_n), \ldots, \mathbf{y}(t_1))] \quad (9.3\text{-}6)$$

By assumption of the relation of Equation 9.1-9,

$$p(\mathbf{y}(t_i) \mid \boldsymbol{\alpha}, \mathbf{y}_0) = p(\boldsymbol{\epsilon}(t_i)) \quad (9.3\text{-}7)$$

and specifically,

$$p(\boldsymbol{\epsilon}(t_i)) = \frac{1}{(2\pi)^{v/2}|\boldsymbol{\Gamma}(t_i)|^{1/2}} \exp [-\tfrac{1}{2}\boldsymbol{\epsilon}^T(t_i)(\boldsymbol{\Gamma}(t_i))^{-1}\boldsymbol{\epsilon}(t_i)] \quad (9.3\text{-}8)$$

where $|\boldsymbol{\Gamma}|$ is the det $\boldsymbol{\Gamma}$, and $\boldsymbol{\Gamma}$ is the covariance matrix of $\boldsymbol{\epsilon}$, i.e., of the responses.

After Equation 9.3-7 is substituted into Equation 9.3-6, and L^* is differentiated with respect to each of the estimates and $\boldsymbol{\epsilon}(t_i)$, and the resulting expression is equated to zero, we get

$$\frac{\partial L^*}{\partial \mathbf{y}_0} = \frac{\partial}{\partial \mathbf{y}_0} \ln p(\boldsymbol{\alpha}, \mathbf{y}_0) - \sum_{i=1}^{n} \boldsymbol{\lambda}^T(t_i) \frac{\partial}{\partial \mathbf{y}_0} \boldsymbol{\Psi}(\boldsymbol{\alpha}, \mathbf{y}_0, t_i) = \mathbf{0}^T \quad (9.3\text{-}9)$$

$$\frac{\partial L^*}{\partial \boldsymbol{\alpha}} = \frac{\partial}{\partial \boldsymbol{\alpha}} \ln p(\boldsymbol{\alpha}, \mathbf{y}_0) - \sum_{i=1}^{n} \boldsymbol{\lambda}^T(t_i) \frac{\partial}{\partial \boldsymbol{\alpha}} \boldsymbol{\Psi}(\boldsymbol{\alpha}, \mathbf{y}_0, t_i) = \mathbf{0}$$

$$\frac{\partial L^*}{\partial \boldsymbol{\epsilon}(t_i)} = \frac{\partial}{\partial \boldsymbol{\epsilon}(t_i)} \ln p(\boldsymbol{\epsilon}^T(t_i)) - \boldsymbol{\lambda}^T(t_i) = \mathbf{0}^T$$

Substitution of Equation 9.3-8 into the last equation of 9.3-9 makes it possible to solve for $\boldsymbol{\lambda}(t_i)$:

$$\begin{aligned}\boldsymbol{\lambda}(t_i) &= -(\boldsymbol{\Gamma}(t_i))^{-1}\boldsymbol{\epsilon}(t_i) \\ &= -(\boldsymbol{\Gamma}(t_i))^{-1}[\mathbf{Y}(t_i) - \boldsymbol{\Psi}(\boldsymbol{\alpha}, \mathbf{y}_0, t_i)]\end{aligned}$$

and to eliminate $\boldsymbol{\lambda}(t_i)$ from the first two equations of 9.3-9.

For convenience we shall define a new column vector $\boldsymbol{\alpha}^*$ in which all the elements of $\boldsymbol{\alpha}$ are arranged as follows:

$$\boldsymbol{\alpha}^* = \begin{bmatrix} \alpha_{11} \\ \alpha_{12} \\ \vdots \\ \alpha_{1v} \\ \alpha_{21} \\ \vdots \\ \alpha_{2v} \\ \vdots \\ \alpha_{vv} \end{bmatrix} \quad (9.3\text{-}10)$$

Finally, we assume that $\mathbf{y}_0$ and $\boldsymbol{\alpha}^*$ are distributed by a joint normal distribution and that the prior distributions of $\boldsymbol{\alpha}^*$ and $\mathbf{y}_0$ are, respectively,

$$p(\boldsymbol{\alpha}^*) = \frac{1}{(2\pi)^{(v/2)}|\boldsymbol{\Omega}_{\boldsymbol{\alpha}^*}|^{1/2}} \cdot \exp [-\tfrac{1}{2}(\boldsymbol{\alpha}^* - \boldsymbol{\alpha}^{*(0)})^T\boldsymbol{\Omega}_{\boldsymbol{\alpha}^*}^{-1}(\boldsymbol{\alpha}^* - \boldsymbol{\alpha}^{*(0)})] \quad (9.3\text{-}11a)$$

$$p(\mathbf{y}_0) = \frac{1}{(2\pi)^{v/2}|\boldsymbol{\Omega}_{\mathbf{y}_0}|^{1/2}} \cdot \exp [-\tfrac{1}{2}(\mathbf{y}_0 - \mathbf{y}_0^{(0)})^T\boldsymbol{\Omega}_{\mathbf{y}_0}^{-1}(\mathbf{y}_0 - y_0^{(0)})] \quad (9.3\text{-}11b)$$

where the $\boldsymbol{\Omega}$'s are the respective covariance matrices for $\boldsymbol{\alpha}^*$ and $\mathbf{y}_0$, and the superscript (0) designates the prior estimates of $\boldsymbol{\alpha}^*$ and $\mathbf{y}_0$. If we assume that $\boldsymbol{\alpha}^*$ and $\mathbf{y}_0$ are independent,

$$\ln p(\boldsymbol{\alpha}^*, \mathbf{y}_0) = \ln p(\boldsymbol{\alpha}^*) + \ln p(\mathbf{y}_0)$$

Introduction of the prior distributions, Equations 9.3-11a and 9.3-11b, plus the expression for $\boldsymbol{\lambda}(t_i)$ into the first two equations of 9.3-9 gives the final equations from which the estimators of $\boldsymbol{\alpha}^*$ and $\mathbf{y}_0$ can be obtained:

$$-(\hat{\mathbf{y}}_0 - \mathbf{y}_0^{(0)})^T\boldsymbol{\Omega}_{\mathbf{y}_0}^{-1} + \sum_{i=1}^{n} [\mathbf{Y}(t_i) - \boldsymbol{\Psi}(\hat{\boldsymbol{\alpha}}, \hat{\mathbf{y}}_0, t_i)]^T\boldsymbol{\Gamma}^{-1}(t_i) \cdot \frac{\partial}{\partial \mathbf{y}_0} \boldsymbol{\Psi}(\hat{\boldsymbol{\alpha}}, \hat{\mathbf{y}}_0, t_i) = \mathbf{0}^T \quad (9.3\text{-}12)$$

$$-(\hat{\boldsymbol{\alpha}}^* - \boldsymbol{\alpha}^{*(0)})^T\boldsymbol{\Omega}_{\boldsymbol{\alpha}^*}^{-1} + \sum_{i=1}^{n} [\mathbf{Y}(t_i) - \boldsymbol{\Psi}(\hat{\boldsymbol{\alpha}}, \hat{\mathbf{y}}_0, t_i)]^T \cdot \boldsymbol{\Gamma}^{-1}(t_i) \frac{\partial \boldsymbol{\Psi}}{\partial \boldsymbol{\alpha}^*} (\hat{\boldsymbol{\alpha}}, \hat{\mathbf{y}}_0, t_i) = \mathbf{0}^T$$

where the overlay caret denotes estimated parameter.

Note that under the assumption that the elements of $\boldsymbol{\Omega}$ are essentially infinite (prior knowledge is diffuse), the elements of $\boldsymbol{\Omega}^{-1}$ are zero, the equations for the maximum likelihood estimates coincide with those for the least squares estimates, and the same calculations for precision in the estimates apply. Maximum likelihood estimates and least square estimates are compared in Example 9.4-1.

9.4 SEQUENTIAL ESTIMATION

Sequential estimation involves using prior observations together with the latest observation to estimate the model parameters and initial conditions and/or responses. We shall write Equation 9.1-8 in a slightly different notation:

$$\mathbf{Y}(t_i) = \mathbf{h}(t_i)\boldsymbol{\eta}(\boldsymbol{\beta}, t_i) + \boldsymbol{\epsilon}(t_i) \quad (9.4\text{-}1)$$

where $\boldsymbol{\eta}$ will be used to represent the model solution. To avoid confusion in the notation, $\boldsymbol{\beta}$ is defined as

$$\boldsymbol{\beta} = \begin{bmatrix} \mathbf{y}_0 \\ \boldsymbol{\alpha}^* \end{bmatrix}$$

and $\boldsymbol{\alpha}^*$, the column vector of the elements of $\boldsymbol{\alpha}$, has been previously defined by Equation 9.3-10. The essential feature of Equation 9.4-1 that makes it tractable for further analysis is that $Y_r(t_i)$ is a *linear* combination of the elements of $\boldsymbol{\eta}$. In signal processing, $\boldsymbol{\eta}$ is regarded as the unknown input to a linear system, and the major emphasis is on the estimation of the value of $\boldsymbol{\eta}$ through observations of $\mathbf{Y}$. But here we regard $\boldsymbol{\eta}$ as a function of known form and we want to estimate the parameters $\boldsymbol{\beta}$ in the function.

The elements of $\boldsymbol{\beta}$ can be estimated from a series of observations by a method variously known as the Wiener-Kalman method, the Kalman-Bucy method, Schmidt's method, and other names (see references at the end of the chapter), depending upon the particular derivation and the computational algorithm. One of the easiest developments, but not the only development, of the estimator equations is to use Bayes' theorem as outlined in Section 3.1-3.

We assume that the following information is known (corresponding to the list in Section 3.1-3):

1. A set of observations of $\mathbf{Y}$ at successive times $t_1, t_2, \ldots, t_i$, all of which together will be denoted by $\mathbf{Y}(t_i)$.
2. A functional relationship between the observations, $\boldsymbol{\eta}$, and $\boldsymbol{\epsilon}$, namely Equation 9.4-1.
3. The joint density function of $\boldsymbol{\eta}$ and $\boldsymbol{\epsilon}$, $p(\boldsymbol{\eta}(t_i), \boldsymbol{\epsilon}(t_i))$; here $\boldsymbol{\eta}(t_i)$ and $\boldsymbol{\epsilon}(t_i)$ are independent so $p(\boldsymbol{\eta}(t_i), \boldsymbol{\epsilon}(t_i)) = p(\boldsymbol{\eta}(t_i))p(\boldsymbol{\epsilon}(t_i))$. Furthermore

(1) $p(\boldsymbol{\eta}(t_i))$ is Gaussian with

$$\mathscr{E}\{\boldsymbol{\eta}(t_i)\} = \boldsymbol{\mu}_\eta$$
$$\text{Covar}\{\boldsymbol{\eta}(t_i)\} = \mathscr{E}\{\boldsymbol{\eta}(t_i)\boldsymbol{\eta}^T(t_i)\} = \boldsymbol{\Omega}_\eta$$

(2) $p(\boldsymbol{\epsilon}(t_i))$ is Gaussian with

$$\mathscr{E}\{\boldsymbol{\epsilon}(t_i)\} = \mathbf{0}$$
$$\text{Covar}\{\boldsymbol{\epsilon}(t_i)\} = \mathscr{E}\{\boldsymbol{\epsilon}(t_i)\boldsymbol{\epsilon}^T(t_i)\} = \boldsymbol{\Gamma}$$

and the $\mathscr{E}\{\boldsymbol{\eta}(t_i)\boldsymbol{\epsilon}^T(t_i)\} = \mathbf{0}$.

In the above list the functional dependence of certain of the matrices on time has been denoted explicitly; in what follows the argument of time is suppressed to save space but the dependence still holds. We now proceed through the steps given in Section 3.1-3 to obtain the posterior density $p(\boldsymbol{\eta}, \mathbf{y})$.

1. Obtain the density function $p(\mathbf{y})$ for the random variable $\mathbf{Y}$. Since $\mathbf{Y} = \mathbf{h}\boldsymbol{\eta} + \boldsymbol{\epsilon}$, and $\boldsymbol{\eta}$ and $\boldsymbol{\epsilon}$ are both Gaussian, $\mathbf{Y}$ is Gaussian; also

$$\mathscr{E}(\mathbf{Y}) = \mathbf{h}\mathscr{E}\{\boldsymbol{\eta}\} = \mathbf{h}\boldsymbol{\mu}_\eta$$
$$\text{Covar}\{\mathbf{Y}\} = \mathscr{E}\{\mathbf{Y}\mathbf{Y}^T\} = \mathscr{E}\{(\mathbf{h}\boldsymbol{\eta} + \boldsymbol{\epsilon})(\mathbf{h}\boldsymbol{\eta} + \boldsymbol{\epsilon})^T\} = \mathbf{h}\boldsymbol{\Omega}_\eta\mathbf{h}^T + \boldsymbol{\Gamma}$$

Therefore, the probability density function $p(\mathbf{y})$ is

$$p(\mathbf{y}) = k_1 \exp\{-\tfrac{1}{2}(\mathbf{Y} - \mathbf{h}\boldsymbol{\mu}_\eta)^T(\mathbf{h}\boldsymbol{\Omega}_\eta\mathbf{h}^T + \boldsymbol{\Gamma})^{-1}(\mathbf{Y} - \mathbf{h}\boldsymbol{\mu}_\eta)\} \tag{9.4-2}$$

where k_1 is a normalizing factor that is not needed here.

2. Obtain the density function $p(\mathbf{y} \mid \boldsymbol{\eta})$ from the relation given in Section 3.1-3:

$$p(\mathbf{y} \mid \boldsymbol{\eta}) = \frac{p(\mathbf{y}, \boldsymbol{\eta})}{p(\boldsymbol{\eta})} = p(\boldsymbol{\epsilon}) = p(\mathbf{y} - \mathbf{h}\boldsymbol{\eta}) = k_2 \exp[-\tfrac{1}{2}(\mathbf{Y} - \mathbf{h}\boldsymbol{\eta})^T\boldsymbol{\Gamma}^{-1}(\mathbf{Y} - \mathbf{h}\boldsymbol{\eta})] \tag{9.4-3}$$

3. Obtain the posterior density $p(\boldsymbol{\eta} \mid \mathbf{y})$ from Bayes' theorem:

$$\begin{aligned} p(\boldsymbol{\eta} \mid \mathbf{y}) &= \frac{p(\mathbf{y} \mid \boldsymbol{\eta})p(\boldsymbol{\eta})}{p(\mathbf{y})} = \frac{p(\boldsymbol{\epsilon})p(\boldsymbol{\eta})}{p(\mathbf{y})} \\ &= k_3 \exp\{-\tfrac{1}{2}[(\mathbf{Y} - \mathbf{h}\boldsymbol{\eta})^T\boldsymbol{\Gamma}^{-1}(\mathbf{Y} - \mathbf{h}\boldsymbol{\eta}) \\ &\quad + (\boldsymbol{\eta} - \boldsymbol{\mu}_\eta)^T\boldsymbol{\Omega}_\eta^{-1}(\boldsymbol{\eta} - \boldsymbol{\mu}_\eta) \\ &\quad - (\mathbf{Y} - \mathbf{h}\boldsymbol{\mu}_\eta)^T(\mathbf{h}\boldsymbol{\Omega}_\eta\mathbf{h}^T + \boldsymbol{\Gamma})^{-1}(\mathbf{Y} - \mathbf{h}\boldsymbol{\mu}_\eta)]\} \end{aligned} \tag{9.4-4}$$

By completing the squares in the [ ], Equation 9.4-4 simplifies to

$$p(\boldsymbol{\eta} \mid \mathbf{y}) = k_3 \exp\{-\tfrac{1}{2}[(\boldsymbol{\eta} - \tilde{\boldsymbol{\eta}})^T\boldsymbol{\Pi}^{-1}(\boldsymbol{\eta} - \tilde{\boldsymbol{\eta}})]\} \tag{9.4-5}$$

in which we have let

$$\boldsymbol{\Pi}^{-1} \equiv \boldsymbol{\Omega}_\eta^{-1} + \mathbf{h}^T\boldsymbol{\Gamma}^{-1}\mathbf{h} \tag{9.4-6}$$

or

$$\boldsymbol{\Pi} = \boldsymbol{\Omega}_\eta - \boldsymbol{\Omega}_\eta\mathbf{h}^T(\mathbf{h}\boldsymbol{\Omega}_\eta\mathbf{h}^T + \boldsymbol{\Gamma})^{-1}\mathbf{h}\boldsymbol{\Omega}_\eta \tag{9.4-6a}$$

and

$$\tilde{\boldsymbol{\eta}} = \boldsymbol{\mu}_\eta + \boldsymbol{\Pi}\mathbf{h}^T\boldsymbol{\Gamma}^{-1}(\mathbf{Y} - \mathbf{h}\boldsymbol{\mu}_\eta) \tag{9.4-7}$$

Maximization of $p(\boldsymbol{\eta} \mid \mathbf{y})$ is equivalent to minimization of the expression in the square brackets in Equation 9.4-5. Minimization of the expression in the [ ] with respect to $\boldsymbol{\eta}$ by differentiating and equating the resulting expression to zero, as in Section 5.1, gives

$$-2\boldsymbol{\Pi}(\hat{\boldsymbol{\eta}} - \tilde{\boldsymbol{\eta}}) = \mathbf{0}$$

Consequently, the best estimate of $\boldsymbol{\eta}$, $\hat{\boldsymbol{\eta}}$, is equal to $\tilde{\boldsymbol{\eta}}$ as given by Equation 9.4-7.

To obtain a relationship to calculate $\hat{\boldsymbol{\eta}}$ recursively and save effort, that is, to compute $\hat{\boldsymbol{\eta}}$ at t_{i+1} from the previous information at t_i plus one new observation at t_{i+1}, we proceed as follows. The subscript index i or $i + 1$ will designate both time dependence and the specific time. We assume that the following information is known at t_{i+1} (corresponding to the list in Section 3.1-3):

1. A set of observations of $\mathbf{Y}$ at successive times t_1 to t_{i+1}. The observations up to and including t_i will be

designated by $\mathbf{Y}_i$, and the observations taken solely at t_{i+1} will be designated by $\mathbf{Y}_{i+1}$.

2. A functional relationship between the observations, $\boldsymbol{\beta}$, and $\boldsymbol{\epsilon}$. It is assumed still that at any t_i

$$\mathbf{Y}_i = \mathbf{h}_i\boldsymbol{\eta}_i + \boldsymbol{\epsilon}_i \tag{9.4-8}$$

and that, in addition,

$$\boldsymbol{\eta}_{i+1} = \mathbf{h}^*\boldsymbol{\eta}_i + \boldsymbol{\Theta}\boldsymbol{\epsilon}_i^* \tag{9.4-9}$$

where $\mathbf{h}^*$ and $\boldsymbol{\Theta}$ are given matrices. Equation 9.4-9 relates the value of the function $\boldsymbol{\eta}$ from one time period to another.

3. The density function $p(\boldsymbol{\eta}_i \mid \mathbf{y}_i)$ as Gaussian, and $\boldsymbol{\epsilon}^*$ and $\boldsymbol{\epsilon}$ as independent unobservable errors. Also,

$$\mathscr{E}\{(\boldsymbol{\eta}_i \mid \mathbf{y}_i)\} = \hat{\boldsymbol{\eta}}_i$$

$$\text{Covar}\,\{\boldsymbol{\eta}_i \mid \mathbf{y}_i\} = \mathscr{E}\{(\boldsymbol{\eta}_i \mid \mathbf{y}_i)(\boldsymbol{\eta}_i \mid \mathbf{y}_i)^T\} = \boldsymbol{\Omega}_{\eta_i}$$

The joint density function between $\boldsymbol{\epsilon}_{i+1}$ and $\boldsymbol{\epsilon}_i^*$ is

$$p(\boldsymbol{\epsilon}_i^*, \boldsymbol{\epsilon}_{i+1} \mid \boldsymbol{\eta}_i, \mathbf{y}_i) = p(\boldsymbol{\epsilon}_i^*)p(\boldsymbol{\epsilon}_{i+1})$$

and

$$\mathscr{E}\{\boldsymbol{\epsilon}_i\} = \mathscr{E}\{\boldsymbol{\epsilon}_{i+1}\} = \mathbf{0}$$

$$\text{Covar}\,\{\boldsymbol{\epsilon}_i^*\} = \mathscr{E}\{\boldsymbol{\epsilon}_i^*\boldsymbol{\epsilon}_i^{*T}\} = \boldsymbol{\Gamma}$$

$$\text{Covar}\,\{\boldsymbol{\epsilon}_{i+1}\} = \mathscr{E}\{\boldsymbol{\epsilon}_{i+1}\boldsymbol{\epsilon}_{i+1}^T\} = \boldsymbol{\Gamma}$$

We proceed as before except that the posterior density function we are seeking is now

$$p(\boldsymbol{\eta}_{i+1} \mid \mathbf{y}_{i+1}) = \frac{p(\boldsymbol{\eta}_{i+1}, \mathbf{y}_{i+1}, \mathbf{y}_i)}{p(\mathbf{y}_{i+1}, \mathbf{y}_i)} \tag{9.4-10}$$

By integration of the numerator and denominator of the right-hand side of Equation 9.4-10 over all the observations except $\mathbf{Y}_{i+1}$, i.e., integration over $\mathbf{Y}_i$, Equation 9.4-10 reduces to

$$p(\boldsymbol{\eta}_{i+1} \mid \mathbf{y}_{i+1}) = \frac{p(\boldsymbol{\eta}_{i+1}, \mathbf{y}_{i+1} \mid \mathbf{y}_i)}{p(\mathbf{y}_{i+1} \mid \mathbf{y}_i)}$$

$$= \frac{p(\mathbf{y}_{i+1} \mid \boldsymbol{\eta}_{i+1}, \mathbf{y}_i)p(\boldsymbol{\eta}_{i+1} \mid \mathbf{y}_i)}{p(\mathbf{y}_{i+1} \mid \mathbf{y}_i)} \tag{9.4-11}$$

Consequently, we need to obtain the three probability density functions on the right-hand side of Equation 9.4-11.

1. Obtain $p(\mathbf{y}_{i+1} \mid \mathbf{y}_i)$. As before, $p(\mathbf{y}_{i+1} \mid \mathbf{y}_i)$ is Gaussian with

$$\mathscr{E}\{(\mathbf{Y}_{i+1} \mid \mathbf{Y}_i)\} = \mathbf{h}\mathbf{h}^*\hat{\boldsymbol{\eta}}_i$$

$$\text{Covar}\,\{\mathbf{Y}_{i+1} \mid \mathbf{Y}_i\} = \mathscr{E}\{(\mathbf{Y}_{i+1} \mid \mathbf{Y}_i)(\mathbf{Y}_{i+1} \mid \mathbf{Y}_i)^T\}$$

$$= \mathbf{h}\boldsymbol{\Omega}_{\eta_{i+1}}\mathbf{h}^T + \boldsymbol{\Gamma}$$

2. Obtain the density function $p(\boldsymbol{\eta}_{i+1} \mid \mathbf{y}_i)$. The density is Gaussian and independent of $p(\boldsymbol{\epsilon}_{i+1})$ with

$$\mathscr{E}\{(\boldsymbol{\eta}_{i+1} \mid \mathbf{Y}_i)\} = \mathbf{h}^*\hat{\boldsymbol{\eta}}_i$$

$$\text{Covar}\,\{\boldsymbol{\eta}_{i+1} \mid \mathbf{Y}_i\} = \mathscr{E}\{(\boldsymbol{\eta}_{i+1} \mid \mathbf{Y}_i)(\boldsymbol{\eta}_{i+1} \mid \mathbf{Y}_i)^T\}$$

$$= \mathbf{h}^*\boldsymbol{\Omega}_{\eta_i}\mathbf{h}^{*T} + \boldsymbol{\Theta}\boldsymbol{\Gamma}\boldsymbol{\Theta}^T \equiv \mathbf{M}_{i+1}$$

3. Obtain the density function $p(\mathbf{y}_{i+1} \mid \boldsymbol{\eta}_{i+1}, \mathbf{y}_i)$. The density is Gaussian with

$$\mathscr{E}\{\mathbf{Y}_{i+1} \mid \boldsymbol{\eta}_{i+1}, \mathbf{Y}_i\} = \mathbf{h}\boldsymbol{\eta}_{i+1}$$

$$\text{Covar}\,\{\mathbf{Y}_{i+1} \mid \boldsymbol{\eta}_{i+1}, \mathbf{Y}_i\} = \mathscr{E}\{(\mathbf{Y}_{i+1} \mid \boldsymbol{\eta}_{i+1}, \mathbf{Y}_i) \cdot (\mathbf{Y}_{i+1} \mid \boldsymbol{\eta}_{i+1}, \mathbf{Y}_i)^T\} = \boldsymbol{\Gamma}$$

Introducing the appropriate quantities for the density functions into Equation 9.4-11, we obtain

$$\begin{aligned} p(\boldsymbol{\eta}_{i+1} \mid \mathbf{y}_{i+1}) &= k_1 \exp\{-\tfrac{1}{2}[(\boldsymbol{\eta}_{i+1} - \mathbf{h}^*\hat{\boldsymbol{\eta}}_i)^T\mathbf{M}_{i+1}^{-1}(\boldsymbol{\eta}_{i+1} - \mathbf{h}^*\hat{\boldsymbol{\eta}}_i) \\ &\quad + (\mathbf{Y}_{i+1} - \mathbf{h}\boldsymbol{\eta}_{i+1})^T\boldsymbol{\Gamma}^{-1}(\mathbf{Y}_{i+1} - \mathbf{h}\boldsymbol{\eta}_{i+1}) \\ &\quad - (\mathbf{Y}_{i+1} - \mathbf{h}\mathbf{h}^*\hat{\boldsymbol{\eta}}_i)^T(\mathbf{h}\mathbf{M}_{i+1}\mathbf{h}^T + \boldsymbol{\Gamma})^{-1} \\ &\quad \cdot (\mathbf{Y}_{i+1} - \mathbf{h}\mathbf{h}^*\hat{\boldsymbol{\eta}}_i)]\} \end{aligned} \tag{9.4-12}$$

where k_1 is a normalizing factor. Completing the squares in the [ ] reduces Equation 9.4-12 to

$$p(\boldsymbol{\eta}_{i+1} \mid \mathbf{y}_{i+1}) = k_1 \exp\{-\tfrac{1}{2}[(\boldsymbol{\eta}_{i+1} - \hat{\boldsymbol{\eta}}_{i+1})^T\boldsymbol{\Omega}_{\eta_{i+1}}^{-1}(\boldsymbol{\eta}_{i+1} - \hat{\boldsymbol{\eta}}_{i+1})]\} \tag{9.4-13}$$

where

$$\hat{\boldsymbol{\eta}}_{i+1} = \mathbf{h}^*\hat{\boldsymbol{\eta}}_i + \mathbf{M}_{i+1}\mathbf{h}^T(\mathbf{h}\mathbf{M}_{i+1}\mathbf{h}^T + \boldsymbol{\Gamma})^{-1}(\mathbf{Y}_{i+1} - \mathbf{h}\mathbf{h}^*\hat{\boldsymbol{\eta}}_i) \tag{9.4-14}$$

and

$$\boldsymbol{\Omega}_{\eta_{i+1}}^{-1} = \mathbf{M}_{i+1}^{-1} + \mathbf{h}\boldsymbol{\Gamma}^{-1}\mathbf{h} \tag{9.4-15}$$

or

$$\boldsymbol{\Omega}_{\eta_{i+1}} = \mathbf{M}_{i+1} - \mathbf{M}_{i+1}\mathbf{h}^T(\mathbf{h}\mathbf{M}_{i+1}\mathbf{h}^T + \boldsymbol{\Gamma})^{-1}\mathbf{h}\mathbf{M}_{i+1} \tag{9.4-16}$$

and $\mathbf{M}_{i+1}$ is defined above in terms of $\boldsymbol{\Omega}_i$. As before, maximization of $p(\boldsymbol{\eta}_{i+1} \mid \mathbf{y}_{i+1})$ with respect to $\boldsymbol{\eta}_{i+1}$ leads to $\boldsymbol{\eta}_{i+1} = \hat{\boldsymbol{\eta}}_{i+1}$, so Equation 9.4-14 is the estimator equation. Equation 9.4-14 is often termed the discrete "Wiener-Kalman filter."

To obtain specifically the equations for estimating $\boldsymbol{\beta}$, it is necessary to use a recursive technique employing the linearized (in $\boldsymbol{\beta}$) solution to the model. Suppose we linearize $\boldsymbol{\eta}_{i+1}$ by a truncated Taylor series about the reference values $\tilde{\beta}_1, \tilde{\beta}_2, \ldots$, where the overlay tilde represents the reference value:

$$\begin{aligned} \boldsymbol{\eta}_i &= \boldsymbol{\eta}_i(\tilde{\boldsymbol{\beta}}) + \frac{\partial\boldsymbol{\eta}_i(\tilde{\boldsymbol{\beta}})}{\partial\beta_1}(\beta_1 - \tilde{\beta}_1) + \frac{\partial\boldsymbol{\eta}_i(\tilde{\boldsymbol{\beta}})}{\partial\beta_2}(\beta_2 - \tilde{\beta}_2) + \cdots \\ &= \boldsymbol{\eta}_i(\tilde{\boldsymbol{\beta}}) + \nabla_{\boldsymbol{\beta}}\boldsymbol{\eta}_i(\tilde{\boldsymbol{\beta}})\,\delta\boldsymbol{\beta} \end{aligned} \tag{9.4-17}$$

where $\delta\boldsymbol{\beta} = \boldsymbol{\beta} - \tilde{\boldsymbol{\beta}}$ and $\nabla_{\boldsymbol{\beta}} \equiv \partial/\partial\beta_1 + \partial/\partial\beta_2 + \cdots$. Then, after substituting Equation 9.4-17 into Equation 9.4-8, assuming that $\mathbf{h}^* = \mathbf{I}$, the identity matrix, and that $\boldsymbol{\Theta}$ is zero in Equation 9.4-9, we obtain

$$\mathbf{Y}_i = \mathbf{h}_i[\boldsymbol{\eta}_i(\tilde{\boldsymbol{\beta}}) + \nabla_{\boldsymbol{\beta}}\boldsymbol{\eta}_i(\tilde{\boldsymbol{\beta}})\,\delta\boldsymbol{\beta}] + \boldsymbol{\epsilon}_i$$

or

$$\delta \mathbf{Y}_i \equiv \mathbf{Y}_i - \mathbf{h}_i \boldsymbol{\eta}_i(\tilde{\boldsymbol{\beta}}) = [\mathbf{h}_i \nabla_{\boldsymbol{\beta}} \boldsymbol{\eta}_i(\tilde{\boldsymbol{\beta}})]\, \delta\boldsymbol{\beta} + \boldsymbol{\epsilon}_i \quad (9.4\text{-}18)$$

Note that Equation 9.4-18 corresponds to Equation 9.4-8 with $\delta\mathbf{Y}_i$ corresponding to $\mathbf{Y}_i$ and $\delta\boldsymbol{\beta}$ corresponding to $\boldsymbol{\eta}_i$. Consequently, $\delta\boldsymbol{\beta}$ can be estimated through use of Equation 9.4-14 with $\delta\hat{\boldsymbol{\beta}}_{i+1}$ replacing $\hat{\boldsymbol{\eta}}_{i+1}$ if the assumptions made about the probability densities involving $\boldsymbol{\eta}$ apply to $\delta\boldsymbol{\beta}$, and we shall assume that they do.

Consequently, if the reference state $\tilde{\boldsymbol{\beta}}$ is assigned as the previously estimated $\boldsymbol{\beta}$, i.e., let $\tilde{\boldsymbol{\beta}} = \hat{\boldsymbol{\beta}}_i$, $\hat{\boldsymbol{\beta}}_{i+1}$ is formed from Equation 9.4-14:

$$\begin{aligned}\delta\hat{\boldsymbol{\beta}}_{i+1} = \delta\hat{\boldsymbol{\beta}}_i &+ \boldsymbol{\Omega}_{\boldsymbol{\beta}_i}[\mathbf{h}_{i+1}\nabla_{\boldsymbol{\beta}}]^T \\ &\cdot[(\mathbf{h}_{i+1}\nabla_{\boldsymbol{\beta}}\boldsymbol{\eta}_i(\hat{\boldsymbol{\beta}}_i))\boldsymbol{\Omega}_{\boldsymbol{\beta}_i}(\mathbf{h}_{i+1}\nabla_{\boldsymbol{\beta}}\boldsymbol{\eta}_i(\hat{\boldsymbol{\beta}}_i))^T + \boldsymbol{\Gamma}]^{-1} \\ &\cdot[\delta\mathbf{Y}_{i+1} - \mathbf{h}_{i+1}(\nabla_{\boldsymbol{\beta}}\boldsymbol{\eta}_i(\hat{\boldsymbol{\beta}}_i))\, \delta\hat{\boldsymbol{\beta}}_i]\end{aligned}$$

Because $\delta\hat{\boldsymbol{\beta}}_i = \hat{\boldsymbol{\beta}}_i - \hat{\boldsymbol{\beta}}_i = \mathbf{0}$, and

$$\delta\mathbf{Y}_{i+1} - \mathbf{h}_{i+1}\nabla_{\boldsymbol{\beta}}\boldsymbol{\eta}_i(\hat{\boldsymbol{\beta}}_i)\, \delta\hat{\boldsymbol{\beta}}_i \cong \mathbf{Y}_{i+1} - \mathbf{h}_{i+1}\boldsymbol{\eta}_i(\hat{\boldsymbol{\beta}}_i)$$

we find

$$\hat{\boldsymbol{\beta}}_{i+1} = \hat{\boldsymbol{\beta}}_i + \mathbf{K}_{i+1}[\mathbf{Y}_{i+1} - \mathbf{h}_{i+1}\boldsymbol{\eta}_{i+1}(\hat{\boldsymbol{\beta}}_i)] \quad (9.4\text{-}19)$$

where

$$\begin{aligned}\boldsymbol{\eta}_{i+1}(\hat{\boldsymbol{\beta}}_i) &= \exp\,[\hat{\boldsymbol{\alpha}}(t_i)t_{i+1}]\hat{\mathbf{y}}_0(t_i) \\ &\quad + \int_0^{t_i} \exp\,[(t_i - \tau)\hat{\boldsymbol{\alpha}}(t_i)]\mathbf{x}(\tau)\, d\tau\end{aligned}$$

$$\begin{aligned}\mathbf{K}_{i+1} = \boldsymbol{\Omega}_{\boldsymbol{\beta}}(t_{i+1})&[\mathbf{h}(t_{i+1})\nabla_{\boldsymbol{\beta}}\boldsymbol{\eta}(\hat{\boldsymbol{\beta}}_i, t_{i+1})]^T \\ &\cdot\{[\mathbf{h}(t_{i+1})\nabla_{\boldsymbol{\beta}}\boldsymbol{\eta}(\hat{\boldsymbol{\beta}}_i, t_{i+1})]\boldsymbol{\Omega}_{\boldsymbol{\beta}}(t_{i+1}) \\ &\qquad\cdot[\mathbf{h}(t_{i+1})\nabla_{\boldsymbol{\beta}}\boldsymbol{\eta}(\hat{\boldsymbol{\beta}}_i, t_{i+1})]^T + \boldsymbol{\Gamma}(t_{i+1})\}^{-1}\end{aligned}$$

and the recursion relationship for the covariance matrix of $\boldsymbol{\beta}$, $\boldsymbol{\Omega}_{\boldsymbol{\beta}}$, from Equation 9.4-16 is

$$\boldsymbol{\Omega}_{\boldsymbol{\beta}}(t_{i+1}) = \boldsymbol{\Omega}_{\boldsymbol{\beta}}(t_i) - \mathbf{K}(t_i)[\mathbf{h}(t_i)\nabla_{\boldsymbol{\beta}}\boldsymbol{\eta}(\hat{\boldsymbol{\beta}}_i, t_i)]\boldsymbol{\Omega}_{\boldsymbol{\beta}}(t_i)$$

The matrix $\boldsymbol{\Omega}_{\boldsymbol{\beta}}$ can be used to estimate the precision of $\hat{\boldsymbol{\beta}}$ for the linearized model as described in Section 6.4.

Example 9.4-1 Comparison of Estimation Techniques

Suppose we take the scalar model, Equation 9.1-1:

$$\frac{dy}{dt} = \alpha y + x_0 \qquad y(0) = y_0 \quad \text{(a)}$$

$$Y = y + \epsilon \quad \text{(b)}$$

where x_0 is a constant input, and find the estimation equations for each of the three main procedures described in Sections 9.2, 9.3, and 9.4. The initial state y_0 and the parameter α are assumed to follow a Gaussian distribution with

$$\mathscr{E}\{y_0\} = \tilde{y}_0 \qquad \mathscr{E}\{\alpha\} = \tilde{\alpha}$$

$$\mathscr{E}\{(y_0 - \tilde{y}_0)^2\} = \sigma_{Y_0}^2 \qquad \mathscr{E}\{(\alpha - \tilde{\alpha})^2\} = \sigma_{\tilde{\alpha}}^2$$

and the variance of the unobservable error is assumed to be constant with

$$\mathscr{E}\{\epsilon(t_i)\} = 0$$

$$\mathscr{E}\{\epsilon(t_i)\epsilon(t_j)\} = \begin{Bmatrix}\sigma_\epsilon^2 \text{ for } i = j \\ 0 \quad \text{for } i \neq j\end{Bmatrix}$$

The *least squares estimation equations*, already given in Section 9.2, are

$$\sum_{i=1}^{n}\left[Y(t_i) - \hat{y}_0\, e^{\hat{\alpha}t_i} + \frac{x_0}{\hat{\alpha}}(1 - e^{\hat{\alpha}t_i})\right] e^{\hat{\alpha}t_i} = 0 \quad \text{(c)}$$

$$\sum_{i=1}^{n}\left[Y(t_i) - \hat{y}_0\, e^{\hat{\alpha}t_i} + \frac{x_0}{\hat{\alpha}}(1 - e^{\hat{\alpha}t_i})\right] \cdot\left\{t_i\hat{y}_0\, e^{\hat{\alpha}t_i} - \frac{x_0}{\hat{\alpha}^2}[e^{\hat{\alpha}t_i}(1 - \hat{\alpha}t_i) - 1]\right\} = 0 \quad \text{(d)}$$

The *maximum likelihood estimation equations* are quite similar and can be obtained directly from Equations 9.3-11:

$$(\hat{y}_0 - y_0^{(0)})\frac{\sigma_\epsilon^2}{\sigma_{y_0}^2} + \text{Equation (c)} = 0 \quad \text{(e)}$$

$$(\hat{\alpha} - \alpha^{(0)})\frac{\sigma_\epsilon^2}{\sigma_\alpha^2} + \text{Equation (d)} = 0 \quad \text{(f)}$$

Equations (c) through (f) are nonlinear in the estimated parameters but can be solved by an iterative technique such as the Newton-Raphson method.

The *sequential estimation equations* can be obtained from Equation 9.4-19:

$$\hat{\boldsymbol{\beta}}(t_i) = \begin{bmatrix}\hat{y}_0(t_i) \\ \hat{\alpha}(t_i)\end{bmatrix} \qquad \mathbf{h} = \mathbf{I}$$

$$\boldsymbol{\Gamma} = \sigma_\epsilon^2\mathbf{I} \qquad \boldsymbol{\Omega}(0) = \begin{bmatrix}\sigma_{y_0}^2 & 0 \\ 0 & \sigma_\alpha^2\end{bmatrix}$$

$$\hat{\boldsymbol{\beta}}(t_{i+1}) = \hat{\boldsymbol{\beta}}(t_i) + \mathbf{K}(t_{i+1})[Y(t_{i+1}) - \eta(\hat{\boldsymbol{\beta}}_i, t_{i+1})] \quad \text{(g)}$$

in which

$$\begin{aligned}\eta(\hat{\beta}_i, t_{i+1}) &= \hat{y}_{0,i}\, e^{\hat{\alpha}_i t_{i+1}} + x_0\int_0^{t_{i+1}} e^{\hat{\alpha}_i(t_{i+1} - \tau)}\, d\tau \\ &= \hat{y}_{0,i}\, e^{\hat{\alpha}_i t_{i+1}} + \frac{x_0}{\hat{\alpha}_i}(e^{\hat{\alpha}_i t_{i+1}} - 1)\end{aligned}$$

$$\begin{aligned}\mathbf{K}_{i+1} = \boldsymbol{\Omega}_{\boldsymbol{\beta},i+1}&[\nabla_{\boldsymbol{\beta}}\eta(\hat{\boldsymbol{\beta}}_i, t_{i+1})]^T \\ &\cdot\{[(\nabla_{\boldsymbol{\beta}}\eta(\hat{\boldsymbol{\beta}}_i, t_{i+1}))\boldsymbol{\Omega}_{\boldsymbol{\beta},i+1}(\nabla_{\boldsymbol{\beta}}\eta(\hat{\boldsymbol{\beta}}_i, t_{i+1}))^T] + \sigma_\epsilon^2\}^{-1}\end{aligned}$$

$$\begin{aligned}\nabla_{\boldsymbol{\beta}}\eta(\hat{\boldsymbol{\beta}}_i, t_{i+1}) &= e^{\hat{\alpha}_i t_{i+1}} + \hat{y}_{0,i}t_{i+1}\, e^{\hat{\alpha}_i t_{i+1}} \\ &\quad - \frac{x_0}{\hat{\alpha}_i}[e^{\hat{\alpha}_i t_{i+1}}(1 - \hat{\alpha}_i t_{i+1}) - 1]\end{aligned}$$

$$\boldsymbol{\Omega}_{\boldsymbol{\beta},i+1} = \boldsymbol{\Omega}_{\boldsymbol{\beta},i} - \mathbf{K}_{i+1}\nabla_{\boldsymbol{\beta}}\eta(\hat{\boldsymbol{\beta}}_i, t_{i+1})\boldsymbol{\Omega}_{\boldsymbol{\beta},i}$$

In the above expressions the subscript index i designates the time.

A direct comparison of the results of estimation of real or simulated data may not be meaningful since one estimation procedure may be more sensitive to the particular error configuration of the data than another. Consequently, Carney

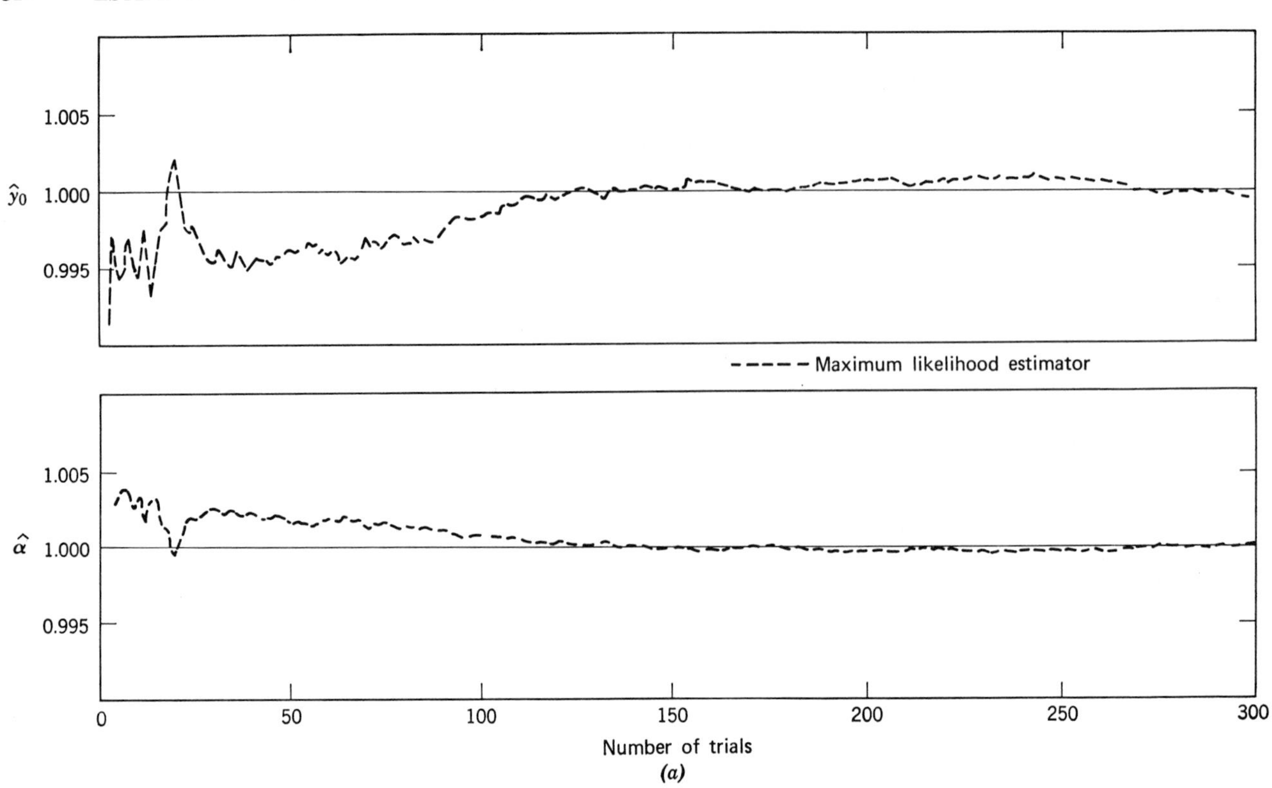

FIGURE E9.4-1a Evolution of estimates for the unstable process model as a function of the number of trials.

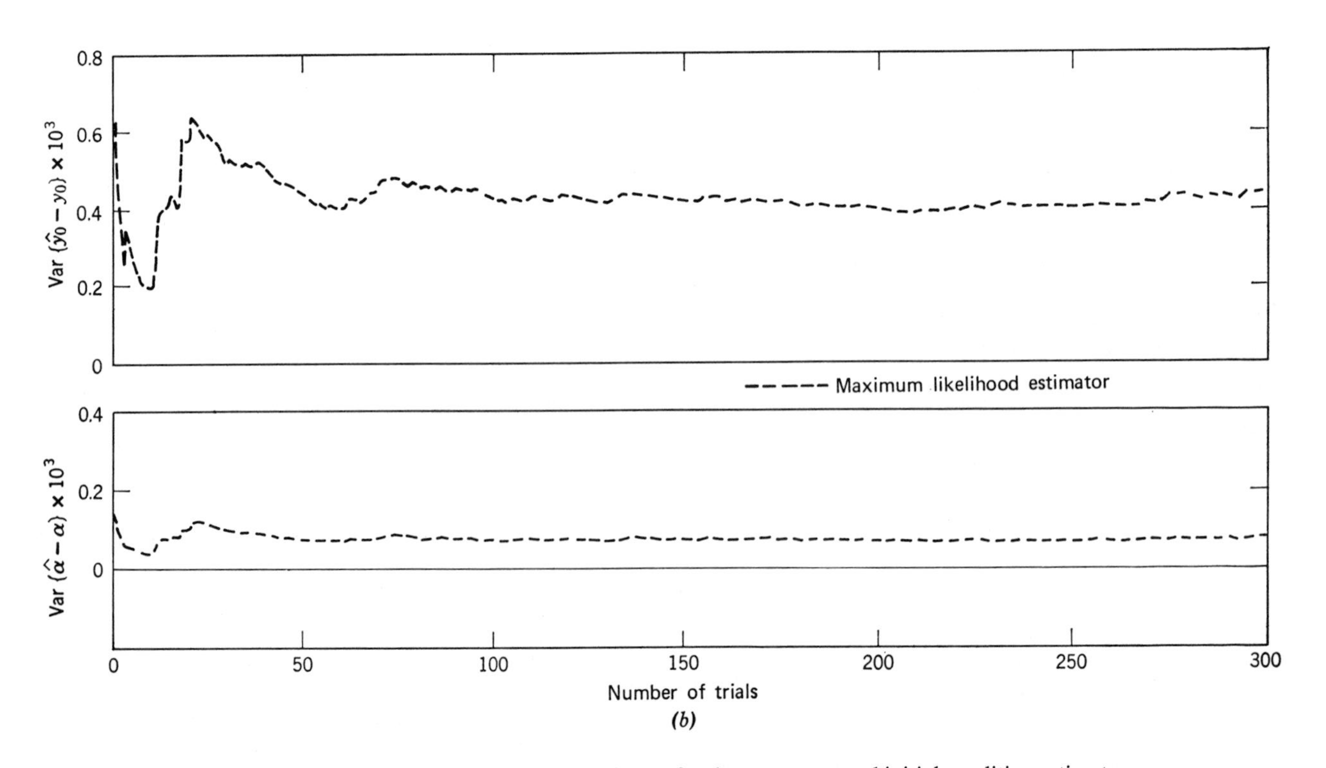

FIGURE E9.4-1b Evolution of estimated variances for the parameter and initial condition estimates with number of trials (unstable model).

and Goldwyn† employed Monte Carlo simulation to evaluate the three estimation procedures. Monte Carlo simulation uses a pseudorandom number generator to provide the values of errors to be added to the assumed initial condition (initial state) and the parameters for an assumed Gaussian distribution of errors. Then the deterministic process model is solved in time repetitively, using the simulated values of y_0 and α. To the generated deterministic outputs are added the observation errors $\epsilon(t_i)$ drawn from a random number generator. Finally, the simulated stochastic outputs are introduced as $Y(t_i)$ into the estimator equations, and estimates of y_0 and α are obtained. Because a large number (about 200–300) of values of Y at any time were calculated, the estimated variances of $\hat{y}_0$ and $\hat{\alpha}$ could also be determined.

Values of the parameter, initial conditions, etc. for the simulation were

$$x_0 = 1 \qquad \sigma_{y_0} = 0.1$$
$$t_f = 2 \qquad \sigma_\alpha = 0.1$$
$$n = 20 \qquad \sigma_\epsilon = 0.1$$

	α	y_0
Stable process	-1	-1
Unstable process	1	1

Figures E9.4-1a and E9.4-1b illustrate the estimates and the variances of the estimates for 300 trials, i.e., 300 different pairs of $\{(y_0 + \epsilon_{y_0})$ and $(\alpha + \epsilon_\alpha)\}$, for the unstable model which exhibits behavior typical of both the stable and unstable cases. All three methods of estimation yielded essentially the same curves. For clarity, the least squares and sequential estimates have been omitted. In Figure E9.4-1b the vertical axes are respectively, the variance of the deviation $\tilde{\alpha} = (\hat{\alpha} - \alpha)$, not the Var $\{\hat{\alpha}\}$ itself, and the variance of $\tilde{\beta} = (\hat{y}_0 - y_0)$, not the Var $\{\hat{y}_0\}$ itself.

Some aspects of changing certain of the assumed "true" quantities were also examined. Figure E9.4-1c illustrates the influence of changing σ_ϵ from 0.0001 to 0.5, all the other quantities remaining constant, for 200 trials. For low observation error (noise), the sequential procedure for the stable model is poorer. But as the observation error becomes larger, its relative position with respect to least squares improves because least squares estimation does not use the *a priori* information that becomes relatively more accurate as the noise increases. For the unstable model (not shown), the same relative trends hold true; however the ratio of Var $\{\hat{a}\}/\sigma_\epsilon^2$ or Var $\{\hat{b}\}/\sigma_\epsilon^2$ is of the order of 100 smaller.

Figure E9.4-1d shows the effect of changing the prior

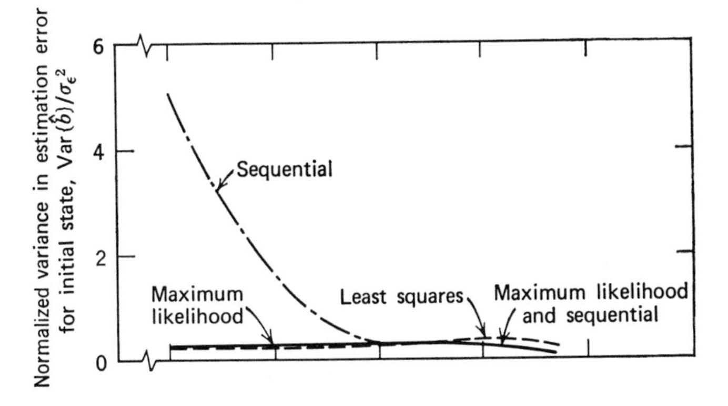

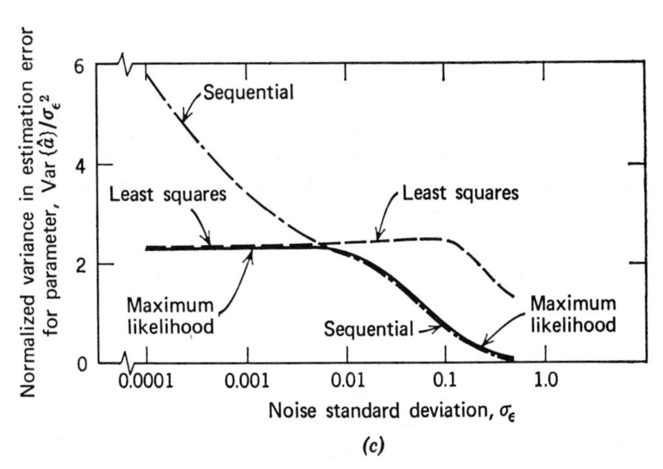

(c)

FIGURE E9.4-1c Variation of estimation error variance with measurement noise (stable process model).

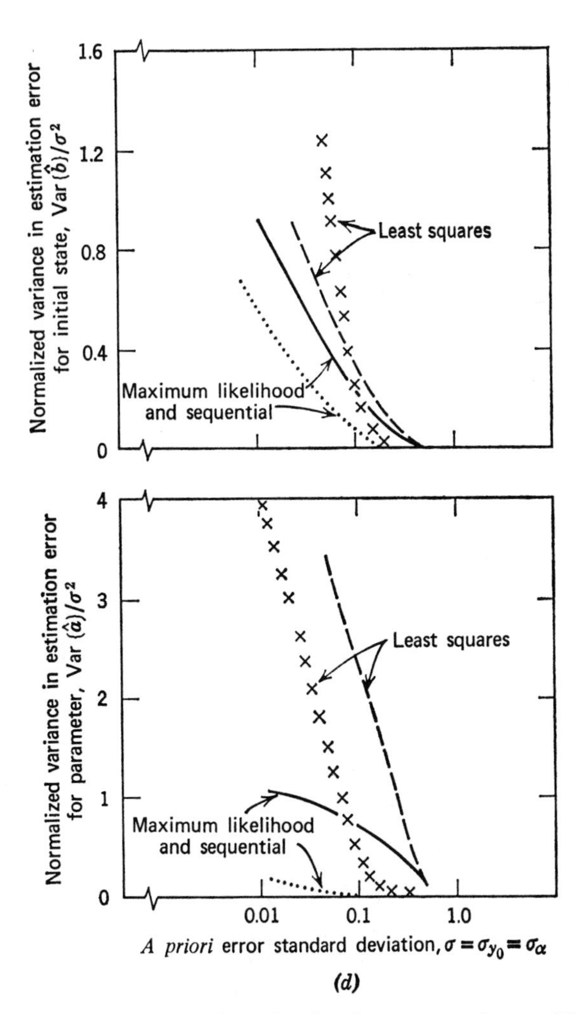

(d)

FIGURE E9.4-1d Variation of estimation error variance with *a priori* standard deviation.

† T. M. Carney and R. M. Goldwyn, *J. Opt. Theory and Applications* 1, 113, 1967.

standard deviations of the initial condition and the parameter for the stable model. For very small *a priori* errors, circa $\sigma_{y_0} = \sigma_\alpha = 0.01$, the maximum likelihood estimator and the sequential estimators give comparable results, while the least squares estimate is degraded as a natural result of not using the rather accurate *a priori* information. As the standard deviation in the *a priori* errors grows, the least squares estimate improves relative to the others.

A third effect studied was that of data length. Sampling intervals of 0.1, 0.2, and 0.4 were used for the same total number of data points (20). As shown in Figure E9.4-1e, for the stable model the signal-to-noise ratio grows worse with time because the deviations from the steady-state solution damp out. However, because the steady-state solution is $-(x_0/\alpha)$, the stable model enables improved estimates of α to be obtained as the data length grows. The maximum likelihood and sequential estimation methods give comparable results, while the least squares estimation procedure gives poorer precision in the estimates, mainly because the *a priori* information is quite important when the measurement information is degraded with time.

For the unstable model, Figure E9.4-1f, all the estimators are essentially the same and show improvement in precision as the length of data increases because the measurement noise has a constant standard deviation whereas the response is growing with time. Consequently, the signal-to-noise ratio is continually improving.

The times required to execute the estimation phase of the study proved to be almost the same for each of the three methods. This result can be explained by the fact that in sequential estimation the data still have to be operated on a number of times equal to the number of measurements at each cycle. Also the covariance matrix has to be updated. The nonlinear estimators operate on the entire data set simultaneously, but iteration is still required because the equations are nonlinear.

It can be concluded that the maximum likelihood estimation procedure in general is best in the sense that it provides the most precise estimates. It also gives estimates which are not any more biased than the other procedures. Sequential estimation compared favorably with maximum likelihood estimation. Least squares estimation is adequate except when precise *a priori* estimates of the parameter and initial state errors are assumed to be known.

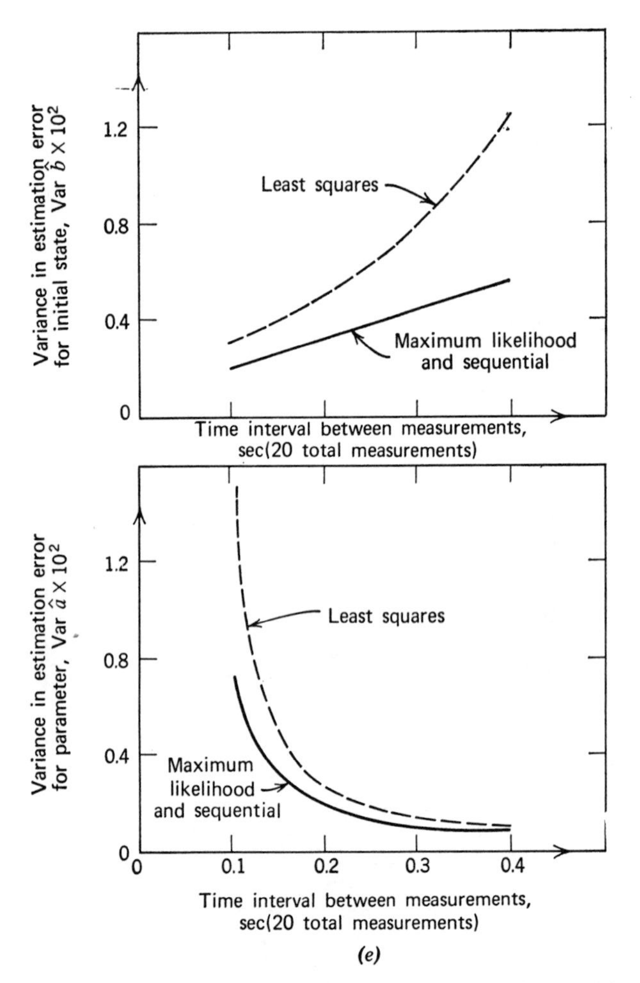

FIGURE E9.4-1e Variation of estimation error variance with measurement interval (stable model).

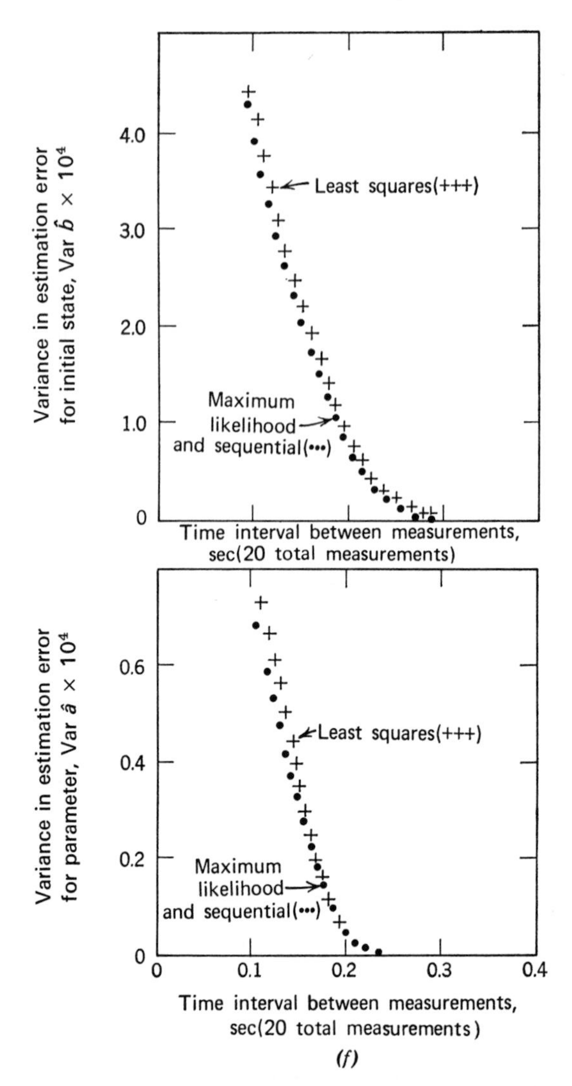

FIGURE E9.4-1f Variation of measurement error with estimation interval (unstable model).

Through use of a limiting process, the discrete sequential estimator equations can be converted to continuous estimation equations. In the most general case the process model is

$$\frac{d\mathbf{y}(t)}{dt} = \mathbf{f}(t)\mathbf{y}(t) + \mathbf{g}(t)\mathbf{x}(t) \tag{9.4-20}$$

and the unobservable error is added as follows:

$$\mathbf{Y}(t) = \mathbf{h}(t)\mathbf{y}(t) + \boldsymbol{\epsilon}(t)$$

The elements of the $\boldsymbol{\epsilon}(t)$ and $\mathbf{x}(t)$ matrices are then assumed to be random variables ($\mathbf{X}(t)$ is a random input) with the covariance matrix

$$\mathscr{E}\left\{\begin{bmatrix}\mathbf{X}(t)\\ \boldsymbol{\epsilon}(t)\end{bmatrix}[\mathbf{X}^T(\tau)\boldsymbol{\epsilon}^T(\tau)]\right\} = \begin{bmatrix}\mathbf{Q}(t)\mathbf{S}(t)\\ \mathbf{S}^T(t)\boldsymbol{\Gamma}(t)\end{bmatrix}\delta(t-\tau)$$

where $\delta(t)$ is the Dirac delta function.

The equations to estimate $\mathbf{y}(t)$ were first derived by Kalman.†

$$\frac{d\hat{\mathbf{y}}}{dt} = \mathbf{f}(t)\hat{\mathbf{y}}(t) + \mathbf{K}(t)[\mathbf{Y}(t) - \mathbf{h}(t)\hat{\mathbf{y}}(t)] \tag{9.4-21}$$

where

$$\mathbf{K}(t) = [\boldsymbol{\Omega}(t)\mathbf{h}^T(t) + \mathbf{g}(t)\mathbf{S}(t)]\boldsymbol{\Gamma}^{-1}(t)$$

The covariance matrix $\boldsymbol{\Omega}(t)$ of $\mathbf{y}(t)$ given $\mathbf{Y}(t)$ is computed from the matrix Riccati equation (the argument (t) is suppressed):

$$\frac{d\boldsymbol{\Omega}}{dt} = \mathbf{f}\boldsymbol{\Omega} + \boldsymbol{\Omega}\mathbf{f}^T + \mathbf{g}\mathbf{Q}\mathbf{g}^T - [\boldsymbol{\Omega}\mathbf{h}^T + \mathbf{g}\mathbf{S}]\boldsymbol{\Gamma}^{-1}[\mathbf{S}^T\mathbf{g}^T + \mathbf{h}\boldsymbol{\Omega}] \tag{9.4-22}$$

$$\boldsymbol{\Omega}(0) = \boldsymbol{\Omega}_0$$

If $\boldsymbol{\Omega}_0$ is singular, the equations are valid. But if $\boldsymbol{\Gamma}$ is singular, modifications are needed. Refer to the references at the end of this chapter.

9.5 METHOD OF QUASILINEARIZATION COMBINED WITH LEAST SQUARES

The method of quasilinearization is essentially a technique for solving nonlinear (in the dependent variable) differential equations in either initial value models or in models where the boundary conditions are split. The latter model has been termed a "boundary value model" in Section 9.1. Quasilinearization as applied to parameter estimation offers a proven strategy that can be coupled with any criterion for optimal estimation, although it is usually used in conjunction with least squares.

† R. E. Kalman, "New Methods and Results in Linear Prediction and Filtering Theory" in *Proceedings of the First Symposium on Engineering Application of Random Function Theory and Probability*, John Wiley, New York, 1963.

As an introduction to the method of quasilinearization, consider the problem of determining a solution of a set of simultaneous deterministic equations:

$$f_i(y_1, y_2, \ldots, y_v) = 0 \qquad i = 1, 2, \ldots, v \tag{9.5-1}$$

In matrix notation, Equation 9.5-1 is

$$\mathbf{f}(\mathbf{y}) = 0 \tag{9.5-2}$$

Equation 9.5-2 can be expanded in a truncated Taylor series about some initial approximation $\mathbf{y}^{(0)}$ as follows:

$$\mathbf{f}(\mathbf{y}) \cong \mathbf{f}(\mathbf{y}^{(0)}) + \mathbf{J}(\mathbf{y}^{(0)})(\mathbf{y} - \mathbf{y}^{(0)}) \tag{9.5-3}$$

where $\mathbf{J}(\mathbf{y}^{(0)})$ is the Jacobian matrix. The argument of $\mathbf{J}$ indicates that the elements are evaluated at $\mathbf{y}^{(0)}$:

$$\mathbf{J}^{(0)} = \mathbf{J}(\mathbf{y}^{(0)}) = \begin{bmatrix}\frac{\partial f_1}{\partial y_1} & \frac{\partial f_1}{\partial y_2} & \cdots & \frac{\partial f_1}{\partial y_v}\\ \vdots & \vdots & & \vdots\\ \frac{\partial f_v}{\partial y_1} & \frac{\partial f_v}{\partial y_2} & \cdots & \frac{\partial f_v}{\partial y_v}\end{bmatrix}_{\mathbf{y}=\mathbf{y}^{(0)}}$$

In the Newton-Raphson method of solving Equations 9.5-1, the next approximation for $\mathbf{y}$, after $\mathbf{y}^{(0)}$, is obtained by equating the right-hand side of Equation 9.5-3 to zero and solving for $\mathbf{y}$:

$$\mathbf{y}^{(1)} = \mathbf{y}^{(0)} - [\mathbf{J}(\mathbf{y}^{(0)})]^{-1}\mathbf{f}(\mathbf{y}^{(0)})$$

In general, the Newton-Raphson iteration scheme uses the recursion relation:

$$\mathbf{y}^{(n)} = \mathbf{y}^{(n-1)} - (\mathbf{J}^{(n-1)})^{-1}\mathbf{f}(\mathbf{y}^{(n-1)}) \tag{9.5-4}$$

Suppose now that the model is Equation 9.1-10, a set of nonlinear differential equations in which the parameters in $\boldsymbol{\alpha}$ are to be estimated as well as a set of initial conditions:

$$\frac{d\mathbf{y}^*}{dt} = \mathbf{f}(\boldsymbol{\alpha}, \mathbf{y}^*, \mathbf{x}, t) \qquad \mathbf{y}^*(0) = \mathbf{y}_0 \quad (9.5\text{-}5) \equiv (9.1\text{-}10)$$

(The * is used to distinguish the model responses here from related variables in subsequent equations.) If the right-hand side of Equation 9.5-5 is replaced with a linearized form of $\mathbf{f}(\boldsymbol{\alpha}, \mathbf{y}^*, \mathbf{x}, t)$, then a method of successive approximations can be used to find $\mathbf{y}^*(t)$, because the right-hand side of Equation 9.5-5 will be linear on each cycle of iteration. Bellman and Kalaba ‡ showed that this technique of "quasilinearization" has the property of quadratic convergence.

We transform the problem of estimating the coefficients in the differential equations and the initial conditions into a problem of estimating only initial conditions

‡ R. E. Bellman and R. E. Kalaba, *Quasilinearization and Nonlinear Boundary-Value Problems*, American Elsevier, New York, 1965, Chapter 1.

by letting the constant coefficients in the model be considered as functions of time. Then

$$\frac{d\alpha_1}{dt} = 0 \qquad \alpha_1(0) = \alpha_1$$

$$\frac{d\alpha_2}{dt} = 0 \qquad \alpha_2(0) = \alpha_2$$

$$\vdots$$

$$\frac{d\alpha_m}{dt} = 0 \qquad \alpha_m(0) = \alpha_m$$

become a set of supplementary differential equations added to Equations 9.5-5 to form an initial value model:

$$\frac{d\mathbf{y}}{dt} = \mathbf{f}(\mathbf{y}, \mathbf{x}, t) \qquad \mathbf{y}(0) = \mathbf{y}_0 \tag{9.5-6}$$

where

$$\mathbf{y} = \begin{bmatrix} y_1^* \\ y_2^* \\ \vdots \\ y_v^* \\ \alpha_1 \\ \vdots \\ \alpha_m \end{bmatrix} \qquad \mathbf{x} = \begin{bmatrix} x_1 \\ x_2 \\ \vdots \\ x_v \\ 0 \\ \vdots \\ 0 \end{bmatrix} \qquad \mathbf{f} = \begin{bmatrix} f_1(\mathbf{y}^*, \mathbf{x}, t) \\ f_2(\mathbf{y}^*, \mathbf{x}, t) \\ \vdots \\ f_v(\mathbf{y}^*, \mathbf{x}, t) \\ 0 \\ \vdots \\ 0 \end{bmatrix}$$

$$\mathbf{y}_0 = \begin{bmatrix} y_{01}^* \\ y_{02}^* \\ \vdots \\ y_{0v}^* \\ \alpha_1 \\ \vdots \\ \alpha_m \end{bmatrix}$$

Notice that $\mathbf{f}(\mathbf{y}^*, \mathbf{x}, t)$ is not explicitly a function of the parameters, because in evaluating $\mathbf{f}$ the α's and the y_0's are presumed known from the previous iteration.

By replacing the right-hand side of Equation 9.5-6 with the linearized form of the function $\mathbf{f}(\mathbf{y}, \mathbf{x}, t)$, we obtain the recursive relation needed for the calculations:

$$\frac{d\mathbf{y}^{(n+1)}}{dt} = \mathbf{J}^{(n)}\mathbf{y}^{(n+1)} + \mathbf{f}(\hat{\mathbf{y}}_0^{(n)}, \mathbf{x}, t) - \mathbf{J}^{(n)}\mathbf{y}^{(n)} \tag{9.5-7}$$

To start, an initial estimate of $\mathbf{y}_0$ is selected, $\hat{\mathbf{y}}_0^{(0)}$, comprised of estimates of the initial conditions in Equations 9.5-5 and estimates of the parameters. It is also necessary to obtain a starting approximate functional relationship for $\mathbf{y}^{(0)}(t)$ to use in Equation 9.5-7, obtained either by assumption or by integrating some (approximate) form of Equation 9.5-6 on the interval $0 \le t \le t_f$.

An improved solution to the model, $\mathbf{y}^{(1)}(t)$, can be obtained by using Equation 9.5-7:

$$\frac{d\mathbf{y}^{(1)}}{dt} = \mathbf{f}(\hat{\mathbf{y}}^{(0)}, \mathbf{x}, t) + \mathbf{J}^{(0)}(\mathbf{y}^{(1)} - \mathbf{y}^{(0)}) \tag{9.5-8}$$

where $\mathbf{J}^{(0)}$ presumably exists and has the elements $\partial f_i(\hat{\mathbf{y}}^{(0)}, \mathbf{x}, t)/\partial y_j$.

Because Equation 9.5-8 is a linear differential equation, a particular solution, $\mathbf{y}_p^{(1)}(t)$, can be numerically determined on the interval $0 \le t \le t_f$ with the unknown initial conditions set at some convenient values, say zero if otherwise unknown:

$$\frac{d\mathbf{y}_p^{(1)}}{dt} = \mathbf{J}^{(0)}\mathbf{y}_p^{(1)} + \mathbf{f}(\hat{\mathbf{y}}^{(0)}, \mathbf{x}, t) - \mathbf{J}^{(0)}\mathbf{y}^{(0)}$$

$$\mathbf{y}_p^{(1)}(0) = \mathbf{0}$$

A general solution of Equation 9.5-8 is the sum of the particular solution and the solution to the homogeneous equation (Equation 9.5-7 with the last two right-hand terms zero):

$$\mathbf{y}^{(1)}(t) = \mathbf{y}_p^{(1)}(t) + \sum_{j=1}^{v+m} c_j^{(1)}\mathbf{h}_j^{(1)}(t) \tag{9.5-9}$$

where $\mathbf{h}_j^{(1)}(t)$ are the $v + m$ linearly independent solutions (also obtained numerically) of the homogeneous set of differential equations:

$$\frac{d\mathbf{h}^{(1)}}{dt} = \mathbf{J}^{(0)}\mathbf{h}^{(1)} \qquad \mathbf{h}_j(0) = \boldsymbol{\delta}_j \tag{9.5-10}$$

with the initial conditions given by $\boldsymbol{\delta}_j$, a $(v + 1) \times 1$ vector with all elements equal to zero except for the jth which is unity.

All that remains to complete the solution for $\mathbf{y}^{(1)}(t)$ is to select the coefficients $\mathbf{c}^{(1)}$. The statistical aspects of the estimation procedure enter here, because the c's are selected so that the least squares criterion ϕ in Equation 9.2-1 or Equation 9.2-2 or some other criterion is minimized. If we substitute Equation 9.5-9 for $\boldsymbol{\Psi}$ in Equation 9.2-1 with $t = t_i$, and minimize ϕ analytically by placing $\partial\phi/\partial c_j = 0$, $j = v + 1, v + 2, \ldots, v + m$, a set of linear algebraic equations is obtained that can be solved for the c_j's:

$$\sum_{j=1}^{v+m} \gamma_{ij}c_j + w_j = 0 \qquad j = 1, \ldots, v + m \tag{9.5-11}$$

where

$$\gamma_{ij} \equiv \int_0^{t_f} (\mathbf{h}_i^T\mathbf{w}\mathbf{h}_j + \mathbf{h}_j^T\mathbf{w}\mathbf{h}_i)\, dt$$

$$w_j \equiv \int_0^{t_f} [\mathbf{h}_j^T\mathbf{w}(\mathbf{y}_p - \mathbf{Y}) + (\mathbf{y}_p - \mathbf{Y})^T\mathbf{w}\mathbf{h}_j]\, dt$$

(For Equation 9.2-2, we would replace the integral by a sum over $i = 1$ to n.)

The c's so chosen make it possible to calculate a new approximation for the initial vector $\mathbf{y}_0$. If $y^*_{01}, y^*_{02}, \ldots, y^*_{0v}$ are all known fixed values, then Equation 9.5-11 will be just a set of m linear equations in the m parameters in the model, and the initial condition vector at the end of the first iteration will be

$$\mathbf{y}_0^{(1)} = \begin{bmatrix} y^*_{01} \\ y^*_{02} \\ \vdots \\ y^*_{0v} \\ \hat{\alpha}_1^{(1)} \\ \vdots \\ \hat{\alpha}_m^{(1)} \end{bmatrix}$$

The entire procedure can be repeated to obtain $\hat{\boldsymbol{\alpha}}^{(2)}$ and $\mathbf{y}_0^{(2)}$, and so on until the change in the estimators (all or some) falls below some prefixed number. The precision in the estimators is calculated in the same way as described in Section 9.1.

Because $\mathbf{J}^{(n)}$ contains many zero entries and has a special structure, the computational effort can be reduced on a digital computer by taking advantage of this structure.† By any computational method, quasilinearization has quadratic convergence if the procedure converges. But it also exhibits the ills of the Newton-Raphson procedure described in Section 6.2-3, such as convergence toward a local rather than global optimum and oscillation. The remedies discussed in Section 6.2 can be applied to overcome these difficulties.

Example 9.5-1 Estimation of Kinetic Coefficients by Quasilinearization

Bellman, Jacquez, Kalaba, and Schwimmer illustrated the results of the quasilinearization procedure as applied to the gas-phase reaction of nitrogen oxide with oxygen.‡ Bodenstein and Linder suggested for the reaction

$$2NO + O_2 \rightleftarrows 2NO_2$$

the following model:

$$\frac{dy}{dt} = k_1(126.2 - y)(91.9 - y)^2 - k_2y^2 \qquad \text{(a)}$$

$$y(0) = 0$$

The observed values of Y were

t	Y
0	0
1	1.4
2	6.3
3	10.5
4	14.2
5	17.6
6	21.4
7	23.0
9	27.0
11	30.5
14	34.4
19	48.8
24	41.6
29	43.5
39	45.3

The coefficients k_1 and k_2 were estimated by the method of quasilinearization by minimizing $\phi = \sum_{i=1}^{14} [Y(t_i) - \hat{y}_i]^2$. The matrix differential equations corresponding to Equation 9.5-6 were

$$\begin{aligned} \frac{dy_1}{dt} &= k_1(126.2 - y_1)(91.9 - y_1)^2 - k_2y_1^2 \equiv f_1 \\ \frac{dk_1}{dt} &= 0 \\ \frac{dk_2}{dt} &= 0 \end{aligned} \qquad \text{(b)}$$

Initial values assumed for k_1 and k_2, respectively, were 10^{-6} and 10^{-4}. The equations corresponding to Equation 9.5-8 were

$$\begin{aligned} \frac{dy_1^{(1)}}{dt} = {} & 10^{-6}(126.2 - y_1^{(0)})(91.9 - y_1^{(0)})^2 - 10^{-4}(y_1^{(0)})^2 \\ & - [2 \times 10^{-6}(126.2 - y_1^{(0)})(91.9 - y_1^{(0)}) \\ & + 10^{-6}(91.9 - y_1^{(0)})^2 - 2 \times 10^{-4}y_1^{(0)}][y_1 - y_1^{(0)}] \end{aligned}$$

$$\frac{dk_1}{dt} = 0$$

$$\frac{dk_2}{dt} = 0$$

For this special case it was easy to obtain $y_0^{(0)}$ by integrating Equation (a) with $k_1 = 10^{-6}$ and $k_2 = 10^{-4}$ as a function of time. Table E9.5-1 lists the estimated coefficients at the end of each cycle of iteration.

TABLE E9.5-1

Cycle	k_1	k_2	ϕ
0	1×10^{-6}	1×10^{-4}	
1	0.3413×10^{-5}	0.2554×10^{-2}	
2	0.4859×10^{-5}	0.3683×10^{-3}	
3	0.4578×10^{-5}	0.2808×10^{-3}	
4	0.4577×10^{-5}	0.2797×10^{-3}	0.21×10^{-2}

† J. K. Donnelly and D. Quon, preprint 19F, Second Joint AIChE-IIQPR Meeting, Tampa, Fla., May 1968.

‡ R. E. Bellman, J. Jacquez, R. E. Kalaba, and S. Schwimmer, Rand Memorandum RM-4721-NIH, Aug. 1965.

Subsequent iterations produced no significant changes. The values after cycle 4 can be compared with those obtained by Bodenstein and Lindner of $k_1 = 0.53 \times 10^{-5}$, $k_2 = 0.41 \times 10^{-3}$, and a sum of squares of deviations of 0.55×10^{-2}.

The variances of the estimates can be obtained as described in Section 6.4; an approximate joint confidence region for them can be set up if an approximate analytical solution for the model is linearized.

Quasilinearization also can be employed for two point or multipoint boundary value models. As explained in Section 9.1, when the boundary conditions are split in models which describe the special dependence of the dependent variables, different numerical procedures must be substituted for direct integration of the differential equations. The method of quasilinearization can be employed as follows. Consider the two differential equations

$$\frac{dy_1}{dt} = f_1(y_1, y_2, t) \qquad y_1(0) = y_{01} \qquad (9.5\text{-}12a)$$

$$\frac{dy_2}{dt} = f_2(y_1, y_2, t) \qquad y_2(L) = y_{L2} \qquad (9.5\text{-}12b)$$

We assume that the starting profiles for y_1 and y_2 can be approximated. The right-hand side of Equation 9.5-12a can be linearized about $y_1^{(0)}(t)$ and $y_2^{(0)}(t)$:

$$f_1(y_1, y_2, t) = f_1(y_1^{(0)}, y_2^{(0)}, t) + \frac{\partial f_1(y_1^{(0)}, y_2^{(0)}, t)}{\partial y_1}(y_1 - y_1^{(0)}) + \frac{\partial f_1(y_1^{(0)}, y_2^{(0)}, t)}{\partial y_2}(y_2 - y_2^{(0)})$$

and a similar equation can be written for the right-hand side of Equation 9.5-12b.

The linearized set of equations is

$$\begin{aligned} \frac{dy_1}{dt} &= f_{11}y_1 + f_{12}y_2 + f_{10} - f_{11}y_1^{(0)} - f_{12}y_2^{(0)} \\ \frac{dy_2}{dt} &= f_{21}y_1 + f_{22}y_2 + f_{20} - f_{21}y_1^{(0)} - f_{22}y_2^{(0)} \end{aligned} \qquad (9.5\text{-}13)$$

where the first subscript on f refers to the function number and the second subscript refers to the dependent variable with respect to which the function is being differentiated. Equations 9.5-13 can be solved by, say, Runge-Kutta integration for a set of initial conditions:

$$y_1(0) = y_{01}$$
$$y_2(0) = a_1 \qquad \text{(assumed)}$$

to give a solution at $t = L$ of $y_1^+(L)$ and $y_2^+(L)$. A second profile can be found for the initial conditions

$$y_1(0) = y_{01}$$
$$y_2(0) = a_2 \qquad \text{(assumed)}$$

to give $y_1^{++}(L)$ and $y_2^{++}(L)$.

The superposition property of linear equations states that any solution of y_1 can be obtained as a linear combination of other solutions of y_1 or, here,

$$y_1 = w_1 y_1^+(L) + w_2 y^{++}(L)$$

We can find the relative weights from the initial conditions

$$y_{01} = w_1 y_{01} + w_2 y_{01}$$

or $w_2 = 1 - w_1$. Consequently,

$$y_2(L) = w_1 y_2^+(L) + (1 - w_1) y_2^{++}(L)$$

or

$$w_1 = \frac{y_2(L) - y_2^{++}(L)}{y_2^+(L) - y_2^{++}(L)} \qquad (9.5\text{-}14)$$

The functional relation between y_1 and t and y_2 and t for any cycle of calculation after the zeroth cycle can be obtained by using the weights given by Equation 9.5-14 and from

$$y_r(t) = w_1 y_r^+(t) + w_2 y_r^{++}(t) \qquad r = 1, 2 \qquad (9.5\text{-}15)$$

New estimates of $y_2(0)$ can be made sequentially from Equation 9.5-15 with $t = 0$ until $y_r(t)$ from stage to stage is less than a preselected number. The point at which the boundary condition at L matches corresponds only to *one* cycle of the initial value problem; hence the complete computations prove quite extensive.

9.6 ESTIMATION USING THE EQUATION ERROR (MODEL RESIDUE)

Many process models consist of an nth order ordinary differential equation:

$$\alpha_q \frac{d^{(q)}y}{dt^{(q)}} + \cdots + \alpha_1 \frac{dy}{dt} + \alpha_0 y = x(t) + \beta_1 \frac{dx(t)}{dt} + \cdots + \beta_m \frac{d^{(m)}x}{dt^{(m)}} \qquad (9.6\text{-}1)$$

in which y is the process output, x is the process input, and the coefficients are not necessarily constant but may be functions of y, x, their derivatives, or t. To simplify the notation, we shall let

$$Y_k = \frac{d^{(k)}Y}{dt^{(k)}} = \frac{d^{(k)}y}{dt^{(k)}} + \epsilon_k \qquad k = 0, 1, \ldots, q$$

$$X_j = \frac{d^{(j)}X}{dt^{(j)}} = \frac{d^{(j)}X}{dt^{(j)}} + \epsilon_j \qquad j = 0, 1, \ldots, m$$

be the stochastic variables which are observed (the process input does not have to be stochastic). In Equation 9.6-1 the coefficient of $x(t)$ has been made 1 by prior division of every term by the original coefficient.

We want to find the best estimates of α_k and β_j (given at least $m + q + 2$ observations), best in the sense of minimizing the square of the "equation error" ("model residue," "satisfaction error"), $\acute{\epsilon}$, defined as follows:

$$\acute{\epsilon} = \sum_{k=0}^{q} \alpha_k Y_k - \sum_{j=1}^{m} \beta_j X_j - X \tag{9.6-2}$$

The equation error is a weighted sum of the errors ϵ_k and $(-\epsilon_j)$, the weights being the model parameters themselves. The square of $\acute{\epsilon}$ in Equation 9.6-2 defines an $m + q + 2$ dimensional hypersurface with coordinates $\acute{\epsilon}^2$, α_k, and β_j. The hyperplane $\acute{\epsilon} = 0$, which is tangent to the hypersurface, can be sought (as a function of time) by using steepest descent or another optimization technique.

As a simple example of the equation error, for the model

$$\beta_1 \frac{dy}{dt} + \beta_0 y = x$$

the corresponding equation error is

$$\acute{\epsilon} = \beta_1 \frac{dY}{dt} + \beta_0 Y - X$$

and the surface $\acute{\epsilon}^2$ versus β_1 and β_0 is shown in Figure 9.6-1. The surface $\acute{\epsilon}^2$ is tangent to the $\beta_0 - \beta_1$ plane in a line denoted by $A - A'$. As the values of dY/dt, Y, and X change with time, the line $A - A'$ rotates about the point (b_1, b_0) and $\acute{\epsilon}^2 \to 0$.

Let us rewrite the generalized error given by Equation 9.6-2 in a slightly different way so that some of the parameters may be negative:

$$\acute{\epsilon} = \sum_{k=0}^{q} \alpha_k Y_k + \sum_{j=1}^{m} \beta_j X_j + X \tag{9.6-3}$$

If *discrete* values of Y_k, X_j, and X are measured, because $\acute{\epsilon}$ is a linear function of the parameters, minimization of $\sum_{i=1}^{n} \acute{\epsilon}_i^2$ with respect to the parameters yields a set of equations analogous to those described in Section 5.6.

If Y_k, X_j, and X are *continuous* functions of time, then by imitating the least squares approach of Section 9.2, we can minimize

$$\psi = \tfrac{1}{2}\left(\int_0^{t_f} \acute{\epsilon}^2 \, dt\right) \tag{9.6-4}$$

by the method of steepest descent, computing $(k > 0)$:

$$\begin{aligned} \frac{d\alpha_k}{dt} &= -k_k \frac{\partial \psi}{\partial \alpha_k} = -k_k \int_0^{t_f} \acute{\epsilon} \frac{\partial \acute{\epsilon}}{\partial \alpha_k} dt \\ \frac{d\beta_j}{dt} &= -k_j \frac{\partial \psi}{\partial \beta_j} = -k_j \int_0^{t_f} \acute{\epsilon} \frac{\partial \acute{\epsilon}}{\partial \beta_j} dt \end{aligned} \tag{9.6-5}$$

To simplify the notation further, we let

$$\int_0^{t_f} \acute{\epsilon} \frac{\partial \acute{\epsilon}}{\partial \alpha_k} dt = \int_0^{t_f} \acute{\epsilon} Y_k \, dt \equiv \langle \acute{\epsilon}, Y_k \rangle \qquad k = 0, \ldots, q$$

and

$$\int_0^{t_f} \acute{\epsilon} \frac{\partial \acute{\epsilon}}{\partial \beta_j} dt = \int_0^{t_f} \acute{\epsilon} X_j \, dt \equiv \langle \acute{\epsilon}, X_j \rangle \qquad j = 1, \ldots, m$$

The restriction that the coefficients are constant in the interval 0 to t_f is not a severe one since the data accumulation time can be reduced to as short as a few system time constants.

Expansion of $\acute{\epsilon}$ in Equations 9.6-5 into individual terms, introduction of the estimates a_i and b_j for the respective parameters α_i and β_j, and organization as a matrix yield

$$\begin{bmatrix} \left(-\frac{1}{k_1}\right) \frac{da_0}{dt} \\ \vdots \\ \left(-\frac{1}{k_{q+1}}\right) \frac{da_q}{dt} \\ \left(-\frac{1}{k_{q+2}}\right) \frac{db_1}{dt} \\ \vdots \\ \left(-\frac{1}{k_{q+m+1}}\right) \frac{db_m}{dt} \end{bmatrix} = \begin{bmatrix} \langle Y_0, Y_0 \rangle & \cdots & \langle Y_q, Y_0 \rangle & \langle X_1, Y_0 \rangle & \cdots & \langle X_m, Y_0 \rangle \\ \vdots & & \vdots & \vdots & & \vdots \\ \langle Y_0, Y_q \rangle & \cdots & \langle Y_q, Y_q \rangle & \langle X_1, Y_q \rangle & \cdots & \langle X_m, Y_q \rangle \\ \langle Y_0, X_1 \rangle & \cdots & \langle Y_q, X_1 \rangle & \langle X_1, X_1 \rangle & \cdots & \langle X_m, X_1 \rangle \\ \vdots & & \vdots & \vdots & & \vdots \\ \langle Y_0, X_m \rangle & \cdots & \langle Y_q, X_m \rangle & \langle X_1, X_m \rangle & \cdots & \langle X_m, X_m \rangle \end{bmatrix} \begin{bmatrix} a_0 \\ \vdots \\ a_q \\ b_1 \\ \vdots \\ b_m \end{bmatrix} + \begin{bmatrix} \langle X, Y_0 \rangle \\ \vdots \\ \langle X, X_q \rangle \\ \langle X, X_1 \rangle \\ \vdots \\ \langle X, X_m \rangle \end{bmatrix} \tag{9.6-6}$$

Equation 9.6-6 can be solved on an analog or hybrid computer in real time (or another selected time scale), assuming that the estimated parameters are constant or change slowly. At the minimum of ψ, when t_f is large, $\psi \to 0$,

$$\frac{\partial \psi}{\partial a_0}, \frac{\partial \psi}{\partial a_1}, \cdots \to 0, \qquad \frac{da_0}{dt}, \frac{da_1}{dt} \cdots \to 0$$

in which case if the matrix of time-averaged quantities does not prove to be singular, the estimated coefficients

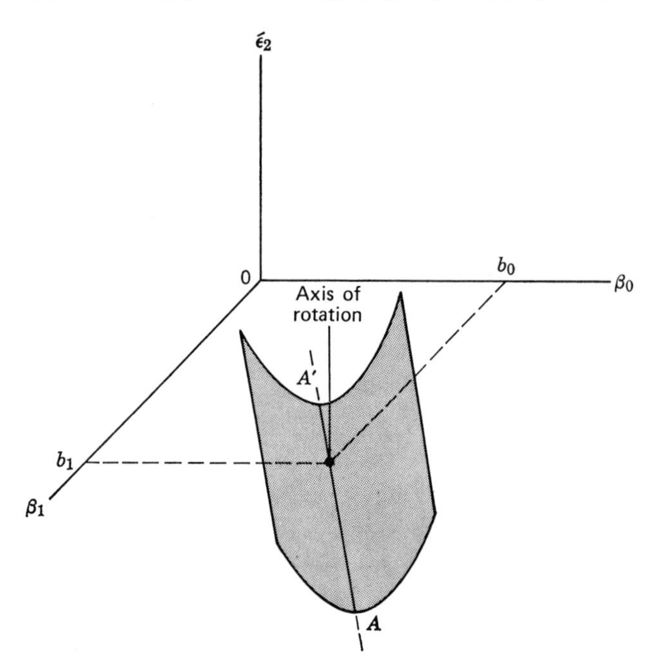

FIGURE 9.6-1 Geometric representation of equation error.

can be obtained after equating the right-hand side of Equation 9.6-6 to zero. How well the estimation procedure converges depends on the shape of the surface ψ in parameter space.

Figure 9.6-2 illustrates the progress of the steepest descent search for two models when the error $\acute{\epsilon}$ is not explicitly a function of time.

Calculation of the bias and precision of the estimates a_k and b_j is uncertain inasmuch as the statistics associated with the terms in Equation 9.6-6 are unlikely to be known.

Aström† showed that for a particular model of a dynamic system a *lower bound* can be placed on the

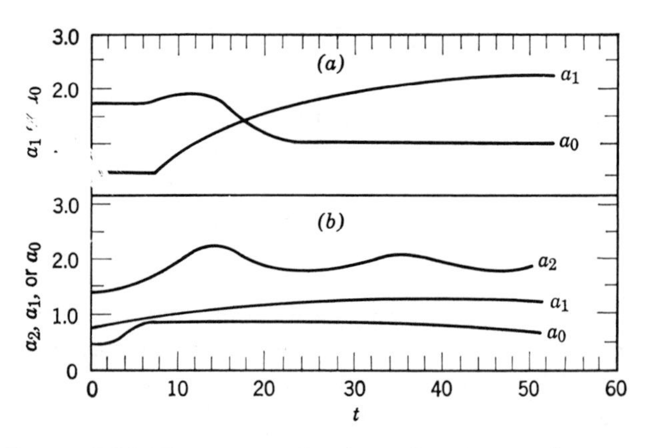

FIGURE 9.6-2 Parameter estimation using the equation error criterion. Models are: (*a*) first-order and (*b*) second-order ordinary differential equations with step inputs.

† K. J. Aström, Preprints of the International Federation of Automatic Control, Paper I.8, Prague, Czechoslovakia, June, 1967.

elements of the covariance matrix of the estimated parameters for any estimation procedure involving the equation error. Equation 9.6-2 can be rewritten as

$$D_k[Y(t)] = D_j X(t) + \lambda\acute{\epsilon}(t) \tag{9.6-7}$$

where

$$D_k[Y(t)] = \alpha_q \frac{d^{(q)}Y(t)}{dt^q} + \cdots + \alpha_0$$

$$D_j[X(t)] = \beta_m \frac{d^{(m)}X(t)}{dt^m} + \cdots + X(t)$$

$$\lambda = \text{a constant}$$

If the $\mathscr{E}\{\acute{\epsilon}(t)\} = 0$, the $\mathscr{E}\{\acute{\epsilon}(t)\acute{\epsilon}(t)\} = 1$, and the Laplace transform of D_j does not have real negative roots (i.e., Equation 9.6-7 represents a stable process), we can proceed as follows.

Observations $Y(t)$ and $X(t)$ are made of $y(t)$ and $x(t)$, and Aström gives the logarithm of the likelihood function of the vector of parameters $\boldsymbol{\theta}$ (λ is not included in $\boldsymbol{\theta}$) as

$$\ln L = -\frac{1}{2\lambda^2}\int_0^{t_f} \acute{\epsilon}^2(t)\,dt - t_f \ln \lambda + \text{constant} \tag{9.6-8}$$

where $\acute{\epsilon}$ is $\acute{\epsilon}(t) = D_k[Y(t)] - D_j[X(t)]$. He applied the Cramér-Rao theorem,‡ which states that if $\hat{\boldsymbol{\theta}}$ is an unbiased estimate of $\boldsymbol{\theta}$ and $\ln L$ is continuous in Y and X and twice differentiable with respect to $\boldsymbol{\theta}$,

$$\mathscr{E}\{(\hat{\boldsymbol{\theta}} - \boldsymbol{\theta})(\hat{\boldsymbol{\theta}} - \boldsymbol{\theta})^T\} \geq \mathbf{J}^{-1} \tag{9.6-9}$$

where $\mathbf{J}$ is the "information matrix" defined by

$$\mathbf{J} = \mathscr{E}\left\{\left(\frac{\partial \ln L}{\partial \boldsymbol{\theta}}\right)\left(\frac{\partial \ln L}{\partial \boldsymbol{\theta}}\right)^T\right\}$$

$$= -\mathscr{E}\left\{\begin{bmatrix} \dfrac{\partial^2 \ln L}{\partial\theta_1\,\partial\theta_1} & \cdots & \dfrac{\partial^2 \ln L}{\partial\theta_1\,\partial\theta_m} & \dfrac{\partial^2 \ln L}{\partial\theta_1\,\partial\lambda} \\ \vdots & & & \vdots \\ \dfrac{\partial^2 \ln L}{\partial\theta_m\,\partial\theta_1} & \cdots & \dfrac{\partial^2 \ln L}{\partial\theta_m\,\partial\theta_m} & \dfrac{\partial^2 \ln L}{\partial\theta_m\,\partial\lambda} \\ \dfrac{\partial^2 \ln L}{\partial\lambda\,\partial\theta_1} & \cdots & \dfrac{\partial^2 \ln L}{\partial\lambda\,\partial\theta_m} & \dfrac{\partial^2 \ln L}{\partial\lambda^2} \end{bmatrix}\right\} \tag{9.6-10}$$

The following quantities are required:

$$\frac{\partial^2 \ln L}{\partial\lambda^2} = -\frac{3}{\lambda^4}\int_0^{t_f} \acute{\epsilon}^2(t)\,dt + \frac{t_f}{\lambda^2}$$

$$\frac{\partial^2 \ln L}{\partial\lambda\,\partial\theta_i} = \frac{2}{\lambda^3}\int_0^{t_f} \acute{\epsilon}(t)\,\frac{\partial\acute{\epsilon}(t)}{\partial\theta_i}\,dt$$

$$\frac{\partial^2 \ln L}{\partial\theta_i\theta_j} = -\frac{1}{\lambda^2}\int_0^{t_f} \frac{\partial\acute{\epsilon}(t)}{\partial\theta_i}\,\frac{\partial\acute{\epsilon}(t)}{\partial\theta_j}\,dt - \frac{1}{\lambda^2}\int_0^{t_f} \acute{\epsilon}(t)\,\frac{\partial^2\acute{\epsilon}(t)}{\partial\theta_i\,\partial\theta_j}\,dt$$

‡ H. Cramér, *Mathematical Methods of Statistics*, Princeton Univ. Press, Princeton, N.J., 1946.

Each second derivative of $\acute{\epsilon}$ vanishes; hence the following expected values result:

$$\mathscr{E}\left\{\frac{\partial^2 \ln L}{\partial \lambda^2}\right\} = -\frac{2t_f}{\lambda^2} \tag{9.6-11a}$$

$$\mathscr{E}\left\{\frac{\partial^2 \ln L}{\partial \lambda\, \partial \theta_i}\right\} = 0 \tag{9.6-11b}$$

$$\mathscr{E}\left\{\frac{\partial^2 \ln L}{\partial \theta_i\, \partial \theta_j}\right\} = -\frac{1}{\lambda^2}\int_0^{t_f} \mathscr{E}\left\{\frac{\partial \acute{\epsilon}(t)}{\partial \theta_i}\cdot\frac{\partial \acute{\epsilon}(t)}{\partial \theta_j}\right\} dt \tag{9.6-11c}$$

Specifically, the expected values in Equation 9.6-11c will be of the following type:

$$\mathscr{E}\left\{\frac{d^{(k)}Y}{dt^k}\cdot\frac{d^{(j)}Y}{dt^j}\right\} \quad \text{or} \quad \mathscr{E}\left\{\frac{d^{(k)}Y}{dt^k}\cdot\frac{d^{(m)}X}{dt^m}\right\}$$

that have been previously related in Example 2.2-4 to the auto- and crosscorrelation functions of X and Y.

Instead of using Equation 9.6-4 as the criterion to be minimized, another technique calls for minimizing $\frac{1}{2}\acute{\epsilon}^2$ instantaneously. Then the equivalent of Equations 9.6-5 are

$$\begin{aligned} \frac{d\alpha_k}{dt} &= -k_k\frac{\partial \acute{\epsilon}^2}{\partial \alpha_k} = -k_k\acute{\epsilon}Y_k \\ \frac{d\beta_j}{dt} &= -k_j\frac{\partial \acute{\epsilon}^2}{\partial \beta_j} = -k_j\acute{\epsilon}X_j \end{aligned} \tag{9.6-12}$$

The rationale for these equations can be seen by assuming that $\acute{\epsilon}$ is also explicitly a function of time, because then

$$\frac{d(\frac{1}{2}\acute{\epsilon}^2)}{dt} = \acute{\epsilon}\left[\frac{\partial \acute{\epsilon}}{\partial \alpha_k}\frac{d\alpha_k}{dt} + \frac{\partial \acute{\epsilon}}{\partial \beta_j}\frac{d\beta_j}{dt} + \frac{\partial \acute{\epsilon}}{\partial t}\right]$$

Introduction of Equations 9.6-12 for the time derivatives of the coefficients yields

$$\frac{d(\frac{1}{2}\acute{\epsilon}^2)}{dt} = -\acute{\epsilon}^2\left[\sum_{k=0}^{q} k_k Y_k^2 + \sum_{j=1}^{m} k_j X_j^2\right] + \acute{\epsilon}\frac{\partial \acute{\epsilon}}{\partial t} \tag{9.6-13}$$

Although the first term on the right-hand side of Equation 9.6-13 is always negative (and follows the path of steepest descent with respect to time), it can be outweighed by the term $\acute{\epsilon}(\partial\acute{\epsilon}/\partial t)$. Hence, convergence is obtained by choosing k_k and k_j such that the sum of the two terms is always negative. Convergence can be proved for cases in which α_k and β_j are constant, and it has been found empirically for certain other special cases. All that is required to prove convergence is to show that Equation 9.6-6 is asymptotically stable, either through the use of linear systems theory or a Lyapunov function as described in texts on systems analysis. The rate of convergence cannot be arbitrarily increased by increasing k. As the gain k is increased or decreased from its optimal value, the time required for the parameter value to converge increases because the surface ψ is not positive-definite with closed contours.

Example 9.6-1 Continuous Estimation Using the Equation Error

The deterministic model for a well-mixed reactor with flow in and out and a reaction (first order to make the model linear) is (V = volume, c = concentration, and F = flow rate):

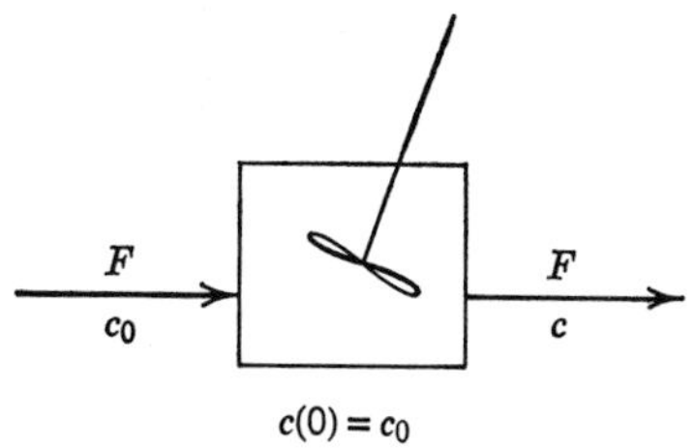

$c(0) = c_0$

$$V\frac{dc}{dt} = Fc_0 - Fc - Vk_rc \tag{a}$$

or

$$\alpha_1\frac{dc}{dt} + \alpha_0 c = c_0 \tag{b}$$

where

$$\alpha_1 = \frac{V}{F} \quad \text{and} \quad \alpha_0 = \frac{F + k_rV}{F}$$

The coefficients α_1 and α_0 are to be estimated from observations of c_0, dc/dt, and c; we let $a_1 = \hat{\alpha}_1$ and $a_0 = \hat{\alpha}_0$.

In what follows the observations of dc/dt, c, and c_0 will be stochastic variables.

The equation error from Equation 9.6-3 is

$$\acute{\epsilon} = \alpha_1 Y_1 + \alpha_0 Y_0 + X \tag{c}$$

where $Y_1 = dc/dt + \epsilon_1$, $Y_0 = c + \epsilon_2$, and $X = c_0 + \epsilon_3$.

The matrices corresponding to Equation 9.6-6 for this problem are

$$\begin{bmatrix} -\dfrac{1}{k_0}\dfrac{da_0}{dt} \\ -\dfrac{1}{k_1}\dfrac{da_1}{dt} \end{bmatrix} = \begin{bmatrix} \langle Y_0, Y_0\rangle & \langle Y_1, Y_0\rangle \\ \langle Y_0, Y_1\rangle & \langle Y_1, Y_1\rangle \end{bmatrix}\begin{bmatrix} a_0 \\ a_1 \end{bmatrix} + \begin{bmatrix} \langle X, X_0\rangle \\ \langle X, Y_1\rangle \end{bmatrix} \tag{d}$$

If we let $k_0 = k_1 = 1$ and

$$\boldsymbol{\xi} = \begin{bmatrix} \langle Y_0, Y_0\rangle & \langle Y_1, Y_0\rangle \\ \langle Y_0, Y_1\rangle & \langle Y_1, Y_1\rangle \end{bmatrix} \qquad \boldsymbol{\zeta} = \begin{bmatrix} \langle X, X_0\rangle \\ \langle X, Y_1\rangle \end{bmatrix}$$

then Equation (d) becomes

$$\frac{d\mathbf{a}}{dt} = -\boldsymbol{\xi}\mathbf{a} - \boldsymbol{\zeta} \qquad \mathbf{a}(0) = \begin{bmatrix} a_{00} \\ a_{10} \end{bmatrix} \tag{e}$$

where a_{00} and a_{10} are the selected initial values of the coefficients. The solution of Equation (e) is

$$\mathbf{a} = e^{-\boldsymbol{\xi}t}\mathbf{a}_0 - \boldsymbol{\xi}^{-1}\boldsymbol{\zeta}$$

or

$$\begin{bmatrix} a_0 \\ a_1 \end{bmatrix} = e^{-\boldsymbol{\xi}t}\begin{bmatrix} a_{00} \\ a_{10} \end{bmatrix} - \boldsymbol{\xi}^{-1}\begin{bmatrix} \langle X, Y_0\rangle \\ \langle X, Y_1\rangle \end{bmatrix} \tag{f}$$

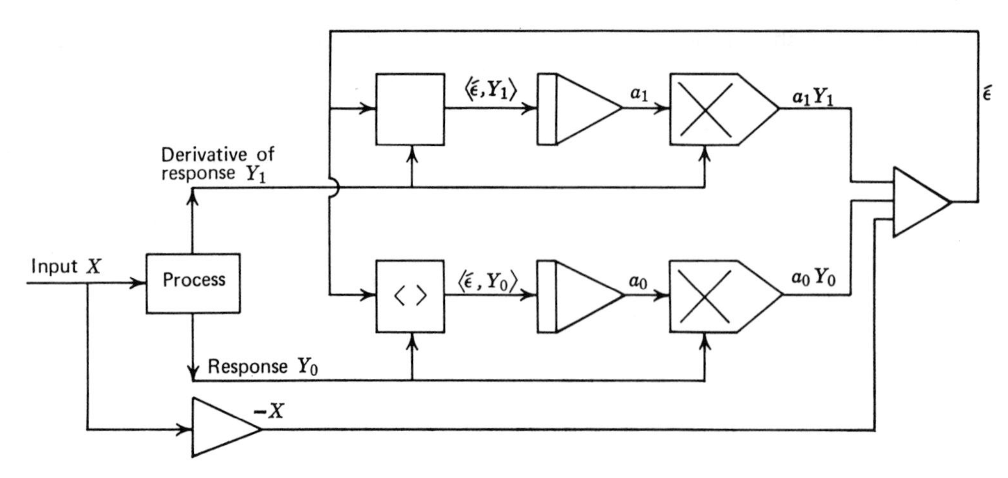

FIGURE E9.6-1

The stability of the calculations for a_0 and a_1 depends on the signs of the eigenvalues of $\boldsymbol{\xi}$.

Figure E9.6-1 illustrates a possible analog computational scheme for the coefficients. Figure 9.6-2a illustrates the estimates of α_0 and α_1 as a function of time for one run in which C and C_0 are measured continuously. The bias and precision of the estimates are not known.

The specific calculations of $\langle Y_i, Y_j \rangle$ are

$$\langle Y_0, Y_0 \rangle = \int_0^{t_f} C^2\, dt$$

$$\langle Y_0, Y_1 \rangle = \int_0^{t_f} (C)\left(\frac{dC}{dt}\right) dt$$

$$\langle Y_1, Y_1 \rangle = \int_0^{t_f} \left(\frac{dC}{dt}\right)^2 dt$$

$$\langle X, Y_0 \rangle = \int_0^{t_f} C_0 C\, dt$$

$$\langle X, Y_1 \rangle = \int_0^{t_f} C_0 \frac{dC}{dt}\, dt$$

Because $\langle Y_0, Y_1 \rangle = \langle Y_1, Y_0 \rangle = 0$,

$$\boldsymbol{\xi}^{-1} = \begin{bmatrix} \langle Y_0, Y_0 \rangle^{-1} & 0 \\ 0 & \langle Y_1, Y_1 \rangle^{-1} \end{bmatrix}$$

and

$$a_0 = a_{00}\, e^{-\langle Y_0, Y_0 \rangle t} - \frac{\langle X, Y_0 \rangle}{\langle Y_0, Y_0 \rangle}$$
$$a_1 = a_{10}\, e^{-\langle Y_1, Y_1 \rangle t} - \frac{\langle X, Y_1 \rangle}{\langle Y_1, Y_1 \rangle} \tag{g}$$

At the minimum of ψ when $\partial\phi/\partial a_0 = \partial\phi/\partial a_1 = 0$,

$$a_0 = \frac{\langle X, Y_0 \rangle}{\langle Y_0, Y_0 \rangle} \quad \text{and} \quad a_1 = \frac{\langle X, Y_1 \rangle}{\langle Y_1, Y_1 \rangle}$$

Example 9.6-2 Comparison of Equation Error and Least Squares Estimates

Estimates of two parameters α_0 and β_1 in the second-order model

$$\frac{d^2y}{dt} + \alpha_1 \frac{dy}{dt} + \alpha_0 y = \beta_0 x + \beta_1 \frac{dx}{dt} \tag{a}$$

were compared in a simulation study under equivalent conditions of additive noise for the derivatives and y. The gains (k's) were the same in each test, and the input $x(t) = \sin t$ was the same in both cases. By increasing the gain the response error became unstable; hence by decreasing the gain the time for $\psi \rightarrow \psi_{\min}$ can actually be reduced. Typical paths of the estimates using Equations 9.2-4 and 9.6-12, respectively, are shown in Figure E9.6-2. The trajectories shown are characteristic of Equation (a) for the selected parameters only; other combinations of parameter values will yield different simulated responses and, hence, different trajectories in parameter space.

As long as the parameters to be estimated enter the model equation linearly, a model nonlinear in the dependent variable or its derivatives is treated by the equation error criterion in exactly the same way as the linear (in the independent variable) model. For example, in the model

$$\alpha_1 \frac{dc}{dt} + \alpha_0 c + \beta c^2 = c_0$$

there is no reason why the term involving c^2 should not be treated in exactly the same fashion as the one containing c or dc/dt; consequently the estimation procedure remains unchanged.

Models that incorporate several dependent variables in the form of independent ordinary differential equations require the use of a weighted objective function such as

$$\psi_1 = \int_0^{t_f} (w_1\epsilon_1^2 + w_2\epsilon_2^2 + \cdots)\, dt$$

and ψ_1 can be minimized by any of the previously described methods. The weights can be unity or selected according to one of the characteristics listed in Section 5.5 for multiresponse models.

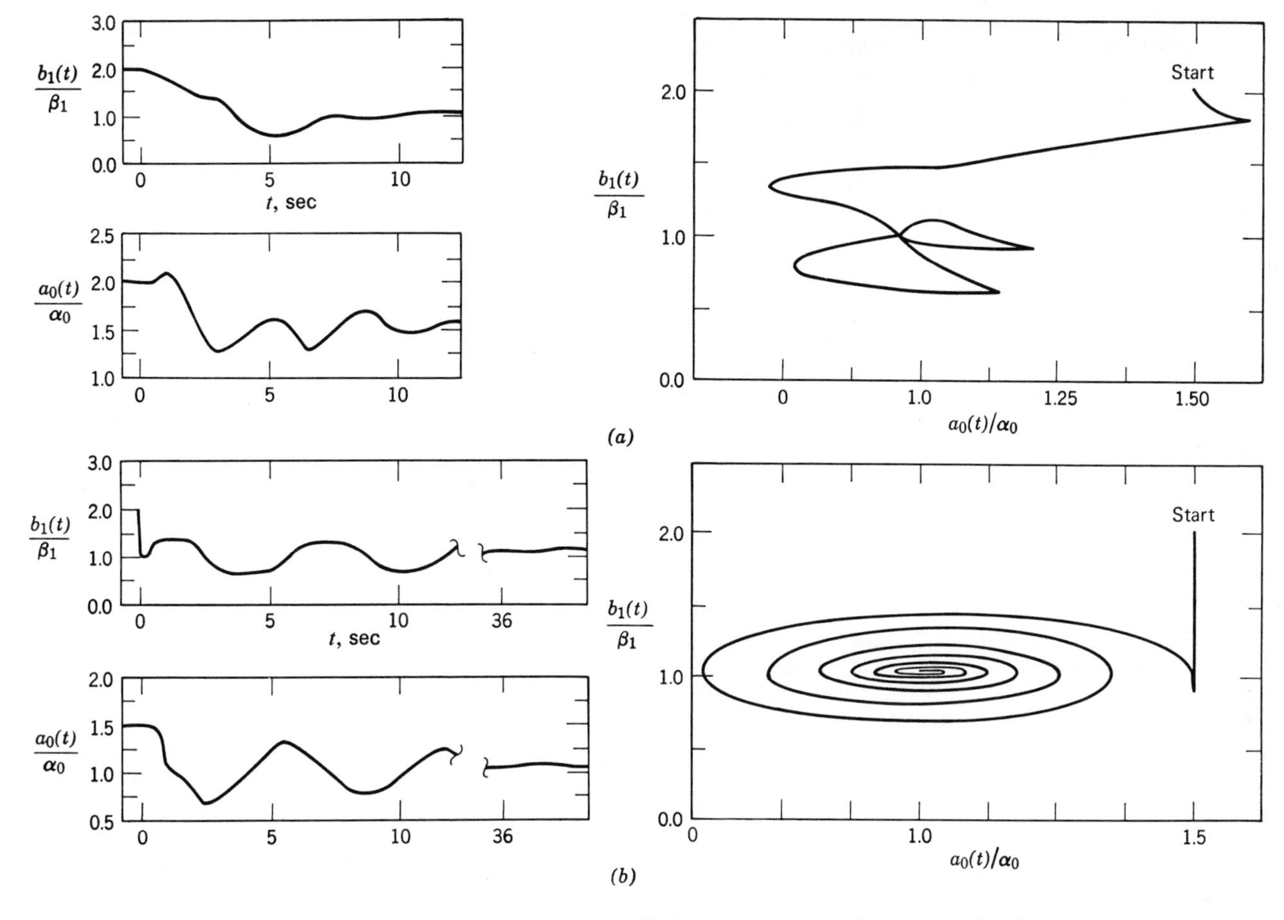

FIGURE E9.6-2 (*a*) Least squares estimation. (*b*) Instantaneous equation error estimation.

Supplementary References

General

Cuenod, M. and Sage, A. P., "Comparison of Some Methods Used for Process Identification," *Automatica* **4**, 235, 1968.

Deutsch, R., *Estimation Theory*, Prentice-Hall, Englewood Cliffs, N.J., 1965.

Eschenroeder, A. Q., Boyer, D. W., and Hall, J. G., *Phys. Fluids* **5**, 615, 1962.

Eykhoff, P., *IEEE Trans. Auto. Control* **AC8**, 347, 1963.

Eykhoff, P., "Process Parameter and State Estimation," *Automatica* **4**, 205, 1968.

Mayne, D. Q., "Parameter Estimation," *Automatica* **3**, 245, 1966.

Stevenson, P. C., *Processing of Counting Data*, National Academy of Sciences, NRC, 1966, Chapter 6.

Least Squares Estimation

Aberbach, L. B., "Estimation of Parameters In Differential Equations," Ph.D. Dissertation, Princeton University, Princeton, N.J., 1967.

Bekey, G. A. and McGhee, R. B., "Gradient Methods for the Optimization of Dynamic System Parameters by Hybrid Computation" in *Computing Methods for Optimization Problems*, ed. by A. V. Balakrishnan and L. W. Neustadt, Academic Press, New York, 1964, p. 305.

Box, G. E. P., "Use of Statistical Methods in the Elucidation of Basic Mechanisms," *Bull. Inst. de Statistique* **36**, 215, 1957.

Box, G. E. P. and Coutie, G. A., "Application of Digital Computers in the Exploration of Functional Relationships," *Proceed. Inst. Elec. Eng.* (London) **103B**, Supl. 1, 100, 1956.

Box, G. E. P. and Hunter, W. G., "A Useful Method for Model Building," *Technometrics* **4**, 301, 1962.

Hartley, H. O., "The Estimation of Nonlinear Parameters by 'Internal Least Squares'," *Biometrika* **35**, 32, 1948.

Meissinger, H. F. and Bekey, G. A., "An Analysis of Continuous Parameter Identification Methods," *Simulation* **6**, 94, Feb., 1966.

Rubin, A. I., Driban, S., and Miessner, W. W., "Regression Analysis and Parameter Identification," *Simulation* **7**, 39, July, 1967.

Equation Error Estimation

Graupe, K. K., "The Analog Solution of Some Functional Analysis Problems," *Trans. AIEE Comm. and Electronics* **79**, 793, 1960.

Hoberock, L. I. and Kohr, R. H., "An Experimental Determination of Differential Equations to Describe Simple Nonlinear Systems," Joint Automatic Control Conference Preprints, 1966, p. 616.

Lion, P. M., "Rapid Identification of Linear and Nonlinear Systems," Joint Automatic Control Conference Preprints, 1966, p. 605.

Sequential Estimation

Davenport, W. B. and Root, W. L., *An Introduction to the Theory of Random Signals and Noise*, McGraw Hill, New York, 1958.

Detchmendy, D. M. and Sridhar, R., "Sequential Estimation of States and Parameters in Noisy Nonlinear Dynamic Systems," *Trans. ASME, J. Basic Eng.* **88D**, 362, 1966.

Friendlander, B. and Bernstein, I., "Estimation of a Nonlinear Process in the Presence of Non Gaussian Noise and Disturbances," *J. Franklin Inst.* **281**, 455, 1966.

Ho, Y. C. and Lee, R. C. K., "A Bayesian Approach to Problems in Stochastic Estimation and Control," *IEEE Trans.* **AC-9**, 333, 1964.

Ho, Y. C. and Lee, R. C. K., "Identification of Linear Dynamic Systems," *Inf. and Control* **8**, 93, 1965.

Kalman, R. E. and Bucy, R. S., "New Results in Linear Filtering and Prediction Theory," *Trans. ASME, J. Basic Eng.* **83D**, 95, 1961.

Rosenbrock, H. H. and Storey, C., "Computation Techniques for Chemical Engineers," Pergamon Press, Oxford, 1966, Chapter 8.

Smith, G. L., Schmidt, S. F., and McGee, L. E., "Application of Statistical Filter Theory to Optimal Estimation of Position and Velocity on Board a Circumlunar Vehicle," N.A.S.A. TR No. R-135, 1962.

Other

Allison, J. S., "On the Comparison of Two Methods of Off-Line Parameter Identification," *J. Math. Anal. Applns.* **18**, 229, 1967.

Howland, J. L. and Vaillancourt, R., "A Generalized Curve-Fitting Procedure," *J. Soc. Ind. Appld. Math.* **9**, 165, 1961.

Lee, E. S., *Quasilinearization and Invariant Imbedding: With Applications to Chemical Engineering and Adaptive Control*, Academic Press, New York, 1967.

Levadi, V. S., "Design of Input Signals for Parameter Estimation," *IEEE Trans.* **AC-11**, 205, 1966.

Mowery, V. O., "Least Squares Recursive Differential-Correction Estimation in Nonlinear Problems," *IEEE Trans.* **AC-10**, 399, 1965.

Schneider, H., "Multidimensional Parameter Estimation by the Summed Weighted Least Squares Minimization of Remainders," *J. Astronau. Sci.* **11**, 61, 1964.

Voronova, L. I. and Krementulo, Yu. V., *Automatika* (2) 3, 1966.

Problems

9.1 A continuous flow, stirred tank reactor (illustrated in Figure P9.1) can be described by the following deterministic model for the component designated by c:

$$V\frac{dc}{dt} = F(c_F - c) + VR$$

$$c(0) = 0$$

where

F = liquid flow rate
V = reactor volume
c = concentration in the reactor
c_F = concentration in the feed
R = loss by reaction = $-kc^n$
t = time

All the variables are deterministic.

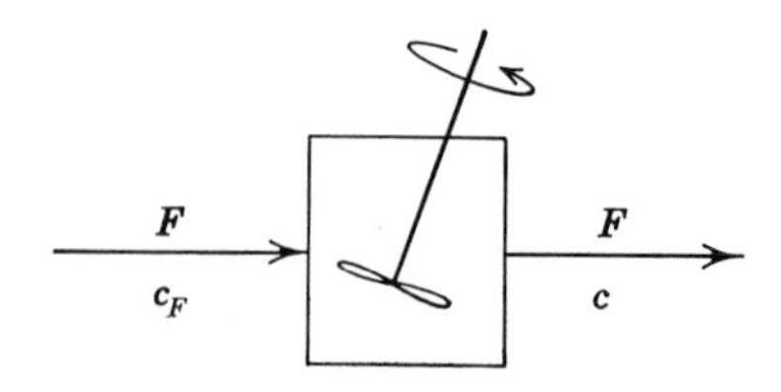

FIGURE P9.1

(a) Explain why each of the dependent variables is stochastic in a real process.
(b) Is this model a boundary value or initial value model?
(c) Are: (1) the model and (2) the solution to the model linear or nonlinear in the parameters k and n?
(d) What criterion might be used to estimate k and n from measurements of c versus t?

9.2 A tank overflows from a notch as shown in Figure P9.2. The height of the water, H, above the notch, is given by the differential equation

$$\frac{dH}{dt} = F(t) - kH^{3/2}$$

$$H(0) = 0$$

where $F(t)$ is the flow in, k is a coefficient, and t is the time. If F is a random variable and t a deterministic variable:

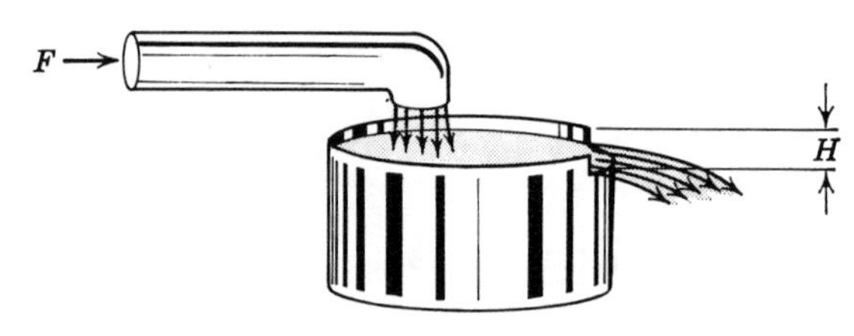

FIGURE P9.2

(a) What is the expected value of H in terms of the expected value of $F(t)$?
(b) Is the equation linear or nonlinear in k?
(c) What is the "equation error" for the model?

9.3 A surge tank damps out the oscillations of pressure in a water line (see Figure P9.3). The unsteady-state material balance is

$$a\frac{dH}{dt} = F_1 - F_2$$

where

a = crosssectional area of the tank
H = height of the water
$F_1 = c_A\sqrt{p_1 - p_2}$
$F_2 = c_B\sqrt{p_2 - p_3}$
$p_2 - p_0 = H\rho$
p_0 = atmospheric pressure, a known constant
p = pressure at indicated point
ρ = density of water, a known value
c = valve coefficient, an unknown constant
t = time, a deterministic variable

Suppose that p_1 is a random variable.

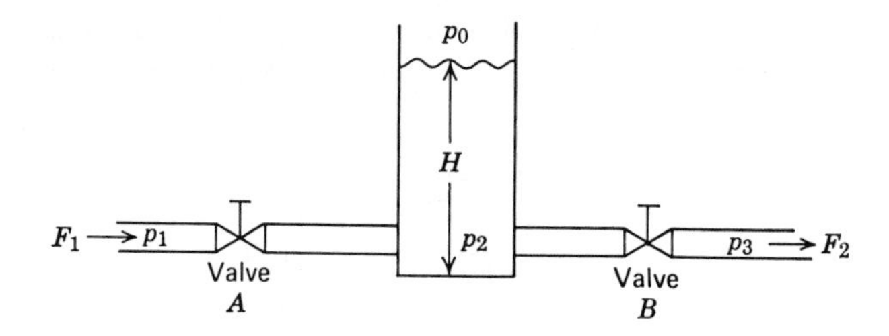

FIGURE P9.3

(a) Are p_2 and p_3 also random variables?
(b) Is H a random variable?
(c) Is the differential equation that represents the process linear or nonlinear? Explain in what sense.
(d) What criterion might be used to estimate c_A and c_B from values of H versus t?
(e) Can this model be decomposed for experimentation into simpler submodels? Give an example or explain why not.

9.4 An nth-order differential equation can be expressed as n first-order differential equations. For example, in the equation

$$\frac{d^2y}{dt^2} + a_1\frac{dy}{dt} + a_2 y = x(t) \qquad \text{(a)}$$

if

$$\frac{dy}{dt} = w$$

$$\frac{d^2y}{dt} = \frac{dw}{dt}$$

and then Equation (a) becomes

$$\frac{dy}{dt} = w$$
$$\frac{dw}{dt} = -a_1 w - a_2 y - x(t) \qquad \text{(b)}$$

Suppose that y is now considered to be a stochastic variable, $Y = y + \epsilon$. What can be said about the stochastic variable w in terms of the expected value and the variance of ϵ? What kind of criterion would be most useful for estimating a_1 and a_2 in: (1) Equation (a) and (2) Equation (b) if Y is the observed variable? Assume that $x(t)$ and t are deterministic variables. Keep in mind that two initial conditions are required for both Equations (a) and (b).

9.5 The steady-state dispersion equation for a packed bed with a first-order reaction is (in dimensionless form)

$$\alpha\frac{d^2y}{dz^2} - \frac{dy}{dz} - \beta y = 0 \qquad \text{(a)}$$

where

$\alpha = \tilde{D}/vL$
$\tilde{D}$ = dispersion coefficient
y = dimensionless concentration
z = dimensionless axial length, $0 \le z \le 1$
$\beta = k\bar{t}$
v = velocity
L = length
k = kinetic rate constant
$\bar{t}$ = mean holding time $= L/v$

See Figure P9.5.

Table P9.5 lists the solutions to Equation (a) for various possible boundary conditions. Plot $\ln y$

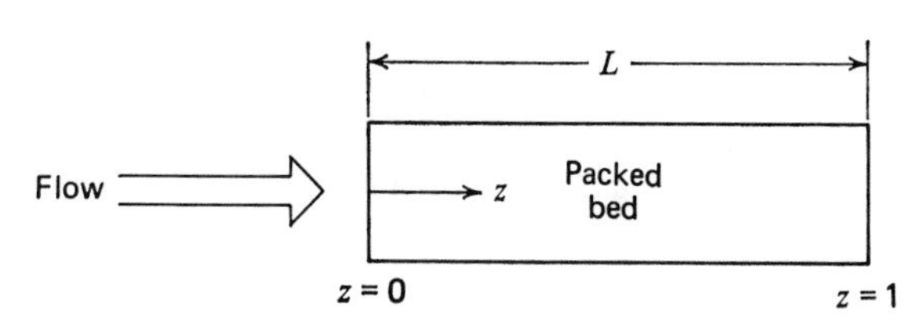

FIGURE P9.5

TABLE P9.5

Solution Number	Boundary Conditions: Entrance ($z = 0$)	Boundary Conditions: Exit or Other	Solution to Model
1	$y = 1.0$	$z \to \infty,\ y = 0$	$y = e^{m_1 z}$
2	$y = 1.0$	$z = 1.0, \frac{dy}{dz} = 0$	$y = \frac{m_2 e^{m_2} e^{m_1 z} - m_1 e^{m_1} e^{m_2 z}}{m_2 e^{m_2} - m_1 e^{m_1}}$
3	$y = 1 + \alpha\frac{dy}{dz}$	$z = 1.0, \frac{dy}{dz} = 0$	$y = \frac{m_2 e^{m_2} e^{m_1 z} - m_1 e^{m_1} e^{m_2 z}}{(1 - \alpha m_1)m_2 e^{m_2} - (1 - \alpha m_2)m_1 e^{m_1}}$
4	$y = 1 + \alpha\frac{dy}{dz}$	$z \to \infty,\ y = 0$	$y = \frac{e^{m_1 z}}{1 - \alpha m_1}$

$$m_1 = \frac{1}{2\alpha}(1 - \sqrt{1 + 4\alpha\beta})$$

$$m_2 = \frac{1}{2\alpha}(1 + \sqrt{1 + 4\alpha\beta})$$

versus β on semilogarithmic paper and answer the following questions:

(a) Under what experimental conditions will it be important to control the boundary conditions, and under what circumstances will it be unimportant?
(b) How can k and α be obtained from separate experimental set-ups?
(c) How can k and α be obtained from the same experimental set-up?
(d) Is it possible to distinguish between a plug flow model (omit dispersion term—the second derivative) and the dispersion model?
(e) Repeat (a) through (c) for a continuous stirred tank reactor model. See Problem 9.1.
(f) If the reaction term were βy^2, how could k be estimated?

9.6 Thaller and Thodos† measured the initial rates (i.e., time derivative at $t = 0$) of the dehydrogenation of sec-butyl alcohol at 600°F:

R, Initial Rate (lb moles alcohol/(hr)(lb catalyst))	F, Feed Rate (100% alcohol)	p_A, Partial Pressure of Alcohol (atm)
0.0392	0.01359	1.0
0.0416	0.01366	7.0
0.0416	0.01394	4.0
0.0326	0.01367	10.0
0.0247	0.01398	14.5
0.0415	0.01389	5.5
0.0376	0.01384	8.5
0.0420	0.01392	3.0
0.0295	0.01362	0.22
0.0410	0.01390	1.0

Estimate the coefficients k_H, k_R, and K_A in the model

$$r = \left[k_H + \frac{k_H^2}{2k_p}\frac{(1 + K_A p_A)^2}{K_A p_A}\right] - \left\{\left[k_H + \frac{k_H^2}{2k_R}\frac{(1 + K_A p_A)^2}{K_A p_A}\right]^2 - k_H^2\right\}^{1/2}$$

by minimization of $\sum_{i=1}^{10} (r_i - R_i)^2$ where $R_i = r_i + \epsilon_i$.

9.7 In many fields the solution to the differential equation

$$\frac{dy}{dt} = \frac{b_1}{b_2}(y - a)[b_2 - (y - a)]$$

is known as the logistic function

$$y - a = \frac{b_2}{1 + c\, e^{b_1 t}}$$

where c is a constant of integration (related to the initial condition), y is the dependent variable, t is time, and b_1, b_2, a, and c are constant parameters to be estimated.

† L. H. Thaller and G. Thodos, *AIChE J.* **6**, 369, 1960.

Given the following data, where Y is a random variable, $Y = y + \epsilon$, determine the best estimates of a, b_1, b_2, and c.

Y Observed	t
195	72
377	144
542	216
687	288
783	346
911	432

9.8 A proposed model is

$$\frac{dC_A}{dt} = -k_1 C_B - k_2 C_E$$

$$\frac{dC_B}{dt} = -k_1 C_B$$

$$\frac{dC_D}{dt} = k_1 C_B - k_2 C_D$$

$$\frac{dC_E}{dt} = k_2 C_D$$

The initial conditions for C_D and C_E are zero, but the initial conditions for C_A and C_B are unknown. Several values of C_D for five runs are shown in Table P9.8.

TABLE P9.8

	Time (sec)				
Run	80	160	320	640	1280
1	14.7	23.4	34.3	34.6	20.3
2	3.72	3.81	17.2	20.0	23.9
3	13.3	27.1	43.0	58.0	49.0
4	30.8	44.4	46.7	24.9	2.94
5	62.6	88.0	89.5	43.4	5.80

Describe two ways in which the coefficients k_1 and k_2 ($k_1 > 0$; $k_2 > 0$) can be estimated from the data. Can the initial conditions for C_A and C_B in each run be estimated? Draw an approximate confidence region in k_2 versus k_1 space for each run. What can you conclude?

9.9 Svirbely and Blaner‡ modelled the reactions

$$A + B \xrightarrow{k_1} C + F$$

$$A + C \xrightarrow{k_2} D + F$$

$$A + D \xrightarrow{k_3} E + F$$

‡ W. J. Svirbely and J. A. Blaner, *J. Amer. Chem. Soc.* **83**, 4118, 1961.

as follows:

$$\frac{dA}{dt} = -k_1AB - k_2AC - k_3AD$$

$$\frac{dB}{dt} = -k_1AB$$

$$\frac{dC}{dt} = k_1AB - k_2AC$$

$$\frac{dD}{dt} = k_2AC - k_3AD$$

$$\frac{dE}{dt} = k_3AD$$

Estimate the coefficients k_1, k_2, and k_3 (all positive) from the following experimental data:

$$C(0) = D(0) = 0$$

$$A(0) = 0.02090 \text{ mole/liter}$$

$$B(0) = (\tfrac{1}{3})A(0)$$

Time (min)	$A \times 10^3$ (mole/liter)
4.50	51.40
8.67	14.22
12.67	13.35
17.75	12.32
22.67	11.81
27.08	11.39
32.00	10.92
36.00	10.54
46.33	9.780
57.00	9.157
69.00	8.594
76.75	8.395
90.00	7.891
102.00	7.510
108.00	7.370
147.92	6.646
198.00	5.883
241.75	5.322
270.25	4.960
326.25	4.518
418.00	4.075
501.00	3.715

The estimates reported in the article were

$$k_1 = 14.7$$

$$k_2 = 1.53$$

$$k_3 = 0.294$$

Could estimates be obtained if the initial conditions were unknown?

9.10 Eakman (private communication) prepared some simulated data to be used in fitting the model:

$$\frac{dC_A}{dt} = -k_1C_AC_B - k_2C_AC_D \qquad \text{(a)}$$

$$\frac{dC_B}{dt} = -k_1C_AC_B \qquad \text{(b)}$$

$$\frac{dC_C}{dt} = k_1C_AC_B \qquad \text{(c)}$$

$$\frac{dC_D}{dt} = k_1C_AC_B - k_2C_AC_D \qquad \text{(d)}$$

$$\frac{dC_E}{dt} = k_2(C_AC_D)^{1/2} \qquad \text{(e)}$$

Assumed values of $k_1 = 0.3$, $k_2 = 0.01$, $C_A(0) = 1.5$, and $C_B(0) = 1.0$ were used to generate the data labeled "true" in Table P9.10. Experimental error was simulated by adding a normal random error to the true values with a standard deviation of from 0.01

TABLE P9.10 SIMULATED CONCENTRATION DATA

Time	A	B	C	D	E	Characterization of Data
0.00	1.5000	1.0000	0.0000	0.0000	0.0000	True
1.00	1.1529	0.6747	0.3252	0.3035	0.0434	True
	1.1652	0.6795	0.3185	0.3056	0.0430	Good
	0.1147	0.6198	0.3111	0.3033	0.0492	Bad
2.00	0.9333	0.4944	0.5055	0.4444	0.1221	True
	0.9251	0.4887	0.5060	0.4442	0.1201	Good
	0.9465	0.5279	0.5352	0.4372	0.1321	Bad
3.00	0.7806	0.3828	0.6171	0.5148	0.2044	True
	0.7941	0.3771	0.6224	0.5136	0.2090	Good
	0.7022	0.4256	0.6145	0.5058	0.2048	Bad
4.00	0.6675	0.3083	0.6916	0.5508	0.2815	True
	0.6803	0.3106	0.6927	0.5570	0.2806	Good
	0.6495	0.2993	0.7034	0.5479	0.2770	Bad
5.00	0.5801	0.2558	0.7441	0.5684	0.3513	True
	0.5724	0.2547	0.7869	0.5688	0.3536	Good
	0.5133	0.2468	0.6495	0.5916	0.3294	Bad
6.00	0.5104	0.2173	0.7826	0.5758	0.4136	True
	0.5103	0.2215	0.7863	0.5790	0.4103	Good
	0.5383	0.1855	0.7144	0.5376	0.4688	Bad
7.00	0.4535	0.1881	0.8118	0.5772	0.4691	True
	0.4528	0.1897	0.8237	0.5750	0.4707	Good
	0.4501	0.2103	0.8014	0.5738	0.4837	Bad
8.00	0.4060	0.1653	0.8345	0.5752	0.5186	True
	0.4072	0.1626	0.8264	0.5766	0.5135	Good
	0.3689	0.1979	0.8312	0.5874	0.4671	Bad
9.00	0.3658	0.1473	0.8526	0.5711	0.5628	True
	0.3705	0.1494	0.8624	0.5736	0.5699	Good
	0.3961	0.1489	0.9286	0.5725	0.6203	Bad
10.00	0.3314	0.1327	0.8672	0.5659	0.6024	True
	0.3276	0.1327	0.8744	0.5613	0.6028	Good
	0.3397	0.1061	0.8383	0.5756	0.6599	Bad

to 0.09. $\sigma_\epsilon = 0.01$ represents approximately a 2-percent error in the value of C_i. The "good" data represent $\sigma_\epsilon = 0.01$ and the "bad" data $\sigma_{\epsilon A} = 0.07$, $\sigma_{\epsilon B} = 0.08$, $\sigma_{\epsilon C} = 0.05$, $\sigma_{\epsilon D} = 0.02$, and $\sigma_{\epsilon E} = 0.09$.

(a) Are all the differential equations in the model independent?
(b) Can k_1 and k_2 be estimated solely from the concentration of one component? If so, which one?
(c) Estimate k_1 and k_2 for each of the three sets of data labeled "true," "good," and "bad."

9.11 Simulate experimental data by using the following set of differential equations, adding normal random error to the dependent variables R, S, T, U, W, X, and Y.

$$\frac{dR}{dt} = -k_1RS - k_2RS - k_3RS + k_9T + k_{10}W$$

$$\frac{dS}{dt} = -k_1RS - k_2RS - k_3RS - k_6TS - k_7US - k_8US + k_9T + k_{12}W + k_{11}X + k_{12}Y$$

$$\frac{dT}{dt} = k_1RS - k_5T - k_6TS - k_9T$$

$$\frac{dU}{dt} = k_2RS + k_5T - k_7US - k_8US + k_{11}X + k_{12}Y$$

$$\frac{dW}{dt} = k_3RS - k_{12}W$$

$$\frac{dX}{dt} = k_6TS + k_7US - k_{11}X$$

$$\frac{dY}{dt} = k_8US - k_{12}Y$$

Design a suitable set of experiments to estimate k_1 through k_{12}, and carry out the estimation.

Each k can be represented as follows:

$$k_i = \rho_i \exp b_i\left[\frac{1}{T^*} - \frac{1}{433}\right]$$

where T^* is the absolute temperature. For the following values of ρ_i and b_i, again generate simulated experimental data but this time estimate the ρ's and b's:

i	ρ_i	b_i
1	6.3×10^{-3}	17,800
2	2.1×10^{-3}	18,000
3	9.0×10^{-3}	11,100
4	1.1×10^{-3}	12,300
5	2.8×10^{-2}	13,100
6	5.0×10^{-4}	14,000
7	4.9×10^{-4}	20,800
8	5.6×10^{-4}	4,330
9	6.7×10^{-2}	28,900
10	2.8	16,000
11	1.03×10^{-1}	14,600
12	3.5×10^{-1}	10,800

MIMIC can be used for the simulation if a large-sized digital computer is available, or the simulation can be executed on a hybrid computer.

9.12 Numerically differentiate the good and bad data of Problem 9.10. Estimate the variance of the derivatives in terms of variances of each observed variable. Select a scheme of differentiation for which the deterministic numerical error in the derivative is less than 10 percent of the estimated stochastic error in the derivative.

9.13 Two methods of obtaining derivatives to be used in ordinary differential equations have been suggested. One is to differentiate the experimental data numerically. The other is to fit a polynomial to the data first and then to differentiate analytically the polynomial. Which technique will lead to the smallest variance for the derivatives? Explain.

9.14 Newton's forward interpolation formula gives the derivatives in terms of forward differences:

$$\left(\frac{dy}{dx}\right)_{x=0} = \frac{1}{h}\left[\Delta y_0 - \tfrac{1}{2}\Delta^2 y_0 + \tfrac{1}{3}\Delta^3 y_0 - \tfrac{1}{4}\Delta^4 y_0 + \cdots\right]$$

The following data were taken from a drying experiment:

Y (lb H_2O/lb dry solid)	t (min)
0.1834	0.9
0.1634	1.0
0.1460	1.1
0.1313	1.2
0.1198	1.3
0.1117	1.4

Starting with increments of ΔY at $t = 0.9$ min, compute ΔY, $\Delta^2 Y$, etc., and compute $(dY/dt)_{t=0.9}$. Then compute the variance of the derivative, assuming the variance of each observed Y is 1 percent of the value in the table.

9.15 Apply the method of steepest descent to obtain least squares estimates of the parameters in the model $\eta = \beta_0 + \beta_1 x_1 + \beta_2 x_1^2$ for continuous data. Draw the analog computer diagram for the circuits which will be required to carry out the computation.

9.16 Use the method of the "equation error" to estimate the coefficients in the model given in Problem 9.5 for a continuous response. Use a hybrid computer or an analog simulator such as MIMIC on a digital computer to carry out the calculations. Select known values for the parameters in the differential equation and add a known normal random error to the response Y prior to feeding the simulated response incorporating the error to the estimation phase of the calculations.

9.17 Repeat Problem 9.16 but use the techniques of sequential estimation. It may be necessary to use the final estimated parameters from the procedure as the initial guesses for a second pass through the estimation procedure.

9.18 In Table P9.18 are yields (concentrations) of the desired component P formed in the following reaction:

$$A + B \rightarrow P$$

$$A + P \rightarrow R$$

Because the initial concentration of A was very much larger than B, the amount of A used up in the reaction was small and its concentration could be considered essentially constant.

Suppose a proposed model of the reaction system is as follows (the initial concentrations are known values):

$$\frac{dc_A}{dt} = -k_1 c_B - k_2 c_P \qquad c_A(0) = c_{A0}$$

$$\frac{dc_B}{dt} = -k_1 c_B \qquad c_B(0) = c_{B0}$$

$$\frac{dc_P}{dt} = k_1 c_B - k_2 c_P \qquad c_P(0) = 0 \qquad \text{(a)}$$

$$\frac{dc_R}{dt} = k_2 c_P \qquad c_R(0) = 0$$

(a) From the data in Table P9.18, estimate the values of k_1 and k_2.

(b) Plot the joint confidence region for k_1 and k_2 for contours of 80 percent and 95 percent.

(c) Plot and analyze the residuals for each run.

TABLE P9.18 CONCENTRATION OF P (MOLES/LITER) × 10^2 AT 160°F

Run	c_{B0} (mole/liter)	$t =$ 1 hr	$t =$ 2 hr	$t =$ 4 hr	$t =$ 8 hr	$t =$ 16 hr
1	1	3.17	5.39	8.66	15.9	22.6
2	1	14.7	23.4	34.3	34.6	20.3
3	2	4.80	10.8	22.5	34.6	42.0
4	2	23.2	39.0	55.6	63.4	41.6
5	1	3.72	3.81	17.2	20.0	23.9
6	1	17.9	28.3	40.5	34.2	21.6
7	2	8.60	13.3	25.9	39.8	50.8
8	2	30.9	51.4	72.2	76.4	38.9
9	1	7.48	9.93	20.0	30.9	24.9
10	1	25.3	35.3	39.1	28.4	7.50
11	2	13.3	27.1	43.0	58.0	49.4
12	2	50.8	75.6	84.2	57.0	11.5
13	1	9.15	15.8	27.5	33.9	23.0
14	1	30.8	44.4	46.7	24.9	2.94
15	2	22.8	37.2	57.9	69.1	53.9
16	2	62.6	88.0	89.5	43.4	5.80

CHAPTER 10

Parameter Estimation in Models Containing Partial Differential Equations

Differential equations that contain derivatives with respect to more than one independent variable are classified as partial differential equations. A wide variety of processes can be represented in terms of such equations such as flow in a porous medium, dispersion of pollutants in air or a river, packed-bed chemical reactors, and transmission lines. Partial differential equations can provide a more detailed description of a process than can ordinary differential equations. But, as indicated in Figure 1.1-2, obtaining either an analytical or numerical solution to a model written in terms of partial differential equations is more difficult than obtaining corresponding solutions for a model written in terms of ordinary differential equations.

Whereas the integration of ordinary differential equations gives rise to arbitrary constants of integration, the general solution of a partial differential equation involves n arbitrary functions for the nth-order case. Consequently, except for first-order equations and a few other special cases, it is seldom possible or necessary to seek a general solution; instead one seeks a particular solution for the specific boundary and initial conditions.

One standard method for the solution of partial differential equations is by separation of variables. A form of the solution is set up such that the dependent variable(s) is (are) equal to the product of the solutions of two or more ordinary differential equations, each of which has a known solution. The initial and boundary conditions for a given problem are used to evaluate the arbitrary functions so that a unique solution can be obtained for the model.

Because the boundary conditions play such an important role in the formulation and solution of the process model, the experimenter tries to select boundary conditions that make the model solution as simple as possible and yet are feasible to achieve experimentally. Naturally, poor results can be expected from a lack of control over the experimental conditions, because then the model and the experimental conditions do not jibe.

In this chapter we shall first examine process inputs and their related outputs that can be used to evaluate the model parameters. We shall then describe how to arrange the experimentation in order to simplify the analytical solution to the process model. Finally, we shall discuss the use of so-called deterministic moments to evaluate the model parameters.

10.1 PROCESS INPUTS

Certain process inputs are more effective than others from the combined viewpoint of experimental execution and model solution. The three deterministic inputs most often used in the evaluation of model parameters are the *step*, *impulse* (*delta*), and *sinusoidal* inputs; refer to Figure 10.1-1. The stochastic input most often used is the random square wave (random binary sequence); sometimes the process input itself can be employed. A discussion of stochastic inputs is deferred until Chapter 12.

The main reason for using deterministic step, impulse, and sinusoidal inputs in *models* of a process is mathematical convenience. It is relatively easy to establish the output for a given model when these functions are used as inputs. The outputs of linear (in the dependent variables) models for sinusoidal inputs are most often interpreted directly in Laplace transform space, whereas the outputs for impulse or step inputs are analyzed in the time domain. In practice, the use of an output from a sinusoidal input requires considerable algebraic manipulation and becomes awkward for modestly complicated models as compared to impulse inputs. However, certain well-established process control system design techniques are based upon the responses to sinusoidal inputs in transform space; hence, it often proves convenient to work with subsystem dynamics with superimposed sinusoidal inputs. In the time domain, any of the three inputs is effective for estimation of the coefficients in linear systems. However, when a sinusoidal input is introduced into nonlinear models, the frequencies of the input and amplitudes are shifted in a nonadditive fashion when they appear in the output. The equivalence of the information provided by step, impulse, and sinusoidal

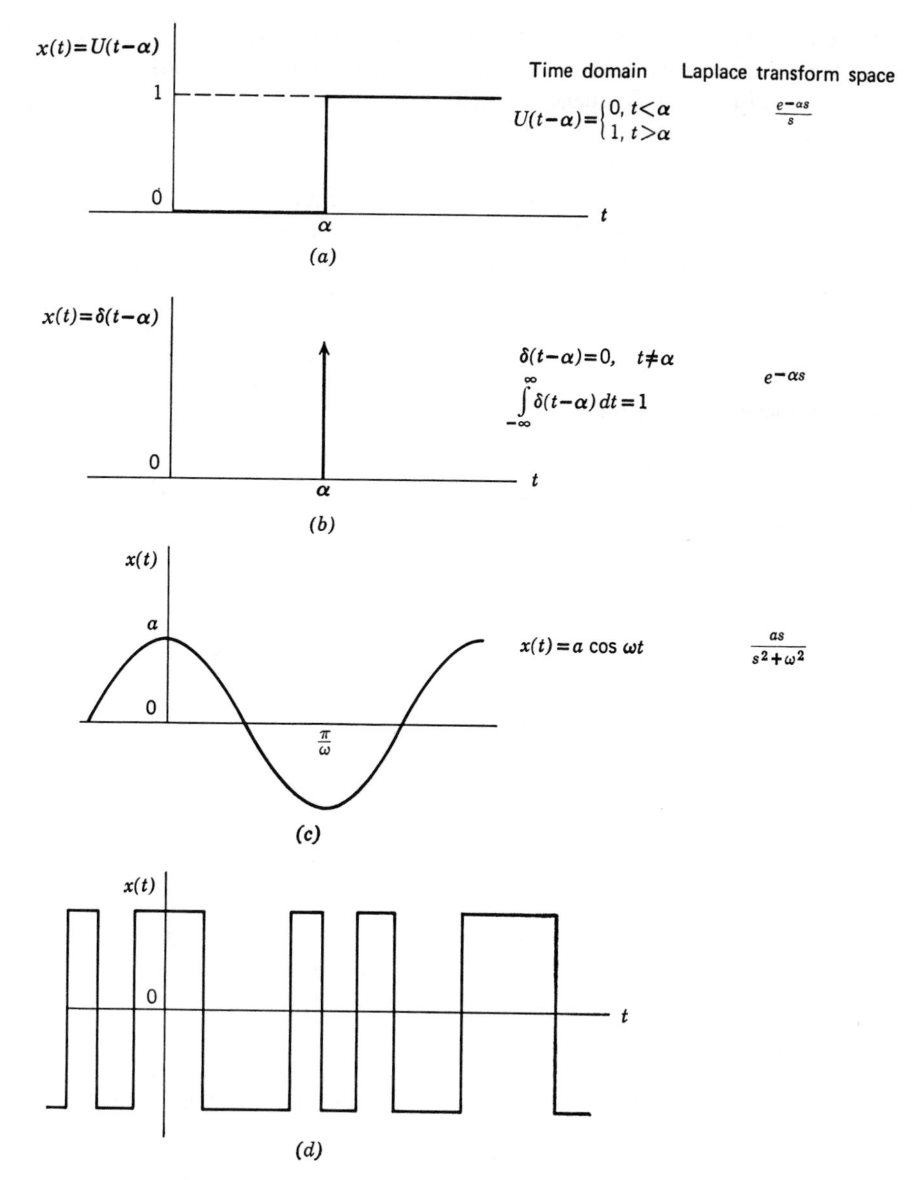

FIGURE 10.1-1 Typical input functions: (*a*) unit step, (*b*) impulse, (*c*) sinusoidal, and (*d*) random square wave (random binary sequence).

inputs for linear systems was described by Nyquist *et al.*†

As a matter of interest at this point, it is worthwhile to mention some of the advantages and disadvantages involved in the use of deterministic step functions, impulse functions, and sine waves as inputs for probing the nature of an actual process (in contrast to a mathematical model of the process).

Experimentally it is impossible to produce an exact step function, but in many studies an input with a fast rise time compared to the process response time can be produced, so the step function can be reasonably approximated. An advantage of the use of a step input is that all the information about the process dynamics is contained in the response to a single step input; hence, the experimentation is economical. But herein lies the major disadvantage in the step input—all the information is packed into a small amount of record. If some noise is present, then much of the fine detail present in the record will be obscured.

Whereas a step function involves moving the process from one steady-state value of the input to another, and therefore requires considerable input material or tracer, the impulse input requires only a relatively small amount of material. But the engineer must consider the possible disruption of the process by a pulse input, and, of course, he has less outlet material to measure than for a step input.

In execution, a sinusoidal input to a real process requires more complicated input equipment to be built

† J. K. Nyquist *et al.*, *Chem. Eng. Progr.*, *Symposium Ser.* No. 46, 98, 1963.

and operated than step or impulse inputs and is very time consuming, since several frequencies are needed and steady state must be achieved for each frequency (which may require hours to achieve). A sinusoidal input does permit only small perturbations to be introduced into the process. However, as mentioned before, if the process is nonlinear, the frequency of the output is shifted from that of the input. Hougen† provided additional information on some of the practical aspects of choosing a process input.

10.2 RESPONSES TO PROCESS INPUTS

The *step response* for models represented by partial differential equations cannot be formulated in general but must be determined separately for each individual case. The examples in Table 10.2-1 illustrate typical analytical solutions for models linear in the dependent variable in which a step input is introduced. Note the highly nonlinear (in the parameters) character of the model solutions.

The impulse input to a process model can appear either as a boundary condition in the model or as a source term in the differential equation itself. For example, an impulse input at $z = 0$ (the start of the axial coordinate z) and $t = 0$ can be introduced into the model as a source term:

$$\frac{\partial y}{\partial t} + v\frac{\partial y}{\partial z} = \alpha\,\delta(t)\,\delta(z) \qquad y(0^-, z) = 0 \qquad (10.2\text{-}1)$$

where the term $\alpha\,\delta(t)\,\delta(z)$ is interpreted from the properties of the delta function as meaning that the pulse takes place at $t = 0$ and $z = 0$. Or the input can be introduced as a boundary condition:

$$\frac{\partial y}{\partial t} + v\frac{\partial y}{\partial z} = 0 \qquad \begin{aligned} y &= (0^-, z) = 0 \\ y &= (t, 0) = \beta\,\delta(t) \end{aligned}$$

In Equation 10.2-1, v represents the velocity of the flowing fluid and y is the dependent variable. Because in the definition of $\delta(x)$, $\int_{\infty}^{-\infty} \delta(x)\,dx \equiv 1$, $\delta(t)$ must have the units of time^{-1}, $\delta(z)$ has the units of length^{-1}, and the units of α and β must be properly assigned to make the differential equation dimensionally consistent. Suppose that a fixed quantity of tracer material m, uniformly spread over the cross-section A of the duct or channel (in which the flow takes place), is introduced into the duct at a fixed location ($z = 0$) at $t = 0$. Then the coefficient in source term specifically becomes $\alpha = m/A$.

Equation 10.2-1 can be solved by taking successive Laplace transforms of both sides on t and z and collecting like terms, followed by inversion

$$y(t, z) = \frac{m}{Av}\,\delta\!\left(t - \frac{z}{v}\right) \qquad (10.2\text{-}2)$$

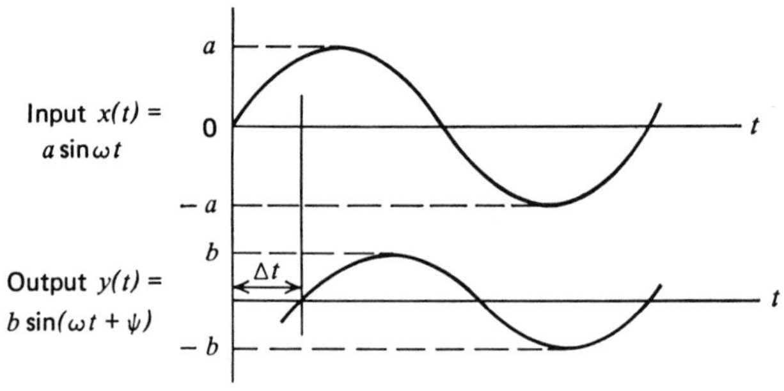

FIGURE 10.2-1 Definitions of gain (amplitude ratio) and phase angle (lag). Amplitude ratio $= |\breve{g}(\omega)| = b/a$. Phase angle $= \angle g(\omega) = \psi = \omega\,\Delta t$.

† J. O. Hougen, "Experiences and Experiments with Process Dynamics," *Chem. Eng. Progr.*, Monograph Ser. No. 4, **60**, 1964.

TABLE 10.2-1 RESPONSES TO STEP INPUTS FOR DETERMINISTIC DISTRIBUTED MODELS*

Process	Model of Process	Model Solution
Heat transfer to a pipe with flowing fluid	$\frac{\partial T(t, z)}{\partial t} + v\frac{\partial T(t, z)}{\partial z} = h[T_a - T(t, z)]$	$T(t, z) = T_a[1 - e^{-(h/v)z}] + T_0 U\left(t - \frac{z}{v}\right) e^{-(h/v)z}$
	$T(0, z) = T_a[1 - e^{-(h/v)z}]$	
	$T(t, 0) = T_0 U(t)$	
Fixed bed adsorber with flowing fluid	Liquid: $\frac{\partial c_L}{\partial t} + v_L\frac{\partial c_L}{\partial z} = -(k)(c_L - c_{L,e})$	$\frac{c_L}{C_{L0}} = 1 - \int_0^{z^*} e^{-(t^* + \tilde{z}^*)} I_0(2\sqrt{t^*\tilde{z}^*})\,d\tilde{z}^*$
	Solid: $\frac{\partial c_s}{\partial t} = \left(\frac{k}{m}\right)(c_L - c_{L,e})$	$z^* = \frac{(k)z}{v_L}$
	$c = 0,\ t < \frac{z}{v_L} \quad z \leq 0$	I_0 = Bessel function
	$c(t, 0) = c_{L0} U(t)$	$\sim$ = dummy variable

* U = unit step function; T = temperature; c = concentration; z = axial direction; v = velocity; t = time; h = interphase heat transfer coefficient; k = interphase mass transfer coefficient.

The interpretation of Equation 10.2-2 is that the input pulse appears at time $t = z/v$ at the point z, still retaining its impulse shape. Since there is no provision for dispersion in the model, this interpretation seems quite reasonable. The impulse proceeds as a front down the duct without spreading and can be measured at the process exit.

A more comprehensive model includes axial dispersion:

$$\frac{\partial y}{\partial t} + v\frac{\partial y}{\partial z} = \tilde{D}\frac{\partial^2 y}{\partial z^2} + \text{source term} \qquad (10.2\text{-}3)$$

where $\tilde{D}$ is an effective dispersion coefficient and the source term is the same as before if the material is uniformly spread over the duct or channel at the location $z = z_0$:

$$\text{source} = \frac{m}{A}\,\delta(z - z_0)\,\delta(t) \qquad (10.2\text{-}4)$$

where

m = amount of tracer injected

$\delta(z - z_0)$ = delta function needed to indicate that the

TABLE 10.2-2 FREQUENCY RESPONSE IN THE TIME DOMAIN FOR SELECTED DISTRIBUTED PARAMETER MODELS

Model	Frequency Response
$\frac{\partial y}{\partial t} = K\frac{\partial^2 y}{\partial z^2}$; $y(0, z) = 0$; $y(t, 0) = e^{i\omega t}$; $\frac{\partial y(t, 0)}{\partial z} = 0$	$y(t, z) = \exp\left[-\left(\frac{\omega}{2K}\right)^{1/2} z\right]\sin\left[\omega t - \left(\frac{\omega}{2K}\right)^{1/2} z\right]$
$\frac{\partial y}{\partial t} + v\frac{\partial y}{\partial z} = K\frac{\partial^2 y}{\partial z^2} - ky$; $y(0, z) = 0$; $y(t, 0) = e^{i\omega t}$; $\frac{\partial y(t, 0)}{\partial z} = 0$	$y(t, z) = \exp\left[\frac{vz}{2K}\left(1 - \sqrt{1 + \frac{4K(k + i\omega)}{v^2}}\right)\right]\exp(i\omega t)$
$\frac{\partial y}{\partial t} + v\frac{\partial y}{\partial z} = K_1\left(\frac{\partial^2 y}{\partial r^2} + \frac{1}{r}\frac{\partial y}{\partial r}\right) + K_2\frac{\partial^2 y}{\partial z^2}$; $y(0, r, z) = 0$; $\frac{\partial y(t, 0, z)}{\partial r} = 0$; $y(t, r, 0) = y_0 e^{i\omega t}$; $y(t, R, z) = 0$; $\frac{\partial y(t, r, 0)}{\partial z} = 0$	$y(t, r, z) = y_0 \sum_{n=1}^{\infty} \frac{2J_0(\lambda_n r)}{\lambda_n R J_1(\lambda_n R)}\exp[-C_1 z + i(\omega t - C_2 z)]$; λ^n are the roots of $J_0(\lambda R) = 0$; $C_1 = \frac{\omega}{2K_2}\left(\frac{1}{C_2} - \frac{v}{\omega}\right)$; $C_2 = \left[-\left(\frac{v^2}{8K_2^2}\right) + \left(\frac{K_1\lambda_n^2}{2K_2}\right) + \sqrt{\left(\frac{v^2}{8K_2^2}\right) + \left(\frac{K_1\lambda_n^2}{2K_2}\right) + \left(\frac{\omega^2}{4K_2^2}\right)}\right]^{1/2}$
$H\frac{\partial y}{\partial t} + V\frac{\partial y}{\partial z} = -ka(y - y^*)$; $h\frac{\partial x}{\partial t} + L\frac{\partial x}{\partial z} = ka(y - y^*)$; $y^* = Kx$; $y(0, z) = 0$; $x(0, z) = 0$; $y(t, z_0) = e^{-i\omega t}$; $x(t, 0) = e^{i\omega t}$	$y(t, z) = \frac{\lambda_1\lambda_2}{Q}[\exp(q_2 z_0 + q_1 z) - \exp(q_1 z_0 + q_2 z)]e^{i\omega t}$; $x(t, z) = \frac{1}{Q}[\lambda_2\exp(q_2 z_0 + q_1 z) - \lambda_1\exp(q_1 z_0 + q_2 z)]e^{i\omega t}$; $Q = \lambda_2 e^{q_2 z_0} - \lambda_1 e^{q_1 z_0}$; $\lambda_{1,2} = \frac{Kka}{vq_{1,2} + Hi\omega + ka}$; $q_{1,2} = -\left(\frac{\beta}{2\alpha}\right) \pm \sqrt{\left(\frac{\beta}{2\alpha}\right)^2 - \left(\frac{\alpha}{\gamma}\right)}$; $\alpha = VL$; $\gamma = Hhi\omega + (h + HK)ka$; $\beta = (Vh + LH)i\omega + (VK + L)ka$

tracer input takes place at position $z = z_0$ only

$\delta(t)$ = delta function needed to indicate that the tracer input takes place at $t = 0$ only

Table 10.4-1 lists known solutions for Equation 10.2-3 for zero initial conditions.

The portion of the *response function for a sinusoidal input* that remains after the transients die out is called the *frequency response* of the system. Since the response of a *linear* (in the dependent variable) system to a sinusoidal input is a sinusoidal output, by comparing the output and input it will be found that they differ (usually) by a displacement in time (phase) and an increase or reduction in amplitude. The former is termed the phase angle (phase shift) of the response while the latter is known as the gain (magnitude ratio, amplitude ratio) of the response. The relationships involved are shown in Figure 10.2-1. The frequency response can be obtained by solving the differential equation(s) for a sinusoidal input and then letting $t \to \infty$. But it is most commonly obtained from the process transfer function $\breve{g}(s)$ (refer to Section 12.1) by letting the real part of the complex parameter s be zero and the complex part be $i\omega$. Then the frequency response can be written in terms of the ratio of the amplitude of the output divided by the input which is equivalent to $\|\breve{g}(\omega)\|$, the absolute value (modulus) of $\breve{g}(\omega)$, and in terms of the phase shift ψ, the argument (angle) of $\breve{g}(\omega)$. Table 10.2-2 lists the frequency response in the time domain for selected models taken from Schiesser.† The frequency ω becomes a controllable independent variable in the experimentation in addition to the time and axial coordinate.

In the remainder of this Chapter we shall use a least squares criterion in estimation and make the same assumptions as in Section 9.1 concerning the unobservable error, namely that

$$Y_r = y_r + \epsilon_r \tag{10.2-5}$$

for each response r. We shall assume that the model adequately represents the process and want to estimate the model parameters from experimental observations Y. For the cases of the step and impulse responses, the Y in Equation 10.2-5 is simply the process response to a deterministic step, or impulse, input. However, for the frequency response, usually two measurements are made, neither of which is directly the process response. The gain (amplitude ratio) may be measured as a single random variable, the output amplitude. Or it may be the ratio of two random variables if the input amplitude is stochastic, in which case Equation 10.2-5 is applied both to the numerator and denominator of the ratio. Refer to Example 10.3-2. The phase angle is usually measured as a single random variable.

† W. E. Schiesser, Preprint, Joint Automatic Control Conference, 1964.

10.3 DESIGN OF EXPERIMENTS TO SIMPLIFY MODEL SOLUTIONS

The major emphasis in the estimation of parameters in partial differential equations and/or in the accompanying boundary conditions is to: (1) arrange the experimental set-up so as to isolate, insofar as possible, the evaluation of each parameter from its fellow parameters, and (2), for a given model and solution, reduce the solution to as simple a form as possible by selecting appropriate values of the independent variables.

As an example of the first strategy, consider the nonlinear model for chemical reaction in a packed bed:

$$\frac{\partial c_A}{\partial t} + \beta_0 \frac{\partial c_A}{\partial z} = \beta_1 \frac{\partial^2 c_A}{\partial z^2} + \beta_2 c_A^2$$

$$c_A(0, z) = c_e$$

$$c_A(t, 0) = c_{A_0}$$

$$\frac{\partial c_A(t, L)}{\partial z} = 0$$

where c_A is the concentration of component A, t is the time, z is the axial coordinate ranging from 0 to L, β_2 is a reaction coefficient, β_1 is a dispersion coefficient, and β_0 is a velocity of flow—a known value. The determination of the coefficient β_2 could be carried out separately and in a different apparatus from the determination of β_1. Measurements to calculate the reaction coefficient might be made in a well-stirred tank using the model

$$\frac{dc_A}{dt} = \beta_2 c_A^2$$

$$c_A(0) = c_e$$

whereas β_1 might be estimated from a steady-state experiment in a packed bed without reaction using the linear model

$$\beta_0 \frac{dc_A}{dz} = \beta_1 \frac{d^2 c_A}{dz^2}$$

$$c_A(0) = c_{A_0}$$

$$\frac{dc_A(L)}{dz} = 0$$

Methods of estimating the coefficients in these much simpler models have been described in Chapter 9. Hopefully, the separate experimental set-ups would faithfully represent the simpler models, and the estimated coefficients from each series of experiments could be validly combined in the full model without impairing the predictive ability of the full model.

As an example of the second strategy mentioned above, we examine models of heat and mass transfer that can be represented by the diffusion equation

$$\frac{\partial y}{\partial t} = \beta \nabla^2 y$$

because for certain boundary conditions these models have solutions in the form of an infinite series:

$$y = \sum_{i=0}^{\infty} a_i u_i \, e^{-m_i \beta t} \qquad (10.3\text{-}1)$$

where the a_i are constants depending upon the initial conditions, and the u_i and m_i are functions of the coordinates whose parameters are determined by known properties of the system.

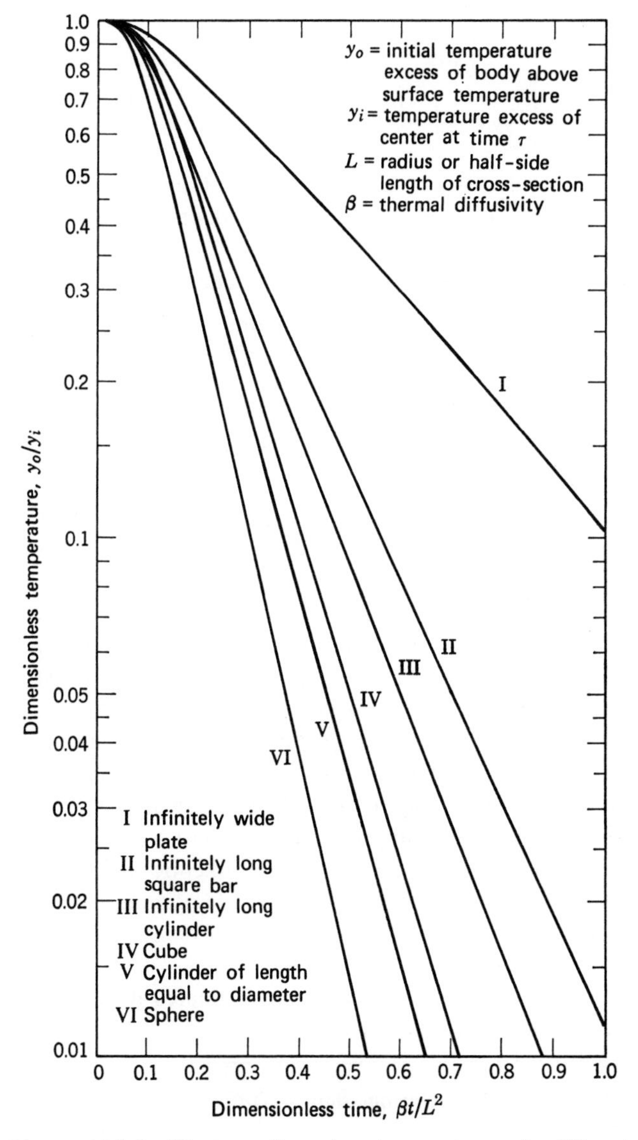

FIGURE 10.3-1 Change of center temperature of different bodies caused by sudden change of surface temperature. (From M. Jakob, *Heat Transfer*, Vol. 1, John Wiley, New York, 1949, p. 266, with permission.)

With increasing time the quantities $e^{-m_i \beta t}$ tend to very small values. After a sufficiently long period of time, the variable y becomes independent of the initial conditions. Some authors have termed this state the "regular" or "quasistationary" condition. In particular, when the first term in Equation 10.3-1 becomes much larger than any of the succeeding terms, Equation 10.3-1 reduces to

$$y \simeq a_0 u_0 \, e^{-m_0 \beta t} \qquad (10.3\text{-}2)$$

As usual, a logarithmic transformation reduces Equation 10.3-2 to an equation linear in β:

$$\ln y = -\beta m_0 t + \text{constant} \qquad (10.3\text{-}3)$$

Figure 10.3-1 illustrates solutions of the diffusion equation (as applied to heat transfer) at the center of various shaped bodies for certain boundary conditions and different geometries as a function of dimensionless time (Fourier number). Note the linear or quasilinear (in the case of the infinitely wide plate) form the functions on the semilogarithmic plot at the longer times.

An important fact to keep in mind when using Equations 10.3-2 and 10.3-3 and their analogs is that the absolute changes in y become vanishingly small with longer times so that increasingly large relative error accompanies their use. Consequently, both long and short times are less favorable for measurement than are intermediate times.

By selection of a suitable measuring point and boundary conditions, considerable simplification can be made in the solution to the diffusion equation used to estimate β. The solution, that is the temperature distribution itself, in an infinite flat plate of thickness $2L$, which is initially at a uniform temperature $y(z, 0) = y_0$ and is heated on both sides by a constant heat flux q, i.e., for the boundary conditions

$$y(z, 0) = y_0$$

$$-k \frac{\partial y(L, t)}{\partial z} = q = h[y_a - y(L, t)] \qquad (10.3\text{-}4)$$

$$\frac{\partial y(0, t)}{\partial z} = 0$$

(where h is the interphase heat transfer coefficient, z is the direction perpendicular to the plate, k is the thermal conductivity, and y_a is the ambient temperature) is

$$y(z, t) - y_0 = \frac{qL}{k}\left[\frac{\beta t}{L^2} - \frac{L^2 - 3z^2}{6L^2} + \sum_{n=1}^{\infty} (-1)^{n+1} \frac{2}{\lambda_n^2} \cdot \cos\left(\lambda_n \frac{z}{L}\right) e^{-\lambda_n^2 \beta t / L^2}\right] \qquad (10.3\text{-}5)$$

where λ_n is an eigenvalue.

Fortunately, the series in Equation 10.3-5 converges quite rapidly. Thus, for values of the Fourier number

$(\beta t/L^2)$ greater than 0.5, the summation can be neglected in comparison with the first two terms in the brackets with an error of less than 0.5 percent. Hence, the temperature distribution in the plate is a parabolic function of the distance z, and it can be estimated as described in Chapter 5 assuming Equation 10.2-5 applies.

In addition, there exists a plane in the plate at which the temperature is exactly equal to the mean temperature of the plate, say at z_e. There

$$y(z_e, t) = \bar{y}(t)$$

From Equation 10.3-5, for $F_0 > 0.5$ it can be shown that

$$z_e = \frac{\sqrt{3}}{3} L$$

A similar unique point exists for other configurations as well. (In a cylinder, $r_e = (\sqrt{2}/2)R$; in a sphere, $r_e = (\sqrt{15}/5)R$.)

Now the average rate of heat transfer to the slab can be calculated by

$$q = \frac{\rho C_p L[y(z_e, t) - y(z_e, 0)]}{(t - 0)}$$

where $\bar{y}$ has been replaced by $y(z_e)$, and the expression for q can be evaluated and introduced into the truncated Equation 10.3-5. For example, if the temperature is also measured at the center of the slab (at $z = 0$), a little manipulation gives β explicitly as

$$\frac{y(z_e, t) - y_0}{y(z_e, t) - y(0, t)} = \frac{6\beta}{L^2} t \tag{10.3-6}$$

A distinct difference exists between Equations 10.3-5 and 10.3-6. In Equation 10.3-5, the left-hand side is the response; hence the unobservable error can be conceptually added as shown in Equation 10.2-5. But in Equation 10.3-6, the left-hand side is a rational fraction of responses, and ϵ is not added directly to the deterministic left-hand side.

Suppose now that this fact is ignored, that Equation 10.2-5 is assumed to hold, and that a least squares estimate of β is calculated by minimizing

$$\sum_{i=1}^{n} \left[\frac{Y(z_e, t_i) - y_0}{Y(z_e, t_i) - Y(0, t_i)} - \frac{6\beta}{L^2} t_i \right]^2$$

(To simplify the notation, we shall let $Y(z_e, t_i) = Y_{1i}$ and $Y(0, t_i) = Y_{2i}$; the capital letters indicate random variables.) Equation 4.3-7a indicates that

$$\hat{\beta} = \frac{\sum_{i=1}^{n} \left(\frac{Y_{1i} - y_0}{Y_{1i} - Y_{2i}} \right) t_i}{\sum_{i=1}^{n} t_i^2} \tag{10.3-7}$$

Unfortunately, the expected value of $\hat{\beta}$ is not β, that is, $\hat{\beta}$ is a biased estimator, and the calculation of the extent of the bias is by no means easy.

If we seek $\mathscr{E}\{\hat{\beta}\}$, with $\hat{\beta}$ calculated by Equation 10.3-7,

$$\mathscr{E}\{\hat{\beta}\} = \left(\frac{1}{\sum_{i=1}^{n} t_i^2} \right) \sum_{i=1}^{n} t_i \mathscr{E}\left\{ \frac{Y_{1i} - y_0}{Y_{1i} - Y_{2i}} \right\} \tag{10.3-8}$$

we can assume that Y_1 and Y_2 are normal random variables. We need to know what the distribution of the quantity $(Y_{1i} - y_0)/(Y_{1i} - Y_{2i})$ is. Let us assume that $\mathscr{E}\{Y_{1i}\} = \mu_1$, $\mathscr{E}\{Y_{2i}\} = \mu_2$, and that $\text{Var}\{Y_{1i}\} = \text{Var}\{Y_{2i}\} = \sigma^2$. Physically these assumptions mean that $y(z_e, t)$ and $y(0, t)$ are different, but the error in their measurement is the same.

We also need an expression for the expected value of the ratio of two normally distributed random variables. An approximate expression for the probability density of $Z = X/Y$, good when the ratio μ_X/σ_X is large, is†

$$p(z) \simeq \frac{\sigma_Y(\mu_X\sigma_Y - \rho_{XY}\mu_Y\sigma_X) - z\sigma_X(\mu_Y\sigma_X - \rho_{XY}\mu_X\sigma_Y)}{(\sigma_X^2 - 2\rho_{XY}z\sigma_X\sigma_Y + z^2\sigma_X^2)^{3/2}}$$

$$\cdot \frac{1}{\sqrt{2\pi}} \exp\left[-\frac{1}{2} \frac{(z\mu_X - \mu_Y)^2}{\sigma_Y^2 - 2\rho_{XY}\sigma_X\sigma_Y z + z^2\sigma_X^2} \right] \tag{10.3-9}$$

where ρ_{XY} is the correlation coefficient between X and Y. Because Y_{1i} and Y_{2i} are independent and normally distributed, $(Y_{1i} - y_0)$ and $(Y_{1i} - Y_{2i})$ are normally distributed with the following parameters:

	Expected Value	Variance
$(Y_{1i} - y_0)$	$(y_1 - y_0)$	σ^2
$(Y_{1i} - Y_{2i})$	$(y_1 - y_2)$	$2\sigma^2$

These quantities (or their estimates) may be substituted for the expected values and standard deviations in Equation 10.3-9 together with ρ_{XY} or its estimate to get $p(z)$. The expected value (and variance) of Z can be evaluated numerically from the moments of Z as described in Chapter 2, and the bias in $\hat{\beta}$ can be evaluated for any specific experiment at each t_i. Each expected value of $(Y_{1i} - y_0)/(Y_{1i} - Y_{2i})$ can be substituted into Equation 10.3-8, and the bias in the estimate of β can be approximated. Clearly, the evaluation of the bias in a least squares estimator when the ratio of two random variables is arbitrarily used as the dependent variable is arduous and only approximate. Equation 10.3-9 in principle can also be used to form a likelihood function so that a maximum likelihood estimate can be made as described in Section 9.3, but the numerical computations would be lengthy.

The precision in the parameter estimates must come from replicate experiments.

Example 10.3-1 Determination of Diffusion Coefficients in a Gas Mixture

Isothermal gaseous diffusion coefficients of ethane in methane are to be measured by use of a long vertical

† R. C. Geary, *J. Royal Stat. Soc.* **93**, 442, 1930.

cylindrical tube. The model assumed for diffusion is

$$\frac{\partial c}{\partial t} = \mathscr{D}\frac{\partial^2 c}{\partial z^2} \tag{a}$$

$$\begin{aligned} c &= 0 \qquad \text{at } t = 0,\ z \geq 0 \\ c &= 0 \qquad \text{at } z = 0,\ t > 0 \\ c &= c_0 \qquad \text{at } z = L,\ t > 0 \end{aligned} \tag{b}$$

where c is the concentration of ethane in moles/liter and c_0 is a known value. Physically, the experimental set-up calls for one gas to be initially placed in the tube and then, at $t = 0$, one end of the tube is opened to the second gas. The denser gas should be underneath the less dense gas to avoid convective mixing engendered by density differences; see Figure E10.3-1a.

The solution to Equation (a) and (b) is†

$$\frac{c}{c_0} = \frac{z}{L} + \frac{2}{\pi}\sum_{n=1}^{\infty}\frac{(-1)^n}{n}e^{-(n\pi/L)^2\mathscr{D}t}\sin\frac{n\pi z}{L} \tag{c}$$

Several possible methods of measuring the concentration of ethane continuously or at discrete intervals (depending upon the analytical equipment available) can be suggested. As one example, suppose two samples are taken simultaneously at several time intervals with hypodermic syringes at sampling points 1 and 2, each located a distance a from the ends of the diffusion tube. The concentration of ethane in methane in the samples can be measured with a gas chromatograph or by other methods. Equation (c) gives, for points 1 and 2,

$$\frac{c_1}{c_0} = \frac{a}{L} + \frac{2}{\pi}\left\{-\exp\left[-\left(\frac{\pi}{L}\right)^2\mathscr{D}t\right]\sin\frac{\pi a}{L} + \frac{1}{2}\exp\left[-\left(\frac{2\pi}{L}\right)^2\mathscr{D}t\right]\sin\frac{2\pi a}{L} + \cdots\right\}$$

$$\frac{c_2}{c_0} = \frac{L-a}{L} + \frac{2}{\pi}\left\{-\exp\left[-\left(\frac{\pi}{L}\right)^2\mathscr{D}t\right]\sin\frac{\pi(L-a)}{L} + \frac{1}{2}\exp\left[-\left(\frac{2\pi}{L}\right)^2\mathscr{D}t\right]\sin\frac{2\pi(L-a)}{L} + \cdots\right\}$$

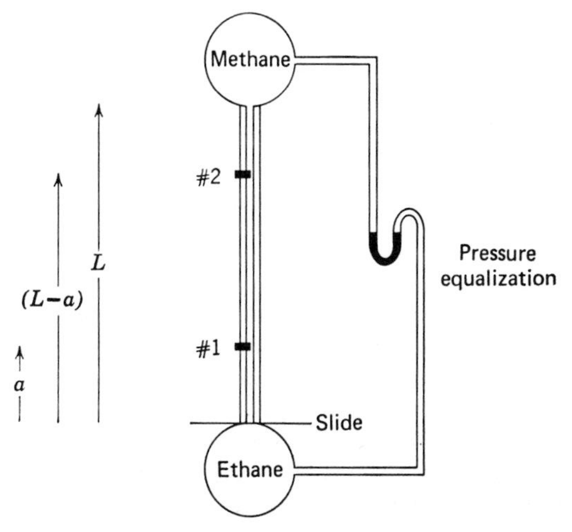

FIGURE E10.3-1a

Addition of c_1 to c_2 gives

$$\frac{c_1+c_2}{c_0} = 1 + \frac{2}{\pi}\left\{-2\exp\left[-\left(\frac{\pi}{L}\right)^2\mathscr{D}t\right]\sin\frac{\pi a}{L} + \exp\left[-\frac{2}{3}\exp\left[-9\left(\frac{\pi}{L}\right)^2\mathscr{D}t\right]\right]\sin\frac{\pi a 3}{L} + \cdots\right\} \tag{d}$$

Note that all the even terms cancel and that the odd terms combine during the addition of c_1 to c_2 because

$$\sin\frac{\pi(L-a)}{L} = \sin\frac{\pi a}{L}$$

$$\sin\frac{2\pi(L-a)}{L} = -\sin\frac{2\pi a}{L}$$

etc.

An additional reduction in the complexity of Equation (d) can be effected by proper choice of the location of the two sampling points. Suppose that the distance a is selected so that

$$\sin\frac{3\pi a}{L} = 0 \qquad \text{or} \qquad a = \frac{L}{3}$$

Then the third term in Equation (d) vanishes, and for any reasonable times the coefficients of the exponents in the expression $\exp[-25(\pi/L)^2\mathscr{D}t]$ and in higher order terms are so small that effectively

$$\frac{c_1+c_2}{c_0} = 1 - \frac{4}{\pi}\exp\left(-\frac{\pi^2\mathscr{D}t}{L^2}\right)\left(\sin\frac{\pi}{3}\right) \tag{e}$$

If only a single measuring location is used, the distance a can be made equal to $L/2$. Then Equation (c) becomes

$$\frac{c_3}{c_0} = \frac{1}{2} + \frac{2}{\pi}\left\{-\exp\left[-\left(\frac{\pi}{L}\right)^2\mathscr{D}t\right] + \frac{1}{3}\exp\left[-9\left(\frac{\pi}{L}\right)^2\mathscr{D}t\right] - \cdots\right\} \tag{f}$$

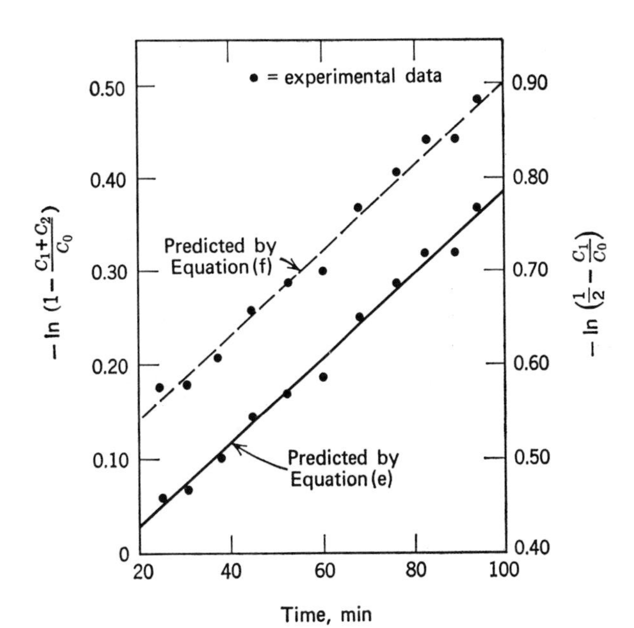

FIGURE E10.3-1b

† H. S. Carslaw and J. C. Jaeger, *Conduction of Heat in Solids* (2nd. ed.), Oxford Univ. Press, Oxford, 1959, p. 313.

Estimation by least squares of $\mathscr{D}$ from Equation (f), including the first and perhaps second exponentials, is straightforward since $c_3 + \epsilon = C_3$. Similarly, in Equation (e), C_1 and C_2 are independent and $\mathscr{E}\{C_1 + C_2\} = c_1 + c_2$, so that $(c_1 + c_2) + (\epsilon_1 + \epsilon_2) = (c_1 + c_2) + \epsilon'$. Figure E10.3-1b illustrates the predicted and experimental values for one experiment in which $L = 100$ cm. The estimated $\mathscr{D}$ was 0.076 cm²/sec, and from replicate experiments $s_{\mathscr{D}} =$ 3.1×10^{-3} cm²/sec. Two exponential terms of Equation (f) were used.

Example 10.3-2 Parameter Estimation Using Frequency Response

If a deterministic sinusoidal input is used as the process input, the steady-state output, termed the frequency response, can be used to estimate the coefficients in a *linear* (in the independent variables) model of the process. For example, if the model for heat transfer in an infinite slab is expressed in rectangular coordinates as (Figure E10.3-2a)

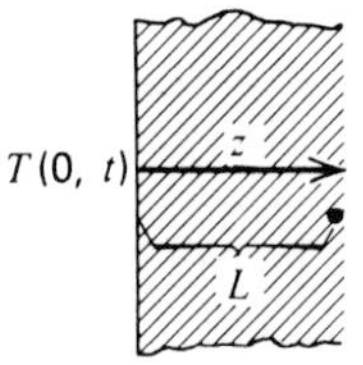

$$\frac{\partial \tilde{T}}{\partial t} = \alpha \frac{\partial^2 \tilde{T}}{\partial z^2} \qquad \text{(a)}$$
$$\tilde{T}(0, t) = a_0 \cos \omega t$$

FIGURE E10.3-2a

where

$\tilde{T}$ = deviation from the mean temperature, a deterministic variable
z = distance measured from face of the slab
t = time
a_0 = amplitude of the maximum temperature deviation at $z = 0$, a deterministic variable
α = thermal diffusivity
ω = frequency of the temperature fluctuations

the time-dependent solution of Equation (a) is quite complicated, but the solution at the point $z = L$ as t becomes large is

$$\tilde{T}(L, t) = a_0 e^{-L\sqrt{\omega/2\alpha}} \cos\left(\omega t - L\sqrt{\frac{\omega}{2\alpha}}\right) \qquad \text{(b)}$$

Although the parameters α (and a_0 if necessary) in Equation (b) can be estimated by nonlinear estimation techniques from temperature measurements, the phase angle (lag) and amplitude ratio both can be measured and used to estimate α by much simpler expressions. The phase angle can be calculated graphically or analytically from measured data, as indicated in Figure 10.2-1, as can the amplitude ratio:

$$\text{Phase angle:} \quad \psi = \omega\, \Delta t = -L\sqrt{\frac{\omega}{2\alpha}} \qquad (\text{c}_1)$$

$$\text{Amplitude ratio:} \quad \frac{A_L}{a_0} = e^{-\sqrt{L\omega/2\alpha}} \qquad (\text{c}_2)$$

The phase angle and amplitude ratio act as dependent variables, and the frequency, ω, can be varied as the independent variable for least squares estimation.

We shall observe next that experimentation in a rectangular geometry may be more difficult than in a cylindrical geometry (particularly for a gas), but that the estimation of α is simpler in the rectangular geometry. Heat transfer in an infinite cylinder can be modeled by

$$\frac{\partial \tilde{T}}{\partial t} = \alpha\left(\frac{\partial^2 \tilde{T}}{\partial r^2} + \frac{1}{r}\frac{\partial \tilde{T}}{\partial r}\right)$$
$$\tilde{T}(R, t) = a_R \cos \omega t \qquad \text{(d)}$$
$$\frac{\partial \tilde{T}(0, t)}{\partial r} = 0$$

where

r = radial direction measured from the center of the cylinder
R = radius of the cylinder, a constant

and where the sinusoidal temperature fluctuation is applied at $r = R$. The solution of Equation (d) for $\tilde{T}(r, t)$ is quite complex, but the frequency response for the center of the cylinder where a fine thermocouple might be placed is

$$\tilde{T}(0, t) = a_R \frac{\cos \omega t[\text{ber}(R\sqrt{\omega/\alpha}) + \sin \omega t[\text{bei}(R\sqrt{\omega/\alpha})]}{[\text{ber}(R\sqrt{\omega/\alpha})]^2 + [\text{bei}(R\sqrt{\omega/\alpha})]^2} \qquad \text{(e)}$$

Both ber and bei are tabulated Bessel functions similar to sine and cosine functions.

If Equation (e) is rewritten as

$$\tilde{T}(0, t) = \frac{a_R}{[\text{ber}^2(R\sqrt{\omega/\alpha}) + \text{bei}^2(R\sqrt{\omega/\alpha})]^{1/2}} \cdot \cos\left[\omega t - \tan^{-1}\left(\frac{\text{bei}(R\sqrt{\omega/\alpha})}{\text{ber}(R\sqrt{\omega/\alpha})}\right)\right]$$

the phase lag and amplitude ratio are clearer:

$$\text{Phase angle:} \quad \psi = \omega\, \Delta t = \tan^{-1}\left(\frac{\text{bei}(R\sqrt{\omega/\alpha})}{\text{ber}(R\sqrt{\omega/\alpha})}\right) \qquad (\text{f}_1)$$

Amplitude ratio:

$$\frac{A_0}{a_R} = \frac{1}{[\text{ber}^2(R\sqrt{\omega/\alpha}) + \text{bei}^2(R\sqrt{\omega/\alpha})]^{1/2}} \qquad (\text{f}_2)$$

It is interesting to note from Figure E10.3-2d that ψ becomes a linear function of $R\sqrt{\omega/\alpha}$ over certain ranges of ω:

$$\psi = b_1\sqrt{\frac{\omega}{\alpha}} - b_0 \qquad \text{(g)}$$

where b_0 and b_1 are empirical coefficients. If Equation (g) is not used, then (e) or (f) must be fitted by nonlinear estimation.

In a typical experiment, a 2.20 cm O.D. ($R = 1.00$ cm) stainless steel tube 1 meter long, containing oxygen, was heated by an electrical current flowing in the tube wall. Temperatures were measured by platinum resistance thermometers at the center and the inner wall of the tube. A syncrogenerator provided the desired sinusoidal input from 0 to 150 amperes.

Figure E10.3-2b illustrates a typical temperature-time recording from which the phase shift and amplitude ratio were obtained as follows. The frequency ω was calculated

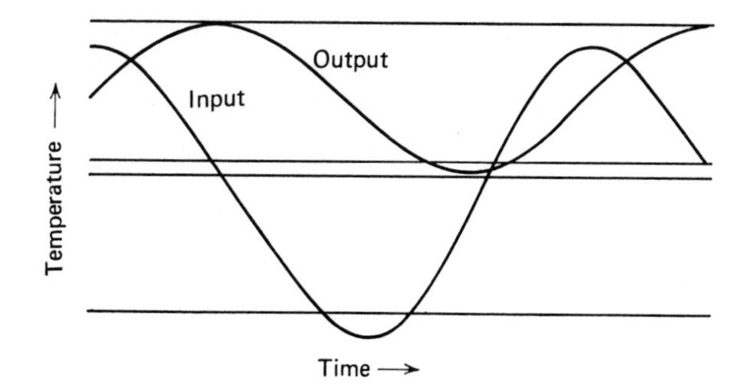

FIGURE E10.3-2b Time record of input and response on the same chart.

from the period of the temperature cycle (both input and output frequencies were the same; hence the process was proved linear) as recorded, and inasmuch as the frequency was controlled, it was regarded as a deterministic variable. To calculate the phase angle ψ, the frequency was multiplied by the measured phase shift from the time record. To calculate the amplitude ratio, the amplitudes from the chart were measured and adjusted for the proper scaling factor.

As mentioned in connection with Equation 10.3-9, the amplitude ratio may give a biased estimate of α; the extent of the bias would be difficult to determine inasmuch as Equation (f_2) is highly nonlinear. Consequently, α was estimated from Equation (f_1). The additional advantage of using Equation (f_1) was that it gave a linear relation between ψ and $\sqrt{(\omega/\alpha)}R$ for $\sqrt{(\omega/\alpha)}R > 3$ that could be approximated within 0.7 percent by the relation

$$\psi = 0.772\sqrt{\frac{\omega}{\alpha}}\,R - 0.631 \qquad \text{(h)}$$

Typical values of the precision for α were obtained from replicate experiments at 150°F and 1 atm, yielding $s_e = 0.065\ \text{cm}^2/\text{sec}$. Figures E10.3-2c and E10.3-2d illustrate the experimental measurements and the predicted amplitude ratio and phase angle for 150°F and $\hat{\alpha} = 0.275\ \text{cm}^2/\text{sec}$ with

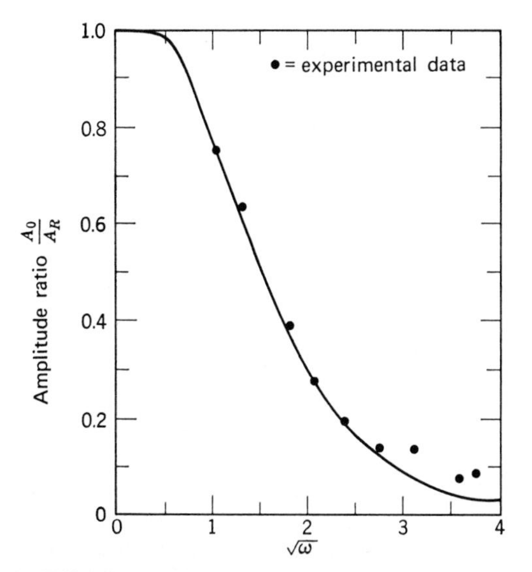

FIGURE E10.3-2c Ratio of amplitude at center of cylinder and amplitude at radius R.

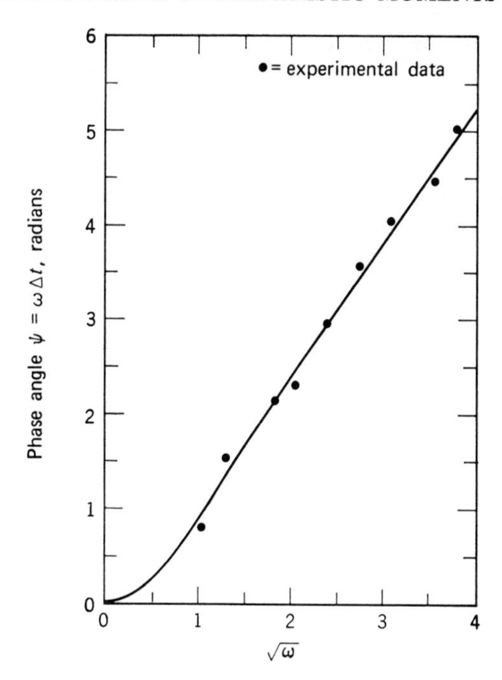

FIGURE E10.3-2d Phase angle between signal at radius R and response at center of cylinder.

$\hat{\alpha}$ calculated by least squares using Equation (g) but with the predictions being made using Equations (f_1) and (f_2). Even though the amplitude ratio was not used to estimate α, the predictions were good at the lower frequencies.

10.4 PARAMETER ESTIMATION USING DETERMINISTIC MOMENTS

Deterministic moments provide an alternate method of estimating the parameters in a distributed model. The term "deterministic moment" will be used to distinguish the moments of this section from the moments of a stochastic variable described in Chapter 2. The major advantage of employing deterministic moments is that the moments of the process responses are related to the model parameters by much simpler equations than the full solution to the process model. Furthermore, no analytical solution can be obtained for some models, whereas the moments can be found analytically. Deterministic moments can be computed only for linear (in the dependent variable) differential equations and have been applied extensively to models involving axial and radial dispersion.

10.4-1 Models of Axial Dispersion

If a pulse of tracer is injected into an actual flowing stream such as in a pipe, porous medium, or open channel, the pulse spreads out as it moves with the fluid downstream because of dispersion. For a fixed distance between the injection point and the measurement point of the response, the amount of spread depends on the intensity of the dispersion in the process. Conversely, the spread can be used to characterize quantitatively the dispersion phenomena, that is, to evaluate the dispersion coefficient in the process model of Equation 10.2-3, written here in dimensionless form:

$$\frac{\partial c^*}{\partial t^*} + \frac{\partial c^*}{\partial z^*} = \frac{1}{P}\frac{\partial^2 c^*}{\partial z^{*2}} + \delta(z^* - z_0^*)\,\delta(t^*) \quad (10.4\text{-}1)$$

where

$t^* = \frac{Vt}{L} = \frac{qt}{V} = \frac{t}{\bar{t}}$, dimensionless time

$c^* = \frac{c}{c_{av}}$, a dimensionless deterministic dependent variable

$P = \frac{vL}{D}$, a dimensionless coefficient called the Peclet number

$z^* = \frac{z}{L}$, a dimensionless axial direction

q = volumetric flow rate of fluid
V = volume of vessel or channel, a deterministic variable
L = length between the injection point and the measurement point $= (z_m - z_0)$
v = velocity of fluid
$c_{av} = m/V$; that is, c_{av} is the concentration of injected tracer if the quantity of tracer m were evenly distributed throughout the vessel
$\bar{t} = (V/q)$, mean residence time of fluid

Note that Model 10.4-1 assumes that a pulse input is introduced into the vessel or channel so that c^* is the dimensionless deterministic impulse response.

We define two deterministic moments as follows:

$$m_1 = \int_0^\infty t^* c^*\, dt^* \quad (10.4\text{-}2)$$

$$m_2^{\#} = \int_0^\infty (t^* - m_1)^2 c^*\, dt^* \quad (10.4\text{-}3)$$

Notice that the definition of m_1 is actually

$$m_1 = \frac{\int_0^\infty t^* c^*\, dt^*}{\int_0^\infty c^*\, dt^*}$$

but we have normalized the denominator, that is, required that

$$\int_0^\infty c^*\, dt^* = 1$$

Consequently, introducing the definitions of c^* and dt^* into the normalization equation, we find

$$\frac{1}{\bar{t}}\int_0^\infty c\, dt = c_{av}$$

and

$$q\int_0^\infty c\, dt = Vc_{av} = m$$

Levenspiel and Smith† first showed that $m_2^{\#}$ can be conveniently related to the dispersion coefficient. Van der Laan‡ and Aris§ pointed out that in using Laplace transforms to solve Equation 10.4-1, if the kth moment is finite,

$$m_k = \frac{\int_0^\infty t^k c^*(z^*, t)\, dt}{\int_0^\infty c^*(z^*, t)\, dt}$$

then

$$\frac{\lim\limits_{s\to 0} \dfrac{d^k \check{c}^*(z^*, s)}{ds^k}}{\lim\limits_{s\to 0} \check{c}^*(z^*, s)} = (-1)^k m_k$$

in which $\check{c}^*(z^*, s)$ is the Laplace transform of $c^*(z^*, t)$ and s is the usual complex parameter. Hence the m_1 and $m_2^{\#}$ can be found from

$$m_1 = -\lim_{s\to 0} \frac{\partial \check{c}^*(z^*, s)}{\partial s} \quad (10.4\text{-}4)$$

$$m_2^{\#} + m_1^2 = \lim_{s\to 0} \frac{\partial^2 \check{c}^*(z^*, s)}{\partial s^2} \quad (10.4\text{-}5)$$

Keeping in mind that the dispersion coefficient is contained in P, one would like to relate m_1 and $m_2^{\#}$ to P for various boundary conditions. Levenspiel and Bischoff‖ gave a number of such relations and appropriate references. Table 10.4-1 is a brief summary of a few of the more useful relationships. Observe that the explicit solutions for the concentration are quite complex or just not available; it would be very hard to evaluate P from c^* versus t^* data, whereas it would be easier to obtain P from m_1 and $m_2^{\#}$.

Because the deterministic moments are actually estimated by using a stochastic variable C through relations:

$$I_0 = \int_0^\infty C\, dt \quad (10.4\text{-}6a)$$

$$I_1 = \int_0^\infty tC\, dt \quad (10.4\text{-}6b)$$

$$I_2 = \int_0^\infty t^2 C\, dt \quad (10.4\text{-}6c)$$

$$M_1 = \frac{1}{\bar{t}}\frac{I_1}{I_0} = \frac{q^2}{mV} I_1 \quad (10.4\text{-}7a)$$

$$M_2^{\#} = \frac{1}{\bar{t}^2}\left[\frac{I_2}{I_0} - \left(\frac{I_1}{I_0}\right)^2\right] = \frac{q}{m}\left(\frac{q}{V}\right)^2\left[I_2 - 2\frac{q}{m}I_1^2 + \left(\frac{q}{m}\right)^2 I_1^2 I_0\right] \quad (10.4\text{-}7b)$$

† O. Levenspiel and W. K. Smith, *Chem. Eng. Sci.* **6**, 227, 1957.
‡ E. T. Van der Laan, *Chem. Eng. Sci.* **7**, 187, 1958.
§ R. Aris, *Proc. Royal Soc.* (London) **A245**, 268, 1958.
‖ O. Levenspiel and K. B. Bischoff, *Advan. Chem. Eng.* **4**, 95, 1963.

TABLE 10.4-1 RELATION BETWEEN MOMENTS AND PECLET NUMBER FOR MODELS WITH AXIAL DISPERSION ONLY

Configuration	Solution	Moments
$-\infty$ $+\infty$ L δ input Response (1) Doubly infinite pipe (single measurement)	$c^* = \frac{1}{2}\left(\frac{P}{\pi t^*}\right)^{1/2} \exp\left[-\frac{P(1-t^*)^2}{4t^*}\right]$	$m_1 = 1 + \frac{2}{P}$ $m_2^{\#} = \frac{2}{P} + \frac{8}{P^2}$
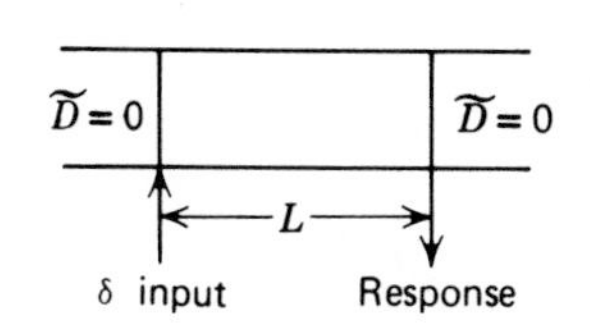 (2) Closed pipe (single measurement)	$c^* = e^{P/2} \sum_{n=1}^{\infty} \frac{(-1)^{n+1} 8\alpha_n^2}{4\alpha_n^2 + 4P + P^2} e^{-\alpha_n t^*}$ $\alpha_n = \frac{P^2 + 4\alpha_n^2}{4P}$ $\tan \alpha_n = \frac{4P\alpha_n}{4\alpha_n^2 - P^2}$	$m_1 = 1$ $m_2^{\#} = \frac{2}{P} - \frac{2}{P^2}(1 - e^{-P})$
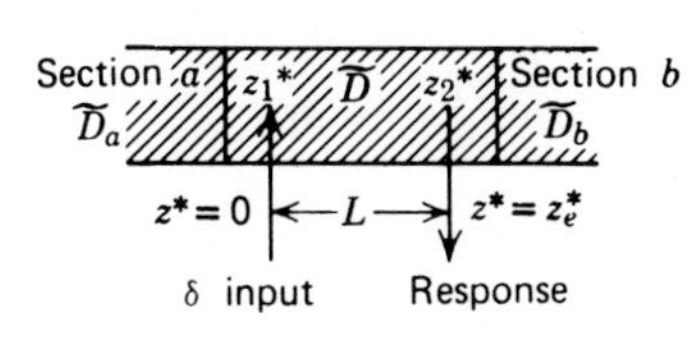 $\alpha = \frac{\tilde{D}_a}{\tilde{D}}$ $\beta = \frac{\tilde{D}^b}{\tilde{D}}$ (3) General case	None available	$m_1 = 1 + \frac{1}{P}[2\cdot(1-\alpha)e^{-Pz_1^*} - (1-\beta)e^{-P(z_e^*-z_2^*)}]$ $m_2^{\#} = \frac{2}{P} + \frac{1}{P^2}$ $\times \{8 + 2(1-\alpha)(1-\beta)e^{-Pz_1^*} - (1-\alpha)e^{-Pz_1^*}$ $\times [4z_1^*P + 4(1+\alpha) + (1-\alpha)e^{-Pz_1^*}]$ $-(1-\beta)e^{-P(z_e^*-z_2^*)}$ $\times [4(z_e^*-z_2^*)P + 4(1+\beta) + (1-\beta)e^{-P(z_e^*-z_2^*)}]\}$
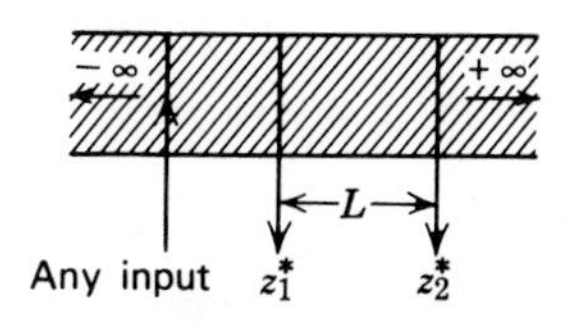 (4) Infinite pipe conditions (double measurement)		$\Delta m_1 = 1$ $\Delta m_2^{\#} = \frac{2}{P}$
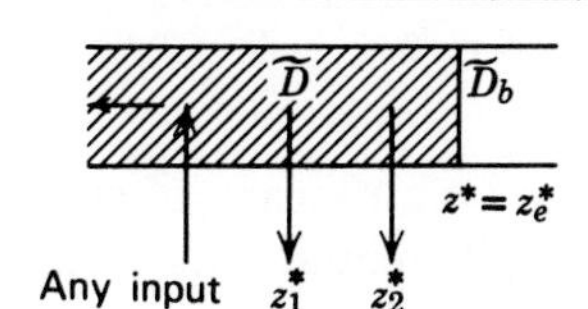 (5) Double measurement within test section $\beta = \frac{\tilde{D}^b}{\tilde{D}}$	None available	$\Delta m_1 = 1 - \frac{1-\beta}{P}[1 - \exp(-P)]\exp[P(z_2^* - z_e^*)]$ $\Delta m_2^{\#} = \frac{2}{P} + \frac{1-\beta}{P^2}\exp[P(z_2^* - z_e^*)]$ $\times \{4(1+\beta)[\exp(-P) - 1] + 4P(z_2^* - z_e^*)$ $+ (1-\beta)[\exp(-2P) - 1]\exp[P(z_2^* - z_e^*)]$ $+ 4P(z_e^* - z_1^*)\exp(-P)\}$

P = Peclet number; * indicates dimensionless variable.

where C is the observed response, a function of time, M_1 is the estimate of m_1, and $M_2^{\#}$ is the estimate of $m_2^{\#}$, we need to investigate the statistical properties of I_0, I_1, and I_2. The integrals I_0, I_1, and I_2 are statistically independent.

Consider the integral I:

$$I = \int_a^b \varphi(t)X(t)\, dt$$

where $\varphi(t)$ is a deterministic function and $X(t)$ is a nonstationary stochastic function. For a given $\varphi(t)$ and integration limits, I is a stochastic variable. The probability distribution of I depends on the chosen function $\varphi(t)$, the probability distribution of $X(t)$ and the integration limits. For the specific case in which $X(t)$ is a normal random function, the integral is also a normal random function, and

$$\mathscr{E}\{I\} = \int_a^b \varphi(t)\mathscr{E}\{X(t)\}\, dt \qquad (10.4\text{-}8)$$

$$\text{Var}\,\{I\} = \int_a^b \varphi(t)\,\text{Var}\,\{X(t)\}\, dt \qquad (10.4\text{-}9)$$

Given the expected value and variance of $X(t)$, Equations 10.4-8 and 10.4-9 can be used to calculate the expected value and variance of I. Even if $X(t)$ is not a normal variable, if the integral is approximated by a sum and the individual observations $X_i(t)$ are independent, by the central limit theorem I is approximately normally distributed.

Inasmuch as the function $X(t)$ is nonstationary, its statistical properties such as expected value and variance are not invariant with respect to a translation in time as described in Section 2.1. Consequently, the expected value of $X(t)$ will be a function of time, and an unbiased estimate of $\mathscr{E}\{X(t)\}$ can be computed from n time records $i = 1, \ldots, n$ at any fixed time t by

$$\hat{\mu}_X(t) = \frac{1}{n}\sum_{i=1}^{n} X_i(t) \qquad (10.4\text{-}10)$$

Unfortunately, large sample sizes are required to reduce the error in the estimates to a reasonable magnitude.

The variance of $\hat{\mu}_X(t)$ at any fixed time t is

$$\text{Var}\,\{\hat{\mu}_X(t)\} = \frac{1}{n^2}\sum_{i=1}^{n}\sum_{k=1}^{n} \text{Covar}\,\{X_i(t)X_k(t)\}$$

$$= \frac{\sigma_X^2(t)}{n} + \frac{1}{n^2}\sum_{\substack{i=1\\ i\neq k}}^{n}\sum_{k=1}^{n} \cdot\mathscr{E}\{[X_i(t) - \mu_X(t)][X_k(t) - \mu_X(t)]\} \qquad (10.4\text{-}11)$$

If $X_i(t)$ and $X_k(t)$ are independent, Equation 10.4-11 reduces to

$$\text{Var}\,\{\hat{\mu}_X(t)\} = \frac{\sigma_X^2(t)}{n} \qquad (10.4\text{-}12)$$

As we let $\Delta t \to 0$ with $n = t_f/\Delta t \to \infty$, the sample data average converges to that of the continuous variable. If $X_i(t)$ and $X_k(t)$ are correlated, the double sum on the right-hand side of Equation 10.4-11 has to be evaluated, and the variance of $\hat{\mu}_X(t)$ may or may not decrease to an acceptably small level with increasing sample size n, depending upon the nature of the double sum. If $r_{XX}(k, t)$ decreases as k increases for large n, Equation 10.4-11 reduces to

$$\text{Var}\,\{\hat{\mu}_X(t)\} = \frac{\sigma_X^2(t)}{n} + \frac{2}{n}\sum_{k=1}^{n-1}\left(1 - \frac{k}{n}\right)[r_{XX}(k, t) - \mu_X^2(t)]$$

$$\approx \frac{\hat{\sigma}_X^2(t)}{n} + \frac{2}{n}\sum_{k=1}^{n-1} R_{XX}(k, t) - \hat{\mu}_X^2(t) \qquad (10.4\text{-}13)$$

where $R_{XX}(k, t)$ is the estimated autocorrelation function at time t.

Application of Equation 10.4-8 to I_0, I_1, and I_2 gives

$$\mathscr{E}\{I_0\} = \int_0^\infty \mathscr{E}\{C\}\, dt = \int_0^\infty c\, dt$$

$$\mathscr{E}\{I_1\} = \int_0^\infty t\mathscr{E}\{C\}\, dt = \int_0^\infty tc\, dt$$

$$\mathscr{E}\{I_2\} = \int_0^\infty t^2\mathscr{E}\{C\}\, dt = \int_0^\infty tc^2\, dt$$

Consequently, if M_1 is calculated from $(q^2/mV)I_1$, M_1 is an unbiased estimate:

$$\mathscr{E}\{M_1\} = \frac{q^2}{mV}\mathscr{E}\{I_1\} = \frac{q^2}{mV}\int_0^\infty tc\, dt = m_1$$

But if M_1 is calculated from the ratio $(I_1/(\bar{t}I_0))$:

$$\mathscr{E}\{M_1\} = \frac{1}{\bar{t}}\mathscr{E}\left(\frac{I_1}{I_0}\right)$$

it would be necessary to use Equation 10.3-9 to evaluate the bias in the estimate of m_1.

Rather than calculate $\hat{\mu}_C(t)$ and $\sigma_{\hat{\mu}_C}^2(t)$, and from those quantities determine the estimates of $\mathscr{E}\{I_k\}$, it may prove simpler, depending upon the data processing equipment available, to evaluate I_k for several different experiments and to calculate a sample average $\bar{I}_k$:

$$\bar{I}_k = \frac{1}{n}\sum_{i=1}^{n} I_{ik} \qquad k = 0, 1, 2$$

Very little can be said about $M_2^{\#}$ as a means of estimating the coefficient in the differential equation. Because the relation between $M_2^{\#}$ and the integrals I_0, I_1, and I_2 is nonlinear, it is difficult to evaluate the bias in $\mathscr{E}\{M_2^{\#}\}$. The precision in $M_2^{\#}$ can be estimated from replicate experiments.

No matter what the probability distribution of $C(t)$, I_k, or M_k, Chebyshev's inequality, Equation 3.3-8, can be used to ascertain the precision in $\hat{\mu}_C(t)$, I_k, or M_k, respectively.

As an example of approximating the precision of the dimensionless parameter P from the deterministic moments, let us take the first entry in Table 10.4-1:

$$m_1 = 1 + \frac{2}{P}$$

After

$$\overline{M}_1 = \frac{q^2}{mV}\frac{1}{n}\sum_{i=1}^{n} I_{i1}$$

has been calculated for independent experiments,

$$\widehat{\mathrm{Var}}\{M_1\} = \frac{\sum_{i=1}^{n} (M_{i1} - \overline{M}_1)^2}{n - 1}$$

and thus

$$\bar{P} = \frac{2}{\overline{M}_1 - 1}$$

$$\widehat{\mathrm{Var}}\{P\} \cong \left(\frac{\bar{P}^2}{2}\right)^2 \widehat{\mathrm{Var}}\{M_1\}$$

Both Levenspiel and Smith's and Van der Laan's work depended on being able to represent the tracer injection by a delta function, a mathematical idealization which physically can only be approximated since it requires a finite amount of tracer be injected in zero time. To circumvent this difficulty, Aris,† Bischoff,‡ and Bischoff and Levenspiel§ described a method that does not require a perfect delta function input. The method involves taking concentration measurements at two points, both within the test section, rather than at only one point as described above. The deterministic moments of the concentration curves at the two points are calculated as before and then the difference between them found. This difference can be related to the parameter P and thus to the dispersion coefficient. It does not matter where the tracer is injected into the system as long as it is upstream of the two measurement points. The injection may be any type of pulse input, not necessarily just a delta function.

Since the position of the injection point is not important, it is convenient to base the dimensionless quantities on the length between the two measuring points. Therefore, let z_1 designate the first measuring point and z_2 the second measuring point. Let

$$\Delta m_1 = (m_1)_{z_2} - (m_1)_{z_1} \tag{10.4-14}$$

$$\Delta m_2^{\#} = (m_2^{\#})_{z_2} - (m_2^{\#})_{z_1} \tag{10.4-15}$$

The expected value and precision in ΔM_1 (or $\Delta M_2^{\#}$) are

$$\mathscr{E}\{\Delta M_1\} = \mathscr{E}\{(M_1)_{z_2}\} - \mathscr{E}\{(M_1)_{z_1}\} \tag{10.4-16}$$

$$\mathrm{Var}\{\Delta M_1\} = \mathrm{Var}\{(M_1)_{z_2}\} + \mathrm{Var}\{(M_1)_{z_1}\} \tag{10.4-17}$$

† R. Aris, *Chem. Eng. Sci.* **9**, 266, 1959.
‡ K. B. Bischoff, *Chem. Eng. Sci.* **12**, 69, 1960.
§ K. B. Bischoff and O. Levenspiel, *Chem. Eng. Sci.* **17**, 245, 1962.

Bischoff and Levenspiel also presented solutions that relate Equations 10.4-14 and 10.4-15 to the dimensionless parameter P for probes both inside and outside the test section. Two such relations are shown in Table 10.4-1. In experimental work it is highly desirable to make measurements far enough away from the ends of the vessels so that end effects become negligible, in which case the extremely simple expressions for the infinite tube can properly be used. Another even more important reason for using the infinite tube expressions is that the end effects cannot be exactly accounted for in real systems because of the complex flow patterns at these locations. Bischoff and Levenspiel presented design charts which allow estimation of the position of the measuring point sufficiently far from the end of the system to neglect end effects.

Example 10.4-1 Estimation of an Axial Dispersion Coefficient from Experimental Data

The response at two measuring points for an experimental set-up corresponding to configuration 4 in Table 10.4-1 is qualitatively shown in Figure E10.4-1a, assuming the dependent variable C is itself recorded. The major source of distortion in the calculation of the estimates of the deterministic moments is in the tail of the C curve where small errors in concentration contribute unduly to the moment. By marking a reference time on the two output charts at the instant when the input pulse of tracer is injected (upstream) into the system, the times on the two output charts can be associated as follows:

$$t_2 = t_1 + (\delta_2 - \delta_1)$$

where the subscript 1 signifies the first measuring point and 2 the second. In computing I_0, I_1, and I_2, the integrals are actually evaluated from the first noticeable breakthrough for each curve, where t_1 and t_2 are placed equal to zero, up to some t_1' and t_2' which represent the time at which the curve again reaches the horizontal axis.

Table E10.4-1 lists some actual concentration-time data (the concentrations were calculated from observed intensities by means of Beer's law and the times from the known chart speeds) for dispersion in a bed packed with 6 mm glass beads. Other measured data were (again calculated from calibration curves, etc.):

Length between points 1 and 2: 60.96 cm
Interstitial velocity, v: 0.7537 cm/sec
$\Big\}\ \bar{t} = \dfrac{60.96}{0.7537}$ sec
Fluid temperature 27.4°C
Flow rate: 5.586 cc/sec
δ_1: 32.88 sec
δ_2: 78.84 sec

If continuous recorder output is to be used directly to calculate the moments, some subsidiary electronic circuitry is required. Otherwise, successive points can be sampled visually or digitally from the recorder output, as listed in

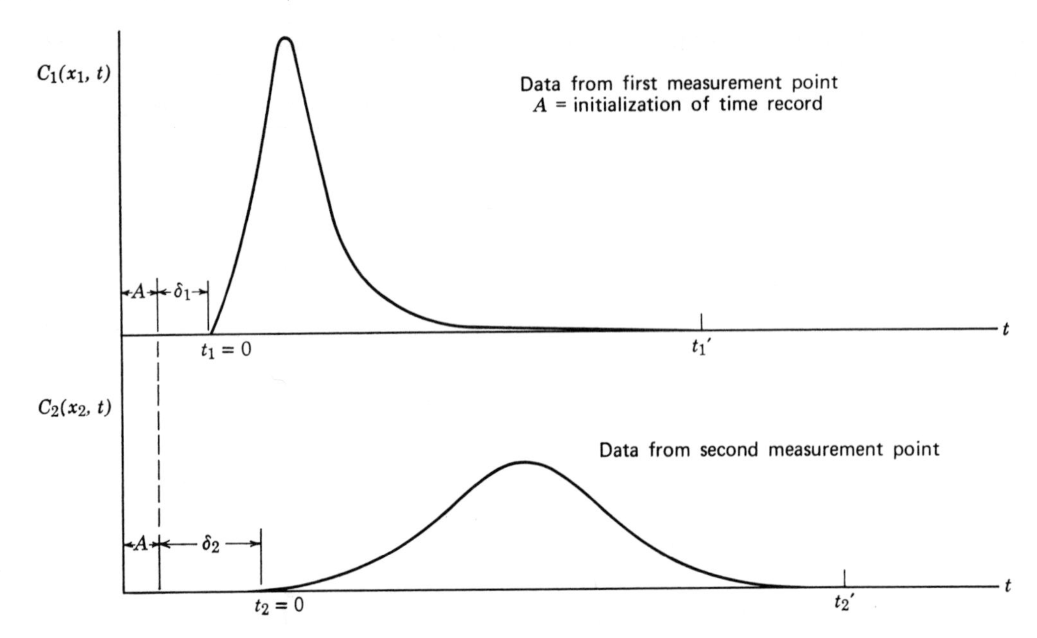

FIGURE E10.4-1a

Table E10.4-1, at times suitable for use in the numerical quadrature scheme to be employed. The integrals I_0, I_1, and I_2, calculated in real time at z_i $(i = 1, 2)$ from Equations 10.4-6, are

$$I_{0i} = \int_0^{t_i'} C_i \, dt_i$$

$$I_{1i} = \int_0^{t_i'} t_i C_i \, dt_i$$

$$I_{2i} = \int_0^{t_i'} t_i^2 C_i \, dt_i$$

and the values of M_{1i} and $M_{2i}^{\#}$ can be calculated from Equation 10.4-7. The desired moments ΔM_1 and $\Delta M_2^{\#}$ in real time are

$$\Delta M_1 = (M_1)_2 - (M_1)_1 + (\delta_2 - \delta_1) \qquad (a_1)$$

$$\Delta M_2^{\#} = (M_2^{\#})_2 - (M_2^{\#})_1 \qquad (a_2)$$

Notice that the shift in the time origin carried out in calculating I_{0i}, I_{1i}, and I_{2i} must be adjusted for in *subtracting* $(M_1)_1$ from $(M_1)_2$, but that $M_2^{\#}$ is independent of the computing origin since it is calculated about M_1. A very practical feature about the method of moments is that the units of C are not important. An arbitrary multiplier of C (to convert from one set of units to another) cancels out if the ratios of the integrals I are used (perhaps at the expense of some bias in the estimate of M_1 and $M_2^{\#}$).

The integrals I_{0i}, I_{1i}, and I_{2i} can be evaluated from the data in Table E10.4-1 by means of the following quadrature formulas based on the trapezoidal rule (the subscript i is suppressed for simplicity):

$$I_0 = \sum_{j=0}^{n-1} \left[C_j t_{j+1} - C_{j+1} t_j + \frac{C_{j+1} - C_j}{2} (t_{j+1} + t_j) \right] \qquad (b)$$

$$I_1 = \sum_{j=0}^{n-1} \left[(C_j t_{j+1} - C_{j+1} t_j)\left(\frac{t_{j+1} + t_j}{2}\right) + \frac{C_{j+1} - C_j}{3} (t_{j+1}^2 + t_{j+1} t_j + t_j^2) \right] \qquad (c)$$

$$I_2 = \sum_{j=0}^{n-1} \left[(C_j t_{j+1} - C_{j+1} t_j)(t_{j+1}^2 + t_{j+1} t_j + t_j^2) + \frac{C_{j+1} - C_j}{4} (t_{j+1}^3 + t_{j+1}^2 t_j + t_{j+1} t_j^2 + t_j^3) \right] \qquad (d)$$

The first data point in the sum is at $t = 0$, $j = 0$; the upper limit of the sum is at $t = t'$ and $j = n - 1$; and $n + 2$ data points are needed. The values of the integrals are

	Point 1	Point 2
I_0 (ppm)(sec)	112.18	112.22
I_1 (ppm)(sec)^2	1.905×10^3	5.833×10^3
I_2 (ppm)(sec)^3	3.923×10^4	3.358×10^5

From these integrals the deterministic moments were found to be

	Point 1	Point 2
$\bar{t}M_1$, sec	16.99	51.98
$\bar{t}^2 M_2^{\#}$, sec^2	349.7	2992.7

which gave

$$\Delta M_1 = [(51.98 - 16.99) + (78.84 - 32.88)]\left(\frac{0.7537}{60.96}\right) = 1.001$$

$$\Delta M_2^{\#} = (2992.7 - 349.7)\left(\frac{0.7537}{60.96}\right)^2 = 0.0351$$

Since I_{01} and I_{02} should represent the same quantity of tracer, we observe that the loss of tracer was negligible.

TABLE E10.4-1 CALCULATED CONCENTRATIONS AND TIMES FOR INPUT PULSE RUN FOR 6 MM GLASS BEADS

First Measuring Point		Second Measuring Point	
Concentration, C (ppm)	Time, t (sec)	Concentration, C (ppm)	Time, t (sec)
0.000	0.0	0.000	0.0
0.012	0.6	0.006	3.6
0.024	1.2	0.018	7.2
0.083	1.8	0.050	9.6
0.282	2.4	0.094	12.0
0.836	3.6	0.142	14.4
1.461	4.8	0.224	16.8
2.061	6.0	0.325	19.2
2.814	7.2	0.468	21.6
3.772	8.4	0.728	25.2
4.832	9.6	1.073	28.8
5.314	10.2	1.470	32.4
5.772	10.8	1.809	36.0
6.189	11.4	2.327	39.6
6.492	12.0	2.412	40.8
6.746	12.6	2.511	42.0
6.913	13.2	2.618	43.2
6.999	13.8	2.701	44.4
6.991	14.4	2.775	45.6
6.913	15.0	2.800	46.8
6.754	15.6	2.820	48.0
6.563	16.2	2.831	49.2
6.304	16.8	2.820	50.4
6.025	17.4	2.786	51.6
5.730	18.0	2.749	52.8
5.078	19.2	2.685	54.0
4.416	20.4	2.622	55.2
3.785	21.6	2.425	57.6
3.220	22.8	2.193	60.0
2.719	24.0	1.965	62.4
2.263	25.2	1.608	66.0
1.900	26.4	1.276	69.6
1.556	27.6	0.991	73.2
1.271	28.8	0.749	76.8
1.056	30.0	0.619	79.2
0.871	31.2	0.505	81.6
0.715	32.4	0.410	84.0
0.590	33.6	0.326	86.4
0.479	34.8	0.260	88.8
0.400	36.0	0.212	91.2
0.330	37.2	0.166	93.6
0.268	38.4	0.133	96.0
0.227	39.6	0.103	98.4
0.194	40.8	0.084	100.8
0.150	42.0	0.068	103.2
0.125	43.2	0.052	106.8
0.109	44.4	0.040	110.4
0.089	45.6	0.035	114.0
0.073	46.8	0.027	117.6
0.063	48.0	0.019	121.2
0.053	49.8	0.015	127.2
0.039	51.6	0.008	132.0
0.031	53.4	0.000	134.0
0.026	55.2		
0.020	57.0		
0.016	60.0		
0.010	66.0		
0.004	72.0		
0.000	84.0		

The dimensionless coefficient P can be approximated from the relationships listed in entry 4 of Table 10.4-1: $\Delta m_1 = 1$, which appears to be satisfied, and $\Delta m_2^\# = 2/P$. Three replicate experiments yielded the following results for $\Delta M_2^\#$ from Equations 10.4-16 and 10.4-17:

$$\overline{\Delta M_2^\#} = 0.0376$$

$$[\widehat{\mathrm{Var}}\{\Delta M_2^\#\}]^{1/2} = 0.00351$$

$$\bar{P} \simeq \frac{2}{\overline{\Delta M_2^\#}} = \frac{2}{0.0376} = 53.2$$

$$[\widehat{\mathrm{Var}}\{P\}]^{1/2} \simeq \left(\frac{\bar{P}^2}{2}\right)[\widehat{\mathrm{Var}}\{\Delta M_2^\#\}]^{1/2} = 4.96$$

Because the distribution of $M_2^\#$ and, hence, P is not known, we apply the Chebyshev inequality, Equation 3.3-8, to determine the approximate confidence limits on P:

$$53.2 - 3(4.96) < P < 53.2 + 3(4.96)$$

with a probability of $1 - \frac{1}{9} = 0.89$.

10.4-2 Models of Axial and Radial Dispersion

The experimental methods for measuring radial-dispersion coefficients involve injection of tracer into the process and measurement of its concentration at some point downstream from the injection point. However, the tracer should not be injected over a plane nor measured over a plane. Instead, to study the radial movement of tracer and thereby gain information on the radial mixing that occurs in the subsystem, an experimental method must be used in which the concentration varies with radial position. Usually the tracer is injected at the axis of the tube ("point source input"); the axis is chosen so that there is radial symmetry about the tube axis which simplifies the mathematics. As the tracer moves down the tube, it spreads radially by dispersion. At long distances (theoretically infinite) from the injection point, the tracer completely mixes with the flowing fluid. As a consequence, the measurement point must not be so far from the injection point that the existing concentration differences cannot be detected.

The mathematical developments are based on the dispersion model

$$\frac{\partial c}{\partial t} + v_z \frac{\partial c}{\partial z} = \tilde{D}_L \frac{\partial^2 c}{\partial z^2} + \frac{\tilde{D}_R}{r}\frac{\partial}{\partial r}\left(r\frac{\partial c}{\partial r}\right) + m\,\delta(z - z_0)f(r) \tag{10.4-18}$$

with all the coefficients assumed to be constant. The last term represents the "source" (injection) of tracer at a specific point. For simplification, Equation 10.4-18 is put in dimensionless form as follows:

$$t^* = \frac{v_z t}{R}$$

$$z^* = \frac{z}{R}$$

$$r^* = \frac{r}{R}$$

$$P_L = \frac{v_z R}{\tilde{D}_L}$$

$$P_R = \frac{v_z R}{\tilde{D}_R}$$

$$c^* = \frac{c}{c_{av}}$$

The mean concentration of the tracer at a point sufficiently downstream from the point of the steady-state tracer injection of rate m is c_{av}. By a mass balance that equates the rates of injection to the rate of flow out of tube, we have

$$m = \pi R^2 v_z c_{av}$$

After introduction of these dimensionless quantities into Equation 10.4-18, we obtain

$$\frac{\partial c^*}{\partial z^*} - \frac{1}{P_L}\frac{\partial^2 c^*}{\partial z^{*2}} - \frac{1}{P_R}\frac{1}{r^*}\left[\frac{\partial}{\partial r^*}\left(r^*\frac{\partial c^*}{\partial r^*}\right)\right] = \delta(z^* - z_0^*)f(r^*) \tag{10.4-19}$$

Solutions to Equation 10.4-19 can be used to estimate dispersion coefficients from experimental data.

The general solution to Equation 10.4-19 is quite cumbersome to use. Table 10.4-2 lists various less complex cases involving simplifications in the differential equation or the boundary conditions. Bischoff and Levenspiel† examined the conditions under which some of the less rigorous expressions would be justified and presented design charts which allow evaluation of the approximation errors in the calculated dispersion coefficients for various conditions. They found that the end-effect approximation errors are smaller for radial than for axial-dispersion coefficients. Even though the error may be large for measurement right at the end of the vessel, the error decreases very rapidly when the detection probe is moved into the bed. Hence, in most typical packed beds, taking measurements one or two particle depths into the bed is often sufficient to make the end-effect approximation errors negligible.

Table 10.4-3 summarizes a number of commonly used models and gives their deterministic moments, as defined in this section, as well as their corresponding frequency response functions. The models in Table 10.4-3 are for the experimental arrangement shown in Figure 10.4-1.

Inspection of Table 10.4-3 might lead one to conclude that the deterministic moments are simpler to use than the frequency response in obtaining estimates of the coefficients. But this conclusion is true only for simple models. Further examination brings out the fact that a model with p coefficients requires the calculation of p deterministic moments if an adequate number of independent equations are to be used. Moments of order greater than two become quite inaccurate because the inherent errors in the "tail" of the response curve are greatly magnified, causing the higher moments to be of little practical use.

In the estimation of parameters in models represented by linear (in the dependent variable) partial differential equations with more than two coefficients, use of the frequency response is recommended. The drawbacks, of course, in the use of the frequency response are that the functions of A_r and ψ are highly nonlinear and that the coefficients in the phase shift, ψ, are quite insensitive to the type of model used.

† K. B. Bischoff and O. Levenspiel, *Chem. Eng. Sci.* **17**, 245, 1962.

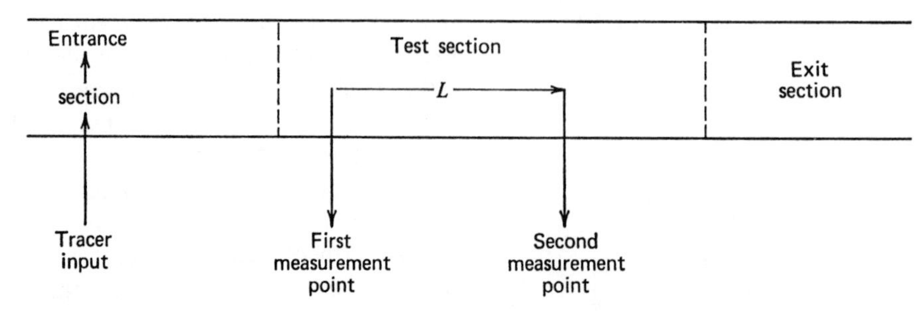

FIGURE 10.4-1 Experimentation for axial dispersion.

TABLE 10.4-2 RESPONSE FOR POINT INPUT IN MODELS WITH RADIAL DISPERSION*

Experimental Scheme	Response to Input

Tracer in (point source) Measurement point

$$c^* = \frac{P_R}{4}\frac{\exp\left[-(P_R/2)(\sqrt{z^2+r^2}-z)\right]}{\sqrt{z^2+r^2}} \quad (1)$$

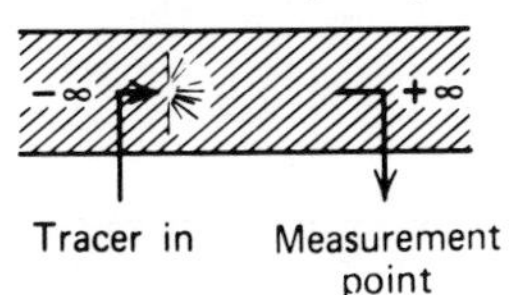

$$c^* = 1 + \frac{1}{2}\sum_{a_i>0}\frac{1}{q_R}\exp\left[(\tfrac{1}{2}-q_R)P_R z\right]\frac{J_0(a_i r)}{J_0^2(a_i)}$$

with $\qquad$ (2)

$$J_i(a_i) = 0, \qquad q_R = \sqrt{\frac{1}{4}+\frac{a_i^2}{P_R^2}}$$

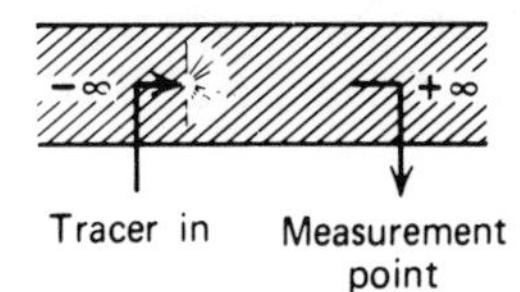

$$c^* = 1 + \frac{1}{2}\sum_{a_i>0}\frac{1}{q}\exp\left[(\tfrac{1}{2}-q)P_L z\right]\frac{J_0(a_i r)}{J_0^2(a_i)}$$

with $\qquad$ (3)

$$J_1(a_i) = 0, \qquad q - \sqrt{\frac{1}{4}+\frac{a_i^2}{P_L P_R}}$$

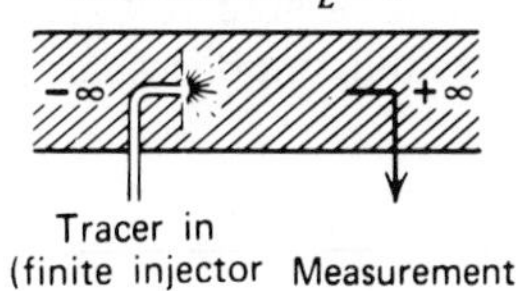

$$c^* = 1 + \frac{2}{e}\sum\frac{J_0(a_i r)}{J_0^2(a_i)}\frac{J_i(a_i e)}{a_i}\exp\left[-\frac{a_i^2 z}{P_R}\right]$$

with

radius of injector $= E$; $e = E/R$ $\qquad J_1(a_i) = 0 \qquad$ (4)

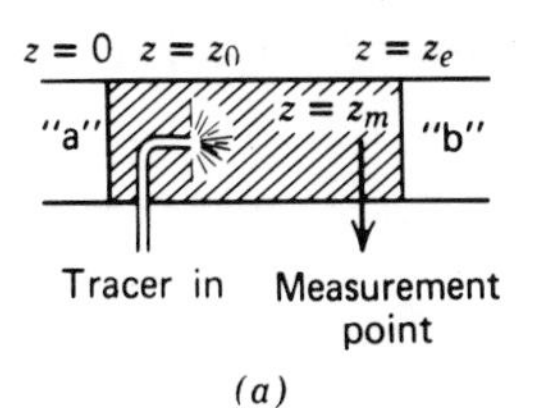

For (a):

$$c^* = 1 + \sum_{a_i>0}\frac{J_0(a_i r)}{J_0^2(a_i)}\frac{J_1(a_i e)}{q a_i e}N_i$$

$$N_i = \exp\left[(\tfrac{1}{2}-q)P_L(z_m-z_0)\right]\left[\frac{\begin{array}{l}(q-q_a)(q-q_b)\exp\left[-2P_L q(z_e-z_m+z_0)\right]\\ \quad +(q-q_a)(q+q_b)\exp\left[-2qP_L z_0\right]\\ \quad +(q+q_a)(q-q_b)\exp\left[-2qP_L(z_e-z_m)\right]\\ \quad +(q+q_a)(q+q_b)\end{array}}{(q+q_a)(q+q_b)-(q-q_a)(q-q_b)\exp\left[-2P_L q z_e\right]}\right] \quad (5)$$

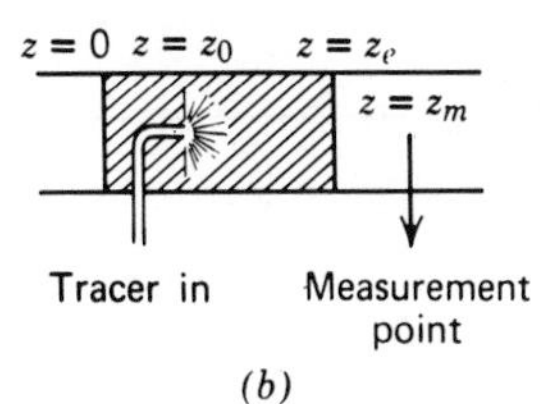

For N_i for (b), refer to reference 5

* References for Table 10.4-2:
1. W. L. Towle and T. K. Sherwood, *Ind. Eng. Chem.* **31**, 457, 1939.
2. R. A. Bernard and R. H. Wilhelm, *Chem. Eng. Progr.* **46**, 233, 1950.
3. A. Klinkenberg, H. J. Krajenbrink, and H. A. Lauwerier, *Ind. Eng. Chem.* **45**, 1202, 1953.
4. R. W. Fahien and J. M. Smith, *AIChE J.* **1**, 28, 1955.
5. K. B. Bischoff and O. Levenspiel, *Chem. Eng. Sci.* **17**, 245, 1962.

TABLE 10.4-3 LINEAR MULTIPLE GRADIENT (DISPERION) MODELS*

Model	Differential Equations	Boundary Conditions	Model Coefficients
1. 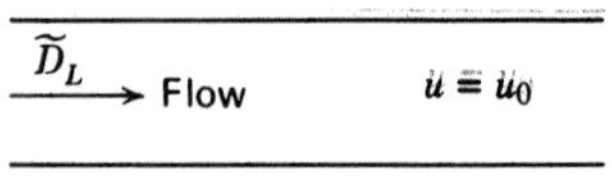 $\tilde{D}_L$ → Flow, $u = u_0$ Dispersion only	$\frac{\partial c}{\partial t} + \frac{\partial c}{\partial z} = \frac{1}{P}\frac{\partial^2 c}{\partial z^2}$	$c(z, 0) = 0$ $c(\infty, 0) =$ finite $c(z_1, t) = 1$	P Peclet Number
2. 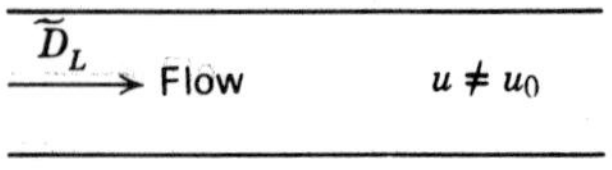 $\tilde{D}_L$ → Flow, $u \neq u_0$ Dispersion only	$\frac{\partial c}{\partial t} + \frac{1}{h}\frac{\partial c}{\partial z} = \frac{1}{P}\frac{\partial^2 c}{\partial z^2}$	Same as Model 1	P $h = u/u_0$
3. 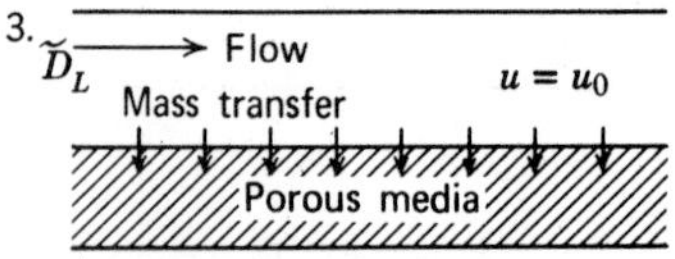 $\tilde{D}_L$ → Flow, $u = u_0$, Mass transfer, Porous media Dispersion plus accumulation by equilibrium adsorption	$\gamma\frac{\partial c}{\partial t} + \frac{\partial c}{\partial z} = \frac{1}{P}\frac{\partial^2 c}{\partial z^2}$ $\frac{1}{\epsilon}\frac{\partial q}{\partial t} = \frac{k_1}{\epsilon}\frac{\partial c}{\partial t}$	Same as Model 1	P $\gamma = [1 + (k_1/\epsilon)] > 1$ $k_1 =$ distribution coefficient for equilibrium adsorption
4. 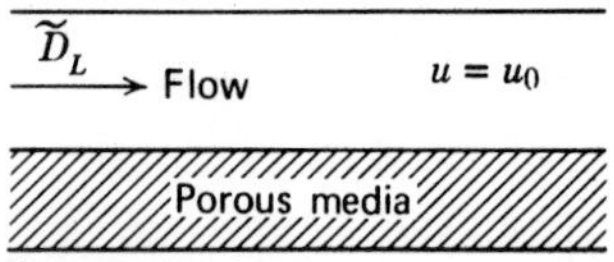 $\tilde{D}_L$ → Flow, $u = u_0$, Porous media Dispersion plus stagnant region	$f\frac{\partial c}{\partial t} + \frac{\partial c}{\partial z} = \frac{1}{P}\frac{\partial^2 c}{\partial z^2}$	Same as Model 1	P $f =$ fraction of void volume containing mobile fluid; $f < 1$
5. 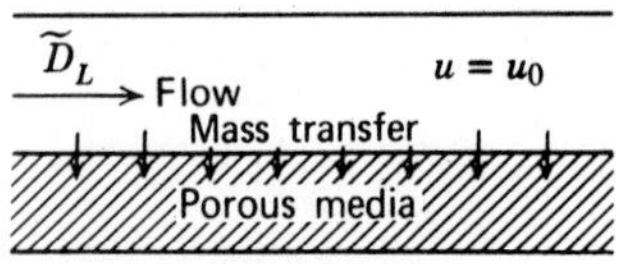 $\tilde{D}_L$ → Flow, $u = u_0$, Mass transfer, Porous media Dispersion plus accumulation by finite rate adsorption	$\frac{\partial c}{\partial t} + \frac{\partial c}{\partial z} = \frac{1}{P}\frac{\partial^2 c}{\partial z^2} - \frac{1}{\epsilon}\frac{\partial q}{\partial t}$ $\frac{\partial q}{\partial t} = k_2(c - c^*) - k_2\left(c - \frac{q}{k_1}\right)$	Same as Model 1 plus $q(z, 0) = 0$	P k_1 $k_2 =$ mass transfer coefficient
6. 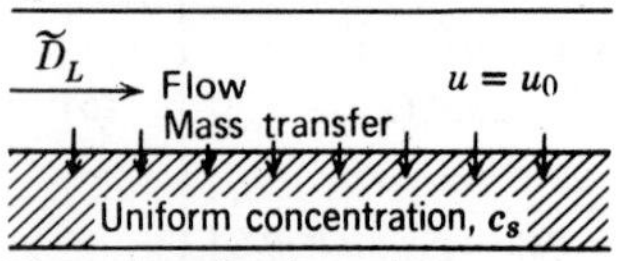 $\tilde{D}_L$ → Flow, $u = u_0$, Mass transfer, Uniform concentration, c_s Dispersion plus interphase mass transfer to a porous stagnant region	$f\frac{\partial c}{\partial t} + \frac{\partial c}{\partial z} = \frac{1}{P}\frac{\partial^2 c}{\partial z^2} - (1 - f)\frac{\partial c_s}{\partial t}$ $(1 - f)\frac{\partial c_s}{\partial t} = k_3(c - c_s)$	Same as Model 1 plus $c_s(z, 0) = 0$	P f $k_3 =$ mass transfer coefficient
7. Dispersion plus accumulation by finite rate adsorption plus mass transfer to a stagnant region	$f\frac{\partial c}{\partial t} + \frac{\partial c}{\partial z} = \frac{1}{P}\frac{\partial^2 c}{\partial z^2}$ $-(1 - f)\left(1 + \frac{k_1}{f\epsilon}\right)\frac{\partial c_s}{dt} - \frac{1}{\epsilon}\frac{\partial q_m}{\partial t} = 0$ $(1 - f)\frac{\partial c_s}{\partial t} = k_3(c - c_s)$ $\frac{\partial q_m}{\partial t} = k_2\left(c - \frac{q_m}{k_1}\right)$	Same as Model 1 plus $c_s(z, 0) = 0$ $q_m(z, 0) = 0$	P k_1 k_2 k_3 f

* For notation, refer to Table 10.4-4.

Deterministic Moments		Frequency Response Parameters for A_r and ψ	
Δm_1	$\Delta m_2^{\#}$	λ_1	λ_2
	$\frac{2}{P}$	$\frac{1}{4}$	$\frac{\omega}{P}$
h	$\frac{2h^3}{P}$	$\frac{1}{4}$	$\frac{\omega h^2}{P}$
$1 + \frac{k_1}{\epsilon} > 0$	$\frac{2\gamma^2}{P}$	$\frac{1}{4}$	$\frac{\omega\gamma}{P}$
$f < 0$	$\frac{2f^2}{P}$	$\frac{1}{4}$	$\frac{\omega f}{P}$
$1 + \frac{k_1}{\epsilon} > 0$	$\frac{2k_1^2}{\epsilon k_2} + \frac{2}{P}\left[1 + \frac{k_1}{\epsilon}\right]^2$	$\frac{1}{4} + \frac{\omega^2 k_2}{\epsilon P[\omega^2 + (k_2/k_1)^2]}$	$\frac{\omega}{P}\left[1 + \frac{k_2^2}{\epsilon k_1[\omega^2 + (k_2/k_1)^2]}\right]$
1	$\frac{2(1-f)^2}{k_3} + \frac{2}{P}$	$\frac{1}{4} + \frac{\omega^2 k_3}{P[\omega^2 + (k_3/(1-f))^2]}$	$\frac{\omega}{P}\left[f + \frac{k_3^2}{(1-f)[\omega^2 + (k_3/(1-f))^2]}\right]$
$1 + \frac{k_1}{f\epsilon} > 1$	$\frac{2(1-f)^2[1 + (k_1/f\epsilon)]}{k_3} + \frac{2k_1^2}{\epsilon k_2} + \frac{2}{P}\left(1 + \frac{k_1}{f\epsilon}\right)^2$	$\frac{1}{4} + \frac{\omega}{P}\left[\frac{k_3\omega(1-f)^2[1 + (k_1/f\epsilon)]}{k_3^2 + \omega^2(1-f)^2} + \frac{k_2\omega}{\epsilon[\omega^2 + (k_2/k_1)^2]}\right]$	$\frac{\omega}{P}\left[f + \frac{k_3^2(1-f)[1 + (k_1/f\epsilon)]}{k_3^2 + \omega^2(1-f)^2} + \frac{k_2^2}{\epsilon k_1[\omega^2 + (k_2/k_1)^2]}\right]$

TABLE 10.4-4 NOTATION FOR TABLE 10.4-3

A_r = amplitude ratio, the ratio of the amplitude at z_2 to that at z_1; $\ln(A_r) = (P/2)\{1 - [2\lambda_1 + \sqrt{(2\lambda_1)^2 + (2\lambda_2)^2}]^{1/2}\}$; λ_i is given in the table
c = dimensionless concentration
c_1 = concentration at z_1 (dimensionless)
c_s = concentration in stagnant region (hypothetical value)
$\tilde{D}$ = axial-dispersion coefficient
f = fraction of the void volume filled with mobile fluid
h = u/u_0
k_1 = adsorption equilibrium coefficient (Henry's law coefficient) between q and c
k_2 = dimensionless interphase mass transfer coefficient in model with adsorption
k_3 = dimensionless interphase mass transfer coefficient in model with capacitance effect
L = length of test section
P = Peclet number $u_0L/\tilde{D}$
q = quantity of tracer adsorbed per unit volume of porous media; $q = k_1c/\epsilon$ so that the accumulation rate is $(dq/dt) = (k_1c/\epsilon)(dc/dt)$
q_m = quantity of tracer adsorbed per unit volume of solid on surfaces that confines the mobile fluid
t = dimensionless time = $\theta u_0/L$
u = mean interstitial velocity
u_0 = open tube velocity divided by the porosity ϵ
z = dimensionless axial distance = $\acute{z}/L$
z_1 = first measuring point (dimensionless)
z_2 = second measuring point (dimensionless)
ϵ = porosity (void fraction)
θ = time
ω = dimensionless frequency = $\acute{\omega}L/u_0$
ψ = phase shift = $(P/2)[-2\lambda_1 + \sqrt{(2\lambda_1)^2 + (2\lambda_2)^2}]^{1/2}$; λ_i is given in the table
Superscript $'$ designates dimensional quantity

Even though the frequency response is used for coefficient estimation, a pulse input is preferred in experimentation because a Fourier analysis of a single pulse response can be run to get the desired frequency information. A single experiment with a pulse input thus can take the place of a number of experiments in which the input is varied sinusoidally at different frequencies.

Supplementary References

Astarita, G., *Mass Transfer with Chemical Reaction*, Elsevier Co., Amsterdam, 1967.

Bellman, R., Detchmendy, D., Kagiwada, H., and Kalaba, R., "On the Identification of Systems and Unscrambling of Data III: One-Dimensional Wave and Diffusion Processes," *J. Math. Anal. Applns.* **23**, 173, 1968.

Crank, J., *The Mathematics of Diffusion*, Oxford Univ. Press, London, 1955.

Jones, F. R., "Various Methods for Finding Unknown Coefficients in Parabolic Differential Equations," *Comm. Pure Appld. Math.* **16**, 33, 1963.

Kudryavtsev, Y. V., *Unsteady State Heat Transfer*, Iliffe Books Ltd., London, 1966.

Perdreauville, F. J. and Goodson, R. E., "Identification of Systems Described by Partial Differential Equations," *J. Basic Eng.* (Trans. ASME) **D88**, 463, 1966.

Tyrell, H. J., *Diffusion and Heat Flow in Liquids*, Butterworths, London, 1961.

Problems

10.1 A packed bed through which a fluid passes in steady flow is a common piece of chemical processing equipment and also has significance in oil reservoir engineering. If heat transfer from the fluid to the bed is of interest, the process can be described in a number of ways. Suppose that a multiple gradient balance is selected with one adjustable parameter in the following form:

$$\rho C_p v_z \frac{\partial T}{\partial z} = \tilde{k}\left[\frac{1}{r}\left(\frac{\partial T}{\partial r}\right) + \frac{\partial T^2}{\partial r^2}\right]$$

where

ρ = density
C_p = heat capacity
v_z = velocity
z = axial direction
T = temperature, the random variable
r = radial direction

Suggest a series of experiments (i.e., prescribe the initial and boundary conditions for the actual process represented by the model) that will be satisfactory to estimate the effective thermal conductivity $\tilde{k}$. What observations should be made? What will be the estimated error in T at $z = 0$ (bed entrance) and at $z = L$ (bed exit)?

10.2 A tubular reactor can be represented by the following model:

$$\frac{\partial C}{\partial t} + v_z \frac{\partial C}{\partial z} = -k_0 C \exp\left(-\frac{\Delta E}{RT}\right)$$

$$\rho C_p\left[\frac{\partial T}{\partial t} + v_z \frac{\partial T}{\partial z}\right] = \frac{2h}{R}(T_w - T) + \Delta H_{Rxn}\left[-k_0 C \exp\left(-\frac{\Delta E}{RT}\right)\right]$$

where

C = concentration, a random variable
ΔH_{Rxn} = heat of reaction, a constant (known)
ΔE = energy of activation, a constant
R = ideal gas constant (known)
k_0 = a constant
h = interphase heat transfer coefficient, a constant
T_w = wall temperature (known)

and the remaining notation is the same as in Problem 10.1.

By what type of experiments can estimates of k_0, h, and ΔE be obtained independently? What will be the respective boundary conditions? Note: T and C are quite sensitive to small changes in L/v_z and T_w in certain ranges of these variables for the model as formulated above.

10.3 For incompressible laminar flow, the Navier Stokes equations are

$$\rho \frac{D\mathbf{v}}{Dt} = -\nabla(p + \rho g z) + \mu \nabla^2 \mathbf{v}$$

where

ρ = density, a constant (known)
$\mathbf{v}$ = velocity, a vector
p = pressure
g = acceleration of gravity, a constant (known)
z = height above a reference level
μ = viscosity

$$\frac{D}{Dt} = \text{substantial derivative} = \frac{\partial}{\partial t} + \mathbf{v} \cdot \nabla$$

Assume that $\mathbf{v}$ is stochastic (it has three components) as is p and that the other variables are deterministic. How can μ be estimated most effectively from velocity measurements? Describe the experiment in detail, giving equations, boundary conditions, and the estimation criterion.

10.4 The dimensionless diffusion equation for absorption or desorption from a finite slab is

$$\frac{\partial c^*}{\partial t^*} = \frac{\partial^2 c^*}{\partial z^*}$$

where

$$c^* = \frac{\omega_A - \omega_{A\infty}}{\omega_{A0} - \omega_{A\infty}}$$

$$z^* = \frac{z}{L}$$

$$t^* = \frac{Dt}{L^2}$$

z = diffusion direction
L = slab width
ω_A = mass fraction of A

For the boundary conditions:

$$\begin{array}{lll} c^* = 1, & t^* < 0, & 0 \le z^* \le 1 \\ c^* = 0, & t^* \ge 0, & z^* = 0 \\ & & z^* = 1 \end{array}$$

how could D be estimated from measurements of ω_A and t?

Hint: Let the fraction of the original material in the slab at time t be

$$M = \frac{\int_0^1 c^* \, dz^*}{\int_0^1 c^*(0) \, dz^*}$$

10.5 A convenient method of determining diffusion coefficients of liquids is the capillary method. A "labelled" (usually with radioactive tracer) liquid is introduced into a uniform capillary tube, roughly 1 mm in diameter and 2 cm long. One end of the tube is sealed. The open end of the tube is placed in contact with a large quantity of the same liquid containing no radioactive tracer. After a selected time t (seconds), the tube is removed and the fraction f of the initial radioactive material left is determined. The diffusion coefficient D of the labelled liquid can then be calculated from the solution to the diffusion equation for a cylinder initially at a uniform concentration C_0:

$$f = \frac{C}{C_0} = \frac{8}{\pi^2}\left[\exp(-\psi) + \tfrac{1}{9}\exp(-9\psi) + \tfrac{1}{25}\exp(-25\psi) + \cdots\right]$$

where $\psi = \pi^2 Dt/4l^2$ and l = length of tube.

How can D be estimated from the ratio f which is a stochastic variable? Assume l and t are deterministic variables.

10.6 Two cylinders, a standard moist cylinder and a cylinder with unknown moisture, are being dried under the same conditions. Solution of the deterministic mathematical model for the temperature profile by Laplace transforms yields an equation which cannot be inverted except into imaginary (rather than real) temperatures. What might be done to enable the experimenter to obtain a solution in terms of real variables? The form of the solution in Laplace transform space is

$$\tilde{T}(x, s) = \left(a \cosh \sqrt{\frac{s}{a}}\, x\right)\left(b \sinh \sqrt{\frac{s}{a}}\, x\right)$$

where x is the distance from the end of the cylinder, s is the Laplace transform parameter, and T is the temperature.

10.7 The differential equation governing steady, laminar, isothermal, horizontal, fully developed flow of an incompressible Newtonian fluid in a rectangular conduit of constant cross-sectional area is

$$\frac{\partial^2 V}{\partial x^2} + \frac{\partial^2 V}{\partial y^2} = \frac{1}{\mu}\frac{\partial p}{\partial z} \tag{a}$$

where V is the velocity in the axial (z) direction, x (horizontal) and y (vertical) are orthogonal Cartesian coordinates also orthogonal to the z direction, $\partial p/\partial z$ is the pressure gradient in the axial direction, and μ is the viscosity, a constant. If Equation (a) is integrated for a conduit having zero surface velocity, a width $2a$ and a depth $2b$, V is as follows:

$$V = -\frac{16b^2}{\pi^3 \mu}\frac{\partial p}{\partial z} \sum_{n=1,3,\ldots}^{\infty} \left[\frac{(-1)^{\frac{n-1}{2}}}{n^3} \frac{\cosh \dfrac{n\pi x}{2b}}{\cosh \dfrac{n\pi a}{2b}} \cos \frac{n\pi y}{2b}\right] - \frac{1}{2\mu}\frac{\partial p}{\partial z}(b^2 - y^2)$$

The volumetric flow rate Q is as follows:

$$Q = -\frac{4ab^3}{3\mu}\frac{\partial p}{\partial z}\left\{1 - \frac{192b}{\pi^5 a}\sum_{n=1,3,\ldots}^{\infty}\left[\frac{1}{n^3}\tanh\frac{\pi n a}{2b}\right]\right\}$$

Can μ best be estimated from measurements of $\partial p/\partial z$ and V or from measurements of $\partial p/\partial z$ and Q? Keep in mind that V is a point velocity, a function of x and y, whereas Q is the rate of flow in the total volume of fluid in the duct, and that the precision in Q may not be the same as the precision in V. Assume x and y are deterministic variables.

10.8 A pulse input to a fluidized bed yielded the following output. The concentration is a coded variable of fraction of full scale of the chart, and the time is in seconds. Select one or more appropriate models and ascertain their parameters by use of the moments of the output curve.

t	Response × 10^2	t	Response × 10^2
1.90	0.000	55.11	0.486
3.80	0.000	57.01	0.420
5.70	0.356	58.91	0.363
7.60	1.245	60.81	0.314
9.50	2.194	62.71	0.271
11.40	2.759	64.61	0.241
13.30	3.229	66.51	0.215
15.20	3.431	68.41	0.192
17.10	3.454	70.31	0.171
19.00	3.346	72.21	0.154
20.90	3.197	74.11	0.142
22.80	2.996	76.01	0.131
24.71	2.761	77.92	0.120
26.61	2.527	79.82	0.111
28.50	2.303	81.72	0.094
30.40	2.089	83.62	0.081
32.30	1.876	85.52	0.068
34.20	1.667	87.42	0.0583
36.10	1.480	89.32	0.0496
38.10	1.315	91.22	0.0422
40.00	1.169	93.12	0.0359
41.81	1.040	95.92	0.0305
43.71	0.926	96.92	0.026
45.61	0.821	98.82	0.022
47.51	0.724	100.72	0.018
49.41	0.644	102.62	0.016
51.31	0.586	104.52	0.014
53.21	0.534		

10.9 Residence time curves such as in Figure P10.9 can be used to obtain the parameters in models of tubular chemical reactors. The figure represents octadecane tagged with C-14 for a pulse input. For two or three models of the reactor, estimate the coefficients in the models. Which model represents the reactor the best? Use either the method of moments and/or the solutions of the differential equations.

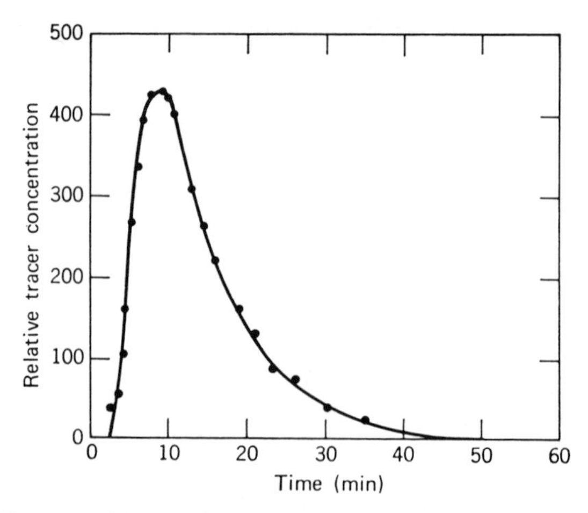

FIGURE P10.9 Typical residence time distribution curve.

10.10 Phillips† used a pulse injection of KCl solution into an orifice plate mixing column to help construct a model for the column. The orifice plate mixer was a column 6 in in diameter and either 6, 12, or 18 in long with 1 or 2 interior plates. Various perforations were used in the plates. Flow of gas and water was concurrent. The output concentration of KCl was measured by a conductivity probe; the millivolt output was a linear function of concentration.

TABLE P10.10

No. plates	1	Water rate	0.94 ft/sec
No. holes	1	Air rate	0.40 ft/sec
Column height	18 in	Volume fraction liquid	0.8

(c mv)	t (sec)	c (mv)	t (sec)	c (mv)	t (sec)
0	1.22	350	2.12	200	2.92
85	1.52	355	2.22	128	3.32
200	1.72	340	2.32	73	3.72
310	1.92	282	2.52	50	4.12
				0	6.12

Water and air rates are total volumetric flows into the column divided by the column cross-sectional area.

Based on Table P10.10, determine how well a:

(a) maximum gradient model

$$\frac{\partial c}{\partial t} + v_z\frac{\partial c}{\partial z} = \frac{I}{A}\,\delta(x)\,\delta(t)$$

(b) or a multiple gradient model

$$\frac{\partial c}{\partial t} + v_z\frac{\partial c}{\partial x} = \tilde{D}_L\frac{\partial^2 c}{\partial x^2} + \frac{I}{\pi R^2}\,\delta(x)\,\delta(t)$$

represents the data. Estimate the value of $\tilde{D}_L$ for the run. A probe correction, which should be included in the analysis, can be ignored, to simplify the calculations.

† J. B. Phillips, M.S. Thesis, Univ. of Texas, 1965.

The notations are:

c = concentration, g/liter
I = tracer injection, g
R = tube radius, cm

10.11 Schiesser and Lapidus† used both step and pulse response techniques on a 4 in. diameter column, 3 ft long, packed with spheres. Water was introduced through a flat distributor at a uniform flow rate, and a step or pulse input of tracer NaCl was introduced by suitable values. Details of the apparatus and procedure can be found in the article.

Data for $\frac{1}{4}$ in. porous and nonporous alumina spheres are listed in Table P10.11 for a water flow rate of 4.06 gal/min. The internal holdup in the porous spheres, as calculated from the difference in step reponses, was 0.215 ft³/ft³ bed, or 79 percent of the pore volume. Determine how well a: (1) macroscopic model, (2) maximum gradient model, and (3) multiple gradient model fit the process. See Problem 10.10 for these latter two models. Explain any discrepancies you note in the conclusions drawn from the respective step and pulse data or type of packing. Estimate the dispersion coefficient in the multiple gradient model.

† W. E. Schiesser and L. Lapidus, *AIChE J.* 7, 163 (1961).

TABLE P10.11*

Pulse Response cQ/m^0, sec^{-1}			Step Response c/c_0		
Time (sec)	Porous	Non-porous	Time (sec)	Porous	Non-porous
12.5	0.0015	0.0015	1.0	1.000	1.000
13.5	0.0050	0.0050	3.0	1.000	1.000
14.5	0.0080	0.0100	5.0	1.000	1.000
15.5	0.0160	0.0190	7.0	1.000	1.000
16.5	0.0220	0.0380	9.0	1.000	1.000
17.5	0.0350	0.0620	11.0	1.000	1.000
18.5	0.0460	0.0850	13.0	1.000	1.000
19.5	0.0590	0.1100	15.0	0.990	—
20.5	0.0650	0.1030	17.0	0.950	0.950
21.5	0.0670	0.1010	19.0	0.820	0.880
22.5	0.0660	0.0910	21.0	0.630	0.750
23.5	0.0630	0.0780	23.0	0.620	0.450
24.5	0.0580	0.0650	25.0	0.500	0.290
25.5	0.0510	0.0530	27.0	0.410	0.190
26.5	0.0470	0.0420	29.0	0.340	0.120
27.5	0.0400	0.0330	31.0	0.290	0.070
29.0	0.0310	0.0240	33.0	0.260	0.050
31.0	0.0220	0.0140	35.0	0.220	0.040
33.0	0.0170	0.0090	37.0	0.210	0.030
35.0	0.01250	0.0070	39.0	0.200	0.020
37.0	0.0100	0.0050	41.0	0.190	0.020
39.0	0.0075	0.0040	43.0	—	0.010
41.0	0.0060	0.0030	45.0	0.180	0.010
43.0	0.0050	0.0020			
45.0	0.0030	0.0020			

* Notation:
Q = volumetric flow rate, ft³ liquid/sec
c = concentration
$m^{\circ}Q$ = moles injected in a pulse input
t = time elapsed since injection of input

CHAPTER 11

Parameter Estimation in Transfer Functions

All the process models described in Chapters 9 and 10 incorporated differential equations in the time domain, either unsteady-state models or steady-state models involving spacial derivatives. For the special case of models represented by nth-order *linear* (in the dependent variable) *ordinary* differential equations with *constant* coefficients and zero initial conditions:

$$\frac{d^n y}{dt^n} + a_{n-1}\frac{d^{n-1}y}{dt^{n-1}} + \cdots + a_1\frac{dy}{dt} + a_0 y = x(t)$$
$$y(0) = \frac{dy(0)}{dt} = \frac{d^2y(0)}{dt^2} = \cdots = \frac{d^{(n-1)}y(0)}{dt^{(n-1)}} = 0 \qquad (11.0\text{-}1)$$

an alternate way to represent the relation between the model output and input is by means of the transfer function defined below by Equation 11.1-1.

We shall describe how the coefficients in Equation 11.0-1, which also appear in the transfer function, can be estimated in both the time domain and in the Laplace transform domain when the transfer function itself is specified as the model.

11.1 THE TRANSFER FUNCTION AS A PROCESS MODEL

The *transfer function*, which will be defined as the ratio of the Laplace transform of the output divided by the Laplace transform of the input of Model 11.0-1, can be found by taking Laplace transforms (refer to Appendix B) of each side of Equation 11.0-1:

$$s^n\check{y}(s) + a_{n-1}s^{n-1}\check{y}(s) + \cdots + a_1 s\check{y}(s) + a_0\check{y}(s) = \check{x}(s) \qquad (11.1\text{-}1)$$

where s (in this chapter) is the complex parameter and the overlay ($\vee$) on the dependent variable signifies it is in the transform domain rather than the time domain. If we rearrange Equation 11.1-1 into the following form:

$$\frac{\check{y}(s)}{\check{x}(s)} = \frac{1}{s^n + a_{n-1}s^{n-1} + \cdots + a_1 s + a_0} \qquad (11.1\text{-}2)$$

we obtain the transfer function corresponding to Equation 11.0-1. If the input function $x(t)$ is generalized to include derivatives of $x(t)$ of order m, then a polynomial in s will appear in both the numerator and denominator of the transfer function:

$$\frac{\check{y}(s)}{\check{x}(s)} = \frac{b_m s^m + b_{m-1}s^{n-1} + \cdots + b_1 s + b_0}{s^n + a_{n-1}s^{m-1} + \cdots + a_1 s + a_0} \qquad (11.1\text{-}3)$$

If the input to the subsystem is an impulse (delta) function, $x(t) = \delta(t)$, the Laplace transform of the impulse input is

$$\check{x}(s) = \mathscr{L}[x(t)] = \int_0^\infty e^{-st}\,\delta(t)\,dt = 1$$

For the impulse input, Equation 11.1-3 becomes

$$\frac{\check{y}(s)}{1} = \check{g}(s) = \frac{b_m s^m + b_{m-1}s^{n-1} + \cdots + b_1 s_1 + b_0}{s^n + a_{n-1}s^{n-1} + \cdots + a_1 s + a_0}$$

where $\check{g}(s) = \mathscr{L}[g(t)]$ is the Laplace transform of the impulse response, usually denoted by $g(t)$. Therefore, the transfer function proves to be just the Laplace transform of the impulse response function, and Equation 11.1-3 can be written in general as

$$\frac{\check{y}(s)}{\check{x}(s)} = \check{g}(s)$$

or

$$\check{y}(s) = \check{g}(s)\check{x}(s) \qquad (11.1\text{-}4)$$

Equation 11.1-4 states that the output can be found from the product of the input times the impulse response, all evaluated in Laplace transform space. When the transfer function is known, the output of a linear (in the dependent variable) model *for any type of input* can be found in Laplace transform space from Equation 11.1-4.

If we now take the inverse Laplace transform of both sides of Equation 11.1-4 to obtain the response in the time domain for any input, we obtain

$$y(t) = \mathscr{L}^{-1}[\check{y}(s)] = \mathscr{L}^{-1}[\check{g}(s)\check{x}(s)]$$
$$= \int_0^t g(t-\alpha)x(\alpha)\,d\alpha \qquad (11.1\text{-}5)$$

Integrals of the type given in Equation 11.1-5 arise in many different applications; they are known in transform theory as *convolution integrals* and in classical mathematics as *Duhamel's integrals.* The impulse response function $g(t)$ is also known as a *weighting function* from its role in Equation 11.1-5, because $g(t - \alpha)$ acts as a weighting function on the input $x(t)$. The response $y(t)$ is equivalent to a weighting of the input values at various times from zero to t. For a model with constant coefficients, the shape of the response to an input applied at any instant depends only on the shape of the input and not on the time of application.

Since Model 11.0-1 in the time domain and the transfer function 11.1-3 have been shown to be directly related, one might inquire as to why transfer functions are used as models at all. One use of transfer functions is in control systems analysis where rather complicated sets of subsystems must be analyzed and combined. Because Laplace transformation reduces differential Equation 11.0-1 to an algebraic equation, historically the study of stability and sensitivity was easier to execute analytically or graphically in Laplace transform space than in the time domain. Then, too, Equation 11.1-4 makes the relation of one subsystem to another quite easy in Laplace space. For example, consider the packed column in Figure 11.1-1. The actual measurements in the flowing stream are made by thermal conductivity cells placed in the inlet and outlet streams. However, to measure dispersion in the packed section, corrections are necessary to exclude instruments and column effects. Figure 11.1-1 shows the information flow of the experimental set-up in the form of a block diagram. The model transfer function for the total process, including the packed section and instruments, is

$$\check{g}_t(s) = \frac{\check{y}_c(s)}{\check{x}_c(s)} \tag{11.1-6}$$

whereas the desired transfer function for the packed section, excluding end effects, is

$$\check{g}_p(s) = \frac{\check{y}_p(s)}{\check{x}_p(s)} \tag{11.1-7}$$

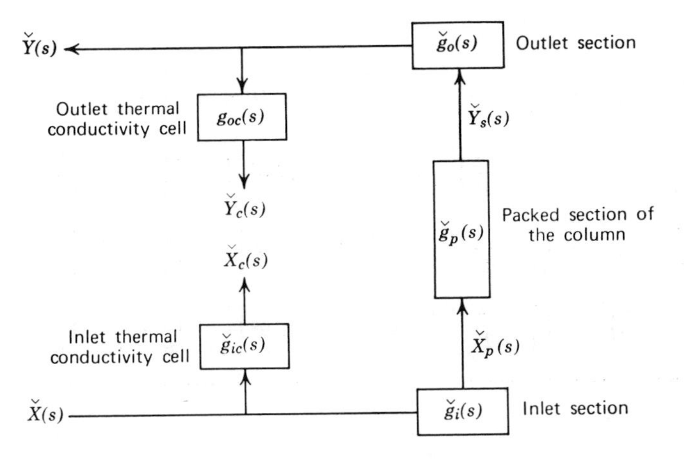

FIGURE 11.1-1 Block diagram of a real process.

The subsystem transfer functions are related (as can be verified by tracing the information flow) as follows:

$$\check{x}_c(s) = \check{g}_{ic}(s)\check{x}(s)$$
$$\check{x}_p(s) = \check{g}_i(s)\check{x}(s)$$
$$\check{y}_c(s) = \check{g}_{oc}(s)\check{y}(s)$$
$$\check{y}(s) = \check{g}_o(s)\check{y}_p(s)$$

so that with the packed section present,

$$\check{g}_t(s) = \check{g}_p(s)\left[\frac{\check{g}_o(s)\check{g}_i(s)\check{g}_{oc}(s)}{\check{g}_{ic}(s)}\right]$$

Without the packed section present (designated by the superscript #), $\check{g}_p(s) = 1$ and

$$\check{g}_t^{\#}(s) = \frac{\check{g}_o(s)\check{g}_i(s)\check{g}_{oc}(s)}{\check{g}_{ic}(s)}$$

Consequently,

$$\check{g}_p = \frac{\check{g}_t(s)}{\check{g}_t^{\#}(s)} \tag{11.1-8}$$

Equation 11.1-8 can be interpreted as relating the transfer function for the nonobservable packed section of the column to the observable transfer function for the entire column with and (in the denominator) without the packed section being present. By experimenting with and without the packed section, the instrumental end effects can be eliminated.

If the frequency response is desired, it can be obtained from the impulse response and/or the step response by the methods described in Hougen,† Nyquist *et al.*,‡ and Schechter and Wissler.§ The frequency response can more easily be obtained directly from the transfer function by replacing the parameter s by $i\omega$ and separating the real and imaginary parts of $\check{g}(i\omega)$. Figure 11.1-2 shows the relation between the time domain, the Laplace transform domain, and the frequency domain. It is because the inversion process indicated in the figure is so difficult that techniques of process analysis have arisen that make direct use of the parameters in the transfer function, and hence make estimation of the parameters in the transfer function itself of significance.

The general set of first-order linear equations given by Equation 9.1-5 with constant $\boldsymbol{\alpha}$ and zero initial conditions:

$$\frac{d\mathbf{y}}{dt} + \boldsymbol{\alpha}\mathbf{y} = \mathbf{x}(t) \qquad \mathbf{y}(0) = \mathbf{0} \tag{11.1-9}$$

† J. O. Hougen, "Experiences and Experiments with Process Dynamics," *Chem. Eng. Progr.*, Monograph Ser. No. 4, **60**, 1964.
‡ J. K. Nyquist *et al.*, *Chem. Eng. Progr.*, Symp. Ser. No. 46, 98, 1963.
§ R. S. Schechter and E. H. Wissler, *Ind. Eng. Chem.* **51**, 945, 1959.

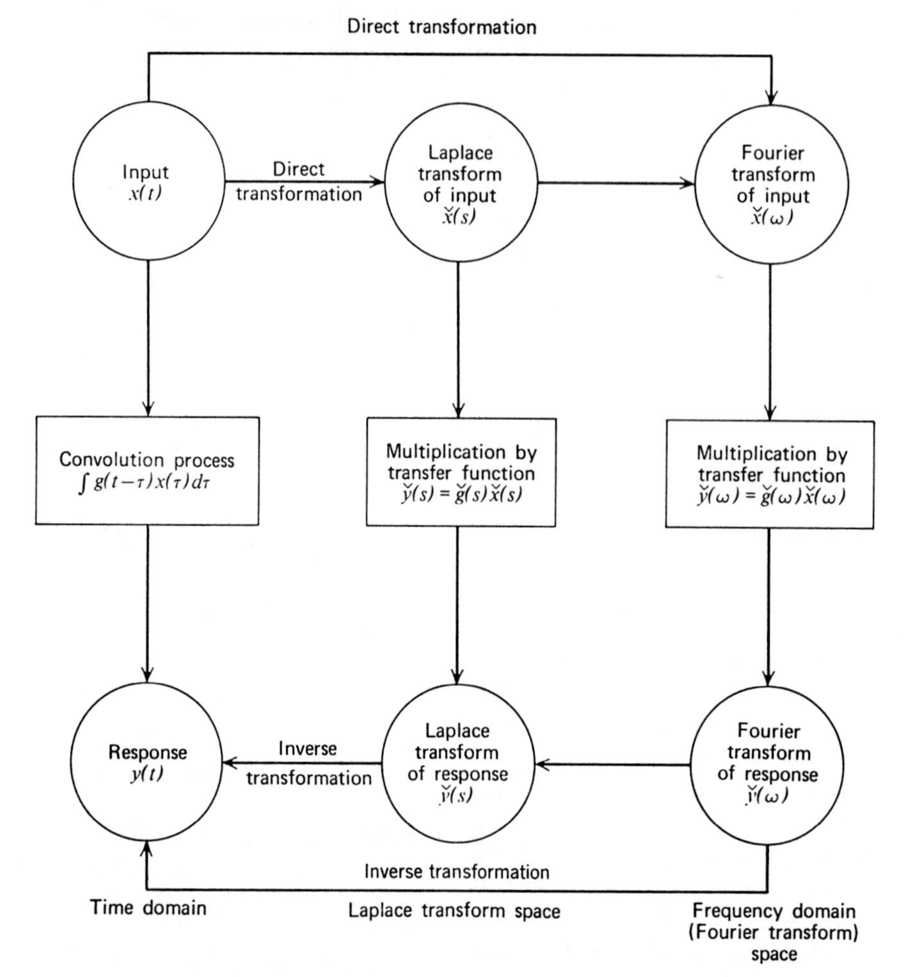

FIGURE 11.1-2 Representation of relationships between input and output of a linear subsystem.

TABLE 11.1-1 TRANSFER FUNCTION FOR A DISTRIBUTION MODEL OF DISPERSION IN A PIPE WITH TWO MEASURING POINTS[a]

Model		Transfer Function
	Test section:	
$\dfrac{\partial c^*}{\partial t^*} + \dfrac{\partial c^*}{\partial z^*} = \dfrac{1}{P}\dfrac{\partial^2 c^*}{\partial z^{*2}}$	$0 \le z^* \le z_e^*$	$\breve{g}(s) = \dfrac{A_1 + A_2}{A_3 + A_4}$
		where
	Exit section:	
$\dfrac{\partial c_b^*}{\partial t^*} + \dfrac{\partial c_b^*}{\partial z^*} = \dfrac{1}{P_b}\dfrac{\partial^2 c_b^*}{\partial z^{*2}}$	$z^* \ge z_e^*$	$A_1 = \left[\left(\frac{1}{4} + \frac{s}{P}\right)^{1/2} - \left(\frac{1}{4} + \frac{s}{P_b}\right)^{1/2}\right] \exp\left\{(z_2^* - z_e^*)P\left[\frac{1}{2} + \left(\frac{1}{4} + \frac{s}{P}\right)^{1/2}\right]\right\}$
$c^*(z^*, 0) = c_b^*(z^*, 0) = 0$		$A_2 = \left[\left(\frac{1}{4} + \frac{s}{P}\right)^{1/2} + \left(\frac{1}{4} + \frac{s}{P_b}\right)^{1/2}\right] \exp\left\{(z_2^* - z_e^*)P\left[\frac{1}{2} - \left(\frac{1}{4} + \frac{s}{P}\right)^{1/2}\right]\right\}$
$c^*(z_e^{*-}) = c_b^*(z_e^{*+})$		
$c^*(z_0, t) = c_0^*$		$A_3 = \left[\left(\frac{1}{4} + \frac{s}{P}\right)^{1/2} - \left(\frac{1}{4} + \frac{s}{P_b}\right)^{1/2}\right] \exp\left\{(z_1^* - z_e^*)P\left[\frac{1}{2} + \left(\frac{1}{4} + \frac{s}{P}\right)^{1/2}\right]\right\}$
$c_b^*(\infty, t)$ is finite		
$c^*(z_e^{*-}) - \dfrac{1}{P}\dfrac{\partial c^*(z_e^{*-})}{\partial z^*} = c_b^*(z_e^{*+}) - \dfrac{1}{P_b}\dfrac{\partial c_b^*(z_e^{*+})}{\partial z^*}$		$A_4 = \left[\left(\frac{1}{4} + \frac{s}{P}\right)^{1/2} + \left(\frac{1}{4} + \frac{s}{P_b}\right)^{1/2}\right] \exp\left\{(z_1^* - z_e^*)P\left[\frac{1}{2} - \left(\frac{1}{4} + \frac{s}{P}\right)^{1/2}\right]\right\}$

[a] P = Peclet number
$*$ = dimensionless variable
z_1^* = location of first measuring point in test section
z_2^* = location of second measuring point in test section
z_e^* = length of test section

can be shown by taking Laplace transforms of both sides:

$$\check{\mathbf{y}}(s) = (s\mathbf{I} + \boldsymbol{\alpha})^{-1}\check{\mathbf{x}}(s)$$

to have the *transfer matrix*

$$\check{\mathbf{g}}(s) \equiv (s\mathbf{I} + \boldsymbol{\alpha})^{-1}$$

with elements $\check{g}_{ij}$. Thus, the relationship

$$\check{\mathbf{y}}(s) = \check{\mathbf{g}}(s)\check{\mathbf{x}}(s) \qquad (11.1\text{-}10)$$

still applies. If Equation 11.1-10 is written in terms of its elements:

$$\check{y}_i(s) = \sum_{j=1}^{n} \check{g}_{ij}(s)\check{x}_j(s) \qquad i = 1, 2, \ldots, n$$

we see that the effect of $\check{x}_k$ on $\check{y}_i$ is given by $\check{g}_{ik}(s)$. Each of the elements $\check{g}_{ij}(s)$ is called the transfer function of $\check{y}_i$ with respect to $\check{x}_j$.

A transfer function can be obtained for models represented by linear partial differential equations, but each model provides a different form for the transfer function; examine Table 11.1-1 for a typical example. A point of interest is that the transfer functions for distributed models include an exponential term containing s as an exponent. Reference to a table of Laplace transforms brings out the fact that in the related time domain the output is delayed by some factor from the input. This delay is characteristic of models represented by partial differential equations which contain the convective term $v(\partial c/\partial z)$ or its equivalent and physically represents the transit time for the input to move through the subsystem. Models with only a dispersion term, such as is found in the diffusion equation $(\partial c/\partial t) = \mathscr{D}(\partial c^2/\partial z^2)$, have instantaneous responses, although the initial magnitude of the response is nil.

Now that we have seen how transfer functions originate and what they look like, we shall turn to consideration of methods of estimating the parameters in a transfer function of known form.

11.2 LEAST SQUARES ESTIMATION OF PARAMETERS

In this section we shall describe how the parameters in a transfer function can be estimated both in the time domain and in the Laplace transform domain by least squares. We shall assume that the form of the transfer function is given and that we want to estimate the coefficients therein. Section 11.3 will treat the situation in which the form of the transfer function is unknown, and both the form and coefficients must be determined.

We shall assume that the coefficients in $\check{g}(s)$ are constants or at least vary insignificantly during the time required to carry out the estimation. The class of functions $g(t)$ in the time domain corresponding to $\check{g}(s)$ must be limited to those that are zero for all times less than zero and that tend to zero as $t \to \infty$, i.e., stable processes. This latter requirement is not particularly restrictive since, for example, if $g(t) \to c_0$ as $t \to \infty$, by redefinition of $g(t)$, $[g(t) - c_0] \to 0$ as $t \to \infty$. In a like manner, a function which tends to a periodic function as $t \to \infty$ can be made to approach zero by subtracting the periodic term, and so forth. Uniqueness of $\mathscr{L}^{-1}[\check{g}(s)]$ also is assumed.

Because experimental data are collected in the time domain, we turn to time domain estimation first.

11.2-1 Estimation in the Time Domain by Inversion of the Transfer Function

The criteria for "ordinary" least squares estimation for continuous data:

$$\phi = \int_0^{t_f} [G(t) - g(t, \boldsymbol{\beta})]^2\, dt \qquad (11.2\text{-}1)$$

or for discrete data:

$$\phi = \sum_{i=1}^{n} [G(t_i) - g(t_i, \boldsymbol{\beta})]^2 \qquad (11.2\text{-}2)$$

where $G(t)$ is the observed (empirical) impulse response, a random variable, $g(t, \boldsymbol{\beta})$ is the model impulse response, the inverse Laplace transform of the transfer function, and $\boldsymbol{\beta}$ is the vector of parameters to be estimated, have been discussed in Section 9.2. The sufficient conditions for ϕ to be minimized are that:

1. $$\frac{\partial \phi}{\partial \beta_j} = 0, \qquad j = 1, 2, \ldots, m$$

2. The Hessian matrix of ϕ:

$$\begin{bmatrix} \dfrac{\partial^2\phi}{\partial\beta_1\,\partial\beta_1} & \dfrac{\partial^2\phi}{\partial\beta_1\,\partial\beta_2} & \cdots & \\ \dfrac{\partial^2\phi}{\partial\beta_2\,\partial\beta_1} & \dfrac{\partial^2\phi}{\partial\beta_2\,\partial\beta_2} & \cdots & \\ & & \ddots & \\ & & & \dfrac{\partial^2\phi}{\partial\beta_m\,\partial\beta_m} \end{bmatrix}$$

is positive-definite (or some equivalent condition to ensure a minimum).

Let us expand the model

$$\check{g}(s, \boldsymbol{\beta}) = \frac{\check{n}(s)}{\check{d}(s)}$$

into partial fractions† as described in most books on

† If

$$\check{g}(s) = \frac{\check{n}(s)}{\check{d}(s)} e^{-f(s)}$$

as in some distributed models, the parameters in the exponential term are usually evaluated separately from the elements of $\boldsymbol{\beta}$ from data based on the time it takes for an impulse input to give the first noticeable response ("breakthrough time").

process control); other types of expansions can also be used

$$\check{g}(s, \boldsymbol{\beta}) = \frac{a_1}{s + s_1} + \frac{a_2}{s + s_2} + \cdots \qquad (11.2\text{-}3)$$

where $-s_1, -s_2, \ldots$ are the roots of the denominator $\check{d}(s) = (s + s_1)(s + s_2)\cdots$ i.e., are the *poles* of $\check{g}(s, \boldsymbol{\beta})$, and may be both real and complex, and where $a_1, a_2, \ldots$ contain the real and imaginary parts of several of the poles. For example, if

$$\check{g}(s, \boldsymbol{\beta}) = \frac{1}{(s + \alpha_1)(s + \alpha_2 + i\beta)(s + \alpha_2 - i\beta)}$$

in which $\boldsymbol{\beta}$ has the elements α_1, α_2, and β, the partial fraction expansion can be shown to be

$$\check{g}(s, \boldsymbol{\beta}) = \frac{a_1}{s + \alpha_1} + \frac{a_2}{s + \alpha_2 + i\beta} + \frac{a_3}{s + \alpha_2 - i\beta}$$

where

$$a_1 = \frac{1}{(\alpha_2 - \alpha_1)^2 + \beta^2} \qquad s_1 = \alpha_1$$

$$a_2 = \frac{1}{2\beta[i(\alpha_2 - \alpha_1) - \beta]} \qquad s_2 = \alpha_2 + i\beta$$

$$a_3 = \frac{1}{-2\beta[i(\alpha_2 - \alpha_1) + \beta]} \qquad s_3 = \alpha_2 - i\beta$$

As a matter of interest, in the time domain the model is

$$g(t, \boldsymbol{\beta}) = \mathscr{L}^{-1}[\check{g}(s, \boldsymbol{\beta})] = \frac{e^{-\alpha_1 t}}{(\alpha_2 - \alpha_1)^2 + \beta^2}$$

$$- \frac{e^{-\alpha_2 t}}{\beta[(\alpha_2 - \alpha_1)^2 + \beta^2]} \sin\left(\beta t + \tan^{-1}\frac{\beta}{\alpha_2 - \alpha_1}\right)$$

It can be shown that the condition $\partial\phi/\partial\beta_j = 0$ can be replaced by

1′. $$\frac{\partial \phi}{\partial a_j} = 0 \qquad \frac{\partial \phi}{\partial s_j} = 0$$

regardless of whether a_j and s_j are real or of a complex conjugate pair.

Then (for continuous data):

$$\frac{\partial \phi}{\partial a_j} = 0 = (-2)\int_0^{t_f}\left\{G(t) - \mathscr{L}^{-1}\left[\sum_{j=1}^{m}\frac{a_j}{s + s_j}\right]\right\} \cdot \left\{\mathscr{L}^{-1}\left[\frac{1}{s + s_j}\right]\right\}$$

$$\frac{\partial \phi}{\partial s_j} = 0 = (2)\int_0^{t_f}\left\{G(t) - \mathscr{L}^{-1}\left[\sum_{j=1}^{m}\frac{a_j}{s + s_j}\right]\right\} \cdot \left\{\mathscr{L}^{-1}\left[\frac{a_j}{(s + s_j)^2}\right]\right\}$$

and the estimation equations are

$$\int_0^{t_f}\left[G(t) - \sum_{j=1}^{m}\hat{a}_j e^{-\hat{s}_j t}\right] e^{-\hat{s}_j t}\, dt = 0 \qquad j = 1, \ldots, m \qquad (11.2\text{-}4)$$

$$\int_0^{t_f}\left[G(t) - \sum_{j=1}^{m}\hat{a}_j e^{-\hat{s}_j t}\right]\hat{a}_j t\, e^{-\hat{s}_j t}\, dt = 0 \qquad j = m + 1, \ldots, 2m$$

Similar equations can be obtained for discrete data with the integral replaced by a summation.

Unfortunately, quite a few vectors $\boldsymbol{\beta}$ satisfy the highly nonlinear Equations 11.2-4; in fact, Deex† showed that the upper bound on the number of solutions is $(4d - 3)$ where d is the order of the denominator of the transfer function used as the model. Thus, the use of Equations 11.2-4 for estimation can lead to quite biased estimates, because of the existence of multiple local extrema for ϕ. Iterative methods of minimizing ϕ in Equations 11.2-1 or 11.2-2 in the time domain, as described in Chapter 6, provide a more favorable route to parameter estimation by least squares.

A more satisfactory procedure for estimating the parameters in Equation 11.1-3 is as follows. Suppose we divide Equation 11.0-1 by a_0 and introduce an appropriate $x(t)$ so that the right-hand side of Equation 11.0-1 has unity in front of the zero-order term for x. Then the transfer function can be written in the revised form:

$$\check{g}(s) = \frac{1 + b_1 s + b_2 s^2 + \cdots + b_m s^m}{1 + a_1 s + a_2 s^2 + \cdots + a_n s^n} \qquad (11.2\text{-}5)$$

Next, define the following transfer functions:

$$\check{g}_1(s) = \frac{1}{s}[1 - \check{g}(s)]$$

$$\check{g}_2(s) = \frac{1}{s}[y_1^* - \check{g}_1(s)] \qquad (11.2\text{-}6)$$

$$\vdots$$

$$\check{g}_{m+n}(s) = \frac{1}{s}[y_{m+n-1}^* - \check{g}_{m+n-1}(s)]$$

where y_i^* is the limiting value of $y_i(t)$ defined as follows as $t \to \infty$:

$$y_1(t) = \int_0^t [1 - y(t)]\, dt$$

$$y_2(t) = \int_0^t [y_1^* - y_1(t)]\, dt \qquad (11.2\text{-}7)$$

$$\vdots$$

$$y_{m+n}(t) = \int_0^t [y_{m+n-1}^* - y_{m+n-1}(t)]\, dt$$

† A. J. Deex, Ph.D. Dissertation, Univ. of Texas, Austin, 1965.

In other words,

$$y_1^* = \int_0^\infty [1 - y(t)]\, dt$$

$$y_2^* = \int_0^\infty [y_1^* - y_1(t)]\, dt$$

$$y_{m+n}^* = \int_0^\infty [y_{m+n-1}^* - y_{m+n-1}(t)]\, dt$$

The values of y_1^* can be related to the parameters in Equation 11.2-5 as follows. Because

$$\check{y}_1(s) = \check{g}_1(s)\check{x}_1(s)$$

$$= \frac{1}{s}[1 - \check{g}(s)]\check{x}_1(s)$$

if we assume $\check{x}_1(s) = 1/s$ (i.e., $x_1(t)$ is unity for $t \geq 0$), then by making use of the final value theorem for Laplace transforms given in Appendix B, we can state that

$$\lim_{s\to 0} s\check{y}_1(s) = \lim_{s\to 0}[1 - \check{g}(s)]\frac{1}{s} = \lim_{t\to\infty} y_1(t) = y_i^*$$

With a little algebraic manipulation, it is easy to show that

$$\frac{[1 - \check{g}(s)]}{s} = \frac{(a_1 - b_1) + (a_2 - b_2)s + \cdots}{1 + a_1 s + a_2 s^2 + \cdots}$$

so that as $s \to 0$, $s\check{y}_1(s) \to (a_1 - b_1)$. Consequently,

$$y_1^* = a_1 - b_1$$

Similarly, with $x_2(t)$ unity,

$$\check{y}_2(s) = \frac{1}{s}[y_1^* - \check{g}_1(s)]\check{x}_2(s)$$

and, after some more algebraic manipulation,

$$s\check{y}_2(s) = \frac{[y_1^* a_1 - (a_2 - b_2)] + [y_1^* a_2 - (a_3 - b_3)]s + \cdots}{1 + a_1 s + a_2 s^2 + \cdots}$$

Then

$$\lim_{s\to 0} s\check{y}_2(s) = \lim_{t\to\infty} y_2(t) = y_2^* = y_1^* a_1 - (a_2 - b_2)$$

Successive continuation of this type of analysis leads to the following matrix equation:

$$\begin{bmatrix} 1 & 0 & 0 & 0 & 0 & \cdots \\ y_1^* & 1 & 0 & 0 & 0 & \cdots \\ y_2^* & y_1^* & 1 & 0 & 0 & \cdots \\ y_3^* & y_2^* & y_1^* & 1 & 0 & \cdots \\ y_4^* & y_3^* & y_2^* & y_1^* & 1 & \cdots \\ \vdots & \vdots & \vdots & \vdots & \vdots & \end{bmatrix} \begin{bmatrix} a_1 \\ -a_2 \\ a_3 \\ -a_4 \\ a_5 \\ \vdots \end{bmatrix} = \begin{bmatrix} y_1^* + b_1 \\ y_2^* - b_2 \\ y_3^* + b_3 \\ y_4^* - b_4 \\ y_5^* + b_5 \\ \vdots \end{bmatrix} \tag{11.2-8}$$

Equation 11.2-8 is a set of linear equations in **a** and **b**, but the elements of the left-hand matrix (other than 0 and 1) are random variables if the y^*'s are computed from experimental data. In that case, the estimation procedure that should be applied is not ordinary least squares but one of the methods discussed in Sections 5.4 and 5.6.

If all the b's are zero, as would be the case in an experiment in which no derivatives appear on the right-hand side of Equation 11.0-1, a set of nonlinear equations in the a's and y^*'s results if each y^* in sequence is replaced by its appropriate a:

$$a_1 = Y_1^*$$

$$a_1^2 - a_2 = Y_2^*$$

$$a_1^3 - 2a_1a_2 + a_3 = Y_3^*$$

$$\vdots$$

By replication of the experiment, the ensemble mean and variance of a_1 can be estimated by the sample mean and variance of Y_1^*. Then, given the (normal) distribution of a_1 and the sample values of Y_2^*, the mean and expected value of a_2 can be approximated, and so on. It is assumed that the numerical error introduced in the integrations in Equations 11.2-7 is unbiased and negligible.

Simulation studies have indicated that if each y_i^* in Equation 11.2-8 is replaced by its sample average $\bar{Y}_i^*$ obtained from replicate experiments for the case in which all the b's are zero, then the solution of Equation 11.2-8 leads to only slightly biased estimates of the parameters in the transfer function and the variances of the coefficients can be approximated from Equation 11.2-8 if solved for $a_1, a_2, \ldots$ as follows:

$$\hat{a}_1 = \bar{Y}_1^*$$

$$\hat{a}_2 = (\bar{Y}_1^*)^2 - \bar{Y}_2^*$$

$$\vdots$$

etc.

so that

$$\text{Var}\{\hat{a}_1\} \approx \text{Var}\{\bar{Y}_1^*\}$$

$$\text{Var}\{\hat{a}_2\} \approx \text{Var}\{\bar{Y}_1^*)^2\} + \text{Var}\{\bar{Y}_2^*\}$$

$$\vdots$$

etc.

Example 11-2.1 Estimation of Parameters in a Transfer Function

A simulated experiment is described in this example to illustrate the use of Equation 11.2-8 in parameter estimation. A transfer function

$$\check{g}(s) = \frac{1}{1 + a_1 s + a_2 s^2} \tag{a}$$

was assigned the values of $a_1 = 0.200$ and $a_2 = 0.050$. The response to a unit step input in the Laplace transform domain is

$$\check{y}(s) = \check{g}(s)\check{x}(s) = \frac{\check{g}(s)}{s}$$

In the time domain the corresponding response with $\check{g}(s)$ given by Equation (a) is

$$y(t) = 1 - \sqrt{1.25}\, e^{-2t} \sin(4t + 1.107) \qquad \text{(b)}$$

A normal random error was added to $y(t)$:

$$Y(t) = y(t) + \epsilon \qquad \text{(c)}$$

with $\mathscr{E}\{\epsilon\} = 0$ and $\mathrm{Var}\{\epsilon\} = 5 \times 10^{-4}$. The generation of both ϵ and $y(t)$ was carried out by the digital analog simulator MIMIC. Figure E11.2-1a illustrates a typical simulated response $Y(t)$.

Equations 11.2-7 were evaluated (also by MIMIC) with $y_i(t)$ being replaced by $Y_i(t)$, the random variable. Figure

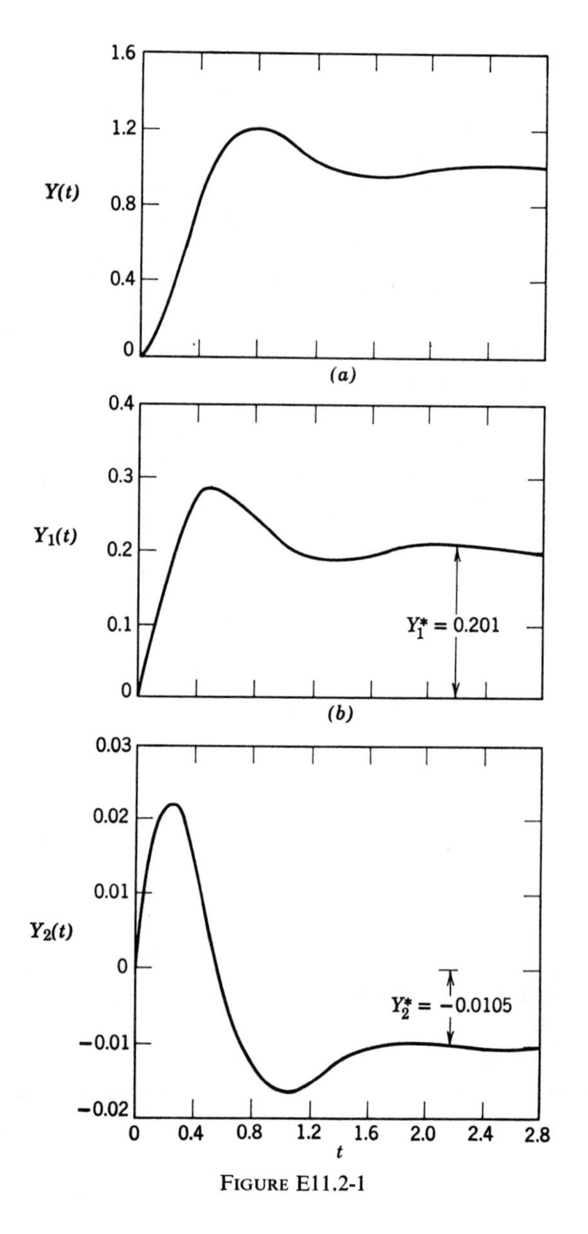

FIGURE E11.2-1

E11.2-1 shows the asymptotic values of $Y_i(t)$ as t becomes large. The numerical integration error was less than 10^{-6} of Y. Four replicate runs were completed from which the following sample values were computed:

$$\bar{Y}_1^* = 0.207 \qquad s^2_{\bar{Y}_1^*} = 7.61 \times 10^{-4}$$

$$\bar{Y}_2^* = -0.0111 \qquad s^2_{\bar{Y}_2^*} = 9.83 \times 10^{-6}$$

$$s^2_{(\bar{Y}_1^*)^2} = 6.41 \times 10^{-6}$$

Equation 11.2-8 gave

$$\begin{bmatrix} 1 & 0 \\ 0.207 & 1 \end{bmatrix} \begin{bmatrix} \hat{a}_1 \\ -\hat{a}_2 \end{bmatrix} = \begin{bmatrix} 0.207 \\ -0.0111 \end{bmatrix} \qquad \text{(d)}$$

or

$$\hat{a}_1 = 0.207$$

$$\hat{a}_2 = (0.207)^2 + 0.0111 = 0.0539$$

The approximate variances were

$$\widehat{\mathrm{Var}}\{\hat{a}_1\} \approx s^2_{\bar{Y}_1^*} = 7.61 \times 10^{-4}$$

$$\widehat{\mathrm{Var}}\{\hat{a}_2\} \approx s^2_{(\bar{Y}_1^*)} + s^2_{\bar{Y}_2^*} = 1.62 \times 10^{-5}$$

11.2-2 Estimation in Laplace Transform Space by Transforming Observations

Some authors have suggested that instead of transforming the transfer function into the time domain, the experimental observations should be transformed into the Laplace transform domain. Two difficulties exist in connection with this suggestion. First, of course, is the problem of how to transform continuous and especially discrete data without significant numerical error being introduced into the results. Second, if least squares is used in Laplace transform space, that is, if

$$\check{\phi}(s) = \sum_s [\check{G}(s) - \check{g}(s)]^2 \qquad (11.2\text{-}9)$$

or some similar expression is minimized to estimate the parameters in $\check{g}(s)$, it is not possible to estimate analytically the bias in the parameters (assuming that time domain least squares would yield unbiased estimates). Simulation studies have indicated that considerable bias exists when either

$$\check{\phi}(s) = \sum_s \left[\check{G}(s) - \frac{\check{n}(s)\check{x}(s)}{\check{d}(s)}\right]^2$$

or

$$\check{\phi}(s) = \sum_s [\check{G}(s)\check{d}(s) - \check{n}(s)\check{x}(s)]^2$$

is minimized by the methods of Section 6.2. Unfortunately, it is not possible in general to answer the question: What criterion should be used in the Laplace transform domain for the estimates to be equivalent to the least squares estimates in the time domain? Nor is it possible generally to answer the question: What criterion in the time domain proves to be equivalent to least squares in the Laplace transform domain?

The first mentioned problem, numerical transformation of data, is less serious. Bellman and others† suggested that the experimental data for the impulse response be collected at specially spaced times (listed in the first reference) so that a special quadrature formula can be used in the transformation. However, this may not prove convenient nor sound from the viewpoint of an appropriate experimental design. Much of the information in the data may be lost between the required sampling times.

Probably a better approach is to take the observations at the most appropriate times and, thereafter, to approximate the empirical response $G(t)$ by piecewise linear segments. If we let the slope of any segment for $t_i \leq t \leq t_{i+1}$ be

$$b_i = \frac{G(t_{i+1}) - G(t_i)}{t_{i+1} - t_i}$$

then the response function $G(t)$ for the same time interval is

$$G_{1i}(t) = \left[\frac{G(t_{i+1}) - G(t_i)}{t_{i+1} - t_i}(t - t_i) + G(t_i)\right] U(t - t_i) \quad (11.2\text{-}10)$$

where $U(t - t_i)$ is the unit step function which is zero for $t < t_i$. Also it is necessary to terminate the contribution of $G(t)$ after t_{i+1} by subtracting the following expression from Equation 11.2-10:

$$G_{2i}(t) = \left[\frac{G(t_{i+1}) - G(t_i)}{t_{i+1} - t_i}(t - t_{i+1}) + G(t_{i+1})\right] U(t - t_{i+1})$$

After adding $G_{1i}(t)$, subtracting $G_{2i}(t)$, and taking the Laplace transform of the sum, we find for the period $t_i \leq t \leq t_{i+1}$

$$\check{G}_i(s) = e^{-st_i}\left(\frac{b_i}{s^2} + \frac{\check{G}_i(s)}{s}\right) - e^{-st_{i+1}}\left(\frac{b_i}{s^2} + \frac{\check{G}_{i+1}(s)}{s}\right) \quad (11.2\text{-}11)$$

where $\mathscr{L}[G(t_i)] \equiv \check{G}_i(s)$. The empirical impulse response in Laplace transform space for any time $t > 0$ is the sum of the segments given by Equation 11.2-11:

$$\check{G}(s) = \sum_{i=1}^{N-1} e^{-st_i}\left(\frac{b_i}{s^2} + \frac{\check{G}_i(s)}{s}\right) - e^{-st_{i+1}}\left(\frac{b_i}{s^2} + \frac{\check{G}_{i+1}(s)}{s}\right) \quad (11.2\text{-}12)$$

Continuous data can be transformed by analog or hybrid computer or by numerical quadrature. If in

$$\check{g}(s) = \int_0^\infty e^{-st} f(t)\, dt$$

we substitute $st = z$:

$$\check{g}(s) = s^{-1} \int_0^\infty e^{-z} f\left(\frac{z}{s}\right) dz$$

A power of z in $f(z/s)$ can be removed to give increased accuracy in the calculations:

$$\check{g}(s) = s^{-1} \int_0^\infty z^q e^{-z} \psi(z)\, dz \qquad q > -1$$

The integral can be approximated by a linear combination of functions involving Laguerre polynomials:

$$\int_0^\infty z^q e^{-z} \psi(z)\, dz \cong \sum_{k=1}^{n} A_k \psi(t_k)$$

Detailed tables of A_k and $\psi(t_k)$ are in Salzer and Zucker.‡

In Equation 11.2-9 the sum on s requires some comment. Because s is complex, s can take any path in the complex plane that does not contain the poles of the transformed solution. For convenience, one usually lets s take on real integer values from $s = 1$ to $s = N$ after making sure that the poles of $g(s)$ do not occur on the real axis. The suggested values of s give more weight to the earlier observations and less weight to the later observations because $\mathscr{L}[t] = 1/s^2$.

We can conclude that estimation in Laplace transform space can be carried out by least squares if the observations are transformed, but that the parameter estimates will be biased.

11.3 ORTHOGONAL PRODUCT METHODS

We shall now describe a method of estimating the coefficients in transfer functions, a method originally proposed by Puri and Weygant.§ Let

$$\check{g}(s, \boldsymbol{\beta}) = c\frac{\check{n}(s)}{\check{n}(s)} \quad (11.3\text{-}1)$$

be the process model where

c = a constant

$$\check{n}(s) = b_n s^n + b_{n-1} s^{n-1} + \cdots + b_1 s + 1$$

$$\check{d}(s) = a_d s^d + a_{d-1} s^{d-1} + \cdots + a_1 s + 1 \qquad d \geq n + 1$$

and the a's and b's are parameters to be estimated. For each of the $p = n + d + 1$ unknown parameters in

† R. E. Bellman, H. H. Kagiwada, R. E. Kalaba, and M. C. Prestrud, *Invariant Imbedding and Time-Dependent Transport Processes*, American Elsevier, New York, 1964; R. Bellman, H. H. Kagiwada, and R. E. Kalaba, *IEEE Trans. Auto. Control* **AC-10**, 111, 1965.

‡ H. E. Salzer and R. Zucker, *Bull. Amer. Math. Soc.* **55**, 1004, 1949.

§ N. N. Puri and C. N. Weygant, "Transfer Function Tracking of a Linear Time Varying System by Means of Auxiliary Simple Lag Networks," Preprints, Fourth Joint Automatic Control Conference, 1963, p. 200.

$\check{g}(s, \boldsymbol{\beta})$, we define an identification constant computed as follows:

$$z_j = \int_0^\infty g(t, \boldsymbol{\beta}) f_j(t)\, dt \qquad j = 0, 1, \ldots, p$$

where $f_j(t)$ are arbitrary functions to be described shortly.

By Parseval's theorem,

$$z_j = \int_0^\infty g(t, \boldsymbol{\beta}) f_j(t)\, dt = \frac{1}{2\pi i} \int_{-i\infty}^{i\infty} \check{g}(s, \boldsymbol{\beta}) \check{f}_j(-s)\, ds$$

If the order of $\check{f}_j(s)$ is such that the denominator of the product $[\check{g}(s, \boldsymbol{\beta}) \check{f}_j(-s)]$ is two or more degrees higher than the numerator, then

$$z_j = \frac{1}{2\pi i} \oint \check{g}(s, \boldsymbol{\beta}) \check{f}_j(-s)\, ds \tag{11.3-2}$$

where the contour encloses the right half-plane in the clockwise direction.

For the functions $f_j(t)$, Puri and Weygant chose the following:

j	$f(t)$	$\check{f}(s)$
$j = 0$	$f_0(t) = U(t)$	$\check{f}_0(s) = \dfrac{1}{s}$
$j \geq 1$	$f_j(t) = e^{-\alpha_j t}$	$\check{f}_j(s) = \dfrac{1}{s + \alpha_j}$

in which the α_j's are arbitrary and $U(t)$ is the unit step function. Deex† proposed using the following functions:

$f_j(t)$	$\check{f}_j(s)$
$f_j(t) = (e^{-\alpha_j t} - e^{-\gamma t}),\ \alpha_j < \gamma$	$\check{f}_j(s) = \dfrac{\gamma - \alpha_j}{(s + \gamma)(s + \alpha_j)} = \dfrac{1}{s + \alpha_j} - \dfrac{1}{s + \gamma}$
$f_\gamma(t) = e^{-\gamma t}$	$\check{f}_\gamma(s) = \dfrac{1}{s + \gamma}$
$f_0(t) = U(t)$	$\check{f}_0(s) = \dfrac{1}{s}$

Then, from Equation 11.3-2:

$$z_j = \frac{1}{2\pi i} \oint \left[\frac{-\check{g}(s, \boldsymbol{\beta})}{s - \alpha_j}\right] ds - \frac{1}{2\pi i} \oint \left[\frac{-\check{g}(s, \boldsymbol{\beta})}{s - \gamma}\right] ds$$

$$\equiv \check{g}(\alpha_j) - \check{g}(\gamma) \qquad j = 1, 2, \ldots, p \tag{11.3-3}$$

$$z_\gamma = \frac{1}{2\pi i} \oint \frac{-\check{g}(s, \boldsymbol{\beta})}{s - \gamma}\, ds \equiv \check{g}(\gamma) \tag{11.3-4}$$

$$z_0 = \int_0^\infty g(t, \boldsymbol{\beta}) f_0(t)\, dt = \lim_{s \to 0} \left[\frac{s}{s} \check{g}(s, \boldsymbol{\beta})\right]$$

$$= c \frac{\check{n}(0)}{\check{d}(0)} = c \tag{11.3-5}$$

† A. J. Deex, Ph.D. Dissertation, Univ. of Texas, Austin, 1965.

If Equations 11.3-3, 11.3-4, and 11.3-5 are combined as follows:

$$z_j = \check{g}(\alpha_j) - \check{g}(\gamma) = c \frac{\check{n}(\alpha_j)}{\check{d}(\alpha_j)} - z_\gamma$$

$$= z_0 \frac{\check{n}(\alpha_j)}{\check{d}(\alpha_j)} - z_\gamma$$

we find

$$z_0 \check{n}(\alpha_j) = (z_j + z_\gamma) \check{d}(\alpha_j)$$

or

$$\check{n}(\alpha_j) = h_j \check{d}(\alpha_j) \tag{11.3-6}$$

where

$$h_j = \frac{z_j + z_\gamma}{z_0}$$

For a slightly more general transfer function than Equation 11.3-1, namely

$$\check{g}(s, \boldsymbol{\beta}) = c s^m \frac{\check{n}(s)}{\check{d}(s)} \qquad m \geq 1 \tag{11.3-7}$$

Deex showed that if we define

$$\mathscr{I}[g(t)] \equiv \int_0^t g(t)\, dt$$

and $\mathscr{I}^k$ as k successive integrations from 0 to t, then we can define new identification constants:

$$z_{0k} = \int_0^\infty \mathscr{I}^k[g(t, \boldsymbol{\beta})] f_0(t)\, dt = 0 \qquad \text{for } k < m$$

$$z_{0k} = \int_0^\infty \mathscr{I}^k[g(t, \boldsymbol{\beta})] f_0(t)\, dt = c \frac{\check{n}(0)}{\check{d}(0)} = c \qquad \text{for } k = m$$

where f_0 is still $U(t)$. The procedure is to evaluate z_{0k} for $k = 1, 2, 3, \ldots$ until one finds the first nonzero z_{0k}, which becomes the z_0 used. Thus

$$z_j = z_0 \alpha_j^m \frac{\check{n}(\alpha_j)}{\check{d}(\alpha_j)} - z_\gamma$$

and the equivalent of Equation 11.3-6 is

$$\alpha_j^m \check{n}(\alpha_j) = h_j \check{d}(\alpha_j) \tag{11.3-8}$$

Equation 11.3-8 leads to the matrix equation that can be used to solve for the parameters in the transfer function Equation 11.3-7:

$$\begin{bmatrix} \alpha_1^{m+1} & \alpha_1^{m+2} & \cdots & \alpha_1^{m+n} & -h_1\alpha_1 & \cdots & -h_1\alpha_1^d \\ \alpha_2^{m+1} & \alpha_2^{m+2} & \cdots & \alpha_2^{m+n} & -h_2\alpha_2 & \cdots & -h_2\alpha_2^d \\ \vdots & \vdots & & \vdots & \vdots & & \vdots \\ \alpha_{n+d}^{m+1} & \alpha_{n+d}^{m+2} & \cdots & \alpha_{n+d}^{m+n} & -h_{n+d}\alpha_{n+d} & \cdots & -h_{n+d}\alpha_{n+d}^d \end{bmatrix} \begin{bmatrix} b_1 \\ b_2 \\ \vdots \\ a_1 \\ \vdots \\ a_d \end{bmatrix} = \begin{bmatrix} h_1 - \alpha_1^m \\ h_2 - \alpha_2^m \\ \vdots \\ h_{n+d} - \alpha_{n+d}^m \end{bmatrix} \tag{11.3-9}$$

An example will clarify the details of the execution of the technique for deterministic variables.

Example 11.3-10 Estimation of Coefficients in a Transfer Function

Suppose the transfer function is

$$\breve{g}(s, \boldsymbol{\beta}) = cs^m \frac{b_1 s + 1}{a_3 s^3 + a_2 s^2 + a_1 s + 1} \quad \text{(a)}$$

We illustrate the application of Equation 11.3-9 in the absence of error by using analytical rather than numerical integration simply to make the procedure clearer. For this example we shall assume that the experimental impulse response in the time domain is the deterministic function

$$g(t) = 2e^{-2t} - e^{-t} \quad \text{(b)}$$

which is equivalent to

$$\breve{g}(s, \boldsymbol{\beta}) = \frac{s}{s^2 + 3s + 2} \quad \text{(c)}$$

Consequently, we know we should find that $c = \frac{1}{2}$, $a_1 = \frac{3}{2}$, $a_2 = \frac{1}{2}$, $a_3 = 0$, $b_1 = 0$, and $m = 1$ by solving Equation 11.3-9.

The first step is to calculate z_{0k}'s by introducing the values for the impulse response in the time domain until the first nonzero z_{0k} is obtained:

$$k = 0: \quad z_{00} = \int_0^\infty \mathscr{I}^0[g(t)]\, dt = \int_0^\infty (2e^{-2t} - e^{-t})\, dt = 0$$

$$k = 1: \quad z_{01} = \int_0^\infty \mathscr{I}^1[g(t)]\, dt$$

$$= \int_0^\infty \int_0^t (2e^{-2\tau} - e^{-\tau})\, d\tau\, dt$$

$$= \int_0^\infty (-e^{-2t} + e^{-t})\, dt = \tfrac{1}{2}$$

From $z_{01} = \frac{1}{2}$, we know that $c = \frac{1}{2}$, and because $k = 1$, $m = 1$.

The next step is to choose four values of α_j (for four h's) and one value of γ:

$$\alpha_1 = \tfrac{1}{2} \qquad \alpha_4 = \tfrac{1}{4}$$

$$\alpha_2 = 1 \qquad \gamma = 2$$

$$\alpha_3 = \tfrac{3}{2}$$

and to compute the z_j's:

$$z_\gamma = \int_0^\infty g(t) f_\gamma(t)\, dt = \int_0^\infty (2e^{-2t} - e^{-t})\, e^{-2t}\, dt = \tfrac{1}{6}$$

$$z_1 = \int_0^\infty g(t) f_1(t)\, dt$$

$$= \int_0^\infty (2e^{-2t} - e^{-t})(e^{-\frac{1}{2}t} - e^{-2t})\, dt = -\tfrac{1}{30}$$

$$z_2 = \int_0^\infty (2e^{-2t} - e^{-t})(e^{-t} - e^{-2t})\, dt = 0$$

$$z_3 = \int_0^\infty (2e^{-2t} - e^{-t})(e^{-\frac{3}{2}t} - e^{-2t})\, dt = \tfrac{1}{210}$$

$$z_4 = \int_0^\infty (2e^{-2t} - e^{-t})(e^{-\frac{1}{4}t} - e^{-2t})\, dt = \tfrac{7}{90}$$

Equation 11.3-9 then is

$$\begin{bmatrix} \frac{1}{4} & -\frac{2}{15} & -\frac{1}{15} & -\frac{1}{30} \\ 1 & -\frac{1}{3} & -\frac{1}{3} & -\frac{1}{3} \\ \frac{9}{4} & -\frac{18}{35} & -\frac{27}{35} & -\frac{81}{70} \\ \frac{1}{16} & -\frac{2}{45} & -\frac{1}{90} & -\frac{1}{320} \end{bmatrix} \begin{bmatrix} b_1 \\ a_1 \\ a_2 \\ a_3 \end{bmatrix} = \begin{bmatrix} -\frac{7}{30} \\ -\frac{2}{3} \\ -\frac{81}{70} \\ -\frac{13}{180} \end{bmatrix} \quad \text{(d)}$$

Equation (d) has the solution

$$\begin{bmatrix} b_1 \\ a_1 \\ a_2 \\ a_3 \end{bmatrix} = \begin{bmatrix} 0 \\ \frac{3}{2} \\ \frac{1}{2} \\ 0 \end{bmatrix}$$

Note especially that $b_1 = 0$ and $a_3 = 0$ as in the assumed experimental response.

Next we shall inquire into the effect of process error on the use of Equation 11.3-9 as an estimation procedure if the experimental values $G(t)$ are to be used for the impulse response in lieu of the deterministic $g(t)$. We shall assume that the error is added to the deterministic process output in the time domain:

$$G(t) = g(t, \boldsymbol{\beta}) + \epsilon(t)$$

where $\epsilon(t)$ is stationary noise normally distributed with respect to amplitude, having zero mean and variance σ_ϵ^2, uncorrelated with $g(t, \boldsymbol{\beta})$.

Suppose that $f_j(t)$ is $e^{-\alpha_j t}$. Then (Z_j is now a random variable)

$$Z_j = \int_0^\infty [g(t, \boldsymbol{\beta}) f_j(t) + \epsilon(t) f_j(t)]\, dt$$

and the expected value of Z_j is

$$\mathscr{E}\{Z_j\} = \mathscr{E}\left\{\int_0^\infty [g(t, \boldsymbol{\beta})\, e^{-\alpha_j t} + \epsilon(t)\, e^{-\alpha_j t}]\, dt\right\}$$

$$= \breve{g}(\alpha_j) \quad \text{(11.3-10)}$$

The variance of Z_j is computed as follows:

$$\operatorname{Var}\{Z_j\} = \sigma_{Z_j}^2 = \mathscr{E}\{Z_j^2\} - [\mathscr{E}\{Z_j\}]^2$$

$$= \mathscr{E}\left\{\left[\int_0^\infty g(t, \boldsymbol{\beta})\, e^{-\alpha_j t}\, dt + \int_0^\infty \epsilon(t)\, e^{-\alpha_j t}\, dt\right]^2\right\} - \breve{g}^2(\alpha_j)$$

$$= \mathscr{E}\left\{\breve{g}^2(\alpha_j) + 2\breve{g}(\alpha_j)\int_0^\infty \epsilon(t)\, e^{-\alpha_j t}\, dt + \int_0^\infty \epsilon(t)\, e^{-\alpha_j t}\, dt \int_0^\infty \epsilon(t)\, e^{-\alpha_j t}\, dt\right\} - \breve{g}^2(\alpha_j)$$

$$= \breve{g}^2(\alpha_j) + 2\breve{g}(\alpha_j)\mathscr{E}\left\{\int_0^\infty \epsilon(t)\, e^{-\alpha_j t}\, dt\right\} + \mathscr{E}\left\{\int_0^\infty \int_0^\infty \epsilon(t)\epsilon(\tau)\, e^{-\alpha_j t}\, e^{-\alpha_j \tau}\, dt\, d\tau\right\} - \breve{g}^2(\alpha_j)$$

$$= \int_0^\infty \int_0^\infty \mathscr{E}\{\epsilon(t)\epsilon(\tau)\, e^{-\alpha_j t}\, e^{-\alpha_j \tau}\, dt\, d\tau\}$$

$$= \int_0^\infty \sigma_\epsilon^2 U(t-\tau)\, e^{-\alpha_j t}\, e^{-\alpha_j \tau}\, dt\, d\tau$$

$$= \sigma_\epsilon^2 \int_0^\infty e^{-2\alpha_j t}\, dt = \frac{\sigma_\epsilon^2}{2\alpha_j} \tag{11.3-11}$$

If $f_j(t) = e^{-\alpha_j t} - e^{-\gamma t}$, then by a similar calculation it can be shown after some extensive manipulation that when Z_γ is computed separately from Z_j as follows:

$$Z_\gamma = \int_0^\infty [g(t, \boldsymbol{\beta}) + \epsilon_1(t)]\, e^{-\gamma t}\, dt$$

that

$$\mathscr{E}\{Z_j\} = \breve{g}(\alpha_j)$$

$$\text{Var}\,\{Z_j\} = \sigma_\epsilon^2 \left[\frac{\gamma^2 - 2\gamma\alpha_j + 2\alpha_j^2}{2\alpha_j\gamma(\gamma + \alpha_j)}\right] \tag{11.3-12}$$

where $\epsilon(t)$ and $\epsilon_1(t)$ may be different but must be uncorrelated,

$$\mathscr{E}\{\epsilon(t)\epsilon(\tau)\} = 0$$

$$\mathscr{E}\{\epsilon(t)\epsilon_1(\tau)\} = 0$$

and $\sigma_\epsilon^2 = \sigma_{\epsilon_1}^2$. The choice of $f_j(t) = e^{-\alpha_j t} - e^{-\gamma t}$ is an improvement over the choice

$$f_j = e^{-\alpha_j t}$$

if $\alpha_j < \gamma$, as originally required. For maximum variance reduction, Deex recommended

$$2 \le \frac{\gamma}{\alpha_j} \le 6$$

By replicate experiments the variance of Z_j can also be reduced; Equation 11.3-9 would become a set of overdetermined equations which could be solved by some type of least squares procedure, or perhaps the estimated values of the parameters could be an average of those obtained from each experiment. Even though the Var $\{Z_j\}$ can be calculated, the variances of the coefficients in $g(s, \hat{\boldsymbol{\beta}})$ cannot be directly calculated although the variances can be approximated by linearizing Equation 11.3-9 in terms of the Z_j's.

Example 11.3-2 Estimation in the Presence of Noise

The examples given here have been taken from Deex and illustrate the influence of normally distributed noise with zero mean as well as the effect of the selection of the values of Δt, α_j, etc. The necessary integrations were carried out on a digital computer by the trapezoidal rule; matrix inversion of Equation (11.3-9) for the parameter vector was by Gaussian elimination.

EFFECT OF CHANGING THE TIME INCREMENT IN THE TRAPEZOIDAL RULE. Simulated data were generated to fit the model

$$\breve{g}(s, \boldsymbol{\beta}) = c\,\frac{(b_1 s + 1)}{(a_2 s^2 + a_1 s + 1)} \tag{a}$$

by adding normal random error to $g(t) = 8e^{-0.4t} - 7e^{-0.6t}$, a function in the time domain which corresponds to a function in the Laplace domain of $\breve{g}(s) = (s+2)/(s^2+s+0.24)$. Hence the "true" values of the parameters are: $c = 8.333$, $b_1 = 0.500$, $a_1 = 4.167$, and $a_2 = 4.167$. Table E11.3-2a demonstrates the effect of the time increments Δt in the trapezoidal rule. The estimates were made by using Equation 11.3-9 with the simulated data used to calculate the Z_j's. It can be concluded that small time steps are more effective.

TABLE E11.3-2a EFFECT OF TIME INCREMENTS ON ESTIMATION

Δt (sec)	$\hat{c}$	$\hat{b}_1$	$\hat{a}_1$	$\hat{a}_2$
0.01	8.327	0.504	4.163	4.191
0.10	8.326	0.509	4.166	4.217
0.50	8.306	0.617	4.259	4.770

$t_f = 20$ sec, $\alpha_1 = 0.3$, $\alpha_2 = 0.6$, $\alpha_3 = 0.9$

EFFECT OF CHANGING THE TERMINAL TIME FOR INTEGRATION, t_f. Simulated data were generated by using $g(t) = 0.5e^{-t} + 0.5e^{-3t}$ which corresponds to $\breve{g}(s) = (s+2)/(s^2+4s+3)$. Then the "true" values of the coefficients are: $c = 0.667$, $b_1 = 0.500$, $a_1 = 1.333$, and $a_2 = 0.333$. Table E11.3-2b illustrates the effect of changing t_f. Values of t_f less than 5 or 6 times the largest time constant of the "process" yielded relatively poor estimates for models such as this one in which the matrix Equation 11.3-9 is ill conditioned.

TABLE E11.3-2b EFFECT OF CHANGING THE TERMINAL TIME FOR ESTIMATION

t_f (sec)	$\hat{c}$	$\hat{b}_1$	$\hat{a}_1$	$\hat{a}_2$
6	0.665	0.096	0.918	0.021
10	0.666	0.471	1.304	0.315
20	0.667	0.500	1.333	0.333

$\Delta t = 0.01$, $\alpha_1 = 0.05$, $\alpha_2 = 0.10$, $\alpha_3 = 0.15$

EFFECT OF SELECTION OF α'S. Simulated data were generated from $g(t) = -12e^{-0.4t} + 13e^{-0.6t}$, which corresponds to $\breve{g}(s) = (s-2)/(s+0.4)(s+0.6)$, from which one finds the "true" coefficients: $c = -8.333$, $b_1 = -0.500$, $a_1 = 4.167$, and $a_2 = 4.167$. Table E11.3-2c shows the effect of the selection of the α's on the estimates. For those choices of α's for which the matrix Equation 11.3-9 was better conditioned (in the sense that the value of the square determinant on the left-hand side of the equation was larger), the estimation tended to be better. The best estimates were obtained when the α's had values approximately equal

TABLE E11.3-2c EFFECT OF SELECTION OF THE α'S ON ESTIMATION

α_1	α_2	α_3	$\hat{c}$	$\hat{b}_1$	$\hat{a}_1$	$\hat{a}_2$	Absolute Value of the Determinant
Exact solution			−8.333	−0.500	4.166	4.166	—
0.4	0.6	2	−8.333	−0.500	4.157	4.166	3.23×10^{-3}
0.3	0.9	2.7	−8.333	−0.500	4.156	4.164	1.35×10^{-2}
0.3	1.5	2.7	−8.333	−0.500	4.157	4.163	1.47×10^{-2}
3	6	9	−8.333	−0.501	4.226	4.155	1.69×10^{-2}
3	5.196	9	−8.333	−0.500	4.215	4.154	1.77×10^{-2}
0.3	0.5196	0.90	−8.333	−0.500	4.155	4.174	5.01×10^{-4}
0.3	0.6	0.9	−8.333	−0.500	4.155	4.173	5.35×10^{-4}
0.001	5	10	−8.333	−0.499	4.146	4.146	1.40×10^{-3}
0.001	0.1	10	−8.333	−0.505	4.141	4.210	9.48×10^{-5}
0.4	0.566	0.8	−8.333	−0.500	4.156	4.172	1.67×10^{-4}
0.4	0.6	0.8	−8.333	−0.500	4.156	4.171	1.72×10^{-4}
0.4	0.489	0.6	−8.333	−0.499	4.156	4.176	2.25×10^{-5}
0.4	0.5	0.6	−8.333	−0.499	4.156	4.176	2.27×10^{-5}
0.03	0.06	0.09	−8.333	−0.464	4.182	4.425	2.85×10^{-8}
0.01	0.02	0.03	−8.333	−0.448	4.198	4.508	6.18×10^{-11}
0.001	0.002	0.003	−8.333	−0.388	4.258	4.791	7.71×10^{-17}

$t_f = 20$ sec $\quad \Delta t = 0.01$ sec

to the magnitude of the real part of the poles and zeros of the process model (the poles are −0.4 and −0.6, the same as in the first case).

From these and other studies, Deex concluded that:

1. The α's should be separated as much as possible.
2. The α's should have magnitudes of the same order as the poles and zeros of the process transfer function.

11.4 ESTIMATION FOR SAMPLED DATA

11.4-1 The z-Transform

The z transform is the basis of a calculus analogous to the Laplace transform, a calculus that is quite useful for sampled, i.e., discrete, data. We shall briefly examine the pertinent relations needed for estimation before describing how the estimation can take place. If a deterministic process impulse response is sampled at discrete instants of time, $t = 0$, $t = 1, \ldots$, each separated by a constant interval τ as in Figure 11.4-1, we can designate the sampled sequence of values of $g(t)$ by $g^*(t)$ and the number of samples as $n = t/\tau$. The z-transform to be denoted by $\mathscr{Z}$, of a time domain function with satisfactory properties (fulfilled by $g(t)$) is defined as

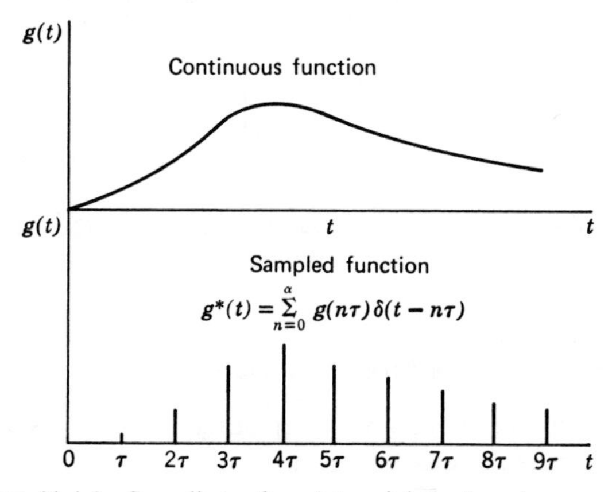

FIGURE 11.4-1 Sampling of a deterministic impulse response function.

$$\breve{g}(z) = \mathscr{Z}[g(n\tau)] = \sum_{n=0}^{\infty} g(n\tau)z^{-n} \tag{11.4-1}$$

and the inversion relation

$$g(n\tau) = \mathscr{Z}^{-1}[\breve{g}(z)] \tag{11.4-2}$$

The Laplace transform corresponding to Equation 11.4-1 is

$$\breve{g}(s) = \sum_{n=0}^{\infty} g(n\tau)\, e^{-n\tau s}$$

from which we observe that the quantity z^{-n} in Equation 11.4-1 has replaced $e^{-n\tau s}$ in the Laplace transform. Table 11.4-1 lists a few z-transforms, the related time domain functions, and the related Laplace transforms; refer to Jury† for more complete tables and for formulas to obtain the z-transform from the Laplace transform.

Unfortunately, from the viewpoint of estimation, transformation from the Laplace transform to the

† E. I. Jury, *Theory and Application of the z-Transform Method*, John Wiley, New York, 1964.

TABLE 11.4-1 LAPLACE AND z-TRANSFORMS

Time Domain Function	Laplace Transform	z-Transform
$\delta(t)$	1	1
1	$\frac{1}{s}$	$\frac{z}{z-1}$
t	$\frac{1}{s^2}$	$\frac{\tau z}{(z-1)^2}$
t^2	$\frac{1}{s^3}$	$\frac{\tau^2}{2}\frac{z(z+1)}{(z-1)^3}$
e^{-t}	$\frac{1}{s+1}$	$\frac{z}{z-e^{-\tau}}$
$f(t-n\tau)U(t-n\tau)$		$z^{-n}\check{f}(z)$

z-transform proves to be of very little assistance. As an example of the complicated nonlinear relationship between the parameters in the Laplace and z-transforms, consider the transfer function

$$\check{g}(s) = \frac{b_0 + b_1 s}{a_0 + a_1 s + s^2}$$

for which the equivalent z-transform is

$$\check{g}(z) = \frac{p_0 + p_1 z^{-1}}{1 + q_1 z^{-1} + q_2 z^{-2}}$$

The following relationships exist between the coefficients:

$$p_0 = b_1$$

$$p_1 = (e^{-a_1\tau/2})[\gamma_3 \sin(\gamma_1\tau) - b_1 \cos(\gamma_1\tau)]$$

$$q_1 = -2(e^{-a_1\tau/2}) \cos(\gamma_1\tau)$$

$$\gamma_1 = \sqrt{a_0 + \frac{a_1^2}{4}}$$

$$\gamma_2 = b_0 + \frac{a_1 b_1}{2}$$

$$\gamma_3 = \frac{\gamma_2}{\gamma_1}$$

$$\tau = \text{sampling period}$$

$$= \frac{2\pi}{\text{(samples cycle)(frequency)}}$$

Thus, the primary value of using z-transforms is not to estimate the coefficients in z-space and then determine the equivalent parameters in the transfer function in s-space but to analyze a dynamic system represented by a *difference equation*. The difference equation may, of course, be the difference analog of a differential equation, or it may just represent discrete sampled data as a distinct model.

Suppose the process is to be represented by a difference equation of known form:

$$y(t_k) + a_1 y(t_k - \tau) + a_2 y(t_k - 2\tau) + \cdots + a_n y(t_k - n\tau)$$
$$= b_0 x(t_k) + b_1 x(t_k - \tau) + \cdots + b_m x(t_k - m\tau) \quad (11.4\text{-}3)$$

where the sequence of y's are the sampled outputs, the sequence of x's are the sampled inputs, and t_k is the kth sample time. Taking the z-transform of Equation 11.4-3 gives

$$\check{y}(z) + a_1 z^{-1}\check{y}(z) + \cdots + a_n z^{-n}\check{y}(z)$$
$$= b_0\check{x}(z) + b_1 z^{-1}\check{x}(z) + \cdots + b_m z^{-m}\check{x}(z)$$

The transfer function in z-space is

$$\check{g}(z) = \frac{\check{y}(z)}{\check{x}(z)} = \frac{b_0 + b_1 z^{-1} + \cdots + b_m z^{-m}}{1 + a_1 z^{-1} + \cdots + a_n z^{-n}} \quad (11.4\text{-}4)$$

Contrast Equation 11.4-4 with 11.1-3, but note that the parameters in Equation 11.4-4 will not be equal to the same symbols in Equation 11.1-3 but to very complicated functions of several of the symbols as mentioned before.

11.4-2 Introduction of Error

We shall designate the vector of the $(m + n + 1)$ coefficients in Equations 11.4-3 and 11.4-4 as usual by $\boldsymbol{\beta}$ in order to express Equation 11.4-3 in matrix notation as

$$\boldsymbol{\beta}^T \mathbf{y}_k = 0 \quad (11.4\text{-}5)$$

And, also as usual, we shall add the error, which incorporates all effects in the process response not attributable to the process input, to the deterministic response

$$\mathbf{Y}_k = \mathbf{y}_k + \boldsymbol{\epsilon}_k \quad (11.4\text{-}6)$$

where

$$\boldsymbol{\beta} = \begin{bmatrix} 1 \\ a_1 \\ \vdots \\ a_n \\ -b_0 \\ -b_1 \\ \vdots \\ -b_m \end{bmatrix} \quad \mathbf{y}_k = \begin{bmatrix} y(t_k) \\ y(t_k - \tau) \\ \vdots \\ y(t_k - n\tau) \\ x(t_k) \\ x(t_k - \tau) \\ \vdots \\ x(t_k - m\tau) \end{bmatrix} \quad \boldsymbol{\epsilon}_k = \begin{bmatrix} \epsilon_Y(t_k) \\ \epsilon_Y(t_k - \tau) \\ \vdots \\ \epsilon_Y(t_k - n\tau) \\ \epsilon_X(t_k) \\ \epsilon_X(t_k - \tau) \\ \vdots \\ \epsilon_X(t_k - m\tau) \end{bmatrix}$$

The additive errors are assumed to have the known covariance matrix $\boldsymbol{\Gamma}$ and individually to have expected values of zero.

11.4-3 "Equation Error" Estimation

If in Equation 11.4-5 we replace $\mathbf{y}_k$ by $(\mathbf{Y}_k - \boldsymbol{\epsilon}_k)$, we can define an "equation error":

$$\acute{\epsilon}_k = \boldsymbol{\beta}^T \boldsymbol{\epsilon}_k = \boldsymbol{\beta}^T \mathbf{Y}_k$$

The elements of the vector $\boldsymbol{\beta}$ can be chosen to minimize the sum of the squares of the equation errors ϵ'_k for all the $\mathbf{Y}_k$ vectors, as in Chapter 5:

$$\text{Minimize} \sum_{k=1}^{K} \epsilon_k'^2 \qquad \begin{array}{l} k = 1, 2, \ldots, K \\ K > n + m + 1 \end{array}$$

to yield

$$\hat{\tilde{\boldsymbol{\beta}}} = -\left[\sum_{k=1}^{K} \tilde{\mathbf{Y}}_k \tilde{\mathbf{Y}}_k^T\right]^{-1} \left[\sum_{k=1}^{K} \tilde{\mathbf{Y}}_k Y(t_k)\right] \tag{11.4-7}$$

where $\tilde{\mathbf{Y}}_k$ is the matrix $\mathbf{Y}_k$ with the first element $Y(t_k)$ deleted and $\hat{\tilde{\boldsymbol{\beta}}}$ is the estimate of the matrix β in which the first element, 1, is deleted. The inverse matrix in Equation 11.4-7 will exist if the input sequence is not the solution of a homogeneous difference equation of order less than m.

Estimation using the equation error does not yield unbiased nor best estimates for the parameters. Its major appeal is that it is simple and may yield reasonably good estimates.

11.4-4 Maximum Likelihood Estimates

Levin † developed a maximum likelihood estimate for the parameters in Equations 11.4-3 and 11.4-4. To obtain independent errors in all the observations, the Y and X sequences must be obtained from *nonoverlapping* sequences in time. Also, there must be more sets of observations than parameters to be estimated. The probability density function for all the observations $\mathbf{Y}_k$ (both Y and X) is assumed to be a multivariate Gaussian density,‡ and the likelihood function is

$$p(\mathbf{y}_1, \ldots, \mathbf{y}_K) = \frac{1}{(2\pi)^{(-K(n+m+2))/2}} \exp\left[-\tfrac{1}{2} \sum_{k=1}^{K} (\mathbf{Y}_k - \mathbf{y}_k)^T \boldsymbol{\Gamma}^{-1} (\mathbf{Y}_k - \mathbf{y}_k)\right]$$

Maximization of $p(\mathbf{y}_1, \ldots, \mathbf{y}_K)$ can be achieved by minimizing

$$\tilde{\phi} = \sum_{k=1}^{K} (\mathbf{Y}_k - \mathbf{y}_k)^T \boldsymbol{\Gamma}^{-1} (\mathbf{Y}_k - \mathbf{y}_k) \tag{11.4-8}$$

with respect to the $\mathbf{y}_k$, where the $\mathbf{y}_k$ are constrained to satisfy Equation 11.4-3. Rogers and Steiglitz§ considered the related maximum likelihood estimation in z-transform space.

Clearly if the errors are correlated for closely spaced samples, the spacing must be extended to allow sufficient time to make the errors uncorrelated. Hence a good share of the available data may not be used in the estimation procedure.

Smith and Hilton‖ and others showed that the constrained minimization of $\tilde{\phi}$ defined by Equation 11.4-8 is equivalent to the minimization of

$$\phi = \frac{1}{K} \sum_{k=1}^{K} \frac{\boldsymbol{\beta}^T \mathbf{Y}_k \mathbf{Y}_k^T \boldsymbol{\beta}}{\boldsymbol{\beta}^T \boldsymbol{\Gamma} \boldsymbol{\beta}} \tag{11.4-9}$$

with respect to $\boldsymbol{\beta}$. The minimum ϕ is the smallest value of λ that satisfies the equation

$$\left(\frac{1}{K} \sum_{k=1}^{K} \mathbf{Y}_k \mathbf{Y}_k^T - \lambda \boldsymbol{\Gamma}\right) \boldsymbol{\beta} = 0 \tag{11.4-10}$$

and $\hat{\boldsymbol{\beta}}$, the estimated coefficient vector, is the eigenvector corresponding to $\lambda = \phi_{\min}$.

Although the estimated parameters are usually biased, Levin stated that if the errors are small in comparison to the observations, the bias is small compared to the standard deviations of the parameters. Levin also stated that, when overlapping sets of observations are used, the bias is no more than when nonoverlapping vectors are used. An approximation for the variances of the estimated coefficients was given as

$$\widehat{\text{Covar}}\,\{\hat{\tilde{\boldsymbol{\beta}}}\} \approx \frac{\hat{\boldsymbol{\beta}}^T \boldsymbol{\Gamma} \hat{\boldsymbol{\beta}}}{K} \left[\frac{1}{K} \sum_{k=1}^{K} \hat{\tilde{\mathbf{Y}}}_k \hat{\tilde{\mathbf{Y}}}_k^T\right]^{-1} \tag{11.4-11}$$

where the overlay ($\sim$) signifies that the first element in the matrix is deleted.

Example 11.4-1 Estimation of Parameters in Difference Equations

Consider the following difference equation:

$$y(t_k) + a_1 y(t_k - \tau) + a_2 y(t_k - 2\tau) = b_0 x(t_k) + b_1 x(t_k - \tau) \tag{a}$$

for which the z-transform is

$$\check{g}(z) = \frac{b_0 + b_1 z^{-1}}{1 + a_1 z^{-1} + a_2 z^{-2}} \tag{b}$$

where τ = sampling period = 2π/(samples per cycle)·(frequency). For brevity, we shall let the first pair of observations at the time equal to 1τ be X_1 and Y_1 for the input and output respectively, X_2 and Y_2 for the observations at $t = 2\tau$, and so forth.

Equation (a) with

$$a_1 = 1.96664 \qquad b_0 = 0$$

$$a_2 = 0.969072 \qquad b_1 = 0.002428$$

corresponds to a transfer function in the Laplace domain of

$$\check{g}(s) = \frac{40}{s^2 + 4s + 40} \tag{c}$$

† M. J. Levin, *IEEE Trans. Auto. Control* **AC-9**, 229, 1964.

‡ The probability distribution is $P\{\mathbf{Y}_1 \le \mathbf{y}_1; \mathbf{Y}_2 \le \mathbf{y}_2; \ldots; \mathbf{Y}_K \le \mathbf{y}_K\}$.

§ A. E. Rogers and K. Steiglitz, *IEEE Trans. Auto. Control* **AC-12**, 594, 1967.

‖ F. W. Smith and W. B. Hilton, *IEEE Trans. Auto. Control* **AC-12**, 568, 1967.

For a process input of

$$x(t) = 1 - e^{-10t} \tag{d}$$

the "true response" in the time domain is

$$y(t) = \frac{800c_1}{c_1^2 + c_2^2}(e^{-2t})\cos(6t) - \frac{800c_2}{c_1^2 + c_2^2}(e^{-2t})\sin(6t) - 4e^{-10t} + 1 \tag{e}$$

where

$$c_1 = -432$$

$$c_2 = 624$$

Figure E11.4-1a illustrates the deterministic input and the deterministic output.

A normal random error was added to $x(t)$ and $y(t)$ to simulate a real process, and the simulated input and response were periodically sampled. The noise added to $x(t)$ and $y(t)$ when sampled gave a normal distribution for the frequency of samples versus amplitude with a mean of 0 and a variance of σ^2. The 3σ limit corresponded to 5 percent of the steady-state value of 1, or $\sigma = (0.005/3) = 0.0167$. Figure E11.4-1b shows the percent error in the largest coefficient among the a's and b's defined as

$$\text{percent error} = \left|\frac{\text{estimated value} - \text{true value}}{\text{true value}}\right| 100$$

for various estimated schemes.

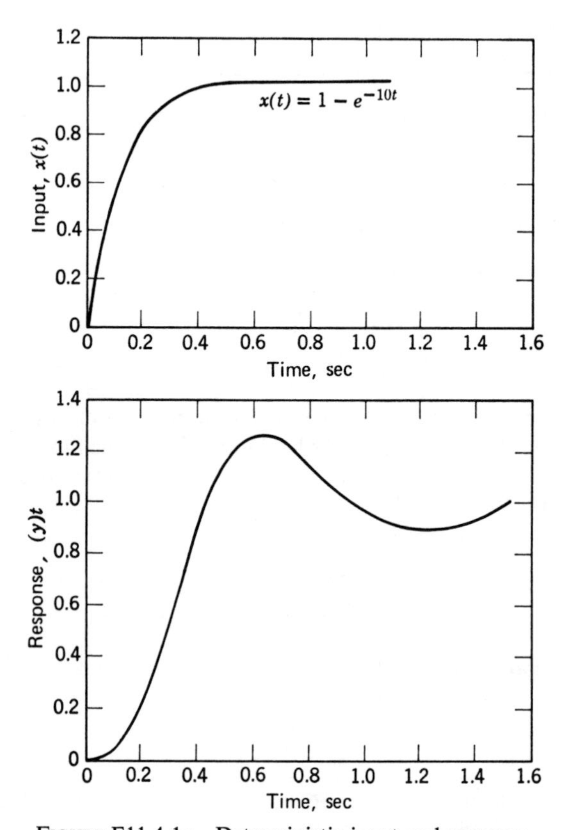

FIGURE E11.4-1a Deterministic input and response.

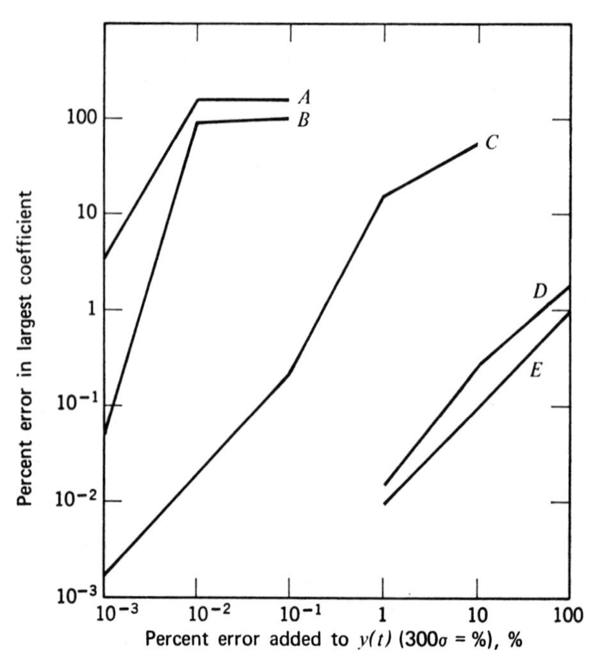

FIGURE E11.4-1b

A—Consecutive equations, Equation (f)
B—Intermittent collocation, Equation (g)
C—Least squares of equation error, Equation 11.4-7
D—Maximum likelihood, Equation 11.4-10
E—Maximum likelihood

USE OF CONSECUTIVE SAMPLES. Consecutive samples were introduced into Equation (a) to replace y and x, and the resulting matrix equation was solved:

$$\begin{bmatrix} X_{n+2} & X_{n+1} & Y_{n+1} & Y_n \\ X_{n+3} & X_{n+2} & Y_{n+2} & Y_{n+1} \\ X_{n+4} & X_{n+3} & Y_{n+3} & Y_{n+2} \\ X_{n+5} & X_{n+4} & Y_{n+4} & Y_{n+3} \end{bmatrix} \begin{bmatrix} b_0 \\ b_1 \\ -a_1 \\ -a_2 \end{bmatrix} = \begin{bmatrix} Y_{n+2} \\ Y_{n+3} \\ Y_{n+4} \\ Y_{n+5} \end{bmatrix} \tag{f}$$

Equation (f) represents only 5 consecutive samples of the input and 6 consecutive samples of the output; it therefore contains less information about $\check{g}(z)$ than does Equation (g) below.

INTERMITTENT COLLOCATION. Instead of using consecutive data sets, every tenth data set was used as shown in Equation (g):

$$\begin{bmatrix} X_{n+2} & X_{n+1} & Y_{n+1} & Y_n \\ X_{\alpha+2} & X_{\alpha+1} & Y_{\alpha+1} & Y_\alpha \\ X_{\beta+2} & X_{\beta+1} & Y_{\beta+1} & Y_\beta \\ X_{\gamma+2} & X_{\gamma+1} & Y_{\gamma+1} & Y_\gamma \end{bmatrix} \begin{bmatrix} b_0 \\ b_1 \\ -a_1 \\ -a_2 \end{bmatrix} = \begin{bmatrix} Y_{n+2} \\ Y_{\alpha+2} \\ Y_{\beta+2} \\ Y_{\gamma+2} \end{bmatrix} \tag{g}$$

Curve B in Figure E11.4-1b illustrates the error for n, α, β, and γ, each differing by 10. With both Equations (f) and (g), the error in the largest estimated coefficient increased rapidly with the error representing the process noise.

LEAST SQUARES OF EQUATION ERROR. If Equation (11.4-7) is used in the estimation, curve C in Figure E11.4-1b shows the results for 100 consecutive samples.

DATA SMOOTHING FOLLOWED BY LEAST SQUARES OF EQUATION ERROR. A polynomial was fit to the input and output observations, the smoothed continuous time functions were sampled at the original sampling times, and the resulting values of X and Y were introduced into Equation 11.4-7. The smoothing, using a sixth-order polynomial, was not satisfactory because the polynomial deviated too far from the true response curve to be of use in estimation. Direct smoothing by a fifth-order formula followed by least squares of the equation error gave estimates no better than those shown by curves A and B in Figure E11.4-1b.

MAXIMUM LIKELIHOOD ESTIMATION. Curves D and E in Figure E11.4-1b illustrates maximum likelihood estimation for two different assumed covariance matrices $\mathbf{\Gamma}$.

Example 11.4-2 Sampling Times and Bias

Smith and Hilton† provided some information, by means of a Monte Carlo simulation study, on the effect of sample times and process error on the bias and variance of the estimates of parameters in a difference equation.

A deterministic step response was introduced into a process model whose transfer function was

$$\frac{1}{s^2 + 0.5s + 1} \tag{a}$$

† F. W. Smith and W. B. Hilton, *IEEE Trans. Auto Control* **AC-12**, 568, 1967.

A normal random error was added to the deterministic response (illustrated in Figure E11.4-1a). A z-transform with 4 parameters

$$\breve{g}(z) = \frac{b_1 z^{-1} + b_2 z^{-2}}{1 + a_1 z^{-1} + a_2 z^{-2}} \tag{b}$$

was estimated from data sampled at 4 different delay times ($\tau = 0.25$, 0.50, 1.0, and 2.0 seconds) for a duration of $-\tau \leq t \leq 8.0$ seconds. The signal-to-noise ratio was 100.

Figure E11.4-2 shows the bias in the parameters as estimated by two different methods: (1) the equation error, and (2) maximum likelihood, for the average of 25 different estimations. The correct values of the parameters in Equation (b) were

$$a_1 = -1.824 \qquad b_1 = 0.030$$
$$a_2 = 0.882 \qquad b_2 = 0.030$$

Table E11.4-2a gives the standard deviations calculated by Equation 11.4-11.

Figure E11.4-2 and Table E11.4-2a show that the bias for the maximum likelihood estimates for all τ is much less than the standard deviation. In contrast, the bias of the parameter estimates in the denominator by the equation error for $\tau \leq 0.50$ is two or more standard deviations in magnitude. Note also that for each τ the estimated coefficient by the equation error with the largest bias magnitude is more biased than the most biased coefficient for the

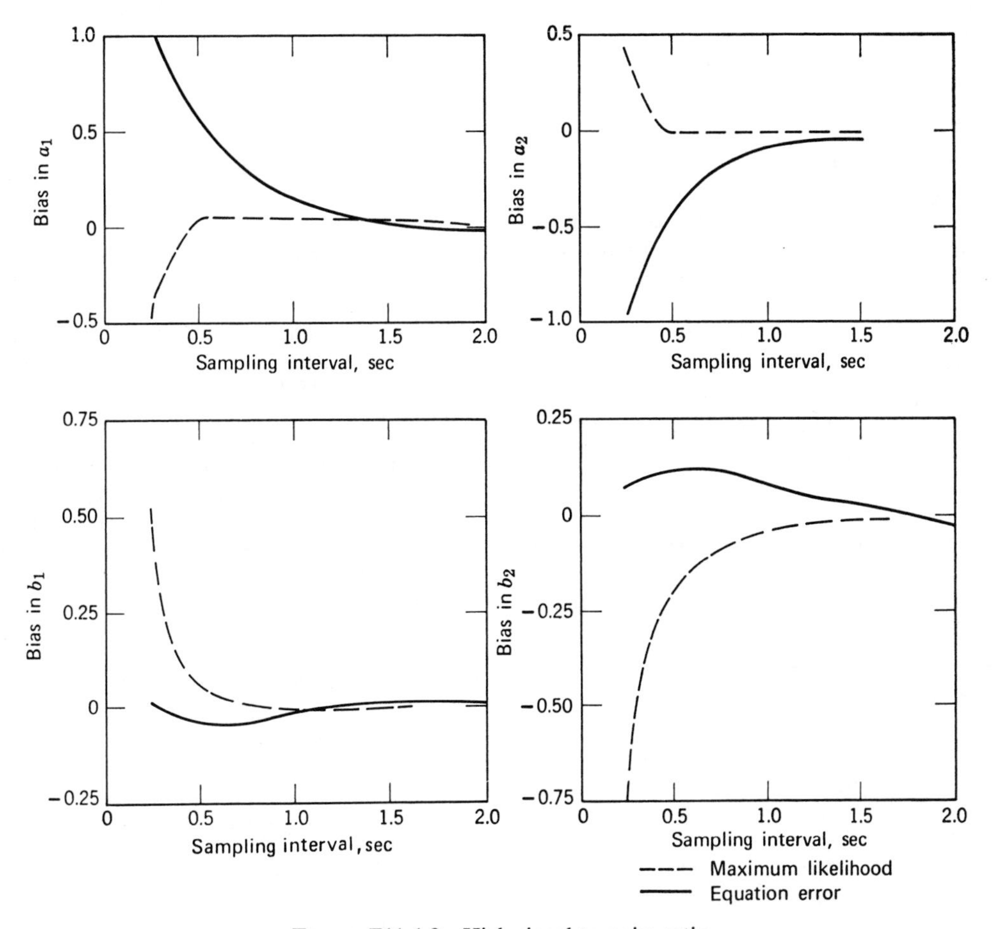

FIGURE E11.4-2 High signal-to-noise ratio.

TABLE E11.4-2a STANDARD DEVIATIONS BASED ON 25 ESTIMATES (APPROXIMATE TRUE VALUE IN PARENTHESES)*

Sampling Interval τ	a_1 I	a_1 II	a_2 I	a_2 II	b_1 I	b_1 II	b_2 I	b_2 II
0.25	0.109 (0.22)	1.825	0.125 (0.25)	2.133	0.172 (0.55)	1.457	0.152 (0.50)	1.191
0.50	0.108 (0.20)	0.221	0.151 (0.25)	0.312	0.187 (0.37)	0.175	0.154 (0.31)	0.103
1.00	0.116 (0.15)	0.137	0.169 (0.24)	0.223	0.205 (0.25)	0.252	0.138 (0.17)	0.148
2.00	0.173 (0.17)		0.578 (0.61)		0.485 (0.47)		0.159 (0.17)	

* I: equation error estimates; II: maximum likelihood estimates.

TABLE E11.4-2b BIAS AND DISPERSION AT LOW SIGNAL-TO-NOISE RATIOS*

	a_1 I	a_1 II	a_2 I	a_2 II	b_1 I	b_1 II	b_2 I	b_2 II
			Overlapping Sequences					
Normalized bias	32.92	−0.184	−28.56	0.186	−0.311	0.123	3.801	−0.166
Sample Standard Deviation	0.040	0.348	0.042	0.333	0.030	0.123	0.032	0.148
			Nonoverlapping Sequences					
Normalized Bias	16.76	0.205	−14.49	−0.203	0.711	−0.091	2.37	0.107
Sample Standard Deviation	0.078	7.49	0.081	7.438	0.043	0.705	0.056	2.25

* $\tau = 0.25$; signal-to-noise ratio = 4.25 sequences.

maximum likelihood estimate. Although the estimates by the equation error appear less biased at short sampling intervals, quite the opposite is true for low signal-to-noise ratios as shown in Table E11.4-2b. The normalized bias is the bias divided by the sample standard deviation.

Table E11.4-2c compares the square root of the covariance matrix elements calculated by Equation 11.4-11 with those obtained from 25 samples for $\tau = 1.0$ and a signal-to-noise ratio of 100. The matrix is symmetric. The standard deviations (on the main diagonal) are reasonably well estimated, but the elements on the off diagonal are quite different.

TABLE E11.4-2c (COVARIANCES)$^{1/2}$ CALCULATED BY EQUATION 11.4-11 (SAMPLE VALUES IN PARENTHESES)

	a_1	a_2	b_1	b_2
a_1	0.0648 (0.0711)			
a_2	0 (−0.481)	0.0902 (0.117)		
b_1	0 (−0.346)	0.696 (0.933)	0.160 0.199	
b_2	0 (0.276)	−0.591 (−0.906)	−0.849 −0.971	0.156 0.191

Supplementary References

Bigelow, S. C. and Ruge, H., "An Adaptive System Using Periodic Estimation of the Pulse Transfer Function," *IRE Nat. Convention Rec.*, Part 4, 24, 1961.

Ellington, J. P. and McCallion, H., "The Determination of a Control System Characteristics from a Transient Response," *Proc. IRE* **105**, Part C, 370, 1958.

Kalman, R. E., "Design of a Self-Optimizing Control System," *Trans. ASME* **80**, 468, 1958.

Levin, M. J., "Optimal Estimation of Impulse Response in the Presence of Noise," *IRE Trans.* **CT7**, 50, 1960.

Rutman, R. C., "Self-Adaptive Systems with Adjustment by Dynamic Characteristics," *Automatic and Remote Control* **23** (5), 1962.

Sanathan, C. K. and Koerner, I., "Transfer Function Synthesis as a Ratio of Two Complex Polynomials," *IEE Trans.* **AC-9**, 56, 1963.

Senf, B. and Stobel, H., "Methods for Determining Transfer Functions of Linear Systems from Measured Values of the Frequency Response," *Measurement, Control and Regulation* **10**, 411, 1961.

Shinbrot, M., "A Description and Comparison of Certain Nonlinear Curve Fitting Techniques with Applications to Analysis of Transient Data," *NACA Tech. Note* 2622, 1952.

Westcott, J. H., "The Problem of Parameter Estimation," *Proc. First Conf. International Fed. in Auto. Control* **3**, Butterworths, London, 1961.

Problems

11.1 As an accelerometer moves over the earth's surface, the deflection angles along a geodetic arc, Y_1, and transverse to the arc, Y_2, are random variables. Typical data are:

Distance, km	Y_1, sec	Y_2, sec
10	30.5	31.0
25	27.4	29.2
50	23.5	27.0
75	19.9	24.3
100	15.9	22.1
150	10.6	17.3
200	9.3	13.7
250	8.4	12.4
300	8.0	11.5
350	7.5	11.1
400	7.3	10.8
450	7.1	10.4
500	6.6	10.0
550	6.4	9.7

It has been suggested that Y_1, or perhaps Y_2, can best be represented by a transfer function containing a polynomial of the first to third degree in the complex parameter s in the denominator and a constant in the numerator. The independent variable is the distance.

Determine which order transfer function represents the data best and estimate the parameters in the transfer function. Three different techniques can be applied: (1) least squares, Equations 11.2-4, (2) Equation 11.2-8, and (3) orthogonal product method, Equation 11.3-9. Compare these methods with Laplace transformation of the data followed by least squares using Equations 11.2-10 and 11.2-12.

11.2 Figure P11.2 is a schematic representation of a well-mixed tank; hence the bulk temperature is the same as T in the tank. Process fluid enters at T_{in} at flow rate F. The fluid in the tank heats the wall (at T_w) which has resistance to heat transfer. In turn, the wall gives up heat to the exterior flowing fluid at T_f.

In the list below are the process response pen readings in mm from a strip chart at one-second intervals for a step down in T_{in} (the equilibrium value is 10 mm).† The chart ratio is 0.10°F/mm.

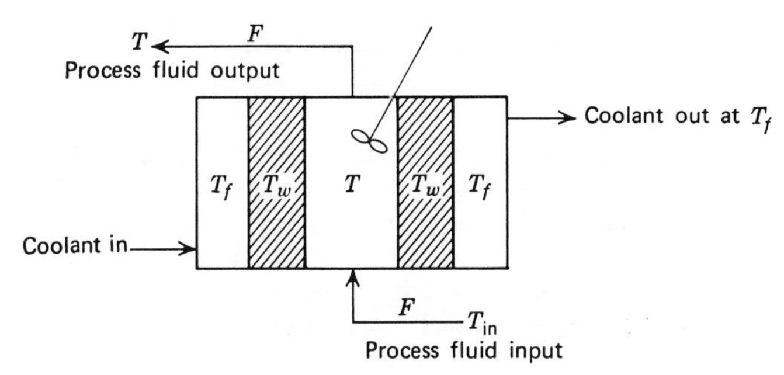

FIGURE P11.2

No.	mm	No.	mm
1	48.5	51	14.7
2	46.7	52	14.5
3	44.9	53	14.4
4	43.3	54	14.2
5	41.8	55	14.0
6	40.3	56	13.9
7	39.0	57	13.7
8	37.7	58	13.6
9	36.4	59	13.4
10	35.4	60	13.3
11	34.2	61	13.2
12	33.1	62	13.1
13	32.1	63	12.9
14	31.2	64	12.8
15	30.3	65	12.7
16	29.4	66	12.6
17	28.6	67	12.5
18	28.0	68	12.4
19	27.1	69	12.5
20	26.5	70	12.2
21	25.7	71	12.5
22	25.1	72	12.3
23	24.5	73	12.0
24	23.9	74	11.9
25	23.2	75	11.8
26	22.8	76	11.8
27	22.3	77	11.7
28	21.8	78	11.6
29	21.3	79	11.6
30	20.9	80	11.5
31	20.4	81	11.5
32	20.0	82	11.4
33	19.6	83	11.4
34	19.3	84	11.3
35	18.5	85	11.1
36	18.0	86	11.1
37	18.0	87	11.0
38	17.8	88	11.0
39	17.6	89	11.0
40	17.3	90	11.0
41	17.0	91	11.0
42	16.7	92	11.0
43	16.5	93	11.0
44	16.2	94	10.9
45	16.0	95	10.8
46	15.7	96	10.8
47	15.5	97	10.7
48	15.3	98	10.5
49	15.1	99	10.5
50	14.9	100	10.6

Because the actual tank may not be well mixed, a transfer function is to be used as the process model. What transfer function represents the data satisfactorily?

† K. A. Bishop, Ph.D. Dissertation, Univ. of Oklahoma, 1965.

11.3 The transfer function for a liquid extraction column† is

$$\check{G}(s) = \frac{T_G^N\sqrt{(T_LT_G - fg + 1)^2 - 4T_LT_G}}{T_LT_G(D_2^N - D_1^N - D_2^{N-1} + D_1^{N-1}) - fg(D_2^N - D_1^N)}$$

where

$$T_L = \frac{h_Ls}{L} + \frac{k_Ls}{L} + 1$$

$$T_G = \frac{h_Gs}{G} + \frac{k_L}{mG} + 1$$

$$f = \frac{k_L}{Lm}$$

$$g = \frac{k_L}{G}$$

s = Laplace transform parameter
L = 13.04 grams water/(min)(cm)^2
G = 11.0 grams organic/(min)(cm)^2
m = 0.766
k_L = interphase mass transfer coefficient, grams/(min)(cm^3)
h_G, h_L = total organic, water phase holdup, grams/cm^3, reported to be 0.075 and 0.5080, respectively
N = number of hypothetical lumped mixing cells
$D_1 = T_LA_1$
$D_2 = T_LA_2$

where A_1 and A_2 are the + and − roots, respectively, of

$$A_{1,2} = \frac{(T_LT_G - fg + 1) \pm \sqrt{(T_LT_G - fg + 1)^2 - 4T_LT_G}}{2T_L}$$

Estimate k_L and N from the given data for the pulse response. Repeat but estimate k_L, N, h_L, and h_G. Use the orthogonal products method. The impulse response is in terms of acid concentration in weight percent:

t (sec)	C × 10³	t (sec)	C × 10³
0	0	400	2.250
20	0.250	420	1.870
40	0.451	440	1.575
60	0.750	460	1.298
80	1.250	480	1.068
100	1.750	500	0.875
120	2.500	520	0.726
140	3.054	540	0.599
160	3.750	560	0.474
180	4.580	580	0.376
200	4.967	600	0.291
220	5.050	620	0.223
240	5.063	640	0.169
260	5.001	660	0.124
280	4.850	680	0.076
300	4.586	700	0.045
320	4.349	720	0.020
340	3.851	740	0.008
360	3.510	760	0
380	2.726		

† J. E. Doninger, Ph.D. Thesis, Northwestern Univ., Evanston, Ill., 1965.

11.4 Tseng‡ evaluated the dynamic response of a finned-tube heat exchanger. See Figure P11.4.

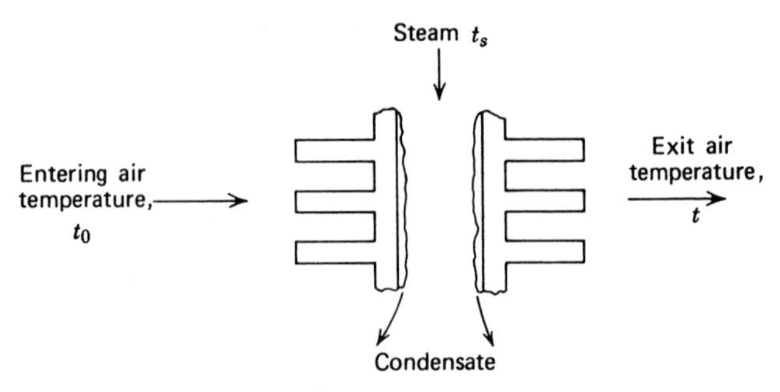

FIGURE P11.4

For the zero initial conditions, a general second-order transfer function (assuming the tube and fin temperatures were the same) was

$$\frac{\Delta t}{\Delta t_s}(s) = \left\{\tfrac{1}{2}C_mC_g\frac{1}{h_1h_2A_1A_2}s^2 + \left[\frac{1}{2}\left(\frac{1}{h_1A_1} + \frac{1}{h_2A_2}\right)C_g + K_aC_m\frac{1}{h_1A_1h_2A_2} + \tfrac{1}{2}C_m\frac{1}{h_1A_1}\right]s + \left(\frac{1}{h_1A_1} + \frac{1}{h_2A_2}\right)K_a + \tfrac{1}{2}\right\}^{-1} \quad \text{(a)}$$

where

Δ = change in temperature
C_m = 0.61 Btu/°F (metal total heat capacity)
C_g = 0.0021 Btu/°F (air total heat capacity)
A_1 = 0.905 ft^2 (inside heat transfer surface)
A_2 = 19.7 ft^2 (outside heat transfer surface)
K_a = 0.0194 × 60 × Q Btu/(°F)(hr)
Q = flow rate, ft^3/min
h_1, h_2 = heat transfer coefficients

A simplified first-order model transfer function is

$$\frac{\Delta t}{\Delta t_s}(s) = \left\{\tfrac{1}{2}C_g\left(\frac{1}{h_1A_1} + \frac{1}{h_2A_2}\right)s + K_a\left(\frac{1}{h_1A_1} + \frac{1}{h_2A_2}\right) + \tfrac{1}{2}\right\}^{-1} \quad \text{(b)}$$

while another hypothetical model transfer function is

$$\frac{\Delta t}{\Delta t_s}(s) = \frac{1}{\dfrac{1}{h_2A_2}C_ms + 1} \quad \text{(c)}$$

Inasmuch as both h_1 and h_2 are hard to determine individually, estimate h_1 for two models and h_2 for all three models from the following results obtained for a sinusoidal input with the air at 14.0 psia, Q = 864 scfm, steam pressure = 11 psig, t_0 = 96.0°F, and t = 126.0°F (average).

‡ Y. M. Tseng, Ph.D. Thesis, Univ. of Rhode Island, 1965.

Frequency (cycles/min), ω	Normalized Amplitude Ratio	Phase Angle (degrees)
0	1.00	0.0
1.17	0.99	5.5
2.26	0.93	15.3
3.24	0.87	26.2
4.62	0.82	41.6
5.72	0.74	41.1
6.80	0.68	46.3
7.90	0.59	50.3
8.10	0.67	58.7
10.66	0.56	58.9
12.50	0.48	73.9

Data have been corrected for a measurement lag of 1 sec. Reported values of h_1 and h_2 (from the literature) are

h_1	h_2
1900	11.1

11.5 Lamb† used a frequency response technique to evaluate the rate of absorption of sulfur dioxide gas into stagnant and turbulent water through a flat interface. The frequency response to sinusoidal pulsations in gas pressure was measured in terms of the difference in the gas pressures in two similar chambers, one with liquid and the other without.

For a stagnant model of the liquid, the approximate result for the amplitude of the difference of the two chambers was

$$\|G(\omega)\| = \frac{P_d - P_w}{P_d} = Q(1 - \sqrt{2}\,Q + Q^2)^{1/2}$$

where

$$Q = \frac{H\gamma S\sqrt{D/\omega}}{V + bS}$$

H = Ostwald absorption coefficient = 42.25 g-mole/cc liquid/g-mole/cc in gas at 25°C and (42.25)(1.1815) at 20°C

γ = heat capacity ratio, C_p/C_v, dimensionless

S = stagnant surface area, cm^2

D = diffusivity of gas in liquid, cm^2/sec

ω = frequency, radians/sec

V = average volume of gas chamber, cm^3

b = height of gas chamber above liquid, cm

$\dfrac{P_d - P_w}{P_w}$ = amplitude of difference signal in terms of pressures of dry (d) and wet (w) chambers

From the following data and Table P11.5, estimate D.

Surface—stagnant	S—668 sq cm
Surfactant—none	t—19.78°C
b—3.077, 3.070	γ—1.292
V—1108 cc.	

TABLE P11.5

Frequency (cps)	P_d (in H_2O)	P_w (in H_2O)	$\Delta P/P_w$
0.159	10.11	9.83	0.03398
	10.11	9.86	0.03173
0.250	10.21	9.98	0.02695
	10.21	10.01	0.02517
0.388	10.27	10.06	0.02404
	10.27	10.09	0.02271
0.612	10.31	10.14	0.01945
	10.31	10.17	0.01730
0.636	10.31	10.18	0.01539
	10.31	10.16	0.01809
0.999	10.33	10.21	0.01360
	10.33	10.21	0.01569
1.557	10.35	10.23	0.01233
	10.35	10.25	0.01265
2.445	10.36	10.28	0.00949
	10.36	10.26	0.01260

† W. B. Lamb, Ph.D. Thesis, Univ. of Delaware, Newark, 1965.

CHAPTER 12

Estimation in the Frequency Domain

Frequency domain estimation is employed in lieu of time domain estimated for two primary reasons, aspects of which have been brought out in connection with the use of moments and the transfer function:

1. The analytical solution of the model may be simpler in the frequency domain than the related solution of the model in the time domain.
2. For some models, analytical solutions can be obtained for the frequency response but not for the response in the time domain. Or, the model response may be in the form of a complicated series which is hard to evaluate numerically. Because of the character of the transformations carried out on both the experimental data and the model, the types of models suitable for analysis are restricted to those corresponding to *linear*, constant coefficient, ordinary or partial differential equations or to their equivalent in the frequency domain.

Impulse or pulselike inputs are the most convenient to use in experimentation because the pulse provides information at all process frequencies. Of course, the pulse may prove to be a significant disturbance to the process, large enough to invalidate the linearity assumption. Therefore, tests should be carried out to verify the linearity of the process by checking, say, with the aid of a sinusoidal input to see that the frequency of the output is not shifted from that of the input.

Along with estimation in the frequency domain, we shall also describe how the coefficients in the process model can be obtained from an applied stochastic input. For random inputs, it is important to ensure that the variations in the input cover the desired frequency range and that they have sufficient amplitude.

Two types of data processing can be employed: analog and digital. The data collected at the testing site can be stored on charts, magnetic or paper tape, etc., either as continuous or discrete functions of time, and processed elsewhere. Or the data can be processed in real time as taken. As a general rule-of-thumb, continuous data are collected when an analog or a hybrid computer is to be used in the analysis; individual frequencies are isolated by harmonic analysis (filtering). Discrete data, on the other hand, are analyzed by a digital computer, often by autocorrelation and cross-correlation methods.

12.1 THE PROCESS MODEL IN THE FREQUENCY DOMAIN

In Section 11.1, in connection with Equation 11.1-5, it was pointed out that the impulse response (also known as the weighting function) for a constant coefficient linear deterministic model corresponding to Equation 11.0-1 can be used to determine the response for any input:

$$y(t) = \int_0^\infty g(t - \tau)x(\tau)\, d\tau \qquad (12.1\text{-}1a)$$

Equation 12.1-1a can be shown to be identical to

$$y(t) = \int_0^\infty g(\tau)x(t - \tau)\, d\tau \qquad (12.1\text{-}1b)$$

by letting $t - \tau = \tau'$ and carrying out the change of variable. If we take the Fourier transform of $g(t)$, we obtain the *frequency response* function:

$$\check{g}(\omega) = \mathscr{F}[g(t)] = \int_{-\infty}^{\infty} g(t)\, e^{-i\omega t}\, dt = \int_0^\infty g(t)\, e^{-i\omega t}\, dt$$

where $\omega = 2\pi f$ has the units of radians per unit time and f has the units of cycles per unit time. The lower limit on the integral can be zero instead of $-\infty$ because $g(t) = 0$ for $t < 0$. Only modest restrictions are placed on the nature of $g(t)$, restrictions that can be found in any advanced calculus text. In effect, we have replaced s in the transfer function by $i\omega$, as mentioned in Chapter 11 in describing the relation between the frequency response and the transfer function.

The transfer function (frequency response) can be characterized in two ways in the frequency domain. Because $\check{g}(\omega)$ is complex, one way of representing the transfer function is

$$\check{g}(\omega) = \mathscr{R}[\check{g}(\omega)] + i\mathscr{I}[\check{g}(\omega)] \qquad (12.1\text{-}2)$$

where $\mathscr{R}[\check{g}(\omega)]$ is the real part of $\check{g}(\omega)$ and $\mathscr{I}[\check{g}(\omega)]$ is the imaginary part (Figure 12.1-1). In polar form, the absolute value or modulus of $\check{g}(\omega)$ is

$$\|\check{g}(\omega)\| = \sqrt{\mathscr{R}^2[\check{g}(\omega)] + \mathscr{I}^2[\check{g}(\omega)]} = r$$

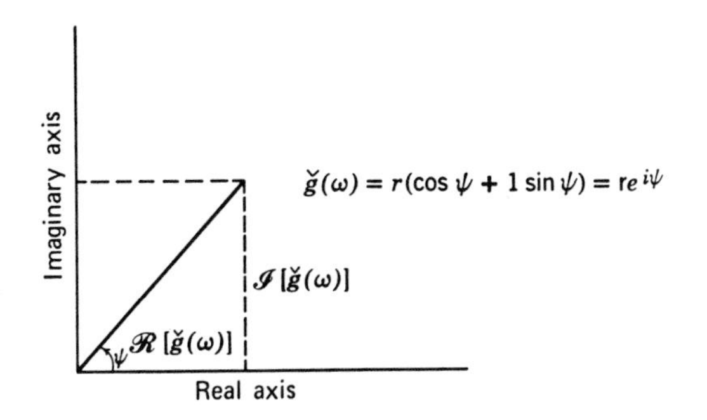

FIGURE 12.1-1 Polar representation of a complex number.

and the argument (angle) of $\check{g}(\omega)$ is ψ, where

$$\tan\psi = \frac{\mathscr{I}[\check{g}(\omega)]}{\mathscr{R}[\check{g}(\omega)]}$$

Thus a second method of representing the transfer function arises from its polar notation

$$\check{g}(\omega) = r(\cos\psi + i\sin\psi) = r\,e^{i\psi} = \|\check{g}(\omega)\|\,e^{i\psi(\omega)} \quad (12.1\text{-}3)$$

For any input $x(t)$, the model response is

$$\check{y}(\omega) = \check{g}(\omega)\check{x}(\omega) \quad (12.1\text{-}4)$$

Consequently, if $\check{g}(\omega)$ and $\check{x}(\omega)$ are expressed in polar form:

$$\check{y}(\omega) = \|\check{g}(\omega)\|\,e^{i\psi}\|\check{x}(\omega)\|\,e^{i\theta}$$

the magnitude of $\check{y}(\omega)$ is seen to be the product of the magnitudes of $\check{y}(\omega)$ and $\check{x}(\omega)$, and the angle of $\check{y}(\omega)$ is the sum of the respective angles

$$\check{y}(\omega) = \|\check{g}(\omega)\|\;\|\check{x}(\omega)\|\;e^{i(\psi+\theta)}$$

As a matter of interest, the frequency response in the *time domain*, i.e., the steady-state response for a sinusoidal input of frequency ω, is the product of the amplitude ratio (which is equal to $\|\check{g}(\omega)\|$ and the sin $(\omega t + \psi)$:

$$y(t) = \|\check{g}(\omega)\|\,\sin(\omega t + \psi)$$

as can be demonstrated by taking the Laplace or Fourier transform of $y(t)$, by taking the inverse Laplace or Fourier transform of Equation 12.4-1, or by applying the convolution theorem mentioned in Section 12.3-1.

In what follows, the observed output, $Y(t)$, will be regarded as being composed of the true response plus a stochastic error, $\epsilon(t)$ (see Figure 9.1-2), whose source may be the process itself or the data collection equipment. In any case the stochastic component will be assumed to be essentially stationary:

$$Y(t) = y(t) + \epsilon(t) \quad (12.1\text{-}5)$$

Figure 12.1-2 compares the confidence region in the time domain with the related confidence region in the frequency domain for a process response.

Least squares estimation of the parameters calls for minimizing

$$\phi = \int_0^\infty [Y(t) - y(t)]^2\,dt \quad (12.1\text{-}6)$$

with respect to the parameters. Because of the upper limit of Equation 12.1-6, $y(t)$ should preferably decrease with time as in a pulse response or decrease at the end of a period as in a sine wave. Parseval's equality:

$$\int_0^\infty [f(t)]^2\,dt = \frac{1}{\pi}\int_0^\infty \|\check{f}(\omega)\|^2\,d\omega \quad (12.1\text{-}7)$$

applied to Equation 12.1-6 gives, in the frequency domain,

$$\phi = \frac{1}{\pi}\int_0^\infty \|\check{Y}(\omega) - \check{y}(\omega)\|^2\,d\omega \quad (12.1\text{-}8a)$$

$$= \frac{1}{\pi}\int_0^\infty [\{\mathscr{R}[\check{Y}(\omega)] - \mathscr{R}[\check{y}(\omega)]\}^2 + \{[\mathscr{I}[\check{Y}(\omega)] - \mathscr{I}[\check{y}(\omega)]\}^2]\,d\omega \quad (12.1\text{-}8b)$$

where $\mathscr{R}[\;]$ is the real part of the Fourier transform in the argument enclosed by [ ] and $\mathscr{I}[\;]$ is the imaginary part. Equation 12.1-8b can be interpreted as calling for minimization of a criterion composed of two terms in the frequency domain, a combination of real and imaginary deviations squared.

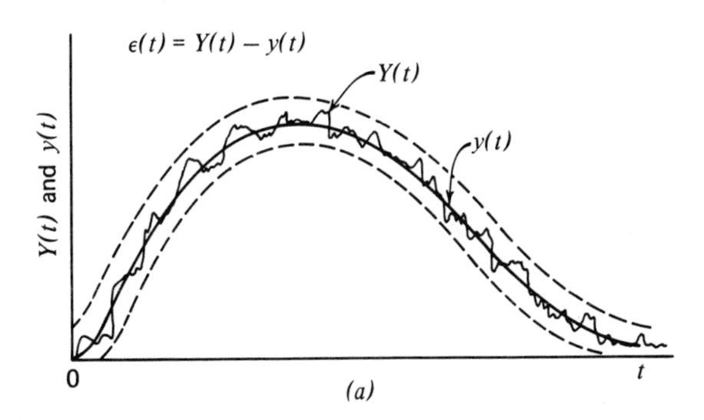

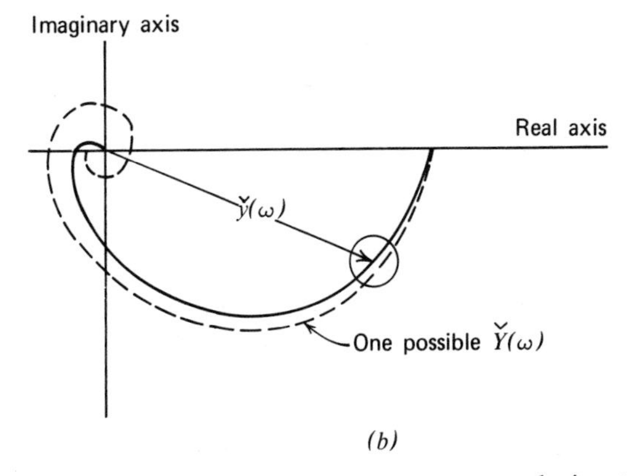

FIGURE 12.1-2 Process response represented in the time domain and in the frequency domain: (*a*) time domain, indicating the confidence region by dashed lines for stationary error, and (*b*) frequency domain, indicating the circle for the confidence region containing the head of the vector $\check{y}(\omega)$ at one frequency.

Equation 12.1-8a can be interpreted more clearly if both the observed and model responses in the frequency domain are put into polar form as follows (the ω is suppressed to save space):

$$\check{Y}(\omega) = V[\cos W + i \sin W]$$

$$\check{y}(\omega) = v[\cos w + i \sin w]$$

where $V(\omega)$ and $v(\omega)$ are the magnitudes of $\check{Y}(\omega)$ and $\check{y}(\omega)$, respectively, and $W(\omega)$ and $w(\omega)$ are the angles, respectively, all real. Substitution of these expressions into Equation 12.1-8a gives

$$\phi = \frac{1}{\pi}\int_0^{\infty} \|V(\cos W + i \sin W) - v(\cos w + i \sin w)\|^2 \, d\omega$$

$$= \frac{1}{\pi}\int_0^{\infty} \|(V \cos W - v \cos w) + i(V \sin W - v \sin w)\|^2 \, d\omega$$

$$= \frac{1}{\pi}\int_0^{\infty} [V^2 + v^2 - 2Vv \cos(w - W)] \, d\omega \qquad (12.1\text{-}9)$$

Inspection of Equation 12.1-9 indicates that the optimum choice of w should be such that $[2Vv \cos(w - W)]$ is a maximum or $w = W$. If so, the resulting equation is expressed only in terms of the amplitude ratios

$$\phi = \frac{1}{\pi}\int_0^{\infty} (V - v)^2 \, d\omega$$

and leads to the conclusion that $V = v$ for ϕ to be zero.

Equation 12.1-8a can also be related to the transfer function if we recall that the absolute value of a product is equal to the product of the individual absolute values. Then

$$\check{Y}(\omega) - \check{y}(\omega) = \check{x}(\omega)\left[\frac{\check{Y}(\omega)}{\check{x}(\omega)} - \frac{\check{y}(\omega)}{\check{x}(\omega)}\right]$$

$$= \check{x}(\omega)[\check{G}(\omega) - \check{g}(\omega)]$$

so that Equation 12.1-8a becomes

$$\phi = \frac{1}{\pi}\int_0^{\infty} [\|\check{x}(\omega)\| \, \|\check{G}(\omega) - \check{g}(\omega)\|]^2 \, d\omega \qquad (12.1\text{-}10)$$

In one sense, $\check{x}(\omega)$ is a "weighting function," because although the Fourier transform of a delta function is just 1, if the input is other than a delta function, its Fourier transform "weights" the expression $\|\check{G}(\omega) - \check{g}(\omega)\|$.

12.2 ESTIMATION USING DETERMINISTIC INPUTS

Simple pulselike inputs or triangular inputs with known Fourier transforms are the most appropriate to use, both because of their mathematic simplicity and because of their ease of execution as process inputs. Table 12.2-1 illustrates typical deterministic inputs and their Fourier transforms. The input pulse excites a wide range of frequencies of the process, a range whose exact characterization depends on the shape of the pulse. Hougen† discussed the advantages and disadvantages of various types of pulse shapes.

Certain problems exist in transforming the experimental data collected in the time domain into the frequency domain without introducing excessive numerical error. A number of algorithms are available for digital processing of *continuous* data; processing on an analog or hybrid computer, if available, is straightforward. Usually the Fourier transformation is truncated before ∞ is reached so that

$$\check{Y}(\omega) \simeq \int_0^{t_y} Y(t)\, e^{-i\omega t}\, dt$$

$$= \int_0^{t_y} Y(t) \cos(\omega t)\, dt + i \int_0^{t_y} Y(t) \sin(\omega t)\, dt \qquad (12.2\text{-}1)$$

where t_y is the termination time. The key to the successful transformation is to evaluate accurately these integrals for use in Equation 12.1-8. Several digital computer programs exist for executing what is known as the Fast Fourier Transform.‡

Hays *et al.*§ suggested that instead of using Equation 12.2-1 and a quadrature formula to evaluate the integrals in the equation for *discrete data* or for sampled continuous data, the following method is more flexible and efficient. The function $x(t)$ or $Y(t)$ in the time domain is characterized through a series of discrete points by a piecewise series of polynomial segments:

$$Y(t) \simeq \sum_{k=1}^{n} U(t - t_k)[a_k + b_k(t - t_k) + c_k(t - t_k)^2 + \cdots] \qquad (12.2\text{-}2)$$

where $U(t - t_k)$ is the unit step function, t_k is the time location of the observation, and a_k, b_k, etc., are coefficients selected to best represent the response between two observations. If only a_k and b_k are used, the data are represented by a piecewise series of straight line segments: if c_k is added, the curves are segments of parabolas, and so forth.

The Fourier transform of Equation 12.2-2 is

$$\check{Y}(\omega) \simeq \sum_{k=1}^{n} e^{-i\omega t_k}\left[\frac{a_k}{i\omega} + \frac{b_k}{(i\omega)^2} + \frac{c_k}{(i\omega)^3} + \cdots\right] \qquad (12.2\text{-}3)$$

† J. O. Hougen, "Experiences and Experiments with Process Dynamics," Monograph Ser. No. 4, AIChE, New York, 1964.

‡ J. W. Cooley, R. E. Miller, and S. Winograd, *Harmonic Analyzer*, IBM Watson Research Center, Yorktown Heights, N.Y., 1963; G. D. Bergland and H. W. Hale, "Digital Real-Time Spectral Analysis," *IEE Trans.* **EC-16**, 180, 1967.

§ J. R. Hays, W. C. Clements, and T. R. Harris, *AIChE J.* **13**, 374, 1967.

TABLE 12.2-1 TYPICAL DETERMINISTIC PULSELIKE INPUTS*

Type of Input	$x(t)$	$\check{x}(\omega)$
Impulse at $t = 0$	$\delta(t)$	1
Impulse at $t = t_0$	$\delta(t - t_0)$	$e^{-i\omega t_0}$
Triangular input initiated at $t = 0$	$c_0 t U(t) + c_1(t - t_1)U(t - t_1) + c_2(t - t_2)U(t - t_2)$	$-\frac{1}{\omega^2}[c_0 + c_1 \cos(t_1\omega) + c_2 \cos(t_2\omega)] + \frac{i}{\omega^2}[c_1 \sin(t_1\omega) + c_2 \sin(t_2\omega)]$

* $U(t)$ is the unit step function.

and $\check{Y}(\omega)$ can be split into its real and imaginary parts for introduction into Equation 12.1-8b

$$\mathscr{R}[\check{Y}(\omega)] \simeq \sum_{k=1}^{n} \left[-\frac{a_k}{\omega} + \frac{2c_k}{\omega^3} + \cdots\right] \sin \omega t_k + \left[-\frac{b_k}{\omega^2} + \frac{6d_k}{\omega^4} - \cdots\right] \cos \omega t_k$$

$$\mathscr{I}[\check{Y}(\omega)] \simeq \sum_{k=1}^{n} \left[-\frac{a_k}{\omega} + \frac{2c_k}{\omega^3} + \cdots\right] \cos \omega t_k - \left[-\frac{b_k}{\omega^2} + \frac{6d_k}{\omega^4} - \cdots\right] \sin \omega t_k$$

Truncation of the summation in Equation 12.2-3 with n too small a number will lead to a poor approximation of $Y(t)$ in the time domain. It can easily be shown by using Parseval's equality that the representation of $Y(t)$ by the truncated series is the best approximation in the least squares sense of $Y(t)$ that can be made with the truncated frequency spectrum.

Once Equation 12.1-8 or 12.1-10 has been formulated, the estimation problem corresponds exactly to the nonlinear estimation problem discussed in Chapter 6. Minimization of ϕ can be carried out by one of the numerical methods described in Section 6.2. The interpretation of the expected value and variance of the estimated parameters is the same as given in that chapter, even though the actual computations are executed in the frequency domain.

Example 12.2-1 Estimation in the Frequency Domain

A typical dispersion equation that was described in Section 10.4 is

$$\frac{\partial y}{\partial t} + v\frac{\partial y}{\partial z} = \tilde{D}\frac{\partial^2 y}{\partial z^2} \quad \text{(a)}$$

where y is the dependent variable, t is the time, v is the velocity, z is the axial direction, and $\tilde{D}$ is a dispersion coefficient to be estimated from observed values of the dependent variable, Y. Equation (a) applies to flow in packed beds, porous media, pollution in rivers, etc. It is assumed that $Y = y + \epsilon$. Only certain boundary conditions can be used in conjunction with Equation (a) if an analytical solution to the process model is to be written. For the initial and boundary conditions:

$$y(z, 0) = 0 \quad \text{(b)}$$

$$y(0, t) = x(t) \quad \text{(b)}$$

$$\lim_{z \to \infty} y(z, t) = 0$$

an analytical solution to Equations (a) and (b) cannot be formulated conveniently. However, a solution in the frequency domain can be formulated.

After taking the Laplace transform of Equation (a) and the boundary conditions in Equation (b),

$$\tilde{D}\frac{d^2\check{y}(z, s)}{dz^2} - v\frac{d\check{y}(z, s)}{dz} - s\check{y}(z, s) = 0 \quad \text{(c)}$$

$$\check{y}(0, s) = \check{x}(s)$$

$$\lim_{z \to \infty} \check{y}(z, s) = 0 \quad \text{(d)}$$

we can solve for $\check{y}(z, s)$ and evaluate the two arbitrary coefficients in the solution by using Equations (d) to obtain

$$\check{y}(z, s) = \check{x}(s) \exp\left[\frac{v - \sqrt{v^2 - 4\tilde{D}s}}{2\tilde{D}} z\right] \quad \text{(e)}$$

At some distance $z = L$, i.e., at the observation point, the dependent variable in the time domain is $y(L, t)$ and in the Laplace transform domain is

$$\check{y}(L, s) = \check{x}(s) \exp\left[\frac{vL}{2\tilde{D}}\left(1 - \sqrt{1 - \frac{4s\tilde{D}}{v^2}}\right)\right] \quad \text{(f)}$$

Note that the transfer function is the exponential quantity. The Fourier transform can be obtained by replacing s by $(i\omega)$:

$$\check{y}(L, \omega) = \check{x}(\omega) \exp\left[\frac{vL}{2\tilde{D}}\left(1 - \sqrt{1 - \frac{4(i\omega)\tilde{D}}{v^2}}\right)\right] \quad \text{(g)}$$

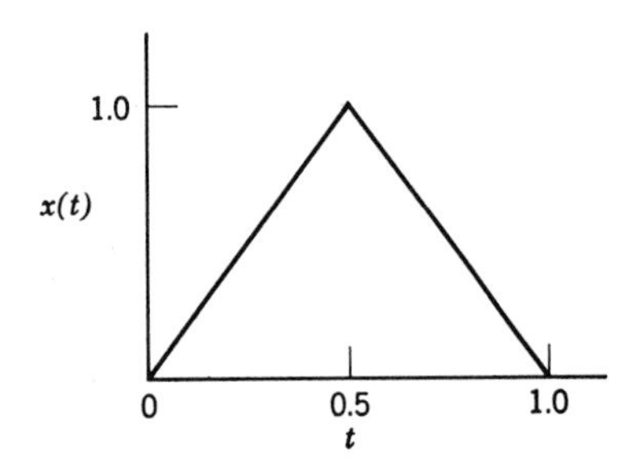

FIGURE E12.2-1

Equation (g) can be introduced into Equation 12.1-8b, or the transfer function itself can be introduced into Equation 12.1-10.

Suppose that a triangular pulse input, as illustrated in Figure E12.2-1, is introduced into the process tube or channel. The time domain representation of $x(t)$ is

$$x(t) = 2tU(t) - 4(t - \tfrac{1}{2})U(t - \tfrac{1}{2}) + 2(t - 1)U(t - 1)$$

and in the frequency domain

$$\check{x}(\omega) = -\frac{1}{\omega^2}[2 - 4\cos(0.5\omega) + 2\cos(\omega)] + \frac{i}{\omega^2}[-4\sin(0.5\omega) + 2\sin(\omega)] \qquad \text{(h)}$$

The value of $\check{x}(\omega)$ at $\omega = 0$ cannot be calculated from Equation (h) but must be obtained from the area under the $x(t)$ relation or $\check{x}(0) = \frac{1}{2}$. (A similar remark applies to $\check{Y}(\omega)$ at $\omega = 0$ and the area under the $Y(t)$ curve.)

Before Equations (g) and (h) are substituted into Equation 12.1-8a, it is necessary to write the transfer function as

$$\exp\left[\frac{vL}{2\tilde{D}}\right]\exp\left[-\sqrt{1 - \frac{4i\omega\tilde{D}}{v^2}}\right]$$

and to expand the term in the square root

$$\sqrt{1 - \frac{4i\omega\tilde{D}}{v^2}} \cong 1 - \frac{2i\omega\tilde{D}}{v^2} + \frac{2\omega^2\tilde{D}^2}{v^4} + \frac{3i\omega^3\tilde{D}^3}{v^6} - \cdots \qquad \text{(i)}$$

and to use the Euler identity

$$e^{i\theta} = \cos\theta + i\sin\theta$$

to separate the product $\check{x}(\omega)\,\check{g}(\omega)$ into real and imaginary parts. In any case, if too many terms are required in Equation (i), the estimation of $\tilde{D}$ by least squares is best undertaken by some other experimental arrangement (so that the boundary conditions can be revised) than that given by Equation (b). Clearly, a true delta function input would be mathematically preferable to a triangular input because $\check{x}(\omega)$ would then be 1.

Rather than integrate to ∞ on ω in Equation 12.1-8a, the integral can be truncated at ω_f where ω_f is selected so that the contribution of $\check{Y}(\omega_f)$ to the value of the integral is negligible. If analytical integration is not feasible in Equation 12.1-8a, a numerical quadrature scheme can be employed.

12.3 ESTIMATION USING STOCHASTIC INPUTS

This section indicates how stochastic inputs can be used to estimate model parameters in the frequency domain. For a single input and response, we combine Equations 12.1-1b and 12.1-5 to obtain

$$Y(t) = \int_0^\infty g(\lambda)x(t - \lambda)\,d\lambda + \epsilon(t) \qquad (12.3\text{-}1)$$

where $\epsilon(t)$ is the random error conceptually added to the response and λ is a dummy variable. Because sinusoidal or step inputs may disrupt the process, the possibility exists of using random inputs for $x(t)$, namely $X(t)$, either deliberately generated random inputs with small amplitudes or possibly the actual process inputs. We shall assume that $\epsilon(t)$ and $X(t)$ are weakly stationary in the sense defined by Equation 2.2-7, because if we do not make this assumption, extra terms will be left in the equations to be developed below that preclude a simple solution for the transfer function. Furthermore, Equation 12.3-1 is not compatible with dead time or lags in the process model; these must be determined separately.

As a criterion for estimation of the impulse reponse $g(t)$ for a *finite* length of record of the input and output, we can minimize

$$\phi = \int_0^{t_f} \epsilon^2(t)\,dt$$

or the mean square error

$$\phi = \frac{1}{t_f}\int_0^{t_f} \epsilon^2(t)\,dt \qquad (12.3\text{-}2)$$

From Parseval's equality, Equation 12.1-7, we can see that when $\epsilon(t)$ is white noise, that is in the frequency domain $\check{\epsilon}(\omega)$ is a constant, each frequency will be weighted equally in the least squares estimation. Weiner† applied the calculus of variations to minimize ϕ in Equation 12.3-2 and showed that

$$r_{XY}(\tau) = \int_0^\infty g(\lambda)r_{XX}(\tau - \lambda)\,d\lambda \qquad \text{for } \tau \geq 0 \quad (12.3\text{-}3)$$

where

$$r_{XY}(\tau) = \mathscr{E}\{X(t)Y(t + \tau)\}$$

$$r_{XX}(\tau - \lambda) = \mathscr{E}\{X(t)X(t + \tau - \lambda)\}$$

There is no guarantee that the estimate of $g(t)$ found by solving Equation 12.3-3 is good in the sense that it has a small variance compared to other possible estimates, but if the $\mathscr{E}\{\epsilon(t)\} = 0$, the estimate is unbiased. For the very special case in which $\epsilon(t)$ is "white noise" described in Section 12.3-1), the estimate obtained by further development of Equation 12.3-3 does have minimum variance among all linear estimates. For many other random

† N. Weiner, *The Extrapolation, Interpolation and Smoothing of Stationary Time Series with Engineering Applications*, John Wiley, New York, 1949.

errors, the least squares estimate asymptotically approaches the minimum variance estimate as the time record is extended.

Equation 12.3-3 can be obtained in a different fashion as follows. Replace t in Equation 12.3-1 by $t + \tau$, multiply both sides by $X(t)$, and then take the expected value of both sides of the resulting equation:

$$\mathscr{E}\{X(t)Y(t+\tau)\} = \mathscr{E}\left\{X(t)\int_0^\infty g(\lambda)X(t+\tau-\lambda)\,d\lambda\right\} + \mathscr{E}\{X(t)\epsilon(t+\tau)\} \qquad (12.3\text{-}4)$$

If the random input and $\epsilon(t)$ are uncorrelated, the second term on the right-hand side of Equation 12.3-4 vanishes. The operations of expected value and integration in time in the first term on the right-hand side can be reversed to obtain

$$r_{XY}(\tau) = \int_0^\infty g(\lambda) r_{XX}(\tau - \lambda)\,d\lambda \qquad (12.3\text{-}5)$$

which is the same as Equation 12.3-3 except for the restriction in 12.3-3 on τ which is encompassed by Equation 12.3-5. Therefore, the impulse response in Equation 12.3-5 is an optimum linear representation of the process. (For a nonlinear process, Equation 12.3-5 is still the best—in the mean square sense—linear representation.)

12.3-1 Power Spectrum

The *power spectrum* or *power spectral density function*,† $\check{s}_{XX}(\omega)$, is the Fourier transform of the autocorrelation function $r_{XX}(\tau)$; the *crosspower spectrum* $\check{s}_{XY}(\omega)$ is the Fourier transform of the crosscorrelation function $r_{XY}(\tau)$. Figure 12.3-1 illustrates the power spectral density functions corresponding to Figures 2.1-3 and 2.2-1. A reciprocal relation exists between the time and frequency domains through use of the Fourier inversion relation, namely

$$\check{s}_{XX}(\omega) = \int_{-\infty}^{\infty} e^{-i\omega\tau} r_{XX}(\tau)\,d\tau \qquad (12.3\text{-}6a)$$

$$r_{XX}(\tau) = \frac{1}{2\pi}\int_{-\infty}^{\infty} e^{i\omega\tau}\check{s}_{XX}(\omega)\,d\omega \qquad (12.3\text{-}6b)$$

and

$$\check{s}_{XY}(\omega) = \int_{-\infty}^{\infty} e^{-i\omega\tau} r_{XY}(\tau)\,d\tau = \check{s}_{YX}(\omega) \qquad (12.3\text{-}7a)$$

$$r_{XY}(\tau) = \frac{1}{2\pi}\int_{-\infty}^{\infty} e^{i\omega\tau}\check{s}_{XY}(\omega)\,d\omega \qquad (12.3\text{-}7b)$$

$\check{s}_{XX}(\omega)$, being the transform of an even function, always has zero phase angle; it is $\check{s}_{XY}(\omega)$ that contains the phase information. Figure 12.3-2 presents some autocorrelation functions for commonly used process inputs and their corresponding power spectral densities.

† If $X(t)$ is a voltage across a device and $Y(t)$ is the resulting input current, then $r_{XY}(0)$ is the average value of the power delivered. The spectrum of a nonperiodic function is obtained by taking its Fourier transform.

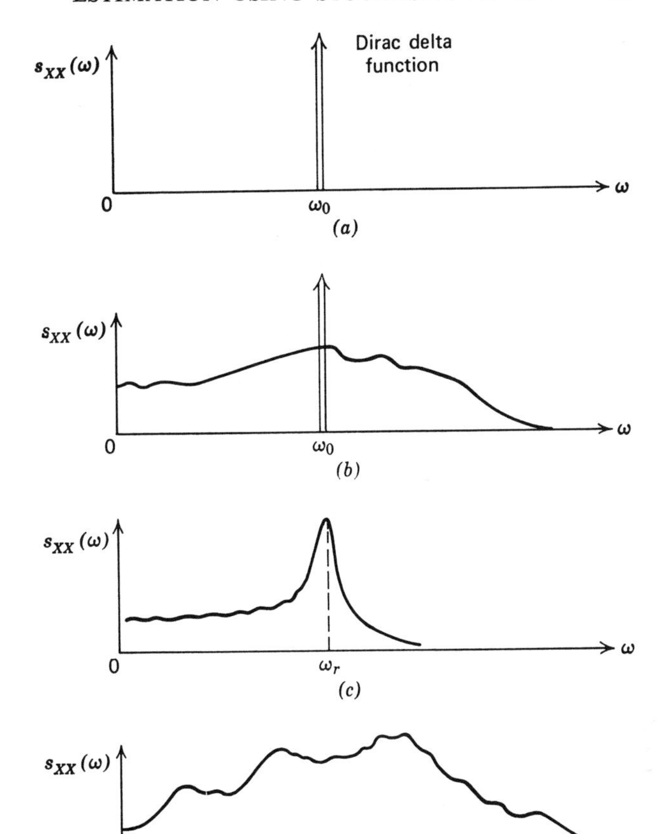

FIGURE 12.3-1 Power spectral density function plots (for $\omega \geq 0$) for the autocorrelation functions shown in Figure 2.2–1: (*a*) sine wave, (*b*) sine wave plus random noise, (*c*) narrow-band random noise, and (*d*) wide-band random noise. (From J. S. Bendat and A. G. Piersol, *Measurement and Analysis of Random Data*, John Wiley, New York, 1966, p. 24, with permission.)

The special case in which the power spectral density is a constant, the diagram in the fourth row of Figure 12.3-2, is known as *white noise*. White noise corresponds to an autocorrelation function for a stationary ensemble of $r(\tau) = k\cdot\delta(\tau)$ where k is a constant and $\delta(\tau)$ is the Dirac delta function. Because the average power becomes infinite for $\check{s}_{XX}(\omega) =$ constant, that is,

$$\text{average power} = r_{XX}(0) = \frac{1}{2\pi}\int_{-\infty}^{\infty}\check{s}_{XX}(\omega)\,d\omega \to \infty$$

a more useful concept is that of white noise over a limited frequency range, or *bandwidth limited white noise*:

$$\check{s}_{XX}(\omega) = k \qquad \left(-\frac{b}{2} \leq \omega \leq \frac{b}{2}\right)$$

$$= 0 \qquad \omega > \frac{b}{2}$$

where b is the *bandwidth* and ω is the frequency.

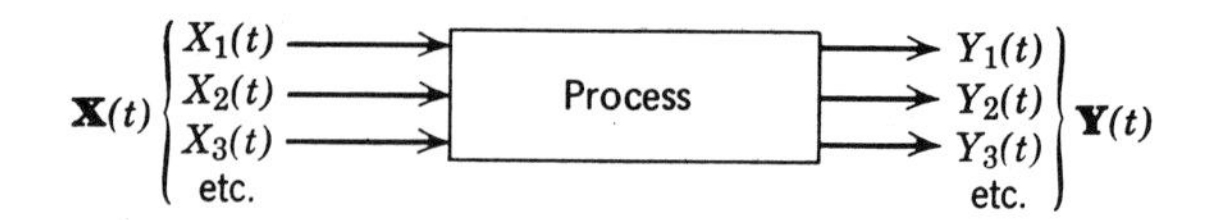

FIGURE 12.3-3 Multiple input-output process.

Type of input	Autocorrelation function $r_{XX}(\tau)$		Spectral density function $\check{s}_{XX}(\omega)$	
Random telegraph signal	$e^{-\alpha\|\tau\|}$		$2/\alpha$	
Random binary sequence	$1 - \frac{\|\tau\|}{t_m}$ if $\|\tau\| < t_m$; 0 if $\|\tau\| > t_m$			$\frac{4\sin^2(\omega t_m/2)}{\omega^2 t_m}$
Constant	c^2			$c^2\,\delta(\omega)$
White noise	$k\,\delta(t)$			k
Sinusoidal	$\cos(\omega_0\tau)$			$\delta(\omega + \omega_0) + \delta(\omega - \omega_0)$

FIGURE 12.3-2 Autocorrelation functions and their corresponding power spectral density functions.

By taking the Fourier transform of Equation 12.3-5, we obtain† a simple expression relating the transfer function in the frequency domain to the power spectrum and crosspower spectrum:

$$\check{s}_{XY}(\omega) = \check{g}(\omega)\check{s}_{XX}(\omega) \tag{12.3-8}$$

Also, by multiplying Equation 12.3-1 by $Y(t + \tau)$ and following the steps outlined before, it can be shown that

$$\check{s}_{YY}(\omega) = \|\check{g}(\omega)\|^2\check{s}_{XX}(\omega) \tag{12.3-9}$$

However, Equation 12.3-9 will not yield any information about the phase of the transfer function which is lost when the absolute value of $\check{g}(\omega)$ is taken; hence the equation has limited use.

With a few additional definitions, we can write an equation like (12.3-8) for the matrix transfer function $\check{\mathbf{g}}(\omega)$ defined in Equation 11.1-7 for a lumped process (or a set of lumped processes) with multiple inputs. See Figure 12.3-3. We shall let the vector of inputs $\mathbf{X}(t)$ contain n inputs and the vector of outputs $\mathbf{Y}(t)$ have m outputs. Also, three spectral density matrices must be defined:

$\check{\mathbf{s}}^X(\omega)$ = input spectral density matrix with $\check{s}_{ii}^X(\omega)$ the power spectral density of input i and $\check{s}_{ij}^X(\omega)$ the cross-spectral density of input i and input j

$\check{\mathbf{s}}^Y(\omega)$ = output spectral density matrix with $\check{s}_{ii}^Y(\omega)$ the power spectral density of output i and $\check{s}_{ij}^Y(\omega)$ the cross-spectral density of output i and output j

$\check{\mathbf{s}}^{XY}(\omega)$ = input-output spectral density matrix with $\check{s}_{ii}^{XY}(\omega)$ the cross-spectral density of input i and output i and $\check{s}_{ij}^{XY}(\omega)$ the cross-spectral density of input i and output j

Then a simple extension of the single variable analysis leads to

$$\check{\mathbf{s}}^Y(\omega) = \check{\mathbf{g}}^*(\omega)\check{\mathbf{s}}^{XY}(\omega) = \check{\mathbf{g}}^*(\omega)\check{\mathbf{s}}^X(\omega)\check{\mathbf{g}}^T(\omega) \tag{12.3-10}$$

$$\check{\mathbf{s}}^{XY}(\omega) = \check{\mathbf{s}}^X(\omega)\check{\mathbf{g}}^T(\omega) \tag{12.3-11}$$

where $\check{\mathbf{g}}^*(\omega)$ is the complex conjugate of $\check{\mathbf{g}}(\omega)$ and T indicates transpose. The transfer matrix $\check{\mathbf{g}}(\omega)$ is calculated from

$$\check{\mathbf{g}}(\omega) = \{[\check{\mathbf{s}}^X(\omega)]^{-1}\check{\mathbf{s}}^{XY}(\omega)\}^T \tag{12.3-12}$$

12.3-2 Experimental Design for Random Process Inputs

Random process inputs in general are selected so as to simplify the solution for the transfer function (frequency response) in the frequency domain. For the special case in which the autocorrelation function of the input is $r_{XX}(\tau) = k\,\delta(t)$, the spectral density is white noise, $\check{s}_{XX}(\omega) = k$, and the transfer function becomes just

$$\check{g}(\omega) = \frac{1}{k}\,\check{s}_{XY}(\omega) \tag{12.3-13}$$

The estimation of the coefficients in the empirical transfer function, the estimate of $\check{g}(\omega)$, would be carried out as described in Chapter 6 under nonlinear estimation if the data are discrete.

To create white noise as a physically realizable input in a flow process is exceedingly difficult; generation of white noise in a computer simulation is much easier. Consequently, it proves easier to use inputs that are not random but only pseudorandom yet whose autocorrelation functions approximate the delta function.

One pseudorandom sequence with useful properties can be defined as follows:

1. The input assumes values of $+a$ or $-a$ with equal probability.

† The *convolution* of two functions $f_1(t)$ and $f_2(t)$:

$$f(t) = \int_{-\infty}^{\infty} f_1(\lambda)f_2(t - \lambda)\,d\lambda$$

has the Fourier transform

$$\check{f}(\omega) = \int_{-\infty}^{\infty} e^{-i\omega t}f(t)\,dt = \int_{-\infty}^{\infty} e^{-i\omega t}\int_{-\infty}^{\infty} f_1(\lambda)f_2(t - \lambda)\,d\lambda\,dt$$

$$= \int_{-\infty}^{\infty} f_1(\lambda)\int_{-\infty}^{\infty} e^{-i\omega t}f_2(t - \lambda)\,dt\,d\lambda$$

With $t - \lambda = \tau$, the last integral on the right-hand side becomes

$$\int_{-\infty}^{\infty} f_1(\lambda)\,e^{-i\omega\lambda}\int_{-\infty}^{\infty} e^{-i\omega\tau}f_2(\tau)\,d\tau\,d\lambda = \check{f}_1(\omega)\check{f}_2(\omega)$$

where $\check{f}_1(\omega)$ and $\check{f}_2(\omega)$ are the Fourier transforms of $f_1(t)$ and $f_2(t)$, respectively.

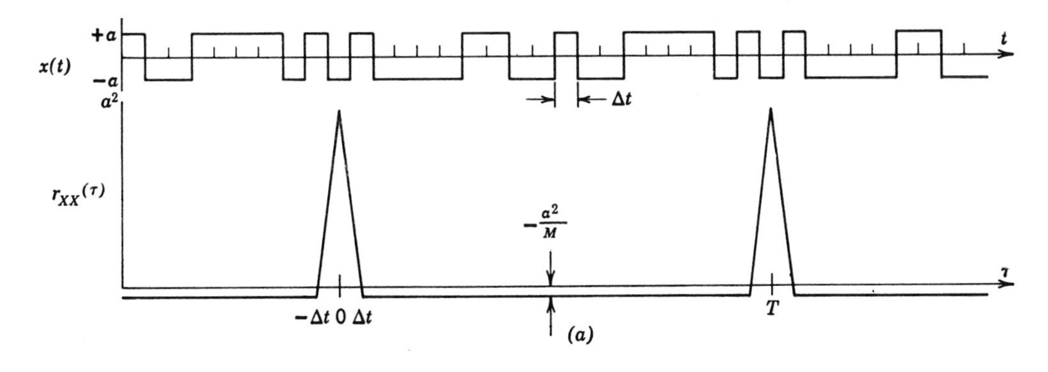

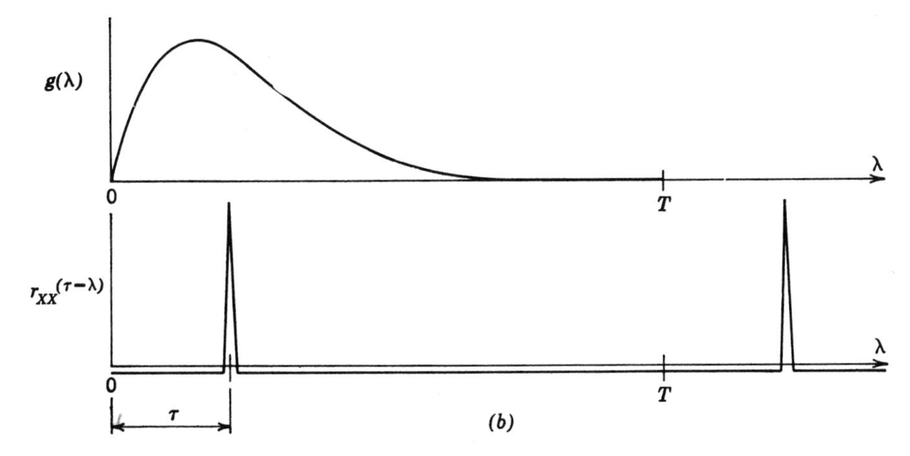

FIGURE 12.3-4 A pseudorandom input with a desirable autocorrelation function. (*a*) Idealized periodic binary signal input for $M = 19$ and the corresponding autocorrelation function. (*b*) Graphical representation of Equation 12.3-5 for the periodic binary input.

2. The input is periodic with a period $T = M\,\Delta t$, where M is the number of basic intervals for switching from $+a$ to $-a$ (bits in a binary code).

Figure 12.3-4a illustrates a typical input and its autocorrelation function, which is somewhat akin to a delta function in each interval 0 to T but is triangular in shape. The input† can be considered to be an ordered sequence of binary elements $c_0, \ldots, c_{M-1}$ such that there is one more $-a$ than $+a$. Consequently,

$$\sum_{i=0}^{M-1} c_i = -a$$

and

$$\sum_{i=0}^{M-1} c_i c_{(i+j)\,\text{modulo}\,M} = \begin{cases} M, j = 0 \\ -a, j \neq 0 \end{cases}$$

The second sum ensures that the autocorrelation function of the input is sufficiently "spiky" and can approximate a delta function of strength equal to the area of the "spike." The sequence of c_i's that determine whether the input is $+a$ or $-a$ for the next Δt can be calculated from a linear recursion relationship given in Zierler and also in Peterson.‡

Table 12.3-1 lists a FORTRAN program for calculating pseudorandom inputs as $+1$ and -1 for the sequence shown Figure 12.3-4. Another possible input is the pseudorandom telegraph input which appears similar to the input in Figure 12.3-4 but cycles from $+1$ to -1 at randomly occurring time intervals prescribed by a Poisson distribution. The ensemble autocorrelation function is illustrated in the first row of Figure 12.3-2. Another possible input is the random binary sequence which again is similar to the input shown in Figure 12.3-4 except that it cycles from $+1$ to -1 at discrete intervals Δt randomly selected; the autocorrelation function and spectral density are in the second row of Figure 12.3-2. King and Woodburn§ described how to choose the values of the period of the input T, the sampling interval, and the basic binary interval Δt, i.e. design the experiment, so as to optimize the determinant described in Section 8.4.

† N. Zierler, *J. Soc. Ind. Appld. Math.* 7, 31, 1959.

‡ W. W. Peterson, *Error Correcting Codes*, MIT Press and also John Wiley, New York, 1961.

§ R. P. King and R. P. Woodburn, *Ind. Eng. Chem.*, a personal communication, 1968.

TABLE 12.3-1 FORTRAN PROGRAM TO CALCULATE PSEUDO-RANDOM INPUTS

```
      PROGRAM RANDOM(INPUT, OUTPUT)
C $$$ NX IS THE NUMBER OF DIGITS IN THE SERIES
      DIMENSION IA(3000)
    4 READ 100,NX
  100 FORMAT(I5)
      IF(NX)2,2,3
    3 PRINT 101,NX
  101 FORMAT(23H TOTAL NO. OF ELEMENTS =,I5//)
      IZ = -1
      DO 5 M = 1,NX
      IA(M) = IZ
    5 CONTINUE
      Y = 0.5*FLOAT(NX)
      LA = 1
      IZ = 1
      A = 1.0
    1 B = A*A
      X = FLOAT(NX)
      D = AMOD(B,X)
      ID = IFIX(D + 0.5)
      IA(ID) = IZ
      LA = LA + 1
      A = FLOAT(LA)
      IF(Y - A - 0.001)6,6,1
    6 PRINT 102, (IA(M),M = 1,MX)
  102 FORMAT(10(10X,15I5/)/)
      PUNCH 103, (IA(M),M = 1,NX)
  103 FORMAT(15I5)
      GO TO 4
    2 CALL EXIT
      END
```

Finally, the following question can be posed: Can the transfer function be estimated from the normal plant operating records? While in principle such estimation is possible, in practice it has not often proved feasible for the following reasons. First, the normal process input may not be a stationary random variable. Second, because of the nature of the data processing that must take place, normal plant operating records usually do not prove accurate enough to provide suitable data. The introduction of adequate instrumentation can overcome this difficulty but may prove expensive. A third problem is the long length of operating record required for the correlation between the process noise and the deterministic portion of the process input to die out.

Finally, a more serious drawback is that the process input will not only be random but will include controlled changes that must be taken into account in the correlation analysis. For example, consider the controlled process in Figure 12.3-5. The process input contains a component of $\epsilon(t)$ by virtue of the connecting link through the controller. Consequently, the term $\mathscr{E}\{X(t-\tau)\epsilon(t)\}$ in Equation 12.3-4 no longer vanishes and must be re-

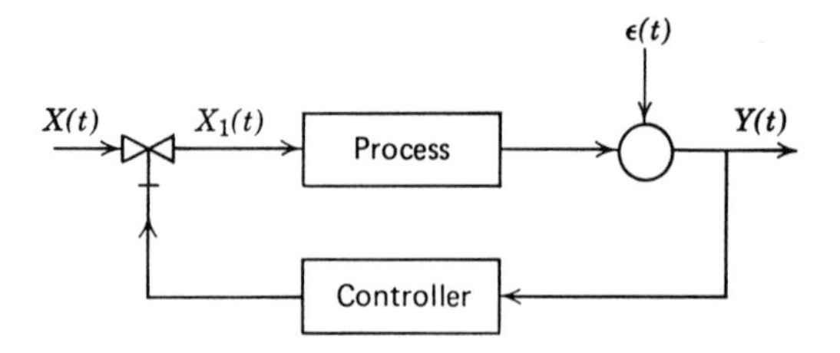

FIGURE 12.3-5 A controlled process with a random input.

tained. Under certain special assumptions about $\epsilon(t)$, it has proved possible to divide the crosscorrelation function $R_{XY}(\tau)$ into two parts, one of which is disturbance free.† But how to treat the process data for more realistic $\epsilon(t)$ is as yet unresolved.‡

12.3-3 Estimation of Spectral Densities

To estimate the transfer function using either Equation 12.3-8, 12.3-9, 12.3-12, or 12.3-13, it is necessary to have adequate estimates of the ensemble average spectral density $\check{s}_{XY}(\omega)$ and perhaps $\check{s}_{XX}(\omega)$. Although the estimates might be determined from sample averages as described in Chapter 2, they are usually calculated from time averages. Time averages make use of a single time record rather than a collection of observations from different time records; they will be designated by $\langle\ \rangle$ unless the meaning is otherwise clear from the text. Time averages are a less powerful method of estimating ensemble averages than sample averages because the latter involve repetitive experiments. We shall be primarily interested in the following two time averages, which are random variables and hence designated by capital letters, for continuous variables:

1. time autocorrelation function

$$R_{XX}(\tau) = \langle X(t)X(t+\tau)\rangle = \frac{1}{t_f}\int_0^{t_f} X(t)X(t+\tau)\,dt$$

2. time crosscorrelation function

$$R_{XY}(\tau) = \langle X(t)Y(t+\tau)\rangle = \frac{1}{t_f}\int_0^{t_f} X(t)Y(t+\tau)\,dt$$

where t_f is the end of the time record. For discrete variables, the time averages are:

1. time autocorrelation function

$$R_{XX}(\tau) = \frac{1}{n}\sum_{k=1}^{n} X(t_k)X(t_k+\tau)$$

† T. P. Goodman and J. B. Reswick, *Trans. Amer. Soc. Mech. Eng.* **78**, 259, 1956.

‡ J. H. Westcott, *Proceed. First Int. Congress of Int. Fed. Auto. Control*, Vol. 2, ed. by J. F. Coales, Butterworths, London, 1961, p. 779; N. R. Goodman and S. Katz, *Math. Comp.* **13**, 289, 1959.

2. time crosscorrelation function

$$R_{XY}(\tau) = \frac{1}{n}\sum_{k=1}^{n} X(t_k)Y(t_k + \tau)$$

where n is the number of data samples.

A real stationary ensemble is termed an *ergodic* ensemble if the time averages equal the corresponding ensemble averages with a probability of one. A process can be ergodic with respect to certain selected parameters or with respect to all of them. It is not enough that the expected value of the time average equal the expected value of the ensemble average; ergodicity *also* requires that the *variance of the time average* tends to zero as the time interval $t_f \to \infty$. Without the latter qualification we cannot state that the expected value of the time average equals the ensemble expected value with a probability of one. The time averages yield estimates which are random variables whereas the ensemble averages do not, hence the necessity of requiring the variance of the random variable to be zero to ensure that the expected values of the former averages are indeed equal to the ensemble averages with a probability of one. For a stationary process (but not necessarily ergodic), it can be shown that the expected value of the correlation functions computed either for continuous or discrete data is equal to the ensemble correlation function:

$$\mathscr{E}\{R_{XX}(\tau)\} = r_{XX}(\tau)$$

$$\mathscr{E}\{R_{XY}(\tau)\} = r_{XY}(\tau)$$

meaning that the respective time averages are unbiased estimates of the ensemble averages. The computation of the variances of $R_{XX}(\tau)$ and $R_{XY}(\tau)$ is quite difficult because fourth-order moments are involved.

Fourier transformation of the time average estimates of the correlation functions gives estimates of the ensemble spectral densities:

$$\check{S}_{XX}(\omega) = \int_{-\infty}^{\infty} e^{-i\omega\tau}R_{XX}(\tau)\,d\tau$$

$$\check{S}_{XY}(\omega) = \int_{-\infty}^{\infty} e^{-i\omega\tau}R_{XY}(\tau)\,d\tau$$

But these estimates prove to be biased estimates of the ensemble spectral densities. Furthermore, the dispersion of the spectral density estimates does *not* decrease with increasing length of the time record (or number of data points for discrete data).

It then becomes necessary to smooth either the correlation function estimates or the spectral density estimates by some appropriate weighting function in order to obtain unbiased estimates of the ensemble spectral densities with reasonable variance. The expected values of the smoothed correlation functions will usually prove to be poor estimates of the ensemble correlation functions. Nevertheless the corresponding Fourier transforms of the smoothed correlation functions, i.e., the spectral densities, will prove to be good estimates of the ensemble spectral densities. Figure 12.3-6 illustrates the information flow in the data processing.

To smooth the autocorrelation functions suitably, they can be multiplied by a weighting function, termed the *window lag*, in the time domain. Alternately, the spectral densities can be multiplied by the Fourier transform of the window lag, termed the spectral window; hence, one refers to window pairs. Lag windows average the correlation function in the neighborhood of a given time giving greater weight to near values and none to remote values. Each time serves in turn as a center for the window so that the entire time range is covered. Lag windows $w(\tau)$ are suitable even functions of τ, such as shown in Figure 12.3-7, subject to the restriction that $w(0) = 1$ and $w(\tau) = 0$ for $|\tau| > t_m$. After multiplication by the window lag, the modified correlation functions are

$$\tilde{R}_{XX}(\tau) = w(\tau)R_{XX}(\tau)$$

$$\tilde{R}_{XY}(\tau) = w(\tau)R_{XY}(\tau)$$

where $R(\tau)$ is defined only for $|\tau| \leq t_m$ but $\tilde{R}(\tau)$ is defined for all τ. As a result, the modified correlation functions are defined for all τ and have valid Fourier transforms:

$$\tilde{\check{S}}_{XX}(\omega) = \check{w}(\omega)\check{S}_{XX}(\omega)$$

$$\tilde{\check{S}}_{XY}(\omega) = \check{w}(\omega)\check{S}_{XY}(\omega)$$

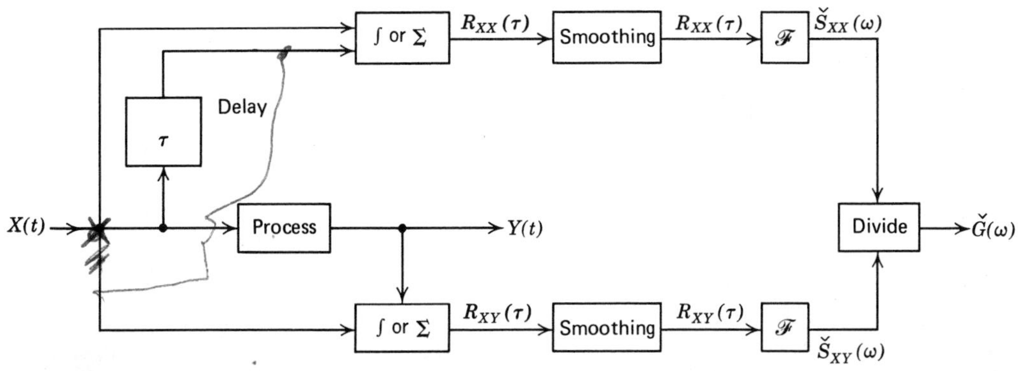

FIGURE 12.3-6 Information flow to calculate the transfer function.

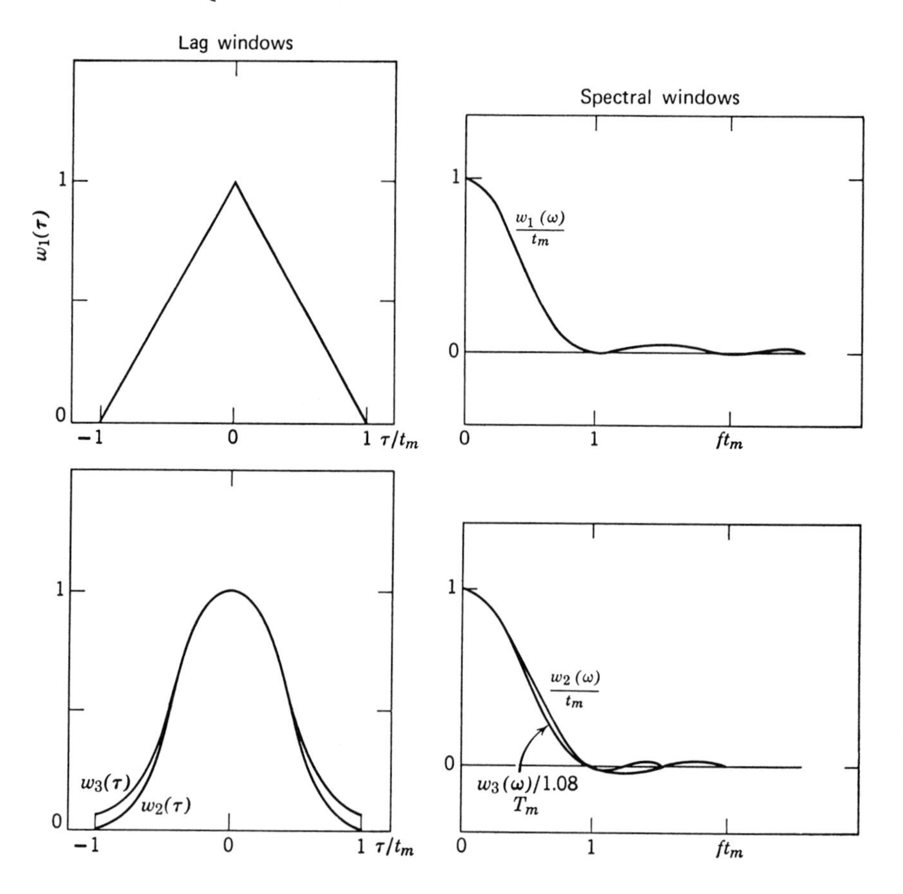

FIGURE 12.3-7 Three frequently used window pairs:

1 Bartlett

$$w_1(r) = 1 - \frac{\|r\|}{t_m} \qquad \text{if } \|\tau\| \le t_m$$
$$= 0 \qquad \text{if } \|\tau\| > t_m$$
$$w_1(\omega) = t_m\left(\frac{\sin \frac{1}{2}\,\omega t_m}{\frac{1}{2}\omega t_m}\right)^2$$

2 Hanning

$$w_2(r) = \tfrac{1}{2}\left(1 + \cos\frac{\pi\tau}{t_m}\right) \qquad \text{if } \|\tau\| \le t_m$$
$$= 0 \qquad \text{if } \|\tau\| > t_m$$
$$w_2(\omega) = \tfrac{1}{2}Qo(\omega) + \tfrac{1}{2}Qo\left(\omega + \frac{\pi}{t_m}\right) + Qo\left(\omega - \frac{\pi}{t_m}\right)$$
$$Qo(\omega) = 2t_m\frac{\sin \omega t_m}{\omega t_m}$$

3 Hanning

$$w_3(\tau) = 0.54 + 0.46\cos\frac{\pi\ \tau}{t_m} \qquad \text{if } \|\tau\| \le t_m$$
$$= 0 \qquad \text{if } \|\tau\| > t_m$$
$$w_3(\omega) = 0.54Qo(\omega) + 0.23\left[Qo\left(\omega + \frac{\pi}{t_m}\right) + Qo\left(\omega - \frac{\pi}{t_m}\right)\right]$$

More important than the specific window shape among the windows in Figure 12.3-7, all of which yield estimates that are barely distinguishable, is the choice of the factor that regulates the width of the window in the time domain, either t_m or the number of data points encompassed by the window m, or its equivalent, the bandwidth b, in the frequency domain. The bandwidth of a window actually refers to the width of a rectangular window. If the window is not rectangular, then some equivalent bandwidth, b_e, is implied such as $b_e = 2\pi/m$. In general, an empirical approach to the selection of m seems to be the most successful. A large bandwidth is initially chosen by making m small, and the effect of decreasing the bandwidth is observed (usually in the frequency domain) by increasing m. Provided that a value of m is reached beyond which no additional fine detail is revealed, the value of m can be fixed. On the other hand, m may prove to be too large in comparison with the number of available data points in the interval 0 to t_f so that the spectrum becomes erratic; i.e., the variance of the spectrum (which is inversely proportional to the bandwidth) becomes unduly large. The choice of a window must rest on judgment and the nature of the correlation function for unusual correlation functions. If the spectrum has a sharp resonant peak, and a broad window is used, the peak is spread thinly over a wide frequency range and may pass unnoticed. To obtain high accuracy for a spectral function having marked changes in amplitude over a small frequency range, a narrow window is needed, but less smoothing takes place and greater variance results for a given length of record. Thus, for a given accuracy, a greater length of record is required than would be needed if the spectral density had been smoother. The minimum frequency that can be computed, $f_{\min} = 1/2\tau_m$, depends on the length of the maximum lag, τ_m, used in calculating the correlation functions.

If continuous data processing equipment or suitable continuous recording equipment is available at the test site, the data can be processed by analog or hybrid computers as continuous data. However, if the experimental data are to be analyzed by a digital computer, the data are usually sampled and stored in digital form. The smoothed spectral density can be computed from a series of n equally spaced values of X and Y in time:

$$\tilde{\check{S}}_{XY}(\lambda\,\Delta\omega) = \sum_{j=-N}^{N} w(j\,\Delta t)R_{XY}(j\,\Delta t)\exp(-i\lambda\,\Delta\omega j\,\Delta t) \tag{12.3-14}$$

$$R_{XY}(j\,\Delta t) \cong \frac{1}{N+1-j}\sum_{k=0}^{N-j} X(k\,\Delta t)Y(k\,\Delta t + j\,\Delta t) \qquad j \geq 0 \tag{12.3-15}$$

$$\cong \frac{1}{n+1+j}\sum_{k=0}^{N+j} X(k\,\Delta t - j\,\Delta t)Y(k\,\Delta t) \qquad j < 0$$

where

$i = \sqrt{-1}$

j = integer multiplier of Δt used to designate a lag, τ

N = number of time intervals between samples; $N + 1 = n$ = number of data samples

Δt = basic binary interval for switching = sampling interval if one sample is taken in each interval; $N\,\Delta t = t_f$ where t_f is the end of the time record

$w(j\,\Delta t)$ = lag window

λ = integer used with the frequency interval to denote the frequency

A corresponding pair of equations can be written for $\tilde{\check{S}}_{XX}(\omega)$ and $R_{XX}(\tau)$ and $\check{S}_{YY}(\omega)$ and $R_{YY}(\tau)$ if the indexes are exchanged.

The accuracy of the estimation of the transfer function, and by implication the coefficients in the transfer function, depends upon the selected values of the maximum lag τ_m, the window width t_m, the switching interval Δt, and the length of the time record t_f, as well as the time constant of the lumped process. As a general rule-of-thumb, correlation functions are not computed for lags greater than 5 to 10 percent of the total record length. For the type of input shown in Figure 12.3-4 and a white noise random error ($R_{\epsilon\epsilon}(t) = a_0^2\,\delta(t)$), Hughes† approximated the mean square error in the impulse response function as

$$\overline{[G(\tau)]^2} \simeq \frac{a_0^2}{a^2(\Delta t)^2 qT} = \frac{a_0^2}{a^2(\Delta t)^3 qM} \tag{12.3-16}$$

where a, Δt, M, and T are defined in connection with Figure 12.3-4, and q is a positive integer representing the number of periods of length T used in the estimate of $\check{g}(\omega)$. Equation 12.3-16 indicates that, all other things being equal, increasing a, Δt, q, and T all decrease the dispersion of $G(\tau)$. Because the dispersion varies inversely with Δt to a large power, the bandwidth of the input should be as small as feasible.

Although the input is constant during each interval Δt so that the input can be sampled at any time, the output changes continuously during the interval Δt and it does make a difference as to when the output is sampled. If we assume that k output samples are taken per interval, they should be taken at times equal to

$$\frac{2n+1}{2k}\Delta t \qquad n = 0, 1, 2$$

where k is the number of samples per interval. Thus, for one sample per Δt, the sample should occur at $\frac{1}{2}\Delta t$, $\frac{3}{2}\Delta t$, etc., that is in the middle of each successive interval.

Two practical limitations which exist in the estimation of the transfer function will now be considered briefly:

1. *Folding* or *aliasing* of information at higher frequencies into information at lower frequencies.
2. Process nonlinearities

If the continuous data are periodic and are sampled at intervals of Δt, no information can be obtained from frequencies higher than those with a period of $2\Delta t$, corresponding to a frequency of $f_{\max} = 2/2\Delta t$ or $\omega_{\max} = 2\pi f_{\max} = \pi/\Delta t$, termed the *turnover*, *folding*, or *Nyquist*

† M. I. G. Hughes, *Inst. Elec. Eng.* **109** (Part B), 77, 1962.

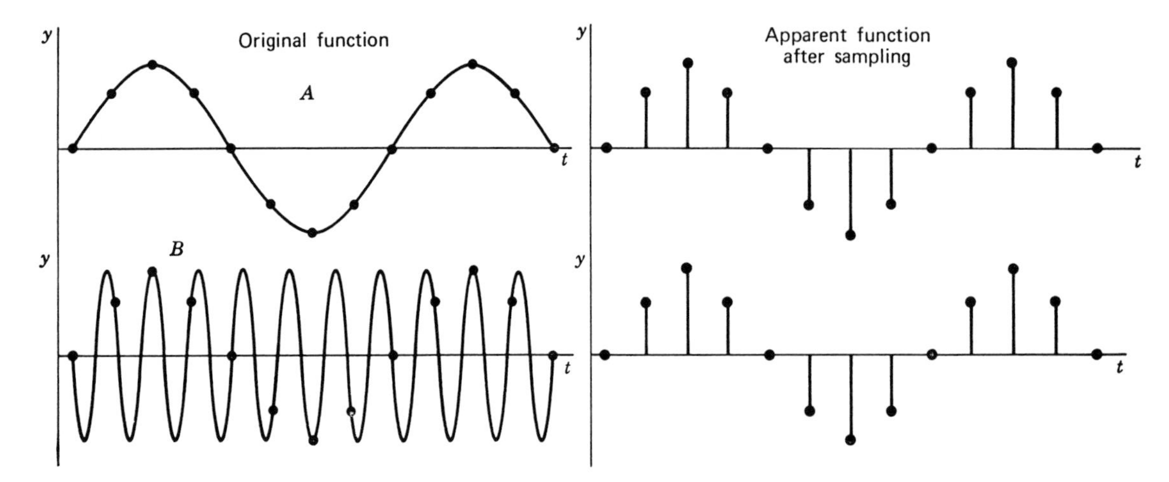

FIGURE 12.3-8 Example of folding (dots indicate samples).

frequency. Higher frequencies are not ignored but are "folded" back into the lower frequencies and confused with data at the lower frequencies, resulting in incorrect estimates of $\check{s}(\omega)$. Figure 12.3-8 demonstrates the concept of folding. Functions A and B are sampled at the same rate, but the sampling frequency for A is well below $\omega_{\max}$ whereas the sampling frequency for B is much higher than $\omega_{\max}$. Function A is unchanged by the sampling, but Function B now appears to be essentially the same as A. The phenomenon of wagon wheels turning backward in movies at slow speeds is an example of aliasing. As the speed of revolution approaches and exceeds the frame repetition rate, the spokes on the wheels appear to change from rotating at a very low speed backward to a slow speed forward. Consequently, although as few samples as possible are desired to reduce the data processing, the sampling frequency must be at least twice and preferably five to ten times the highest frequency of interest. If the process exhibits nonlinearities, the assumption that ϵ and X are uncorrelated will be violated and the estimates of $\check{g}(\omega)$ distorted. In any actual experiment, it is desirable, insofar as feasible, to examine the output amplitude as the input amplitude is changed at some given frequency, or to see if the output frequency is the same as the input frequency. Such tests disclose any significant nonlinearities present. A step input can be used to get a rough idea of the process time constants if the process is modeled as a lumped system. From these values the frequency range of interest can be selected. A record of at least four times the largest time constant of the process should suffice to let $G(t)$ settle to its final value.

12.3-4 Precision of Transfer Function Estimates

Both systematic and random errors can arise and be merged into the calculations used to estimate the transfer function. *Systematic errors* include such factors as:

1. Improper initial conditions.
2. Periodicity in the input.
3. Bias in the input.
4. Improper approximation of a delta function input, which is equivalent to bandwidth limitations on the test signal.
5. Interference from other inputs.
6. Instrument error.
7. Quantization error.

Such errors can be kept small by appropriate choice of the test input and proper control of the experiment. The error in the transfer function due to initial conditions and/or periodicity of the test signal can be made of the order of 1 percent or less by applying the input at least one signal period prior to observation and by making this period at least five times the dominant time constant of the system.

The bias error can be shown to be inversely proportional to M. Unfortunately, mean square errors due to wide-band noise are proportional to M, so the bias error cannot, in general, be made suitably small by choosing M large. As a result, a correction must be made based on measurement of $R_{XY}(\tau)$ over regions of its argument where the transfer function is expected to be negligible. In the presence of significant amounts of noise, the estimation of the bias would have to be made by using several independent values of $R_{XY}(\tau)$.

The error due to the delta function approximation can be shown to be related to the ratio of the input bandwidth to the output bandwidth. When this ratio is of the order of 14:1, the fractional error introduced is of the order of 0.5 percent. Instrument limitations, either caused by nonlinearity or distortion of the record at high and low frequencies, must be avoided by calibration and choice of proper instruments. A sinusoidal input passed into the instruments to check bandwidth and linearity validates instrument performance. If the process record is quantized, the discrete intervals chosen must be small enough to make the error negligible. Florentin *et al.*† gave a comprehensive account of the possible sources of systematic error and the possible methods of error reduction.

Let us now turn to the *random errors* $\epsilon(t)$ of observation. While it is possible to estimate the variances of the estimated autocorrelation function, crosscorrelation function, power spectral density, and cross-spectral density for certain types of random error, these variances are of lesser interest than the variance of the estimated transfer function itself and the variances of the estimated coefficients in the estimated transfer function. We shall assume that random uncorrelated errors are added to the deterministic input and output as shown in Figure 12.3-9:

$$X(t) = x(t) + \epsilon_0(t) \qquad (12.3\text{-}17)$$

$$Y(t) = y(t) + \epsilon(t) \qquad (12.3\text{-}18)$$

By multiplying each equation by itself with the argument of $(t + \tau)$, taking the expected value of both sides to get the correlation functions, and then taking the Fourier transforms, we obtain

$$\check{s}_{XX}(\omega) = \check{s}_{xx}(\omega) + \check{s}_{\epsilon_0\epsilon_0}(\omega)$$

$$\check{s}_{YY}(\omega) = \check{s}_{yy}(\omega) + \check{s}_{\epsilon\epsilon}(\omega)$$

$$\check{s}_{XY}(\omega) = \check{s}_{xy}(\omega)$$

The correlation functions between x and ϵ_0, y and ϵ, and ϵ and ϵ_0 are assumed to be zero. We also assume that the random errors $\epsilon(t)$ and $\epsilon_0(t)$ are stationary so that the observed process input $X(t)$ is stationary. The tests described in Chapter 3 can be used to verify these assumptions.

The coherence function between the input and output $\check{\gamma}^2_{XY}(\omega)$, defined as

$$\check{\gamma}^2_{XY}(\omega) = \frac{\|\check{s}_{XY}(\omega)\|^2}{\check{s}_{XX}(\omega)\check{s}_{YY}(\omega)} \qquad 0 \le \check{\gamma}^2_{XY}(\omega) \le 1 \qquad (12.3\text{-}19)$$

is a measure of the linear dependence between X and Y in the frequency domain corresponding to the square of the correlation coefficient in the time domain. To observe the effect of the additive errors, the coherence function can also be expressed as follows:

$$\begin{aligned}
\check{\gamma}^2_{XY}(\omega) &= \frac{\|\check{s}_{xy}(\omega)\|^2}{[\check{s}_{xx}(\omega)+\check{s}_{\epsilon_0\epsilon_0}(\omega)][\check{s}_{yy}(\omega)+\check{s}_{\epsilon\epsilon}(\omega)]} \\
&= \frac{\|\check{s}_{xy}(\omega)\|^2}{\check{s}_{xx}(\omega)\check{s}_{yy}(\omega)\left[1+\dfrac{\check{s}_{\epsilon\epsilon}(\omega)}{\check{s}_{yy}(\omega)}+\dfrac{\check{s}_{\epsilon_0\epsilon_0}(\omega)}{\check{s}_{xx}(\omega)}+\dfrac{\check{s}_{\epsilon_0\epsilon_0}(\omega)\check{s}_{\epsilon\epsilon}(\omega)}{\check{s}_{xx}(\omega)\check{s}_{yy}(\omega)}\right]} \\
&= \frac{1}{1+\dfrac{\check{s}_{\epsilon\epsilon}(\omega)}{\check{s}_{yy}(\omega)}+\dfrac{\check{s}_{\epsilon_0\epsilon_0}(\omega)}{\check{s}_{xx}(\omega)}+\dfrac{\check{s}_{\epsilon_0\epsilon_0}(\omega)\check{s}_{\epsilon\epsilon}(\omega)}{\check{s}_{xx}(\omega)\check{s}_{yy}(\omega)}} \qquad (12.3\text{-}20)
\end{aligned}$$

For the special case of no error in the input, making use of Equation 12.3-9,

$$\check{\gamma}^2_{XY}(\omega) = \frac{1}{1+\dfrac{\check{s}_{\epsilon\epsilon}(\omega)}{\check{s}_{yy}(\omega)}} = \frac{1}{1+\dfrac{\check{s}_{\epsilon\epsilon}(\omega)}{\check{s}_{xx}(\omega)\|\check{g}(\omega)\|^2}} \qquad (12.3\text{-}21)$$

If $\check{s}_{\epsilon\epsilon}(\omega) \ll \check{s}_{yy}(\omega)$, then

$$\check{\gamma}^2_{XY}(\omega) \approx 1 - \frac{\check{s}_{\epsilon\epsilon}(\omega)}{\check{s}_{yy}(\omega)}$$

Goodman‡ made use of the coherence function to obtain the following approximate confidence statement for the gain and phase angle when $\widehat{\check{G}(\omega)}$ is an unbiased estimate of $\check{g}(\omega)$ and sampled data are used:

$$\begin{aligned}
&\text{Probability}\left\{\left\|\frac{\|\widehat{\check{G}(\omega)}\| - \|\check{g}(\omega)\|}{\|\check{g}(\omega)\|}\right\| < \sin\theta \right. \\
&\qquad \left. \text{and } \|\widehat{\psi(\omega)} - \psi(\omega)\| < \theta\right\} \\
&\qquad \simeq 1 - \left[\frac{1-\check{\gamma}^2_{XY}(\omega)}{1-\check{\gamma}^2_{XY}(\omega)\cos^2\theta}\right]^{k/2} \qquad (12.3\text{-}22)
\end{aligned}$$

where θ is the limiting error (in radians). The number of degrees of freedom per spectral calculation point, k, for sampled data is calculated as follows. If $N = t_f/\Delta t$ is defined as the number of sample intervals, and N_m is defined as the number of intervals in the maximum lag:

$$N_m = \begin{cases} \dfrac{\tau_m}{\Delta t}, & \tau_m < t_m \\ \dfrac{t_m}{\Delta t}, & t_m < \tau_m \end{cases}$$

then $k \approx 2N/N_m$ for all frequencies $0 < \omega < \omega_{\max}$.§ At $\omega = 0$, $k = t_f/\tau_m$. For continuous data, $k = 2b_e t_f$ where b_e is the equivalent bandwidth of the input and window.

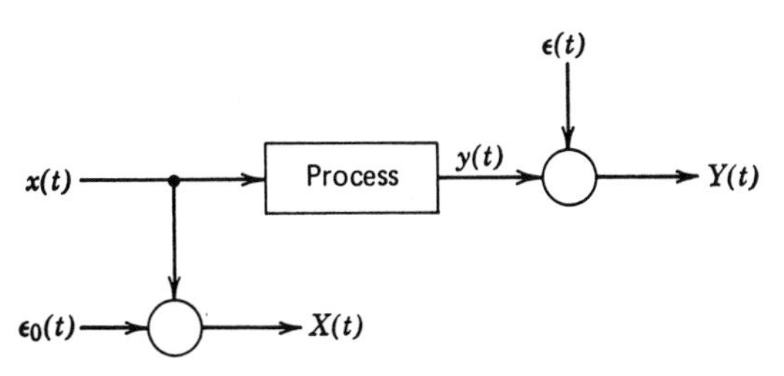

FIGURE 12.3-9 Additive measurement errors.

† J. J. Florentin, B. D. Hainsworth, J. B. Reswick, and J. H. Wescott, *Joint Symp. on Instr. and Comp.*, Inst. of Chem. Eng., London, 1959, p. 18.

‡ N. R. Goodman, *Technometrics* 3, 245, 1961.

§ R. B. Blackman and J. W. Tukey, *The Measurement of Power Spectra*, Dover, New York, 1959.

Because $\check{\gamma}^2_{XY}(\omega)$ is not known, a conservative estimate must be substituted; hence the confidence statement will be only approximate. Once a first estimate of the coherence function has been made, the approximate number of degrees of freedom needed to measure the frequency response to the desired accuracy can be calculated from Equation 12.3-22. Then $\check{g}(\omega)$ is estimated for the indicated degrees of freedom and $\widehat{\check{\gamma}^2_{XY}(\omega)}$ is reevaluated.

As an example of the application of Equation 12.3-22, consider an experiment in which

$$\omega = 0.100 \text{ radian}$$

$$\mathscr{R}[\check{S}_{XY}(0.100)] = 0.01287$$

$$\mathscr{I}[\check{S}_{XY}(0.100)] = 0.04822$$

$$\check{S}_{XX}(0.100) = 0.09875$$

$$\check{S}_{YY}(0.100) = 0.003153$$

Number of data points = 240

Time of record = 1 hr

Window lag time = 10 min

Maximum time lag for correlation = 20 min

Hence,

$$\widehat{\check{\gamma}^2_{XY}(0.100)} = 0.800$$

$$\widehat{\check{\gamma}_{XY}(0.100)} = 0.89$$

and

$$k = \frac{2(60)}{10} = 12$$

For a confidence coefficient of 0.90,

$$0.90 = 1 - \left[\frac{1 - 0.80}{1 - 0.80\cos^2\theta}\right]^6$$

from which $\theta = 0.349$ radian and $\sin\theta = 0.342$. Hence,

$$\left\|\frac{\widehat{\|\check{G}(\omega)\|}}{\|g(\omega)\|} - 1\right\| < 0.342 \quad \text{and} \quad \|\widehat{\psi(\omega)} - \psi(\omega)\| < 20.0°$$

To ascertain the maximum and minimum limits on the frequency, note that $\Delta t = \frac{60}{240} = \frac{1}{4}$ second so that $f_{\max} = 1/[2(\frac{1}{4})] = 2$ cycles/min or $\omega_{\max} = 2(2\pi) = 12.56$ radians/min. The minimum frequency is $f_{\min} = 1/[2(20)] = 0.025$ cycle/min or 0.157 radian/min.

Jenkins† gave approximate expressions for the variances of $\widehat{\check{\gamma}^2_{XY}(\omega)}$, $\|\widehat{\check{G}(\omega)}\|$, and $\widehat{\psi(\omega)}$ if large samples are taken:

$$\text{Var}\{\widehat{\check{\gamma}_{XY}(\omega)}\} \approx \frac{a_k}{2}[1 + \check{\gamma}^2_{XY}(\omega)] \tag{12.3-23}$$

$$\text{Var}\{\|\widehat{\check{G}(\omega)}\|\} \approx \frac{a_k}{2}|\check{g}(\omega)|^2\left[\frac{1}{\check{\gamma}^2_{XY}(\omega)} - 1\right] \tag{12.3-24}$$

$$\text{Var}\{\widehat{\psi(\omega)}\} \approx a_k\left[\frac{1}{\check{\gamma}^2_{XY}(\omega)} - 1\right] \tag{12.3-25}$$

where

$$a_k = \tfrac{1}{2}\sum_{k=-N}^{+N} w_k^2$$

where w_k is the window in the time domain with $k = \tau/\Delta t$, or $\tau = k\,\Delta t$, and $m = t_m/\Delta t$, or $t_m = m\,\Delta t$, replacing τ and t_m. The particular quadrant for the phase angle has to be obtained by independent means. Equation 12.3-24 indicates that as $\check{\gamma}^2_{XY}(\omega) \to 0$, the Var $\{\|\widehat{\check{G}(\omega)}\|\} \to \infty$, and as $\check{\gamma}^2_{XY}(\omega) \to 1$, the Var $\{\|\widehat{\check{G}(\omega)}\|\} \to 0$. Jenkins also showed that to a first order of approximation the Covar $\{\|\widehat{\check{G}(\omega)}\|, \widehat{\psi(\omega)}\} = 0$, indicating that the gain and phase angle can be treated separately. If $p(\|\widehat{\check{G}(\omega)}\|)$, the probability density of the estimated gain, is represented approximately by a χ^2 probability density with the number of degrees of freedom defined by

$$\nu = \frac{4}{a_k\left[\frac{1}{\check{\gamma}^2_{XY}(\omega)} - 1\right]}$$

the confidence limits for a confidence coefficient of $(1 - 2\alpha)$ for the log $[\|\check{g}(\omega)\|]$ are

Lower	Upper
$\log \frac{\nu\vert\widehat{\check{G}(\omega)}\vert}{\chi^2_{1-\alpha}}$	$\log \frac{\nu\vert\widehat{\check{G}(\omega)}\vert}{\chi^2_{\alpha}}$

where χ^2_α and $\chi^2_{1-\alpha}$ are, respectively, the lower and upper α percentage points of the χ^2 density.

For multiple inputs and outputs, the *coherency matrix*, taking into account the correlation which may exist between inputs, is

$$\check{\boldsymbol{\gamma}}^2 = [\{[\check{\mathbf{s}}^X(\omega)]^{-1}\check{\mathbf{s}}^{XY}(\omega)\}^T]^*\check{\mathbf{s}}^{XY}(\omega)[\check{\mathbf{s}}^Y(\omega)]^{-1} \tag{12.3-26}$$

where the * designates conjugate. In the absence of error, $\check{\boldsymbol{\gamma}}^2 = \mathbf{I}$. Each element of $\check{\boldsymbol{\gamma}}^2$ can be used in Equation 12.3-22 at a given frequency.

Example 12.3-1 Estimation of the Transfer Function by Correlation Analysis

Gallier, Sliepcevich, and Puckett‡ used a simple lumped (well-mixed) tank as a heat exchanger to estimate the process transfer function by the correlation technique. Figure E12.3-1a illustrates the experimental arrangement. Both the

† G. M. Jenkins, Chapter 18 in *Time Series Analysis*, ed. by M. Rosenblatt, John Wiley, New York, 1963.

‡ P. W. Gallier, C. M. Sliepcevich, and T. H. Puckett, *Chem. Eng. Progress Symp.* (Ser. No. 36) **57**, 59, 1961.

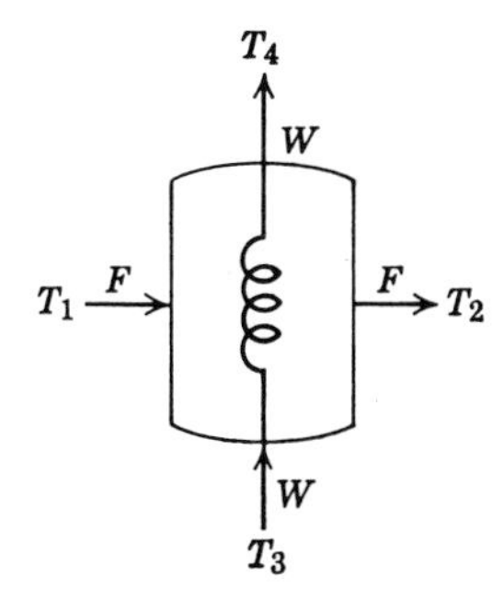

FIGURE E12.3-1a

heating and cooling fluids were water. The mathematical model for this process is†

1. Hot outside fluid

$$\rho_2 C_{p_2} V \frac{dT_2}{dt} = \rho_1 C_{p_1} F T_1 - \rho_2 C_{p_2} F T_2 - UA(T_2 - T_4) \quad \text{(a1)}$$

Accumulation input output interphase transfer

2. Cold inside fluid

$$C_{p_4} M \frac{dT_4}{dt} = W C_{p_3} T_3 - W C_{p_4} T_4 + UA(T_2 - T_4) \quad \text{(a2)}$$

Accumulation input output interphase transfer

where

A = outside area of coolant coils, ft^2
C_p = heat capacity, Btu/(lb)(°F)
F = volumetric flow rate, ft^3/min
M = mass of coolant inside coils, lb
T = temperature, °F
t = time, min
U = interphase heat transfer coefficient, Btu/(min)(ft^2)(°F)
V = volume of fluid in the tank, ft^3
W = mass flow rate of fluid through coils, lb/min
ρ = density of fluid, lb/ft^3

Equations (a) can be rearranged as follows:

$$\frac{dT_2}{dt} + \left[\frac{F}{V} + \frac{UA}{\rho_2 C_{p_2} V}\right] T_2 - \left[\frac{UA}{\rho_2 C_{p_2} V}\right] T_4 = \frac{\rho_1 C_{p_1} F T_1}{\rho_2 C_{p_2} V} \quad \text{(b1)}$$

$$\frac{dT_4}{dt} + \left[\frac{W}{M} + \frac{UA}{M C_{p_4}}\right] T_4 - \left[\frac{UA}{M C_{p_4}}\right] T_2 = \frac{W C_{p_3} T_3}{M C_{p_4}} \quad \text{(b2)}$$

The product $T_4 W/M$ was approximately zero because of the small amplitude variations in W, as demonstrated by tests on an analog computer, and hence the product was neglected in order to linearize Equations (b). For the special case in which the heat capacities and densities of the two fluid streams were the same, a random input was introduced in the form of fluctuations in W while F, F_1, and T_3 were held constant. The random input was physically generated by setting a solenoid valve at a high or low flow rate at time intervals of 10 seconds based on a table of random numbers depending upon whether the last digit was odd or even, respectively. Figure E12.3-1b illustrates a record of the input function which is essentially a random square wave (random binary sequence) with amplitude from $+a$ to $-a$. The input function $W(t)$ can be expressed as

$$W(t - Z) = \begin{cases} +a & \text{odd digit} \\ -a & \text{even digit} \end{cases} \quad \text{with } (n-1)t_m < t < nt_m$$

where Z is a random variable designating the time interval between $t = 0$ and the very first switching time. All subsequent switching decisions were made at $n\,\Delta t$ with n an integer.

† D. M. Himmelblau and K. B. Bischoff, *Process Analysis and Simulation*, John Wiley, New York, 1968, Chapter 2.

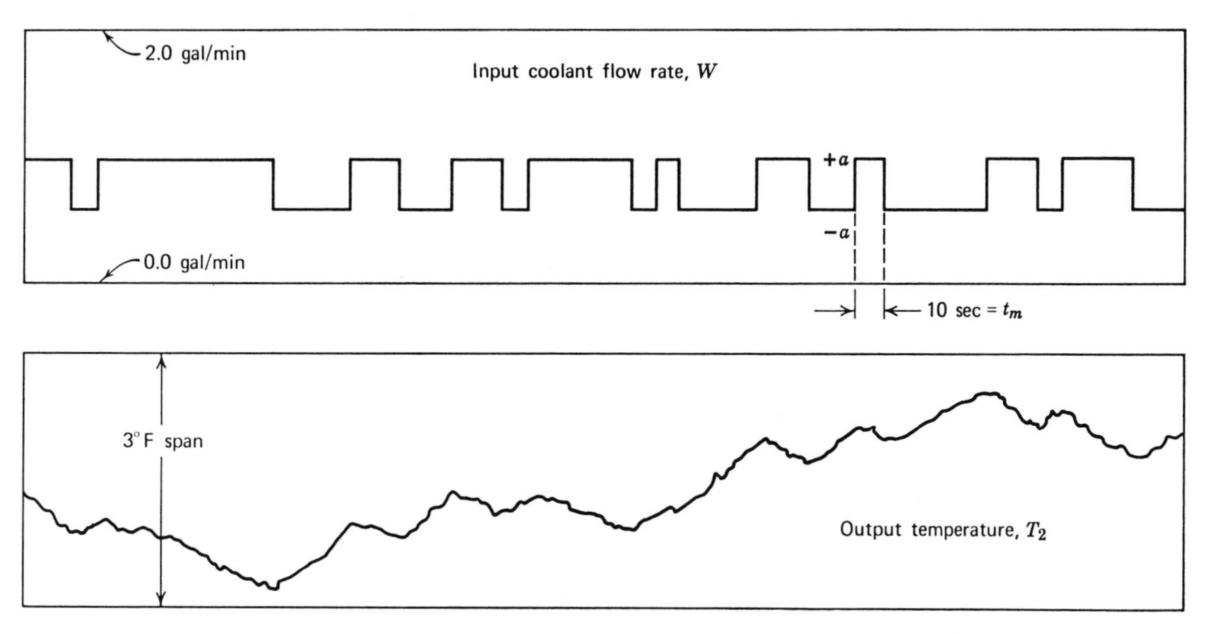

FIGURE E12.3-1b Input and output record of the tank.

The autocorrelation function of the random binary sequence (square wave) with a switching period t_m (refer to Figure 12.3-2) is

$$r_{WW}(\tau) = \mathscr{E}\{W(t)W(t+\tau)\}$$
$$= \begin{cases} a^2\left(1 - \dfrac{\|\tau\|}{t_m}\right) & \text{for } 0 \le \|\tau\| \le t_m \\ 0 & \text{for } \|\tau\| > t_m \end{cases}$$

The autocorrelation function has a triangular shape similar to the diagram in the second row of Figure 12.3-2 except that the peak is at a^2. As t_m approaches zero, $r_{WW}(\tau)$ approaches an impulse function, and the crosscorrelation function r_{WT} between the input and output approaches that of an impulse response function. The input power spectral density is determined by taking the Fourier transform of $r_{WW}(\tau)$:

$$\check{s}_{WW}(\omega) = \int_{-\infty}^{\infty} r_{WW}(\tau)\, e^{-i\omega\tau}\, d\tau$$
$$= a^2 \int_{-t_m}^{t_m} \left(1 - \frac{\|\tau\|}{t_m}\right) e^{-i\omega\tau}\, d\tau$$
$$= 2a^2 \int_0^{t_m} \left(1 - \frac{\|\tau\|}{t_m}\right) \cos \omega\tau\, d\tau$$
$$= a^2 t_m \left(\frac{\sin \dfrac{\omega t_m}{2}}{\dfrac{\omega t_m}{2}}\right)^2 \tag{c}$$

It is necessary that the input contain enough power at frequencies of interest so that the response can be measured at those frequencies. Based on the assigned values (with t in minutes, temperature in °F, and flow in lb/min) listed in Table E12.3-1, the transfer function of the linear model was

$$\frac{\check{T}_2(s)}{\check{W}(s)} = \frac{4.215}{s^2 + 13.082s + 9.413} \tag{d}$$

and the corresponding impulse response for the model in the time domain was

$$g(t) = 0.3646(e^{-0.764t} - e^{-12.32t})$$

Consequently, the time constants for the equipment were approximately $1/0.764 = 1.3$ and $1/12.32 = 0.08$ minute. A comparison of the frequency response of the model (omitting the T_4W/M term) obtained from the transfer function of Equation (d) by placing $s = i\omega$ and putting $\check{g}(\omega)$ in polar form and the input power spectral density given by Equation (c) led to the choice of t_m as 10 seconds. The power spectral density of the random square wave has a repeating zero value at $(\omega t_m/2) = n\pi$, $n = 1, 2, 3, \ldots$. Selection of $t_m = 10$ makes the first zero occur at 37.7 radians/min where

TABLE E12.3-1

$V = 0.7671$ ft³
$F = 32.0$ lb/min
$\rho_1 = \rho_2 = \rho = 62.4$ ft³
$C_{p_2} = C_{p_4} = C_p = 1.0$ Btu/(lb)(°F)
$U = 39.7$ Btu/(min)(°F)(ft²)
$A = 1.84$ ft²
$M = 1.682$ lb
$T_1 = 140$°F
$T_3 = 70$°F

the system response was attenuated by a factor of more than 100.

To process the input-output data on a digital computer, the time record was sampled at 1.93 second intervals, corresponding to 1 millimeter of record length and to a folding frequency of 97.5 radians/min, a frequency well above that which could be obtained from the equipment. The mean values of T_2 and W were evaluated from the entire time record and subtracted from the respective variables in order to remove bias from the correlation functions; $+a$ corresponded to 1.0 gal/min and $-a$ to 0.6 gal/min, so that $|a| = 0.2$ gal/min as shown in Figure E12.3-1b.

The estimated crosscorrelation function was calculated from the adjusted discretized data by

$$R_{WT}(\tau) = \frac{1}{n-m}\sum_{i=1}^{n-m} W_i(t)T_{i+\tau} - \left[\frac{1}{n-m}\sum_{i=1}^{n-m} W_i(t)\right]\left[\frac{1}{n-m}\sum_{i=m+1}^{n} T_i\right] \tag{e}$$

where $T \equiv T_2$, m is the maximum number of time delay increments used, and n is the total number of discrete data points abstracted from the time record. The autocorrelation function can be computed similarly. The terms subtracted on the right-hand side of Equation (e) were used to remove any possible bias caused by working with fractions of the total time record. The time required for the computations was approximately proportional to $(n - m)m$.

Estimates of the power spectral density were determined by representing $R_{WT}(\tau)$ by straight-line segments between the discrete values as described in Section 12.2-2. To test the magnitude of the error involved in representing $R_{WT}(\tau)$ by straight-line segments, some simulation studies were carried out for assumed autocorrelation functions. At frequencies of 0.032 radian/min, the numerical error was in the fourth significant figure; at 1 radian/min the error was in the third significant figure; at 500 radians/min the error was in the second significant figure in the real part of the spectral density.

Sums of the terms resulting from analytical integration over each increment in the expression for $R_{WT}(\tau)$ were added together to obtain the Fourier transform of $R_{WT}(\tau)$:

$$\check{S}_{WT}(\omega) = \sum_{j=-N}^{N-1} \int_j^{j+1} w_l(\tau)[\alpha_j + \beta_j(\tau - \tau_j)]\, e^{-i\omega j\,\Delta t}\, d\tau \tag{f}$$

where

$w_l(t)$ = the lag window = $[1 - (\tau/t_m)^2]^l$; $l = 2$

α_j = the value of a correlation function at time $n\,\Delta t$

β_j = the average slope of the correlation function between time $j\,\Delta t$ and $(j+1)\,\Delta t$, or

$$\frac{\alpha_j((j+1)\,\Delta t - \alpha_j(j\,\Delta t))}{\Delta t}$$

j = the number of the time increments

N = the total number of time intervals in the time record

Equation (f) avoids inaccuracies at high frequencies which might result if the product $\alpha_j(j\,\Delta t)\, e^{-i\omega j\,\Delta t}$ were integrated by the trapezoidal rule.

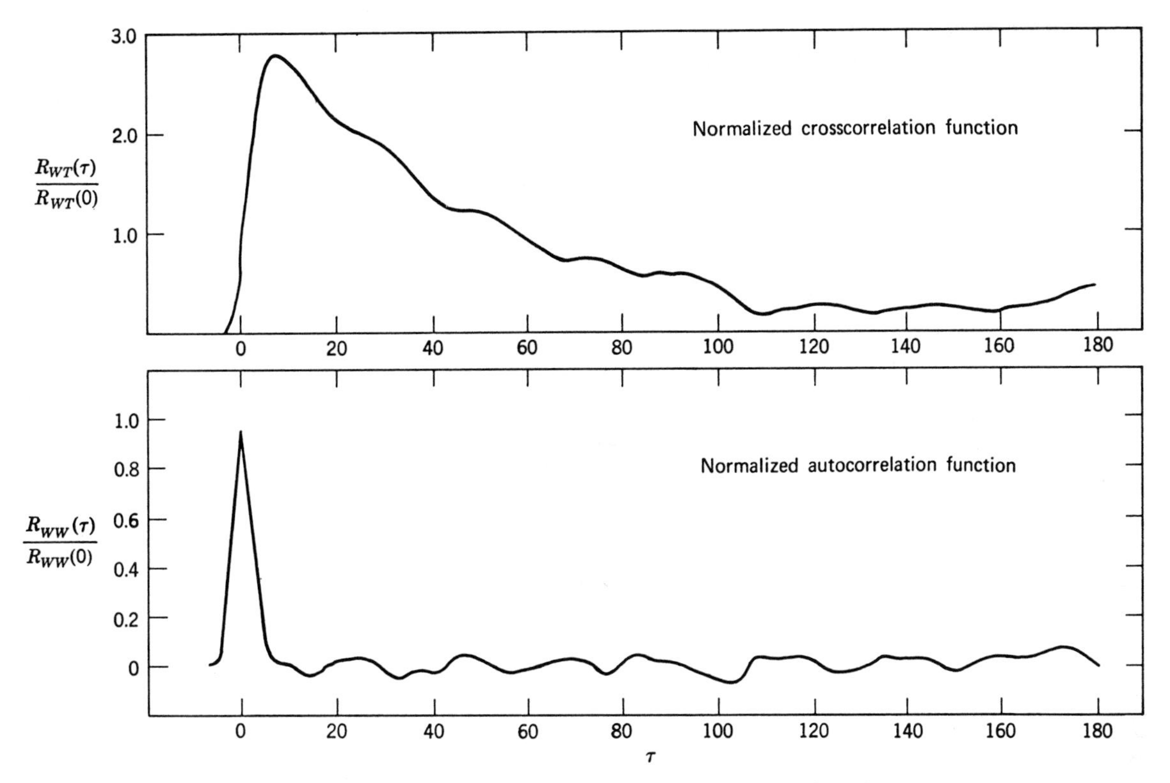

FIGURE E12.3-1c Correlation estimates for 3000 lagged products.

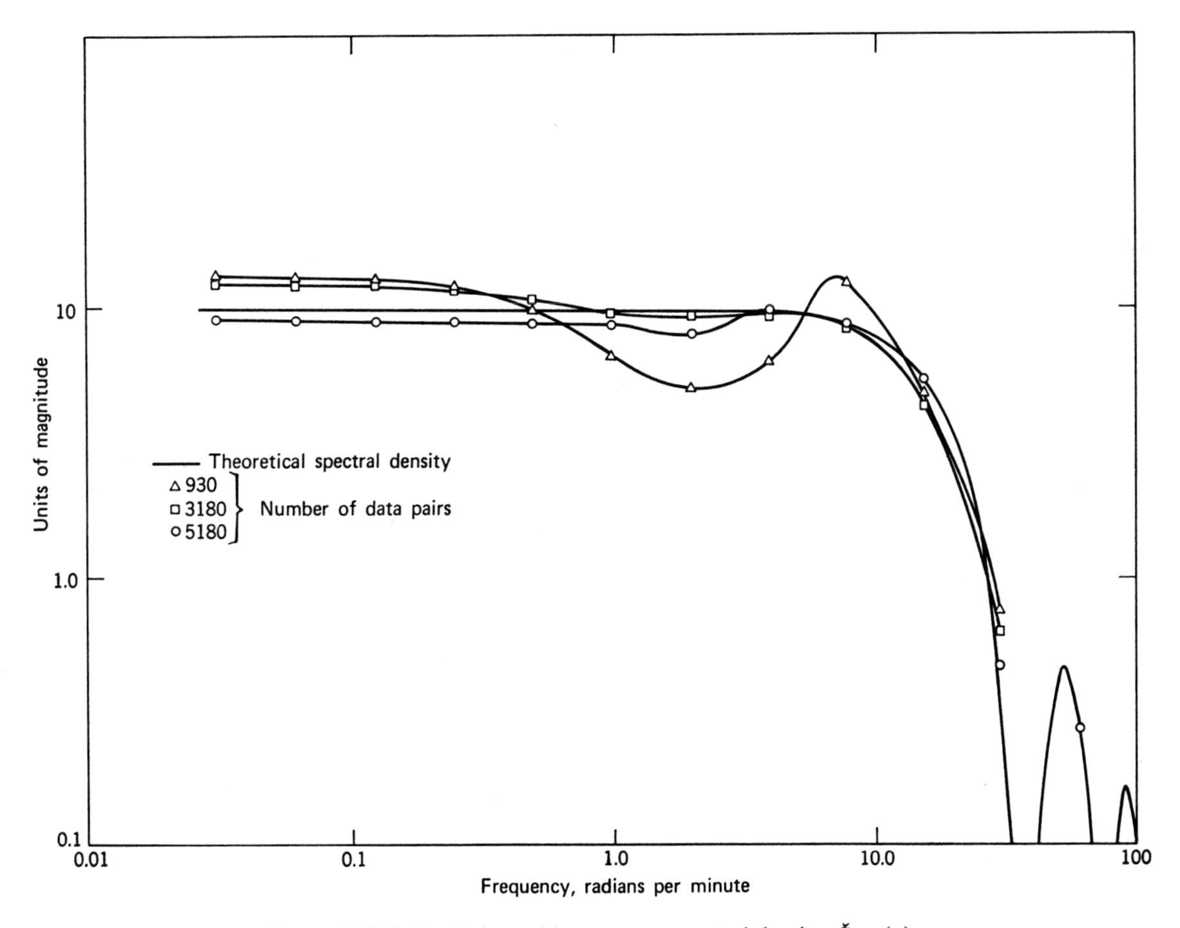

FIGURE E12.3-1d Estimated input power spectral density, $\check{S}_{WW}(\omega)$.

After integration, Equation (f) can be placed in a form more suitable for digital computation:

$$
\begin{aligned}
\check{S}_{WT}(\omega) = {} & \frac{1}{\omega}\left\{f_N \sin \omega N\,\Delta t + \frac{1}{\omega\,\Delta t}\left[(f_j - f_{j-1}) \cos \omega N\,\Delta t - (f_1 - f_0) + \sum_{j=1}^{N-1} (2f_j - f_{j+1} - f_{j-1}) \cos \omega j\,\Delta t\right]\right\} \\
& + \frac{1}{\omega}\left\{f_{-N} \sin \omega N\,\Delta t + \frac{1}{\omega\,\Delta t}\left[f_{-N} - f_{-(N-1)} \cos \omega N\,\Delta t - (f_{-1} - f_0) + \sum_{j=-1}^{-(N-1)} (2f_j - f_{j+1} - f_{j-1}) \cos \omega j\,\Delta t\right]\right\} \\
& + \frac{i}{\omega}\left\{(f_N \cos \omega N\,\Delta t - f_0) - \frac{1}{\omega\,\Delta t}\left[(f_N - f_{N-1}) \sin \omega N\,\Delta t + \sum_{j=1}^{N-1} (2f_j - f_{j+1} - f_{j-1}) \sin \omega j\,\Delta t\right]\right\} \\
& + i\left\{-(f_N \cos \omega N\,\Delta t - f_0) - \frac{1}{\omega\,\Delta t}\left[-(f_{-N} - f_{N-1}) \sin \omega N\,\Delta t - \sum_{j=1}^{N-1} (2f_j - f_{j+1} - f_{j-1}) \sin \omega j\,\Delta t\right]\right\} \qquad \text{(g)}
\end{aligned}
$$

where $f_j \equiv w(j\,\Delta t) \cdot R_{WT}(j\,\Delta t)$.

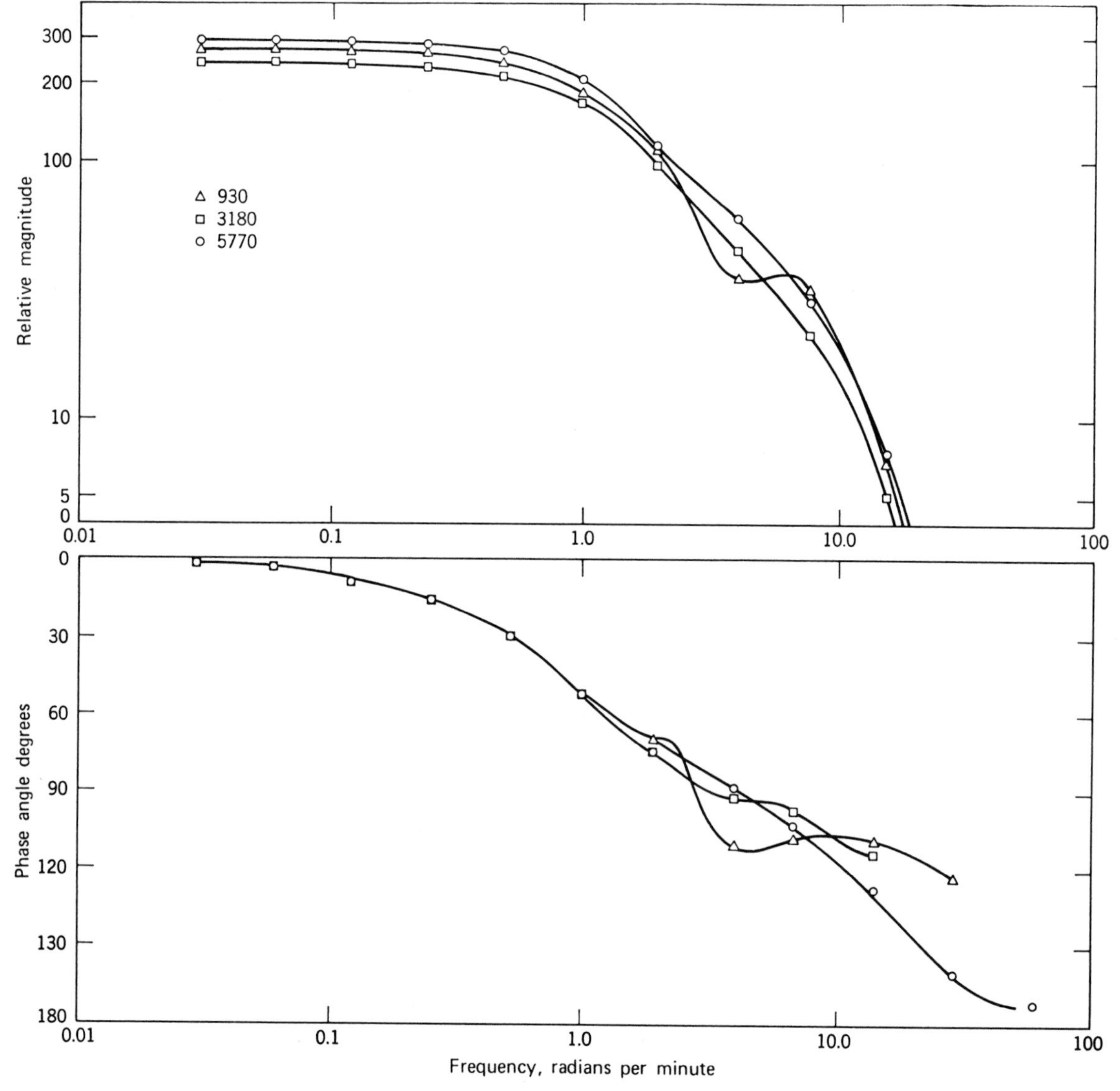

FIGURE E12.3-1e Estimated crosspower spectral density.

Figure E12.3-1c illustrates typical values of $R_{WT}(\tau)$. Also, contrast the expected value of the autocorrelation function, an isosceles triangle at the origin as shown in Figure 12.3-2 with $R_{WW}(\tau)$ in Figure E12.3-1c. As the record length increases, the fluctuations to the right of the triangle die out. The estimated autocorrelation and crosscorrelation function values have been made dimensionless by dividing by the value of $R(\tau)$ at $\tau = 0$.

Figure E12.3-1d compares the input power spectral density $\check{S}_{WW}(\omega)$ for increasingly long time records with the theoretical density given by Equation (c). Figure E12.3-1e shows the crosspower spectral density. The transform calculations were truncated at a lag of 180 sampling increments.

Finally, the estimated transfer function was computed by division as

$$\check{G}(\omega) = \frac{\check{S}_{WT}(\omega)}{\check{S}_{WW}(\omega)} \tag{h}$$

for various record lengths consisting of 930 to 5770 pairs of $[W_i(t), T_i]$. Since the largest time constant of the experimental equipment was 1.31 minutes, the maximum time delay of (180)(1.93)/60 or 4.4 time constants permitted the response to settle to roughly one percent of its initial value. Figure E12.3-1f compares the gain for $\check{G}(\omega)$ (the gain has been normalized by dividing by the highest value of $|\check{G}(\omega)|$) from Equation (h) with the corresponding quantities determined: (1) theoretically from the parameters given in Table E12.3-1 and (2) by frequency response experiments on the same equipment using a sinusoidal input. Division of $\check{S}_{WT}(\omega)$ in Equation (g) by $\check{s}_{WW}(\omega)$, the theoretical input spectral density, yielded improved predictions of the gain (not shown). Estimates of the gain also improved as the record length increased.

One reason for the departures from the theoretical and frequency response curves, in addition to the truncation of the time record, was believed to be the correlation of the process noise with the input as mentioned in connection with Equation 12.3-4. Equation 12.3-8 would become

$$\check{s}_{XY}(\omega) = \check{g}(\omega)\check{s}_{XX}(\omega) + \check{s}_{X\epsilon}(\omega)$$

or

$$\check{g}(\omega) = \frac{\check{s}_{XY}(\omega) - \check{s}_{X\epsilon}(\omega)}{\check{s}_{XX}(\omega)}$$

It was not possible to isolate $\epsilon(t)$ so that $\check{s}_{X\epsilon}(\omega)$ could be evaluated, but the direction of the correction to improve $\check{g}(\omega)$ is downward. The recorded data showed evidence of low amplitude, high frequency variations associated with temperature eddies in the fluid in the tank as a source of the possible correlation.

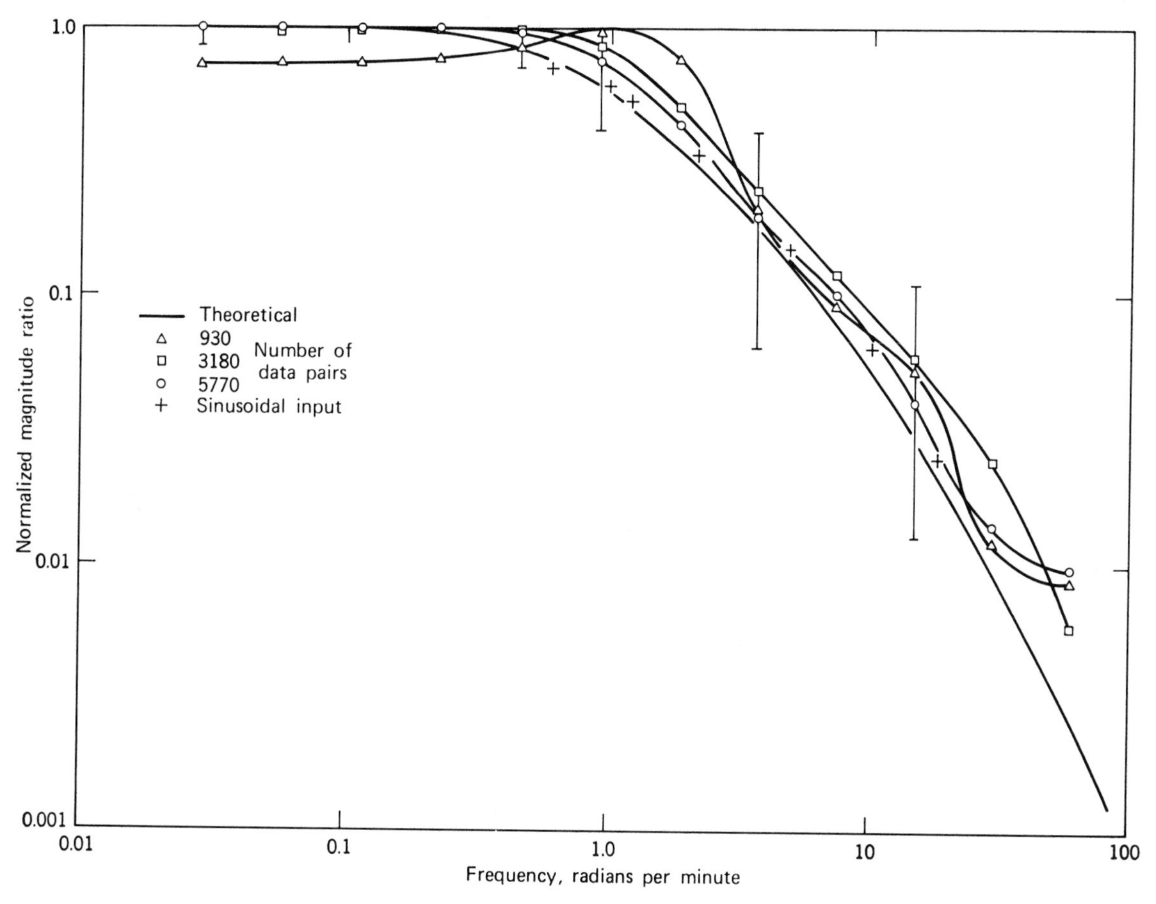

FIGURE E12.3-1f Normalized magnitude of the transfer function in the frequency domain using the estimated input power spectral density (80-percent confidence limits from Equation 12.3-22 shown by Ī).

Supplementary References

Angus, R. M. and Lapidus, L., "Characterization of Multiple Variable Linear Systems from Random Inputs," *AIChE J.* **9**, 810, 1963.

Aris, R. and Amundson, N. R., "Statistical Analysis of a Reactor—Linear Theory," *Chem. Eng. Sci.* **9**, 250, 1959.

Cowley, P. E. A., "The Application of an Analog Computer to the Measurement of Process Dynamics," *Trans. Amer. Soc. Mech. Eng.* **79**, 823, 1957.

Goodman, T. P., "Determination of the Characteristics of Multi-Input and Nonlinear Systems from Normal Operating Records," *Trans. Amer. Soc. Mech. Eng.* **79**, 567, 1957.

Goodman, T. P. and Reswick, J. B., "Determination of System Characteristics from Normal Operating Records," *Trans. Amer. Soc. Mech. Eng.* **78**, 259, 1956.

Goodman, N. R., Katz, S., Kramer, B. H., and Kuo, M. T., "Frequency Response from Stationary Noise: Two Case Histories," *Technometrics* **3**, 245, 1961.

Homan, C. J. and Tierney, J. W., "Determination of Dynamic Characteristics of Processes in the Presence of Random Disturbances," *Chem. Eng. Sci.* **12**, 153, 1960.

Levin, M. J., "Optimum Estimation of Impulsive Response in the Presence of Noise," *IRE Trans. PGCT* **7**, 50, 1960.

Kerr, R. B. and Surber, W. H., Jr., "Precision of Impulse-Response Identification Based on Short, Normal Operating Records," *IRE Trans. PGAC* **6**, 173, 1961.

Qvarnstrom, B., "An Uncertainty Relation for Linear Mathematical Models" in *Proceed. Congress of Int. Fed. Auto. Control*, Basle, Switz., ed. by V. Broida, Butterworths, London, 1964, p. 634.

Woodrow, R. A., "On Finding a Best Linear Approximation to System Dynamics from Short Duration Samples of Operating Records," *J. Electronics and Control* **7**, 176, 1959.

Problems

12.1 The frequency response of an electropneumatic transducer is measured by feeding a sinusoidal voltage into the transducer and recording the output pressure produced. The output pressure is measured with another transducer (pressure pickup) which converts the pressure signal to a voltage. This voltage is recorded on a Sanborn Recorder, thus showing the electropneumatic transducer's output pressure. The measured data are shown in Table P12.1.

A proposed transfer function for the transducer is

$$g(s) = \frac{k}{\tau s + 1}$$

where τ is the time constant in seconds and k is in psia/volt. (Note: an 18-volt change (± 9 volts) from the midscale voltage is equivalent to a $1\frac{1}{2}$-psia change.) Estimate τ. What is the precision of τ?

TABLE P12.1 DATA FOR THE FREQUENCY RESPONSE TEST OF THE ELECTROPNEUMATIC TRANSDUCER

Input Frequency (cycles/sec)	Magnitude Ratio	Phase Lag (degrees)
0.01	1.00	0
0.04	1.00	3
0.07	0.98	6
0.10	0.97	9
0.20	0.94	13
0.40	0.90	23
0.70	0.84	32
1.00	0.75	36
2.00	0.54	55
4.00	0.34	63
7.00	0.23	64

12.2 A frequency response test of an essentially frictionless control valve is conducted by applying a sinusoidal pressure input to the valve and recording the change in position of the valve stem. The amplitude of the input signal is 1.50 psia as in Problem 12.1. Data for the frequency response test of the control valve are shown in Table P12.2.

TABLE P12.2 DATA FOR THE FREQUENCY RESPONSE TEST OF THE CONTROL VALVE

Input Frequency (cycles/sec)	Magnitude Ratio	Phase Lag (degrees)
0.01	1.0	11
0.02	1.0	13
0.04	1.0	15
0.07	0.99	19
0.10	0.98	27
0.20	0.94	48
0.40	0.69	87
0.70	0.35	131
1.00	0.17	164

Determine the two time constants, τ_1 and τ_2, as well as k in the transfer function model of the valve:

$$g(s) = \frac{k}{(\tau_1 s + 1)(\tau_2 s + 1)}$$

12.3 The same tank as described in Problem 11.2 was tested with a sinusoidal input. The figure to the immediate right of each data column represents scale factor adjustments to be added to the data to obtain the proper response.† What frequency response function represents the data satisfactorily? (Refer to Problem 11.2 for additional information.)

† K. A. Bishop, Ph.D. Dissertation, Univ. of Oklahoma, 1965.

Number	Data	Number	Data
1	18.7 + 66.7	51	27.2
2	13.1	52	26.9
3	8.0	53	26.6
4	3.3	54	26.3
5	43.1 + 22.6	55	26.1
6	39.1	56	25.8
7	35.4	57	25.6
8	32.0	58	25.4
9	28.9	59	25.1
10	26.0	60	24.9
11	23.3	61	24.7
12	20.8	62	24.5
13	18.4	63	24.4
14	16.2	64	24.2
15	14.2	65	24.0
16	12.3	66	23.8
17	10.5	67	23.8
18	8.9	68	23.5
19	7.3	69	23.4
20	5.9	70	23.3
21	46.9 − 19.8	71	23.1
22	45.6	72	23.0
23	44.4	73	23.0
24	43.3	74	22.9
25	42.2	75	22.8
26	41.2	76	22.8
27	40.2	77	22.6
28	39.3	78	22.4
29	38.4	79	22.3
30	37.6	80	22.2
31	36.8	81	22.0
32	36.1	82	21.9
33	35.4	83	21.8
34	34.7	84	21.8
35	34.1	85	21.7
36	33.5	86	21.6
37	32.9	87	21.5
38	32.4	88	21.5
39	31.9	89	21.5
40	31.4	90	21.3
41	30.9	91	21.4
42	30.4	92	21.4
43	30.0	93	21.2
44	29.6	94	21.3
45	29.2	95	21.2
46	28.8	96	21.1
47	28.5	97	21.0
48	28.1	98	21.0
49	27.8	99	20.9
50	27.5	100	21.0

12.4 Compute the finite time autocorrelation function and the finite time averaged poweral spectral density for the dependent variable in the process model

$$\frac{dY}{dt} + aY = X(t)$$

if $X(t)$ is periodic and is given by:

(a) $X(t) = A\cos(\omega t + \psi)$, where A is a random variable.

(b) $X(t) = a\cos(\omega t + \phi)$, where ϕ is a random variable.

12.5 Suppose a process output is the stationary random variable $Y(t) = a\sin(\omega t + \psi)$. Estimate the ensemble autocorrelation function $r_{YY}(\tau)$ by calculating $R_{YY}(\tau)$ for the period 0 to t_f and then let $t_f \to \infty$.

12.6 Show that the estimated autocorrelation coefficient for the stationary random input

$$X(t) = a_0 + \sum_{k=1}^{n} a_k \sin(\omega_k t + \psi_k)$$

is

$$R_{XX}(\tau) = a_0^2 + \sum_{k=1}^{n} \frac{a_k^2}{2}\cos\omega_k\tau$$

as t_f becomes large.

12.7 The structure of turbulence can be analyzed by calculating the crosscorrelation function for two fluid velocities in a given coordinate direction observed at different positions. Assume the velocity fluctuations $V_A(t)$ and $V_B(t)$ at positions A and B are ergodic, the estimated crosscorrelation function is $R_{AB}(\tau) = \langle V_A(t)V_B(t)\rangle$, and the standardized estimated crosscorrelation function is

$$R^* = \frac{\langle V_A(t)V_B(t)\rangle}{\sqrt{\langle V_A^2\rangle\langle V_B^2\rangle}}$$

Let

$$K = \sqrt{\frac{(V_A + V_B)^2}{(V_A - V_B)^2}}$$

i.e., the rms of ratio of the sum and difference of the velocities. If $\langle V_A^2\rangle = \langle V_B^2\rangle$, show that

$$R^* = \frac{K^2 - 1}{K^2 + 1}$$

Also show that R^* is relatively insensitive to the changes in the magnitude of V_A and V_B by letting $(\langle V_B^2\rangle)^{1/2} = (1 + \delta)(\langle V_A^2\rangle)^{1/2}$, where δ is a perturbation, and showing that R^* is computed correctly even if $\langle V_A^2\rangle \neq \langle V_B^2\rangle$. What is the error in R^* for a 20-percent difference in the V's?

APPENDIX A

Concepts of Probability

The term probability has a wide variety of meanings, ranging from "the probability of getting a head on the next toss of a penny is $\frac{1}{2}$" to "probably our team will win the next game." For convenience, theories of probability can be roughly classified in one of the following three ways:

1. Frequency theories (objectivistic theories).
2. Subjective theories (personalistic theories).
3. Logical theories (axiomatic theories).

The frequency theories of probability have many nuances and refinements that are described in the references at the end of this appendix, but a common interpretation is:†

> Whenever we say that the probability of an event E with respect to an experiment ϵ is equal to P, the concrete meaning of this assertion [is]: In a series of repetitions of ϵ, it is practically certain that the frequency of E will be approximately equal to P.

Briefly, a probability space (Ω, E, P) consists of a physical *sample space* Ω (list of all events), a class E of subsets of Ω termed *events* or *outcomes*, and a *probability* measure P defined on E. Suppose that an experiment is carried out many times, each time determining whether E occurs or does not occur. After many experiments ("in the long run") the ratio of the number of times E occurs to the sum of number of times E occurs and does not occur is P. Presumably the event E can be expressed as the value of a random variable. The probability of an event E is *independent* of the beliefs or expectations of the experimenter. Formulated in the above fashion, probability statements have a dual nature. Given a probability statement, the analyst can roughly predict the result of a long series of experiments. Given the results of the series of experiments, he can decide what probability statement to accept for some given decision criterion.

When probability is defined by the frequency theory as above, certain natural uses of the word probability cannot be accommodated. For example, statements of the type "it is probable that there is no life on Mars" cannot be given a frequency interpretation. Also, statements such as "the probability that the next toss of a penny will be a head is $\frac{1}{2}$" is hard to explain, since only one unique experiment takes place with a frequency either or 0 or 1. Nevertheless, the frequency theory approach to probability serves adequately in engineering and science.

Subjective theories of probability rest on statements concerning actual degrees of belief. To interpret the statement "it will rain tomorrow" in terms of quantitative probabilities, the analyst must ask himself whether he would prefer to bet on the occurrence of an event E or on the lack of occurrence of E. A person has certain degrees of confidence in given postulates, and he modifies his beliefs as he gains additional information. It is quite plausible that two rational people will arrive at different estimates of the probability of an event using subjective theories of probability, although as their information increases they will tend to reach like estimates. Thus, according to the subjective theory of probability, the probability that an event has occurred or that an event will occur is a measure of one's belief in its occurring.

The Bayesian approach to estimation makes use of prior information. Such prior knowledge can come from theoretical considerations, from the results of previous experiments, or from assumptions by the experimenter. Typically, a Bayesian approach assumes a prior probability distribution of an unknown parameter θ in some parameter space $\boldsymbol{\theta}$. The distribution is updated by using Bayes' rule to obtain the posterior probability distribution.

Consider a set of events or outcomes, $A_1, A_2, \ldots, A_n$, and some other event, B. Bayes' theorem states that the probability that event A_i will occur, given that event B has already occurred, which will be denoted by $P\{A_i \mid B\}$, is equal to the product of the probability that A_i will occur regardless of whether B will take place and the probability that B will occur, given that A_i has already taken place, divided by the probability of the occurrence of B:

$$P\{A_i \mid B\} = \frac{P\{B \mid A_i\}P\{A_i\}}{P\{B\}}$$

† H. Cramér, *Mathematical Methods of Statistics*, Princeton Univ. Press, Princeton, N.J., 1946, p. 149.

Further, if all events comprising the set $\{A_i\}$ are included in $A_1, A_2, \ldots, A_n$, then

$$P\{A_i \mid B\} = \frac{P\{B \mid A_i\}P\{A_i\}}{\sum_{i=1}^{n} P\{B \mid A_i\}P\{A_i\}} \quad \text{(A-1)}$$

We can interpret these symbols as follows:

1. $P\{A_i\}$ is a measure of our degree of belief that event A_i will occur or that hypothesis A_i is true prior to the acquisition of additional evidence that may alter the measure. $P\{A_i\}$ is denoted the *prior probability*.
2. $P\{A_i \mid B\}$ is a measure of our degree of belief that event A_i will occur or that hypothesis A_i is true, given additional evidence B pertinent to the hypothesis. $P\{A_i \mid B\}$ is termed the *posterior probability*.
3. $P\{B \mid A_i\}$ denotes the likelihood that event B will occur, given that event A_i or hypothesis A_i is true. $P\{B \mid A_i\}$ is a conditional probability, interpreted in the Bayesian framework as a likelihood, $L(A_i \mid B)$.

For continuous variable, Bayes' theorem can be more conveniently expressed in terms of the probability density functions rather than the probabilities themselves. Equation A-1 can be expressed in terms of a set of observed values of the random variable X, $\mathbf{x}$, and an unknown parameter $(s)\theta$ as

$$p(\theta \mid \mathbf{X} = \mathbf{x}) = p(\theta \mid \mathbf{x}) = \frac{L(\theta \mid \mathbf{x}) \cdot p(\theta)}{\int_{-\infty}^{\infty} L(\theta \mid \mathbf{x}) \cdot p(\theta)\, d\theta} \quad \text{(A.2)}$$

where

$p(\theta \mid \mathbf{x})$ = the posterior probability density function for θ; it includes knowledge of the possible values of θ gained from the experimental data $\mathbf{x}$

$p(\theta)$ = the prior probability density function for θ (before the experiment in which $\mathbf{x}$ was observed)

$L(\theta \mid \mathbf{x}) = p(\mathbf{x} \mid \theta)$ = the probability density function termed the likelihood function of θ given $\mathbf{x}$ (described in more detail in Section 3.2-1)

The denominator in Equation A-2 is a normalizing factor chosen so that the integration of the posterior distribution is unity, i.e., $\int_{-\infty}^{\infty} p(\theta \mid \mathbf{x})\, d\theta = 1$. From Section 2.1 we find that

$$\int_{-\infty}^{\infty} p(\mathbf{x} \mid \theta) p(\theta)\, d\theta = p(\mathbf{x})$$

If the prior distribution is a uniform distribution, that is the prior distribution is a constant, then Equation A-2 reduces to

$$p(\theta \mid \mathbf{x}) = \frac{L(\theta \mid \mathbf{x})}{\int_{-\infty}^{\infty} L(\theta \mid \mathbf{x})\, d\theta}$$

If prior knowledge concerning a postulated event or hypothesis is poor, the posterior probability is largely or entirely determined by the likelihoods, that is, by the additional accumulated evidence for which the likelihood function acts as a mathematical expression. If prior knowledge outweighs recent evidence, however, then the posterior probability is determined almost solely by the prior probability.

The third concept of probability treats probability axiomatically. According to Keynes,† who first explicitly promulgated the use of the axiomatic approach:

> All propositions are true or false, but the knowledge we have of them depends on our circumstances; and while it is convenient to speak of propositions as certain or probable, this [statement only] expresses strictly a relationship in which they stand to a corpus of knowledge, actual or hypothetical. . . . A proposition is not probable because we think it so.

In other words, the probability of a statement is a real number determined on logical grounds only, and the degree of belief which a rational person can have about a statement, given certain evidence, is also determined on logical grounds alone.

It is not possible in the brief space permitted here to illustrate the interaction of these concepts of probability nor the ramifications of them. The interested reader should refer to the works of Kyburg, Körner, Lees, or Papoulis, listed as references at the end of this appendix, for authoritative and detailed nonmathematical discussions.

In the application of tests and the design of experiments, certain definitions and rules concerning probability are needed and are listed below.

1. It follows from the frequency theory of probability that

$$0 \le P \le 1 \quad \text{(A-3)}$$

2. If the probability of occurrence of one event A depends on whether or not event B has occurred, the two events are termed *dependent*: if the probability of occurrence of event A does not depend on the occurrence of B, or the reverse, then the two events are *independent*.

3. ADDITION RULE. If $A_1, A_2, \ldots, A_n$ are *mutually exclusive* events, i.e., cannot occur at the same time, the probability of occurrence of just one of the events is equal to the sum of the probabilities of each A_k:

$$P(A_1, \text{or } A_2, \ldots, \text{or } A_n) = \sum_{k=1}^{n} P(A_k) \quad \text{(A-4)}$$

Very often we let

$$\sum_{k=1}^{n} P(A_k) = 1 \quad \text{(A-5)}$$

† J. M. Keynes, *A Treatise on Probability*, MacMillan, London, 1921, pp. 3–4.

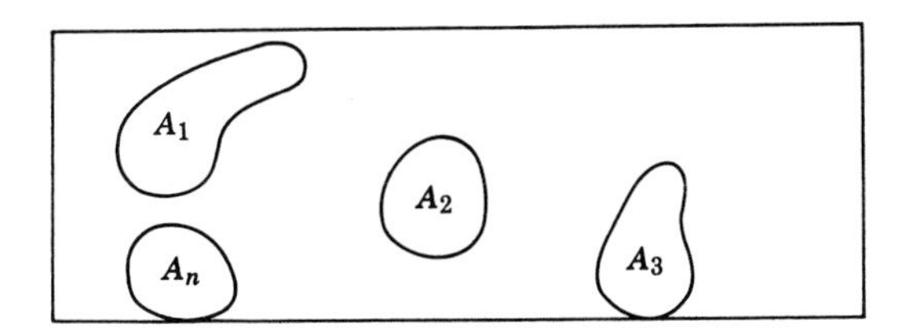

FIGURE A.1 Mutually exclusive events (sets).

Also, if each event is equiprobable so that $P(A_k) = q$,

$$\sum_{k=1}^{n} q = nq = 1 \quad \text{or} \quad q = \frac{1}{n} = P(A_k) \quad \text{(A-5a)}$$

In set theory, mutually exclusive events have no points in common. See Figure A.1. The union of the sets which represents the set of all elements that belong to A_1 or A_2 or ...(union is designated by the symbol $\cup$) is $A_1 \cup A_2 \cup \cdots \cup A_n$ and

$$P(A_1 \cup A_2 \cup \cdots \cup A_n) = P(A_1) + P(A_2) + \cdots + P(A_n) \quad \text{(A-4a)}$$

If the space is completely divided up into sets, then Equation A-5 holds. See Figure A.2.

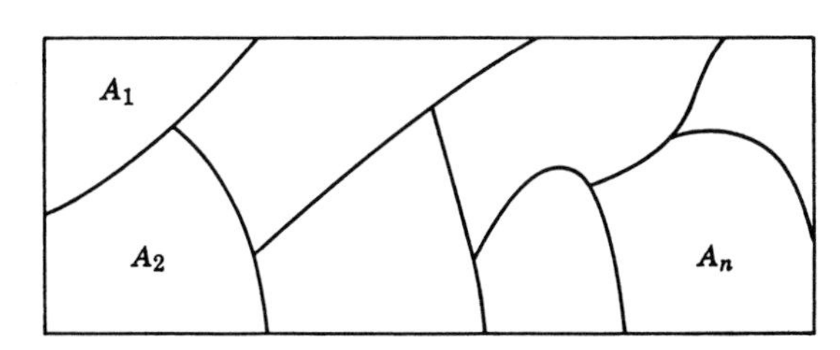

FIGURE A.2 A partition of the sample space into n events (sets).

4. MULTIPLICATION RULE. If A and B are *independent* events,

$$P(A \text{ and } B) = P(A) \cdot P(B) \quad \text{(A-6)}$$

In set theory the intersection of A and B is the set of all elements that belong to A and B and is designated by the symbol $\cap$. Thus,

$$P(A \cap B) = P(A) \cdot P(B) \quad \text{(A-7)}$$

as illustrated in Figure A.3 by the shaded area. (Keep in mind that mutually exclusive does not mean that the events (sets) are independent.)

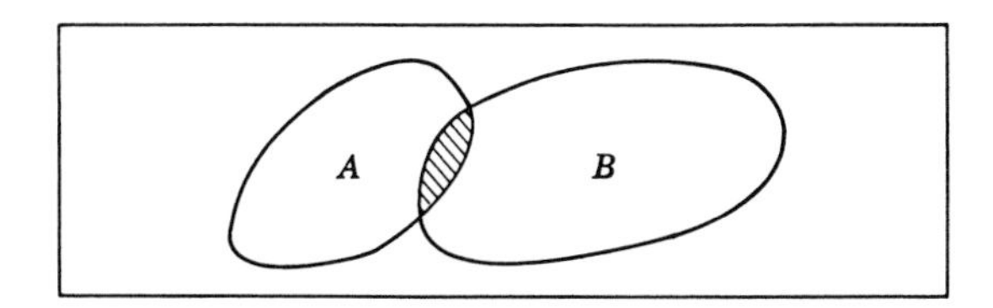

FIGURE A.3 Intersection of A and B.

If A and B are *dependent* events,

$$P(A \mid B) = \frac{P(A \cap B)}{P(B)}; \qquad P(B) \neq 0 \quad \text{(A-8)}$$

where the symbol $P(A \mid B)$ means the "probability of A given B." As a corollary,

$$P(A \cap B) = P(B)P(A \mid B) \quad \text{(A-8a)}$$

$$= P(A)P(B \mid A) \quad \text{(A-8b)}$$

Two kinds of probabilities enter Equation A-8a (or A-8b): the absolute probability of event B (or A) irrespective of whether or not A (or B) has occurred, and the conditional probability of event A (or B) computed on the assumption that B (or A) has occurred. It is easy to see that Equation A-6 or Equation A-7 is a special case of Equation A-8, because if the events are independent, $P(A \mid B) = P(A)$.

For the case of many events, Equation A-6 can be expanded to

$$P(A_1 \text{ and } A_2 \text{ and} \cdots \text{and } A_n) = P(A_1) \cdot P(A_2) \cdot \ldots \cdot P(A_n)$$

$$= \prod_{k=1}^{n} P(A_k) \quad \text{(A-9)}$$

5. Another useful relationship for events which are not mutually exclusive is

$$P(A) + P(B) - P(A \cap B) = P(A \cup B) \quad \text{(A-10)}$$

Example A.1 Application of Addition and Multiplication Rules

The toss of two dice, one red and one blue, provides a simple illustration which makes the above rules more meaningful. When the two dice are tossed, because each has six sides and can yield one number (as the up number), the events which can occur can be portrayed in a two-dimensional array termed the "sample space." For fair dice it is possible to predict in advance the sample space; in other cases the same information must be obtained by experimentation. In Table EA.1, each box represents one possible outcome.

TABLE EA.1

Upper Face (Outcome)

	Blue → Red ↓	1	2	3	4	5	6
Upper Face (Outcome)	1	1, 1	1, 2	1, 3	1, 4	1, 5	1, 6
	2	2, 1	2, 2	2, 3	2, 4	2, 5	2, 6
	3	3, 1	3, 2	3, 3	3, 4	3, 5	3, 6
	4	4, 1	4, 2	4, 3	4, 4	4, 5	4, 6
	5	5, 1	5, 2	5, 3	5, 4	5, 5	5, 6
	6	6, 1	6, 2	6, 3	6, 4	6, 5	6, 6

The general procedure to determine the probabilities of various events is to:

1. Set up the sample space of all possible outcomes in a table (above) or equation, if possible.
2. Assign probabilities to each element ($\sum P = 1$).
3. Obtain the probabilities of an event by adding the probabilities assigned to elements of the subset comprising the event.

For two fair dice, each outcome in the table is equally probable, so that Equation A-5a applies and $P = \frac{1}{36}$.

Now we can pose and answer some specific questions:

1. What is the probability of throwing the same number on each die, i.e., (1, 1), or (2, 2), or (3, 3), etc.? Six elements make up the event, each mutually exclusive. Applying Equation A-4a yields

$$P[(1, 1), \text{ or } (2, 2), \text{ or } (3, 3), \text{ or } (4, 4), \text{ or } (5, 5), \text{ or } (6, 6)]$$
$$= \tfrac{1}{36} + \tfrac{1}{36} + \tfrac{1}{36} + \tfrac{1}{36} + \tfrac{1}{36} + \tfrac{1}{36} = \tfrac{1}{6}$$

2. What is the probability that the sum of two tosses is 10? Three elements (on a diagonal), each mutually exclusive, comprise the event. The probability of each is $\frac{1}{36}$ so that the sum of three $\frac{1}{36}$'s is $\frac{1}{12}$ according to Equation A-4a:

$$P[B + R = 10] = P[(6, 4), \text{ or } (5, 5), \text{ or } (4, 6)] = \tfrac{1}{12}$$

3. What is the probability that the blue toss is ≤ 3 or the red toss is ≤ 2? These are *not* mutually exclusive events so that one must avoid double counting of overlapping outcomes. Inspection of the table yields the following events:

$$\left.\begin{array}{c} B \leq 3 \\ \hline (1, 1), (1, 2), (1, 3) \\ (2, 1), (2, 2), (2, 3) \\ \cdots \\ (6, 1), (6, 2), (6, 3) \end{array}\right\} 18 \qquad \left.\begin{array}{c} R \leq 2 \\ \hline (1, 1), (1, 2), \ldots, (1, 6) \\ (2, 1), (2, 2), \ldots, (2, 6) \end{array}\right\} 12$$

Enumeration of the duplicated outcomes:

$$\left.\begin{array}{c} (1, 1), (1, 2), (1, 3) \\ (2, 1), (2, 2), (2, 3) \end{array}\right\} 6$$

in the counting gives

$$P[(B \leq 3), \text{ or } (R \leq 2)] = 18(\tfrac{1}{36}) + 12(\tfrac{1}{36}) - 6(\tfrac{1}{36}) = \tfrac{24}{36} = \tfrac{2}{3}$$

Direct application of Equation A-10 gives

$$P(A) + P(B) - P(A \cap B) = P(A \cup B)$$
$$\tfrac{18}{36} + \tfrac{12}{36} - \tfrac{6}{36} = \tfrac{24}{36}$$

4. What is the probability that the blue toss is ≤ 3 and the red toss is ≤ 2? By counting outcomes, only six events represent $B \leq 3$ *and* $R \leq 2$:

$$(1, 1), (1, 2), (1, 3)$$
$$(2, 1), (2, 2), (2, 3)$$

so that $P(B \leq 3 \text{ and } R \leq 2) = \frac{6}{36} = \frac{1}{6}$. Because the tosses are independent of each other, Equation A-7 applies:

$$P(B \leq 3 \cap R \leq 2) = P(B \leq 3) \cdot P(R \leq 2)$$
$$= (\tfrac{18}{36})(\tfrac{12}{36}) = \tfrac{1}{6}$$

5. What is the probability that the sum of the dice is < 4 given that the blue dice is 1? In this example we are dealing with a conditional probability (dependent events) so Equation A-8 applies:

$$P[(B + R < 4 \mid (B = 1)] = \frac{P[(B + R < 4) \cap (B = 1)]}{P(B = 1)}$$

From the table one can count

$$P(B = 1) = 6(\tfrac{1}{36}) = \tfrac{1}{6}$$
$$P[(B + R < 4) \cap (B = 1)] = 2(\tfrac{1}{36}) = \tfrac{1}{18}$$

so that

$$P[(B + R < 4) \cap (B = 1)] = \frac{\frac{1}{18}}{\frac{1}{6}} = \tfrac{1}{3}$$

as might be observed directly.

6. What is the probability that the first toss of the dice yields 4 and the next toss of the dice yields 4? Presumably the tosses are independent events so Equation (A-7) applies:

$$P(B + R = 4) \cap P(B + R = 4) = P(B + R = 4)$$
$$\cdot P(B + R = 4)$$
$$= (\tfrac{3}{36})(\tfrac{3}{36}) = \tfrac{1}{144}$$

Supplementary References

Ayer, A. J., "Chance," *Scientific Amer.* **211**, Nov. 1964, p. 44.

Fisher, R. A., "Mathematical Probability in the Natural Sciences," *Technometrics* **1**, 21, 1959.

Körner, S. (ed.), *Observation and Interpretation*, Butterworths, London, 1957.

Kyburg, H. E., *Probability and Logic of Rational Belief*, Wesleyan Univ. Press, Middletown, Conn., 1961.

Lees, S., in *Joint Auto. Control Conference* (Paper 5, Session 16), p. 36, 1965.

Papoulis, A., "The Meaning of Probability," *IEEE Trans.* **E-7**, 45, 1964.

APPENDIX B

Mathematical Tools

This appendix is designed to provide the reader with a summary of certain mathematical principles employed in the main body of the text. These principles are usually encountered after a study of calculus. *This appendix is only a summary* and is not intended to be of sufficient scope or detail to be adequate preparation for those who have not encountered these principles before.

B.1 PROPERTIES AND CHARACTERISTICS OF LINEAR EQUATIONS

A combination of vectors or scalars is said to be *linear* if it can be assembled in the following form:

$$c_1x_1 + c_2x_2 + \cdots + c_nx_n \tag{B.1-1}$$

The x's in Equation B.1-1 may be vectors:

$$\mathbf{x}_1 \equiv a_{11}\,\boldsymbol{\delta}_1 + a_{12}\,\boldsymbol{\delta}_2 + \cdots \tag{B.1-2}$$

$$\mathbf{x}_n \equiv a_{n1}\,\boldsymbol{\delta}_1 + a_{n2}\,\boldsymbol{\delta}_2 + \cdots \tag{B.1-3}$$

or they may be polynomials

$$x_1 = P_1(x) \tag{B.1-4}$$

$$x_n = P_n(x) \tag{B.1-5}$$

or other functions.

The x's are said to be *linearly dependent* if for some set of the c_i's (assuming the c_i's are not all zero) the following is true:

$$c_1x_1 + c_2x_2 + \cdots + c_nx_n = \sum_{i=1}^{n} c_ix_i = 0 \tag{B.1-6}$$

On the other hand, if

$$\sum_{i=1}^{n} c_ix_i = 0$$

only if the c_i's are all zero, then the x_i's are said to be *linearly independent*.

Example B.1-1 Linear Independence

The polynomials

$$P_1(x) = 2x_1 - x_2$$

$$P_2(x) = x_1 + 2x_2$$

$$P_3(x) = x_1 + 4x_2$$

are linearly dependent over the field of rational numbers since

$$c_1(2x_1 - x_2) + c_2(x_1 + 2x_2) + c_3(x_1 + 4x_2) = 0 \quad \text{(a)}$$

holds for at least one set of c_i's, namely, $c_1 = 1$, $c_2 = -\frac{9}{2}$, and $c_3 = \frac{5}{2}$. On the other hand the vectors

$$\mathbf{x}_1 = \boldsymbol{\delta}_1 + 2\boldsymbol{\delta}_2 + 3\boldsymbol{\delta}_3$$

$$\mathbf{x}_2 = 3\boldsymbol{\delta}_1 + 2\boldsymbol{\delta}_2 + \boldsymbol{\delta}_3$$

are linearly independent since there is no set of c_i's that satisfies the following equation:

$$c_1(\boldsymbol{\delta}_1 + 2\boldsymbol{\delta}_2 + 3\boldsymbol{\delta}_3) + c_2(3\boldsymbol{\delta}_1 + 2\boldsymbol{\delta}_2 + \boldsymbol{\delta}_3) = 0 \quad \text{(b)}$$

This latter can be seen to be true if we write three scalar equations in place of Equation (b):

$$\begin{aligned} \boldsymbol{\delta}_1&: \; c_1 + 3c_2 = 0 \\ \boldsymbol{\delta}_2&: \; 2c_1 + 2c_2 = 0 \\ \boldsymbol{\delta}_3&: \; 3c_1 + c_2 = 0 \end{aligned} \quad \text{(c)}$$

There is no nontrivial ($c_i \neq 0$) solution to this set of equations as can be seen by inspection.

B.2 LINEAR AND NONLINEAR OPERATORS

Given that $\mathscr{H}$ is an operator, by linear operator we mean:

1. Additivity (superposition):

$$\mathscr{H}(f_1 + f_1) = \mathscr{H}(f_1) + \mathscr{H}(f_2) \tag{B.2-1}$$

2. Proportionality:

$$\mathscr{H}(kf) = k\mathscr{H}(f) \tag{B.2-2}$$

where f_1 and f_2 are functions. A well-known example of a linear operator is $D^2 = d^2/dx^2$; observe that

$$D^2(y_1 + y_2) = \frac{d^2y_1}{dx^2} + \frac{d^2y_2}{dx^2}$$

But on the other hand, $(d/dx)^2$ is not a linear operator since

$$\left(\frac{dy_1}{dx}\right)^2 + \left(\frac{dy_2}{dx}\right)^2 \neq \left(\frac{dy_1}{dx} + \frac{dy_2}{dx}\right)^2$$

Other typical examples of linear and nonlinear operators are listed in Table B.2-1.

TABLE B.2-1 EXAMPLES OF LINEAR AND NONLINEAR OPERATORS

Linear	Nonlinear
$\mathscr{H}(u) \equiv \frac{du}{dx}$	$\mathscr{N}(u) \equiv R(x)u^2$
$\mathscr{H}(u) \equiv Q(x)u$	$\mathscr{N}(u) \equiv \left(\frac{du}{dx}\right)u$
$\mathscr{H}(u) \equiv \frac{d^n u}{dx^n}$	$\mathscr{N}(u) \equiv \int_u^b H(x, s)u(s)u(s + x)\, ds$
$\mathscr{H}(u) \equiv \frac{\partial^2 u}{\partial x^2} + \frac{\partial^2 u}{\partial y^2} + \frac{\partial^2 u}{\partial z^2}$	$\mathscr{N}(u) \equiv P(x)\, e^u$
$\mathscr{H}(u) \equiv A(x) + B(x)u = C(x)\frac{du}{dx}$	$\mathscr{N}(u) \equiv V(u)\frac{\partial^n u}{\partial x^n}$

u = a continuous variable in a given interval
x, y, z = independent variables

From the above definitions—Equations B.2-1 and B.2-2—it is easy to interpret a nonlinear equation or a linear equation in a general way as

$$\mathscr{H}(f) = \psi(x) \tag{B.2-3}$$

(for a single independent variable x) in which Equation B.2-3 is linear if $\mathscr{H}(f)$ is a linear operator and is nonlinear if it is not. The *solution* of such an equation, if it exists, will be a function

$$f(x, y) = F(x) \tag{B.2-4}$$

involving one or more arbitrary constants or parameters, which satisfies Equation B.2-3.

B.3 LINEAR SYSTEMS

The ease and accuracy of predicting system outputs are significantly dependent upon whether the system is linear or nonlinear. We shall call a system linear if the operator representing the input-output relation for the system is linear. For example, given the responses

$$y_1(t) = f[x_1(t)] \tag{B.3-1}$$

$$y_2(t) = f[x_2(t)] \tag{B.3-2}$$

the following must be true for the system to be linear:

$$y_1(t) + y_2(t) = f[x_1(t) + x_2(t)] \tag{B.3-3}$$

A linear system also exhibits the property of proportionality or homogeneity: if $f[x_1(t)]$ is actually $kf[x_1(t)]$, then the response is $ky_1(t)$. The real significance of the principle of superposition is that linear transformations cannot be applied in an exact manner to nonlinear systems. For example, a "square law" device which has the input-output relation $y(t) = [x^2(t)]$ is not a linear system since

$$y_1(t) + y_2(t) = [x_1^2(t) + x_2^2(t)] \neq [x_1(t) + x_2(t)]^2$$

B.4 MATRIX ALGEBRA

Matrix algebra is widely used whenever large numbers of linearly combined variables must be handled. Familiarity with some of the notations, methods, limits, and applications of matrix theory is essential to the understanding of how to solve important classes of linear problems and how to simplify complex notation. Matrix operations are particularly adaptable to manipulation at high speeds on digital computers, so whoever uses these techniques is relieved of an immense amount of tedious repetitive detail. We shall summarize here the important properties of and operations on matrices and then illustrate some typical applications.

A *matrix* is an array of elements†

$$\mathbf{a} = \begin{bmatrix} a_{11} & a_{12} & \cdots & a_{1n} \\ a_{21} & a_{22} & \cdots & a_{2n} \\ \cdot & \cdot & \cdot & \cdot \\ a_{m1} & a_{m2} & \cdots & a_{mn} \end{bmatrix} \tag{B.4-1}$$

in a definite order. A *square matrix* is one in which the number of rows and the number of columns are equal. For example, an $n \times n$ matrix with $n = 3$ is

$$\mathbf{a} = \begin{bmatrix} 1 & 2 & 3 \\ 2 & 3 & 4 \\ 3 & 4 & 5 \end{bmatrix} \quad \text{a } 3 \times 3 \text{ matrix} \tag{B.4-2}$$

† In texts, matrices are identified by bold-faced letters.

In the *identity matrix* (given the special notation **I**), the elements on the main diagonal are 1 and the rest of the elements are 0:

$$\mathbf{I} = \begin{bmatrix} 1 & 0 & 0 \\ 0 & 1 & 0 \\ 0 & 0 & 1 \end{bmatrix} \quad \text{a } 3 \times 3 \text{ identity matrix} \tag{B.4-3}$$

For two matrices to be equal, each and every element in the corresponding position in the two matrices must be equal.

The *transpose of a matrix* is obtained by interchange of rows and columns:

$$\mathbf{a}^T = \begin{bmatrix} a_{11} & a_{21} & \cdots & a_{m1} \\ a_{12} & a_{22} & \cdots & a_{m2} \\ \cdot & \cdot & \cdot & \cdot \\ a_{1n} & a_{2n} & \cdots & a_{mn} \end{bmatrix} \tag{B.4-4}$$

For example, if

$$\mathbf{a} = \begin{bmatrix} 2 & 0 & -1 \\ 1 & 1 & 4 \end{bmatrix}, \qquad \mathbf{a}^T = \begin{bmatrix} 2 & 1 \\ 0 & 1 \\ -1 & 4 \end{bmatrix} \tag{B.4-5}$$

A *symmetric matrix* is one in which $\mathbf{a} = \mathbf{a}^T$:

$$\mathbf{a} = \begin{bmatrix} 0 & 1 & 2 \\ 1 & 2 & 3 \\ 2 & 3 & 4 \end{bmatrix}, \qquad \mathbf{a}^T = \begin{bmatrix} 0 & 1 & 2 \\ 1 & 2 & 3 \\ 2 & 3 & 4 \end{bmatrix} \tag{B.4-6}$$

An *adjoint matrix* ($\mathscr{A}$) is a transposed matrix of cofactors. Given the square matrix **a** of order n:

$$\mathbf{a} = \begin{bmatrix} a_{11} & a_{12} & \cdots & a_{1n} \\ a_{21} & a_{22} & \cdots & a_{2n} \\ \cdot & \cdot & \cdot & \cdot \\ a_{n1} & a_{n2} & \cdots & a_{nn} \end{bmatrix}$$

the elements of the cofactor a_{11} are outlined by the dashed lines:

$$\begin{bmatrix} a_{22} & \cdots & a_{2n} \\ \cdot & \cdot & \cdot \\ a_{n2} & \cdots & a_{nn} \end{bmatrix}$$

By definition the cofactor A_{ij} of a_{ij} in the matrix is $[(-1)^{i+j}]$ times [the determinant obtained by deleting the ith row and jth column from the matrix]. Then,

$$\mathscr{A} = \text{adjoint matrix} = \begin{bmatrix} A_{11} & A_{21} & \cdots & A_{n1} \\ A_{12} & A_{22} & \cdots & A_{n2} \\ \cdot & \cdot & \cdot & \cdot \\ A_{1n} & A_{2n} & \cdots & A_{nn} \end{bmatrix} \tag{B.4-7}$$

For example,

$$\mathbf{a} = \begin{bmatrix} x & 1 & y \\ 1 & 2 & 1 \\ 0 & 3 & 2 \end{bmatrix}$$

$$A_{11} = (1)(4 - 3) = 1$$

$$A_{12} = (-1)(2 - 0) = -2$$

$$A_{13} = (1)(3 - 0) = 3$$

etc.

$$\mathscr{A} = \begin{bmatrix} 1 & -(2 - 3y) & (1 - 2y) \\ -2 & 2x & -(x - y) \\ 3 & -3x & (2x - 1) \end{bmatrix}$$

The inverse of a matrix ($\mathbf{a}^{-1}$) is defined as

$$\mathbf{a}^{-1}\mathbf{a} = \mathbf{I} \tag{B.4-8}$$

One way that an inverse may be calculated is

$$\mathbf{a}^{-1} = \frac{\mathscr{A}}{\det \mathbf{a}} \quad \text{if } \det \mathbf{a} \neq 0 \tag{B.4-9}$$

An illustration is

$$\mathbf{a} = \begin{bmatrix} 8 & 4 & 2 \\ 2 & 8 & 4 \\ 1 & 2 & 8 \end{bmatrix} \qquad \text{cofactor matrix } [A_{ij}] = \begin{bmatrix} 56 & -12 & -4 \\ -28 & 62 & -12 \\ 0 & -28 & 56 \end{bmatrix}$$

$\det \mathbf{a} = 392$

$$\mathbf{a}^{-1} = \begin{bmatrix} \frac{56}{392} & \frac{12}{392} & -\frac{4}{392} \\ \frac{-28}{392} & \frac{62}{392} & \frac{-12}{392} \\ \frac{0}{392} & -\frac{28}{392} & \frac{56}{392} \end{bmatrix}^T = \begin{bmatrix} \frac{1}{7} & -\frac{1}{14} & 0 \\ -\frac{3}{98} & \frac{31}{196} & -\frac{1}{14} \\ -\frac{1}{98} & -\frac{3}{98} & \frac{1}{7} \end{bmatrix}$$

If the determinant of **a** is 0, then **a** is said to be *singular*, and $\mathbf{a}^{-1}$ does not exist. If $\mathbf{a}^T = \mathbf{a}^{-1}$, **a** is said to be *orthogonal*. A matrix consisting of a single column is called a *column vector* while a matrix consisting of a single row is a *row vector*:

$$\begin{bmatrix} 1 \\ 2 \\ 3 \\ 4 \end{bmatrix} \text{ Column vector} \qquad [1 \quad 2 \quad 3 \quad 4] \text{ Row vector}$$

To multiply two matrices together, **ab**, they must be *conformable*; that is, the number of columns of the first matrix **a** must equal the number of rows of the second matrix **b** (**a** is the *premultiplier* and **b** the *postmultiplier*). Notice that **ab** does not equal **ba** except in unusual cases.

To multiply the two matrices **a** and **b** together, take the first element in the first row of **a** and multiply it into the first element of the first column of **b**. Take the second element of the first row of **a** and multiply it into the second element of the first column of **b**. Continue until each element of the first row of **a** has been multiplied into the corresponding element of the first column of **b**, and then sum the products. This sum forms the new element c_{11} of $\mathbf{ab} = \mathbf{c}$:

$$c_{11} = \sum_{j=1}^{n} a_{1j} b_{j1} \tag{B.4-10}$$

Next, multiply in a similar fashion the first row of **a** into the second column of **b**, and sum the products; this becomes c_{12}. Repeat until the first row of **a** has been multiplied into each column of **b**. This completes the first row of the product **c**. Then repeat the entire process by using the second row of **a**; multiply it into the first, second, etc., columns of **b** to form the second row of the product **c**. The sequence of steps is continued until all rows of **a** have been accounted for. We now illustrate the method for two 3 by 3 matrices.

$$\mathbf{a} = \begin{bmatrix} 1 & 0 & 2 \\ 2 & 1 & 1 \\ 0 & 1 & 2 \end{bmatrix} \qquad \mathbf{b} = \begin{bmatrix} 0 & 1 & 3 \\ 2 & 1 & 0 \\ 3 & 2 & 1 \end{bmatrix}$$

$$\mathbf{ab} = \begin{bmatrix} (0+0+6) & (1+0+4) & (3+0+2) \\ (0+2+3) & (2+1+2) & (6+0+1) \\ (0+2+6) & (0+1+4) & (0+0+2) \end{bmatrix} = \begin{bmatrix} 6 & 5 & 5 \\ 5 & 5 & 7 \\ 8 & 5 & 2 \end{bmatrix}$$

If $\mathbf{a}(x)$ is a matrix or a vector with elements that are functions of x, the *derivative* of $\mathbf{a}(x)$ with respect to x is obtained by differentiating *each element* of $\mathbf{a}(x)$ with respect to x:

$$\frac{d\mathbf{a}(x)}{dx} = \frac{d[a_{ij}(x)]}{dx} \tag{B.4-11}$$

Then the *integral* of $\mathbf{a}(x)$ is

$$\int_{x_1}^{x_2} \mathbf{a}(x)\, dx = \left[\int_{x_1}^{x_2} [a_{ij}(x)]\, dx\right] \tag{B.4-12}$$

that is, integrate each element of $\mathbf{a}(x)$.

The *rank* of a matrix is the order of the highest nonzero determinant contained in the matrix.

B.4-1 Solution of Algebraic Equations

One important application of matrix algebra is in the solution of sets of simultaneous linear equations. For a set of linear equations to have a solution, the rank of the matrix of coefficients **a** and the rank of the augmented matrix $[\mathbf{a}, \mathbf{b}]$, as described below, must be the same. With the use of a digital computer, it is possible to handle hundreds of sets of simultaneous independent equations of the form

$$\begin{aligned} a_{11}x_1 + a_{12}x_2 + \cdots + a_{1n}x_n &= b_1 \\ a_{21}x_1 + a_{22}x_2 + \cdots + a_{2n}x_n &= b_2 \\ \cdots\cdots\cdots\cdots\cdots\cdots & \\ a_{n1}x_1 + a_{n2}x_2 + \cdots + a_{nn}x_n &= b_n \end{aligned} \tag{B.4-13}$$

in which the a_{ij}'s are constant coefficients and the x_j's are unknowns. In compact matrix notation with

$$\mathbf{a} = \begin{bmatrix} a_{11} & a_{12} & \cdots & a_{1n} \\ a_{21} & a_{22} & \cdots & a_{2n} \\ \cdot & \cdot & \cdot & \cdot \\ a_{n1} & a_{n2} & \cdots & a_{nn} \end{bmatrix}$$

$$\mathbf{x} = \begin{bmatrix} x_1 \\ x_2 \\ \vdots \\ x_n \end{bmatrix} \qquad \mathbf{b} = \begin{bmatrix} b_1 \\ b_2 \\ \vdots \\ b_n \end{bmatrix}$$

Equations B.4-13 become

$$\mathbf{ax} = \mathbf{b} \tag{B.4-14}$$

which has the solution

$$\mathbf{x} = \mathbf{a}^{-1}\mathbf{b} \tag{B.4-15}$$

if $\det \mathbf{a} \neq 0$.

Example B.4-1 Solution of a Set of Simultaneous Independent Linear Equations

Find the solution of

$$\begin{aligned} 2x_1 + 3x_2 + 4x_3 + 5x_4 &= 1 \\ 3x_1 + 7x_2 + 5x_3 + 4x_4 &= 1 \\ x_1 + 4x_2 + 9x_3 + 2x_4 &= 1 \\ 5x_1 + 2x_2 + 7x_3 + x_4 &= 1 \end{aligned}$$

Solution:

We know that **a** and **b** are

$$\mathbf{a} = \begin{bmatrix} 2 & 3 & 4 & 5 \\ 3 & 7 & 5 & 4 \\ 1 & 4 & 9 & 2 \\ 5 & 2 & 7 & 1 \end{bmatrix} \qquad \mathbf{b} = \begin{bmatrix} 1 \\ 1 \\ 1 \\ 1 \end{bmatrix}$$

By use of some suitable iterative method (such methods are discussed in detail in books on numerical analysis), the inverse of **a** can be calculated as

$$\mathbf{a}^{-1} = \begin{bmatrix} -0.007246 & 0.050725 & -0.188406 & 0.201045 \\ -0.176812 & 0.237681 & 0.002899 & -0.072464 \\ 0.013043 & -0.091304 & 0.139130 & 0.021739 \\ 0.298551 & -0.089855 & -0.037681 & -0.057971 \end{bmatrix}$$

and by Equation B.4-15

$$\mathbf{x} = \begin{bmatrix} x_1 \\ x_2 \\ x_3 \\ x_4 \end{bmatrix} = \begin{bmatrix} 0.065217 \\ -0.008696 \\ 0.082609 \\ 0.113043 \end{bmatrix}$$

B.4-2 Eigenvalues and Eigenvectors

If $\mathbf{a}$ is an $n \times n$ matrix and $\mathbf{x}$ is a column vector of order n, we can generate a new column vector $\mathbf{y}$ by multiplication:

$$\mathbf{ax} = \mathbf{y} \tag{B.4-16}$$

Now we pose the question: Is $\mathbf{y}$ in the same direction as $\mathbf{x}$? If so, we can think of $\mathbf{y}$ as being some scalar λ times $\mathbf{x}$, or

$$\mathbf{y} = \mathbf{ax} = \lambda\mathbf{x} \tag{B.4-17}$$

or

$$(\mathbf{a} - \lambda\mathbf{I})\mathbf{x} = 0 \tag{B.4-18}$$

For Equation B.4-18 to be true, it is necessary for either det $(\mathbf{a} - \lambda\mathbf{I})$ to be zero or for $\mathbf{x}$ to be zero, but the latter is a trivial solution. Hence,

$$\det(\mathbf{a} - \lambda\mathbf{I}) = \begin{vmatrix} (a_{11} - \lambda) & a_{12} & \cdots & a_{1n} \\ a_{21} & (a_{22} - \lambda) & \cdots & a_{2n} \\ \cdot & \cdot & \cdot & \cdot \\ a_{n1} & a_{n2} & & (a_{nn} - \lambda) \end{vmatrix} = 0 \tag{B.4-19}$$

$(\mathbf{a} - \lambda\mathbf{I})$ is called the *characteristic matrix* of $\mathbf{a}$, the det $(\mathbf{a} - \lambda\mathbf{I})$ is called the *characteristic* (or secular) *function* of $\mathbf{a}$, and Equation B.4-19 is termed the *characteristic equation* of $\mathbf{a}$ which can be expanded as a *characteristic polynomial* $P(\lambda)$:

$$\det(\mathbf{a} - \lambda\mathbf{I}) = P(\lambda) = \lambda^n + P_1\lambda^{n-1} + \cdots + P_{n-1}\lambda + P_n = 0 \tag{B.4-20}$$

The scalar multiplier λ we sought is one of the n roots (real or complex) of Equation B.4-20; each of these roots is called an *eigenvalue* (or characteristic value or latent root) of $\mathbf{a}$ and, in general, can be obtained by iterative methods.

If λ_1 is an eigenvalue of $\mathbf{a}$, then for this value of λ Equation B.4-19 is satisfied, and has a nontrivial solution. Momentarily, let us assume that Equation B.4-19 does not have multiple roots so that there are n distinct λ's. Associated with each one of these λ_i's is a column vector $\mathbf{x}_i$ that satisfies the equation

$$(\mathbf{a} - \lambda_i\mathbf{I})\mathbf{x}_i = 0 \tag{B.4-21}$$

These vectors are called *eigenvectors* or *principal axes* of the matrix $\mathbf{a}$. The elements in the columns of $\mathbf{x}_i$ are thus directly proportional to each other. If $\mathbf{x}_i$ is regarded as a vector in n-dimensional space, only its direction is uniquely determined, not its length.

For multiple roots, only one eigenvector is found for each distinct root.

Example B.4-2 Eigenvalues and Eigenvectors

Let

$$\mathbf{a} = \begin{bmatrix} 1 & 2 \\ 2 & 1 \end{bmatrix}$$

Then

$$P(\lambda) = \begin{bmatrix} 1 - \lambda & 2 \\ 2 & 1 - \lambda \end{bmatrix} = \lambda^2 - 2\lambda - 3 = 0 \tag{a}$$

Equation (a) has the roots (the eigenvalues) $\lambda_1 = -1$ and $\lambda_2 = 3$. Usually the roots are not integers—as used in this example for illustrative purposes—but can be obtained by numerical methods (such as Newton's method) for nonlinear equations as described in texts on numerical analysis. For the root $\lambda = -1$, we can find an eigenvector which satisfies the equation $(\mathbf{a} - (-1)\mathbf{I})\mathbf{x}_1 = 0$; that is,

$$\begin{bmatrix} 1 - (-1) & 2 \\ 2 & 1 - (-1) \end{bmatrix} \begin{bmatrix} x_1 \\ x_2 \end{bmatrix} = 0 \tag{b}$$

which yields

$$\begin{aligned} 2x_1 + 2x_2 &= 0 \\ 2x_1 + 2x_2 &= 0 \end{aligned} \tag{c}$$

or

$$x_1 + x_2 = 0 \tag{d}$$

The second eigenvalue gives

$$\begin{bmatrix} 1 - 3 & 2 \\ 2 & 1 - 3 \end{bmatrix} \begin{bmatrix} x_1 \\ x_2 \end{bmatrix} = 0 \tag{e}$$

or

$$x_1 - x_2 = 0 \tag{f}$$

From Equations (d) and (f) the eigenvectors associated with the eigenvalues -1 and 3 are those values of x_1 and x_2 that satisfy Equations (d) and (f); one (of many) pair of eigenvectors is

$$\begin{bmatrix} 1 \\ -1 \end{bmatrix} \quad \text{and} \quad \begin{bmatrix} 1 \\ 1 \end{bmatrix} \tag{g}$$

respectively. Any scalar multiple of each of the vectors in Equation (g) is equally suitable.

Example B.4-3 Eigenvectors

If

$$\mathbf{a} = \begin{bmatrix} 5 & 2 & 0 & 0 \\ 2 & 2 & 0 & 0 \\ 0 & 0 & 5 & -2 \\ 0 & 0 & -2 & 2 \end{bmatrix}$$

$\lambda_1 = \lambda_2 = 1$; $\lambda_3 = \lambda_4 = 6$. For the eigenvalue 1, we find

$$4x_1 = 2x_2 + 0 + 0 = 0$$
$$2x_1 + x_2 + 0 + 0 = 0$$
$$0 + 0 + 4x_3 - 2x_4 = 0$$
$$0 + 0 - 2x_3 + x_4 = 0$$

which yields

$$2x_1 + x_2 = 0$$
$$2x_3 - x_4 = 0$$

or

$$\mathbf{x}_1 = \begin{bmatrix} -1 \\ 2 \\ 0 \\ 0 \end{bmatrix} + \begin{bmatrix} 0 \\ 0 \\ 1 \\ 2 \end{bmatrix} = \begin{bmatrix} -1 \\ 2 \\ 1 \\ 2 \end{bmatrix}$$

For $\lambda = 6$, it can be shown that an eigenvector is

$$\mathbf{x}_2 = \begin{bmatrix} 2 \\ 1 \\ 0 \\ 0 \end{bmatrix} + \begin{bmatrix} 0 \\ 0 \\ -2 \\ 1 \end{bmatrix} = \begin{bmatrix} 2 \\ 1 \\ -2 \\ 1 \end{bmatrix}$$

B.4-3 Normalization

The *norm* of a real vector is defined as

$$\text{Norm} = (\mathbf{x}^T\mathbf{x})^{1/2} = \sqrt{\sum_{i=1}^{n} x_i^2} \qquad \text{(B.4-22)}$$

Normalization of a vector $\mathbf{x}$ is the process of dividing every component of $\mathbf{x}$ by the length of the vector (to yield a unit vector). For example,

$$\text{Vector } \mathbf{x} = [1, 2, -3, 0]$$

$$\text{Norm } \mathbf{x} = \sqrt{1^2 + 2^2 + (-3)^2 + 0^2} = \sqrt{14}$$

$$\left.\begin{matrix}\text{Normalized vector} \\ \text{(unit vector)}\end{matrix}\right\} \tilde{\mathbf{x}} = \left[\frac{1}{\sqrt{14}}, \frac{2}{\sqrt{14}}, \frac{-3}{\sqrt{14}}, 0\right]$$

An *orthonormal* vector is a normalized orthogonal vector (orthogonal unit vector) as described below, and can be constructed by the Gram Schmidt process or other techniques given in the references at the end of the Appendix.

Example B.4-4 Normalized Vector

The eigenvectors from Example B.4-2 $\begin{bmatrix} 1 \\ -1 \end{bmatrix}$ and $\begin{bmatrix} 1 \\ 1 \end{bmatrix}$, are normalized as follows:

$$\text{Norm} = \sqrt{1^2 + (-1)^2} = \sqrt{2}$$
$$\text{Norm} = \sqrt{1^2 + 1^2} = \sqrt{2}$$

Normalized vectors are

$$\begin{bmatrix} \frac{1}{\sqrt{2}} \\ -\frac{1}{\sqrt{2}} \end{bmatrix} \quad \text{and} \quad \begin{bmatrix} \frac{1}{\sqrt{2}} \\ \frac{1}{\sqrt{2}} \end{bmatrix}$$

The normalized eigenvectors from Example B.4-3 are

$$\begin{bmatrix} -\frac{1}{\sqrt{5}} \\ \frac{2}{\sqrt{5}} \\ 0 \\ 0 \end{bmatrix}, \begin{bmatrix} 0 \\ 0 \\ \frac{1}{\sqrt{5}} \\ \frac{2}{\sqrt{5}} \end{bmatrix}, \begin{bmatrix} \frac{2}{\sqrt{5}} \\ \frac{1}{\sqrt{5}} \\ 0 \\ 0 \end{bmatrix}, \text{ and } \begin{bmatrix} 0 \\ 0 \\ -\frac{2}{\sqrt{5}} \\ \frac{1}{\sqrt{5}} \end{bmatrix}$$

B.4-4 Transformation to Canonical Form†

A *quadratic form* is defined as

$$q = \sum_{i=1}^{n} \sum_{j=1}^{n} a_{ij}x_ix_j = \mathbf{x}^T\mathbf{a}\mathbf{x} \qquad \text{(B.4-23)}$$

To avoid confusion in connection with terms such as $a_{ij}x_ix_j$ and $a_{ji}x_jx_i$ (i.e., $a_{12}x_1x_2$ and $a_{21}x_2x_1$), which contain the same independent variables and would be ordinarily combined, we shall agree to eliminate any ambiguity by replacing each member of every pair of coefficients by their mean:

$$\frac{a_{ji} + a_{ji}}{2}$$

By this rule, the expansion of $\mathbf{x}^T\mathbf{a}\mathbf{x}$ yields a symmetric matrix $(\mathbf{a} + \mathbf{a}^T)/2$, the symmetry property retaining definite advantages as will appear below. To illustrate the rule, if

$$q = x_1^2 - 4x_1x_2 + x_2^2 + x_1x_3 \qquad \text{(B.4-24a)}$$

by agreement q is also in full expansion:

$$q = x_1^2 - 2x_1x_2 - 2x_2x_1 + x_1^2 + \tfrac{1}{2}x_1x_3 + \tfrac{1}{2}x_3x_1 \qquad \text{(B.4-24b)}$$

† N. V. Yefinov, *Quadratic Forms and Matrices*, Academic Press, New York, 1964.

In matrix notation, Equation B.4-24b is

$$\mathbf{x}^T\mathbf{ax} = [x_1x_2x_3]\begin{bmatrix} 1 & -2 & \frac{1}{2} \\ -2 & 1 & 0 \\ \frac{1}{2} & 0 & 0 \end{bmatrix}\begin{bmatrix} x_1 \\ x_2 \\ x_3 \end{bmatrix}$$

where $\mathbf{a}$ is called the *symmetric bilinear form.*

Of particular interest in interpreting empirical models is a method of reducing the general quadratic form, which may contain crossproduct terms, to the so-called *canonical form* which does not contain crossproduct terms. For example, if

$$q_1 = 29x_1^2 + 24x_1x_2 + 5x_2^2 = \mathbf{x}^T\begin{bmatrix} 29 & 12 \\ 12 & 5 \end{bmatrix}\mathbf{x}$$

$$= \mathbf{x}^T\mathbf{ax}$$

by a suitable transformation introducing a special matrix $\mathbf{b}$ as follows

$$\mathbf{x} = \mathbf{by} \tag{B.4-25}$$

or in particular here

$$\mathbf{x} = \mathbf{by} = \begin{bmatrix} 1 & -2 \\ -2 & 5 \end{bmatrix}\mathbf{y}$$

we can compute $\mathbf{x}^T$:

$$\mathbf{x}^T = (\mathbf{by})^T = \mathbf{y}^T\mathbf{b}^T = \mathbf{y}^T\begin{bmatrix} 1 & -2 \\ -2 & 5 \end{bmatrix}^T = \mathbf{y}^T\begin{bmatrix} 1 & -2 \\ -2 & 5 \end{bmatrix}$$

and the canonical form of q_1 is evolved as

$$q_1 = \mathbf{y}^T\mathbf{b}^T\mathbf{aby} = \lambda_1y_1^2 + \lambda_2y_2^2 + \cdots + \lambda_ny_n^2$$

or specifically here

$$q_1 = \mathbf{y}^T\begin{bmatrix} 1 & -2 \\ -2 & 5 \end{bmatrix}\begin{bmatrix} 29 & 12 \\ 12 & 5 \end{bmatrix}\begin{bmatrix} 1 & -2 \\ -2 & 5 \end{bmatrix}\mathbf{y} = \mathbf{y}^T\begin{bmatrix} 1 & 0 \\ 0 & 1 \end{bmatrix}\mathbf{y}$$

$$= y_1^2 + y_2^2$$

To effect such a transformation, the major question is: How, in general, can one obtain the appropriate matrix for the transformation, such as $\mathbf{b} = \begin{bmatrix} 1 & -2 \\ -2 & 5 \end{bmatrix}$ above? A number of methods exist to reduce a quadratic form to equivalent forms (see the references at the end of this appendix), but the method described below is fairly simple and quite effective. It is known as *reduction by orthogonal transformation,* and it employs a unitary matrix $\mathbf{U}$ for the matrix $\mathbf{b}$ above, i.e., $\mathbf{x} = \mathbf{U}\tilde{\mathbf{x}}$, where the vector $\mathbf{x}$ represents the old coordinates and $\tilde{\mathbf{x}}$ represents the new coordinates.

Certain new nomenclature and properties of matrices that are needed are:

DEFINITION 1. A $n \times n$ real *unitary* matrix is one with the property $\mathbf{U}^T\mathbf{U} = \mathbf{U}\mathbf{U}^T = \mathbf{I}_n$.

DEFINITION 2. A real unitary matrix is an *orthogonal matrix,* i.e., $\mathbf{U}^T = \mathbf{U}^{-1}$.

PROPERTY 1, If $\mathbf{a}$ is a real symmetric matrix, there always exists an orthogonal matrix $\mathbf{U}$ such that $\mathbf{U}^T\mathbf{aU}$ is a diagonal matrix whose diagonal elements are the eigenvalues (characteristic roots) of $\mathbf{a}$:

$$\mathbf{U}^T\mathbf{aU} = \begin{bmatrix} \lambda_1 & & & 0 \\ & \lambda_2 & & \\ & & \ddots & \\ 0 & & & \lambda_n \end{bmatrix} \tag{B.4-2}$$

PROPERTY 2. The eigenvectors associated with distinct eigenvalues of a real symmetric matrix are orthogonal.

The matrix $\mathbf{U}$ can be obtained by finding the eigenvalues of $\mathbf{a}$ and then forming the set of normalized (orthonormal) eigenvectors associated with the eigenvalues. The set of orthonormal vectors form the unitary matrix $\mathbf{U}$. (In passing it should be noted that not every square matrix can be diagonalized by a nonsingular transformation such as Equation B.4-26; only Hermitian matrices can be.) Finally, the orthogonal transformation can be accomplished:

$$q = \mathbf{x}^T\mathbf{ax} = (\mathbf{Uy})^T\mathbf{a}(\mathbf{Uy}) = \mathbf{y}^T(\mathbf{U}^T\mathbf{aU})\mathbf{y} = \lambda_1y_1^2 + \lambda_2y_2^2 + \cdots \tag{B.4-27}$$

Example B.4-5 Unitary Matrix

By using the normalized vectors from Example B.4-4 for $\lambda_1 = 3$ and $\lambda_2 = -1$,

$$\mathbf{U} = \begin{bmatrix} \frac{1}{\sqrt{2}} & -\frac{1}{\sqrt{2}} \\ \frac{1}{\sqrt{2}} & \frac{1}{\sqrt{2}} \end{bmatrix}$$

Then, by multiplication,

$$\mathbf{U}^T\mathbf{aU} = \begin{bmatrix} \frac{1}{\sqrt{2}} & \frac{1}{\sqrt{2}} \\ -\frac{1}{\sqrt{2}} & \frac{1}{\sqrt{2}} \end{bmatrix}\begin{bmatrix} 1 & 2 \\ 2 & 1 \end{bmatrix}\begin{bmatrix} \frac{1}{\sqrt{2}} & -\frac{1}{\sqrt{2}} \\ \frac{1}{\sqrt{2}} & \frac{1}{\sqrt{2}} \end{bmatrix}$$

$$= \begin{bmatrix} 3 & 0 \\ 0 & 1 \end{bmatrix}$$

In the other case, with $\lambda_1 = 1, \lambda_2 = 1, \lambda_3 = 6$, and $\lambda_4 = 6$,

$$\mathbf{U} = \begin{bmatrix} \frac{1}{\sqrt{5}} & 0 & \frac{1}{\sqrt{5}} & 0 \\ -\frac{2}{\sqrt{5}} & 0 & \frac{1}{\sqrt{5}} & 0 \\ 0 & \frac{1}{\sqrt{5}} & 0 & -\frac{2}{\sqrt{5}} \\ 0 & \frac{2}{\sqrt{5}} & 0 & \frac{1}{\sqrt{5}} \end{bmatrix}$$

and it can be shown that, with **a** from Example B.4-3,

$$\mathbf{U}^T\mathbf{a}\mathbf{U} = \begin{bmatrix} 1 & 0 & 0 & 0 \\ 0 & 1 & 0 & 0 \\ 0 & 0 & 6 & 0 \\ 0 & 0 & 0 & 6 \end{bmatrix}$$

A geometric interpretation of an orthogonal transformation is that of rigid rotation of Cartesian axes about the origin. Consider two sets of coordinates (see Figure B.4-1):

$$Ox_1, Ox_2, Ox_3$$

and

$$O\tilde{x}_1, O\tilde{x}_2, O\tilde{x}_3$$

Let u_{ij} be the cosine of the angle between Ox_i and $O\tilde{x}_j$.

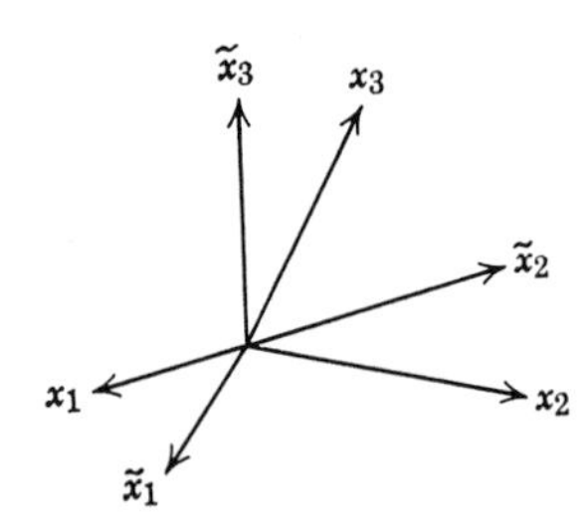

FIGURE B.4-1

The following equations give the relation between the two coordinate systems:

$$\begin{aligned} x_1 &= u_{11}\tilde{x}_1 + u_{12}\tilde{x}_2 + u_{13}\tilde{x}_3 \\ x_2 &= u_{21}\tilde{x}_1 + u_{22}\tilde{x}_2 + u_{23}\tilde{x}_3 \\ x_3 &= u_{31}\tilde{x}_1 + u_{32}\tilde{x}_2 + u_{33}\tilde{x}_3 \end{aligned} \qquad \text{(B.4-27)}$$

or

$$\mathbf{x} = \mathbf{U}\tilde{\mathbf{x}}$$

Note that u_{11}, u_{12}, and u_{13} (the first row) are the cosines of the angles Ox_1 makes with $O\tilde{x}_1$, $O\tilde{x}_2$, and $O\tilde{x}_3$, respectively, while u_{11}, u_{21}, and u_{31} (the first column) are the angles that $O\tilde{x}_1$ makes with Ox_1, Ox_2, and Ox_3, respectively.

One can show that **U** is an orthogonal matrix ($\mathbf{U}^{-1} = \mathbf{U}^T$) since

$$u_{1j}^2 + u_{2j}^2 + u_{3j}^2 = 1, \qquad j = (1, 2, 3) \qquad \text{(B.4-28a)}$$

and since

$$u_{1i}u_{1j} + u_{2i}u_{2j} + u_{3i}u_{3j} = 0 \qquad (i \neq j) \qquad \text{(B.4-28b)}$$

Also one can solve for $\tilde{\mathbf{x}}$ in terms of **x** and thus relate the new coordinates to the old ones:

$$\tilde{\mathbf{x}} = \mathbf{U}^{-1}\mathbf{x} = \mathbf{U}^T\mathbf{x} \qquad \text{(B.4-29)}$$

Example B.4-6 Transformation to Canonical Form

Reduce $q = 7x_1 - 4x_1x_2 + 2x_1x_3 + 10x_2^2 - 4x_2x_3 + 7x_3^2$ to canonical form.

Solution:

Set up q in matrix notation:

$$q = \mathbf{x}^T\mathbf{a}\mathbf{x} = [x_1 \quad x_2 \quad x_3]\begin{bmatrix} 7 & -2 & 1 \\ -2 & 10 & -2 \\ 1 & -2 & 7 \end{bmatrix}\begin{bmatrix} x_1 \\ x_2 \\ x_3 \end{bmatrix}$$

Next, find the eigenvalues of **a** from $\det(\mathbf{a} - \lambda\mathbf{I}) = 0$:

$$\det\begin{bmatrix} 7-\lambda & -2 & 1 \\ -2 & 10-\lambda & -2 \\ 1 & -2 & 7-\lambda \end{bmatrix} = 0$$

or

$$\begin{aligned} &(7-\lambda)[(10-\lambda)(7-\lambda) - (-2)(-2)] + 2[(-2)(7-\lambda) + 2] \\ &\quad + 1[4 - (10-\lambda)] = \lambda^3 - 24\lambda^2 + 180\lambda - 432 = 0 \end{aligned} \qquad \text{(a)}$$

The eigenvalues which satisfy Equation (a) are 6, 6, and 12. Consequently, the canonical form of q is

$$\begin{aligned} q &= \mathbf{x}^T\mathbf{a}\mathbf{x} = \tilde{\mathbf{x}}^T(\mathbf{U}^T\mathbf{a}\mathbf{U})\tilde{\mathbf{x}} = \lambda_1\tilde{x}_1^2 + \lambda_2\tilde{x}_2^2 + \lambda_3\tilde{x}_3^2 \\ &= 6\tilde{x}_1^2 + 6\tilde{x}_2^2 + 12\tilde{x}_3^2 \end{aligned} \qquad \text{(b)}$$

where the $\tilde{x}_i$'s are the new axes.

To relate the new coordinate system to the old coordinate system, use Equation B.4-29:

$$\tilde{\mathbf{x}} = \mathbf{U}^T\mathbf{x}$$

The matrix **U** is obtained as follows. For $\lambda = 6$, the eigenvectors are obtained as described in Section B.4-2. First introduce $\lambda = 6$ into $(\mathbf{a} - \lambda\mathbf{I}) = 0$:

$$\begin{bmatrix} 1 & -2 & 1 \\ -2 & 4 & -2 \\ 1 & -2 & 1 \end{bmatrix}\begin{bmatrix} u_1 \\ u_2 \\ u_3 \end{bmatrix} = 0$$

and then choose two sets (because of the repeated eigenvalues) of x_i that satisfy Equation (c), i.e., satisfy $u_1 - 2u_2 + u_3 = 0$, and also form orthogonal vectors. As the orthogonal pair, select, say,

$$\mathbf{U}_{\lambda=6} = \begin{bmatrix} 1 \\ 1 \\ 1 \end{bmatrix} \quad \text{and} \quad \mathbf{U}_{\lambda=6} = \begin{bmatrix} 1 \\ 0 \\ -1 \end{bmatrix}$$

For $\lambda = 12$, select x's that satisfy

$$-5u_1 - 2u_2 + u_3 = 0$$

or, for example,

$$\mathbf{U}_{\lambda=12} = \begin{bmatrix} 1 \\ -2 \\ 1 \end{bmatrix}$$

Next, the three vectors are normalized which leads to the following result for **U**:

$$\mathbf{U} = \begin{bmatrix} \frac{1}{\sqrt{3}} & \frac{1}{\sqrt{2}} & \frac{1}{\sqrt{6}} \\ \frac{1}{\sqrt{3}} & 0 & -\frac{2}{\sqrt{6}} \\ \frac{1}{\sqrt{3}} & -\frac{1}{\sqrt{2}} & \frac{1}{\sqrt{6}} \end{bmatrix}$$

It is not difficult to show that

$$\mathbf{U}^T\mathbf{a}\mathbf{U} = \begin{bmatrix} 6 & 0 & 0 \\ 0 & 6 & 0 \\ 0 & 0 & 12 \end{bmatrix}$$

as follows:

$$\mathbf{U}^T\mathbf{a}\mathbf{U} = \begin{bmatrix} \frac{1}{\sqrt{3}} & \frac{1}{\sqrt{3}} & \frac{1}{\sqrt{3}} \\ \frac{1}{\sqrt{2}} & 0 & -\frac{1}{\sqrt{2}} \\ \frac{1}{\sqrt{6}} & -\frac{2}{\sqrt{6}} & \frac{1}{\sqrt{6}} \end{bmatrix} \begin{bmatrix} 7 & -2 & 1 \\ -2 & 10 & -2 \\ 1 & -2 & 7 \end{bmatrix}$$

$$\cdot \begin{bmatrix} \frac{1}{\sqrt{3}} & \frac{1}{\sqrt{2}} & \frac{1}{\sqrt{6}} \\ \frac{1}{\sqrt{3}} & 0 & -\frac{2}{\sqrt{6}} \\ \frac{1}{\sqrt{3}} & -\frac{1}{\sqrt{2}} & \frac{1}{\sqrt{6}} \end{bmatrix}$$

$$= \begin{bmatrix} 6 & 0 & 0 \\ 0 & 6 & 0 \\ 0 & 0 & 12 \end{bmatrix}$$

The relation between the new and old coordinates then is

$$\tilde{\mathbf{x}} = \begin{bmatrix} \frac{1}{\sqrt{3}} & \frac{1}{\sqrt{3}} & \frac{1}{\sqrt{3}} \\ \frac{1}{\sqrt{2}} & 0 & -\frac{1}{\sqrt{2}} \\ \frac{1}{\sqrt{6}} & -\frac{2}{\sqrt{6}} & \frac{1}{\sqrt{6}} \end{bmatrix} \mathbf{x} \tag{d}$$

B.4-5 Solution of Nonlinear Sets of Equations

The solution of a single nonlinear algebraic or transcendental equation, or sets of nonlinear equations, is a much more formidable task than the solution of one or more linear equations. In steady-state processes, one or more material and energy balances can be nonlinear, in which case iterative techniques are required to evaluate the unknowns. Unfortunately, it is neither possible to determine in advance if the set of equations has a unique solution nor to assure that the iterative method will find the solution if it exists. But in working with equations based on real physical processes, this handicap may not prove to be of much practical significance.

One basic technique to solve sets of nonlinear equations is to linearize them by a Taylor series expansion, guess a set of values for the solution, solve the approximate linear problem, and improve the guesses by iteration. Such a procedure is known as the Newton-Raphson technique.†

B.5 SOLUTIONS OF SINGLE ORDINARY DIFFERENTIAL EQUATIONS

Many problems in unsteady-state heat, mass, and momentum transfer reduce to the solution of one or more ordinary differential equations. The engineer is usually interested in carrying out the integration of these equations in a formal analytical fashion, if possible. Sometimes the forms of the analytical solutions are rather complex and the determination of input-output *numbers* is tedious or even impractical. In such instances, it may be faster to solve the equation(s) by graphical or numerical methods with or without the assistance of an analog or digital computer. On the other hand, the generality of a formal solution has many desirable features.

The *order* of a differential equation is the order of the highest derivative; that is, $d^3y/dx^3 + d^2y/dx^2 = 3x$ is of order 3. The *general solution* of any ordinary differential equation contains as many *arbitrary constants* as the order of the differential equation; consequently, we need as many initial or boundary conditions as there are constants. The general solution of an nth-order differential equation essentially consists of a relation between the independent and dependent variables (involving also n arbitrary constants) which, when introduced into the differential equation, satisfies it.

The *degree* of a differential equation is the highest power to which the highest-order derivative is raised. As a consequence of this definition, all equations of higher degree than 1 are nonlinear.

In the limited space here, we can only list solutions for a few ordinary differential equations which are widely used in process analysis. Murphy‡ and Kamke§ listed solutions for more than 2000 ordinary differential

† L. Lapidus, *Digital Computers for Chemical Engineers*, McGraw-Hill, New York, 1962, p. 288; K. S. Kunz, *Numerical Analysis*, McGraw-Hill, New York, 1957, p. 10.

‡ G. M. Murphy, *Ordinary Differential Equations*, D. Van Nostrand, New York, 1960.

§ E. Kamke, *Differential Equations*, Edward Bros., Ann Arbor, Mich., 1945.

equations and outlined the general solution techniques; other references which cover many special cases will be found at the end of this appendix.

B.5-1 Single Second- and Higher-Order Linear Equations

Second- and higher-order linear equations are encountered in many process models. A linear equation of order n has the form

$$a_n(t)\frac{d^n y}{dt^n} + a_{n-1}(t)\frac{d^{n-1}y}{dt^{n-1}} + \cdots + a_1(t)\frac{dy}{dt} + a_0(t)y = x(t) \quad \text{(B.5-1)}$$

If $x(t) = 0$, Equation B.5-1 is called a *homogeneous equation* (all the terms are of the first degree in y and its derivatives). Sometimes Equation B.5-1 is known as the *complete* equation or, with $x(t) = 0$, as the *reduced equation.*

Solutions of the reduced equation are known as *complementary functions*; a particular solution of Equation B.5-1 is known as the *particular integral*, $y_p(t)$. A general solution to the complete equation consists of the sum of n linearly independent homogeneous solutions plus a particular integral:

$$y = \sum_{i=1}^{n} c_i y_i + y_p(t) \quad \text{(B.5-2)}$$

To evaluate the arbitrary constants, we need to be given y and its derivatives (n specifications) at $t = 0$ (the Cauchy problem), or y at n values of t (the LaGrange problem), or some combination of y and its derivatives.

A particularly important equation is the second-order equation with constant coefficients

$$\frac{d^2y}{dt^2} + a\frac{dy}{dt} + by = f(t) \quad \text{(B.5-3)}$$

or, in operator notation,

$$(D^2 + aD + b)y = f(t)$$

We make use of the roots of the auxiliary equation

$$r^2 + ar + b = 0, \qquad r = \frac{-a \pm \sqrt{a^2 - 4b}}{2}$$

to establish three categories of solutions shown in Table B.5-1. The particular integrals can be obtained by: (1) the method of undetermined coefficients, (2) the method of variation of parameters, or (3) operator methods, the details of which can be found in texts on differential equations.

Example B.5-1 Solution of a Linear Second-Order Equation

In determining the thermal conductivity of a metal from the electrical conduction and temperature drop in a rod, the resulting differential equation was evolved:

$$k\frac{d^2T}{dx^2} = \frac{(h)(2D)T}{A} - Q$$

where

T = temperature
x = distance from center
A = cross-section of rod
D = diameter of rod
Q = energy generated/volume of rod

TABLE B5.1 SOLUTIONS OF LINEAR SECOND-ORDER DIFFERENTIAL EQUATIONS WITH CONSTANT COEFFICIENTS

Case	Solution
I. Homogeneous equations:	
Real roots (r_1, r_2)	
$a^2 - 4b > 0$	$y = c_1 e^{r_1 t} + c_2 e^{r_2 t}$
Real (equal) roots $(r_1 = r_2 = r)$	
$a^2 - 4b = 0$	$y = c_1 e^{rt} + c_2 t e^{rt}$
Complex roots $\begin{Bmatrix} r_1 = \alpha + \beta i \\ r_2 = \alpha - \beta i \end{Bmatrix}$	
$a^2 - 4b < 0$	$y = e^{\alpha t}(c_1 \cos \beta t + c_2 \sin \beta t)$ $= A e^{\alpha t} \sin \beta(t + \delta)$
II. Nonhomogeneous equations	
As above	$y = y_c + y_p(t)$ (y_c shown above in I)

The boundary conditions are

$$T = T_0 \qquad \text{at } x = 0$$

$$\frac{dT}{dx} = 0 \qquad \text{at } x = 0$$

What is the relation between T and x?

Solution:

Place the equation in the form

$$\frac{d^2T}{dx^2} - m^2T = -\frac{Q}{k} = -q \tag{a}$$

where

$$m^2 = \frac{(h)(2\pi D)}{Ak}$$

The solution of the homogeneous equation is

$$T = A\,e^{-mx} + B\,e^{mx} \tag{b}$$

while a particular solution is (by inspection)

$$T_p = \frac{q}{m^2} \tag{c}$$

so that the general solution is

$$T = A\,e^{-mx} + B\,e^{mx} + \frac{q}{m^2} \tag{d}$$

Introducing the boundary conditions and evaluating A and B result in

$$T = \left(T_0 - \frac{q}{m^2}\right)\frac{e^{-mx} + e^{mx}}{2} + \frac{q}{m^2} \tag{e}$$

The techniques described above can be extended to nth-order linear differential equations with constant coefficients. Such equations often arise when first-order effects occur in series. The auxiliary equation becomes a polynomial of the nth order and has n roots. If none of the roots is equal, the complementary function is

$$y = c_1\,e^{r_1 t} + c_2\,e^{r_2 t} + \cdots + c_n\,e^{r_n t} \tag{B.5-4}$$

For multiple roots, independent solutions of the form $y = f(t)\,e^{rt}$ can be employed under certain conditions.

Methods of solution using Laplace transforms will be discussed later.

B.5-2 Special Linear Equations with Variable Coefficients

Several forms of linear equations with variable (in the independent variable) coefficients occur so frequently that they have been given special names; their solutions (although expressed in terms of converging infinite series) have become standard tabulated functions, such as Bessel functions and Legendre functions. Bessel functions have appeared with great frequency in chemical engineering models because the processes being examined frequently occur in cylindrical vessels or tubes.

Many equations arise that may not appear to be Bessel equations until a judicious change of variable is made. We shall show here only a generalized Bessel equation that is quite widely used:†

$$t^2\frac{d^2y}{dt^2} + t(a + bt^r)\frac{dy}{dt}$$
$$+ [c + ht^{2s} - b(1 - a - r)t^r + b^2t^2]]y = 0 \tag{B.5-5}$$

B.5-3 Nonlinear Equations

Only a small number of the nonlinear differential equations that are developed in process analysis can be solved by exact analytic methods. Much time and effort have been devoted to finding means of obtaining approximate solutions to nonlinear equations. A third method of attack is to use digital- or analog-computer solutions which are the usual tools of the engineer in treating nonlinear ordinary differential equations.

Refer to the references at the end of this appendix for particular techniques.

B.6 SOLUTION OF SETS OF LINEAR ORDINARY DIFFERENTIAL EQUATIONS WITH CONSTANT COEFFICIENTS

A multitude of diverse processes from every field can be described by a set of ordinary differential equations of the following form:

$$\frac{dy_1}{dt} = f_1(y_1, y_2, \ldots, y_n, t)$$
$$\frac{dy_2}{dt} = f_2(y_1, y_2, \ldots, y_n, t) \tag{B.6-1}$$
$$\cdots\cdots\cdots$$
$$\frac{dy_n}{dx} = f_n(y_1, y_2, \ldots, y_n, t)$$

or, in matrix notation,

$$\frac{d\mathbf{y}}{dt} = f(\mathbf{y}, t) \tag{B.6-1a}$$

Such equations have solutions of the form

$$y_1 = F_1(t)$$
$$\vdots \tag{B.6-2}$$
$$y_n = F_n(t)$$

or

$$\mathbf{y} = \mathbf{F}(t) \tag{B.6-2a}$$

if certain initial conditions are met and if certain other conditions (which are unimportant to us) are satisfied. The problem of finding the solution of the set of Equations B.6-1, given the initial conditions, is called the Cauchy integration problem. At the other extreme, given the

† For solutions, see H. S. Mickley, T. K. Sherwood, and C. E. Reed, *Applied Mathematics in Chemical Engineering*, McGraw-Hill, New York, 1957, p. 174.

values of F_i at various t's, the problem is called the Lagrange integration problem. If the set of Equations B.6-1 does not include an explicit dependence on t, then the set can be called an *autonomous system*.

To state a general solution of the form of Equations B.6-2 is merely an exercise in the handling of notation; to find a specific closed-form solution for a given set of Equations B.6-1 may prove to be most difficult.

In another form, a set of differential equations can exist as a single equation of nth order:

$$\frac{d^n y}{dt^n} = g\left(y, \frac{dy}{dt}, \frac{d^2y}{dt^2}, \ldots, \frac{dy^{n-1}}{dt^{n-1}}, t\right) \quad \text{(B.6-3)}$$

This can be transformed to the canonical form by use of the following transformations:

$$\left.\begin{aligned} y &= z_1 \\ \frac{dy}{dt} &= \frac{dz_1}{dt} = z_2 \\ &\cdots \\ \frac{dy^{n-1}}{dt^{n-1}} &= \frac{dz_{n-1}}{dt} = z_n \end{aligned}\right\} \quad \text{(B.6-4)}$$

The equivalent system of differential equations is

$$\begin{aligned} \frac{dz_1}{dt} &= z_2 \\ \frac{dz_2}{dt} &= z_3 \\ &\cdots \\ \frac{dz_{n-1}}{dz} &= z_n \\ \frac{dz_n}{dt} &= G[z_1, z_2, \ldots, z_n, t] \end{aligned} \quad \text{(B.6-5)}$$

Thus an nth order differential equation can always be reduced to a system of differential equations equivalent to Equations B.6-1. (The converse is not true—that a system such as Equations B.6-1 can always be transformed into a single nth-order equation.)

Example B.6-1 Reduction of Higher-Order Equations to First-Order Equations

Let us put

$$\ddot{u} + 3\dot{u} - 4u + \dddot{v} + \dot{v} + 3v = 0 \quad \text{(a)}$$

$$\ddot{u} + 3\dot{u} \qquad - \dddot{v} + 3\dot{v} + v = 0 \quad \text{(b)}$$

in the form of Equation B.6-4. Let

$$x_1 = u, \quad x_2 = \dot{u}, \quad x_3 = v, \quad x_4 = \dot{v}, \quad x_5 = \ddot{v}$$

Then,

$$\begin{aligned} \dot{x}_1 &= x_2 \\ \dot{x}_3 &= x_4 \\ \dot{x}_4 &= x_5 \end{aligned}$$

and Equations (a) and (b) become

$$\dot{x}_2 + \dot{x}_5 - 4x_1 + 3x_2 + 3x_3 + x_4 = 0$$

$$\dot{x}_2 - \dot{x}_5 \qquad + 3x_2 + x_3 + 3x_4 = 0$$

or

$$\dot{x}_2 = 2x_1 - 3x_2 - 2x_3 - 2x_4$$

$$\dot{x}_5 = 2x_1 \qquad - x_3 + x_4$$

In matrix notation:

$$\begin{bmatrix} \dot{x}_1 \\ \dot{x}_2 \\ \dot{x}_3 \\ \dot{x}_4 \\ \dot{x}_5 \end{bmatrix} = \begin{bmatrix} 0 & 1 & 0 & 0 & 0 \\ 2 & -3 & -2 & -2 & 0 \\ 0 & 0 & 0 & 1 & 0 \\ 0 & 0 & 0 & 0 & 1 \\ 2 & 0 & -1 & 1 & 0 \end{bmatrix} \begin{bmatrix} x_1 \\ x_2 \\ x_3 \\ x_4 \\ x_5 \end{bmatrix}$$

Before discussing techniques of solution of sets of simultaneous differential equations, a few remarks are in order about the difference between *initial-value* problems and *boundary-value problems*.

1. Initial-value problem: The values of the dependent variables and their derivatives at $t = 0$ are given; that is, $y_1(0) = y_{10}$, $dy_2/dt = 0$, etc.
2. Boundary-value problem: The values of the dependent variables at the ends of an interval, distance, or perhaps time are specified; that is, $y_1 = y_{10}$ at $t = 0$; $y_1 = y_{1f}$ at $t = t_f$.

The solution of boundary-value problems is more complicated than initial-value problems, except when the exact solution to the differential equation is known and the boundary conditions are simply used to evaluate the arbitrary constants in the solution.

Given the set of Equations B.6-1 and the initial conditions, it should be possible even in the worst cases to start at the initial conditions and numerically construct the curves representing $\mathbf{y}$ versus t. Some error may accumulate, of course, but at least an approximate set of curves may be established. On the other hand, if the problem is stated in terms of boundary conditions, then even an approximate construction of the curve at $t = 0$ is awkward because to know the value of y at another t is not very useful.

Sets of linear simultaneous differential equations *with constant coefficients* can be treated by first reducing them to matrix notation and then using matrix or numerical methods to effect a solution. Computer programs are available to carry out the detailed steps. A set of equations of the form

$$\begin{aligned} \dot{y}_1 &= a_{11}y_1 + a_{12}y_2 + \cdots + a_{1n}y_n \\ \dot{y}_2 &= a_{21}y_1 + a_{22}y_2 + \cdots + a_{2n}y_n \\ &\cdots \\ \dot{y}_n &= a_{n1}y_1 + a_{n2}y_2 + \cdots + a_{nn}y_n \end{aligned} \quad \text{(B.6-6)}$$

reduces to the *homogeneous* form

$$\dot{\mathbf{y}} = \frac{d\mathbf{y}}{dt} = \mathbf{a}\mathbf{y} \tag{B.6-7}$$

As long as $\mathbf{a}$ is continuous for $t \geq 0$, Equation B.6-7 has a unique solution for the initial conditions $\mathbf{y}(0) = \mathbf{y}_0$. Recalling that for the scalar case

$$\frac{dy}{dt} = ay \qquad y(0) = y_0$$

has a solution of the form $y = (e^{at})y_0$, we look for an analogous solution of Equation B.6-7 in the form

$$\mathbf{y} = (e^{\mathbf{a}t})\mathbf{y}_0 \tag{B.6-8}$$

To do this we need to make use of the matrix exponential function which is defined as

$$e^{\mathbf{a}t} = \mathbf{I} + \mathbf{a}t + \cdots + \frac{\mathbf{a}^n t^n}{n!} + \cdots \tag{B.6-9}$$

analogous to the scalar expansion

$$e^x = 1 + x + \frac{x^2}{2!} + \cdots$$

It can be demonstrated by substitution and subsequent differentiation that Equation B.6-8 is the solution of Equation B.6-7:

$$\frac{d\mathbf{y}}{dt} = \frac{d}{dt}(e^{\mathbf{a}t})\mathbf{y}_0$$

$$= \frac{d}{dt}\left(\mathbf{I} + \mathbf{a}t + \frac{1}{2!}\mathbf{a}^2t^2 + \frac{1}{3!}\mathbf{a}^3t^3 + \cdots\right)\mathbf{y}_0$$

$$= \left(\mathbf{a} + \mathbf{a}^2t + \frac{1}{2!}\mathbf{a}^2t^2 + \cdots\right)\mathbf{y}_0$$

$$= \mathbf{a}\left(\mathbf{I} + \mathbf{a}t + \frac{1}{2!}\mathbf{a}^2t^2 + \cdots\right)\mathbf{y}_0$$

$$= \mathbf{a}(e^{\mathbf{a}t})\mathbf{y}_0 = \mathbf{a}\mathbf{y}$$

Next consider an inhomogeneous set of equations where $\mathbf{x}(t)$ is the forcing function (or set of inputs) for the set of equations

$$\frac{d\mathbf{y}}{dt} + \mathbf{a}\mathbf{y} = \mathbf{x}(t) \tag{B.6-10}$$

As with a single scalar differential equation, an integrating factor $e^{\mathbf{a}(t-t_0)}$ can be introduced such that

$$[e^{\mathbf{a}(t-t_0)}]\frac{d\mathbf{y}}{dt} + [e^{\mathbf{a}(t-t_0)}]\mathbf{a}\mathbf{y} = [e^{\mathbf{a}(t-t_0)}]\mathbf{x}(t)$$

Then

$$\frac{d}{dt}[e^{\mathbf{a}(t-t_0)}\mathbf{y}] = [e^{\mathbf{a}(t-t_0)}]\frac{d\mathbf{y}}{dt} + \left[\frac{d}{dt}e^{\mathbf{a}(t-t_0)}\right]\mathbf{y}$$

$$= [e^{\mathbf{a}(t-t_0)}]\frac{d\mathbf{y}}{dt} + \mathbf{a}\,e^{\mathbf{a}(t-t_0)}\mathbf{y}$$

$$= [e^{\mathbf{a}(t-t_0)}]\mathbf{x}(t) \tag{B.6-11}$$

Each side of Equation B.6-11 may now be integrated from t_0 to t, using the condition that at $t = t_0$, $\mathbf{y} = \mathbf{y}_0$:

$$[e^{\mathbf{a}(t-t_0)}\mathbf{y}]_{y_0}^{y} = \int_{t_0}^{t} [e^{\mathbf{a}(t'-t_0)}]\mathbf{x}(t')\,dt'$$

or

$$e^{\mathbf{a}(t-t_0)}\mathbf{y} - \mathbf{I}\mathbf{y}_0 = \int_{t_0}^{t} [e^{\mathbf{a}(t'-t_0)}]\mathbf{x}(t')\,dt' \tag{B.6-12}$$

Premultiplying both sides of Equation B.6-12 by $e^{-\mathbf{a}(t-t_0)}$ yields

$$\mathbf{y} = e^{-\mathbf{a}(t-t_0)}\mathbf{y}_0 + e^{-\mathbf{a}(t-t_0)}\int_{t_0}^{t} [e^{\mathbf{a}(t'-t_0)}]\mathbf{x}(t')\,dt' \tag{B.6-13}$$

For the special case of $\mathbf{y} = \mathbf{0}$ at $t = 0$, Equation B.6-13 becomes

$$\mathbf{y} = e^{-\mathbf{a}t}\int_0^t e^{\mathbf{a}t'}\mathbf{x}(t')\,dt'$$

$$= \int_0^t e^{\mathbf{a}(t'-t)}\mathbf{x}(t')\,dt' \tag{B.6-14}$$

To obtain closed-form scalar solutions of Equation B.6-7 explicitly is somewhat lengthy. A number of procedures are available of which we show one.

If the $n \times n$ matrix $\mathbf{a}$ in its canonical form is nonsingular and has n distinct eigenvectors (latent vectors) $\mathbf{h}_1, \mathbf{h}_2, \ldots, \mathbf{h}_n$, let $\mathbf{h}$ be the square matrix formed by the column vectors $\mathbf{h}_i$

$$\mathbf{h} = [\mathbf{h}_1, \mathbf{h}_2, \ldots, \mathbf{h}_n] \tag{B.6-15}$$

Then

$$\mathbf{y} = e^{\mathbf{a}t}\mathbf{y}_0$$

$$= \left[\mathbf{I} + \mathbf{a}t + \frac{(\mathbf{a}t)^2}{2} + \cdots\right]\mathbf{y}_0$$

Since $\mathbf{h}\mathbf{I} = \mathbf{h}$, we know that

$$\mathbf{I} = \mathbf{h}(\mathbf{I})\mathbf{h}^{-1}$$

$$\mathbf{a} - \lambda\mathbf{I} = 0 \quad \text{or} \quad \mathbf{a} = \lambda\mathbf{I}$$

$$\mathbf{a} = \mathbf{h}(\lambda\mathbf{I})\mathbf{h}^{-1}$$

$$\mathbf{a}^2 = (\mathbf{h}(\lambda\mathbf{I})\mathbf{h}^{-1})(\mathbf{h}(\lambda\mathbf{I})\mathbf{h}^{-1}) = \mathbf{h}(\lambda\mathbf{I})^2\mathbf{h}^{-1}$$

etc.

Finally,

$$\mathbf{y} = \left[\mathbf{h}(\mathbf{I})\mathbf{h}^{-1} + \mathbf{h}(\lambda\mathbf{I})\mathbf{h}^{-1}t + \mathbf{h}(\lambda\mathbf{I})^2\mathbf{h}^{-1}\frac{t^2}{2} + \cdots\right]\mathbf{y}_0$$

$$= \left[\mathbf{h}\left(\mathbf{I} + \lambda\mathbf{I}t + (\lambda\mathbf{I})^2\frac{t^2}{2} + \cdots\right)\mathbf{h}^{-1}\right]\mathbf{y}_0$$

$$= [\mathbf{h}\,e^{\lambda\mathbf{I}t}\mathbf{h}^{-1}]\mathbf{y}_0 \tag{B.6-16}$$

In the above, all n roots are distinct:

$$\lambda\mathbf{I} = \begin{bmatrix} \lambda_1 & 0 & \cdots & 0 \\ 0 & \lambda_2 & \cdots & 0 \\ \cdot & \cdot & \cdot & \cdot \\ 0 & 0 & \cdots & \lambda_n \end{bmatrix}$$

(The eigenvalues $\lambda_1, \lambda_2, \ldots$ may be real or complex.)

Now we can identify from the above that

$$e^{\mathbf{a}t} = \mathbf{h}\, e^{\lambda\mathbf{I}t}\mathbf{h}^{-1} \tag{B.6-17}$$

so that

$$e^{\mathbf{a}t} = \mathbf{h}\begin{bmatrix} e^{\lambda_1 t} & 0 & \cdots & 0 \\ 0 & e^{\lambda_2 t} & \cdots & 0 \\ \cdot & \cdot & \cdot & \cdot \\ 0 & 0 & \cdots & e^{\lambda_n t} \end{bmatrix}\mathbf{h}^{-1} \tag{B.6-18}$$

and if $\mathbf{h}^{-1}\mathbf{y}_0$ is defined to be equal to $\mathbf{b}$:

$$\mathbf{y}(t) = [\mathbf{h}_1\mathbf{h}_2\cdots\mathbf{h}_n]\begin{bmatrix} e^{\lambda_1 t} & 0 & \cdots & 0 \\ 0 & e^{\lambda_2 t} & \cdots & 0 \\ \cdot & \cdot & \cdot & \cdot \\ 0 & 0 & & e^{\lambda_n t} \end{bmatrix}\begin{bmatrix} b_1 \\ b_2 \\ \vdots \\ b_n \end{bmatrix}$$

$$= \mathbf{h}_1\, e^{\lambda_1 t}b_1 + \mathbf{h}_2\, e^{\lambda_2 t}b_2 + \cdots + \mathbf{h}_n\, e^{\lambda_n t}b_n \tag{B.6-19}$$

If the roots λ_i are complex, as long as they occur in conjugate pairs, Equation B.6-19 can be arranged to contain only real numbers including sines and cosines.

Example B.6-2 Matrix Solution of Sets of Differential Equations

Solve $\ddot{y} - 3\dot{y} + 2y = e^{-t}$ with initial conditions $y = 0$ and $\dot{y} = 1$ at $t = 0$.

Solution:

Let

$$y_1 = y, \qquad y_2 = \dot{y}, \qquad y_3 = e^{-t}$$

Then

$$\dot{y}_1 = y_2 \qquad\qquad y_{1,0} = 0$$

$$\dot{y}_2 = -2y_1 + 3y_2 + y_3 \qquad\qquad y_{2,0} = 1$$

$$\dot{y}_3 = -y_3 \qquad\qquad y_{3,0} = 1$$

or

$$\dot{\mathbf{y}} = \begin{bmatrix} 0 & 1 & 0 \\ -2 & 3 & 1 \\ 0 & 0 & -1 \end{bmatrix}\mathbf{y}$$

The determinant $(\mathbf{a} - \lambda\mathbf{I})$ is

$$\begin{vmatrix} (0-\lambda) & 1 & 0 \\ -2 & (3-\lambda) & 1 \\ 0 & 0 & (-1-\lambda) \end{vmatrix} = (-1-\lambda)(\lambda-2)(\lambda-1) = 0$$

and the eigenvalues of A are $\lambda = 1, 2$, and -1, respectively. The latent vectors are found from

$$\begin{bmatrix} -1 & 1 & 0 \\ -2 & 2 & 1 \\ 0 & 0 & -2 \end{bmatrix}\mathbf{h}_1 = 0 \quad \text{so} \quad \mathbf{h}_1 = \begin{bmatrix} 1 \\ 1 \\ 0 \end{bmatrix}$$

$$\begin{bmatrix} -2 & 1 & 0 \\ -2 & 1 & 1 \\ 0 & 0 & -3 \end{bmatrix}\mathbf{h}_2 = 0 \quad \text{so} \quad \mathbf{h}_2 = \begin{bmatrix} 1 \\ 2 \\ 0 \end{bmatrix}$$

$$\begin{bmatrix} 1 & 1 & 0 \\ -2 & 4 & 1 \\ 0 & 0 & 0 \end{bmatrix}\mathbf{h}_3 = 0 \quad \text{so} \quad \mathbf{h}_3 = \begin{bmatrix} 1 \\ -1 \\ 6 \end{bmatrix}$$

This means

$$\mathbf{h} = \begin{bmatrix} 1 & 1 & 1 \\ 1 & 2 & -1 \\ 0 & 0 & 6 \end{bmatrix} \quad \text{and} \quad \mathbf{h}^{-1} = \tfrac{1}{6}\begin{bmatrix} 12 & -6 & -3 \\ -6 & 6 & 2 \\ 0 & 0 & 1 \end{bmatrix}$$

and

$$\lambda\mathbf{I} = \mathbf{h}^{-1}\mathbf{a}\mathbf{h} = \begin{bmatrix} 1 & 0 & 0 \\ 0 & 2 & 0 \\ 0 & 0 & -1 \end{bmatrix}$$

Finally, the solution is

$$\mathbf{y}(t) = \mathbf{h}\, e^{\lambda\mathbf{I}t}\mathbf{h}^{-1}\mathbf{y}_0$$

$$= \tfrac{1}{6}\begin{bmatrix} 1 & 1 & 1 \\ 1 & 2 & -1 \\ 0 & 0 & 6 \end{bmatrix}\begin{bmatrix} e^t & 0 & 0 \\ 0 & e^{2t} & \\ 0 & 0 & e^{-t} \end{bmatrix}\begin{bmatrix} 12 & -6 & -3 \\ -6 & 6 & 2 \\ 0 & 0 & 1 \end{bmatrix}\begin{bmatrix} 0 \\ 1 \\ 1 \end{bmatrix}$$

or

$$\begin{bmatrix} y_1 \\ y_2 \\ y_3 \end{bmatrix} = \begin{bmatrix} -\frac{3}{2}e^{+t} + \frac{4}{3}e^{2t} + \frac{1}{6}e^{-t} \\ -\frac{3}{2}e^{t} + \frac{8}{3}e^{2t} - \frac{1}{6}e^{-t} \\ e^{-t} \end{bmatrix}$$

If the matrix $\mathbf{a}$ has one or more roots that are not distinct, Equation B.6-18 can be formally generalized to include this situation by making each exponent of e a matrix instead of a scalar:

$$\lambda_r t \to \boldsymbol{\lambda}_r t$$

in which $\boldsymbol{\lambda}_r$ is a $q \times q$ matrix (where q is the multiplicity):

$$\boldsymbol{\lambda}_r = \begin{bmatrix} \lambda_r & 1 & 0 & \cdots & 0 & 0 \\ 0 & \lambda_r & 1 & \cdots & 0 & 0 \\ \cdot & \cdot & \cdot & \cdot & \cdot & \cdot \\ 0 & 0 & 0 & \cdots & \lambda_r & 1 \\ 0 & 0 & 0 & \cdots & 0 & \lambda_r \end{bmatrix} \tag{B.6-20}$$

If each root is distinct, $\boldsymbol{\lambda}_r$ reduces to $\lambda_r\mathbf{I}$. It can be shown from the properties of matrices that

$$e^{\lambda t} = \begin{bmatrix} 1 & t & t^2/2 & \cdots & t^{q-1}/(q-1)! \\ 0 & 1 & t & \cdots & t^{q-2}/(q-2)! \\ \cdot & \cdot & \cdot & \cdot & \cdot \\ 0 & 0 & 0 & \cdots & 1 \end{bmatrix} \qquad \text{(B.6-21)}$$

Examination of Equations B.6-18 and B.6-21 shows that the solution of Equation B.6-8 will now consist of linear combinations of products of polynomials with exponentials.

Example B.6-3 Case of Multiple Eigenvalues

Solve

$$\dot{y}_1 = y_2 - y_3 \quad \text{with} \quad y_1(0) = 1$$
$$\dot{y}_2 = 2y_2 + y_3 \quad \text{with} \quad y_2(0) = 1$$
$$\dot{y}_3 = 4y_1 - 2y_2 + 5y_3 \quad \text{with} \quad y_3(0) = -2$$

Solution:

$$\mathbf{a} = \begin{bmatrix} 0 & 1 & -1 \\ 0 & 2 & 1 \\ 4 & -2 & 5 \end{bmatrix}$$

The $\det(\mathbf{a} - \lambda\mathbf{I}) = (2-\lambda)^2(3-\lambda) = 0$; the eigenvalues are 2, 2, and 3.

The eigenvector $\mathbf{h}_1$, corresponding to $\lambda = 2$, is found from $(\mathbf{a} - 2\mathbf{I})\mathbf{h}_1 = 0$:

$$\begin{bmatrix} -2 & 1 & -1 \\ 0 & 0 & 1 \\ 4 & -1 & 3 \end{bmatrix}\mathbf{h}_1 = 0 \quad \text{so} \quad \mathbf{h}_1 = \begin{bmatrix} 1 \\ 2 \\ 0 \end{bmatrix}$$

Since there is a repeated root, we need to find a second vector, not an eigenvector, from the equation

$$(\mathbf{a} - 2\mathbf{I})\mathbf{h}_2 = \mathbf{h}_1$$

$$\begin{bmatrix} -2 & 1 & -1 \\ 0 & 0 & 1 \\ 4 & -1 & 3 \end{bmatrix}\mathbf{h}_2 = \begin{bmatrix} 1 \\ 2 \\ 0 \end{bmatrix} \quad \text{so} \quad \mathbf{h}_2 = \begin{bmatrix} 0 \\ 3 \\ 2 \end{bmatrix}$$

From $(\mathbf{a} - 3\mathbf{I})\mathbf{h}_3 = 0$, we find

$$\mathbf{h}_3 = \begin{bmatrix} 0 \\ 1 \\ 1 \end{bmatrix}$$

Then

$$\mathbf{h} = \begin{bmatrix} 1 & 0 & 0 \\ 2 & 3 & 1 \\ 0 & 2 & 1 \end{bmatrix}, \quad \mathbf{h}^{-1} = \begin{bmatrix} 1 & 0 & 0 \\ -2 & 1 & -1 \\ 4 & -2 & 3 \end{bmatrix}$$

$$\lambda I = \mathbf{h}^{-1}\mathbf{a}\mathbf{h} = \begin{bmatrix} 2 & 1 & 0 \\ 0 & 2 & 0 \\ 0 & 0 & 3 \end{bmatrix}$$

The solution is

$$\mathbf{y}(t) = e^{\mathbf{a}t}\mathbf{y}_0 = \mathbf{h}\, e^{\lambda t}\mathbf{h}^{-1}\mathbf{y}_0$$

$$= \begin{bmatrix} 1 & 0 & 0 \\ 2 & 3 & 1 \\ 0 & 2 & 1 \end{bmatrix}\begin{bmatrix} e^{2t} & te^{2t} & 0 \\ 0 & e^{2t} & 0 \\ 0 & 0 & e^{3t} \end{bmatrix}\begin{bmatrix} 1 & 0 & 0 \\ -2 & 1 & -1 \\ 4 & -2 & 3 \end{bmatrix}\begin{bmatrix} 1 \\ 1 \\ -2 \end{bmatrix}$$

or

$$\begin{bmatrix} y_1(t) \\ y_2(t) \\ y_3(t) \end{bmatrix} = \begin{bmatrix} e^{2t} + te^{2t} \\ 5e^{2t} + 2te^{2t} - 4e^{3t} \\ 2e^{2t} - 4e^{3t} \end{bmatrix}$$

If the equations are not homogeneous but the *nonhomogeneous terms themselves satisfy homogeneous equations*, they can be reduced to a homogeneous set by introducing extra equations satisfied by the nonhomogeneous parts. Consider a set of equations of the form

$$\dot{\mathbf{y}} = \mathbf{b}\mathbf{y} + \mathbf{f}(x) \qquad \text{(B.6-22)}$$

Suppose $\mathbf{f}(x)$ can be written as a linear combination $\mathbf{cz}$ of m functions $\mathbf{z}$ which satisfy the differential equation $\dot{\mathbf{z}} = \mathbf{dz}$. If we replace Equation B.6-22 by the two sets

$$\dot{\mathbf{y}} = \mathbf{b}\mathbf{y} + \mathbf{c}\mathbf{z}$$

and

$$\dot{\mathbf{z}} = \mathbf{d}\mathbf{z}$$

Equation B.6-22 can be written in the so-called irreducible form

$$\begin{bmatrix} \dot{\mathbf{y}} \\ \dot{\mathbf{z}} \end{bmatrix} = \begin{bmatrix} \mathbf{b}\mathbf{c} \\ \mathbf{0}\mathbf{d} \end{bmatrix}\begin{bmatrix} \mathbf{y} \\ \mathbf{z} \end{bmatrix} \qquad \text{(B.6-23)}$$

the same form as Equation B.6-7, or

$$[\dot{\mathbf{y}}^*] = [\mathbf{a}][\mathbf{y}^*] \qquad \text{(B.6-23a)}$$

where

$$\dot{\mathbf{y}}^* = \begin{bmatrix} \dot{\mathbf{y}} \\ \dot{\mathbf{z}} \end{bmatrix} \qquad \mathbf{y}^* = \begin{bmatrix} \mathbf{y} \\ \mathbf{z} \end{bmatrix}$$

and

$$\mathbf{a} = \begin{bmatrix} \mathbf{b}\mathbf{c} \\ \mathbf{0}\mathbf{d} \end{bmatrix}$$

Example B.6-4 Solution of Simultaneous Equations

Solve the simultaneous equations:

$$\dot{y}_1 = 3y_1 + y_2 + e^{2t} + \sin t + 3 \qquad \text{(a)}$$
$$\dot{y}_2 = -y_1 + 2y_2 + te^{2t} - 2\sin t - 2 \qquad \text{(b)}$$

Solution:

Let

$$z_1 = t\,e^{2t}; \quad z_2 = \dot{z}_1 = 2t\,e^{2t} + e^{2t}$$
$$z_3 = \sin t; \quad z_4 = \dot{z}_3 = \cos t; \quad z_5 = -1$$

Other selections might be adopted. Then

$$\ddot{z}_1 - 4\dot{z}_1 + 4z_1 = 0 \quad \text{or} \quad \dot{z}_2 = 4z_2 - 4z_1$$

$$\ddot{z}_3 + z_3 = 0 \quad \text{or } \dot{z}_4 = -z_3$$

$$\dot{z}_5 = 0$$

In matrix notation,

$$\begin{bmatrix} \dot{z}_1 \\ \dot{z}_2 \\ \dot{z}_3 \\ \dot{z}_4 \\ \dot{z}_5 \end{bmatrix} = \begin{bmatrix} 0 & 1 & 0 & 0 & 0 \\ -4 & 4 & 0 & 0 & 0 \\ 0 & 0 & 0 & 1 & 0 \\ 0 & 0 & -1 & 0 & 0 \\ 0 & 0 & 0 & 0 & 0 \end{bmatrix} \begin{bmatrix} z_1 \\ z_2 \\ z_3 \\ z_4 \\ z_5 \end{bmatrix}$$

$$\mathbf{d} = \begin{bmatrix} 0 & 1 & 0 & 0 & 0 \\ -4 & 4 & 0 & 0 & 0 \\ 0 & 0 & 0 & 1 & 0 \\ 0 & 0 & -1 & 0 & 0 \\ 0 & 0 & 0 & 0 & 0 \end{bmatrix}$$

$$\mathbf{c} = \begin{bmatrix} -2 & 1 & 1 & 0 & 3 \\ 1 & 0 & -2 & 0 & -2 \end{bmatrix}$$

and

$$\begin{bmatrix} \dot{y}_1 \\ \dot{y}_2 \\ \hline \dot{z}_1 \\ \dot{z}_2 \\ \dot{z}_3 \\ \dot{z}_4 \\ \dot{z}_5 \end{bmatrix} = \left[\begin{array}{cc|ccccc} 3 & 1 & -2 & 1 & 1 & 0 & 3 \\ -1 & 2 & 1 & 0 & -2 & 0 & -2 \\ \hline 0 & 0 & 0 & 1 & 0 & 0 & 0 \\ 0 & 0 & -4 & 4 & 0 & 0 & 0 \\ 0 & 0 & 0 & 0 & 0 & 1 & 0 \\ 0 & 0 & 0 & 0 & -1 & 0 & 0 \\ 0 & 0 & 0 & 0 & 0 & 0 & 0 \end{array}\right] \begin{bmatrix} y_1 \\ y_2 \\ \hline z_1 \\ z_2 \\ z_3 \\ z_4 \\ z_5 \end{bmatrix}$$

B.7 SOLUTION OF SINGLE PARTIAL DIFFERENTIAL EQUATIONS

Differential equations that contain derivatives of more than one independent variable are classified as partial differential equations. We have already seen the wide variety of problems that are described in terms of such equations. The definitions of order, degree, and linearity retain the same meaning for partial differential equations as for ordinary differential equations.

Although the integration of ordinary differential equations gives rise to arbitrary constants of integration, the *general solution* of a partial differential equation involves n arbitrary functions for the nth-order case. Except for first-order equations and a few other special cases, it is seldom possible or necessary to seeks a general solution; instead the engineer must seek a particular solution for the specific conditions of the problem.

A large number of problems in transport phenomena and mathematical physics can be described by one of the following partial differential equations:

1. Laplace's equation $\quad \nabla^2\phi = 0$
2. Poisson's equation $\quad \nabla^2\phi = \begin{cases} -K(x^1, x^2, x^3) \\ \text{or a constant} \end{cases}$
3. Diffusion equation $\quad \nabla^2\phi = \dfrac{1}{h^2}\dfrac{\partial\phi}{\partial t}$
4. Wave equation $\quad \nabla^2\phi = \dfrac{1}{c^2}\dfrac{\partial^2\phi}{\partial t^2}$
5. Damped-wave equation (telegraph equation) $\quad \nabla^2\phi = \dfrac{1}{c^2}\dfrac{\partial^2\phi}{\partial t^2} + R\dfrac{\partial\phi}{\partial t}$
6. Helmholtz's equation $\quad \nabla^2\phi + \gamma^2\phi = 0$

In the above list, h, c, and γ are constants.

One conventional method for the solution of these equations is by *separation of variables*. A solution of the form $\phi = U^1(u^1)U^2(u^2)U^3(u^3)$ is used which permits the partial differential equation to be separated into three ordinary differential equations, each of which has a U solution. The initial and boundary conditions for a given problem are used to evaluate the arbitrary constants so that a unique solution can be obtained. Refer to the references at the end of this appendix for a discussion of particular techniques to carry out the separation of variables solution.

B.8 LAPLACE TRANSFORMS

In this section we shall describe the use of operational mathematics to solve *linear* ordinary and partial differential equations and, in particular, the use of Laplace transforms.

The *Laplace transform* $\check{f}(s)$ of a function $f(t)$ is defined by

$$\mathscr{L}[f(t)] = \check{f}(s) = \int_0^\infty e^{-st}f(t)\,dt \tag{B.8-1}$$

The inverse Laplace transform is denoted by $\mathscr{L}^{-1}[\check{f}(s)]$ so that

$$f(t) = \mathscr{L}^{-1}[\check{f}(s)] \tag{B.8-2}$$

The following properties of the Laplace transform are of interest to us.

1. *Linearity.* If

$$\check{f}_1(s) = \mathscr{L}[f_1(t)]$$

and

$$\check{f}_2(s) = \mathscr{L}[f_2(t)]$$

then

$$\mathscr{L}[c_1f_1(t) + c_2f_2(t)] = c_1\mathscr{L}[f_1(t)] + c_2\mathscr{L}[f_2(t)] = c_1\check{f}_1(s) + c_2\check{f}_2(s) \tag{B.8-3}$$

2. *Transform of a derivative.*

$$\mathscr{L}\left[\frac{df(t)}{dt}\right] = L[f'(t)] = s\check{f}(s) - f(0) \quad \text{(B.8-4)}$$

where $f(0)$ = value of $f(t)$ as $t \to 0$.

3. *Transform of an integral.*

$$\mathscr{L}\left[\int_0^\infty f(\tau)\,d\tau\right] = \frac{1}{s}\check{f}(s) \quad \text{(B.8-5)}$$

4. *Complex translation.* If

$$\check{f}(s) = \mathscr{L}[f(t)]$$

then

$$\check{f}(s-a) = \mathscr{L}[e^{at}f(t)] \quad \text{(B.8-6)}$$

That is, multiplying $f(t)$ by e^{at} results in substituting $(s - a)$ for s in its transform.

5. *Derivative of a transform.*

$$\frac{d_n\check{f}(s)}{ds^n} = \mathscr{L}[(-t)^n f(t)] \quad \text{(B.8-7)}$$

6. $$\lim \check{f}(s) = 0 \quad \text{as} \quad s \to \infty \quad \text{(B.8-8)}$$

7. *Integration of a transform.*

$$\int_s^\infty \check{f}(x)\,dx = \mathscr{L}\left[\frac{f(t)}{t}\right] \quad \text{(B.8-9)}$$

That is, division of $f(t)$ by t corresponds to integration of the transform of $f(t)$ from s to ∞.

8. *Transform of a step function.* If $U(t)$ is a unit step function:

$$U(t) = 0 \quad t < 0 \quad \text{(B.8-10)}$$
$$= 1 \quad t > 0$$

and if $U(t - \tau)$ is a unit step function starting at $t = \tau$ (see Figure B.8-1):

$$U(t-\tau) = 0 \quad t < \tau \quad \text{(B.8-11)}$$
$$= 1 \quad t > \tau$$

then

$$\mathscr{L}[U(t)] = \frac{1}{s}$$

$$\mathscr{L}[U(t-\tau)] = \frac{e^{-s\tau}}{s}$$

$$\mathscr{L}[f(t-\tau)] = e^{-\tau s}\check{f}(s) \quad \text{if} \quad f(t-\tau) = 0 \quad \text{for} \quad 0 < t < \tau$$

$$\mathscr{L}[f(t-\tau)U(t)] = e^{-\tau s}\check{f}(s)$$

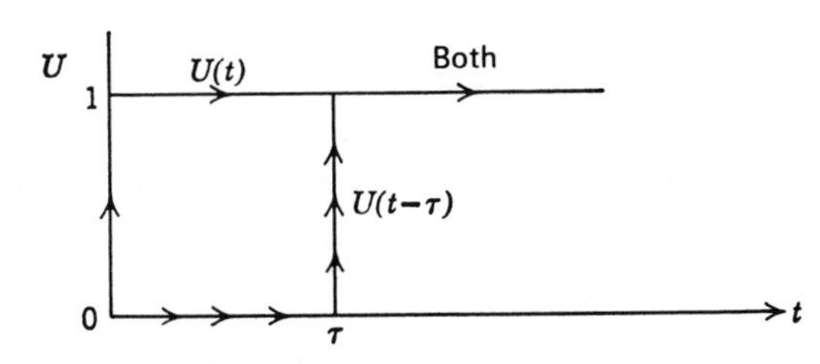

FIGURE B.8-1 Unit step functions $U(t)$ and $U(t - \tau)$.

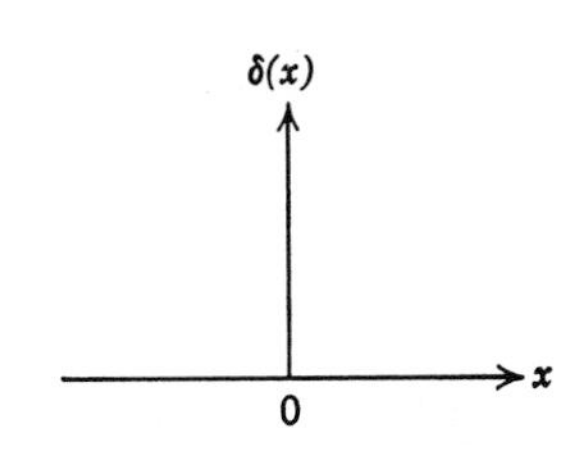

FIGURE B.8-2 Unit impulse function.

9. *Transform of a unit impulse function.* If $\delta(t)$ is the unit impulse function (Dirac delta function) shown in Figure B.8-2

$$\delta(x) = 0, \quad x \neq 0$$

$$\int_{-\infty}^\infty \delta(x)\,dx = 1$$

$$\mathscr{L}[\delta(t)] = 1 \quad \text{(B.8-12)}$$

$$\mathscr{L}[\delta(t-\tau) = e^{-\tau s}$$

10. *Transform of the convolution of two functions.* The integral

$$\int_0^t f_1(t-\tau)f_2(\tau)\,d\tau$$

is called the convolution of the functions f_1 and f_2:

$$\mathscr{L}\left[\int_0^t f_1(t-\tau)f_2(\tau)\,d\tau\right] = \check{f}_1(s)\check{f}_2(s) \quad \text{(B.8-13)}$$

11. *Initial and final values.*

Initial value: $\lim_{s\to\infty} s\check{f}(s) = \lim_{t\to 0} f(t)$

Final value: $\lim_{s\to 0} s\check{f}(s) = \lim_{t\to\infty} f(t)$

Example B.8-1 Laplace Transforms

Find the Laplace transforms of the following functions

1. $f(t) = U(t) = \begin{Bmatrix} 0 & t < 0 \\ 1 & t > 0 \end{Bmatrix}$
2. $f(t) = e^{at}$
3. $f(t) = \cos at$

Solution:

1. $$\mathscr{L}[U(t)] = \int_0^\infty e^{-st}(1)\,dt = -\left[\frac{1}{s}e^{-st}\right]_0^\infty$$
$$= -\frac{1}{s}[-1] = \frac{1}{s} \qquad s > 0$$

2. $$\mathscr{L}[e^{at}] = \int_0^\infty e^{-st}e^{at}\,dt = \int_0^\infty e^{(a-s)t}\,dt$$
$$= \frac{1}{a-s}[e^{(a-s)t}]_0^\infty = \frac{1}{a-s}(-1) = \frac{1}{s-a}$$

3. $\mathscr{L}[\cos at] = \int_0^\infty e^{-st} \cos at \, dt$

$$= \tfrac{1}{2}\int_0^\infty e^{-st}(e^{-iat} + e^{-iat})\, dt$$

$$= \frac{1}{2}\left[\frac{e^{(ia-s)t}}{(ia-s)} + \frac{e^{(-ia-s)t}}{(-ia-s)}\right]_0^\infty$$

$$= \frac{1}{2}\left[\frac{1}{s-ia} + \frac{1}{s+ia}\right] = \frac{s}{s^2+a^2}$$

Example B.8-2 Laplace Transforms of Derivatives

Find the Laplace transform of $f'(t)$ and $f''(t)$.

Solution:

$$\mathscr{L}[f'(t)] = \mathscr{L}\left[\frac{df(t)}{dt}\right] = \int_0^\infty \frac{df(t)}{dt} e^{-st}\, dt$$

Integrating by parts, let

$$u = e^{-st}$$

$$dv = f'(t)\, dt$$

$$\mathscr{L}[f'(t)] = [e^{-st}f(t)]_0^\infty + s\int_0^\infty e^{-st}f(t)\, dt$$

$$= -f(0) + s\check{f}(s)$$

Next,

$$\mathscr{L}[f''(t)] = \int_0^\infty f''(t)\, e^{-st}\, dt$$

$$= [e^{-st}f'(t)]_0^\infty + s\int_0^\infty f'(t)\, e^{-st}\, dt$$

$$= s^2\check{f}(s) - sf(0) - f'(0)$$

Evaluation of the inverse of Laplace transforms of functions (a topic of considerable importance) may be exceedingly difficult to accomplish. In determining a function $f(t)$ from a function $\check{f}(s)$, as a practical matter we usually first seek a table of Laplace transforms and try to match the transform of interest with one in the table. If this is successful, the inverse is immediately found. Other ways to evaluate $\mathscr{L}^{-1}$ are by:

1. The Heaviside expansion theorem.
2. Partial fractions.
3. The methods of poles and residues.
4. The convolution integral.
5. Numerical calculation.†

In view of the large detailed tables of transforms available today,‡ if one is unsuccessful in finding the proper transform, methods 1 to 4 will no doubt involve extensive complex manipulations. Table B.8-1 is a brief list of transforms, some of which are used in the examples below and others in the main body of the text.

† See H. L. Salzer, *J. Math. Phys.* **36**, 89, 1958, for a convenient set of tables to aid in numerical calculations.

‡ A. Erdélyi *et al.*, *Tables of Integral Transforms*, McGraw-Hill, New York, 1954.

Example B.8-3 Solution of an Ordinary Differential Equation

Solve $y''(t) + k^2y(t) = 0$ by Laplace transforms for the initial conditions $y(0) = c_1$ and $y'(0) = c_2$.

Solution:

First take the Laplace transform of both sides of the given equation with the help of Table B.8-1:

$$L\{y''(t) + k^2y(t)\} = 0$$

$$s^2\check{y}(s) - sy(0) - y'(0) + k^2\check{y}(s) = 0$$

Next introduce the initial conditions

$$s^2\check{y}(s) - sc_1 - c_2 + k^2\check{y}(s) = 0$$

and solve for $\check{y}(s)$

$$\check{y}(s) = \frac{sc_1 + c_2}{s^2+k^2} = \frac{sc_1}{s^2+k^2} + \frac{c_2}{s^2+k^2}$$

Finally, take the inverse transform of each side, again with the help of Table B.8-1:

$$y(t) = c_1 \cos(kt) + \frac{c_2}{k}\sin(kt)$$

Example B.8-4 Solution by Transformation of a Partial Differential Equation into an Ordinary Differential Equation

Solve the diffusion equation in a semi-infinite media:

$$\frac{\partial c}{\partial t} = \mathscr{D}\frac{\partial^2 c}{\partial x^2}$$

for the following boundary conditions:

(1) $c = 0, \quad t = 0, \quad x \geq 0$

(2) $\mathscr{D}\left(\frac{\partial c}{\partial x}\right)_{x=0} = \text{constant} = k, \quad t > 0, \quad x = 0$

(3) $c = 0, \quad t \geq 0, \quad x \to \infty$

Solution:

Take Laplace transforms of both sides of the equation with respect to the variable t. (In doing this, we assume it is permissible to interchange the order of differentiation with respect to x and the taking of the Laplace transform.)

$$s\check{c}(x,s) - c(x,0) = \mathscr{D}\frac{d^2\check{c}(x,s)}{dx^2} \tag{a}$$

Equation (a) is an ordinary differential equation which has the solution (with $c(x, 0) = 0$)

$$\check{c}(x,s) = A_1 e^{\left(\sqrt{\frac{s}{\mathscr{D}}}\right)x} + A_2 e^{-\left(\sqrt{\frac{s}{\mathscr{D}}}\right)x} \tag{b}$$

Note that the transformed boundary conditions become

(2′) $\mathscr{L}\left[\mathscr{D}\left(\frac{\partial c(x,t)}{dx}\right)_{x=0}\right] = \mathscr{D}\left(\frac{d\check{c}(x,s)}{dx}\right)_{x=0} = \mathscr{L}(k) = \frac{k}{s}$

(3′) $\mathscr{L}[\underset{x\to\infty}{c(x,t)}] = \underset{x\to\infty}{\check{c}(x,s)} = 0$

TABLE B.8-1 SELECTED LAPLACE TRANSFORMS

Function $f(t)$	Transform $\bar{f}(s) = \int_0^\infty e^{-st} f(t)\, dt$
1	$\frac{1}{s}$
$\frac{t^{n-1}}{(n-1)!}$	$\frac{1}{s^n}$
$\frac{1}{(\pi t)^{1/2}}$	$\frac{1}{s^{1/2}}$
$\frac{t^{n-1} e^{at}}{(n-1)!}$	$\frac{1}{(s-a)^n}$
$\frac{e^{at} - e^{bt}}{a - b}$	$\frac{1}{(s-a)(s-b)} \quad a \neq b$
$1 + \frac{1}{b-a}[a\, e^{-\frac{t}{a}} - b\, e^{-\frac{t}{b}}]$	$\frac{1}{s(1+as)(1+bs)}$
$\frac{1}{a} \sin at$	$\frac{1}{s^2 + a^2}$
$\cos at$	$\frac{s}{s^2 + a^2}$
$\frac{1}{a} \sinh at$	$\frac{1}{s^2 - a^2}$
$\cosh at$	$\frac{s}{s^2 - a^2}$
$J_0(at)$	$\frac{1}{(s^2 + a^2)^{1/2}}$
$erfc\left(\frac{k}{2\sqrt{t}}\right)$	$\frac{e^{-k\sqrt{s}}}{s}$
$\frac{1}{\sqrt{\pi t}} \exp\left(-\frac{k^2}{4t}\right)$	$\frac{e^{-k\sqrt{s}}}{\sqrt{s}}$
$2\left(\frac{t}{\pi}\right)^{1/2}\left[\exp\left(-\frac{k^2}{4t}\right)\right] - (k) erfc\left(\frac{k}{2\sqrt{t}}\right)$	$\frac{e^{-k\sqrt{s}}}{(s)^{3/2}}$
$f(t)$	$\bar{f}(s)$
$f(t-b)U(t-b)$	$e^{-bs}\bar{f}(s)$
$f'(t)$	$s\bar{f}(s) - f(0)$
$f^{(n)}(t)$	$s^n\bar{f}(s) - s^{n-1}f(0) - s^{n-2}f'(0) \cdots - f^{(n-1)}(0)$
$t^n f(t)$	$(-1)^n \bar{f}^{(n)}(s)$
$\delta(t)$	1
$U(t)$	$\frac{1}{s}$

By making use of condition (3′), $A_1 = 0$; then

$$\check{c}(x, s) = A_2 e^{-\left(\sqrt{\frac{s}{\mathscr{D}}}\right)x} \tag{c}$$

Now

$$\mathscr{D}\left(\frac{d\check{c}(x, s)}{dx}\right)_{x=0} = \mathscr{D}A_2\left(-\sqrt{\frac{s}{\mathscr{D}}}\right)\left(e^{-\left(\sqrt{\frac{s}{\mathscr{D}}}\right)x}\right)_{x=0} + \frac{k}{s}$$

so that

$$A_2 = -\left(\frac{k}{s\mathscr{D}}\right)\left(\frac{\mathscr{D}}{s}\right)^{1/2} \tag{d}$$

Finally

$$\check{c}(x, s) = -\frac{k}{s\mathscr{D}}\left(\frac{\mathscr{D}}{s}\right)^{1/2} e^{-\left(\sqrt{\frac{s}{\mathscr{D}}}\right)x} \tag{e}$$

The inverse Laplace transform of Equation (e) is

$$c(x, t) = -\frac{k}{\mathscr{D}^{1/2}}\mathscr{L}^{-1}\left[\frac{1}{s^{3/2}} e^{-\left(\sqrt{\frac{s}{\mathscr{D}}}\right)x}\right]$$

$$= -\frac{k}{\mathscr{D}^{1/2}}\left[2\left(\frac{t}{\pi}\right)^{1/2}\exp\left(-\frac{x^2}{4\mathscr{D}t}\right) - \frac{x}{\sqrt{\mathscr{D}}}\, erfc\left(\frac{x}{2\sqrt{\mathscr{D}t}}\right)\right] \tag{f}$$

By changing boundary condition (2) to

$$c = c_0, \qquad t > 0, \qquad x = 0$$

we would have

$$\mathscr{L}[c(x, t)] = \check{c}(x, s) = \ll[c_0] = \frac{c_0}{s}, \qquad x = 0$$

so that

$$\frac{c_0}{s} = A_2\left(e^{-\left(\sqrt{\frac{s}{\mathscr{D}}}\right)x}\right)_{x=0} = A_2 \tag{g}$$

Then

$$\check{c}(x, s) = \frac{c_0}{s} e^{-\left(\sqrt{\frac{s}{\mathscr{D}}}\right)x} \tag{h}$$

and

$$c(x, t) = c_0\mathscr{L}^{-1}\left(\frac{1}{s} e^{-\left(\sqrt{\frac{s}{\mathscr{D}}}\right)x}\right)$$

$$= c_0\, erfc\left(-\frac{x}{2\sqrt{\mathscr{D}t}}\right) \tag{i}$$

This sample problem can also be solved by use of multiple Laplace transforms.†

B.9 GENERALIZED FUNCTIONS

The engineer is often called upon to deal with system responses to sudden step changes in input or to "pulse" types of input. Such inputs can be represented by discontinuous types of functions as illustrated in Figures B.9-1, B.9-2, and B.9-3. In this section we shall consider certain properties of these so-called "generalized functions" which are useful in process analysis and the solution of mathematical models.

† T. A. Estrin and T. J. Higgins, *Quart. Appl. Math.* **9**, 153, 1951.

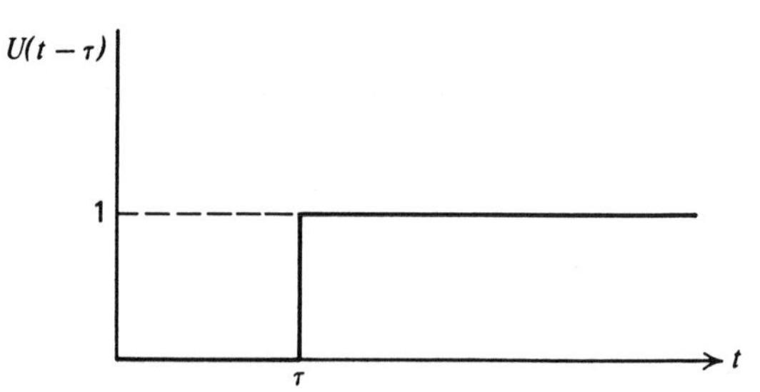

FIGURE B.9-1 Representation of a unit step function.

A *unit step function* is defined by

$$U(t - \tau) = \begin{cases} 0, & t < \tau \\ 1, & t > \tau \end{cases} \tag{B.9-1}$$

and its graph appears as in Figure B.9-1. Thus, $U(t - \tau)$ is continuous (a constant) for $t < \tau$ and $t > \tau$ and has a unit jump discontinuity at $t = \tau$. Combinations of step functions with other functions can be used to represent certain kinds of discontinuous behavior as shown in Figure B.9-2.

The second type of generalized function mentioned above is called the *unit impulse function* or the *Dirac delta function* (Figure B.9-3) and is defined by

$$\delta(t - \tau) = 0, \qquad t \neq \tau$$

$$\int_{-\infty}^{\infty} \delta(t - \tau)\, dt = 1 \tag{B.9-2}$$

This is the mathematical idealization of what might be described as a sudden "jolt" in input to the system. The delta function assumes that the "jolt" occurs in zero time which is, of course, physically not strictly possible. The limiting case is convenient mathematically, however, and closely approximates many physical situations.

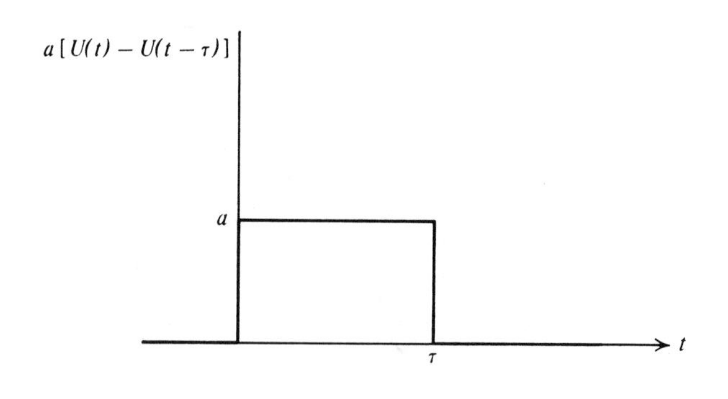

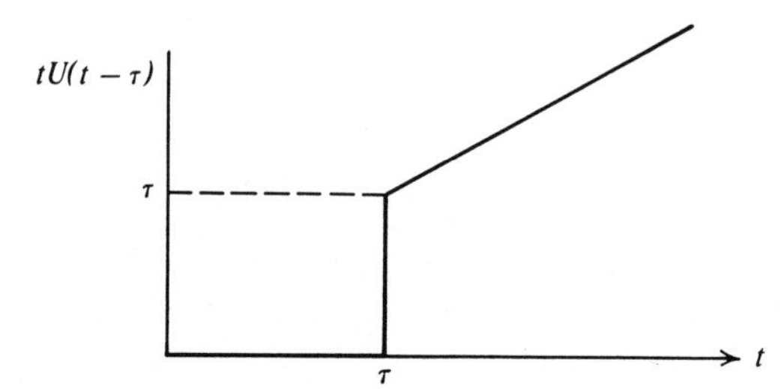

FIGURE B.9-2 Examples of functions constructed by combinations of the step function and ordinary functions.

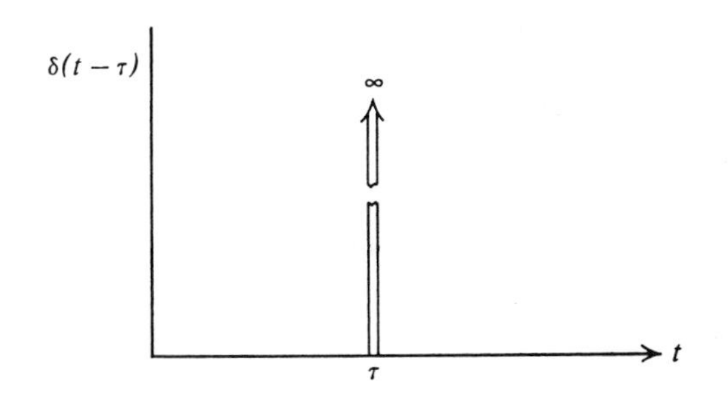

FIGURE B.9-3 The unit impulse function.

Of the general class of generalized functions, we shall only be concerned with the unit step function $U(t - \tau)$ and the unit impulse function $\delta(t - \tau)$. The unit impulse function can be taken to be the derivative of the step function

$$\delta(t - \tau) = \frac{d}{dt} U(t - \tau) \tag{B.9-3}$$

heuristically verified by comparing Figures B.9-1 and B.9-3.

B.10 USE OF LAGRANGIAN MULTIPLIERS TO DETERMINE AN OPTIMUM

A convenient way to find the extremum (not on a boundary) of a function subject to equality constraints is to use Lagrangian multipliers. In calculating the minimum (or maximum) of a function of several independent variables, $f(x_1, x_2, \ldots, x_n)$, we know that the solution of

$$\frac{\partial f}{\partial x_1} = 0$$
$$\frac{\partial f}{\partial x_2} = 0 \tag{B.10-1}$$
$$\text{etc.}$$

yields values of x_i which are at the extremum. Suppose, however, that the values of x_i are joined by some equality relationships (constraints) which can be expressed in general as

$$\Psi_1(x_1, x_2, \ldots, x_n) = 0$$
$$\Psi_2(x_1, x_2, \ldots, x_n) = 0 \tag{B.10-2}$$
$$\text{etc.}$$

For example, suppose that the sum of the mole fractions in a mixture equals unity. Attempting to solve each one of the Equations B.10-2 for a single x_i so that it can be eliminated by substitution into Equation B.10-1 does not usually prove fruitful. Instead, we proceed as follows.

At an extremum the total derivative of $f(x_1, x_2, \ldots, x_n)$ vanishes even if the variables $x_1, x_2, \ldots, x_n$ are not independent:

$$df = 0 = \frac{\partial f}{\partial x_1} dx_1 + \frac{\partial f}{\partial x_2} dx_2 + \cdots + \frac{\partial f}{\partial x_n} dx_n \tag{B.10-3}$$

Also, since $\Psi_i(x_1, x_2, \ldots, x_n) = 0$,

$$\frac{\partial \Psi_1}{\partial x_1} dx_1 + \frac{\partial \Psi_1}{\partial x_2} dx_2 + \cdots + \frac{\partial \Psi_1}{\partial x_n} = 0$$
$$\frac{\partial \Psi_2}{\partial x_1} dx_1 + \frac{\partial \Psi_2}{\partial x_2} dx_2 + \cdots + \frac{\partial \Psi_2}{\partial x_n} = 0 \tag{B.10-4}$$
$$\cdot \quad \cdot \quad \cdot \quad \cdot \quad \cdot \quad \cdot$$
$$\frac{\partial \Psi_p}{\partial x_1} dx_1 + \frac{\partial \Psi_p}{\partial x_2} dx_2 + \cdots + \frac{\partial \Psi_p}{\partial x_n} = 0$$

If Equations B.10-4 are multiplied, respectively by $\lambda_1, \lambda_2, \ldots, \lambda_p$, and the resulting products added to Equation B.10-3, we obtain

$$\left(\frac{\partial f}{\partial x_1} + \lambda_1 \frac{\partial \Psi_1}{\partial x_1} + \lambda_2 \frac{\partial \Psi_2}{\partial x_1} + \cdots + \lambda_p \frac{\partial \Psi_p}{\partial x_1}\right) dx_1 + \left(\frac{\partial f}{\partial x_2} + \lambda_1 \frac{\partial \Psi_1}{\partial x_2} + \lambda_2 \frac{\partial \Psi_2}{\partial x_2} + \cdots + \lambda_p \frac{\partial \Psi_p}{\partial x_2}\right) dx_2 + \cdots = 0 \tag{B.10-5}$$

If $x_1, x_2, \ldots$ are considered to be the independent variables, and $x_m, \ldots, x_n$ (a total of p variables) are those eliminated as independent variables because of the p constraining equations, and if the Jacobian of the derivatives of $\Psi_1, \Psi_2, \ldots, \Psi_p$ with respect to the x_m to x_n designated x's does not vanish:

$$\mathbf{J}(x_m, \ldots, x_n) = \begin{vmatrix} \frac{\partial \Psi_1}{\partial x_m} & \cdots & \frac{\partial \Psi_1}{\partial x_n} \\ \frac{\partial \Psi_2}{\partial x_m} & \cdots & \frac{\partial \Psi_2}{\partial x_n} \\ \vdots & & \vdots \\ \frac{\partial \Psi_p}{\partial x_m} & \cdots & \frac{\partial \Psi_p}{\partial x_n} \end{vmatrix} \neq 0$$

we can fix the λ's such that at the extremum

$$\left(\frac{\partial f}{\partial x_m} + \lambda_1 \frac{\partial \Psi_1}{\partial x_m} + \lambda_2 \frac{\partial \Psi_2}{\partial x_m} + \cdots\right) = \mathrm{p}$$
$$\cdot \quad \cdot \quad \cdot \quad \cdot \quad \cdot \quad \cdot \tag{B.10-6}$$
$$\left(\frac{\partial f}{\partial x_n} + \lambda_1 \frac{\partial \Psi_1}{\partial x_n} + \lambda_2 \frac{\partial \Psi_2}{\partial x_n} + \cdots\right) = 0$$

Because of Equations B.10-6, Equation B.10-5 reduces to a truncated expression in which each of the x_i's are truly independent. Consequently, each of the terms in the parentheses must vanish, or

$$\frac{\partial f}{\partial x_1} + \lambda_1 \frac{\partial \Psi_1}{\partial x_1} + \lambda_2 \frac{\partial \Psi_2}{\partial x_1} + \cdots + \lambda_p \frac{\partial \Psi_p}{\partial x_1} = 0$$
$$\frac{\partial f}{\partial x_2} + \lambda_1 \frac{\partial \Psi_1}{\partial x_2} + \lambda_2 \frac{\partial \Psi_2}{\partial x_2} + \cdots + \lambda_p \frac{\partial \Psi_p}{\partial x_2} = 0 \tag{B.10-7}$$
$$\text{etc.}$$

If we simultaneously solve Equations B.10-7 together with Equations B.10-4 and B.10-6, we can find the values of the x's at the extremum and also the λ's. If $\mathbf{J}(x_m, \ldots, x_n) = 0$, it may prove feasible to interchange the role of some of the independent variables with those in $\mathbf{J}(x_m, \ldots, x_n)$, but if not, the method fails.

Example B.10-1 Extremum of a Constrained Function

Find the maximum and minimum distances from the origin to the surface $5x^2 + 6xy + 5y^2 = 8$.

Solution:

The distances (function to be optimized), as shown in Figure EB10.1, are

$$d = f(x, y) = \sqrt{x^2 + y^2} \tag{a}$$

while the constraining equation is

$$\Psi(x, y) = 5x^2 + 6xy + 5y^2 - 8 = 0 \tag{b}$$

It is just as satisfactory, and easier, to find the extremum of

$$f(x, y) = x^2 + y^2 \tag{a'}$$

as for Equation (a) itself.

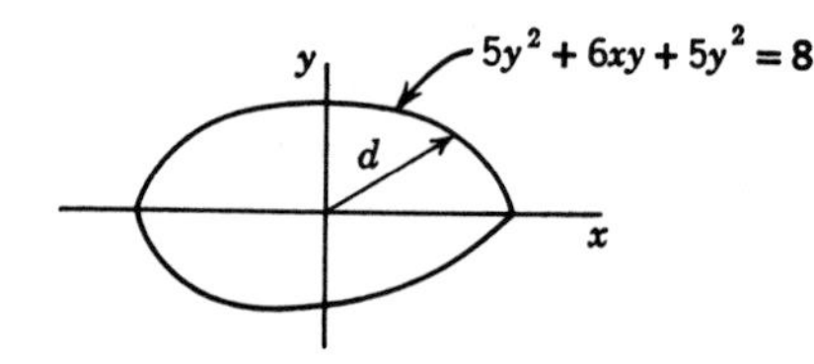

FIGURE EB10.1

Equations B.10-6 and B.10-7 are

$$\frac{\partial f}{\partial x} + \lambda \frac{\partial \Psi}{\partial x} = 0 \quad \text{or} \quad 2x + \lambda(10x + 6y) = 0 \tag{c}$$

$$\frac{\partial f}{\partial y} + \lambda \frac{\partial \Psi}{\partial y} = 0 \quad \text{or} \quad 2y + \lambda(6x + 10y) = 0 \tag{d}$$

Equations (b), (c), and (d) are solved together. Multiplying Equation (c) by y and Equation (d) by x and then subtracting, we obtain

$$6\lambda(y^2 - x^2) = 0$$

or

$$y = \pm x \tag{e}$$

Introduction of Equation (e) into Equation (b) gives

$$\begin{aligned} x^2 &= \tfrac{1}{2} \\ x^2 &= 2 \end{aligned} \tag{f}$$

Consequently,

$$f(x, y) = x^2 + y^2 = \tfrac{1}{2} + \tfrac{1}{2} = 1; \qquad d = 1(\text{min})$$

$$f(x, y) = x^2 + y^2 = 2 + 2 = 4; \qquad d = 2(\text{max})$$

It is now possible to state a general rule for the use of Lagrangian multipliers. To determine the extreme values of a function

$$f(x_1, x_2, \ldots, x_n)$$

whose variables are subjected to p constraining relations:

$$\Psi_i(x_1, x_2, \ldots, x_k) = 0 \qquad i = 1, 2, \ldots, p \tag{B.10-8}$$

form the function

$$F = f + \sum_{i=1}^{p} \lambda_i \Psi_i \tag{B.10-9}$$

and determine the parameters λ_i and the values of x_1, x_2, $\ldots$, x_n from the n equations:

$$\frac{\partial F}{\partial x_j} = 0 \qquad j = 1, 2, \ldots, n \tag{B.10-10}$$

and the p Equations B.10-8.

Supplementary References

Matrix Methods

Amundson, N. R., *Mathematical Methods in Chemical Engineering*, Prentice-Hall, Englewood Cliffs, N.J., 1966.

Bellman, R., *Introduction to Matrix Analysis*, McGraw-Hill, New York, 1960.

Frazer, R. A., Duncan, W. J., and Collar, A. R., *Elementary Matrices*, Cambridge Univ. Press, 1960.

Hohn, F. E., *Elementary Matrix Algebra*, MacMillan, New York, 1958.

Ordinary Differential Equations

Ayres, F., *Theory and Problems of Differential Equations*, Schaum, New York, 1952.

Davis, H. T., *Introduction to Nonlinear Differential Equations*, Govt Printing Office, Washington, D.C., 1960.

Kaplan, W., *Ordinary Differential Equations*, Addison-Wesley, Reading, Mass., 1958.

Murphy, G. M., *Ordinary Differential Equations and Their Solutions*, D. Van Nostrand, New York, 1960.

Struble, R. A., *Nonlinear Differential Equations*, McGraw-Hill, New York, 1962.

Partial Differential Equations

Greenspan, D., *Introduction to Partial Differential Equations*, McGraw-Hill, New York, 1961.

Moon P. and Spencer, D. M., *Field Theory for Engineers*, D. Van Nostrand, New York, 1961.

Sagan, H., *Boundary and Eigenvalue Problems in Mathematical Physics*, John Wiley, New York, 1961.

Sneddon, I. N., *Elements of Partial Differential Equations*, McGraw-Hill, New York, 1957.

Operational Calculus

Brown, B. M., *The Mathematical Theory of Linear Systems*, John Wiley, New York, 1961.

Churchill, R. V., *Operational Mathematics* (2nd ed.), McGraw-Hill, New York, 1958.

Kaplan, W., *Operational Methods for Linear Systems*, Addison-Wesley, Reading, Mass., 1962.

Spiegel, M. R., *Laplace Transforms*, Schaum, New York, 1965.

Generalized Functions

Liverman, T. P. G., *Generalized Functions and Direct Operational Methods*, Prentice-Hall, Englewood Cliffs, N.J., 1964.

Difference Equations

Levy, H. and Lessman, F., *Finite Difference Equations*, MacMillan, New York, 1961.

APPENDIX C

Tables

TABLE C.1 THE NORMAL PROBABILITY DISTRIBUTION

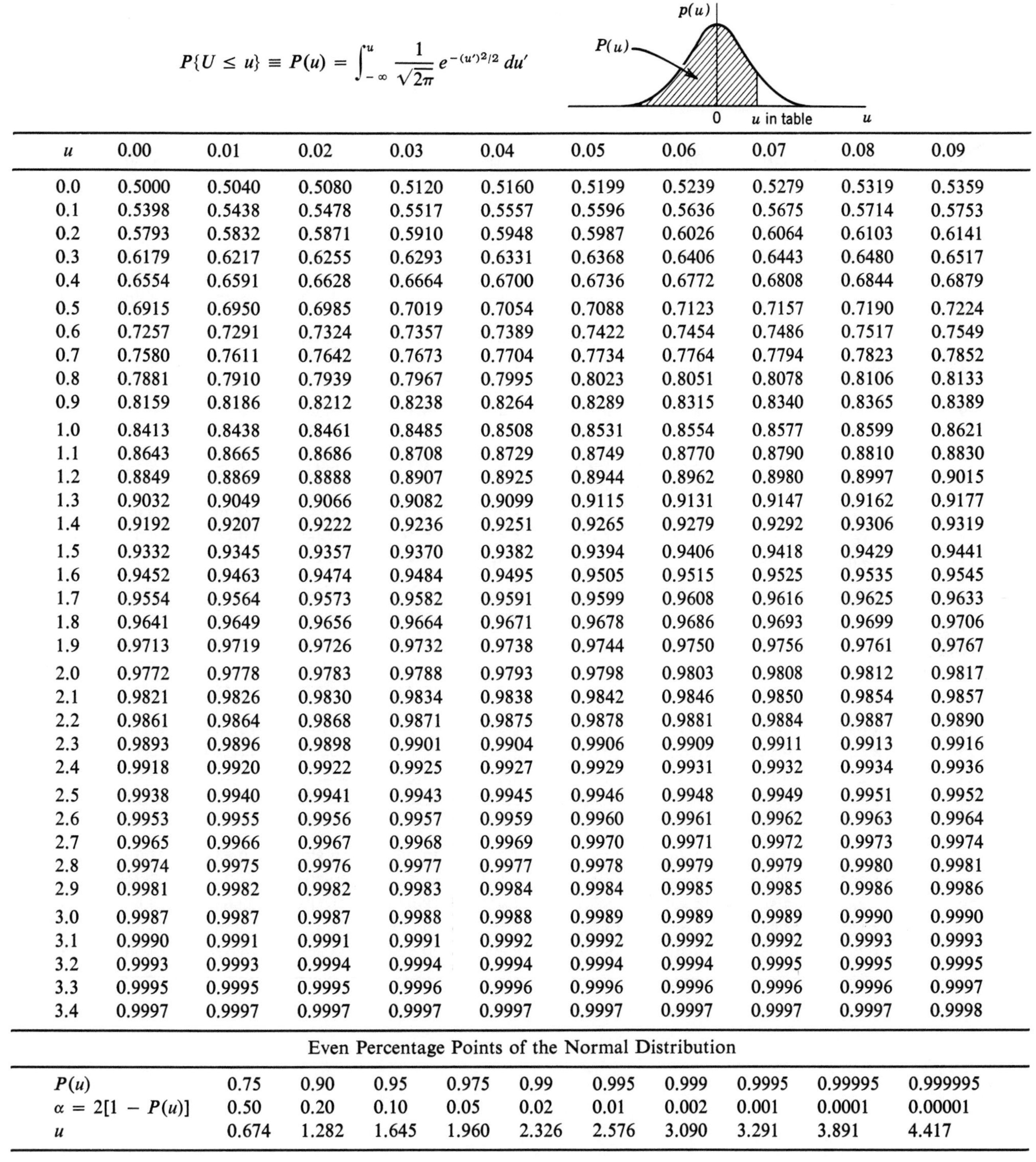

$$P\{U \leq u\} \equiv P(u) = \int_{-\infty}^{u} \frac{1}{\sqrt{2\pi}} e^{-(u')^2/2} \, du'$$

u	0.00	0.01	0.02	0.03	0.04	0.05	0.06	0.07	0.08	0.09
0.0	0.5000	0.5040	0.5080	0.5120	0.5160	0.5199	0.5239	0.5279	0.5319	0.5359
0.1	0.5398	0.5438	0.5478	0.5517	0.5557	0.5596	0.5636	0.5675	0.5714	0.5753
0.2	0.5793	0.5832	0.5871	0.5910	0.5948	0.5987	0.6026	0.6064	0.6103	0.6141
0.3	0.6179	0.6217	0.6255	0.6293	0.6331	0.6368	0.6406	0.6443	0.6480	0.6517
0.4	0.6554	0.6591	0.6628	0.6664	0.6700	0.6736	0.6772	0.6808	0.6844	0.6879
0.5	0.6915	0.6950	0.6985	0.7019	0.7054	0.7088	0.7123	0.7157	0.7190	0.7224
0.6	0.7257	0.7291	0.7324	0.7357	0.7389	0.7422	0.7454	0.7486	0.7517	0.7549
0.7	0.7580	0.7611	0.7642	0.7673	0.7704	0.7734	0.7764	0.7794	0.7823	0.7852
0.8	0.7881	0.7910	0.7939	0.7967	0.7995	0.8023	0.8051	0.8078	0.8106	0.8133
0.9	0.8159	0.8186	0.8212	0.8238	0.8264	0.8289	0.8315	0.8340	0.8365	0.8389
1.0	0.8413	0.8438	0.8461	0.8485	0.8508	0.8531	0.8554	0.8577	0.8599	0.8621
1.1	0.8643	0.8665	0.8686	0.8708	0.8729	0.8749	0.8770	0.8790	0.8810	0.8830
1.2	0.8849	0.8869	0.8888	0.8907	0.8925	0.8944	0.8962	0.8980	0.8997	0.9015
1.3	0.9032	0.9049	0.9066	0.9082	0.9099	0.9115	0.9131	0.9147	0.9162	0.9177
1.4	0.9192	0.9207	0.9222	0.9236	0.9251	0.9265	0.9279	0.9292	0.9306	0.9319
1.5	0.9332	0.9345	0.9357	0.9370	0.9382	0.9394	0.9406	0.9418	0.9429	0.9441
1.6	0.9452	0.9463	0.9474	0.9484	0.9495	0.9505	0.9515	0.9525	0.9535	0.9545
1.7	0.9554	0.9564	0.9573	0.9582	0.9591	0.9599	0.9608	0.9616	0.9625	0.9633
1.8	0.9641	0.9649	0.9656	0.9664	0.9671	0.9678	0.9686	0.9693	0.9699	0.9706
1.9	0.9713	0.9719	0.9726	0.9732	0.9738	0.9744	0.9750	0.9756	0.9761	0.9767
2.0	0.9772	0.9778	0.9783	0.9788	0.9793	0.9798	0.9803	0.9808	0.9812	0.9817
2.1	0.9821	0.9826	0.9830	0.9834	0.9838	0.9842	0.9846	0.9850	0.9854	0.9857
2.2	0.9861	0.9864	0.9868	0.9871	0.9875	0.9878	0.9881	0.9884	0.9887	0.9890
2.3	0.9893	0.9896	0.9898	0.9901	0.9904	0.9906	0.9909	0.9911	0.9913	0.9916
2.4	0.9918	0.9920	0.9922	0.9925	0.9927	0.9929	0.9931	0.9932	0.9934	0.9936
2.5	0.9938	0.9940	0.9941	0.9943	0.9945	0.9946	0.9948	0.9949	0.9951	0.9952
2.6	0.9953	0.9955	0.9956	0.9957	0.9959	0.9960	0.9961	0.9962	0.9963	0.9964
2.7	0.9965	0.9966	0.9967	0.9968	0.9969	0.9970	0.9971	0.9972	0.9973	0.9974
2.8	0.9974	0.9975	0.9976	0.9977	0.9977	0.9978	0.9979	0.9979	0.9980	0.9981
2.9	0.9981	0.9982	0.9982	0.9983	0.9984	0.9984	0.9985	0.9985	0.9986	0.9986
3.0	0.9987	0.9987	0.9987	0.9988	0.9988	0.9989	0.9989	0.9989	0.9990	0.9990
3.1	0.9990	0.9991	0.9991	0.9991	0.9992	0.9992	0.9992	0.9992	0.9993	0.9993
3.2	0.9993	0.9993	0.9994	0.9994	0.9994	0.9994	0.9994	0.9995	0.9995	0.9995
3.3	0.9995	0.9995	0.9995	0.9996	0.9996	0.9996	0.9996	0.9996	0.9996	0.9997
3.4	0.9997	0.9997	0.9997	0.9997	0.9997	0.9997	0.9997	0.9997	0.9997	0.9998

Even Percentage Points of the Normal Distribution

$P(u)$	0.75	0.90	0.95	0.975	0.99	0.995	0.999	0.9995	0.99995	0.999995
$\alpha = 2[1 - P(u)]$	0.50	0.20	0.10	0.05	0.02	0.01	0.002	0.001	0.0001	0.00001
u	0.674	1.282	1.645	1.960	2.326	2.576	3.090	3.291	3.891	4.417

TABLE C.2 THE χ DISTRIBUTION*

$$P\{\chi^2 \le \chi_*^2\} \equiv P(\chi_*^2) = \int_0^{\chi_*^2} p(\chi^2)\, d\chi^2$$

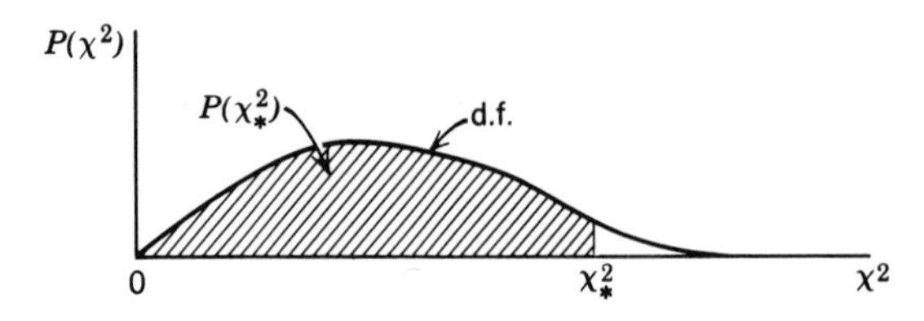

d.f. \ $P(\chi_*^2)$	0.005	0.01	0.02	0.025	0.05	0.10	0.20	0.25	0.30	0.50
1	0.0^4392	0.0^3157	0.0^3628	0.0^3982	0.00393	0.0158	0.0642	0.101	0.148	0.455
2	0.010	0.0201	0.0404	0.051	0.103	0.211	0.466	0.575	0.713	1.386
3	0.072	0.115	0.185	0.216	0.352	0.584	1.005	1.213	1.424	2.366
4	0.207	0.297	0.429	0.484	0.711	1.064	1.649	1.923	2.195	3.357
5	0.412	0.544	0.752	0.831	1.145	1.610	2.343	2.675	3.000	4.351
6	0.676	0.872	1.134	1.237	1.635	2.204	3.070	3.455	3.828	5.348
7	0.989	1.239	1.564	1.690	2.167	2.833	3.822	4.255	4.671	6.346
8	1.344	1.646	2.032	2.180	2.733	3.490	4.594	5.071	5.527	7.344
9	1.735	2.088	2.532	2.700	3.325	4.168	5.380	5.899	6.393	8.343
10	2.156	2.558	3.059	3.247	3.940	4.865	6.179	6.737	7.267	9.342
11	2.603	3.053	3.609	3.816	4.575	5.578	6.989	7.584	8.148	10.341
12	3.074	3.571	4.178	4.404	5.226	6.304	7.807	8.438	9.034	11.340
13	3.565	4.107	4.765	5.009	5.892	7.042	8.634	9.299	9.926	12.340
14	4.075	4.660	5.368	5.629	6.571	7.790	9.467	10.165	10.821	13.339
15	4.601	5.229	5.985	6.262	7.261	8.547	10.307	11.037	11.721	14.339
16	5.142	5.812	6.614	6.908	7.962	9.312	11.152	11.912	12.624	15.338
17	5.697	6.408	7.255	7.564	8.672	10.085	12.002	12.792	13.531	16.338
18	6.265	7.615	7.906	8.231	9.390	10.865	12.857	13.675	14.440	17.338
19	6.844	7.633	8.567	8.906	10.117	11.651	13.716	14.562	15.352	18.338
20	7.434	8.260	9.237	9.591	10.851	12.443	14.578	15.452	16.266	19.337
21	8.034	8.897	9.915	10.283	11.591	13.240	15.445	16.344	17.182	20.337
22	8.643	9.542	10.600	10.982	12.338	14.041	16.314	17.240	18.101	21.337
23	9.260	10.196	11.293	11.689	13.091	14.848	17.187	18.137	19.021	22.337
24	9.886	10.856	11.992	12.400	13.848	15.659	18.062	19.037	19.943	23.337
25	10.520	11.524	12.697	13.120	14.611	16.473	18.940	19.939	20.867	24.337
26	11.160	12.198	13.409	13.844	15.379	17.292	19.820	20.843	21.792	25.336
27	11.808	12.879	14.125	14.573	16.151	18.114	20.703	21.749	22.719	26.336
28	12.461	13.565	14.847	15.308	16.928	18.939	21.588	22.657	23.647	27.336
29	13.121	14.256	15.574	16.047	17.708	19.768	22.475	23.567	24.577	28.336
30	13.787	14.953	16.306	16.791	18.493	20.599	23.364	24.478	25.508	29.336

(continued)

TABLE C.2 (*continued*)

d.f. \ $P(\chi^2_*)$	0.70	0.75	0.80	0.90	0.95	0.975	0.98	0.99	0.995	0.999
1	1.074	1.323	1.642	2.706	3.841	5.024	5.412	6.635	7.879	10.827
2	2.408	2.772	3.219	4.605	5.991	7.378	7.824	9.210	10.597	13.815
3	3.665	4.108	4.642	6.251	7.815	9.348	9.837	11.345	12.838	16.268
4	4.878	5.385	5.989	7.779	9.488	11.143	11.668	13.277	14.860	18.465
5	6.044	6.626	7.289	9.236	11.070	12.833	13.388	15.086	16.750	20.517
6	7.231	7.841	8.558	10.645	12.592	14.449	15.033	16.812	18.548	22.457
7	8.383	9.037	9.803	12.017	14.067	16.013	16.622	18.475	20.278	24.322
8	9.524	10.219	11.030	13.362	15.507	17.535	18.168	20.090	21.955	26.125
9	10.656	11.389	12.242	14.684	16.919	19.023	19.679	21.666	23.589	27.877
10	11.781	12.549	13.442	15.987	18.307	20.483	21.161	23.209	25.188	29.588
11	12.899	13.701	14.631	17.275	19.575	21.920	22.618	24.725	26.757	31.264
12	14.011	14.845	15.812	18.549	21.026	23.337	24.054	26.217	28.299	32.909
13	15.119	15.984	16.985	19.812	22.362	24.736	25.472	27.688	29.819	34.528
14	16.222	17.117	18.151	21.064	23.685	26.119	26.873	29.141	31.319	36.123
15	17.322	18.245	19.313	22.307	24.996	27.488	28.259	30.578	32.801	37.697
16	18.418	19.369	20.465	23.542	36.296	28.845	29.633	32.000	34.267	39.252
17	19.511	20.489	21.615	24.769	27.587	30.191	30.995	33.409	35.719	40.790
18	20.601	21.605	22.760	25.989	28.869	31.526	32.346	34.805	37.156	42.312
19	21.689	22.718	23.900	27.204	30.144	32.852	33.687	36.191	38.582	43.820
20	22.775	23.828	25.038	28.412	31.410	34.170	35.020	37.566	39.997	45.315
21	23.858	24.935	26.171	29.615	32.671	35.479	36.343	38.932	41.401	46.797
22	24.939	26.039	27.301	30.813	33.924	36.781	37.659	40.289	42.796	48.268
23	26.018	27.141	28.429	32.007	35.172	38.076	38.968	41.638	44.181	49.728
24	27.096	28.241	29.553	33.196	36.145	39.364	40.270	42.980	45.559	51.179
25	28.172	29.339	30.675	34.382	37.652	40.647	41.566	44.314	46.928	52.620
26	29.246	30.435	31.795	35.563	38.885	41.923	42.856	45.642	48.290	54.052
27	30.319	31.528	32.912	36.741	40.113	43.194	44.140	46.963	49.645	55.476
28	31.391	32.621	34.027	37.916	41.337	44.461	45.419	48.278	50.993	56.893
29	32.461	33.711	35.139	39.087	42.557	45.722	46.693	49.588	52.336	58.302
30	33.530	34.800	36.250	40.256	43.773	46.979	47.962	50.892	53.672	59.703

* Adapted from Table IV of R. A. Fisher and F. Yates, *Statistical Tables for Biological, Agricultural and Medical Research*, Oliver & Boyd, Ltd., Edinburgh and London, 1953, by permission of the authors and publishers.

d.f. = degrees of freedom = ν. For $30 < \nu < 100$, linear interpolation where necessary will give four significant figures. For $\nu > 100$, take $\chi^2_{\nu,\alpha} = \frac{1}{2}(t_\alpha + \sqrt{2\nu - 1})^2$.

TABLE C.3 THE t-DISTRIBUTION

$$P\{t \le t_*\} \equiv P(t_*) = \int_{-\infty}^{t_*} p(t)\, dt$$

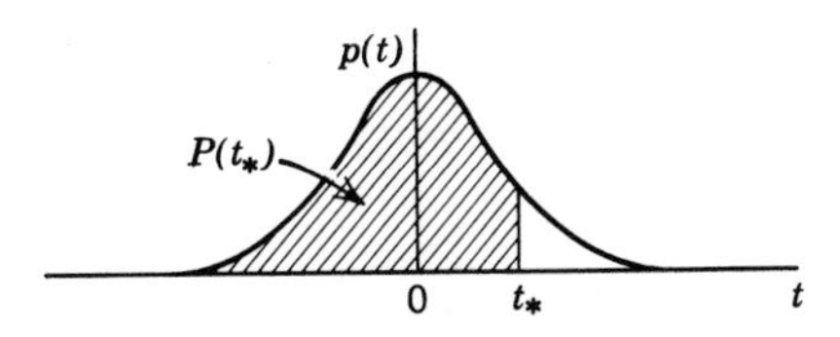

d.f. \ $P(t^*)$	0.55	0.60	0.65	0.70	0.75	0.80	0.85	0.90	0.95	0.975	0.99	0.995	0.9995
1	0.158	0.325	0.510	0.727	1.000	1.376	1.963	3.078	6.314	12.706	31.821	63.657	636.619
2	0.142	0.289	0.445	0.617	0.816	1.061	1.386	1.886	2.920	4.303	6.965	9.925	31.598
3	0.137	0.277	0.424	0.584	0.765	0.978	1.250	1.638	2.353	3.182	4.541	5.841	12.941
4	0.134	0.271	0.414	0.569	0.741	0.941	1.190	1.533	2.132	2.776	3.757	4.604	8.610
5	0.132	0.267	0.408	0.559	0.727	0.920	1.156	1.476	2.015	2.571	3.365	4.032	6.859
6	0.131	0.265	0.404	0.553	0.718	0.906	1.134	1.440	1.943	2.447	3.143	3.707	5.959
7	0.130	0.263	0.402	0.549	0.711	0.896	1.119	1.415	1.895	2.365	2.998	3.499	5.405
8	0.130	0.262	0.399	0.546	0.706	0.889	1.108	1.397	1.860	2.306	2.896	3.355	5.041
9	0.129	0.261	0.398	0.543	0.703	0.883	1.100	1.383	1.833	2.262	2.821	3.250	4.781
10	0.129	0.260	0.397	0.542	0.700	0.879	1.093	1.372	1.812	2.228	2.764	3.169	4.578
11	0.129	0.260	0.396	0.540	0.697	0.876	1.088	1.363	1.796	2.201	2.718	3.106	4.437
12	0.128	0.259	0.395	0.539	0.695	0.873	1.083	1.356	1.782	2.179	2.681	3.055	4.318
13	0.128	0.359	0.394	0.538	0.694	0.870	1.079	1.350	1.771	2.160	2.650	3.012	4.221
14	0.128	0.258	0.393	0.537	0.692	0.868	1.076	1.345	1.761	2.145	2.624	2.977	4.140
15	0.128	0.258	0.393	0.536	0.691	0.866	1.974	1.341	1.753	2.131	2.602	2.947	4.073
16	0.128	0.258	0.392	0.535	0.690	0.865	1.071	1.337	1.746	2.120	2.583	2.291	4.015
17	0.128	0.257	0.392	0.534	0.689	0.863	1.069	1.333	1.740	2.110	2.567	2.898	3.965
18	0.127	0.257	0.392	0.534	0.688	0.862	1.067	1.330	1.734	2.101	2.552	2.878	3.922
19	0.127	0.257	0.391	0.533	0.688	0.861	1.066	1.328	1.729	2.093	2.539	2.861	3.883
20	0.127	0.257	0.391	0.533	0.687	0.860	1.064	1.325	1.725	2.086	2.528	2.845	3.850
21	0.127	0.257	0.257	0.532	0.686	0.859	1.063	1.323	1.721	2.080	2.518	2.831	3.819
22	0.127	0.256	0.390	0.532	0.686	0.858	1.061	1.321	1.717	2.074	2.508	2.819	3.792
23	0.127	0.256	0.390	0.532	0.685	0.858	1.060	1.319	1.714	2.069	2.500	2.807	3.767
24	0.127	0.256	0.390	0.531	0.685	0.857	1.059	1.318	1.711	2.064	2.492	2.797	3.745
25	0.127	0.256	0.390	0.531	0.684	0.856	1.058	1.316	1.708	2.060	2.485	2.787	3.725
26	0.127	0.256	0.390	0.531	0.684	0.856	1.058	1.315	1.706	2.056	2.479	2.779	3.707
27	0.127	0.256	0.389	0.531	0.684	0.855	1.057	1.314	1.703	2.052	2.473	2.771	3.690
28	0.127	0.256	0.389	0.530	0.683	0.855	1.056	1.313	1.701	2.048	2.467	2.763	3.674
29	0.127	0.256	0.389	0.530	0.683	0.854	1.055	1.311	1.699	2.045	2.462	2.756	3.659
30	0.127	0.256	0.389	0.530	0.683	0.854	1.055	1.310	1.697	2.042	2.457	2.750	3.646
40	0.126	0.255	0.388	0.529	0.681	0.851	1.050	1.303	1.684	2.021	2.423	2.704	3.551
60	0.126	0.254	0.387	0.527	0.679	0.848	1.046	1.296	1.671	2.000	2.390	2.660	3.460
120	0.126	0.254	0.386	0.526	0.677	0.845	1.041	1.289	1.658	1.980	2.358	2.617	3.373
∞	0.126	0.253	0.385	0.524	0.674	0.842	1.036	1.282	1.645	1.960	2.326	2.576	3.291

* Adapted from Table III of R. A. Fisher and F. Yates, *Statistical Tables for Biological, Agricultural and Medical Research*, Oliver & Boyd, Ltd., Edinburgh and London, 1963, by permission of the authors and publishers.

TABLE C.4a THE F-DISTRIBUTION*—$P(F_*) = 0.50$

$$P\{F \leq F_*\} \equiv P(F_*) = \int_0^{F_*} p(F)\, dF$$

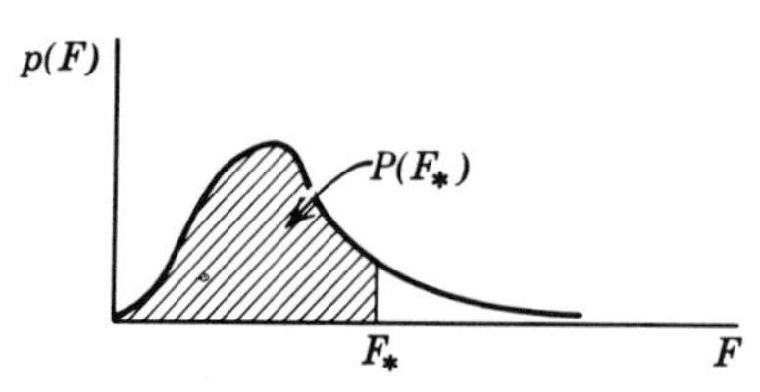

ν_2 \ ν_1	1	2	3	4	5	6	7	8	9
1	1.0000	1.5000	1.7092	1.8227	1.8937	1.9422	1.9774	2.0041	2.0250
2	0.66667	1.0000	1.1349	1.2071	1.2519	1.2824	1.3045	1.3213	1.3344
3	0.58506	0.88110	1.0000	1.0632	1.1024	1.1289	1.1482	1.1627	1.1741
4	0.54863	0.82843	0.94054	1.0000	1.0367	1.0617	1.0797	1.0933	1.1040
5	0.52807	0.79877	0.90715	0.96456	1.0000	1.0240	1.0414	1.0545	1.0648
6	0.51489	0.77976	0.88578	0.94191	0.97654	1.0000	1.0169	1.0298	1.0398
7	0.50572	0.76655	0.87095	0.92619	0.96026	0.98334	1.0000	1.0126	1.0224
8	0.49898	0.75683	0.86004	0.91464	0.94831	0.97111	0.98757	1.0000	1.0097
9	0.49382	0.74938	0.85168	0.90580	0.93916	0.96175	0.97805	0.99037	1.0000
10	0.48973	0.74349	0.84508	0.89882	0.93193	0.95436	0.97054	0.98276	0.99232
11	0.48644	0.73872	0.83973	0.89316	0.92608	0.94837	0.96445	0.97661	0.98610
12	0.48369	0.73477	0.83530	0.88848	0.92124	0.94342	0.95943	0.97152	0.98097
13	0.48141	0.73145	0.83159	0.88454	0.91718	0.93926	0.95520	0.96724	0.97665
14	0.47944	0.72862	0.82842	0.88119	0.91371	0.93573	0.95161	0.96360	0.97298
15	0.47775	0.72619	0.82569	0.87830	0.91073	0.93267	0.94850	0.96046	0.96981
16	0.47628	0.72406	0.82330	0.87578	0.90812	0.93001	0.94580	0.95773	0.96705
17	0.47499	0.72219	0.82121	0.87357	0.90584	0.92767	0.94342	0.95532	0.96462
18	0.47385	0.72053	0.81936	0.87161	0.90381	0.92560	0.94132	0.95319	0.96247
19	0.47284	0.71096	0.81771	0.86987	0.90200	0.92375	0.93944	0.95129	0.96056
20	0.47192	0.71773	0.81621	0.86830	0.90038	0.92210	0.93776	0.94959	0.95884
21	0.47108	0.71653	0.81487	0.86688	0.89891	0.92060	0.93624	0.94805	0.95728
22	0.47033	0.71545	0.81365	0.86559	0.89759	0.91924	0.93486	0.94665	0.95588
23	0.46965	0.71446	0.81255	0.86442	0.89638	0.91800	0.93360	0.94538	0.95459
24	0.46902	0.71356	0.81153	0.86335	0.89527	0.91687	0.93245	0.94422	0.95342
25	0.46844	0.71272	0.81061	0.86236	0.89425	0.91583	0.93140	0.94315	0.95234
26	0.46793	0.71195	0.80975	0.86145	0.89331	0.91487	0.93042	0.94217	0.95135
27	0.66744	0.71124	0.80894	0.86061	0.89244	0.91399	0.92952	0.94126	0.95044
28	0.46697	0.71059	0.80820	0.85983	0.89164	0.91317	0.92869	0.94041	0.94958
29	0.46654	0.70999	0.80753	0.85911	0.89089	0.91241	0.92791	0.93963	0.94879
30	0.46616	0.70941	0.80689	0.85844	0.89019	0.91169	0.92719	0.93889	0.94805
40	0.46330	0.70531	0.80228	0.85357	0.88516	0.90654	0.92197	0.93361	0.94272
60	0.46053	0.70122	0.79770	0.84873	0.88017	0.90144	0.91679	0.92838	0.93743
120	0.45774	0.69717	0.79314	0.84392	0.87521	0.89637	0.91164	0.92318	0.93218
∞	0.45494	0.69315	0.78866	0.83918	0.87029	0.89135	0.90654	0.91802	0.92698

(continued)

TABLE C.4a (*continued*)

ν_2 \ ν_1	10	12	15	20	24	30	40	60	120	∞
1	2.0419	2.0674	2.0931	2.1190	2.1321	2.1452	2.1584	2.1716	2.1848	2.1981
2	1.3450	1.3610	1.3771	1.3933	1.4014	1.4096	1.4178	1.4261	1.4344	1.4427
3	1.1833	1.1972	1.2111	1.2252	1.2322	1.2393	1.2464	1.2536	1.2608	1.2680
4	1.1126	1.1255	1.1386	1.1517	1.1583	1.1649	1.1716	1.1782	1.1849	1.1916
5	1.0730	1.0855	1.0980	1.1106	1.1170	1.1234	1.1297	1.1361	1.1426	1.1490
6	1.0478	1.0600	1.0722	1.0845	1.0907	1.0969	1.1031	1.1093	1.1156	1.1219
7	1.0304	1.0423	1.0543	1.0664	1.0724	1.0785	1.0846	1.0908	1.0969	1.1031
8	1.0175	1.0293	1.0412	1.0531	1.0591	1.0651	1.0711	1.0771	1.0832	1.0893
9	1.0077	1.0194	1.0311	1.0429	1.0489	1.0548	1.0608	1.0667	1.0727	1.0788
10	1.0000	1.0116	1.0232	1.0349	1.0408	1.0467	1.0526	1.0585	1.0645	1.0705
11	0.99373	1.0052	1.0168	1.0284	1.0343	1.0401	1.0460	1.0519	1.0578	1.0637
12	0.98856	1.0000	1.0115	1.0231	1.0289	1.0347	1.0405	1.0464	1.0523	1.0582
13	0.98421	0.99560	1.0071	1.0186	1.0243	1.0301	1.0360	1.0418	1.0476	1.0535
14	0.98051	0.99186	1.0033	1.0147	1.0205	1.0263	1.0321	1.0379	1.0437	1.0495
15	0.97732	0.98863	1.0000	1.0114	1.0172	1.0229	1.0287	1.0345	1.0403	1.0461
16	0.97454	0.98582	0.99716	1.0086	1.0143	1.0200	1.0258	1.0315	1.0373	1.0431
17	0.97209	0.98334	0.99466	1.0060	1.0117	1.0174	1.0232	1.0289	1.0347	1.0405
18	0.96993	0.98116	0.99245	1.0038	1.0095	1.0152	1.0209	1.0267	1.0324	1.0382
19	0.96800	0.97920	0.99047	1.0018	1.0075	1.0132	1.0189	1.0246	1.0304	1.0361
20	0.96626	0.97746	0.98870	1.0000	1.0057	1.0114	1.0171	1.0228	1.0285	1.0343
21	0.96470	0.97587	0.98710	0.99838	1.0040	1.0097	1.0154	1.0211	1.0268	1.0326
22	0.96328	0.97444	0.98565	0.99692	1.0026	1.0082	1.0139	1.0196	1.0253	1.0311
23	0.96199	0.97313	0.98433	0.99558	1.0012	1.0069	1.0126	1.0183	1.0240	1.0297
24	0.96081	0.97194	0.98312	0.99436	1.0000	1.0057	1.0113	1.0170	1.0227	1.0284
25	0.95972	0.97084	0.98201	0.99324	0.99887	1.0045	1.0102	1.0159	1.0215	1.0273
26	0.95872	0.96983	0.98099	0.99220	0.99783	1.0035	1.0091	1.0148	1.0205	1.0262
27	0.95779	0.96889	0.98004	0.99125	0.99687	1.0025	1.0082	1.0138	1.0195	1.0252
28	0.95694	0.96802	0.97917	0.99036	0.99598	1.0016	1.0073	1.0129	1.0186	1.0243
29	0.95614	0.96722	0.97835	0.98954	0.99515	1.0008	1.0064	1.0121	1.0177	1.0234
30	0.95540	0.96647	0.97759	0.98877	0.99438	1.0000	1.0056	1.0113	1.0170	1.0226
40	0.95003	0.96104	0.97211	0.98323	0.98880	0.99440	1.0000	1.0056	1.0113	1.0169
60	0.94471	0.95566	0.96667	0.97773	0.98328	0.98884	0.99441	1.0000	1.0056	1.0112
120	0.93943	0.95032	0.96128	0.97228	0.97780	0.98333	0.98887	0.99443	1.0000	1.0056
∞	0.93418	0.94503	0.95593	0.96687	0.97236	0.97787	0.98339	0.98891	0.99445	1.0000

* Reproduced by permission of E. S. Pearson from "Tables of Percentage Points of the Inverted Beta (F) Distribution," *Biometrika* **33**, 73–88, 1943, by Maxine Merrington and Catherine M. Thompson.

Where necessary, interpolation should be carried out using the reciprocals of the degrees of freedom. The function $120/\nu$ is convenient for this purpose, ν_1 = numerator, ν_2 = denominator.

TABLE C.4b THE F-DISTRIBUTION*—$P(F_*) = 0.75$

$$P\{F \leq F_*\} \equiv P(F_*) = \int_0^{F_*} p(F)\, dF$$

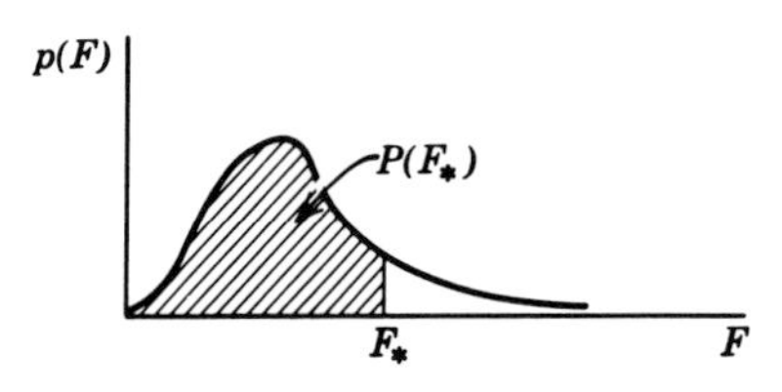

ν_2 \ ν_1	1	2	3	4	5	6	7	8	9
1	5.8285	7.5000	8.1999	8.5810	8.8198	8.9833	9.1021	9.1922	9.2631
2	2.5714	3.0000	3.1534	3.2320	3.2799	3.3121	3.3352	3.3526	3.3661
3	2.0239	2.2798	2.3555	2.3901	2.4095	2.4218	2.4302	2.4364	2.4410
4	1.8074	2.0000	2.0467	2.0642	2.0723	2.0766	2.0790	2.0805	2.0814
5	1.6925	1.8528	1.8843	1.8927	1.8947	1.8945	1.8935	1.8923	1.8911
6	1.6214	1.7622	1.7844	1.7872	1.7852	1.7821	1.7789	1.7760	1.7733
7	1.5732	1.7010	1.7169	1.7157	1.7111	1.7059	1.7011	1.6969	1.6931
8	1.5384	1.6569	1.6683	1.6642	1.6575	1.6508	1.6448	1.6396	1.6350
9	1.5121	1.6236	1.6315	1.6253	1.6170	1.6091	1.6022	1.5961	1.5909
10	1.4915	1.5975	1.6028	1.5949	1.5853	1.5765	1.5688	1.5621	1.5563
11	1.4749	1.5767	1.5798	1.5704	1.5598	1.5502	1.5418	1.5346	1.5284
12	1.4613	1.5595	1.5609	1.5503	1.5389	1.5286	1.5197	1.5120	1.5054
13	1.4500	1.5452	1.5451	1.5336	1.5214	1.5105	1.5011	1.4931	1.4861
14	1.4403	1.5331	1.5317	1.5194	1.5066	1.4952	1.4854	1.4770	1.4697
15	1.4321	1.5227	1.5202	1.5071	1.4938	1.4820	1.4718	1.4631	1.4556
16	1.4249	1.5137	1.5103	1.4965	1.4827	1.4705	1.4601	1.4511	1.4433
17	1.4186	1.5057	1.5015	1.4873	1.4730	1.4605	1.4497	1.4405	1.4325
18	1.4130	1.4988	1.4938	1.4790	1.4644	1.4516	1.4406	1.4312	1.4230
19	1.4081	1.4925	1.4870	1.4717	1.4568	1.4437	1.4325	1.4228	1.4145
20	1.4037	1.4870	1.4808	1.4652	1.4500	1.4366	1.4252	1.4153	1.4069
21	1.3997	1.4820	1.4753	1.4593	1.4438	1.4302	1.4186	1.4086	1.4000
22	1.3961	1.4774	1.4703	1.4540	1.4382	1.4244	1.4126	1.4025	1.3937
23	1.3928	1.4733	1.4657	1.4491	1.4331	1.4191	1.4072	1.3969	1.3880
24	1.3898	1.4695	1.4615	1.4447	1.4285	1.4143	1.4022	1.3918	1.3828
25	1.3870	1.4661	1.4577	1.4406	1.4242	1.4099	1.3976	1.3871	1.3780
26	1.3845	1.4629	1.4542	1.4368	1.4203	1.4058	1.3935	1.3828	1.3737
27	1.3822	1.4600	1.4510	1.4334	1.4166	1.4021	1.3896	1.3788	1.3696
28	1.3800	1.4572	1.4480	1.4302	1.4133	1.3986	1.3860	1.3752	1.3658
29	1.3780	1.4547	1.4452	1.4272	1.4102	1.3953	1.3826	1.3717	1.3623
30	1.3761	1.4524	1.4426	1.4244	1.4073	1.3923	1.3795	1.3685	1.3590
40	1.3626	1.4355	1.4239	1.4045	1.3863	1.3706	1.3571	1.3455	1.3354
60	1.3493	1.4188	1.4055	1.3848	1.3657	1.3491	1.3349	1.3226	1.3119
120	1.3362	1.4024	1.3873	1.3654	1.3453	1.3278	1.3128	1.2999	1.2886
∞	1.3233	1.3863	1.3694	1.3463	1.3251	1.3068	1.2910	1.2774	1.2654

(*continued*)

TABLE C.4b (*continued*)

ν_2 \ ν_1	10	12	15	20	24	30	40	60	120	∞
1	9.3202	9.4064	9.4934	9.5813	9.6255	9.6698	9.7144	9.7591	9.8041	9.8492
2	3.3770	3.3934	3.4098	3.4263	3.4345	3.4428	3.4511	3.4594	3.4677	3.4761
3	2.4447	2.4500	2.4552	2.4602	2.4626	2.4650	2.4674	2.4697	2.4720	2.4742
4	2.0820	2.0826	2.0829	2.0828	2.0827	2.0825	2.0821	2.0817	2.0812	2.0806
5	1.8899	1.8877	1.8851	1.8820	1.8802	1.8784	1.8763	1.8742	1.8719	1.8694
6	1.7708	1.7668	1.7621	1.7569	1.7540	1.7510	1.7477	1.7443	1.7407	1.7368
7	1.6898	1.6843	1.6781	1.6712	1.6675	1.6635	1.6593	1.6548	1.6502	1.6452
8	1.6310	1.6244	1.6170	1.6088	1.6043	1.5996	1.5945	1.5892	1.5836	1.5777
9	1.5863	1.5788	1.5705	1.5611	1.5560	1.5506	1.5450	1.5389	1.5325	1.5257
10	1.5513	1.5430	1.5338	1.5235	1.5179	1.5119	1.5056	1.4990	1.4919	1.4843
11	1.5230	1.5140	1.5041	1.4930	1.4869	1.4805	1.4737	1.4664	1.4587	1.4504
12	1.4996	1.4902	1.4796	1.4678	1.4613	1.4544	1.4471	1.4393	1.4310	1.4221
13	1.4801	1.4701	1.4590	1.4465	1.4397	1.4324	1.4247	1.4164	1.4075	1.3980
14	1.4634	1.4530	1.4414	1.4284	1.4212	1.4136	1.4055	1.3967	1.3874	1.3772
15	1.4491	1.4383	1.4263	1.4127	1.4052	1.3973	1.3888	1.3796	1.3698	1.3591
16	1.4366	1.4255	1.4130	1.3990	1.3913	1.3830	1.3742	1.3646	1.3543	1.3432
17	1.4256	1.4142	1.4014	1.3869	1.3790	1.3704	1.3613	1.3514	1.3406	1.3290
18	1.4159	1.4042	1.3911	1.3762	1.3680	1.3592	1.3497	1.3395	1.3284	1.3162
19	1.4073	1.3953	1.3819	1.3666	1.3582	1.3492	1.3394	1.3289	1.3174	1.3048
20	1.3995	1.3873	1.3736	1.3580	1.3494	1.3401	1.3301	1.3193	1.3074	1.2943
21	1.3925	1.3801	1.3661	1.3502	1.3414	1.3319	1.3217	1.3105	1.2983	1.2848
22	1.3861	1.3735	1.3593	1.3431	1.3341	1.3245	1.3140	1.3025	1.2900	1.2761
23	1.3803	1.3675	1.3531	1.3366	1.3275	1.3176	1.3069	1.2952	1.2824	1.2681
25	1.3750	1.3621	1.3474	1.3307	1.3214	1.3113	1.3004	1.2885	1.2754	1.2607
25	1.3701	1.3570	1.3422	1.3252	1.3158	1.3056	1.2945	1.2823	1.2698	1.2538
26	1.3656	1.3524	1.3374	1.3202	1.3106	1.3002	1.2889	1.2765	1.2628	1.2474
27	1.3615	1.3481	1.3329	1.3155	1.3058	1.2953	1.2838	1.2712	1.2572	1.2414
28	1.3576	1.3441	1.3288	1.3112	1.3013	1.2906	1.2790	1.2662	1.2519	1.2358
29	1.3541	1.3404	1.3249	1.3071	1.2971	1.2863	1.2745	1.2615	1.2470	1.2306
30	1.3507	1.3369	1.3213	1.3033	1.2933	1.2823	1.2703	1.2571	1.2424	1.2256
40	1.3266	1.3119	1.2952	1.2758	1.2649	1.2529	1.2397	1.2249	1.2080	1.1883
60	1.3026	1.2870	1.2691	1.2481	1.2361	1.2229	1.2081	1.1912	1.1715	1.1474
120	1.2787	1.2621	1.2428	1.2200	1.2068	1.1921	1.1752	1.1555	1.1314	1.0987
∞	1.2549	1.2371	1.2163	1.1914	1.1767	1.1600	1.1404	1.1164	1.0838	1.0000

* Reproduced by permission of E. S. Pearson from "Tables of Percentage Points of the Inverted Beta (*F*) Distribution," *Biometrika* **33**, 73–88, 1943, by Maxine Merrington and Catherine M. Thompson.

Where necessary, interpolation should be carried out using the reciprocals of the degrees of freedom. The function $120/\nu$ is convenient for this purpose. ν_1 = numerator, ν_2 = denominator.

TABLE C.4c THE F-DISTRIBUTION*—$P(F_*) = 0.90$

$$P\{F \le F_*\} \equiv P(F_*) = \int_0^{F_*} p(F)\, dF$$

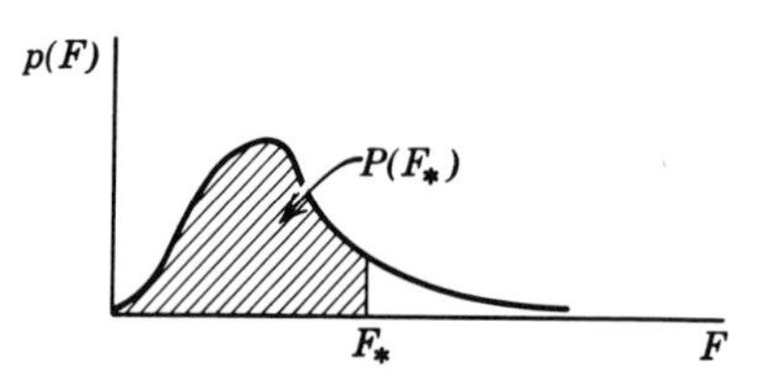

ν_1 \ ν_1	1	2	3	4	5	6	7	8	9
1	39.864	49.500	53.593	55.833	57.241	48.204	58.906	59.439	59.858
2	8.5263	9.0000	9.1618	1.2434	9.2926	9.3255	9.3491	9.3668	9.3805
3	5.5383	5.4624	5.3908	5.3427	5.3092	5.2847	5.2662	5.2517	5.2400
4	4.5448	4.3246	4.1908	4.1073	4.0506	4.0098	3.9790	3.9549	3.9357
5	4.0604	3.7797	3.6195	3.5202	3.4530	3.4045	3.3679	3.3393	3.3163
6	3.7760	4.4633	3.2888	3.1808	3.1075	3.0546	3.0145	2.9830	2.9577
7	3.5894	3.2574	3.0741	2.9605	2.8833	2.8274	2.7849	2.7516	2.7247
8	3.4579	3.1131	2.9238	2.8064	2.7265	2.6683	2.6241	2.5893	2.5612
9	3.3603	3.0065	2.8129	2.6927	2.6106	2.5509	2.5053	2.4694	2.4403
10	3.2850	2.9245	2.7277	2.6053	2.5216	2.4606	2.4140	2.3772	2.3473
11	3.2252	2.8595	2.6602	2.5362	2.4512	2.3891	2.3416	2.3040	2.2735
12	3.1765	2.8068	2.6055	2.4801	2.3940	2.3310	2.2828	2.2446	2.2135
13	3.1362	2.7632	2.5603	1.4337	2.3467	2.2830	2.2341	2.1953	2.1638
14	3.1022	2.7265	2.5222	2.3947	2.3069	2.2426	2.1931	2.1539	2.1220
15	3.0732	2.6952	2.4898	2.3614	2.2730	2.2081	2.1582	2.1185	2.0862
16	3.0481	2.6682	2.4618	2.3327	2.2438	2.1783	2.1280	2.0880	2.0553
17	3.0262	2.6446	2.4374	2.3077	2.2183	2.1524	2.1017	2.0613	2.0284
18	3.0070	2.6239	2.4160	2.2858	2.1958	2.1296	2.0785	2.0379	2.0047
19	2.9899	2.6056	2.3970	2.2663	2.1760	2.1094	2.0580	2.0171	1.9836
20	2.9747	2.5893	2.3801	2.2489	2.1582	2.0913	2.0397	1.9985	1.9649
21	2.9609	2.5746	2.3649	2.2333	2.1423	2.0751	2.0232	1.9819	1.9480
22	2.9486	2.5613	2.3512	2.2193	2.1279	2.0605	2.0084	1.9668	1.9327
23	2.9374	2.5493	2.3387	2.2065	2.1149	2.0472	1.9949	1.9531	1.9189
24	2.9271	2.5383	2.3274	2.1949	2.1030	2.0351	1.9826	1.9407	1.9063
25	2.9177	2.5283	2.3170	2.1843	2.0922	2.0241	1.9714	1.9292	1.8947
26	2.9091	2.5191	2.3075	2.1745	2.0822	2.0139	1.9610	1.9188	1.8841
27	2.9012	2.5106	2.2987	2.1655	2.0730	2.0045	1.9515	1.9091	1.8743
28	2.8939	2.5028	2.2906	2.1571	2.0645	1.9959	1.9427	1.9001	1.8652
29	2.8871	2.4955	2.2831	2.1494	2.0566	1.9878	1.9345	1.8918	1.8568
30	2.8807	2.4887	2.2761	2.1422	2.0492	1.9803	1.9269	1.8841	1.8490
40	2.8354	2.4404	2.2261	2.0909	1.9968	1.9269	1.8725	1.8289	1.7929
60	2.7914	2.3932	2.1774	2.0410	1.9457	1.8747	1.8194	1.7748	1.7380
120	2.7478	2.3473	2.1300	1.9923	1.8959	1.8238	1.7675	1.7220	1.6843
∞	2.7055	2.3026	2.0838	1.9449	1.8473	1.7741	1.7167	1.6702	1.6315

(continued)

TABLE C.4c (*continued*)

ν_2 \ ν_1	10	12	15	20	24	30	40	60	120	∞
1	60.195	60.705	61.220	61.740	62.002	62.265	62.529	62.794	63.061	63.328
2	9.3916	9.4081	9.4247	9.4413	9.4496	9.4539	9.4663	9.4746	9.4829	9.4913
3	5.2304	5.2156	5.2003	5.1845	5.1764	5.1681	5.1597	5.1512	5.1425	5.1337
4	3.9199	3.8955	3.8689	3.8443	3.8310	3.8174	3.8036	3.7896	3.7753	3.7607
5	3.2974	3.2682	3.2380	3.2067	3.1905	3.1741	3.1573	3.1402	3.1228	3.1050
6	2.9369	2.9047	2.8712	2.8363	2.8183	2.8000	2.7812	2.7620	2.7423	2.7222
7	2.7025	2.6681	2.6322	2.5947	2.5753	2.5555	2.5351	2.5142	2.4928	2.4708
8	2.5380	2.5020	2.4642	2.4246	2.4041	2.3830	2.3614	2.3391	2.3162	2.2926
9	2.4163	2.3789	2.3396	2.2983	2.2768	2.2547	2.2320	2.2085	2.1843	2.1592
10	2.3226	2.2841	2.2435	2.2007	2.1784	2.1554	2.1317	2.1072	2.0818	2.0554
11	2.2482	2.2087	2.1671	2.1230	2.1000	2.0762	2.0516	2.0261	1.9997	1.9721
12	2.1878	2.1474	2.1049	2.0597	2.0360	2.0115	1.9861	1.9597	1.9323	1.9036
13	2.1376	2.0966	2.0532	2.0070	1.9827	1.9576	1.9315	1.9043	1.8759	1.8462
14	2.0954	2.0537	2.0095	1.9625	1.9377	1.9119	1.8852	1.8572	1.8280	1.7973
15	2.0593	2.0171	1.9722	1.9243	1.8990	1.8728	1.8454	1.8168	1.7867	1.7551
16	2.0281	1.9854	1.9399	1.8913	1.8656	1.8388	1.8108	1.7816	1.7507	1.7182
17	2.0009	1.9577	1.9117	1.8624	1.8362	1.8090	1.7805	1.7506	1.7191	1.6856
18	1.9770	1.9333	1.8868	1.8368	1.8103	1.7827	1.7537	1.7232	1.6910	1.6567
19	1.9557	1.9117	1.8647	1.8142	1.7873	1.7592	1.7298	1.6988	1.6659	1.6308
20	1.9367	1.8924	1.8449	1.7938	1.7667	1.7382	1.7083	1.6768	1.6433	1.6074
21	1.9197	1.8750	1.8272	1.7756	1.7481	1.7193	1.6890	1.6569	1.6228	1.5862
22	1.9043	1.8593	1.8111	1.7590	1.7312	1.7021	1.6714	1.6389	1.6042	1.5668
23	1.8903	1.8450	1.7964	1.7439	1.7159	1.6864	1.6554	1.6224	1.5871	1.5490
24	1.8775	1.8319	1.7831	1.7302	1.7019	1.6721	1.6407	1.6073	1.5715	1.5327
25	1.8658	1.8200	1.7708	1.7175	1.6890	1.6589	1.6272	1.5934	1.5570	1.5176
26	1.8550	1.8090	1.7596	1.7059	1.6771	1.6468	1.6147	1.5805	1.5437	1.5036
27	1.8451	1.7989	1.7492	1.6951	1.6662	1.6356	1.6032	1.5686	1.5313	1.4906
28	1.8359	1.7895	1.7395	1.6852	1.6560	1.6252	1.5925	1.5575	1.5198	1.4784
29	1.8274	1.7808	1.7306	1.6759	1.6465	1.6155	1.5825	1.5472	1.5090	1.4670
30	1.8195	1.7727	1.7223	1.6673	1.6377	1.6065	1.5732	1.5376	1.4989	1.4564
40	1.7627	1.7146	1.6624	1.6052	1.5741	1.5411	1.5056	1.4672	1.4248	1.3769
60	1.7070	1.6574	1.6034	1.5435	1.5107	1.4755	1.4373	1.3952	1.3476	1.2915
120	1.6524	1.6012	1.5450	1.4821	1.4472	1.4094	1.3676	1.3203	1.2646	1.1926
∞	1.5987	1.5458	1.4871	1.4206	1.3832	1.3419	1.2951	1.2400	1.1686	1.0000

* Reproduced by permission of E. S. Pearson from "Tables of Percentage Points of the Inverted Beta (*F*) Distribution," *Biometrika* **33**, 73–88, 1943, by Maxine Merrington and Catherine M. Thompson.

Where necessary, interpolation should be carried out using the reciprocals of the degrees of freedom. The function $120/\nu$ is convenient for this purpose. ν_1 = numerator, ν_2 = denominator.

TABLE C.4d THE F-DISTRIBUTION*—$P(F_*) = 0.95$

$$P(F \leq F_*) \equiv P(F_*) = \int_0^{F_*} p(F)\, dF$$

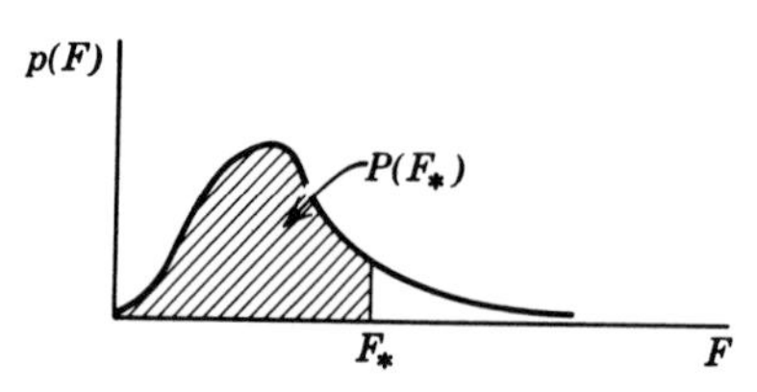

ν_2 \ ν_1	1	2	3	4	5	6	7	8	9
1	161.45	199.50	215.71	224.58	230.16	233.99	236.77	238.88	240.54
2	18.513	19.000	19.164	19.247	19.296	19.330	19.353	19.371	19.385
3	10.128	9.5521	9.2766	9.1172	9.0135	8.9406	8.8868	8.8452	8.8123
4	7.7086	6.9443	6.5914	6.3883	6.2560	6.1631	6.0942	6.0410	5.9988
5	6.6079	5.7861	5.4095	5.1922	5.0503	4.9503	4.8759	4.8183	4.7725
6	5.9874	5.1433	4.7571	4.5337	4.3874	4.2839	4.2066	4.1468	4.0990
7	5.5914	4.7374	4.3468	4.1203	3.9715	3.8660	3.7870	3.7257	3.6767
8	5.3177	4.4590	4.0662	3.8378	3.6875	3.5806	3.5005	3.4381	3.3881
9	5.1174	4.2565	3.8626	3.6331	3.4817	3.3738	3.2927	3.2296	3.1789
10	4.9646	4.1028	3.7083	3.4780	3.3258	3.2172	3.1355	3.0717	3.0204
11	4.8443	3.9823	3.5874	3.3567	3.2039	3.0946	3.0123	2.9480	2.8962
12	4.7472	3.8853	3.4903	3.2592	3.1059	2.9961	2.9134	2.8486	2.7964
13	4.6672	3.8056	3.4105	3.1791	3.0254	2.9153	2.8321	2.7669	2.7144
14	4.6001	3.7389	3.3439	3.1122	2.9582	2.8477	2.7642	2.6987	2.6458
15	4.5431	3.6823	3.2874	3.0556	2.9013	2.7905	2.7066	2.6408	2.5876
16	4.4940	3.6337	3.2389	3.0069	2.8524	2.7413	2.6572	2.5911	2.5377
17	4.4513	3.5915	3.1968	2.9647	2.8100	2.6987	2.6143	2.5480	2.4943
18	4.4139	3.5546	3.1599	2.9277	2.7729	2.6613	2.5767	2.5102	2.4563
19	4.3808	3.5219	3.1274	2.8951	2.7401	2.6283	2.5435	2.4768	2.4227
20	4.3513	3.4928	3.0984	2.8661	2.7109	2.5990	2.5140	2.4471	2.3928
21	4.3248	3.4668	3.0725	2.8401	2.6848	2.5727	2.4876	2.4205	2.3661
22	4.3009	3.4434	3.0491	2.8167	2.6613	2.5491	2.4638	2.3965	2.3419
23	4.2793	3.4221	3.0280	2.7955	2.6400	2.5277	2.4422	2.3748	2.3201
24	4.2597	3.4028	3.0088	2.7763	2.6207	2.5082	2.4226	2.3551	2.3002
25	4.2417	3.3852	2.9912	2.7587	2.6030	2.4904	2.4047	2.3371	2.2821
26	4.2252	3.3690	2.9751	2.7426	2.5868	2.4741	2.3883	2.3205	2.2655
27	4.2100	3.2541	2.9604	2.7278	2.5719	2.4591	2.3732	2.3053	2.2601
28	4.1960	3.3404	2.9467	2.7141	2.5581	2.4453	2.3593	2.2913	2.2360
29	4.1830	3.3277	2.9340	2.7014	2.5454	2.4324	2.3463	2.2782	2.2229
30	4.1709	3.3158	2.9223	2.6896	2.5336	2.4205	2.3343	2.2662	2.2107
40	4.0848	3.2317	2.8387	2.6060	2.4495	2.3359	2.2490	2.1802	2.1240
60	4.0012	3.1504	2.7581	2.5252	2.3683	2.2540	2.1665	2.0970	2.0401
120	3.9201	3.0718	2.6802	2.4472	2.2900	2.1750	2.0867	2.0164	1.9588
∞	3.8415	2.9957	2.6049	2.3719	2.2141	2.0986	2.0096	1.9384	1.8799

(*continued*)

TABLE C.4d (*continued*)

ν_2 \ ν_1	10	12	15	20	24	30	40	60	120	∞
1	241.88	243.91	245.95	248.01	249.05	250.09	251.14	252.20	253.25	254.32
2	19.396	19.413	19.429	19.446	19.454	19.462	19.471	19.479	19.487	19.496
3	8.7855	8.7446	8.7029	8.6602	8.6385	8.6166	8.5944	8.5720	8.5494	8.5265
4	5.9644	5.9117	5.8578	5.8025	5.7744	5.7459	5.7170	5.6878	5.6581	5.6281
5	4.7351	4.6777	4.6188	4.5581	4.5272	4.4957	4.4638	4.4314	4.3984	4.3650
6	4.0600	3.9999	3.9381	3.8742	3.8415	3.8082	3.7743	3.7398	3.7047	3.6688
7	3.6365	3.5747	3.5108	3.4445	3.4105	3.3758	3.3404	3.3043	3.2674	3.2298
8	3.3472	3.2840	3.2184	3.1503	3.1152	3.0794	3.0428	3.0053	2.9669	2.9276
9	3.1373	3.0729	3.0061	2.9365	2.9005	2.8637	2.8259	2.7872	2.7475	2.7067
10	2.9782	2.9130	2.8450	2.7740	2.7372	2.6996	2.6609	2.6211	2.5801	2.5379
11	2.8536	2.7876	2.7186	2.6464	2.6090	2.5705	2.5309	2.4901	2.4480	2.4045
12	2.7534	2.6866	2.6169	2.5436	2.5055	2.4663	2.4259	2.3842	2.3410	2.2962
13	2.6710	2.6037	2.5331	2.4589	2.4202	2.3803	2.3392	2.2966	2.2524	2.2064
14	2.6021	2.5342	2.4630	2.3879	2.3487	2.3082	2.2664	2.2230	2.1778	2.1307
15	2.5437	2.4753	2.4035	2.3275	2.2878	2.2468	2.2043	2.1601	2.1141	2.0658
16	2.4935	2.4247	2.3522	2.2756	2.2354	2.1938	2.1507	2.1058	2.0589	2.0096
17	2.4499	2.3807	2.3077	2.2304	2.1898	2.1477	2.1040	2.0584	2.0107	1.9604
18	2.4117	2.3421	2.2686	2.1906	2.1497	2.1071	2.0629	2.0166	1.9681	1.9168
19	2.3779	2.3080	2.2341	2.1555	2.1141	2.0712	2.0264	1.9796	1.9302	1.8780
20	2.3479	2.2776	2.2033	2.1242	2.0825	2.0391	1.9938	1.9464	1.8963	1.8432
21	2.3210	2.2504	2.1757	2.0960	2.0540	2.0102	1.9645	1.9165	1.8657	1.8117
22	2.2967	2.2258	2.1508	2.0707	2.0283	1.9842	1.9380	1.8895	1.8380	1.7831
23	2.2747	2.2036	2.1282	2.0476	2.0050	1.9605	1.9139	1.8649	1.8128	1.7570
24	2.2547	2.1834	2.1077	2.0267	1.9838	1.9390	1.8920	1.8424	1.7897	1.7331
25	2.2365	2.1649	2.0889	2.0075	1.9643	1.9192	1.8718	1.8217	1.7684	1.7110
26	2.2197	2.1479	2.0716	1.9898	1.9464	1.9010	1.8533	1.8027	1.7488	1.6906
27	2.2043	2.1323	2.0558	1.9736	1.9299	1.8842	1.8361	1.7851	1.7307	1.6717
28	2.1900	2.1179	2.0411	1.9586	1.9147	1.8687	1.8203	1.7689	1.7138	1.6541
29	2.1768	2.1045	2.0275	1.9446	1.9005	1.8543	1.8055	1.7537	1.6981	1.6377
30	2.1646	2.0921	2.0148	1.9317	1.8874	1.8409	1.7918	1.7396	1.6835	1.6223
40	2.0772	2.0035	1.9245	1.8389	1.7929	1.7444	1.6928	1.6373	1.5766	1.5089
60	1.9926	1.9174	1.8364	1.7480	1.7001	1.6491	1.5943	1.5343	1.4673	1.3893
120	1.9105	1.8337	1.7505	1.6587	1.6084	1.5543	1.4952	1.4290	1.3519	1.2539
∞	1.8307	1.7522	1.6664	1.5705	1.5173	1.4591	1.3940	1.3180	1.2214	1.0000

* Reproduced by permission of E. S. Pearson from "Tables of Percentage Points of the Inverted Beta (F) Distribution," *Biometrika* **33**, 73–88, 1943, by Maxine Merrington and Catherine M. Thompson.

Where necessary, interpolation should be carried out using the reciprocals of the degrees of freedom. The function $120/\nu$ is convenient for this purpose. ν_1 = numerator, ν_2 = denominator.

TABLE C.4e THE F-DISTRIBUTION*—$P(F_*) = 0.975$

$$P(F \leq F_*) \equiv P(F_*) = \int_0^{F_*} p(F)\, dF$$

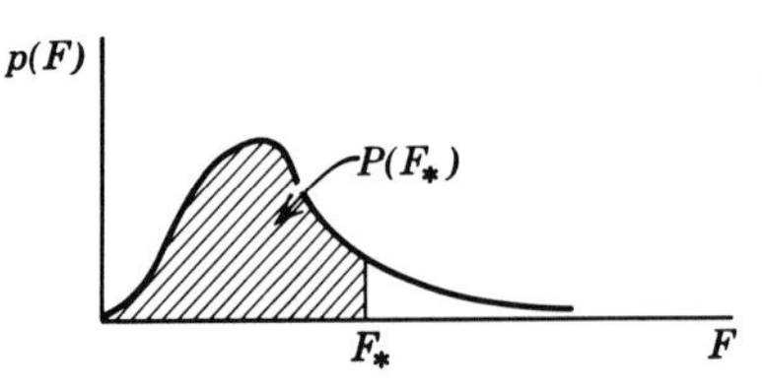

ν_2 \ ν_1	1	2	3	4	5	6	7	8	9
1	647.79	799.50	864.16	899.58	921.85	937.11	948.22	956.66	963.28
2	38.506	39.000	39.165	39.248	39.298	39.331	39.355	39.373	39.387
3	17.443	16.044	15.439	15.101	14.885	14.735	14.624	14.540	14.473
4	12.218	10.649	9.9792	9.6045	9.3645	9.1973	9.0741	8.9796	8.9047
5	10.007	8.4336	7.7636	7.3879	7.1464	6.9777	6.8531	6.7572	6.6810
6	8.8131	7.2598	6.5988	6.2272	5.9876	5.8197	5.6955	5.5996	5.5234
7	8.0727	6.5415	5.8898	5.5226	5.2852	5.1186	4.9949	4.8994	4.8232
8	7.5709	6.0595	5.4160	5.0526	4.8173	4.6517	5.5286	4.4332	4.3572
9	7.2093	5.7147	5.0781	4.7181	4.4844	4.3197	4.1971	4.1020	4.0260
10	6.9367	5.4564	4.8256	4.4683	4.2361	4.0721	3.9498	3.8549	3.7790
11	6.7241	5.2559	4.6300	4.2751	4.0440	3.8807	3.7586	3.6638	3.5879
12	6.5538	5.0959	4.4742	4.1212	3.8911	3.7283	3.6065	3.5118	3.4358
13	6.4143	4.9653	4.3472	3.9959	3.7667	3.6043	3.4827	3.3880	3.3120
14	6.2979	4.8567	4.2417	3.8919	3.6634	3.5014	3.3799	3.2853	3.2093
15	6.1995	4.7650	4.1528	3.8043	3.5764	3.4147	3.2934	3.1987	3.1227
16	6.1151	4.6867	4.0768	3.7294	3.5021	3.3406	3.2194	3.1248	3.0488
17	6.0420	4.6189	4.0112	3.6648	3.4379	3.2767	3.1556	3.0610	2.9849
18	5.9871	4.5597	3.9539	3.6083	3.3820	3.2209	3.0999	3.0053	2.9291
19	5.9216	4.5075	3.9034	3.5587	3.3327	3.1718	3.0509	2.9563	2.8800
20	5.8715	4.4613	3.8587	3.5147	3.2891	3.1283	3.0074	2.9128	2.8365
21	5.8266	4.4199	3.8188	3.4754	3.2501	3.0895	2.9686	2.8740	2.7977
22	5.7863	4.3828	3.7829	3.4401	3.2151	3.0546	2.9338	2.8392	2.7628
23	5.7498	4.3492	3.7505	3.4083	3.1835	3.0232	2.9024	2.8077	2.7313
24	5.7168	4.3187	3.7211	3.3794	3.1548	2.9946	2.8738	2.7791	2.7027
25	5.6864	4.2909	3.6943	3.3530	3.1287	2.9685	2.8478	2.7531	2.6766
26	5.6586	4.2655	3.6697	3.3289	3.1048	2.9447	2.8240	2.7293	2.6528
27	5.6331	4.2421	3.6472	3.3067	3.0828	2.9228	2.8021	2.7074	2.6309
28	5.6096	4.2205	3.6264	3.2863	3.0625	2.9027	2.7820	2.6872	2.6106
29	5.5878	4.2006	3.6072	3.2674	3.0438	2.8840	2.7633	2.6686	2.5919
30	5.5675	4.1821	3.5894	3.2499	3.0265	2.8667	2.7460	2.6513	2.5746
40	5.4239	4.0510	3.4633	2.1261	2.9037	2.7444	2.6238	2.5289	2.4519
60	5.2858	3.9253	3.3425	3.0077	2.7863	2.6274	2.5068	2.4117	2.3344
120	5.1524	3.8046	3.2270	2.8943	2.6740	2.5154	2.3948	2.2994	2.2217
∞	5.0239	3.6889	3.1161	2.7858	2.5665	2.4082	2.2875	2.1918	2.1136

(continued)

TABLE C.4e (*continued*)

ν_2 \ ν_1	10	12	15	20	24	30	40	60	120	∞
1	968.63	976.71	984.87	993.10	997.25	1001.4	1005.6	1009.8	1014.0	1018.3
2	39.398	39.415	39.431	39.448	39.456	39.465	39.473	39.481	39.490	39.498
3	14.419	14.337	14.253	14.167	14.124	14.081	14.037	13.992	13.947	13.902
4	8.8439	8.7512	8.6565	8.5599	8.5109	8.4613	8.4111	8.3604	8.3092	8.2573
5	6.6192	6.5246	6.4277	6.3285	6.2780	6.2269	6.1751	6.1225	6.0693	6.0153
6	5.4613	5.3662	5.2687	5.1684	5.1172	5.0652	5.0125	4.9589	4.9045	4.8491
7	4.7611	4.6658	4.5678	4.4667	4.4150	4.3624	4.3089	4.2544	4.1989	4.1423
8	4.2951	4.1997	4.1012	3.9995	3.9472	3.8940	3.8398	3.7844	3.7279	3.6702
9	3.9639	3.8682	3.7694	3.6669	3.6142	3.5604	3.5055	3.4493	3.3918	3.3329
10	3.7168	3.6209	3.5217	3.4186	3.3654	3.3110	3.2554	3.1984	3.1399	3.0798
11	3.5257	3.4296	3.3299	3.2261	3.1725	3.1176	3.0613	3.0035	2.9441	2.8828
12	3.3736	3.2773	3.1772	3.0728	3.0187	2.9633	2.9063	2.8478	2.7874	2.7249
13	3.2497	3.1532	3.0527	2.9477	2.8932	2.8373	2.7797	2.7204	2.6590	2.5955
14	3.1469	3.0501	2.9493	2.8437	2.7888	2.7324	2.6742	2.6142	2.5519	2.4872
15	3.0602	2.9633	2.8621	2.7559	2.7006	2.6437	2.5850	2.5242	2.4611	2.3953
16	2.9862	2.8890	2.7875	2.6808	2.6252	2.5678	2.5085	2.4471	2.3831	2.3163
17	2.9222	2.8249	2.7230	2.6158	2.5598	2.5021	2.4422	2.3801	2.3153	2.2474
18	2.8664	2.7689	2.6667	2.5590	2.5027	2.4445	2.3842	2.3214	2.2558	2.1869
19	2.8173	2.7196	2.6171	2.5089	2.4523	2.3937	2.3329	2.2695	2.2032	2.1333
20	2.7737	2.6758	2.5731	2.4645	2.4076	2.3486	2.2873	2.2234	2.1562	2.0853
21	2.7348	2.6368	2.5338	2.4247	2.3675	2.3082	2.2465	2.1819	2.1141	2.0422
22	2.6998	2.6017	2.4984	2.3890	2.3315	2.2718	2.2097	2.1446	2.0760	2.0032
23	2.6682	2.5699	2.4665	2.3567	2.2989	2.2389	2.1763	2.1107	2.0415	1.9677
24	2.6396	2.5412	2.4374	2.3273	2.2693	2.2090	2.1460	2.0799	2.0099	1.9353
25	2.6135	2.5149	2.4110	2.3005	2.2422	2.1816	2.1183	2.0517	1.9811	1.9055
26	2.5895	2.4909	2.3867	2.2759	2.2174	2.1565	2.0928	2.0257	1.9545	1.8781
27	2.5676	2.4688	2.3644	2.2533	2.1946	2.1334	2.0693	2.0018	1.9299	1.8527
28	2.5473	2.4484	2.3438	2.3224	2.1735	2.1121	2.0477	1.9796	1.9072	1.8291
29	2.5286	2.4295	2.3248	2.2131	2.1540	2.0923	2.0276	1.9591	1.8861	1.8072
30	2.5112	2.4120	2.3072	2.1952	2.1359	2.0739	2.0089	1.9400	1.8664	1.7867
40	2.3882	2.2882	2.1819	2.0677	2.0069	1.9429	1.8752	1.8028	.7242	1.6371
60	2.2702	2.1692	2.0613	1.9445	1.8817	1.8152	1.7440	1.6668	1.5810	1.4822
120	2.1570	2.0548	1.9450	1.8249	1.7597	1.6899	1.6141	1.5299	1.4327	1.3104
∞	2.0483	1.9447	1.8326	1.7085	1.6402	1.5660	1.4835	1.3883	1.2684	1.0000

* Reproduced by permission of E. S. Pearson from "Tables of Percentage Points of the Inverted Beta (F) Distribution," *Biometrika* **33**, 73–88, 1943, by Maxine Merrington and Catherine M. Thompson.

Where necessary, interpolation should be carried out using the reciprocals of the degrees of freedom. The function $120/\nu$ is convenient for this purpose. ν_1 = numerator, ν_2 = denominator.

TABLE C.4f THE *F*-DISTRIBUTION*—$P(F_*) = 0.99$

$$P(F \leq F_*) \equiv P(F_*) = \int_0^{F_*} p(F)\, dF$$

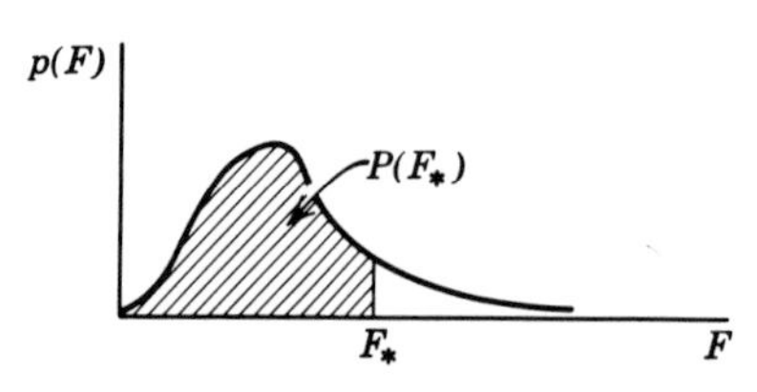

ν_2 \ ν_1	1	2	3	4	5	6	7	8	9
1	4052.2	4999.5	5403.3	5624.6	5763.7	5859.0	5928.3	5981.6	6022.5
2	98.503	99.000	99.166	99.249	99.299	99.332	99.356	99.374	99.388
3	34.116	30.817	29.457	28.710	28.237	27.911	27.672	27.489	27.345
4	21.198	18.000	16.694	15.977	15.522	15.207	14.976	14.799	14.659
5	16.268	13.274	12.060	11.392	10.967	10.672	10.456	10.289	10.158
6	13.745	10.925	9.7795	9.1483	8.7459	8.4661	8.2600	8.1016	7.9761
7	12.246	9.5466	8.4513	7.8467	7.4604	7.1914	6.9928	6.8401	6.7188
8	11.259	8.6491	7.5910	7.0060	6.6318	6.4707	6.1776	6.0289	5.9106
9	10.561	8.0215	6.9919	6.4221	6.0569	5.8018	5.6129	5.4671	5.3511
10	10.044	7.5594	6.5523	5.9943	5.6363	5.3858	5.2001	5.0567	4.9424
11	9.6460	7.2057	6.2167	5.6683	5.3160	5.0692	4.8861	4.7445	4.6315
12	9.3302	6.9266	5.9526	5.4119	5.0643	4.8206	4.6395	4.4994	4.3875
13	9.0738	6.7010	5.7394	5.2053	4.8616	4.6204	4.4410	4.3021	4.1911
14	8.8616	6.5149	5.5639	5.0354	4.6950	4.4558	4.3779	4.1399	4.0297
15	8.6831	6.3589	5.4170	4.8932	4.5556	4.3183	4.1415	4.0045	3.8948
16	8.5310	6.2262	5.2922	4.7726	4.4374	4.2016	4.0259	3.8896	3.7804
17	8.3997	6.1121	5.1850	4.6690	4.3359	4.1015	3.9267	3.7910	3.6822
18	8.2854	6.0129	5.0919	4.5790	4.2479	4.0146	3.8406	3.7054	3.5971
19	8.1850	5.9259	5.0103	4.5003	4.1708	3.9386	3.7653	3.6305	3.5225
20	8.0960	5.8489	4.9382	4.4307	4.1027	3.8714	3.6987	3.5644	3.4567
21	8.0166	4.7804	4.8740	4.3688	4.0421	3.8117	3.6396	3.5056	3.3981
22	7.9454	5.7190	4.8166	4.3134	3.9880	3.7583	3.5867	3.4530	3.3458
23	7.8811	5.6637	4.7649	4.2635	3.9392	3.7102	3.5390	3.4057	3.2986
24	7.8229	5.6136	4.7181	4.2184	3.8951	3.6667	3.4959	3.3629	3.2560
25	7.7698	5.5680	4.6755	4.1774	3.8550	3.6272	3.4568	3.3239	3.2172
26	7.7213	5.5263	4.6366	4.1400	3.8183	3.5911	3.4210	3.2884	3.1818
27	7.6767	5.4881	4.6009	4.1056	3.7848	3.5580	3.3882	3.2558	3.1494
28	7.6356	5.4529	4.5681	4.0740	3.7539	3.5276	3.3581	3.2259	3.1195
29	7.5976	5.4205	4.5378	4.0449	3.7254	3.4995	3.3302	3.1982	3.0920
30	7.5625	5.3904	4.5097	4.0179	3.6990	3.4735	3.3045	3.1726	3.0665
40	7.3141	5.1785	4.3126	3.8283	3.5138	3.2910	3.1238	2.9930	2.8876
60	7.0771	4.9774	4.1259	3.6491	3.3389	3.1187	2.9530	2.8233	2.7185
120	6.8510	4.7865	3.9493	3.4796	3.1735	2.9559	2.7918	2.6629	2.5586
∞	6.6349	4.6052	3.7816	3.3192	3.0173	2.8020	2.6393	2.5113	2.4073

(*continued*)

TABLE C.4f (*continued*)

ν_2 \ ν_1	10	12	15	20	24	30	40	60	120	∞
1	6055.8	6106.3	6157.3	6208.7	6234.6	6260.7	6268.8	6313.0	6339.4	6366.0
2	99.399	99.416	99.432	99.449	99.458	99.466	99.474	99.483	99.491	99.501
3	27.229	27.052	26.872	26.690	26.598	26.505	26.411	26.316	26.221	26.125
4	14.546	14.374	14.198	14.020	13.929	13.838	13.745	13.652	13.558	13.463
5	10.051	9.8883	9.7222	9.5527	9.4665	9.3793	9.2912	9.2020	9.1118	9.0204
6	7.8741	7.7183	7.5590	7.3958	7.3127	7.2285	7.1432	7.0568	6.9690	6.8801
7	6.6201	6.6591	6.3143	6.1554	6.0743	5.9921	5.9084	5.8236	5.7372	5.6495
8	5.8143	5.6668	5.5151	5.3591	5.2793	5.1981	5.1156	5.0316	4.9460	4.8588
9	5.2565	5.1114	4.9621	4.8080	4.7290	4.6486	4.5667	4.4831	4.3978	4.3105
10	4.8492	4.7059	4.5582	4.4054	4.3269	4.2469	4.1653	4.0819	3.9965	3.9090
11	4.5393	4.3974	4.2509	4.0990	4.0209	3.9411	3.8596	3.7761	3.6904	3.6025
12	4.2961	4.1553	4.0096	3.8584	3.7805	3.7008	3.6192	3.5355	3.4494	3.3608
13	4.1003	3.9603	3.8154	3.6646	3.5868	3.5070	3.4253	3.3413	3.2548	3.1654
14	3.9394	3.8001	3.6557	3.5052	3.4274	3.3476	3.2656	3.1813	3.0942	3.0040
15	3.8049	3.6662	3.5222	3.3719	3.2940	3.2141	3.1319	3.0471	2.9595	2.8684
16	3.6909	3.5527	3.4089	3.2588	3.1808	3.1007	3.0182	2.9330	2.8447	2.7528
17	3.5931	3.4552	3.3117	3.1615	3.0835	3.0032	2.9205	2.8348	2.7459	2.6530
18	3.5082	3.3706	3.2273	3.0771	2.9990	2.9185	2.8354	2.7493	2.6597	2.5660
19	3.4338	3.2965	3.1533	3.0031	2.9249	2.8442	2.7608	2.6742	2.5839	2.4893
20	3.3682	3.2311	3.0880	2.9377	2.8594	2.7785	2.6947	2.6077	2.5168	2.4212
21	3.3098	3.1729	3.0299	2.8796	2.8011	2.7200	2.6359	2.5484	2.4568	2.3603
22	3.2576	3.1209	2.9780	2.8274	2.7488	2.6675	2.5831	2.4951	2.4029	2.3055
23	3.2106	2.0740	2.9311	2.7805	2.7017	2.6202	2.5355	2.4471	2.3542	2.2559
24	3.1681	3.0316	2.8887	2.7380	2.6591	2.5773	2.4923	2.4035	2.3099	2.2107
25	3.1294	2.9931	2.8502	2.6993	2.6203	2.5383	2.4530	2.3637	2.2695	2.1694
26	3.0941	2.9579	2.8150	2.6640	2.5848	2.5026	2.4170	2.3273	2.2325	2.1315
27	3.0618	2.9256	2.7827	2.6316	2.5522	2.4699	2.3840	2.2938	2.1984	2.0965
28	3.0320	2.8959	2.7530	2.6017	2.5223	2.4397	2.3535	2.2629	2.1670	2.0642
29	3.0045	2.8685	2.7256	2.5742	2.4946	2.4118	2.3253	2.2344	2.1378	2.0342
30	2.9791	2.8431	2.7002	2.5487	2.4689	2.3860	2.2992	2.2079	2.1107	2.0062
40	2.8005	2.6648	2.5216	2.3689	2.2880	2.2034	2.1142	2.0194	1.9172	1.8047
60	2.6318	2.4961	2.3523	2.1978	2.1154	2.0285	1.9360	1.8363	1.7263	1.6006
120	2.4721	2.3363	2.1915	2.0346	1.9500	1.8600	1.7628	1.6557	1.5330	1.3805
∞	2.3209	2.1848	2.0385	1.8783	1.7908	1.6964	1.5923	1.4730	1.3246	1.0000

* Reproduced by permission of E. S. Pearson from "Tables of Percentage Points of the Inverted Beta (*F*) Distribution," *Biometrika* **33**, 73–88, 1943, by Maxine Merrington and Catherine M. Thompson.

Where necessary, interpolation should be carried out using the reciprocals of the degrees of freedom. The function $120/\nu$ is convenient for this purpose. ν_1 = numerator, ν_2 = denominator.

TABLE C.5 CRITICAL VALUES OF r FOR THE SIGN TEST*

(α = confidence level)

	α for Two-Sided Test					α for Two-Sided Test			
	0.01	0.05	0.10	0.25		0.01	0.05	0.10	0.25
	α for One-Sided Test					α for One-Sided Test			
n	0.005	0.025	0.05	0.125	n	0.005	0.025	0.05	0.125
1	—	—	—	—	46	13	15	16	18
2	—	—	—	—	47	14	16	17	19
3	—	—	—	0	48	14	16	17	19
4	—	—	—	0	49	15	17	18	19
5	—	—	0	0	50	15	17	18	20
6	—	0	0	1	51	15	18	19	20
7	—	0	0	1	52	16	18	19	21
8	0	0	1	1	53	16	18	20	21
9	0	1	1	2	54	17	19	20	22
10	0	1	1	2	55	17	19	20	22
11	0	1	2	3	56	17	20	21	23
12	1	2	2	3	57	18	20	21	23
13	1	2	3	3	58	18	21	22	24
14	1	2	3	4	59	19	21	22	24
15	2	3	3	4	60	19	21	23	25
16	2	3	4	5	61	20	22	23	25
17	2	4	4	5	62	20	22	24	25
18	3	4	5	6	63	20	23	24	26
19	3	4	5	6	64	21	23	24	26
20	3	5	5	6	65	21	24	25	27
21	4	5	6	7	66	22	24	25	27
22	4	5	6	7	67	22	25	26	28
23	4	6	7	8	68	22	25	26	28
24	5	6	7	8	69	23	25	27	29
25	5	7	7	9	70	23	26	27	29
26	6	7	8	9	71	24	26	28	30
27	6	7	8	10	72	24	27	28	30
28	6	8	9	10	73	25	27	28	31
29	7	8	9	10	74	25	28	29	31
30	7	9	10	11	75	25	28	29	32
31	7	9	10	11	76	26	28	30	32
32	8	9	10	12	77	26	29	30	32
33	8	10	11	12	78	27	29	31	33
34	9	10	11	13	79	27	30	31	33
35	9	11	12	13	80	28	30	32	34
36	9	11	12	14	81	28	31	32	34
37	10	12	13	14	82	28	31	33	35
38	10	12	13	14	83	29	32	33	35
39	11	12	13	15	84	29	32	33	36
40	11	13	14	15	85	30	32	34	36
41	11	13	14	16	86	30	33	34	37
42	12	14	15	16	87	31	33	35	37
43	12	14	15	17	88	31	34	35	38
44	13	15	16	17	89	31	34	36	38
45	13	15	16	18	90†	32	35	36	39

* Adapted with permission from W. J. Dixon and F. J. Massey, Jr., *Introduction to Statistical Analysis* (2nd. ed.), McGraw-Hill, New York, 1957.

† For values of n larger than 90, approximate values of r may be found by taking the nearest integer less than $(n - 1)/2 - k\sqrt{n + 1}$, where k is 1.2879, 0.9800, 0.8224, 0.5752 for the 1, 5, 10, 25 percent values, respectively.

Table C.6 Mann and Whitney test
(Probability of Obtaining a U^* not Larger than that Tabulated in Comparing Samples of n and m)

$m = 3$

U^* \ n	1	2	3
0	0.250	0.100	0.050
1	0.500	0.200	0.100
2	0.750	0.400	0.200
3		0.600	0.350
4			0.500
5			0.650

$m = 4$

U \ n	1	2	3	4
0	0.200	0.067	0.028	0.014
1	0.400	0.133	0.057	0.029
2	0.600	0.267	0.114	0.057
3		0.400	0.200	0.100
4		0.600	0.314	0.171
5			0.429	0.243
6			0.571	0.343
7				03.44
8				0.557

$m = 5$

U^* \ n	1	2	3	4	5
0	0.167	0.047	0.018	0.008	0.004
1	0.333	0.095	0.036	0.016	0.008
2	0.500	0.190	0.071	0.032	0.016
3	0.667	0.286	0.125	0.056	0.028
4		0.429	0.196	0.095	0.048
5		0.571	0.286	0.143	0.075
6			0.393	0.206	0.111
7			0.500	0.278	0.155
8			0.607	0.365	0.210
9				0.452	0.274
10				0.548	0.345
11					0.421
12					0.500
13					0.579

$m = 6$

U^* \ n	1	2	3	4	5	6
0	0.143	0.036	0.012	0.005	0.002	0.001
1	0.286	0.071	0.024	0.010	0.004	0.002
2	0.428	0.143	0.048	0.019	0.009	0.004
3	0.571	0.214	0.083	0.033	0.015	0.008
4		0.321	0.131	0.057	0.026	0.013
5		0.429	0.190	0.086	0.041	0.021
6		0.571	0.274	0.129	0.063	0.032
7			0.357	0.176	0.089	0.047
8			0.452	0.238	0.123	0.066
9			0.548	0.305	0.165	0.090
10				0.381	0.214	0.120
11				0.457	0.268	0.155
12				0.545	0.331	0.197
13					0.396	0.242
14					0.465	0.294
15					0.535	0.350
16						0.409
17						0.469
18						0.531

(continued)

TABLE C.6 (*continued*)

$m = 7$

U^* \ n	1	2	3	4	5	6	7
0	0.125	0.028	0.008	0.003	0.001	0.001	0.000
1	0.250	0.056	0.017	0.006	0.003	0.001	0.001
2	0.375	0.111	0.033	0.012	0.005	0.002	0.001
3	0.500	0.167	0.058	0.021	0.009	0.004	0.002
4	0.625	0.250	0.092	0.036	0.015	0.007	0.003
5		0.333	0.133	0.055	0.024	0.011	0.006
6		0.444	0.192	0.082	0.037	0.017	0.009
7		0.556	0.258	0.115	0.053	0.026	0.013
8			0.333	0.158	0.074	0.037	0.019
9			0.417	0.206	0.101	0.051	0.027
10			0.500	0.264	0.134	0.069	0.036
11			0.583	0.324	0.172	0.090	0.049
12				0.394	0.216	0.117	0.064
13				0.464	0.265	0.147	0.082
14				0.538	0.319	0.183	0.104
15					0.378	0.223	0.130
16					0.438	0.267	0.159
17					0.500	0.314	0.191
18					0.562	0.365	0.228
19						0.418	0.267
20						0.473	0.310
21						0.527	0.355
22							0.402
23							0.451
24							0.500
25							0.549

(*continued*)

TABLE C.6 (*continued*)

$m = 8$

U^* \ n	1	2	3	4	5	6	7	8	t	normal
0	0.111	0.022	0.006	0.002	0.001	0.000	0.000	0.000	3.308	0.001
1	0.222	0.044	0.012	0.004	0.002	0.001	0.000	0.000	3.203	0.001
2	0.333	0.089	0.024	0.008	0.003	0.001	0.001	0.000	3.098	0.001
3	0.444	0.133	0.042	0.014	0.005	0.002	0.001	0.001	2.993	0.001
4	0.556	0.200	0.067	0.024	0.009	0.004	0.002	0.001	2.888	0.002
5		0.267	0.097	0.036	0.015	0.006	0.003	0.001	2.783	0.003
6		0.356	0.139	0.055	0.023	0.010	0.005	0.002	2.678	0.004
7		0.444	0.188	0.077	0.033	0.015	0.007	0.003	2.573	0.005
8		0.556	0.248	0.107	0.047	0.021	0.010	0.005	2.468	0.007
9			0.315	0.141	0.064	0.030	0.014	0.007	2.363	0.009
10			0.387	0.184	0.085	0.041	0.020	0.010	2.258	0.012
11			0.461	0.230	0.111	0.054	0.027	0.014	2.153	0.016
12			0.539	0.285	0.142	0.071	0.036	0.019	2.048	0.020
13				0.341	0.177	0.091	0.047	0.025	1.943	0.026
14				0.404	0.217	0.114	0.060	0.032	1.838	0.033
15				0.467	0.262	0.141	0.076	0.041	1.733	0.041
16				0.533	0.311	0.172	0.095	0.052	1.628	0.052
17					0.362	0.207	0.116	0.065	1.523	0.064
18					0.416	0.245	0.140	0.080	1.418	0.078
19					0.472	0.286	0.168	0.097	1.313	0.094
20					0.528	0.331	0.198	0.117	1.208	0.113
21						0.377	0.232	0.139	1.102	0.135
22						0.426	0.268	0.164	0.998	0.159
23						0.475	0.306	0.191	0.893	0.185
24						0.525	0.347	0.221	0.788	0.215
25							0.389	0.253	0.683	0.247
26							0.433	0.287	0.578	0.282
27							0.478	0.323	0.473	0.318
28							0.522	0.360	0.368	0.356
29								0.399	0.263	0.396
30								0.439	0.158	0.437
31								0.480	0.052	0.481
32								0.520		

Reproduced by permission of the publisher from H. B. Mann and D. R. Whitney, *Annals Math. Stat.* **18**, 52–54, 1947.

TABLE C.7 CRITICAL VALUES FOR THE RUN DISTRIBUTION

(Values of $U_\alpha^\dagger$ such that Prob $\{U^\dagger \geq U_\alpha^\dagger\} = \alpha$, for the case of $n_1 = n_2 = \frac{n_1 + n_2}{2}$)

	$1 - \alpha$: 0.01	0.025	0.05	0.95	0.975	0.99
$n/2$	α: 0.99	0.975	0.95	0.05	0.025	0.01
5	2	2	3	8	9	9
6	2	3	3	10	10	11
7	3	3	4	11	12	12
8	4	4	5	12	13	13
9	4	5	6	13	14	15
10	5	6	6	15	15	16
11	6	7	7	16	16	17
12	7	7	8	17	18	18
13	7	8	9	18	19	20
14	8	9	10	19	20	21
15	9	10	11	20	21	22
16	10	11	11	22	22	23
18	11	12	13	24	25	26
20	13	14	15	26	27	28
25	17	18	19	32	33	34
30	21	22	24	37	39	40
35	25	27	28	43	44	46
40	30	31	33	48	50	51
45	34	36	37	54	55	57
50	38	40	42	59	61	63
55	43	45	46	65	66	68
60	47	49	51	70	72	74
65	52	54	56	75	77	79
70	56	58	60	81	83	85
75	61	63	65	86	88	90
80	65	68	70	91	93	96
85	70	72	74	97	99	101
90	74	77	79	102	104	107
95	79	82	84	107	109	112
100	84	86	88	113	115	117

Reproduced from J. S. Bendat and A. G. Piersol, *Measurement and Analysis of Random Data*, John Wiley, New York, 1966, with permission.

TABLE C.8 CRITICAL VALUES FOR THE SUM OF SQUARED LENGTHS TEST FOR EQUAL SAMPLE SIZE, n: $P\{N \geq N_\alpha\} \leq \alpha$

n	α			
	0.10	0.05	0.025	0.01
3	18	18	18	18
4	26	32	32	32
5	34	38	42	50
6	38	44	50	58
7	46	52	60	68
8	54	60	68	80
9	62	70	78	90
10	68	78	86	100
11	76	86	96	108
12	84	94	104	118
13	92	102	112	128
14	98	110	122	136
15	106	118	130	146

TABLE C.9 CRITICAL VALUES FOR THE REVERSE ARRANGEMENT DISTRIBUTION

(Values of T^* such that Prob $\{T^* > T_\alpha\} = \alpha$ where n = total number of measurements)

n	α					
	0.99	0.976	0.95	0.05	0.025	0.01
10	9	11	13	31	33	35
12	16	18	21	44	47	49
14	24	27	30	60	63	66
16	34	38	41	78	81	85
18	45	50	54	98	102	107
20	59	64	69	120	125	130
30	152	162	171	263	272	282
40	290	305	319	460	474	489
50	473	495	514	710	729	751
60	702	731	756	1013	1038	1067
70	977	1014	1045	1369	1400	1437
80	1299	1344	1382	1777	1815	1860
90	1668	1721	1766	2238	2283	2336
100	2083	2145	2198	2751	2804	2866

Reproduced from J. S. Bendat and A. G. Piersol, *Measurement and Analysis of Random Data*, John Wiley, New York, 1966, with permission.

TABLE C.10 * DURBIN-WATSON STATISTIC FOR SERIAL CORRELATION—THE DISTRIBUTION OF

$$D = \frac{\sum_{2}^{n} (E_t - E_{t-1})^2}{\sum_{1}^{n} E_t^2}$$

Sample Size n	Probability in Upper Tail†	Values of D_L and D_U from Durbin and Watson									
		$K = 1$		$K = 2$		$K = 3$		$K = 4$		$K = 5$	
		D_L	D_U	D_L	D_U	D_L	D_U	D_L	D_U	D_L	D_U
	0.01	0.81	1.07	0.70	1.25	0.59	1.46	0.49	1.70	0.39	1.96
15	0.025	0.95	1.23	0.83	1.40	0.71	1.61	0.59	1.84	0.48	2.09
	0.05	1.08	1.36	0.95	1.54	0.82	1.75	0.69	1.97	0.56	2.21
	0.01	0.95	1.15	0.86	1.27	0.77	1.41	0.68	1.57	0.60	1.74
20	0.025	1.08	1.28	0.99	1.41	0.89	1.55	0.79	1.70	0.70	1.87
	0.05	1.20	1.41	1.10	1.54	1.00	1.68	0.90	1.83	0.79	1.99
	0.01	1.05	1.21	0.98	1.30	0.90	1.41	0.83	1.52	0.75	1.65
25	0.025	1.18	1.34	1.10	1.43	1.02	1.54	0.94	1.65	0.86	1.77
	0.05	1.29	1.45	1.21	1.55	1.12	1.66	1.04	1.77	0.95	1.89
	0.01	1.13	1.26	1.07	1.34	1.01	1.42	0.94	1.51	0.88	1.61
30	0.025	1.25	1.38	1.18	1.46	1.12	1.54	1.05	1.63	0.98	1.73
	0.05	1.35	1.49	1.28	1.57	1.21	1.65	1.14	1.74	1.07	1.83
	0.01	1.25	1.34	1.20	1.40	1.15	1.46	1.10	1.52	1.05	1.58
40	0.025	1.35	1.45	1.30	1.51	1.25	1.57	1.20	1.63	1.15	1.69
	0.05	1.44	1.54	1.39	1.60	1.34	1.66	1.29	1.72	1.23	1.79
	0.01	1.32	1.40	1.28	1.45	1.24	1.49	1.20	1.54	1.16	1.59
50	0.025	1.42	1.50	1.38	1.54	1.34	1.59	1.30	1.64	1.26	1.69
	0.05	1.50	1.59	1.46	1.63	1.42	1.67	1.38	1.72	1.34	1.77
	0.01	1.38	1.45	1.35	1.48	1.32	1.52	1.28	1.56	1.25	1.60
60	0.025	1.47	1.54	1.44	1.57	1.40	1.61	1.37	1.65	1.33	1.69
	0.05	1.55	1.62	1.51	1.65	1.48	1.69	1.44	1.73	1.41	1.77
	0.01	1.47	1.52	1.44	1.54	1.42	1.57	1.39	1.60	1.36	1.62
80	0.025	1.54	1.59	1.53	1.62	1.49	1.65	1.47	1.67	1.44	1.70
	0.05	1.61	1.66	1.59	1.69	1.56	1.72	1.53	1.74	1.51	1.77
	0.01	1.52	1.56	1.50	1.58	1.48	1.60	1.46	1.63	1.44	1.65
100	0.025	1.59	1.63	1.57	1.65	1.55	1.67	1.53	1.70	1.51	1.72
	0.05	1.65	1.69	1.63	1.72	1.61	1.74	1.59	1.76	1.57	1.78

* From C. F. Christ, *Economic Models and Methods*, John Wiley, 1966, with permission, based on J. Durbin and G. S. Watson, *Biometrika* **37**, 409, 1950; **38**, 159, 1951.

† The probability shown in the second column is the area in the upper tail. K is the number of independent variables in addition to the constant term.

TABLE C.11 CRITERIA FOR REJECTION OF OUTLYING OBSERVATIONS*

Statistic	Number of Observations, n	$1-\alpha$						
		0.70	0.80	0.90	0.95	0.98	0.99	0.995
	3	0.684	0.781	0.886	0.941	0.976	0.988	0.994
	4	0.471	0.560	0.679	0.765	0.846	0.889	0.926
r_{10}	5	0.373	0.451	0.557	0.642	0.729	0.780	0.821
	6	0.318	0.386	0.482	0.560	0.644	0.698	0.740
	7	0.281	0.344	0.434	0.507	0.586	0.637	0.680
	8	0.318	0.385	0.479	0.554	0.631	0.683	0.725
r_{11}	9	0.288	0.352	0.441	0.512	0.587	0.635	0.677
	10	0.265	0.325	0.409	0.477	0.551	0.597	0.639
	11	0.391	0.442	0.517	0.576	0.638	0.679	0.713
r_{21}	12	0.370	0.419	0.490	0.546	0.605	0.642	0.675
	13	0.351	0.399	0.467	0.521	0.578	0.615	0.649
	14	0.370	0.421	0.492	0.546	0.602	0.641	0.674
	15	0.353	0.402	0.472	0.525	0.579	0.616	0.647
	16	0.338	0.386	0.454	0.507	0.559	0.595	0.624
	17	0.325	0.373	0.438	0.490	0.542	0.577	0.605
	18	0.314	0.361	0.424	0.475	0.527	0.561	0.589
r_{22}	19	0.304	0.350	0.412	0.462	0.514	0.547	0.575
	20	0.295	0.340	0.401	0.450	0.502	0.535	0.562
	21	0.287	0.331	0.391	0.440	0.491	0.524	0.551
	22	0.280	0.323	0.382	0.430	0.481	0.514	0.541
	23	0.274	0.316	0.374	0.421	0.472	0.505	0.532
	24	0.268	0.310	0.367	0.413	0.464	0.497	0.524
	25	0.262	0.304	0.360	0.406	0.457	0.489	0.516

* Adapted by permission from W. J. Dixon and F. J. Massey, Jr., *Introduction to Statistical Analysis* (2nd. ed.), McGraw-Hill, New York, 1957.

APPENDIX D

Notation

GENERAL CONVENTIONS

1. Vectors and matrices are boldface.
2. Random variables are capital letters, primarily from the end of the alphabet, plus certain other commonly accepted symbols such as t and s.
3. Laplace transforms are indicated by both an overlay ˇ and (s).
4. Fourier transforms are indicated by both an overlay ˇ and (ω).
5. Greek letters represent expected values.
6. Square brackets are used for matrices; large vertical parallel lines are used for determinants.
7. Operators are script letters.

Symbol	Meaning
a	deterministic constant in general
a_{ij}	element in the matrix $\mathbf{a}$
$\mathbf{a}$	the matrix $\mathbf{x}^T\mathbf{w}\mathbf{x}$
A	area
A_{ij}	element of the matrix $\mathbf{A}$
A_{ij}^*	scaled element in the matrix $\mathbf{A}$
$\mathbf{A}^{(\ )}$	identical to $[(\mathbf{X}^{(\ )})^T\mathbf{w}\mathbf{X}^{(\ )}]$
ARL	average run length
b	estimated parameter in a linear (in the parameters) model, random variable
b	deterministic constant in general
b	bandwidth (range of frequency)
b_e	equivalent bandwidth, $2\pi/m$
b_k	element of the matrix $\mathbf{b}$
b_0	estimated intercept in a linear (in the parameters) model $\eta=\beta_0+\beta_1(x-\bar{x})$, random variable
b_0'	estimated intercept in model $\eta = \beta_0' + \beta_1 x$
$b_{i,j}$	element of $\mathbf{b}_i$
$b^{(n)}$	element of the matrix $\mathbf{b}^{(n)}$
$\mathbf{b}$	matrix of estimated model parameters
$\mathbf{b}_i$	vector of parameters specifying the coordinates of a vertex of a simplex
$\mathbf{b}_p$	$p \times p$ submatrix of b's
$\mathbf{b}_1$	matrix of coefficients defined in connection with Equation 8.2-4
$\mathbf{b}_{11}$	matrix of coefficients defined in connection with Equation 8.2-4
$\mathbf{b}^{**}$	coordinates of an expansion or contraction vertex in a simplex
$\mathbf{b}^*$	submatrix of $\mathbf{b}_1$ with one element deleted
$\mathbf{b}^*$	coordinates of a vertex in a simplex
$\mathbf{b}^*$	selected matrix of $\mathbf{b}$ for a Taylor series expansion of a model in Section 8.4
$\mathbf{b}^{(n)}$	vector of estimated parameters at the nth stage in calculations
$\mathbf{B}^{(n)}$	column vector of the elements $(\Delta b_1^{(n)} \cdots \Delta b_m^{(n)})$ at the nth stage in calculations
c	concentration, deterministic variable
c	sample coefficient of variation; see Equation 2.4-4
c	constant
c_A	concentration of component A
$c_A(0)$	initial concentration of component A
c_{ij}	element in the matrix $\mathbf{c} = (\mathbf{x}^T\mathbf{w}\mathbf{x})^{-1}$
c_{ij}	element of the matrix $\mathbf{c}$
c^*	dimensionless concentration
$\mathbf{c}$	coordinates of the centroid of a simplex
$\mathbf{c}$	the matrix $(\mathbf{x}^T\mathbf{w}\mathbf{x})^{-1}$
$\mathbf{c}_p$	submatrix of $\mathbf{c}$
Covar	Covariance
C	discrimination criterion for models in Section 8.5-3
C	coefficients, random variables
C	concentration, random variable
C_p	heat capacity
C_q	test statistic given by Equation 7.3-1
C_{ij}	element of the matrix $\mathbf{C}$
$C^{(\ )}$	identical to $(\mathbf{A}^{(\ )})^{-1}$
$\check{d}(s)$	polynomial in denominator of $\check{g}(s)$
d.f.	degrees of freedom
D	Durbin-Watson statistic in the test for serial correlation
D	difference, random variable
D	ratio of $K_v/K_{v,\max}$ in Section 8.5-3
D_i	difference, random variable
D_k	kth derivative operator

$\tilde{D}$	dispersion coefficient in axial direction; subscripts L and R refer to axial and radial direction, respectively
$D^k_{m+1}(y)$	approximation for the kth ordinary derivative
$\mathscr{D}$	diffusion coefficient
$\mathbf{D}$	diagonal matrix used in Equation 6.2-20 composed of the elements on the main diagonal of $\mathbf{A}$
E	defined in connection with Equation 8.5-4
E	activation energy
E	event
E_i	the ith residual $(Y_i - \hat{Y}_i)$, also $[Y_i - (\eta_i)]$ for a nonlinear model
$E^\dagger$	residual for a suspected outlier
$\mathscr{E}$	expectation
$\mathbf{E}$	matrix of residuals
f	function in general
$f(x_i)$	relative frequency function
f_{max}	folding or Nyquist frequency
f_{min}	$\frac{1}{2}\tau_m$
$\mathbf{f}$	Covar $\{\mathbf{X}_i\mathbf{X}_j\}$, the covariance matrix of Equation 2.3-6
$\mathbf{f}(\boldsymbol{\alpha}, \mathbf{y}, t)$	nonlinear function of $\boldsymbol{\alpha}$, $\mathbf{y}$, t
F	flow rate of fluid
F	the Fisher F used in the variance ratio test, random variable
$F_{1-\alpha}(m, n)$	the Fisher F for the significance level α and for m degrees of freedom in the numerator and n degrees of freedom in the denominator in the variance ratio
$\mathscr{F}$	Fourier transform
$\mathscr{F}^{-1}$	inverse Fourier transform
$g(t)$	general function of time
$g(t)$	deterministic impulse response or weighting function
$g_i(\ )$	the ith general function
$\check{g}(s)$	deterministic transfer function $(\mathscr{L}[g(t)])$
$\check{g}(\omega)$	deterministic transfer function (frequency response) in frequency domain
$\check{g}(z)$	z-transform of impulse response (i.e. of $g[t_k]$)
$\check{\mathbf{g}}(s)$	transfer matrix
$G(t)$	empirical impulse response (weighting function)
$\check{G}(s)$	stochastic transfer function, Laplace transform of $G(t)$
$\mathbf{G}$	the matrix $\mathbf{x}^T\mathbf{w}\mathbf{Y}$
h	interphase heat transfer, a constant; hold-up, a constant
h_{jk}	element from a Hessian matrix
h_j	$(Z_j + Z_\alpha)/Z_0$ in Section 11.3
h_j	step length in an iterative least squares procedure
$\mathbf{h}$	Hessian matrix
$\mathbf{h}(t)$, $\mathbf{h}(t_i)$	known matrix in Equation 9.1-6
$\mathbf{h}^*$	matrix defined in connection with Equation 9.4-9
$\mathbf{h}^{(n)}$	solution of Equation 9.5-10 at the nth stage of calculation
H_0	null hypothesis
$H_1, H_2, \ldots$	alternate hypotheses
$\mathscr{H}$	general linear operator
i	$\sqrt{-1}$
I_0, I_1, I_2	integrals of experimental data in Section 10.4
$I(r:s)$	Kullback criterion for model discrimination in Section 8.5
I^*	number of inversions
$\mathscr{I}(\ )$	imaginary part
$\mathbf{I}$	identity matrix
$J(1, 2)$	Kullback criterion for model discrimination in Section 8.5
$J_1(\lambda_n r)$, $J_0(\lambda_n r)$	Bessel functions in Tables 10.2-2 and 10.4-2
$\mathscr{J}$	successive integrations in Section 11.3
$\mathbf{J}$	information matrix defined in connection with Equation 9.6-9
$\mathbf{J}^{(0)}$, $\mathbf{J}(y^{(0)})$	Jacobian matrix defined in a Taylor series expansion in Section 9.5
k	constant in general, chemical reaction rate constant
k	preexponential factor, constant
k_n^*, k_n	normalizing constant in a probability density in Sections 8.4 and 9.4
K	constant in general
K_v	discrimination criterion for different models in Section 8.5
$\mathbf{K}_{i+1}$	matrix defined in connection with Equation 9.4-19
l_j	lower bound on parameter b_j or variable x_j
$\mathbf{l}$	a matrix of lower bounds
L	likelihood function in general
L	length of a piece of equipment
$L(\boldsymbol{\beta} \mid \mathbf{y}_n)$	likelihood function of the elements of the matrix $\boldsymbol{\beta}$ given the $n \times 1$ vector of observations, $\mathbf{Y}$
$\mathscr{L}$	Laplace transform
$\mathscr{L}^{-1}$	inverse Laplace transform

Symbol	Meaning
m	the number of binary intervals encompassed by the spectral window
m	amount of tracer injected in Chapter 10
m	number of parameters in nonlinear models
m_i	function defined in connection with Equation 10.3-1
m_1	deterministic first raw time moment in Chapter 10
$m_2^{\#}$	deterministic second central time moment in Chapter 10
M	number of basic intervals in switching the input in Section 12.3
M_X	finite time continuous data average of $X(t)$
M_1	calculated (experimental) estimate of m_1
$M_2^{\#}$	calculated (experimental) estimate of $m_2^{\#}$
$\mathscr{M}_i$	ith central moment
$\mathscr{M}_{ij}$	central moment; i and j indicate the respective orders
$\mathbf{M}$	matrix defined in connection with Equation 8.5-9
$\mathbf{M}_{i+1}$	covariance matrix defined in Section 9.4
n	total number of observations or data sets
n_i	number of pulses present in a sample
n_{t_1}	number of times a value of $X(t) \le x$ in a group of time records
$\check{n}(s)$	polynomial in numerator of $\check{g}(s)$
N	number of time intervals between samples in Section 12.3
N	total number of time records
N	total number of pulses
N	sum of squared run lengths in Equation 3.7-9
p	partial pressure of a component denoted by a subscript
p_i	number of replicate measurements of the dependent variable for a given x_i
$p(x)$	probability density of X in general
$p(x, t)$	the first-order probability density function of $X(t)$
$p(x_1 \mid x_3)$	probability density of X_1 given X_3
$p_n(\boldsymbol{\beta} \mid \mathbf{y}_n)$	probability density for the elements of the matrix $\boldsymbol{\beta}$ given the $n \times 1$ vector of observations of Y, $\mathbf{Y}_n$, i.e., the density after n observations
$p(x_1, x_2; t_1, t_2)$	the second-order probability density function of $X(t)$
P	probability in general
P	dimensionless parameter termed the Peclet number in Chapter 10; subscripts designate axial (L) and radial (R) directions
$P(x)$	probability distribution function of X in general
$P(E)$	probability of event E
$P(x; t)$	the first-order probability distribution function $P\{X(t) \le x\}$ or the probability that the random variable $X(t)$ is less than or equal to the value of a deterministic variable x
$P(x_1; t)$	marginal probability distribution function of $X(t_1)$
$P(x_k; t)$	the first-order probability function for a discrete variable $X(t_k)$
$P(x_1, x_2; t_1, t_2)$	the second-order probability distribution function defined by Equation 2.1-2
$P_{m,n}(x)$	orthogonal polynomial term defined by Equation 5.1-21
$P_r^{(n)}$	probability of model r being the correct model after n experiments have been carried out
$P\{A \mid B\}$	probability of A given B
q	volumetric flow rate in Chapter 10
q	number of independent variables in an empirical model
q	$(\mathbf{x} - \mathbf{u})^T\mathbf{f}^{-1}(\mathbf{x} - \mathbf{u})$ in Equation 2.3-6
r	radius
r	magnitude (modulus) of a complex number in Chapter 12
r_i	reaction rate, deterministic variable
r_{ij}	Dixon criterion in the test of outliers
r^*	dimensionless radial coordinate
$r_{XX}(t_1, t_2)$	ensemble autocorrelation function for the nonstationary random variable $X(t)$
$r_{XX}(\tau)$	ensemble autocorrelation function for the stationary random variable $X(t)$
$r_{XY}(t_1, t_2)$	ensemble crosscorrelation function for the nonstationary random variables $X(t_1)$ and $Y(t_2)$
$r_{XY}(\tau)$	ensemble crosscorrelation function for the stationary random variables $X(t)$ and $Y(t)$
R	ideal gas constant
R	range for a random variable
R_i	reaction rate random variable
$R_{XX}(\tau)$	finite time or empirical continuous variable estimate of the ensemble autocorrelation function

$R_{XY}(\tau)$	finite time or empirical continuous variable estimate of the ensemble crosscorrelation function
$\mathscr{R}(\ )$	real part
s	distance in Euclidean space in Section 9.2
s	complex parameter in the Laplace transform
s_X	sample standard deviation
s_{b_k}	standard error of estimate for the estimated parameter b_k
s_j	root of $\check{d}(s)$ in Chapter 11
s_{XY}	sample crosscovariance
$\check{s}_{XX}(\omega)$	power spectral density function
$\check{s}_{XY}(\omega)$	crosspower spectral density function
s_X^2	sample variance
s_e^2	error mean square in regression analysis
s_p^2	pooled sample variance
s_r^2	residual mean square in regression analysis
$s_{\bar{Y}_i}^2$	estimated variance of $\bar{Y}_i$ (in regression analysis)
S	sum in general
S^*	$S^* = T^* - I^*$
$\check{S}_{XX}(\omega)$	finite time power or empirical spectral density, random variable
$\check{S}_{XY}(\omega)$	finite time or empirical crosspower spectral density, random variable
SS	sum of squares
t	time
t	the Student-t used in the t-test, random variable
t	relative temperature (°C or °F)
t^*	dimensionless time
t_f	time at the end of a time record
t_0	initial time
T	period of input in Section 12.3
T	temperature, absolute temperature
T	Wilcoxon T, the rank sum
T_x, T_y	Wilcoxon T for x or y
T^*	number of times a larger number is followed by a smaller number in a sequence
$\mathscr{T}$	transformation operator
u_j	upper bound for a parameter b_j or a variable x_j
$\mathbf{u}$	sensitivity coefficient matrix
$\mathbf{u}$	matrix of upper bounds
U	standardized normal random variable defined in Equation 2.3-2
$U(t)$	unit step function
$U(x)$	unit step function
U^*	the number of y's preceding an x in a ranked list, the Mann-Whitney statistic
$U^\dagger$	the Wald-Wolfowitz statistic for the total number of runs in a time record
$\mathbf{U}$	unitary matrix
v	velocity of flow
v_j	parameter used in exploration of the surface of ϕ in parameter space
v_j^*	value of v_j at the minimum ϕ
v_{ij}	element in $\mathbf{V}$
V	volume of process equipment
V	Thompson statistic in the test of outliers
$\mathbf{V}$	matrix in Wilks' test among estimated regression equations
$V_{rs}^{(n)}$	summation defined in connection with Equation 8.4-11
Var	variance
w_i	weight
$w(\tau)$	lag window
$\check{w}(\omega)$	spectral window
$\mathbf{w}$	matrix of weights
$W(t)$	a random variable in general
$\mathbf{W}_r^{(n+1)}$	compact notation for $\mathbf{X}_r^{(n+1)}\mathbf{M}^{-1}\cdot(\mathbf{X}_r^{(n+1)})^T$ in Section 8.5
x	general deterministic independent variable
x_i	ith independent variable
x^*	additional value of x (in regression analysis)
$x(t)$	deterministic process or model input
$x_{r,iq}$	element of $\mathbf{x}_{r,i}$; r denotes the model number, i the experimental run number, and q the independent variable number
$\check{x}(s)$	Laplace transform of $x(t)$
$\check{x}(z)$	z-transform of $x(t_k)$
$\check{x}(\omega)$	process or model input in the frequency domain
$\mathbf{x}$	a vector of deterministic variables, of independent deterministic variables
$\mathbf{x}_i$	vector composed of all the independent variables in a model evaluated on the ith run
$\mathbf{x}(t)$	matrix of deterministic process or model inputs
$\mathbf{x}_e$	the matrix $\mathbf{x}$ at an extremum
$\mathbf{x}^\dagger$	the alias matrix defined in Equation 5.1-17
$\mathbf{x}^*$	submatrix of $\mathbf{x}$ (one column is deleted)

$\mathbf{x}^{(n)}$	matrix of elements of the independent variables for the nth experiment in Section 8.5
$\mathbf{x}_{rj}^{(n)}$	matrix of elements of the independent variables in the jth equation in the rth model on the nth experiment in Section 8.5
$\mathbf{x}_{ri}, \mathbf{x}_{r,i}$	vector of elements $(x_{r,i1} \cdots x_{r,iq})^T$
$\check{\mathbf{x}}(s)$	Laplace transform of $\mathbf{x}(t)$
$\acute{\mathbf{x}}$	difference between $\mathbf{x}$ and $\mathbf{x}_e$
X	general stochastic variable or stochastic independent variable
$X(t)$	general stochastic variable that is explicitly a function of time; a subscript on t designates $X(t)$ at a particular time
$X_i(t)$	the ith stochastic variable that is a function of time, occasionally the ith time record
$\bar{X}$	sample mean
X^*	additional value of X, a random variable (in regression analysis)
X_j	jth-order derivative of X in Section 9.6
$X_{ij}^{(n)}, X_{ij}$	notation for $\partial\eta_i(\mathbf{x}_i, \mathbf{b})/\partial\beta_j$; the superscript n denotes the stage of the calculations or the experimental run number
$X_{r,ij}$	notation for $\partial\eta_r(\mathbf{b}, \mathbf{x}_{r,i})/\partial\beta_j$ in Section 8.4
$X_{rj,k}^{(n)}$	compact notation for $[\partial y_{rj}(\hat{\boldsymbol{\beta}}_r^{(n)}, \mathbf{x}_{rj}^{(n)})/\partial\beta_{rk}]$ in Section 8.5
$\check{X}(s)$	Laplace transform of $X(t)$
$\mathbf{X}$	a vector of random variables X_1, X_2, ..., a matrix of partial derivatives evaluated from the most recent data in which the rows represent successive experimental runs
$\mathbf{X}^{(n)}$	matrix of elements of $X_{ij}^{(n)}$; the superscript denotes the nth stage or cycle of calculations
$\mathbf{X}_{rj}$	matrix of elements of $X_{rj,k}^{(i)}$ with $i = 1, \ldots, n$ in Section 8.5
$\mathbf{X}_r$	matrix of elements of $X_{r,ij}$
y	general deterministic dependent variable
$y(t)$	general deterministic dependent variable which is explicitly a function of time
$y(t_i)$	model response for a discrete time
$y(0)$	initial condition for y
y_{rj}	response of the jth equation in the rth model in Section 8.5
y_n^*	defined in connection with Equation 11.2-7
y_r	response of model r
$\check{y}(\omega)$	process or model output in the frequency domain
$\check{y}(s)$	Laplace transform of $y(t)$
$\check{y}(z)$	z-transform of $y(t_k)$
$\mathbf{y}$	matrix of model responses
$\mathbf{y}_r$	matrix of elements y_{rj}
$\mathbf{y}_p$	particular solution of a matrix differential equation
$\mathbf{y}(t_i)$	matrix of model responses (for discrete times)
$\mathbf{y}(0)$	matrix of initial conditions for $\mathbf{y}$
$\mathbf{y}_0$	initial condition for $\mathbf{y}$
$\mathbf{y}^*$	defined in connection with Equation 9.5-5
$\mathbf{y}^{(n)}$	matrix of model responses on the nth cycle of calculation
$\mathbf{y}_k$	defined in Section 11.4
$\check{\mathbf{y}}(s)$	Laplace transform of $\mathbf{y}(t)$
Y	general stochastic dependent variable
Y_i	ith observation of Y
Y_j	observation for jth equation in Section 8.5
Y_k	kth-order derivative of Y with respect to time in Section 9.6, otherwise the observation at t_k
Y_{ij}	jth observation of Y at x_i
Y_{ij}	new observation of Y or of $\bar{Y}_i$
Y_{ri}	ith observation for model r in Section 8.4
$\bar{Y}_i$	sample average of observed Y's at x_i
$\hat{Y}_i$	predicted value of Y_i at x_i
$\hat{Y}_e$	$\hat{Y}$ at an extremum (for a response surface)
$\bar{Y}$	sample mean of Y_i, grand mean of Y in a regression model
$Y(t_i)$	discrete observation at t_i
$Y(t)$	general stochastic dependent variable that is explicitly a function of time
Y_{n+1}^*	one additional observation of Y
$\check{Y}(s)$	Laplace transform of $Y(t)$
$\check{Y}(\omega)$	Fourier transform of $Y(t)$
$Y \mid x$	value of Y given the value of x
Y^*	new observed value of Y (in regression analysis)
$Y^{(n)}$	value of the nth response in Section 8.5
$\hat{Y}_r^{(n)}$	predicted reponse for model r after n data points have been collected in Section 8.5
$\mathbf{Y}$	matrix of observations (dependent variables)

Symbol	Meaning
$\hat{\mathbf{Y}}$	matrix of values of Y predicted from the estimated regression equation
$\mathbf{Y}_i$	matrix of observations at t_i in Section 9.4
$\mathbf{Y}(t_i)$	matrix of observations at t_i
z	parameter in the z-transform
z	axial coordinate
z_j	integral in orthogonal product method
z^*	dimensionless axial coordinate
Z	random variable in general
$\mathscr{Z}$	z-transform operator
Z_j	integral in orthogonal product method, random variable
Z^*	$Z^* = \tanh^{-1} \rho_{XY}$
Z^*_{ij}	scaled element in the matrix $\mathbf{Z}$
$\mathbf{Z}$	identical to $[(\mathbf{X}^{(\ )})^T \mathbf{w} \mathbf{E}^{(\ )}]$

Greek

Symbol	Meaning
α	a parameter in a probability distribution in general
α	significance level = 1 − confidence coefficient
α	parameter in a model
α_{rs}	coefficient
$\boldsymbol{\alpha}$	matrix of coefficients
$\boldsymbol{\alpha}^*$	column vector of parameters given by Equation 9.3-10
β	parameter in a probability distribution in general
β	operating characteristic of a test = 1 − power of a test
β	parameter in an empirical model
β_k	element of the matrix $\boldsymbol{\beta}$
β_0	intercept in a linear (in the parameters) empirical model
$\boldsymbol{\beta}$	matrix of model parameters
$\boldsymbol{\beta}^*$	submatrix of β with one element deleted
γ	constant
$\gamma(t)$	blunder in Section 4.7
γ_c	contraction coefficient
γ_e	expansion coefficient
γ_r	reflection coefficient
$\gamma_X(t)$	ensemble coefficient of variation of the stochastic variable $X(t)$
$\boldsymbol{\gamma}^2(\omega)$	coherency matrix
$\tilde{\gamma}^2_{XY}(\omega)$	coherence function
$\Gamma(x)$	gamma function
$\boldsymbol{\Gamma}$	covariance matrix among models in Chapter 8
$\boldsymbol{\Gamma}$	covariance matrix among observations
δ	a difference
$\delta(x)$	delta function
δ_{kl}	Dirac delta function: $\delta_{kl} = 0$ if $k \neq l$; $\delta_{kl} = 1$ if $k = l$
Δ	difference, as $\Delta X = X_{i+1} - X_i$ or $\Delta b_j = \beta_j - b_j$
$\Delta^{(n)}, \Delta, \Delta_1, \Delta_r$	criteria for discrimination among models in Sections 8.4 and 8.5; the superscript n denotes the stage of calculations
ΔSS	difference in the sum of squares
ϵ	stochastic unobservable error, error in general
$\acute{\epsilon}$	equation error defined in Equation 9.6-2
ϵ_{ri}	unobservable error for rth model on ith experimental run in Section 8.4
$\epsilon^{(n)}_{rj}$	unobservable error for the jth equation in the rth model for the nth experiment
$\boldsymbol{\epsilon}$	matrix of unobservable errors
$\boldsymbol{\epsilon}^*_i$	matrix of unobservable errors defined in connection with Equation 9.4-9
$\boldsymbol{\epsilon}_i, \boldsymbol{\epsilon}(t_i)$	column vector of unobservable errors (at t_i for discrete times)
ζ	outcome of an experiment
η	dependent variable in empirical model, the expected value of Y
η_i	model response evaluated with the vector of independent variables used in the ith experiment
η_{ri}	ensemble response from rth model on ith experimental run in Section 8.4
$\boldsymbol{\eta}$	matrix of dependent variables for a model
$\boldsymbol{\eta}_i$	matrix of model responses at t_i in Section 9.4
θ	parameter in a probability distribution in general
$\boldsymbol{\theta}$	vector of parameters in a probability distribution
$\boldsymbol{\Theta}$	matrix defined in connection with Equation 9.4-9
κ	constant
λ	eigenvalue of a matrix
λ	Lagrangian multiplier
λ	parameter in the Marquardt least squares method, a parameter in general
Λ	random variable for Bartlett's test (Equation 3.6-2)

Symbol	Meaning
μ_n	raw moment; n indicates order
μ_X	expected value of X
$\mu_X(t)$	expected value of $X(t)$
μ_{ij}	joint moment of two random variables; i and j indicate the respective orders
μ_X^*	expected value of X^*
ν	degrees of freedom in general
ξ	known value for a model parameter
ξ_i	scale factor $= (A_{ii})^{-1/2}$
$\prod$	multiple product
$\mathbf{\Pi}$	matrix defined by Equation 9.4-6a
ρ	density in Chapter 10
ρ_{XY}	ensemble correlation coefficient for the stationary random variables $X(t)$ and $Y(t)$
$\hat{\rho}_{XY}$	sample correlation coefficient
$\hat{\rho}_\pi$	estimated multiple correlation coefficient
$\rho_{12\cdot 3}$	partial correlation coefficient between variables 1 and 2 eliminating variable 3
$\hat{\rho}_{Yx_j\cdot x_k}$	estimated partial correlation coefficient between Y and x_j eliminating the effect of all the other variables
σ	ensemble standard deviation of the random variable X
$\sigma_X(t)$	ensemble standard deviation of the random variable $X(t)$
σ_{ij}	ensemble covariance for X_i and X_j in general
$\sigma_{XX}(t_1, t_2)$	ensemble autocovariance function of the nonstationary random variable $X(t)$
$\sigma_{XX}(\tau)$	ensemble autocovariance function of the stationary random variable $X(t)$
σ_{rs}	element of the covariance matrix $\mathbf{\Gamma}$ in Section 8.4
$\sigma_{XY}(t_1, t_2)$	ensemble crosscovariance function for the nonstationary random variables $X(t)$ and $Y(t)$
$\sigma_{XY}(\tau)$	ensemble crosscovariance function for the stationary random variables $X(t)$ and $Y(t)$
$\breve{\sigma}_{XY}(\omega)$	ensemble cospectrum function for the stationary random variables $X(t)$ and $Y(t)$
σ^{rs}	element of the inverse of the matrix $\mathbf{\Gamma}$
σ_X^2	ensemble variance of the random variable X
$\sigma_X^2(t)$	ensemble variance of the random variable $X(t)$
$\sigma_{\bar{Y}_i}^2$	ensemble variance of $\bar{Y}_i$ (in regression analysis)
$\mathbf{\Sigma}_r$	compact notation for $\mathbf{\Sigma_Y} + \mathbf{W}_r^{(n+1)}$
$\mathbf{\Sigma_Y}$	covariance matrix for multiresponse models in Section 8.5
τ	difference in two times
τ_m	maximum lag in correlation
$\mathbf{\Gamma}$	covariance matrix of $\boldsymbol{\epsilon}_i^*$
ϕ	sum of squares of deviations (or integral as in Equation 9.2-1)
ϕ_i	value of ϕ at vertex i of a simplex
ϕ_l	lowest value of ϕ at a simplex vertex
ϕ_q	sum of squares for a model with q parameters
ϕ_u	highest value of ϕ at a simplex vertex
$\phi_{\min}$	sum of squares of residuals, the minimum ϕ
$\bar{\phi}$	average value of ϕ
$\breve{\phi}(s)$	Laplace transform of $\phi(t)$
ϕ^*, ϕ^{**}	values of ϕ at vertices of a simplex in Section 6.2-1
χ^2	chi-square, random variable
χ_α^2	chi-square for a significance level of α
$\tilde{\chi}^2$	approximate chi-square random variable
ψ	phase angle
$\psi \varphi$	functions in general
$\boldsymbol{\psi}$	matrix of functions in general
ω	frequency in general
ω_{ij}	element of the matrix $\mathbf{\Omega}$
ω^*	frequency in cycles/time
$\omega_{\max}$	Nyquist frequency in radians
$\mathbf{\Omega}$	covariance matrix for the model parameters, $\boldsymbol{\beta}$; subscripts designate a covariance matrix for other parameters
$\mathbf{\Omega}_{\eta_{i+1}}$	covariance matrix defined in Equation 9.4-16

OVERLAYS

Symbol	Meaning
^	estimated
ˇ	Fourier or Laplace transform
‾	sample average
~	in canonical coordinates
~	incorrectly estimated, approximate

SUPERSCRIPT

Symbol	Meaning
(n)	intermediate estimate (0 = initial guess); also indicates order of derivative (1 = first derivative); also indicates in Section 8.4 the sequence number of the data set

*	differs in some way from the usual definition of the parameter or variable
T	transpose of a matrix
$V^{ij} = [V_{ij}]^{-1}$	inverse element of a matrix $\mathbf{V}$
$\mathbf{V}^{-1}$	inverse of matrix $\mathbf{V}$

OTHER

$\lvert\mathbf{x}\rvert$	determinant of the matrix $\mathbf{x}$
$\lVert\mathbf{f}\rVert$	norm, i.e., the square root of the sum of the squares of the components of the vector $\mathbf{f}$; also the absolute value
∇	gradient of a function or matrix
$\oint$	contour integral
$\cap$	intersection
$\cup$	union
$\langle\ \rangle$	finite time average for continuous random variables
$\langle\ \rangle_n$	sampled data finite time average for random variables
$\mid$	symbol for given, as $Y \mid x$

Author Index

Subject Index

QUEEN MARY
COLLEGE
LIBRARY